Longevity, Senescence, and the Genome

Longevity, Senescence, and the Genome

Caleb E. Finch

The University of Chicago Press
Chicago and London

Caleb E. Finch is the ARCO and William F. Kieschnick Professor in the Neurobiology of Aging and professor of gerontology and biological sciences at the University of Southern California.

The University of Chicago Press, Chicago 60637
The University of Chicago Press, Ltd., London
© 1990 by The University of Chicago
All rights reserved. Published 1990
Printed in the United States of America

99 98 97 96 95 94 93 92 91 90   5 4 3 2 1

The University of Chicago Press gratefully acknowledges a subvention from the John D. and Catherine T. MacArthur Foundation in partial support of the costs of production of this volume.

Library of Congress Cataloging-in-Publication Data

Finch, Caleb Ellicott.
  Longevity, senescence, and the genome / Caleb E. Finch.
    p. cm.—(The John D. and Catherine T. MacArthur Foundation series on mental health and development)
  Includes bibliographical references and index.
  ISBN 0-226-24888-7 (alk. paper)
  1. Longevity. 2. Longevity—Genetic aspects. 3. Aging. 4. Physiology, Comparative. I. Title. II. Series.
QP85.F47 1990
574.3′74—dc20
                                                 90-43971
                                                 CIP

To my mother, Faith Stratton Finch Campbell,
A grand and very lively Yankee lady and se'er
Who nurtured my curiosity and love of reading.

Contents

Preface

This project really originated in a literature review for my Ph.D. dissertation, *Cellular Activities during Aging in Mammals* (Finch, 1969), in which I considered the then heretical positions that senescence might not be pervasive in cells throughout the body; that many cellular changes during senescence in mammals were driven by changes in hormones and other regulatory factors, rather than by random degeneration of the genome in somatic cells or random disturbances of macromolecular biosynthesis; and that there might be fundamental differences between species in the mechanisms of senescence.

The goal of this book was to pursue questions raised in my laboratory research on the cellular selectivity of aging in the rodent and human brain into the literature on organisms with different life history patterns. In particular, I sought to assemble a body of informtion on life history patterns that would give fresh insight on the role of the genome in regulating physiological mechanisms in senescence; on relationships between reproductive modes and senescence; and on relationships between clonal senescence of cells and organismic senescence. A set of questions is posed as an Agenda in Chapter 1 that is reflected on throughout the book. Thereby I went far outside my primary areas of expertise as a molecularly and developmentally oriented physiologist, often finding myself a stranger in literatures in which I had little knowledge and few collegial contacts. To my great pleasure, most fellow scholars treated me generously, making thoughtful suggestions and often spending hours in correcting early drafts. I began the writing in January 1987, and since then have spent all the time I could spare at this task.

I tried not to dodge controversial or difficult issues, but also tried not to stray unproductively far into areas so uncertain that the issues could not be discerned. There may be howlers: despite intense concern, a project of this scope written by one person cannot escape errors and misconceptions. It is my hope that this book will draw attention to a new set of intellectual issues that I believe are ripe and tractable for experimental analysis.

I cannot express enough gratitude to three great colleagues, who read several drafts of the entire book and steered me right many times: Roger Gosden, Thomas Johnson, and Robert Sapolsky. I also greatly appreciate key input and critiques from William DiMichele, Richard Grosberg, Larry Noodén, and Michael Rose. Michael Rose and I exchanged drafts of our books; I benefited from knowing early on that our approaches differed in their intellectual structures and gestures to our chosen scholarly communities. Two readers for the University of Chicago Press allowed me to recognize their valuable critiques, John Eisenberg and David Prescott. Special recognition is due the John D. and Catherine T. MacArthur Foundation and the leadership of Dennis Prager and John Rowe that created the intellectual environment in the Network on Successful Aging which so encouraged this project.

While the number of scholars that I contacted for specific information is too numerous to list, I am greatly indebted to the following for their written comments on specific chapters: Marilyn Ader (Chapters 3, 10); Robert Arking (6); Steven Austad (2, 11); David Bardack (11); John M. Baynes (7); Robert J. Behnke (2, 9, 11); Lisa F. Berkman (1, 9); David Bottjer (11); Edward Brothers (2, 11); Leo Bustad (3); Bennett Cohen (3); Vincent J. Cristofalo (7); Donald Cummings (2, 6); Kelvin J. Davies (11); Jonathan Day (4, 11); William DiMichele (2, 4, 11); Serge Doroshov (3); George M. Dunnet (3); Mark W. Ferguson (3); James E. Fleming (7); Paul Gallop (7); Murray B. Gardner (6); John J. Gilbert (2, 8, 9); Barbara Gilchrest (3, 8); Wendy Gilmour (6); Myron Goodman (8); Richard Grosberg (2, 4); Michael G. Hadfield (2, 3, 9); David Hammerman (3); John M. Harris (11); David E. Harrison (6, 8, 10); Peter W. Hochachka (5); Terrance P. Hughes (4); Michael S. Jazwinski (2); Steven A. Johnson (7); J. Kukalova-Peck (11); Conrad Labandeira (2, 11); Calvin A. Lang (2, 6, 7); Bruce M. Leaman (1, 4); Monte Lloyd (2, 6); Jaakko Luume (6); Robert M. McDowall (2, 11); Margaret McFall-Ngai (1, 7, 11); Armand V. Maggenti (4); George M. Martin (3, 6, 7, 8); Edward J. Masoro (3, 10); Richard A. Miller (3, 10); Vincent M. Monnier (7); Hamish Munro (5); James S. Murphy (9); Joseph Nelson (2, 11); John Nesselroade (1, 11); R. Bruce Nicklas (8); Marcel Nimni (7); Larry Noodén (2, 4, 11); John Ogden (4); Mary Ann Ottinger (3); Peter Plagemann (6); Edward G. Platzer (4); Robert E. Ricklefs (3, 11); George S. Roth (7); Frank G. Rothman (2); Stephen E. Scheckler (11); Edward L. Schneider (3); Arthur G. Schwartz (7); Virginia Scofield (2); James R. Smith (7); Joan Smith-Sonneborn (2–4); Stephen Spindler (7); Alan R. Templeton (6); A. C. Triantaphyllou (4); Angelo Turturro (2); Frederick vom Saal (9); Roy L. Walford (1, 3); Richard Weindruch (3, 10); Mark Winston (2); David and Phyllis Wise (2); Matthew Witten (1); Carl Woese (11); Vernon R. Young (7).

Technical help in the book involved many individuals. Kathleen McElhinney became indispensable to the project by her knack of obtaining copies of obscure reports and her skill in computer graphics. The tedious job of typing the references was done with good cheer by Carmen Washington. Others who gave notable help were Chris Anderson, photography and graphics; Lalli Medina, drawing; Marc Gallardo, Frank Vargas, and Antonius Van Haagen, assistance in calculations; Shu-gwan Xu and Julie Moore, library assistance. At the University of Chicago Press, Joann Hoy and Julie McCarthy did an outstanding job of editing the manuscript, while the overall project was skillfully organized by Senior Editor T. David Brent. My administrative assistant Amy Weinstein deserves special thanks for overseeing the daily business of the Alzheimer Disease Research Consortium of Southern California and of my lab while I was so preoccupied. With me throughout was Doris Jane, whose moral support and hearthside arts brought serenity and restoration.

This preface, written as the book was being edited, pulls a string of memories that course my mind, much like the sense of motion continuing after a strenuous journey when real motion has ceased. Moreover, certain changes occurred during this long project: I have passed 50 years and now need spectacles, true to the canon of aging in humans; our dog Romulus passed dodderingly into his last sleep; and my first PC became defunct after recording about 10^7 bytes. On to the book!

Altadena, California
29 May 1990

Synopsis

The nature of genetic controls on senescence and lifespan constitutes one of the most intriguing, but difficult, problem areas in modern biology. This book addresses evidence and issues surrounding the basic questions: *Is somatic senescence an obligatory feature in the life of eukaryotes?* And, *what roles does the genome have in mechanisms that may be pacemakers for senescence and lifespan?* We know at the outset that lifespans range nearly a millionfold among different species of higher animals and plants and must be species characteristics under considerable genetic control.

But there are also important variations among and within populations which reflect myriad environmental or epigenetic mechanisms. Some readers may be startled by the twofold or more differences of lifespan between genetically identical members of the same birth cohort of laboratory mice, flies, or roundworms. These variations implicate various types of random and epigenetic influences during development or adult life. While some influences come through random traumas ("wear and tear") via the external environment, other random events occur during development without specific links to the outside.

The subject of gerontology then must recognize various complex interactions between the genome and the environments, both internal and external, across the lifespan, from which evolve important differences in patterns of aging and potential lifespan among individuals and species. The natural history of senescence reveals a vast variety in temporal organization, both among species as well as among and within populations, that may vary over as large ranges of scale and qualitative characteristics as do morphological and biochemical variations. This variety, I believe, gives invaluable insights into the mechanisms of senescence within and among species.

The material ahead is often controversial and challenges generalizations about the universality of senescence that were based on limited numbers of species. Some theoretical approaches with a narrow perspective have been embarrassed by the varieties of senescence. Efforts are made to identify important gaps of information and to critically review presumptions that I think have slowed progress. Many basic facts about senescence in even the best-studied species (humans) are still being revised as the multifarious confoundings of age-related diseases are better understood. For example, the dark view that 100,000 neurons are inevitably lost each successive day throughout the lifespan (Burns, 1958) is being revised by studies from better-controlled populations of laboratory animals and human specimens that are more exactingly classified to identify subjects according to the presence of vascular diseases, strokes, and Alzheimer's disease (Chapter 3.6.10).

Another issue concerns the large scope of environmental influences, particularly in long-lived organisms. Environmental influences that affect lifespan and senescence generally differ widely among individuals in any population on a statistical basis. Perhaps the potential for environmental diversification in the phenotypes of senescence scales with the lifespan. This diversity argues against a single or universal mechanism in organismic senescence. Rather, I propose that there are many different mechanisms. These may be ultimately traced to diverse arrays of genes that could differ among organisms according to their evolutionary histories and even within a species according to the environment and natural history of its populations.

The structure and language of the book are intended to bring this now rapidly moving subject to the attention of researchers in behavioral, developmental, evolutionary, and molecular biology. Although gerontology as a subject area is often considered closer to clinical than to basic research, there are important connections to areas throughout basic biological research. In particular, I hope to extend gerontology beyond its usual focus on biological, medical, and social phenomena to an array of basic questions concerning the various outcomes of life history that are shared with developmental, evolutionary, and reproductive biology. It is also noteworthy that many diseases of adult humans, such as diabetes, hypertension, and vascular disease, have been historically considered in isolation from the larger manifold of changes in the age group; this narrow view may have limited progress. As pertinent, I note historical precedents of gerontological questions from developmental and evolutionary biology.

Although the book considers many topics that are not usually discussed in treatments of gerontology, it is not a comprehensive review of all major areas of data or theory. As described on the first page of Chapter 1, I address an agenda of select questions that are taken up throughout the book. These questions are illuminated by examples that are discussed in different chapters. It is my intent to provide examples that will expand our present experimental and theoretical approaches to senescence and the lifespan. The reader should thus expect to find many patches and gaps, which, hopefully, will stimulate my colleagues in a number of disciplines. I should also state an attitude, that I do not expect a simple unifying explanation to account for the manifest diversity of senescence. There may be, nonetheless, domains of phenomena and manifestations in some evolutionary radiations that reflect a predominating mechanism of senescence.

The book has twelve chapters in two major sections: *Section I,* the comparative biology of senescence, and *Section II,* genomic mechanisms. In *Section I, Chapters 1–6* illustrate species differences in the natural history of senescence that indicate various typologies of senescence. *Chapter 1* sets the agenda and defines terms and concepts with an effort to minimize vagaries that have often biased thought in this field. A classification of senescence is proposed which defines a spectrum of senescence according to whether organismic involutional changes are scaled ranging from rapid to gradual to negligible. I also provide the most extensive multiphyletic calculations of mortality rate constants according to the Gompertz model of exponentially accelerating mortality (summarized in *Appendix 1*). Because many species have short lifespans owing to high overall mortality rates, I argue that lifespan is often an unreliable basis for comparing mechanisms or rates of senescence. *Chapters 2–4* present the astonishing variety of pathophysiological changes, which argues against a universal cause of organismic senescence in eukaryotes. This case is strongest for species that die after their first and only bout of reproduction. As described in *Chapter 4,* the widespread occurrence of vegetative or asexual reproduction from somatic cells (somatic cell cloning) differs importantly from the limited ability of some mammalian cells to proliferate in culture (clonal senescence, also known as the Hayflick phenomenon). *Chapter 5* discusses correlations of lifespan with body size and other attributes (the allometric relationships). *Chapter 6* describes genetic variants that influence the patterns of senescence and lifespan, including effects of gender. These perspectives on the natural history of senescence set the context for analyzing genomic mechanisms.

Section II analyzes various genomic influences at the molecular, cell, and physiological levels. *Chapters 7 and 8* address macromolecular synthesis and other genomic functions during senescence. Most changes in macromolecular biosynthesis appear to represent selective, rather than random, changes in gene regulation. *Chapters 9 and 10* analyze environmental and experimental influences on senescence and the lifespan during development and adult life, which indicate a remarkable plasticity in phenotypes throughout the life course. This plasticity also argues against the intrinsicality of senescence in higher cells. *Chapter 11* discusses evolutionary aspects of senescence and lifespan, concluding that lifespans have been both shortened and lengthened as judged from the fossil record. *Chapter 12* closes the book by revisiting the agenda. There is a recapitulation of major themes, a synopsis of the plasticity of senescence and its possible limits, and a perspective on the subject of aging as the biology of extended time.

A major conclusion is that mechanisms by which genes influence lifespan and the particulars of senescence can be identified in some organisms. Many examples show an important role of alternative patterns of gene expression or of changes in gene activity as underlying causes of differences in lifespans between related organisms. Other examples indicate changes in the genome itself, for which the best examples are somatic cell chromosomal abnormalities. However, these mechanisms in most other species remain obscure. From the vantage of our present scant knowledge, genomic mechanisms in senescence appear much more complex and phyletically diverse than genetic mechanisms in development. So far, no new biological mechanisms are yet required to account for senescence beyond those which operate during development or in pathogenesis during the early adult years. Moreover, there is an impressive number of examples of molecular and cell functions that remain unimpaired during long lifespans. The conclusion from developmental biology that most somatic cells of young adults are genetically totipotent may be provisionally extended to later ages. If so, many aspects of senescence should be strongly modifiable by interventions at the level of gene expression. This proposal is supported by modifications of senescence through manipulations of the external or internal (physiological) environments.

Readers may be assisted by the *Glossary,* in which special terms are defined and abbreviations are listed. The *Index* includes topics as listed in the table of contents, as well as the outlines that precede each chapter. Numerous other items, such as species names, are listed in the *Index;* where the common name is listed, reference is also given to the Linnean.

Section I

THE COMPARATIVE BIOLOGY OF SENESCENCE

This analysis of how the genome influences lifespan and senescence in higher organisms is based on species comparisons across many levels of biological organization. I have sought out the life histories of many species besides the usual mammals, flies, and nematodes to expand the view of mechanisms that limit lifespans. A comparative approach is crucial to the biology of senescence, just as for any other aspect of the life history: virtually all fields of life sciences have had phases in their history when a few capriciously favored organisms dominated the literature and limited progress because of features that were not shared by other species (Beach, 1950). Chapters 1–6 describe the diversity of species differences in the patterns of senescence and lifespans of animals and plants. This diversity enriches discussions of mechanisms and provides a framework for analyzing the genomic roles in senescence. Chapter 6 discusses genetic variants that are associated with influences on longevity and senescence, as observed in natural and laboratory populations. Emphasis is given to age-related changes in reproduction, which I believe are central to understanding the evolution of senescence.

1

Introduction

1.1. The Agenda

Analysis of the following issues is key to learning how the genome influences the temporal organization of life histories, especially in regard to the causes of senescence that influence the lifespan of individual organisms. In essence, I am inquiring about the dimensions, constraints, and plasticity in the temporal features of eukaryotic life histories.

(a) *How valid is the lifespan as a measure of the rate of senescence?* Comparisons of biochemical and other attributes of species on the basis of the age of the oldest reported survivor may not always be appropriate because of high overall mortality or because of different natural histories.

(b) *When does senescence begin?* That is, how should we deal with the duration of developmental stages? For example, consider the 17-year life cycle of slowly growing cicadas with adult phases of about 1 month versus insect queens with adult phases of 5–15 years versus flies with total lifespans of 1 month.

(c) *At what levels of organization do genes influence senescence?* Age-related dysfunctions can occur in macromolecules, cells, organs, etc., all of which are directly specified by the genome at some time during the life cycle.

(d) *What is the relative contribution of selective (nonrandom) versus random changes in gene expression and other cell functions during senescence?* While many studies and theories emphasize random damage to genomic functions, most changes as reviewed in this book appear to be selective. The importance of selective gene regulation throughout development and differentiation is clear, but the role of changes of gene regulation in senescence is not generally known.

(e) *How directly do genes operating during development or adult life specify senescence?* Genes may be direct determinants or alternatively may set indirect constraints on the lifespan. Direct effects are epitomized by early onset hereditary diseases. Genes with indirect effects include those governing the replacement of vital cells (e.g. neurons) or organs (e.g. insect wings).

(f) *How do species vary in reproductive senescence, including the total production of gametes and incidence of developmental abnormalities?* Reproductive age changes are important to the selection of genes that influence senescence.

(g) *How much plasticity in senescence is there from environmental (i.e. nongenetic) sources?* This question is asked for variations among populations of the same genotype, as well as among outbred populations. These variations give clues about the limits of genetic control on lifespan and senescence. Plasticity in senescence is seen virtually everywhere.

(h) *How universal are age-related degenerative changes at the organismic, cellular, and molecular levels as may be judged by species comparisons?* While some species suffer rapid senescence through specific dysfunctions, organisms with slower senescence may still have more general cell changes. Yet other species show few signs of senescence. The comparative approach gives many clues about vulnerabilities that limit lifespan under certain conditions.

These issues, discussed throughout the book, are recapitulated in Chapter 12.

1.2. Definitions and Concepts

1.2.1. The Lexicon

First the word. The term aging, like many words in ordinary usage, is haunted by assumptions that can bias scientific thinking. Assumptions about aging that are held by all sentient beings must influence the choice of experimental model and the types of experiments done, no less than they influence societal attitudes about providing resources for the rehabilitation or long-term care of the elderly. Aging, as Medawar (1952) pointed out, is used to describe virtually all time-dependent changes to which biological entities, from molecules to ecosystems, are subject, though the mechanisms and consequences to function may be vastly different. We read, for example, about the aging of macromolecules such as collagen or heterogeneous nuclear RNA; aging of the ovum, embryo, placenta; aging of diploid cells in culture and of erythrocytes in the circulation; aging of populations and societies; aging of genes and species during evolution. All these share only the trait that changes occur during the passage of physical time, for which there need be no common mechanism. Because many age-related changes in adult organisms have little or no adverse effect on vitality or lifespan, I avoid the word **aging** in this book. You won't miss it!

The lexicon is chosen to stay close to the empirical phenomenology. Changes during the lifespan seem better described as age-related, which deemphasizes the sense that time itself has caused a biological change or that a change is necessarily adverse. The literature of gerontology until fairly recently has been permeated with the prejudicial presumption that age-related changes are all adverse to some degree and that most components of an organism should decay as the end of the lifespan is approached. Evidence contrary to these important assumptions is presented throughout the book. Senescence is mainly used to describe age-related changes in an organism that adversely affect its vitality and functions, but most importantly, increase the mortality rate as a function of time. Senility represents the end stage of senescence, when mortality risk is approaching 100%.

Although it is common practice to speak of senescence in molecules and cells, I have minimized this usage to avoid confusion between the phenomena leading to organismic death and numerous phenomena of cell death. Cell death is a normal feature of morphogenesis and of the maintenance of proliferating tissues in the bone marrow, gut, skin, etc., and is also ongoing throughout adult human life. For example, epithelial cell death and turnover has no obvious relationship to neuronal involution during Alzheimer's disease, in which there is highly selective neuronal involution (cell senescence). A converse example to these uses of senescence is given by the growth of malignant cells. While malignant cells might be considered very healthy by the standards of cell culture and might be used to establish immortal cell lines, their growth kills the organism. Whether a cell grows vigorously or dies is therefore distinct from the viability or senescence of the organism.

Nonetheless, senescence can arise at virtually any level of organization, from the molecular upwards. Long-lived macromolecules can become irreversibly damaged through racemization of amino acids or through glycation of proteins (Chapter 7.5.4). At the tissue level, blood vessels can become narrowed from atherogenesis, while joints and appendages can mechanically wear out. All these may contribute to organismic dysfunctions that increase the mortality risk. The onset of deteriorative changes need not be simultaneous at any level of organization. As will be discussed throughout the book, senescent changes generally show high selectivity for particular structures, independent of their level of organization.

Another distinction concerns proximal versus distal causes of senescence. For example, the cause of death may be listed on a coroner's report as heart failure from a myocardial infarction. This proximal cause of death, however, often results from a slow process of atherosclerosis of the coronary vessels that might have a distal cause in organismic factors related to blood lipid abnormalities. In the cases of familial lipid disorders that cause premature death from atherosclerotic lesions, the ultimate distal cause would be a particular genotype, as discussed in Chapter 6. As has been noted (Medawar, 1952; Williams, 1957), senility, whether or not associated with diseases, is usually the end state of a long-term process that might have either a few or many distal causes which can be traced to earlier stages in life.

The approach used here converts the use of time in the analysis of senescence from an independent variable to a dependent variable (Finch, 1988; Schroots and Birren, 1988; Arking and Dudas, 1989). Event-dependent changes during senescence result from specific proximal causes that are extrinsic to the molecule, cell, organ, or organism. Biological time thus becomes equivalent to trains of specific physical or chemical events, which is a very different concept than that of an intrinsic clock based on sidereal or calendar time. This philosophical model gives release from the general intuition that time itself inevitably causes damage and disorder in some entropic sense. In fact, a remarkable array of repair mechanisms have evolved to cope with damage at the molecular and other levels of organization. These repair and regulatory mechanisms allow most, if not all, aspects of the life course to be determined by specific events that are independent of absolute time.

In practice, most data are collected with reference to particular ages; accordingly the figures in this book display age-related changes against time according to the usual convention that statistically time is the independent variable with which changes in various parameters may be correlated. Many reports present data as scatter plots with regression lines, which I believe often bias their reading. When age-related trends are strong, no further information is gained, for example Figure 1.13, which is graphed as originally presented. Some figures were redrawn by removing lines as indicated in the legend. In other cases, the curves were retained as originally drawn to help discriminate between different data sets.

Some time-dependent changes in nonreplaced proteins are sufficiently general and predictable to be biological chronometers. For example, there is slow racemization of aspartic acid in lens proteins and other proteins that is thermodynamically driven and epitomizes "intrinsic senescence" at a molecular level (Chapter 7.5.3). Very slowly replaced molecules are at risk for accumulating damage through racemization or glycation (Chapter 7.5). Molecules having time-dependent changes raise questions about how to view the relationships of cell lifespans to organismic senescence. Circulating erythrocytes, for example, manifest a well-characterized sequence of changes in surface molecules (Chapter 7.5) which has been described as "aging." Erythrocytes are among the few cells of mammals that totally lack RNA and protein synthesis. They are therefore subject to unique types of age-related changes because of the limited ability to replace damaged or altered molecules, though even erythrocytes have some enzyme-based molecular repair mechanisms. Because the production and lifespan of circulating erythrocytes changes little during the organismic lifespan (Chapters 3, 7), the erythrocyte lifespan should have no bearing on organismic lifespan, unless it can be shown (a dubious prospect) that the same molecular changes also occur on an age-related basis throughout the organism. The finite lifespans of certain cells or molecules is pertinent to discussions of the organismic lifespan only insofar as there may be age-related changes in the renewal of these molecules and cells.

In contrast to these examples are the permanently postreplicative cells, that is, permanently nondividing cells, such as brain neurons of mammals and nearly all cells of adult flies and soil nematodes (Chapters 2–4). While permanently postreplicative cells may be nearly as old as the organism, most of their molecules are probably replaced within days to weeks in mammals. A relationship to senescence could arise if there were age-related slowing of molecular renewal in postreplicative cells that decreased viability of the organism. Many scientists, but few neurobiologists, have implicitly considered the postreplicative status as a sufficient cause for cell senescence. In my view, diploid fibroblast cultures of mammals are valuable experimental models for clonal senescence, specifically, limited proliferative capacity. Postreplicative fibroblasts may, nonetheless, serve as models for particular aspects of cell involution during organismic senescence.

Note, however, that some bone marrow stem cells show no evidence of finite replicative capacity. Moreover, many lower species of animals and most plants can propagate asexually from somatic cells—by fission, fragmentation, or budding—through vegetative reproduction (Chapter 4). In most cases, these species also have sexual phases. The examples of somatic cell cloning show no evidence for clonal senescence and furthermore show that clonal senescence is not an obligate characteristic of all differentiated somatic cell lineages. Vegetative reproduction involves somatic cell redifferentiation and depends on genomic totipotency, that is, the preservation in these somatic cell lineages of a complete set of genes that are equivalent to those in a germ line of gametically reproducing species.

The major focus on genes draws on special terms. Genes refer to nucleotide sequences that may reside in chromosomal DNA; in mitochondrial or chloroplast DNA; or in viral nucleotide sequences, either RNA or DNA, depending on the type of virus. The genome is the ensemble of all genes. Gene expression refers to the activities of a gene as represented by some cell characteristic, mostly messenger RNA, but also the protein encoded. The number of coding sequences or genes in the genome may be much less than the total amount of DNA. Mammals have an estimated 50,000–200,000 genes (Bantle and Hahn, 1976; Kaplan and Finch, 1982; Sutcliffe, 1988), of which less than 10% have been cloned and sequenced. Differential gene expression refers to the selective regulation of genes, such that only part of the genome is expressed in any cell type; this selectivity is usually the result of different rates of transcription (RNA synthesis), but may also result from the control of translation (protein synthesis). The genotype is defined comparatively, usually within a population, by heritable DNA or RNA sequences that are distinguished from those in other members of the species or taxonomic grouping. Reference to an organism's genotype does not mean that all the genes have been located or that their nucleotide sequence is known. Besides these familiar concepts, certain unconventional infectious agents, prions, are associated with neurodegenerative diseases such as scrapie and Creutzfeldt-Jacob disease (Chapter 6.6). Prions are small proteinaceous particles with infectivity that resists modification by treatments that usually destroy RNA and DNA, and with an abnormal form (isoform) of a protein also found in host cells (Prusiner, 1990; Westaway et al., 1987).

The phenotype of an organism represents any of its physical characteristics. Readers who are not close to modern genetics may be confused by statements indicating that the same genotype may have alternate phenotypes, depending on differential gene expression. For example, fetal exposure to androgens through "virilizing" adrenal tumors of the mother results in a male genital phenotype, even though the Y chromosome may be absent. Numerous examples are given of variations in the phenotype during senescence from environmental or experimental influences. Ultimately, these variations in phenotype at any stage of the life cycle are constrained or controlled by the genome.

A major question in interpreting age-related changes concerns whether they arise from causes that are extrinsic or intrinsic to the organism. Degenerative changes and increased mortality can arise from extrinsic—that is, environmental or ecological—hazards that may not be experienced by all populations nor by all within a population. For example, neonatal infections can cause long-lasting damage to the heart or immune system that shortens lifespan (Chapter 9.3). Exposure to stress, carcinogens, or other traumas that may vary among individuals can also be superposed on more ubiquitous changes, such as the accumulation of lipofuscins, or aging pigments (Chapter 7.5.6). Ecological hazards may, however, determine the lifespans of groves of trees (Chapter 4.3). We must be concerned with adventitious as opposed to universal aspects of senescence, whatever the role of the genome. Concern about extrinsic or ecological factors in senescence is considerable for the many species that have been observed only in field populations, that is, under suboptimal and uncontrolled circumstances. While the environment can be made nearly uniform for simple, nonsocial aquatic organisms like rotifers, most other laboratory species will encounter diverse and haphazard risks from their confinement (e.g. flies may suffer mechanical damage to irreplaceable appendages through collisions with each other or the cage walls). Some types of extrinsic senescence can be considered ecological, such as may result from damage by meteorological exposure or by exposure to ambient pathogens.

The variable duration of development, or time to maturation, necessitates more clearly defining what is meant by lifespan. Although lifespan in the gerontological literature commonly means the time between birth (or maturation) and death, many species characteristically have prolonged development (e.g. 17-year cicadas) or prolonged existence as free-living but immature organisms (e.g. bamboo species which grow for decades before maturation and rapid senescence). I propose to use lifespan in reference to the total life duration of an individual organism, from its earliest developmental phase, whether it is derived from an egg or from a vegetatively propagating clone, to its death in the adult phase that ordinarily culminates its life cycle. In vegetatively reproducing organisms, the lifespan of the clone may vastly exceed the lifespan of the individual. Harper (1977) coined the term genet in reference to all members of a clone with the same genotype. Individual, physiologically distinct organisms in a clone are called ramets. Clonal senescence of a genet would be equivalent to the death of that genetic strain.

Before discussing the rates of senescence in the next section, other concepts that are fundamental to discussions of senescence are the reproductive schedule that describes the age when reproduction begins, the rate of reproduction of viable adults, and the duration of the reproductive phase, from maturation through reproductive senescence. In some species reproductive senescence is absent or may be restricted to females (Chapters 3, 4, 8). A classification of life history according to reproductive schedules was introduced by Cole (1954) and is used by biogerontologists (Kirkwood, 1985) and population biologists. Semelparous species reproduce only once, in contrast to iteroparous species. Similar terms traditionally describe plant life histories as annual or perennial, monocarpic or polycarpic. One must then speak of perennial monocarps—for example, bamboo species that propagate vegetatively for many decades before flowering and dying—a terminology that seems awkward. Specifically, semelparous organisms confine reproduction to a single event or season that marks the end of their lifespan, while iteroparous organisms may reproduce for many years. Some semelparous species have very long total lifespans as the result of extended development before they mature and suddenly die (e.g. the 17-year cicadas or the 100-year bamboos). The terminology is misleading, however, when referring to very short-lived species like flies and rotifers. The multiple batches of eggs laid during a reproductive lifespan of a few weeks or months could qualify flies or rotifers as iteroparous, with age classes designated by days or weeks. However, in the context of annual cycles that define perennial reproduction, these short-lived species have only one bout of reproduction that occurs shortly after hatching; they might as well be considered semelparous.

1.2.2. A Continuum of Senescence: Rapid, Gradual, and Negligible

To help discuss the diverse phenomena associated with senescence, I find it useful to characterize senescence using a continuum with three general subdivisions according to the rate of degenerative changes: rapid, gradual, and negligible (Chapters 2, 3, 4, respectively). These subdivisions of senescence are not absolute categories and do not imply specific mechanisms. Similar terms have been used by Comfort (1979) and others, but not as a general characterization. Many examples show that environmental variables like temperature and nutrition can shift the rate of senescence of some species, from rapid to gradual or even towards the negligible, in association with major effects on mortality rate, specific pathologic changes, and lifespan (Chapters 1–4, 9, 10). The description of senescence as a continuum of changes subject to environmental effects gives a supple framework for analyzing shifts in the intensity of senescent changes. Also, better than might a more rigid typology, it facilitates discussions about the plasticity of these phenomena. The comparison of species on this basis gives a compelling picture of diversity in the mechanisms of senescence within all time frames.

Rapid senescence (Chapter 2) is the rapid onset of major pathophysiologic changes at some time after maturation in most or all members of a birth cohort which precipitate exponential increases in mortality rates, causing death, typically within a year's time. Examples include species that die during or after their first reproduction, such as certain annual plants, marsupial mice, and Pacific salmon. However, rapid senescence may occur at the end of a decades-long juvenile phase, as in the bamboo. Thus rapid senescence can occur in species with long lifespans as free-living organisms as well as in the very short-lived. Most short-lived invertebrates—such as rotifers, nematodes, and flies, whose entire lives span a few days, weeks, or months, respectively—are also considered to manifest rapid senescence. Few species of mammals do so, with the exceptions of marsupial mice and rodents, which show seasonal die-off. The ability of marsupial mice to exceed their usual lifespans if isolated from reproductive stress shows that rapid senescence need not be obligate. Aphagous moths and some other insects, however, show obligate rapid senescence that resists natural variations or experimental interventions. There is thus reason to consider subdivisions of rapid senescence according to whether the changes are obligate or conditional. Many species showing rapid senescence are also semelparous.

Gradual senescence (Chapter 3) characterizes populations manifesting a slower onset of pathophysiologic disorders. The disorders emerge over many months or years after maturation and are associated with age-related increases of mortality rate, but are not synchronized to any life course event. Gradual senescence occurs in all placental mammals for which data are available. Although small rodents have much shorter lifespans than humans, rodent senescence is also gradual, since its manifestations are not synchronous in the population. Even when senescence is gradual, individuals may nonetheless undergo a rapid pre-morbid involution lasting days to months, but these changes are usually not manifested synchronously across the birth cohort. In general, species showing gradual senescence manifest progressive dysfunctions during one or more years or even decades, as in humans.

Negligible (or extremely gradual) senescence (Chapter 4) is at the other extreme from rapid senescence and describes species in which dysfunctional changes have so far eluded detection. This position in the continuum of senescence is hypothetical at present, since with sufficient time senescence might be manifested as increased mortality or as sterility. The duration of adult phases may span decades or centuries, as in anemones, clams, trees, and maybe also fish and reptiles that have shown no evidence of progressive dysfunctions before death, in contrast to that observed in humans or other gradually senescing mammals. In yet other species, the high overall mortality rate may yield a characteristic maximum lifespan in finite populations (Section 1.4), even without allowing manifestations of senescence that might have had time to develop under more permissive conditions. Because many pathophysiological changes of senescence in mammals can be greatly ameliorated by caloric restriction (Chapter 10.3), it is possible that more optimum diet or other environmental factors can shift senescence in some genotypes towards the more gradual. The still unknown extent to which species fail to show senescence has major bearing on population genetics theories of senescence, as discussed in Section 1.5.

The diverse lifespans and patterns of senescence of closely related species as found in many phyla, when considered with the evidence for the absence of clonal senescence in vegetative reproduction, make the case that senescent changes leading to finite lifespans generally result from degeneration of particular cell types or organs. The role of the genome in senescence and lifespan thus may be devolved to the genetic programming of vulnerability in particular organ systems.

1.3. Sources of Information and Experimental Approaches

This book assembles information from the scientific literatures of many fields that are not usually identified as "gerontology," including veterinary medicine and toxicology, as well as developmental biology, evolutionary biology, ecology, and wildlife management. Many studies have valuable data about senescence that are not explicitly shown in their title, abstract, or key words. There is thus a huge and interstitial literature on the natural history of senescence.[1]

I have come to believe that the genomic basis for senescence, as established for a few laboratory models, cannot be understood without knowing the different natural histories of senescence in related species, as well as the range of variations owing to different extrinsic factors, such as diet, population density, and social conditions. Most current experimental studies in gerontology deal with a few organisms (domestic mice and rats, fruit flies, soil nematodes, and soybeans) and human diploid cell lines. Although increasingly the same might be said for current developmental biology and genetics, these subjects have a much longer history of examining diverse models (Davidson, 1990). Consequently, the range of variations in developmental and genetic mechanisms among species is much better recognized than for most models in gerontology. Thus, I made a special effort to identify species that are not usually considered in gerontology, so as to better appreciate species variation in the range of senescent traits and life history scheduling. Moreover, comparisons between populations exposed to varying vicissitudes of nature and comparisons between species within an evolutionary radiation give many insights into the genomic basis for the plasticity of senescence. Besides the description of different types of senescence, mortality rate schedules by age group and survival curves are used to calculate age-related changes in mortality rates, which are emphasized as a major index of senescence.

Several experimental approaches have been used to analyze mechanisms of senescence. Transplantation of cells and organs between differently aged organisms (heterochronic transplantation) helps identify the intrinsicality of age changes. In rodents, for example, it is used to analyze the role of the ovary in reproductive senescence or the capacity of bone marrow for erythropoiesis. Gland ablation is used widely to identify the role of hormones in the senescence of plants and animals. Transplantation and ablation techniques have a classical position in this subject, much as in developmental biology. Treatment with hormones is used to evaluate the role of hormonal deficits during senescence as well as to accelerate some aspects of senescence that are thought to be hormone dependent. Diet has important effects on organismic senescence and the degenerative changes

---

1. Computerized literature databases were often helpful, particularly for recent biomedical publications. However, much of the older literature that I used is becoming progressively less accessible. Enormous space problems are faced by libraries, requiring storage of older volumes. Not infrequently, it took me months to get major works from earlier in this century that had been removed off site to less-expensive storage. While many libraries agree in principle to share materials with off-campus users, in reality there is no widely working system and access is erratic. Moreover, electronic cataloguing, recording on microfiche, or orderly relocation is rarely given high priority for infrequently requested archives. Within a generation, it seems, the library staff who know where they themselves located little-used materials will have passed on to other postings or into retirement. Many libraries are already storing volumes more than 25 years old. Besides early journals, many major books originally contained information not repeated in later editions (e.g. E. B. Wilson's *The Cell in Development and Heredity* [1896, 1925] and E. H. Davidson's *Gene Activity in Early Development* [1968, 1976, 1986]). Thus, much of the older literature may become generally unavailable. This loss of access to a most valuable literature from our predecessors portends a huge tragedy that future scholars may regard with the same sense of waste as we do the destruction of the libraries of ancient Alexandria.

of macromolecules and cells, as do drugs that influence metabolism and neurotransmission. These experimental approaches indicate great plasticity and few temporal constraints in the genomic control of life history changes.

While characterization of genes that influence senescence and lifespan is new, a sizable literature on inbred lines of many species show genotypic differences in senescence and lifespan. Recently, artificial selection for increased lifespan has succeeded in flies (Chapter 6.3.2). Much may be expected from ongoing transformational genetic studies, for example, the increased lifespans of flies transfected with plasmids containing additional copies of the gene for EF–1α (Shepherd et al., 1989; Chapter 7.4.3).

1.4. Mortality Rates and Lifespans

Senescence can be quantified using age-related changes in mortality rates in a population. Exponential increases of mortality rates as a function of age after maturation are widely observed in human and animal populations, and can be *the* major determinant of maximum lifespan. Because the incidence of most cancers and other age-related progressively degenerative diseases also increases exponentially with age, age-related accelerations of mortality rates give a basic measure of the rate of senescence. However, if the population average mortality rates (AMRs) are very high, an exponential increase of mortality rates may be obscured. Frequently in natural populations few adults survive to ages when senescence may be manifested. Thus, variations in lifespan are not necessarily statistically linked to variations in the rate of senescence. Moreover, because reproduction may cause increased mortality, characterizations of senescence require knowledge of the mortality rates across all adult ages. One cannot conclude that experimental manipulations of mean or maximum lifespan altered senescence without analyzing the age-related mortality rate acceleration throughout the lifespan.

Thus, I argue that the mean or maximum lifespan cannot by itself indicate the characteristics of senescence. Increases in mortality rate from senescence may be difficult to detect in wild populations under suboptimum conditions where overall mortality rates are very high because of malnutrition, infectious disease, social conflict, or predation. Such conditions often prevent enough animals from surviving to advanced ages for their cohort to manifest exponential increases of mortality. For example, adult birds in nature have high AMRs that often exceed 0.5/year (Lack, 1954, 1966; Botkin and Miller, 1974; Chapter 3.5) and severely limit the number that could survive long enough to manifest senescence (Section 1.4.8). In the robin (*Erithacus rubecula*) with an AMR of 0.6/year, as determined from banding and recapture studies, only 1/60,000 will survive to the recorded (apparent) maximum lifespan of 12 years (Lack, 1943a; Botkin and Miller, 1974).

Since other long-lived birds do not manifest age-related increases in mortality until twice that age (Botkin and Miller, 1974; Ollason and Dunnet, 1988), little can be said about the role of senescence in limiting the robin's apparent lifespan maximum. This view weighs against using maximum lifespans recorded for a few specimens living in zoos or recaptured in the wild after banding and release. However, such data are often used for species comparisons of lifespan potential and have figured prominently in many theories correlating physical and biochemical characteristics to lifespan (Chapters 5, 7, 8). Certainly, the age of the oldest survivor gives more insight about the rapidity of senescence than the mean lifespan (Simms, 1945; Sacher, 1959), but this still may be misleading.

An alternative to the use of mean or maximum lifespan in evaluating senescence is using mortality rate parameters according to the Gompertz and other power function models, which give a deeper quantitative basis for comparing different genotypes and species. The several different mortality parameters described below also can be used to evaluate whether the statistical lifespan has been altered without changing the characteristics of senescence. While mortality rates in a population are ultimately under genetic influence, much work remains to resolve how the genome influences the different mortality rate parameters. The next section describes graphic representations of mortality changes with age in human populations, followed by a simple mathematical treatment.

1.4.1. *The Gompertz Model for Mortality Rates*

A widely used population model for mortality rates and lifespan derives from classic actuarial observations by Benjamin Gompertz (1825, 514) that "the number of living corresponding to ages increasing in arithmetical progression, decreased in geometrical progression." This statement of survivorship as a negative power function of time indicates that the mortality rate accelerates with age in most populations that live long enough to show senescence. The Gompertz model has been the major mortality rate model in gerontology for more than 60 years.[2]

The Gompertz model is usually expressed as an exponential function, because the logarithm of the age-group-specific mortality rate (fraction of survivors that die in the next time interval) increases linearly with age over most of the adult life phase of humans and many other species. Graphs of the mortality rate against age on a semi-log scale (ln mortality on age) give a straight line from the ages of sexual maturity through the average lifespan (e.g. data on women from the 1980 U.S. Census shown in Figure 1.1, top). The slope of the line represents the acceleration of mortality rate with age (Gompertz coefficient G), which is an estimate of the rate of senescence. The corresponding equations are developed below. Figure 1.1, bottom, shows survival as a function of age by a direct plot of the data on a linear scale, giving a curve that approximates a rectangular shape. Similar graphs for the patterns of other human populations are shown in Figures 1.2 and 1.3.

Besides the slope given by the ln mortality rate plotted against age (Gompertz slope or Gompertz coefficient), other parameters are indicated. Inspection of human mortality rate changes shows families of curves displaced by constant increments at all adult ages, as well as decreases in the slopes of mortality in some populations during mid-life (Figure 1.2, top). Different birth cohorts show strong historical trends for a right shift of the ln mortality rates versus age within the last 150 years (Figure 1.2, bottom). Changes in slope during the adult phase may be smaller in populations with lower overall mortality, for example, U.S. (Figure 1.1) and Swedish populations (Figure 1.2). Figure 1.2, top, also

---

2. For historical and recent discussions of mortality statistics, see Brody, 1924, Pearl, 1928, Simms, 1945, Rosenberg et al., 1973, Sacher, 1977, Strehler, 1977, Comfort, 1979, Economos, 1982, Hirsch and Peretz, 1984, Kirkwood, 1985, Witten, 1983, 1984, 1985, 1988, and Hibbs and Walford, 1989. Other mortality models are discussed later in this section.

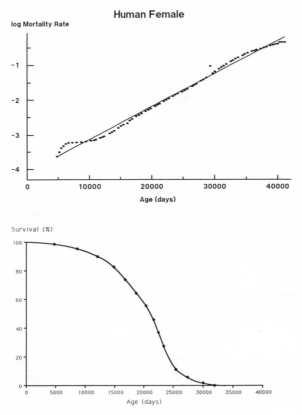

Fig. 1.1. *Top,* human mortality rate data plotted on a semi-log scale against age, showing that the logarithm (ln) of the age-specific mortality rate (fraction of survivors that died in the next time interval) increases linearly with increasing age (the Gompertz model, equation [1.2]). *Bottom,* graph of survival by age (survivorship plot, equation [1.4]). Mortality rate coefficients are given in Appendix 1. Data from the U.S. Census of 1980.

shows sharp decreases of mortality in all populations between birth and 10–15 years (puberty) that are treated separately below.

Description of mortality rates across the adult life phase requires parameters besides the Gompertz slope that are incorporated in the two age-independent mortality rate coefficients of the Gompertz-Makeham equation (Comfort, 1979; Sacher, 1977; Strehler, 1977; Witten, 1985; Finch et al., 1990):

$$\mathrm{m}(t) \;=\; A_0 e^{Gt} \;+\; M_0, \tag{1.1}$$

where $\mathrm{m}(t)$ is the mortality rate function of age, t, such that $\mathrm{m}(t)dt$ is the fraction of the surviving population that dies between t and $t + dt$;

G is the exponential (Gompertz) mortality rate coefficient, the slope of the line in Figure 1.1, top;

A_0 is a constant estimated by extrapolation to an early age, which the literature usually gives at the intercept of $t = 0$ (i.e. extrapolated to birth). In this formulation, A_0 is constant across age groups and is referred to as an age-independent mortality, and is sometimes called the "intrinsic vulnerability" (Sacher, 1977). However, A_0 can also be increased by environmental hazards. For reasons described below, I calculate A_0 at the age of puberty and refer to it in subsequent chapters for simplicity as the initial mortality rate (IMR);

and M_0 is the age-independent mortality rate constant of Makeham (1867), which is distinct from A_0 and which may also represent environmental influences.[3]

In regard to the Gompertz-Makeham equation, there is no *a priori* reason why a single exponential coefficient, G, should describe mortality rate changes across the adult lifespan, as presumed by the Gompertz model. Deviations of slope are often seen, particularly for survivors to very advanced ages, when mortality rates may decrease (Economos, 1979, 1982; Witten, 1988). While one could argue that the rate of senescence decreases at later ages, another view is that the survivors to advanced ages are a special subpopulation, with different mortality statistics. Thus, equation (1.1) is a statistical hypothesis that

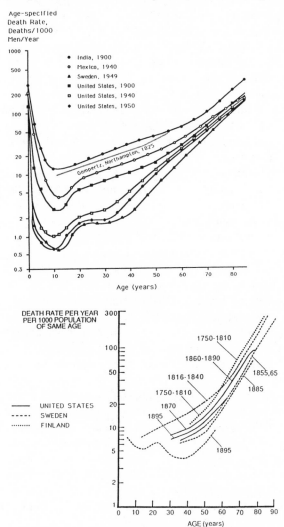

Fig. 1.2. *Top,* mortality rates as a function of age for different human populations, plotted on a semi-log scale. The curve designated Gompertz, Northhampton, 1825, represents the classic study of Gompertz (1825). The accelerations of slope in the earlier populations may represent larger values of the Makeham coefficient (M_0, equation [1.1]; Figure 1.4). *Bottom,* mortality rates as a function of age for different cohorts (presumably both genders) with the same approximate birth date. Both panels show that, despite large variations among populations in the initial mortality rates (IMR), the slope of the mortality rate varies little after mid-life (40 years) and is equivalent to a mortality rate doubling time (MRDT) of about 8 to 9 years. Redrawn from Jones, 1956.

3. Letters chosen for parameters in the Gompertz-Makeham equation vary among authors. I chose capital letters for the age-independent coefficients to avoid confusion with symbols preferred by population geneticists in the Euler-Lotka equation, equation (1.11) (Rose, 1991; Charlesworth, 1980, 1990a, 1990b).

accounts for most of the variance of mortality in large populations and within defined age limits. With few exceptions, organisms that manifest pathophysiological senescence in association with a characteristic maximum lifespan have exponential increases of mortality rate.

The Gompertz model is empirical and is not based on any principle or law that requires a particular relationship between mortality rate parameters and age. Occasionally, a linear model for age-related mortality rate increases has been used (Botkin and Miller, 1974; Miller, 1988), but the data scatter does not exclude an exponential function. Alternative to the Gompertz model, power functions like the Weibull model are also used to describe population senescence: $m(t) = At^{c-1}$. As shown for several invertebrates, the fit of the Gompertz and Weibull models varies among populations (Slob and Janse, 1988; Hirsch and Peretz, 1984). Graphs of $\ln(m[t])$ against age give a straight line with the Gompertz model, but a concave curve with the Weibull that represents slowing of the mortality rate at advanced ages. The Gompertz analysis is emphasized here because of precedence, through the present, in analysis of animal mortality data, and the virtues of a simpler graphic representation with directly obtained parameters of mortality that represent averages across the adult lifespan. Over the adult range the fitted Gompertz straight line closely approximates the fitted Weibull, at least for present data sets. The choice of mortality functions is also discussed by Rosenberg and Juckett (1987) and Witten (1987, 1988). Hopefully the present analysis of mortality data will stimulate a thorough inquiry on the generality of mortality rate accelerations, using the vast and often unpublished archive of mortality schedule data from field studies. As discussed in Section 1.4.7, the extent to which the maximum reported lifespan is less than that predicted if the mortality rate remained constant throughout the lifespan gives an approach to characterizing whether mortality rates accelerate by the Gompertz or other models.

Data on mortality rates by age are of several types. The most desirable are lifelong mortality data for particular birth cohorts that give individual lifespans up to the last survivor as a function of age, from which mortality statistics can be directly calculated by maximum likelihood estimates. Such data are rarely published because of space constraints in the journals and are hard to obtain for most species; there is a major need for an on-line database containing primary mortality data for a wide range of human and animal populations. The next best source is mortality schedule or actuarial table data of mortality rate by age group in the cohort. Even these are available for few species besides humans and laboratory animals such as rodents, flies, and soil nematodes. Total population mortality (cross-sectional) data by age group are more widely available (Figure 1.3), but are less desirable because cohorts are susceptible to important fluctuations in infections, stress, etc. Lacking these sources, mortality rates must be in general extracted from survival curves like Figures 1.1 (bottom) and 1.3 through graphic partitioning of the survival curves in arbitrary intervals. Because survivorship curves are often based on small samples, the curves may not be smooth, and the resulting mortality rate estimates are not very precise. Even so, in general the derived regressions of mortality rate by age are strongly linear, with slopes that are remarkably constant between diverse populations of the same species (Appendix 1).[4]

4. For details on the mortality rate estimates, see footnotes to each species in Tables 2.1 and 3.1. Because mortality rate acceleration is such a powerful trend with age, the correlation of ln mortality on age *must* be strong; the particular values of correlation coefficients are therefore not informative.

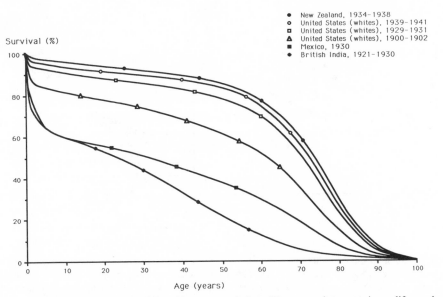

Fig. 1.3. Survival plotted by age for different human populations. The approach to a maximum lifespan is not shown, since the authentication of advanced ages is uncertain. Nonetheless, it seems likely that all populations have exceptional survivorship into the range 90–110 years. Redrawn from Comfort, 1979, 6.

1.4.2. Age-independent Mortality Coefficients of the Gompertz-Makeham Equation

Next I consider the two nonexponential coefficients of equation (1.1). The age-independent mortality rate coefficients A_0 and M_0 have distinctly different impacts on mortality rates at any age. As modelled in Figure 1.4, top, variations in A_0 (holding G and M_0 constant) yield a set of parallel lines with different intercepts at $t = 0$. In contrast, variations in M_0 (holding G and A_0 constant) yield a fanlike set of curves that converge towards a single line at later ages (Figure 1.4, bottom). According to this model, the Makeham constant M_0 has a diminishing impact on the mortality rate function at later ages. This is because the accelerating mortality rate from advancing senescence progressively dominates the contribution from environmental risk factors. In contrast, the intrinsic vulnerability constant A_0 multiplies the exponential effect of mortality throughout adult life, so that variations of A_0 yield the same shift of mortality on semi-log plots at early or later ages. Figure 1.2 shows examples from human populations that fit these cases. Various national populations show remarkably parallel mortality rate lines at later ages, despite a threefold range of absolute mortality rates. This suggests threefold variations in A_0 like those modelled in Figure 1.4. There are also increases of mortality rate slopes (e.g. in colonial India in Figure 1.2, top), so that after 50–60 years of age the curve parallels the other populations. This decrease of slope could represent a relatively large Makeham constant M_0, with a diminishing impact on total mortality rate at later ages as modelled in Figure 1.4, bottom.

Despite the recognition for decades of A_0 and M_0 as distinct parameters, there has not been any analysis of real data to resolve their relative contribution in different populations. While A_0 has been estimated for a few mammals by back extrapolation to $t = 0$ (Sacher, 1977), this value would also represent contributions from M_0 (Figure 1.2). Such data are

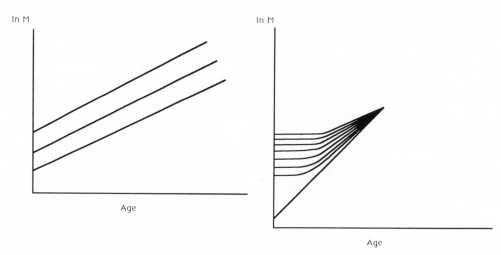

Fig. 1.4. Influences of mortality rate coefficients on the mortality rate functions, modelled according to the Gompertz-Makeham equation (equation [1.1]). *Left,* with different values for the age-independent mortality rate coefficient, A_0, holding constant the Makeham age-independent coefficient M_0 and the Gompertz mortality rate coefficient G. *Right,* with different values for M_0, holding constant A_0 and G. Redrawn from Witten, 1985.

complicated by genetic diversity as well as myriad environmental factors, which in humans, for example, may differ among various ethnic and socioeconomic subgroups. The values for A_0 and M_0 in defined genotypes should be studied with inbred animals exposed to specific pathogens and diets at different times in development and after maturation. As described in Chapter 10 (Table 10.1), restricted diets that considerably increase mean and maximum lifespans in rodents influence both the Gompertz slope and the age-independent mortality rate.

It is hard to imagine an environment that eliminates all extrinsic factors of risk. For example, cesarean-derived mice and rats can be kept in a "germ-free" environment without bacterial infections or parasites. The incidence of kidney disease (Gordon et al., 1966) or hepatoma (Pollard et al., 1989) at later ages may not be different from controls, but prostatic abscesses were eliminated (Pollard et al., 1989). However, germ-free mice acquire a new mortality risk, because their flaccid intestines often become fatally twisted (intestinal volvulus; Gordon et al., 1966); also, cesarean derivation does not necessarily exclude germ-line transmissible viruses that in some mouse strains cause a senescent-like syndrome (Chapter 6.5.1.7.2). The different mortality rate curves in Figure 1.2 suggest that meaningful estimation by extrapolation of the minimum A_0 values requires evaluation of a number of populations living in different environments.

To summarize, Gompertz plots, on a semi-log scale, of adult mortality rate versus age indicate the extent of senescence in a population. When the mortality rate curve increases to approach a straight line with a positive slope after maturity, then manifestations of senescence may be expected in individuals of the population. The slope of the line is the Gompertz exponential coefficient G (equation [1.1]). When the mortality rate line is horizontal, senescence may not yet be manifested; this condition is like radioactive decay. When the mortality rate line has a negative slope such that mortality decreases with age, this may indicate subpopulations with different rates of maturation, since mortality risks usually decrease between birth and maturity and then again decrease during adult life as senescence becomes manifest. The age-independent mortality coefficients A_0 and M_0 modify the intercept of the line at age 0, while M_0 has the additional effect of reducing the slope (making the line more horizontal) at younger ages. Whatever the contributions of the age-independent mortality coefficients may be at younger ages, they do not change the slope at later ages when senescence is generally manifested. Thus, the slope of the ln mortality rate versus age is a key parameter for analyzing the impact of experimental manipulations on senescence.

1.4.3. Issues of Subgroups with Different Mortality Patterns

At the extreme ages (before maturation or beyond the average lifespan at the tip of the survival curve), the picture of age-related mortality rates becomes more complex. Neither in nature nor in the most carefully maintained laboratory colony does the mortality rate change monotonically across the postnatal lifespan, as is implied by equation (1.1). In all human populations, the observed mortality rate is large at early postnatal stages, then decreases to a minimum at about the age of sexual maturity (Figure 1.2). Analogous bidirectional changes in the mortality rate function with age also are seen in populations of fish (Egami, 1971; Beverton and Holt, 1959), birds (Lack, 1943a, 1943b), and numerous mammals (Deevey, 1947; Spinage, 1972; Caughley, 1977; Ralls et al., 1980). The initial, age-independent mortality rates that are estimated by extrapolating back to $t = 0$ thus are usually much smaller than the observed mortality rates of neonates. At advanced ages beyond the mean lifespan, the mortality rate acceleration may slow slightly, as described for humans and *Drosophila* (Economos, 1979, 1982; Witten, 1988). Multiple exponential functions can represent these complex changes of mortality rates with age (Witten, 1986). Moreover, an ongoing analysis of human populations distinguished by different fatal diseases suggest subgroup differences in the age-related disease incidence (Juckett and Rosenberg, 1988). Thus the Gompertz model is best regarded as an approximation which does not preclude a diversity of subpopulations with different changes in mortality risks as a function of age.

Two major factors may account for differences in mortality rate changes with age between populations: (1) physiological changes; and (2) effects of subgroups that differ by genetic polymorphisms, by exposure to various traumas, or by developmental deviations which need not be under genetic control. The same biological mechanisms cannot be assumed to determine the risk of mortality across all stages of life. The age-group specificity of diseases is shown by multiphasic changes in the incidence of specific infectious and malignant diseases, from childhood through old age. Thus, on the basis of only physiological changes it is not surprising that the mortality risk would change in complex ways.

Superimposed on such changes in the mortality rate slopes may be influences traced to various subgroups that can be differentiated by genotype (e.g. gender, polymorphisms) or by environmental factors (e.g. exposure to local infections). In some species, mortality rates may increase sharply during reproductive activities. For example, in bighorn rams

(*Ovis canadensis*) mortality is stable at about 0.04/year between sexual maturity at 2 years of age and age 7–8, when rams reach their maximum size and begin to fight more intensely (Geist, 1971). After this "behavioral maturation," mortality increases at least fivefold, but for causes that have nothing to do with pathophysiological senescence. For it is the oldest rams, the survivors to 13 years or more, that are the dominant leaders and the most active breeders. Their mortality rates at later ages have not been characterized, but presumably show further accelerations as in all other mammals. Female bighorns do not show as marked age-related changes in mortality during their adult years, although mortality in breeding females may be greater than in nonreproducers because of the nutritional drain from bearing and nursing (reproductive costs are discussed further in Section 1.5). The female mortality rates are slightly higher than for young males. Males and females have similar maximum lifespans in nature, about 20 years (Geist, 1971). This example of early increases in mortality that are unrelated to functional senescence emphasizes the need for detailed mortality schedules for both genders, as well as knowledge of the natural history in order to interpret mortality patterns in species and gender subgroups. In many other species, the mortality rate also changes during reproduction and other activities that are under neural and endocrine control and that influence life expectancy and maximum lifespan (Chapters 2, 3).

Another type of subgroup may originate from childhood diseases that inflict permanent organ damage and increase adult mortality risk. A classic case is the rheumatoid fever of children and the subsequent rheumatic heart disease of adults which is associated with increased adult mortality risk (Chapter 9.3.5). Correlations between A_0 and G for various national groups (Strehler and Mildvan, 1960) could also reflect subgroups with different effects of early environment on adult mortality risks. The relationships between neonatal and adult mortality rate parameters have not been studied in detail and could give much insight about environment-genotype interactions during development that may influence adult longevity. There is no general approach to the identification of subgroups with distinct mortality rates at advanced ages, and they pose a major enigma. This problem is particularly great in longer-lived, gradually senescing species such as mammals, which differ widely among individuals in type and extent of dysfunctions, even within inbred rodent populations (Chapter 3).

Certain highly adverse conditions during adult life have little impact on the slope of the mortality curve and mostly increase the constant A_0. Figure 1.5, top, shows the astonishing conclusions of Jones (1959) and Bergman (1948) that Australian prisoners in Japanese concentration camps during World War II had Gompertz curves with slopes that were indistinguishable from two civilian populations. Despite the constant slope, the mortality rates range thirtyfold between these populations at any adult age shown. Similar parallelism in mortality curves is shown by Dutch civilian populations during and just after the Nazi occupation (Figure 1.5, bottom). Graphs of this larger data set suggest an increased slope of the mortality curve before mid-life as well as the upward shift of the whole curve, which could be interpreted as a larger contribution of A_0 than seen in the populations of Figure 1.5 (Appendix 1). The parallelism of these lines on semi-log plots requires that the basal mortality rate remain fixed in proportion to the acceleration of mortality. A provisional conclusion from the World War II data is that dire stress lasting several years does not influence the rate of senescence during that time; subsequent effects on longevity have

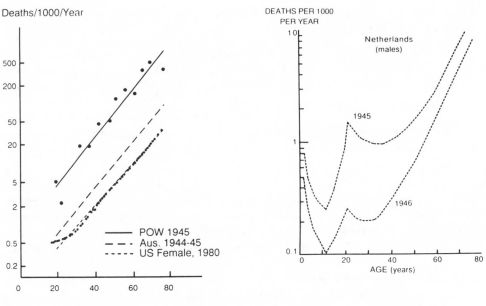

Fig. 1.5. Mortality rates as a function of age in human populations subjected to prolonged dire stress duress, as analyzed by the Gompertz mortality rate model (equation [1.2]), in which the ln age-specific mortality rate is plotted against age. The slope of this line estimates the Gompertz mortality rate coefficient G, from which is calculated the MRDT (equation [1.3]). The extrapolated value of the line at age 15 years estimates the IMR, an age-independent mortality rate term. The mortality coefficients are given in Appendix 1. The MRDTs of the five samplings are indistinguishable (range, 7.6–8.9 years), while IMR ranges 35-fold (0.04×10^{-3} to 1.4×10^{-3}/ year). *Left, POW,* Australian prisoners of war held in concentration camps by the Japanese Army during 1945; *Aus,* civilians in Australia, 1944–1945; *U.S. female,* white women in 1980 census. The POW data of Bergman (1948) as analyzed by Jones (1959) are replotted here for ages up to 75 years, and show a major increase in overall mortality due to an increased IMR, without influencing the mortality rate slope. *Right,* shifts in IMR without effect on the mortality rate slope occurred in Netherlands male civilians in 1945 versus 1946 during and after World War II. Redrawn from Jones, 1959.

not been analyzed. While little is known about the impact of various environmental factors during development on subsequent mortality rates, the parallelism of mortality rate curves for different national populations (Figure 1.2) suggests that adverse environments throughout the lifespan do not strongly influence mortality rate accelerations.

The independence of the coefficients A_0 and G suggested above is also indicated in mouse strains which have wide variations in age-related degenerative diseases and in the values of A_0, in contrast to G (Table 6.2). In view of the correlation of A_0 with G among different human populations (Strehler and Mildvan, 1960), further studies are needed to reveal the situations where such correlations may occur. A major question concerns the genetic determination of the different mortality rate coefficients (Chapter 6). The relative stability of the Gompertz mortality coefficient G in different environments for a variety of species gives a provisional basis for comparing data from environments that cannot be described in detail.

1.4.4. The Working Model: The Initial Mortality Rate (IMR) and the Mortality Rate Doubling Time (MRDT)

A_0 and M_0 have not been resolved for any species or population, and the data may be inadequate to do so in most cases. Therefore, I have chosen as a simpler working model the Gompertz equation without the Makeham coefficient M_0, for which a single age-independent coefficient, A_0, is calculated at the age of puberty and is designated as the initial mortality rate (IMR). By puberty, the neonatal mortality component has sharply declined. The IMR represents contributions from both A_0 and M_0 and is calculated from the linear regression of ln $(m[t])$ on t, with values of t calculated at the age of puberty:

$$m(t) = Ae^{Gt}. \tag{1.2}$$

This simpler and widely used approximation of equation (1.1) (Strehler, 1977; Sacher, 1977; Johnson, 1987) facilitates further derivations and calculations. No conclusions can be drawn about the relationship between IMR and G, which will be treated as empirical parameters. The pathophysiological substrata need not be the same for any of these coefficients.

Comparison of species using the Gompertz model is helped by calculating the mortality rate doubling time (MRDT). In preference to G, which varies inversely with lifespan potential, the MRDT is a more natural unit for mortality rate acceleration since it is in the same direction as the lifespan and is measured in the same units of time (Sacher, 1977; Kirkwood, 1985). Equation (1.2) is easily solved for MRDT:

$$\text{MRDT} = \frac{\ln 2}{G} = \frac{0.693}{G}. \tag{1.3}$$

Representative species values of IMR, MRDT, and the observed maximum lifespan are given in Table 1.1. A summary by phylum of all values throughout the book is in Appendix 1. Humans have the slowest MRDT among species with a defined *tmax*, while that of short-lived flies, nematodes, and rotifers is about a thousandfold greater. The IMR often varies widely among populations, which makes it less specific for the species than the MRDT, which is far less variable. The range of IMR among populations may nonetheless be biologically significant, for example, to the force of selection (Section 1.5).[5]

The wide occurrence of exponential increases of mortality rate throughout the phyla has long been recognized as a remarkably general biological phenomenon and has led to many proposals of universal or general mechanisms in senescence. How can the pan-phyletic generality of exponential increases in mortality be explained *except* by a shared fundamental cellular change, one might ask? A widely discussed hypothesis proposes that senescence is caused by an accumulation of somatic mutations or other informational errors (Szilard, 1959; Orgel, 1963, 1970; Holliday and Tarrant, 1972; Burnet, 1974). But, as described in Chapters 7 and 8, there is little evidence that accumulated point mutations

---

5. In reference to the choice between the Gompertz and Weibull distributions, the mortality rate constant G increases slightly at later ages in the Weibull but remains constant in the Gompertz. Therefore, the MRDT decreases slightly at later ages in the Weibull, which would make species comparisons more awkward than with the Gompertz, which gives a single value for MRDT. Use of the Weibull would require calculating an average MRDT (Finch et al., 1990).

Table 1.1 Representative Values of Mortality Rate Coefficients

Species	IMR/year (initial mortality rate)	MRDT, years (mortality rate doubling time)	*tmax*, years (maximum lifespan)
Human[T3.1]	0.0002	8	>110
Domestic dog[T3.1]	0.02	3	20
Laboratory mouse[T3.1, T6.1]	0.01	0.3	4–5
Laboratory rat[T3.1]	0.02	0.3	5–6
Pipestrelle bat[T3.1, A2]	0.36	3–8	≥11
Herring gull[T3.1, A2]	0.2	6	49
Brush turkey[T3.1]	0.05	3.3	12.5
Japanese quail[T3.1]	0.07	1.2	5
Fruit fly[T2.1]	1	0.02	0.3
Aphagous moth[T2.1] (Automeris)	10	0.005	0.03
Soil nematode[T2.1]	2	0.02	0.15
Rotifer[T2.1] (Lecane, female amictic)	6	0.005	0.10

Note: Further information on these species is in the notes to tables 2.1, 3.1, 6.1, and Appendix 2, as indicated by the superscripts T2.1, T3.1, T6.1, and A2.

or other types of errors in nuclear genomic information flow are a basic mechanism in senescence.

A major thesis of this book is that the diverse but characteristic species differences in the pathophysiologic mechanisms of senescence restrict the generalizability of random genomic damage as a major mechanism of senescence. Organisms with similar mortality rate coefficients can have radically different pathophysiological changes during senescence (Chapters 2, 3, 4), particularly in species manifesting rapid senescence (Chapter 2). Eukaryotes differ at least as much in their patterns of senescence as they do in development. Thus, comparisons of age-related changes in species that differ widely in development, physiology, or other major characteristics need careful justification.

The difficulties in making inferences about mechanisms of senescence from mortality data are amusingly shown by machines that have rectangular survival curves and exponential increases of dysfunction ("mortality"), such as the electrical relays shown in Figure 1.6 (Haviland, 1964). On the other hand, the survival of glass tumblers approximates an age-independent mortality rate with exponential loss, an example used by Medawar (1952) and others (Brown and Flood, 1947; Comfort, 1979). Interestingly, if tumblers were toughened by heat annealing, they showed a senescent survival curve, because abrasions at the lip of the tumbler increased the likelihood of subsequent cracking. Stochastic theories of mortality based on random damage or wearing out of "critical elements," as in parts of machines, have been derived with Gompertzian properties (Abernethy, 1979; H. A. Johnson, 1985; Witten, 1985; Hibbs and Walford, 1989).

These inanimate models suggest an insight about the universality of exponential increases in mortality with age. The connectedness of mechanically integrated systems, either through distinct parts, as in the relay of Figure 1.6, or through materials with differ-

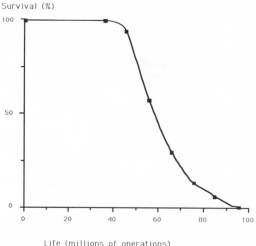

Survival (%)

Fig. 1.6. Survival of functions in electrical relays plotted as a function of the number of operations (events). Data of W. H. Lesser, *Electronic Design* (October 24, 1958).

Life (millions of operations)

ent properties, as in the glass tumblers, would appear to amplify the effects of subthreshold lesions as they accumulate. This emergent trait may be similar in physiological interactions of multicellular organisms. Effects of subthreshold injury or lesions might also be looked for in complex subcellular organizations, such as occur in neurons and in unicellular organisms like ciliates. Intracellular critical elements for cascading injuries might involve various polymerized connections that transmit mechanical forces and compartments that differ in their molecular and particulate transport properties.

1.4.5. *Survival Curves and the Maximum Lifespan*

Next I consider the shapes of the survival curves (Figures 1.1, bottom, 1.3). The proportion of the population surviving to age t, $s(t)$, is obtained by integrating the survival function

$$\frac{ds(t)}{dt} = m(t)s(t), \tag{1.4a}$$

where $m(t)$ is equation (1.2). The result is the Gompertz survival model,

$$s(t) = \exp\left[\left(\frac{A}{G}\right)(1 - e^{Gt})\right] \tag{1.4b}$$

If the initial age-independent mortality rate constant A is low, so that A is less than G, mortality is low during most of adult life. Survival then decreases sharply only towards the end of the lifespan, resulting in nearly rectangular survival curves, as observed for populations under optimal conditions such as in developed countries (Figure 1.3) or in healthy populations of laboratory animals. A low mortality rate until after maturity with subsequent exponential increases in mortality rate is the most reliable indication for an age-related process of senescence. Influences of these variations on survivorship curves are modelled in Figure 1.7, top, where the age-independent mortality coefficient A (equation [1.1]) is varied without changing G. Note that as A becomes greater than G, survival decreases nearly exponentially as a function of age (Figure 1.7, bottom). In feral populations with high natural attrition (high IMR), it is often difficult to detect age-related increases in mortality rate, as observed in long-lived birds. (Another approach to this is given in Section 1.4.7.) Moreover, the apparent maximum lifespan for small populations of a negligibly or slowly senescing species that has a high value of IMR in nature even under optimum conditions can easily be shorter than for more rapidly senescing species, as modelled in Figure 1.7. Others have noted the importance of the age-independent IMR to the lifespan potential (Simms, 1945; Edney and Gill, 1968).

In conclusion, *both* the MRDT and the IMR must be calculated to establish whether natural or experimentally induced differences in lifespan result from alterations in the rate of senescence. Few report these or equivalent mortality rate coefficients or give data in a format such that they could be easily calculated. Most merely provide mean lifespan with some variance estimates. Given the data on humans and on the snail *Lymnaea*, the IMR might vary widely among studies, even those from the same laboratory, and it would be helpful to know how widely. Without information about possible changes in IMR, it cannot be concluded that experimental manipulations of lifespan have altered senescence.

1.4.6.   *Evidence for Delays in the Onset of Senescence*

I propose another factor that may be useful in describing the age-related increases of mortality. Populations may vary in the age when they manifest a given rate of mortality.

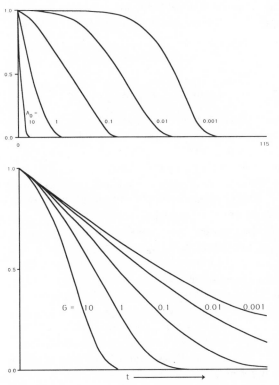

Fig. 1.7. Survival curves according to the Gompertz model (computed from equation [1.4]). *Top*, survivorship vs. age, plotted with different values of the age-independent mortality coefficient A_0 and with the Gompertz mortality coefficient G held constant. The linear ordinate scale does not show clearly the asymptotic approach to zero survival. *Bottom*, survivorship vs. age, with different values of G and with A_0 held constant. Redrawn from Witten, 1985.

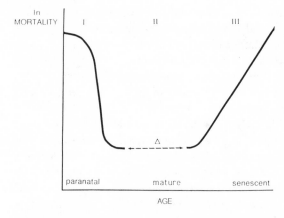

Fig. 1.8. A model of mortality rate changes showing three phases of mortality: *I*, paranatal mortality decreases sharply to a low value. *II*, a hypothetical phase lasting Δ with stable mortality rates. *III*, a senescent phase with rapidly increasing mortality. This model fits certain plants and invertebrates that have prolonged free-living phases before maturation and sudden death, e.g. monocarpic perennials (Chapter 2.3.2), 17-year cicadas (2.2.3.3), and insect queens (Chapter 2.2.3.1). The mortality rate curves in Figure 1.2 might also be interpreted by this model.

The comparisons of different cohorts (Figure 1.2, bottom) suggests another mortality parameter (delta, Δ): the interval between maturation and the onset of the Gompertz exponential mortality increase. Delta would constitute a latent phase in senescence before senescence-related increases of mortality (Figure 1.8). Although high values of the Makeham coefficient M_0 (equation [1.1]) can simulate a delay before the acceleration of mortality is detected (Figure 1.4, bottom), indefinite maintenance of low mortality rates is not readily modelled without another parameter that separates the mortality function into time domains. Clear examples of delays are the sudden deaths of social insect queens after many years of fertile adult life (regicide by worker castes, apparently upon exhaustion of sperm stores) and the multi-decades-long juvenile phase of bamboos, which is followed by flowering and rapid senescence (Chapter 2.3.2). Possible latent phases before the onset of senescent increases of mortality in adult human populations are purely speculative.

1.4.7.   *Estimates of Mortality Rates from the Maximum Lifespan*

A major frustration is the dearth of age-group-specific mortality rate data needed to calculate MRDT, particularly for feral species like birds where only the oldest surviving age or maximum lifespan is reported (Clapp et al., 1982; Klimkiewicz and Futcher, 1987; Zammuto, 1986). Several note the imperceptibility of age-related increases in mortality rate in some birds with long maximum lifespans (Lack, 1954; Buckland, 1982; Dunnet and Ollason, 1979; Ollason and Dunnet, 1988). Yet some short-lived birds clearly show Gompertzian mortality rate accelerations, for example, the Japanese quail (*Coturnix*) and the brush turkey (*Alectura;* Figure 1.9). In a pioneering analysis, Botkin and Miller (1974) showed that the maximum reported lifespan in very large bird populations is less than predicted from constant mortality rates. Extending these studies, my colleagues Malcolm Pike and Matthew Witten and I showed how to estimate the Gompertz mortality rate coefficients from the average (adult) mortality rate, population size, and maximum reported lifespan when mortality schedule data by age group are lacking. Our approach, derived from equation (1.4) (Finch et al., 1990), required no assumptions about body size, in contrast to that of Calder (1983a), which is otherwise similar.

For a population of size *N*, the age when the population has diminished to one survivor (s[t] = 1/*N*) approximates the maximum lifespan (*tmax*). Thus,

$$s(tmax) = \frac{1}{N} = \exp\left[\left(\frac{A_0}{G}\right)(1 - e^{Gtmax})\right], \qquad (1.5)$$

or,

$$tmax = \ln\left[1 + G(\ln N)/A_0\right]/G \qquad (1.6)$$

From Gumbel (1958), the average mortality rate, AMR, of a steady-state population subject to the age-group-specific mortality rates of equation (1.2) is known to be

$$AMR = 1/\int_0^\infty s(t)dt . \qquad (1.7)$$

For a given AMR, *tmax,* and *N,* equations (1.6) and (1.7) can be numerically solved for IMR and *G.* The MRDTs were then calculated for a range of *N* (Appendix 2).

By these calculations, the pipestrelle bat (*Pipistrellus*) has a long MRDT that approaches values for humans (Table 1.1, Appendix 2) and that would not be predicted from its 90% smaller apparent maximum lifespan of 11 years. Bats in general have *tmax*'s that are severalfold higher than those of other mammals of the same size (Jürgens and Prothero, 1987; Austad and Fischer, in press); the absence of degenerative joint disease at 19 years of age in another feral bat (also a vespertilionid; Sokoloff, 1969) also indicates slow senescence. Using the data sets on feral birds from Botkin and Miller (1974), at least five species have long MRDTs (Table 3.1; Appendix 2), close to those for humans, the pipe-

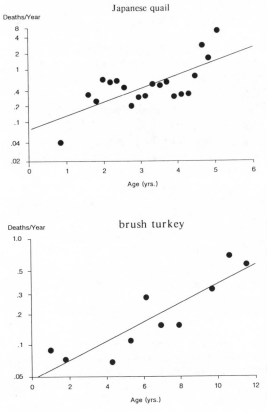

Fig. 1.9. Mortality rates graphed on a semi-log scale as a function of age for two short-lived species of birds, Japanese quail (*Coturnix coturnix japonica,* laboratory population, males) and brush turkey (*Alectura lathami,* zoo population), which show good fits to the Gompertz model (equation [1.2]). Consistent with the rapid accelerations of mortality rate, Japanese quail have numerous manifestations of senescence (Chapter 3.5). The extent of senescence in brush turkeys is unknown. Mortality coefficients are in Tables 1.1 and 3.1. Data of Cherkin and Eckhardt, 1977, Woodard and Abplanalp, 1971, and Comfort, 1962, as analyzed by Finch, Pike, and Witten, 1990.

strelle bat, and the rhesus monkey. In the case of herring gulls, this approach could be validated from data on mortality by age group using equation (1.3), which gave similar MRDT and IMR (Table 3.1; Appendix 1). Consistent with these slow MRDTs is the slow rate of reproductive senescence in many marine birds, some retaining fertility for several decades (Chapter 3.5). These calculations are sensitive to uncertainties about population size and the age of the oldest survivor. Nonetheless, the MRDT is probably underestimated because the *tmax* will almost certainly increase further.[6]

This analysis gives two major conclusions. First, short *tmax*'s do not rule out long MRDTs. This conclusion is important for predictions from population genetics theory (Section 1.5.2) that high mortality will permit the accumulation of germ-line mutations with delayed effects that should cause fast rather than slow rates of pathophysiological senescence. Long MRDTs are incompatible with rapid pathophysiological senescence. This conclusion reemphasizes concerns about predicting the rates of senescence from *tmax* in the absence of data on IMR and MRDT. Secondly, small size does not preclude slow MRDTs: adult pipestrelle bats weigh less than twenty grams (Nowak and Paradiso, 1983). Slow MRDTs are also seen in some small but long-lived birds that survive beyond 20 years (Table 2.1). Two analyses show that *tmax*'s of bats (Jürgens and Prothero, in prep.; Austaa and Fischer, in press) and birds (Prothero and Jürgens, 1987) are much longer than for other small-sized species (mostly rodents). This result departs from correlations between body size and *tmax* in many birds and mammals (Chapter 5).

1.4.8. Initial Mortality Rates (IMR) and the Lifespan Potential

The greater mean lifespan of humans (75 years) versus laboratory mice (2.5 years) or flies (0.1 year) under good conditions results from species (genotypic) differences in both IMR and MRDT. The IMR in humans is typically about a hundredfold less than in mice and a further fiftyfold less than in flies (Table 1.1). The typically high vulnerabilities of the short-lived invertebrates require that even if senescence failed to occur, the *tmax* must be short. The strong impact of IMR on *tmax* was shown by Simms's (1945) thought-provoking calculation. Assuming that a constant fraction of the population dies per unit time ($G = 0$, equation [1.2]), then the survival curve is given by the familiar zeroeth-order reaction rate equation (analogous to exponential decay of an unstable radioisotope):

$$dN = -NR dt. \tag{1.8a}$$

By integration,

$$N(t) = N_0 e^{-Rt}, \tag{1.8b}$$

where N(t) is the number surviving at age t; N_0 is the initial population size; and R is the initial mortality rate coefficient (IMR, Section 1.4.4).

---

6. Some values of AMR assumed by Botkin and Miller (1974) did not permit the reported *tmax*. For example, the AMR = 0.2/year for the arctic tern (not shown) did not give the observed survival to *tmax* of 31.2 years with populations up to 10^6 in size. A lower estimate, AMR = 0.1/year, gave the reported *tmax* and may be more realistic. This analysis gives a new validation of AMR and *tmax* from banding studies (Finch et al., 1990).

The time, $t_{0.5}$, required for loss (decay) of 50% of the population (i.e. median lifespan) is

$$\frac{N}{N_0} = 0.5 = e^{-Rt}.$$

Then, solving for $t_{0.5}$,

$$t_{0.5} = \frac{\ln 0.5}{-R} = \frac{0.693}{R}. \qquad (1.9)$$

Assuming a constant IMR for the rest of life (i.e. without senescence) at the minimum value observed in the most-favored human population at the age of 10–15 years (5×10^{-4} deaths/year; Figure 1.2), then the age at median lifespan (50% dead) would be about 1,200 years. This is only tenfold greater than the present maximum human lifespan. Similar calculations for mice show that in a typical laboratory colony in the absence of senescence, the median lifespan would be about 30 years, also about tenfold longer than the usual maximum lifespan. For *Drosophila* and probably many other invertebrates, the median adult lifespan in the absence of senescence would be about 0.5 year. This result is important to certain evolutionary theories of senescence, as discussed in Section 1.5, because most short-lived organisms are subject to mortality risks that severely restrict their mean lifespans even if there were no senescent increases of mortality rate. The relatively few individuals that statistically survive to later ages thus weaken the effects of natural selection on phenomena manifested at later ages, according to the analysis of Medawar and others.

Another form of equation (1.8b) estimates the minimum lifespan of the last survivor in a population of size N_0 (Simms, 1945). Solving for t yields

$$t = -R \left(\ln \frac{N}{N_0} \right). \qquad (1.10)$$

For the present world population of 4 billion, with the above value of R for a favored population, the last survivor would live at least 25,000 years in the absence of senescence. Calculating for a population of 1 billion, the last surviving mouse would live beyond 900 years and the last fly beyond 6 years. Thus, in the hypothetical absence of senescence, the environmental risk factors for most species are great enough to increase the median lifespan by a smaller multiple than the present differences between mice and humans.

The high values of age-independent mortality risks in natural populations have many implications for theories of senescence. Under conditions where normally $A > G$ (equation [1.3]) and few survive to advanced ages, senescence and the maximum lifespan could not be subject to strong selection. This perspective also emphasizes that any genetically determined effects of senescence are superimposed on age-independent mortality risks that must also be genetically determined. For example, the IMR of rhesus monkeys raised under apparently optimal conditions is nonetheless at least tenfold higher than that of humans subjected to dire conditions during war (Appendix 1). Possible exceptions are plant and animal species that do not manifest reproductive senescence or mortality rate increases (negligible senescence, Chapter 4). The hypothesis that a crucial aspect of the genetic influence on potential lifespan arises from determinants of IMR is considered in Chapter 6.

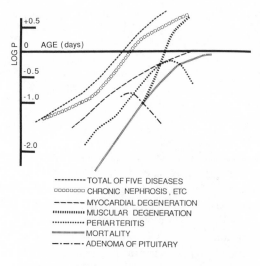

Fig. 1.10. The age-related increases of mortality and incidence of specific chronic diseases in male Sprague-Dawley rats that have closely similar exponential accelerations with age. Ln *P* is probability of occurrence of diseases or death. The diseases reported at necropsy in this early study continue to be found in roughly similar incidence in Sprague-Dawley rats. Also see Figure 3.15. Redrawn from Simms and Berg, 1957.

---------- TOTAL OF FIVE DISEASES
□□□□□□□□ CHRONIC NEPHROSIS, ETC
– – – – – MYOCARDIAL DEGENERATION
""""""""""" MUSCULAR DEGENERATION
············ PERIARTERITIS
""""""""""""" MORTALITY
–·–·–·– ADENOMA OF PITUITARY

1.4.9. Mortality Rates and Age-related Diseases as Measures of Senescence

The constancy of the species-characteristic MRDT or Gompertz mortality rate coefficient in human populations throughout the world and in different laboratory rodents poses a major puzzle in view of the very different incidence of age-related diseases that might be expected to differentially affect mortality. As noted above, despite major differences among genotypes in age-related diseases, most inbred mouse strains have indistinguishable MRDTs (Table 6.1). In some cases, the age-related incidence of many diseases associated with senescence parallels the acceleration of mortality rate (Simms and Berg, 1957; Figure 1.10). Another example is laboratory populations of the snail *Lymnaea stagnalis,* some of which had infections that elevated the overall mortality rate and caused major differences in mean and maximum lifespan; nonetheless the MRDTs were relatively constant (Slob and Janse, 1988). Similarly, older human populations across the world differ in incidences of malignant diseases, diabetes, hypertension, strokes, etc. (Jones, 1956, 1959; Silverberg and Lubera, 1986), yet have indistinguishable MRDTs. Deaths attributed to cardiovascular disease continue to increase exponentially and in parallel with total mortality (Figure 1.11). However, other diseases may not conform to this pattern; for example, recent data from the United States indicate that the overall incidence of cancer does not increase exponentially at advanced ages and may even decrease after 80 years (Brody, 1983).

The question of competing risk in age-related mortality also enters this discussion. Brody (1983) cogently argues that decrease or prevention of any one cause of mortality among the elderly will increase the relative contribution to mortality from other causes, but will not increase their *absolute* contribution. At present in the United States, cardiovascular disease is the prime cause of mortality after 75 years. Thus, during the impressive recent decrease in cardiovascular mortality (a negative age-related slope, -1.7%/year, 1970–1977), *no* other major sources of mortality showed absolute increases among the elderly, including cancer. The age-related increase of cancer is not exponential throughout the lifespan. Unlike cardiovascular disease, cancer decreases as a cause of mortality—by 65 years, 30% of deaths are attributable to cancer, whereas by 80 years the contribution is less than 12% in North America (Brody, 1983).

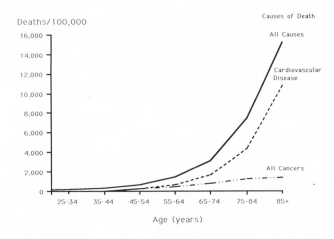

Deaths/100,000

Causes of Death

Fig. 1.11. Death rates by cause and age group in the U.S., 1976 (National Center for Health Statistics). Redrawn from Brody, 1983.

Age (years)

Prevention or reduction of some age-related diseases may increase human longevity, although this is contested by Fries (1980, 1988), who predicts that older age groups will show a compression of morbidity as causes of premature death, such as heart disease (myocardial infarction), are further diminished. Brody (1983) makes the important point that these various age-related risk factors do not necessarily compete at the same age. Therefore, one cannot conclude that the maximum lifespan is fixed in humans simply because other diseases are coexistent in the same age group. Thus far, the demographic data continue to indicate trends for increased survival in *all* older age groups, contrary to Fries's prediction (Guralnik et al., 1988). The mortality curves of Figure 1.2, moreover, suggest an increasing delay over the last two centuries in the age when mortality rates reach a stable exponential acceleration; possibly this trend can be modelled according to Figure 1.8 (Finch, 1969). Time will tell within a few decades if there is a strict biological limit on the human lifespan.

Relationships between the age-related incidence of various diseases and individual mortality risk at advanced ages are also sensitive to psychosocial variables. For example, men show increased mortality during the first few years after the death of their spouses, whereas women do not (Jacobs and Ostfeld, 1977); other examples are given in Berkman and Breslow, 1983, and Levav et al., 1988. A converse case is the decreased mortality at major social occasions, when "death takes a holiday" (Phillips and King, 1988). Moreover, mice may live longer if caged alone than in groups (Muhlbock, 1959), probably because of fighting and other social stresses. The behavioral interactions of social insects will be discussed as key to the long lifespans of insect queens (Chapter 2). Thus, numerous examples indicate that the environment of socially responsive animals is a key factor in the potential plasticity of senescence and lifespan.

1.4.10. *The Biomarker Problem*

The prediction of lifespan for an individual is another frustrating issue in the analysis of population mortality risks. Many attempts have been made to predict individual lifespan from age-related changes or from physiological or molecular characteristics at a particular age, the elusive biomarkers of aging (Reff and Schneider, 1982). The following brief discussion introduces this subject; a more detailed treatment is given in Chapter 10.9.

Few age-related changes are good predictors of lifespan in humans, but this is not surprising given the widely differing lifestyles, diet, genotypes, exposures to disease, etc., that characterize individuals in any population. The best predictors are related to premature death from cardiovascular disease, as shown in the famous longitudinal Framingham Study (Kannel et al., 1981; Kannel, 1985a, 1985b). Cardiovascular diseases and their attendant mortality risk are linked less strongly to chronological age than to hypertension. Small lung volumes (measured as the forced vital capacity) also predict early mortality (Figure 10.27). However, lung volume is no better than chronological age in predicting mortality at all adult ages (Kannel et al., 1980). Premature mortality can be predicted from particular diseases, but unusual longevity cannot.

One might have hoped to find better biomarkers for the lifespans of inbred animals under laboratory environments. Efforts using a range of behavioral, physiological, and biochemical parameters that are sensitive to certain changes have so far failed to predict individual lifespans for laboratory populations of the nematode *Caenorhabditis* or various inbred mouse strains (Chapter 10.9, Figure 10.25). The extent of these differences between genetically identical individuals in *C. elegans* (Figure 1.12) is made more dramatic by the very short lifespans which give little time for the accumulation of individual experiences. There may be many subtle differences between individuals that influence mortality risk after mid-life. Even differences between genetically identical individuals may arise through random variations of cell migration and cell death during development. Differences in neuron number, for example, could constitute cryptic defects of little consequence early in life that might have a multiplicative effect later in life with subsequent lesions (Chapter 9.3.3).

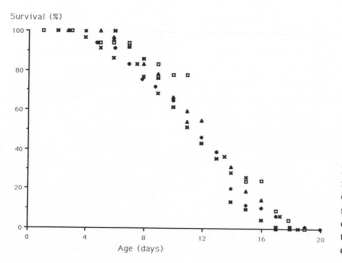

Fig. 1.12. Survivorship as a function of age in soil nematodes, *Caenorhabditis elegans*. The symbols represent four replicates of genetically identical populations. Redrawn from Bolanowski et al., 1981.

1.5. Evolution and the Relationships between Reproduction and Senescence

This section briefly identifies issues about the evolution of senescence, with emphasis on defining the information needed for discussions that will continue throughout the book. Evolutionary shifting in senescence and longevity immediately become obvious as one considers the life histories of familiar organisms, for example, short-lived flies versus long-lived insect queens; mice versus elephants; or annual flowering plants versus millennia-old bristlecone pines. Population geneticists often view the evolution of senescence as an aspect of the more general problem of genetic selection for life history patterns. It is intuitively presumed that senescence is a highly labile phenomenon during evolution. Accordingly, shortening or lengthening of the lifespan, with acceleration or deceleration of senescence, may be supposed to have occurred on innumerable occasions. There have, however, been few analyses of the fossil record for such trends (Chapter 11). The following reviews first-reproduction schedules; these give the basis for evolutionary hypotheses about senescence that are based on premises from population genetics.

1.5.1. *Reproductive Schedules and Darwinian Fitness*

The forces driving evolutionary changes of life history are sometimes ascribed to the optimization, with respect to mortality rates, of *reproductive schedules,* that is, the age when reproduction begins, the rate of reproduction, and the duration of the reproductive phase. The actions of natural selection on existing genetic polymorphisms are thought by population geneticists to allow for rapid alterations in the distribution of reproduction by age group. Thus, reproductive schedules over the lifespan are fundamental to life history theory, as well as to a major focus of this book: the nature of *genotypic* influences on senescence and the lifespan and the mechanisms of *genomic expression* through which a given set of genes may influence the characteristics of senescence.

A considerable literature has developed during the past 30 years from major efforts by population geneticists to describe the distribution of genes in populations that influences the reproductive schedule (Murray, 1979; Hamilton, 1966; Bell, 1980; Charlesworth, 1980, 1990a, 1990b; Rose and Graves, 1989; Rose, 1991; Sibly and Calow, 1986; Stearns, 1977; Williams, 1957).[7] Many mathematical treatments of the relationships between mortality and reproduction have been based on Fisher's (1930) concept of fitness, defined according to the population growth rate in the Euler-Lotka equation, given here in its integrated form:

$$\int_0^\infty e^{-rt} s(t) f(t) \, dt = 1 \, , \tag{1.11}$$

where r is the Malthusian parameter of population growth for which the equation may be solved, given either the function or a range of values for $s(t)$ and $m(t)$; $s(t)$ is the proportion

---

7. I cannot do justice in this short section to the highly developed literature on the evolutionary theory of senescence that draws from population genetics models. My intention is to give the gist of arguments and their predictions for the outcomes of senescence from a selection of major papers. Comprehensive treatments are given in books by Murray (1979), Charlesworth (1980, especially his chapter 5), and Rose (1991). A. R. Wallace's early statement on the evolution of senescence is given in Appendix 3.

of the population surviving from birth to age t; and $f(t)$ is the fecundity per individual at age t.

This equation determines the net population growth rate that would counterbalance the net reproductive output as specified by the parameters of mortality and fecundity. Thus the equation allows many mechanisms at different times in the life history for responding to demographic pressures. The relationships between mortality and reproduction can vary with particular trade-offs between different age groups. Many specific cases were worked out by Hamilton (1966) and Charlesworth (1980, 1990a, 1990b), and are reviewed by Rose (1991). The population genetics models to date have not used specific mortality rate functions; the Gompertz functions for $s(t)$ in equation (1.4) could be used here for $s(t)$.

The concept of Darwinian fitness as the basis for natural selection relies on the relative net reproductive output of individuals in a population who are assumed to be competing for limited resources. At once the theorist confronts questions about the lifespan reproductive capacity, a value that represents the total (integrated) reproduction across the survival distribution of the population from maturity to the statistical lifespan or until reproductive senescence. Key parameters are the number of offspring that can be produced at different ages, as well as their viability, including developmental abnormalities at different parental ages; such information is given in the discussion of life histories in Chapters 2–4 and 8. In general, population genetics models indicate that natural selection often acts to maximize the intrinsic rate of increase in a population, the Malthusian r in equation (1.11). Moreover, the optimization of reproduction against finite resources presumes constraints, against which genes are selected on the basis of their impact on the internal (physiological) distribution of resources between those drawn by somatic repair and those drawn for the completion of reproduction.

Natural selection is thought to operate on reproductive schedules through two main determinants that must also influence the total lifespan: (1) the age at first reproduction, which varies over a huge range in most phyla (Chapters 2–4); and (2) the life expectancy, which includes effects of mortality rates after first reproduction and which estimates the reproductive potential. Numerous comparative studies of iteroparous species indicate negative correlations between the age of first reproduction and the statistical life expectancy, such that an early onset of reproduction is correlated with shorter statistical adult life phases, as shown for mammals in Figure 1.13 (Harvey and Zammuto, 1985) and in birds (Lack, 1966). These general trends do not apply as neatly to semelparous species that may delay their event of reproduction for years or decades (Chapter 2). A further complication is that some genders show greater trends towards semelparity.

For a generation, population geneticists and ecologists have discussed these issues in terms of "r-selection versus K-selection," which describes selection for reproductive scheduling and lifespan on the basis of mortality rates and environmental stability.[8] Under r-selection, genotypes are favored that have rapid development, early reproduction, and

8. Pianka (1970) traces this concept to Dobzhansky's (1950) proposal that natural selection operates differently in the relatively constant environment of the tropics, where mortality results more from competition; by contrast, the fluctuations in temperate zones purportedly favor higher fecundity and more rapid development. MacArthur and Wilson (1967) proposed that variable climate was associated with fluctuating mortality rates that were independent of population size and favored r-selection, or selection on the basis of the Malthusian rate parameter of population increase, r, in equation (1.11).

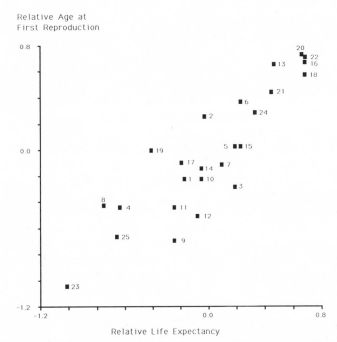

Relative Age at
First Reproduction

Relative Life Expectancy

Fig. 1.13. Relative age at first reproduction plotted against relative life expectancy at birth for natural populations of mammals. Relative values refer to deviations from logarithmic regression lines of age at females' first breeding, or expectation of life at birth, on adult female body size. The correlation coefficient of $r = 0.98$ is only slightly decreased to 0.89 ($p < 0.001$) by removing the effects of body size through partial correlation. The numbers refer to different mammalian genera. Artiodactyla: *1, Syncerus* (buffalo); *2, Hippopotamus; 3, Aepyceros* (impala); *4, Sus* (pig); *5, Cervus* (deer); *6, Ovis* (sheep); *7, Hemitragus* (tahr); *8, Phacochoerus* (warthog); *9, Kobus* (kob); *10, Connochaetes* (gnu). Carnivora: *11, Taxidea* (American badger); *12, Felis lynx; 13, Lutra* (river otter); *14, Mephitis* (skunk). Lagomorpha: *15, Sylvilagus* (cottontail rabbit); *16, Ochotona* (pika). Perissodactyla: *17, Equus* (zebra). Proboscidea: *18, Loxodonta* (African elephant). Rodentia: *19, Castor* (beaver); *20, Tamias* (chipmunk); *21, Spermophilus* (ground squirrel); *22, Sciurus* (gray squirrel); *23, Clethrionomys* (vole); *24, Tamisciurus* (red squirrel); *25, Peromyscus* (mouse). Redrawn and the legend quoted from Harvey and Zammuto, 1985.

semelparity; these life history traits are often associated with high adult mortality rates. The other extreme is stable environments where stable mortality rates permit populations to approach the carrying capacity of the environment, *K*-selection; these conditions favor genotypes with slower development, iteroparity, larger size, and longer statistical lifespans; these traits are often associated with much lower adult mortality rates. Overall, insects are considered to be *r*-selected, while most birds and mammals are *K*-selected. However, numerous exceptions weaken the dichotomy (Pianka, 1970; Templeton, 1982); for example, the very long-lived cicadas with adult life phases that are less than 1% of the total 17-year lifespans (Chapter 2.2.3.3). Moreover, semelparity is not obligate in some genotypes nor in populations of fish and angiosperms that show conditional semelparity (Chapters 2, 11). There is no doubt that ecological conditions are important in the selection for different reproductive schedules (Bull and Shine, 1979; Leutenegger, 1979; Promislow and Harvey, 1990; Stearns, 1984). However, I will make minimal use of the *r*- and *K*-selection classification to avoid prejudging the situation. There may always be unknown factors that allow variations in reproductive schedules that are viable in a particular environment but were not specifically selected for by parameters in the present environment.

Differences in mortality are hypothesized to select for the age at maturation (Williams, 1966a, 1966b; Gadgil and Bossert, 1970; Charlesworth, 1980; Sibly and Calow, 1986). These trends are supported by artificial selection for reproduction at late ages in *Drosophila,* which increased lifespans and delayed the onset of reproduction (Chapter 6.3.2). Thus, genotypes may be selected for trade-offs between early maturation with higher mortality rates, and slow growth with delayed maturation and greater survival to later ages when reproduction occurs. Semelparous species do not necessarily have short total lifespans, however, as shown by the free-living animals and plants that have very delayed maturation (Chapter 2).

Life history theories also consider the costs of reproduction, a concept deriving from Williams' (1966b) prediction that species with greater seasonal reproductive efforts should be shorter-lived. In general, theories on the costs of reproduction presume that natural selection cannot maximize reproduction in all adult ages (Partridge and Harvey, 1985). A considerable literature describes how adult-phase mortality rates increase in diverse species in association with the acts of reproduction (e.g. Bell, 1980; Murray, 1979; Reznick, 1985; Clutton-Brock et al., 1983; Clutton-Brock et al., 1988) and that mortality rates tend to be higher in species with greater seasonal fecundity (e.g. Tinkle et al., 1970; Stearns, 1984).

1.5.2. Population Genetics Theories on the Evolution of Senescence

Evolutionary theories of senescence generally emphasize how natural selection can be exerted on a gene pool despite the high rates of postmaturational mortality that characterize many species. As described in Section 1.4.6, the aggregate age-independent mortality rate (IMR, equation [1.4]) may be so high from natural hazards, irrespective of any senescence, that only a small fraction can survive much beyond the first season of reproduction. Such situations are found, for example, in rapidly senescing insects with short life cycles of a month or so (Chapter 2). Few populations of mammals in nature allow survival of a substantial fraction to advanced ages, whether or not senescence is observed (Chapter 3). For example, a review of studies on wild *Mus musculus* and *Peromyscus leucopus* indicated that less than 10% survive to ages permitting coexistence of more than three generations (Phelan and Austad, 1989); these conclusions are tentative because of the relatively small sampling and because it is difficult to evaluate the contribution of migration. However, certain slowly reproducing and long-lived vertebrates, including many birds and fish, have greater survival to multigenerational age groups (Nesse, 1988; Leaman and Beamish, 1984; Chapter 4.2).

Haldane (1941) founded a major line of reasoning by considering that mouse strains with different hereditary diseases at mid-life are models for the escape of genotypes from natural selection when adverse effects are late in onset. Subsequently, Medawar (1952), Williams (1957), and Hamilton (1966) argued in seminal papers that selection is weak against genotypes for chronic diseases and other dysfunctions that arise long after the start of reproduction. These quantitative arguments are based on the population model of equation (1.11). Thus, the extent of senescence according to this theory should be closely linked to adult-phase mortality rates. These proposals are cornerstones in a major theoretical prediction that the high natural mortality rates of most populations allow the accumulation of genes in older age groups that cause age-related deteriorations or senescence, because so few individuals survive to ages when these adverse genetic effects are manifested (Charlesworth, 1980, Chapter 5; Rose, 1991). The outcome of this prediction is a major issue, and conclusions are by no means straightforward or settled, in my view.

In a mathematical treatment based on equation (1.11), Hamilton (1966, 25) concluded that "increases of the rate of growth of fertility will always reduce senescence." The maintenance of stable population sizes with exponential increases of reproduction and constant mortality rates *requires* diminution of reproduction after some age. However, even if "individuals increase their fecundity exponentially and do so indefinitely . . . phenomena of senescence will tend to creep in." A similar conclusion was reached by A. R. Wallace about a century before (Appendix 3).

Williams (1957) proposed the "negative or antagonistic pleiotropy" hypothesis, in which genes might be selected for on the basis of advantages conferred to reproductive fitness early in life but that also had delayed harmful characteristics. Williams' hypothetical example was a gene selected because it enhanced bone calcification during development, but had a delayed adverse consequence by allowing calcification of arterial connective tissue. Moreover, some genes could be relatively neutral in their effects on the young adult, while crucial to earlier development. Haldane (1941) also proposed that genotypes with delayed adverse effects could be subject to selection for modifier genes that delay the onset age of particular hereditary diseases. Subsequently, selection for modifiers was considered unlikely on the basis that mutant alleles are predicted to exert effects over a wide distribution of ages (Charlesworth, 1980, 218).

Few negatively pleiotropic genes or alleles are characterized. An example is the *aa* polymorphism in natural populations of *Drosophila mercatorum*. Although *aa* shortens the lifespan, it allows earlier reproduction and responds dynamically to environmental fluctuations that select for rapid reproduction just after hatching (Chapter 6.3.3). However, one may not conclude from artificial selection studies that a suite of traits for reproductive scheduling (age at first reproduction, age-group-specific fecundity, etc.) are controlled by single alleles. Genes may be linked, as observed for the gene clusters that determine histocompatibility as well as aspects of reproduction (Chapter 6.5.1.4). Williams (1957) and Charlesworth (1990a, 1990b) propose that the accumulation of genes with delayed adverse consequences that contribute to senescence would be favored by very high mortality rates in young adults. This question is examined in birds and mammals (Chapter 3). Postmaturational changes in gene expression also bear on this hypothesis, as discussed below.

Another important genetically influenced trait concerns reproductive senescence. As noted by Fisher (1930), Williams (1957), and Charlesworth (1980) among others, human and animal populations show a reciprocal relationship between mortality rates and fecundity. While this trait is observed in innumerable animal species, other evidence suggests that not all species show reproductive senescence (Chapter 4.1). Survivorship to advanced ages can decrease *independently* of changed fecundity in that age group, as allowed in equation (1.11). Thus, the diminishing force of selection with age does not depend on reproductive declines. However, if reproduction diminishes with increasing age in long-lived species, then I deduce from the evolutionary theories of senescence that genotypes with reproductive senescence should more rapidly accumulate genes with adverse effects later in their lifespan than those with smaller or negligible declines in reproduction.

The implication that genes differ in the manifestations of adverse effects immediately raises questions about the timing of gene expression. The simplest alternatives are that (1) the gene was expressed from development onwards or (2) that the gene was not expressed until later in adult life. The converse also may hold, that other genes may exert negative effects by being turned off at various times. The production of erythrocytes with hemoglobin S that causes sickle cell anemia illustrates a gene with adverse consequences that is expressed early in life. An example of a gene with later onset expression that probably results from increased transcription is the gene for hairy pinna (Figure 7.2), though in this case no harm results.[9] A crucial feature in evolutionary changes of senescence concerns the timing of changes in gene expression during the lifespan. Chapter 7 reviews evidence on gene expression during adult life that represents a number of conceivable cases: increases, decreases, and no change, according to the cell type and the age. Since many genes are controlled by hormones and other physiological factors, and since there are many age-related changes in the regulation of levels and timing of hormones and other physiological regulators of gene activity (Chapters 2, 3), we may anticipate numerous age-related changes in gene expression. Some may have adverse effects; some may not.

A corollary to the proposition that the force of selection diminishes with age, emphasized by Haldane, Medawar, Williams, and Rose, is that genetically based manifestations of senescence should vary widely among species and within species. Here it is useful to note the already large variations among taxa in regard to developmental patterns, for example, cell lineage determination, the onset of the zygote genomic functions, and so on (Davidson, 1986, 1990). Superimposed on the species-characteristic developmental differences, we should then expect even greater variations in senescence. The following chapters demonstrate this variety at all levels of biological organization, although the changes of senescence in a particular species may be quite restricted.

An open question concerns the partitioning of mortality rates in feral populations between the age-independent mortality rates (coefficients A_0 and M_0, equation [1.1]) and the two different age-related mortality changes associated with the neonatal and adult (senescent) phases (Figures 1.1, 1.2). The population genetics models often give separate terms for these mortality rates, but few of the available data have been analyzed for the relative contributions from age-dependent and age-independent mortality in the adult phase. The relative impact of these mortality parameters may not be equal throughout species of a given body size or construction.

A little explored question is the extent to which co-adapted traits become disordered during senescence. During development and maturation, numerous aspects of fitness are selected for co-adapted effects in a "suite of characters," such as evolved in the shapes of the teeth and muscles of early mammals that allowed them to chew their food (Chapter 11.3.5.2). Genes with late effects that damage one of such co-adapted characters might

---

9. The genotypes of Huntington's disease and familial Alzheimer's disease, while having delayed manifestations, have not been traced to a specific gene product, so that we cannot say which pattern these genotypes represent at the level of transcription.

then have cascading effects that I will refer to as "dysorganizational senescence." I suggest that recently evolved aspects of senescence might differ from those of longer standing because of the occasions when the adult phase was selected for lengthening. Selection for longer adult phases would tend to eliminate genotypes that promote rapid senescence. Of these, the most labile should be genotypes that are not deeply embedded in a co-adapted suite of characters. Examples may be found in the Diptera that differ according to their need to feed as adults (autogeny), a trait that is under neuroendocrine regulation and that may be rapidly reversible as judged from the geographic clines in autogeny of pitcher-plant mosquitos (Chapter 2.2.1.1.2).

The evolutionary theory of senescence has led to studies of artificial selection for different lifespan and reproductive schedules in *Drosophila* (Chapter 6.3.2). While this experimental focus is laudable, it may have furthered the neglect of information on lifespan and reproduction from field studies. As reviewed in Chapter 4.2, some organisms have extremely slow, or possibly negligible, senescence. A presumption that has been widely held by biologists of many fields is that senescence must be ubiquitous in somatic cells. For example, Williams (1957, 406), in considering iteroparous organisms with declining reproduction and increasing mortality, concluded that "senescence should always be generalized deterioration, and never due largely to changes in a single system." To evaluate this presumption, I give many examples of the specificity of senescence in different cells and organ systems.

Population geneticists, however, allow an exception for organisms that reproduce vegetatively by fission, since in this case the distinction between germ line and soma is lost. Clones need not senesce (e.g. Williams, 1957; Bell, 1984b; Rose, 1991; Chapter 4.4), although many vegetatively reproducing plants and animals do, in fact, show rapid senescence in association with sexual reproduction (Chapters 2, 4). The coexistent potentials for somatic cell immortality through vegetative reproduction and for somatic senescence under other circumstances with the same genome epitomize the role of selective gene regulation in senescence.

However, the arguments of population genetics do not always demand the acquisition of genes that cause senescence. Special cases of the theory in its present form allow natural selection against genotypes programming for senescence, either by eliminating those alleles or by selecting for other alleles that suppress particular phenotypes of senescence. For example, Williams (1957) and Hamilton (1966) predicted that the trait of continuously increasing reproductive capacity as a function of age would select against senescence. Such a circumstance of net population growth might prevail during population fluctuations that arise spontaneously (Charlesworth, 1980, 216) or from climatic shifts and geologic changes that opened new territory for invasions. Even in the absence of major climatic or geologic changes, sudden population expansions may occur through adventitious immigration of new species from elsewhere. Population expansions might give windows of opportunity for the evolution of very extended reproductive schedules with high fecundity that might purge some radiations of genes with delayed adverse effects by eliminating these alleles or by selecting for modifiers.

Under some conditions, it might be difficult to detect delayed adverse effects of genes. Perhaps the senescence-retarding effects of lowered temperature in invertebrates and of caloric restriction in mammals and many invertebrates (Chapter 10) has revealed an important gene/environment interaction, in which antagonistic pleiotropic effects are environmentally dependent. From this perspective, senescence in many species should be extensively modifiable but by routes that are not generalizable and thus are taxon specific.

We can consider two parameters that might enhance the strength of selection of genotypes with slow or negligible senescence: (1) initial mortality rates that, if sufficiently low, may permit a substantial fraction to survive to multigenerational ages; and (2) maintenance of reproduction. The latter may depend on continuing *de novo* gametogenesis during adult life, that is, in contrast to female mammals, which have a fixed oocyte stock that is lost irreversibly (Figure 3.10). Species with sustained or increased fecundity with age that thereby augments their representation in the population gene pool at later ages should therefore have protection against accumulated germ-line mutations with adverse effects. This question will be examined through examples of long-lived animals and plants that have progressive increases of fecundity as age advances.

In all these cases, the nature of changes of reproduction with age have great bearing on the force of selection, independently of how mortality rates may change. The changes of reproductive performance with age across the lifespan are also pertinent to the potential for varying the reproductive schedule, especially in species and genders where gametogenesis is unlimited. While life history theory has emphasized these parameters in abstract population models, relatively little attention has been given to specific biological factors that might constrain the flexibility of reproductive schedules, such as the limitation of gametogenesis that is characteristic of female mammals and many soil nematodes. The distribution of senescence and mortality in relation to reproduction are given particular attention throughout the book. While none of these theories make assumptions about particular mechanisms of cell or organ senescence, nonetheless the types of senescence and the extent of disorders that may occur need careful evaluation.

An important role of senescence in selection for life history alternatives is implied by correlations between mortality rates and reproductive scheduling. The "disposable soma" theory (Kirkwood and Cremer, 1982; Kirkwood, 1985) considers the optimum allocation of energy to somatic repair processes according to the reproductive schedule. Species with short adult lifespans, early maturation, and high reproductive yield are predicted to "invest" fewer resources in somatic repair than those with longer lifespans and slower rates of reproduction. A major question concerns the relations of the various mortality rate terms to repair capacity, which are genetically coded. As noted above, statistically short lifespans in species with early maturation and high fecundity need not be consequent to senescence. The relative size of the two major adult mortality rate terms will be analyzed where possible in relation to the different patterns of senescence (Chapters 2–4).

A related question is the onset of senescence. Some species have lengthy developmental stages or arrested development at ambient temperatures (Chapter 9) that implicitly require a wide range of repair mechanisms, from molecular repair (e.g. thymidine dimer excision, nonenzymatic glycation) to cell replacement and organ regeneration. Optimization of the resources committed to repair may be stage specific, and hence could be as much a question of the regulation of the associated genes as the presence or absence of repair mechanisms. An interesting comparison will be the social insects, in which both short- and long-lived castes develop from the same genome (Chapter 2.2.3.1). The capacity for somatic cell repair and regeneration will be considered in relation to the potential variations of reproductive scheduling.

Because of the life histories of the semelparous, in which maturation or reproduction is the trigger for rapid senescence (Chapter 2), it is important to consider if reproduction in iteroparous species also triggers somatic senescence in more gradual forms. In some species, reproduction increases both the age- dependent and age-independent mortality rates, and accelerates certain aspects of senescence beyond that seen in virgins (Chapter 10.6). Another source concerns the impact of castration and hypophysectomy on lifespans (Chapters 10.4, 10.8).

The distribution of senescence in postreplicative somatic cells and the capacity for somatic cell proliferation have long figured in theories of senescence. Somatic cell senescence and limited somatic cell replicative capacity were presumed to result from natural selection in Weismann's concept of the immortal "germ plasm" (Williams, 1957; Kirkwood and Cremer, 1982). However, there is abundant evidence for vegetatively reproducing organisms that are essentially immortal somatic cell clones (Chapter 4). I will also review how somatic cell replacement and regeneration vary between taxa in relation to lifespan and the patterns of senescence. Understanding the particular mechanisms of senescence and the limits of selection, both natural and artificial, requires treatment of the developmental and pathophysiological characteristics of each taxon. The diverse phylogenetic origins of life histories give an approach to these complex, elusive issues and help devolve them into tractable problems.

To close this prologue on evolution and senescence, I briefly discuss issues that are fundamental to mechanisms in the theories of evolutionary biologists about life history and reproductive scheduling, as well as to mechanisms of senescence as seen by gerontologists. An implicit assumption in theories of reproductive scheduling and life history is that nearly all temporal variations are evolutionarily possible during development and growth phases, as well as in the length of adult life as needed to propagate the population gene pool. In turn, this requires important assumptions concerning (1) the maintenance of macromolecules by somatic cell gene expression, and (2) the efficacy of gametogenesis, including the stability of irreplaceable oocyte pools.

Firstly, the extent of somatic cell genome maintenance across the adult phase is a question of very broad significance to evolutionary change in reproductive scheduling. That is, we wish to know the extent of irreversible molecular change or damage to the somatic genome during the interval from fertilization (or the initiation of development) to the end of the reproductive phase. Although complete germ-line gene sets (genomic totipotency) may not be needed in each differentiated somatic cell, the differentially active gene sets must remain intact throughout the phase of fitness selection, that is, during development and into reproductive life. Specific molecular repair mechanisms must remain effective throughout this often very prolonged duration. These mechanisms of repair are required not only for DNA, but for long-lived proteins which are at risk for post-translational damage, like glycation, racemization, and deamidation that cause age-related dysfunctions at the molecular level. A question of major interest concerns whether or not these mechanisms then fail, thereby causing organismic senescence.

Secondly, are there time limitations in the efficacy of gametogenesis that might arise, for example, through clonal limitations in the proliferation of gametogenic stem cells? The fixed numbers of ovarian oocytes in mammals illustrate this case, but are usually not considered to result from mechanisms like those that limit diploid somatic cell proliferation *in vitro,* as will be discussed in Chapter 7.5.4. Moreover, we wish to know what limitations there may be in the viability of gametes stored over prolonged times at various stages of differentiation, from primordial oocytes to mature sperm. The questions about somatic cell genome stability and gamete viability are obviously interrelated by the exposure of tissues throughout the body to hazards. These questions about mechanisms of life history variations are shared by evolutionary biologists and gerontologists.

1.6.   Summary

This chapter introduced the Gompertz mortality model as a descriptor of senescence at the population level. Two mortality rate parameters limit the lifespan, the MRDT and IMR. The age-related acceleration of mortality rate is measured as the mortality rate doubling time, MRDT, and is held to be a fundamental measure of senescence. The age-independent mortality risk is measured as the initial mortality rate, IMR, and sets limits on the statistical lifespan that may prevent most individuals from achieving ages when senescence could be manifested. In characterizing species according to their type of senescence in the following chapters, the MRDT places them in the continuum from rapid to negligible senescence more clearly than the values of IMR. For example, no species with a maximum lifespan over 4 months has an MRDT higher than 0.2 year, while occasional populations of long-lived animals have values of IMR that approach those of flies. Similarly with regard to lifespan, the herring gull and the quail have similar IMRs, while their MRDTs are fiftyfold apart and their lifespans differ sevenfold.

According to population genetics models, the force of natural selection generally diminishes with increasing age because of the usually high values of mortality in nature. Consequently, there should be an increased likelihood that genes with adverse and delayed effects will be permitted to accumulate in the germ line. The accelerations of mortality rates with age that are manifested by many species can be viewed as resulting from genotypes with delayed adverse effects. If reproduction also declines with age, then the strength of selection against senescence will be further weakened. But, as described in Chapter 4, some species do not manifest a marked senescence, as judged by mortality rates and reproductive performance.

The natural history of senescence as described in the next chapters gives an approach to the difficult problem of identifying the types of mechanisms in senescence, the biological contexts within which they may arise, and the role of the genome. A major question posed by evolutionary theories of senescence concerns the extent to which age-related changes during adult life vary among species and taxa. Can we find canonical patterns in the phenotypes of senescence, that is, changes during senescence that are universal phenotypes, or nearly so, in a group of species or an evolutionary radiation, as in canonical patterns of cleavage or cell movement during development that occur throughout a group of species (Davidson, 1990). Moreover, is there a relaxation or breakdown of the selection for co-adapted traits upon which the integrated cellular and physiological functions depend during development and early adult life? In organisms with slow senescence, are all vital physiological systems affected? Does the evolution of senescence then present a picture of chaos or of order? Examples of species that show both outcomes are described in Chapters 2–4, which should help set the stage for more critical evaluations of the extent of degenerative changes at various levels of the organism later in the lifespan.

2

Rapid Senescence and Sudden Death

2.1. Introduction

In this chapter I discuss widely differing processes that lead to rapid senescence or sudden death in association with maturation or gametic reproduction. Most species showing rapid senescence have only one bout of reproduction, while a few others die suddenly after long and fertile lifespans. Thus, the *total* lifespan may be virtually any length. In this classification of life histories, the involutional phase preceding death is brief, lasting less than a year regardless of the total lifespan of the free-living organism. Most individuals in populations of species showing rapid senescence or sudden death undergo similar and generally synchronous involutional changes that may be considered canonical. In contrast, the characteristics that I propose to use for gradual senescence are rarely synchronous in an age group or birth cohort and show much greater individual variability. This point was brought out in a discussion on senescence in *Drosophila* and mice by Miquel, Bensch, Philpott, and Atlan (1972), which noted the near universality of multiple degenerative changes in flies at the end of the average lifespan. By contrast, in healthy laboratory mice that had survived beyond the average lifespan, most tissues lacked pathologic lesions, while the degenerative changes that did occur did not arise with any universality or synchrony.[1] Subsequent chapters describe other species with gradual (Chapter 3) or negligible senescence (Chapter 4).

  Rapid senescence and sudden death are found in organisms with a wide range of lifespans, from flies and nematodes that live for 1 month to bamboos that live beyond a century. The combination of rapid senescence or sudden death with long lifespan may at first seem contradictory. However, sudden death from rapid senescence can terminate long lifespans. This life history pattern is exemplified by the numerous plants with prolonged vegetative growth phases lasting for one or more decades and followed by fruiting, rapid senescence, and death within a few months. Other examples, discussed in detail below,

---

  1. Miquel, Bensch, Philpott, and Atlan (1972) quote from several other papers that are authoritative for that time and still valuable to recall. The following is verbatim:

  In regard to mammalian aging, Alexander (1966) has noted that "The most striking factor is that even very old mice (e.g. more than two and one-half years) when killed while still fit have remarkably few pathologies and are almost indistinguishable from young animals." In *Drosophila* the situation was strikingly different, since all the 70–100-day-old flies showed degenerative changes in the tissues, which, in our opinion, fulfilled the criteria set forth by Casarett (1963): "Many highly inbred, or genetically more homogeneous, aging populations of animals tend to show a lesser variety of causes of death and usually have a high incidence of specific causes of death for which they have a strong genetic susceptibility. In general, the age-dependent diseases or the so-called diseases of aging are essentially either degenerative or neoplastic in character." . . . The greater severity of the degenerative changes observed in *Drosophila* in comparison with those seen in healthy men or experimental animals may be due to the fact that all tissues of the fly except for the gonads are made of postmitotic cells.

are the insect queens which die suddenly after long adult lifespans. Each organism discussed below merits much deeper study than is possible in this overview. I hope my vignettes stimulate further discussion of these complex phenomena.

During rapid senescence, the mortality rate sharply increases, with acceleration as an exponential function of age as observed in mammals, but at much greater rates. The mortality rate doubling times (MRDT)[2] are in the range of 0.005–0.1 year (Table 2.1), whereas mammals and other species classified as showing gradual senescence have MRDTs that range from 0.3 year to more than 10 years (Table 3.1; Appendix 1). In graphic form, there is no difference other than in scale between the mortality rate curves on age shown by the species discussed below, and there is no reason so far to consider other mortality models besides the Gompertz or Weibull formulae.

A subset of species showing rapid senescence and sudden death have only one bout of reproduction or season of fertility, and are called semelparous to distinguish them from iteroparous species with repeated reproductive cycles (Cole, 1954; Chapter 1.2.1). Similarly, botanists have long distinguished plants as annuals or perennials (usually polycarps), to which must be added the often overlooked perennial monocarps. These plants grow vegetatively for extended times before suddenly flowering and dying (Watkinson and White, 1985). The phenomenon of semelparity is of much interest to theories on the evolution of reproductive strategies (Chapters 1.5.1, 11.4.3).

An important distinction is that rapid senescence and semelparity are not obligatory in many species, which may survive under favorable conditions for further reproduction. Thus, rapid senescence and sudden death can be classified by whether rapid senescence or semelparity are obligate (universal) versus conditional, or triggered by environmental or physiological signals during the adult stage. In many cases, major extensions of lifespan result from hormonal or other physiological manipulations that postpone or modify senescence. This plasticity in the schedule of senescence and in the lifespan is seen again and again, and depends on the capacity of cells in organisms with allegedly fixed lifespans for far greater potential longevities than the usual (Chapters 7, 10–12).

The causes of rapid senescence and sudden death differ widely among phylogenetic lineages. The following presentations will examine the levels of organization that are causal in rapid senescence and sudden death—that is, the relative contribution to mortality and morbidity from physiological and structural changes, as may be distinguished from more intrinsic forms of cell involution. The causes of rapid senescence and sudden death include wearing out of irreplaceable parts, alterations of hormones, impairments in feeding, or viral proliferation. Some of these harmful changes resemble the hereditary progeroid syndromes of humans, in which a single specific condition causes premature debilitation and death (Martin, 1982; Chapter 6.5.3.3). Although the causes of rapid senescence are unknown for many other species, adults of more slowly senescing species seem to develop a broad range of age-related disorders that appear to have multifocal origins. I keep open the possibility of some shared mechanisms in rapid and gradual senescence.

---

2. The MRDT measures the acceleration of the mortality rate with adult age and is derived from the Gompertz mortality coefficient (equation [1.2]). As described in Chapter 1.4.4, MRDT is a useful measure of the rate of senescence in a population.

Table 2.1 Mortality Rates in Species with Rapid Senescence

Phylum	IMR/year	MRDT, years	*tmax*, years
Animalia			
ARTHROPODA			
Crustacea			
Daphnia longispina[1]	0.7	0.02	0.1
Insecta			
Diptera			
Calliphora erythrocephala[2]	0.4	0.07	0.3
Drosophila melanogaster[3]	0.01–4	0.02–0.04	0.3
Musca domestica[4]	4–12	0.02–0.04	0.3
Hymenoptera			
Apis mellifera[5]			
Workers (June)	0.2	0.02	0.2
Workers (winter, > 240 days)	< 0.001	0.03	0.9
Queens	< 0.1		> 5
Lepidoptera[6]			
Automeris boucardi	10	0.005	0.03
A. junonia	20	0.005	0.03
Dirphia eumedide	7	0.005	0.05
D. hircia	20	0.01	0.05
Lonomia cynira	45	0.007	0.02
MOLLUSCA			
Gastropoda			
Aplysia juliana[7]	0.007	0.13	0.9
NEMATODA			
Caenorhabditis elegans[8]	1–7	0.02–0.04	0.16
ROTIFERA			
Lecane inermis[9]			
Female, amictic	6	0.005	0.10
Female, mictic	20	0.03	0.08
Male, aphagic	0.4	0.002	0.02
Fungi			
Saccharomyces cerevisiae[10]		0.004	

Notes: Estimates of mortality rates were calculated according to the Gompertz model: $\ln m = \text{IMR} + Gt$, where m is the mortality rate as a function of age (the fraction of survivors dying in arbitrary intervals, usually deciles of *tmax;* IMR is the initial mortality rate calculated at maturation; and G is the mortality rate exponential coefficient (equation [1.2]). The MRDT is the mortality rate doubling time: MRDT = $(\ln 2)/G$ (equation [1.3]). Data on mortality rate as a function of age were obtained from mortality schedule data where available (*Apis, Aplysia, Lymaea, Caenorhabditis,* and *Drosophila*) or were extracted from survivorship graphs. The regressions usually had correlation coefficients $r^2 > 0.8$. Use of primary mortality schedule data would generally improve the precision of these estimates. The numbers in the table should be considered as first approximations within a two-fold range.

[1] Water flea. Mortality rates were estimated from graph of Ingle et al., 1937.

[2] Male blowfly. Mortality rates were estimated from graph of Comfort, 1979, which was based on data of Feldman-Muhsam and Muhsam, 1946. Females showed similar MRDT and *tmax*.

[3] Male and female fruit fly. Mortality rates were estimated from graphs of Rockstein & Lieberman, 1958. The large range of IMR indicates variations in local conditions. These figures are particularly open to refinement using the abundant life table data that have recently been gathered but not analyzed for mortality rate changes.

[4] Male and female housefly. Mortality rates were estimated from graph of Feldman-Muhsam and Muhsam, 1946.

[5] Honeybee worker, female. Mortality rates were calculated from mortality schedule tables of Sakagami and Fukuda, 1968. The winter worker bees were born in the summer and showed little mortality during the winter

Table 2.1   (*continued*)

when activity is less; the mortality rates of winter bees are calculated at ages > 245 days, when mortality rates accelerate at about the same rate as in summer bees. For maximum lifespans of queens, see text.
[6]Male saturniid hemileucine moths. The adults are aphagic because of degenerate mouthparts. Mortality rates were extracted from survivorship graph of Blest, 1963. These indications of accelerating mortality rates with respectable correlation coefficients (r^2 of 0.8–0.87 according to equation [1.2]) oppose the conclusion of Blest, 1963, 6, that "the rate of dying remains constant"; no basis was given for this statement.
[7]Opisthobranch mollusk; laboratory *Aplysia*. From calculations of Hirsch and Peretz, 1984.
[8]Soil nematode, range of genotypes. Mortality rates calculated by Johnson, 1987 and 1990, from mortality schedule for genotypes shown in Figures 6.1 and 6.2. Replicates varied over a two-fold range.
[9]Monogonont. Males lack an intestine and are aphagic as adults. Mortality rates based on survivorship graphs of H. M. Miller, 1931.
[10]Eumycota; yeast. Mortality rates of yeast mother cell during successive budding cycles was based on the calculation of Jazwinski et al., 1989; the exponential mortality rate per bud generation was then converted to calendar time by assuming one generation lasts an average of 2 hours.

2.2.   Developmental Determinants of Rapid Senescence

The first major aspect of rapid senescence concerns its occurrence as a specific consequence of development, through which adults have characteristics that doom them to short lifespans, unless restrained from normal activity by environmental perturbations or experimental intervention. It is obvious that most characteristics of senescence, or at least the limitations on the type of senescence, must be set during development when most characteristics of the future adult are predetermined. Such features as irreplaceable body parts that are subject to mechanical wear (exoskeletal appendages of insects; most teeth of mammals) or irreplaceable cells (most cells of adult insects; oocytes or neurons of mammals) are part of the characteristic adult body-plan that arises during development.[3] Any features of adult senescence that may be attributable to specific events in development need not, on the other hand, have been explicitly selected for on the basis of the later dysfunction. That is, we cannot say without much inquiry if a particular aspect of senescence is programmed genetically, whatever the role of genes may be in its occurrence. Moreover, organisms differ widely in the rigidity of these predestining developmental events, and alternative developmental pathways can lead to adults having the same genome but very different phenotypes of senescence. Social insects are discussed as an example of different developmental pathways that the same genome can direct, leading from the same batch of eggs to very long-lived queens or to short-lived worker bees. Peculiarities of development in some species cause the absence of mouth or gut organs, which leads to the inability to ingest a complete diet as adults (adult aphagy).

Another characteristic established during differentiation is the inability to regenerate vital organs of adults that are thence subject to irreversible damage from mechanical effects of usage: mechanical senescence, which is a recurrent canonical characteristic of senescence, whether rapid or gradual. Besides the wings and other appendages of insects that are considered in this chapter, the erosion of irreplaceable molars in adult mammals will be considered in Chapter 3.6.11.4. These characteristics of the body-plan implicate species differences in gene regulation during organogenesis that limit the potential lifespan. DNA sequences may be identified that determine the absence of particular organs

---

3. *Body-plan* has the same meaning as *Bauplan* in Gould (1977).

in the adult or the capacity for cell division that is important for functions of some organs, particularly those that must continuously shed cells, like the exfoliating epithelia of the gut.

Many other aspects of senescence must also be developmentally influenced, but the influence on the mortality rate need not be either strong or direct. The irreplaceability of adult molars in most mammals is likewise determined developmentally, although total tooth loss is not rigidly predestined (Section 2.2.2). The DNA sequences that prevent the regeneration of teeth in adult mammals are only a weak factor of mortality risk, rather than the major mortality risk factor associated with the genes that determine the absence of mouthparts in insects. The absence of a gall bladder in adult mice and its presence in rats shows that not all anatomic deficiencies have important effects on lifespan or other aspects of senescence, since laboratory-bred mice and rats are so similar in most of these aspects. On the other hand, the numbers of irreplaceable ovarian oocytes in mammals, while a weak determinant of longevity, are a major determinant of the age-related loss of fertility (Chapters 3, 8).

One more feature of development is pertinent to the examples that follow. Developmental plasticity that is environmentally dependent is widely documented for many species and considered a major potential source of evolutionary variation (West-Eberhard, 1986). Such alternatives in development are also known as polyphenism (Mayr, 1961; Shapiro, 1976). Moreover, closely related species can have very different developmental stages yet very similar adults. For example, the sea urchin, *Heliocidaris erythrogramma,* develops directly from the embryo to an adult, and bypasses the feeding larval stage that is typical of others in the genus and of sea urchins in general (Raff, 1987; Williams and Anderson, 1975). In this example, the alternative development arises from a different temporal programming of gene activities. The widespread phenomenon of plasticity in development gives a natural basis for inquiring about the developmentally influenced alternatives in senescence, which can also be regarded as alternative phenotypes. This theme will be followed in the book, particularly in Chapters 9 and 10, which summarize influences on alternative phenotypes of senescence.

2.2.1. *Anatomic Deficiencies in Mouth or Gut Organs*

Many insects and some other invertebrates are aphagous as adults because they lack mouthparts or digestive organs, this lack being a genetically programmed characteristic of organogenesis in the species. Other animals characteristically cease to eat during their adult life, as described later for Pacific salmon and spiders, but do so for very different causes that do not involve anatomic deficiencies. The inability to ingest a complete diet as free-living adults poses a fundamental limitation on lifespan and sets the animals on a course of obligatory rapid senescence, differing radically from the age-related changes that may accrue in most other organisms. In the absence of ingesting a complete diet, the adult must draw on the nutrient deposits accumulated during development, these deposits often becoming depleted during adult life from gametogenesis.

An important subtype of aphagy concerns differences in adult food requirements among closely related species, between genders, and sometimes even among populations. Those that require food after hatching to reproduce are called anautogenous, while those that do not are autogenous. As discussed in Chapters 6 and 11, autogeny with adult aphagy has arisen numerous times during evolution and may be favored when reproduction must be rapidly consummated in environments with uncertain food resources.

2.2.1.1. Aphagy in Adult Insects

2.2.1.1.1. Nutritional Deficits and Senescence

Aphagy from defective mouthparts or digestive organs is very common during the adult phases of insects (Weismann, 1889b; Metchnikoff, 1915; Norris, 1934; Brues, 1946; Wigglesworth, 1972; Dunlap-Pianka et al., 1977) and is *the* limiting factor in the adult lifespan of many short-lived species. This phenomenon is, inarguably, programmed senescence. Nearly all insects with deficient mouthparts have adult phases of 1 year or less.

The mouthparts of insects are among the most widely varying anatomical feature (Richards and Davies, 1977; Labandeira, 1990). Thousands of species of adult insects from at least nine orders cannot eat or drink at all or cannot digest a complete diet after hatching (Table 2.2). Most such species have degenerate nonfunctional mouthparts or lack them altogether. The larval stages of these (holometabolous) insects are the trophic phase. The first examples to be discussed are mayflies, followed by Diptera and other orders with sporadic occurrence of adult aphagy according to the species or gender.

All adults in extant species of mayflies (227 genera of the order Ephemeroptera) are completely aphagous and have long been known for their ultra-short adult phases that last a few minutes or hours to a few weeks (Edmunds, 1965; Brittain, 1982; Edmunds and McCafferty, 1988; Peters and Peters, 1988). Exceptionally long-lived mayflies have fully active adult phases of a few weeks at best. The total lifespan, however, is typically several years, mostly spent in developmental stages (naiads and nymphs), when feeding is active. The number of nymphal instars is unusually large (ten or more) and varies among species. Mayflies have a unique, winged subadult stage, the subimago, which is generally shorter than the immediately ensuing adult (imaginal) phase. Most species then undergo a complete molt (ecdysis) that may include further anatomic reductions. Gametogenesis is complete by the subimaginal stage, which allows some species to eliminate the adult phase completely. This precocious maturation of gonads is an example of neoteny (Chapter 11.4.2).

As the last molt approaches, the nymph stops feeding and the gut epithelium is phagocytosed by blood cells (Pickles, 1931); the cellular mechanisms have not been studied and may be an active process, as determined for other occurrences of cell death during development (Chapter 7.7). At hatching, the mouth is always degenerate and the abdomen is commonly inflated with air. Besides the reduction of mouth and digestive tract during maturation, males (e.g. in *Dolonia americana*) lose major portions of their legs, which is thought to speed their hatching or flight (Peters and Peters, 1986). The ultra-short adult phase of mayflies is further divided according to two patterns (both aphagous), in which the subimaginal and imaginal phases are less than a few hours or may last a few days to two weeks.

Hatching of a generation is often synchronous, within a few minutes, on virtually the same day each year (e.g. in *Dolonia americana*); this precise termination of the relatively prolonged developmental phase is thought to be under environmental control by light and temperature (Peters et al., 1987; Brittain, 1982). Males typically hatch first and immedi-

Table 2.2 Aphagy in Adult Insects

Order	Adult Lifespan	Degenerate Adult Mouth	Type of Feeding Deficit in Adults	
			No Feeding	Only Liquids
Coleoptera[1]				
Bruchidae (seed beetles)		yes		
Drilidae		yes		
Lampyridae		yes		
Phenogodidae		yes		
Rhipidiinae (wedge-shaped beetles)[2]		yes (males)		
Scarabaeoidae				
Lucanidae (stag beetles)		nectar at most		
Lichthnanthe		yes	yes	
Pleocomidae (*Pleocoma*)[3]	< 1 yr (males)	yes	yes	
Diptera[4,5]				
Nymphomyiidae		vestigial mouthparts; atrophied gut	yes	
Oestridae (botflies)		vestigial mouthparts	yes	
Syrphidae (hoverflies)		some	some species	
Ephemeroptera (mayflies)[7]	< 7 days	yes		sometimes water; otherwise complete
Homoptera				
Coccoidea (scale insects)[8]	1–10 days (males)	vestigial or absent mouthparts	yes	
Hymenoptera				
Ceraphronidae[9]				yes
Eupelmidae[10]				yes
Figitidae[11]				yes
Mymaridae[12]				yes
Platygasteridae[13]				yes
Proctotrupidae[12]				yes
Lepidoptera		functional mandibles rare		
Bombycidae (silk moths)[14]	< 30 days	yes	yes	
Eriocranioidea (dacnonyphan moths)[15]		vestigial mandibles		
Eucleidae (bagworm moths)[5]		Females only	yes females	
Hepialidae (swifts)[16]		vestigial mandibles		little or no solids
Lescocampidae[17]		yes	yes	
Saturniidae (giant silkworm moths)[18]	< 20 days	yes	yes	
Plecoptera (stoneflies)[19]	short	vestigial or absent in some species	yes	
Strepsiptera (twisted-wing parasites)[20]		female endoparasitic, without mouth or mouthparts; male has rudimentary		yes yes

Table 2.2 (*continued*)

Order	Adult Lifespan	Degenerate Adult Mouth	Type of Feeding Deficit in Adults	
			No Feeding	Only Liquids
		mouthparts that are sensory and not trophic		
Trichoptera (caddisflies)[5,21]		Some species have degenerate mouthparts; no functional mandible in 16 families; others in order have complete mouthparts		

Notes: The table refers to aphagy in the adult phase that arises through anatomical deficiencies in mouthparts or gut that impair or prevent ingestion of a compete diet, as indicated. This table does not represent a complete survey. Many more examples are scattered throughout the orders of insects and will be turned up by sifting through the enormous literature on insect anatomy. Only in mayflies are all members of an order aphagous. Insect phylogeny is shown in Figure 11.7. Valuable additions to this table were provided by Conrad Labandeira.

[1]Crowson, 1981; Lawrence, 1982.
[2]Silvestri, 1905; Grandi, 1936.
[3]Ellerston and Ritcher, 1959; Hovore, 1979.
[4]Bickel, 1982.
[5]Comstock, 1950.
[6]Holloway, 1976; Gilbert, 1981; Schuhmacher and Hoffmann, 1982.
[7]Edmunds, 1965; Richards and Davis, 1964.
[8]Kosztarab, 1982.
[9]Haviland, 1921.
[10]Clancy, 1946.
[11]James, 1928.

[12]Jackson, 1961.
[13]Hill and Emery, 1937.
[14]Norris, 1934; Rockstein and Miquel, 1973.
[15]Davis, 1978.
[16]Essig, 1947; Nielsen and Scoble, 1986.
[17]Rouchy, 1964.
[18]Balázs and Burg, 1963; Borror and Delong, 1954; Blest, 1963.
[19]Bauman, 1982.
[20]Kathirithamby, 1989.
[21]Crichton, 1957; but see Porsch, 1958.

ately swarm to mate with the females as they emerge. After one or more matings, *Dolonia* soon become too weak to fly and fall into the water (Peters and Peters, 1977, 1988). The degenerative changes that precede death have not been characterized. It would be valuable to know about changes in glycogen and other energy reserves. In the absence of mortality rate data for the very brief adult phases of mayflies, we do not know whether mortality rates accelerate progressively or increase in one huge step.

Unlike mayflies, many other insects can digest a partial diet. The best studied are the butterflies and moths (order Lepidoptera), which usually ingest only carbohydrates (Norris, 1934). Other Lepidoptera can feed as adults on nitrogen-containing pollen, but the relative prevalence of these two diet patterns is unknown (Conrad Labandeira, pers. comm.). The giant silkworm moths (family Saturniidae) and the domesticated silk moth (*Bombyx mori*) have adult phases of 6–8 weeks, while some silk moth strains have very short adult phases of a few days (Murakami et al., 1985). Other saturniid species typically live less than 2 weeks (Blest, 1963; Table 2.1). The age-related decrease of some amino acids in *Bombyx* (Osani and Yonezawa, 1984) is consistent with aphagy and degenerate mouthparts. Metchnikoff himself (1915) noted that adult *Bombyx* had few intestinal bacteria, indicative of aphagy. The great wax moth *Galleria mellonella* also appears to die from nutrient exhaustion and shows major loss of body weight (40%), nitrogen (10–20%), and lipids (30–50%) (Balázs et al., 1962). Adult butterflies and moths show extremely rapid accelerations of mortality; MRDTs are on the order of a few days, which is among the shortest in animals and about 25% of those in laboratory flies (Table 2.1). The mortality rate curves are well fitted by the Gompertz model.

An instructive variant of aphagy in moths with short adult phases is the lymantrid *Orygia dubia judaea* (Hafez and El-Said, 1969). Both genders are aphagous as adults; males have a shorter adult phase (< 2 days) than females (< 2 weeks). The wingless female never emerges from her cocoon, where the male finds her for mating. If mating fails to occur, eggs develop parthenogenetically. Here we see further shortening of the lifespan by elimination of the adult phase. Although the evolution of parthenogenesis in this species is incomplete, many other insect orders show parthenogenetic reproduction (Bell, 1982).

However, other adult butterflies continue to feed and maintain total body fat content during their adult phase, for example, in *Lysandra* (Gere, 1978). The long-lived butterfly *Heliconius charitonius* eats pollen as a source of amino acids and continues to lay eggs throughout a 6-month adult phase (Gilbert, 1972; Dunlap-Pianka et al., 1977). Despite advanced ages, pollen-feeding heliconiines show no shrinkage of their fat body, an organ for storage of reserve nutrients. The presence of all stages of oogenesis and numerous mitotic figures in the *Heliconius* ovarioles indicates continued *de novo* formation of oocytes and the possible absence of ovarian senescence due to oocyte exhaustion. In contrast, the related *Dryas julia* feeds only on nectar and dies at about 1 month with depleted ovaries. *Heliconius,* if fed only carbohydrates, has a much reduced lifespan and prematurely stops laying eggs to become a "phenocopy" of *Dryas.* Another heliconiine that cannot feed on pollen, *Agraulis vanillae,* also shows ovarian depletion. Dunlap-Pianka et al. (1977) suggest that ovarian decline in *Dryas* is not programmed at the level of the ovary, but rather is determined by "programmed starvation." The shutdown of ovarian function during protein starvation in *H. charitonius* or during usual senescence in *Dryas* could be regulated by juvenile hormone or by availability of yolk proteins such as vitellogenin, which are present in hemolymph. Yolk protein production might decrease during nutrient depletion. Mammals also show decreased ovarian activity during diet restriction, as well as slowed ovarian oocyte loss (Chapter 10.3.2.8.2).

Two special adaptations are indicated. The need for nitrogenous nutrients to complete oocyte maturation in the pierid butterfly *Colias erytheme* is met by the male, which transfers amino acids into the female along with his sperm (Boggs and Watt, 1981). Moreover, some female flies (e.g. *Probezzia flavonigra*) consume the male during mating by puncturing his head and liquefying his tissues (Downes, 1971). Presumably, these nutrients are used to complete oocyte maturation.

Senescence is attributable to aphagy from developmentally programmed anatomical deficiencies in other insects as well (Table 2.2). Among beetles, the aphagous *Callosobruchus maculatus* shows a more than 50% loss of total body lipids during its short lifespan of under 2 weeks (Sharma et al., 1982). However, not all aphagous beetles are short-lived (see below). Among the Diptera, all botflies (Oestridae) and certain hover flies (Syrphidae) have degenerate mouths and mouthparts. However, many dipteran families retain powerful mandibles, for example, mosquitos (Culicidae). Related dipteran genera show major differences in oral anatomy. Among the few remaining species of the archaic Tanyderidae, the Australian *Radinoderus* has well-developed mandibles while the North American *Protoplasa* has nonfunctional vestiges (Downes and Colless, 1967). These scattered examples hint at a widely distributed phenomenon that has not been systematically analyzed.

In light of these examples, how should we regard the laboratory Diptera that are used as models for senescence? As described in the next section on autogeny and anautogeny, many ultra-short-lived flies and mosquitos are aphagous as adults. However, the role of nutrient reserves in the 5- to 10-week lifespan of the fruit fly *Drosophila melanogaster* has not been well characterized. The shrinking fat body (Krumbiegel, 1929; Miquel, Bensch, Philpott, and Atlan, 1972; Rockstein and Miquel, 1973) and smaller glycogen content (Williams et al., 1943) strongly suggest that nutrient reserves are depleted or exhausted during senescence in *Drosophila*.

A possible early biochemical marker for impending inanition in *D. mojavensis* is the decrease of the larval isozyme ADH-1 (alcohol dehydrogenase) during the first week of the adult phase, which represents the histolysis of the larval fat body (Figure 2.1; Batterham et al., 1983); ADH-2, which persists into adult life, is found in the adult fat body and would be interesting to study at later ages. Clearly, adult *Drosophila* species depend on continued feeding, because starvation with access to water sharply decreased lifespan from 39 days to 3 days (Kopec, 1928; Chapter 10.3.1.3). *Drosophila* lines that were selected for extended reproductive schedules and 30% longer lifespans also had greater resistance to starvation and desiccation (Service et al., 1985; Chapter 6.3.2), implying greater nutrient reserves with a possible role in longevity in the absence of these stressors. Failure of the digestive and excretory systems during senescence has also been considered (Miquel, Bensch, Philpott, and Atlan, 1972); no direct tests of these systems have been reported across adult ages. Major questions remain about the effects of aging on the bioavailability of nutrient reserves and the efficiency of ingestion and digestion in *Drosophila* and other short-lived insects with functional mouthparts. Senescence in flies is discussed further in Section 2.2.2.1.

Gender differences in oral anatomy are numerous in the Diptera. In many biting Diptera, such as mosquitos (Culicidae), blackflies (Simuliidae), and midges (Ceratopogonidae), the adult females are equipped with strong, toothed mandibles for obtaining blood meals from vertebrates or even other insects. In contrast, males usually feed only on carbohydrates in nectar and often have reduced mouthparts (Gad, 1951; Downes, 1958a, 1958b, 1971; Davies and Peterson, 1956; Peterson, 1981; Anderson, 1987). A striking example of gender dimorphism in mouthparts is the biting midge *Dicrobezzia venusta* (Figure 2.2). Sex differences in adult feeding are a hallmark of nematocerous Diptera. However, in the common laboratory brachyceran Diptera, *Drosophila* and *Musca*, gender differences in oral anatomy are not prominent.[4] Nonfeeding (or autogenous) Diptera gen-

---

4. The "higher" Diptera, or suborder Brachycera that includes the Cyclorrhapha, feed by sponging surface fluids with their labellum. Most of the species that display gender dimorphism in mouthparts are in the suborder Nematocera and have stylate mouthparts for piercing and sucking subdermal fluids from animals and plants (Labandeira, 1990).

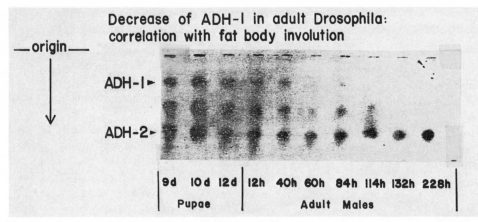

Fig. 2.1. Involution of the larval fat body after hatching in *Drosophila mojavensis,* as detected on the electrophoretic profiles of the larval isozyme ADH-1 (alcohol dehydrogenase), which decreases during the first week of the imaginal phase (Batterham et al., 1983), when the larval fat body degenerates. ADH-2 persists into adult life in the adult fat body and would be interesting to study at later ages. Rephotographed from Batterham et al., 1983.

erally die within a few days after hatching (Downes, 1965, 1971; Edwards, 1929; Wood, 1978). Gender differences of oral anatomy outside the Diptera are considered less common (Downes, 1971). One example is the bagworm moths (family Eucleidae), in which only females lack mouthparts (Table 2.2).

A prediction for aphagous species is that starvation should have little effect on their adult lifespan. Life-shortening effects of starvation from aphagy were shown in the short-lived moths *Ephestia cautella, E. elutella,* and *E. kühniella,* which usually live less than 2 weeks if given only water. Longevity was increased 30% in both sexes by providing sucrose solutions (Norris, 1934). The lack of effect on longevity of albumin solutions is consistent with the absence of proteases in the lepidopteran gut. The fat body was exhausted at death in moths fed only on water, but was not exhausted in most moths fed on sucrose solutions. It is interesting that egg production in *Ephestia* was not influenced by diet (Norris, 1934); the extension of lifespan occurs in the postreproductive phase which is analogous to menopause. While it is unknown if *Ephestia* or other moths survive after egg laying ceases in natural populations, these laboratory observations show this possibility.

Laboratory studies of mosquitos encounter a different aspect of this problem. Adult female yellow fever mosquitos (*Aedes aegypti*), which are anautogenous, can ingest protein obtained from whole blood, but live longer if fed only on sucrose (Clements, 1963; Putnam and Shannon, 1934; Pena and Lavoipierre, 1960a, 1960b; Lavoipierre, 1961). Their ovaries do not develop at all on a sucrose diet. Sucrose-fed mosquitos are a preferred model by some experimentalists, because the ovarian protein synthesis required for egg maturation has a major impact on systemic physiology and metabolism (Lang et al., 1972; Chapter 7.4.3). Despite the absence of nitrogen intake on a diet of only sucrose, female *Aedes* do not show loss of total body protein or nucleic acids (Lang et al., 1965). However, there are major decreases of cysteine and glutathione during the 4-week adult life-

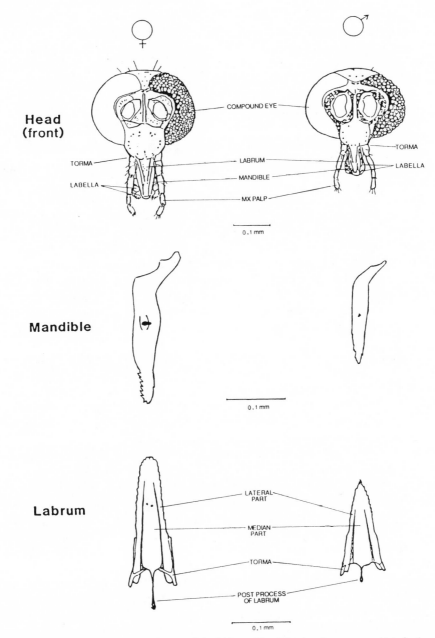

Fig. 2.2. Gender differences in mouthparts of the biting midge *Dicrobezzia venusta,* which feeds on other insects in Scotland. Females have serrated mandibles for piercing that are threefold larger than the nearly toothless male mandible. The labus, through which females draw blood, is also smaller in males. Redrawn from Gad, 1951, by Lali Medina.

span (Richie and Lang, 1988). Because starvation on diets of water did not alter levels of cysteine or glutathione within 3 days, the slow age-related decreases of cysteine and glutathione on the sucrose diet are interpreted as independent of starvation. The extent of recycled catabolized nitrogen is not known.

2.2.1.1.2.   Autogeny and Anautogeny

In general, Diptera whose eggs are immature in newly hatched adults require a blood meal (anautogeny), while those with mature eggs at hatching do not, at least for their first clutch of eggs (autogeny) (Downes, 1971; Spielman, 1971; Anderson, 1987). Within numerous genera of blackflies and mosquitos, some species were autogenous at hatching, while others had immature eggs that required a blood meal for ripening (Spielman, 1971; Davies, 1978; Anderson, 1987). Anautogenous species are characterized by five or more cycles of egg laying, production of larger numbers of small eggs, stronger mouthparts, and longer lifespans (Anderson, 1987). An extreme example is the palearctic *Anopheles maculopensis*, which lays up to thirteen batches of eggs during its adult phase, which can extend across three seasons because of overwintering (diapause; Gillies, 1964). Gillies' review also graphs survivorship as a function of the number of egg-laying cycles for natural populations of this and other mosquitos. There is little or no evidence for accelerating mortality; the graphs resemble the exponentially declining survival of Figure 1.7 when $R \gg G$.

Many genera show remarkable differences in autogeny among species. For example, females of the primitive blackfly *Cnephia eremites* have unserrated mouthparts that are unsuited for biting and drawing blood, while *C. mutata* is well equipped for this (Davies and Peterson, 1956; Downes, 1971). Corresponding to the ability to feed on a protein-rich source (blood) as adults, the eggs of the autogenous and weak-mouthed *C. eremites* are already mature by hatching, while eggs of the fiercer anautogenous *C. mutata* are smaller and appear to require a blood meal for maturation. A common association of aphagy is the resorption of oocytes in some species upon failure to obtain a blood meal; the resorbed proteins are used to complete the maturation of the remaining few (Corbet, 1967; Downes, 1971; Anderson, 1987). There is an age-related increase in the resorption of oocytes in mosquitos, as well as in species in other insect orders, according to scattered references assembled by Bell and Bohm (1975). A similar process may be important to oocyte maturation in nonparasitic lampreys, which do not feed in their short adult phase (Section 2.3.1.4.1).

A neuroendocrine basis for autogeny was revealed by Lea's (1970) classic work on the mosquito *Aedes taeniorhynchus*. Ablation and grafting of neuroendocrine glands (corpora allata and medial neurosecretory cells of the procerebrum) showed the necessity of a blood meal to the neurosecretory cascade that induces the synthesis of vitellogenic proteins by the fat body that are needed for egg maturation. However, even without a blood meal, the nutrient reserves suffice to mature 60% of the normal egg clutch, as shown by grafting supplemental neurosecretory cells. Both juvenile hormone and ecdysone are needed for vitellogenesis (formation of the vitelline layer of the next batch of eggs) after blood meals (Borovsky et al., 1985; Lea, 1982).

An intriguing geographical cline of autogeny is found in the pitcher-plant mosquito (*Wyeomyia smithii*). From New Jersey north (at or above 40°N), this species does not take blood meals and is obligately autogenous; its eggs are mature and yolky at hatching (Brad-shaw and Lounibos, 1977; O'Meara and Lounibos, 1981). Southern populations (below 36°N), in contrast, are transitionally autogenous and can produce one clutch of eggs au-togenously, though at eclosion, eggs have little yolk and require more time before the eggs are mature; subsequent batches of eggs require blood meals. These races are interfertile and appear to differ in gene polymorphisms that determine the time when a gonotropic hormone is secreted from the head. The difference is at a neuroendocrine level, since precocious oocyte maturation can be induced in the more southerly pupae by a juvenile hormone analogue. Genetic influences on the incidence of autogeny in mosquitos are discussed in Chapter 6.3.3.

An open and important question is the capacity of Diptera for continued oogenesis, particularly the long-lived anautogenous species with numerous cycles of oviposition. *Anopheles maculopensis* showed major decreases in fertility with age because fewer ova-rioles in each mosquito produce mature ova (Detinova, 1955). These less productive ova-rioles might merely be quiescent, lacking neuroendocrine signals, or might be depleted of cell oocytes. Quiescent ovarioles are found in young mosquitos. Thus, the age-related loss of fertility could derive from ovarian and/or neuroendocrine changes, as observed in *Drosophila* (Section 2.2.2.1).

The lifespan of biting flies and the distribution of autogeny has great significance to the transmission of parasitic diseases to humans and the vertebrates upon which humans de-pend. The capacities of different biting flies as vectors depend both on the need for blood meals and on whether the fly lives long enough to acquire the parasite (Gillies, 1964). For example, even under favorable conditions, only a minority ($< 10\%$) of *Anopheles* in East Africa will have enough blood meals and live long enough to transmit malaria in their short lifespan of 1 to 2 weeks. Tsetse flies (e.g. *Glossina morsitans*) are among the longest-lived Diptera, with continuously active adult life phases up to 5 months, during which there can be at least five cycles of oogenesis, each triggered by a blood meal (Saun-ders, 1962a, 1962b). These characteristics give the tsetse its fearsome efficacy as a vector, because the protozoan trypanosomes that cause sleeping sickness in humans and other diseases in domestic animals must reside in the tsetse for 2 to 3 weeks before they can be transmitted. Shorter-lived species would not be effective vectors. Many other examples show how variations in lifespan influence the life cycles of associated organisms. Such temporal factors in life cycle coordination bear on little-discussed questions in the evolu-tion of life history (Chapter 11).

2.2.1.2. *More Examples: Other Invertebrates and* Tokophrya

Developmental defects that influence the adult phase also occur in other phyla, includ-ing the Protista. One of the best examples is given by male rotifers, which are much shorter-lived than females, surviving only hours to a few days (Hyman, 1951, 123), with clear exponential increases of mortality (Figure 2.3, top). In contrast to females, male rotifers lack an intestine, anus, or other excretory and digestive organs (Figure 2.3, bot-tom) (Tannreuther, 1919; Beauchamp, 1956; Thane, 1974; Hyman, 1951; Remane, 1932; Gilbert, 1968, 1988b; Miller, 1931). In the two monogononts, both genders have similar development initially, but then select degeneration occurs so that adult males lack the stomach and other organs. Thus, as in mayflies, at least some of the tissues missing in

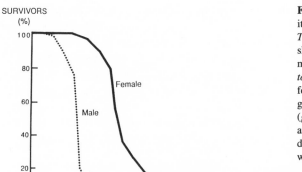

Fig. 2.3. Gender differences in longevity and adult morphology of two rotifers. *Top, Lecane inermis* survivorship curves, showing the markedly shorter lifespans of males. Redrawn from Miller, 1931. *Bottom, Asplanchna brightwelli gosse. Left,* females have well-developed stomach (*s*), gastric glands (*g*), and germovitellarium (*gv*). *Right,* the males have only degenerate remnants (*r*) of these organs and lack a digestive tract; (*t*), testis. Rephotographed with permission from Gilbert, 1968.

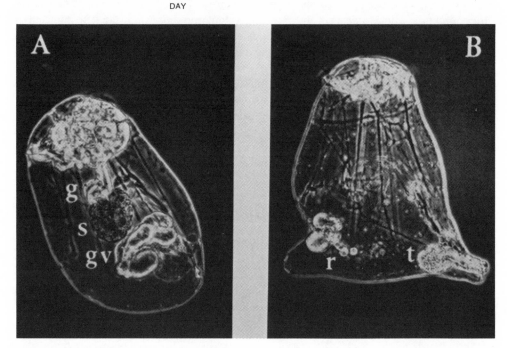

adults are present in an early form. While male rotifers cannot feed, their lack of excretory organs also could cause death by autointoxication from metabolic wastes. Most rotifer populations consist of parthenogenetically reproducing females, and males are described in only a few species, particularly in Monogononta (Wesenberg-Lund, 1923; Bell, 1982). While adult female rotifers feed actively, they nonetheless lose digestive functions during senescence and may die from malnutrition as do the males (Section 2.2.3.2).

Here are a few other examples. The dog tick *Rhipicephalus sanguineus* (an arachnid) usually dies 3–5 days after laying its only clutch of eggs. The increase of lifespan up to 31 days after injection of a sugar solution into the body cavity (Achan, 1961) suggests that exhaustion of nutrient reserves is the pacemaker of its lifespan. Dicyemid parasites of octopuses are simple organisms (probably degenerate flatworms) with a brief free-living larval stage during which they cannot eat. Upon infecting their hosts, the somatic cells deteriorate, while the germ cells proliferate to form the next generation (Stunkard, 1954). The orthonectid parasites of brittle stars have similar life cycles (Caullery, 1952; Stunkard, 1954). In the water flea *Daphnia* (crustacean), which lives 1–2 months, the fat body and intestinal epithelium deteriorate (Schulze-Robbecke, 1951). In the laboratory, limitations of vitamins or other nutrients accelerate daphnid senescence (Comfort, 1979, 121).

In regard to the evolutionary potential for modifying lifespan, it is important that some parasitic invertebrates lacking a digestive system nonetheless are thought to have considerable longevity, because they can absorb nutrients directly from their hosts, for example, the degenerate crustaceans, *Sacculina* and *Xenocoeloma* (Caullery, 1952). Through parasitism, an anatomically degenerate species could still access the evolutionary opportunities of a longer lifespan.

A different life history is shown by the protozoans *Tokophrya infusionum* and *T. lemnarum* in the family Dendrosomatidae. These are single-celled, sessile suctorians with tentacles for catching and ingesting small ciliates like *Tetrahymena* (Colgin-Bukovsan, 1979; Rudzinska, 1961a, 1961b, 1962). *Tokophrya,* unlike most other protozoans, lacks the cytophage, a cytoplasmic organelle for eliminating waste products (Rudzinska, 1962; Smith-Sonneborn, 1981). This species reproduces asexually by internal budding. The free-swimming bud attaches itself and metamorphoses into a stalked adult with a short lifespan, lasting less than 2 weeks. Adults reproduce by budding for 4 to 5 days and will feed continuously. Cultures that are propagated vegetatively eventually show clonal senescence (Chapter 3.2). If an endless supply of living ciliates is available, they enlarge to 120 times their normal size. This gluttony shortens the lifespan to less than 4 days, whereas food restriction lengthens it (Rudzinska, 1951, 1952, 1962). Budding ceases, most tentacles are lost, and feeding becomes ineffective. The cytoplasm of senescent or overfed young *Tokophrya* becomes filled with vacuoles containing 300 Å particles and small, dense lipoidal bodies that resemble lipofuscins (aging pigments). These accumulations appear to be insoluble waste products, possibly partially degraded proteins and oxidized lipids. Lipofuscins are often associated with, and may derive from, mitochondria. The inevitable senescence of *Tokophrya* is thus a consequence of autointoxication from digestive wastes that cannot be eliminated because of species-characteristic congenital defects in exocytosis. The distribution of other species in the family that lack a cytophage is unknown. It would be interesting to examine *Dendrosoma:* this monster among ciliates has ramified branches up to three millimeters high that do not suggest a life history terminated by autointoxication as in *Tokophrya.*

2.2.1.3. *Aphagy without Rapid Senescence*

In contrast to the short-lived aphagous insects, other invertebrates may have extended lives as aphagous, but active, adults. Several different cases are represented in beetles. In rain beetles *Pleocoma* (Scarabaeoidea), adults of both sexes are completely aphagous from reduced oral anatomy. Adult females can survive for at least a year (Ellertson and Ritcher, 1959; Hovore, 1979). Their large size (3–4 centimeters) and relatively greater nutrient stores may permit this unusually long adult aphagous phase. Female *Pleocoma* are semelparous, unlike many other beetles (Chapter 3.3.5), and die within days or weeks after laying fifty or so large eggs the size of peas (Fellin, 1981; Frank Hovore, pers.

comm.). *Pleocoma,* with a prolonged development of 9 to 13 years (Ellertson and Richter, 1959), have abundant fat stores at hatching, which became depleted by death in one specimen kept by Hovore (pers. comm.). Another scarabaeid, *Lichthnanthe,* also has aphagous adults (Crowson, 1981, 401).

Other beetles may starve slowly despite the presence of mouthparts at hatching. The short-lived *Pterohelaeus darlingensis* (Tenebrionidae) shows a nutritional imbalance from the progressive diminution of abdominal fat in both sexes during a 2- to 3-month adult phase (Allsopp, 1979). The decreasing presence of fat is used by agriculturalists to estimate the age of this pest, which feeds as an adult on wheat sprouts. A final example among beetles illustrates incidental aphagy from mechanical senescence. In *Carabus* (Carabidae), which is carnivorous and definitely not aphagous (Thiele, 1977), the fat body often shrinks during a 1- to 5-year adult lifespan (Krumbiegel, 1929; Houston, 1981). This is a result of mechanical wear on the mandibles, which impairs feeding and leads to death by starvation (Houston, 1981). *Carabus* specimens with severely worn mandibles also have small ovaries. Ovarian maturation, oviposition, and even the fat reserves are restored to ripeness by macerated liver, which the senescent beetles still can ingest (Houston, 1981).

The Mermithidae, nematode parasites of insects and snails, are long-lived, although the adults are aphagous because of nonfunctional mouths. Adults draw on nutrient stores that they acquire during an intermediate parasitic stage (Hyman, 1951, 277; Maggenti, 1982, 894), for example *Agamermis decaudata* (Christie, 1929). After hatching from eggs laid in the earth, juvenile worms enter their hosts, grasshopper nymphs. *Agamermis* remain in the hemocoele up to 3 months, where food reserves accumulate in intestinal trophosomes. After returning to the soil, they overwinter. The trophosomal food stores give the young adults an opaque appearance (Christie, 1929). No food is ingested by mermithids after the parasitic phase because the intestine becomes disconnected from the pharynx and anus (Hyman, 1951, 239). During the prolonged phase of egg laying, the increasing transparency of the worms is thought to indicate exhaustion of their food stores (Christie, 1929). However, if kept in the laboratory, unfertilized female worms do not appear to rapidly exhaust their food reserves and live more than 2 years. While no mortality rate data are available, these descriptions suggest that adult aphagy is not incompatible with relatively long lifespans and more gradual senescence.

Nutritional deficiencies may arise because insects as a class cannot synthesize *de novo* the sterol ring that is needed for the important ecdysteroid hormones. Species differ widely in the particular sterols they must obtain from their diet or symbionts (Norris et al., 1986; Svoboda and Thompson, 1983). The long-lived ambrosia beetle *Xyleborus ferrugineus* requires a dietary source of Δ^5, Δ^7 sterols in order to make the ecdysteroids that are transferred into each egg during oogenesis; this species ceases to lay eggs about a month before death at 6–9 months (Norris and Moore, 1980). Although most other insects can use cholesterol (Δ^5 sterol) as a precursor for sterols, *Xyleborus* and a few others cannot. If cholesterol is substituted for the required sterol, then the duration of reproduction, locomotor activities, and lifespan are shortened by about 30%, and the ovarian follicular cells become structurally abnormal (Norris et al., 1983). Nonetheless, the whole body content of ecdysteroids was neither altered nor changed with age in cholesterol-fed beetles or controls. This result is equivalent to the life-shortening effects of vitamin deficiencies and raises questions about irreplaceable steroids or other nutrient stores in the senescence of other insects.

In conclusion, the various extents of dietary insufficiency in aphagous adults result in some cases from developmentally determined deficiencies of mouthparts in the adult. Developmental determinants of the capacity for replacement and regeneration are no less a factor in other species that are fully equipped at hatching with mandibles, teeth, etc., that are worn out adventitiously and cannot be replaced. The end result in both conditions is the exhaustion of metabolic reserves, such as glycogen in insect flight muscles, and the major reserve depots of fat body cells for protein, carbohydrates, and lipids. Rapid senescence in autogenous species is clearly obligatory and one of the best illustrations of the "natural death" that was well noted by early biologists (Metchnikoff, 1915; Weismann, 1889a, 1889b; Wallace, 1870 [date approximate, see Appendix 3]). Aphagy is also observed in spawning Pacific salmon, eels, and lampreys (Section 2.3). Aphagy can occur in most vertebrates as a consequence of loss of irreplaceable teeth owing to injury or erosion from use; loss of teeth is incidental, as in the erosion of mandibles in beetles. The aphagous death of these species certainly underestimates the potential of their cells for continued function. For example, in a further twist to the aphagy tale, decapitation of the blood-sucking bug *Rhodnius prolixus* prevented molting after a meal, but also increased the lifespan manyfold to beyond a year (Wigglesworth, 1934). Decapitation also increases the lifespan of butterflies (Camboué, 1926).

2.2.2. Lack of Cell Replacement and Mechanical Senescence

The capacity for cell replacement and regeneration in adults varies widely, and is a major determinant of longevity and the rapidity of senescence in certain species. The extent of cell proliferation in adults is acquired during development as one of the characteristics of differentiation. On one extreme are soil nematodes, rotifers, and many insects in which adults are virtually postmitotic (having few or no cells that are capable of dividing). On the other are sponges and other asexually reproducing organisms in which most or all cells continue proliferation indefinitely and give the capacity for extensive regeneration of damaged organs; such species tend to have indefinite lifespans (Chapter 4). Mammals and many invertebrates with rapid or gradual senescence have intermediate powers of cell replacement and regeneration.

The relationship of cell proliferation to senescence is a long-standing theme in theories of senescence. Weismann, in his famous essay "The Duration of Life" (1889a), was the first to discuss possible relationships between the number of generations in somatic cells, senile changes, and the lifespan of the organism. There is no doubt that the cessation of mitosis increases the risk of organismic mortality, since postmitotic cells often accumulate damage during the lifespan that compromises organ function. Many short-lived and rapidly senescing species have few or no dividing cells, and little or no capacity for regeneration as adults. However, we cannot conclude that the postmitotic state immediately or inevitably brings senescence or loss of other functions, since full functions are retained by postmitotic myocardial cells and many types of neurons in mammals for many decades, maybe even beyond a century, since their last division in humans (Chapter 3.6.9). Moreover, the loss of proliferation by diploid human fibroblasts *in vitro* (a popular model for cell senescence) does not result in diminished capacity for many other biosynthetic activities (Chapter 7.2.5, 7.6.4). Thus, the postmitotic status should not be equated with cell senescence.

It helps to classify irreplaceable (postmitotic) cells and cell organizations according to whether age-related losses occur through programmed or incidental cell death. Programmed cell death concerns losses that always occur in all individuals of all populations of a species. The most clear case is the programmed cell death that occurs within a defined time during development, for example, in the morphogenesis of vertebrate limbs and insect wings, and in the nervous system. The mechanism in some cases involves changes in gene activity that are controlled by trophic factors (Chapter 7.7). During the adult phase, there are also cell populations that, being postmitotic, are subject to loss. In some cases, the loss occurs over extended time and is hormonally influenced. For example, the fixed stock of ovarian oocytes in mammals is slowly lost at a relatively constant rate, somewhat like radioactive decay during adult life (Figure 6.11). Most oocyte loss represents atresia, or death during oocyte maturation that is distinct from the minority of oocytes that are eventually ovulated (vom Saal and Finch, 1988; Gosden, 1985). The rate of loss is, however, under hormonal control, such that removal of pituitary gonadotropins by hypophysectomy of young rodents greatly slows atresia (Chapter 10.8); conversely, atresia of the remaining follicles may be accelerated just before menopause, possibly as a consequence of chronic elevations of gonadotropins (S. J. Richardson et al., 1987). While the molecular mechanisms of atresia and oocyte loss are unknown, the regularity of this phenomenon in all mammals qualifies it as an example of programmed cell death over extended time.

On the other hand, some neurons, myocardial cells, and other postmitotic cells are not regularly lost in all humans during adult life. Any loss of postmitotic cells can then be considered as incidental, occurring as a consequence of events that vary among individuals, like stroke, mechanical trauma (Chapters 3.6.7, 3.6.10), or perhaps chronic stress (Chapter 10.5). Incidental cell death thus results from factors that may vary widely among individuals and that are often strongly influenced by the environment. Nonetheless, genetics strongly influences incidental cell death, because the capacity to replace lost cells or tissues is determined during development as an aspect of cell differentiation. While the timing of incidental cell death is not immediately under genetic control, the patterns of differential genomic expression clearly set constraints on the extent of incidental cell death during adult life. In many examples from organisms that are postmitotic as adults, the loss of cells appears to be incidental rather than programmed.

2.2.2.1. Insects

Mechanisms of senescence in short-lived insects are likely to be very diverse. Besides the aphagy described above, another key factor must be the postmitotic status of virtually all cells in most adult insects. This places the insects at risk for debilitating damage to irreplaceable cells and organs, particularly to their fragile wings and exoskeleton, which Comfort (1979) has called mechanical senescence. Such damage is irreversible. While phenomena of mechanical senescence are sometimes disdained as being trivial, they represent a major feature of senescence for which insects apparently have developed a range of adaptations, most of which are behavioral. This discussion emphasizes flies, whose senescent changes are described in the most detail, but other species will be mentioned.

The housefly (*Musca domestica*) and the fruit fly (*Drosophila melanogaster*) have adult lifespans of 1 to 3 months with exponential accelerations of mortality (Table 2.1) associated with marked degeneration of wings and other organs, as discussed in detail below.

Few winged insects of similar construction have much longer adult lifespans (Comfort, 1979; Corbet et al., 1960), except for dragonflies, worker bees, and others with diapause or quiescent phases (overwintering) that extend total lifespan. Beetles (Section 2.2.1.1) and worker ants (Section 2.2.3.1), however, which also have fragile exoskeletons, have adult phases of several years without replacement of their cuticles. At the far extreme are the insect queens, which live over 5 years (see below).

The capacity for cell proliferation and macromolecular biosynthesis in adult insects will be reviewed before considering the evidence for mechanical senescence. Adult fruit flies and mosquitos incorporate [^{3}H]-thymidine into gonadal cells, but also into cells of visceral organs. In adult *Drosophila,* most cells are postmitotic, with the exception of gonadal cells and gut (Bozcuk, 1972; Maynard-Smith, 1962; Mayer and Baker, 1985). The incorporation of [^{3}H]-thymidine into gut cell nuclei even in senescent flies is thought to be due to continuing endoreplication of their amitotic, polytene chromosomes (Bozcuk, 1972); occasionally neurons are also [^{3}H]-thymidine labelled just after eclosion (Technau, 1984). Mosquitos (*Culex pipiens*), another short-lived dipteran, also incorporate [^{3}H]-thymidine into gut and fat cells up to 3–4 days after hatching, and possibly longer (limited mention was given of nongonadal tissues); incorporation into spermatogonia and oocytes continued for about 2 weeks (Sharma et al., 1970). Although mitotic figures have not been described in somatic cells of adult flies or mosquitos, worker bees show mitotic figures in the intestinal crypts (Snodgrass, 1956, 189) and in brains (observations of P. Mobbs, cited in Technau, 1984). Cytogeneticists studying grasshoppers (Acrididae) also rely on finding mitotic figures in the gonads and cecum of adults (e.g. John and Hewitt, 1966; John and Freeman, 1976). Although blood cells of the cockroach *Blaberus giganteus* rarely show mitotic figures, occasionally 6-month-old cockroaches had increased numbers of prohemocytes (Arnold, 1959a), suggesting possible resurgence of hematopoeisis. Thus, it is likely that many adult insects, even short-lived species, have continuing cell replacement of some soft tissues. A related question concerns the occurrence of tumors in adult insects, which, for example, have not been reported in *Drosophila* (Miquel, Bensch, Philpott, and Atlan, 1972), but are described in insect queens (Haydak, 1957). I also note a phylogenetic difference: no adult insect can regenerate or replace components of its exoskeleton; spiders can do so, but they are in a different arthropod subphylum (Chelicerata) than the insects (Uniramia).

Little is known about cell loss in any arthropod. *Drosophila* show a small loss of neurons in the mushroom bodies of the brain after 4 weeks (mid-life; Technau, 1984). Neuron atrophy also occurs (Miquel and Philpott, 1986; Kern, 1986) and may be more consequential than cell loss.

In *Drosophila,* the bulk of proteins may not be replaced during the lifespan (Clarke and Maynard- Smith, 1966; Baker et al., 1985). There are also suggestions of marked impairments of protein synthesis (Webster, 1986; Niedzwiecki and Fleming, 1990; Chapter 7.3.3). The role of nutrient availability in these changes is unknown (e.g. through fat body shrinkage; see Krumbiegel, 1929; Rockstein and Miquel, 1973; Miquel, Bensch, Philpott, and Atlan, 1972; Miquel and Philpott, 1986). The mitochondria and some macromolecules may not be renewed in flight muscle (Maynard-Smith et al., 1970; Tribe and Ashhurst, 1972); ribosomes are decreased during senescence in nerve and gut cells, but not in flight muscle (Miquel et al., 1976; Miquel and Philpott, 1986). Nonetheless, when challenged by heat shock, senescent flies can make new proteins (Figures 7.11, 7.12).

Senescent flies also accumulate lipofuscin-like aging pigments in many cell types (Miquel et al., 1976) with which no dysfunction has been associated; the slowed turnover of proteins and the accumulation of lipofuscins may be related phenomena. These and other molecular changes are discussed in Chapter 7.

The wings, cuticular hair, and other brittle irreplaceable parts of the exoskeleton accumulate wear-and-tear damage from friction and collisions that is a marker of age in field studies (Tyndale-Biscoe, 1984). The frayed wings of flies within several weeks of hatching is common and epitomizes mechanical senescence. In the laboratory, flying capacity progressively decreases after the first week in *Drosophila* (Figure 2.4, top) and many other insects (Rowley and Graham, 1968; Collatz and Wilps, 1986).

Cooler temperature slows the age-related loss of flight and damage to wings in *Musca* and increases lifespan (15°C vs. 23°C; Figure 2.4, bottom) (Ragland and Sohal, 1968). The importance of flight activity to lifespan is indicated by the 6-month average lifespan of flies at 11°C (sixfold greater than at 30°C), a temperature at which there is no flight and only walking movements (Massie and Williams, 1987). The inverse relationship of ambient temperature and lifespan shown in this study and others (e.g. Loeb and Northrop, 1917) could be the result of a slowed mechanical senescence. Glycogen content is also reduced in most tissues of *Drosophila,* in close parallel with the decline of flight capacity (Williams et al., 1943; Baker et al., 1985). Similar glycogen loss occurs in mosquitos (*Aedes aegypti;* Rowley and Graham, 1968). Although the density of mitochondrial cristae in flight muscles decreases notably (Fleming, 1986; Turturro and Shafiq, 1979), activities of enzymes in the inner and outer mitochondrial membranes did not change across the *Drosophila* lifespan (Massie and Kogut, 1987). Another factor in mitochondrial dysfunctions is the major loss of mitochondrial DNA (Massie and Williams, 1987). The specific role of flight activity in causing these changes is unknown.

Another critical appendage is the front tarsi, particularly in male flies, where accumulated damage may result in death. The loss of front tarsi, with their taste receptors, predictably leads to cessation of feeding and to starvation (Lockshin and Zimmerman, 1983). Damage to the front tarsi may be especially germane to the early death of male houseflies. Autopsies showed that almost all males, far more than females, had incurred extensive damage to their front tarsi (Angelo Turturro, pers. comm.). Experimental lesioning of front legs with a needle, equivalent to depriving flies of food and water, caused death within 3 days. The irreplaceable and vulnerable tarsi are thus at risk for accumulating activity-related injuries.

Isolation of male houseflies in small bottles allows them to live as long as females (Angelo Turturro, pers. comm.; Figure 2.4 legend). Male flies are well-known fighters in the field as well as the laboratory (Partridge et al., 1987; Partridge, 1988), which contributes to their shorter adult phases. Conditions that favor mating also shorten lifespan and increase the number of flightless flies with damaged wings in *Musca* (Ragland and Sohal, 1973) and *Drosophila* (Partridge and Farquhar, 1981). For example, introduction of males to a female *Drosophila* population dramatically accelerated female mortality after a lag of several weeks, while removal had the opposite effect (Partridge et al., 1986).

The vestigial-winged mutant of *Drosophila* gives another approach to evaluating contributions from mechanical senescence to lifespan. This mutant, which cannot fly at all, has a short lifespan (Baker, 1975), but still shows age-related decreases of the muscle enzyme arginine phosphokinase. Removal of wings lessens the loss of mitochondrial DNA in

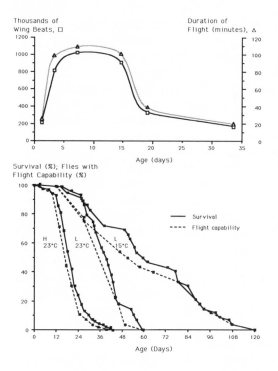

Fig. 2.4. Senescence and the capacity to fly in two dipterans. *Top,* the fruitfly *Drosophila melanogaster.* Relationships between flight ability and age: upper curve (*triangle*), average duration of flight as a function of age; lower curve (*square*), average total number of wing beats as a function of age. Redrawn from Williams et al., 1943. *Bottom,* the housefly *Musca domestica.* Effects of ambient temperature (23°C and 15°C) and behavior on age-related flight capacity through wing damage (loss of > 35% of either wing) and on survivorship. *H,* high flight activity, induced by caging five males with one female; females generally mate once and their resistance to further approaches stimulates flight activity in males. *L,* low flight activity; males singly housed in small vials, which virtually suppresses flight. Redrawn from Ragland and Sohal, 1968.

Drosophila (Massie and Williams, 1987). Such drastic manipulations must also cause major disturbances in other physiological functions, since removal of wings shortens lifespan (Ragland and Sohal, 1973) and increases the number of small mitochondria (Sohal, 1975).

Besides the wings, tarsi, and other appendages, which are obvious targets of mechanical senescence, it would be important to know about the respiratory apparatus, particularly whether the spiracles become clogged. These vital conduits from the cuticle pass oxygen into the tracheolar system inside. Lockshin and Zimmerman (1983) note that the circulatory system of insects is vulnerable to occlusion by small embolisms, which would then disable sensory or motor functions in that appendage. Observations of blood flow in the cockroach *Blaberus giganteus* (Arnold, 1959, 1964) showed that 6-month-old (senescent) specimens had slowed flow. Gradually, most veins became partly blocked by hemocytes and, as death approached, degenerating blood cells became common. Blood cells also became more vacuolated and adhesive, which may be a factor in the cell clumping (Arnold, 1961). Other insects also showed circulatory blockage with age, including *Apis mellifera,* presumably a lower caste, and a dragonfly (Arnold, 1964). There is no information on the ability of the adult insect for collateral vascularization, a phenomenon which is crucial in the recovery of humans from myocardial infarction. Arnold's observations give a basis for systematic studies of age-related damage to the circulatory and respiratory systems of insect models, which might have some parallels to occlusive vascular changes in mammals (Chapter 3.6.7).

Progressive loss of fertility in *Drosophila* cannot be attributed to an absence of germ cell proliferation in adults. Despite decreased numbers of eggs, reduced size (David,

1959a, 1962), decreased viability, and increased developmental abnormalities (Chapter 8), there is no depletion of oocyte precursors in their ovarioles (Wattiaux, 1967, 1968) and no depletion of primary spermatogonia (Miquel, 1971). Even 3-month-old flies can copulate, which is long after they become sterile (Miquel, Bensch, Philpott, and Atlan, 1972). Neuroendocrine or other neural (extra-gonadal) changes could contribute to reproductive senescence (Baker et al., 1985). In young flies, feeding triggers secretion of hormones just before vitellogenesis, but the extent of impairments in the process with age is not clear.

An important example relating hormones to the intensity of wing usage and lifespan is the African migratory locust (*Locusta migratoria migratorioides;* Orthoptera). The lifespan of this and three other acridids is doubled by removal of the corpora allata (Joly, 1965; Wajc and Pener, 1971; Pener, 1976), the endocrine gland that produces juvenile hormone. Allatectomy also reduced flight activity by 70% in this species. Conversely, lifespan was greatly shortened by implantation of extra glands or confinement under crowded conditions (Pener, 1976). A specific relationship between flight activity and lifespan is suggested by the shorter lifespans of allatectomized locusts that are allowed to fly compared to a flightless subgroup (Wajc and Pener, 1971). The data are consistent with (but do not prove) a lifetime maximum number of wing beats. Nonetheless, such a maximum flight capacity is indicated for the worker bee (Neukirch, 1982; Section 2.2.3.1). The widely differing lifespans of worker bees depend on the season of birth, which influences the onset of intense foraging flights. Another example may be the progressively frayed appearance of wings during the 6-week adult phase of periodical cicadas (*Magicicada*) (Monte Lloyd, pers. comm.; Section 2.2.3.3, Chapter 6.3.3).

Hormonal influences that stimulate the flying phase are indicated for locusts and worker bees, and may be a trigger for the mechanical senescence of irreplaceable structures. The inability to replace most cells and having fragile exoskeletal structures, like wings, tarsi, and setae, are candidates as major causes of their short lifespans and rapid senescence. This intrinsic vulnerability does not, however, preclude prodigious use of irreplaceable wings, as shown by the monarch butterflies, which make roundtrips of up to four thousand miles from Canada to California or Mexico where they overwinter. Another counterexample is the scorpion fly *Panorpa vulgaris:* despite major muscle atrophy by its average lifespan of 2 months, flight mating behavior remains intact and is not lost except in rare survivors to 3 months (Collatz and Wilps, 1986).

In moths, the mechanical senescence of wings and cuticles may increase the risk for predation and accidents (Dunlap-Pianka et al., 1977). Damage to wings and antennae occurred during breeding in the aphagous great wax moth (*Galleria mellonella*), but did not necessarily shorten lifespan (Balázs and Burg, 1963). These changes may also occur during the exhaustion of nutrient reserves from adult aphagy (Section 2.2.1). That is, new sources of potential dysfunction will arise if an organism survives beyond the age of its natural death. This is a general phenomenon that may be viewed as sequential (not competing) risk (Chapter 1). Beetles, as described in Section 2.2.1.3, show progressive wear on their mandibles, as in *Carabus glabratus,* for which the mandibular wear was useful as a marker for age that distinguished up to three successive annual generations (Houston, 1981). Ultimately, the maximum lifespan is determined by the irreversible mechanical senescence, which by impairing prey capture and feeding then leads to nutritional imbalance and the cessation of reproduction.

2.2.2.2. Nematodes and Rotifers

Two major invertebrate phyla, Rotifera and Nematoda, contain species that as adults are virtually postmitotic. The absence of vegetative (asexual) reproduction in either phylum is consistent with the generally postmitotic status. In particular, soil and plant nematodes have few, if any, dividing somatic cells as adults, as shown for the short-lived (< 1 month) and rapidly senescing soil nematodes such as *Caenorhabditis briggsae* and *C. elegans* (Russell and Jacobson, 1985; Hyman, 1951). The DNA synthesis inhibitor FUdR even slightly increases their lifespan (Ghandi et al., 1980). In contrast to somatic cells in *Caenorhabditis,* germ-line cells continue to proliferate during adult life (Kimble and Ward, 1988). Moreover, mating can stimulate oogenesis in hermaphrodites. After spontaneous cessation of oogenesis, "old" worms if mated can produce additional fertile eggs, up to about fourfold the usual number of 350.

Some parasitic nematodes live for many months to more than 10 years, during which they lay prodigious numbers of eggs. This hundredfold range of lifespans and growth implies somatic cell division in the intestine and musculature (Chapter 4.2.2). There is no implication that nematode senescence is linked to mechanical factors as described above for flies.

Rotifers are also postmitotic as adults, and the somatic cell number is generally fixed during development (Hyman, 1951, 59; Gilbert, 1968, 1980a, 1980b, 1983b). The continuing synthesis of DNA in the postmitotic gastric gland of *Asplanchna* is an exception (Jones and Gilbert, 1977; Gilbert, 1983b) that might be equivalent to the polyploidization of hepatocytes that is observed postmaturationally in mammals. All species apparently have short adult phases, from a few days to about 2 months (Comfort, 1979). The hard shells of rotifers may be a basis for mechanical senescence, but there is no evidence that this is a factor in their short lifespans. Rapid senescence in nematodes and rotifers is discussed in more detail below.

2.2.3. Alternate Developmental Pathways and the Lifespan

Social insects and rotifers can develop into adults with very different lifespans and body characteristics, depending on the developmental pathway chosen. These examples add force to the argument that manifestations of senescence are highly subject to influences throughout the lifespan and that the characteristics of senescence can be widely varied as a function of development, even within the same species.

2.2.3.1. Social Insects

Social insects are wonderful examples of how the same nuclear genome can be programmed during development to yield very short- or very long-lived adults, which represent extremes in the spectrum of rapid to negligible senescence. In the perennials described below, queens can live 5 or even 15 years, which is a hundredfold longer than short-lived castes, yet both can develop from the same clutch of fertilized eggs. At the other extreme in queen lifespans is the Brazilian social wasp *Mischocyttarus drewseni,* where both workers and queens have a maximum longevity of 3 months (R. L. Jeanne, cited in Wilson, 1971, 429). Caste differentiation, with the implication for major differ-

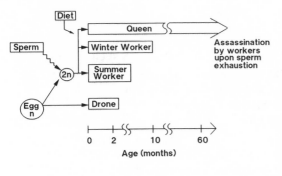

Fig. 2.5. Castes of the honeybee *Apis mellifica* have greatly differing lifespans. Male drones live 1 to 2 months and arise parthenogenetically from unfertilized eggs; death occurs just after mating. Female workers live 1 to 2 months if born in the summer or 6 to 8 months if born in the fall, and arise from fertilized eggs. Queens may live more than 5 years. The development of either queens or workers from larvae of the same clutch of fertilized eggs depends on the food provided to individual larvae by nurse bees (workers; see text). The lifespan of queens appears to be set by the depots of sperm that are stored during their nuptial flights; when the sperm depots are exhausted after some years, queens are killed by the attending worker bees. Original figure by the author.

ences in longevity and type of senescence, is thought to have evolved independently in the Hymenoptera many times during the last 100 million years (Chapter 11.3.3). Honeybees are the best studied in regard to senescence and longevity. The large and sophisticated literature on social insects contains many other fascinating examples which cannot be discussed here.

Castes of the honeybee (*Apis mellifera* L.) have greatly differing adult lifespans, which arise as a consequence of epigenetic factors that influence the choice of alternate developmental pathways (Figure 2.5). Male drones and workers live 1–2 months, though female workers may live up to 12 months if they overwinter. In great contrast, queens are widely acknowledged to live at least 5 years (Ribbands, 1953; Maurizio, 1946, 1959, 1961; Snodgrass, 1956, 302; Winston, 1987) and even 8 years, according to Bozina (1961). Worker honeybee populations show clear accelerations of mortality (Wilson, 1971; Sakagami and Fukuda, 1968), with MRDTs like those of *Drosophila* and *Musca*, but much longer than in aphagous moths (Table 2.1, Appendix 1). Genetic influences on the lifespan are indicated for workers (Kulincevic and Rothenbuhler, 1982; Winston and Katz, 1981) and are also suspected for queens.

In honeybees, the drones arise parthenogenetically if the haploid egg is not fertilized, whereas either queens or workers may develop from fertilized eggs (Figure 2.5). This haplodiploid mode of sex determination is not found in termites (Wilson, 1971). The honeybee queen is larger than her workers, but is much smaller than ant and termite queens that may live even longer (see below). The age of the larvae normally influences the potential for forming queens, so that larvae older than 3 days, when specialized feeding commences, usually become workers rather than queens (Winston, 1987). A complex regulation by nutrition and hormones determines this developmental branch point. Queen-destined larvae are fed ten times more often by the nurse bees (female workers) with a diet that is threefold higher in fructose and glucose. The volume and sugar content of the diet provided by nurse bees regulates the secretion of juvenile hormone through the corpora allata, presumably through stretch receptors in the gut (Dietz et al., 1979; Asencot and Lensky, 1984; Winston, 1987). Exogenous juvenile hormone can restore to larvae as old as 4.5 days the ability to induce queens and prevent the degeneration of the ovarioles. Removal of a queen induces development of ovaries in worker bees within a few weeks if a new queen is not reared. Commonly, however, replacement of a new queen occurs within 16 days after the egg is laid, that is, 6 days faster than needed for development of workers. Queen-worker intermediates can be formed by diet and hormonal manipulations (Taber and Poole, 1973; Winston, 1987), but their lifespan is not described. Thus, the vastly longer lifespan of queens than of workers arises from neuroendocrine influences during development that, in turn, derive from the nurturing behavior of workers.

If senescence occurs in the long-lived honeybee queens, it is very gradual; little mortality schedule data are available. Queen bees slowly develop pigmentation and oily globules in the mandibular and other glands during their lifespans; tumors may also arise occasionally (Haydak, 1957). Old queens also develop a yellow ring around the base of their ovaries that represents the accumulation of debris from the incomplete resorption of unovulated eggs (Flanders, 1957). While these changes might indicate a gradual senescence, nonetheless queens aged 2 or more years continue laying at least as many eggs (up to 3,000/day) as at 1 year of age (Braun, 1942; Flanders, 1957). Even greater fertility is documented below for ant queens. There is no evidence that queens in any social insect have a limited number of oocytes. Whatever senescent changes may occur in queens emerge much more gradually than in the shorter-lived, rapidly senescing workers and drones.

The lifespan of honeybee queens, however, is ultimately limited by the exhaustion of sperm stores. During their famous nuptial flights, young queen bees are inseminated by a succession of drones, which explosively pump sperm into the queens' spermatheca and then fall away, leaving part of their endophallus behind. The spent drones then die almost immediately from this gross injury (Winston, 1987), which can be considered a form of mechanical senescence. Bee queens must fertilize all eggs from this store of up to 5 million sperm, which suffice for many years of fertilizing 175,000 to 200,000 eggs per year. This survival of sperm is as astounding as the longevity and fertility of the queens, and by far exceeds these capacities in domestic fowl and rodents. However, not all sperm are conserved or utilized: by 2 to 4 years the sperm stores are usually depleted, and the queen will then be killed by the workers and replaced (Winston, 1987, 45).[5] The supersedure, or elimination of senile queens, appears to be triggered by diminished pheromone production or by some other mechanism that enables the constantly attending workers to recognize the insufficiency of fertilized eggs (Butler, 1957, 1960; Winston, 1987, 197; Michener, 1974, 112). This mortality pattern fits the model of rapid senescence characterized by an extensively delayed onset (Figure 1.8).

Hormones have a major role in determining the lifespan of workers and its variations according to the season of birth and the duration of specialized phases of labor, as was hypothesized by Maurizio (1959). Several temporal castes (age-group polyethism) are observed among worker bees (Wilson, 1971; Winston, 1987). Young hive worker bees, which attend the queen and nurse the brood, feed mostly on protein-rich pollen; and older field worker bees, which forage for food, feed mainly on carbohydrates. The role transition from hive to field bees occurs by 3 weeks after hatching and is accompanied by major physiological changes. These include tenfold increases of juvenile hormone (Figure 2.6), major decreases of vitellogenin levels in the hemolymph, atrophy of wax and hypopharyngeal glands, and the onset of foraging flights (Haydak, 1957; Fluri et al., 1982). These

---

5. Skaife (1954, 264) described this extraordinary regicide: "When the queen is failing for any reason, the workers put her to death by licking her. This has happened several times in my artificial nests. She is surrounded by a crowd of workers, all with their mouth-parts applied to her skin, and this goes on for three or four days, her body shrinking until little more than the shrivelled skin is left."

changes appear to involve juvenile hormone secreted by the corpora allata and are prematurely induced in young workers by injection of juvenile hormone or implantation of corpora allata (Rutz et al., 1977; Robinson and Ratnieks, 1987; Robinson, 1987a, 1987b). The age of onset of foraging is also under genetic influence (Kolmes et al., 1989).

Workers born in the summer have lifespans of 1 to 2 months (Ribbands, 1952), whereas workers born in the fall survive through the winter and are physiologically active for 6 to 8 months in nursing the queen, the drones, and the larvae with brood food. The onset of foraging and the degeneration of hypopharyngeal glands in winter worker bees is postponed for many months (Haydak, 1957), as is the elevation of juvenile hormone (Fluri et al., 1982). The lifespan of summer bees can be extended at least threefold, to 6 months, by placing summer bees in a queenless colony without larvae or queen to feed, a situation that arises naturally in the fall and winter, when little or no nursing is needed (Maurizio, 1954, 1959). Under these conditions summer bees also have low juvenile hormone levels and high levels of vitellogenin, as found in winter bees (Fluri et al., 1982). Summer worker bees show a progressive loss of brain cells (ca. 30%) during their short lifespan (Hodge, 1894; Rockstein, 1950). The surviving neurons are markedly deteriorated with loss of cytoplasm and nuclear chromatin (Pixell-Goodrich, 1919). No comparisons have been made between neuron loss in summer and winter bees; bees and other insects should be valuable for studying the influences of diet and endocrine functions on neuronal loss. In the moth *Manduca,* the steroid-like hormone ecdysone protects against neuronal loss during metamorphosis (Bennett and Truman, 1985). In parallel with longevity, the hypopharyngeal glands degenerate earlier in breeding than in queenless colonies (Maurizio, 1954). Another major difference between summer and winter bees is in the fat body, which is undeveloped throughout the life of summer worker bees, but which becomes engorged with nutrient stores in winter bees (Ribbands, 1952; Maurizio, 1954, 1959). Brood rearing by summer bees appears to exhaust the protein stores that derived from pollen eaten after hatching. Winter bees feed mainly on honey, drawing on protein reserves from the fat body and elsewhere, and cluster to maintain temperature, with few foraging flights until late winter when larval rearing begins.

The onset of foraging appears to be a crucial determinant of worker bee lifespan and is associated with major increases of juvenile hormone as the female workers cease their nursing activities in the hive and begin foraging flights as field bees (Figure 2.6). The lifetime maximum amount of flight may be fixed at about eight hundred kilometers in worker bees, as determined by varying the duration of daily availability of a sugar source (Neukirch, 1982); the average lifetime flight, however, varied considerably with the daily demand. These results generally concur with the well-known minimization of foraging flights by winter bees until spring and the earlier onset of foraging flights after application of juvenile hormone; such treated bees also had shorter lifespans (Jaycox et al., 1974; Rutz et al., 1977). Thus senescence and death are not strictly determined by age, but are more linked to the onset of foraging with its demands for intense flight that causes mechanical senescence. This gives another branch point in alternate pathways of senescence, in which elevations of juvenile hormone may be the crucial trigger for the onset of mechanical senescence. Presumably, worker bee lifespans could be extended even more by further postponing the onset of foraging. The reader may recall the retarding of wing senescence in flies by the lowering of temperature (Figure 2.4). The apparently constant amount of potential flight may be due to limitations in carbohydrate metabolism (Neu-

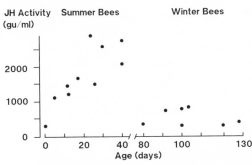

Fig. 2.6. Age-related changes in juvenile hormone (JH) levels in the hemolymph of worker bees. Summer workers show major increases of JH during their transition from hive to field bees. Winter bees maintain much lower JH during their longer lifespan. The short lifespans of summer bees is associated with foraging flights and can be provoked prematurely by injecting summer bees with JH. Conversely, depriving summer bees of a queen yields the winter bee phenotype with low JH. Redrawn from Fluri et al., 1982.

kirch, 1982) or to mechanical senescence of the irreplaceable wings; definitive experiments are needed.

How can honeybee queens live so much longer than the workers of generally similar construction? I suspect that their few flights outside of the nest spares them from as severe mechanical abrasion. Other factors could include the steady feeding by workers (Mark Winston, pers. comm.) or the constant grooming and licking by the workers, which could serve to cleanse and keep open the spiracles, the tiny holes in the cuticle through which insects breathe, as well as its best-known function of maintaining pheromonal contact with the queen. There is no information on the patency of spiracles with age in bees. It is also interesting to consider the role of cell replacement in the greater longevity of worker bees than of laboratory flies. In contrast to the shorter-lived *Drosophila,* intestinal epithelial cells of workers are replaced from progenitor cells in the intestinal crypts that contain mitotic figures (Snodgrass, 1956, 189), apparently as in the vertebrates. Mitotic chromosomes are seen in brains of adult bees, presumably workers (P. Mobbs, cited in Technau, 1984), as in *Drosophila* (Section 2.2.2.1).

Many other social insects have short-lived lower castes and very long-lived queens (Michener, 1974; Wilson, 1975). The great longevity of their queens as documented here may also apply to the tens of thousands of ants, bees, termites, and wasps. There is little cause to suspect the typicality of the ten examples of species whose queens have lifespans greater than 5 years (summarized by Wilson, 1971). The record is held by an ant queen (*Stenamma westwoodi*) that lived at least 16 years and continued to lay eggs until 3 months before death (Donisthorpe, 1936). Fertile eggs were also laid by a 14-year-old *Formica fusa* (Lubbock, 1888) and by a 12-year-old *Aphaenogaster picea* (Haskins, 1960). In the laboratory, queens of *A. picea* remain fertile at least through 13 years until sperm stores are depleted; ten individuals lived 7 or more years (Haskins, 1960). As in honeybees, the lifespan of queen ants is attributed to the exhaustion of the sperm stores that are acquired during a single mating a decade or more earlier (Haskins, 1960; Donisthorpe, 1936). However, termite queens are inseminated intermittently during development of the colony by termite kings (Wilson, 1971, 434) and thus appear to have a continuing source of sperm; whenever reproductive failure occurs, the queens are killed by their attending workers (Skaife, 1954), apparently as in bees. The greater reported lifespans of ant and termite queens (over 10 years) than bee queens (8 years or less) may be more apparent than real.

Worker ants may also live many years in some species, for example, 3 years in *Aphaenogaster picea* (Fielde, 1904) and 6 years in *Messor semirufus* (Tohmé and Tohmé, 1978). Worker castes thus may vary in longevity between social insect species over a 25-fold range: worker bees and wasps characteristically have lifespans much shorter than worker and soldier ants or termites (Wilson, 1971). This major difference could derive from the absence of wings and the associated risk of mechanical senescence in these longer-lived lower castes, though species differences in the grooming of workers by each other need consideration.

In summary, the widely ranging lifespans of the social insects are striking examples of how the developmental pathways and the natural environment can determine different patterns of senescence and longevity from the same set of chromosomal genes. Thus the presence of genes that determine rapid senescence and short lifespans does not preclude the possibility that evolutionary changes in gene expression could yield closely related species or even phenotypically different organisms in the same population that have vastly longer lifespans and negligible senescence. Neural and endocrine mechanisms are manifestly important in the lifespans of all castes. Whether a particular larva will become a worker or queen is controlled by juvenile hormone, and the age-related polyethism of adults is also controlled by this substance. Although the lifespan of the queen is terminated by the grooming behavior of the workers when egg production fails, the feeding by the workers allows the queen to avoid foraging for the nutrients required to maintain her vast egg production, and thereby minimizes mechanical senescence. I further hypothesize that the workers' grooming behavior may cleanse and maintain patency of the spiracles that are crucial for breathing. The short-lived and largely asocial Diptera have none of these advantages. Other important questions concern the relationships between mortality rates and the capacity for cell replacement and regeneration. These examples set the stage for further discussions of the role of gene regulation in longevity and senescence.

2.2.3.2.   *Rotifers*

Rotifers also have developmental choices that influence the lifespan, although the lifespan variations are much less than in the social insects. Rotifers generally have short adult phases, lasting a few days, but some live for months (Comfort, 1979). Unlike nematodes and other phyla with both short- and long-lived species, no rotifer is known to have a long adult phase. Species in two orders (Monogononta and Bdelloidea) have been studied since early in this century as models of senescence. In monogononts (e.g. *Asplanchna, Euchlanis, Lecane, Proales*), males are found (Figure 2.3), whereas in bdelloids (e.g. *Philodina*), males have never been observed (Bell, 1982). As noted in Section 2.2.1.2, males lack excretory and digestive organs at birth (Figure 2.3), which explains why their adult phase is always shorter than in females. Consistent with their truncated existence, spermatogenesis begins early; for example, monogonont males are fertile at hatching (Wesenberg-Lund, 1923; Gilbert, 1983b). The duration of development, unless diapause occurs, is short, taking about a week. The number of sperm is limited (40–200), though in some species spermatogonial proliferation may continue a short while after hatching; males nonetheless can mate repeatedly, despite their limited sperm (Gilbert, 1983b).

During oogenesis in monogononts, obscure mechanisms determine if the eggs are to be mictic, yielding both sexes and capable of being fertilized, or amictic, yielding only females (Bell, 1982; Margulis and Schwartz, 1988; Gilbert, 1983b). If mictic eggs are not fertilized, males are formed, as in *Asplanchna*. Because this developmental switch has been characterized in only a few monogonont species, the many differences among genera in rotifer reproduction (Bell, 1982) make generalizations unwise. Environmental factors influence whether a given female will produce mictic eggs. In *A. sieboldi,* vitamin E (tocophorol) and high population density (Gilbert and Thompson, 1968; Gilbert, 1980a, 1980b) stimulate production of mictic eggs, while photoperiod is important in *Notommata copeus* (Pourriot and Clément, 1975). Others note effects of adverse environments (Hyman, 1951, 136; Thane, 1974, 478). Females of *Lecane inermis* (Miller, 1931) and other species not named by Hyman (1951) live longer if they are mictic, while *Euchlanis dilatata* does not show this effect (King, 1970). Mictic females are thought to live longer because they may be less exhausted by the fewer and smaller eggs they produce (Miller, 1931); rotifer eggs are large relative to the adult. These different controls on the outcomes of oogenesis show the importance of environmental factors in the alternative life phases of rotifers.

Rotifers show many somatic degenerative changes and accelerations of mortality that constitute a rapid senescence and that are fitted by the Gompertz model (Comfort, 1979, 114; Lansing, 1947; Jennings and Lynch, 1928a, 1928b). The ultra-short lifespans of aphagous males with an extremely short MRDT (Figure 2.3; Table 2.1) implies nutrient depletion, but there are no data. The absence of feeding in males might alter sperm viability later in the short lifespan when nutrient reserves may be reduced (Gilbert, 1983b).

Major manifestations of senescence as characterized in female monogononts and bdelloids include decreased body movement, accumulation of cellular pigments and lysosome-like inclusion bodies, and decreased egg production. An increase of free ribosomes is consistent with reduced protein synthesis (Szabó, 1935; Comfort, 1979; Herold and Meadow, 1970). Although regeneration of lost cells does not occur, some cytoplasmic appendages, such as the coronal lobes, can be replaced in young adults (Hyman, 1951), but this response is lost in a few days (Pai, 1934). The gut in females may be the first organ system to deteriorate, which Comfort (1979, 115) suggests may be a pacemaker for senescence. Advanced age in some species reduced digestion and defecation (Szabó, 1935). Spemann's (1924) figures show a dramatic body shrinkage just before death that suggests starvation. Thus, starvation may be the proximal cause of senescence in both sexes.

Reproduction also influences the lifespan of females. Mortality in *Philodina* (Meadow and Barrows, 1971a) and *Asplanchna* (Enesco et al., 1989) increased sharply at the cessation of egg laying. This major effect was seen throughout experimental manipulations using cooler temperature and diet restriction (Meadow and Barrows, 1971a, 1971b) and vitamin E (Enesco et al., 1989) which extended the duration of reproduction by up to 100%. While diet restriction reduced the number of offspring by half in *Philodina,* vitamin E and cortisone increased fecundity by 50% in *Asplanchna* (Enesco et al., 1989). There could be common mechanisms, possibly in gut function, that might cause failure of reproduction and increased mortality. A continued need for feeding is implied by the large yolk volume provided to the large eggs during the oocyte maturation that continues into adult life. The size of the residual oocyte stock when reproduction has ceased is not known.

Maternal age has several major effects on the characteristics of rotifer eggs, which are compared with those of other phyla in Chapter 9. Mictic eggs from older female *Asplanchna sieboldi* cannot be fertilized and yield only males with ultra-short lifespans, whereas mictic eggs from young adults can be fertilized (Buchner and Kiechle, 1967). Maternal age also can influence the lifespan of offspring, the "Lansing effect" (Lansing, 1947), although this remarkable phenomenon has not been observed in all species (Chapter 9.3.2.2). Rotifers have a short postreproductive stage that ranges from less than 1 day up to 3 days (King, 1967; Jennings and Lynch, 1928a, 1928b; Szabó, 1935; Enesco et al., 1989). After reproduction ceases, mictic females live several days longer before death than amictic females. In *A. brightwelli,* the duration of the postreproductive period was not linked to the length of the reproductive phase, whether the variations were spontaneous among individuals in the population or whether the reproductive phase was lengthened by vitamin E or cortisone (Enesco et al., 1989).

In conclusion, rotifers are models for very short phases of development and adult life. The modes of gender determination could be valuable for further study of how oogenesis can be scheduled in relation to age changes in fecundity. The Lansing effect, while controversial, is intriguing and deserves further attention. One drawback is that little is known about the molecular biology and reproductive physiology of rotifers.

2.2.3.3. Different Total Lifespans That Arise through Varying Durations of Development

Chronological variations in life history of rotifers include extended diapause during development or as adults. Adverse conditions can arrest development and induce a hypometabolic state that delays hatching for many months (Gilbert, 1968; Hyman, 1951, 140). Adult bdelloid rotifers can also shrivel during dry periods into a "tun," which is highly resistant to temperature extremes and desiccation for years or decades (Hyman, 1951, 144). Through diapause, rotifers and many other species can reschedule their natural history in accord with environmental opportunities. Numerous other invertebrates and vertebrates can enter extended diapause at particular stages of development, including insects, annual fish, and mammals before preimplantation (Chapter 9.2). There are probably a vast number of other life history variations with major consequences to total lifespan that arise as switch points during development. Here are several more examples.

The flatworm *Polystoma integerrimum,* a parasite of amphibia, has two very different life histories, lasting alternatively 2–3 months (branchial form) or over 6 years (bladder form), at least a tenfold difference (Williams, 1961; Caullery, 1952; Gallien, 1935; Kaestner, 1968). The life cycle is prolonged when these worms develop in the bladder of the tadpole, because maturation is delayed in synchrony with the metamorphosis (maturation) of the host, which in temperate zones may not occur before 3 years. The bladder form appears to be continuously fertile, and there is no mention of senescence in the literature. However, the shorter-lived branchial form matures on the exterior gills, develops gonads precociously (neotenously), and dies at the metamorphosis of its host. The branchial form otherwise resembles immature bladder worms.

The duration of development in many invertebrates is quite flexible, so that the numbers of larval instars can vary widely, depending on temperature and nutrition. A classic case of this indeterminacy is the beetle *Trogoderma.* Food and water deprivation can extend the duration of larval life phases by years. Wodsedalek (1917) noted that in the absence of food for 5 years, there occurred gradual shrinkage to less than 0.02% of the original size during "retrogressive molting"; growth resumed with refeeding. A shorter study of starvation for 1 year recorded sixteen supranumerary molts during an extended larval period more than ten times the usual length; the immature ovaries persisted in the larvae throughout, and the imaginal discs for adult structures did not differentiate until commitment to pupation (Beck, 1971a, 1971b, 1972); Lockshin and Zimmerman, 1983). The temporal plasticity of imaginal discs is also documented by their capacity for extended serial culture for 5 or more years (Hadorn, 1966).

Cicadas also have widely varying durations of larval stages, with the total lifespan ranging at least fivefold. Three species of periodical cicadas (*Magicicada*) in the eastern United States also have extremely prolonged, yet precisely timed development. Each species has broods which take either 13 or 17 years to develop, depending on the duration of the second larval instar (Alexander and Moore, 1962; Lloyd, 1987). The 17-year broods grow very slowly as subterranean nymphs during their first 4 years, but grow thereafter as fast as the 13-year broods (White and Lloyd, 1975). Each has five instars despite 4-year differences in development. The periodical cicadas are the only insects known to feed on stem xylem fluid, which is an unlikely food source because of its low caloric content, being 99.9% water (Cheung and Marshall, 1973; White and Strehl, 1978). The rate of development may be limited by the scarcity of nutrients in xylem fluid (White et al., 1979; White and Strehl, 1978). An advantage inferred for the underground habitat of cicada nymphs is the relative freedom from birds and other predators encountered above ground.

Emergence of the nymphs from the ground appears to irreversibly initiate maturation, as manifested by rapid hardening and darkening of the exoskeleton. Under optimal conditions, eclosion soon follows. However, if unable to find a suitably uncrowded vertical perch, eclosion is blocked; exoskeletal hardening continues in such uneclosed nymphs, who fail to become free of their nymphal skin, fall to the ground, and slowly die (White et al., 1979). This premature death is thus a result of behavioral cues.

The causes of death in the short (6-week) adult phase are unclear, but include mechanical senescence, as judged by the progressive fraying of wings, but also probably through slow starvation (Monte Lloyd, pers. comm.). Although adults of both sexes feed daily on the watery xylem fluid, it is likely that the adult lifespan is ultimately limited by nutrient stores, as in certain butterflies that feed only on nectar (e.g. *Dryas;* Section 2.2.1.1.1).

For the 13- and 17-year broods, the alternate durations of development cause total lifespan to vary by about 25%, without differences in the length of their adult phases; genetic aspects of these different developmental schedules are discussed in Chapter 6.3.3. *Tibicen,* a larger cicada also found in North America, may have an even longer life cycle than 17 years: more time should be needed at *Magicicada*'s growth rate for *Tibicen* to reach the same size at maturity (M. Lloyd, 1984). At the shorter extreme is *Melampsalta cruentata,* a New Zealand species with the same number of instars as *Magicicada,* but which takes just 3 years to develop from egg to adult (Cumber, 1952).

These examples of variably extended development in invertebrates suggest that the clock for adult senescence is not started in some species until commitment to the adult phase. This circumstance appears to fit *Caenorhabditis* (Chapter 9.3.3). A similar conclusion is indicated for species that die in association with maturation and reproduction (Section 2.3). Further relationships between development and the adult phase are discussed in Chapter 9.

2.3. Reproduction-related Rapid Senescence and Sudden Death

Sudden death, often preceded by dramatic rapid senescence, commonly occurs as the specific result of reproduction and has long been discussed (Weismann, 1889b; Korschelt, 1924). In many cases, under special conditions individuals can survive for subsequent breeding and thus are not obligatorily semelparous. The trigger for rapid senescence often involves physiological feedback (Chapter 10), and the lifespan can be greatly extended by postponing maturation or reproduction.

2.3.1. Neuroendocrine Mechanisms in Animals

2.3.1.1. Insects

The examples of autogeny in flies and mosquitos described in Section 2.2.1 show a major neuroendocrine role in regulating the timing of egg maturation and the need for feeding in the adult stage. Autogeny thus commits an insect to an earlier demise than in the closely related genotypes which must survive for days or weeks as feeding adults to complete the reproductive quota required for the perpetuation of their genotype. The role of hormones in regulating insect development is pertinent to recall here, since again, hormones program the clonal lifespan of cells in the imaginal discs that form the adult. For example, if metamorphosis of *Drosophila* is prevented by removing the corpora allata, the imaginal discs are prevented from differentiating into adult tissues. In the presence of sufficient levels of juvenile hormone, undifferentiated imaginal discs proliferate continuously for at least 6 years during serial transplantation so that the clonal lifespan exceeds the adult lifespan by thirtyfold, as shown in Hadorn's classic studies (1966, 1968). Another role of the corpora allata was described above in the lifespans of locusts (Section 2.2.2.2) and worker bees (2.2.3.1), where hormonal secretions influence the onset or intensity of flight and associated mechanical senescence. These varied roles of hormone establish their fundamental place in regulating the life history and total lifespan of flies and the other insects which use these same mechanisms.

2.3.1.2. Arachnids

Most examples of rapid senescence in arachnids concern the spiders, and several examples show links to reproduction. The "bowl and doily" spider (*Frontinella pyramitela*) is extremely short-lived—less than 3 weeks after sexual maturity—with no evidence of age-related mortality acceleration, presumably because of environmental hazards such as predators and parasites (Figure 2.7, top; Austad, 1989). In the laboratory, however, its lifespan is extended for up to several months, with survival curves that demonstrate an acceleration of mortality indicative of senescence (Figure 2.7, bottom). This species, like most spiders, deposits yolk into the eggs after copulation, and this requires continued eating before oviposition. Austad clearly showed a link between the stress of oviposition and increased mortality. Access to increasing amounts of food (flies) decreased the age of oviposition and decreased the latent period when mortality increased sharply. As the volume of food was decreased, the egg clutch decreased, while the fraction of barren females also increased sharply. Associations of delayed reproduction and increased lifespan with restricted diets are widely observed (Chapter 10).

While Austad (1989) did not report on the survival of male *Frontinella* in relation to mating, male spiders in many species die soon after their first mating, whereas females survive them by months or years (Edgar, 1971; Foelix, 1982, 207; Vlijm et al., 1963). Males of some species clearly die from starvation in association with reproduction. W. J. Baerg made valuable observations of tarantulas in the field and during prolonged captivity

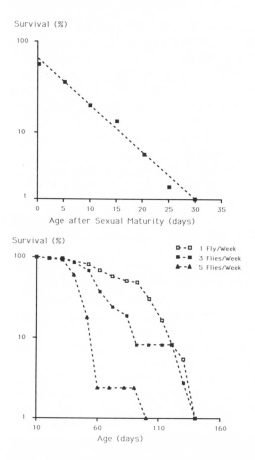

Fig. 2.7. *Top,* the lifespan of the "bowl and doily" spider (*Frontinella pyramitela*) is extremely short in nature, less than 3 weeks after sexual maturity, with no evidence for age-related mortality acceleration. *Bottom,* in the laboratory, survival was extended for up to several months, with survival curves that show accelerating mortality rates that indicate senescence. Access to increasing amounts of food (1, 3, or 5 flies/week) caused earlier onset of oviposition and decreased the latent period before mortality increased sharply. Redrawn from Austad, 1989.

that are the best record of gender differences in the lifespan of any long-lived spider. As noted above for *Frontinella,* males in other families cease to eat and drink after mating. The best characterized example is the male tarantula (*Eurypelma californica*) that matures at 8 to 10 years and shows marked abdominal shrinkage during the month or so before death (Baerg, 1920). Similar death within the first season of mating is probably general in the Mygalomorpha (Steven Austad, pers. comm.). Nonetheless, male tarantulas can mate with at least four different females during the month before death (Baerg, 1963). The close relationship to mating suggests neuroendocrine mechanisms in senescence, as in the octopus (Section 2.3.1.3). It would be of much interest to know if male tarantulas that were not given access to a mature female could survive longer than usual.

Males are killed by the female during mating in only a few species (*Araneus, Argiope, Crytophria;* Foelix, 1982, 197). Contrary to popular belief, male black widows (*Latrodectus mactans*) usually survive mating. Their maximum total lifespan is about 7 months, with a mean adult phase lasting 21 days. Female black widows, however, live up to 30 months, with mean adult phases of 7 months (Kaestner, 1968).

Female tarantulas may also have a sudden death, but from different causes. Unlike males, females survive to breed in successive seasons. Soon after the spiderlings exit the cocoon, the adult females molt, thereby replacing their entire exoskeleton. Molting typically occurs every other year, which gives protection against mechanical senescence. At some age, at least a decade after maturing, the molting schedule becomes irregular and the act of molting less effective, so that females may bleed to death through rips in their delicate skin (Baerg, 1958, 1963; Baerg and Peck, 1970). Female tarantulas represent the far end of arachnid lifespans, exceeding 25 years (Baerg, 1920, 1958, 1963; Baerg and Peck, 1970; Gertsch, 1949, 51; Foelix, 1982, 231; Savory, 1977, 57). It is unclear whether this event is senescent, perhaps through altered neuroendocrine regulation of molting, or whether it is statistically distributed across all adult ages.

Other long-lived species include spitting spiders (*Scytotidae*) (Bristowe, 1958; Gertsch, 1949, 51) and purse web spiders (*Atypidae*) (Bristowe, 1958, 69), which both can live 7–10 years. Purse web spiders reach maturity after 4 years (78). The female purse web spider ceases to feed, and its abdomen shrinks before death (109), but the relationship to reproduction is not stated by Bristowe. At least one small spider, *Siro rubens* (cyphophthalmid), also lives nearly a decade, of which 4 years is required for its remarkably slow development (Savory, 1977, 194; Juberthie, 1967); the characteristics of death are not reported.

Environmental factors also influence the lifespan. As noted above, lifespan is increased by limiting food, but also by cold (Savory, 1977, 57). In nature, the same species may have different lifespans depending on location and mean temperature. The wolf spider *Lycosa lugubris* takes 2 years (2 winters) to reach maturity in Scotland, but has a 1-year life cycle in the slightly warmer Netherlands and England (Edgar, 1971). Male wolf spiders die during their first breeding season. Ancestral spiders may have had much longer lifespans (Gertsch, 1949, 51; Bristowe, 1958, 78).

Mites, which are also arachnids, have a wide range of lifespans between species. At one extreme is the 6- to 10-day adult phase of *Parasitus coleoptratorum;* this is so short as to imply rapid senescence and sudden death, though no description is available (Kaestner, 1968). The host of this mite species is a beetle (*Geotrupes*) that lives at least 2 years (Crowson, 1981). Other mites have much longer adult phases, for example, 3 years in *Arrenurus globator* (Kaestner, 1968). Thus, arachnids have the same range of lifespans as shown by insects, from days to years.

2.3.1.3. Mollusks

Many cephalopod mollusks (e.g. squid, octopus) are thought to be semelparous and live only 1–3 years (Van Heukelem, 1979; Kaestner, 1968). Spawning death would be a spectacular event in the giant squid *Architeuthis,* the largest invertebrate, which is thought to live several years (Kaestner, 1968). Rapid senescence in association with spawning has been described for only a few species, particularly in the genus *Octopus*. Both sexes of most small species of *Octopus* die by 1–2 years (Van Heukelem, 1979; Calow, 1983). In general, octopus do not eat after spawning and lose weight before death (Wodinsky, 1977; Van Heukelem, 1976, 1977), with marked atrophy of their digestive tracts (Sakaguchi, 1968). In *O. vulgaris* the salivary glands and hepatopancreas lose more than 90% of their proteolytic digestive enzymes (Sakaguchi, 1968). Other changes in *O. maya* include lethargy, loss of the ability for protective color changes, and impaired wound healing (Van Heukelem, 1977, 1978, 1979). In contrast to these species, *O. chierchiae* is iteroparous and continues to feed after spawning (Rodaniche, 1984); other unnamed species also are iteroparous. *Nautilus* is a major exception, showing iteroparity over a longer lifespan (Arnold and Carlson, 1986).

Neuroendocrine mechanisms probably determine the age of sexual maturity and repro-duction as well as lifespan. Wodinsky (1977) prevented the postspawning involution of female *O. hummelincki* by removing the optic gland just after spawning; the lifespan was increased by two- to threefold, and most animals continued to feed and grow. Males also had increased lifespans after optic gland removal. Thus, secretions of the optic gland influence longevity as well as control reproductive behavior and feeding. Starvation is not likely to be the only cause of postspawning senescence, because some females can survive without feeding after gland removal for up to 4 months, an interval that exceeds by three-fold the normal postspawning interval to death of about 1 month. Thus different mecha-nisms are involved in the control of longevity and cessation of feeding by the optic gland (Wodinsky, 1977) Although Van Heukelem (1978) did not find a similar life-extending effect of optic gland removal in *O. maya,* it seems likely that neuroendocrine controls are generally involved in octopus senescence. High light intensity, long days, and cool tem-peratures can also extend lifespan by delaying maturation (Van Heukelem, 1979).

Some gastropod mollusks (sea hares, Opisthobranchia) also senesce rapidly over 2–4 months with decreased food intake, loss of body weight, prolonged inactivity, loss of muscle tone, depigmentation, and exponential increases of mortality, as observed in at least three species: *Aplysia juliana* and *A. dactylomela* (Figure 2.8), and *Stylocheilus longicauda* (Switzer-Dunlap and Hadfield, 1979; Hadfield and Switzer-Dunlap, 1990). Importantly, sea hares with senescent appearance have been found among natural field populations in Hawaii (Hadfield and Switzer-Dunlap, 1990). The mean lifespans in the laboratory have a tenfold range, from 2–3 months (*S. longicauda*) to 5 years (*Dolabella auricularia*) (Hadfield and Switzer- Dunlap, 1984). The neurobiologists' favorite, *A. cal-ifornica,* has an intermediate lifespan (Hirsch and Peretz, 1984). Short lifespans occur in both large and small species (Calow, 1983). Some other marine gastropods, however, live at least 10 years, with few signs of senescence (Chapter 4.3.2).

The coincident cessation of growth, egg laying, feeding, and depigmentation suggests

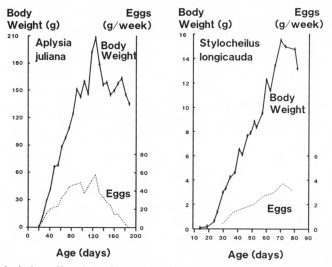

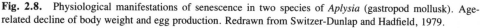

Fig. 2.8. Physiological manifestations of senescence in two species of *Aplysia* (gastropod mollusk). Age-related decline of body weight and egg production. Redrawn from Switzer-Dunlap and Hadfield, 1979.

a neuroendocrine involvement in the rapid senescence of sea hares. A clue comes from one individual in a laboratory population of *A. californica* that failed to lay eggs and also lived 30% longer than the rest which died by 8–9 months (Hirsch and Peretz, 1984). This occurrence is consistent with the linkage between egg laying and senescence in the octopus. Evidence for altered neural function in *Aplysia* includes impaired memory of habituation responses and decreased arousal behavior, as measured by cardiac acceleration in response to food stimuli (Bailey et al., 1983). Another clue comes from influences of temperature and light in laboratory studies of *Aplysia*. When grown at 28°C, *A. juliana* lived 35% longer than when kept at 20°C (maximum of 10 months vs. 6 months) (Hadfield and Switzer-Dunlap, 1990). Egg production also peaked and terminated in scale to the lifespan. Constant darkness or blinding caused greater growth to larger sizes; feeding may be increased with constant darkness, because *Aplysia* tends to feed during dark phases on ordinary light-dark cycles; the lifespans of dark-reared animals are not known. Because temperature and light effects are probably both transduced through neural pathways, they imply a neuroendocrine role in *Aplysia* senescence. If neuroendocrine causes could be established for *Aplysia* senescence, powerful physiological and genetic manipulations might be possible.

Among snails, closely related species may be semelparous or iteroparous (Calow, 1978a, 1983; Hunter, 1961). Death soon after first spawning is very common among annual freshwater snails, but for unknown causes. Depending on the location, populations have very different lifespans. *Lymnaea peregra,* if born in the spring, dies by 5 months after spawning; but if born in the late summer, *Lymnaea* can overwinter to spawn in the spring with typical lifespans of 9 months (Hunter, 1961). Because light and temperature have major effects on the adult lifespan of *Aplysia juliana,* it seems likely that the seasonal effects on *Lymnaea* life history are also mediated by light and temperature. Mortality data on natural populations suggest that *Lymnaea* do not have a rapid senescence in association with increased mortality rates and that individuals rarely survive beyond 15 months (Calow, 1978a). Thus, rapid senescence is conditional, unlike that in octopus or, probably, in *Aplysia*. Gradual senescence in other snails is described in Chapter 3.3.6.

2.3.1.4. Lampreys, Eels, and Salmon

Proceeding to aquatic vertebrates, the lampreys, eels, and Pacific salmon share a characteristic death at first spawning in which neuroendocrine mechanisms are clearly indicated. I describe later in this section how death at first reproduction also occurs in other teleosts in association with a common evolutionary history of migration between fresh and salt water.

2.3.1.4.1. Lampreys

Lampreys, one of the two surviving classes of Agnatha, are among the most primitive fish, lacking jaws. Most surprisingly, lampreys also appear to lack corticosteroids (Buus and Larsen, 1975; Weisbart et al., 1978; Kime and Rafter, 1981; Larsen, 1980; Hardisty, 1979). Of the forty-one species, the life history is known for only a few, particularly the northern hemispheric *Ichthyomyzon, Lampetra,* and *Petromyzon,* and also *Geotria* from the southern hemisphere.

Some lamprey species are anadromous, migrating from the sea to mature and spawn in fresh water. The sea-run lampreys are external parasites of fish in estuaries or the ocean from which they obtain blood and other tissues before maturation causes them to cease feeding. Other species known as "brook lampreys" are nonparasitic, landlocked, and much smaller at maturity. Nonparasitic lampreys do not grow larger than the premeta-morphic larval or ammocoete stage. They are aphagous throughout their short 6- to 9-month-long adult phase (Hubbs and Trautman, 1937; Hardisty and Potter, 1971; Hardisty, 1979). The parasitic and nonparasitic forms within a genus are referred to as paired species and are nearly identical before the adult stage. Their evolutionary relationships are discussed in Chapter 11.3.4.1. The adult (postlarval) phase lasts several years in the parasitic species and is shorter in nonparasitic lampreys.

Metamorphosis of sea lampreys (*Petromyzon marinus*) typically occurs between 3 and 7 years, but may be delayed up to 18 years (Manion and Smith, 1978). Metamorphosis in nonparasitic lampreys may occur between 1 and 5 years (William et al., 1988). Factors in the age of metamorphosis include temperature, population density, and body fat content (Purvis, 1980; Mormon, 1987), while the seasonal timing of metamorphosis varies inversely with latitude among populations in the northern hemisphere (William et al., 1988). The validation of ages has not been certain, but may be improved by estimation from the rings of the statoliths, which are formed annually under some circumstances; lampreys lack most hard tissues used to estimate age in teleosts (Medland and Beamish, 1987; Volk, 1986).

In general, feeding ceases in nonparasitic lampreys at metamorphosis from the free-living larval or ammocoete stage and leads to extensive visceral atrophy. However, the arctic nonparasitic lamprey (*Lampetra japonica kessleri*) continues feeding during meta-morphosis (Poltorykhina, 1971). During and after metamorphosis, there is a general deterioration of internal organs except for the heart and gonads. The intestine collapses to a thread; the liver shrinks and becomes green from the biliverdin degradation products of hemoglobin; there is a general loss of fat and protein, both of which are needed as energy for locomotion as well as to complete oocyte maturation (Larsen, 1980; Bentley and Follett, 1965). In a few well-studied species of parasitic lampreys, feeding never occurs after spawning and death soon ensues in both sexes (Hagelin and Steffner, 1958; Hardisty et al., 1970).

Lampreys have a fixed number of primordial germ cells (Hardisty and Cosh, 1966; Davidson, 1968, 174), which is in contrast to the other surviving agnathan (see below) and also to most teleosts (Chapter 3.4.3). Another notable feature is the extensive follicular atresia or spontaneous death of the bulk of maturing oocytes (Hardisty et al., 1970; Weissenberg, 1927; Hardisty, 1963). Semelparity with aphagous rapid senescence and a limited oocyte stock appear to be ubiquitous in lampreys, and are the only concurrence of the traits in any vertebrate family.

Nonparasitic species differ in the timing of some developmental events: oogenesis and sexual differentiation occur earlier, while vitellogenesis is later (Figure 2.9). Because the nonparasitic species take 2 years longer to mature, the total lifespan is about the same in both types of life history. It seems plausible that the prolonged larval stage during which feeding continues is important to providing nutrients for egg maturation. The failure of nonparasitic adults to feed is associated with a much smaller production of mature ova, about 80% fewer in *L. planeri* (Hardisty 1979, 287; 1970). Hardisty and Potter (1971) hypothesize that the burrowing into mud of adult nonparasitic lampreys is an adaptation

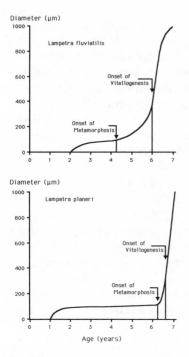

Fig. 2.9. Ovarian oocyte development and age in lampreys, comparing the anadromous (sea-run) *Lampetra fluviatilis,* which parasitizes fish in its adult phase, with *L. planeri,* a freshwater species that is entirely aphagous in its adult phase. Such examples of very different life histories in the same genus are called "paired species." Oocyte growth begins about 1 year earlier in the freshwater *L. planeri,* although vitellogenesis and the final stages of oocyte maturation are delayed by a year or more. The total duration of oogenesis and the lifespan are the same in both species. Redrawn from Hardisty and Potter, 1971.

to reduce mortality and compensate for lower fecundity as compared with the parasitic forms.

Neuroendocrine mechanisms are implicated as regulators of the lamprey lifespan, although environmental influences and other details are not well understood. Atrophy of the intestines and other viscera is tightly linked to gonadal developments (reviewed in Larsen, 1980; Youson, 1980). Gonadectomy just after the upstream migration begins prevents intestinal atrophy (Larsen, 1974; Pickering, 1976), as well as the development of the urogenital papillae and other secondary sex characteristics (Larsen, 1974). Correspondingly, testosterone and estradiol implants accelerate intestinal atrophy in intact lampreys and gonadectomized lampreys, but in the latter only during the onset of the spawning migration (Pickering, 1976). Curiously, estradiol prolonged the survival of intact females (Larsen, 1974).

Studies were also done with hypophysectomy, in this case selective extirpation of the pro- and meso-adenohypophysis, which correspond to the pars distalis or anterior pituitary of higher vertebrates, but sparing the meta-adenohypophysis and brain. Hypophysectomy before maturation extended lifespan by up to 11 months in *Lampetra fluviatilis* (Larsen, 1965, 1969). While ovulation, spermiation, and the development of secondary sex characteristics were blocked, nonetheless vitellogenesis and spermatogenesis continued; moreover, target tissue sensitivity to sex steroids also increased on its usual schedule (Larsen, 1987). These studies show that the proximal pacemaker for senescence in lampreys is gonadal maturation, but do not yet identify the triggers for such maturation or the extragonadal mechanisms that may involve the nervous system or extrinsic environmental cues. Larsen (1985) comments that manipulations prolonging survival also slow the depletion of body tissues, giving further weight to the conclusion that death after spawning is caused by effects of starvation.

Some discussion is warranted about the lamprey's unusual endocrinology. Lampreys differ from higher vertebrates, in which hypophysectomy completely blocks gonadal activity, and, moreover, do not appear to have corticosteroids (Larsen, 1980; Buus and Larsen, 1975; Weisbart et al., 1978). Although there is some steroidogenic capacity in the interrenal gland, an analogue of the corticosteroid-producing pronephric tissues of teleosts (Buus and Larsen, 1975; Weisbart et al., 1978), these steroids have not been identified (Kime and Larsen, 1987). Neither cortisol, cortisone, corticosterone, 11-deoxycortisol, nor 11-deoxycorticosterone were formed from incubations of [1,2–3H]cholesterol with gonadal and interrenal tissues of lampreys (Weisbart et al., 1978). The familiar sex steroids of higher vertebrates, however, are present, including estradiol, testosterone, and DHEA. On the other hand, the neuroendocrine regulation is unique, since plasma levels of estradiol and testosterone are *increased* by hypophysectomy (Kime and Larsen, 1987). These findings differ greatly from the teleosts, which arose about 200 million years later (Chapter 11) and whose corticosteroids are under pituitary control (Henderson et al., 1975). The interrenal mass increases sharply during metamorphosis in *Petromyzon planeri* (Sterba, 1955), which implies increased production of osmoregulatory hormones that are physiologically adaptive in the shift from sea to fresh water. Sea-run lampreys in addition cease drinking (Larsen, 1980).

The other surviving jawless fish, the hagfish (Myxinidoidae), apparently differs greatly from lampreys in its continuing growth, *de novo* oogenesis, and iteroparity (Chapter 4.2.2).

2.3.1.4.2. Salmon, Eels, and Other Teleosts

A small fraction ($<$ 1%) of the ca. 22,000 teleost species are semelparous and die suddenly after spawning (Tables 2.3, 2.4). My survey of this phenomenon indicates that nearly all of this fraction share another unusual trait with lampreys: they or their likely ancestors are diadromous, migrating between fresh water and the ocean for reproduction (Chapter 11.3.4.4). Diadromy itself is unusual, occurring in less than 10% of the 445 families of fish. McDowall (1988) emphasizes that diadromy occurs widely in only the most primitive teleost families and is sporadic elsewhere. Anadromy (migration from ocean to fresh water for spawning), as done by Pacific salmon, is far more common than the reverse route, catadromy, which is mainly represented in the family Anguillidae, in some gobies, and in galaxiids. Clearly, not all diadromous fish are semelparous.

The Pacific salmon (genus *Oncorhynchus*) are the best-understood examples of death at spawning in teleosts. However, semelparity also occurs in ten out of ninety genera of Salmoniformes (see below). The Pacific salmon species are briefly described in Table 2.4, which also gives the numerous common names of these fish. As explained in Chapter 11.3.4.4, ancestral oncorhynchids probably were repeat spawners. The perennially spawning Pacific *Salmo gairdneri* (steelhead trout, anadromous; rainbow, landlocked) was recently renamed *O. mykiss*. Whether anadromous or landlocked, six out of seven species in this genus show a spectacular senescence at their first spawning, as described in the classic studies of O. H. Robertson and his student B. C. Wexler.[6] The pathophysiology of hyperadrenocorticism and aphagy at spawning have been characterized only in *O. nerka*

---

6. Oswald H. Robertson (d. 1966) was a leading clinical researcher who created the first blood bank during World War I. Taking early retirement from the University of Chicago, he set up a laboratory near Stanford and began his classic work on the pathophysiology of spawning in salmon.

Table 2.3   Semelparity in Fish

Taxon	Semelparity
Class Agnathy (jawless fish) Order Petromyzontiformes (lampreys) Families **Petromyzontidae, Geotriidae, Mordacidae**	Obligate semelparity is indicated for several species in each family; the extent of obligate semelparity is unknown for others. Lifespans, typically 7 years but may be > 20 years if metamorphosis is delayed.[1]
Class Osteichthyes (bony fish) Order Anguilliformes Infraorder Anguilloidea Family **Anguillidae** (freshwater eels)	16 species. Obligate semelparity generally indicated; widely varying lifespans, 8–88 years.
Family **Nemichthyidae** (snipe eels)	9 species. Semelparity is indicated but not rigorously shown for *Nemichthys scolopaceus;*[2] status is unknown for other species.
Infraorder Congroidea Family **Congridae** (conger eels)	109 species. Semelparity is indicated for *Conger vulgaris;*[3] the extent of semelparity is unknown for other species.
Order Clupeiformes Family **Clupeiformidae** (herrings)	190 species. Most are iteroparous.
Alosa pseudoharengus (alewife)	Cease feeding at migration from seas to rivers; mortality at spawning is 39–57%; a few live to spawn the next year.[4]
Alosa sapidissima (American shad)	Complete semelparity below 32°N; semelparity decreases greatly at northerly latitudes; see Figure 2.14.[5] Most shad species are iteroparous.[6]
Order Cypriniformes Family **Cyprinidae** (minnows and carps)	2,070 species. Most are iteroparous. In the genus *Pimephales,* which is native to the Midwest, at least 3 of the 4 species are semelparous: *P. promelas* (fathead minnow), *P. notatus* (bluntnose minnow), and *P. vigilax* (bullhead minnow); some survive spawning and overwinter but die before the next breeding season; repeat spawning is atypical.[7] Lifespan 1 year. There is no histological or hormonal data to indicate the adrenal status or the cause of death. The life history of most minnows native to North America is incompletely known.
Order Salmoniformes Suborder Salmonoidei Superfamily Galaxioidea Family **Galaxiidae** Subfamily **Aplochitoninae** (Tasmanian whitebait)	4 species.
Lovettia sealii	Only species with obligate semelparity; may cease feeding at migration from sea; dies at spawning in rivers;[8] lifespan, 1 year.
Aplochiton zebra and *Aplochiton taeniatus*	*A. zebra* and *A. taeniatus*[9] of Chilean rivers and lakes are probably iteroparous; lifespans, 2–3 years.[10]
Subfamily **Galaxiinae**	4 species. Semelparity to varying degrees.
Brachygalaxias bullocki	Apparently annual.[11]
Galaxias divergens (dwarf galaxias)	Mature at 1–2 years; semelparity is implied by the lifespan of 2–3 years.[12]
Galaxias maculatus (Inanga)	Famous for spawning on tidal banks; nearly all die as the high tide recedes;[13] some may survive spawning;[14] semelparity

Table 2.3 (*continued*)

Taxon	Semelparity
	here would appear to be consequent to behavior rather than pathophysiology; essentially an annual fish, although some may delay maturation to 2 or 3 years.
Galaxiella pusilla (dwarf galaxiid)[14]	R. M. McDowall (pers. comm.) considers that all species in the genus are semelparous. Most of the other 45 galaxiids are perennial and iteroparous: e.g. *G. fasciatus* (banded kokopu), lifespan, > 10 years; *G. argenteus* (giant kokopu), lifespan, > 21 years.[15]
Family **Retropinnidae** (southern smelts)	4 species. *Retropinna retropinna* and *R. tasmanica* are semelparous; lifespan, 2 years. *Stokellia anisodon* is considered semelparous; some may mature 1–2 years later than the norm of 1 (*Retropinna*) or 2 years (*Stokellia*). The cause of death is unknown; *Stokellia* is eaten by gulls at spawning. Lifespan, 1–2 years.[16]
Superfamily Osmeroidea	
Family **Osmeridae** (smelts)	
Hypomesus olidus (= *transpacificus*)	Conditionally semelparous in Japan.[17] A-type: all spring migrants die at spawning; natural history like *Oncorhynchus masou*. B-type: autumn anadromous migrants and landlocked populations are iteroparous.
Thaleichthys pacificus (eulachon)	Most die at first spawning, but a few survive to spawn 2–3 times.[18]
Mallotus villosus (capelin)	Most die at first spawning, by 4–5 years.
Family **Plecoglossidae** (Japanese ayu)	1 species.
Plecoglossus altevis (dwarf ayu)	Semelparous with adrenocortical hypertrophy during gonadal maturation; lifespan, 1 year.[20] The landlocked ko-ayu (dwarf ayu) has higher postspawning survival than the ayu. Also shows adrenocortical hypertrophy during gonadal maturation.[21]
Family **Salangidae** (ice-noodle fish)	*Salangichthys ishikawae* and *S. microdon* are considered to die at first spawning.[22]
Family **Sundasalangidae**	Smallest salmonid (< 2 cm); 2 species of *Sundasalanx*; lifespan, 1–2 years.[23]
Superfamily Salmonoidea	
Family **Salmonidae**	68 species. Most are long-lived and iteroparous.
Subfamily **Salmoninae**	
Oncorhynchus	5 species. The only genus with semelparity, which is prevalent or universal. *O. gorbuscha*, *O. keta*, *O. nerka*, *O. kisutch*, *O. tschawytscha*, lifespans, 3–8 years; *O. mykiss*; (steelhead) tends to semelparity at more northerly latitudes.[24]
Suborder Lepidogalaxioidei	
Lepidogalaxia salamandroides	Smallest salmoniform;[25] lifespan, 3–4 years; semelparity uncertain.
Order Atheriniformes	
Family **Atherinidae** (silversides)	160 species.
Labidesthes sicculus	Semelparous; lifespan, 1 year.[26]
Leuresthes tenuis (grunion)	Iteroparous despite its risky spawning on beaches.[27]
Order Gasterosteiformes	
Family **Gasterosteidae** (sticklebacks)	7 species with complex taxonomy.
Spinachia spinachia	Semelparous; lifespan, 1 year.[28]

Table 2.3 *(continued)*

Taxon	Semelparity
Order Perciformes	
Family **Callionymidae** (dragonets)	130 species.
Callionymus lyra	Males mature by 3–5 years and show rapid weight loss after the start of the breeding season, suggesting aphagy; there is no evidence for survival to the next year; lifespan, ≤ 5 years. Females may spawn perennially and live ≤ 7 years.[29] The distribution of semelparity is not reported for other dragonets.
Family **Gobiidae** (gobies)	2,000 species. 6 species with proven or suspected semelparity; most are probably iteroparous.
Leucopsarion petersi (Japanese ice goby)	Aphagy during maturation; obligate semelparity. The interrenal gland does not hypertrophy during maturation; the thymus atrophies extensively;[30] lifespan, 1 year.
Pomatoschistus microps (common goby)	Semelparous; growth resumes after breeding during the first season; but mortality increases for unknown causes; about 20% survive to the second year and are fertile; lifespan, 2.25 years. Miller remarked on "ovarian exhaustion in a minute proportion of females, although no detailed examination of wild fishes was attempted."[31]
Pseudaphya ferreri	Semelparous;[32] empty gut at spawning;[33] lifespan, 1 year.
Aphia minuta, Crystallogobius linearis, Pomatoschistus minutus, Pseudaphya pelagica	Stated to be semelparous; lifespan, 1 year.[34]
Family **Mugilidae** (mullets)	95 species. Most are iteroparous.
Myxus capensis	South African; aphagic during maturation and semelparous;[35] lifespan, 2–5 years.[36]

Notes: Families and subfamilies are indicated in bold. Table 11.1 shows the relationship to migration; Figures 11.8–11.10 show phylogeny. The sequence of families is based on the cladistics of Nelson, 1984.

[1] Manion and Smith, 1978.
[2] Nielsen and Smith, 1978.
[3] Cunningham, 1891; Wheeler, 1969, 229.
[4] Durbin et al., 1979.
[5] Leggett and Carscadden, 1978.
[6] McDowall, 1988.
[7] Markus, 1934; Larry Page, Illinois Natural History Survey, pers. comm.
[8] McDowall, 1988; Blackburn, 1950.
[9] Campos, 1969; McDowall, 1971a.
[10] R. M. McDowall, pers. comm.; H. Campos, Universidad Austral de Chile, Valdivia, pers. comm.
[11] McDowall, 1988.
[12] Hopkins, 1971; McDowall, 1978, 77.
[13] McDowall, 1978, 72.
[14] McDowall, 1971b.
[15] R. M. McDowall, pers. comm.; Hopkins, 1979a, 1979b.
[16] McDowall, 1978, 43, 46; R. M. McDowall, pers. comm.

[17] Hamada, 1961.
[18] McDowall, 1988; Scott and Crossman, 1973.
[19] Leggett and Frank, 1990; Frank and Carscadden, 1984.
[20] Honma, 1960, 1959.
[21] Honma and Tamura, 1963.
[22] McDowall, 1988.
[23] Roberts, 1981; Nelson, 1984.
[24] Withler, 1966; McDowall, 1988, 113.
[25] Allen and Berra, 1989; McDowall, 1988.
[26] Hubbs, 1921; Nelson, 1968.
[27] Walker, 1952.
[28] Johnsen, 1944.
[29] Chang, 1951.
[30] Tamura and Honma, 1969, 1970; McDowall, 1988.
[31] Miller, 1975, 444; Healey, 1972.
[32] Miller, 1973.
[33] Fage, 1910.
[34] Miller, 1973, 1975, 1979; Swedmark, 1957.
[35] Bok, 1979; McDowall, 1988, 88.
[36] Bruton et al., 1987.

and *O. tschawytscha,* but there is agreement that all species, including the Asiatic *O. masou,* show the same generalized pathophysiology during spawning.

A hallmark of maturation is a set of rapid degenerative changes associated with progressive hyperplasia of interrenal cells leading to fivefold or greater elevations of circulating corticosteroids (Hane and Robertson, 1959; Idler et al., 1959; Robertson and Wexler, 1960, 1962; Oguri, 1960; Robertson et al., 1961a; Love, 1970; Figure 2.10). Nonetheless, maturation and spawning alone do not cause sustained cortisol elevations in *O.*

Table 2.4 Pacific Salmon (*Oncorhynchus*)

O. gorbuscha (pink salmon): invariant 2-year lifespan in Asia and America; in a recently landlocked population in Lake Superior, some fish do not mature until 3 years (Kwain, 1987; Kwain recently died, but these studies are continued by Paul McMann, Lake Superior Fisheries Unit, Guelph, Ontario.

O. keta (chum): Asia and America; most live 3 to 6 years, but maturation may be delayed to ≥ 7 years (Hanamura, 1966).

O. kisutch (coho, silver salmon): America and Asia; 3-year lifespan, of which half is spent in freshwater, half in the ocean.

O. masou (cherry salmon, masu, yamame): both anadromous and landlocked populations; mature at 3 to 4 years, but as late at 6 years (Machidori and Kato, 1984). Asian side of North Pacific only, in lands near Sea of Japan, including Korea and Siberia, but also Taiwan. While death at spawning is nearly universal, some exceptions are noted for males (Tanaka, 1965; Ono, 1933, cited in Tanaka; Oshima, 1934), as for *O. tschawytscha.* Smith and Stearly (1989) cite reports that some populations are iteroparous: Masuda et al., 1984; Jordan and Evermann, 1896. Landlocked variants with red spots found only on the Asian side are sometimes designated as a separate species.

O. mykiss (steelhead, anadromous; rainbow trout, landlocked; formerly known as *Salmo gairdneri*):[1] high mortality at spawning but iteroparous, unlike others in the genus. Lifespan of the rainbow increases at more northerly latitudes.

O. nerka (sockeye, migratory; kokanee, red salmon, nonmigratory): landlocked in northern Asia and America; lifespan of ≤ 8 years; longer lifespans in Alaska. Sockeyes introduced to New Zealand have maintained semelparity (McDowall, 1978). Kokanee occasionally survive spawning. Several North American subspecies cited in early literature are no longer recognized: *O. nerka nerka* (sockeye); *O. nerka kennerlyi* (freshwater kokanee, kickaninny).

O. rhodurus (amago, streams; biwa-masu, lakes): *O. masou* and *O. rhodurus* are interfertile; *O. rhodurus* is sometimes classified as a subspecies of *O. masou.* Precociously mature males (jacks) survive spawning (Oshima, 1934; Ono, 1933, cited in Tanaka, 1965).

O. tschawytscha (chinook, king, or spring salmon); America and Asia; introduced to New Zealand (locally known as the Quinnat salmon) where semelparity persists (McDowall, 1978). Robertson (1957) notes that jacks can survive spawning. Precocious maturation provides an alternative life history strategy with different mortality risks (see Chapter 9.2.2.2.1).

Notes: Common names in parentheses. America = northern Pacific coast of the United States and Canada; Asia = northern Pacific and all shores of the Sea of Japan. Figure 11.9 shows the evolutionary relationships.
[1]For general descriptions of salmon life cycles, see Vladykov, 1963, Hoar, 1976, McDowall, 1988, scholarly reports of the International North Pacific Fisheries Commission (Vancouver, British Columbia, especially numbers 16, 18, 23, 43, 46; sent free on request), and J. Van Dyk, *Natl. Geog.* (July 1990, 3–38). Also see Dadswell et al., 1987, with valuable articles besides those cited here. Salmon names have been recently revised by the American Fisheries Society (Behnke, 1990b; Smith and Stearly, 1989).

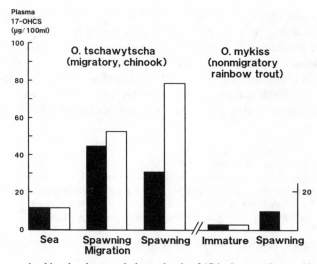

Fig. 2.10. Senescence in chinook salmon and plasma levels of 17-hydroxycorticosteroids (17-OHCS). *Black bars* are females; *white bars* are males. The chinook salmon (*Oncorhynchus tschawytscha*), a migratory Pacific salmon that dies at first spawning, is compared with nonmigratory rainbow trout (*O. mykiss*, formerly named *Salmo gairdneri* that usually survives first spawning. Redrawn from Hane and Robertson, 1959.

nerka, without additional stress (Fagerlund, 1967). The hyperadrenocorticism is probably caused by gonadal steroids during maturation, since it can be induced in castrated immature sockeyes (*O. nerka*) by androgens and estrogens (van Overbeeke and McBride, 1971).

Spawning Pacific salmon appear to die from a shared set of pathophysiologic changes. Spawning fish do not ingest food, the villi of their intestinal epithelium almost completely disappear, and their fat reserves are depleted. This aphagy has a completely different basis than that in insects, as there is no anatomic deficit. Gross pathologic changes occur in many tissues, but are selective. In regard to the question of gene activities during senescence, it is of interest that during spawning, the interrenal cells show conspicuous enlargement of their cell nuclei and nucleoli (McBride and van Overbeeke, 1969). Nucleolar enlargement is generally considered to indicate increased transcription of ribosomal RNA genes. Presumably, interrenal cell steroid synthesis requires increased amounts of enzymes that, in turn, depend on increased macromolecular biosynthesis. The coronary arteries, but also renal and other arteries, show an extensive proliferation of endothelial cells (Figure 2.11) that strikingly resembles coronary artery disease in humans (Robertson et al., 1961b). Similar coronary lesions are found in spawning Atlantic salmon (*Salmo salar*) and persist for at least 5 months after spawning in this perennial species (Saunders and Farrell, 1988). The pancreatic β-cells and most skeletal muscles retain normal cytology, while other cells show atrophic changes that may be attributed to toxic effects of elevated steroids, as described next.

The liver, kidney, spleen, and thymus degenerate markedly, with pycnosis and destruction of cell nuclei (Figure 2.12), in contrast to the burgeoning interrenal cells. (As described in Chapter 7.7, elevated glucocorticoids selectively kill thymic and other cell types in mammals through a process that is dependent on dynamic changes in gene regulation.) The heart shows extensive myofibrillar degeneration. Fungal infections are com-

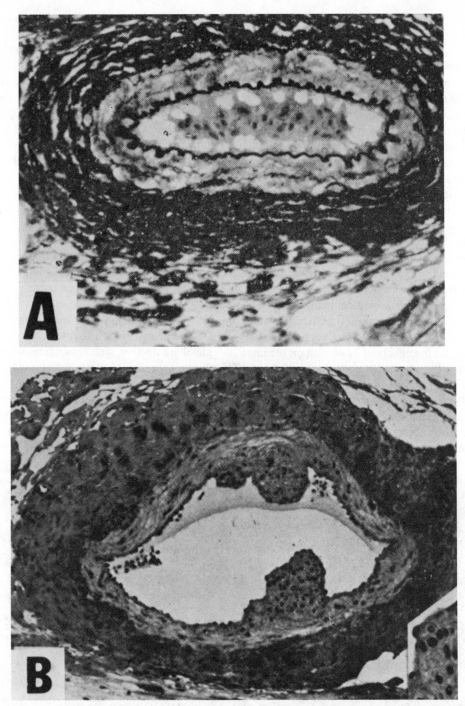

Fig. 2.11. Coronary artery degeneration in chinook salmon (*Oncorhynchus tschawytscha*) during spawning. Enlargened to equivalent magnification (× 50) from Robertson et al., 1961. *Top*, a normal artery of a migratory fish before major senescent involution. *Bottom*, a degenerating artery of a spawning fish showing proliferation of intimal cells that project into the vessel lumen. The adventitial layer of the wall is much thicker. The insert shows a nodule of intimal cells (× 350). Rephotographed from Robertson et al., 1961b.

mon on the skin, and there is an extensive involution of immune functions. *O. kisutch* shows reduced resistance to a bacterial pathogen (*Vibrio anguillarum*) and fewer splenic antibody-producing cells in response to an antigen from *V. anguillarum* (Maule et al., 1987). This immune depression illustrates the trade-offs of reproduction, by which elevated corticosteroids may be physiologically adaptive to mobilize stored energy for reproduction but at the expense of many natural defense mechanisms. This pathophysiological syndrome strongly resembles the consequences of Cushing's disease in humans (hyperadrenocorticism). There are also parallels between these changes and certain age-related changes in mammals (Chapters 3.6.10, 10.5).

Several lines of experiments showed that these pathophysiological changes are linked to the adrenocortical hypertrophy and elevated blood cortisol (hydrocortisone). Senescent-like changes can be prematurely induced by hydrocortisone implants in rainbow trout (*O. mykiss*) (Robertson et al., 1963), a species that usually does not show the extreme hyperadrenocortical changes of the Pacific salmon and which can spawn repeatedly. The findings were one of the earliest examples of a bidirectional gerontological manipulation, in which senescence can be delayed to advanced ages or prematurely induced in the young by specific physiological manipulations of steroids and diet.

Conversely, castration before spawning prevents interrenal gland hypertrophy and the hypercorticism of *O. nerka* and *O. tschawytscha* (Robertson, 1961). The role of the pituitary in the effects of castration on the interrenal gland are unclear, since castration did not alter the histologic appearance of the ACTH-containing cells (McBride and van Overbeeke, 1969). Castrated fish escape from having senescence occur at the usual age; instead they continue to grow and may even double their lifespans (Figure 2.13) without interrenal gland hypertrophy if completely castrated (Robertson and Wexler, 1962; McBride and van Overbeeke, 1969). These castrated fish eventually lost weight and showed some of the usual corticosteroid-dependent changes. Unfortunately, data on plasma steroids are

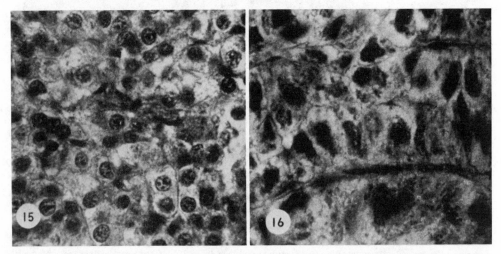

Fig. 2.12. Cell degeneration during spawning in Pacific salmon. Hematoxylin and eosin staining, × 900. *Left,* liver of male chinook salmon (*Oncorhynchus tschawytscha*) that would probably spawn within 1–2 months and showing early degenerative changes. *Right,* liver of postspawning kokanee (*O. nerka*) showing marked degeneration. Rephotographed from Robertson and Wexler, 1960.

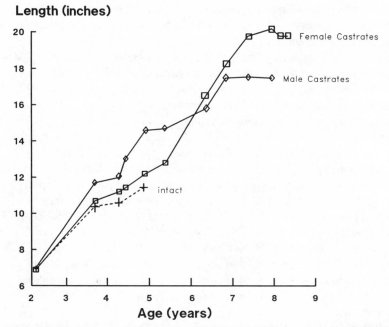

Fig. 2.13. Castration before sexual maturity increases the lifespan of kokanee salmon (*Oncorhynchus nerka*), a nonmigratory variant that normally does not live beyond 4 years and rarely survives spawning. Fish were castrated when six inches long (25 months old). Most castrates died by 4 years, but some survived to greater ages. The initial number of fish was 473; except for the last four ages, the data points represent means of five or more fish. Redrawn from Robertson, 1961.

not available to indicate if there was cryptic regeneration of the gonad; the causes of death were not characterized and may or may not be related to those of normal fish.

The gastrointestinal atrophy of postspawning *O. nerka* can also be greatly reduced by providing zooplankton, which the fish will eat (McBride et al., 1965). While the importance of elevated corticosteroids is indisputable in postspawning senescence, food reserves and diet may have independent contributions (Larsen, 1985).

Even greater superannuation of salmon may be realized through attempts to increase the size of the chinook by extending their lifespan through sterilization (Schmidt, 1986).[7] Sterile triploids are being produced by heat-shocking eggs just after fertilization. The resulting fish are expected to grow normally, but should have impaired gonadal development, based on triploidy of landlocked Atlantic salmon (Benfey and Sutterlin, 1984). These fish were tagged before release in 1986, and some should survive the predations of sports fishermen to become giants with lifespans far beyond the normal 4 years.

Next, I describe some variations in the natural history of salmon which indicate the great plasticity of these phenomena. Contrary to the general rule of postspawning mortality, there is some postspawning survival of precociously mature parr (juvenile males) of *O. tshawytscha* (Robertson, 1957) and *O. masou* (Oshima, 1934; Ono, 1933, cited in Tanaka, 1965). Precocious maturation of parr is common in iteroparous salmon (Chapter 9). Thus, an escape from mortality in usually semelparous *Oncorhynchus* species is con-

---

7. Donald Garling and Howard Tanner of the State of Michigan's Department of Natural Resources.

sistent with life history variants in closely related species, and underlines the preservation of plasticity and the potential for rapid evolutionary changes in life history through variants of reproductive scheduling.

The migratory anadromous pattern itself must be a major factor in the (near) universal death at spawning, as judged by comparing the survival of some migratory steelheads with that of landlocked rainbow trout after spawning; both are *O. mykiss,* formerly *S. gairdneri.* Only a small fraction of anadromous steelheads survive, but these can spawn again, while the nonmigratory rainbows have much greater survival to second spawning (Robertson, Krupp, Thomas, Favour, Hane, and Wexler, 1961). Moreover, the extent of death at spawning increases with the length of the upstream run, as observed for *O. mykiss* (Robert Behnke, pers. comm.). Tanaka (1965) comments that freshwater *O. masou* often occur in the same streams as the anadromes, which suggests a facile transition or possibly a reproductively isolated sympatric population. A landlocked stock of *O. gorbuscha* was recently established in Lake Superior (Kwain, 1987).[8] A notable change in these *O. gorbuscha* is the delay of maturity to 3 years, which departs from the invariant 2-year life cycles of anadromous populations. This change in entire populations is so rapid as to suggest a phenotype with alternate reproductive scheduling, rather than selection of genetic variants. Future variants of these or other landlocked species should be followed to see if some individuals begin to escape death at spawning, allowing reestablishment of the ancestral perenniality.

Of particular interest to the genetic basis for semelparity are phenotypes of conditional semelparity. A striking example is that of the American shad *Alosa sapidissima;* shad are clupeiforms, a different order than salmon (Table 2.3). The most southerly populations that spawn in Florida rivers are semelparous, while iteroparity increases linearly with the degree latitude, exceeding 70% in the St. John River of Canada (Figure 2.14). In the iteroparous populations, shad occasionally spawn up to seven times during lifespans of at least 13 years (McDowall, 1988, 69).

Steelhead trout (*O. mykiss*) also show conditional semelparity. A minority (1–20%, depending on location) of sea-run steelhead trout survive to spawn a second season, while landlocked (nonmigratory) rainbow trout usually survive to spawn the next season (Robert Behnke, pers. comm.). The elevation of corticosteroids and the extent of tissue deterioration at spawning show similar rankings: other *Oncorhynchus* > steelheads > rainbows (Hane and Robertson, 1959; Robertson et al., 1961a). Also, rainbows eat more actively during the spawning season than steelheads and other *Oncorhynchus* species, and may have less severe nitrogen imbalances that could favor resistance to adverse effects from elevated corticosteroids. Moreover, mortality in spawning steelheads increases at more northerly latitudes, so that the most northerly steelhead populations are virtually semelparous (Withler, 1966; McDowall, 1988, 113). Consistent with this trend, the lifetime egg production was relatively constant at 250,000 to 400,000 per fish at all latitudes. Other salmoniforms that show conditional semelparity include pond smelt (*Hypomesus olidus*), various galaxiids, and *Pimephales promelas* (Cypriniformes) (Table 2.3).

These examples of conditional semelparity contrast with the obligate semelparity of some lampreys and Anguillidae, and most Pacific salmon. The extent to which semelpar-

---

8. Kwain recently died, but the studies are continuing under Paul McMann of the Lake Superior Fisheries Unit in Guelph, Ontario.

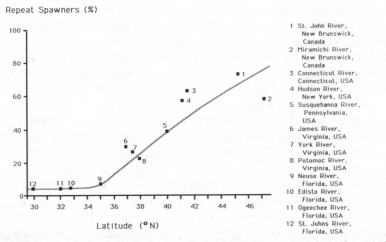

Repeat Spawners (%)

1 St. John River,
   New Brunswick,
   Canada
2 Miramichi River,
   New Brunswick,
   Canada
3 Connecticut River,
   Connecticut, USA
4 Hudson River,
   New York, USA
5 Susquehanna River,
   Pennsylvania,
   USA
6 James River,
   Virginia, USA
7 York River,
   Virginia, USA
8 Potomac River,
   Virginia, USA
9 Neuse River,
   Florida, USA
10 Edisto River,
    Florida, USA
11 Ogeechee River,
    Florida, USA
12 St. Johns River,
    Florida, USA

Latitude (°N)

Fig. 2.14. Variable semelparity is a function of geographic latitude in the American shad (*Alosa sapidissima*). In samples of populations from rivers below 32°N on the eastern American shore, the scale rings of shad indicated semelparity, while further north, the fraction that survived to spawn another year increased progressively beyond 50%. Redrawn from Leggett and Carscadden, 1978.

ity is conditional is not known for most of the other species in Table 2.3. Conditional semelparity could be an intermediate to obligate semelparity, in conjunction with genetic changes that may be readily reversible (polymorphisms) or not (mutations). The variable survival after spawning in these species shows that semelparity coupled with rapid senescence is not under rigid genetic programming.

Eels, an entirely distinct radiation (order Anguilliformes, Figure 11.8) are another major example of semelparity. Among the nineteen families in this order, at least a few species in three families die universally or with high frequency at first spawning: the Anguillidae, Nemichthyidae, and Congridae. The best known are catadromous freshwater eels, the European *Anguilla anguilla,* which die at spawning in association with an aphagous adult phase after their return from fresh water. Nearly all species (14 out of 16) of *Anguilla* spawn in the Sargasso Sea (Moriarity, 1987; Post and Tesch, 1982). European eels are generally presumed universally to die after first spawning. Although there are no descriptions of spent eels at the spawning grounds, the otolith rings indicate only one marine phase (Todd, 1980). Two other catadromous anguillids are also found offshore from southeastern Africa and in the North and South Pacific, but some Atlantic species live there as well.

Anguilla usually die at age 10–15 years upon first spawning (Bertin 1956; Tesch, 1977; Williams and Koehn, 1984), with a rapid senescence from elevated corticosteroids that is suggestively like that of Pacific salmon. Little is known of the pathophysiology of eel senescence, beyond the atrophy of the digestive tract during gonadal maturation. In the laboratory, juvenile eels withstand starvation for at least a year, losing up to 70% of body weight before death (Boëtius and Boëtius, 1967). The transformation from the juvenile (yellow or green eel) to the adult (silver eel) stage occurs just before the remarkable migration of eels from European freshwater lakes and rivers to the Atlantic shores off North America (Bertin, 1956). During this change, hydrocortisone (cortisol) increased twofold (Leloup-Hatey, 1964, 319; Henderson et al., 1974). It is unknown why preventing migration extends lifespan; certainly the ability of such eels to continue feeding is crucial. Male eels, unlike females, appear to survive spawning and maturation when induced by human chorionic gonadotropin in the laboratory (Boëtius and Boëtius, 1967, 1980). Moreover, spermiation can be induced more than once in successive studies during the same year, whereas females died within 2 weeks of induced ovulation. The male eel thus may occupy a shifting position in the spectrum of senescence and may not be obligately semelparous.

Anguillids appear to have the same life cycle phases, with wide variations in their duration. Nutrient reserves are accumulated during slow maturation in fresh water, followed by migration for thousands of kilometers to spawn and die. The immature leptocephalus larvae and the metamorphosed glass eels must then make the long return to fresh water (Nelson, 1984, 408; Bruton, Bok, and Davies, 1987; Jellyman, 1987). The age at migration varies widely, and maturation is not tightly linked to body size, unlike in most other vertebrates. One *A. dieffenbachii* had sixty "annual" growth rings from its freshwater phase (Todd, 1980), while others of unknown age grew to huge lengths, nearly two meters (McDowall, 1978). This enormous plasticity in the duration of the growth phase before maturation and the cessation of feeding is entirely consistent with the many-decade-long lifespans of numerous other teleosts (Chapters 3, 4), and makes the extended lifespans of aquarium eels less strange. If eels are prevented from returning to the ocean, they can achieve remarkable lifespans without further growth (Roule, 1933; Gandolfi-Hornyold, 1935; Lekholm, 1939; Wundsch, 1953; Tesch, 1977). "Putte" resided at the Halsingborgs Museum (Sweden), where its finicky feeding amused the museum staff for decades (Lekholm, 1939). She died at 88 years (Vladykov, 1956), which is about six times the normal lifespan. Tesch (1977) gives other examples. These lifespans appear to extend far beyond the much shorter-lived lampreys and are the best vertebrate equivalent of the very long-lived semelparous and rapidly senescing plants like the bamboo (Section 2.3.2). The extended growth phases in eels suggest that there must be continuing proliferation of intestinal and other exfoliating cell types.

Indications of semelparity are found in two families distinct from the anguillids, the conger eel and snipe eel, which are both entirely marine. Conger eels (*Conger vulgaris,* one of 109 species in the family Congridae) cease feeding for 6 months during gonadal maturation, with atrophy of the stomach and intestines, and loss of teeth in both sexes (Cunningham, 1891). They are widely considered semelparous (Wheeler, 1969, 229). From these findings and other accounts, Cunningham concluded that congers breed only once, arguing that they could not regain feeding owing to the loss of teeth. Snipe eels, family Nemichthyidae with nine species, are also candidates for semelparity. In *Nemichthys scolopaceus* both sexes lose teeth at maturation, but only males lose them all (Nielsen and Smith, 1978); this implies impairments of feeding, but I have not found further information. Nothing is known about semelparity, obligate or conditional, in the other sixteen families of eels that are entirely marine.

Other teleosts that commonly die at first spawning are also listed with references in Table 2.3. The anadromous and semelparous ice goby (*Leucopsarion petersi*) also shows adrenal enlargement and splenic degeneration at spawning (Tamura and Honma, 1969, 1970). Most examples have unknown pathophysiology, including Tasmanian whitebait (*Lovettia sealii*), Japanese ayu (*Plecoglossus altevis*), and the capelin (*Mallotus villosus*). An unusual example is the gender-specific semelparity of the common dragonet, *Callionymus lyra* (Perciformes): males disappear after first spawning and are considered semelparous, while females spawn perennially (Chang, 1951). The poor and wasted condition after spawning in most of these fish suggests the effects of stress and starvation. As noted above, the semelparity of many of these species is not obligatory.

As will be discussed in Chapter 11.3.4, there is a strikingly disproportionate incidence of semelparity in fish species that are—or whose ancestors were—migratory between fresh water and the seas (diadromy; Table 11.1). Nearly all of the eleven orders with at least one semelparous species are diadromous or have diadromous ancestors. Most of these belong to two major orders, the Anguilliformes and the Salmoniformes, which are found in temperate zones of both hemispheres. In the salmoniform genera, *Galaxias* and *Retropinna,* semelparity prevails in most species, while other genera may have a more scattered incidence. The orders Atheriniformes, Clupeiformes, Cypriniformes, Gasterosteiformes, and Perciformes each have at least one semelparous species. The associations of diadromy and semelparity may help find other examples of semelparity in related species whose natural history may not be known, as well as in identifying links of semelparity to ancestral diadromy (Chapter 11). We can be sure that there is no *obligate* link between semelparity and diadromy, as shown in numerous *Salmo* that are repeat spawners. Similarly, many species of sturgeon, a primitive teleost, are anadromous and survive to spawn repeatedly despite elevations of corticosteroids and aphagy (Chapter 3.4.1.2.1). Their large size and nutrient reserves probably make them better protected than smaller fish. The distribution of aphagy during reproduction in nonmigratory fish is indicated in scattered references and might also lead to further discoveries of semelparity. In contrast to these examples are most other teleosts, which spawn on successive years and retain fertility in both sexes throughout lifespans that occasionally exceed the average human life (Chapters 3, 4).

2.3.1.5. *Marsupial "Mice"*

About ten small dasyurid marsupials in the genera *Antechinus* and *Phascogale* are usually semelparous and show a rapid senescence linked to adrenal corticosteroids (Lee and Cockburn, 1985a; Woolley, 1966, 1981; Lee et al., 1982; Diamond, 1982; Calaby and Taylor, 1981; Braithwaite and Lee, 1977). The mouse-sized males of *A. stuartii, P. tapoatafa,* and other dasyurids generally die within 2 to 3 weeks of mating before they reach 1 year. Females of these monoestrous species usually survive beyond 1 additional year. Depending on location, the proportion having a second litter varies from 0 to 0.67% (Cockburn et al., 1983). Other species have different postmating survivals according to whether they forage in trees or in soil litter.

The postmating mortality of males may be stress related (Barnett, 1973; Bradley et al., 1980; Barker et al., 1978) and consequent to repeated bouts of prolonged copulation that in the laboratory last up to 12 hours (Lee et al., 1982). If males of several semelparous species are captured before breeding and kept in the lab, they will survive to the age of at least 3 years, more than 2 years longer than in nature (Moore, 1974; Woolley, 1971). Their escape from "natural death" under optimum conditions and their capacity to more than double their natural lifespan caution against overemphasizing lifespan and mortality rates as a basic index of cellular "aging." Moreover, males of the closely related *A. bilarni* are perennial and live for at least 32 months (Begg, 1961). The DNA satellite distributions (Timms et al., 1982) indicate that four species whose males die at mating (*A. swainsonii,*

A. bellus, A. flavipes, and *A. stuartii*) are more closely related than the perennial *A. bilarni.*

The pathophysiology of postmating rapid senescence in *Antechinus* is similar in many ways to that of Pacific salmon. In the month before death, the adrenal weight abruptly increases by 50%, with hyperplasia in the zona fasciculata (Barnett, 1973; Moore, 1974). Free-circulating corticosteroids (principally cortisol) increase more than fivefold because of two factors: increases in total corticosteroids; and larger decreases of corticosteroid-binding globulins in *A. stuartii,* as shown in Figure 2.15 (Bradley et al., 1976, 1980; Bradley, 1982; Moore, 1974), and in *A. flavipes* and *A. swainsonii* (Bradley, 1980). Thus, the net increase of free corticosteroids (the most biologically active) is much greater than the total blood levels would indicate. Androgens, principally testosterone, also increase sharply (Bradley et al., 1980; Moore, 1974) and trigger the decrease of corticosteroid-binding globulins, as shown by experimental studies with castration and androgen replacement (Lee and Cockburn, 1985a). No *Antechinus* species has a sex-steroid-binding globulin, unlike some other marsupials (Bradley, 1982).

By the time *Antechinus* mate, the negative feedback of corticosteroids on ACTH secretion becomes greatly diminished (Lee and Cockburn, 1985b). The sustained corticosteroid elevations imply that the brain-pituitary-adrenocortical axis has been desensitized to negative feedback. How this may occur is a matter of conjecture. A plausible mechanism suggested by Sapolsky (1990) involves a decrease of hippocampal corticosteroid receptors as part of the physiological changes during the reproductive season. Rats treated to lower their hippocampal corticosteroid receptors show impaired feedback, possibly by diminished inhibition of the hippocampus on the hypothalamic output of corticotropin-releasing hormone (CRH; Sapolsky et al., 1984a, 1984b; Chapter 9.3.4.2).

The males rapidly become sick from a range of lesions and stop grooming their fur, a mark of serious illness in most mammals. They lose weight and have a negative nitrogen balance, despite the presence of food in their stomachs; feces become watery and contain but few fragments of insects, a main staple (Lee and Cockburn, 1985b). The testes involute, with dramatic loss of the germinal epithelium. Other important involutionary changes occur in the spleen, and there are marked decreases of gamma globulin-A and -G and increased parasitic and microbial infections. These are hallmarks of glucocorticoid-induced immunosuppression in placental mammals (Munck et al., 1984). *Antechinus* shows a remarkable ability to consummate reproduction despite these numerous pathophysiological changes associated with elevated glucocorticoids that, by comparison with most placental mammals, would cause reproductive collapse. The schedule and extent of testicular involution must be delicately adjusted so that there is sufficient sperm at copulation.

Bleeding gastric and duodenal ulcers cause anemia (Moore, 1974; Woolley, 1971; Cheal et al., 1976; Barker et al., 1978; Bradley et al., 1980; Calaby and Taylor, 1981). The absence of lipofuscins (aging pigments) during these dramatic changes (Barnett, 1973) argues against a generalized acceleration of the usually gradual senescence that is characteristic of placental mammals (Chapter 3.6). Many of these pathophysiological changes can be induced in laboratory *Antechinus* by cortisol injections (Barker et al., 1978; Bradley et al., 1975, 1980). In contrast to males, female *Antechinus* maintain a healthy appearance with sleek coats during this time and do not have marked increases of free corticosteroids.

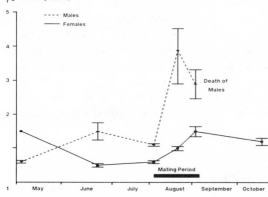

Free Corticosteroids
(µg/100 ml plasma)

- - - - Males
——— Females

Mating Period

May   June   July   August   September   October

Death of Males

Fig. 2.15. Plasma elevations of corticosteroids in association with mating-related death in *Antechinus stuartii*, a mouse-sized marsupial. The assay represents total free corticosteroids (principally cortisol) that are not bound to corticosteroid-binding globulin. The males, which die shortly after mating, have greater corticosteroid levels during the mating season than do females that survive to breed another year. Redrawn from Bradley et al., 1980.

A triggering role of the gonads is established for *Antechinus*: males castrated before the mating season did not show adrenocortical hyperplasia, cortisol elevations, weight loss, or major organ degeneration (Moore, 1974), and their fur remained sleek (Inns, 1976). Androgens suppress corticosteroid-binding globulins in placental mammals, which could set up a precondition for further changes (Sapolsky, 1990). Even in the field, castrated males had a greater survival (Moore, 1974; Lee and Cockburn, 1985a). The external stress associated with breeding is widely thought to cause the postmating mortality (Diamond, 1982; Woolley, 1971; Lee et al., 1982). Male-male aggression, as tested in captivity, becomes progressively more intense during each successive encounter (Braithwaite, 1974), which suggests a positive behavioral-feedback cycle. However, breeding stress cannot be the only factor, since even unmated captive males show seasonal involution. Although captives can survive at least 3 years under favorable conditions (see above), they are permanently sterile in three species (Woolley, 1981). The limited data for other marsupials (Lee and Cockburn, 1985a) suggest important endocrine variations at breeding that influence survival.

2.3.1.6. *Rodents*

Besides these Australian marsupials, bush rats (*Rattus fuscipes*) in the same locales have stress-related mortality during reproduction, concurrent with elevated androgens and lowered corticosteroid-binding globulin (Lee and Cockburn, 1985a). These striking similarities indicate the potential for parallel evolution in the same locale of the stress-related semelparous pathophysiological syndrome. However, fluctuations in food availability are crucial as well, as shown for the smokey mouse (*Pseudomys fumeus*), also Australian: males have over 50% mortality outside of their preferred habitat, while nearly all survive in locales where there is a plentiful supply of their food, a migratory moth and a fungus (Lee and Cockburn, 1985a). In Canada, voles (*Microtus townsendii*) often have population crashes in the spring; the extent of spring declines varies widely between years, possibly as a result of differences in food or population pressures. Survivors have many rump wounds, presumably from fighting within their species (Krebs and Boonstra, 1978). Seasonal population crashes are also common in deer mice (*Peromyscus maniculatus*), where losses are highest in breeding females (Fairbairn, 1977), as well as in lemmings and many other rodents (Krebs and Myers, 1974).

Thus, in many species of rodents that are short-lived (5–10 years) under optimum circumstances, few will usually survive to the next year because of high mortality rates (AMRs) that range from 3.6/year (1%/day) observed in *Microtus agrestis* during most of the year (Krebs and Myers, 1974) to 18/year in *M. townsendii* during its spring decline (Krebs and Boonstra, 1978). The stress hypothesis of Christian (1950) that adrenal hyperfunction is the major cause of population cycles does not explain seasonal mortality of all species (Krebs and Myers, 1974). The stress from repeated breeding that accelerates senescence in laboratory rats (Chapter 10.6) and the postmating mortality of *Antechinus* and the Pacific salmon are extremes in a continuum of adrenocortical involvement in the senescence of vertebrates.

2.3.2. Hormonally Triggered Senescence in Plants

Rapid senescence has a well-defined hormonal basis in vascular plants, particularly for the monocarpic angiosperms (equivalent to semelparous animals) in which senescence and death of the whole plant rapidly follow the reproductive phase of flowering and fruiting (Leopold, 1961; Thimann, 1980; Thimann and Giese, 1981; Noodén, 1980a, 1988c; Noodén and Thompson, 1985; Woolhouse, 1978, 1983). The timing of these events ranges from 5 weeks in the ephemera (e.g. *Stellaria media*) to more than 100 years for some bamboos (see below). Plants exhibit an anatomical spectrum of senescence, from overall rapid senescence in which the entire plant dies, to weaker localized changes in which only select organs die. At the extreme in the spectrum are numerous examples of limitless vegetative (asexual) growth and proliferation (Chapter 4). Many botanists emphasize that plant senescence is an orderly and active process (Leopold, 1961; Noodén, 1988a, 1988b, 1988c), which is a very different view than that held by most investigators of animal senescence. A major emphasis in the literature of plant physiology is given to correlation effects, probably mediated by hormones, through which roots, flowers, and fruits coordinate the development, maturation, and senescence of other, sometimes distant parts.

The hormonal mechanisms that cause whole-plant senescence are also thought to cause organ senescence. Besides death of the entire plant, as in monocarps, the senescence and abscission of shoots, leaves, petals, flowers, thorns, and root nodules occur regularly during growth and development (e.g. Millington and Chaney, 1973; Thimann, 1980). For example, the ripening of fruit is considered to be a senescence process, and involves many of the same metabolic changes as in senescence of leaves and petals, including yellowing (loss of chlorophyll) and decreased protein synthesis (Leopold, 1961; Woolhouse, 1978, 1983; Gepstein, 1988). During leaf senescence, leaf chloroplasts of the flowering plant *Perilla fructescens*, for example, show greater loss of polyribosomes than does the cytoplasm (Callow et al., 1972). Structural and biochemical studies show the great selectivity of senescence at cellular and subcellular levels, such that neighboring cells of different types are differentially affected (Noodén and Thompson, 1985). In many regards, the curling of leaves and other degenerative changes in plant organs resemble the cyclic involution of reproductive tissues during the menstrual cycle, such as the involution of uterine cells or the corpus luteum. Further examples from the sophisticated literature on the biochemistry of organ senescence in higher plants are not discussed since my focus is on whole plants.

Fig. 2.16. Delay of senescence in soybean plants (*Glycine max*) by removing flowers before polination. *Right,* controls died at 119 days (mean). *Left,* early deflorated plants died at 179 days. Effects in delaying senescence are lessened if flowers are ablated at later ages. From Leopold et al., 1959. Photograph kindly provided by Carl Leopold.

Many classic studies show how the senescent death of monocarps is triggered by secretions from specific reproductive structures. Senescence can be delayed or prevented altogether by removing these organs (Molisch, 1938; Noodén, 1988c). As examples of these correlation effects, experimental removal of flowers from soybeans (*Glycine max;* Figure 2.16) or spinach (*Spinacia oleraca*) delayed the senescence that normally occurs at the end of the reproductive phase (Leopold et al., 1959); the effects diminished progressively as defloration was done nearer to maturation. However, maintenance of soybean plants at short photoperiods that prevented the induction of flowering, in turn promoted indefinite vegetative growth (Noodén, 1980b). Monocarps exhibit many species differences in the influence on whole-plant senescence of removing various organs, for example, the seed, the flower, or the roots (Noodén, 1980a). In some plants, defloration does not retard whole-plant senescence. For example, soybeans stop growing early in reproductive development and die when the seeds ripen. But in contrast to some other plants, removal of their flowers does not influence the rate of growth or the age when it ceases. In others, deflorated plants retain green leaves longer, for example the cocklebur (*Xanthium pennsylvanicum;* Krizek et al., 1966). Leaf senescence can also be delayed by factors resulting from fungi, insect larvae, and insect-induced galls (Molisch, 1938). In yet other species, experimental defruiting accelerates senescence, for example, in some varieties of corn (*Zea mays;* Allison and Weinmann, 1970).

A small number of major hormones are involved in plant senescence (Table 2.5). Both in plant development and senescence, these hormones appear to work in various combi-

Table 2.5 Major Hormones in Vascular Plant Senescence

Abscisic acid (ABA): A substituted cylohexenone that promotes senescence by inhibiting protein synthesis and accelerating chlorophyll loss in the leaves of many species. ABA may also stimulate protease activity. Endogenous ABA may increase before or during senescence in different tissues. ABA also interacts with ethylene and other plant hormones, as shown by treatment of organs at different phases of senescence.

Cytokinins (CK): A diverse class of substituted adenines that includes kinetin (synthetic) and zeatin, among other naturally occurring compounds. CK treatment characteristically delays senescence of excised leaves by slowing the breakdown of chlorophyll (yellowing) and maintaining RNA and protein levels. CK are produced by roots and generally decline in a senescing plant organ.

Ethylene: Treatment with ethylene promotes fruit senescence (ripening), leaf senescence, and abscission, and increases hydrolylic enzymes. Endogenous ethylene increases in many tissues during senescence.

Gibberellic acid (GA): GA and other giberellins are four-ring carbon compounds that promote shoot elongation and retard senescence in leaves and petals in certain species. Endogenous GA decreases during senescence in many tissues.

Indoleacetic acid (IAA): Indoleacetic acid is a growth regulator formed at the growing tip (apical meristem), as well as by leaves and other parts. IAA has multiple effects that retard senescence in different parts of a plant. IAA is transported downwards, where it promotes root growth. Another effect is to promote the growth of fruit. Endogenous IAA decreases during leaf senescence and abscission, while exogenous IAA delays leaf senescence. IAA may also promote senescence by inducing protoplast breakdown during xylem differentiation.

Polyamines: Putrescine and other polyamines have antisenescent effects and inhibit ethylene production. Strictly speaking, the polyamines are not hormones since they are not transported.

Sources: Noodén and Leopold, 1988, and Raven et al., 1986, in consultation with Larry Noodén.
Notes: There are numerous species variations, exceptions, and apparent contradictions. No species have been studied systematically for the major hormones throughout the entire life cycle, and few hormonal systems have been worked out in detail.

nations, and their relative importance seems to vary widely among organs, stage of development, and species (Jacobs, 1979; Noodén, 1988a, 1988c). For example, decreased cytokinin in the sap of fruiting plants is widely implicated in shoot senescence in some plants (e.g. Davey and Van Staden, 1978; Van Staden et al., 1988), but not in all species (Noodén, 1980a, 1988c; Chapter 12). The senescence of excised leaves can be retarded by auxins and cytokinins, whereas it is often accelerated by abscissic acid, ethylene, and gibberellic acid (J. Sacher,[9] 1957, 1965; Noodén, 1988a; Thimann, 1980). In species with whole-plant senescence and death, the mechanisms are thought to be similar to organ senescence. A role of nuclear genes in leaf senescence is partly known (Chapter 7.4)—an intriguing correlation between the minimum generation time (minimum lifespan), nuclear DNA, and time in meiosis in some surveys (Chapter 5.4.1) suggests a constraint in life histories.

9. There were two Sachers, brothers who each made pioneering contributions to gerontology. Joseph A. (d. 1985) characterized hormonal mechanisms in leaf senescence and showed alterations in RNA and protein synthesis while at California State University, Los Angeles. George A. (d. 1983) investigated effects of radiation on mammalian lifespans and the allometric correlations of lifespan with various physical characteristics, at Argonne National Laboratories (Chapter 5.3). George was the more widely recognized, as a theorist and promoter of comparative approaches.

Natural variations within species in the timing of senescence show that senescence is not rigidly linked to chronological age *per se,* but is triggered by flowering. The following are examples of perennial but monocarpic plants; Watkinson and White (1985) and Wangerman (1965) give others. The century plant (*Agave americana*) grows vegetatively for many years, to then terminate its lifespan after formation of a large mass of fruit and flowers, typically at 8–10 years in Mexico. However, in the Mediterranean *Agave* may grow vegetatively for up to 100 years before flowering and rapid senescence (Molisch, 1938). The most extreme example of a perennial monocarp is *Puya raimondii,* a pineapple relative that may mature as late as 150 years, its lifespan (Raven et al., 1986, 447). In yet others, flowering and vegetative growth can coexist: the morning glory *Ipomoea caerulea,* when grown under 16 hours of light daily, produces flowers from lateral buds, while the apical shoot grows vegetatively; on 8 hours of daylight, the apical bud becomes a flower (Ashby, 1950). In *Xanthium* (see above), long days greatly extend lifespan with continued vegetative growth (Krizek et al., 1966). In some species, the capacity for vegetative production may decline with age, but there is no evidence for general clonal senescence (Chapters 4, 7).

Closely related species manifest a range of lifespans that parallel the range in mammals. Various species of the thick-stemmed bamboos (*Phyllostachys*) have prolonged phases of vegetative growth that last for many years or decades (7, 30, 60, or 120 years) according to the species, before suddenly flowering and dying (perennials, yet monocarps; Janzen, 1976; Soderstrom and Calderon, 1979). Their death is often "gregarious," such that local populations show synchrony. In contrast, other bamboo species flower perennially (perennial polycarps). These lifespans are stable species characteristics, despite different climates. The 120-year flowering cycles of *P. bambusoides* and *P. henonis* are documented in Japan, going back 1,000 years (Kawamura, 1927). The environmental or internal pacemaker for these epidemics of bamboo flowering and senescence is unknown (Janzen, 1976). The recently achieved induction of flowering *in vitro* (Nadgauda et al., 1990) should elucidate the mechanisms of bamboo monocarpy.

Other closely related plants have annual and perennial forms. There is the genus *Lobelia,* an herbaceous angiosperm that grows on Mount Kenya (Young, 1984). *L. telekii,* which grows on arid slopes, is annual, while *L. keniensis* is a perennial in moist valleys at the same altitude. Both species have similar minimum sizes at flowering, as is widely observed in plants. The branch that contained flowers in the perennial *L. keniensis* dies back, while other branches that have not yet matured do not die back. The annual *L. telekii* has a much higher mortality before flowering, in general accord with other semelparous species. Both bamboo and *Lobelia* again show that the termination of vegetative growth is closely linked to flowering and fruiting. Other important examples are the numerous strains (races) of rice (*Oryza perennis*) that may be exclusively annual, exclusively perennial, or conditionally annual; these constitute an interfertile continuum of annuals and perennials (Morishima et al., 1961; Chu et al., 1969; Oka, 1976).

Adaptive aspects of plant senescence have long been discussed (Leopold, 1961). The ripening of fruits, for example, enhances seed release, whereas the export of nitrogen from senescing leaves back to the plant allows many plants to recover valuable nutrients for dormant phases in their life cycles. Leaf senescence or "leaf fall" may serve many functions, especially in seasonally deciduous plants. Possible functions include the elimination of toxins, release of allelopathic agents that influence interspecific competition, and contributions of nutrients to the local environment (Hardwick, 1986). Leaf senescence is thus subject to many points of selection in relation to the whole plant. Environmental cues from light, temperature, and moisture can trigger senescence of either the whole plant or of select organs, depending on the species.

A different selective mechanism is proposed for a neotropical tree *Tachigalia versicolor.* After growing to huge heights (30–40 meters), individual trees flower and produce fruit, followed by death within a year (Foster, 1977). A few individuals were observed to produce flowers and fruits on only a few branches; although the branches died, these trees survived, suggesting that local hormonal influences associated with reproduction on individual branches might summate to a threshold before death of the whole tree ensues. Such semelparity in a large, branched dicotyledon is rare. The death of *Tachigalia* is proposed to favor the survival of its progeny by opening a "light gap" in the forest canopy. Chapter 4.3 gives examples in which such light gaps arise from external causes, rather than from reproduction- related senescence.

Noodén (1980b) emphasizes the importance of human influences in the selection of monocarpy for field crops, where short lifespans, with complete and synchronous desiccation before harvest, would have many advantages. The diversity of senescent patterns suggests that monocarpy evolved independently many times, possibly with different control mechanisms (Noodén, 1980b; Chapter 11.3.2). Many plant families have both annual and perennial species. Evolutionary selection for annuals appears to include unstable environments, as also suspected for mammals with the seasonal die-offs described in Section 2.3.1.2.

These examples indicate that the variable phenotypes of senescence in higher plants can be resolved in terms of environmental and hormonal influences on selective regulation of genes. Moreover, they show the great plasticity of reproductive scheduling in plants that is fundamental to the variations in lifespan and the onset of senescence. Senescence from endogenous causes in the perennial plants (polycarps) is not demonstrated, and their recorded lifespan maxima could arise from many different causes, some of which may be external (Chapter 4.3).

2.3.3.   Birth-related Sudden Maternal Death

Sudden maternal death (also called endotokic matricide) is the immediate result of birth in a number of invertebrates. This phenomenon is not senescence in the usual sense, but is no less real than the death at spawning that is commonly accepted as senescence in Pacific salmon. These species still conform to the same tautology that governs all phenomena of senescence: any genetically influenced lethal change is permitted if the next generation has already been established. The following examples illustrate these bizarre phenomena.

One of the first descriptions was of the nematode *Ascaris nigrovenosa,* in which the young kill their mother by boring through her body wall (Weismann, 1889b). Similarly, in rhabdocoele flatworms (family *Typhloplanidae*), eggs are fertilized internally; the young escape by rupturing the body wall and kill their parent (Hyman, 1951, 135). The Orthonectida (phylum Mesozoa; primitive invertebrate parasites that may be degenerate flatworms) also hatch from within to kill the female (Weismann, 1889b; Stunkard, 1954).

Numerous polychaete worms (Annelida), especially the nereids, die at spawning when the ripe eggs are shed before fertilization by bursting through the body wall (Kaestner, 1968). An even stranger example is given by the beetle *Micromalthus debilis,* in which male embryos are produced by haploid parthenogenesis from unfertilized eggs within the ovaries (Crowson, 1981). In a new twist to the Oedipus tale, Scott (1937, 1941) described how one of these embryos is shed, but remains stuck to the mother's body; within a week the developing male embryo has cannibalized her, apparently for nourishment, followed by pupation and then hatching. No other beetle has this reproductive pattern. In general, closely related species do not show these phenomena, which suggests that the phenomena have arisen independently and sporadically during evolution (Chapter 11.3.3).

Matricidal hatching can also be induced in the laboratory nematode *Caenorhabditis* by starvation of larvae which is used to induce diapause and extend total lifespan (Chapter 9.2.2.1). When fourth-stage larvae or fecund adults are starved, the developing eggs hatch within, causing maternal death (Johnson et al., 1984). This demonstrates another alternate pattern of life history under epigenetic regulation and suggests that some of the above examples may be conditional to diet or temperature.

2.3.4. Finite Budding Capacity and Lifespan of Yeast Mother Cells

The budding yeast (*Saccharomyces cerevisiae*), a single-celled organism with all the organelles and genetic mechanisms of multicellular organisms, presents another type of rapid senescence that is attractive as a model for consideration of particular questions about unicellular organismic senescence, especially because its molecular biology is so highly developed (Strathern et al., 1982). The senescence of yeast is manifested during asexual reproduction, which occurs through budding (for general reviews, Mortimer and Johnston, 1959; Jazwinski et al., 1989). A daughter cell is formed from a small zone on the mother cell every 1–2 hours at room temperature or above. When the bud grows to a certain size and a nucleus has been passed from the mother cell, the daughter cell breaks away. A characteristic bud scar (2μ diameter, 1% cell area) remains on the mother cell wall (Bartholomew and Mittwer, 1953; Barton, 1950). A major component of this scar is chitin, a polysaccharide (Cabib et al., 1974). Bud scars are not repaired and accumulate to a maximum number, depending on the genotype, for an average of twenty to thirty, accounting for up to 50% of the cell surface area (Barton, 1950; Mortimer and Johnston, 1959; Muller, 1971; Egilmez et al., 1990; Figure 2.17). During this growth phase, the generation time (budding cycle) gradually lengthens (Egilmez and Jazwinski, 1989). As the limiting number of bud generations (cell divisions) is approached, about one to three budding cycles before death, the mother cells become wrinkled and lose turgor, and the budding cycle lengthens precipitantly by fivefold (Mortimer and Johnston, 1959). The ability to mate also diminishes (Muller, 1985). Cytolysis ensues frequently; otherwise the cells become granular in appearance. The lifespan, measured as the number of generations to mother cell death, differs among strains, but is on the order of fifteen to fifty maximum (Muller et al., 1980; Egilmez and Jazwinski, 1989).

The calendar lifespan is sensitive to temperature and nutrition. Growth at 10°C, by slowing the budding cycle severalfold, caused a doubling of the lifespan (from about 2 to 4 days), but did not alter the number of daughters produced; nutrient restriction and re-

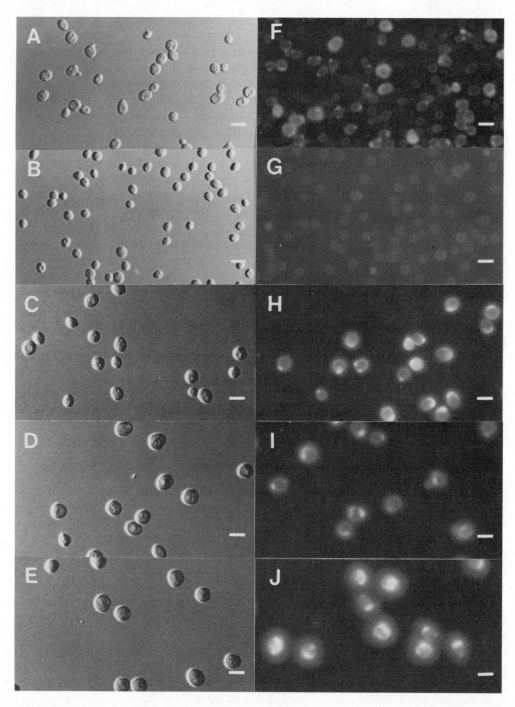

Fig. 2.17.   Senescence in the budding yeast (*Saccharomyces cerevisiae*) at different stages of growth in culture. *A–E*, Nomarski-DIC optics; *F–J*, epifluorescent staining of bud scars with Cellufluor. *A* and *F*, logarithmic culture; *B* and *G*, virgin cells; *C* and *H*, second-generation cells; *D* and *I*, eleventh-generation cells; *E* and *J*, seventeenth-generation cells. The growth of bud scars is seen as increasing amounts of lighter areas in *H–J*. The *bar* represents 7.5 μ, except in panels *A* and *F*, where the *bar* represents 9.4 μ. From Egilmez et al., 1990. Photograph kindly provided by Michael Jazwinski.

placement gave similar results (Muller et al., 1980). These results indicate that the number of buds produced is the pacemaker of senescence, rather than chronological age. Graphs of the mortality rate (mother cell deaths per bud generation) against the number of generations demonstrate exponential increases of mortality (Pohley, 1987; Jazwinski et al., 1989). Although this analysis is not based on a metric of calendar time, an equivalent MRDT of 0.004 years can be calculated from the data of Jazwinski et al. (1989) by assuming one generation lasts an average of 2 hours; this value is among the fastest MRDTs in Table 2.1.

The mechanism of yeast senescence is under renewed study after long neglect. Early workers suggested the accumulation of bud scars on the maternal cell wall as a source of pathophysiology through disturbances of the physiologically active surface/volume ratio (Muller, 1971). However, subsequent studies appear to rule this out. Cell membranes or wall components have also been proposed as distal or originating causes of senescence. This view gains plausibility from the observation that 50% more buds can be formed if ethanol is substituted for glucose as a carbon source (Muller et al., 1980). This conclusion must be tempered by the increased mitochondrial respiration with ethanol as a substrate. Recent work indicates other mechanisms as well and is based on a temperature-sensitive mutant *cdc24,* which cannot form buds at nonpermissive temperatures but does accumulate chitin, the major component of bud scars (Egilmez and Jazwinski, 1989). Despite the twofold increase of chitin—equivalent to five normal budding cycles—when the *cdc24* mutants were returned to the growth-permitting temperature, the subsequent lifespan (number of generations) was normal. This experiment thus argues against the chitin component of bud scars as the pacemaker of senescence, but by itself does not rule out the role of membrane changes.

Other studies indicate a diffusible factor (Egilmez and Jazwinski, 1989). The budding cycles of newborn daughter cells are slower and in synchrony with the mother cells. However, the daughter cells return to the "young" rate of multiplication within two to three budding cycles. Dominant genotypic influences are shown in the diploid progeny formed from parents with different "lifespans": the diploid progeny retained the "lifespan" of the longer-lived parent, as discussed further in Chapter 7.

Sequences for nuclear genes are being isolated from cDNA libraries that show altered expression as a function of the number of budding cycles. One mRNA (O30) shows major increases at later generations; the increase occurs independently of the cell-cycle stage (Egilmez et al., 1989; Jazwinski et al., 1990). Yet other mRNA decrease progressively. The emerging picture then is of selective changes in transcription during the life history of this simple unicellular organism.

There is no evidence for changes in the structure of the genome itself, a possibility suggested by other fungi that show marked genetic instability (Section 2.4). Because the daughters of senescing mother cells grow as well as the firstborn and do not have increased incidence of petites or auxotrophs (Muller, 1971; Johnston, 1966), there is no reason to assume an age-related change in the maternal genome or a transmissible senescence involving episomes. Also, there is no preferential segregation of sister chromatids in mother or daughter cells (Williamson and Fennell, 1981). This observation is pertinent to theoretical discussions of chromosome modification, such as are important in mammalian embryogenesis. Gender differences in genomic "imprinting" during gametogenesis alter the pattern of DNA methylation between maternal and paternal genomes and may also provide a barrier to parthenogenesis (e.g. McGrath and Solter, 1984).

While numerous protistans and other unicellular eukaryotes show gradual clonal senescence during asexual reproduction (Chapter 3), yeast senescence would appear to be a different phenomenon because the daughter and mother cells are clearly distinct throughout each generation; their cell nuclei, however, are equivalent as far as is known. As discussed in Chapter 7, the clonal senescence of human diploid fibroblasts *in vitro* is driven by the number of cell generations rather than calendar time; this phenomenon is also considered separately, because fibroblasts are only one cell of the organism, whereas yeast mother cells are *the* organism.

2.3.5. Botryllus: *A Colonial Ascidian with Semelparity*

Colonial ascidians (sea squirts) have complex life histories that can include two divergent characteristics: indefinite asexual reproduction from somatic cells (vegetative reproduction), yet rapid senescence associated with reproduction. While these phenomena may be found in plants and invertebrates, no vertebrate chordate can propagate both by sexual and vegetative reproduction. Ascidians are protochordates with a two-chambered beating heart, intestine, notochord, and a complex neural network (Figure 2.18, top). Some species are solitary and grow to a determinate size before sexual reproduction, while other ascidians are colonial, with ongoing vegetative and sexual reproduction. While most familiar to developmental biologists, ascidians have only recently been used to study mechanisms in senescence, in particular the *Botryllus schlosseri*. The diversity of life histories in colonial ascidia is large, genera differing in the seasonality and synchrony of the changes.

Originating from mosaic eggs, sexually produced tadpole larvae settle and metamorphose into an oözooid (Berrill and Karp, 1976; Berrill, 1951; Brunetti, 1974; Milkman, 1967; Grosberg, 1988a; Harp et al., 1988; Figure 2.18, bottom). In colonial and social ascidians, the oözooid vegetatively produces daughter zooids, which may or may not remain connected to other daughter zooids. These zooids, like those of solitary ascidians, are hermaphroditic; each is capable of either vegetative or sexual reproduction. *Botryllus*, like other colonial ascidians, grows into a colony of one or more thousand individuals through cycles of asexual budding (cyclic blastogenesis; Figure 2.18). During each blastogenic cycle, the oözooid grows a set of one to four buds that ultimately differentiate into adult zooids (about 2 millimeters long and 0.5 millimeter in diameter). The life cycle of oözooids lasts only 6 days, and is terminated by a dramatic cellular degeneration and resorption that occurs at the same time throughout a *Botryllus* colony. In synchrony with this degeneration is a maturation of the buds that then form the next cohort of zooids. Simultaneously, the primary buds begin growth to replace the previous generation of adult zooids, a process called "takeover" which perpetuates the colony during its asexual phase. The botryllids are unlike most colonial ascidians in the synchronous degeneration of their zooids and synchronous sexual maturation. Also, *Botryllus* does not usually propagate clonally by fragmentation (Brunetti, 1974), again unlike most colonial ascidians. However, vegetative reproduction can be perpetuated by fragmentation in the laboratory, apparently indefinitely (Grosberg, 1982; Sabbadin, 1979; Chapter 4.4.3).

Throughout these processes, the zooids connect to others of the colony by a vascular system that contains a diverse array of cells. Some of the circulating blood cells have cell-cell recognition with cytotoxic mechanisms, an allorecognition, like that in vertebrate

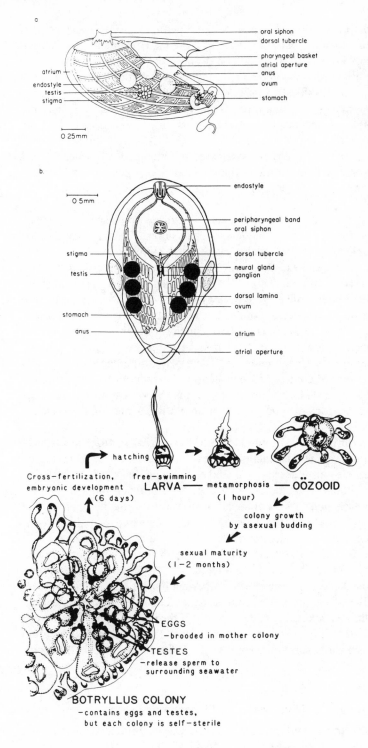

Fig. 2.18. *Botryllus schlosseri*, a colonial ascidian. *Top*, anatomy of the mature zooid. Drawing kindly provided by Richard Grosberg. *Bottom*, the two alternate modes of reproduction: asexual generation by budding from somatic cells and sexual generation by self- or cross-fertilization. Drawing kindly supplied by Virginia Scofield.

a

oral siphon
dorsal tubercle
pharyngeal basket
atrial aperture
anus
ovum
stomach

atrium
endostyle
testis
stigma

0 25mm

b.

endostyle

peripharyngeal band
oral siphon

0 5mm

stigma

testis

dorsal tubercle
neural gland
ganglion
dorsal lamina
ovum

stomach

anus

atrium

atrial aperture

hatching

Cross−fertilization,
embryonic development
(6 days)

free−swimming
LARVA — metamorphosis — OÖZOOID
(1 hour)

colony growth
by asexual budding

sexual maturity
(1−2 months)

EGGS
−brooded in mother colony

TESTES
−release sperm to
surrounding seawater

BOTRYLLUS COLONY
−contains eggs and testes,
but each colony is self−sterile

histocompatibility, which prevents neighboring colonies with incompatible genotypes from fusing through their vascular systems (Grosberg, 1988b; Hildemann, 1979; Bancroft, 1903). The mechanisms of cell degeneration during each asexual takeover cycle also involve autoreactive blood cells that increase during the senescent phase and that can be studied *in vitro* (Harp et al., 1988). This programmed degeneration of zooid cells may involve histocompatibility-like phenomena through changes in self-recognition epitopes that arise during maturation or degeneration of the zooids. This process resembles the regression of hydranths in certain cnidarians (Section 2.5.2) and does not kill the individual organism. While these processes may be considered programmed cell senescence, I prefer to use the word *senescence* according to the population genetics concept, for circumstances when the entire organism dies, thereby removing its genome from the population.

After five to ten of these asexual cycles, a pair each of mature ovaries and testes differentiate from somatic cells, followed by ovulation and fertilization. The eggs then develop rapidly (5 days) into tadpole larvae which are liberated just as regression begins; some of these then settle to start new colonies. Grosberg (1982, 1988a) distinguished by their reproductive schedule two types of colonies in *B. schlosseri* that coexist in the same environment (Eel Pond at the Woods Hole Marine Biology Laboratory). A semelparous type grows rapidly, produces a single large clutch of fertilized eggs, and then dies. In contrast, another is iteroparous and forms up to ten successive clutches of embryos before senescence and colony death ensue. However, the iteroparous colonies grow more slowly, and produce fewer and smaller offspring per zooid type. These major differences in reproductive scheduling are under genetic control, as indicated by the stability of these traits in laboratory populations. The hybrids formed from iteroparous and semelparous crosses did not show dominance, the progeny from F_1 clutches presenting a mix of reproductive schedules (Grossberg, 1988a).

Senescent degeneration during sexual reproduction appears to arise from mechanical factors, rather than from programmed degeneration at a cellular level. In his Ph.D. thesis, Grosberg (1982) described elegant microsurgical studies showing that death is related to the presence of embryos. Removal of the embryos prevented death in the semelparous type, but had no effect on early generations of the iteroparous race. If the iteroparous type was given supplemental embryos, however, mortality greatly increased. Another experiment indicated that the embryo itself was not the key factor, since substitution with an equivalent-sized glass bead (0.25 millimeter) also led to death in the semelparous race. Grosberg hypothesized that the embryo bulk interfered with respiration, which would constitute a mechanical distal cause of senescence, somewhat like the myocardial or brain lesions that are caused, albeit on a more haphazard basis, in humans by vascular stenosis. No studies were done with later sexual generations to test if the colony lifespan of iteroparous races could be extended. It is unknown if the cytotoxic blood cell reactions during cell degeneration of the asexual generations (Harp et al., 1988) have a role in the reproduction-related death.

2.4. Virus- and Episome-induced Senescence in Filamentous Fungi

Many strains of the filamentous ascomycete *Podospora anserina* become senescent and die (Figure 2.19) after a phase of vegetative growth that lasts 1–15 weeks, depending on

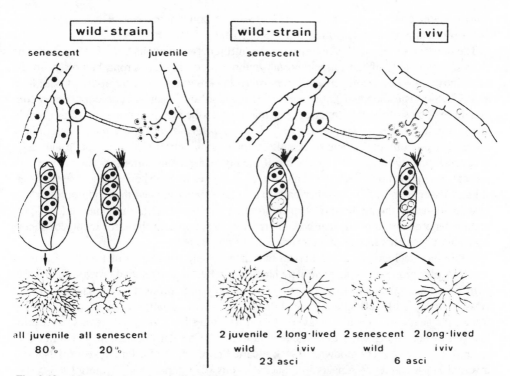

Fig. 2.19. Maternally (cytoplasmically) inherited senescence in the filamentous fungus *Podospora anserina*. Senescence is transmitted cytoplasmically by a plasmid. *Left,* cross between female senescent mycelia with juvenile (nonsenescent) male mycelium, both wild type. The progeny are predominantly juvenile, showing the absence of segregation as would be expected for a nuclear factor. *Right,* cross between the senescent wild type and a juvenile male mycelial strain containing the mutations *i* and *viv* that suppress the formation of the infectious plasmid. Rephotographed from Tudzynski and Esser, 1977.

the strain (Marcou, 1961; Smith and Rubenstein, 1973; Kuck et al., 1985). A novel infective cytoplasmic agent was discovered by Rizet (1953) to rapidly transfer senescence after fusion of senescent hyphae to juvenile cells. Maternal inheritance of senescence is consistent with the infectivity of this type of senescence. This phenomenon differs fundamentally from that of the single-cell ascomycete, yeast (see above), since the diffusible factor associated with yeast senescence is not infectious.

Senescence in *Podospora* is associated with decreased cytochrome oxidase activity (Belcour and Begel, 1978) and major increases in a 2.6-kilobase plasmid that is referred to as α-senDNA (Wright et al., 1982; Jamet-Vierny et al., 1980) or p1DNA (Stahl et al., 1978). This plasmid can infect (transform) juvenile protoplasts and rapidly induces senescence (Tudzynski et al., 1980). The free plasmid is found in low amounts in juvenile cells and is also integrated in mitochondrial DNA like a provirus. The plasmid arises spontaneously in *Podospora* strains that undergo senescence (Kuck et al., 1981). Several nonsenescing strains and long-lived mutants of senescing strains lack the mitochondrial sequence that gives rise to α-senDNA (Belcour and Vierny, 1986; Cummings et al., 1986). Other strains displaying delayed senescence (Vierny et al., 1982) or a temperature-sensitive senescent phenotype (Turker et al., 1987a) do not manifest the free plasmid during nonsenescent conditions. Nuclear DNA does not contain the α-senDNA sequence in either senescing or nonsenescing strains (Koll, 1986).

The α-senDNA plasmid is identical to the intron between exons I and II of cytochrome oxidase, subunit I, of the mitochondrial genome, and also has strong homology to the same intron of other ascomycetes (*Saccharomyces cerevisiae* and *Neurospora crassa;* Koll et al., 1984; Osiewacz and Esser, 1984). Mitochondria of rapidly senescing strains contain more copies of the free (excised) plasmid than do the slowly senescing strains (Wright and Cummings, 1983). Senescence in *Podospora anserina* can be postponed indefinitely by growth on sublethal levels of inhibitors for mitochondrial DNA or protein synthesis (Tudzynski and Esser, 1979). Moreover, if the nuclear gene mutations *gv* and *viv* are both present, free plasmid is not formed and these genotypes do not show senescence (Tudzynski and Esser, 1982). On the other hand, Holliday (1969) showed that addition of amino acid analogues to the medium (D,L-p-fluorophenylalanine or D,L-ethionine) induced premature senescence, a result that was interpreted as supporting the error theory of senescence but needs reexamination in terms of these plasmids.

While the mechanisms are unknown that lead to major increases of free infectious α-senDNA plasmids after a characteristic latent period during vegetative growth, the *Podospora* intron has several features that may contribute to the production of plasmids. It is a class II intron with self-splicing capability (Lambowitz, 1989). It has an open reading frame, the sequence of which suggests that it is coding for a viral-type reverse transcriptase (Osiewacz and Esser, 1984; Michel and Lang, 1985). This protein may also function as an RNA maturase (Lambowitz, 1989; Picard-Bennoun, 1985). Finally, α-senDNA has inverted DNA sequence repeats in adjacent exons (Cummings and Wright, 1983; Osiewacz and Esser, 1983) that are characteristic of proviruses and other mobile genetic elements and that could favor horizontal transmission. Under some conditions yet other plasmids are seen besides α-senDNA. A temperature-sensitive mutant of *Podospora* was found which showed senescence only at elevated growth temperatures (34°C). No α-senDNA was formed at the nonsenescing temperature (27°C), but at 34°C several plasmids were produced in addition to α-senDNA, most of which originated from a recombinational "hot spot" on the mitochondrial genome (Turker et al., 1987a, 1987b).

Senescence in the *kalilo* strains of *Neurospora intermedia,* another filamentous ascomycete, has also been linked to a plasmid. A 9-kilobase linear extrachromosomal DNA sequence, which appears to be associated with nuclei in up to two hundred copies of senescent and pre-senescent strains, is absent in nuclei of stable, nonsenescing strains (Bertrand et al., 1986). This nuclear plasmid appears to initiate senescence by insertion into one of several sites in the mitochondrial genome, of which at least two are in the large subunit rRNA genes. Insertion causes loss of mitochondrial cytochromes and other functions; ultimately the culture dies (Bertrand et al., 1985). The free-plasmid sequences may not normally be found in mitochondria, but they can be maternally inherited.

The novel mechanisms described above have not so far been observed in mitochondria of higher species and represent the firmest example of senescence in association with organelle genomic instability. As discussed in Chapter 8.4.1, recent evidence indicates the possibility that mammalian mitochondrial DNA accumulate deletions. The usual mode of reproduction for *Neurospora* in nature is asexual, which Holliday (1969) noted would strongly select for indefinite growth and against genotypes programmed for senescence.

Apparently similar senescent phenomena can also be found in vegetative clones of the filamentous ascomycete *Aspergillus glaucus* (Holliday, 1969; Jinks, 1959). The course of "vegetative death" obtained from laboratory stocks is much slower than in the above strains of *Podospora,* taking 5–24 months, during which sexual reproduction is lost, growth slows, and intense brown pigments accumulate. Jinks (1959) described lethal cytoplasmic mutations, presumably in mitochondrial DNA, that were "infectious" and observed to be transferred by hyphal fusion (anastomoses of adjacent cylindrical hyphae that permit exchange of cytoplasmic organelles). The incidence of these changes varied extensively among clones and, although universal in this species (Holliday, 1969), appears to be less programmed than in *Podospora*. Another age-related change in *A. ornatus* was observed in apparently healthy vegetative cultures. As they approached their asymptotic maximum growth in a single Petri dish within 7 to 10 days, induction of the enzyme o-pyrocatechuic acid carboxylase became slowed and the net incorporation of [^{14}C]-tryptophan into protein decreased (Wilson and Coursen, 1987). This phenomenon appears unrelated to the cytoplasmic changes associated with vegetative death, since the authors give no indication that vigorous growth would not be renewed on subculture.

Holliday (1969) suggested that *Podospora anserina* and *Aspergillus glaucus,* being self-fertile and homothallic (i.e. produces both mating types in one mycelium), should therefore have continual sexual cycles permitting the accumulation of genes predisposed to senescence during vegetative growth. This is a variant of the population genetics hypothesis that genes with delayed adverse effects are not strongly selected against (Chapter 1.5.2). An open question concerns whether heterothallic fungi like *Neurospora*, which require two separate mycelia for a sexual cycle, have a lower incidence of vegetative death mutants than in the homothallic fungi, like *Saccharomyces,* which is generally homothallic in the wild. This result might be predicted if the heterothallic fungi used sexual reproduction less frequently.

2.5.   Rapid Senescence and Sudden Death of Unknown Origin

The above examples may encourage readers to expect that all causes of rapid senescence can be easily established. However, the mechanisms remain unknown for many other species in which lifespan is terminated from internal degenerative causes. There is no reason to preclude more complex causes involving numerous genes, such as probably underlie gradual senescence (Chapter 3).

2.5.1.   Algae

The green alga *Volvox carteri* undergoes rapid and uniform senescence of its somatic cells during a life cycle of about 8 days. During the asexual cycle, *Volvox* gives rise to a spheroid of two thousand somatic cells in a single layer surrounding about ten gonidia (single reproductive cells). After ten synchronous divisions during the first 4 days, the somatic cells grow to their maximum size and divide no more. Then about 2 days after the last division, somatic cells become senescent and die in association with major reduction of protein synthesis (Pommerville and Kochert, 1982; Hagen and Kochert, 1980), disorganization of chloroplast structure, and accumulation of lipids (Pommerville and Kochert, 1981). In some mutants, somatic cells escape senescence by redifferentiating into reproductive cells (Sessoms and Huskey, 1973). The mechanisms of somatic cell senescence are not known, but seem approachable at the level of gene regulation in view of the selective changes in proteins synthesized as revealed by two-dimensional gel electrophoresis patterns at various stages of senescence (Hagen and Kochert, 1980). Growth at lower temperatures or treatment with chloramphenicol can delay somatic cell senescence (Pommerville and Kochert, 1982), whereas removal of the gonidia (Hagen and Kochert, 1980) or inhibition of gonidial growth by puromycin (Pommerville and Kochert, 1982) does not. The finding that neuron death from withdrawal of nerve growth factor can be prevented by inhibiting protein synthesis (Chapter 7.7) suggests a similar mechanism, in which proteins (thanatins) could consummate a cell autolysis cascade.

The blue-green alga *Acetabularia mediterranea* is a giant single cell (3–5 centimeters) that when mature has a cap (1-centimeter diameter) containing thousands of spores. Maturation requires several months of growth, but the organism disintegrates after release of spores. During stalk elongation, the single zygotic nucleus increases its volume by a millionfold and then disintegrates to form a secondary nucleus, which divides mitotically and from which the spores are derived (Franke et al., 1974; Berger and Schweiger, 1975, 1986; Hammerling, 1963). If the cap is amputated, disintegration of the primary nucleus can be prevented. A new cap will regenerate as long as the primary nucleus remains, but if these new caps are removed in succession, the plant continues to make them for at least 3 years without signs of "irreversible ageing" (Hammerling, 1963).

2.5.2. Cnidaria

The Cnidaria, which include the hydra and jellyfish, are among the simplest multicellular invertebrates and are widely known for their asexual reproduction (Chapter 4). The hydranths of some colonial Cnidaria undergo a cellular regression that is viewed as a form of senescence (Huxley and de Beer, 1923; Strehler, 1977, 67; Brock, 1970). For example, in the hydrozoans *Bougainvillea carolinensis* and *Campanularia flexuosa,* the individual hydranth has a "lifespan" of 1 to 2 weeks and regresses with an exponentially increasing risk according to the Gompertz mortality model (Brock and Strehler, 1963). The autolyzed cellular debris returns through the stalk, where it is thought to nourish the rest of the colony (Strehler and Crowell, 1961; Brock et al., 1968, Strehler, 1977, 67). Hydranths of the hydrozoan *Clytia johnstone* have a very different mortality pattern: hydranth regression maintained a constant mortality risk over its short lifespan and did not senesce (Brock and Strehler, 1963). Because in neither example does the parent organism die, hydranth regression may be considered more similar to the senescence of leaves and branches in perennial trees or of tissues that slough off and are continuously regenerated in vertebrates. A hint of senescence in the solitary scyphozoan *Nausithoe* may be the declining growth rate as the length of its chitinous body tube approaches seven centimeters; nonetheless, this species reproduces asexually and without limit in the laboratory (Werner, 1979). Numerous others in this phylum do not show any evidence of senescence.

2.5.3. Flatworms

The acoelomate flatworms (phylum Platyhelminthes) have wide variations in patterns of senescence and longevity, including indefinite propagation by asexual reproduction

(Chapter 4). The rapid senescence of *Planaria velata* occurs over 1 to 2 months, when feeding ceases; the head, eyes, and pharynx may disintegrate; dark body pigments are lost; and movements slow. This rapid senescence signals the onset of an asexual reproductive cycle involving fragmentation and encystment, followed by the hatching of miniature rejuvenated adults (Chapter 4.2). Many small turbellarians also have short lifespans of several weeks (*Stenostomum leucops, S. ignavum,* and *S. fasciatum*), whereas larger species live much longer and show a more gradual senescence (Haranghy and Balázs, 1964; Hyman, 1951, 195). In sexually reproducing species, lifespans sometimes exceed 4 years (Chapter 3.3.1).

2.5.4. Nematodes

Several free-living soil nematodes are widely studied because of their very short life cycle and rapid senescence, altogether 1 month or less (Johnson, 1983; Russell and Jacobson, 1985), as observed in the soil species *Caenorhabditis elegans* and *C. briggsae* and in the vinegar eelworm *Turbatrix (= Anguillula) aceti.* Maupas (1900) described sixteen other short-lived rhabditoid nematodes, of which *Cephalobus dubius* had the longest lifespan of 5 months and showed a senile decline. Many physiological, cellular, and biochemical changes of senescence in *Caenorhabditis elegans* are known in detail. These species can be grown under axenic conditions and with close synchrony of age groups (Wood 1988). Although the causes of senescent death remain unknown (Russell and Jacobson, 1985), recent genetic variants with nearly doubled lifespan suggest that a limited number of causes may be identified (Johnson, 1987; Friedman and Johnson, 1988a, 1988b; Chapter 6). As discussed in Section 2.3.2, the postmitotic status of *Caenorhabditis* does not account for its short lifespan. The total lifespan of *C. elegans* can be extended severalfold by the formation of dauer larvae, a semiquiescent state that larvae can enter if starved at the end of the second larval stage (Figure 9.1). Nematode phytoparasites display another variant of metabolic arrest and can survive desiccation for decades (Hyman, 1951, 405). Reproduction in *Caenorhabditis* sharply declines during the second week and few remain fertile after 10 days, mainly because of the exhaustion of sperm stores (Johnson, 1987). Laboratory nematodes have postreproductive phases lasting up to 10 days. Cytogenetic abnormalities in oocytes are described in Chapter 8.5.3.2.

Most plant parasitic nematodes are short-lived, for example, the genera *Heterodera* and *Meloidogyne* live 30–40 days at 25°C, but up to 4 months at lower temperatures. They have limited production of gametes. Female *Meloidogyne* produce 300–800 eggs within 20–30 days; while mitoses are seen in the oogonial ridge of young adults, mitotic activity soon ceases. At death, females show depletion of lipid and protein globules, which suggests inability to feed. Males do not feed as adults and die with a transparent appearance soon after inseminating a female (A. C. Triantaphyllou, pers. comm.).

In *Caenorhabditis* (Figure 2.20) and *Turbatrix,* body movement and defecation frequency decline linearly with age, which may be consequent to the smaller amount of food ingested (^{35}S-labeled *Escherichia coli;* Johnson and McCaffrey, 1985). Moreover, longitudinal studies show the cessation of pharyngeal pumping and feeding 3 days before death; the ability of serotonin to reactivate pumping in about 50% of senescent worms suggests involvement of this neurotransmitter (Rothman and Lewis, in prep.). The syn-

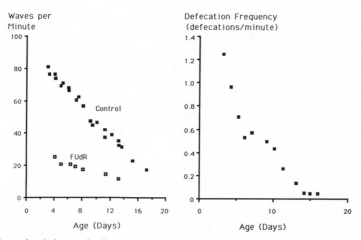

Fig. 2.20. Age-related changes in *Caenorhabditis elegans*. *Left,* whole-body (sinusoidal) movements in control and populations treated with FUdR, a halogenated nucleoside that blocks all DNA synthesis and cell division; despite major effects of FUdR on body movement, the lifespans were not altered. *Right,* defecation frequency shows nonmonotonic changes with age. The regression lines in the original report are not shown. Redrawn from Bolanowski et al., 1981.

thesis rate of enolase and bulk proteins also decreases (Sharma et al., 1979; Karey and Rothstein, 1986; Chapter 7), but the two hundred most abundant proteins continue to be synthesized at advanced ages (Johnson and McCaffrey, 1985). In senescent *T. aceti,* several enzymes have decreased specific activity and accumulate inactive forms with altered conformations (Gershon and Gershon, 1970; Gupta and Rothstein, 1976), as also occurs in senescent rats (Chapter 7.5.1); however, not all other enzymes of *T. aceti* show these changes (Gupta and Rothstein, 1976). Nematodes also have age-related accumulations of autofluorescent, intracellular lipofuscins (aging pigments), but only in the intestinal cells, where they are identified as secondary lysozomes (Clokey and Jacobsen, 1986). The recently discovered longevity mutants of *Caenorhabditis* that increase maximum lifespan as a result of slowed MRDT (Chapter 6.3.1) have not been analyzed for most details of senescence.

2.5.5. *Crustacea*

Age-related changes of two isopod Crustacea suggest rapid senescence. In the isopod crustacean *Asellus aquaticus,* growth slows and limb regeneration approaches zero during its 3- to 5-month lifespan; a rapid decline of growth rates and a decrepit appearance before death implied pathologic changes of senescence (Needham, 1950). Similar changes in growth were described for another isopod, *Ligia oceanica* (Inagaki and Berreur-Bonnenfant, 1970). On the other hand, barnacles (a different crustacean class) show no signs of senescence during long lifespans (Chapter 4).

2.6. Summary

Most of the organisms discussed above are semelparous and occur in numerous phyla (Table 2.6). The pathophysiological changes associated with rapid senescence and sudden

Table 2.6   Distribution of Semelparity or Sudden Death in Animals: A Survey

Taxonomic Summary by Phylum	Species, % (No. in Family)
Arthropoda	

Arachnida	
Araneae (spiders; 35,000 species)	
Theraphosidae (tarantulas in North America)	100? males (600)
Lycosidae (wolf spiders)	≥ 1 males (3,000)
Insecta (600 families; > 1,000,000 species)	
Coleoptera (beetles)	
Drilidae	≥ 1 (1/80)
Lampyridae (fireflies)	≥ 1 (2,000)
Phenogodidae	≥ 1 (200)
Rhipidiinae (Pleocomidae)	100? males (35)
Stylopidae	100? males (200)
Diptera (flies)	
Nymphomyiidae	100 (4 genera)
Oestridae (botflies)	100? (65)
Syrphidae (hoverflies)	100? (5,000)
Ephemeroptera (mayflies)	100 (2,100)
Lepidoptera (butterflies and moths)	?
Bombycidae (includes silk moth)	≥ 1 (100)
Eriocraniidae	≥ 1 (20)
Eucleidae (bagworm moths)	≥ 1 (800)
Hepialidae (swifts)	≥ 1 (500)
Megalopygidae	≥ 1 (220)
Saturniidae	≥ 1 (1,000)
Plecoptera (stoneflies)	most (1,500)
Trichoptera (caddisflies)	100? (7,000)

Mollusca	

Cephalopoda (44 families, 600 species)	
Octopus Vulgaris	< 100%, females

Nematomorpha (horsehair worms)	

Life histories are known for few species.[1]	100%? males (≥ 200)

Chordata	

Agnatha	
Monorhini (1 family, 41 species)	
Petromyzontidae (lampreys)	100? (≥ 10/41); (no known exceptions, but many species are uncharacterized)
Gnathostomata	
Osteichthyes (418 families, 20,857 species)	
Anguilliformes (22 families, 600 species)	
Anguillidae (freshwater eels)	100? (16); (probably all)
Congridae (conger eels)	(≥ 1/100)
Nemichthyidae (snipe eels)	≥ 1/9
Atheriniformes (16 families, 894 species)	
Atherinidae (silversides)	< 1 (1/160)

(continued)

Table 2.6   (*continued*)

Taxonomic Summary by Phylum	Species, % (No. in Family)
Chordata	

Clupeiformes (4 families, 330 species)	
Clupeidae (shad)	1 (2/190)
Cypriniformes (25 families, 3,000 species)	
Cyprinidae (minnows)	< 1 (1/2,000)
Gasterosteiformes (3 families, 10 species)	
Gastereidae (sticklebacks)	10 (1/10)
Perciformes (150 families, 6,800 species)	
Callionymidae (dragonets)	< 1 (1/130)
Gobiidae (gobies)	< 1 (6/2,000)
Mugilidae (mullets)	1 (1/95)
Salmoniformes (24 families, 508 species)	
Galaxiidae	10 (5/45)
Osmeridae (smelts)	20 (2/10)
Plecoglossidae (ayu)	100 (1)
Retropinnidae	100 (4)
Salangidae (ice-noodle fishes)	15 (2/14)
Salmonidae (*Oncorhynchus;* Pacific salmon)	85 (6/7)
Marsupialia (16 families; 258 species)	
Dasyuridae (*Antechinus;* marsupial mice)	< 5 (9/258)

Estimates of Semelparity or Sudden Adult Death by Major Taxa	
Insecta	1%
Mollusca	1?
Nematomorpha	1?
Fish	0.5
Marsupialia	< 5

Notes: This survey represents a minimal estimate of semelparity (i.e. one season of breeding in species) or sudden death from anatomical deficiencies in mouthparts, as represented mostly in the insects that have defects in adult mouthparts that render them partly or completely aphagous, and have necessarily short adult life phases (Table 2.2). Numerous other examples may be realized with more systematic surveys. No examples of obligate semelparity have come to my attention for placental mammals, birds, or reptiles. In general, the references are in Tables 2.2–2.4 or in the text. The numbers of species are from Parker, 1982, and Margulis and Schwartz, 1988. It is often unclear how many species show these pheomena, as indicated by question marks.
[1]Hyman, 1951; Margulis and Schwartz, 1988.

death vary widely and in most cases represent supracellular and dysorganizational distur-bances, rather than insidious cellular defects (Table 2.7). The failure or inability to ingest a complete diet during all or some of adult life is a recurrent phenomenon during rapid senescence that is discussed again in the next chapter. Yet, starvation during rapid senes-cence arises from very different causes in mayflies, tarantulas, and Pacific salmon. A broad variety of endocrine changes that arise during rapid senescence differ among auto-genous dipterans, lampreys, Pacific salmon, octopus, and flowering plants. Various inver-tebrates die from mechanical causes during ovulation or birth that are probably only dis-tally linked to the hormonal controls of reproduction. The fungal genotypes with episomal-related senescence are exceptions in this discussion, since these clearly have an intracellular defect. For some of these species, the causes of rapid senescence and sudden death are unimodal and, like the unimodal progeroid syndromes in humans (Martin, 1982; Chapter 6.5.3), may eventually be traced to a limited number of genes that are key to the trait. For others, a multiplicity of separate genetic contributions to various pathological changes may prevail (polymodal senescence).

A polygenic basis may be suspected for rapid senescence of other short-lived species, such as flies, which wear out their irreplaceable wings and have deficits of energy metabolism despite their continued feeding. As flies and others with irreplaceable parts age into prolonged time, a wide range of damage is observed which must derive from numerous genetic determinants. Even so, the very long lifespans of insect queens with similar construction demonstrate adaptations that permit long lifespans despite a fragile body. This is accomplished for the insect queens by behavioral and endocrine modifications, again supracellular and dysorganizational.

Taken together, the variations of rapid senescence and sudden death fit well with the population genetics view that the weakening of natural selection in populations where older organisms produce a dwindling minority of the offspring will allow the penetrance of diverse genetic defects with delayed adverse effects that are not selected against so long as sufficient progeny are produced to propagate the next generation (Chapter 1.5). There would seem to be more than one such adverse trait in the rapid senescence of insects that

Table 2.7   Typology in Sudden Death and Rapid Senescence: A Sampling

	Organism	Text section
Obligate Rapid Senescence		
Aphagy due to congenital defects of mouth or gut organs	mayflies, moths	(2.2.1.1)
Mechanical senescence due to wearing out of wings or other irreplaceable organs	winged insects	(2.2.2.2, 2.2.3.1)
Birth-related maternal death	orthonectids, a few nematodes and flatworms	(2.3.3)
Induced by virus or episome	some fungal genotypes	(2.5)
Hormonally triggered by sexual maturation	many plants, Pacific salmon lampreys(?), eels(?)	(2.3.1, 2.3.2)
Aphagy triggered by mating	some male spiders some octopods some Aplysia(?)	(2.3.1.2) (2.3.1.3) (2.3.1.3)
Conditional Senescence		
Stress-induced during breeding	marsupial mice; species that can be either iteroparous or semelparous	(2.3.1.2)
Seasonally induced	many plants can be annuals or perennials	(2.3.2)
Geographic variations	shad, steelhead	(2.3.1.4.2)

are aphagous and also wear out their wings. The number of adverse genetic traits during rapid senescence may be related to the time elapsed since the evolutionary origin of rapid senescence in that taxon (Chapter 11). Williams (1957) predicted that ultra-short-lived genotypes should accumulate genes in their populations that allow numerous concurrent dysfunctions; this may be demonstrated by the short-lived insects and rotifers. When organisms show a conditional semelparity, as in the geographic populations of shad and salmon, alleles or a gene for multiple postreproductive defects may not accumulate in the gene pools. The evolutionary history may be looked to in many species for insight into the genetic basis of its patterns of senescence and longevity. Little has been done along these lines.

The varieties of rapid senescence and sudden death may be considered in context with particular ecological constraints, such as unpredictable environments at hatching that favor rapid reproductive schedules and permit sudden death thereafter to have no impact on future population stability (*r*-selection, Chapter 1.5.1). A theoretical prediction (Chapter 1.5.2) is that short adult phases should show great species variations in the details of senescence within organisms of the same radiation. Careful study of the mayflies and dipterans for species differences in senescence is warranted. Such comparative studies might also reveal a theoretically important but undemonstrated life history pattern—organisms that are semelparous, yet retain robust health and live for extended time after their only bout of reproduction.

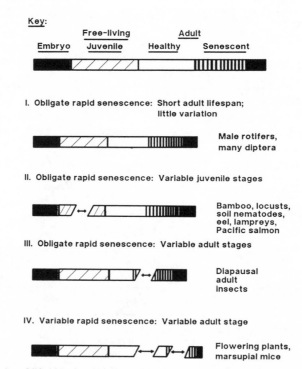

Fig. 2.21. Typologies of life histories showing rapid senescence or sudden death. Examples discussed in the text.

Developmental events clearly determine rapid senescence of those invertebrates that lack mouth or gut parts as adults or that wear out wings and other irreplaceable parts. Such species are born with ticking time bombs for their own self-destruction that are triggered by hatching and are little influenced by the environment. On the other hand, there are many species in which hormonal changes during maturation trigger senescence (Pacific salmon and annual plants), while in others changes at maturation increase the risk of death, though they may survive in favorable environments (marsupial mice). These diverse phenomena suggest that rapid senescence and sudden death have originated many times during evolution (Chapter 11). The great extent to which rapid senescence and sudden death can be manipulated or influenced by the environment is consistent with the many points in the life course that are subject to major changes in scheduling (Figure 2.21).

As described in the next chapter, many longer-lived species that manifest more gradual forms of senescence also are subject to extensive variations in the timing and intensity of age-related degenerative changes. We will continue to ask about the importance of supracellular dysorganizational defects relative to cellular defects in organismic senescence.

3

Gradual Senescence with Definite Lifespan

3.1. Introduction

This chapter concerns organisms manifesting gradual forms of senescence during lifespans in the range of 2 to 100 years. Like rapid senescence (Chapter 2), populations of gradually senescing organisms show age-related exponential accelerations of mortality rate (G; equations [1.2], [1.3]) at some time in their adult lives, from months to decades after sexual maturity. I will continue to use the MRDT in place of the Gompertz exponential coefficient G. For comparison, in gradually senescing organisms MRDT ranges at least a hundredfold: 0.1 year (short-lived fish), to 0.3 year (laboratory rodents), to 8 years or longer (humans, rhesus monkeys, and long-lived birds; Table 3.1; Appendix 1).

In contrast to rapid senescence, which is often nearly synchronous in a population, gradual senescence generally varies extensively among individuals, in the time of onset and rate of progression. Age-related dysfunctions emerge gradually and need not be smoothly progressive after the onset of reproduction. Most gradually senescing species are iteroparous and have the potential for more than one annual reproductive cycle, in distinction to the semelparous animals and plants that show rapid senescence (Chapter 2). By definition, species that show no evidence for age-related dysfunctions from endogenous causes and no exponential increases of mortality rate after the onset of reproduction are classified under negligible senescence. The differences in age-related changes between rapid, gradual, or negligible senescence (Chapter 4) may be subtle and subject to debate, especially for species with very long lifespans. This uncertainty is an expected outcome of choosing subdivisions in phenomena that I believe are best viewed as continual.

The mechanisms of gradual senescence are much less understood than the manifestations of rapid senescence or sudden death, which could be grouped according to a predominant cause of senescence (Chapter 2). For this reason, I return to taxonomic groupings in this and the next chapter. There is good indication of canonical patterns of senescence throughout mammalian radiations, particularly for reproductive senescence in females, but very sketchy evidence for the presence of these patterns elsewhere in vertebrates. I have tried to find information about senescence of many types of organisms that are rarely discussed in the gerontological literature. Readers may expect some patchiness as a result of insufficient information.

Throughout these examples, some of the same mechanisms described in Chapter 2 can also be observed, including mechanical damage to irreplaceable parts and alterations in humoral regulatory factors with cascading effects upon other functions. Both mechanical senescence and humoral cascades share the attribute of focal changes with organizational consequences to organismic functions. Does more gradual senescence merely represent slower versions of the same phenomena seen in Pacific salmon or in flies? This is a major and open question. I briefly describe cellular and molecular mechanisms that may be independent of the "dysorganizational" mechanisms of rapid senescence and that may be consequent to the longer adult phases, for example, genetic changes that may be important in cancers or the accumulation of aging pigments (lipofuscins). Nonetheless, extended developmental phases were encountered in species with rapid senescence and raise continued questions about possible similarities and differences in mechanisms that permit long total lifespans through varying the length of developmental versus adult phases. The major discussion of mechanisms in cell changes during organismic senescence is in Chapters 7–10, which draw on the portraits of organismic senescence developed in this chapter.

Table 3.1 Mortality Rate Coefficients in Gradual Senescence (in most cases, the IMR and MRDT should be regarded as first approximations within a two-fold range; see notes to table)

		IMR/year	MRDT, years	*tmax*, years
ANNELIDA				
Lumbricus terrestris[1]	common earthworm	0.1		> 6
MOLLUSCA				
Lymnaea stagnalis[2]	pond snail	0.15	0.25	2
ARTHROPODA				
Coleoptera				
Carabus coerulescens, *C. glabratus*[3]	carabid beetles			3–4
Macronychus glabratus[4]	elmid beetle			> 9
Microcylloepus pussilus[4]	elmid beetle			> 9
Tribolium castaneum[5]	flour beetle	0.12		2–3
CHORDATA				
Teleosta				
Acipenser fulvescens[6]	lake sturgeon	0.013	10	> 150
Betta splendens[7]	Siamese fighting fish			1.5–3
Cynolebias adolffi[8]	annual fish (male)	0.07	0.1	1–2
Lebistes reticulatus[9]	guppy	0.07	0.8	5
Oryzias latipes[10]	rice fish, or medaka			3–6
Perca fluviatilis[11]	river perch	0.07	27	25
Xiphophorus maculatus[12]	platyfish			2–3
Aves				
Apodiformes:				
Apus apus[13]	common swift	0.1	5	21
Selasphorus platycerus[14]	broad-tailed hummingbird	0.25		> 8
Charadriiformes:				
Vanellus vanellus[15]	lapwing	0.2	6	16
Larus argentatus[16]	herring gull	0.004	6	49
Galliformes:				
Alectura lathami[17]	brush turkey	0.05	3.3	12.5
Coturnix coturnix[18]	Japanese quail (male)	0.07	1.2	5
Pavo cristatus[19]	peafowl	0.06	2.2	9.2
Syrmaticus reevesi[20]	Reeves pheasant	0.02	1.6	9.2
Passeriformes (perching birds):				
Lonchura striata[21]	Bengal finch (female)	0.1	2.5	10
Erithacus rubecula[22]	European robin	0.5	8	12
Sternus vulgaris[23]	starling	0.5	> 8	20
Mammalia				
Artiodactyla (hoofed with even numbers of toes):				
Hippopotamus amphibius[24]	hippopotamus	0.01	7	> 45
Ovis dalli[25]	Dall mountain sheep	0.05	1.5	15–20
Chiroptera (bats):				
Pipistrellus pipistrellus[26]	pipestrelle bat	0.36	3–8	≥ 11
Myotis lucifugis[26]	little brown bat			> 32

Table 3.1 (continued)

		IMR/year	MRDT, years	*tmax*, years
Carnivora:				
Canis familiaris[27]	domestic dog	0.02	3	20
Perissodactyla (hoofed with odd numbers of toes):				
Equus caballus[28]	horse	0.0002	4	> 45
Rhinosceros unicornis[29]	Indian rhinoceros			> 50
Proboscidea:				
Loxodonta africana[30]	African elephant	0.002	8	> 70
Primates:				
Macaca mulatta[31]	rhesus monkey	0.02	8	> 35
Homo sapiens[32]	humans (female)	0.0002	8	> 110
Rodentia:				
Mus musculus[33]	mouse (female)	0.01	0.3	4–5
Rattus norvegicus[34]	rat (male)	0.002	0.3	5–6
Peromyscus leucopus[35]	white-footed mouse	0.001	0.5	7–8

Notes: Two methods of calculation were used. Estimates of mortality rates were calculated according to the Gompertz model by linear regression (LR): $\ln m = \text{IMR} + Gt$, where m is the mortality rate as a function of age (the fraction of survivors dying in arbitrary intervals, usually deciles of *tmax*); IMR is the initial mortality rate calculated at maturation; and G is the mortality rate exponential coefficient (equation [1.2]). The MRDT is the doubling time of the mortality rate acceleration: $\text{MRDT} = \ln 2/G$ (equation [1.3]). Data on mortality rate as a function of age were obtained from mortality schedule data where available or extracted from survivorship graphs. The regressions had correlation coefficients $r^2 > 0.9$. Use of primary mortality schedule data would generally improve the precision of these estimates. The numbers in the table should be considered as first approximations within a twofold range.

Using the best-fit method (BF), where survivorship plots or mortality schedule data were not available, mortality coefficients were estimated as described in Chapter 1.4.7 and Appendix 2. Calculations of IMR and MRDT are from reported average annual population mortality rates (AMR, the mean across all ages) and maximum lifespans (*tmax;* equations [1.6] and [1.7]; Appendix 2). IMR was extrapolated to age at maturation, equation (1.2) and Table 1.1. Calculations are based on populations of 10^5; values for other population sizes are in Appendix 2. Items 13–23, 25–28, 31–34 are from Finch et al., 1990.

[1]*Lumbricus terrestris*. Calculated by LR from survivorship plot of a population that was "reared outdoors" (Satchell, 1967, 266).

[2]*Lymnaea stagnalis*. Laboratory-bred pond snail. From calculations for Figure 1 of Slob and Janse, 1988. Comparisons of seven populations showed a twofold range for MRDT and a hundredfold range for IMR.

[3]*Carabus coerulescens, C. glabratus*. Field and laboratory specimens. van Dijk, 1979; Houston, 1981.

[4]*Macronychus glabratus, Microcylloepus pussilus*. Maximum lifespans in the laboratory from Brown, 1973.

[5]*Tribolium castaneum*. Calculated for laboratory populations by LR from survivorship plots of Mertz, 1975.

[6]*Acipenser fulvescens*. Calculated for little-exploited lake population that was caught by sports fishing. LR from survivorship plot of Beverton and Holt, 1959, shown in Figure 3.4. Based on data of Probst and Cooper, 1954. *tmax* from "One hundred fifty-two year old lake trout," 1954, and Tsepkin and Sokolov, 1971.

[7]*Betta splendens*. Lifespan from Woodhead, 1974.

[8]*Cynolebias adolffi*. Based on incomplete survivorship plot of laboratory populations from Liu and Walford, 1969.

[9]*Lebistes reticulatus*. Calculated for laboratory populations by LR from mortality schedule data of Comfort, 1963, shown in Figure 3.6.

[10]*Oryzias latipes*. *tmax* from Egami, 1971, and Takamoto et al., 1984.

[11]*Perca fluviatilis*. Calculated for populations by LR from survivorship plot of Beverton and Holt, 1959, shown in Figure 3.4. Based on data of Alm, 1952, for marked fish in a protected population.

[12]*Xiphophorus maculatus*. Laboratory population *tmax* from Schreibman and Margulis-Nunno, 1989.

Table 3.1. (Continued)

[13]*Apus apus.* Calculated for feral populations by BF from data given by Botkin and Miller, 1974, AMR, 0.189; *tmax* from Ollason and Dunnet, 1988, 263–278; body weight, 40 gm (Calder, 1984).

[14]*Selasphorus platycerus.* Feral populations, (estimated from Calder, 1985).

[15]*Vanellus vanellus.* Calculated for feral populations by BF from data given by Botkin and Miller, 1974, AMR, 0.34; body weight, 220 gm (Calder, 1984).

[16]*Larus argentatus.* Calculated for feral populations by BF from data of Ollason and Dunnet, 1988. AMR, 0.04; *tmax* from Pearson, 1935; body weight, 950 gm (Calder, 1984). Data on mortality rates by age group calculated by LR agree closely with these results: MRDT, 5 years.

[17]*Alectura lathami.* Calculated for zoo populations by LR from survivorship plots of Comfort, 1962.

[18]*Coturnix coturnix.* Calculated for laboratory populations by LR from survivorship plots of Cherkin and Eckhardt, 1977. Other quail also have similar short *tmax* in the wild (Clapp et al., 1982): *Lophortyx californica* (California quail), 6.8 years; *L. gambeli* (Gambel's quail), 7.4 years.

[19]*Pavo cristatus.* Calculated for zoo populations by LR from survivorship plots of Comfort, 1962.

[20]*Syrmaticus reevesi.* Calculated for zoo populations by LR from survivorship plots of Comfort, 1962.

[21]*Lonchura striata.* Calculated for zoo populations by LR from mortality schedule data of Eisner, 1962.

[22]*Erithacus rubecula.* Calculated for feral populations by BF from AMR, 0.64, given by Botkin and Miller, 1974; body weight, 16 gm (Calder, 1984).

[23]*Sternus vulgaris.* Calculated for feral populations by BF from AMR, 0.52, given by Botkin and Miller, 1974; body weight, 75 gm (Calder, 1984). The MRDTs in Appendix 2 may be high, and are represented conservatively as > 8 years.

[24]*Hippopotamus amphibius.* Calculated by LR for two populations averaged for table: Queen Elizabeth Park (IMR, 0.002/year; MRDT, 5 years) and Parc National Albert (0.014; 8.8). From survivorship plots of Laws, 1968.

[25]*Ovis dalli.* Calculated by LR from life table data of Deevey, 1947.

[26]*Pipistrellus pipistrellus, Myotis lucifugis.* For pipestrelle bat, calculated by BF assuming AMR, 0.36, from data of Thompson, 1987; body weight, 20 gm, from Nowak and Paradiso, 1983. The 11-year *tmax* reported from Thompson (1987) may be a considerable underestimate. For comparison, *P. subflavus* lives $\geq$ 14.8 years, whereas other small vespertilionids are reported to live longer, e.g. *Myotis lucifugis* (8 g), $\geq$ 32 years (E. B. Hitchcock, cited in Thompson, 1987). The table entry shows a range of MRDTs from Appendix 2 calculated on the basis of *tmax*s of 11 and 15 years, which represent the present uncertainty. Data for mortality by age group (Thompson, 1983) were also analyzed by LR; although the regression analysis did not achieve significance, the estimated mortality coefficients were similar to those given by BF: IMR, 0.25/year; MRDT, 8 years. Provisionally, pipestrelle bats have MRDTs that approach the upper range in mammals.

[27]*Canis familiaris.* Calculated by LR from survivorship plots of Comfort, 1960b. Average of values for the following breeds: cocker spaniel (IMR, 0.004/year; MRDT, 2.4 years); Pekinese (0.012; 3.1); mastiff (0.012; 2); wolfhound (0.044; 4).

[28]*Equus caballus.* Calculated by LR from survivorship plots of Comfort, 1959, for two breeds. Average of thoroughbred mare (IMR, 0.0002/year, MRDT, 4.5 years); Arabian mare (0.0002; 3.8).

[29]*Rhinosceros unicornis.* No life table data are available. *tmax* from Nowak and Paradiso, 1983.

[30]*Loxodonta africana.* Calculated by LR from data extracted from survivorship plots of Laws, 1966. Average of populations in Queen Elizabeth Park (IMR, 0.0016/year; MRDT, 7.4 years) and Murchison Falls (0.003; 8.7).

[31]*Macaca mulatta.* Calculated by LR from mortality schedule data from two populations aged 4.5 to 24.5 years. Monkeys born at the Wisconsin Regional Primate Center: data from Dyke et al., 1984, Table 2, column q(x), which represents spontaneous deaths and excludes monkeys removed for experimentation (Bennett Dyke, pers. comm. to Matthew Witten); data were combined for males and females: IMR, 0.02/year; MRDT, 18 years. Feral-born monkeys brought when about 2 years old to the Yerkes Regional Primate Center: mortality schedule data Tigges et al., 1988, which presents the combined genders: IMR, 0.02/year; MRDT, 12 years. The table gives a lower value for MRDT, the same as for humans, because a longer MRDT seems misleading in view of the many manifestations of senescence by 25 years. *tmax* from Tigges et al., 1988, may be a considerable underestimate.

[32]*Homo sapiens.* U.S. white female humans in 1980. Calculated by LR from USDHHS life tables. Jones's

Table 3.1. (Continued)

(1956) analysis of populations throughout the world during the nineteenth to mid-twentieth centuries indicates the remarkable consistency of MRDT, 7 to 8.5 years. The human *tmax* is uncertain. Indisputable birth records and lifelong identification are not available before about 1880.

[33]*Mus musculus.* Average of values calculated by LR for six long-lived inbred mouse strains at the Jackson Laboratory. Data were extracted from survivorship plots of Storer, 1978, which represented 7–9 adult age groups and > 600 mice for each strain: A/J (IMR, 0.024/year; MRDT, 0.26 year); CBA/J (0.027; 0.26); C57BL/6J (0.025; 0.30); DBA/2J (0.008; 0.26); SJL/J (0.073; 0.26); 129/J (0.02; 0.28). The values for MRDT were statistically indistinguishable. Overall mean ± SD (*n* = 6 strains): IMR, 0.03 ± 0.025/year; MRDT, 0.27 ± 0.02 year.

[34]*Rattus norvegicus.* Calculated by LR from mortality schedule data for *ad libitum*–fed Sprague-Dawley and Fischer 344 laboratory rats in seven studies shown in Table 10.1. The grand mean MRDT was 0.26 ± 0.06/year (± SD).

[35]*Peromyscus leucopus.* Calculated by LR for laboratory populations from survivorship plots of Smith and Walford, 1990.

3.2. Clonal Senescence in Ciliates

Clonal senescence in ciliates was recognized more than a century ago through the studies of Butschli (1876), Maupas (1889), and others reviewed by E. B. Wilson in the third edition of his classic, *The Cell in Development and Heredity* (1925, 238–247) and most recently by Bell (1988). These studies and subsequent ones by Jennings (1944a, 1944b, 1944c, 1945) and Sonneborn (1954) gave focus early in the era of cell biology to the possible roles of cell senescence in the somatic senescence of organisms. In many cases, the senescence of ciliates develops slowly during serial passage and terminates in the loss of proliferation after a characteristic number of divisions, as observed in *Euplotes*, *Paramecium* (Smith-Sonneborn, 1985, 1990; Takagi, 1987), and a few species of *Tetrahymena* (Nanney, 1974). Depending on the clone of *Tetrahymena*, for example, the number of fissions ranges from 40 to 1,500 before clonal extinction. Many strains of *Paramecium* and *Tetrahymena*, among other ciliates, also manifest an analogue of adolescence, in which sexual competence is not acquired for up to a hundred generations after conjugation (Jennings, 1945; Nanney, 1974; Rogers and Karrer, 1985). The period of sexual maturity may occupy many hundreds of generations over several months or even several years before senescence. A general analogy can be made to the rhythm of growth and fecundity, senescence, and death in the life histories of multicellular organisms.

During senescence in *Paramecium*, clones manifest slowing of proliferation, increased morphological abnormalities, accumulation of pigments, and increased death after conjugation, and then stop dividing permanently. Macronuclear DNA is lost after 300 fissions in *P. caudatum* (Takagi and Kanazawa, 1982) and after 100 fissions in *P. aurelia* (Klass and Smith-Sonneborn, 1976). For comparison, DNA loss does not happen during clonal senescence of diploid human fibroblasts (Stanulis-Praeger, 1987). If conjugation occurs, some of the daughter cells can be clonally rejuvenated.

The molecular basis for clonal senescence in the free-swimming ciliates is not known. Nuclear transplantation studies point to various roles for both the cytoplasm and the macronucleus (reviewed in Smith-Sonneborn, 1990). In *Paramecium* young macronuclei extend the clonal lifespan of older recipients of a short-lived strain (Aufderheide and

Schneller, 1985; Aufderheide, 1984, 1987). Cytoplasm from immature clones increased the sexual capacity of later stage clones that were sexually waning, but did not restore the proliferative rate (Aufderheide, 1984; Haga and Karino, 1986). The mating of paramecia from young and senescent clones, a type of micronuclear transplantation, indicates that "senescent" cytoplasm often damages the young micronucleus so that few progeny survive; those that do, however, are rejuvenated (Sonneborn and Schneller, 1960). The success in detecting changes of proteins synthesized during clonal adolescence (Rogers and Karrer, 1985) suggests the possibility of molecular analysis of clonal senescence at the level of gene expression. The increased mortality of exconjugants from parents mated after many generations of vegetative reproduction is interpreted by Bell (1988, 71–73) as the result of accumulated lethal recessive mutations. There is no direct evidence for the mutational hypothesis, however.

The rapid individual senescence of the sessile ciliate *Tokophrya* is attributed to a very different mechanism, insufficient removal of waste products (Chapter 2.2.1.2). *Tokophrya* also shows clonal senescence during propagation by budding. Budding, or vegetative reproduction (Chapter 4.4), in *Tokophrya* involves an internal morphogenetic process, through which a limited number of embryos are formed. Upon release the embryos swim about for minutes to hours before settling and metamorphosing into a smaller adult with tentacles that permit feeding (Millecchia and Rudzinska, 1970, 1971; Hascall and Rudzinska, 1970). Vegetative propagation through repeated transfer is limited with clonal lifespans that range from 1 to 33 days and wide variations among clones (Karakashian et al., 1984). Remarkably, clones from older individuals gave rise to individual cells with shorter lifespans and reduced capacity to form buds. Thus, *Tokophrya* shows both clonal and individual cell senescence leading to death.

In contrast to these examples, numerous other species and strains have no clonal senescence or apparent limit to asexual replication (Chapter 4).

3.3. Plants and Invertebrates

The distribution of gradual senescence in plants and invertebrates is more difficult to establish than dramatic rapid senescence, because few mortality rate data are available for the few species documented to live beyond 3–6 months. A few vascular plants and quite a number of flatworms, annelids, and arthropods show gradual losses of function and nominal maximum lifespans that suggest gradual senescence.

3.3.1. Plants

In distinction to the annual vascular plants, many perennials show no signs of whole-plant senescence (Noodén, 1988c). Indeed, the list of very long lived trees is impressive (Chapter 4.2.1). Nonetheless, many of these have maximum lifespans that separate them into distinct groups. Thus, few deciduous trees exceed 1,000 years, while numerous conifers live beyond 1,500 years. It is well known that apple, orange, and other fruit trees commonly lose capacity before they reach 100 years, which is suggestive of a slow senescent process (Wangerman, 1965; Leopold, 1981). Removal of buds, however, may restore the ability of cuttings from these branches to form roots (Wellensick, 1952). The slowed growth of mature trees can often be restored by pruning, cutting, or grafting (Leopold and Kriedemann, 1975; Leopold, 1981). Citrus trees showing age-related declines of fruit yield can also be rejuvenated by vegetative propagation through cuttings (Frost, 1938). These phenomena recall the correlation effects, and the role of the hormones produced by roots or growing tips, that are known to be crucial in the prevention of senescence in annuals (Chapter 2.3.2). No studies have been done of plant hormones in the slowly senescing examples. Because there have been no systematic studies of the vigor, as a function of age, of cuttings and grafts from species like fruit trees, which have some indications of slow senescence, strong conclusions would be premature.

3.3.2. Flatworms

Flatworms are the simplest of the invertebrates to show slow senescence leading to organismic death. Many species that live several years combine senescence with asexual reproduction and rejuvenation (Hyman, 1951, 195). The sexually reproducing *Dugesia*

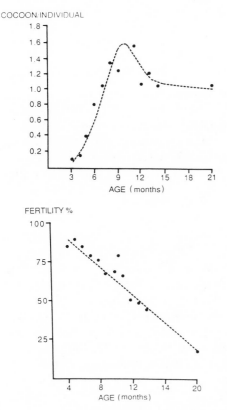

Fig. 3.1. Senescence in the flatworm *Dugesia lugubris*. *Top,* age-related changes in cocoon production. *Bottom,* age-related decreases in the fertility of cocoons. Redrawn from Balázs and Berg, 1962.

(= *Planaria) lugubris* (Haranghy and Balázs, 1964) has a 3-year lifespan, at least fiftyfold greater than short-lived, rapidly senescing species (Chapter 2.5.3). These fresh-water triclads have been studied for many decades because of their widely varying life histories. Some species can reproduce asexually and escape senescence (Chapter 4.2), while others are semelparous and show rapid senescence (Chapter 2.5.3), and yet others have intermediate characteristics.

In the laboratory, *D. lugubris* reaches sexual maturity and maximum size by 2 to 5 months (Balázs and Berg, 1962; Calow et al., 1979). Egg production peaks by 10 months and continues with modest decreases throughout life (Figure 3.1, top). Moreover, the fertility of the cocoons decreases linearly with age after maturity (Figure 3.1, bottom); the size and shape of the cocoons also become variable. Spermatogenesis, however, is main-tained. Concurrently, the dark pigmentation of the young adult worm gradually fades, and morphologic abnormalities arise that include swellings and wounds. Taken together, these changes indicate that senescence is gradual. However, regeneration of *D. lugubris* occurs as readily at 30 months as at 6 (Haranghy and Balázs, 1964; Balázs, 1966), which shows maintenance of somatic cell genomic totipotency. In this regard, the outcome of regener-ation and rejuvenation of *D. lugubris* is similar to the rejuvenation of *Planaria velata* after fragmentation and encystment (Chapter 2.5.3). *Dugesia polychroa,* however, failed to regenerate when 2 years old (Lindh, 1957). These examples indicate extensive species differences in the timing and manifestations of senescence. No life table data are available for mortality rate calculations.

3.3.3. Nematodes

In contrast to the short-lived soil nematodes (Chapter 2.5.4), some parasitic species have much longer lifespans. The mermithid parasites of grasshoppers, which are apha-gous as free-living adults, nonetheless have a gradual senescence and adult life phases lasting more than 2 years (Chapter 2.2.1.3). Other parasitic nematodes live even longer in human hosts and may not show any senescence (Chapter 4.1.2). These examples show that the postmitotic status of nearly all cells in adult nematodes is not incompatible with extended lifespans.

3.3.4. Annelids

Annelid earthworms, in contrast to the long-lived flatworms described above, have an extensive postreproductive phase within their even longer lifespans. Earthworm lifespans are documented to exceed 10 years for *Allolobophora (= Helodrilus) longa* (Korschelt, 1914), which may include 4 months of annual diapause (Edwards and Lofty, 1972). Oth-ers live 4–6 years (Korschelt, 1914) and even survive 4 years as hosts for larval or juvenile stages of nematode parasites (Keilin, 1925; Otter, 1933; Hyman, 1951, 348). Develop-ment from the cocoon to sexual maturity takes 4 to 24 months, depending on the species and its environment (moisture and temperature); slow growth usually continues after ma-turity (Murchie, 1960; Edwards and Lofty, 1972; Satchell, 1967). The active phase of

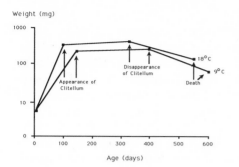

Fig. 3.2. Life history of the earthworm *Dendrobaena subrubicunda*. The disappearance of the external reproductive organ (clitellum) is followed by a phase of sterility lasting almost 1 year. The life history is slightly prolonged at cooler temperatures. Data of Michon, 1953, 1954. Redrawn from Satchell, 1967.

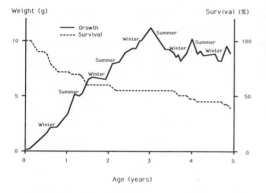

Fig. 3.3. Growth and survival in the earthworm *Lumbricus terrestris,* showing that maximum body size is achieved at 3 years of age but without subsequent changes in mortality rate. Redrawn from Satchell, 1967.

cocoon laying in *Dendrobaena subrubicunda* lasts 8–12 months, as marked by the disappearance of the clitellum (Michon, 1953, 1954). Yet this species has a postreproductive phase of up to 8 months, so that the adult phase lasts about 2 years (Figure 3.2). The occurrence of a postreproductive phase in earthworms is particularly interesting because of the ability of annelids to reproduce by fission (vegetatively; Chapter 4.4.3); it is unknown if vegetative reproduction is altered during sexual reproductive senescence.

Weight loss before death is another manifestation of senescence in *Dendrobaena* and in the common earthworm *Lumbricus terrestris,* where it is associated with fungal infections (Satchell, 1967). *Lumbricus* lives 6–10 years (Korschelt, 1914; Edwin Cooper, pers. comm.) but does not show marked age-related increases in mortality (Figure 3.3); the IMR is about 0.1/year, which is low among invertebrates (Appendix 1). Worms that survived to the age of 3 years showed no signs of senescence. It would be important if *Lumbricus* also had an extensive postreproductive phase as seen in *Dendrobaena.* Another species of interest is the largest earthworm, *Microchaetus,* up to 7 meters long and 7.5 centimeters in diameter, which must be very long lived to achieve this size. Earthworms may provide the first examples of species with long postreproductive lifespans but without increases of mortality, whether exponential or otherwise.

3.3.5. Arthropods

The arthropods, with the largest number of species of the animal phyla, may well represent the broadest range of senescent phenomena among the animals. Chapter 2 describes the extremely long lived insect queens, with negligible senescence until sudden death; aphagous mayflies; etc. Some social insects may have gradual senescence, particularly the lower castes of ants and termites that live several years; no mortality data are reported.

The best examples of gradual senescence among insects are found in the Coleoptera, many species of which have adult life phases of 2 or more years (Comfort, 1979). The record lifespans are at least 9 years for two beetles, *Macronychus glabratus* and *Microcylloepus pussilus* (family Elmidae): several adults of each species showed no loss of vitality in an aquarium during this time (Brown, 1973). In the absence of mating and under protected conditions, senescence in these beetles approaches the negligible end of the spectrum and recalls the great longevity of insect queens (Chapter 2.2.3.1).

In contrast to these examples of longevity, numerous beetles have shorter lifespans and well-defined gradual senescence. Flour beetles, *Tribolium castaneum,* are used in selection studies to investigate how different reproductive schedules may influence lifespan (Chapter 6.3.2), and are favored because of their rapid development and maturation (1 month, ca. 6 instars) and long adult phases (*tmax* of 2–3 years; Mertz, 1975; Good, 1936). While female fertility decreases sharply after 6 months (Mertz, 1975; Good, 1936), male fertility continues at least to 3 years (Good, 1936). The cellular basis for this major gender difference is unknown and implies ovarian or neuroendocrine senescence in females only. Mortality rate and life table data are not available. A mean lifespan of 130 days (Lints and Soliman, 1977) against an eightfold greater *tmax* (Good, 1936) implies an *AMR* of 0.9/ year, which would permit about 0.1% of mature beetles to reach 3 years, even without accelerating mortality. *Carabus* live longer than *Tribolium,* up to 4 years. Egg production in *Carabus* females gradually wanes after several years (Thiele, 1977; Houston, 1981; van Dijk, 1979). The absence of correlations between egg production and survival to the next season (van Dijk, 1979) indicates that these beetles were able to feed adequately during the first several years. Van Dijk concluded that, in contrast to some other species, the majority of *Carabus coerulescens* in natural populations were 2 or more years old. The very long adult phases of beetles implies cell replacement in the intestinal epithelium or elsewhere, but I have not found documentation. A general factor in beetle mortality must be the irreplaceability of the cuticle and its appendages, which slowly become worn and so serve as markers of age in field studies. Described in Chapter 2.2.2.1, wear of the mandibles can impair feeding and cause starvation, for example, in 2–3-year-old *Carabus glabratus* (Houston, 1981).

Cockroaches (Orthoptera) also give evidence for gradual senescence. In a major survey of fifteen species that represent different patterns of oviposition, Willis et al. (1958) observed adult lifespans of 5 to 15 months. Egg production and viability gradually declined in *Blatella germanica* (lifespan 180 days) and most other species, but did not in *Supella aupellectilium* (lifespan 141 days). These and longer-lived species (*Diploptera punctata,* lifespan 423 days) had postreproductive phases of a few days to 5 months before death. In general, the most fecund individuals in a species lived longest. These examples span the shorter end of the spectrum, verging into rapid senescence.

3.3.6. Mollusks

This large phylum of more than 100,000 species presents as extensive a range of lifespans (more than fiftyfold) as the huge arthropod phylum and may also prove to have as much diversity in the patterns of senescence, judging by the few species studied. At one extreme are certain clams, with lifespans beyond a century and no definable senescence (Chapter 4.2.2). At the other extreme are sea hares, with lifespans that may be as short as 6 months (Chapter 2.3.1). This section discusses snails with intermediate lifespans.

Some terrestrial mollusks grow slowly and show high mortality before maturation. As marked by cessation of growth at 20 mm length, maturation in the Hawaiian tree snails *Achatinella mustellina* and *Partulina proxima* occurs at ages up to 7 years (Hadfield and Mountain, 1980; Hadfield and Miller, 1989; Hadfield, 1986), which is surprisingly old for such small animals. Mortality then decreases sharply after maturation, for unknown reasons. Calculations based on estimates of the age structure by snail size indicate that, to maintain stable populations, the lifespan could be 11 to 19 years for these two species, respectively (Hadfield and Miller, 1989). Four species of *Conus* also have late maturation and lifespans in this range (Perron, 1982). One suspects that lifespans could be even longer. These long lifespans suggest very gradual or negligible pathophysiological senescence.

Other mollusks, including certain octopods and squid (cephalopod mollusks), are shorter-lived and die at first spawning (Chapter 2.3.1.3). *Aplysia* is also a candidate for rapid senescence. However, semelparity is not characteristic of all *Octopus* (Rodaniche, 1984). *Nautilus,* which is considered to be the most primitive living cephalopod, can produce eggs in captivity for at least 30 months (Arnold and Carlson, 1986). The examples of *Lottia* and *Nucella* (Chapter 4.3) show that closely related marine gastropod mollusks within each genus have manyfold differences in mortality (AMR), lifespan, and fecundity in nature that are not the result of any known differences in rates of senescence. Freshwater gastropods also show a range of life histories, from semelparous annuals to iteroparous perennials (Hunter, 1961; Calow, 1978a).

The pulmonate snails *Lymnaea* and *Planorbis* show well-defined accelerations of mortality (MRDT = 0.2 years) and maximum lifespans of about 2 years, which are in the same range as small rodents (Table 3.1; Appendix 1). Laboratory-bred adults of *L. stagnalis* showed accelerations of mortality when 6 months old. Growth, measured as shell lengthening, ceased as death neared (Janse et al., 1988; Janse, Wildering, and Popelier, 1989; Janse, ter Maat, and Pieneman, 1990). Egg laying ceased typically by 1 year, which implies a postreproductive phase of about 0.5 years. Egg laying also gradually declined in *Planorbis contortus* (Calow, 1973).

A series of studies from Holland points to oviductal and neuronal changes in *L. stagnalis* (Janse, van der Roest, Bedaux, and Slob, 1986; Janse, Beek, Van Oorschot, and van der Roest, 1986; Janse et al., 1988; Janse, ter Maat, and Pieneman, 1990; Janse, Wildering, and Popelier, 1989; Janse and Joosse, 1989). The caudodorsal cells of the cerebral ganglion control oviposition through secretion of caudodorsal-cell ovulation hormone and other neuropeptides during prolonged, synchronized electrical activity. Induced ovulation by synthetic hormone was 50% less effective in causing egg laying in senescent snails that had stopped laying, as compared with others of the same age or young snails that were laying; this indicates a decreased oviduct responsiveness, which may or may not be an all-or-none phenomenon. There were no differences between old layers or nonlayers in the stimulation of ovulation in young adults by injections of cerebral ganglion homogenates. However, the caudodorsal cells showed electrophysiological changes in after discharges, which lasted longer in senescent nonlayers than in young layers and nonlayers. Membrane biophysical properties in the parietal ganglia of *L. stagnalis* show alterations (Frolkis et al. 1984, 1989). Moreover, the parietal ganglion showed decreased RNA, thinning of endoplasmic reticulum, and increased lipoidal pigments (probably lysosomes with pigments like lipofuscins). The slower recovery of the tentacle withdrawal reflex after nerve injury in 10-month-old *Lymnaea* suggests impaired neuronal regeneration (Janse, Beek, Van Oorschot, and van der Roest, 1986), that is, before reproductive senescence. Many changes in neurons and egg laying are reported in the even shorter lived *Aplysia* (opisthobranch gastropod), which has a more rapid senescence (Chapter 2. Although present data cannot show if the basic mechanisms are different, there is no hint that senescence in *Lymnaea* is triggered by reproduction, as may be true for *Aplysia*.

An open question is whether *de novo* oogenesis continues in adult *Lymnaea*, or in other snails that maintain fertility to ages five- to tenfold longer, like *Achatinella*. In *Helix aspersa*, the primordial germ cells are segregated early in development. A fixed oocyte pool in adult *Helix* does not permit gonadal regeneration at ages older than 6 months (see literature cited in Hogg and Wijdenes, 1979; Griffond and Bride, 1987). It would be valuable to make comparisons of reproductive age changes of pulmonate snails with different life histories, for example, annuals and perennials with different seasonal trends in oviposition (Calow, 1978a) under laboratory conditions.[1]

Neuronal age-related changes are also indicated in bivalve mollusks (class Pelecypoda), but no mortality data are available. The mussel *Mytilus edulis*, which lives up to 10 years (Comfort, 1957), has decreased neuronal receptors for opioid ligands (etorphine and naloxone) and decreased dopamine-sensitive adenylate cyclase in its pedal ganglia by 4 years (Stefano, 1982; Stefano et al., 1982; Chapman et al., 1984); these specimens were collected in suburban New York. Some of these changes resemble those of the mammalian brain (Makman and Stefano, 1984). While such comparisons stimulate discussion about possible generalizations in senescence that may transcend great evolutionary divergences, it seems unlikely that mechanisms in senescence have persisted, since the lines leading to snails and humans separated long before the Cambrian (Chapter 11.2).

Freshwater mussels (*Anodonta cygnea* and *Unio pictorum*) continue to produce fertile ova during lifespans of 10 years or more, though some older females had no ripe oocytes throughout the year and appeared to be degenerating; males did not seem to degenerate (Haranghy et al., 1964). These reports suggest gradual senescence that might be impli-

---

1. Some age-related changes may result from husbandry that was adequate for development and maturation, but that might be suboptimal for long-term studies. Another general concern is pollution in the environment from which the parental generation is collected, which could influence the quality of eggs and thus carry over effects into the next generation. Some laboratory-bred populations of *Lymnaea* were afflicted with undefined waterborne infections that shortened lifespan and accelerated overall mortality, but did not alter the MRDT (Slob and Janse, 1988). This example warns against mean lifespan as the sole criterion for senescence and indicates the need for replicate studies over a span of time to indicate confounds from physiological and pathological fluctuations that may occur in the same laboratory. How frustrating that such great efforts are needed to establish new models for gerontological studies, even when the development and adult neurobiology, for example in *Lymnaea*, are so well studied.

cated in the apparently short lifespan of some bivalves, but be cautioned: some bivalve species live beyond 100 years (Chapter 4.2.2). Lacking data on the age-dependent and age-independent mortality coefficient (equation [1.3]), we cannot tell if the age-related changes impact on mortality rates.

3.4. Fish and Lower Vertebrates

The lifespans and patterns of senescence in the vertebrates with well-developed brains (the Craniata) range as widely as in most invertebrate phyla. The evidence developed here and in Chapter 2 suggests that numerous radiations developed particular patterns of senescence (Chapter 11). There are few clear suggestions of senescence outside of certain tetrapods.

3.4.1. Chondrichthyes

3.4.1.1. Sharks and Rays (Elasmobranchs)

Sharks and rays have the following differences from teleosts: (1) long gestation times in viviparous species, (2) long lifespans compared with most ray-finned fish, and (3) tentatively, a fixed number of ovarian oocytes. Toothed sharks also have a remarkable capacity for continuous tooth replacement (Roberts, 1967), which protects them against accumulations of dentitional damage shown in many mammals (Section 3.6.11.4). While there is as yet no information about senescence in sharks, these characteristics predict female reproductive senescence specifically, and also bear on Williams' (1957) prediction that organisms with decreasing reproductive capacity will manifest senescence (Chapter 1). There is no evidence in elasmobranchs for semelparity or increased mortality in association with spawning (Beverton, 1987).

Age determination in elasmobranchs is studied to determine the effects of the recently intensified gill net and recreational fishing on the age stratification of shark populations. Although sharks lack the scales and otoliths that are widely used to age teleosts, tetracycline marking studies established the annual production of bands in the dorsal fins (S. E. Smith, 1984; Beamish and McFarlane, 1985). Counting of bands in conjunction with X-ray and radiometric techniques (Cailliet et al., 1986; Cailliet, Martin, Kusher, Wolf, and Welden 1983; Cailliet, Martin, Harvey, Kusher, and Welden, 1983; Martin and Cailliet, 1988a, 1988b; Welden et al., 1987) shows the rapid growth of large sharks, with maturation at 60–90% of the maximum body size: 5 years/0.8 meters for bat rays (*Myliobatis californica*); 6 years/2 meters for blue sharks (*Prionace glauca*); 10 years/3 meters for white sharks (*Carcharodon carcharias*) (Holden, 1977; Cailliet and Bedford, 1983); but more than 20 years/0.7 meter for the more slowly growing dogfish (*Squalus acanthias;* Ketchen, 1975). Dogfish have the greatest lifespans (70 years) documented so far in elasmobranchs, as determined from annual rings on the dorsal spine (Beamish and McFarlane, 1985). Recapture studies of the Australian school shark (*Galeorhinus australis*) show lifespans of at least 28 years (Hansen, 1963), whereas aquarium records do not exceed 25 years (Clark, 1963). Captured specimens rarely exceed 15 years (Cailliet et al., 1985), which suggests that much older sharks are very rare. By contrast, numerous records show that eels (Chapter 2.3.1.4.2), sturgeon, and teleosts (discussed below) can live 80 years or longer. Still, these thin data do not preclude elasmobranch lifespans equal to those of sturgeon and teleosts.

Female elasmobranchs are believed to acquire a fixed ovarian oocyte stock during development, as the result of cessation of mitosis in the oogonia of the germinal epithelium. Meiosis then begins in the newly formed oocytes (Franchi et al., 1962; Nieuwkoop and Sutasurya, 1979). There are, however, few studies. Matthews (1950) reported that adult basking sharks (*Cetorhinus maximus*) showed "very few . . . in the earlier stages of development"; the small number is not necessarily in conflict with the above conclusions, since the ages of the specimens were unknown. Similarly, cessation of oogonial proliferation and finite oocyte stocks are believed to be characteristic of nearly all birds and mammals. In marked contrast, there is a continuing renewal of primary oocyte pools by oogonial division in most teleosts and reptiles, but probably not in all amphibia (Franchi et al., 1962; Nieuwkoop and Sutasurya, 1979; Tokarz, 1978). Overall, the annual birthrate is low, particularly in viviparous sharks, where most species produce fewer than 20 offspring per year (Cailliet and Bedford, 1983). In bat rays, the number of ova increased progressively with body size, from 30 ova at 6 years to 250 at 13 years; the few older specimens collected had fewer ova at 14–17 years (Martin and Cailliet, 1988b), possibly indicating ovarian senescence. The right ovary of bat rays is inactive, as is typical of viviparous rays (Dodd, 1983; Wourms, 1977). The possible reactivation of the dormant right ovary might be looked for at later ages; this occasionally happens in birds (Section 3.5). The kinetics of elasmobranch oocyte maturation and follicular growth are not known as a function of adult age.

3.4.1.2.   Ray-finned Fish

The ray-finned fish (actinopterygians) generally have well-ossified skeletons. An exception to this is the sturgeon, whose scant calcification is a derived characteristic from ray-finned ancestors in the Paleozoic (Chapter 11.3.4). Among the ray-finned fish, there is an indication of an evolutionary divergence in continuing *de novo* oogenesis in adults.

3.4.1.2.1.   Sturgeon

Sturgeon are of particular interest because of their great longevity and size (occasionally 5 meters long) and early appearance in the fossil record (Chapter 11; Figure 11.8), but also because, like sharks, they may not continue *de novo* oogenesis into adult life. The record lifespans are held by the sturgeon *Acipenser fulvescens,* 152 years ("One-hundred-fifty-two-year-old lake sturgeon," 1954) and *Huso,* 118 years (Tsepkin and Sokolov, 1971). Maturation varies among species and is determined by size more than chronological age. Larger species may delay maturation to 20 or more years, while 10 years is more common (Doroshov, 1985).

Limited data indicate that sturgeon have a fixed ovarian oocyte stock. In sturgeon as throughout the Chordata, primordial oocytes are derived from oogonial cells. In several acipenserids, active gonial cell mitoses were seen at 3–7 months, an age long before maturation (Persov, 1975). During the first spawning season in *Acipenser stellatus* and *A. güldenstädti* at 15 years, the germinal epithelial layer of the ovary retained many clusters of primordial oocytes or gonial cells (Raikova, 1976). Immature oocytes that are present during the spawning season are in a previtellogenic stage and are still increasing in size (F. Chapman and S. Doroshov, in prep.; Kornienko et al., 1988, translated for me by Serge Doroshov). When the second spawning season occurs after the age of 20 years in *A. güldenstädti,* the clusters of primordial oocytes have dwindled to single cells (Kornienko et al., 1988). No data are available on the population census of oogonia, oocytes, or follicles at later ages, or on the possibility of gonial mitoses that might replenish the primordial oocyte pool. Nonetheless, data of Kornienko et al. suggest that the ovarian oocyte population is limited. If so, oocyte exhaustion and reproductive senescence are predicted.

The intervals between spawning vary from 2 to 8 years; sturgeon are presumed to spawn several times (Doroshov, 1985). Spawning in some species is associated with aphagy (McDowall, 1988, 38) and lipid depletion (Krivobok and Tarkovskaya, 1970). Cortisol, which is the main corticosteroid in these as in other teleosts (Idler and Sangalang, 1970), increases during experimentally induced ovulation (Lutes et al., 1987). Dramatic cytological changes in hypothalamic neurosecretory cells also occur during spawning (Polenov and Pavlovic, 1978).

Unlike many other anadromous fish with extended migrations (Chapter 2.3.1.4), sturgeon have low mortality after spawning. Growth continues asymptotically long after maturity; males are generally smaller and live less long than females (Probst and Cooper, 1954; Doroshov, 1985). Some evidence for gradual senescence is the increased mortality in *A. fulvescens* after 30 years (Figure 3.4), with an estimated MRDT that is at least as long as the better-documented 8-year MRDT of humans. The age distribution of *A. fulvescens* also indicates that few (ca. 1%) in the population survive beyond 40 years (Probst and Cooper, 1954). Little postmortem data are available from which to judge senescence. A male sterlet (*A. ruthenus*) that lived 70 years began to lose its balance a week before death; infections were found at autopsy (D. Dekker, Amsterdam Aquarium, pers. comm.). That this lifespan is severalfold longer than expected for sterlets (Doroshov, 1985) indicates how much lifespans in nature may underestimate the potential when individuals are protected.

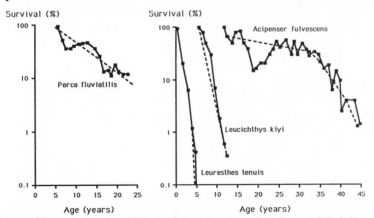

Fig. 3.4. Survivorship by age in natural fish populations that were relatively unexploited. Three species did not show detectible age-related increases in mortality: the goby (*Leuresthes tenuis*), lifespan ≤ 10 years; the chub (*Leucichthys kiyi*), lifespan ≥ 10 years; and the perch (*Perca fluviatilis*), lifespan ≥ 22 years. In contrast, there is a hint of increased mortality in the Great Lakes sturgeon (*Acipenser fulvescens*), lifespan > 80 years and estimated MRDT 10 years. Other mortality coefficients are in Table 3.1. Redrawn from Beverton and Holt, 1959.

The extent of senescence in sturgeon is a question of great interest. If sturgeon in general have a fixed stock of oocytes yet can live to ages beyond humans without signs of senescence as is indicated, this information would suggest that a finite gametogenic capacity can be dissociated from the necessity of senescence. The pursuit of this question rests on obtaining more data on whether oogenesis is limited in all species, on the extent of pathophysiology at later ages, and on the acquisition of better life table data to evaluate the nature of age-related increases of mortality rate.

3.4.1.2.2. Teleosts

The bony ray-finned fish (teleosts) are highly diverse and represent a fiftyfold range of lifespans and diverse patterns of senescence. The role of senescence in the lifespan of iteroparous teleosts is clearest for a few short-lived species, but is virtually unknown for most others. Teleost lifespans vary fifty- to a hundredfold, a wider range than in mammals, and a range that continues to increase as analysis of annual rings on scale and otoliths becomes more developed (Beverton, 1987; Leaman and Beamish, 1984). Several species of rockfish (*Sebastes*) appear to live at least 120 years (Leaman and Beamish, 1984; Chapter 4.2.2), in the range of the oldest sturgeon, turtles, and humans. The rockfish and many other very long lived fish do not show any indications of ovarian senescence, which makes these species candidates for negligible senescence (Chapter 4).

Shorter (2- to 15-year) lifespans are indicated for some teleosts, but these reports must be considered as representing a meager and uneven sample of the 20,000 species. In many cases, these short lifespans (Beverton and Holt, 1959; Beverton, 1987) probably result from high age-independent mortality risks (IMR), but the age-related mortality patterns needed to establish the rate of senescence are generally lacking for long-lived populations. Little can be said about the rates of senescence from the survival records of fish in aquaria (Hinton, 1962; Comfort, 1979). Mortality schedules are available for relatively few species, mostly the shortest-lived (Table 3.1; Appendix 1). A few long-lived and iteroparous species appear to show age-related increases of mortality in natural populations: the Pacific herring *Clupea pallasi* (Woodhead, 1978), the rock bass (*Ambloplites rupestris*), and the whitefish (*Coregonus clupeaformis*) (Ricker, 1949); the present data do not show whether the mortality rate increases are definitively exponential.

Little is known about senescence in fish with intermediate lifespans, ranging from 5 to 30 years, and the literature is fragmentary. Gerking (1959) suggested, based on a few specimens, that fecundity decreased at later ages in haddock (*Melanogrammus aeglefinis*). The interoparous anchovies (*Engraulis mordax;* Fitch and Lavenburg, 1971) and many species of trout (*Salmo trutta;* Behnke, 1989) rarely live longer than 7 years and might show senescence. Although there is an increased lipid peroxidation by the sarcoplasmic reticulum of flounders (*Pseudopleuronectes americanus*) between 1 and 9 years (Luo and Hultin, 1986), it should not be presumed that such changes are debilitating and increase the mortality risk. The Atlantic salmon (*Salmo salar*) have high, but not complete, mortality at spawning, with life histories that are different only in degree from the Pacific salmon (Chapter 2.3.1.4.2). Some spawn a second time, and rare exceptions spawn up to six times (Ducharme, 1969). A recent study shows that spawning Atlantic salmon have extensive coronary arteriosclerosis (Saunders and Farrell, 1988) that resembles that of Pacific salmon (Figure 2.11). The coronary lesions had not regressed by 5 months after spawning. The role of the coronary lesions in mortality is unknown. Corticosteroids are implicated.

Some species of very short-lived fish present characteristics of gradual senescence, including values for MRDT like those of small rodents (Table 3.1; Appendix 1), which are accompanied by pathological changes such as cellular hyperplasia and pre-morbid weight loss (Craig, 1985; Comfort, 1979; Gerking, 1959). Specific examples are given below. The degenerative changes do not arise rapidly or synchronously in the population. Together, the variety of lesions and their asynchronous occurrence in individual fish discount the occurrence of pathophysiologic changes like the hyperadrenocorticism of rapidly senescing Pacific salmon and eels (Chapter 2.3.1.4).

Perhaps the most is known about gradual senescence in cyprinodonts, a large order with 845 species (Nelson, 1984). The "annual" *Cynolebias* and *Nothobranchius* (briefly mentioned in Chapter 2.4) live for 2–3 years in the laboratory and show a well-defined senescence with exponential accelerations of mortality and heterogeneous pathological changes (Figure 3.5; Liu and Walford, 1966, 1969; Markofsky and Perlmutter, 1972, 1973). Similar mortality accelerations are shown by the guppy (*Lebistes reticulatus;* not an "annual"), with lifespans in the range of 2 to 6 years (Comfort, 1961, 1963). Comfort's classic study gave the first characterization of senescence in a fish by anatomical and actuarial criteria. The acceleration of mortality becomes apparent at about the same age in the population (1.5 years), when body size is approaching the maximum (Figure 3.6).

While small fish are often thought to have the shortest lifespans, consider the following evidence about gobies, which are part of the huge perciform order (8,700 species). One of the smallest adult vertebrates is the goby *Pandaka pygmea;* males at maturity can be less than one centimeter long, weighing less than twenty milligrams, and are the shortest freshwater fish (Bruun, 1940; Miller, 1979; Nelson, 1984). *Pandaka* captured as adults

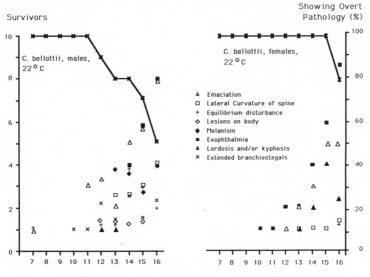

Age (months)

Fig. 3.5. Age-related pathological lesions in the short-lived (annual) fish, *Cynolebias bellottii*, males and females. Survival is indicated by the *solid line*. Redrawn from Liu and Walford, 1969.

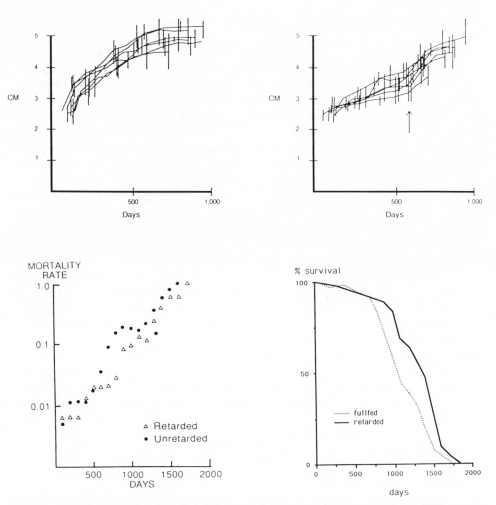

Fig. 3.6. Age-related growth and survival of the guppy *Lebistes reticulatus* maintained on *ad libitum* or re-stricted feeding. *Top left,* body length (cm) when fed *ad libitum; top right,* body length when retarded by diet restriction until fully fed at 600 days *(arrow); bottom left,* mortality rates; *bottom right,* survivorship. Retarded fish were diet-restricted from 100 to 600 days. There were slight decreases in the IMR with no effect on MRDT (Table 3.1). Redrawn from Comfort, 1963.

survived for 14 months in the laboratory (Liu and Walford, 1970). There are also two gobies with similar maximum sizes of seven to nine centimeters with lifespans that differ at least fivefold in nature: Fries' goby (*Lesueurigobius friesii*), 11 years (Gibson and Ezzi, 1978), versus the sand goby (*Pomatoschistus minutus*), 2.25 years (Fonds, 1973; Lee, 1974; Miller, 1975; Table 2.2G). Fries' goby breeds perennially, as is characteristic of most gradually senescing species, while most (80%) sand gobies breed only for one sea-son. Miller lucidly discusses the continuum between semelparity and iteroparity in these and other fish. Further arguments against a strong constraint of body size on longevity are given in Section 3.6.1 and Chapter 5.2.2.

Although *Lebistes* are considered to have indeterminate growth (Comfort, 1963), they may not live to manifest this potential because of pathophysiological changes during senescence. Indeterminate growth, or continued axial growth after maturation that approaches an asymptotic maximum, is widely noted in teleosts (Beverton and Holt, 1959; Figure 5.9); the growth rate constants show inverse correlations with the population mortality rate (Figure 5.10). Another aspect of growth-mortality interactions is given by a diet-restricted group that grew much more slowly until *ad libitum* feeding was restored at 1.65 years: an ensuing growth spurt let them nearly catch up after 2 years, but mortality accelerations then overwhelmed the population (Comfort, 1963). These observations bear on Bidder's (1932) hypothesis that indeterminate growth protects against senescence. While Comfort (1963) concluded that his data on *Lebistes* were incompatible with Bidder's hypothesis, these data do not seem to resolve the issue, since the cessation of growth and the accelerations of mortality are more or less concurrent. The data do show that senescence can occur when axial growth continues beyond sexual maturation. Bidder's hypothesis is discussed further in Chapter 4.5.3.

The thymus gland shows a striking shrinkage and involution in some species during maturation, for example, the medaka or rice fish (*Oryzias latipes*), which lives 3–6 years (Egami, 1971; Takimoto et al., 1984). Atrophy continues during the rest of life; cysts form and connective tissue increases (Ghoneum and Egami, 1982). Similar thymic shrinkage occurs during or after maturation in other short-lived fish (Rasquin and Hafter, 1951; Deansley, 1927; Robertson and Wexler, 1960). These changes at least superficially resemble events during maturation in mammals.

Senescence is abundantly shown at the anatomical level in these short-lived fish, as a type of gradual, but not rapid, senescence. *Cynolebias* frequently become thin or emaciated and eat less before death (Myers, 1952; Liu and Walford, 1969; Walford and Liu, 1965). Among other degenerative changes, the body may also become twisted or hunched (lordosis, kyphosis), the cornea clouded, the scales disordered (Walford and Liu, 1965; Liu and Walford, 1969). Egg production continues through 1 year, but ovaries become atretic and contain fewer mature ova (Walford and Liu, 1965).

Abnormal growths are common at later ages. *Cynolebias bellottii* shows them in the thyroid, in the walls of the gas bladder (reticuloendothelial cell types), and in the kidney (Liu and Walford, 1969). The platyfish (*Xiphophorus maculatus*) also develops malignant melanomas, at an exponential rate influenced by several genes (Anders et al., 1981; Ozato and Wakamatsu, 1983). *Xiphophorus* is a small viviparous fish with a mean lifespan of 2.5 years and a maximum lifespan of 5 years. Melanomas also increase with age in other colored species (Takayama et al., 1981). Ovarian tumors occur by sexual maturity (4–6 years) in long-lived domestic carp *Cyprinus carpio* (Ishikawa et al., 1976). In general, bony and cartilaginous fish are thought to have fewer spontaneous malignancies than mammals, but as in mammals some genotypes are relatively susceptible (Ponten, 1977).

Regeneration becomes impaired in the guppy (*Lebistes reticulatus*). Growth continues after maturity and maximum size is approached by 1.5 to 2 years, close to the lifespan. Amputation of the tail web induces a vigorous regeneration in young fish, with replacement of dermal bone and a complex sequence of cell differentiation that is complete by 20 days (Comfort and Doljanski, 1958). Regeneration of the tail web declines with age; most of the decrease occurs by 1 year, with more gradual decreases up through the oldest age examined, 3 years. If tails of adults were amputated during food restriction that had limited growth, the regeneration rate increased when more food was given and growth was transiently resumed. A related phenomenon is the increase of thymidine labelling in livers of senescent *Oryzias,* which was induced by transfer to warmer water (Shima et al., 1978); the basal labelling rate had decreased progressively with age (Shima and Egami, 1978). Unlike most mammals (Chapter 5.2.2), hepatocytes do not become increasingly polyploid with age in *Oryzias.* The evidence for impaired regeneration and cell proliferation during senescence is limited to studies on short-lived fish; long-lived species that reach puberty at manyfold greater ages would appear to maintain proliferative and regenerative capacities.

Oogenesis and primordial germ cell proliferation leading to *de novo* formation of primary oocytes is thought to continue into adult life in many teleosts (Tokarz, 1978; Nieuwkoop and Sutasurya, 1979), although the evidence allows only provisional conclusions. While many fish maintain fertility throughout their lifespans and may even increase their yield of ripe ova during continued growth, these observations do not preclude the depletion of a fixed oocyte store. Despite gradual age-related decreases in reproduction and histological changes in the gonads, some guppies (*Lebistes reticulatus*) of both sexes remain fertile even at the advanced age of 5.5 years, when more than 95% of the population was dead (Comfort, 1961; Woodhead and Ellett, 1969a, 1969b). While *Xiphophorus* gonads accumulate fat and connective tissue that might be markers for degeneration, spermatogenesis continues beyond 3–4 years (Schreibman et al., 1983), as does oocyte maturation, as judged by the presence of yolky oocytes in the oldest females (Schreibman and Margolis-Nunno, 1989). These observations, however, do not inform us of *de novo* oogenesis or indicate if there are any age-related changes in the reserve oocyte pool.

Oryzias also retains fertility throughout its lifespan (Egami, 1971). *De novo* oogenesis may possibly continue in *Oryzias* (Yamamoto, 1962), as indicated by the presence of numerous mitotic figures in small oogonial cells in a minority (30%) of ovaries taken from a sample of spawning fish. Ovaries after the spawning season showed numerous small oocytes that subsequently grew in size; this next oocyte cohort could have come either from newly formed oogonia or from residual stores of primary oocytes. A complete census of oocytes and follicles is needed to determine this question in fish of known age; unfortunately the age of these fish was known only by the season.

In the hake (*Merluccius*), oogonial proliferation ceases before maturation, so that the young ovaries contain a finite but huge store of immature resting oocytes (Hickling, 1936). Age changes in fertility are not reported in hake, a much larger fish than *Xiphophorus* or *Oryzias*. Schreibman and Margolis-Nunno (1989) conclude that age-related decreases in reproduction are not obligatory in teleosts. The question of *de novo* oogenesis during adult life of teleosts is not generally resolved by these old reports.

An extended postreproductive state is also reported. Siamese fighting fish, *Betta splendens* (lifespan 1.5–3 years) cease spawning after 1.5 years and their testes gradually regress, so that some fish at 2 years completely lack germinal cells (Woodhead, 1974). This gonadal senescence was considered distinct from the effect of global pituitary failure, since hypophysectomy arrests spermatogenesis at the spermatogonial stage (Atz, 1957), whereas all stages were present except in the most senescent fish.

The mosquito fish (*Gambusia affinis*) and other poeciliids also manifest an age-related decrease in brood size, which is followed by a definite postreproductive stage during their short (1–2-year) lifespan; no ova were found by histologic studies (Krumholz, 1948), but it is unknown if any primordial ovarian follicles remain (discussed in Gerking, 1959).

There is also evidence that natural populations have genetic polymorphisms that influence size and age at maturation and that might also influence senescence. Stearns (1983a, 1983b) studied populations derived from 150 founders introduced for mosquito control in Hawaii early in this century. Their habitats varied in the annual fluctuations of water levels and thus might select for different growth characteristics. About 150 generations later, wild-caught fish were sampled from six locations that differed in water-level stability and were bred for two subsequent generations under standard conditions. The six stocks differed nearly twofold in their fecundity, with smaller differences in age at maturity, size at maturity, and the interval between broods. However, the mean lifespan of about 9 months did not differ significantly. The survival of some fish to 19 months (the cut-off age for the study and double the mean lifespan) implies a nonrectangular survival curve and large age-independent mortality rates; the overall mortality rates were too high to observe if there were differences in the rates of senescence.

Data on the neurobiology of senescence in fish are virtually restricted to the *Xiphophorus* (see above). There are genetic influences on age of maturation in *Xiphophorus*, an event under neuroendocrine control in all vertebrates. Five alleles of the *P* gene (X and Y chromosomes, also near loci for body pigmentation that identify these alleles) are associated with discrete and nonoverlapping ages at maturation, which occurs from 2 to 24 months (Schreibman and Kallman, 1977). In studies of different genotypes of *Xiphophorus*, those of both genders that matured earliest (*P*1 allele) lived longest (Schreibman and Margolis-Nunno, 1989). These *P* alleles are also associated with different and complex fluctuations of age changes in gonadotropin hormone-releasing hormone (GnRH), arginine vasotocin, and other hypothalamic neuropeptides, catecholamines, and serotonin (Schreibman et al., 1983; Schreibman et al., 1987; Margolis-Nunno et al., 1986, 1987; Schreibman and Margolis-Nunno, 1989). A major change was found in tyrosine hydroxylase, an enzyme that synthesizes catecholamines and that became undetectable in the paraventricular organ of 2.5-year-old fish (Halpern-Sebold et al., 1986; Schreibman and Margolis-Nunno, 1989). None of these changes has been related to specific functional changes in circulating hormones or physiological function. Intriguing preliminary data indicate degeneration of the olfactory epithelium by 2 years (Schreibman and Margolis-Nunno, 1989), which could have cascading effects on closely linked hypothalamic functions. The present data point to dysorganizational mechanisms in senescence, but so far experimental manipulations of diet or hormones in different genotypes have not been reported. In regard to the potential role of biorhythms in lifespan, I also mention a blind goby that lives more than 12 years (*Typhlogobius californiensus*), a crustacean's symbiont (MacGinitie, 1939; Miller, 1979).

In closing this discussion, I give examples of superannuations that are remarkable because they involve apparently opposite effects of diet. The first example supports notions that diet restriction extends lifespan, as discussed in depth in Chapter 10.3. Brook trout (*Salvelinus fontinalis*) rarely live beyond 6 years, although the role of senescence in this population maximum is unknown. However, brook trout introduced to Bunny Lake in the Sierra Nevada lived at least 24 years (Reimers, 1979). The Bunny Lake brook trout lived in a very cold environment with sparse food supply that allowed very slow growth, so that maturation was delayed to 15 years. Autopsies and histological study showed few signs of senescence: "they experienced a form of aging that proceeded almost imperceptibly" (Reimers, 1979, 212). Nonetheless, some older fish showed clouded lenses and impaired photoresponses. Upon return to warmer waters, the Bunny Lake trout showed increased gametogenesis, although normalcy was never regained. In this case, the cold and spare environment may have caused a transition from gradual to negligible senescence. Mortality data were not provided.

Relative longevity, however, is also achieved by brown trout (*Salmo trutta*), the sought-after great lake trout or "ferox trout" of isolated lakes in Ireland and Scotland that many records show can live at least 20 years and reach at least thirty-two pounds and ninety centimeters. In contrast, most others in the local population die by 4–6 years and do not grow longer than ten centimeters (Campbell, 1979; Ferguson and Mason, 1981; Hamilton et al., 1989; Figure 3.7). To become a ferox trout, the trout's early growth apparently must be slow, since ferox are not found in rapidly growing populations. After maturation, a very few trout show accelerated growth; their stomachs often contain small fish that are rarely preyed on by the other brown trout. The accelerated growth is attributed to this richer diet, which only a few learn to catch. In Scotland, ferox only occur in lochs that also have char populations (Campbell, 1979). Similarly, Behnke (1989) reported continued growth in brown trout populations in the Gunnison River, where crayfish and other large food organisms maintained growth and allowed trout to avoid competition with particle-feeding smaller fish. Hardy-Weinberg analysis of enzyme polymorphisms indicates reproductive isolation of the ferox in Lough Melvin and several other lakes (Ferguson and Mason, 1981; Hamilton et al., 1989): the *LDH–5 (105)* allele is very rare in other Lough Melvin *Salmo trutta* subspecies except for ferox that in this lake are recognizable before they achieve unusual size. The perpetuation of sympatric but noninterbreeding populations is consistent with the salmonid characteristic of breeding in their natal grounds. It cannot be concluded that the *LDH–5 (105)* allele alone accounts for the continued growth of ferox trout and potential longevity.

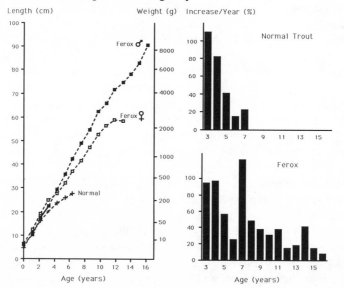

Fig. 3.7. Ferox trout, a biotype (alternate phenotype) of brown trout (*Salmo trutta*), has greatly increased potential lifespan. In certain lakes of Ireland and Scotland, these subpopulations of brown trout achieve huge size and longevity, apparently as the result of their learning to prey upon smaller fish, which gives a richer food source than the drift food that is eaten by younger and smaller fish in the same lakes. *Left*, length increases after 4–5 years, when brown trout usually cease growth. *Right*, the growth increments in normal and ferox trout, as judged by cross-sectional analysis. Redrawn from Campbell, 1979.

While there is no information on age-related mortality statistics or pathology of brown trout, whether in the conventional or exceptional growth patterns, there is little doubt that the factors permitting increased growth after maturation also extend the lifespan. The mechanisms are unknown and pose a tantalizing paradox in view of the extended lifespans of the starved Bunny Lake brook trout and of *Lebistes,* among many other species on diet restriction (Chapter 10.3). The conclusion that prematurational growth must be slow suggests neuroendocrine mechanisms, speculatively, through altered secretions of glucocorticoids and other metabolic hormones that may elevate blood glucose and promote damage to proteins through glycation. Involvement of corticosteroids in the landlocked ferox is suggested by their resemblance to "fresh run salmon" (Campbell, 1979) and their long heads (Ferguson and Mason, 1981). Similar characteristics are seen in prespawning salmon that have just arrived from the sea and should have elevated glucocorticoids (Figure 2.10).

In conclusion, these descriptions of senescence in fish suggest wide species variations in different evolutionary radiations (sturgeon, salmon, etc.) that are genetically as well as environmentally determined. In iteroparous, gradually senescing species like *Cynolebias* and *Xiphophorus,* numerous pathophysiologic changes are strikingly similar to those for mammals (Section 3.6). Unlike mammals, however, the brown trout give startling exceptions to the general belief that *ad libitum* food consumption shortens lifespan. Besides fish, many other very long lived species may have extremely gradual or negligible phenomena of senescence (Chapter 4). This major vertebrate group may contain accessible examples of different patterns of senescence and reveal different types of evolutionary mechanisms that limit lifespan.

3.4.2. *Amphibia and Reptiles*

The occurrence of senescence in amphibia and reptiles is not strongly indicated by the small literature on age-related physiological and pathological changes, or mortality rates by age group. Jacobson (1980) surveyed neoplasms in reptiles; these scattered case studies are not sufficient to indicate any age trends. Some functional data on the frog *Xenopus* suggest that senescence is negligible or very slow, as does the longevity under favorable conditions of some amphibia and reptiles that live for many decades, at least beyond a century for turtles (Chapter 4.2.2). There is no example of rapid senescence from causes associated with reproduction in amphibia or reptiles. Limited data suggest gradual senescence in a few species, but these examples give no basis for generalization.

The garden lizard *Calotes versicolor,* as dated by annual bone growth rings, lives about 4 years in the wild and continues to increase in size (Patnaik and Behera, 1981; Kara and Patnaik, 1985). The potential lifespan of *Calotes,* however, may be much longer, since other Agamidae live at least 10 years in captivity (Slavens, 1988, 116).[2] Older *Calotes* collected from the wild show modest physiological and biochemical decrements, for example, liver oxygen uptake (Kara and Patnaik, 1981), responses to cold stress (Kara and Patnaik, 1985), and slight reductions in bulk brain-RNA concentration (Padhi and Patnaik, 1976). Because no data are available on any species reared throughout its lifespan

2. An annually updated record of the breeding performance and lifespans of captive amphibia and reptiles is published by Frank Slavens, P.O. Box 30744, Seattle, WA 98103.

under controlled laboratory conditions, the differences that are attributed to effects of age might equally well reflect adverse environmental effects or disease rather than senescence.

The extent of age change in reproduction is virtually undocumented in lizards as well as other reptiles. *Calotes,* a seasonal breeder, retains into adulthood a germinal epithelium with oogonia from which new cohorts of growing oocytes and follicles are formed (Varma, 1970). Although Varma's report gives no information about whether or not the oogonial cells continue to proliferate and form new oocytes in adults, oogonial mitosis is described in adult turtles (Altland, 1951; Chapter 4.2.2) and might be expected in adult lizards. While both lizards and turtles are egg-laying tetrapods, they represent radiations that separated toward the end of the Paleozoic era. There is no information on age changes in the number of oogonia or oocytes in lizards.

Alligators, for example *Alligator mississippiensis* with lifespans of at least 50–60 years (Slavens, 1988, 114; Flower, 1937; Bowler, 1977), are also candidates for gradual senescence, as indicated by modest evidence for reproductive declines. Ferguson (1984) noted that after 30 years there were marked declines of egg production and sharp increases of malformed embryos. These trends followed the same trends observed in mammals, where fertility and birth defects vary reciprocally in females from adolescence through late middle age. Fertility was maximum at 15–20 years in these seasonal breeders. Some females (3%) were judged senescent by the absence of ova during the short season of ovarian activity (Joanen and McNease, 1980; Ferguson, 1985), about the same frequency reported for crocodiles in East Africa (Cott, 1961; Graham, 1968). Nonetheless, *de novo* oogenesis through oogonial proliferation may continue into adult life (Ferguson, 1984), as in most reptiles (Nieuwkoop and Sutasurya, 1979; Franchi et al., 1962; Tokarz, 1978). Examination of a few ovaries from alligators older than 30 years indicated continuing oogenesis with *in situ* resorption (Mark Ferguson, pers. comm.). Any conclusions must be regarded as provisional, particularly because a substantial fraction (40%) of females may not become reproductively active in any year (Lance et al., 1983). Moreover, stress can suppress gonadal functions (Lance and Elsey, 1986). Age-related pathology is also poorly documented. One 40-year-old zoo alligator had a large seminoma (Jacobson, 1980; Wadsworth and Hill, 1956).

3.5. Birds

Birds represent the range of gradual to negligible senescence, with lifespans from a few years to many decades. There is no example of semelparity or rapid senescence. More than twenty-five species have proven lifespans of 40 or more years (Comfort, 1979), while field studies continue to extend the maximum lifespans of numerous others (Clapp et al., 1982; Klimkiewicz and Futcher, 1987). Useful data on mortality rate are available from field studies of banded and recaptured birds, conducted by devotees throughout the world. The mortality rate of adults is often high throughout life, for example, either IMR or AMR are typically 0.1–0.5/year for fulmars, gulls, terns, and other seabirds in nature (Botkin and Miller, 1974; Goodman, 1974; Coulson and Horobin, 1976; Ollason and Dunnet, 1988; Lack, 1943a, 1943b, 1954, 1966). These values are in the range of bats but are higher than most long-lived mammals (Table 3.1).

Banding and recapture methods can give erroneous estimates of population age structure, due to unknown effects of migration (Buckland, 1982). Nonetheless, there is general agreement that, in sharp contrast to mammalian populations, age-related accelerations of mortality rates are difficult to demonstrate directly in some long-lived birds (Botkin and Miller, 1974; Ollason and Dunnett, 1988). For example, no age-related increases of mortality were detected up to age 20 in the arctic tern (*Sterna paradisaea;* Coulson and Horobin, 1976) or up to age 30 in the fulmar petrel (*Fulmarus glacialis;* Ollason and Dunnet, 1988). The astonishing data on fulmars come from George Dunnet (pers. comm.), who has studied a colony on the Orkney Islands since 1950. These gull-sized birds are related to the albatrosses and shearwaters and have a low annual mortality (AMR = 0.03). Corresponding to these long lifespans, reproductive changes are extremely modest, as discussed below. Yet we can deduce that there must be senescent processes in some of these very long lived birds.

Botkin and Miller (1974) showed in a pioneering analysis that the annual AMRs of the population cannot give the observed maximum lifespan; some increase of mortality rate is required. Calculations by a new method based on the model of exponentially accelerating mortality (equation [1.2]; Finch et al., 1990; Chapter 1.4.5) indicate that the maximum reported lifespans are consistent with very slow accelerations of mortality. For most of the birds in the Botkin and Miller analysis, we estimate that MRDTs exceed 8 years, which is as found for humans (Table 3.1; Appendix 1). It is reassuring that independent calculations from mortality schedule data for the herring gull gave the same results (Table 3.1, n. 16). These findings support the presence of gradual or delayed senescence in certain birds, whatever their mean lifespans. In any case, the absence of evidence for senescence does not preclude its emergence at later ages.

Reproductive changes of some long-lived birds suggest very gradual or even negligible senescence within the ages so far observed. Adult birds are generally thought to have a finite stock of oocytes (Tokarz, 1978; Franchi et al., 1962). However, adults of few species besides the domestic galliform birds have been studied with modern techniques. Whatever the case may prove to be, birds represent a range of reproductive schedules that is very broad, perhaps as extensive as that in mammals.

The best-documented example of negligible decline in fertility is given for fulmars in Dunnet's ongoing study: even after 30 years fulmars "do not show any obvious decline of fertility in their later years" (Ollason and Dunnet, 1988, 268). During lifespans that last at least 30 years, these slowly reproducing and monogamous birds will produce about eleven fledglings. The longest-lived fulmars produce the most offspring. Breeding typically begins in females at 9–12 years, but may be delayed to 19 years. Reproductive success (egg to fledgling) increases steadily until about the tenth breeding year in fulmars (Figure 3.8). These ages are 15–20 years later than reproduction can continue in domestic quail and hens (see below). Another example without a definite postreproductive phase was a pair of crowned pigeons (*Goura netoria*) that continued to mate and produce eggs throughout a 40-year lifespan; however, no eggs hatched after the pair was 20 years old (Flower, 1938; Comfort, 1979, 86).

Condors have the longest documented lifespans among birds. The Andean condor (*Vultur gryphus*) lives at least 75 years; zoo records have not (yet) indicated lifespans as long for the California condor (*Gymnogyps californianus*), which is considered to have identical life history traits (Table 3.1). Mortality in field populations of condors appears to be among the lowest reported for birds, about 0.05/year (Temple and Wallace, 1989). These

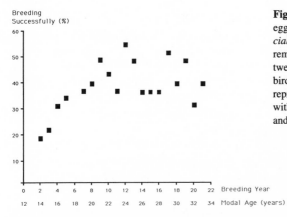

Fig. 3.8. Reproductive success (rearing from egg to fledgling) in female fulmars (*Fulmarus glacialis*) gradually improves with age and apparently remains stable thereafter through at least the twenty-second breeding year, when these banded birds were older than 30 years. The age at first reproduction in females ranges from 9 to 19 years, with a mode of 12 years. Redrawn from Ollason and Dunnet, 1988.

data are consistent with a mathematical analysis of the observed low fecundity (one egg per 2 years), from which it was shown that mortality rates must remain very low and any manifestations of reproductive or actuarial senescence must be postponed for 30 to 50 years after maturation at 5–10 years in order to maintain population size (Mertz, 1971). Senescence has not been described in condors.[3]

Some long-lived species show decreases in the size of eggs or the total clutch (egg) volume, for example, the herring gull (*Larus argentatus;* Davis, 1975) and ruff (*Philomachus pugnax;* Andersen, 1951). However, *Larus californicus* showed no decrease in clutch volume (Pugasek, 1983). It is remarkable that age-related sterility or reproductive senescence has not been documented in long-lived birds (George Dunnet and Bernt-Erik Saerther, pers. comm.; Saerther, review in prep.). Without more mortality rate data or histological evidence for abnormal cell growth and the other common age-related disorders, as described above for fish and which also occur in mammals, it is difficult to place most long-lived birds in the spectrum from gradual to negligible senescence. A major survey is needed of pathological lesions in wild and captive birds from different orders, for example, to establish the incidence of arteriosclerosis, cataracts, and tumors. Moreover, the widespread belief that all birds have fixed oocyte stocks that therefore must decline irreversibly as in mammals needs direct study, particularly in the very long lived birds.

Another remarkable age-related change in long-lived birds is that their success as parents increases for some years, for example, in California gulls (Figure 3.9). Table 3.2 shows the age-related improvement in parenting and long reproductive lifespans of other birds. This phenomenon provides one of the few examples where longevity could be directly selective. The late maturity, small production of offspring, and increasing age-related *increases* of effectiveness in foraging and in rearing young may constitute traits that select positively for long lifespans (extreme *K*-selection).

In contrast, other birds have short lifespans with accelerated senescence, particularly the galliforms. The best-studied example is the Japanese quail (*Coturnix coturnix japonica;* Figure 1.9; Cherkin and Eckhardt, 1977). Analysis of mortality data from domestic-bred populations gave an MRDT of 1 year (threefold longer than in laboratory rodents)

---

3. An Andean condor of unknown age was found with cataracts; this female was still flying but had difficulty in landing (Michael Wallace, pers. comm.).

and an extremely high IMR of 2/year (Table 3.1; Appendix 1). The maximum lifespan of 6 years for Japanese quail is less than most birds. As described below, *Coturnix* has age-related pathophysiologic changes that resemble some changes in laboratory rodents. Other galliform birds with maximum lifespans of less than 20 years and explicit accelerations of mortality include the Reeves pheasant (*Syrmaticus reevesi*), peafowl (*Pavo cristatus*), and brush turkey (*Alectura lathami*) (Table 3.1). Anecdotal reports on domestic fowl (*Gallus gallus domesticus*) suggest lifespans of at least 15 years (Flower, 1938). But there are no mortality schedule data on these most essential birds!

Besides the galliforms, other species in captivity also are short-lived with rapid accelerations of mortality, including the Bengalese finch (*Lonchura striata;* Eisner and Etoh, 1967). Reproduction declined after 4 years in captivity, but most birds continued to produce eggs when mated. Because the 10-year maximum lifespan in this captive population is far less than the 20–30 years recorded for finches (Flower, 1938), this species probably should not be classified as short-lived. Few other life tables are available for birds. Although lifespans of birds show some overall correlation with body size (Chapter 5.3.1.2; Western and Ssemakula, 1982; Calder, 1983a), the constraints of body size on maximum lifespan cannot be strong. Examples of small but long-lived birds include a caged chaffinch (*Fringilla coelobs*), typically twenty-five grams, that survived 29 years (Moltoni, 1947) and Leach's storm petrel (*Oceanodroma leucorhoa*), typically forty-five grams, that

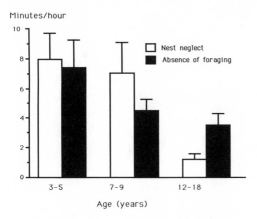

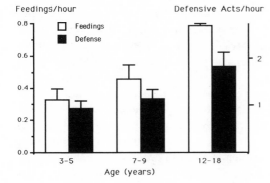

Fig. 3.9. Success as parents increases with age in gulls (*Larus californicus*). *Top, open bars,* duration of nest neglect when neither parent was present and, *shaded bars,* absence of foraging by either parent. *Bottom, open bars,* frequency with which parents fed offspring and, *shaded bars,* defended their territory. Redrawn from Pugasek, 1981.

Table 3.2 Age, Reproductive Success, and Longevity in Female Birds (in years)

Species	Reproductive Trend	Oldest Breeder	Mortality Trend	tmax
Fulmar (*Fulmarus glacialis*)[a]	gradual ↑ to 20; no change > 30	> 40	no change > 30	> 40
Gulls (*Larus californicus*)[b]	gradual ↑ > 12	> 18		> 27[k]
Arctic tern (*Sterna paradisaea*)[c]		> 29		> 34[k]
Kittiwake gull (*Rissa tridactyla*)[d]	gradual ↑ to 5; ↓ > 11	> 19	↑ > 10	> 19[k]
Bewick's swan (*Cygnus columbianus bewickii*)[e]	gradual ↑ to 10; variable > 10	> 15	no change 2–10	36
Sparrow hawk (*Accipiter nisus*)[f]	gradual ↑ to 7; possible ↓ > 8	> 10	↑ > 8	> 11
Andean condor (*Vultur gryphus*)[g]	very slow reproduction, ≤ 1 egg/yr;[i] young condors have an "apprenticeship" of about 7 years while they learn foraging skills from older birds;[j] extent of reproductive decline is unknown			70–80
California condor (*Gymnogyps californianus*)[h]	as for Andean condor			> 45

[a]Ollason and Dunnet, 1988
[b]Pugasek, 1981, 1983; Pugasek and Diem, 1983
[c]Coulson and Horobin, 1976
[d]Thomas and Coulson, 1988
[e]Scott, 1988
[f]Newton, 1988
[g]Milwaukee Zoo, > 70 years (Edward Diebold, pers. comm.); Moscow Zoo, > 75 years (Michael Wallace, pers. comm.)
[h]Mertz, 1975
[i]Temple and Wallace, 1989
[j]Michael Wallace, pers. comm.
[k]Clapp et al., 1982

survives 30 years or longer, according to careful ongoing banding studies (C. E. Huntington, pers. comm.).

Feral populations of some species also show indications of reproductive declines. For example, the sparrow hawk (*Accipiter nisus*) rarely lives more than 10 years and shows increased mortality; older females produce fewer and smaller eggs (Newton, 1988). Moreover, some senescent birds showed swollen, apparently arthritic joints. One presumes that even mild bone or joint impairments would greatly increase mortality risks in feral birds, and so might be difficult to observe generally. No definite postreproductive phase was noted in sparrow hawks. The Florida scrub jay (*Aphelocoma coerulescens coerulescens*) also has indications of gradual reproductive senescence (Fitzpatrick and Woolfenden, 1988). The monogamous habits that are characteristic of most birds could also contribute to decreased fertility: reproduction is noted to decline if a change of mate is necessitated by death of the partner, as compared with other birds of the same age; this may contribute to the average declines of reproduction after 4 years in the great tit (*Parus major;* McCleery and Perrins, 1988). It is striking that none of these show evidence for complete infertility at a characteristic age like menopause. The only examples approaching a mid-life sterility phase like menopause are found in domestic species.

Most laboratory studies of senescence have used two domesticated galliform birds, Japanese quail and hens. The quail *Coturnix* show numerous manifestations of senescence that are in accordance with the accelerations of mortality. Egg laying decreases progressively after 6 months, leading to a postreproductive phase of a year or more, which is equivalent to 30% of the lifespan (Woodward and Abplanalp, 1971). This decline is predicted from the cessation of primordial germ cell replication in females by hatching (Franchi et al., 1962; Tokarz, 1978), which implies a finite number of primary oocytes and their subsequent attrition as clearly shown only for mammals (Section 3.6). However, data on the ovarian oocyte stock as a function of age are not characterized in any adult bird. The oviduct atrophies with major loss (> 80%) of the oviduct-specific proteins ovalbumin and avidin (Bernd et al., 1983). Importantly, treatment with diethylstilbestrol (a potent synthetic estrogen) or progesterone did not restore the levels of avidin or ovalbumin; if confirmed on other specimens, this would be an important example of lost sex-organ target cell responsiveness (Chapter 7.3).

Males show progressive declines in blood testosterone and testis weight between 6 months and 4 years, and elevations of follicle-stimulating hormone (FSH; Balthazart et al., 1984; Ottinger et al., 1987). Elevations of FSH in mammals are linked to reduced gonadal activities in both genders and thus are consistent with gonadal involution. Testicular tumors (Sertoli cell) were found in about 10% of males aged 3–5 years in one colony (Gorham and Ottinger, 1986). Sex behavior in intact males and castrates that were given testosterone implants deteriorates, and an increasing fraction of males are behaviorally inactive (Ottinger et al., 1983; Balthazart et al., 1984; Ottinger and Balthazart, 1986). Several neuroendocrine and neurochemical changes are found in senescent males: smaller gonadotropin secretion in response to GnRH injections (Ottinger and Balthazart, 1986) and decreased testosterone 5β-reductase in the hypothalamus (Balthalzart et al., 1984). Conditioned learning also becomes impaired (Cherkin and Eckhardt, 1977; Woodward and Abplanalp, 1971; Meinecke, 1974).

Similar reproductive changes occur during senescence in domestic chickens. Spermatogenesis gradually declines with age in roosters (Payne, 1952), while hens show progressive decrease in egg laying during their first decade (Figure 3.10; Brody et al., 1923; Clark, 1940; Hall and Marble, 1931; Williams and Sharp, 1978; Bahr and Palmer, 1989) and commonly "go out of production" by 2–3 years, in association with marked atrophy of the oviduct compared with hens of the same age that are still laying (Boyd-Leinen et al., 1982). There is also decreased follicular growth (Williams and Sharp, 1978; Bahr and Palmer, 1989) and decreased follicular responsiveness to luteinizing hormone (LH; Johnson et al., 1986). However, a few hens may continue to produce eggs beyond 10 years, indicating a much broader spread in the ages of reproductive senescence than characteristic of most mammals. While plasma estrogens did not change notably in layers versus nonlayers of the same age, plasma progesterone and cytosolic progesterone receptors decreased markedly in the nonlayers (Boyd-Leinen et al., 1982). The oviductal atrophy and loss of progesterone receptors are puzzling, since both are strongly linked to plasma estrogens which did not change. Another phenomenon of senescence is that hens and other

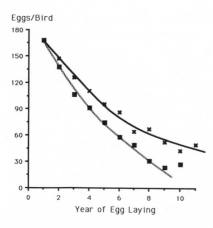

Eggs/Bird

Fig. 3.10. The age-related decline of egg production in the domestic chicken. Based on data of Clark, 1940 (*square*), and Hall and Marble, 1931 (*x*). Redrawn from Comfort, 1979.

Year of Egg Laying

birds occasionally develop masculine plumage and behavior (Forbes, 1947), which probably results from the activation of testicular cells in the normally undeveloped right gonad, as the ovarian follicles become exhausted, and the hypothalamic-pituitary axis responds as at menopause to increase the secretion of LH and FSH.

In the domestic duck (*Anas platyrhynchos,* Anseriformes), a brief report (Tanabe, 1985) indicates that egg production decreased by 4 years without change in plasma estradiol or progesterone. Thus, we have examples from two orders (Galliformes and Anseriformes) for decreased egg production without dramatic decreases in ovarian steroid production. Age changes in the molecular responses to hormones are discussed in Chapter 7.4.

In conclusion, gradual or rapid senescence is demonstrated best for galliform birds. In contrast, birds of several other orders have very long lifespans and maintain reproduction for several decades; their estimated MRDTs are at least as slow as observed in humans. The examples of pathophysiological senescence and accelerated mortality in galliform birds suggest that this radiation departed evolutionarily from more slowly senescing ancestors (Chapter 11.3.5.1). Even in these examples, the most rapidly senescing birds still live longer and have more gradual changes than observed in laboratory rodents.

3.6. Mammals

3.6.1. *General Features of Senescence*

This section sketches a portrait of senescence in mammals that is needed for later discussions of mechanisms (Chapters 7–10). In its brevity it represents only some of the general features and neglects important phenomena and differences between species and genders. Even so, one is impressed by some larger generalities in the pattern of senescence, which imply the evolutionary persistence of similar genetic determinants of senescence during the last 100 million years. The most secure generalizations about senescence in placental mammals are the exponential increase of mortality rate and the depletion of ovarian oocytes during mid-life as seen in laboratory rodents, domestic animals, and humans. No well-studied mammal has failed to show these changes, which can be considered a canonical pattern of senescence.

Concurrently with accelerations of mortality, there are myriad pathophysiological changes. With the exception of seasonal die-offs of feral rodents (Chapter 2.3.1.6), these changes are not synchronous in any population and vary widely between and within the same age groups in populations. Mammals, whether short- or long-lived species, can be considered to manifest gradual senescence in association with numerous pathological changes. This section also discusses the individual differences in senescent changes, which are observed even between individuals of highly inbred rodent strains. I also propose a set of age-related pathophysiological syndromes, in which changes in a limited number of cells can have cascading influences throughout the body and which are another type of dysorganizational mechanism during senescence. Ultimately, the characteristics of different cells that are established during differentiation will dictate their response to changes in regulatory factors in various pathophysiological cascades. An open question is whether the myriad changes of mammalian senescence can be linked as aspects of specific pathophysiological syndromes that may have a small number of cellular or molecular causes.

The patterns of changes in cellular and organ function have been studied in few of the three thousand extant species in the laboratory or in defined populations: about ten laboratory rodents, five domestic animals, a few primates, and humans. Data on many mammalian orders are only fragmentary and anecdotal. The range of maximum lifespans extends from the 100 years of humans and perhaps whales (see below) downwards to 2–5 years for most laboratory rodents. There are few rodents with *tmax* less than 2 years, entirely represented by the laboratory-bred genotypes that die from early onset diseases before 1–2 years (Chapter 6.5.1). The very short lifespans of less than a year that are sometimes attributed to shrews, for example, should be considered minimum estimates: the range of records for the shrew genera *Blarina, Neomys,* and *Sorex,* for example, indicates maximum lifespans of 2 years or more (Nowak and Paradiso, 1983). Improvements of husbandry could well extend these lifespans further into the range of laboratory rodents. Even in laboratory rodents, the maximum lifespan of cohorts may fluctuate by more than 7 months (Hollander et al., 1984).

Some marsupials are very short lived. Readers may recall from Chapter 2.3.1.5 that marsupial mice (*Antechinus*), despite maximum lifespans of some genders and species of 1 year in nature, can live several years if reproductive-related stress is controlled. Other very short lived mammals are the opossum (*Didelphis virginianus* and *D. marsupialis*) and other New World marsupials that generally live less than 5 years (Hunsaker, 1977). Austad (1988) observed cataracts in 2-year-old opossums. In captivity, the mean lifespan of males and females is about 21 months, and the maximum 28 months (Donna Holmes, pers. comm.). I conclude that mammals are like birds and others in tetrapod radiations that characteristically have *minimum* potential lifespans of 2–3 years.

Chapter 2 gave examples of species in the same genus with different life history patterns (e.g. lampreys, salmon, marsupial mice, etc.). Although this approach has not been widely used for mammals, the genus *Peromyscus* (white-footed mice) offers the interesting comparison of *P. maniculatus* and *P. leucopus* (Figure 3.11) which, on the basis of newly established colonies (Smith et al., 1990), appear to live twice as long as laboratory mice and have correspondingly longer MRDTs. Values of these mortality coefficients, like most others, are provisional since they are based on limited populations that have not had as many generations as rodents to adjust to laboratory life. *P. leucopus* maintains ovarian

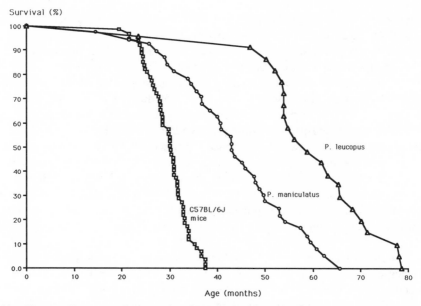

Fig. 3.11. The mortality curves of two species of *Peromyscus,* the white-footed mouse. *P. maniculatus* has a shorter lifespan than *P. leucopus* under laboratory conditions. Mortality coefficients are in Table 3.1. Redrawn from Smith et al., 1990.

follicular activity longer than do laboratory mice (Peluso, Montgomery, Steger, Meites, and Sacher, 1980; Peluso, England-Charlesworth, and Hutz, 1980); none have been available for comparison with *P. maniculatus*.

Certain bats can live at least 30 years (Jürgens and Prothero, 1988; Nowak and Paradiso, 1983; Austad and Fischer, in press), which is three- to fivefold longer than rodents of the same body size (Chapter 5.3) and which suggests a different and much slower senescence. The little brown bat (*Myotis lucifugis*) weighs eight to fourteen grams (Nowak and Paradiso, 1983), yet survives at least 32 years (H. B. Hitchcock, cited in Thompson, 1987). Based on limited evidence, the mortality rate accelerates slowly in female pipestrelles (*Pipistrellus sp.*) (Thompson, 1987). Depending on assumptions about *tmax* and population size (Appendix 2), the pipestrelle MRDT is in the range of 3–8 years in these tiny twenty-gram bats (Table 1.2; Table 3.1; Finch et al., 1990). This MRDT ranks with the horse and is exceeded only by humans and rhesus monkeys (*Macaca mulatta*). The absence of degenerative joint diseases by 19 years in a related vespertilionid bat (Sokoloff, 1969) is also consistent with very gradual senescence. Another small but much longer lived mammal is the naked mole rat (*Heterocephalus glaber,* family Bathyergidae), whose remarkable social system with nonbreeding worker castes has been compared with social insects. Naked mole rats weigh thirty to eighty grams and survive more than 10 years in captivity (Nowak and Paradiso, 1983; Jarvis, 1981; Gamlin, 1987). Thus, small size is not a strict constraint on the lifespan. More on this important point is found in Chapter 5.3.1.

Whales are at the other size extreme to bats. Analysis of annual rings in whale ear plugs indicates that some species may live beyond 100 years: blue whales, *Sibbaldus,* are the largest mammal at more than 10^8 grams; *Balaenoptera* are slightly smaller (Ohsumi,

1979). Data on the racemization of amino acids in dentine and other nonreplaced proteins also indicate very long lifespans (Bada et al., 1980). Life table data, while indicating sharp increases of mortality at later ages, for example, in short-finned pilot whales (*Globicephala macrorhynchus*) after 35 years in males and 50 in females (Kasuya and Marsh, 1984), do not show if mortality rates continue to accelerate according to the usual pattern.

A remarkable feature of mammalian life history is that many postnatal changes occur on a time scale that is in approximate proportion to the lifespan, whatever its length. Under favorable conditions, laboratory rodents and humans have low mortality between puberty and mid-life; by mid-life the age-related mortality rate has begun to increase exponentially (Figure 1.2; Chapter 1). Many of the same characteristic changes of senescence occur in mammals whose lifespans vary more than thirtyfold, yet they occur in the same fraction of the lifespan (Table 3.3). The concept of physiological scaling in relation to lifespan or mortality rate acceleration is traditional (Moulton, 1923; Pearl and Doering, 1923; Asdell, 1946; Brody, 1945; Simms, 1945; Jones, 1956; Fabens, 1965; Calder, 1984). However, many cellular properties do not scale with lifespan or basal metabolic rate, for example, cortical neurons have about the same rate of oxygen consumption in humans and rats (Hofman, 1983; Chapter 5.2.4).

For example, thymic involution and the approach to maximum skeletal length generally occur in the first third of the lifespan, even in rodents which have late epiphyseal plate closure. Contrary to mythology, bones of laboratory rats do not continue to grow throughout their lifespan (McDonald et al., 1986). The continued increase of shoulder height and total size in male elephants for at least 20 years beyond sexual maturity implies continued bone growth (Sikes, 1971), but this is unusual among mammals. Loss of female fertility (Section 3.6.4.1), decreased immune responses (Section 3.6.6), osteoporosis (Section 3.6.5), and decrease of dopamine (D-2) receptors in the striatum (caudate nucleus of the brain; Section 3.6.10) begin by the middle third of the lifespan. Another common age change is male-pattern baldness, which occurs after maturation in many human males and in some primates (Uno et al., 1967, 1969). This is a change of "aging," not senescence, since there is no associated mortality risk. The basis for balding is a change in the activity of hair follicles that represent the interaction of androgens with non-Y-linked genes (Chapter 10.4.1).

Senescence, the last third of the lifespan, is characterized by a generally exponential increase of diverse spontaneous pathophysiological disorders in many organ systems, whose incidence and intensity can differ widely among populations as well as among species. In humans, for example, the reported cause of mortality varies widely among countries, so that heart attacks from ischemia are cited as the cause of 43% of deaths in the age group 65–74 years in Sweden, but only 8% in Japan (MacFadyen, 1990; also see Guralnik et al., 1988). There is no systematic information about coexisting lesions throughout the body at death in any human population; the expense of this major project has not yet been justified. At a physiological level, homeostatic reserves are also often impaired after mid-life, as seen in the higher mortality of elderly mammals during fluctuations of environmental temperature (Shattuck and Hilferty, 1932; Ellis et al., 1976; Collins et al., 1977; Finch et al., 1969; Finch and Landfield, 1985). Other changes during senescence are the decrease of organ reserve or maximum potential for function by the heart, lungs, or kidneys (Shock, 1961, 1985) and the reduced capacity for surviving hemorrhage (Simms, 1942). The net result is the reduced survival of fluctuations, so-called

Table 3.3 Age-related Changes of Mammals Scaled on a Unit Lifespan: A Sampling

	Mid-life	Senescence
Reproduction	menopause or total infertility in females, in association with imminent depletion of ovarian oocytes (C,[3] D,[1] Hm,[2] H,[3] P,[3] M,[3] Op,[3] R,[3] Wh[3])	decreased male fertility (H, P, M, R)[4]
Hair	male-pattern baldness (H, P)[5]	
Tumors	many endocrine-related tumors of reproductive organs (H,[6,7] P,[8] M,[6,7] R,[6,7] D[1])	
Growth hormone		decreased frequency of pulses; smaller amplitude (H, P, R)[9]
Fat	increased fat depots (H, P, R, M)[10]	loss of weight varies widely between individuals (H, M, R)[10]
Plasma DHEA	decline begins (H, P)[9]	very low levels (H, P)[9]
Thermoregulation		impaired responses to cold (H, M, R)[11] impaired febrile response (H, P)[11]
Coronary and cerebral arteries	early atherosclerotic lesions (H, M, R)[1,12]	widespread atherosclerosis (H, P, D, R, Rb, M)[1,12]
Immune functions	slow decline in humoral antibody production (H, P, R, M)[13]	
Bone	onset of osteoporosis in female (H, P, R, Rb, M)[14]	female osteoporosis > male (H, P, R, Rb, M)[14]
Joints		arthritic changes very common (C, D, H, M, R)[15]
Reaction times		generally slowed (H,[16] R[17])
Vision		universally decreased accommodation of eye (presbyopia) (H, P)[18]
Hearing		sporadic loss (H[19])
Striatal dopaminergic	onset of D-2 receptor decline (H, M, P, R, Rb)[20]	Dopamine conc. decline (H, M, R, Rb)[20]
Large neurons		heterogeneous changes; some show increased size and dendritic complexity; others show atrophy and dendritic withering (H, R, M)[21]
cerebrovascular β-amyloid protein		widespread in absence of Alzheimer's disease (H, P, D)[22]

Notes: Onset of senescence in population (usual age, in years):

C, cattle = 15–25; Hm, golden hamster = 2–3; P, (rhesus) monkey = 20–30;
D, dog = 10–15; M, mouse = 2–3; R, rat = 2–3;
H, human = 60–80; Op, opossum = > 2; Rb, rabbit = 4–6;
    Wh, pilot whale = > 40??

Table 3.3.   (Continued)

[1]Anderson, 1970. Continued proliferation in the ovary cortex of adults raises the possibility that *de novo* oogenesis continues throughout life (Anderson 1970, 323).
[2]Thorneycroft and Soderwall, 1969: Blaha, 1964.
[3]Section 3.6.4.1.
[4]Section 3.6.4.2.
[5]Section 3.6.1.
[6]Section 3.6.3.
[7]Table 6.6.
[8]Alsum et al., 1990.
[9]Section 3.6.9; Figure 3.36.
[10]Section 3.6.8.
[11]Chapter 10.3.2.8.4; Finch and Landfield, 1985.
[12]Section 3.6.7.
[13]Section 3.6.6.

[14]Section 3.6.5.1.
[15]Section 3.6.5.2.
[16]Welford, 1977, 1980; Welford et al., 1969; Birren, 1964.
[17]Wallace et al., 1980; Campbell et al., 1980
[18]Section 3.6.1; Figure 3.14.
[19]Section 3.6.11.2; Figure 3.42.
[20]Morgan and May, 1990; Morgan et al., 1987; Wong et al., 1984.
[21]Section 3.6.10.
[22]Selkoe et al., 1987; Ogomori et al., 1988; Yamada et al., 1987; L. J. Martin et al., 1989.

"homeostenosis," or the shrinkage of the homeostatic "envelope" (Fries and Crapo, 1981; F. E. Yates, unpublished lecture, 1985).

These statements are provisional, because the data are dominated by the very few species at the extremes of lifespan. Many issues must be resolved for species with intermediate lifespans, without which generalizations cannot be made. For example, rhesus monkeys showed the same values for MRDTs as humans, which would imply an equally slow rate of senescence, yet the maximum lifespan is much shorter (< 40 years). Moreover, rhesus monkeys aged 20–30 years show many characteristics of senescence found in humans, including degenerative changes in some of the same brain regions (Table 3.3). The shorter lifespans of rhesus monkeys seems mainly attributable to the very high IMRs in these well-cared-for monkeys, about five- to tenfold higher than in nearly all human populations (Finch et al., in prep.; Table 3.1). These observations could mean that many age-related changes are not related to mortality rate accelerations, and their interpretations are far from clear.

The time-scaling of life history events in general proportion to the duration of the adult life phase in diverse short- and long-lived mammals poses perplexing questions to evolutionary theories of senescence (Chapter 1.5.2). For example, if selection weakens with advancing age, then it is difficult to rationalize the apparent similarity in the larger patterns of senescence that are manifested by radiations that diverged more than 70 million years ago (Chapter 11.3.5). The numerous genotype-related diseases of senescence within a species are, however, easier to understand. Although we are still far from understanding the genetic bases for lifespan differences among mammals, some genetic polymorphisms identified in mice and humans are determinants of different age-related diseases (Chapter 6.5). The limits to longevity in mammals, however, can only partly be attributed to specific diseases, as discussed below.

Despite the general manifestations, the changes of senescence do not occur simultaneously in any population, nor do they progress regularly. Early cross-sectional comparisons of healthy young adults with largely infirm older subjects gave the impression that most functions declined linearly with age (Shock, 1961). Untrue! While average declines

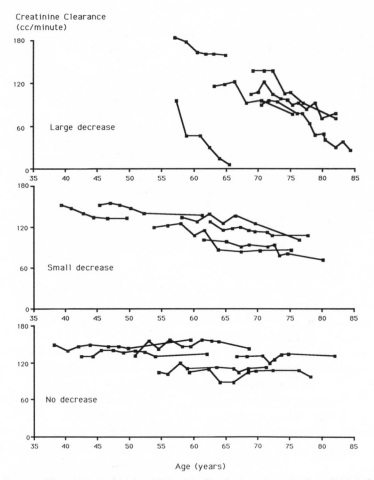

Fig. 3.12. Humans differ widely in age-related kidney dysfunctions, as shown in plots of kidney function from the Baltimore Longitudinal Study. Longitudinal testing of creatinine clearance is an assay of glomerular filtration rate. Three patterns of age-related change that represent subgroups are shown. *Top,* major age-related deterioration; *middle,* modest deterioration; *bottom,* no change. Redrawn from Lindeman et al., 1985.

are often seen as a population average, it is now widely recognized that individuals differ widely in the extent and timing of age-related changes (Shock, 1985). Subsequent longitudinal studies of adults selected for good health clearly showed remarkable maintenance of performance by some individuals at advanced ages. For example, kidney functions (glomerular filtration rate, measured by creatinine clearance), which show linear age-related declines in populations (Shock, 1961), did not change at all in longitudinal studies of healthy men during 20 years of observations that extended beyond the mean lifespan (Figure 3.12; Lindeman et al., 1985). It seems likely that disease-related factors can explain most age-related declines in kidney function. These findings indicate that age-related kidney dysfunctions are not inevitable like menopause, though the risk is high.

Kidney dysfunctions illustrate another recurrent phenomenon of senescence: the cascading and multiplicative consequences of some disorders. Some extent of impaired glomerular function represents a threshold, after which glomerular kidney disease progresses inexorably in humans and laboratory rodents, leading to sclerotic changes in the remaining glomeruli and degeneration of the glomerular basement membrane (Brenner et al., 1982). Kidney disease may be exacerbated by particular protein-rich diets and may be controlled by diet restriction (Chapter 10.3.2.2). Other consequences of age-related kidney disease are disturbances in mineral metabolism (Section 3.6.5).

The circulatory system also demonstrates maintained function despite common age-related increases of diseases. The invariance of hematologic values (e.g. packed erythrocyte volume, hemoglobin concentration) in a healthy group up through the tenth decade indicates that the replacement of circulating erythrocytes is not necessarily impaired (Zauber and Zauber, 1987). Because anemias of many origins are common in the elderly, however, cross-sectional studies could give a misleading impression. The maximal aerobic capacity declines progressively with age in the more sedentary, but can be substantially increased even at advanced ages by exercise (Heath et al., 1981; Hodgson and Buskirk, 1977). Moreover, no age-related changes were found in the maximal aerobic capacity of highly trained endurance athletes, as measured by cardiac stroke volume and arteriovenous difference in oxygen pressure (Hagberg et al., 1985; Pollack et al., 1987).

Although the peak heart rate declined progressively with age even in these highly motivated individuals (Figure 3.13; Pollack et al., 1987), nonetheless training substantially improved the cardiac stroke volume (Rodeheffer et al., 1984). The increased stroke volume, however, requires a greater pulmonary venous pressure, which in turn is predicted to increase the risk of pulmonary congestion; this cascade may contribute to the shortness of breath that is commonly observed during exercise in the elderly (Lakatta, 1990). Even if the heart has suffered major damage from ischemic attacks, remarkable improvements in performance are possible at advanced ages (Section 3.6.5.3).

Another example is the average decline of circulating testosterone levels in older age

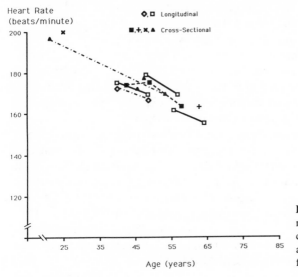

Fig. 3.13. Age-related decreases in maximum heart rates of endurance athletes, a compilation of data from two longitudinal and four cross-sectional studies. Redrawn from Pollack et al., 1987.

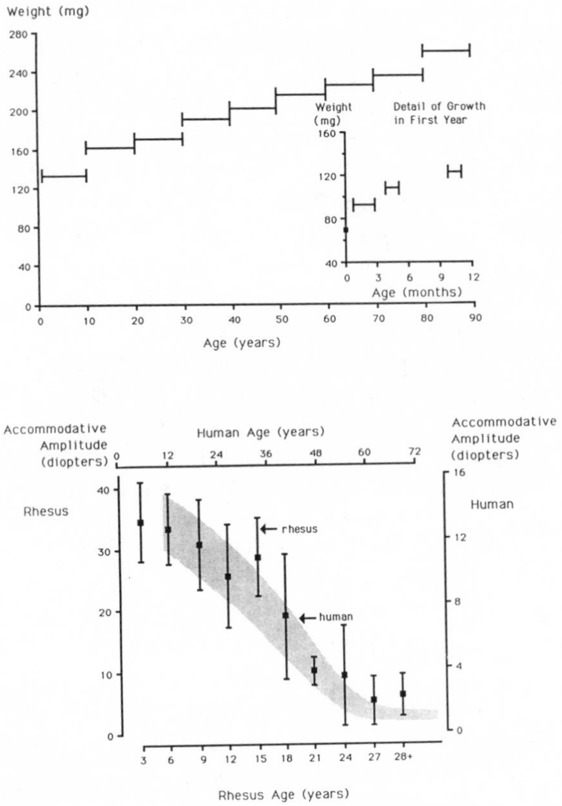

Fig. 3.14. Age-related changes that lead to presbyopia (farsightedness). *Top,* the weight of the lens in humans. Redrawn from Fozard et al., 1985, 503 (based on data of Salmony and Weale, 1961, and Scammon and Hesdorffer, 1937). *Bottom,* the elastic properties of the eye lens of rhesus monkeys (*Macaca mulatta*) reduce the ability to change focal length, as measured in diopters of accommodative amplitude; ranges shown are $\pm$ SD. The normal range for humans is indicated by the *shaded zone,* with a scaling in approximate proportion to the lifespan. Monkeys were resident in the Wisconsin Regional Primate Center. Redrawn from Bito et al., 1982.

groups, excluding a subgroup of men with normal levels at advanced ages (Section 3.6.4.2). It seems fair to conclude that the trend for functional declines in many different systems is a usual feature after mid-life that is revealed by either cross-sectional or longitudinal sampling. Nonetheless most individuals at advanced ages still maintain the potential for performance of many organ functions within the range of young adults.

Certainly genotype has a major role in the early onset degenerative diseases (Chapter 6.5.3.3), but the penetrance of adverse genes in later life may be far more haphazard. If so, considerable success may be achieved in further reducing or postponing age-related degeneration. The limits are not defined, excepting female fertility. The exceptional health of some individuals of advanced ages gives a general basis for optimism.

Many common age-related changes can be attributed to specific diseases, rather than to "aging." The age-related decline of cardiac performance found in most population samples is mostly consequent to coronary artery disease. The age-related increase in blood pressure that is so common in human populations (Section 3.6.7) has many adverse consequences, including loss of kidney functions (Lindeman et al., 1984). A different example is the age-related increase of male impotence (Kinsey et al., 1948; Hegler, 1981) and its association with age-related diseases. For example, 30–60% of adult male diabetics are impotent, while alcoholism and the drugs used to treat hypertension often cause impotence as a side effect (Comfort, 1980; Lipson, 1984a, 1984b).

In addressing these complex age-related changes, it is useful to distinguish changes that are consequent to disease of older age groups as pathogeric, while those changes general to all populations and not caused by disease are eugeric (Finch, 1972a). With the apparently improving health of the elderly in many populations, many usual age-related changes are found to be much smaller or not present at all. Menopause at mid-life and the presbyopia (farsightedness) from decreased elasticity of the lens in both genders are among the few age-related changes that are universal in humans. Both menopause (Bowden and Williams, 1985) and presbyopia (Figure 3.14) have been documented in primates (Table 3.3).

3.6.2. Age-related Diseases and Mortality Risk

The relationship of age-related gross pathologic changes to the exponential increase of mortality has puzzling features. The cause of death at advanced ages is often difficult to determine because multiple pathologic changes become very common in humans (Howell, 1963; Zeman, 1962) and laboratory rodents (Finch and Foster, 1973; Simms and Berg, 1957; Hollander et al., 1984; Zurcher, van Zwieten, Solleveld, van Bezooijen, and Hollander, 1982). Simms' and Berg's classic study of age-related disease in a carefully maintained rat colony showed a close correlation between the risk of new lesions (determined by sampling of living rats) and the mortality rate (Figure 1.10). The onset of five types of lesions increased exponentially after ages that were characteristic for each type of lesion and, moreover, anticipated the increases of mortality rate by several or more months. There was a strong correlation in the colony between the probability of death and the onset of new lesions (Figure 3.15). Although such data suggest links between age-related specific diseases and mortality in populations, the picture is less clear for individuals. Thus, the increasing incidence of lesions as reported does not mean that all organs or tissues degenerate during senescence (Chapter 2, n. 1). The extent of causality between specific age-related diseases and age-related increases of mortality is far from clear even under laboratory conditions.

By convention, gross lesions like cancer, aneurysms, pneumonia, infarcts, diabetes, and hypertension are designated as cause of death. However, Kohn (1982, 2795) suggested "lesions that would not cause death in younger persons are commonly accepted as a cause in the aged . . . . the final disease that goes on the death certificates is arbitrary." This

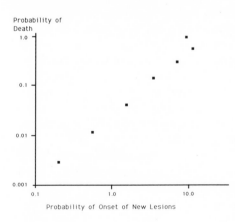

Fig. 3.15. Disease and mortality rates in Sprague-Dawley male rats, showing the correlation between the risk of new degenerative diseases (kidney, lung, and muscular degeneration; abnormal growths) and risk of mortality. Also see Figure 1.10. Redrawn from Simms and Berg, 1957.

often-cited view, however, may not be supported by thorough postmortem analysis. As noted earlier, the type and severity of lesions of the very elderly are poorly documented, because thorough autopsies are infrequently performed, for example, on fewer than 20% of those older than 70 years in contrast to more than 60% at 30 years (*Vital Statistics of the United States,* 1976, cited by Kohn, 1982). Similarly, Fries and Crapo (1981, 99–100) noted that many older humans die of no specific major cause while being treated for medical crises in modern intensive care facilities, as shown by this example.

> An 88 year old woman is admitted to the hospital from a nursing home with an infection of the bladder. When given intravenous fluids she develops heart failure, and the diuretics given to remove the fluid of the heart failure result in the deterioration of the kidney function. She is transferred to the intensive care unit where oxygen is given. . . . She begins to bleed slightly from a stress ulcer in the stomach, and during passage of a stomach tube to remove the blood she vomits and a small amount of vomitus enters the lung, where an aspiration pneumonia develops. Despite oxygen treatment she becomes comatose, and a tracheostomy is performed, and a respirator used to assist her breathing. Several other adverse events occur, the family is eventually told that there is no hope, and she dies in her seventh week in the hospital. This scenario is all too familiar to every physician, where a trivial initial challenge in an individual without organ reserve leads to a sequence of catastrophes despite extraordinary care, and despite extraordinary expenditure of resources. The medical care results in survival of one catastrophe, or even several, but there is always another catastrophe awaiting its turn.

Thus, it is often difficult to assign a *single* major cause of death to an older person where *many* aspects of homeostasis are impaired. Although relatively few individuals past midlife altogether escape some lesion like vascular disease or nonideal regulation of glucose or blood pressure, it does not necessarily follow that the presence of focal age-related degenerative diseases account for *all* of the increased mortality risk, although multiple lesions may interact in subtle ways to lower homeostatic responses.

Two key features of cardiac lesions at advanced ages were revealed by a study of hearts 90–105 years old from patients autopsied at the Mayo Clinic (Lie and Hammond, 1988): lesions are multiple, but some individuals escape particular lesions. In one of the few large-scale studies of pathology in the oldest human age group, death was mainly attributed to cardiovascular disease and stroke. The 237 hearts showed multiple lesions including calcification of plaques in the coronary arteries (93%), mitral annulus (53%), and aortic valve (38%), and also accumulation of amyloid (65%). Calcium deposits that are considered unusual in younger age groups, however, may be characteristic of the oldest old (Waller, 1988; Lie and Hammond, 1988).

Remarkably, most hearts (79%) lacked "critical" occlusions of the coronary arteries. This observation is extremely important, because it suggests that the very general trend for atherosclerosis is, after all, not universal. Lie and Hammond (1988, 563) concluded, "The secret to attaining longevity seems to be an unexplained protection from . . . the development of both significant coronary artery disease and cardiac dysfunction despite a multitude of ageing-related structural changes." Two protective mechanisms are discussed in Chapter 6.5.3.1 (hereditary lipoproteinemias that reduce atherosclerosis and developmental determinants for larger-diameter coronary arteries). While no heart from this age group was untouched by some lesion, the individual differences in type and distribution of lesions strongly suggests that the proverbial lifetime number of heartbeats (3 billion) need not be associated with any specific pathological outcome during the heart's prodigious work.

Further indication of the obscure linkage between age-related mortality and specific diseases of the elderly comes from the comparisons of different populations. Despite major differences in average longevity and in childhood mortality, the MRDT is remarkably similar (± 20%) for different geographic populations or cohorts (Jones, 1956; Figure 1.2). The invariance of the MRDT in different human populations becomes even more remarkable when population differences in disease prevalence are considered, for example, in cancers of the breast and other tissues (Section 3.6.3). Populations also differ widely in the incidence of heart disease and bone fractures to an extent (Section 3.6.5) that is not easily attributed to differences in diagnosis or record keeping. Although systolic blood pressure rises progressively in men and women of most industrial societies (Figure 3.16), a few groups appear to escape this trend, particularly certain traditional people in rural populations and possibly in association with low sodium intake (Maddocks, 1961a; Maddocks, 1961b; Page et al., 1974; Mann et al., 1964; Eaton et al., 1988). The similar age-related mortality rates throughout all human populations despite their wide differences in the incidence of specific diseases suggests the unexpected possibility that certain processes related to mortality risk are not closely linked to specific age-related diseases.

Analysis of inbred mouse strains leads to similar conclusions. Each mouse strain has its characteristic spontaneous age-related diseases. Inbred strains may or may not have malignant lymphoreticular disease, hepatomas, mammotrophic pituitary tumors, or soft-tissue calcification (Chapter 6.5.1). Yet despite the many differing diseases that might be assigned as causes of death, their acceleration of mortality rates (MRDTs) is very similar

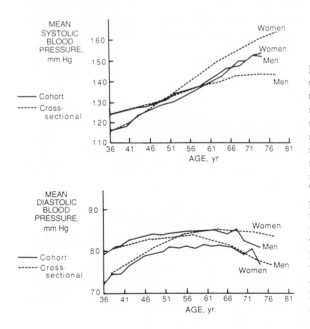

Fig. 3.16. Age-related trend in (*top*) systolic and (*bottom*) diastolic blood pressure in the Framingham Study. Different impressions of age-related changes in blood pressure are given by longitudinal studies of the same cohort compared to cross-sectional samplings of that and other cohorts in the same community, Framingham, Massachusetts (Framingham Study, cohorts 3–10). The progressive elevation of systolic blood pressure is attributed to loss of arterial elasticity and occurs widely, though some individuals escape this trend (Kannel, 1985a). This graph shows a major difference between longitudinal and cross-sectional observations beyond the average lifespan. Diastolic pressure, however, tends to decline by 65 years. Redrawn from Kannel et al., 1961, and Kannel, 1985a.

(Table 3.1). There is another puzzle: according to several colleagues, careful necropsy of diet-restricted rodents sometimes does not reveal gross pathology.

Taken together, the similarity of increases in age-related mortality rate among human or mouse populations with very different patterns of disease suggests that other factors may be more influential than specific diseases in age-related mortality. I propose that decreased homeostatic powers are a major source of age-related mortality and may decrease independently of specific diseases. This would account for the similar mortality accelerations in genotypes and populations with different disease distributions. The presence of a major disease may also decrease the homeostatic reserve and physiological resiliency, however. Age-related decreases of homeostasis are poorly understood and may arise from many subclinical causes that probably vary among individuals, including arteriosclerosis, impaired glucose regulation, and reduced kidney reserve.

The early onset of homeostatic decline is indicated by the sharp increase of mortality from standardized bleeding during maturation in rats (50 versus 100 days; Simms, 1942); subsequent increases of mortality from bleeding were more gradual and in parallel with spontaneous mortality. If we do not take homeostatic changes into account, we must then attribute quantitatively similar contributions to mortality risk for diverse lesions. Homeostatic impairments would decrease resistance to many types of stresses, from specific diseases as well as physiological and environmental fluctuations. The boundaries between impairments in homeostasis and certain of the usual age-related diseases become vague (Dilman, 1984). Even without major diseases, relatively small impairments of organ functions could synergize during stress to exceed the homeostatic limit, and might result in the apocryphal "natural death" without specific disease. Kohn (1971, 117) proposed that "there may be a small number of basic aging processes in tissues which cause or predispose to (all) of these diseases. Aging processes would then constitute a 100% total disease that everyone has. The age-related diseases could then be regarded as complications." The hypothetical age-related processes might then increase individual susceptibility to a wide range of diseases, according to genotype. The roots of age-related mortality may be traced to earlier ages, when homeostatic capacity has begun to change, but before the onset of age-related diseases.

3.6.3. *Abnormal Cell Growths and Metaplastic Syndromes*

The increased incidence of abnormal cell growths is strikingly general by mid-life and later ages in most populations of humans (Figures 3.17 and 3.18) and of rodents (Figure 1.10). With few exceptions, cancers of humans increase exponentially with age during most of adult life (Doll, 1970; Peto et al., 1975; Dix et al., 1980; Dix, 1989). The age distribution of abnormal growths is not known in most other mammals but is expected to show these same phenomena. There is an extensive literature on abnormal growths in domestic pets and zoo animals (McClure, 1980). These reports are difficult to generalize, however, since the local conditions are unknown and the numbers are generally small. The cell types commonly involved in age-related abnormal growths vary widely among species and are strongly influenced by genotypes within a species, as shown in laboratory mice (Ponten, 1977; Chapter 6.5.1). The widespread age-related increase of abnormal cell growths, both benign and malignant, comprise the metaplastic syndromes of mam-

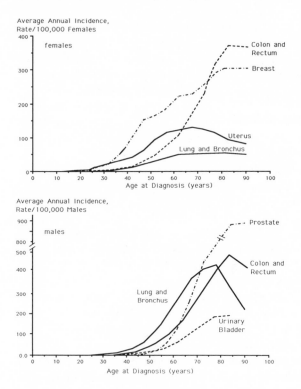

Fig. 3.17. Age-related incidence of cancer in different organs of humans in the Third National Cancer Survey. Nine geographic areas in the U.S. represent 21 million people. Redrawn from Cutler et al., 1974.

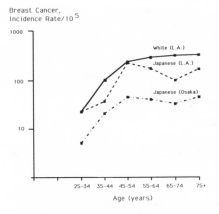

Fig. 3.18. Breast cancer in women as a function of age in three contemporary populations: white, Los Angeles; Japanese, Los Angeles; Japanese, Osaka. Each population has different incidence, but all show slowed rates of increase after menopause. Redrawn from Henderson et al., 1984.

malian senescence. In contrast, select cell subpopulations in the immune system, for example, show decreased proliferation (Section 3.6.6).

The hyperplasias and dysplasias of different cell populations are being considered collectively as syndromes of abnormal proliferative homeostasis, and could reflect various responses of stem cell differentiation mechanisms to common age-related changes in growth factors, which so far have not been described. It is important that not all aspects of cell proliferation show age-related impairments, for example, the capacity for red blood cells (erythropoiesis; Chapter 7.6.1).

The causes of abnormal growths are diverse and include clonal growth of cells that became transformed through one or more changes in nucleotide sequence or through viruses, as well as less drastic changes in proliferative homeostasis that arose through alterations in endocrine or immunological regulation (Section 3.6.6). The smooth muscle cells found in coronary artery plaques of humans may represent benignly growing tumor clones (Benditt and Benditt, 1973; Pearson et al., 1980; Chapter 8.3). Correlations between arteriosclerotic heart disease and cancer in diverse populations (Benditt, 1977; Jones, 1956) imply common factors that might act ultimately as inappropriate mitogens. In many populations, there are specific environmental risk factors that are time-dependent and that interact with manifestations of senescence. The cumulative exposure over time to cigarette smoke, radiation, or exogenous hormones (oral contraceptives, postmenopausal estrogen replacements) may each increase the age-group-specific risk factors for abnormal growths in different cell populations.

A major debate concerns the specific role of biological age changes in the progressive age-related increases of cancer. Arguments have been made that the age-related effect represents exposure to environmental carcinogens (Doll, 1970; Peto et al., 1975). However, a stable correlation in the progression with age for cancers of the stomach and bronchus is not consistent with a simple environmental model, because environmental causes of these cancers are thought to have changed in opposite directions (Dix, 1989).

Both genotypic and environmental factors can influence the age-related risk of abnormal growths. Laboratory mice vary extensively between genders and among strains, in the incidence of abnormal growths (Chapter 6.5.1). Diet is also a major influence (Chapter 10.3). Human populations also differ widely in the incidence and organ location of abnormal growths (Henderson et al., 1984; Silverberg and Lubera, 1986). For example, the incidence of breast cancer in women after mid-life is tenfold higher in North America than in Japan (Figure 3.18). The source of these differences is thought to be environmental. A likely factor is diet, particularly the fat content, as deduced from higher incidence of breast cancer observed in Japanese women who reside in the U.S. (Henderson et al., 1984; Pike et al., 1983). Because virtually all tumors with age-related incidence in adults show some demographic differences (see above), chronological age should be considered as a risk factor, not an absolute determinant, of this aspect of senescence. Genetic risk factors in human malignant disease are also described (Chapter 6.5.3).

Benign prostatic hyperplasia is considered to be almost universal in men (Harman and Talbert, 1985; Siiteri and Wilson, 1970), but not all of the world's populations have been surveyed. Benign prostatic hyperplasia (BPH) also occurs later in the lifespan of domestic dogs, in which it is very common (Zuckerman and Groome, 1937; Huggins and Clark, 1940; Hart, 1970). However, the incidence after mid-life in other species is unclear. Anecdotal reports indicate BPH after mid-life in a squirrel monkey (*Saimiri sciureus;* Adams and Bond, 1979), but there are no indications of extensive spontaneous incidence at later ages in other primates, including rhesus monkeys and baboons. Chronic androgen treatment did induce BPH in baboons (Karr et al., 1984), suggesting that species differences in BPH could arise through tissue sensitivity to androgens. A thorough search for BPH in the tissue banks kept by zoos and primate colonies could indicate if its phylogenetic distribution is as rare as it would seem in primates. In other mammals BPH may also be rare. Huggins (1943) reported BPH in a lion, whereas laboratory rodents do not develop anything like it. The distended seminal vesicles found in 2-year-old C57BL/6J and other mouse strains (Finch and Girgis, 1974; Cohen et al., 1987; Wolf et al., 1988) are not associated with glandular hyperplasia and may represent ductal blockade. Gloyna and Wilson (1969) speculated that age changes in the conversion of testosterone to dihydrotestosterone may be a factor in these species differences. The species distribution of prostatic hyperplasia implies genetic factors at a higher taxonomic level (Chapter 11).

3.6.4. Reproductive Changes

3.6.4.1. Ovariprival Syndromes in Females

Female mammals without exception show a complete or at least major loss of fertility during mid-life, so that under favorable conditions about 35% of the remaining lifespan is postreproductive (Harman and Talbert, 1985; Finch and Gosden, 1986; vom Saal and Finch, 1988). These changes, as documented in laboratory rodents, domestic animals, and several primates (Table 3.3), result mainly from a single cause, the irreversible age-related loss of ovarian oocytes, the number of which was fixed during development.

With possible rare exceptions in a few prosimians (e.g. *Galagoides demidoffi;* Butler and Juma, 1970), the formation of new oocytes through oogonial mitotic division ceases within a few weeks of birth in placental mammals (Franchi et al., 1962; Tokarz, 1978). By mid-life when fertility has greatly decreased, few oocytes remain in most species, for example, mice and humans (Figure 3.19). The depletion of growing follicles results in the reduction of ovarian steroids to castrate levels, and as a consequence many pathophysiological changes may occur, the ovariprival syndromes of mammalian senescence. This loss of ovarian oocytes is one of the few, perhaps the only, examples of major postmaturational loss of cells that occurs in all populations of all species. No such general loss of cells occurs in the brain (Section 3.6.10).

The survival of feral mammals after the onset of age-related sterility is documented in scattered species. An unusual source led to important insights about the short-finned pilot whale (*Globicephala macrorhynchus*). A sample of 373 females caught by commercial

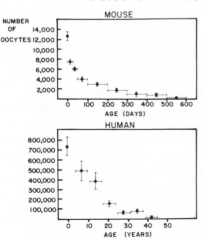

Fig. 3.19. The loss of ovarian oocytes as a function of age in laboratory mice and humans. Redrawn from Jones and Krohn, 1961b, Block, 1952, and Finch et al., 1984.

whalers was analyzed for ovarian histology and annual growth layers in the dentine and cementum of their teeth (Kasuya and Marsh, 1984). The lifespan in this sample exceeded 60 years; mortality rates are discussed in Section 3.6.1. Menopause occurs between 29 and 40 years, based on the absence of a corpus luteum to indicate recent ovulation and the absence of nonatretic follicles. The unexpected conclusion is that *postreproductive females represented 25% of all adult females*! The mean postreproductive phase (menopause to death) was estimated at 14 years. Just as important to population genetics theory (Chapter 1.5) is the observation that 15% of postreproductive females were lactating, implying that they were still being suckled. Kasuya and Marsh (1984) cited for precedent that !Kung women continue to nurse during menopause (Lee, 1980). Wisely, I think, Kasuya and Marsh also proposed that continued suckling of postreproductive females is the *result of,* rather than the cause of, the mother-calf bonds, stable social structure, and the longer-than-usual duration of maternal care observed in pilot whales. The generalizability of these important observations to other whales is unclear. Postmenopausal females are much rarer in the long-finned pilot whale (*G. melaena*) up to 50 years, an age that approximates the lifespan (Sergeant, 1962; Marsh and Kasuya, 1986). Marsh and Kasuya (1986) give a thoughtful survey of information about other whales, indicating that killer whales (*Orcinus orca*) and false killer whales (*Pseudorca crassida*) may also become postreproductive at later ages. Evidence is negative or insufficient to evaluate for other whales.

Other species also can survive in nature beyond their equivalent of menopause, including African elephants, to $\geq$ 60 years (Moss, 1988); the opossum (*Didelphis virginiana marsupialis*), reproductively senescent by 2 years (Austad, 1988); a ringed seal (*Phoca hispida*), 35 years (McLaren, 1958); a chimpanzee (*Pan troglodytes*), > 40 years, that appeared to retain a maternal role (Goodall, 1979; Bowden and Williams, 1985; Hrdy, 1981). These reports, though intriguing, do not give enough information to judge the generality of survival to postreproductive ages, nor the possible benefits of this to the fecundity (fitness) of individual germ lines in the social group, nor benefits to group survival.

Despite the massive age-related loss of oocytes, most species continue to ovulate until the ovary is virtually depleted of oocytes (Harman and Talbert, 1985; Finch and Gosden, 1986). However, fertility decreases sharply with advancing maternal age, and there is a sharp increase of resorbed or aborted fetuses. Characteristic among these are many abnormalities, including Down's syndrome and other aneuploidies as well as morphological abnormalities (Figure 3.20). A different spectrum of age-related maternal birth defects occurs in laboratory mice (Fabricant and Schneider, 1978; Finch and Gosden, 1986). The dwindling supply of oocytes has an implied role in the increase of fetal abnormalities with maternal age, since fetal abnormalities prematurely increase in hemi-ovariectomized mice (Brook et al., 1984; Chapter 8.5.1.1). This important study indicates that the chronological age is less crucial as a factor in the increased fetal abnormalities than is the size of ovarian follicular reserves. The marked irregularities of fertility cycles and delayed ovulation are thought to cause defective ova (Fugo and Butcher, 1971; Page et al., 1983; reviewed by Finch and Gosden, 1986), but the mechanisms are not understood.

The age-related dwindling of ovarian steroids, particularly estrogens, generally result in the atrophy of target cells in the uterus, vagina, and breast (Steger et al., 1981; Finch and Gosden, 1986; Roth and Hess, 1982; Nelson and Felicio, 1985), as found in humans and laboratory rodents. Not all women experience major atrophic changes. Diet may in-

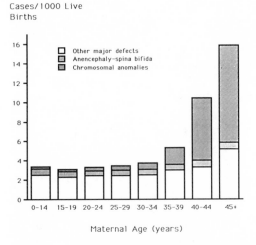

Cases/1000 Live Births

Other major defects
Anencephaly-spina bifida
Chromosomal anomalies

16
14
12
10
8
6
4
2
0

0-14 15-19 20-24 25-29 30-34 35-39 40-44 45+

Maternal Age (years)

Fig. 3.20. The age-related increase of birth defects in humans is mainly due to increasing maternal age. Most defects are associated with chromosomal abnormalities. Redrawn from Goldberg et al., 1979.

fluence the postmenopausal levels of estrogens, which are lower in vegetarians (Armstrong et al., 1981). Many of these atrophic changes are reversed or prevented by replacing ovarian estrogens (Chapters 9.3.4.3, 10.4.2). Some (but not all!) aspects of the age-related loss of bone (osteoporosis) are also associated with decreased secretion of estrogens (Davidson et al., 1982; Section 3.6.5.1) and can be partly prevented by steroid supplements (Chapter 10.4.2). A salutary consequence of menopause is the slowed rate of increase for breast cancer after menopause (Figure 3.18), probably because of decreased estrogens, progesterone, or prolactin. However, early menopause (before 40) is associated with twofold higher risk of mortality in Seventh-Day Adventists who, reliably, do not use tobacco (Snowdon et al., 1989). The ovariprival syndromes of reproductive senescence seem to occur widely in mammals and can be extensively modified by steroid treatments.

Some ovariprival changes of senescence are species-specific. Hypersecretion of the pituitary gonadotropins (LH and FSH) is very common in the ovariprival postreproductive phase of women, monkeys (*Macaca mulatta* and *M. nemestrina;* Short et al., 1987; Dierschke, 1985), and the C57BL/6J mice (Gee et al., 1983; vom Saal and Finch, 1988) and results from the decreased secretion of ovarian hormones (estradiol and inhibin) that regulate pituitary secretions through physiological feedback. The hypersecretion of LH is rarely seen in senescent laboratory rats, probably because most rat strains also have hypersecretion of prolactin in association with age-related pituitary tumors (Huang et al., 1976; Lu et al., 1979). At sufficient elevations, prolactin itself can inhibit LH secretion in rodents and humans (Bohnet et al., 1976; Grandison et al., 1977). Another hallmark of menopause is hot flushes, which are transient episodes of vasodilation under hypothalamic control and which arise after the loss of estrogens to castrate levels (Judd, 1983). Hot flushes from ovarian exhaustion have only been described in humans and rhesus monkeys (Jelinek et al., 1984; Dierschke, 1985).

3.6.4.2. Male Reproductive Senescence

Male mammals generally maintain spermatogenesis and gonadal functions throughout the lifespan, so that only a minority in any population become hypogonadal to the extent of the ovariprival syndromes. Besides examples from our species, spermatogenesis continues into the last quartile of the lifespan of pilot whales (Kasuya and Marsh, 1984) and elephants (Moss, 1988). In humans, paternity is documented to at least 94 years (Seymour et al, 1935), which is 35 years beyond the oldest recorded birth in women (Fergusson et al., 1982). However, the fraction of men with sexual impairments increases rapidly after 50, so that by 80 years only an exponentially decreasing minority is still sexually active, according to data from the first half of this century (Kinsey et al., 1948). Age-related decreases in male sex drive occur in otherwise fully functional men and lower animals (vom Saal and Finch, 1988; Bishop, 1970; Bronson and Desjardins, 1977).

Although levels of testosterone did not decrease in a sample of healthy older men from the Baltimore Longitudinal Study (Harman and Tsitouras, 1980), a trend for decrease is usually seen in general samples (e.g. Vermeulen et al., 1972; Figure 3.21). Contrary to folklore, there is little correlation between sexual activities and the levels of blood testosterone in older men (Tsitouras et al., 1982) or male laboratory rodents (Coquelin and Desjardins, 1982; vom Saal and Finch, 1988). While very low blood testosterone can cause impaired male sexual performance at all ages, the physiological range of adult testosterone is associated with a wide range of male sexual capacity. The number of testosterone-producing Leydig cells appears to decrease progressively with advancing age (Kaler and Neaves, 1978), though at a much slower rate than the exponential age-related loss of follicles in the ovary. There is no evidence to suggest an age-related limit in the formation of sperm or depletion of the spermatogonial stem cell populations.

The age-related reductions of male sex activities appear to have complex and diverse causes. A wide range of impairments at advanced ages appear to develop at loci that vary widely among individuals, even within highly inbred rodent populations which have extremely few genetic variations. Sexual performance may be disrupted by age-related lesions at many different loci in the nervous system or peripheral organs. Remarkable differences in performance are observed among 2-year-old inbred mice, which cannot be attributed to the same cause (Coquelin and Desjardins, 1982; Bronson and Desjardins, 1982).

Impairments in men are also linked to other age-related conditions, such as diabetes, hypertension, genitourinary cancer or infections, and other major disabling diseases with indirect effects on many body functions (Nelson et al., 1975; Lipson, 1984a, 1984b). Drugs used to treat these diseases may also secondarily impair sexual performance, especially β-adrenergic blockers that are used for treating hypertension. Histological studies of testicular biopsies from men of proven fertility, aged 27–89, suggest the importance of gradual damage to the seminiferous tubules from ischemia in association with arteriosclerosis (Paniagua et al., 1987). These diverse changes, though age-related, do not compose a generally definable syndrome. These disruptions are another illustration of dysorganizational changes during senescence, in which lesions at a small number of loci can precipitate major impairments. The age-related impairments of reproduction in both genders bear on the evolutionary potential of older age groups by diminishing their possible contribution to the population gene pool.

3.6.5. Skeletal and Muscular Changes

Two major age-related skeletal changes are documented in humans, domestic animals, and laboratory rodents: osteoporosis, a loss of bone mass which predisposes to breakage,

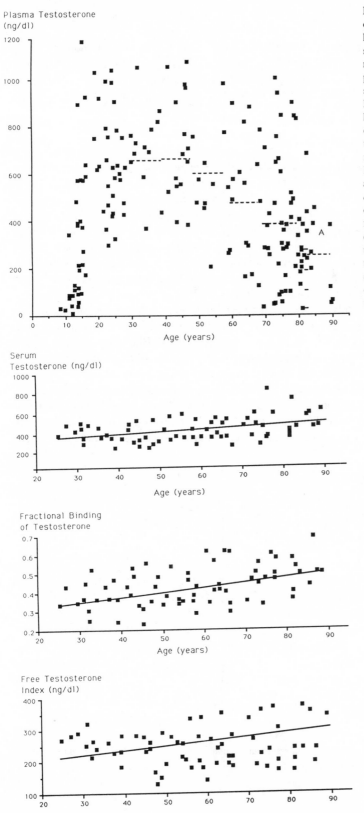

Fig. 3.21. Age-group data on circulating testosterone in humans from cross-sectional samplings. *Top,* marked age-related declines of total plasma testosterone in a sample described as "ambulatory and in general good health . . . [not] taking any drug known to influence testosterone metabolism" (Vermeulen et al., 1972, 730). Redrawn from Vermeulen et al., 1972. *Bottom three panels,* no change in a subgroup of men with privileged health from the Baltimore Longitudinal Study, as assayed by total serum testosterone, fractional binding of testosterone, and free testosterone. Redrawn from Harman and Tsitouras, 1980.

and arthritis, an often painful inflammation of the joints which reduces mobility. Both conditions can drastically curtail life's activities and increase the risk of crippling falls and accidents. Of the 200,000 cases of hip fracture associated with osteoporosis per year in the U.S., up to 20% are fatal for reasons that are largely obscure (Cummings et al., 1985, Riggs et al., 1986).

3.6.5.1. Osteoporotic Syndromes

Osteoporosis is considered a nearly universal age-related change in humans (Figures 3.22–3.24) and is generally more intense in women (Garn et al., 1967; Avioli, 1982; Riggs et al., 1986). Age-related osteoporosis is also documented in 2,000- to 4,000-year-old (nonfossilized) bones from North American Indian graves (Perzigian, 1973). Loss of bone mineral begins insidiously in women by the fourth decade (Parfitt et al., 1983; Riggs et al., 1981), which is about 5–10 years before the menopausal decrease in plasma estradiol. Bone loss is often intensified after menopause (Figure 3.24). Bone loss greatly increases the risk of bone fractures, since the total mass of bone is the primary determinant for the risk of fractures (Newton-John and Morgan, 1970).

Genetic factors are widely suspected in determining bone mass. For example, monozygotic twins have greater concordance in the mass of their radius bone than dizygotic twins (Smith, Nance, Kang, Christian, and Johnston, 1973). Black women have lower risk of age-related fractures than Caucasians, probably because of their larger bones with greater mineral reserves (Cohn et al., 1977), which may be a genetic influence.

Bones have differential susceptibility to osteoporosis. Mineral loss begins earlier and is more intense in trabecular bone (vertebrae, pelvis, ends of long bones) than in cortical bone (shafts of long bones), and is reflected in fewer fractures of the latter. In both sexes, osteoporosis has many causes which can differ among individuals, including deficient

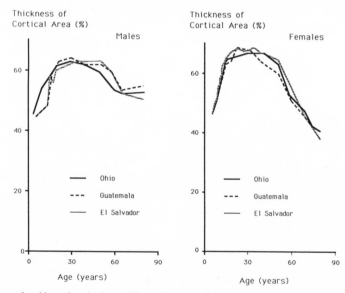

Fig. 3.22. Age-related bone loss in three different human populations, as indicated by decreased thickness of the second metacarpal bone in X-ray images. Redrawn from Garn et al., 1967.

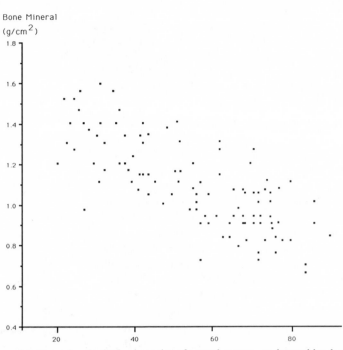

Fig. 3.23. Age-related bone loss in the lumbar spine of normal women, as detected by dual photon absorptiometry. Redrawn from Riggs et al., 1981.

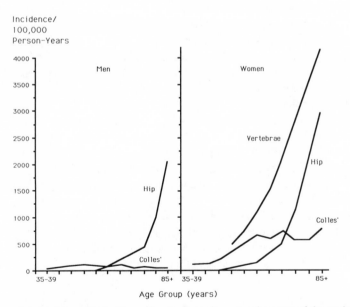

Fig. 3.24. Age-related osteoporotic fractures of major bones in men and women of the predominantly white population of Rochester, Minnesota. Colles' fracture occurs in the forearm (radius) near the wrist joint. Redrawn from Riggs and Melton, 1986.

mineral intake, physical inactivity, drugs and smoking, and complex effects from age-related changes in the regulation of numerous hormones (Avioli, 1982; Raisz, 1982; Zerwekh) et al., 1983; Armbrecht, 1984; Riggs and Melton, 1986).

Age-related elevations in men and women of parathyroid hormone and reductions of 24,25-dihydroxyvitamin D and 25-hydroxyvitamin D are given particular attention (Orwoll and Meier, 1986; Gallagher et al., 1980). Calcium absorption through the intestine is decreased by mid-life in rats (Armbrecht, 1986; Horst et al., 1978) and possibly humans (Alevizaki et al., 1973), and could be an early factor in osteoporosis. Riggs and Melton (1986) propose an endocrine cascade in postmenopausal osteoporosis, involving parathyroid hormone, calcitonin, which ultimately decreases vitamin D and thence decreases calcium absorption. Unlike similar phenomena in rats (see below), the alterations in vitamin D and parathyroid hormone are not strongly associated to age-related kidney dysfunctions in the typical forms of osteoporosis in humans (Orwoll and Meier, 1986; Gallagher et al., 1980), although impaired synthesis of vitamin D by the kidney may still be important (Tsai et al., 1984). There is thus a basis to provisionally consider a renal-osteodystrophic syndrome in human osteoporosis, as well as in rodents.

Laboratory rodents and monkeys of both genders also show age-related osteoporosis (Silberberg and Silberberg, 1941; Krishna Rao and Draper, 1969; Papadaki et al., 1979; Kalu, Hardin, Cockerham, Yu, Norling, and Egan, 1984; Kalu et al., 1988; Short et al., 1987; Zihlman et al., 1989). Besides the possible influence of decreased estrogen secretions in females (Aitkin et al., 1972; Papadaki et al., 1979), the common age- related kidney diseases may play an important role through the large increase of parathyroid hormone that often results from renal insufficiency (Kalu, Hardin, Cockerham, and Yu, 1984; Kalu, Hardin, Cockerham, Yu, Norling, and Egan, 1984). The complex interactions between kidney diseases and osteoporosis typify many aspects of senescence. The age-related trend for impaired kidney functions may prove to have general importance to bone disorders, but the extent of a renal-osteodystrophic syndrome of senescence is not clear outside of laboratory rats and possibly humans. Whatever the relative balance of various mechanisms may prove to be, the major mechanisms of osteoporosis are believed to be endocrine or behavioral (smoking, diet, exercise), although primary changes in bone-forming cells is not ruled out.

At a cell level, osteoporosis may arise from an imbalance in bone formation by osteoblasts and resorption by osteoclasts during the continuously ongoing bone-remodelling processes. Glycated forms of osteocalcin are also being considered in osteoporosis. There is a marked increase of glycated osteocalcin after middle age in humans and cattle (Gundberg et al., 1986). Osteocalcin is the second most common bone protein after collagen; carbohydrate is added nonenzymatically, and like glycated hemoglobin A_{1c} reflects exposure to circulating glucose (Chapter 7.5.2), which may increase with age because of the general trend for fasting and postprandial hyperglycemia (Section 3.6.8). Combinations of mineral supplements, ovarian steroid replacements, and exercise may in the near future minimize one of the traditional scourges of advanced age.

3.6.5.2. Arthritic Syndromes

The arthritic syndromes of senescence are primarily osteoarthritis (osteoarthrosis), in which the cartilage is the main site of degeneration, and rheumatoid arthritis, which involves inflammation and invasion of the synovial membranes by lymphocytes. Osteo-arthritis increases progressively during adult life. Most humans show some manifestations by 70–80 years (Barney and Neukom, 1979; Calkins and Challa, 1985). The arthritides are also common later in life in domestic dogs, cattle, and rodents (Silberberg and Silber-berg, 1965; Sokoloff, 1969; Poole, 1986). Nonetheless, a few individuals have very lim-ited degenerative changes even at advanced ages.

Osteoarthritis leads to destruction of the articular cartilages, in association with mod-erate size degradation of matrix proteoglycans. Collagen is disrupted but also becomes more cross-linked, which increases its brittleness (Chapter 7.5.3). Deposits of calcium pyrophosphate (chondrocalcinosis) commonly increase with age in the human knee joint (Sokoloff and Varma, 1988). Osteoarthritis is accelerated by heavy usage (ankles of danc-ers and fingers of cotton pickers), indicating an example of mechanical senescence. How-ever, osteoarthritis is often also intensified by diabetes (Lee et al., 1974; Kellgren et al., 1963; Calkins and Challa, 1985). Feral animals with such changes would have an in-creased risk of mortality because of impaired mobility. One extensive survey indicates the near universality of arthritic changes in a wide range of field and zoo mammals (Fox, 1939). However, the effects of captivity cannot be underestimated; elephants, for ex-ample, have more degenerative joint diseases in captivity than in the wild (Schmidt, 1978).

Rheumatoid arthritis is considered distinct from osteoarthritis. This disorder is usually a slow inflammatory process involving connective tissue in the joints, with lymphocytic infiltration and the deposition of immune complexes. The inflamed, hypertrophic synovial membrane may then invade bone and cartilage.

The primary causes of osteoarthritis and rheumatoid arthritis are unresolved but are suspected to involve slow and complex modifications in cartilage macromolecules in the former, and immunological responses in the latter. Much attention is given to modifica-tions of the various types of collagen through cross-linking and degradation, to the in-crease of proteases in synovial tissues, and to autoimmune reactions. An idea threading through this literature is that reaction to local antigens, newly expressed during later adult life, could promote inflammatory and autoimmune responses. One source of new antigens could arise from accumulations of degraded collagen or proteoglycans, or from the nonen-zymatic glycation of collagen (Schnider and Kohn, 1980; Kohn et al., 1984), as for osteo-calcin (Section 3.6.5.1). Another mechanism is the association of rheumatoid arthritis with alleles in the main histocompatibility complex, the *HLA* of humans. Particular *HLA*-coded proteins are recognized by antibodies to the Epstein-Barr virus, which may trigger autoimmune responses (Chapter 6.5.3.2).

Neuropeptides and neuroendocrine functions are also being considered in rheumatoid arthritis. Substance P, which is secreted from sensory nerve fibers, has many proinflam-matory actions, including the release of collagenase from synovial cells (Lotz et al., 1987). β_2-adrenergic mechanisms are also implicated in experimental arthritis (Levine et al., 1988). Neural involvement in rheumatoid arthritis is also indicated by clinical obser-vations that hemiplegic patients are usually spared from rheumatoid inflammations on the paralyzed side (Thompson and Bywaters, 1962) and that psychological trauma may provoke or intensify this disease (Baker, 1982). The Lewis rat strain, which has high susceptibility to rheumatoid arthritis, is also deficient in hypothalamic-adrenal cortical responses, possibly through aberrant regulation of the gene that encodes corticotrophin-releasing hormone (CRH; Sternberg et al., 1989). This model for neuroendocrine involvement in rheumatoid arthritis at the level of hypothalamic gene regulation fits within the framework of cascading dysorganizational effects of gene expression as a major mechanism in senescence.

3.6.5.3. Musculature

Age-related loss of skeletal muscle mass is typically 25% between maturity and the mean lifespan and is thought to be a general phenomenon in humans and laboratory rodents (Gutmann, 1977). The role of disuse in these changes is certainly important but may not be the only factor. A loss of muscle fibers (cells) is documented in only a few muscles, for example, the biceps femoris, which loses 30% of its fibers over the rat lifespan (Grimby and Saltin, 1983; Gutmann, 1977). There is also evidence for a loss of slow fibers in other muscles (Hooper, 1981; Rebeitz et al., 1972; Rowe, 1969). The remaining fibers have normal size and appearance in certain human muscles. Moreover, increased exercise can induce considerable increases of strength and oxidative capacity in response to training even late in life (Orlander and Aniansson, 1980; Suominen, Heikkinen, Liesen, Michel, and Hollmann, 1977; Suominen, Heikkinen, and Parkatti, 1977; Goldberg and Hagberg, 1990). Whatever intrinsic causes of senescence skeletal muscle may have, the capacity for reversal of changes is impressive. It seems likely that the age-related atrophic changes have several causes, including reduced physical activity (disuse atrophy), altered trophic stimulation through the innervating motor neurons, and hormonal influences from reduced stimulation by androgens (Section 3.6.4.2) and growth hormone (Section 3.6.9). For example, growth hormone treatment dramatically increased protein synthesis in the diaphragm of senescent rats (Figure 7.9).

3.6.6. *Immunological Changes*

Major changes in immunological cell populations and functions are reported in a huge literature of rodents and humans (Hausman and Weksler, 1985; Miller, 1990; Goidl, 1987; Makinodan et al., 1987; Walford, 1969), but their significance to dysfunctions of senescence and to mortality rates is unclear. Moreover, wide species differences in age changes are recognized, and the evidence is often controversial and difficult to generalize between genotypes and genders (Miller, 1989; Kay et al., 1979; Chapter 6.5.1). As in so many age-related changes, the general state of health can influence immune responses (see below). Beyond these well-known problems, the explosive developments in molecular and cellular immunobiology, the use of heterochronic transplantation, and *in vitro* assays with cell components from donors of different ages give powerful experimental approaches for unravelling the complexities of immunological senescence.

The thymus gland, the major locus for the differentiation of lymphocytes, undergoes a dramatic involution during maturation that is a very general trait in mammals and is also seen in some short-lived fish (Section 3.4.1.2.2). Thymic involution is clearly under endocrine control, possibly through growth hormone (Kelley et al., 1986). During thymic involution, there is a major loss of cells in the thymic cortex; this loss continues at a slower rate throughout adult life. The thymic medulla decreases in size, but much less than the cortex (Kay, 1978a; Makinodan and Hirokawa, 1985; Hausman and Weksler,

1985; Wade and Szewczuk, 1984; these references pertain to most of the following discussion). Although there are some reports of massive decrements of thymic hormones in peripheral blood by mid-life in humans (Lewis et al., 1978) and mice (Goldstein et al., 1974), no changes are seen by others (Weindruch et al., 1988). The massive decreases in blood levels of thymic hormones by mid-life far exceed changes in other immunological functions. Moreover, adult thymectomy has modest effects on immune responsiveness compared to the massive effects of neonatal thymectomy (Simpson and Cantor, 1975; Hirokawa and Hayashi, 1930; Miller, 1989, 1990). Among numerous other factors is the reduced export of new T cells by the thymus, decreasing by 95% between 1 and 6 months in mice (Scollay et al., 1980), and decreased maturation of bone marrow–derived cells into mature T cells in the thymus, between 1 and 2 years as shown by thymus transplantation (Hirokawa and Makinodan, 1975). On the other hand, treatment of senescent mice with thymic hormones partly restored immunologic responses (Weksler et al., 1978; Wade and Szewczuk, 1984). These complex issues cannot be resolved by the present data.

Soon after maturation, the strength of many immunological responses begins to decline. While the process of antigen presentation by macrophages shows few impairments (Miller, 1990), humoral antibody production by B cells (bone marrow–derived lymphocytes) tends to decline by the average lifespan of humans and mice to less than 50% of the peak at puberty (Figure 3.25). Primary antibody responses decrease much more than secondary responses as represented in the classic studies of Makinodan (e.g. Makinodan and Peterson, 1964). However, others report no age changes of certain B-cell functions (Snow, 1987; Kishimoto et al., 1978) or in the size of B-cell populations in humans and rodents (Miller, 1989). The mucosa-associated lymphoid populations in the gut, which are often the first line of defense in infections, show few age-related changes in mice (Wade and Szewczuk, 1984).

The *in vitro* responses of peripheral lymphocytes to mitogens (T cell–specific) give an important experimental model, discussed further in Chapter 7.6.3, that typically shows dramatic declines (Figure 3.26). The evidence generally indicates decreased production of interleukin-2, the growth factor required for T- and B-cell proliferation during immune

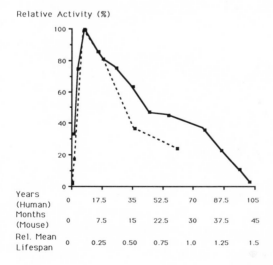

Relative Activity (%)

Fig. 3.25. Comparison of age-related changes in circulating agglutinin titers (related to B-lymphocyte function). *Solid line,* natural anti-A isoagglutinin in humans (data of Thomsen and Kettel, 1929); *dotted line,* agglutinin response to sheep erythrocytes in mice. Redrawn from Makinodan and Adler, 1975.

responses, while production of interleukin-2 per helper T cell may not decrease (Chapter 7.3.1.2).

The diverse subpopulations of T cells show more age-related changes than do B cells or macrophages. Dramatic changes are observed in killer (natural cytotoxic) cells of rodents (Miller, 1989, 1990; Goidl, 1987), while humans do not show as consistent decrements, for example in centenarians (Thompson et al., 1984). Using the Pgp-1 surface marker for memory T lymphocytes, Lerner et al. (1989) found a 2.5-fold increase of this cell population by the average lifespan in mice, which may reflect a lifetime of exposure to environmental antigens. An increasing fraction of memory T cells and other T-cell types become unresponsive in limiting dilution assays *in vitro,* while those that can, respond within the normal range. This important conclusion indicates that immune changes during senescence may be "mosaic" (Miller, 1989, 1990), rather than distributed throughout the entire T-cell populations of any individual. The generality of this finding to human populations will be important to establish as a function of different environments.

Autoantibody production increases during mid-life and later in most humans and rodents (Figure 3.27), as do the numbers of autoreactive cells (Gozes et al., 1978). These changes are without any clear link to pathogenesis, however. The NZB and some other short-lived mouse strains have major diseases by 8–16 months as a consequence of autoantibody production (Chapter 6.5.1.7.3), but these extreme changes are atypical. Another change is the increased incidence of individuals producing high levels of particular immunoglobulins (benign monoclonal gammopathy), which affects 3% of some 70-year age groups (Hallen, 1966). The growing evidence for the importance of autoanti-idiotypic antibodies in regulating clonal expansions of responding immune cells (Jerne, 1975) suggests that many functional age-related changes will be linked to altered regulation at a

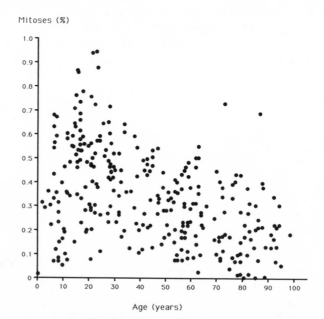

Fig. 3.26. Mitotic responses of peripheral lymphocytes (T cells) decrease with age in normal humans. Cells were cultured with the mitogen phytohemagglutinin and counted for mitotic figures. Age accounts for about 25% of the variance. Redrawn from Pisciotta et al., 1967.

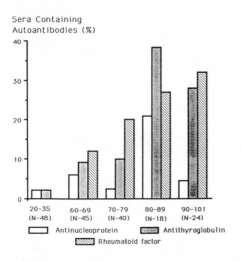

Sera Containing
Autoantibodies (%)

20-35 (N-48) 60-69 (N-45) 70-79 (N-40) 80-89 (N-18) 90-101 (N-24)

☐ Antinucleoprotein ▨ Antithyroglobulin
▨ Rheumatoid factor

Fig. 3.27. The percent of human sera containing autoantibodies increases with age. Based on data of Hallgren et al., 1973. Redrawn from Hausman and Weksler, 1985, 420.

clonal level. The major experimental manipulations of immunological age changes by transplantation of cells between different aged mice (Hausman and Weksler, 1985; Makinodan and Hirokawa, 1985) and by endocrine manipulations (Harrison et al., 1982; Weksler et al., 1978; Kelley et al., 1986) is consistent with this view.

A major controversy is over the functional consequences of altered immune changes, which are generally modest until after mid-life. Moreover, most older individuals clearly resist ordinary infections and are far more functional, for example, than those ravaged by AIDS. The 30% shortening of latency of AIDS in patients older than 60 who were transfused with HIV-infected blood (Medley et al., 1987) nonetheless implies the functional importance of age-related immune changes. Another example is the correlation between low responsiveness in delayed hypersensitivity testing and increased mortality from cardiovascular and other diverse causes during the next several years (Roberts-Thompson et al., 1974; Murasko et al., 1988). Moreover, elderly individuals varied in immune responses to a multivalent pneumococcal vaccine according to whether they resided at home (Roghmann et al., 1987). There may be numerous cryptic relationships between the immune system and manifestations of pathology in other organ systems as well as influence from behavioral states. Immune changes may have greater significance to subgroups with severe bone and joint disease or to subgroups with tumors from uncontrolled growth of immune cell clones. From this viewpoint, age-related changes in immune function may not be strong factors in mortality risk for most individuals, until perhaps very advanced ages. In the next few years, we may expect major findings on the selectivity and root causes of changes in different lymphocyte subpopulations. The present data indicate highly selective and regulated changes of a nature that could arise from subtle shifts in gene regulation during lymphocyte differentiation and from cumulative interactions with environmental antigens.

3.6.7. *Vascular Diseases and Hypertension*

Vascular diseases, particularly arteriosclerosis and atherosclerosis, are the major factors in heart disease and stroke, which together account for the majority of deaths in most human populations (Figure 1.11). The genesis of major changes in the vulnerable coro-

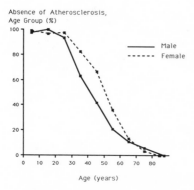

Fig. 3.28. Age-related increase of atherosclerotic changes in cerebral arteries, as shown by the diminishing fraction of the age group that had no atherosclerosis (score of zero) in 5,033 consecutive autopsies in a Minnesota population. Redrawn from Flora et al., 1968.

nary arteries that supply the heart and in other arteries may be traced to childhood. The necropsy of U.S. soldiers killed in Korea and Vietnam revealed that emergent lesions are common in robust young adults (McGill, 1984; McNamara and Malot, 1971; Strong et al., 1978), again emphasizing the early beginnings of many senescent processes. By the age of 65, the majority of North Americans have some degree of arterial disease, for example, in the cerebral arteries (Figure 3.28). The International Atherosclerosis Project (23,207 autopsies from fourteen countries) revealed qualitatively similarly progressive age-related increases of fatty streaks in the aorta and coronary arteries throughout the world, but also important demographic and ethnic differences in the extent to which the vessel wall was covered by the lesion (Eggen and Solberg, 1968).

Atherosclerosis may begin with the accumulation of cholesterol and other lipids in the intima of blood vessels to form visible fatty streaks (Figure 3.29). The process continues with proliferation of smooth muscle cells in the artery wall and the formation of fibrous tissue that causes "raised lesions" that narrow the lumen. Calcification may occur in the smooth muscle cells and also in the leaflets of the aortic valve, particularly in the oldest groups (Section 3.6.2). The widely observed calcification of vascular components is concurrent with osteoporosis in many species. Although linkage between the two phenomena is not widely recognized, both could involve similar changes in hormones that promote calcium mobilization and allow its adventitious deposition in abnormal (ectopic) soft tissues.

Fibrous plaques and adhering blood clots (thromboses) may severely narrow the lumen, causing a chronic insufficiency of oxygen that is often manifested as pain from the heart during stress (angina pectoris). In the worst case the plaques and narrowed vessel may trap circulating blood clots to cause a local block in blood circulation. The resulting ischemia usually causes irreversible damage to nonreplaceable cells in the myocardium or brain, and may immediately precipitate death. However, sudden coronary death in the absence of obstructions in coronary artery blood flow may be more common than once thought, and has been associated with a type of damage to the heart muscle (myocytolysis) that could result from cumulative effects of catecholamines released by the sympathetic nervous system (Chapter 10.5.1; Baroldi et al., 1979). Vascular stenosis can cause the upstream portion of the vessel to develop outpocketing (aneurysms) that may burst and also cause death. A range of other age-related disorders is associated with vascular disease. Gangrene may also result from inadequate blood supply of the extremities. Many of these lesions are found in adult diabetics (Section 3.6.8). The vascular-ischemic syndromes are clearly a major cause of senescence in human populations and can be viewed either as disorders from mechanical senescence (Section 3.6.11.1) or aberrant proliferation.

Risk of ischemic heart disease is influenced by a range of now-familiar risk factors including chronic blood elevations of low-density lipoproteins (LDL), very low-density lipoproteins (VLDL), and lipoprotein(a) (e.g. De Backer et al., 1982; Rifai, 1986; Goldstein and Brown, 1982). Obesity and hypertension are also well-established risk factors (Kannel et al., 1961; Ross, 1986). The distribution of vascular diseases in feral populations is not known for most species. Nonetheless, the presence of atherosclerotic lesions or calcified valves in various large mammals, for example, feral East African elephants (McCullagh, 1972) and captive leopards (Fox, 1939), strongly indicates the generality of these diseases in long-lived mammals. Similar vascular diseases can be induced in laboratory rodents, rabbits, pigs, and monkeys by cholesterol-rich diets (Gerrity et al., 1979; Faggiotto and Ross, 1984), stress (Wexler and True, 1963; Henry et al., 1971; Kaplan et al., 1983), or hypertension (Wexler et al., 1977). Hereditary and environmental influences through diet and stress are discussed later (Chapters 6, 10). For now, I mention that hypertension-prone rat genotypes have intensified arteriosclerosis and a high incidence of stroke (Okamoto et al., 1974).

Another factor may be age-related increases in blood coagulability (increased prothrombin fragment F_{1+2}), which is hypothesized to be a component of "prethrombic" state in healthy elderly men (Bauer et al., 1987). A lower threshold for activation of clotting could generate clots that block the blood supply and contribute to strokes and heart attacks. Lipid peroxides, which are formed by free radical attack, are decreased by diet restriction in rats (Chapter 10.3.2.5) and are a candidate factor in atherogenesis.

A long-suspected mechanism of cellular damage during myocardial ischemia was recently demonstrated: free radicals. These were directly demonstrated during the restoration of blood flow (reperfusion), and free radical damage was prevented by administration of catalase and supraoxide dismutase (SOD; Bolli et al., 1989). The protective effects of SOD and catalase draw attention to the levels of these enzymes. Besides inhibiting free radical production, SOD activities in tissue homogenates show some correlations with the lifespan in species comparisons (Chapter 5.4.3).

Atherosclerosis is no longer thought to be a cause of senile dementia of the Alzheimer's type; recent studies indicate nonatheromatous abnormalities in the small blood vessels and capillaries during this disease, including the accumulation of β-amyloid (Miyakawa et

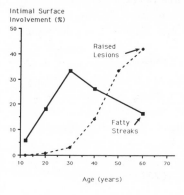

Fig. 3.29. The progression of atherosclerosis in coronary arteries, showing the early onset of fatty streaks before puberty and the development into fibrous plaques and other raised lesions that can trap circulating blood clots and otherwise block the flow of blood. Data represent the fraction of the artery inner surface that is covered by these changes, as determined on 1,013 autopsies of white males in New Orleans. Based on the data of Strong et al., 1978. Redrawn from McGill, 1984.

al., 1982; Section 3.6.9). Cerebrovascular amyloid is commonly deposited at advanced ages in a wide range of species, including dogs, macaques, and neurologically normal humans (Selkoe et al., 1987). Amyloids represent a wide variety of fibrillar proteinaceous aggregates with different amino acid sequences that yield a β-configuration of protein structure that gives Congo red dye a green birefringence under polarizing light (Glenner, 1980; Calkins and Wright, 1989; Husby and Slettin, 1986). The proteins deposited are usually derived from a larger precursor and may also be further modified by covalent attachment of carbohydrates, ubiquitin, and other proteins. The β-amyloid of cerebral vessels mentioned above also occurs within the parenchyma of the brain, particularly during Alzheimer's disease (Section 3.6.10). Amyloids with other amino acid sequences are deposited during other diseases, including scrapie (prion disease; Bruce et al., 1976; Hsiao et al., 1989) and human type 2 diabetes (Cooper et al., 1988), and in the senescence-accelerated mouse (Chapter 6.5.1.7.2).

3.6.8. Diabetic Syndromes

With advancing age, impairments of glucose homeostasis regulation are very common in humans, but the generality of these changes remains to be established in other species. While fasting blood glucose shows only slight age-related increase, much larger increases of blood glucose occur after a meal or after ingestion of glucose alone in the oral glucose tolerance test (Andres and Tobin, 1977; Minaker et al., 1985; Goldberg and Coon, 1987). These changes progress with age after maturity and appear to result largely from a decreasing sensitivity of peripheral tissues to insulin, insulin resistance (DeFronzo, 1979; Chen et al., 1988; Minaker et al., 1985). By 60 years, about 50% of the general adult population would be classified as diabetics if their age were not considered (Andres and Tobin, 1977; Rowe and Minaker, 1985). Nonetheless, the increase of blood glucose may be small, if body weight is accounted for. For example, comparisons of young (20–35-year-old) and older (60–72-year-old) men in good health who were matched for body mass index showed that the integrated (24-hour) blood glucose levels were 10% higher (Gumbiner et al., 1989; Marilyn Ader, pers. comm.). The different tendencies for excercise and activity are difficult to account for in such studies.

Effects of age on the capacity to secrete insulin over a full range of blood glucose are controversial, but most agree that the maximum capacity of pancreatic β-cell for insulin secretion, when stimulated sufficiently, is not greatly impaired (Andres and Tobin, 1977; DeFronzo, 1979). Moreover, circulating insulin is normal or even slightly higher up through advanced ages (Andres and Tobin, 1977; Minaker et al., 1985). The elderly may thus be described as a group that is unable to compensate fully for insulin resistance. In other words, younger healthy age groups would be expected to show greater secretion of insulin to compensate for comparable insulin resistance (Bergman, 1990).

A molecular index of the increased blood glucose is the age-related increase of glycated hemoglobin A_{1c} (Chapter 7.5.2), a covalent glucose adduct that forms nonenzymatically and that indicates the integrated exposure to glucose over the preceding several months in humans and mice. The trend for greater transient glucose elevations is implicated in numerous age-related disorders because of the glycation of certain long-lived proteins, for example, in the kidney basement membrane, bones and joints, and eye lens.

Much evidence suggests that chronological age, while the correlate of these changes in glucose regulation, is not their cause, since age-related trends in glucagon and growth hormone, obesity, exercise, and diet all influence glucose homeostasis (Elahi, 1982; Goldberg and Coon, 1987). Evidence is mounting that many of these age-related changes are reversible. For example, increasing dietary carbohydrates of nonobese healthy elderly males for 3–5 days improved insulin sensitivity, insulin secretion, and glucose tolerance (Chen et al., 1988).

Laboratory rodents also show age-related changes in glucose homeostasis with many parallels to humans, but the literature is difficult to interpret, because rats are prone to obesity under laboratory conditions and because the ages usually used for comparison, for example, 2–3 months (postpubertal) versus 12 months (middle-aged) or older, include major differences in lean body mass due to the extensive growth continuing up to about 9 months. Even so, many studies describe reduced oral glucose tolerance and reduced β-cell secretion of insulin (Reaven and Reaven, 1980; Chaudhuri et al., 1983; Elahi, 1982; Florini and Regan, 1985). Nonetheless, other indices of glucose regulation (fasting blood glucose levels, clearance of a glucose injection) did not deteriorate with age in one study of male C57BL/6J mice up to 25 months (Leiter et al., 1988).

The consistent, if subtle, age-related trend for greater average circulating levels of post-prandial blood glucose in humans described above may increase the risk of many of the same diseases risked by diabetics (Cerami et al., 1987). Clinicians have long noticed that diabetics appear to "age faster" and as a subpopulation have earlier onset of arteriosclerosis, bone and joint diseases, cataracts, and hypertension. Diabetic retinopathy and kidney disease, for example, have increased incidence and severity in general proportion to the duration of diabetes. Diabetes is often associated with multiple degenerative lesions in central and peripheral neurons (polyneuropathy), including the eye (diabetic retinopathy) and other peripheral senses, as well as motor and autonomic dysfunctions (Sidenius and Jakobsen, 1982; Thomas, 1987). Vascular-related ischemia is thought to be an important factor in the multifocal degenerations of nerve fibers commonly found in older diabetics (Thomas, 1987). Modest age-related blood glucose elevations could thus contribute to the degenerative changes in the retina, peripheral nerves, and kidneys of the elderly who would not otherwise be classified as diabetics. Experimentally induced diabetes in rats causes neuronal degeneration in peripheral and hypothalamic neurons within 2 months (Medori et al., 1985; Bestetti et al., 1985). Thus, it seems likely that glucose intolerance intensifies many age-related disorders. Furthermore, glucose intolerance itself is specifically implicated as a risk factor that increases mortality by 50% (Fuller et al., 1980).

A tendency to accumulate fat deposits is common in many mammals during mid-life. The ratio of the waist-hip circumference has been implicated as a risk factor of mortality, and increases progressively with age in men more than in women, as revealed in the Baltimore Longitudinal Study (Shimokata et al., 1989). Rhesus monkeys from two Regional Primate Centers also show gradual increases of body fat after maturity (Kemnitz, 1984; Kemnitz et al., 1989). The origins of this are not well studied and probably include altered locomotor activity and neuroendocrine mechanisms (e.g. growth hormone, Section 3.6.9), which could have common origins in the brain. Rodent genotypes differ in the increase of body fat and adipocyte size with age (Roth and Livingston, 1976). The higher incidence of diabetes and hypertension in obesity are risk factors in cardiovascular disease (Kannel et al., 1961).

3.6.9. Neuroendocrine and Endocrine Syndromes

The importance of the hypothalamus and limbic system in regulating many of the physiological functions that also change with postmaturational age has led to speculation that these brain regions may be centers for the regulation of mammalian senescence (Dilman, 1981; Everitt, 1976b; Finch, 1972a, 1976a; Finch and Landfield, 1985). The crucial role of the brain, particularly the hypothalamus, in regulating the onset of puberty is already well established (Reiter, 1982; Ojeda et al., 1980; Terasawa et al., 1984). A major pending question concerns the role of the peripheral and central nervous system in the numerous age-related changes of hormones.

Besides alterations in the levels of hormones, another influence on target tissues may be through alterations in the temporal pattern of secretion (described below). The temporal pattern can have profound influences on target cells, as illustrated by the critical fre-

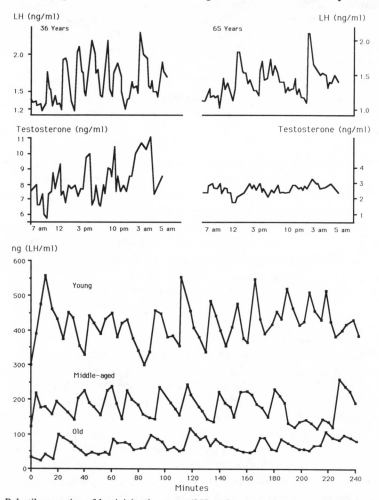

Fig. 3.30. Pulsatile secretion of luteinizing hormone (*LH*) and testosterone slows with age in men, as measured from sampling healthy monks every 20 minutes. *Top,* individual profiles of men aged 36 years and 65 years. Redrawn from Deslypere et al., 1987. *Bottom,* pulsatile LH profiles from three individual castrate male rats aged 3, 18, and 26 months. Redrawn from Karpas et al., 1983.

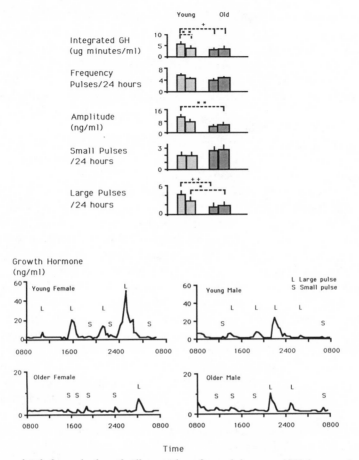

Fig. 3.31. Age-related change in the pulsatile secretion of growth hormone (*GH*) in normal young (18–33 years) and older (56–76 years) men and women. The 24-hour secretory profile of growth hormone was assayed from sampling every 20 minutes. *Top,* groups that are connected by dashed lines and * are significantly different averages of the 24-hour integrated growth hormone levels in serum, the pulse amplitude, the numbers of small and large pulses, and the frequency of growth hormone pulses. *Bottom,* representative profiles. Redrawn from Ho et al., 1987.

quency of LH pulses needed to stimulate ovarian follicle growth (Pohl et al., 1983) and testosterone production (Ellis and Desjardins, 1982). Temporal patterns also influence receptors for other hormones, as shown by the gender differences of prolactin receptors on hepatic membranes, which are controlled by gender differences in the diurnal secretion of growth hormone (Norstedt and Palmiter, 1984). These examples suggest that some subsets of age-related changes of gene regulation may be traced to altered temporal patterns of hormone secretion.

Many hypothalamic neurosecretory changes with age occur after mid-life in humans and laboratory rodents. Men show trends for reduced pulse frequency of ultradian testosterone secretion (Bremner et al.,1983; Deslypere et al., 1987) (Figure 3.30), which implies altered hypothalamic regulation of LH. There are also age-related trends for reduced nocturnal secretion of growth hormone during mid-life (Prinz et al., 1983; Figure 3.31). The influence of diet, obesity, and physical activity needs to be evaluated here, as for

glucose homeostasis. The postmenopausal hot flushes are considered to involve hypothalamic responses to the ovariprival estrogen levels (Section 3.6.4.1 above).

The regulation of vasopressin is altered in a particularly interesting way: with advancing
age, vasopressin is *more readily secreted* in response to increases in blood osmolality
(Figure 3.32), so that the osmoreceptor sensitivity of men aged 65 was fourfold higher
than at 30 years (Helderman et al., 1978; Robertson and Rowe, 1980). Resting plasma
levels of vasopressin increased markedly in 32-month-old rats that did not manifest obvious kidney damage (Fliers and Swaab, 1983); other reports indicate a similar trend for
increased vasopressin secretion (Miller, 1987; Frolkis et al., 1982). The number of vasopressinergic neurons of the suprachaismatic nucleus, however, decreased by 30% in rats
(Roozendaal et al., 1987) and humans (Swaab et al., 1985; Fliers et al., 1985). In senescent rats, the remaining neurons show compensatory increases in size, and their histological appearance shows no evidence of degeneration in rats (Peng and Hsu, 1982; Fliers
and Swaab, 1983). The total neurons of the suprachaismatic nucleus, however, did not
decrease in old rats (Peng and Hsu, 1982; Roozendaal et al., 1987), an observation discussed later in terms of recently described novel changes in the gene expression of vasopressinergic neurons (Chapter 7.3.3).

Much more is known about neuroendocrine changes in rodents. Female mice and rats
have major impairments in the regulation of gonadotropic hormones (Sahu et al., 1988;
Meites, 1990; Scarbrough and Wise, 1990; Finch et al., 1984). By the age when cycles
are lost (12–18 months), the estrogen-inducible preovulatory gonadotropin surge is virtually lost (Figure 3.33). The pulsatile secretion of LH, which is directly linked to hypothalamic neurosecretion of GnRH, is also greatly impaired (Figure 3.34). At least in C57BL/
6J mice, there is no loss of GnRH-containing neurons up through the average lifespan
(Hoffman and Finch, 1986; Don Carlos et al., 1986), although loss of these neurons is
indicated at later ages (Miller et al., 1990). Male rodents share with females an age-related
reduction in the frequency of LH pulsatile secretions (Coquelin and Desjardins, 1982;
Karpas et al., 1983). The mechanism is unknown, but changes in the hypothalamus are
clearly indicated. Alterations in other hypothalamically seated rhythms are described
below.

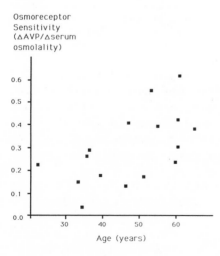

Osmoreceptor
Sensitivity
(ΔAVP/Δserum
osmolality)

Fig. **3.32.** Hypothalamic osmoreceptor sensitivity increases with age in normal humans, assayed by the secretion of arginine vasopressin as serum osmolality is varied.
The sensitivity represents the slope (increment of arginine
vasopressin (AVP) per increment of serum osmolality). The
regression of sensitivity on age, $p < 0.02$; age accounted
for 36% of variance. Regression line not shown. Redrawn
from Robertson and Rowe, 1980.

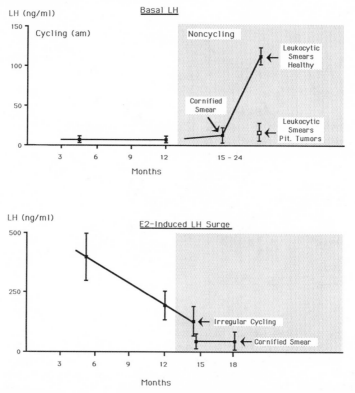

Fig. 3.33. Age-related changes in the regulation of LH in female C57BL/6J mice. *Top,* basal circulating levels of LH showing spontaneous elevations that roughly coincide with the exhaustion of ovarian follicles (see Figure 3.19). Ovarian follicles produce estradiol, whose decreased secretions can be easily detected by the loss of cornified epithelial cells in the vaginal smear and their replacement by leukocytes. The elevation of LH results from the deficits of estradiol, as at menopause in humans. Noncycling mice of ages 15–20 months are combined. *Bottom,* the estrogen-induced LH surge in mice (simulation of the LH surge at proestrus) becomes progressively smaller with advancing age and is a marker for age-related impairments in hypothalamic and pituitary functions. Redrawn from data of Kevin Flurkey, Danny Gee, and Charles Mobbs. From Finch, 1987.

In female rodents, prolactin secretion tends to increase, as noted in Section 3.6.9, and pituitary lactotrope adenomas become increasingly common. Pituitary tumors, though rarely metastatic in rodents, can press disruptively into the brain (Stein and Mufson, 1979). A major species difference is that postmenopausal women rarely experience pituitary tumors or elevated prolactin (reviewed in vom Saal and Finch, 1988).

A major cause of the altered regulation of gonadotropins in female mice and rats has been traced to the cumulative impact of ovarian estradiol on the hypothalamus and pituitary. Nearly all of the age-related reproductive neuroendocrine changes in female rodents can be delayed by ovariectomy without steroid replacement, or can be prematurely induced in young rodents by chronic treatments that elevate estradiol (Chapter 10.4.2; Table 10.6).

Growth hormone secretion shows marked age-related decreases in pulse frequency and amplitude in rats (Figure 3.35) and in rhesus monkeys (Kaler et al., 1986). Because the normal frequency and amplitude of growth hormone and LH secretions can be rapidly

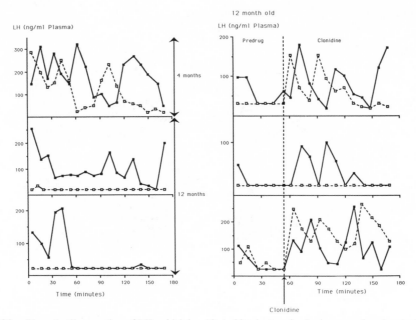

Fig. 3.34. The pulsatile secretion of LH, which has 10- to 20-minute episodes in young female rats. *Top left,* individual plots of serial samplings. *Middle and lower left,* by 12 to 15 months most rats show fewer secretory episodes; at this age, estrous cycles are also fewer and more irregular. *Right,* these age-related neuroendocrine changes are rapidly reversed at 11–12 months by injecting clonidine, an α_2-adrenergic agonist. Redrawn from Estes and Simpkins, 1982.

restored in old rats by catecholaminergic drugs (Sonntag et al., 1982), it seems likely that any neurochemical age changes are distal to the neurons that secrete the releasing factors controlling these pituitary secretions.

In contrast to many of the above examples, the basal circulating levels of adrenal corticosteroids and output of ACTH are not impaired in older humans. Adrenocortical responses after surgery, for example, are at least as large in older patients (Blichert-Toft, 1975; Blichert-Toft and Hummer, 1977). Basal cortisol levels in humans do not decrease (Minaker et al., 1985) and may increase slightly (Weiner et al., 1987). Age-related increases of plasma corticosterone are found in rats of several lines (Landfield et al., 1978; DeKosky et al., 1984; Sapolsky et al., 1986a, 1986b) but do not occur in C57BL/6J mice (Latham and Finch, 1976; Finch et al., 1969). In view of a recent report that 10–15% of Syrian hamsters (*Mesocricetus auratus*) show age-related adrenocortical hyperplasia at age 1–3 years (Deamond et al., 1990), the tendency for progressive adrenocortical hyperactivity might be ubiquitous in certain rodent radiations. In each case, however, the effects of the caging (social) environment need careful consideration. While there are no data on possible age changes in the numbers of hypothalamic neurons that secrete CRH, this evidence indicates sparing of the crucial hypothalamic-pituitary-adrenal axis across the lifespan.

Moreover, secretions of norepinephrine by the sympathetic nervous system are not impaired in humans, as judged by the age-related trend for elevated plasma norepinephrine (basal and in response to postural changes) and the absence of changes in norepinephrine clearance (Rowe and Troen, 1980; Young et al., 1980; Ziegler et al., 1978; Fleg et al., 1985; Lakatta, 1990).

Another adrenal steroid, dehydroepiandrosterone sulfate (DHEA), shows remarkable decreases that begin soon after maturity in humans of both sexes and continue progressively throughout life (Orentreich et al., 1984; Figure 3.36). Similar decreases are reported in rhesus monkeys (Orentreich Foundation for Advancement of Science, 1987). The cause of this major change is unknown and clearly represents a selective regulation of steroid biosynthesis, rather than a general defect in the adrenal cortex. The consequences of decrease in this weak androgen are unknown, though evidence indicates that low production is a risk factor in breast cancer (Bulbrook et al., 1971). DHEA treatment retards some age-related diseases in rodents and may play a role in intermediary metabolism (Chapter 10.3.2.6).

Some evidence indicates that corticosteroids from the adrenal gland may also participate in neuronal damage in the hippocampus. A loss of corticosterone-binding, pyramidal CA_3 neurons was described in 2-year-old rats and a similar loss results from 3 months of elevated corticosterone (Chapter 10.5.1). Correspondingly, certain of these changes are induced by chronic corticosteroids and are prevented by adrenalectomy. These findings suggest that adrenal steroids and certain stressors have cumulative effects on the hippocampus. Moreover, since the hippocampus has an inhibitory influence on secretion of ACTH via the hypothalamus, hippocampal damage caused by the putative cumulative effects of stress could contribute to a vicious cycle (Chapter 10.5.1). The progressive age-related elevations of corticosteroids noted above in older rats and humans suggest such a process. The corticosteroid-related deaths of Pacific salmon and marsupial mice (Chapter 2.3.1) represent extremes in a widely distributed and possibly long-standing vertebrate tendency, which may contribute to gradual senescence in many other species. The hippocampus, crucial to memory, is well known for its vulnerability to ischemia and other insults (Siesjo, 1981; Sloviter, 1987; Zola-Morgan et al., 1986). There is no information on whether chronic stress or sustained elevations of endogenous or therapeutic glucocorticoids or anti-inflammatory steroids can damage the human hippocampus.

The altered pulsatile release of LH and growth hormone is only one manifestation of altered rhythmic activity by the brain. The diurnal rhythm of glucose metabolism, as judged by regional analysis of 2-deoxyglucose uptake, showed marked alteration in senescent rats (Wise et al., 1987). A general change of major importance is the fragmentation

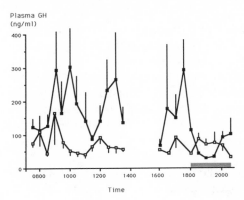

Plasma GH (ng/ml)

Fig. 3.35. Age-related changes in the pulsatile secretion of growth hormone in 4-month-old (young adults, *solid square*) and 20-month-old (late middle-aged, *open square*) male Sprague-Dawley rats, showing the decreased frequency and amplitude of growth hormone pulses. Redrawn from Sonntag et al., 1980.

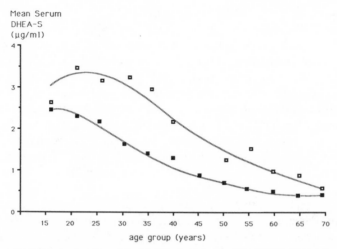

Fig. 3.36. Age-related changes in serum dehydroepiandrosterone sulfate (DHEA-S) in normal humans. *Open square,* men; *solid square,* women. The decrease across the lifespan begins soon after maturity, but there are numerous individual differences in the rates of change in this weak androgen. Redrawn from Orentreich et al., 1984.

of sleep cycles in association with disrupted brain electrical activities and breathing patterns (Richardson, 1990; Dement et al., 1985; Knight et al., 1987). Many species, including humans, cats, and rodents, show increased wakefulness during their sleep phases and increased napping during the phase of usual daily wakefulness. During sleep apnea (interrupted breathing), blood hemoglobin can become considerably depleted of oxygen (Block et al., 1979). These trends are observed in mammals at later ages. Their extent varies widely among individuals.

In summary, age-related changes in neuroendocrine functions show considerable selectivity. While there may be impairments in the regulation of the hypothalamic-pituitary-gonadal axis that vary among species, at least two other systems show no evidence for hypothalamic-level impairments in the capacity to make positive responses: the neural systems regulating ACTH and vasopressin.

3.6.10. Neurodegenerative Syndromes

Neurons, like oocytes, are irreplaceable in the adult brain of rodents, primates, and humans. But, unlike oocytes, neurons are not lost at an exponential rate or even at a regular rate throughout the adult life phase. There is presently no example of a neuronal type that is always lost in all human and rodent populations by the end of the average lifespan. Many neuronal clusters ("nuclei") show no loss at all in the absence of vascular disease (Curcio et al., 1982; Monagle and Brody, 1974), and several studies suggest that shrinkage of cerebral cortex neurons in elderly humans may have been mistaken for neuron loss (Terry et al., 1987; Haug 1984; Coleman and Flood, 1987; Finch and Morgan, 1990).

Nonetheless, there is an apparent age-related trend for loss of large neurons in the cerebral cortex, hippocampus, cerebellum, and especially the pigmented neurons of the locus ceruleus and substantia nigra (reviewed in Curcio et al., 1982; Flood and Coleman, 1990). It is unlikely that extensive neuron loss occurs in all these regions in all individuals. The upper range of neuron loss in the neurologically normal elderly overlaps with the extensive, but pathway-specific, neuron loss in Huntington's, Parkinson's, and Alzheimer's diseases. The loss of large pyramidal neurons in the hippocampus during Alzheimer's disease appears to be much greater than age-related loss (Figure 3.37). Each of these diseases has intensified neuron death in limited brain regions and pathways (Finch and Morgan, 1987; Morgan et al., 1987; Finch, 1980) and occurs in a characteristic quartile of the lifespan (Figure 3.38). Schizophrenia is included as a neurological condition that shows marked spontaneous improvement during its later course in middle age in about 50% of the cases. In view of evidence for excessive dopaminergic activities in schizophrenia, the early decrease of dopamine receptors may help ameliorate the condition or may lower the risk for new cases with advancing age (Finch and Morgan, 1987). These age-related neurological diseases afflict less than 20% of adults during the present human lifespan of 75 years. However, the rapidly increasing size of the 80-year and older age groups in all countries will probably be accompanied by major increases of Alzheimer's disease, since its incidence increases progressively with age.

The general trend for age-related neuron loss and neuron atrophy has raised controversial questions about interactions of age-related changes with genetic and other risk factors for Alzheimer's, Parkinsonism, and the other diseases (Agid et al., 1989; McGeer et al., 1989; Wolters and Calne, 1989; Morgan et al., 1987). For example, the neurocytological changes that are hallmarks of Alzheimer's disease also gradually increase with advancing age in the neurologically normal population: the intraneuronal neurofibrillary tangles and the senile (neuritic) plaques, which contain degenerating axon terminals and are commonly embedded in a core of β-amyloid (Selkoe et al., 1987; Guiroy et al., 1987; Kem-

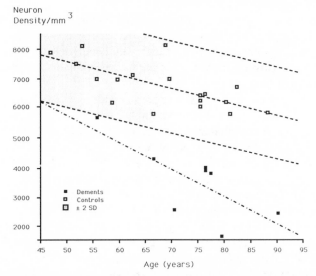

Fig. 3.37. The age-related loss of hippocampal pyramidal neurons is much smaller in neurologically normal elderly than in a sample of demented brains, mostly with Alzheimer's disease. The data indicate a considerable range of hippocampal neuron loss at later ages, with the possibility of considerable overlap between demented and nondemented. Redrawn from Ball, 1977.

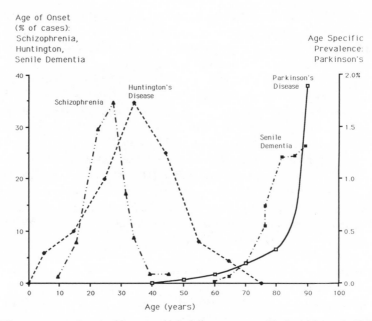

Fig. 3.38. The age-group estimates of four neurological diseases, grouped in 5-year intervals. Epidemiological and case incidence data: schizophrenia (Loranger, 1984), Huntington's disease (Brackenridge, 1973), Parkinson's disease (Mutch et al., 1986), and senile dementia (including Alzheimer's disease; Mortimer et al., 1981; Larsson et al., 1963). Data on Parkinson's disease represent a household survey that estimates prevalence by age group, not age of onset. The distinct age group of onset for each of these conditions suggests hypotheses about how the lesion interacts with age-related neurochemical changes. Modified and redrawn from Finch and Morgan, 1987.

per, 1984). This same β-amyloid is also found in cerebral blood vessels in older age groups of humans (Glenner and Wong, 1984; Goldgaber et al., 1987; Masters et al., 1985; Robakis et al., 1987). By immunological criteria, similar amyloids accumulate in cerebral blood vessels of shorter-lived carnivores and herbivores but have not been found so far in senescent rodents (Selkoe et al., 1987).

The neurocytological changes of Alzheimer's disease are clearly distinguished from normal age groups below 75 years, but increasingly converge thereafter with the incidence of plaques and tangles in the nondemented (Mann, 1985; Figure 3.39). The distinctions between "normal aging" and Alzheimer's disease at ages beyond 80 years may not be clarified until the genetics of Alzheimer's disease is better understood. The β-amyloid deposits in the senile plaques of Alzheimer's disease may be different from those in the nondemented elderly by their content of tau and neurofilament proteins (Arai et al., 1990). Besides those of older humans, amyloid deposits in neuritic plaques and in small blood vessels increase with age in monkeys, dogs, and bears (Wisniewski et al., 1978; Selkoe et al., 1987). These widely distributed changes suggest that some ubiquitous neurodegenerative process selectively afflicts certain neocortical neurons and related subcortical pathways, particularly through the hippocampus. However, neurofibrillary tangles and paired helical filaments have not been reported during senescence in any species besides humans. Concurrent with β-amyloid accumulation during Alzheimer's disease are selective changes in alternatively spliced amyloid mRNA species. There is evidence for a relative increase of mRNA encoding protease inhibitors (APP-751-mRNA) in regions of plaque accumulation (Johnson et al., 1988, 1989, 1990; but see Koo et al., 1990 for different conclusions). Increased proteases could favor the accumulation of abnormal proteins by altering proteolysis (Kitaguchi et al., 1988; Ponte et al., 1988; Tanzi et al., 1988; Chapter 7.4).

Besides neuron loss, another important change of Alzheimer's disease concerns the dendritic arbor. A major ongoing study of dendritic morphology of a well-characterized set of brains from normal and Alzheimer's disease victims by Paul Coleman, Dorothy Flood, and colleagues at the University of Rochester shows a high degree of regional specificity in whether the arbor of a given neuronal type in the cortex or hippocampal formation expands (grows), shrinks, or does not change (Flood and Coleman, 1990; Flood, 1990; Buell and Coleman, 1981). It is clear that, even at advanced stages of Alzheimer's disease or in the tenth decade, healthy neurons remain. These changes in dendritic arbor may represent several different mechanisms that include neuron compensatory responses to changes in afferent fibers from neuron death elsewhere (Pasinetti et al., 1989; see below), changes in the availability of local neurotrophic factors (Hefti et al., 1990), and response to the sociobehavioral environment that can reversibly influence neuron morphology and synaptic density at adult ages in rodents (Rosenzweig et al., 1971; Greenough, 1985).

The loss of neurons may stimulate compensatory responses in certain areas. Both Alzheimer's victims and the normal elderly show synaptic remodelling in the dendritic arbor of neurons in the granule layer of the hippocampal formation (Geddes et al., 1985 Flood and Coleman, 1990). These particular responses are associated with degeneration of the perforant pathway from the entorhinal cortex. The capacity to modify the dendritic field may diminish after 80 years in the normal elderly (Flood and Coleman, 1990).

At the neurochemical level, several general age-related changes are described in laboratory rodents, rabbits, monkeys, and humans. The most generalized involves the progressive loss from maturity onwards of dopaminergic type 2 receptors (D-2) in the neos-

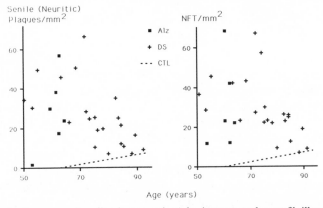

Fig. 3.39. *Left,* the frequency of senile plaques and, *right,* intraneuronal neurofibrillary tangles (NFT) in postmortem specimens of temporal cortex. +, Alzheimer's disease and, *square,* Down's syndrome as a function of age; *dotted line,* average values for ten "normal" controls. These findings suggest that postmortem diagnosis of Alzheimer's disease by histopathological criteria may be more definitive at ages less than 70 years. The criteria for Alzheimer's disease at later ages may need to be revised. Redrawn from Mann et al., 1984.

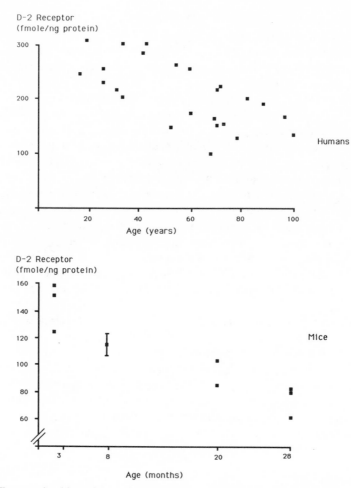

Fig. 3.40. The age-related loss of dopaminergic receptors (D-2 subtype) in the neostriatum. *Top,* neurologically normal humans (from Morgan et al., 1987) and, *bottom,* C57BL/6J male mice. Redrawn from Severson and Finch, 1980.

triatum (Morgan et al., 1987; Morgan and May, 1990; Figure 3.40). Small but reproducible decreases are widely found between puberty and mid-life in mice, rats, rabbits, and humans (Table 3.3). Similar decreases occur in the serotonergic type 2 (S-2) receptors of the neocortex (Marcusson et al., 1984; Morgan et al., 1987). The cause of the loss is unknown and could involve reduced dendritic fields (Cotman, 1990; Flood et al., 1985; Scheibel, 1978; Pentney, 1986) or neuronal loss. The consequences of these substantial biochemical changes are not clear but may be linked to the age-related changes in the precision of motor control (D-2 receptors) or emotional stability (S-2 receptors). The loss of striatal dopamine is more modest in rodents than in humans (Finch, 1973a; Morgan et al., 1987). The trend for gradual age-related loss of striatal dopamine results in much smaller loss than found in Parkinsonism, which is typically more than 90% (Bernheimer et al., 1973; Riederer and Wuketich, 1976). While the age-related risk of Parkinsonism increases strikingly after mid-life (Figure 3.38), there is little insight about the relation of Parkinsonism to basal gangliar neuronal loss or neurochemical changes in the general population (Agid et al., 1989; McGeer et al., 1989; Wolters and Calne, 1989).

In conclusion, neuronal changes show a high degree of selectivity with respect to neuron loss, changes in neuronal morphology, and biochemical changes throughout the lifespan and in Alzheimer's disease. These complex phenomena are ultimately rooted in selective aspects of gene regulation that determine the cellular characteristics during differentiation and that determine the responses of neurons to shifts in the chemistry of their environment as well as the consequences of function over extended time.

3.6.11. Wear-and-Tear Syndromes

During long lifespans of years to decades, many mammals are exposed to diverse risks for damage to irreplaceable cells and tissues that can be viewed as a source of gradual senescence through "wear and tear." At the subcellular level, the accumulation of lipofuscins progresses throughout the lifespan in some cells (Chapter 7.5.6) but without any associated dysfunction. Although mammals have no counterpart to the mechanical senescence shown in insect wings and other irreplaceable appendages during their short lifespans (Chapter 2.2.2.1), mammals do manifest certain cumulative effects from chronic use, the wear-and-tear syndromes. A model for cumulative effects from exogenous or endogenous factors in Figure 3.41 represents individual life histories as trajectories that are defined over time by various physical and chemical agents. These vectors define surfaces in accordance with parameters representing strength and duration of agents having time-dependent impact that collectively can be called gerontogens, a useful term coined by George Martin (1987). This model fits certain cumulative effects of estrogen exposure on the hypothalamic pituitary axis of rodents during female reproductive senescence (Finch, 1988; Lerner and Finch, in prep.; Chapters 3.6.9, 10.4.2). Gerontogens span a wide range of agents, from mechanical transferred energy that abrades a surface like insect wings to chemically transduced signals like those involving steroids or carcinogens. The following are additional examples of gerontogens that mostly involve mechanical transfer of energy.

3.6.11.1. Vascular Lesions

Hemodynamic factors influence the extent of vascular lesions. As described in Section 3.6.7, hypertension promotes atherogenesis. Moreover, lesions tend to develop at points of the most turbulent blood flow (Blumenthal et al., 1954), for example, at the bifurcation of the carotid artery near the carotid sinus (Ku and Giddens, 1983; Zarins et al., 1983). These phenomena could be the basis for predictions about the extent of age-related vascular lesions in species with different blood pressures and vessel configurations.

Another insight comes from the deterioration of saphenous vein bypass grafts, which are used to replace atherosclerotically blocked coronary arteries. Within 1 day after surgery, the intimal layer of the grafted veins developed changes resembling those of early atherogenesis, which can lead to the formation of new fibrous plaques. Some degree of atherogenic changes had occurred in patients who died within 1 month of surgery (Bulkley and Hutchins, 1977). Together with the blood lipid risk factors, these examples emphasize the importance of the blood vessel hemodynamic environment to emergent ather-

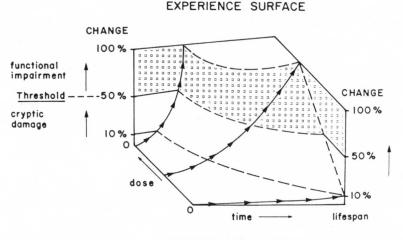

Fig. 3.41. A model for the cumulative impact of cell or organ damage over time that may be incurred by an individual from either endogenous or exogenous factors (toxins, nonenzymatic glycation, mechanical trauma, etc.). This "experience surface" shows two trajectories that differ in the intensity of exposure to damage. The model assumes that damage results from a simple strength-duration relationship that is commutative within certain boundary values. Thus, an intensity of four units per year over 1 year exposure gives the same integrated value as two units per year for 2 years. The damage is assumed to be irreversible. A threshold is depicted to represent a transition from cryptic damage to manifested dysfunction. A more realistic model in relation to senescence might include effects of age on the threshold and changes in sensitivity. From Finch, 1988, and S. Lerner and C. Finch, unpublished.

ogenesis. The irreversible damage done to blood vessels by hypertension and by turbulent flow can be considered a mammalian equivalent to the mechanically abraded and irreplaceable insect wings.

3.6.11.2. Hearing Loss

Age-related hearing loss (presbycusis) is very common, third in rank of chronic conditions of the noninstitutionalized elderly in the U.S. (Feller, 1981; Rees and Duckert, 1990). Presbycusis also occurs in laboratory mice, with marked strain differences (Willott et al., 1985; Henry, 1982), but the distribution in other species is not known. An important factor in age-related hearing loss is the loss of hair cells in the cochlea, as observed in humans (Johnsson and Hawkins, 1972) and guinea pigs (Coleman, 1976; Úlehlová, 1975). Two major causal factors are thought to be vascular disease and cumulative effects from noise. One study suggests that age-related hearing loss is not inevitable, at least in very quiet environments. The Mabaan, a peaceful, isolated tribe in the southern Sudan, have unusually well-preserved hearing in the higher frequency range (> 4,000 Hertz; Figure 3.42; Rosen et al., 1962, 1964). Few noises exceed 80 decibels, except at festivals during the 2-month spring harvest, where group singing transiently reaches 110 decibels,

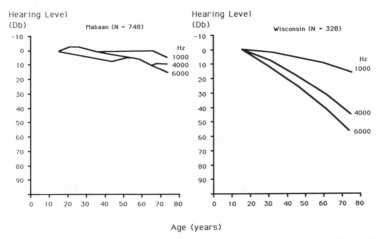

Fig. 3.42. Age-related trends for loss of hearing sensitivity at different frequencies (1,000, 4,000, and 6,000 Hertz) in two populations of men: the Mabaan, a pastoralist tribe in the Sudan, and a Wisconsin population. Redrawn from Rosen et al., 1962.

an intensity that is loud but not painful. The Mabaan eat little fat and show little hypertension, cardiovascular disease, or other signs of atherosclerosis. Their mean blood pressures remain stable at least through the eighth decade, at about 15 mm Hg below North Americans. Although not investigated in the Mabaan, atherosclerotic degeneration and stenosis are common in the vessels of the inner ear in other populations of older humans (Fisch et al., 1972; Bochenek and Jachowska, 1969; Schuknecht, 1974). Populations with extensive hypertension and vascular disease apparently have more hearing loss (Rosen et al., 1964; Bochenek and Jachowska, 1969; Rosen and Rosen, 1971).

Exposure to noise has cumulative effects, particularly for higher frequency sounds, 2,000–6,000 Hertz, which are in the upper third of frequencies encountered in ordinary conversation (1 to 2 octaves above middle C). Acoustical trauma causes the destruction of irreplaceable hair cells in the cochlea (Spoendlin and Brun, 1973). The extent of hearing loss fits well to commutative strength-duration relationships (the product of noise intensity × duration of exposure) above a threshold characterized for occupational exposure (Henderson et al., 1976; Robinson and Shipton, 1973). Similar exposure relationships have been described in laboratory animals (Spoendlin and Brun, 1973). Thus, there may be a lifetime dose of sound exposure, which could be modelled according to Figure 3.41. However, it is not clear what the boundary values are, and it seems plausible that there is a lower threshold for sound intensity below which no damage is done.

3.6.11.3. Skin Changes from Exposure

Changes in skin epitomize many aspects of gradual senescence that are driven by extrinsic gerontogens. The vast majority of humans older than 75 years appear to have at least one cutaneous abnormality (Beauregard and Gilchrest, 1987). Some, but by no means all, of the age-related changes in skin are caused by damage from UV from sunlight ("photo-aging"), particularly in regard to abnormal growths but also for damage to collagen and other connective-tissue macromolecules (Kligman et al., 1985; Gilchrest, 1990). Many degenerative changes are prematurely induced by intense exposure to sunlight

(Kligman, 1969; Forbes et al., 1979; Braverman and Fonferko, 1982; Montagna and Carlisle, 1979; Selmanowitz et al., 1977). Wrinkling is also associated with exposure to sunlight. Elastic fibers in the upper layers of the skin are more damaged than those in deeper layers, and there is evidence of hyperactive fibroblasts in the epidermis (Gilchrest, 1990; Tsuji, 1987). The cumulative exposure to sunlight might conform to the model in Figure 3.41. As for sound exposure in the ear, it is plausible that there is a lower threshold of UV below which damage does not occur. UV-induced DNA repair processes could also be important in protecting against some aspects of solar exposure (Chapter 8.2.6).

However, age-related degenerative changes are also observed in regions of the human skin that have minimal exposure to sunlight. For example, the buttocks also often show a loss of subepidermal elastic fibers and flattening of the rete pegs that help hold the dermis and epidermis together; individuals can vary markedly in these changes (Braverman and Fonferko, 1982; Kligman et al., 1985). Decreased density of veins is also indicated in sun-protected buttock skin (Gilchrest et al., 1983). Although skin changes may be dramatic at advanced ages in unexposed areas, thorough samplings have not been made in individuals of known health status. Possible relationships between changes in connective tissue elements of the skin and menopause or other physiological changes is unknown. The increased glycation of human skin collagen in diabetics in excess of the slower increase for the same ages (Schnider and Kohn, 1980, 1981, 1982) indicates that another source is damage from elevated blood glucose. The role of glycation in the fraying of skin elastic fibers of 2-year-old male rats (Imayama and Braverman, 1989) is unknown.

Despite strong biochemical protective mechanisms, abnormal growths in the skin are strongly associated with exposure to sunlight. The very common senile keratoses cause patches of skin to thicken and discolor and may be premalignant. Fair-skinned people have the highest risk for solar keratoses. The incidence often exceeds 50% by 70 years, apparently as the result of cumulative exposure to sunlight (Forbes et al., 1979; Marks et al., 1988). Squamous cell carcinomas (nonmelanomatous) and melanomas, though rarer, also increase with age and solar exposure (Lee, 1982; Forbes et al., 1979). Nearly all individuals with squamous cell carcinomas also had solar keratoses, an observation that firmly links both to solar exposure (Marks et al., 1988). The epidemiology of squamous cell carcinomas is unclear, in part because this common cancer is not reported to U.S. tumor registries. The incidence elsewhere increases exponentially with age, approximating the acceleration of mortality (MRDT, equation [1.3]); Appendix 1; Glass and Hoover, 1989).

These observations on a common phenomenon also yield the consideration that biochemical protective mechanisms were evolved in certain mammalian radiations to permit humans, elephants, and other sparsely haired diurnal terrestrial organisms to survive "lifetime" doses of sunlight in accord with selection for extended reproductive schedules.

3.6.11.4. Tooth Erosion

The last example of wear-and-tear syndromes is the wearing down of adult teeth. As Williams noted (1957), the senescence of teeth amounts to their nonreplacement when worn down, not tooth erosion itself. With few exceptions, mammals have a fixed number of adult teeth which are irreplaceable. Depending on the species, certain teeth continue to grow throughout the lifespan (rodent incisors and elephant tusks); the molars are generally unable to do so (horses are an exception) and therefore show cumulative erosion.

Survey of a scattered literature indicates that tooth erosion is a major limitation on the maximum potential lifespan of numerous placental mammals. Tooth wear is a particular problem for grazing animals, because of the abrasive phytoliths that are present in many grasses and because of the earth or sand that is chewed along with the vegetation. Although many other sources of senescent debilitation may cause death, the inability to ingest an adequate diet because of worn or damaged teeth is an ultimate and predictable outcome of advanced age. Erosion of teeth to a dysfunctional limit is documented for the hippopotamus by 40 years (Laws, 1966) and the elephant by 60–80 years (Laws and Parker, 1968; Sikes, 1971) even in zoos (Schmidt, 1978). The progression of molar erosion is quantitatively documented for the hippopotamus (Figure 3.43) and the white-tailed deer (*Odocoileus virginiansus*; Severinghaus, 1949; Geist, 1971). Besides the loss of grinding surface in white-tailed deer, the wearing of teeth down to the gum and the exposure of the pulp cavity may increase the risk of infections (Severinghaus, 1949). Extensive erosion of molars by 10–20 years must be an ultimate limiting factor on lifespan in many species, unless non-abrasive foods are available. The particular pattern of tooth erosion varies widely among species because species differ in the teeth that continue growth.

Few quantitative studies of tooth erosion or loss by age group are available for mammals besides the hippopotamus (see above) and the white-tailed deer. Domestic cattle also show age-related erosion of teeth, particularly incisors (Wass et al., 1981). Metal tooth-caps or special diets are sometimes used to prolong the breeding or milk-producing life-

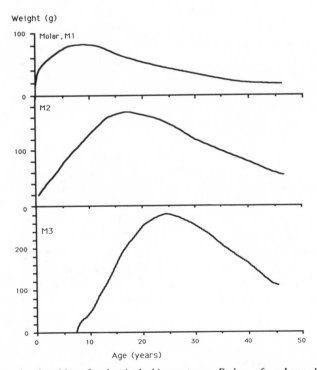

Fig. 3.43.   The age-related attrition of molars in the hippopotamus. Redrawn from Laws, 1966.

span in commercial herds. Anecdotal evidence indicates the importance of tooth loss in feral animals, including kangaroos (Breeden, 1967, 25); chimpanzees by 40 years, where the resulting diminution of food intake may compound osteoporosis (Zihlman et al., 1989); and African lions by 20–25 years (Schaller, 1972, 191). Schaller noted the debilitated appearance of tooth-worn old lions, even though they continue to feed on kills made by others in their pride. The irreversible wear on teeth is also widely used to estimate the ages of feral primates, according to a working assumption that wear proceeds linearly after mid-life (Bowden and Williams, 1985).

Tooth loss in humans increases quite linearly with age. By 65 years, 30–60% have lost all teeth, depending on the survey, socioeconomic group, and gender (Weintraub and Burt, 1985). There is a trend for decreasing tooth loss at later ages, which reflects cohort-specific effects including improved dental care as well as attitudes about tooth removal.[4] Loss of teeth may be underestimated as an adverse factor in the already precarious health of many older people, because of its impact on mental well-being and eating habits, which only rarely were fully correctable by prosthetic teeth.

Elephants are a unique example of evolutionary adaptations in dentition that permitted an extended reproductive schedule. Their lifespans last 70 or more years (Nowak and Paradiso, 1983; Laws, 1966; Sikes, 1971), a mammalian longevity that may be excelled in nature only by humans. While African elephants can reproduce as early as 8 years, limited food or water in bad years may on rare occasions delay maturation until after 20 years (Laws and Parker, 1968; Moss, 1988). Reproduction adds major and prolonged demands on the teeth from increased food intake. After a long gestation of 22 months, the cows must provide milk, usually for 9 months. Intervals between calving are about 8 years. No other large ungulate or carnivore has such an extended reproductive schedule. In fact, few other species are reproductively active beyond 20 years, an age group when some elephants may be calving their first. The vital role of female elephants extends to their later years: old females at 50 or 60 are the herd leaders (Moss, 1988). They may eventually become menopausal, but at ages when their teeth are about worn out. Sikes (1971, 184, 222) emphasizes the connection between dental insufficiency and eventual decrepitude that may be the leading factor in the senescence of those few that survive to advanced ages. However, other factors of senescence lurk behind tooth wear, including sensory impairments and vascular diseases (atherosclerosis, medial calcification, and varicose veins; Sikes, 1971; McCullagh, 1972).

Elephants place huge demands on their teeth during their daily intakes of 100–200 kilograms of foodstuffs that require continuous chewing during most of the waking hours (Eisenberg, 1981; Kingdom, 1979). Elephants appear to have managed this through several evolutionarily adaptive changes in dentition, which postpone the age when tooth wear becomes a vital deficiency. First, the specialized dentition of elephants is used for the horizontal shearing of food against the enamel ridges by anterior-posterior motions of the jaw (Maglio, 1972; Turnbull, 1970). The pacemaker for tooth wear in elephants is the enamel, which is more durable than dentine or cementum (Maglio, 1972). Two specializations favor long use of the molars: their unusual length and their elevated (hypsodont) enamel ridges (Figure 3.44).

---

4. Until recently, health professionals in some parts of the U.S. would commonly advise removing all teeth as a hedge against further dental problems.

Top (Crown)

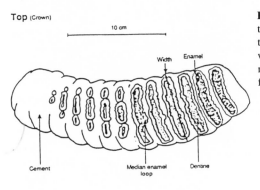

Fig. 3.44. The structure of elephant molars, showing the multiple lamellar plates (*1–13*) and the enamel loops that form the shearing edge. With wear, each tooth wastes away but is simultaneously replaced by another molar that grows in horizontally. Adapted and redrawn from Maglio, 1972, by Lali Medina.

FRONT   REAR

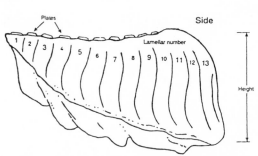

Second, elephants have an unusual scheduling of tooth eruption or emergence, which provides replacement teeth long after maturity. Unlike nearly all other mammals, elephants (*Loxodonta africana*, African; *Elephas maximus*, Asian) have up to six replacements of their adult cheek teeth (Peyer, 1968; Kingdom, 1979; Laws, 1966; Sikes, 1971; Moss, 1988). In present elephants, the teeth on each side of the jaw (in the anatomic position of molars) are replaced continuously, so that a new molar is growing in as the prior molar wears down. Tooth replacement results from sequentially delayed tooth eruption, so that the total number does not exceed the eleven adult teeth per jaw quadrant that characterize placental mammals.

Just one of the elongate molars is fully used at any time in adult African and Asian elephants (Peyer, 1968; Laws, 1966; Maglio, 1972; Sikes, 1971). Each tooth set is identifiable by the number of enamel ridges on the grinding surface, which are formed by the sequential addition of contiguous lamellar plates (Figure 3.44). In contrast, the molars of most other mammals are formed as a unit. The next ingrowing tooth is slightly larger with one or more additional lamellae, which are added sequentially. Tooth wear and tooth growth proceed more or less in synchrony, so that worn teeth are shed as the lengthening molar moves forward. Again this is unusual, because most mammalian teeth are replaced from below, rather than horizontally. Set 4 is lost by 30 years and set 5 after 40. Set 6, which begins to grow in at about 30 years, is usually the last. Occasionally a seventh set may appear (Laws, 1966; Sikes, 1971). For comparison, humans have delayed eruption of just the third molars, or "wisdom teeth," which usually appear between 20 and 30 years but may be delayed further. In elephants, each set of molars lasts about 20 years, the last

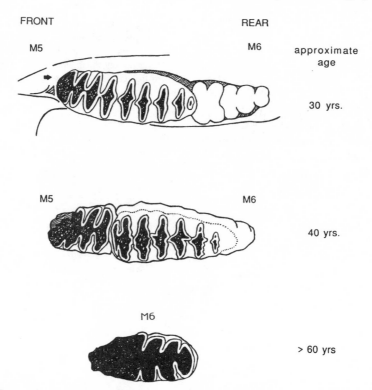

Fig. 3.45. Progressive tooth erosion and replacement in African elephants: the last two molars. Most elephants have a total of six molars on each jaw quadrant. When the last molars are ground down after 50 years, elephants can no longer feed adequately (Sikes, 1971; Moss, 1988; Laws, 1966). This is an example of mechanical senescence but is only one of several major causes limiting the elephant lifespan. Adapted and redrawn from Maglio, 1972, by Lali Medina.

(sixth) set usually grinding down by 60 years (Figure 3.45). It would be interesting to know if the rarely occurring seventh set was a genetic polymorphism and if it conferred a longer lifespan.

The elephant dental pattern evolved recently in gomphothere radiations during the Miocene, about 5–10 million years ago (Figure 11.14). This evolutionary change raises an interesting question: Did the likely increase of lifespan and extension of the reproductive schedule then select in small or large steps for genes like Haldane's (1941) modifiers of senescence (Chapter 1.5.2) that delayed disorders of senescence besides tooth erosion? Alternatively, the gomphotherian stem line might not have yet acquired genotypes for these particular disorders. The increase of lifespan could conceivably reschedule the delayed pleiotropy of numerous genes whose expression was related to reproductive senescence (e.g. osteoporosis) or might purge them from the genome altogether. These questions are considered further in Chapter 11.4.

Continuous replacement of molars is rare in most other ungulates that must also endure tooth wear over extended time. Manatees (*Trichechus*) show a pattern, producing unlimited numbers of molars, at least up to thirty, on each side of their jaw. This most unusual phenomenon occurs throughout the genus: *T. inunguis* of the Amazon basin; *T. manatus* of Florida; and *T. senegalensis* of West Africa (Weber, 1928; Peyer, 1968; Domning, 1982, 1983, 1987; Domning and Hayek, 1984). No other mammal has this degree of polyphyodonty, which may be considered evolutionarily adaptive for extreme tooth wear caused by sand chewed along with seeds and other fodder in the manatees' aquatic environment. Tooth replacement presumably favors their relatively long lifespans, which exceed 30 years in captivity (Nowak and Paradiso, 1983). New teeth are formed at the back of the jaw and enter horizontally, as worn teeth are shed from the front of the mouth. Replacement varies with diet and is stimulated mechanically by chewing fibrous plants (Domning and Hayek, 1984). Another less documented occurrence of continuous tooth replacement is found in the Australian rock wallaby *Petrogale* (Domning, 1983).[5] The 14-year survival record for *Petrogale* is based on one individual (Nowak and Paradiso, 1983) but does not exceed the reported maximum lifespans of other macropods (Jones, 1982). Manatees and rock wallabies are among the few mammals recovering a capacity for *de novo* tooth formation that characterized their reptilian ancestors (Chapter 11.3.5.2). Nonetheless, long lifespans do not always depend on the unusual tooth replacement capacities of elephants or manatees.[6]

In summary, animals that must routinely chew abrasive foods are at particular risk for encountering a debilitating extent of tooth wear. While many specializations in dentition compensate for tooth wear (continuous growth of incisors or molars, delayed tooth emergence, continuous replacement), there is no report of an older animal, except perhaps humans, that can maintain its full adult strength when tooth erosion is encountered later in life and alternate diets may be sought. For many species besides ungulates, tooth erosion is a mechanical form of senescence that serves as a pacemaker for longevity.

The above examples were chosen to represent different types of experience, involving the transfer of small amounts of energy per event, that are cumulative with injurious consequences to the transducing organ. While the trajectories of Figure 3.41 are defined for particular age-related (actually event-dependent) vectors, a further development of such models would define the relation to the trajectory of mortality risk. Numerous other physical, chemical, and behavioral factors bear careful consideration as possible gerontogens. Such models may be made more complex to account for age-related changes in other parameters that may alter responses to a particular agent, for example, the interaction of declining kidney function with steroid clearance. Beyond cumulative effects from gerontogens on specific functions is their relation to mortality risk or, of equal importance to population genetics theories on the evolution of senescence, the risk of reproductive failure (Chapter 1.5).

Future discussions may consider cryptic and cumulative damage from head injury in young adults in relation to later cognitive decline (Corkin et al., 1989); alcohol in relation

5. There are six species of *Petrogale* (macropod), also congeneric with two species of *Percodoras*.

6. In contrast to the manatees, the other surviving sirenian (*Dugong* sp.) has a maximum of six teeth per jaw quadrant and lacks continuous replacement. As the molars erupt, their enamel crowns are soon worn off. The first molar is usually lost by 25 years, but the dentine pegs on the remaining two molars continue to grow axially and provide grinding surface for masticating the dugongs' abrasive diets (Marsh, 1980; Domning 1982, 1983). Dugongs live at least 55 years (Nowak and Paradiso, 1983). The natural history of Sirenia is too poorly known to say if manatees are really shorter-lived than dugongs. However, there is no doubt that the preferred manatee diets would, through tooth erosion, curtail their lifespan potential, were it not for continuing tooth replacement. The dugongs prove that the general habits of Sirenia do not require continuous tooth replacement. Moreover, the recently extinct Steller's sea cow (*Hydrodamalis*) had no adult teeth (Peyer, 1968; Nowak and Paradiso, 1983).

to degenerative changes in the heart and skeletal muscles (Urbano-Marquez et al., 1989); smoking in relation to hypertension as well as abnormal cell proliferation; stress in relation to neurodegeneration (Chapter 10.5.1); numerous toxins and chemical carcinogens from the environment; and occult toxins, such as mercury that is a major component in the amalgams used during the last 150 years to make "silver" tooth fillings (Hahn et al., 1989; Enwonwu, 1987). It is already clear that the hypothalamic-pituitary axis of rodents but not humans is vulnerable to damage from chronic physiological elevations of estrogen (Section 3.6.9; Chapter 10.4.2). Thus, there will be major influences of species and genotypes on the cumulative impact of prospective mechanical and chemical gerontogens.

3.7 Summary

To summarize these observations on gradual senescence is to acknowledge enormous gaps of facts and insights. A range of proximal causes of senescence is observed in multicellular organisms that are semelparous or iteroparous. In contrast to semelparous species that die rapidly in association with reproduction with a cohort-wide canonical pattern of senescence, individuals in an iteroparous population appear to vary more widely in the extent of specific lesions, like arteriosclerosis, cataracts, tumors, or kidney lesions. Also, in contrast to some semelparous species, there is no evidence that senescence is triggered in iteroparous species by the cessation of growth or the onset of reproduction. In the small, short-lived fish *Lebistes*, exponential accelerations of mortality occur despite continued growth. In mammals, skeletal growth usually ceases soon after maturity. If sexual maturity or the acts of reproduction trigger senescence, the adverse consequences must be very delayed; some evidence that castration and hypophysectomy may slow senescence is reviewed in Chapter 10. As for semelparous species, variations among taxa in the types of lesions are consistent with predictions from evolutionary theory (Chapter 1.5) that wide variability of age-related lesions are to be expected because the force of selection on the survivors decreases sharply with advancing age.

Patterns of senescence are known in detail for only a few vertebrates besides laboratory rodents, some domestic animals, and humans, certainly far fewer than those whose patterns of development are known. The only universal indicated for vertebrate senescence is the exponential acceleration of mortality: again, except for mammals this is documented for only a few fish and birds. There are numerous examples of epigenetic variations in the timing of events during development and maturation throughout chordate species, but with the exception of the peculiar growth patterns of ferox trout, these alternate schedules have little impact on adult morphology.

Among the candidates for universals in senescence at an organ or cell level are accumulations of lipofuscins, the risk for damage or loss of irreplaceable cells, and increases of abnormal growths. A neuroendocrine change reported for rodents and humans is the circadian rhythm of growth hormone, which shows deficits in at least one secretory phase. The pulsatile secretion of LH also becomes slowed with lower amplitude. Changes in hormone regulation could alter gene activities in many target cells. Data are insufficient to judge if the age-related increase of tumors and other abnormal growths commonly seen in mammals also generally occurs in populations of long-lived birds, reptiles, or fish. While several organ systems of mammals show evidence for mechanical senescence (teeth, joints, blood vessels), the role of these phenomena in the senescence of lower vertebrates is uncertain. Mechanical senescence is shown, however, in long-lived itero-parous beetles like *Carabus*. Virtually nothing is known about the cellular, physiological, or pathological changes of long-lived invertebrates.

The extent of age effects on the production of abnormal gametes is of great interest for species that have not yet shown declining gamete production at later ages, for example, long-lived birds (Section 3.5) or lower vertebrates described in the next chapter. Some evidence indicates maternal age-related increases of abnormalities in ova and development in alligators, as has been well-documented in rodents and humans. The extent to which *de novo* oogenesis continues during adult life is little investigated in long-lived iteroparous species besides mammals, and bears on evolutionary predictions (Chapter 1.5) that repro-ductive declines will favor the accumulation during evolution of germ-line mutations and gene combinations with delayed adverse impact.

For example, the development of tumors or kidney diseases by 2-year-old laboratory rodents could not be strongly selected against because female rodents have virtually ex-hausted their ovarian oocyte pools by this age (the Haldane-Medawar-Williams view, Chapter 1.5.2). The limited evidence for a few species of elasmobranchs and sturgeon that they too have limited oocyte stocks (Section 3.4) leads to a prediction that manifes-tations of senescence should be more intense in these fish than in teleosts at the same age that may maintain *de novo* oogenesis and increase their production of ova commensurately with their continued growth (Chapter 4.2.2).

The present data suggest that few mammals besides elephants, humans, and whales survive in nature or in domestic protection beyond 50 years. The vast majority of mam-mals propagate their species in much shorter lifespans, with maturation rarely later than 6–10 years. A substantial number of birds, however, continue reproduction into their 20s, some to even later ages, without any signs of reproductive senescence. Thus, mammalian life histories are relatively accelerated by comparison with many long-lived birds for which senescence has yet to be observed.

In mammals, the most universal change is failure of reproductive capacity because of the exhaustion of ovarian oocytes during the middle third of the potential lifespan. Nu-merous other changes in physiological functions also occur in virtually all individuals during the middle phase of adult life of short- and long-lived mammals, including declines in thermoregulation, immune responses, and other homeostatic capacities and neural cir-cuit functions like asynchrony of biorhythms and slowed motor responses (Table 3.3). Together with ovarian oocyte exhaustion, these pathophysiological changes can be con-sidered elements in the canonical patterns of mammalian senescence. However, there are many taxa-specific diseases of senescence: for example, pituitary tumors are far more common during senescence in rodents than in humans, whereas benign prostatic hyperpla-sia is very common in humans, but rarer in rodents. As described in greater detail in Chapter 6, inbred rodent strains characteristically differ in the organ or cell systems that show age-related lesions. This diversity in the manifestation of mammalian senescence matches expectations from evolutionary theory (Chapter 1.5).

However, several features are not consistent with the concept that the germ lines in populations inevitably accumulate random mutations and genetic polymorphisms through combination during evolution that have delayed adverse effects (Chapter 1.5). If so, why do certain manifestations occur so consistently in corresponding fractions of the lifespan?

For example, both genders show progressive losses of striatal dopaminergic receptors and in T-cell responses that begin soon after maturation, and a later onset of bone and joint disorders (osteoporosis and the arthritides). In regard to reproduction, two features stand out. Female mammals have about the same fraction of the lifespan (30–50%) remaining after reproductive failure from oocyte exhaustion. In contrast, few males ever show complete failure of spermatogenesis; I know of no mammal that shows 100% reproductive failure during senescence in males. The proportionate scaling of at least some age-related changes in particular cell and organ systems across the thirtyfold range of lifespans (Table 3.3) is contrary to the population genetics models (Chapter 1.5) that the weakening of selection from the relative rarity of survival to advanced ages will therefore yield accumulated germ-line genes that encode essentially random delayed onset dysfunctions. Conclusions on this major issue are of course tentative, since reliable data are available for only a handful of species.

The survival of occasional females in natural populations to postreproductive ages is also of much interest in regard to the general premise that competition for resources must rapidly eliminate individuals that are not strongly contributing to their germ line. Some pilot whales, and possibly also ringed seals, elephants, and several primates, are reported to survive in field populations into postreproductive years (Section 3.6.1). The potential for continued health and survival after age-related sterility is likely for other species under protected conditions, including earthworms and Siamese fighting fish. Further understanding of these potentially remarkable findings requires much more detailed information on population age structure and reproductive status by age.

The apparent lack of impairment in other cell functions during the lifespans of rodents and humans is of great interest in general, but particularly in predictions from the population genetics arguments that all organismic physiological systems should eventually display impairments during senescence. Provisional examples include the undiminished capacity for elevations of corticosteroids by the hypothalamic-pituitary-adrenocortical axis, the capacity to secrete vasopressin, the sympathetic neurosecretion of norepinephrine, the capacity to make red blood cells, and antigen presentation by macrophages. However, review of molecular studies on genomic activities throughout the lifespan, mostly in mammals (Chapters 7, 8), gives a mixed outcome, with evidence for random as well as selective impairments in genomic functions at later ages and major species differences, for example, in bulk protein synthesis. Further analysis is needed to determine whether or not there are species differences in the cell systems that appear to remain intact throughout the lifespan.

So far, there is no *general* way to decide whether a particular age-related change is truly intrinsic to an organism and would arise in all environments, or whether it represents a time-dependent interaction with environmental risk factors that can be greatly modified. Environmental risk factors of vascular disease, hypertension, and cancer are amply documented in humans and several laboratory mammals. These and other examples of environmental modifications of senescence will be analyzed in Chapters 9 and 10. Many species showing rapid senescence in association with their first and only bout of reproduction, for example, marsupial mice and Pacific salmon (Chapter 2), can live much longer under some conditions and would probably have a more gradual senescence with age-related increases in lipofuscins, cell loss, and abnormal growths. Such phenomena remain to be shown.

The next chapter raises the possibility that pathophysiologic senescence and marked increases of mortality are not universal in all phyla. The evidence will be discussed as a further test of population genetics hypotheses on senescence (Chapter 1.5). Moreover, the absence of senescence and definite lifespans in some species that are anatomically similar to senescing organisms suggests a general underlying plasticity in lifespan outcome. If so, then we can consider the possibility that many mechanisms of senescence are biologically superficial and should be reversible, if somatic cells retain a complete set of chromosomal genes.

4

Negligible Senescence

4.1. Introduction

The first three chapters gave many examples of species with definite maximum lifespans that resulted from progressive age-related accelerations of mortality rate in association with progressive pathophysiological changes. Yet scattered throughout the world are other species at all grades of organization that do not show indications of whole-organism senescence, that do not show age-related increases of mortality risks, and that represent the slow extreme in the continuum of senescence. Depending on the IMR (equation [1.2]), their lifespans may be long or short. Recall from Chapter 1.4.8 calculations that, even without senescence-related increases of mortality, flies would have median lifespans of about 0.5 years. Often in natural populations, relatively high age-independent rates of mortality (IMR) will obscure age-dependent mortality, so that few individuals may survive to ages when biological senescence is expressed, as modelled in Figure 1.7. In nature, the lowest mortality rates are still high enough to prevent nearly all individuals from surviving beyond 1,000 years, even in the absence of senescence. The values of IMR are low enough in numerous other species to permit individual lifespans of hundreds to thousands of years, or up to 10^4 longer. However, maximum lifespans in populations of slowly senescing organisms are more commonly 50 to 100 years. Because mortality rates are relatively stable for these negligibly senescing species, their maximum lifespans are statistically indeterminate. Nonetheless, because population sizes are finite, a maximum lifespan nonetheless will occur, irrespective of the extent of senescence (Chapter 1.4.8).

Important exceptions to finite lifespans are given by organisms that reproduce vegetatively and agametically and that are immortal by clonal definition. Harper (1977) introduced the term *genet* to represent all organisms from the same zygote and *ramet* to represent a demographically distinct individual, whatever its genotype. Vegetative reproduction is discussed in Section 4.4.

When populations show high AMRs, it may be difficult to detect age-related increases in mortality that arise from various exogenous causes in the environment or from endogenous senescence. From one perspective in population genetics, there may be no difference in whether or not senescence contributes to the AMR that sets the statistical lifespan.

Species that do not show age-related increases of mortality challenge premises about the inevitability of senescence at cellular or physiological levels, and such organisms are scattered throughout the animal and plant kingdoms. The occurrence of very slow or negligible senescence at the level of the whole organism suggests that experimental manipulations and natural variations of senescence, particularly in species with rapid senescence and sudden death, have relieved or alleviated specific points of vulnerability. In most species, variations in senescence appear to occur at an epigenetic level, rather than through a hypothetical "intrinsic or basal senescing property" of molecules and cells.

Two major categories of organisms with indeterminate lifespan are discussed. (1) Sexual reproducers have indeterminate lifespans without detectable age-related increase of mortality, and very gradual or negligible senescence. (2) Vegetative reproducers propagate by vegetative (asexual) reproduction from somatic cells to yield ramets that usually have finite lifespans as individuals (ramets), yet have indeterminate lifespans as a genetic clone (genet). Although the individual lifespan may be considered finite, vegetatively reproducing species demonstrate the indeterminate lifespan of some euploid somatic cell clones; this has major theoretical importance. Taken together, these two categories give a basis for the hypothesis that *somatic cell lineages are not inevitably predestined to senescence*. Because many of the examples given in the previous chapters manifest the onset of senescence, rapid or gradual, in association with maturation and the cessation of growth, I also review Bidder's hypothesis (1932) that relates senescence to the cessation of growth, by examining the patterns of growth in species with negligible senescence.

4.2. Indeterminate Lifespans and Negligible Senescence

The concept of indeterminate lifespan arises from the absence of a characteristic maximum lifespan and the absence of a clear-cut senescent degeneration at the level of the organism. Two alternative criteria indicate if lifespan is potentially indeterminate: (1) the age-dependent mortality rate remains relatively constant after maturity; or (2) there are negligible postmaturational functional impairments with advancing age. Nonetheless, within negligibly senescing organisms, particular cell populations may degenerate and turn over regularly, but without detriment to the organism itself as indicated by the absence of increases in mortality rate. To establish the presence of senescence requires good data on age-related mortality rates and/or tests of physiological functions, especially reproduction. Unfortunately, there are only scant data on mortality rates together with assays of function across all age groups for most species represented in this book. Such information is needed to evaluate other species that may prove to have indeterminate lifespans and negligible senescence. Examples of long-lived sexually propagating species are summarized in Table 4.1.

Table 4.1 Examples of Very Long Lived, Sexually Propagating Species

	tmax, years
Negligible Senescence with Indefinite Lifespan	
Vascular plants	
Bristlecone pine	> 5,000[a]
Other conifers and deciduous trees	300–1,500[a]
Invertebrates	
Lobster	probably > 50–100[b]
Quahog	> 200[b]
Vertebrates	
Rockfish	> 120[b]
Tortoise	> 150[b]
Senescence with Major Increases of Mortality Rates	
Human (gradual senescence)	> 110[c]
Vascular plants (rapid senescence, monocarpic perennials)	
Bamboo (*Phyllostachys*)	≤ 120[d]
Century plant (*Agave*)	> 100[d]
Puya raimondii	150[d]

[a]References in Section 4.1.1.   [c]References in Chapter 3.6.1.
[b]References in Section 4.1.2.   [d]References in Chapter 2.3.2.

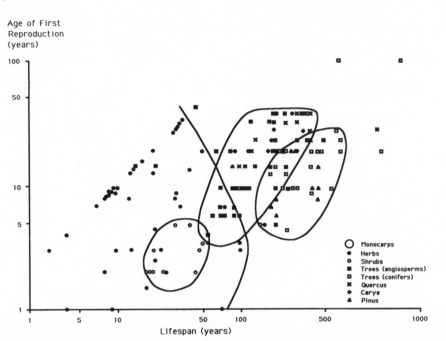

Fig. 4.1. Durations of prereproductive life (or juvenile period) and total lifespan in perennial plants. The data show known normal lifespans but not exceptional cases such as the coast redwoods. For identification and bibliographic references for points on the diagram, see Harper and White, (1974). Redrawn from Harper and White, 1974.

4.2.1. Plants

Certain conifers (gymnosperms) have enormous longevities of many thousand years in nature. Famous examples are the redwoods (*Sequoia sempervirans* and *S. gigantea*), the bristlecone pine (*Pinus aristata*), and the giant Mexican cyprus (*Taxodium distichum*; Stebbins, 1950; Noodén, 1988c). However, many other conifers have lifespans that exceed 1,000 years: Douglas fir (*Pseudotsuga menzeisii*), alerce (*Fitzroya cupressoides*), cedar of Lebanon (*Cedrus libani*), yew (*Taxus baccata*), kauri (*Agathis australis*), and the ginkgo (*Ginkgo biloba*) (Faber, 1897; Molisch, 1938; Schulman, 1954; Ogden, 1985b; Noodén, 1988c).[1] Some Australian conifers (e.g. *Dacrydium franklinii*) also exceed 1,000 years, showing the global reach of this phenomenon (Ogden, 1978). The record for bristlecone pines by dendrochronology is close to 5,000 years (LaMarche, 1969). Even these extraordinary longevities may be underestimates, since up to 5% of the annual growth rings can be absent (Schulman, 1954). Some of these species also propagate vegetatively, for example, redwoods, as discussed in Section 4.4.2.

Many conifers live longest in harsh environments, which are nonetheless conducive for longevity because of fewer fungi and smaller amounts of inflammable underbrush (Schulman, 1954; LaMarche, 1969). Success in these stringent conditions also shows that wide extremes in the physical environment are not incompatible with great longevity. Besides

---

1. Molisch (1938) gives access to a rich and largely neglected literature.

the conifers and the English oak (*Quercus pedunculata*), few angiosperms or other deciduous trees exceed 1,000 years (Molisch, 1938), although lifespans of 200 to 400 years are
not uncommon. Long-lived trees (perennial polycarps) also tend to mature late (Figure
4.1; Molisch, 1938).

Mortality rates are not available for most long-lived perennials (Harper, 1977). Opportunities for study are scarce, since few virgin forests have survived hurricanes, fires, de-
foliating infestations, or depredations by humans. Little is known about the distribution
of tree ages in most forests. Accurate dating requires counting of annual growth rings
from a core sample or a freshly cut end, which is undesirable in the few remaining long-
lived stands. Tree girth can mislead, especially in groves where growth may be stunted by
a high canopy, so that some very old trees are small. In some groves, redwoods mainly
propagate by sprouting, rather than from seedlings (Wensel and Krumland, 1986). Since
robust growths can sprout from the stumps of older fallen trees in the coast redwood and
others (Bosch, 1971; Harper, 1977), giving rise to normal adult trees, these could be very
much younger than the clone from which they came. The limited data on survivorship as
a function of age in the coast redwood (Fritz, 1929; Harper, 1977) suggest no change in
mortality rate between 200 and 1,000 years (Figure 4.2). Similarly constant mortality
rates are also indicated for the adult pencil pine (*Athrotaxis capressoides*), which survives
at least 1,000 years (Odgen, 1985a, 1985b), and for mature oaks (in Hungary) from 20 to
200 years (Szabó, 1931); the upper age in this latter study is less than observed for *Quercus* elsewhere (see above). By contrast, as described in Section 4.3, some stands of long-
lived hardwood trees show synchronized cohort mortality, a form of gradual senescence,
which results from extrinsic ecological factors.

More is known about mortality rates in smaller perennials, particularly from longitudinal studies that C. O. Tamm began in 1943 on a meadow in Sweden. In this population of
initially unknown age structure, the herb *Sanicula europea* is long-lived and has a low and

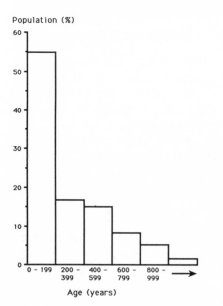

Fig. 4.2. The distribution of ages in a stand of redwood
Sequoia sempervirens. Data did not include trees that were
< 30-cm diameter (1–200 year). In a thirty-acre tract of
primary growth redwoods (Humboldt County, CA), about
3% of trees ≥ 30 cm were > 1,000 years old. Data of Fritz,
1929. Redrawn from Harper, 1977.

constant mortality rate, approximately a half-life of more than 50 years (Harper, 1967; Tamm, 1956) that predicts survival of one plant per thousand to at least 500 years (equation [1.10]). *Sanicula* and other small plants maintain remarkably constant AMRs that are characteristic of each species for at least 14 years and that give no hint of age-related increases or senescence. The threefold or more differences in mortality rates among species of these field herbs suggest that even shorter-lived species should nonetheless be classified as having indeterminate lifespans in the absence of senescence or absence of increases of age-related mortality rates. The manipulations of lifespan by natural selection or in the laboratory thus need not influence senescence and may operate only on the initial mortality rate (IMR, equation [1.2]).

Many botanists conclude that the lifespans of long-lived perennial trees and other vascular plants are not limited by endogenous senescence (Leopold, 1981; Noodén, 1988c; Watkinson and White, 1985). The continuation of growth for hundreds or thousands of years depends on apparently unlimited capacity for cell division in meristematic growth zones (Watkinson and White, 1985; Westing, 1964). Pith cells in cactus plants also continue to increase in size without dividing for at least 100 years (MacDougal and Long, 1927; Steward and Mohan Ram, 1961). Moreover, pith cells from basswood trees (*Tilia americana*) aged about 50 years can proliferate in culture as well as those from trees aged 5 years (Barker, 1953). Present evidence points to exogenous or mechanical factors in limiting the lifespan of hardwoods and conifers, such as insufficient stability of the roothold causing blow-down from episodic wind gusts. Other factors related to continued growth and not derived from cellular limitations on growth are discussed in Section 4.3.

The fecundity of many perennial seed plants progressively *increases* with size with each successive year, so that the largest (and often oldest) trees provide disproportionately more seedlings (Watkinson and White, 1985). Examples include the trees *Quercus crispula, Pentaclethra macrolobata,* and *Symphoricarpos occidentalis* (Figure 4.3). Although seed viability is reported to change with age of redwoods (Bosch, 1971), the local environment of the tree may be at least as much of an influence as age (William J. Libby, University of California School of Forestry, pers. comm.). There are no definitive data on seedling viability as a function of age in any long-lived polycarpic perennial. The coexistence of traits allowing great longevity with constant or increasing fecundity bear on population genetics models of longevity (Harper and White, 1974; Chapters 1, 12). The increase of gamete production may serve to increase the contribution of these senior citizens to the gene pool and, if so, would select against negatively pleiotropic genes and senescence (Chapter 1.5).

Nonetheless, some trees may show some very slow age-related declines, for example, the decreasing annual growth increments of the Ponderosa pine (*Pinus ponderosa;* Figure 4.4). The cambium or other growth zones do not seem to become proliferatively exhausted in very long lived trees, despite slowing growth rates. Considerable reversibility in growth slowing is also well documented. The slowed growth rate in mature trees often can be reversed by pruning, cutting, or grafting (Leopold and Kriedmann, 1975; Leopold, 1981). In old apple trees, removal of the buds restores the ability of cuttings subsequently made from these branches to form roots (Wellensick, 1952). Old citrus trees showing reductions in fruit yield can also be rejuvenated by propagation through cuttings (Frost,

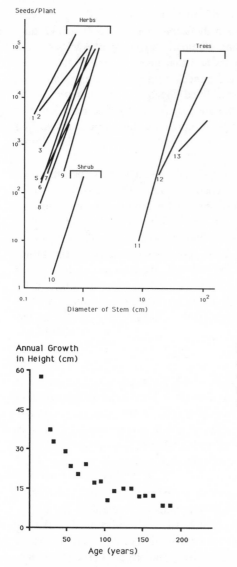

Fig. 4.3. Fecundity as a function of size in angiosperm seed plants. Size represents basal stem diameter for herbs and low-growing plants or diameter at breast height for trees. *1, Erigeron canadensis* (Hayashi and Numata, 1968); *2, E. annuus* and *E. strigosus* (Hayashi, 1984); *3, Oenothera parviflora* (Hayashi and Numata, 1968); *4, Chenopodium album* (Fukuda and Hayashi, 1982); *5, Polygonum persicaria* (Hayashi, 1984); *6, Chenopodium album* (Hayashi and Numata, 1984); *7, Artemisia princeps* (Hayashi, 1984); *8, Daucus carota* (Holt, 1972); *9, Lactuca virosa* (Boorman and Fuller, 1984); *10, Symphoricarpos occidentalis* (Pelton, 1953); *11, Bursera simaruba* (Hubbell, 1980); *12, Quercus crispula* (Kanazawa, 1982); *13, Pentaclethra macroloba* (Hartshorn, 1975). Redrawn from Watkinson and White, 1985.16

Fig. 4.4. Annual growth of Ponderosa pines (*Pinus ponderosa*) showing a gradual decline in the annual-growth increment, which could be due to factors that are either age-related or size-related. Redrawn from Leopold and Kreidmann, 1975.

1938). The absence of systematic senescence in long-lived plants, by contrast with annual plants, strongly suggests that the mortality risk may not increase after a certain age and that random attrition is the major lifespan determinant.

Lichens also are very long lived. These symbionts are formed by Ascomycetes or other fungi with algae. A 25-fold difference in lifespan was deduced within the genus *Lecida*, with *L. parasema* living 20 years and *L. promiscens* probably living 500 years (Wangerman, 1965; Beschel, 1955). The stability of symbiotic relationships during prolonged growth poses interesting questions about how genetic variants are accommodated. Little is known about age-related changes in symbionts.

4.2.2. Animals

Scattered throughout the phyla are sexually reproducing species that show no signs of senescence, with indeterminate lifespans, continuing growth, and no known dysfunctions or increase of mortality. The following examples are judged by the same criteria for senescence used in previous chapters: individual growth patterns, egg production, and mortality rates. The sequence of presentation is protostomates and then deuterostomates. The frequency of species with negligible senescence is uncertain outside of Porifera and Cnidaria, which have innumerable such species, most of which reproduce vegetatively as well as sexually (Section 4.4.2).

Some bivalve mollusks (class Pelecypoda) have exceptional longevities, gauged by annual growth rings in the shells or ligaments (sclerochronology). Among the slowly growing bivalves, the best examples are the ocean quahog (*Arctica islandica*) with 220 annual rings, and the geoduck (*Panope generosa*) (Thompson et al., 1980; Jones, 1983) and the freshwater mussel (*Margeritifera*; Hendelberg, 1960), each showing about 120 annual rings. Other bivalves also exceed 80–100 years (Comfort, 1979; Jones, 1983). Many slowly growing marine gastropods do not show age-related increases of mortality during adult life phases of a decade or more (Comfort, 1979; Frank, 1969).

While there are no data to indicate if the mortality rate increases with age or size, large sizes do not always mean greater longevity. For example, hard-shelled clams (*Mercenaria* (= *Venus*) *mercenaria*) with 40 or more annual growth rings are much smaller than those with the largest shells whose lifespans rarely exceed 20 years (Hopkins, 1930). Furthermore, gonadal mass, an estimate of gamete production, scales with size in *Merceneria* up through 46 years, the record age for this species (Peterson, 1986). There is no evidence for gonadal involution at advanced ages. Consistent with the decoupling of size and age, gonadal size correlates better with shell length ($r^2 = 0.53$), rather than age ($r^2 = 0.11$).

In contrast to these examples, the blue mussel *Mytilus edulis* and other faster-growing bivalves have shorter life expectancies of about 10 years (Figure 4.5). In view of these shorter lifespans, it is of interest that age-related biochemical changes are reported for *Mytilus*, for example, decreased dopamine- and opioid-stimulated adenylate cyclase of pedal ganglia (Stefano et al., 1982; Chapman et al., 1984). *Aplysia* (Chapter 2.3.1.3), *Lymnaea* (Chapter 3.3.6), and some other gastropods that live 3 years or less.

Numerous arthropods show indeterminate growth and no signs of senescence. Examples represent the three major subphyla: Crustacea, Chelicerata, and Uniramia. Several barnacles (in the subphylum Crustacea) have stable mortality rates after maturity: for ≥ 16 years in *Tetraclita squamosa* and ≥ 7 years in *Balanus glandula* (Hines, 1979; Connell, 1970). During these extended lifespans, egg production continued to increase with the size of the animal; growth generally followed an asymptotic pattern (Hines, 1979). The primitive stomatopod crustaceans also grow incrementally with progressive increases of egg production during lifespans estimated to be several decades (Reaka, 1979).

Lobsters (*Homarus americanus*) are decapod crustaceans that probably have a very long lifespan and also show asymptotic growth (Terao, 1928). The replacement of hard tissues at molting minimizes the accumulation of wear-and-tear effects on molted parts, which should give them an escape from mechanical senescence. Molting, however, prevents establishing the age of field-caught specimens, by discarding hard tissues that might

Shell Length (mm)

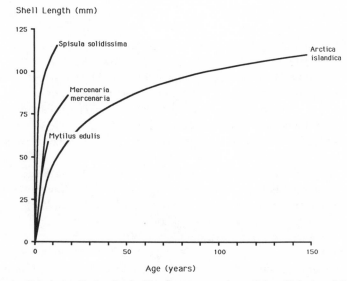

Age (years)

Fig. 4.5. Growth of bivalve mollusks. On the New Jersey coast, the surf clam (*Spisula solidissima*), the hard clam (*Mercenaria = Venus*), and the blue mussel (*Mytilus edulis*) grow more rapidly and have shorter lifespans than the ocean quahog (*Arctica islandica*). As described in text, *Mercenaria* lives at least 46 years and retains gametogenesis; the oldest clams, however, were smaller than predicted by these growth curves (Peterson, 1986). Redrawn from Jones, 1983.

accumulate growth rings. As concluded by Herrick (1896, 1911), and supported by subsequent records, lobsters occasionally grow larger than forty pounds (Wolff, 1978). Growth curves indicate that these giants would take 50 to 100 years to achieve the maximum predicted size (Cooper and Uzmann, 1977; Aiken, 1980).[2] Even the biggest females have abundant eggs, with continued increase as a hypoallometric function of size (Figure 4.6; hypoallometry is discussed in Chapter 5.2.1). Whatever their age, there is no doubt that the largest females make the most eggs. There is no evidence for a postreproductive phase in either sex, nor evidence of cessation of the molting. Molting occurs about every other year in young adults, and up to 4–5 years in larger lobsters (David Dow, University of Maine, pers. comm.; Aiken and Waddy, 1980). The increasing intervals between molting also means that the sperm stores acquired at molting when mating occurs are held for increasing durations before fertilization, up to several years (Aiken and Waddy, 1980; Waddy and Aiken, 1986). There is no evidence, however, for decreased viability of fertilized eggs as a function of female size; tumors are rare in adults and do not show age (size) correlation (David Aiken, pers. comm.). The extent of cell turnover during these long lifespans is unknown. While direct evidence is lacking, some neurobiologists consider it likely that motorneurons are not replaced during or between molts.

Together, this information suggests that lobsters can be very long lived and are candidates for negligible senescence. We must remain cautious, since great longevity and continued growth alone cannot rule out the occurrence of a slow increase of age-specific

2. Regulations setting an upper size limit to the capture of lobsters in the state of Maine have been enforced for more than 40 years, so that some lobsters could be at least this age. However, size may not be a reliable measure of age (Aiken, 1980).

mortality and a gradual senescence. Smaller annual increments of growth, shown by bivalves, lobsters, and isopod Crustacea (Needham, 1950; Inagaki and Berreur-Bonnenfant, 1970), may be a manifestation of gradual senescence. Recall from Chapter 3.2 that clonal senescence develops in some paramecia during hundreds of divisions over months to years. Relationships among body size, growth rate, and lifespan are treated in detail in Chapter 5.3.1.

Female tarantulas (in the subphylum Chelicerata) continue to molt annually or biannually during lifespans of at least 20 years in captivity; their senescence is sporadic and unrelated to reproduction (they continue to mate for many years, unlike the males, which die soon after mating, Chapter 2.3.1.2). Several reliable anecdotal reports associate senescence with difficulties in molting or sudden cessation of feeding (Baerg, 1963; Baerg and Peck, 1970). Insect queens (Uniramia) also are very long lived, with few indications of senescence; their lifespans appear to be limited by an exogenous event, the depletion of sperm stores that triggers their assassination by workers (Chapter 2.2.3.1).

Nematodes that are parasites of long-lived mammals have adult lifespans a hundredfold longer than soil and plant nematodes (Chapter 2.6.4). The hookworms *Ancyclostoma duodenale* and *Necator americanus* are thought to live 5 to 15 years (Chandler and Read, 1961; Noble and Noble, 1971; Faust and Russell, 1964; Palmer, 1955; Sandground, 1936); the filiarid *Loa loa* lives 15 years or more (Sandground, 1936). In a longitudinal study of a voluntary infection with *Necator,* egg production peaked by 2–5 years and gradually declined to zero by 15 years (Palmer, 1955); it is unknown whether the worm infections were eliminated accidentally or through endogenous senescence of the individual nematodes. These studies suppose that reinfection did not occur, which seems reasonable for patients who left the tropics for temperate zones where the intermediate hosts are rare. Some parasitic nematodes are short-lived, however. *Trichinella spiralis,* the cause of trichinosis in humans, has one of the shortest lifespans, in females about 3 months; nonetheless, its cysts can survive at least 35 years (Armand Maggenti, pers. comm.).

Many of these diabolical marvels have a vast daily production of eggs. *Ascaris,* for example, produces about 200,000 eggs a day (Levine, 1980). Although there is no direct evidence from labelling studies of ovarian cell kinetics, it seems likely that *de novo* ga-

Number of eggs

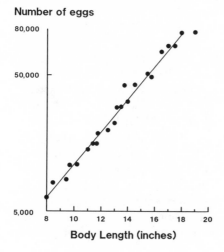

Body Length (inches)

Fig. 4.6. Egg content of feral lobsters (*Homarus americanus*) as a function of size (semi-log plot of number against size). A best-fitting function was calculated by Roger Gosden (pers. comm.): number of eggs = $5.9L^{3.3}$, where L is length (inches). Nineteen-inch lobsters may be 20 years old. Data of Herrick, 1896.

metogenesis continues during these long lives. The picture from *Caenorhabditis* of post-mitotic adult somatic cells and a developmentally fixed number of somatic cells (eutely) probably does not apply to the long-lived nematodes that lengthen extensively after hatching. For example, a proliferation of intestinal and muscle cells is implied by body elongation in *Ascaris* (Maggenti, 1981, 157). Nonetheless, there is no example of vegetative reproduction in any nematode (Table 4.3). Vegetative reproduction would not be expected for *Ascaris* because of chromatin (nuclear DNA) diminution in somatic cells (Davidson, 1986). Whatever the case in regard to somatic cell proliferation may prove to be, there is no proof that infections of the very long lived parasitic nematodes are limited by senescence.

The deuterostomate phylogenetic branch, which includes the Chordata, also has many examples of long-lived species. Many sea urchins, for example, appear to live a decade or more. To the extent that longevity can be judged from the rates of growth and the distribution of size classes, there is no indication of later age changes in mortality rate (Ebert, 1975, 1982, 1985). The mortality rate of juveniles is greater in species that have faster early growth (Ebert, 1975). In general, mortality rates *decrease* with size in adults (Ebert, 1982). The question of mortality rates at later ages in marked individuals has not been addressed, however. The annual mortality rates of *Strongylocentrotus purpuratus* and other species predict survival of about one in a thousand to reach 40 years (Ebert, 1985). Other echinoderms and starfish are discussed in Ebert, 1983.

Among vertebrates, the fact stands out that very few species have lifespans that plausibly exceed a century. Documentation is sparse but shows that a few fish, turtles, perhaps whales, and certainly humans reach these ages. While human longevity is clearly limited by senescence, some teleosts and reptiles may have indeterminate lifespans with negligible senescence. That the lifespan of pike and carp might exceed the human is unproven legend, although plausible (Comfort, 1979). Aquarium data are limited to specimens of only ten species that have been observed for more than 20 years (Hinton, 1962). However, several ray-finned fish (sturgeon, rockfish) have been dated by annual rings as living beyond 100 years, while numerous others live at least 50 years (Leaman and Beamish, 1984; Mulligan and Leaman, in prep.; Fitch and Lavenburg, 1971). The record is held by sturgeon (154 years). However, this great age may not preclude gradual senescence, in view of evidence suggesting finite ovarian oocyte stocks in sturgeon (Chapter 3.4.1.2.1). In species that are of little value to commercial or sports fishing, other record lifespans may well be found.

The jawless fish (Agnatha) show divergent, even opposite, life history patterns. Hagfish or slime eels (Myxinoidae) with continued growth could have very slow senescence. These characteristics differ radically from semelparous and rapidly senescing lampreys. Like lampreys, hagfish lack bone and scales, but unlike lampreys they are exclusively marine and develop directly without a larval stage (Hardisty, 1979; Brodal and Fänge, 1963; Gorbman, 1983). In contrast to adult lampreys, which do not eat, adult hagfish are broad-range scavengers of invertebrates as well as fish (Hardisty, 1979, 64). Furthermore, hagfish ovaries show various stages of gametogenesis, from the formation of primordial germ cells to mature oocytes; ovulation is continuous in some species (Dodd, 1977; Gorbman, 1983; Patzner, 1978). The best studied is *Eptatretus burgeri,* which is found seasonally in Japan (Fernholm, 1974; Tsuneki et al., 1983). The number of maturing ova scales with body size over the range 40–63 centimeters, and most individuals have two generations of postovulatory follicles (Tsuneki et al., 1983). These limited observations are consistent with continuing *de novo* oogenesis in adults. Follicular atresia is seen and accounts for the smaller number of large follicles. *Eptatretus* and other hagfish are thought to spawn repeatedly, since the ovaries of specimens caught by baited traps had corpora lutea and collapsed follicles indicative of recent ovulation (Cunningham, 1891; Tsuneki et al., 1983). The age may be guessed at by growth rates. Banding and recapture studies showed that a 36-centimeter-long *Myxine* grew 2.1 centimeters in 11 months, while another of 27 centimeters grew just 0.5 centimeter during 29 months; the slower growth in the latter might be related to the maturation of large eggs (Foss, 1963). By another estimate, *Eptatretus* grows five centimeters/year (Patzner, 1978). Based on this limited data, large hagfish of eighty to a hundred centimeters could be 40 years old.

Some teleosts live as long as the sturgeon and have not manifested senescence. The possibility of a nonsenescing vertebrate is, of course, of great interest. It must be said forthrightly that the evidence is suggestive but not yet detailed enough to remove all reasonable doubt. Rigorous evidence from measurements of radioisotopes and otolith rings indicates great ages in the rockfish genus *Sebastes,* scorpaenids (Bennett et al., 1982; Leaman and Beamish, 1984; Chilton and Beamish, 1982; Stanley, 1987; Leaman, 1990; Mulligan and Leaman, in prep.). Several species of *Sebastes* regularly live beyond 50 years. Fecundity increases with continued growth and with no indications of reproductive senescence. The record lifespans are 140 years for rougheye rockfish (*S. aleutianus*), 120 years for silver-gray rockfish (*S. borealis*), and 90 years for Pacific ocean perch or deep-water rockfish (*S. alutus*).

A lightly exploited population of *S. alutus* offshore of British Columbia was sampled by net fishing. The age structure analysis showed a population AMR (Chapter 1.4.5) of 0.05/year (Leaman, 1990; Mulligan and Leaman, in prep.). While the data are limited, there is no evidence for increased mortality rate at later ages (Figure 4.7). Deeper waters (200–600 meters), where there are far fewer predators and competitors than near the surface, harbor the oldest fish. Fish with the most otolith rings are not the largest (Mulligan and Leaman, in prep.). Thus, slower growth may be associated with lower mortality risk. Because there were slightly more older females than males, there is no evidence for "reproductive costs" (Chapter 1.5; Leaman, 1988). Necropsy did not disclose age-related gross pathology; rarely ($< 0.1\%$) fish showed ovarian tumors (Bruce Leaman, pers. comm.). These studies have particular importance because there are apparently few remaining populations of fish and other vertebrates that have not been exploited commercially or disturbed through other human encroachments. The cold of the deep waters inhabited may be another factor in these great ages achieved by *Sebastes*. Recall from Chapter 3.4.1.2.2 that, in very cold waters, brook trout lived fivefold longer than in warmer waters and showed negligible, rather than gradual, senescence.

Histological surveys showed the maintenance of ovarian follicular pools in 60- to 64-year-old *S. alutus* (Leaman, 1988, pers. comm.) and in sablefish (*Anoplopoma fimbria*) and rockfish of advanced but unspecified ages (Leaman and Beamish, 1984). There was no indication of ovarian senescence according to the presence of mature follicles in the oldest fish, nor was follicular atresia increased. Moreover, oogonial nests, the source of new oocytes, can also be found in older fish, although analysis is incomplete. Egg production per gram body weight remains relatively constant throughout life, after maturation at about 10 years. Mean egg size, however, decreased slightly with age. As emphasized

Length (cm)

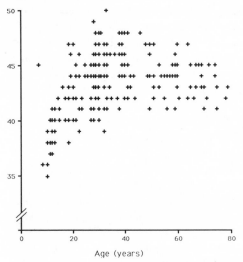

Fig. 4.7. Age structure of an unexploited population of the *Sebastes alutus,* Pacific ocean perch, that indicates a constant mortality rate at advanced ages, AMR = 0.05/year. The oldest fish tended to be smaller, implying that the population contained fish of different growth rates, or survival as a function of size (see text). Redrawn from Leaman, 1990.

for seed production in plants, we need to know the viability of the gametes produced by the oldest fish.

Other teleosts live many decades, though not as long as *Sebastes,* and also continue to produce eggs in proportion to body size and without evidence of limit. For example, cod (*Gadus morhua;* Figure 4.8) produce commensurately more eggs as they grow (Love, 1970), as do iteroparous salmonids, for example, the char *Salvelinus fontinalis* (Rounsefell, 1957). One 60-year-old female halibut (*Hippoglossus;* dated by scale rings) was still fertile and growing upon its capture (Comfort, 1979, 68). Other authorities consider that halibut continue to grow indefinitely (C. L. Hubbs, cited in Hinton, 1962). Calculations from limited data for the perch (*Perca fluviatilis;* Figure 3.4) indicate very slow accelerations of mortality (MRDT of 27 years, which is about twofold slower than in humans or elephants; Table 3.1) and very slow senescence. Lake trout or arctic char *Salvelinus* (= *Cristovomer) namaycush* in northern Canada mature at 14–15 years, live for at least 30–40 years, and continue to grow at advanced ages; egg production continues at least past 20 years, though not in all individuals (Miller and Kennedy, 1948; Sprules, 1952). Individual differences could reflect intermittent (nonannual) cycles of oogenesis and breeding, or adventitious parasitic infections and other exogenous diseases.

While it is often declared that teleosts sustain *de novo* oogenesis during adult life (Franchi et al., 1962; Zuckerman and Baker, 1977), I have not found any quantitative study of ovarian oogonial nests or primary oocyte populations with a series of adult ages in a long-lived species. Such an analysis would be very important for *Sebastes* (see above). There is no evidence that any slowly growing and iteroparous fish eventually becomes sterile. More recent studies support earlier conclusions (Love, 1970; Bidder, 1932; Lankester, 1870; Comfort, 1979) that senescence is unusual in teleosts, particularly in large species that grow slowly.

Further search for unexploited populations of the 500 flatfish, 150 perch, and 300 rockfish species might identify other very long lived species that may be candidates for negligible senescence. Studies of unexploited populations of long-lived fish are highly desirable and could pose major issues about the evolution of senescence, since these complex vertebrates would appear to escape both reproductive and actuarial senescence.

Little is known about senescence or mortality rates in amphibia, but relatively long lifespans are reported, particularly for larger species. The giant salamander *Megalobatrachus japonicus* lives at least 55 years, the toad *Bufo* at least 36 years, the axolotl (*Ambystoma maculata*) 30 years, and the newt *Triturus pyrrhogaster* 25 years. Many others exceed 15 years (Noble, 1931; Freytag, 1975; Flower, 1936; Mertens, 1970a, 1970b). This range of lifespans overlaps with other vertebrates but does not reach the upper range.

In field studies of the salamander *Desmognathus ochrophaeus* from southern Appalachia, fertility did not decline with age. Maturation occurred by 5 years, and lifespan was at least 16 years (Tilley, 1980). Based on the estimated annual mortality (AMR = 0.3, average of two populations), the age-specific mortality rate might decrease at later ages. A simple calculation shows that at this AMR only 0.3^{16}, or one in 250 million, would survive to 16 years, which implies an unreasonably large population.

Age-related increases of fitness are indicated for bullfrogs (*Rana catesbeiana*), since the older females "consistently" mated with the oldest and largest males (Howard, 1978, 1983). The older males were larger and had dominated the smaller males, as well as

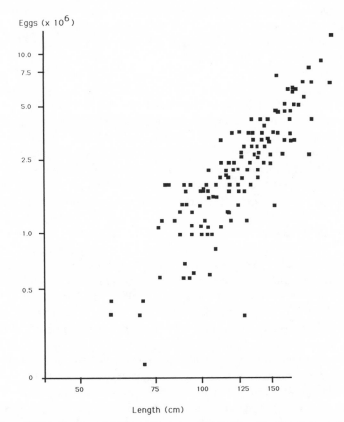

Fig. 4.8. Egg production by the cod *Gadus morhua*, a semi-log plot against length (age). Number of eggs = $0.62L^{3.4}$ where L is length (cm). Data of Mays, 1967, plotted by Roger Gosden (pers. comm.).

having greater success in finding mates. Moreover, the number of hatchlings increased progressively with age and size up through 5 years after metamorphosis (Howard, 1983). These bullfrogs matured by 2 years and lived 5 after metamorphosis. On the basis of the last 2 years, mortality appeared to increase after 5 years, but in view of the continuing reproductive increases of the few survivors to this age, it seems inappropriate to classify this as senescence, and more likely that predation was involved. Captive *R. catesbeiana* have lived 16 years (Flower, 1936).

Laboratory-bred *Xenopus laevis* aged 4 to 13 years (Brocas and Verzar, 1961) also did not show reproductive senescence. *Xenopus* lives at least 15 years (Flower, 1936). At all ages and in both sexes, gonadotropin injections induced normal copulation behavior. The numbers of induced ova were greater in the older, larger females. Females continued to grow after sexual maturity at 2 to 3 years, while males had stable sizes. One age-related change was found, a progressive increase in the tensile strength of finger tendons (Brocas and Verzar, 1961); this is a widely observed age change also seen in mammals, which suggests increased cross-linking of collagen (Chapter 7.5.3). Liver and kidneys had normal weights in the oldest frogs. These limited findings suggest that senescence is very gradual or negligible in *Xenopus*. Taken together, the data on *Desmognathus, Rana,* and *Xenopus* suggest that some amphibia do not show reproductive senescence. Although *de novo* formation of oocytes from oogonial proliferation continues past maturity in some fish and amphibia (Davidson, 1968), there are no reports on later ages.

The greatest claim for tetrapod longevity is the famous specimen of Marion's tortoise (*Geochelone gigantea;* also known as *Testudo sumerii*), which reputedly died older than 150 years at a British fort in Mauritius (Flower, 1937; Comfort, 1979). Apart from this anecdotal claim, which seems plausible, other captive turtles have lifespans documented up to 70 years (Bowler, 1977; Slavens, 1988; Gibbons, 1987). In feral populations, mortality rates of *Geochelone* increase sharply at larger sizes when growth ceases; few exceed 60 years (Gibson and Hamilton, 1984). In view of the greater ages of the captive, the lifespan in nature may be limited by external risks. There is documentation for only a few captives of the box turtle *Terrapene carolina* beyond 50 years (Gibbons, 1987): not every turtle carved with the initials AL was born before 1809. No mortality acceleration with age is shown in three turtles, up to the population maximum ages of 15 to 25 years (Figure 4.9; Gibbons, 1987). Species from eight turtle families do not show age-specific mortality rate increases or senescence (Gibbons, 1976, 1987). The higher mortality of the largest individuals in a field population of giant tortoises as judged by recapture studies (Gibson and Hamilton, 1984) could indicate age-related mortality but does not rule out emigration. The patterns of reproduction with age are little known. Histological studies indicate seasonal formation of new oogonia and ova in box turtles of undetermined age (*Terrapene carolina;* Altland, 1951). While we lack anatomic, physiologic, and demographic data on turtles of advanced age, there is no evidence yet for senescence (Gibbons, 1987).

Snakes may also include species with negligible senescence. Little has been added to Flower's (1925) compilation showing that some species survive beyond 20 years. A captive king snake (*Lampropeltus getulus*) lasted 23 years (Bowler, 1977). Many species continue slow, but diminishing, growth after maturity but lay progressively more eggs with increasing size at later ages, as in many other vertebrates. For example, the clutch size in captive king snakes progressively increased after maturation at 4 years up to 12

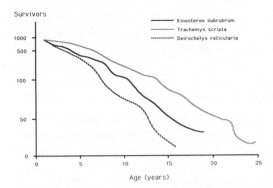

Fig. 4.9. Survival of freshwater turtles as a function of age on a semi-log plot, as determined by recapture from the wild. Redrawn from Gibbons, 1987.15

years, the oldest in the sample (Zweifel, 1980). In the water snake *Natrix natrix helvetica,* which lives at least 20 years, the number of eggs increased progressively with size and age at least beyond 15 years; unlike some other snakes (e.g. *Natrix maura* and *Vipera aspis,* see below), growth continued in the oldest specimens dated by growth rings and was indeterminate (Petter-Rousseau, 1953). The oldest field specimens, and they were few, manifested mild involutional changes in reproductive organs, but their unknown life history precludes characterization of these changes as senescence.

Natrix maura matures at about 5 years and reaches maximum size by 12, as dated by growth rings on the jawbones (Hailey and Davies, 1987). The survivorship data on captured specimens do not show whether mortality increases at later ages; the AMR is about 0.65/year. Few lasted beyond 20 years (Hailey and Davies, 1987), consistent with this high attrition rate. Most specimens of all ages had actively growing ovarian follicles; their occasional absence was not considered to be age-related. Limitations on growth are also described in *Vipera aspis.* After *V. aspis* matures by 3–4 years, its growth slows and stops by 7–8 years. The cessation of growth is not a sign of senescence, since the lifespan may exceed 20 years (Castenet and Naulleau, 1985). Reproductive involution was reported in *Spalerosophis cliffordi* at 13–15 years (Dmi'el, 1967) but the relation to possible gonadal senescence cannot be evaluated from these field specimens. So far, no upper age limit for egg production is reported for any reptile. As for fish and amphibia, we do not know if oogonial proliferation continues in snakes beyond maturity.

There is no evidence that homeothermic tetrapod lineages altogether escape senescence. Although some birds may also have a stable mortality rate for several decades, the observed maximum lifespans generally require that mortality accelerate (Chapters 1.4.8, 3.5). Nonetheless, the calculated accelerations (MRDT) are even slower in some long-lived birds than in humans (Table 3.1). The provisional conclusion that female birds have fixed oocyte stocks also predicts reproductive senescence. However, these arguments cannot preclude that some birds with very long MRDTs are far outside of most species in the slowness of their senescence. Nor can we exclude the possibility of continued *de novo* oogenesis in some species as a recurrence of the trend found in many earlier vertebrate lineages. Such species, if they exist, would be candidates for very gradual senescence.

The preceding conclusions support views of Strehler (1977) and Comfort (1979), that certain animals and plants do not manifest increases of mortality rate or other signs of senescence. Before considering the theoretical implications of this, two more subjects are introduced: the possibility of finite lifespans that are due to causes that are extrinsic to the organism altogether (Section 4.3), and the capacity of numerous species for indeterminate clonal expansion through vegetative reproduction (Section 4.4).

4.3. Finite Lifespans with Negligible Senescence

Here I consider a proposal some may find radical: *that some species have finite lifespans yet do not deteriorate or become senescent from internal causes.* Finite lifespans can occur from two types of factors that may be unrelated to endogenous causes of senescence. (1) Overall mortality rates (AMR) may not increase with age. Yet, the real finiteness of cohort sizes will result in maximum lifespans, whether the AMR is high or low. (2) Exogenous or ecological forces, or other organisms, may cause cumulative damage that increases age-related mortality rates, but without endogenous causes of age-related dysfunctions. In either case, the finite lifespans of individuals in a population may be either short or long.

This recognition of exogenous factors in senescence may appear to depart from the population-based definitions of senescence in Chapter 1. However, the genotype must have a major role in the individual choice of habitats or environments that may cause senescence through exogenous factors. The exogenous risks, whatever their contribution to senescence in a population, must be factors in the balance of reproductive effort that must be achieved for population stability or growth. Thus, we must recognize the contributions from the environment to the endogenous factors of senescence.

In species with small population sizes and high AMR, few individuals will survive very long. An appreciation of this is given by calculations for *Drosophila*, a species with AMRs that are typically so high that, even in a population of 1 billion, only one fly would survive to 6 years, this without the normal onset of senescence and accelerations of mortality rate (Chapter 1.4.8). Similarly, even for the lowest values reported for humans, a species with one of the lowest mortality rates as young adults, the median lifespan would be 1,200 years and the maximum 25,000 years in a population of 1 billion (Chapter 1.4.8); these values are fivefold greater than in conifers. Examples of evidently finite lifespans in nature that are not associated with senescence or loss of function are found throughout the phyla (Comfort, 1979; Charlesworth, 1980). However, detailed studies are needed to ascertain whether the high mortality rates from external causes have obscured senescence-related increases of mortality. Alternatively, senescence may be extremely gradual or even negligible. Readers may recall calculations of MRDTs for some long-lived birds for which physiological evidence of senescence is presently lacking (Chapters 1.4.5, 3.5).

It is important to recognize at the outset that some species with finite lifespans demonstrate senescence under more favorable conditions that reduce external causes of mortality as shown in domestic animals. Another good example is the so-called annual fish, cyprinodonts that live in temporary pools and that have lifespans of 1 year or less (Myers, 1952; Turner, 1964; Hildemann and Walford, 1963; Walford and Liu, 1965). Among about 20 species, the best studied are *Cynolebias* from South America and *Nothobranchius* from sub-Saharan Africa. The eggs are typically laid in the rainy season. As their habitat dries up, development is arrested at one of several stages before hatching, and the embryos survive in diapause for up to 8 months, when rains of the next season permit hatching (Chapter 9.2.1.1). Maturation can occur within 2 months, which is very accelerated relative to most teleosts. The adults die soon after spawning in nature. Even if the ponds do not completely disappear, these fish do not survive long (Myers, 1952). Nonetheless, their short lifespan in nature can be extended by several years (severalfold) in the laboratory, where a gradual senescence is observed (Chapter 3.4.1.2.2). A question of major interest then is whether all other examples of species with short lifespans in nature must also show senescence under protected conditions when low mortality allows the survival of a larger proportion to later ages. The evidence indicates both possibilities.

The first group of examples concerns various mollusks, in which species of the same genus have major differences in life expectancy. In a classic study of gastropod limpets, *Lottia (= Acmaea) insessa* and *L. digitalis*, Choat and Black (1979) showed that *L. insessa* has a statistically short life expectancy of about 1 month, while that of *L. digitalis* is 9 months. The survival curves show no hint of age-related increases in mortality that would imply senescence in the longer-lived species. The data for *L. insessa* indicate *decelerating* mortality rates at later ages; the major factor in its short lifespan is a thirtyfold greater age-independent mortality rate (IMR; Figure 4.10). The extremely short lifespan of *L. insessa* is attributed to its obligate natural substrate, sporophytes of the brown alga *Egregia laevigata* that themselves have a life cycle shorter than a year in the intertidal habitat. The longer-lived limpet *L. digitalis*, however, does not require this alga, and settles on more permanent rocky substrates. Corresponding to its habitat, the shorter-lived *L. insessa* grows faster and reproduces at a much earlier age; laboratory studies are needed to evaluate whether its lifespan potential is greater than seen in nature. Provisionally, these two life cycles would seem to be mainly limited by ecological factors and without evidence of intrinsic senescence. For comparison, some gastropod mollusks have lifespans that are at least 5 to 10 years long (Chapter 3.3.6).

A similar example is the species pair *Nucella (= Thais) emarginata* and *N. lamellosa*. These marine snails occupy nearby but distinct sites in rocky, intertidal habitats, yet their lifespans and reproductive schedules are different (Spight, 1979). *N. emarginata* has a higher AMR of > 0.9/year (effectively, a field lifespan of 1 year) and early maturation. In contrast, *N. lamellosa* has a lower AMR, 0.5/year, that allows delayed maturation until 3

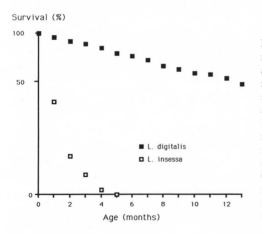

Fig. 4.10. Survival curves for field populations of two marine limpets (gastropod mollusks). *Lottia (= Acmaea) insessa,* which has a very high annual mortality rate (AMR $=$ 0.2/month or 2.5/year) because of the seasonal life cycle of its obligate substrate, the brown alga *Egregia laevigata.* A Gompertz analysis of these limited data (equation [1.2]) indicates that the mortality rate decreased at later ages; there was no indication of mortality rate accelerations. In contrast *L. digitalis* has a much lower and constant mortality rate (AMR $=$ 0.06/month) over the same span, because of the permanence of its substrate of intertidal rocks. The lifespans might be much greater under protected conditions. Redrawn from Choat and Black, 1979.

years. There is no evidence that senescence determines the lifespan of either species in nature. In the laboratory, *N. emarginata* can live at least 5 years and is iteroparous (A. R. Palmer, pers. comm.).

Another marine mollusk reveals an external factor related to size and continued growth that limits the lifespan. *Chiton tuberculatus* shows a major increase of mortality after 4 years, because of its growth to a size that forces it out from its cryptic habitat of inaccessible crevices into more exposed environments where, presumably, it is exposed to predation, that is, lifespan limitations from dislodgement. Even then, growth continued at least to 12 years, the oldest found in the study (Crozier, 1918). The survivorship data indicate accelerating mortality with an MRDT of 1.5 years (Appendix 1). Whether this is due to predation or to endogenous senescence cannot be determined from the field data. Either factor could increase age-specific mortality rates.

In view of hypotheses that statistically short lifespans allow the accumulation of germ-line mutations in a population with deleterious later effects (Chapter 1.5.2), it would be of much interest to study *Lottia* or *Nucella* in the laboratory at later ages than usually achieved in nature. Does reproductive decline and pathophysiological senescence eventually occur? The body-plan and slow locomotion of these snails might favor the detection of genes with delayed harmful effects on cellular or metabolic functions, as distinct from mechanical senescence. Under protected conditions, these snails might show minimal mechanical senescent damage to irreplaceable brittle appendages, compared to the mechanical senescence of fly wings and appendages (Chapter 2.2.2). Population genetics theories predict that the shortest-lived species should accumulate senescence-causing genes at a faster rate (Chapter 1.5.2). Future studies of this question might well consider species, in addition to fragile dipterans, that have different capacities for regeneration and cell replacement, and a different body-plan.

Better-defined ecological factors influence the maximum lifespan in certain plants. The giant saguaro cactus (*Cereus giganteus*, an angiosperm) grows slowly during a juvenile phase lasting for about 30 to 60 years, until reaching a height of about 2 m, when vertical growth decelerates and reproduction begins (Steenburgh and Lowe, 1977, 1983). As described for annual plants in Chapter 2, removal of buds allows continued rapid vertical growth for at least 2 more years. After maturation, these sentinels of the Sonoran desert develop their characteristic up-stretched arms and continue slow growth for at least 200 years, occasionally reaching 15 meters high. While vertical growth slows, as measured from ground to apex, the total extension accounting for growth of the arms continues at about the same rate. According to age distributions that were estimated from growth curves (Steenburgh and Lowe, 1983), there are slow accelerations of mortality after 60 years, with MRDT of 18 years (Appendix 1).

The very gradual senescence of saguaros is considered to result exclusively from accumulations of damage from exogenous causes (Steenburgh and Lowe, 1983). Freezing, wind, and lightning all inflict permanent structural damage, for which the risk must increase with continued growth in the above-ground target size. Similar risks are encountered by other long-lived and continuously growing plants like redwoods. Damage from exogenous organisms is slight by comparison with abiotic causes, because saguaros protect themselves by forming callus tissues that effectively encapsulate the holes made by insects and birds. Even so, large saguaros may have twenty to fifty nesting holes made by woodpeckers. These holes are thought to increase the risk of further mechanical damage by freezing or wind. Ultimately, severely injured (moribund) saguaros become vulnerable to bacterial infections that may be a common proximal cause of death.

A different phenomenon, cohort senescence, limits the lifespans of some trees that form predominantly single-species (monospecific) stands (Mueller-Dombois, 1983; Ogden, 1988). Again, we are dealing with the interactions of slow growth patterns with exogenous sources of damage that limit lifespan and cause slow senescence. Many hardwoods form stands of the same age, as described by Ogden (1988) for New Zealand beeches (*Nothofagus*). Each species also has a characteristic maximum lifespan in nature. Stands of black beech and mountain beech (variants of *N. solandri*) live 300 to 400 years, while stands of silver beech (*N. menziesii*) last 500 to 600 years. As the trees grow, they become more vulnerable to damage from windthrow from storms, or from accumulated snow that breaks upper branches and creates openings in the canopy (Wardle, 1984; Ogden, 1988).

The occurrence of openings in the canopy varies with the severity of winters and may trigger senescence of the stand. Most stands accumulate dead branches and other organic debris on the ground below, which become host to populations of herbivorous and wood-boring beetles, and fungal pathogens carried by beetles. Field observations suggest the working hypothesis that a sudden increase in the mass of fallen branches may increase the beetle population to a critical threshold, beyond which these exogenous agents destroy the stand (Wardle, 1984; Ogden, 1988). Older trees or trees stressed by drought are considered more vulnerable to beetle and fungal infestations (Hosking and Kershaw, 1985). Senescence spreads to other trees, causing die-back of the patch. The opening then is filled by a new cohort of young trees from the previously suppressed seedlings. The death of large trees in a cohort may continue one by one for a decade, spreading into areas that were not originally storm-damaged (Wardle, 1984, 314). The mortality is synchronous and relatively rapid when considered within the context of the total lifespan of these long-lived trees. Similarly, deadfall accumulation under *Eucalyptus* stands may increase risk of damage from fire (John Ogden, pers. comm.). The seedling pool, which was suppressed by the shade of the canopy and which may contain a range of ages, can rapidly expand to form a pole stand of young trees. For other aspects of forest dynamics, see Ogden, 1985a, 1985, Shigo, 1979, and Hallé et al., 1978.

Thus, two major age-related changes limit lifespan of saguaros and beeches through exogenous factors: (1) Continued growth increases vulnerability to exogenous risk factors that increase the risk of vital damage with increasing size. (2) The extended time of exposure to the environment causes the accumulation of structurally weakening injuries from both the biotic and physical environment. These examples from plants represent aspects of mechanical senescence and dysorganizational senescence from biotic and abiotic causes. Other examples of these exogenous phenomena that bear on the extent of endogenous mechanisms of senescence at the cellular or molecular level as a cause of limited lifespan might be sought in plants and animals.

In conclusion, I suspect that there is as much diversity in the causes of finite lifespans through exogenous and ecological factors as there are of more endogenous types of senescence. In some cases, lifespans are statistically finite without endogenous causes of senescence. Yet much more information is needed to determine if mortality rates could be further lowered and lifespans extended without senescence being ultimately manifested as predicted by population genetics theory. The present data merely suggest that this possibility can be seriously considered and that the topic deserves much more attention.

At the least, researchers of senescence must recognize the great importance of environment in analysis of the life cycle. It is important to compare more than one species within the same genus for the patterns of mortality and types of senescence after removal of the ecological hazards that usually terminate lifespan long before mortality might increase from endogenous causes. These comparisons could reveal much about the accumulation of genes in the germ line of populations that have delayed adverse effects as postulated by population genetics theory (Chapter 1.5.2). The extent to which the germ line is derived from somatic cells, as is widely true for plants, might strongly influence this proposed tendency, and future studies should recognize the patterns of germ-line differentiation. The next section deals explicitly with this question.

4.4. Vegetative Reproduction by Somatic Cell Cloning

Senescence is not universal in all eukaryotic somatic cell lineages. This vital fact is proven by the capacity of many plants, animals, and protozoans for indefinite vegetative reproduction by budding, fission, or fragmentation, all of which are equivalent to somatic cell cloning (for major reviews, see Bell, 1982; Jackson et al., 1985; Thimann and Giese, 1981; Noodén, 1980b; Giese and Pearse, 1975, 1977; Vorontsova and Liosner, 1960; Bounoure, 1940; Buss, 1987; Jackson and Coates, 1986; Watkinson and White, 1985). The question about somatic versus germ-line senescence has a long history that includes Weismann's 1881 doctrine of the continuity of the germ cells (Buss, 1987; Kirkwood and Cremer, 1982).

Vegetative reproduction by somatic cell cloning differs fundamentally from the parthenogenetic or unisexual reproduction that proceeds from an egg through the full morphogenetic course. Examples are clonality in parthenogenetic rotifers (Chapter 2.2.3.2), fish (Schultz, 1971), and lizards (Cuellar, 1977), among many others (Bell, 1982). The cell lineages leading to germ cells are segregated early in development in many animals, both in protostomates and deuterostomates, while this is not the case in other scattered phyla as discussed below. In vascular plants, for example, the germ line for sexual reproduction becomes differentiated from the somatic cells of the apical meristems at the growing tips. The widespread occurrence of vegetative reproduction from somatic cell cloning throughout the phyla gives a powerful argument against the proposition that the germ-line cells have some special protection from senescence that is otherwise lost during differentiation of somatic cells.

In view of the patterns of transcription that are unique to oogenesis and to early embryos in many species (Davidson, 1986), it might be thought that reproduction should require recycling of the genome through meiosis and oogenesis. Another consideration could be that DNA methylation and other types of molecular imprinting of parental genomes are barriers to parthenogenesis in mice (e.g. McGrath and Solter, 1984; Reik et al., 1987). However, the universality of this is disproved by the widespread, if scattered, occurrence of unisexual propagation from unfertilized eggs (Bell, 1982) and by the widespread occurrence of vegetative reproduction in diverse organisms. Vegetative reproduction depends on the maintenance of genomic totipotency in the somatic cell lineages that are cell sources for vegetative reproduction. The capacity that all plant phyla and some animals show for vegetative reproduction by somatic cell cloning demonstrates that nuclear totipotency is not a unique feature of germ-line cells. The extensive capacity shown by most vegetative reproducers for regenerating damaged appendages or internal organs is another manifestation of somatic cell nuclear totipotency. The usual emphasis on sexual reproduction in discussing senescence gives a narrow impression about the demonstrably wide occurrence of vegetative reproduction. Further evidence is given in Chapter 7.6.3 that the finite *in vitro* proliferation of diploid mammalian fibroblasts is not generalized to all somatic cells.

The capacity of vegetatively reproducing species to fragment indefinitely raises some issues about the definition of the individual that are philosophically irritating, yet also profound. These terms confront us with issues about modularity in multicellular organisms. An ongoing debate concerns how to cut the distinctions between unitary and modular organisms (Harper and Bell, 1979; Watkinson and White, 1985; Grosberg and Patterson, 1989), but the details do not concern us here. In essence, it is useful to distinguish between organisms that possess internal repeated structures, which can continuously regenerate themselves internally as well as vegetatively by fragmentation (call these modular), and organisms that have a nonrepeating internal structure, which typically do not reproduce vegetatively but which also typically show senescence (call these unitary). Semantically comfortable examples of unitary organisms are flies and fish. Examples of modular organisms are colonial ascidians and solitary organisms like sponges and vascular plants that grow indeterminately and reproduce vegetatively. Often, but not always, modular organisms do not segregate germ cell lines from somatic cell lines. Populations of modular organisms often represent clones of the same genotype; such populations are called ramets to distinguish them from genets, or clonal populations with different genotypes of the same species (Harper, 1977). Modular organisms may not be subject to the same geometric constraints that govern metabolic rate as a function of size in unitary organisms (Chapter 5.2), because of the greater flexibility in growth patterns, for example, as shown by certain bryozoans (Hughes and Cancino, 1985; Grosberg and Patterson, 1989).

The term *lifespan* need not apply to ramets of asexually reproducing species, since lifespan is by convention defined as the interval between fertilization and death. Moreover, ramets usually form from adult genets by budding or fragmentation. Individual organisms of asexually reproducing species (i.e. physiologically independent organisms or daughter colonies) may have finite lifespans that are determined by environmental risk factors, but senescence of these ramets is the exception (Jackson and Coates, 1986). Nonetheless, clones of vegetatively reproducing animals, plants, and protista sometimes eventually fail to thrive, in a possible equivalent of the clonal senescence observed in cultured diploid human fibroblasts. Even when older regions of colonial organisms are degenerating, however, vigorous asexual propagation continues in the younger parts of the colony, as widely observed in cnidarians and tunicates. Thus, some vegetatively reproducing organisms may be immortal and lack the more intense forms of senescence seen in most unitary organisms (Jackson and Coates, 1986; Watkinson and White, 1985; Grosberg and Patterson, 1989). Although vegetative organisms may propagate indefinitely, older modules may degenerate, as illustrated below. Such degeneration should not be considered an organismic senescence, since it does not interrupt the propagation of the individual genotype and has no more impact on the gene pool than the death of circulating erythrocytes or exfoliated intestinal cells.

4.4.1. Protista

Many ciliates and other protista completely lack sexual phases yet proliferate freely with no evidence of clonal senescence. Numerous examples are found in the Amoebina, Cryptomonadina, Phytomonadina, Sarcosporidia, and Radiolaria, which reproduce only asexually, by fission or budding (Hyman, 1940; Weinrich, 1954; Smith-Sonneborn, 1985; Anderson, 1988). Another set of examples comes from ciliates lacking micronuclei, the "vegetative" mutants. Amicronucleate strains are sexually inert, since gamete nuclei do not arise from the macronucleus, yet the amicronucleates can replicate indefinitely. For example, amicronucleate *Tetrahymena* strains have multiplied vigorously in the laboratory for 50 years or more, achieving thousands of generations (Corliss, 1953; Nanney, 1959, 1974). Bell (1988, 41) records that an amicronucleate strain isolated by Andre Lwoff in 1921 is still vigorous. The origin of most amicronucleate strains is not easy to date, because micronuclei may be occasionally lost and amicronucleate *Tetrahymena* have been found in nature (Elliott and Hayes, 1955). Not all amicronucleate *Tetrahymena* lines can replicate indefinitely, and some show clonal senescence after about fifty divisions (Nanney, 1957). There are many examples of related species with both modes of reproduction. The endosymbiotic flagellates in the gut of termites proliferate only asexually (Cleveland 1947a), while closely related endosymbionts of wood-eating roaches have sexual cycles that are controlled by host hormones during molting (Cleveland, 1947b).

A famous attempt to establish the possibility of immortal, nonsenescing clones ultimately proved misleading, giving a sad vignette arising from insufficient knowledge about the life cycles of models for senescence. A culture of *Paramecium aurelia,* "Paramecium Methuselah," was cloned from a single cell and proliferated from 1907 to 1943 during 21,800 calculated generations (Woodruff, 1943). However, unknown to Woodruff, who was a pioneer in ciliate biology, his culture conditions allowed a type of sexual process, autogamy, at regular intervals (Diller, 1936; Bell, 1988; Sonneborn, 1947). Sonneborn (1954) later showed that if autogamy was prevented from occurring, these paramecia clones inevitably slowed in growth. Woodruff's studies thus reconfirmed that, after all, clonal senescence can occur in some unicellular organisms. Other examples of clonal senescence in ciliates were described in Chapter 3.2.

4.4.2. Plants

Asexual, agametic reproduction is a major natural means of propagation in plants through budding, fission, and other derivations of somatic cells (Watkinson and White, 1985). Sexual cycles have never been seen in some robustly growing plants that include the genera *Rubis, Hieracium, Poa,* and *Taraxacum* (Cook, 1983, 1985; Silander, 1985). Other plant clones (Table 4.2) include specimens of the box huckleberry and creosote bush, whose size and growth rates indicate that the colony originated 10,000 years ago or longer. Additional evidence for an ancient colony origin is available for the quaking (Rocky Mountain) aspen (*Populus tremuloides*). The post-Pleistocene climate for more than 10,000 years may not have allowed seed germination of this aspen species (Cottam, 1954; Leopold, 1978). The longevity of individual clones, however, depends on the temperature and soil conditions. Liverworts (phylum Bryophyta) provide even more extreme

Table 4.2 Long-lived Clones of Vascular Plants

Species	Age, years	Diameter, meters	Origin of Growth
Lily of the valley (*Convallaria majalis*)	> 670	850	rhizome[1]
Ground cedar (*Lycopodium complanatum*)	850	250	rhizome[2]
Red fescue (*Festuca rubra*)	> 1,000	220	tillering[3]
Sheep fescue (*Festuca ovina*)	> 1,000	8.25	tillering[4]
Velvet grass (*Holcus mollis*)	> 1,000	880	tillering[5]
Bracken (*Pteridium aquilinum*)	1,400	489	rhizome[6]
Quaking aspen (*Populus tremuloides*)	> 10,000	81,000[7]	root buds[8]
Creosote bush, "King Clone" (*Larrea tridentata*)	> 11,000	7.8	basal[9]
Huckleberry (*Gaylussacia brachycerium*)	> 13,000	1,980	rhizome[10]

Source: All references excerpted from Cook, 1985.

[1]Oinonen, 1969.   [6]Oinonen, 1967a.
[2]Oinonen, 1967b.   [7]In square meters.
[3]Harberd, 1961.   [8]Kemperman and Barnes, 1976.
[4]Harberd, 1962.   [9]Vasek, 1980.
[5]Harberd, 1967.   [10]Wherry, 1972.

examples. These species have never revealed sexual stages (Schuster, 1979) and may have propagated asexually for millions of years (E. F. Warburg, cited in Sax, 1962).

Asexual propagation is an alternate growth mode in some plants that spread vegetatively by rhizomes, underground bulbs, or runners on the surface (Cook, 1985; Patelka and Ashmun, 1985). A related phenomenon is shown by the grass *Elymus*, which forms robustly growing, but completely sterile, interspecies hybrids; these spread so effectively by rhizomal extension that they often grow better than the parental species (Stebbins, 1950, 183).

Asexual growth can be readily induced by cuttings in most plant species and is widely used throughout agriculture (Molisch, 1938; Crocker, 1942; Cook, 1983; Raven et al., 1986). Great clonal longevity is established for the Winter Pearmain apple and the Cabernet Sauvignon grape, both propagated for more than 800 years by serial grafting, according to historical records (Westing, 1964; Noodén and Thompson, 1985). Other fruit cultivars are discussed by Schaffalitzky de Muckadell (1954). Many domesticated plant clones are sterile and can only have propagated asexually during their recorded lineages. The pentaploid garden tulip (*Tulipa clusiana*) is traced back nearly 400 years to Constantinople under the Ottomans (Hall, 1929). The highly sterile saffron crocus (*Crocus sativa*) was cultivated 2,500 years ago by Greeks for its yellow dye. Present saffron clones apparently descended vegetatively from fragments of the bulbous underground stem (corm; Stebbins, 1950, 138). Finally, the banana (*Musa sapientum*) has never formed viable seed during the hundreds, if not thousands, of years of its cultivation in Asia, Africa, and the Americas (Robbins, 1957). The basis for continued propagation of cuttings or shoots is unlimited meristematic cell proliferation in the shoot tips and rhizomes (Watkinson and White, 1985).

In contrast to these examples, however, other species may lack the potential for unlimited meristematic growth. Some crop plants that are propagated asexually by cuttings often lose vigor over a span of years to decades, which may indicate clonal senescence (Silander, 1985; Simmonds, 1979). Limited clonality is observed in cuttings of sugarcane (Trippi and Montaldi, 1960) and citrus (Frost, 1938), among others (Molisch, 1938; Schaffalitzky de Muckadell, 1954). Viral infections may limit clonal propagation (Crocker, 1942; Hartman and Kester, 1968; Wangermann, 1965; Watkinson and White, 1985), but the general mechanisms of clonal senescence in plants are undefined. In view of the rapid progress being made on the mechanisms limiting DNA synthesis during clonal senescence in diploid human fibroblasts (Chapter 7.6.4), it would be of interest to conduct molecular studies of plant clonal senescence.

Asexual growth may be controlled by the photoperiod. During growth under restricted exposure to light, the scarlet pimpernel (*Anagallis arvensis*) propagates asexually by forming secondary plants that arise through the rooting of plagiotropic offshoots from lateral branches (Trippi and Brulfert, 1973). This vegetative growth is indeterminate, and no flowers are formed. Nonetheless, the basal leaves show senescence, the only organs to do so. With longer daily light periods, vegetative growth is arrested, flowering occurs, and the whole plant undergoes rapid senescence and death. Other examples of environmentally regulated vegetative growth and senescence are cited in Noodén, 1980a.

This theme of limitless asexual reproduction is also represented in the potential for cloning a complete plant from a single cultured cell, as was done for carrots (Steward and Mohan Ram, 1961) and tobacco (Braun, 1959). These pioneering studies anticipated the demonstration that nuclei of differentiated cells, when transplanted to oocytes, could give rise to normal adults in some amphibians and gave key evidence for the concept that cell differentiation need not occur at the expense of genomic totipotency (Gurdon, 1962; Davidson, 1986). Moreover, some cultured plant cell lines have been maintained in active proliferation for decades (White, 1934, 1939). In contrast, tobacco callus cells, among other plant cells, gradually lose the ability to form shoots during extended culture (Murashige and Nakano, 1965). This resembles the loss in proliferative capacity of diploid human fibroblasts during serial *in vitro* transfer (Hayflick, 1977; Chapter 7.5.3). In conclusion, unlimited vegetative propagation is very common in higher and lower plants. The capacities to form new tissues and to increase in reproductive capacity during very extended lifespans are prime factors in the negligible senescence of many plants and derive directly from the capacity of somatic cells for unlimited proliferation while maintaining genomic totipotency.

4.4.3. Animals

Vegetative propagation by somatic cell cloning also occurs in many phyla (Table 4.3; Bell, 1982; Buss, 1987; Hughes and Cancino, 1986; Hughes, 1989). In some phyla, vegetative reproduction predominates, while other phyla have only a few such species. Few

Table 4.3 Vegetative Agametic Reproduction in Animals

	Size of Phylum (No. of species)	Regeneration	Prevalence of Vegetative Reproduction
Phyla with at Least One Species Showing Vegetative Reproduction			
Placazoa	1	extensive	present
Porifera	10,000	extensive	universal
Coelenterata	10,000	unknown	common
Mesozoa	small	unknown	common
Platyhelminthes	15,000	extensive	common, but not universal
Nemertina	900	limited	common, but not universal
Entoprocta	150	extensive	common
Ectoprota	5,000	probable	all marine species
Phoronida	unknown	limited	moderately common
Sipuncula	300	unknown	rare
Annelida	9,000	extensive, except leeches	common, except leeches
Mollusca	110,000	unknown	extremely rare, except for planktonic pteropods
Pogonophora	100	unknown	occasional
Echinodermata	6,000		rare
Hemichordata	90		common
Urochordata	unknown	extensive	common, but not universal
Capacity for Appendage Regeneration in Phyla without Examples of Vegetative Agametic Reproduction			
Gnathostomulida	1,000	unknown	
Acanthocephala	600	unknown	
Rotifera	2,000	doubtful	
Gastrotricha	400	unknown	
Kinorhyncha	150	doubtful	
Priapulida	8	limited	
Nematoda	80,000	no	
Gordiacea	small	?	
Brachiopoda	335	doubtful	
Echiura	140	limited	
Tardigrava	unknown	doubtful	
Onychophora	80	?	
Arthropoda	> 1,000,000		
Crustacea		yes	
Uniramia (Insecta)		no	
Chelicerata (Arachnida)		yes	
Chaetognatha	100	unknown	
Cephalochordata	small	unknown	
Vertebrata	45,000	limited regeneration of appendages and eye lens (Chapter 11.4.1.3), exclusive of birds and mammals; regeneration of liver and gut mucosa retained in mammals throughout adult life (Chapter 7.6)	

Table 4.3 (*continued*)

Summary of Vegetative Agametic Reproduction and Regeneration in Animals[1,2]

	Phyla	Percentage
Vegetative reproduction		
Never observed	2	6
Definitive evidence	13	40
Appendage regeneration		
Never observed	3	9
Definitive evidence	15	45

Sources: Bell, 1982; Buss, 1987; Hughes and Cancino, 1985; Margulis and Schwartz, 1988.
[1] Vegetative reproduction occurs throughout plant phyla.
[2] These calculations are based on the thirty-three animal phyla listed in Margulis and Schwartz, 1988.

species are exclusive vegetative reproducers, while many reproduce sexually or parthenogenetically as well. The distribution of vegetative reproduction in Table 4.3 could have serious omissions or errors because the *complete* life cycle is known for few species in most phyla. Closely related species differ in their capacity for vegetative reproduction. The following examples illustrate limitless vegetative reproduction without clonal senescence.

Cnidaria have a great capacity for asexual reproduction by budding, as first recognized by van Leeuwenhoek (1702). In some species of hydroids, both budding and gametogenesis occur simultaneously on the same polyp (Burnett and Diehl, 1964). General impressions indicate continued vigorous budding was maintained by *Hydra* strains over decades (Lenhoff and Loomis, 1961) and in a clone of *Campanularia flexuosa* for fewer years (Strehler, 1977). However, there are few quantitative studies over an extended time. While Turner's (1950) data on *Hydra* show no change in budding rate over eighteen generations, Bell (1988, 85) noted a decline. The production of buds is influenced by size and by extrinsic factors that include temperature and nutrition (Shostak, 1981). Thus, the modest decline of budding observed by Bell need not represent clonal senescence.

Many sea anemones reproduce only by fission (Chia, 1976; Shick et al., 1979). Some have reached great ages in aquaria without evidence of senescence, for example, several specimens of *Cereas pedunculatus* at the University of Edinburgh that maintained active budding for 70 to 90 years but then died suddenly (Comfort, 1979, 110). In view of the evidence for survival of corals to greater ages (see below), the sudden death of these specimens seems more likely to have been the result of extrinsic causes than clonal senescence. Several large natural populations of the anemone *Haliplanella luciae* showed no variation in electrophoretic isoenzyme patterns (Shick and Lamb, 1977), which is consistent with reproduction either by fission only (Shick et al., 1979), or by parthenogensis, which occurs in related species (Terrence Hughes, pers. comm.).

Although individuals of *Hydra* and the scyphozoan *Cyanea capillata* are considered not to show senescence (Brock and Strehler, 1963; Brien, 1953), individual hydranths of the related colonial *Campanularia* manifest a well-defined cycle of growth and rapid involution (Brock, 1984; Brock and Strehler, 1963). The involution does not cause death of the organism itself, loss of the ability to regenerate, or breaking of the clonal lineage. Therefore, this partial or mosaic involution should not, I believe, be considered in the same category as an organismic senescence.

Similarly, the bryozoan *Steginoporella* shows localized degeneration, in which adjacent patches continue to asexually regenerate or expand the colony (Palumbi and Jackson, 1983). The degenerating regions are older in clonal origin and are marked by the accumulation of "brown bodies" that appear to represent nonexcretable wastes within the polypide gut wall. Zooids containing brown bodies feed less actively and are less able to regenerate after experimental excision of tissue. Other bryozoans do not accumulate brown bodies (Terrence Hughes, pers. comm.). *Cellepora pumicosa* shows another type of senescence, in which some "senescent zooids became bleached and died," causing shrinkage of large colonies (Hughes, 1990). Localized degeneration also occurs in the older regions of colonial corals and tunicates, while younger parts continue vigorous growth (Jackson, 1985; Winston and Jackson, 1984). Bryozoan colonies generally have annual mortality of more than 0.2/year, according to limited demographic data (Hughes, 1990). The mechanism of localized degeneration has not been characterized at a cellular or molecular level. Clonal segregation of growth potential might arise through differential nuclear gene expression or through abnormal mitochondrial DNA as described in mammals (Chapter 8.4.1).

Reef corals formed by Cnidaria can grow vegetatively for hundreds of years, as dated from the size (Connell, 1973) and annual rings or other events recorded in their continuously growing skeletons (Jones, 1983; Hughes and Jackson, 1985). Moreover, the larger, older colonies have lower mortality rates than smaller colonies (Hughes and Jackson, 1980, 1985; Hughes and Connell, 1987; Figure 4.11); smaller colonies could be either daughters or new recruits. Many species of coral, as well as bryozoans and ascidians (Jackson, 1985; Hughes and Cancino, 1986) tend to have *increased* fecundity as their colonies grow larger, as in many plants. Nonetheless, reef-building corals may show a marked senescence over several months that cannot be attributed to extrinsic factors (Hildemann, 1978; Palumbi and Jackson, 1983). A detailed study showed declining growth, calcification, and reproduction that preceded colony death in *Stylophora pistilla* (Rinkevich and Loya, 1986). There are also age-by-size effects; for example, the smallest colonies of the oldest specimens in three genera (*Acropora*, *Porites*, and *Pocillopora*) had 50% higher mortality and twofold greater likelihood of shrinkage (negative growth) than similarly sized young colonies (Hughes and Connell, 1987). Even when a coral colony has shrunk, several species demonstrate the ability for continuing sexual reproduction and for regeneration (Szmant-Frohlich, 1985; Wahle, 1983, discussed by Hughes and Connell, 1987). Thus shrinkage, with an apparently higher risk of mortality, may allow redistribution of resources for repair or sexual reproduction. Scleractinian corals commonly have annual reproductive phases, besides ongoing vegetative reproduction (Szmant, 1986).

The next set of examples concern several phyla with species that can reproduce by fission. Flatworms (Platyhelminthes) show extensive asexual reproduction by fission, alternating with sexual reproduction as a function of season or diet (Grasso, 1974; Schroeder and Hermans, 1975; Haranghy and Balázs, 1964; Lassere, 1975; Calow et al., 1979). Cycles of senescence and rejuvenescence by asexual reproduction in *Phagocata velata* occur in its habitat of temporary pools (Child, 1913; Chapter 2.5.3). Vegetative reproduction in turbellarians may be more prevalent in juveniles (Bell, 1982, 193). Nonetheless,

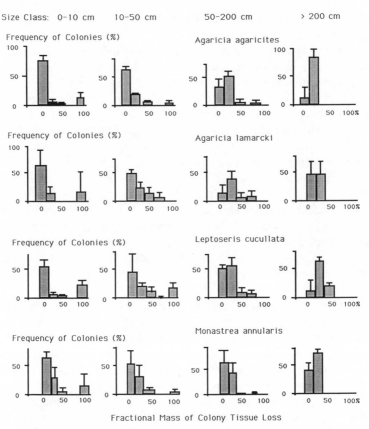

Fig. 4.11. Mortality in reef corals. Mean annual mortality rate ($\pm$ SD) as a function of colony size for four species of foliaceous reef corals (Discovery Bay, Jamaica). The abscissa shows the amount of tissue lost by a colony each year; if 100%, the colony died. Redrawn from Hughes and Jackson, 1985.

age did not influence the capacity of *Dugesia lugubris* to regenerate from fragments (Haranghy and Balázs, 1964; Balázs, 1966). The fragments thrown off by *Phagocata* and other flatworms reorganize to form a tiny, rejuvenated worm, just as though hatched from an egg (Child, 1914). This process can be repeated indefinitely without sign of clonal senescence.

However, both nonsenescing and senescing clones are formed by the flatworm *Stenostomum incaudatum* (Sonneborn, 1930). This classic study showed that worm lines derived from the posterior zooids (fission products) appear to have unlimited potential for asexual reproduction, while anterior lines show limited propagation (Figure 4.12). Turning to other phyla, a similar anterior restriction of regeneration is reported in some sea cucumbers (Holothuroidea; Domontay, 1931). Bell (1988, 91) recently reevaluated fission in the oligochaete *Aelosoma hemprichii*. In contrast to Hammerling (1924), Bell (1984b, 1985) found no clonal decline in either anterior or posterior lines. In view of the glandular functions associated with the cerebral ganglia of so many invertebrates, it is possible that the anterior restrictions on serial regeneration in certain species represent some aspect of neuronal differentiation or of neurohumoral influences on the cell sources for regeneration. Annelids differ from flatworms in their metameric construction of tandemly repeated compartments, each with excretory and reproductive organs. Certain annelids, except for leeches (Huridinea), can reproduce by fission and budding, which gives them complete powers of regeneration. In general, fewer annelids propagate by fission than do flatworms. It is striking that several earthworm species appear to show reproductive senescence and an extensive postreproductive phase, but without accelerations of mortality (Chapter 3.3.4). While others in the family Lumbricidae regenerate, the long-lived *Lumbricus terrestris* shows limited regeneration.

Other worms are prodigies of somatic cell cloning. In the laboratory, the flatworm *Stenostomum tenuicauda* produced a calculated 1,000 asexual generations during 11 years (Haranghy and Balázs, 1964) and the polychaete annelid *Ctenodrilus monostylos* reproduced only by fission during 60 years (Korschelt, 1942). During successive transfers of *Taenia crassiceps* through mice, the capacity of metacestodes for vegetative budding increased continuously (Freeman, 1962). Vast amounts of somatic tissues are produced vegetatively from the proglottids, for example, a 10-year infection of *Diphyllobothrium latum* in humans yields the equivalent of seven kilometers of segments (Stunkard, 1962). There are also reports of parasitic infections of humans that must have proliferated asexually for several decades or more (Sandground, 1936; Moore, 1981). A further example is the tiny hyatid flatworm *Echinococcus granulosus,* whose adult form occur in dogs. In humans, their intermediate hosts, the larvae can reproduce asexually for at least 30 years after the original infection (Faust and Russell, 1964; Moore, 1981; Whitfield and Evans, 1983). The tapeworm *Taeniarhynchus saginatus* (cestode flatworm), whose larval host is cattle, can also live 35 years in humans (Penfield et al., 1937). In the laboratory, another tapeworm (*Hymenolepis diminuta*) was transferred through a series of rats (the intermediate host) for 14 years, without diminishing its production of viable eggs and without any signs of senescence (Read, 1967). One assumes considerable cell turnover through the formation of new proglottids from the stem cells in the anterior germinative zone (Bolla and Roberts, 1971). More examples are discussed by Whitfield and Evans (1983). Although humans endure nearly lifelong duration of many cestode flatworms, the life expectancy of individual adult worms in their human host is probably less than 5 years in most species

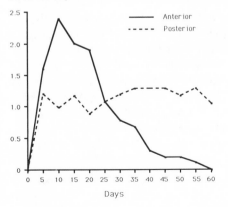

Divisions/Day

Anterior
Posterior

Days

Fig. 4.12. Differential clonal senescence in the flatworm *Stenostomum.* Proliferation of clonal lines from anterior fragments, which show clonal senescence, and from posterior fragments, which are apparently immortal. Redrawn from Sonneborn, 1930.

(Anderson and May, 1985). Lifespans of other cestodes can exceed 10 years (Comfort, 1979, 71; Hyman, 1951, 403) and other long-lived parasites might be sought in worms whose intermediate or final hosts are also long-lived.

Protochordates also reproduce vegetatively. The invertebrate ascidians use fission and budding as a major reproductive mode. Buds form from various somatic cells, depending on the species, redifferentiating to form a miniature adult with heart, neural structures, intestine, and other cells in less time than development from an egg. For example, in *Perophora viridis,* buds are formed from certain circulating lymphocytes that are, in effect, stem cells for the complete repertory of adult cell types, including the gonads (Freeman, 1964; Brien, 1948). Ascidians are thus classified among the organisms that have delayed differentiation of germ cell lineages. In the laboratory, Sabbadin (1979) observed the colonial ascidian *Botryllus schlosseri* for hundreds of vegetative asexual generations, but without any diminution of budding that might have been analogous to a "Hayflick limit" to proliferation as observed for cultured human diploid fibroblasts (Chapter 7.5).[3] *Botryllus* also shows an organismic senescence that is induced by the embryos' bulk during sexual reproduction (Chapter 2.3.5). This example thus resembles the numerous plants that can propagate vegetatively, yet senesce and die during reproduction (Section 4.4.2).

Tunicates may have several cycles of asexual and sexual reproduction in a single season (*Diplosoma, Botryllus*), or budding may occur mainly in the winter with sexual reproduction in the summer (*Clavelina*) (Bell, 1982; Berrill, 1935: Grosberg, 1988b). As in plants, temperature and nutrition regulate the patterns of asexual growth. With the advent of winter or other adverse conditions, temperate-zone ascidian colonies typically regress and their cells dedifferentiate, but this is reversible with the return of favorable conditions (Huxley, 1921; Barth and Barth, 1967). Overlapping (nested) cycles of growth and regression also occur (Berrill, 1951; Birkeland, 1981).

Why some phyla lack vegetative agametic reproduction is unclear. Examples of phyletic exclusion include arthropods, mollusks, nematodes, rotifers, and vertebrate chordates. However, some might argue that polyembryony in armadillos and a few other mammals represents the phenomenon. Although it is thought that the rigid skeletons of such species may be a factor in their lack of vegetative reproduction (Bell, 1982), the capacity for regeneration of eye lens, tail, limbs, or teeth that is shown by particular poikilothermic tetrapods and in others that lack vegetative reproduction argues against an intrinsic limitation. Certain parasitic nematodes, particularly the ascarids, eliminate portions of their nuclear DNA in different somatic cells, that is, chromatin elimination (Davidson, 1986; Wilson, 1925), which would render vegetative reproduction impossible, at least from those cells. Chromatin elimination is uncommon in nematodes, however.

Another intriguing issue is the regulation of module size, which is considered to be larger in aclonal (usually solitary) than clonal (and colonial) species of a wide range of marine invertebrates that use both reproductive modes (Coates and Jackson, 1985). While larger module size is generally correlated with greater life expectancy at maturity, many long-lived plants and animals are quite small (Chapter 5.3). Another difficult question

---

3. Asexual fission is not the usual reproductive mode in *B. schlosseri,* occurring in < 1% colonies, unlike numerous other species (Harvell and Grosberg, 1988).

concerns the selective advantage of sexual versus vegetative propagation in closely related species that show predominantly one or the other reproductive modes. This question has been considered for flatworms and hydras, but without resolution (Bell, 1982).

4.5. Summary of Chapters 2–4

4.5.1. *Vegetative Reproduction and the Issue of Clonal Senescence*

Somatic cell lineages can be immortal in plants, animals, and protozoans, as proven by the unlimited capacity for vegetative reproduction that pervades eukaryotic phyla (Table 4.3). The evolutionary persistence of vegetative reproduction poses major theoretical questions, since vegetative reproduction greatly reduces the capacity to produce new combinations of alleles and to disseminate genetic variants in the absence of syngamy (gametic fusion) and recombination (e.g. Templeton, 1982; S. E. Kelley et al., 1988; Michod and Levin, 1987; Bell, 1982). There is also considerable fossil evidence for the persistence (repeated evolution) of vegetative reproduction in corals, bryozoans, and vascular plants since the Paleozoic. In particular, coral genera that reproduce exclusively sexually have the same phylogenetic lifespan as those that reproduced vegetatively *and* sexually (Figure 11.3; Chapter 11.3.1).

According to traditional arguments, exclusively asexually reproducing taxa should be less competitive and more vulnerable to extinction in fluctuating environments since they cannot produce genetic variants as readily. Moreover, harmful mutations are predicted to accumulate more readily in asexually reproducing populations than during sexual reproduction, because of the absence of recombination. Small populations of asexually reproducing organisms are more vulnerable to the accumulation of harmful mutations than if they reproduced sexually (Muller, 1964; Bell, 1982). In Muller's words, "an asexual population incorporates a kind of ratchet mechanism, such that it can never get to contain, in any of its lines, a load of mutation smaller than that already existing in its at present least-loaded lines" (Muller's ratchet; Muller, 1964, 8). On the other hand, asexual reproduction, either by somatic cell cloning or by self-fertilization, gives the advantage of mating-independent reproduction, which may be advantageous in some environments and life cycles where the population is sparse or favorable conditions are transient. Such conditions are often associated with short adult phases, whether from rapid senescence or from high age-independent mortality rates that statistically permit few to survive beyond one generation in the population.

Individual size (mass) and population size (number) may interact to promote sexual reproduction and recombination. Elaborating on Muller's ratchet, Bell (1988, 148) argues that statistical fluctuations in occurrence of lethal mutations will cause extinction of purely asexually reproducing species, unless the population size is huge; Bell estimates that populations larger than 10^{10} should be safe from extinction, a population size that becomes vanishingly unrealistic in organisms with large body sizes. Smaller populations can maintain themselves, according to this classic argument, by recombination, which allows the maintenance of balanced lethals. In support of this, Bell (1988, 147) showed a correlation between the incidence of sexual forms and colony size in volvocaceans. He also made the interesting proposal that the evolution of larger body sizes depended on the prior evolution of sexual recombination. The incidence of strict agametic asexual reproduction in plants and animals is not known in detail.

What may be the relationships between the lifespan and reproductive modes? The examples given above show that vegetative reproduction occurs independently of total or adult lifespan, from long-lived vascular plants, sponges, and coral to short-lived flatworms. Tunicates may cover both extremes of lifespan. Thus, vegetatively producing plants are probably the longest-lived organisms, definitively releasing the restriction of asexual reproduction to short lifespans. In regard to sexually reproducing species with slow or negligible senescence, few animals (e.g. bivalve mollusks, sturgeon, rockfish, tortoises) but many more plants are documented to live beyond 100 years and can be considered as candidates for indeterminate lifespans. The evidently indeterminate lifespans in these examples from phyla that are not known to have vegetative reproduction (Table 4.3) suggests the important conclusion that indeterminate lifespans without senescence and the capacity for vegetative reproduction in animals are not obligately linked in animals.

Nor do species in phyla with vegetative reproduction necessarily escape senescence, as shown for some vascular plants, flatworms, and tunicates. Because senescence with finite lifespans often occurs in closely related species that escape senescence either by vegetative reproduction or by indeterminate lifespans, neither senescence nor its absence seem obligatory. In plants and animals, clonal senescence is not an obligatory or universal characteristic of somatic cells.

Vegetative propagation absolutely requires the genomic totipotency of the somatic stem cells that give rise through redifferentiation to the next generation. Thus, the role of these somatic stem cells is equivalent to that of the fertile ovum and, like the ovum, must contain a complete set of functional somatic genes equivalent to those of the germ line. Such cases negate Weismann's proposition for fundamental distinctions between senescence in somatic cells and in germ-line cells (Chapter 1), and show that cell differentiation does not confer inevitable senescence on a cell or its progeny.[4] Passaging of genes through oogenesis (whether or not fertilization and gametic fusion occurs, allowing for parthenogenesis) has been considered crucial in propagation. Without some undefined process of rejuvenation through the germ-line passage, species should otherwise succomb to senescence, according to the Weismannian proposals (Kirkwood and Cremer, 1982). However, examples of vegetatively reproducing species with indeterminate longevity that escape from senescence without requiring gametic fusion have long been recognized (Wilson, 1925, 235) and demonstrate that the existence of senescent changes in the individual organism itself is not linked to any specific mode of reproduction.

A general feature of organisms that reproduce vegetatively as adults is that the germ cell lineage is not differentiated during early development (Buss, 1987). In vascular plants

---

4. The linkage between cell senescence and cell differentiation has been debated for more than 100 years by eminent biologists including Child, R. Hertwig, L. Loeb, Minot, and Weismann. These historically important discussions are described by Child (1915, 59), Wilson (1925, 233), and Kirkwood and Cremer (1982). The significance of unlimited asexual, agametic reproduction was also long recognized in regard to theories of senescence (Lankester, 1870; Wilson, 1925; Szabó, 1935; Sonneborn, 1960; Harper, 1977). Cellular senescence and rejuvenescence were of much interest to early biologists in their efforts to understand the relationships of sexual processes to the life cycle.

Table 4.4 Time of Germ-Line Segregation: Examples in Relation to Senescence or Reproduction

	During Development	During Free-Living Adult Stage
Type of senescence		
Rapid	lampreys, Pacific salmon	monocarpic plants, ascidians
Gradual	placental mammals	
Negligible	cod, rockfish	conifers, hardwoods
Type of reproduction		
Gametogenic		
Sexual	mammals, most insects	vascular plants, tunicates
Parthenogenetic	some lizards, some insects, some nematodes, some rotifers	a few angiosperms and ferns
Vegetative		corals, some flatworms; vascular plants

Sources: Bell, 1982; Buss, 1987; Gifford and Foster, 1989.

and tunicates, for example, germinal tissue is generally formed from somatic cells just before reproduction in adult, free-living organisms. If vegetative reproduction can proceed without clonal senescence, then we must ask if the late differentiation of germ cells in certain organisms excludes organismic senescence. Clearly this is not the case, as was also concluded by Williams (1957). As shown in Table 4.4, numerous plants and animals show delayed (late) differentiation of germ cell yet also have finite lifespans in association with organismic senescence. It is also clear that the timing of germ cell differentiation can occur early or late, independently of the mode of reproduction, vegetative or sexual, including parthenogenesis.

Another important point is that the capacity for vegetative reproduction shown by so many diverse species of plants and animals is perfectly consistent with the continuous replication of many cell populations throughout the lifespans of other species that show senescence. The existence of rapidly proliferating cells in the gut and bone marrow that show no depletion during the long lifespans of mammals (Chapter 7.6.1) could represent the persistence of access during differentiation to some of the same gene sets required for unlimited proliferation by vegetative reproduction. The genes that determine the capacity for clonal propagation should be identifiable in closely related organisms with and without the capacity for vegetative reproduction. The small number of point mutations or other genetic changes required for the acquisition of somatic cell immortality (unlimited cell proliferation) in cancer (Newbold et al., 1982; Weinberg, 1985) suggests that such interspecies differences that determine the presence or absence of vegetative reproduction could be minor.

Finally, I note that the phylogenetic lifespan of exclusively asexually reproducing species during evolution is *identical* to the clonal lifespan of somatic cells that are used in vegetative propagation. This situation raises questions about the meaning of speciation in vegetatively reproducing organisms whose diversification cannot arise by recombination from a population gene pool (Dobzhansky, 1941; Sonneborn, 1957). The evolution of senescence in regard to the phylogenetic lifespan of species, the individual lifespan, and patterns of reproduction is discussed in Chapter 11.

4.5.2. Bidder's Hypothesis on the Cessation of Growth and the Onset of Senescence

G. P. Bidder (1932) proposed from species comparisons that the onset of senescence is linked to the cessation of growth. He argued that senescence did not arise from "a weakness inherent in protoplasm of nucleated cells; (senescence) is the unimportant by-product of regulating mechanisms." He postulated that senescence resulted "from the continued action of the regulator after growth was ceased," whereas numerous species that continued growth escape senescence. He was impressed by the continuing growth of a flatfish, the female plaice (*Pleuronectes platessa*): "we have no evidence of her senescence, nor any cause of death except violence." Male plaice, in contrast, are much shorter lived. Besides those discussed by Bidder, further examples from the preceding chapters can now be brought into discussion to support his hypothesis, while other examples indicate that senescence may be initiated despite continuing growth and that senescence may not be manifested until long after growth has ceased.

Organisms that fit Bidder's model include numerous plants that continue to grow and to increase in fecundity (Figure 4.3). Complex and long-lived animals without evidence of senescence may also show continuing growth after maturation, including certain mollusks (bivalves, chitons), crustaceans (barnacles, lobsters), flatfish and rockfish, and possibly amphibia. A presently unique example concerned the acquisition of longer lifespans by ferox trout (brown trout, *Salmo trutta*); after maturation, these individuals show accelerated growth and reach much larger sizes with severalfold longer potential lifespans than the rest of the population, apparently because they learned to capture larger, more nutritious prey (Chapter 3.4.1.2.2). Also, the examples of vegetative propagation in plants, sponges, corals, and tunicates represent indeterminate growth without clonal senescence, although individual modules or organisms may nonetheless die from senescence, depending on the species. The cellular basis for continued postmaturational growth is not known for most species, though there is little doubt that cell numbers increase, even in nematodes (Section 4.2.2). There are also no data to evaluate if species differences in growth capacity represent differently programmed capacities for somatic cell proliferation, and/or different sensitivities to regulators of cell proliferation.

Other organisms in which growth becomes slowed during senescence may not give critical support for Bidder's hypothesis, since the onset of pathophysiological changes during senescence could impair growth. The effects of disease in retarding the growth of children is well known and gives a model for the adverse impact of disease on growth during adult phases in other types of organisms. In the guppy (*Lebistes* (= *Poecilia*) *reticulatus*), growth begins to slow at ages in the population when spontaneous diseases arise and when the mortality rate begins to accelerate exponentially (Chapter 3.4.1.2.2). Moreover, some isopod crustaceans that continue growth after maturation also manifest senescence as their maximum size is asymptotically approached (Chapter 2.5.5). Elephant males, whose shoulder height and body size continue to increase during adult life, may be a special case because of the loss of molars at later ages through tooth erosion, which in turn might prevent further growth through gradual progressive malnutrition (Chapter 3.6.11.4). While sharks and sturgeon are long-lived and continue to grow after maturation, their ovarian histology indicates a finite oocyte stock that predicts eventual reproductive senescence (Chapter 3.4). However, reproductive senescence itself need not trigger more general dysfunctions, as is indicated by the general enhancement of maximum lifespan in spayed domestic animals (Chapter 10).

Plants show another facet of this issue, since monocarps generally cease growth before flowering and the onset of whole-plant senescence (Noodén, 1988c, 400). This cessation of shoot elongation is equivalent to determinate growth. However, the cessation of growth is prerequisite but not causal to senescence in some species. For example, in soybeans removal of the fruit blocks the rapid senescence but does not restore growth. There may be other circumstances under which growth cessation can be experimentally dissociated from senescence. Moreover, even with growth-arrested apical meristems, new growth occasionally can be reactivated, as shown by the formation of new shoots from the arrested meristems in needles of the Monterey pine (*Pinus radiata*); the shoots can then regenerate a whole plant (Raven et al., 1986, 338). Such examples argue against a rigid proliferative limit in plant meristems, even in those with determinate size.

While Bidder's conclusion that some slowly growing teleosts escape senescence appears to be sustained in a few examples, this statement cannot yet be generalized and merits unremitting scrutiny. Detailed study of many more species for possible cellular, genetic, and biochemical changes may reveal if there is a general pattern.

4.5.3. *Gametic Reproduction and Senescence*

Three important issues can now be compared across the range of lifespans and the patterns of senescence: the generality of reproduction as a trigger in senescence, the generality of reproductive senescence in both genders, and the increase of fecundity with size and age.

In many semelparous and rapidly senescing species, the events leading to reproduction also trigger death. Examples occur panphyletically, in the forever-cited Pacific salmon; in monocarpic plants, both the annuals and the perennials that may grow vegetatively for many years before flowering and rapid senescence; and in male tarantulas, among many others discussed in Chapter 2. The range of total lifespans in semelparous species exceeds 100 years in the case of certain bamboos but is not as great as reached by certain iteroparous species, like rockfish and sturgeon (> 120 years) or conifers (> 1,000 years). A dichotomy of semelparity versus iteroparity does not fairly describe numerous plants and animals that can display either phenomenon, depending on geographic location or environmental fluctuations, for example, the American shad (Figure 2.14) and the morning glory (Chapter 2.3.2).

Many species are *statistically* semelparous because of high mortality rates that allow few to survive to the next year or breeding cycle. This is the case for many fish, birds, and rodents, whose short mean lifespans do not indicate anything about the type of senescence or the potential longevity (Chapter 1). Fighting and other changes of behavior during mating can also increase mortality risk independently and cause irreparable body damage, as seen in mountain sheep and flies. The maturation of ova can also exhaust nutrient reserves and increase mortality, as observed in some insects. In numerous cases of semelparity, we find organisms of the same size and construction within a taxon that are iteroparous. This contrast also often occurs between genders of the same species, for example, plaice and tarantulas. I also note the numerous examples of reproduction-related death that involve hormones or the nervous system, either through its control of hormones or through behavior.

The second issue is the widely observed gender differences in reproductive senescence. In male mammals, reproductive senescence is generally much more gradual than the typically complete onset of sterility at mid-life in females. This gender difference derives directly from the continued capacity for spermatogenesis throughout life, whereas female mammals have a fixed oocyte stock that is progressively depleted. An open question concerns the extent of male reproductive senescence in other iteroparous organisms that have fixed oocyte stocks, for example, elasmobranchs and sturgeon for which only a few species have been studied. While some short-lived domestic birds also show marked reproductive declines, many others maintain undiminished fecundity for decades with no observed decline (Chapter 3.5). Yet some degree of reproductive senescence and organismic senescence is ultimately predicted, if elasmobranchs generally prove to have fixed oocyte stocks, as is believed to be general in birds as in mammals. Studies are needed on oogenesis and spermatogenesis in long-lived fish and birds. At the other extreme are species in which both genders have fixed numbers of gametic cells, probably all rotifers and soil nematodes, certainly the short-lived nonparasitic species.

The third issue is a question of the utmost importance to evolutionary theories of senescence, the apparent absence of senescence in some plants and animals that show continued growth and gametogenesis as a function of size. Moreover, some vascular plants, corals, crustaceans (lobsters, barnacles), mussels, and teleosts (cod) even show increasing fecundity as they grow larger (Section 4.2.2). Senescence has not been described in the most long lived of these species. The examples of *Sebastes* (rockfish) and *Mercenaria* (mussels) are of great importance because of data on the age and the gonad. These examples challenge the view that "senescence will tend to creep in" (Chapter 1.5.2).

Just how reproduction changes with age in iteroparous species has a fundamental bearing on the selection of genes that may promote longevity or on the selection against genotypes that shorten lifespan and reduce fecundity. According to the current tenets, selection generally becomes weaker against adverse traits with advancing age, particularly for organisms with high mortality rates and age-related decreases in reproduction that reduce the capacity to transmit genes, even favorable ones; this situation has been investigated by mathematical models (Williams, 1957; Hamilton, 1966; Charlesworth, 1980; Rose, 1991). High mortality rates of adults are argued to favor accumulations of genes with adverse effects on the few survivors to later ages, for example, negatively pleiotropic genes with delayed adverse effects that should contribute to senescence and increased mortality of adults (Williams, 1957; Hamilton, 1966; Charlesworth, 1980; Rose, 1991).

The evidence for this important prediction is mixed. In support is the rapid senescence of most dipterans. However, the very long lived insect queens with the same sets of genes as short-lived worker bees show that genes with delayed adverse effects, whether or not negatively pleiotropic, can be subject to modified expression (Chapter 2). Long-lived corals (> 200 years) also have very high juvenile mortality rates (Hughes and Jackson, 1985). Bats are another example that does not fit the inevitable accumulation of genes with adverse effects, despite high mortality rates of young adults, for example, *Pipestrellus pipestrellus* at 0.36/year (Appendix 1, 2). Under favorable conditions, bats may live for several decades (Chapter 3.6.1). The accelerations of mortality rate, based on limited data, are slow for these small creatures (Appendix 1, 2), and there are few indications of senescence, for example, the degenerative joint diseases that are common in terrestrial mammals were absent in one 19-year-old bat (Sokoloff, 1969). The present evidence thus does not establish the inevitability of rapid senescence in species with high mortality as young adults.

As reviewed in Chapter 1.5.2, Williams (1957) and Hamilton (1966) discussed how organisms with increasing fecundity as a function of age may be protected from senescence by their disproportionate contributions to the population gene pool. In particular, certain plants are noted by Watkinson and White (1985) to meet conditions that Hamilton (1966) set for a nonsenescing organism: (1) mortality rates that do not increase with age and (2) exponential increases of fertility. Some increases of mortality rate could be offset by commensurate increases of fecundity. The extent to which fecundity continues to increase with age or size in the examples of long-lived organisms merits renewed attention.

Each type of organism may have different changes in reproduction and mortality rate at advanced ages or sizes. Larger trees may become more vulnerable to critical damage from meteorological hazards, while larger animals may become less vulnerable to predators. Major efforts are warranted to explore these important questions undisturbed natural populations of long-lived, slowly senescing trees and crustaceans, but probably few fish. Of particular interest would be changes described in mammals during senescence, such as the accumulation of glycated proteins, the pathology of excretory organs, abnormal gametes, and the recently reported mitochondrial DNA deletions (Chapters 7, 8). Such information is needed to test the validity of the popular belief, supported by evolutionary hypotheses, that senescence is inevitable in all multicellular organisms.

The case of vertebrates is of particular importance, since vegetative reproduction is unknown. It is conceivable that mechanisms that were evolved to allow vegetative reproduction at most or any ages would protect other somatic cells from a hypothetical generalized senescence, that is, a type of senescence distinct from the rapid senescence that can often be attributed to a limited number of neuroendocrine or other causes. Vertebrates with apparently unlimited capacity to produce ova, like certain teleosts and amphibia, should have no reproductive senescence. The possibility that some vertebrates lack female reproductive senescence should receive particular attention through studies of oogenesis at later adult ages. The absence of reproductive senescence in species with high mortality rates in field populations would be an important failure of population genetics theories that predict the accumulation of germ-line genes in populations that *should* have delayed adverse effects at ages when individual reproductive probability distributions have dwindled.

In conclusion, there may be two types of events in the acquisition of senescence that allow the accumulation of genes that determine later onset dysfunctions. (1) Mortality rates may be sufficiently high to diminish the genetic contribution of that individual to the gene pool in an age-stratified population. (2) Independently, there may be a determinate capacity for gametogenesis in either gender, such that the genetic contribution of those individuals will diminish further than by their fractional presence in the population. The tremendous diversity of senescence occurs in lifespans that range from a week (rotifers) to more than a century (humans). Certain organisms with very gradual or negligible senescence live even longer. The million-fold range of adult phase lengths within which senescence is manifested argues that time in itself is not the crucial factor and that there are a host of molecular, cellular, and physiological adaptations that allow natural selection for virtually any schedule in reproduction.

4.5.4.   Examples of Senescence in Nature

Some may argue that overall mortality rates are high enough to prevent virtually any individuals in natural populations from reaching ages when senescence is expressed. The population genetics models, for example, would not predict survival to postreproductive ages, on the basis of competition for resources that would eliminate genotypes that were vigorous enough to survive to advanced ages but that became postreproductive. The following brief review from previous chapters emphasizes reproductive senescence and does not do justice to numerous descriptions of senescence from field studies. However, it is not always clear from population age-structure analyses that disproportionate decreases in older age groups are due to endogenous senescence. Age-group representation in a population can decrease from many factors that are distinct from endogenous senescence, such as migration or size-specific mortality risks from niche dislodgement (*Chiton;* Section 4.2), storm damage (trees; Section 4.2), predation (*Physella;* Chapter 9.2.2.1), or male fighting behaviors (sheep; Chapter 1.4.3). Thus, actuarial evidence for senescence may be misleading without specific descriptions of the conditions associated with increased mortality. The mean or maximum lifespans similarly can be misleading (Chapter 1.4). A critical and comprehensive review of senescence in field populations is long overdue, but would far exceed the scope of this volume.

Survival to the age of reproductive senescence in nature is reported for scattered iteroparous animals that include short-lived *Aplysia* (Hadfield and Switzer-Dunlap, 1990) and long-lived *Carabus* beetles (Houston, 1981). Among natural populations of mammals, menopausal females are documented in pilot whales (Kasuya and Marsh, 1984), elephants (Moss, 1988), harbor seal (McLaren, 1958), and several primates (Goodall, 1979; Bowden and Williams, 1985; discussed in Chapter 3.6.1). The extent of survival to postreproductive ages is unknown, but should not be dismissed as insignificant.

4.5.5.   A Phenomenology of Senescence

Despite the interspecies and phyletic diversity in senescence, some general trends can be considered as a phenomenology of senescence. In the semelparous plants and animals where rapid senescence or sudden death occurs after first reproduction, one or another hormonal factor can be clearly implicated as a distal cause of senescence. In animals, stress-inducing behaviors that increase mortality risk are also frequently linked to hormones and may extend these phenomena to more gradual types of senescence in iteroparous species, for example, rodents (Chapters 3, 10). Another recurrent aspect of increased mortality risk during reproduction is the cessation of feeding, seen in octopus, male tarantulas, and most migratory fish (Chapter 2). It is worth remembering that cessation of feeding during reproduction is not necessarily lethal, as shown by the perennially spawning sturgeon (Chapter 3.4.1.2.1).

Death from starvation may also arise from anatomic deficiencies that impair the ability to ingest a complete diet. An extreme example of adult aphagy leading to rapid senescence is the mayflies. However, adult aphagy shows up sporadically in other insects (e.g. autogenous biting flies) and other invertebrates (e.g. male rotifers).

Starvation from a different developmental determinant with delayed manifestation results from the inability to replace molars. This absence of grinding-tooth replacement is characteristic of nearly all mammals and constitutes a form of mechanical senescence. The oldest individuals in many ungulate and carnivore populations, for example, have a high incidence of lost or damaged molars, which may be supposed to increase mortality risk because of difficulties in feeding on their usual diet or in defending themselves (Chapter 3.6.11.4). The unusually long lifespans of elephants can be attributed to a variant developmental schedule in tooth eruption that provides renewal with fresh sets of molars up through the sixth or seventh decades. While other, unrelated senescent changes, for example, arteriosclerosis, are concurrent with tooth erosion and pose additional, independent risk factors, few mammals besides humans have behavioral and dietary options that would permit them to escape the consequences of the use-related mechanical senescence of teeth at later ages.

Mechanical senescence is a recurrent phenomenon in other irreplaceable organs as well. On one hand we have insect wings and exoskeletal appendages that show a wear-and-tear that is faster in flies (Chapter 2) than in beetles (Chapter 3.3.5) that live several years. Crustacea and spiders that continue to replace their exoskeletons at molting should largely escape from this aspect of senescence. On the other hand are those soft tissues in mammals that also show use-related damage: the use-related arthritic syndromes and the vascular damage that is generally caused by hypertension and at foci of turbulent blood flow (Chapter 3.6.7). Vascular damage from glycation of membrane proteins is a different phenomenon that may be related to metabolic shifts (Chapter 7.5.2).

Damage to macromolecules may also accumulate when the turnover or replacement is very slow. This phenomenon is best documented for the accumulation of glycation derivative in lens crystallins and in collagen and other proteins of connective tissue in kidney, bone, and blood vessels (discussed at length in Chapter 7.5.2). Speculatively, the consequences to the lens may be increased rigidity and opacity; connective tissue glycation may impair kidney function and increase the rigidity of connective tissues with consequent acceleration of mechanical senescence in the joints. With the exception of these examples, most other macromolecules of mammals appear to remain intact and functional through regular replacement or repair. The extent to which senescence impairs molecular turnover is controversial (Chapter 7.3).

Provisionally at this stage in the book, I consider these widely ranging phenomena of rapid and gradual senescence as contributing to increased mortality through their effects on homeostasis and integrated physiological functions involving the interrelated nervous, endocrine, and immune systems. I suggest that the most critical aspects of senescence are *dysorganizational,* through their influence on homeostasis at a systems level. That is, we need not invoke a primary or intrinsic senescence of each molecule or cell in the body. According to this view, most cells have potential lifespans that far exceed the organism that they compose. This potential is indicated in very gradually senescing iteroparous species with apparently indeterminate lifespans, but is particularly clear for the semelparous species whose lifespans can be greatly extended by specific physiological (i.e. organizational) manipulations. These examples, sampling over widely different higher and lower eukaryotes, reinforce a long-standing view that "senescence of somatic cells, though real, does not result from an inherent property of living protoplasm as such, but is due to secondary conditions" (Wilson, 1925, 235).

The many varieties of rapid and gradual senescence described in Chapters 2 and 3 suggest that a genetic basis for senescence has occurred independently many times during evolution. A diversity of genetically influenced patterns of senescence is consistent with evolutionary theories based on the weakening of selection in older age groups because of limited survival that proportionately allows older age groups little contribution to the population gene pool (Chapters 1, 11). The particular genes involved have not been identified as having neutral effects on young adults with delayed harmful effects or as having positive effects on young adults with delayed negative consequences (negative pleiotropy; Chapter 6).

Conversely, the numerous examples of plants and animals spanning the range of organizational and genomic complexity for which evidence of senescence is lacking strongly imply that there must be individual genes or coadapted gene sets that protect against senescence. Examples of "antisenescent" genes would include those coding for enzymes that protect against free radical and oxidative damage, renature proteins (heat shock proteins), or protein regulators that maintain mitotic proliferation to replace damaged cells. The involvement of few genes would permit relatively rapid reversion to very slow or negligible senescence from species with more rapid senescence. Alternatively, genetic combinations could cause similar senescent changes but need not bar rapid reversion if they were based on polymorphisms that were subject to intra- or interchromosomal rearrangements through recombination or chromosomal assortment. Possible examples of the potential for major and rapid temporal changes in life history organization are (1) the long-lived queens versus short-lived castes in numerous social insects; (2) the geographically distinct annual and perennial races of rice (Chapter 2.3.2); and (3) selection studies that show considerable increases in the mean and maximum lifespans of flies and nematodes (Chapter 6.3).

The generality of accelerations in mortality during senescence remains unknown. As discussed in Chapter 1.4, the Gompertz mortality model that describes an average mortality rate doubling time (MRDT) may not fit all populations of a particular species; in certain cases the Weibull model may fit better (Hirsch and Peretz, 1984; Slob and Janse, 1988). I chose the Gompertz model for this book because the slowing of the MRDT at later ages in the Weibull model makes it more awkward to compare species and populations. There is no *a priori* reason why mortality rates should conform to simple power functions. Moreover, there may be extensive delays (Figure 1.8) before mortality rates increase during adult life.

Nonetheless, the Gompertz model gives a good working approximation for age-related changes in mortality rates during the adult phases of a remarkable range of species with different mechanisms of senescence. Aphagous moths and male rotifers, feeding female rotifers and flies, short-lived cyprinodont fish and galliform birds, and nearly all mammals that have been studied show mortality rate increases that can be well fitted by the Gompertz model. The populations characterized for age structure are, of course, a minute sample that do not yet establish the generality of any model. A result of these future efforts should be the refinement of criteria by which to judge whether some organisms with increasing fecundity with advancing age do or do not escape senescence.

Chapter 5 considers a different aspect of senescence by comparing certain physiological, cellular, and biochemical characteristics of organisms within the same taxon that have widely different lifespans. Here we confront the question of which suborganismic features, such as metabolic rate and body size, show scalings with lifespan potentials and mortality rates.

5

Rates of Living and Dying: Correlations of Lifespan with Size, Metabolic Rates, and Cellular and Biochemical Characteristics

5.1. Introduction

Chapters 1–4 presented typologies of life histories throughout the eukaryotic phyla and developed a framework that is formed by two axes: the rates of senescence, from rapid to negligible; and the duration of the life cycle, from days to millennia. Short-lived species need not show senescence, nor do all very long lived species escape senescence. At one extreme are species that reproduce just once during short lifespans that may be statistically limited because of high overall mortality (e.g. the limpet *Lottia* (= *Acmaea) insessa*) or may be terminated by senescence in association with rapid increases of mortality rates (e.g. mayflies or Pacific salmon). At the other extreme are species with extremely long lifespans that show few if any indications of senescence (e.g. rockfish, redwoods). The rate or duration of the senescent phase was shown not to be linked to the *total* lifespan. Thus, some plants and animals have decades-long juvenile phases that are terminated by rapid senescence soon after maturation and reproduction (bamboo, 17-year cicadas).

These questions of species comparisons have long permeated thinking about mechanisms in senescence as the "rate-of-living" hypothesis and have lead to presumptions about biological quanta in rates of living (such as, the lifespan limit is related to fixed numbers of heartbeats or of cell divisions, etc.) that have been uncritically transmitted, in my view, from one generation of scientists to the next. It is often useful to compare species of different sizes in terms of metabolic rate (e.g. mice metabolize oxygen per gram body weight twenty times faster than do elephants). However, such calculations do not establish that the "rate of aging" is also scaled twentyfold faster.

This chapter addresses a basic aspect of biological time: How do the rates of metabolism as cellular and physiological levels as *young adults* correspond to the adult lifespan? This recurrent issue in discussions of life history organization is also contained in the "rate-of-living" hypothesis. Information on this long-standing question comes from the examination of metabolic, cellular, and biochemical characteristics of organisms in the same taxon that vary in the duration of their adult phases. While many age-related changes do show a scaling in rate that is proportionate to lifespan (Table 3.3), others that do not may give important clues to the types of parameters that influence lifespan. The mechanisms that may be found to underlie size or metabolic correlations with lifespan also bear on the evolutionary theories of senescence that predict wide variations in senescence among species (Chapter 1.5). If so, then the extent to which senescence or lifespan has strong correlations with size or metabolism, either between or within species, has profound bearing on evolutionary mechanisms in senescence.

This chapter is organized reductionistically, starting with body and organ sizes and proceeding to cell and biochemical characteristics. Attempts are made to understand the correspondences between scaling of metabolism and lifespan at the anatomic-physiological level versus the cellular-molecular level. The evidence does not yet allow us to decide if the cellular and molecular functions vary systematically among species in accord with the numbers of cells, rate of total body metabolism, or adult lifespan. This also bears on the important issue of extrapolating data on carcinogens and environment from rodents to humans. Discussed here and in Chapters 7 and 8, the variations in gene expression during development that determine the numbers of adult cells may also influence the patterns of senescence or limit lifespan. These are difficult questions that might be accessed through transformational genetics. Mammals will dominate the discussion, but not to the exclusion of some material on plants and other animals. The text thus mainly deals with species that are iteroparous and show gradual forms of senescence (Chapter 3). To develop this approach, the next section summarizes topics from a large literature of quantitative comparisons between species that help thinking about the comparative biology of senescence.

A major question is, Why do larger organisms so often have longer lifespans? This is best addressed in mammals, which have wider variations than most other taxa for which data are available, in regard to lifespan (50-fold), brain (8,000-fold), and body mass (40,000,000-fold), and resting metabolic rate (30-fold). These major differences could imply systematic differences in the rates of living and dying. Early biologists recognized that larger plants and animals have longer lifespans (Lankester, 1870, 72; Molisch, 1938, 81) and that animals with bigger brains for a given body size also live longer (Friedenthal, 1910). Higher metabolic activity, on the other hand, can be associated with shorter lifespans. Weismann (1889a, 7) succinctly stated a rate-of-living hypothesis, "[in addition to size] . . . the second factor which influences the duration to life is purely physiological: it is the rate at which the animal lives." A generation later, Rubner (1908) calculated from metabolic rates that most mammals have similar total basal energy production over the lifespan. Subsequent quantifications include the inverse relationships among environmental temperature and longevity and mortality rates in the fruit fly (*Drosophila;* Loeb and Northrup, 1917) and the constant total (lifespan) number of heartbeats of the water flea (*Daphnia*), over a range of temperatures that influence heart rate (MacArthur and Baillie, 1929). Such calculations[1] fostered the often implicit presumption of intrinsic biological limitations on the maximum metabolic or mechanical work over the lifespan. Rate-of-living hypotheses have greatly influenced biogerontology but, in my opinion, are not supported by many examples that include lifespan extensions (Chapters 2, 9, 10) and indefinite lifespans (Chapter 4).

The search among body size and metabolic parameters for the basic principles that determine the lifespan has lead to analysis of various correlations of the characteristics of young animals to the organismic lifespan, which are commonly called "allometric relationships." The term *allometric* was coined by Julian Huxley (1932) to represent relative growth rates of organs that were nonlinear, that is, nonisometric functions of body size. While most correlations of size and other parameters with lifespan are allometric (nonlinear), the word *allometry* has come to include any correlations where size or rate is involved, whether nonlinear or linear. As will be described, the allometric relationships are statistical hypotheses about relationships between size and other parameters, which differ among species. It is tantalizing to consider which parameters might predict characteristics of senescence and lifespans of the many species about which little is known. But there is no reason to expect that such predictions could be as powerful or precise as those drawn, for example, from the periodic table of the elements in predicting chemical properties or from physics in predicting energy transfer. Moreover, we do not know the cellular basis for these variations or their underlying genetic controls.

The discussion of lifespan in relation to species differences in size, metabolism, and biochemistry involves another basic issue in evolution: To what extent do such species or

---

1. Weismann (1889b), Brody (1945), Hill (1950), Bonner (1965), Sohal (1976), and Calder (1984) give other perspectives on the rate-of-living hypothesis and on size-lifespan relationships.

size characteristics result from independent selection versus their occurrence as pleiotropic consequences of other genes? The adaptationist-versus-epigeneticist interpretation (Gould and Lewontin, 1979) bears on the number of mutational changes needed to modify quantitatively so many organismic characteristics during mammalian evolution. In the future, this correlative subject may be made experimentally intriguing through new models from transformational genetics, which will allow examination of functions of the same genes, but placed in hosts of widely different sizes or physiological characteristics. The transgenic giant mice (Palmiter et al., 1982; Shea et al., 1987) are an important first step (Section 5.3.1) in creating a new research topic, experimental allometry, that may give valuable models for analyzing various correlations of the body-plan with senescence.

Section 5.2 develops the mathematical description of allometry in simple terms, which gives a foundation for discussing the correlations of lifespan with size and whole-body metabolic rate (Section 5.3) and with cellular and molecular characteristics (Section 5.4). Many questions remain open, especially about the relationships of macromolecular biosynthesis to body size, for which the controls are unknown. I have tried to proceed with a general organization that identifies major gaps of information, though this may be inconclusive and frustrating to the reader. Those familiar with the physiological literature on allometry may wish to skip to Sections 5.3 and 5.4.[2]

5.2. Correlations of Body Size and Metabolic Rates

5.2.1. Whole-Body Metabolism

Huxley (1932) described the relative growth rates of particular organs during development by the equation

$$Y = aM^b, \tag{5.1}$$

where Y is the size of the differentially growing organ (gm); a is the coefficient determined at $M = 0$; M is the size of the body minus the organ Y (gm); and b is the exponential growth coefficient, usually < 1.

When $b < 1.0$, the equation describes a parabola. Although equation (5.1) does not include time explicitly, it is widely used to predict changes in the growth rate of an organ as a function of body size, with empirically determined coefficients (Section 5.3.2). Most organs in mammals show negative allometry (*hypo*allometric scaling, $b < 1.0$) with body mass (liver, brain), so that as size increases, an organ may represent a progressively smaller fraction of the body mass. For example, in elephants the eyes are a much smaller part of their head than are the eyes in mice. Skeletal muscle, lungs, and blood are isometric ($b = 1.0$) and constant fractions of body mass. Only the skeleton and fat show positive allometry (*hyper*allometric scaling, $b > 1.0$), so that they represent progressively larger fractions of the body mass at larger sizes (Figure 5.1).

Correlations between body mass and energy metabolism are also approximated by par-

---

2. Three important studies dealing with body size, metabolic rate, and lifespan came to my attention as this book was in the final editing process (Read and Harvey, 1989; Promislow and Harvey, 1990; Austad and Fischer, in press). I was able to incorporate only some of this new material.

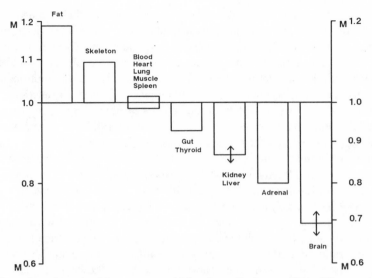

Fig. 5.1. The scaling of organ weights as exponential functions of body mass (M) in different species of adult mammals. Fat and skeleton are an increasing fraction of body mass in the comparison of large and small species (hyperallometric scaling); blood and spleen are nearly constant fractions of body mass (isometric scaling); most other organs are smaller fractions of body mass as size increases (hypoallometric scaling). The *arrows* in the boxes for kidney/liver and brain indicate the range of values. Related taxa have less variation. Data were excerpted from Calder, 1984, Prothero, 1980, Stahl, 1962, 1963, and Schmidt-Nielsen, 1984.

abolas, both within and between species. These relationships are expressed by the Brody-Kleiber equation:

$$P = aM^b \qquad (5.2a)$$

where P is basal metabolism (liters of oxygen consumed/hr); M is body mass (gm); a is the scalar mass constant (kcal); and b is the mass exponential constant.

With few exceptions in placental mammals, the Brody-Kleiber equation has the empirically obtained values of

$$P = 70M^{0.75}. \qquad (5.2b)$$

The conventional values for the constants in equation (5.2b) were determined from a great range of species: the exponent ($M^{0.75}$) fits both mammals and birds (Kleiber, 1932, 1947, 1975; Brody, 1924; Benedict, 1938; Zar, 1969), though Iberall (1972) argues for slightly higher values. As an example, two species that differ in body weights by 10,000-fold would have total basal oxygen consumptions that differ only 1,000-fold, for example, a 20-gram mouse versus a 200-kilogram pig.

A long-standing issue concerns whether the exponential constants relating basal metabolism and other rates to body mass (M) are fundamental numbers of biology in the sense used by physicists and chemists (Hofman, 1983; Boddington, 1978; McMahon, 1973; Stahl, 1962; Calder, 1984; Yates, 1987; Schmidt-Nielsen, 1984; Iberall, 1972; Prothero, 1986). The empirical exponent for basal metabolism ($\approx M^{0.75}$) is closely approximated by fractions formed from integers ($\frac{2}{3}$, $\frac{3}{4}$, etc.) that have been rationalized from dimensional analysis (Stahl, 1962; Feldman and McMahon, 1983; Heusner, 1985). As pointed out by Schmidt-Nielsen (1984), these widely used integers also resulted from rounding off before the advent of pocket calculators.

So far, no argument from first principles has established that either constant should be an *exact* number, which would be a rare occurrence in biology. This lack of universal biological constants for organismic-level functions stands in sharp distinction to the universal constants of physics and chemistry. It seems more likely that the allometric coefficients represent normative tendencies from constraints in anatomic and physiologic characteristics. A familiar example is the huge design problem cows would have in dissipating heat if they metabolized at the same rate as mice, since mice produce about tenfold more watts per gram body weight. Consequently, metabolic rate may closely follow the body area to volume ratio of ⅔ that sets a baseline for heat dissipation from physical objects. Nonetheless, even physically constrained characteristics are open to many genotypic modulations during evolution, as shown by species differences in shape, skin and hair, epidermal fat content, etc., any of which can cause variations in basal metabolism and heat transfer. Prothero (1986) showed how uncertainties as small as 0.03 in the exponents give deviations of more than 50% over the mammalian weight range. Experimental errors and biological fluctuations preclude estimates of the mass exponents much closer than 0.02 (Iberall, 1972). Another source of uncertainty is species differences in response to the anesthesia that is needed to measure basal metabolism. In conclusion, while these relationships are almost always strong statistically if enough species are considered, it seems best to view these as normative approximations, or working values, rather than rigorously established constants.

Another expression relating metabolism per gram body weight (M) is useful in making species comparisons. The *specific metabolic rate, P^**, is simply derived from equation (5.2b):

$$P^* = P/M = 70M^{-0.25}. \tag{5.3}$$

The negative exponent in equation (5.3) derives directly from the property of negative allometry from equation (5.2a), in the case when the exponential coefficient $b < 1$. Continuing the example from above, the basal oxygen consumption *per gram* in a 200-kilogram pig would be 10% of that in a 20-gram mouse, a difference that has implications for the amount of metabolic work done by cells during the lifespan. Since this calculation averages over all body cell types, we should not presume that all cells in a large animal have slower metabolism than in a small one. Neurons, for example, may not (Section 5.2.2).

Many physiological rates also show negative allometry on body mass, with exponents that approximate the specific metabolic rate, $\approx M^{-0.25}$ (Table 5.1). The duration of physiologic cycles (the reciprocal of rate) ranges from $\approx M^{0.20}$ to $\approx M^{0.40}$ for the half-lives of many molecules in circulation, respiratory and cardiac cycles, and age at maturation, as well as the lifespan (Section 5.3). Methotrexate (used in cancer chemotherapy; Dedrick et al., 1970) and inulin (a polysaccharide marker that is injected to estimate extracellular fluid volume; Adolph, 1949; Edwards, 1975) also have similar scaling of the circulatory half-lives (Table 5.1). Since the kidneys are the major locus for clearance of marker molecules like methotrexate and the carbohydrate marker inulin, these similar scalings are consistent (Dedrick et al., 1970) with the scaling for the circulation time of the blood

Table 5.1   Examples of Scaling Relationships with Time

	Allometric Equation	Notes
Cardiac cycle		
Mammals (second)	$0.25M^{0.25}$	1, 2
Birds (second)	$0.39M^{0.23}$	1, 2
Circulation of the blood volume		
Mammals (second)	$21M^{0.21}$	1, 2
Half-lives of Molecules in Circulation		2
Albumin (day)	$4M^{0.31}$	3, 4
Glucose (minute)	$0.2M^{0.25}$	3, 5
Inulin (minute)	$6.5M^{0.27}$	3, 6
Methotrexate (minute)	$58M^{0.20}$	7
Transferrin (day)	$3.8M^{0.23}$	8
Reproductive Maturity		
Mammals (year)	$0.75M^{0.29}$	2
Birds (year)	$2.4M^{0.23}$	2, 9
Lifespans		
Mammals (year)	$12M^{0.20}$	2,10
Birds (year)	$28M^{0.19}$	2,11

[1]Stahl, 1967.
[2]Calder, 1984.
[3]Lindstedt and Calder, 1981.
[4]Calculated from Allison, 1961, and Munro and Downie, 1964.
[5]Calculated from Ballard et al., 1969.
[6]Calculated from Stahl, 1967, and Edwards, 1975.
[7]Dedrick et al., 1970.
[8]Calculated by Jang H. Youn (pers. comm.) from Regoeczi and Hatton, 1980.
[9]Western and Ssemakula, 1982.
[10]Sacher, 1959.
[11]Lindstedt and Calder, 1976.

volume and indicate an overall coherence in the scaling of physiological time in mammals.

In parallel to basal metabolism, key biochemical components of oxidative metabolism and biosynthesis are also functions of body mass, with similar allometric constants. For example, both the total body content of cytochrome c (Drabkin, 1950) and cytochrome oxidase (Jansky, 1961; Munro, 1969) scale as $\approx M^{0.67}$, which is equivalent to $\approx M^{-0.33}/$ gram body weight (equation [5.3]). Thus the basal oxygen consumption and the body content of key enzymes in oxidative metabolism are well correlated, decreasing in concert with increasing body size.

An intriguing correlate, with implications for the rates of RNA turnover and genomic transcription, is that the rates of whole-body protein synthesis also show negative allometry, as determined for the incorporation rates of labelled amino acids into whole-body proteins of humans and other species intermediate between mice and cows (Munro, 1969; Figure 5.2). Urinary nitrogen, which estimates whole-body protein turnover without the possible confounds from precursor pools that always concern tracer studies, gives the same result: daily urinary nitrogen is fivefold greater per gram in rats than in humans (Munro, 1969). These important relationships have been extensively confirmed (Arnal et al., 1983; Waterlow and Jackson, 1981; Garlick et al., 1980; Reeds et al., 1980). The rate of protein synthesis per gram body mass is

$$\text{gm protein synthesized}/M \sim M^{-0.25} \qquad (5.4)$$

and has the same scaling as oxygen consumption per gram body weight (equation [5.3]). Moreover, equation (5.4) describes the rate of protein synthesis from birth onwards throughout the entire postnatal and adult phases (Waterlow and Jackson, 1981). Thus, the rate of protein synthesis is thought to be remarkably constant across a twentyfold range of body weights.

Coulson (1986) also noted the proportionality of protein turnover in vertebrates to the rate of blood flow, which suggests a mechanism for regulation of overall protein synthesis rates (Section 5.2.5). The rates of whole-body macromolecular synthesis and energy metabolism thus show the same coefficients for negative allometry with body weight. This scaling will be extended to the synthesis of several liver proteins (Section 5.2.4). A major open question is the relationship among protein turnover, mRNA turnover, and genomic transcription in the same cell type as compared between species with different body sizes. While there are major differences in the transcription rates of genes during development (Davidson, 1986), the extent of these differences is not established across species. Because of evidence that the capacity for repair of UV-induced DNA damage is higher in the most actively transcribed genes (Bohr et al., 1986), the extent of species differences in transcription rate bears on the scaling of genomic repair mechanisms with lifespan.

Most blood-borne substrates for metabolism vary little with body size among species, in contrast to gross metabolic parameters which scale close to $\approx M^{0.25}$. Arterial blood oxygen (Bartels, 1964), total hemoglobin (Burke, 1966), and hematocrit (the fractional volume of packed red blood cells after centrifugation; Prothero, 1980) are independent of body size in the species studied. Plasma amino acids (Coulson et al., 1977) and 2,3-

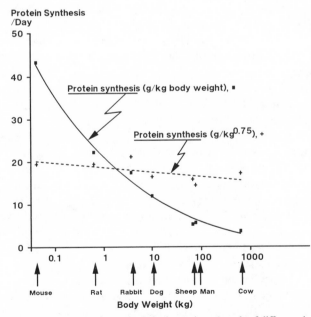

Fig. 5.2. The rates of whole-body protein synthesis in domestic mammals of different sizes, as determined by the incorporation of amino acids. The rates of protein synthesis conform to the canonical hypoallometric formulae. Data of Arnal et al., 1983, Garlick et al., 1980, Munro, 1969, Reeds et al., 1980, and Waterlow and Jackson, 1981. Regraphed by Stephen Lerner.

diphosphoglycerate (Bunn et al., 1974; Nakashima et al., 1985)—a compound, produced during intermediary metabolism by pyruvate kinase, that regulates the rate of oxygen dissociation from hemoglobin (Section 5.2.4)—do not seem to scale with body size.

Blood glucose, however, appears to scale negatively with size ($\approx M^{-0.1}$; Umminger, 1975), over a fourfold range from bats (200 mg/100 ml serum) to cattle (50 mg/100 ml serum). While this exponential constant ($M^{-0.1}$) is smaller than those of oxidative metabolism per gram or protein synthesis per gram ($M^{-0.25}$), there could be a relationship to lifespan through the pathological consequences of formation of glycated adducts to slowly metabolized proteins in basement membranes (Chapters 3.6.8, 7.5.2). No analysis has been made of collagen glycation, for example, in relation to blood glucose, body size, and lifespan.

The above analyses are sensitive to the physiological and behavioral state at sampling and should be taken as first approximations. A constant frustration of outsiders to this literature is that little data are available for comparison of different species under the same physiological conditions. Nonetheless, the data raise an interesting issue about why the rates of metabolism and protein synthesis show such consistent scaling with body size, whereas the blood levels of oxygen, and possibly glucose and amino acids, change much less.

5.2.2. Cell Size and Body Size

Most cells in mammals have very similar dimensions and structures for the same cell type throughout the range of body sizes or lifespans (David, 1977). Important exceptions to this include ovarian cells and neurons (see below). There is a size scaling of erythrocytes and other cells with the nuclear DNA content and nuclear volume in fish and amphibia (Szarski, 1976; Mirsky and Ris, 1950; Commoner, 1964). However, these size scalings do not extend to mammals, which have similar nuclear genome sizes within a narrow range of less than 25% (Section 5.4.1). The size of the enucleated mature erythrocytes in mammals (Schmidt-Nielson, 1984) does not vary systematically with body size. This conclusion also holds for nucleated cells of visceral tissues, such as hepatocytes or tongue epithelia (Teissier, 1939; Munro, 1969; David, 1977). Although hepatocyte polyploidy increases postnatally in rodents and humans (reaching $32C$, where C is the haploid DNA content), polyploid hepatocytes have proportionately increased cytoplasmic volume (Epstein, 1967), which is consistent with DNA–cell size correlations for species comparisons of similar cell types. David's (1977) tabulation of hepatocytes also indicates similar sizes in cows and rats. However, such comparisons are problematic because the size of hepatocytes and some other cells is subject to variable conditions before death, such as the recentness of feeding or stress in the holding pen and abattoir.

In contrast to these examples, the sizes of some ovarian and brain cells increase with larger-sized species. Comparison of twenty-two species from nine orders (mice to horses) showed that the diameter of Graafian (mature) follicles ranged 100,000-fold with body size with *positive* allometry, $\approx M^{1.21}$ (Gosden and Telfer, 1987a). However, primordial oocytes showed only modest correlations with adult body size, with diameters that varied 3-fold over a 100,000-fold range of body size (Gosden and Telfer, 1987a; Austin, 1961). These statistically strong relationships include a few major outliers, for example, the

Graafian follicles of the rhesus monkey and marmoset are tenfold greater than predicted. From these and other deviations, we must conclude that the allometric trends do not *absolutely* constrain variations of specific cell types. In contrast to cell size, the numbers of oocytes and primordial follicles scale *hypo*allometrically with body size, $\approx M^{0.47}$, whereas scaling with lifespan (L) is *hyper*allometric, $\approx L^{1.58}$ (Gosden and Telfer, 1987b); in this equation, the lifespan is the average, not the record. The rate of loss of oocytes with age is discussed in Section 5.4.5.

Neurons are of particular interest in questions of size scaling with lifespan, because some of the largest neurons of humans are vulnerable to age-related damage and loss. Of particular interest are the large pyramidal-shaped neurons of the frontal cortex and the hippocampus that die during Alzheimer's disease (Chapter 3.6.10). However, the size of the Betz cells (among the largest pyramidal cells in the cortex) does not scale with brain size among primates (E. Armstrong, 1989). This confirms analyses of other cortical neurons that found no correlation in a sampling from three orders (mice, rats, guinea pigs; rabbits; humans) (Bok, 1959). Olzewski's (1950) analysis of spinal cord neurons also failed to detect a cell-size correlation in spinal cord neurons. Other neuron types show indications that cell size scales with brain or body size. Tower (1954) considered that cerebral cortical neurons in the fin whale (*Balaenoptera physalus*) are "somewhat larger" than in humans. However, it is often unclear if neurons really have the same metabolism or circuit functions, despite their similar locations in specific brain regions or even cortical layers. To close this discussion, there are two convincing reports of neuron cell size–body size correlations. Cerebellar Purkinje-cell volumes vary fivefold from rodents to cattle, with humans between (Friede, 1963), whereas superior cervical ganglion neurons and their preganglionic cells vary fourfold in volume between mice and rabbits (calculated from Purves et al., 1986).

Few detailed comparisons of neuronal size, number, and cytoarchetectonics in specific cortical regions are available, besides the study of Zilles et al. (1986). Nerve cell bodies should scale in accord with the length of axonal projections and body size, because of an implied a greater axoplasmic volume to be synthesized. Correlations of cortical neuron cell body and cell nuclear sizes in rats (Bok, 1959) suggest that larger neurons have greater rates of RNA synthesis, since cell nuclei generally shrink or expand in accord with the amount of gene activity in a cell. Comparative measurements of transcription rates through nuclear run-on studies would be extremely interesting, for example in cortical neurons of mice versus cows. Other questions are raised in Section 5.2.5 about the allometry of transcription in relation to cell and body size.

Dendritic arborization in some neurons also varies with body size and again may be pertinent to the risk of neuron death because of the huge ion fluxes and energy consumption that may take place on dendritic membranes. In the superior cervical ganglion (Purves and Lichtman, 1985) and submandibular ganglia neurons (Snider, 1987), larger animals had greater total dendritic lengths and dendritic extensions from the cell body, as well as more primary and higher-order dendrites. A generally consistent pattern is seen in many species comparisons, such that the neuronal density varies inversely with body size, for example in superior cervical ganglion (Ebbesson, 1968) and cerebral cortex (Bok, 1959; Jerison, 1973). The cerebral cortex is of particular interest in these comparisons. Analysis of cortical columns (the vertical array of neurons through numerous layers of cells and dendrites that is a basic feature of cerebral cortex organization) showed that the number of neurons per cortical column is remarkably constant among species, whereas the neuron density in the cortical columns varies inversely with cortex thickness (Bok, 1959). The density of cortical neurons varies as brain size, E (gm):

$$\text{neurons/mm}^3 \sim E^{-0.32} \tag{5.5}$$

(Tower, 1954). Moreover, direct measurements of the dendritic tree from cortical neurons of mice, rats, guinea pigs, and rabbits indicated that the length of dendritic segments varied with body size as $\approx M^{0.167}$ (Bok, 1959) or as brain size, $\approx E^{0.33}$ (calculated by Jerison [1973] from Bok, 1959). Bok's 1959 monograph is still the prime source for quantitative comparisons of dendritic length in the mammalian brain. Because some neurons, for example hippocampal granule cells, can sprout additional dendrites even during Alzheimer's disease (Geddes et al., 1985), one suspects a high plasticity throughout development, with epigenetic control by growth factors, synaptic contact, and other factors extrinsic to neurons, from which could arise major species differences in dendritic arbor (Snider, 1987).

The trend for larger cell volumes in some neurons is accompanied by an increased amount of neuronal nuclear DNA through polyploidization. Polyploidization occurs postnatally in rodent cerebellar Purkinje cells, hippocampal pyramidal neurons, and some other neurons (Lapham, 1968; Jacobson, 1970). A similar overall increase of gene copy number results from the formation of binucleate neurons in the superior cervical ganglion of guinea pigs and rabbits; binucleate neurons in this ganglion have not been seen in mice (Purves et al., 1986). From another perspective, these species differences in nuclear DNA content can be considered as homeostatic (epigenetic) responses to maintain similar ratios of nucleus to cytoplasm.

Skeletal muscle cell volumes also increase with body size, as judged by muscle cell mass per milligram of DNA in a few species of intermediate size, mouse, horse, and cow (Munro and Gray, 1969), but these data need detailed extension. The thyroid gland shows another scaling. Whereas the thyroglobulin-storing follicles increase with animal size, the cell size is constant, while the RNA to DNA ratio decreases (Munro, 1969). With the potentially important exceptions of oocytes, some neuron, and possibly skeletal muscle, differences in the rate of age changes between short- and long-lived mammals could not be easily attributed to size or shape. Diffusion rates, however, could vary importantly in those exceptional cells and might influence a variety of age-related changes.

5.2.3.   Body Size and the Proportion of Highly Active Cells

Basic to interpreting size comparisons of whole-body metabolism is the variation among species in the proportion of the most active cells. If large and small animals differed in the proportion of highly metabolically active cells, then false conclusions could be drawn from comparisons of whole-body metabolism. The resting *in vivo* oxygen consumption per gram of brain, liver, kidney, and gut in humans is up to a hundred times more than in skeletal muscle, so that most (> 70%) of the oxygen consumed at rest is used by only 10% of the body mass (Schmidt-Nielsen, 1984; Aschoff et al., 1971). However, several calculations show that this is not the only factor. The five- to tenfold smaller whole-body basal metabolism and protein synthesis per gram body mass in large animals versus mice (Section 5.2.1, equations [5.3] and [5.4]) can be only in part attributed to different fractions of the body mass that are associated with these most metabolically active organs (Holliday et al., 1967; Schmidt-Nielson, 1984).

The most metabolically active tissues show negative allometry on body size, $\approx M^{0.85}$ (average of brain, liver, kidney, and gut from Figure 5.1; Holliday et al., 1967). These tissues contribute little to body mass, about 15% in mice and 8–10% in mammals that weigh 100–1,000 kg (estimated from Calder, 1984, 16). However, Hofman (1983) makes the firm point that species with relatively large brains ("highly encephalized") use more of their energy to fuel the brain, particularly humans, other primates, and toothed whales (Mysticeti). The human brain uses the most oxygen, about 20% of the total at rest. About five- to tenfold more of the body is formed by tissues that scale isometrically with size, particularly skeletal muscle (45% of body weight) and blood (7%). On the other hand, skeleton and fat (low basal metabolism) scale hyperallometrically with body size (Figure 5.1): skeleton ($\approx M^{1.1}$) and fat ($\approx M^{1.2}$). Together these contribute more than 20% of the body mass in the range of body sizes 0.1 kilogram to 100 kilograms (estimated from Calder, 1984, 16). Thus, the hypoallometric scaling of the most metabolically active tissues cannot account for differences as a function of size in basal metabolism or protein synthesis per average body cell (Holliday et al., 1967; Schmidt-Nielson, 1984). As shown next, some cells have different rates of biosynthesis as a function of species differences in body size.

5.2.4. Energy Metabolism in Specific Cells

The next question concerns species variations in the energy metabolism of specific cell types. Much research into cell senescence involves these issues, particularly species differences in oxidative damage and the maximum capacity for biosynthesis over the lifespan in nondividing cells.

Oxygen consumption during short-duration *in vitro* incubation has been long used to relate whole-body metabolism and regional blood flow to oxygen utilization in particular tissues. In general, the *in vitro* oxygen consumption of slices from most tissues decreases progressively with animal size (Elliot, 1948; Krebs, 1950; Martin and Fuhrman, 1955; Davies, 1961). Tissues with the greatest *in vitro* oxygen usage also ranked highest *in vivo:* kidney, brain, liver, heart, and gut (Davies, 1961; Aschoff et al., 1971). Overall, species differences are smaller *in vitro* than *in vivo,* which raises serious technical concerns about *in vitro* studies (Kleiber, 1947; Davies, 1961; Schmidt-Nielson, 1984). Data from Krebs' (1950) and other *in vitro* studies fit regressions on body weight in the range of O_2 consumed/gram body weight/hour $\approx M^{-0.07}$ to $M^{-0.17}$ (Davies, 1961; Smith, 1956), which gives larger values for oxygen consumption per gram (because of the negative exponent) than the $\approx M^{-0.25}$ calculated from whole-body basal metabolism (Schmidt-Nielsen, 1984). Thus *in vitro* measurements may considerably overestimate the resting tissue metabolism.

Neural tissues differ systematically in metabolic rates per gram (Elliot, 1948; Mink et al., 1981; Hofman, 1983). Calculations for brain in seven species, rat to man, gave O_2 consumed/gram brain/hour $\approx M^{-0.13}$. The spinal cord showed a similar scaling (Mink et al., 1981). These trends of oxygen utilization comfortably agree with estimates of local cerebral glucose utilization that were obtained from metabolic brain-imaging techniques by the uptake of 2-deoxyglucose in normal, conscious animals. The utilization of glucose by the brain is tightly coupled to oxygen consumption. In a version of the 2-deoxyglucose method that involves rapid sacrifice followed by autoradiography, rhesus monkeys had two- to threefold more glucose uptake per gram tissue than rats, throughout twenty-five gray- and white-matter subregions (reviewed in Sokoloff, 1981). Human cerebral cortex, assayed for 2-deoxyglucose utilization by positron emission tomography in conscious subjects, gave values that were similar to the rhesus monkey studies (e.g. Duara et al., 1984).

The inverse relation of oxygen consumption per gram tissue to brain size in species comparisons mostly results from lower neuronal density in larger brains (equation [5.7]). However, some differences may still be unexplained. Assuming that neurons account for most of the brain's oxygen consumption (Hess, 1961; Epstein and O'Connor, 1965; Hertz, 1966), Hofman (1983) calculated the important result that the average oxygen consumption per neuron is within 10% for man and rat. Oxygen consumption per gram brain in humans and rats appears to differ threefold at the most (see above), whereas the density of neurons is five- to tenfold different (Tower, 1954). I suggest that the greater than expected cerebral metabolism in larger brains may be partly explained by some neurons that have much larger cell bodies in larger brains (see above). Such large cells might also have greater biosynthesis in association with needs for more axoplasmic flow, or active transport of ions and metabolites. However, detailed data are lacking.

Glial cell densities may not vary with size as do neuronal densities. As judged by biochemical markers of glia (butyrylcholinesterase, carbonic anhydrase, and anaerobic glycolysis), glial density is remarkably independent of brain size over a 3,000-fold range from rats to whales (Tower and Young, 1973). These markers, however, are not always faithful, and the comparative distribution of the various types of glia (several astrocyte classes; oligodendroglia; microglia) should be considered an open question.

Data on subcellular activities from the liver also support these trends. Moreover, we can proceed at least a short way on the path to subcellular analysis of allometry. Cytochrome oxidase decreases per gram liver wet weight or nitrogen, as a function of increasing body size, $\approx M^{-0.24}$/gram body weight (Kunkel and Campbell, 1952). These calculations suggest that aerobic glycolysis per liver cell is inverse to body size, in parallel to the relative decrease of whole-body basal metabolism and protein synthesis, $\approx M^{-0.25}$/gram body weight (Section 5.2.1).

The density of mitochondria per hepatocyte may also scale with size. According to one estimate in postnuclear centrifugal fractions by phase contrast microscopy, the number of mitochondria per gram liver decreased progressively with body size, $\approx M^{-0.1}$/gram body weight (Smith, 1956). However, mitochondria cannot be unambiguously identified from other minute particles in crude suspensions by light microscopy. Unfortunately, there is little electron-microscopic analysis to answer this question. Other studies indicate that the fractional liver cell volume occupied by mitochondria in rats, 21% (Oudea, Collette, Dedieu, and Oudea, 1973) was twice that in humans (Oudea, Collette, and Oudea, 1973). The fraction of the cell volume occupied by the cell nucleus and the area of rough endoplasmic reticulum were about 30% *greater* in rats. In contrast, the area of smooth endoplasmic reticulum was 45% *smaller* in rats than humans. According to another brief survey, the inner mitochondrial membranes that contain the enzymes of oxidative phosphorylation were more concentrated in mice and rats than in humans and cattle (Plattner, 1968). Thus, the numbers of mitochondria per hepatocyte may be scaled with body size, though the data leave much uncertainty.

Skeletal muscle is important in discussions of brain allometry since, like the brain, it consists of mostly postmitotic cells. Skeletal muscle has a very different energy metabolism, however, being less dependent than the brain on glucose as its energy source and less rapidly damaged by ischemia (Harris et al., 1986). A first consideration is the different size allometry between individual muscles. While the total skeletal muscle mass is roughly size-independent in mammals (Figure 5.1; Calder, 1984), individual muscles showed widely differing negative and positive allometry in a set of African mammals ranging a thousandfold in size: the diaphragm varied as $\approx M^{0.7}$ (or $M^{-0.3}$/gram body weight), while the semitendinosus varied as $\approx M^{1.03}$ (or $M^{0.03}$/gram body weight) (Mathieu et al., 1981). It would be interesting to study whether age-related changes in muscles differed among species according to these negative and positive correlations with body size.

At the subcellular level, the density of skeletal muscle mitochondria showed a negative allometry per gram body weight (Gauthier and Padykula, 1966; Mathieu et al., 1981) that differed widely among muscles: diaphragm muscle, $\approx M^{-0.06}$/gram body weight; semitendinosus muscle $\approx M^{-0.23}$/gram body weight, with three other muscles between (Mathieu et al., 1981). These allometric variations in mitochondrial density are generally smaller than for cytochrome oxidase content in the gracilis muscle, which varied as $\approx M^{-0.24}$/gram body weight (Kunkel et al., 1956). The total mitochondrial volume per muscle has a scaling on body size that is close to the maximum oxygen consumption during heavy work (Taylor et al., 1981), when skeletal muscles are the major users of oxygen. These analyses also include outliers that demonstrate the variations outside of normative trends. The different mitochondrial densities in muscle types may give insights about species differences in muscle-cell dysfunctions during senescence, since free radical production might vary with mitochondrial activities.

Muscle enzymes for aerobic and anaerobic glycolysis also vary widely with size. Larger mammals and reptiles appear generally to use more anaerobic mechanisms; glycerol may also be a more important substrate for gluconeogenesis in larger animals (Coulson, 1987). In the gastrocnemius muscle of ten species (shrews to cattle), the enzymatic activity for aerobic metabolism per gram muscle was inverse to body size: β-hydroxybutyryl-CoA dehydrogenase, $\approx M^{-0.21}$/gram body weight (Figure 5.3, top) and citrate synthase, $\approx M^{-0.11}$/gram body weight (Emmett and Hochachka, 1981). In contrast to these inverse trends in size, several enzymes of anaerobic glycolysis had activities per gram muscle that *increased* with body size, lactic dehydrogenase, $\approx M^{0.15}$/gram muscle (Figure 5.3, bottom) and pyruvate kinase, $\approx M^{0.14}$. Ratios of the aerobic to anaerobic enzymes vary a hundredfold between shrews and cows. These variations between species could be valuable in comparing age-related changes in oxidative damage. These findings suggest different needs according to size for anaerobic glycolysis during intense bursts of work (Emmett and Hochachka, 1981).

Oxygen-binding characteristics of blood are valuable in discussing epigenetic and adaptive mechanisms. Because mammals do not vary in the amount of hemoglobin per milliliter blood, all mammals have similar maximum oxygen binding capacity (Section 5.2.1; Schmidt-Nielsen, 1984). However, the oxygen dissociation rate (affinity) varies inversely

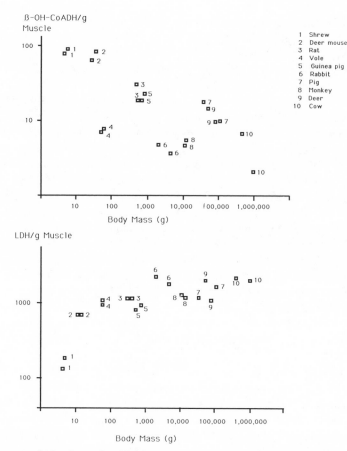

Fig. 5.3. Enzyme activities that scale with body size in the mammalian gastrocnemius muscle. *Top,* β-hydroxybutyryl-CoA dehydrogenase; *bottom,* lactate dehydrogenase. Redrawn from Emmett and Hochachka, 1981.

with size in placental mammals, as measured by the half-saturation pressure (P_{50}) of oxygen in whole blood (Bunn et al., 1974; Scott et al., 1977; Dhindsa et al., 1971, 1972; Schmidt-Nielson and Larimer, 1958; Nakashima et al., 1985). The generalizability of this trend is not clear from the surveys of more than twenty species. For example, the few whales sampled so far do not show a clear trend (Dhindsa et al., 1974). An adaptive rationale for this trend is that oxygen should be released more readily from the blood of small species as the partial pressure of blood oxygen decreases, rendering more oxygen available during bursts of muscular activity. Exceptions include burrowing animals such as the spiny anteater and the armadillo, whose very low P_{50}'s for their size may pertain to a hypothetical need for enduring prolonged hypoxia as they tunnel to escape from predators (Dhindsa et al., 1971). The slower oxygen dissociation of larger animals is also consistent with their high levels of anaerobic enzymes in skeletal muscle (see above). The greater rate of oxygen consumption by smaller mammals predicts inverse relationships with body size and differences of oxygen content between arterial and venous blood (Iberall, 1972).

The species differences for oxygen affinity in whole blood imply differences in the hemoglobin molecule itself. These different affinities may ultimately be attributed to particular amino acid substitutions that were not neutral. The example of species differences in the hemoglobin oxygen affinity indicates that epigenetic controls cannot explain all allometric differences. Overall, the tertiary and quaternary structure of hemoglobin has varied little during vertebrate evolution (Perutz, 1983). Systematic effects from amino acid substitution on hemoglobin are strongly indicated by comparisons of the Bohr effect, in which oxygen is released faster from hemoglobin as blood pH decreases. Sampling of fifteen or so species indicates that smaller mammals release relatively more oxygen as blood pH is decreased (Riggs, 1960; Bartels, 1964; Tomita and Riggs, 1971).

However, studies on whole blood and intact erythrocytes are subject to other factors, such as variations in blood levels of 2,3-diphosphoglycerate, which reduces the affinity of hemoglobin for oxygen by stabilizing the structure of deoxyhemoglobin. Other heterotropic ligands may also be important (Perutz, 1983). Under well-defined conditions and after removal of endogenous diphosphoglycerate, the hemoglobin of mouse, human, and elephant retained different-sized Bohr effects and influences from diphosphoglycerate (Tomita and Riggs, 1971; Riggs, 1971). Provisionally, it appears that more oxygen is available to the faster metabolizers during muscular action that produces lactic acid and CO_2, which in turn lowers blood pH. These species differences in hemoglobin chemistry suggest a basis for species variations in cell functions that can arise from genetic selection, as distinct from epigenetic variations (Gould and Lewontin, 1979). The particular amino acid substitutions responsible for altered binding of protons and diphosphoglycerate have not been identified. In any case, amino acid substitutions in hemoglobin (summarized in Nakashima et al., 1985) are not always selectively neutral (Perutz, 1983).

The content of diphosphoglycerate also varies in erythrocytes—for example, sheep and cattle have levels less than 1% of pigs (Bunn et al., 1974)—and these variations indicate the potential importance of other genetic factors that influence hemoglobin's functions through the physiological environment. For example, the production of 2,3-diphosphoglycerate is regulated by pyruvate kinase. High levels of this enzyme are considered to increase the oxygen affinity of fetal blood in rats and other species that lack fetal hemoglobin, thereby facilitating placental transfer of oxygen (Jelkmann and Bauer, 1980). The amount or activity of pyruvate kinase is presumably under genetic control, but again, it is unknown whether the effects are *direct* (different amino acid sequences that are dictated by DNA differences) or *indirect* (epigenetically and pleiotropically influenced through physiological and homeostatic regulation of gene expression or enzyme activity).

5.2.5. Macromolecular Synthesis in Specific Cells

Body size correlations with energy metabolism raise interesting questions about the rates of macromolecular biosynthesis and give a basis for discussions in relation to lifespan in Sections 5.3 and 5.4. For example, do small and short-lived species have correspondingly scaled-up rates of both RNA and protein synthesis? Data that appear reliable show that the half-lives of some plasma proteins made in the liver scale with body size, $\approx M^{0.3}$ (Table 5.1). For example, plasma albumin has a half-life of 21 days in cows but only 2 days in mice and rats, which is tenfold slower (Allison, 1961; Munro and Downie,

1964). The rate of albumin or transferrin synthesis per gram liver can be estimated by assuming that the rate of plasma clearance (turnover), $\approx M^{0.7}$/day for albumin (Munro and Downie, 1964; Calder, 1984) or transferrin (Regoeczi and Hatton, 1980), is equal to the rate of synthesis and secretion by the liver. Since the liver varies with body size as $\approx M^{0.88}$ (average of values cited in Calder, 1984), I suggest further extensions of these relationships:

$$\text{plasma albumin synthesis/day/gm liver} \sim M^{0.7}/M^{0.88}.$$

This simplifies to synthesis per day per gram liver as a function of body size:

$$\text{plasma albumin synthesis/day/gm liver} \sim M^{-0.18}. \tag{5.6}$$

Calculating from equation (5.6), the liver in a 70-kilogram human should synthesize about one-fifth as many molecules of albumin or transferrin per day per gram liver than in a 35-gram mouse. Assuming that the hepatocyte volume is approximately size-independent (Section 5.2.2), this calculation implies a fivefold difference in rates of synthesis of some plasma proteins between laboratory rodents and humans. So far, however, no single study has directly compared small and large species for the synthesis rates of particular RNA or proteins in liver or other cells from different species. This interpretation also assumes that hepatocytes compose similar fractions of liver mass across species sizes. Although larger animals are thought to have more metabolically inert connective tissue in their livers (Krebs, 1950) and elsewhere (Pettengill and Martin, 1947), it seems unlikely that connective tissue content could account for such major differences in plasma protein synthesis per gram liver.

If hepatocytes of larger species have manyfold less protein synthesis for cells of similar size, then the hepatocytes should also show varying amounts of RNA. According to older data, the bulk RNA content per gram tissue and the RNA to DNA ratio vary inversely with body size over a threefold range in liver and skeletal muscle (Munro and Downie, 1964; Munro and Gray, 1969). In particular, total RNA per liver appeared to decrease with body size more than for skeletal muscle (Munro and Gray, 1969). The twentyfold smaller half-life of plasma albumin is associated with a threefold smaller liver RNA:DNA in cows versus mice. The RNA content per cell corresponds in general to the amount of protein synthesis in tissue comparisons (Munro, 1969) and with changes in physiological state (Waterlow and Jackson, 1981). Because ribosomes are the major contribution (> 80%) to bulk RNA, Munro (1969) predicted that the decrease of RNA:DNA with animal size yields fewer ribosomes per liver cell, but this has not been directly established. In turn, fewer ribosomes would be consistent with less protein synthesis per cell, including that of albumin, which accounts for almost half of liver protein synthesis. Contrary to these arguments, the post hoc comparison of data on hepatocytes from rats and from human biopsy specimens (Oudea, Collette, Dedieu, and Oudea, 1973; Oudea, Collette, and Oudea, 1973) did not show consistent differences in the amount of smooth or rough endoplasmic reticulum (Section 5.2.4). Body size correlations of transcription and translation rates are an interesting future area for studying epigenetic versus DNA-encoded aspects of changes in organ size.

Brain protein synthesis may not fit these trends, though at first glance it does. In three gray-matter brain regions, determinations with corrections for pool size indicated twofold more protein synthesis per gram in rats (Smith et al., 1984; Ingvar et al., 1985) than in rhesus monkeys (Carolyn Smith and Louis Sokoloff, pers. comm.). However, this difference can be attributed to the lower density of neurons in primates versus laboratory rodents. According to equation (5.4), a rat with its 2-gram brain should have about four times more neurons per cubic millimeter than a rhesus monkey with a 300-gram brain. The absolute rates of protein synthesis per neuron remain to be compared between species.

Some proteolytic enzymes may scale allometrically with brain size. Calpains (calcium-dependent, neutral thiol proteases) have activities in major brain regions that decrease five- to tenfold per gram brain protein with increasing brain size (E) from mice to horses. The allometric coefficients are in the range of $\approx E^{-0.17}$/gram (cerebellum) to $\approx E^{-0.38}$/gram (cerebral cortex and pons medulla; Baudry et al., 1986). Because these exponential scalings approximate that of cortical neuron density (equation [5.4]), the calpain content per neuron could be relatively constant across species. The differential susceptibility of cytoskeletal-related proteins to calpain led Lynch et al. (1986) to hypothesize that calpain is involved in synaptic turnover and that there are systematic species differences in the replacement rates of cytoskeletal elements and their associated neuronal processes. Very limited data do not indicate body size correlates of protease concentration in the gut (Coulson, 1986). A possible role of calpain in protecting against damage to neurons is discussed in Section 5.4.3. Data are lacking on species variations in ubiquitin and its half-life and are pertinent here in view of the importance of the ubiquitin pathway to protein degradation.

Mechanisms that could coordinate the rates of cell energy metabolism and macromolecular biosynthesis pose an important puzzle, since the tissue environment (temperature and nutrients) is similar for animals of different sizes. One appealing proposal concerns the rate of blood circulation and the availability of substrates (Coulson et al., 1977; Coulson, 1986). Larger animals have slower circulation times, $\approx M^{0.25}$ sec (Schmidt-Nielsen, 1984; Stahl, 1967). This scaling of circulation time resembles the turnover time for blood glucose, seconds $\approx M^{0.25}$ sec (Ballard et al., 1969). Since resting blood levels of oxygen, glucose, and amino acids are considered to vary little among species (Section 5.2.1), their availability to cells must be determined mainly by the rate of circulation. Because the pumping capacity of the heart is also size-determined (cardiac output, ml/hr, $\approx M^{0.81}$; Stahl, 1967), the allometric variations in physiologic regulation appear to arise from the necessity that cells must, on the average, consume less substrates as species size increases. Otherwise, the capacity of the circulation to deliver substrates would soon be outstripped, the "flow hypothesis" of Coulson (1986). Little is known about the rates of substrate uptake for the same cell types in different species, which could be of great significance (Coulson, 1986). I note that epigenetic variations in cardiac pumping capacity of larger mammals are much smaller than the differences across the species range; for example, maximum cardiac output increases in humans only up to twofold from training (Saltin, 1969; Hagberg et al., 1985) or hyperthyroidism (Guyton, 1963). There is no obvious way that humans or larger mammals could sustain the five- to tenfold greater metabolic rates per gram body weight of the smallest mammals (Bartels, 1964).

In essence, I hypothesize that the circulation sets the boundary values on metabolism that also require commensurate adjustments of gene expression, either through epigenetic (homeostatic) modulation of gene expression or through adaptive changes in DNA sequences, or both. If the slower biosynthetic rates of cells in larger animals represent epigenetic (homeostatic) adjustments by some cells to their environment that down-regulate the number of ribosomes and mitochondria per cell, then a mouse hepatocyte should down-regulate its biosynthetic rates by fivefold or more if placed in a horse. This result is already anticipated by the constant rate of protein synthesis in humans, neonate through adults, when expressed per $M^{0.75}$, despite the more than 25-fold range of total metabolism and body weight (Waterlow and Jackson, 1981). Another approach would be to examine the influence of substrate delivery rates on the rates of transcription and translation *in vitro*, for example, of albumin and other proteins in primary hepatocyte cultures. The generality of metabolic scaling within and among species argues for the importance of epigenetic mechanisms, which would ultimately reduce to a small number of genetic changes among species of different size, with ramifying consequences on selective gene regulation. A further implication is that such epigenetic mechanisms are evolutionarily ancient, allowing relatively rapid changes in body size without major changes in DNA sequence *per se* or sequence organization.

The regulation of metabolism with body size despite similar body temperatures and apparently similar levels of blood oxygen and most substrates (Section 5.2; but these data are not certain) hints at some yet undescribed mechanisms. Are there controls that adjust metabolic rates in cells in accord with the *rate* of substrate availability? Hepatocytes and neurons might respond very differently, in view of the brain's greater dependency on oxygen and glucose. The extent of species and size influences on the hormonal regulation of protein synthesis also bears on these questions. Are there size scaling influences on the secretion of growth hormone and somatomedin-C, which are produced by organs (pituitary and liver, respectively) that scale with the same coefficients of negative size allometry, while another major hormone target, skeletal muscle, scales isometrically?

There could also be body size correlations or adjustments at the level of hormone receptors or postreceptor mechanisms. The importance of body size to hormonal regulation is indicated by the correlation of puberty with body size and fat content in most mammals (Frisch, 1985; Kennedy and Mitra, 1963; Frisch and Revelle, 1971). Insulin is also of great interest here as a regulator of muscle protein synthesis. A major gap is the absence of data on the scaling of pancreas (the source of insulin) with body size. Nor are there useful comparative data relating plasma insulin availability to metabolic stimulation in muscle, liver, or fat cells in species of different size. Genotypic polymorphisms can have major effects on endocrine regulation, as shown in mice by twofold strain differences in thyroidal ^{131}I secretion rates (Amin et al., 1957) and the different estrous-cycle lengths (Lerner and Finch, in prep.) and scheduling of reproductive senescence (Lerner et al., 1988; Chapter 6.5.1.4) that are associated with variants in the *H-2* locus. Such genetic differences, in turn, could determine species differences in the constitutive rates of macromolecular biosynthesis and its sensitivity to physiological regulation.

5.3. Lifespan, Size, and Metabolism

This section addresses the lifespan and physiological time that have long been noted to correlate with body size. While most of the data concern mammals, there is some discussion of fish and birds. As the reader will see, despite a considerable literature on size correlations with lifespan, there are enough exceptions to show that size does not firmly constrain lifespan. Most of the studies discussed use lifespan data for various species from zoos that generally represent the record survival as discussed in Chapter 1.4.5. The sample (population) size can strongly influence the apparent maximum lifespan. Moreover, the maximum lifespan is subject to cohort variation of 10–25% even in inbred mice (Hollander et al., 1984). These inherent variations limit the strength of conclusions.

5.3.1. Body Size and Lifespan

5.3.1.1. Mammals

George Sacher (1959), by drawing an analogy from Huxley's allometric equations (Section 5.2.1), formulated the first statistical relationship between postnatal lifespan (*L*) and adult body mass (*M*) in different mammals:

$$L \sim M^{0.20}. \tag{5.7}$$

This result is supported by more extensive surveys, though the exponent varies a bit (Sacher, 1976, 1978; Economos, 1980a, 1980b; Lindstedt and Calder, 1981; Mallouk, 1975; Eisenberg, 1981; Western and Ssemakula, 1982; Prothero and Jürgens, 1987). Because the interval between birth and sexual maturity is approximately proportionate to the length of the adult "lifespan" in mammals, these relationships *should* also hold for the

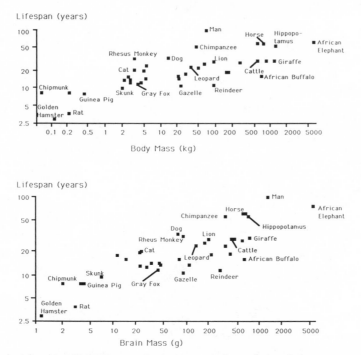

Fig. 5.4. *Top*, relationship of log lifespan to log body mass in forty species of placental mammals. *Bottom*, relationship of log lifespan to brain mass for the same species. Redrawn from Economos, 1980a; based on data from Spector, 1961, and Walker, 1975.

adult life phase. Overall, more than 50% of the statistical variance for total postnatal lifespan is attributable to adult body size, as shown for placental mammals (Figure 5.4, top), primates alone (Figure 5.5, top), and marsupials (Figure 5.6) (Eisenberg, 1981; Economos, 1980a). Among the marsupials, herbivores of all sizes live longer than same-sized carnivores or insectivores (Eisenberg, 1981; Figure 5.6). Nonetheless, the correlation is far from "perfect," and the statistical lifespan varies widely within most body size classes (Eisenberg, 1981). Prothero and Jürgen (1987) present the most comprehensive analysis to date, of 578 species from 130 families of mammals. There was no statistically significant influence on the mass exponent from gender or by taxonomic order; in the aggregate data set, lifespan varied as $\approx M^{0.17}$.

Another indirect argument about body size and lifespan can be developed from body sizes. Eisenberg and Wilson's (1978) survey of body lengths (Figure 5.7) suggests that adults in most mammalian genera are shorter than one meter. Using Jerison's (1973) formula to estimate body weight from body length, Z ($Z = 0.02M^3$), a mammal one meter

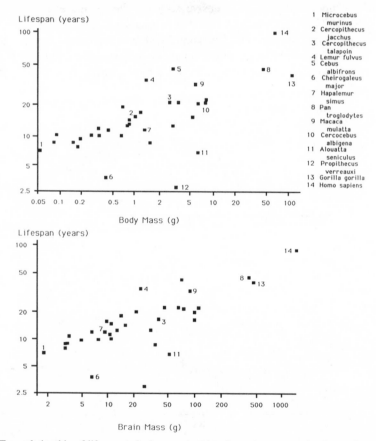

Fig. 5.5. *Top*, relationship of lifespan to body mass in thirty-five primates on a log-log scale. *Bottom*, relationship of lifespan to brain mass in the same species as in Figure 5.4, top. Redrawn from Economos, 1980a; based on data from Spector, 1961, and Walker, 1975.

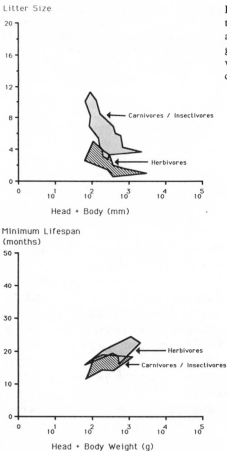

Litter Size

Head + Body (mm)

Carnivores / Insectivores

Herbivores

Minimum Lifespan
(months)

Herbivores

Carnivores / Insectivores

Head + Body Weight (g)

Fig. 5.6. Longevity as a function of body mass and trophic strategies in marsupials. The points are represented as two convex polygons reflecting two broad trophic strategies. The herbivores tend to be longer-lived than the carnivores and insectivores (microtine rodent herbivores are excluded). Redrawn from Eisenberg, 1981, Figure 86.

long should weigh about twenty kilograms. Because the genera with the most species (rodents, bats, and insectivores) are typically small, $Z \geq 30$ centimeters (Eisenberg, 1981), the allometric correlations predict that the maximum lifespan of most species of mammals is less than 30 years.

Nonetheless, caution is needed because the available data on lifespan represent so few ($< 15\%$) of the four thousand mammals and because of anomalies like the much longer lifespans of bats than are predicted by body size. For example, recall from Chapter 3.6.1 the slow accelerations of mortality rate of twenty-gram pipestrelle bats (Table 3.1).

A major issue concerns the contributions in these correlations from the IMR (calculated at puberty) and the MRDT (equation [1.4]). Limited data suggest that the MRDT also scales with size in mammals: $\approx M^{0.27}$ (Calder, 1984, 1982, 1983a). Not surprisingly, this mass exponential coefficient is similar to that in the lifespan–body mass relationships, as are so many other physiological rates. There is also an inverse relation of the IMR to body size across species (Calder, 1983a); this would be expected from the greater prematurational mortality rates of small mammals (Eisenberg, 1981; Figure 5.8), which generally have large litters and higher mortality rates. Because the values of the IMR are more

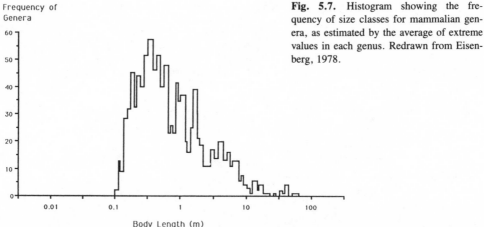

Frequency of
Genera

Body Length (m)

Fig. 5.7. Histogram showing the frequency of size classes for mammalian genera, as estimated by the average of extreme values in each genus. Redrawn from Eisenberg, 1978.

subject to the environment than those of the MRDT (Chapter 1), it is difficult to draw strong conclusions about the relationships between species and environments.

Covariate analyses of size in relation to maximum lifespan, and parameters of growth and reproduction indicate important differences among taxonomic groups, particularly among orders of placental mammals (Stearns, 1983b, 1984; Read and Harvey, 1989; Promislow and Harvey, 1990). These studies indicate that most of the variance in life history parameters is accounted for at higher taxonomic levels, that is, 80% of the overall variance is explained by differences among families and orders. Body size had a major influence on the ranking with respect to age at maturation, gestation length, and number of offspring, and different orderings occurred when body size was controlled for (statistically removed); lifespans differed among orders, after body size was controlled (Stearns, 1983b).[3] Stearns proposed that these differences among orders represent phylogenetic constraints.

Taxonomic effects were also strongly shown in more extensive analyses by Paul Harvey and colleagues. Gestation length, size and frequency of litters, and maximum lifespans form a fast-slow continuum, from lagomorphs to bats and primates, *irrespective of body size* (Read and Harvey, 1989). This gives strong evidence that variations in these life history parameters are not constrained by body size or other size-related scaling principles. Furthermore, the mortality rates of juveniles and adults retained strong correla-

3. The source of data was Eisenberg's (1981) tables, which are mainly based on zoo records from the 1970s, particularly those collected by Marvin Jones (1982). Some lifespan records are now longer by 25% or more, e.g. *Macaca mulatta*, given as 29 years (Eisenberg, 1981), but which more recent data show live at least 35 years (Table 3.1). Except for humans and a few domesticated species where large poplations in good health have been characterized, nearly all other data represent the oldest survivors in zoos (Jones, 1983; Nowak and Paradiso, 1983; Comfort, 1979), where few were born in captivity until recently. Thus, we do not know that these lifespans are the maximum obtainable under optimum conditions for physical or mental (psychological) health. I anticipate further increases of maximum lifespans of most species in captivity, through the continuing efforts to improve health. The maximum lifespan of the C57BL/6J laboratory mouse strain, for example, has increased by 25% during the past 30 years (Finch, 1971). The maximum lifespan for lab mice is now 4.8 years (Table 6.1, footnote 14, a record that would have caused great skepticism 20 years ago). Despite great efforts, the age-independent mortality rates (IMR) of many nondomestic animals are still high in captivity by comparison with privileged laboratory populations, as shown by the absence of rectangular mortality curves for most feral or captive species (e.g. the rhesus monkey; Tigges et al., 1988), which indicates high values of IMR that could considerably reduce the apparent maximum lifespan for small populations, as modelled in Figure 1.7.

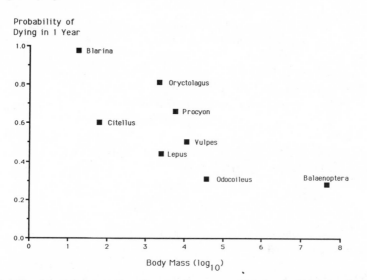

Fig. 5.8. Probability of death for a sampling of mammalian genera within 1 year of birth as a function of body mass. Data for baleen whales (*Balaenoptera*) from an exploited population; short-tailed shrew (*Blarina*); ground squirrels and gophers (*Citellus*); hares and jack rabbits (*Lepus*); domestic rabbit (*Oryctolagus*); raccoons (*Procyon*); and red foxes (*Vulpes*). Redrawn from Eisenberg, 1981, Figure 48.

tions after size was controlled for ($r^2 = 0.64$; Promislow and Harvey, 1990). However, adult mortality rates were correlated only with age at maturity, while maximum lifespan correlated best with annual production of offspring.

These results sharpen the questions about the roles of size and mortality rates as the basis for selection of life history variations. There is now a good rationale for analyzing contributions of overall adult mortality rate (IMR) and age-related accelerations of mortality (MRDT) to life history variations according to the phylogenetic leneage. Although phylogeny by itself may "explain nothing" (Read and Harvey, 1989), comparisons at higher taxonomic levels should give insight about genetic determinants of the numerical ranges of IMR and MRDT (Appendix 1). Examples might include the relative strength of biochemical mechanisms for preventing the accumulation of molecular damage from glycation (Chapter 7.5.2) or free radicals (Chapter 7.5.6), and the extent of proliferation in lymphopoietic cells before clonal senescence (Chapter 7.6). Analyses so far, as discussed in Section 5.4, have emphasized correlations with lifespan rather than with different components of mortality rate.

5.3.1.2. Other Animals

For other animals, the relationships between body size and lifespan draws on very limited data, mostly from field studies. Without reference to mortality constants, the maximum lifespans (*Lmax*) of birds scale with body size, according to exponential coefficients that are similar to mammals (equation [5.7]; Western and Ssemekula, 1982; Lindstedt and Calder, 1976): wild passerines (*Lmax* $\approx M^{0.26}$) and nonpasserines (*Lmax* $\approx M^{0.18}$), and captive birds (*Lmax* $\approx M^{0.19}$). However, these relationships need reevaluation in terms of the separate mortality coefficients, IMR and MRDT. It is already clear that some very long lived birds weighing less than one kilogram show no signs of reproductive senescence at least to 30 years and have MRDTs at least as short as in humans (Chapter 3.5; Tables 1.2, 3.1).

For lower vertebrates, data on mortality rate and lifespans are even scarcer (Chapters 3, 4). A little is known for reptiles, which appear to show very gradual to negligible senescence. The scaling of body size and lifespan from compiled data on reptiles (Mallouk, 1975) may pertain more to the IMR than to the rate of senescence.

Other insights are available for teleostan fish. Most long-lived teleosts continue growing after sexual maturation (Figure 5.9) at a constant rate that fits the negatively allometric Brody-Bertalanffy growth equation (Brody, 1945; Bertalanffy, 1938; Fabens, 1965; Ricker, 1979):

$$Z_t = Z_{max}\,(1 - e^{-gt}),\qquad\qquad(5.8)$$

where Z_t is length at time t; Z_{max} is the asymptotic length, or growth limit; g is a constant for the rate at which growth slows.

Beverton and Holt (1959) showed that the observed *Lmax* was correlated with the maximum size over a wide range of lifespans. Moreover, estimates of the average mortality

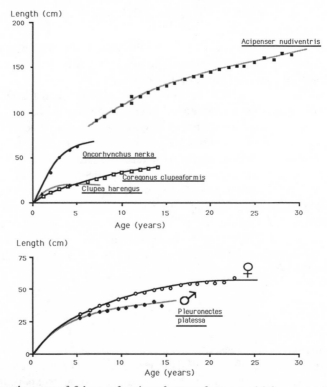

Fig. 5.9. The growth curves of fish as a function of age as fit to a model for asymptotic approach to a maximum size (equation [5.9]). The semelparous salmon (*Onchorynchus nerka,* migratory sockeye) and three iteroparous species: Atlantic herring (*Clupea harengus*), plaice (*Pleuronectes platessa*), and a sturgeon (*Acipenser nudiventris,* schip). Redrawn from Beverton and Holt, 1959.

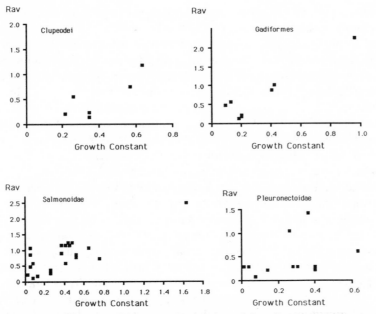

Fig. 5.10. Plot of the population mortality constant in natural populations (Rav/year) and the growth constant, *g* (equation [5.8]) for iteroparous fish: Clupeodei (herrings and sardines); Gadiformes (cod, haddock, hake); Salmonoidae (salmon, smelt, trout, whitefish); and Pleuronectoidae (flounder, halibut, plaice, sole). Redrawn without regression lines from Beverton and Holt, 1959, Figure 6; correlation coefficients were not provided by those authors.

rate, an approximation of IMR, indicated its correlation with the rate of slowing of growth (Figure 5.10). The slope of the regression, maximum lifespan (*Lmax*) on maximum size, varies with the family of fish. Other compilations for flatfish indicate annual mortality rates (*Rav*) of 0.1–0.2/year that are not correlated to lifespan (Roff, 1981).

No slowly growing and iteroparous fish manifests senescence *before* achieving most (75%) of its predicted growth (Beverton and Holt, 1959). In the semelparous Pacific salmon, nearly 90% of the maximum predicted growth occurs before death at spawning. However, castrate Pacific salmon continued to grow for 3 or more years after their escape from rapid senescence (Robertson, 1961; Roberston and Wexler, 1962; Figure 2.13). Therefore, the rapid senescence after first spawning of Pacific salmon is not linked to the *inability* for further growth. The radical endocrine changes and cessation of feeding that are associated with spawning in some semelparous fish may be the cause, not the effect, of truncated growth. Analyses of growth and maturation are subject to confounds from seasonal changes in temperature and nutrients, or from shifts in food preference that occur with growth; for example, some perch *increase* their growth rate when their size allows them to switch their diet from insects to small fish (Ricker, 1979, 690; also Figure 3.7).

Two other examples are briefly considered. Sea urchins are of interest because of their indeterminate growth, their distant evolutionary relationship as deuterostomes to the chordates, and the apparent absence of senescence. An analysis of size in sea urchins suggests that larger, faster-growing species with small values of *g* (equation [5.8]) also have smaller age-independent mortality rate constants (IMR; Ebert 1982, 1985). Long-standing colonies of sea urchins in several laboratories might give direct information on sea urchin mortality schedules.

The final example concerns the strong correlations between body size and individual lifespans in the *same* species; note that this analysis is fundamentally different from the between-species comparisons. For both male and female *Drosophila*, the smallest flies had 50% smaller lifespans than the largest (Partridge and Farquhar, 1981). In view of the evidence that mechanical damage from flying, fighting, and mating is a major factor in fly senescence (Chapter 2.2.2.1), this correlation might be consequent to vulnerabilities to mechanical damage. It would be easy to sort flies by size for an analysis of mortality rate coefficients (IMR and MRDT). Numerous examples from field studies indicate many animals for which larger size protects them against predation; this may be a basis for a trend in many radiations, for size to increase during evolution (Cope's rule; Chapter 11.4.1.2).

A major open question concerns whether growth cessation in general is consequent to endocrine changes or to genetically programmed limitations for growth and cell proliferation. Besides the above example of continued growth in Pacific salmon after castration, transformational genetics gives another. An important ongoing study is examining mice that were injected just after fertilization with rat growth hormone genes under the control of the strong metallothionine promoter and that became genetically integrated into a single chromosome (Palmiter et al., 1982; Shea et al., 1987). The transgenic mice experienced high levels of growth hormone and insulin-like growth factor, growing in some cases to adults of seventy grams, which is two- to threefold the normal size. The growth acceleration became notable by 3 weeks after birth (the usual age of weaning). While most organs were larger than in normal mice, brain size was not effected (Shea et al., 1987). Allometric analysis showed other important differences, so that heart, lungs, and kidneys had identical coefficients (equation [5.1]) to normals, while liver and spleen differed. These results suggest that elevations of two major growth factors (growth hormone, insulin-like growth factor) can stimulate cell proliferation in certain organs during development, but only within limits set by other, still unknown mechanisms. The lifespan of these giant mice is unknown.

5.3.2. Body Size and Rates of Postnatal Development

The timing of changes during development, as well as in senescence, shows a scaling with body size. In general, larger animals mature later ($\approx M^{0.27-0.29}$) and have greater generation times ($\approx M^{0.27}$) (Western, 1979; Eisenberg, 1981; Calder, 1984; Millar and Zammuto, 1983; Figure 5.11). Birds (Western and Ssemekula, 1982), lizards and snakes (Stearns, 1984), and even trees (Figure 4.3; White, 1973) also show strong correlations between body size and the age of maturation. As noted above for mammals, there are important correlations of taxonomic level (family and order) with growth and reproductive parameters in reptiles (Stearns, 1984).

The age of maturation has a fundamental relationship to life expectancy and adult mortality risk, as required to provide adequate numbers of offspring for propagating the next generation. As may be recalled from Chapter 1 (Figure 1.13), the age of first reproduction is balanced against life expectancy in many mammals (Harvey and Zammuto, 1985) and birds (Lack, 1966). Thus, animals with long life expectancy (low IMR and/or short MRDT) tend to be late maturing. This effect holds even when size effects were statistically

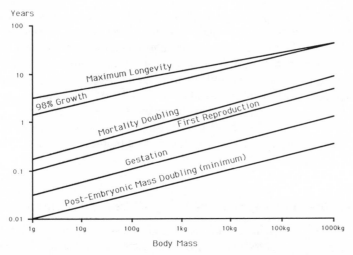

Fig. 5.11. Allometric representations of the life histories of eutherian mammals. Redrawn from Calder, 1984.

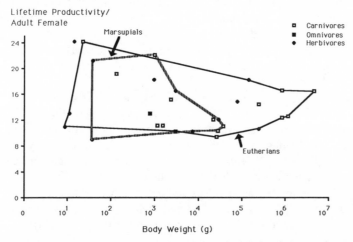

Fig. 5.12. The estimated mean lifetime productivity (births) of adult female mammals according to body weight, showing the remarkably similar reproductivity across a body-weight range of 10^6 in marsupials and placental mammals. Redrawn from Eisenberg, 1981.

removed by partial correlation (Harvey and Zammuto, 1985). Eisenberg (1981) proposed that lifetime number of offspring per female is relatively constrained across marsupials and placental mammals, so that few species have the potential for producing more than ten to twenty-five neonates (Figure 5.12), despite a millionfold range of body weights. Whereas clear exceptions are small rodents and pigs, most mammals produce about twelve neonates during their lifespans. Furthermore, this figure suggests that body weight is not a major factor in the *minimum* lifetime productivity, although the smallest mammals may have the largest numbers of offspring. These remarkable trends beg to be studied in more detail.

The exponential coefficient for age at maturation ($\approx M^{0.27}$) exceeds that for maximum lifespan ($\approx M^{0.20}$). This relationship suggests that maturation occupies relatively more of the lifespan in the longer-lived mammals and birds. Many examples suggest that warm-blooded vertebrates have rates of change during postnatal development that are calibrated in general proportion to their size. Moreover, postmaturational changes are often shared by short- and long-lived mammals and occur in similar fractions of the lifespan, despite thirtyfold difference between species in lifespan (Table 3.3). The similarity of these scalings in postmaturational changes raises evolutionary issues (Chapter 3.6.1). The relationships of the rates of oxidative metabolism and macromolecular biosynthesis to the scaling of postmaturational changes is obscure and remains a problem of great intellectual interest.

5.3.3.   *Brain Size and Lifespan: A Potentially Misleading Correlation*

Brain size (Figure 5.5, bottom), the brain/body mass ratio, and combinations of the brain and body mass show somewhat stronger correlations of maximum lifespans between species ($r^2 > 0.79$, or 79% of the variance) than does body weight alone ($r^2 = 0.63$; Sacher, 1959; Hofman, 1983). If insectivores and bats are excluded, the correlation of lifespan with the compound parameters of body and brain size improves further (Hofman, 1983). Two species of bats lived fourfold longer than predicted from the multivariate allometric equation for body and brain mass, whereas four insectivores lived 50% or less than predicted. These deviations may represent phylogenetic constraints, noted above for body size by Stearns (1983b). Further analysis is needed to resolve influences from taxonomic level (family and order).

The brain has also been proposed for a role as pacemaker of physiological changes during senescence (Dilman, 1971, 1981; Everitt, 1973, 1980; Finch, 1972a, 1976a). Brain size is appealing for consideration as a parameter in relation to lifespan in several regards. (1) Brain size varies about 50% less among individuals than body weight for most species (Calder, 1984, 9). (2) It is also less influenced by seasonal and environmental conditions in mammals than body weight. In the comparison of two body size classes, lifespans were longer for the smaller species across a twentyfold range of brain sizes (Sacher, 1975a; Figure 5.13). Whereas the human lifespan is about fourfold longer than predicted from body size alone, it is more consistent with species correlations with the

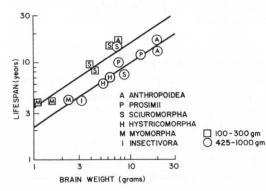

Fig. 5.13. Regressions of lifespan on brain mass for two body-weight classes of small mammals, plotted on a log-log scale. Redrawn from Sacher, 1975a.

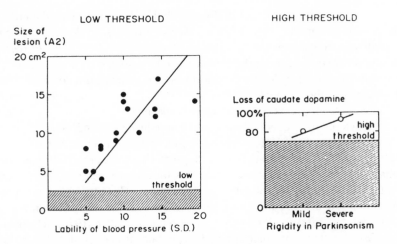

Fig. 5.14. Relationships between brain lesion size and dysfunctional manifestation in two dopaminergic systems of mammals, suggesting major differences among brain regions in the threshold for manifesting damage. *Left,* a low-threshold neural system is the A2 brain stem nucleus: minor lesions are expressed as increased variance (lability) of the blood pressure, which this nucleus influences (Talman et al., 1980). *Right,* a high-threshold neural system is the A9 (substantia nigra): relatively huge (> 80%) loss of neurons can be sustained with little dysfunction (Bernheimer et al., 1973); greater damage (loss of dopamine) is associated with Parkinsonism. These examples of neural systems with low and high thresholds for expressing damage suggest that the relative proportions of different brain regions between mammalian species may be a factor of safety (Meltzer, 1906) in buffering neuron damage accumulated over a lifetime. From Finch, 1982.

ratio of brain to body mass (Hofman, 1983, 1984; Schmidt-Nielsen, 1984; Yates and Kugler, 1986). The brain/body mass ratio may also increase with the fraction of the postnatal lifespan spent in learning from parents or siblings (Eisenberg, 1981, 327). Relatively larger brain size for a body weight class could influence the potential lifespan by providing a greater neuronal reserve for buffering the impact of neuron loss (Finch, 1988). Neuron populations show a critical threshold for manifestations of loss, for example, the substantia nigra in relationship to the extrapyramidal signs of Parkinsonism (Figure 5.14). While speculative, the brain/body mass implies a greater neuron reserve. The relationships between the brain/body mass ratio and the critical thresholds for neurological dysfunction are entirely unknown.

In regard to the pronounced evolutionary changes of brain structure in mammals, the neocortical mass shows the strongest correlations with lifespan among twelve major brain regions. Even so, neocortex accounts for less of the lifespan variance than the whole brain (Sacher, 1975a). Using an index of encephalization that relates the volume of the neocortex to both body and brain size, Hofman (1982, 1983, 1984) attributed the long lifespans of elephants and humans and other primates to the relatively large size of their neocortex. A combination of brain weight and cerebrocortical surface area accounts for 99% of the variance in lifespan (Hofman, 1983). In particular, most of the long human lifespan was attributed to extreme encephalization.

Although Hofman (1984), Sacher (1975), and Cutler (1976a) attribute the greater maximum lifespan of humans, in comparison to other anthropoids, to slower senescence, another possibility is indicated by calculations that the mortality coefficients of rhesus monkeys (Chapter 3.6.2) differ from most human populations in the higher values of IMR. In contrast, estimates of the acceleration of mortality rate with age did not show differences between humans and rhesus monkeys; that is, the MRDT was about as long (Finch et al., in prep.; Table 3.1). From this perspective, the capacity for statistically long life expectancy might be favored by neocortical functions that reduce IMR, independently of the rate of senescence. Reductions of IMR might be caused by less dangerous behaviors in one or both sexes, greater skill in avoiding predators, greater skill in finding adequate diets, etc. These behavioral traits are reasonable candidates for improvements of brain functions at the neocortical level during recent anthropoid evolution. This proposal is a major departure from views that delayed or slowed senescence was a major factor in the putative increase of potential lifespan during primate evolution (Cutler, 1976a; Hofman, 1984; Sacher, 1975a). Pending more mortality schedule data, it may be possible to relate the age-independent and age-dependent mortality rates (Chapter 1) to particular brain structures.

Several caveats should be considered about these and other correlations with lifespan. First, there is a fundamental difference between the allometric *equations* describing organ and body size relationships, and the *correlations* of lifespan with size. Huxley's allometric equations (equation [5.1]) are dimensionally consistent, algebraic identities with the same units (e.g. weight) on both sides, whereas the correlations with lifespan are not equations in the same sense. Any relationship between lifespan and size is only a statistically formulated hypothesis and is not based on any physical principle *requiring* a link between the variables on either side of the equation. Second, the data on maximum lifespan are very limited. Correlations with the maximum apparent lifespans of shorter-lived species may, on further analysis, show influence from the age-independent mortality rate (IMR). The IMRs appear to be smaller in longer-lived and K-selected species (Chapter 1.5.1). Third, even the largest samplings (85 species, Sacher, 1978; 170 species, Economos, 1980a) represent only about half of the mammalian orders and less than 5% of extant species (Honacki et al., 1982; Nowak and Paradiso, 1983). At best, we could hope to discern trends, while accepting that conclusions may be sensitive to the particular sampling. The present coefficients thus cannot be presumed generalizable to species with different body designs and adaptations to different environments. To press for finer distinctions seems premature. Fourth, because nearly all organs increase to some extent with body size, strong correlations with lifespan and most organs are not only expected, but are statistically unavoidable (Calder, 1976). Exceptions would be the relatively smaller appendix in humans versus most carnivores and the absence of a gall bladder in rats but not mice. Fifth (and finally), in regard to brain mass, there are equally good correlations of some visceral organ weights with lifespan: adrenal, liver (Figure 5.15; Economos, 1980a), and spleen (Calder, 1976). Differences in these correlation coefficients are small and sensitive to the species sampled. Moreover, multiple determinants of mortality risk seem likely through involvement of multiple-organ system defects. In conclusion, little insight is given by correlations of lifespan with the sizes of various body components. We need further studies to resolve the relationships of brain mass and lifespan to mortality rate coefficients IMR and MRDT.

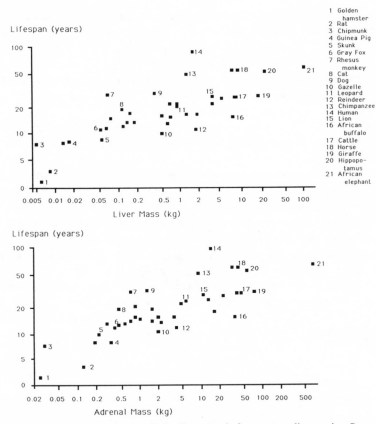

Fig. 5.15. *Top,* correlation of log lifespan with log liver mass in forty mammalian species. *Bottom,* correlation of log lifespan with log adrenal mass in the same species. Redrawn from Economos, 1980a.

5.3.4. Metabolism

Because smaller mammals have relatively higher metabolism and shorter lifespans, it is no surprise that basal metabolism shows a general inverse correlation. Allometric equations for birds and mammals have similar mass exponential coefficients (Calder, 1984). For mammals,

$$\text{cal/gm/lifetime} = 300,000 \ M^{0.07}; \tag{5.9}$$

for birds (passerine),

$$\text{cal/gm/lifetime} = 1,000,000 \ M^{0.07}. \tag{5.10}$$

These correlations suggest that the basal metabolism of some birds consumes three times the energy over their lifespans than that of mammals of the same size does. It does not necessarily follow, however, that the lifespan is *limited* by the metabolic or functional capacity of all or even most cells. As discussed in Chapters 7 and 8, an impressive number of cell functions of mammals remain normal beyond the average lifespan.

A multivariate analysis of metabolic rate, brain weight, and body weight (Sacher 1976,

1978) accounted for 81–85% of the lifespan variance in eighty-five species, somewhat more than body or brain weight alone.

$$L = 0.66E^{0.6}M^{-0.4}P^{-0.5}T^{0.25}, \tag{5.11}$$

where L is maximum lifespan (yrs.); E is brain weight (gm); M is body weight (gm); P is resting specific metabolic rate (oxygen/gm/min); T is body temperature °C.

The small contribution from body temperature to the overall lifespan correlation is consistent with the size independence of body temperature in mammals (e.g. Morrison and Ryser, 1952).

The meaning of the inverse correlation of basal metabolic rate to lifespan and the positive correlation with body size is unclear. It can no longer be concluded (Hofman, 1984) that this is consistent with the prolongation of rodent lifespan by diet restriction, since more recent studies clearly show that the basal metabolism per gram is not influenced by such diets (Chapter 10.3.1.1). Moreover, under some circumstances (exercise, cold exposure), food intake increases without shortening lifespan. The additional maternal heat production per kilogram of fetus during pregnancy increases exponentially with species comparisons of maternal body size, $\approx M^{1.2}$ (Brody, 1945), and predicts a trend opposite to the inverse correlations of basal metabolism and lifespan.

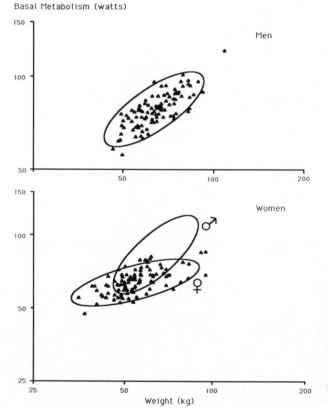

Fig. 5.16. Scatter diagrams of body mass and basal metabolism in men and women with 95% equal-probability ellipses. *Top,* adult men; *bottom,* adult women, with the ellipse for men superimposed for comparison. The effects of menopause on basal metabolism have not been described. Redrawn from Heusner, 1985.

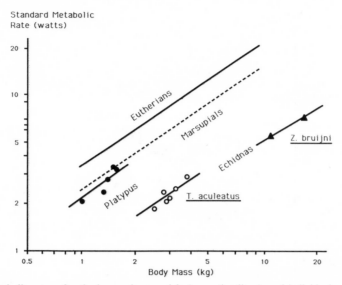

Fig. 5.17. Metabolic rates of eutherians and marsupials (normative lines); and individual specimens of the platypus (*Ornithorhynchus anatinus*) and two echidnas (*Tachyglossus aculeatus* and *Zaglossus bruijni*). Although the egg-laying mammals produce less than one-third of the basal metabolic energy of eutherians, size has essentially the same effect, the $M^{0.75}$ scaling (equation [5.2]). Redrawn from Dawson et al., 1979.

Moreover, the meaning of relationships among basal metabolism and lifespan is colored by variation between related organisms and the sensitivity of basal metabolism to many physiological variables. For example, in gender comparisons of basal metabolism, young adult women have 40% smaller mass exponents (*b* in equation [5.2a]; Figure 5.16; Heusner, 1985). Major differences in the scalar coefficient *a* (equation [5.2a]) are found between marsupials and monotremes (Figure 5.17; Dawson et al., 1979) and laboratory rats and mice (Figure 5.18; Heusner, 1985; Bartels, 1982). Another example of deviations is the soricine shrews, whose resting metabolic rates are up to twofold greater than for rodents of the same size (Eisenberg, 1981). On the other hand, spiny anteaters (*Tachyglossus*) live at least 50 years (Nowak and Paradiso, 1983; Jones, 1982), exceeding the 30-year maximum predicted by their three- to six-kilogram body size alone. This extended lifespan is matched by few other intermediate-sized mammals and has been attributed to their low, 30°C body temperature (Dawson et al., 1978) and slow metabolism (Figure 5.17). The number of egg-laying mammals is too few for statistical analysis. When body size was statistically controlled across different orders of placental mammals in the most comprehensive survey to date, metabolic rate no longer correlated with lifespan or other life history variables (Read and Harvey, 1989).

The effects of metabolic down-regulation during hibernation, aestivation, and torpor also influence the relationships between basal metabolism and lifespan. Hibernation in Turkish hamsters (*Mesocricetus brandti*) commensurately increases lifespan (Lyman et al., 1981; Chapter 10.2.2). Field observations support this view. The 8-year lifespan of the tiny (ten gram) pocket mouse *Perognathus longimembris* (Egoscue et al., 1970) is also attributed to extended daily torpor and low waking metabolism (Sacher, 1978). Recall from above that bats live longer than expected from their body or brain size and that at least one species, the pipestrelle bat (twenty grams), has a statistically short lifespan be-

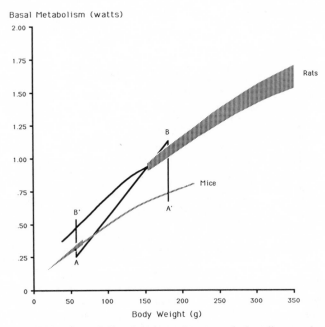

Fig. 5.18. Body mass and basal metabolism in mice and rats, graphed on linear scales according to the allometric equation aM^b. The shadings indicate 95% confidence ranges. Both species show the same allometric mass exponent ($b = 2/3$). However, the mass coefficients differ: $a = 2.3$ in mice but 3.4 in rats. This discrepancy is shown by the extrapolations of values for each species (A' and B') and the line intersecting A and B. Redrawn from Heusner, 1985.

cause of high values of IMR (0.36/year), with a considerable MRDT, as long as that of horses (Table 3.1). Other small bats survive beyond 30 years (Austad and Fischer, in press; Jürgens and Prothero, 1987). The greater-than-predicted lifespan of this and other temperate-zone bats has been thought to result from dramatic reductions of body temperatures by 20°C or more during daily sleep or during the 4–6-month-long hibernation (Bourlière, 1958; Economos, 1980a; Sacher, 1978). However, nonhibernating tropical bats live about as long, despite having an estimated tenfold greater total annual energy expenditure than hibernators (Herreid, 1964). Analysis of homeothermic and heterothermic bats ($N = 50$) and marsupials ($N = 67$) definitively showed that their lifespans are unrelated to the duration of torpor (Austad and Fischer, in prep.).

Together, these results argue against an intrinsic maximum limit to energy consumption, or the number of heartbeats or breaths, etc., as *direct* limits to the lifespan. Rather, it seems more likely that the metabolic work done during the lifespan is not linked strongly, if at all, to the major causes of death during senescence. Rubner's (1908) hypothesis can be laid to rest with full dignity.

5.4. Cellular and Biochemical Correlations with Lifespan

The bulk correlations imply that larger and more long lived species have much less biosynthesis per cell. Although most characteristics of cells do not vary consistently in relation to body size, there are some interesting candidates for cell and molecular scaling with lifespan. Because protein turnover is inversely correlated with body weight (these studies focus on young adults) and because body weight is positively correlated with lifespan, it is inescapable that there are inverse statistical relationships between the rates of protein synthesis and lifespan (Spector, 1974). Similar trends might be predicted for RNA synthesis. I chose not to pursue these generalizations further, until data become available on macromolecular biosynthesis in specific cell types that become generally dysfunctional with vital consequences during senescence across the range of mammalian lifespans. Candidates for such cells could include neurons that show wasting of their dendritic arbors and cell-body atrophy (Chapter 3.6.10), or select lymphocyte lineages that show altered differentiation from stem cells (Chapter 3.6.6). The following material presents candidates for various mechanisms relating subcellular and biochemical parameters to lifespan, but is by no means inclusive.

5.4.1. Genome Size

The nuclear genome size (haploid DNA content) does not vary systematically with size or lifespan in mammals, and clusters narrowly around 3–4 picograms per gamete cell nucleus (Ohno, 1970; Olmo, 1983; Bachmann, 1972; Bachmann et al., 1972; Altman and Katz, 1976; Kato et al., 1980). The genome size of mammals is slightly greater than reptiles, but smaller than most amphibia, and varies much less. For example, birds range twofold in haploid DNA; teleosts tenfold; and there is an astounding hundredfold range in amphibia (Altman and Katz, 1976; Britten and Davidson, 1971; Bachmann et al., 1972; Olmo and Morescalchi, 1975) and in vascular plants (Bennett, 1972). There is no clear understanding of how these variations relate to the number of genes or transcription units. As noted in Section 5.2.2, the size of a given type of cell scales with the nuclear DNA content.

Whereas the haploid DNA content in mammals varies little and in no apparent relation to body size, there are major, if nonsystematic, differences among species in the frequency and organization of repeated sequences; in the presence of various transposable elements, including viral genes; and in the rate of nucleotide divergence in single-copy sequences (Bernardi et al., 1985; Wichman et al., 1985; Schmid and Jelinek, 1982; Britten, 1986). So far, no particular DNA sequence or organizational variant is implicated in lifespan differences among mammals, with the exception of grossly abnormal genotypes, like trisomy of chromosome 21 or Down's syndrome, that have short lifespans and segmental accelerations of senescence (Chapter 6.5.3.3).

The wide range of nuclear DNA content in plants may, however, influence the duration of their life cycles. An intriguing analysis (Bennett, 1972, 1977; Francis et al., 1985) showed that a sample of perennial species of vascular plants, which often take many years to reach maturity, had a greater upper range of nuclear DNA contents (1–400 picograms) than did shorter-lived annual species (0.8–100 picograms), particularly in the monocotyledons. Thus, there is considerable overlap. Ephemeral species that can complete their life cycle in 1 month or less have the smallest haploid DNA, averaging 1.2 picograms, with the lowest value of 0.6 picogram in *Arabidopsis thalania*. Temperature also influences the cell cycle, and species that inhabit zones of low temperatures also appear to have smaller DNA content (Bennett, 1972).

A mechanism is suggested by the correlations of nuclear DNA content with the duration of the somatic cell cycle in root-tip meristems of angiosperms (Van't Hof and Sparrow, 1963) and with the duration of meiosis in diploid angiosperms as well as some insects, over a more than tenfold range (Bennett, 1977). From this perspective, the minimum generation time in a phylogenetic line may be constrained by the time required for DNA synthesis. Here may be an important influence of the genome on the life history that is independent of informational DNA sequences, that is, those coding for specific genes or functions. Thus, the short life cycles of ephemeral plants may have constrained the rate at which the nuclear DNA content could increase during evolution. However, the cell-cycle time could be decreased by increasing the numbers of nuclear DNA replicons, a mechanism that may also favor the frequent polyploidization in plant radiations.

5.4.2.   *Rates of Accumulation of Damaged Molecules and Lipofuscins*

The accumulation of damage to macromolecules during the lifespan is speculated to be involved in the accumulation of lipofuscins or aging pigments, which are considered to be slowly accumulated metabolic by-products, particularly of oxidized lipids (Chapter 7.5.6). To continue the inquiry about scaling with rates of metabolism and the lifespan, by making post hoc comparisons of studies from separate laboratories, it is provocative that lipofuscin accumulates five times faster in dog myocardium (Munnell and Getty, 1968) than in humans (Strehler et al., 1959). This may be related to the frequency of heartbeats. The scaling of lipofuscin accumulation with metabolic activities or organismic lifespan is virtually undescribed.

The accumulation of racemized amino acids (Chapter 7.4.4), which is temperature driven, should not scale with lifespan, because basal temperature varies little with body size in mammals (Morrison and Ryser, 1952). It is unknown if glycation of long-lived proteins scales with lifespan. The small allometric correlations of body size and blood glucose (Section 5.2) suggest weak correlation with lifespan. These and other post-transcriptional modifications are discussed in Chapter 7.5.

In conclusion, there are insufficient data to indicate whether age-related accumulations of damaged molecules fit an allometric pattern.

5.4.3.   *Correlations with Mechanisms Protecting against Oxidation and Other Types of Damage*

Antioxidant mechanisms that protect cells from endogenously made oxygen radicals may also be correlated with lifespan, the free radical theory of aging as propounded by Harman (1962, 1968, 1981). Many studies show how exogenous superoxide dismutase and other antioxidants can protect laboratory rodents or cultured cells against damage by radiation or endogenous free radicals derived from oxygen (Schneider and Reed, 1985; Petkau et al., 1975). Cytotoxic derivatives of oxygen include the free radicals with an unpaired electron (singlet oxygen; the superoxide, hydroxyl, peroxyl, and alkoxyl radicals), but also hydrogen peroxide and hyperperoxides, which readily form free radicals upon degradation (Pryor, 1986).

That mammalian cells produce partially reduced oxygen radicals under ambient conditions is well established. For example, more than 1% of the oxygen used by heart mitochondria is converted to superoxide radicals (Chance et al., 1979). Recent evidence for the *in vivo* importance of free radical damage was obtained by directly showing the increase of free radicals in the myocardium during reperfusion after experimental ischemia and the minimization of cell damage through infusions of superoxide dismutase and catalase (Bolli et al., 1989; Chapter 3.6.7). Although much evidence suggests links between reactive oxygen species and nuclear genomic mutations (e.g. Hsie et al., 1986), direct proof is lacking. Cytoplasmic proteins and membrane lipids appear to be the major target of endogenous free radicals (Mead, 1976; Westerberg et al., 1979; Davies et al., 1990). There is mounting evidence that fluorescent products from oxidized membranes are produced by lymphocytes and other cells during normal functions. Moreover, oxidation of lipoproteins may be important in atherogenesis (Palinski et al., 1989) and some lymphocyte subtypes from senescent mice produce more peroxidized lipid products (Hendricks and Heidrick, 1988). The role of free radicals in the production of lipofuscins is indicated circumstantially.

Some data suggest correlations of free radical production to species lifespans. The reader may recall that all mammals at rest have the same arterial blood oxygen (Bartels, 1964; Section 5.3.4). However, there is little information on the levels of intracellular oxygen in different species. Furthermore, the levels of free radicals in a whole brain decreased slightly with increasing size and lifespan in a sample of eleven mammals and eight birds (Marechal et al., 1973). Singlet electron, paramagnetic signals were measured in fresh tissue, but the types of free radical were not identified. Most recently, Sohal et al. (1989, 1990) reported that the production of superoxide anion radicals by submitochondrial particles ranked in an inverse order to lifespan: mice > rat > rabbit > pig > cow.

Among the numerous natural antioxidants that serve to limit free radical damage are enzymes like superoxide dismutase and glutathione peroxidase. There are also a variety of endogenous chemical free radical scavengers and antioxidants like ureate (uric acid), ascorbic acid (vitamin C), and glutathione in the "liquid" phase of the cell, while α-tocopherol (vitamin E), β-carotene are thought to protect membranes particularly because of their hydrophilic character (Ames et al., 1981). While the chemistry of free radical quenching is being worked out in model reactions (e.g. Cohen et al., 1984), the complex antioxidant mechanisms are less understood.

Some correlations of antioxidant mechanisms are emerging in species comparisons. Humans rank the highest among mammals in tissue and blood levels of several antioxidants, in the limited species sampled. Urate levels in plasma and brain (Ames et al., 1981; Cutler, 1984) were correlated with lifespan in primates (Figure 5.19), but not in nonprimates with 3- to 50-year lifespans (Cutler, 1984). Plasma urate levels in humans at up to 0.6 mM, approach the maximum solubility, and are about a hundredfold greater than in rats or prosimians (Ames et al., 1981). Humans lack the enzyme uricase, which is the usual catabolizer of urate, forming allantoin; urate is mostly formed from purine degradation. Urate itself was recently shown to protect ascorbate from oxidation by divalent metal ions (Sevanian et al., 1985). The loss during human evolution in the capacity to make vitamin C thus may have been compensated by the increased levels of urate (Ames et al., 1981; Hochstein et al., 1986).

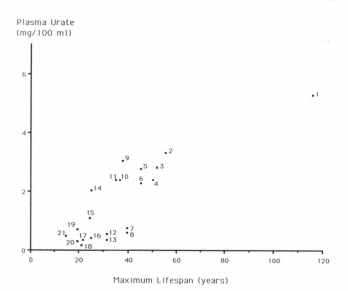

Fig. 5.19.   Plasma urate levels in primates as a function of maximum reported lifespan. *1*, human; *2*, chimpanzee (*Pan troglodytes*); *3*, orangutan (*Pongo pygmaeus*); *4*, gorilla (*Gorilla* sp.); *5*, gibbon (*Hylobates lar*); *6*, capuchin (*Cebus capucinus*); *7*, macaque (*Macaca mulatta*); *8*, baboon (*Papio cynocephalus*); *9*, spider monkey (*Ateles paniscus*); *10*, Siamang monkey (*Symphalangus* [= *Hylobates*] *syndactylus*); *11*, woolly monkey (*Lagothrix lagotricha*); *12*, langur (*Semnopithecus entellus*); *13*, grivet (*Cercopithecus aethiops*); *14*, tamarin (species not provided); *15*, squirrel monkey (*Saimiri sciureus*); *16*, night monkey (*Aotus trivirgatus*); *17*, potto (*Perodicticus potto*); *18*, patas (*Erythrocebus patas*); *19*, galago (*Galago*); *20*, howler monkey (*Alouatta*); *21*, common tree shrew (*Tupaia glis*). Correlation coefficient for linear regression was $r^2 = 0.85$ ($p \leq 0.001$); regression line not shown. Redrawn from Cutler, 1984.

Superoxide dismutase activities of liver, heart, and brain are weakly correlated with lifespan across twelve species (Figure 5.20; Tolmasoff et al., 1980). The activity in crude extracts was slightly lower in tissues of the short-lived A/J mouse strain (16-month mean lifespan) versus the LP/J strain that has a more typical lifespan (25 months; Kellogg and Fridovich, 1976). Cutler (1983, 1984) proposed that, because correlations are improved by expressing superoxide dismutase activity in terms of basal metabolic rate or lifespan energy production, this and other antioxidants may be considered as strong longevity promoters. As discussed in Section 5.2.4, some other enzymes involved in oxygen metabolism that vary with metabolic rate or body size, however, *decrease* per liver or muscle cell of larger animals (e.g. cytochrome oxidase and enzymes of aerobic glycolysis). Superoxide dismutase activity can also vary depending on metabolic state and can be induced in lung cells by oxygen and other treatments (Hass et al., 1982; Freeman et al., 1986) and monocytes (Asayuma et al., 1985), among other cells (Pugh and Fridovich, 1985). It is possible, therefore, that species differences arise epigenetically during development or through homeostatic influences on gene regulation.

Besides urate and superoxide dismutase, a range of antioxidant mechanisms may have relative importance. Other antioxidants that do not correlate with lifespan include ascorbate, glutathione, glutathione peroxidase, and glutathione S-transferase (Cutler, 1984, 1985). Carotenoids, which are powerful scavengers of singlet oxygen and which can protect against membrane lipid peroxidation (Foote et al., 1970), are relatively high in blood and tissues of humans; the correlation with lifespan is weak for shorter-lived species (Cutler, 1984). So far, there is no evidence supporting the age-related accumulation of somatic mutations that are speculatively associated with damage from endogenous free radicals (Chapter 8.2.1).

The hypothesis that senescence is caused by the accumulation of point mutations or other types of DNA damage (Chapter 8.2.1) has led to studies on relations between species lifespan and cell responses *in vitro* to DNA-damaging agents (Tice and Setlow, 1985). A major approach involves UV-irradiation, which causes the formation of covalent links between adjacent pyrimidines, among other lesions (Friedberg, 1985). Pyrimidine dimers are subsequently excised and repaired by local DNA synthesis that occurs independently of the cell cycle (unscheduled DNA synthesis) in patches that can be detected by the incorporation of [^{3}H]-thymidine (Setlow and Setlow, 1972; Friedberg, 1985).

Some comparisons indicate good correlations between species lifespans and the amount of DNA repair, as assayed after UV-irradiation of diploid fibroblast cultures from different mammals (Figure 5.21; Hart and Setlow, 1974; Hart and Daniel, 1980; Francis et al., 1981; Hall et al., 1984). Similar trends are shown by lens epithelial cells (Trenton and Courtois, 1982). Other species do not fit as well, particularly bats (Kato et al., 1980; Figure 5.22), giving another example of the divergence of bats with respect to correlates

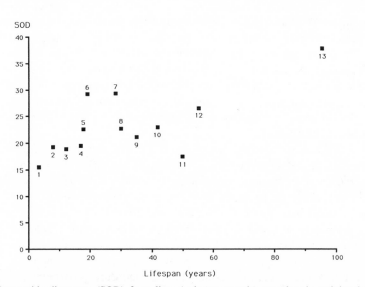

Fig. 5.20. Superoxide dismutase (SOD) from liver (units/mg protein; cytoplasmic activity that represents mixture of Cu/Zn − and Mn-types of SOD) of rodents and primates (young adult males) plotted against lifespan (95% *tmax*). *1*, house mouse (*Mus musculus*); *2*, deer mouse (*Peromyscus maniculatus*); *3*, common tree shrew (*Tupaia glis*); *4*, squirrel monkey (*Saimiri scuireus*); *5*, bush baby (*Galago crassicaudatus*); *6*, mustached tamarin (*Seguinus mystak*); *7*, lemur (*Lemur macaco fulvus*); *8*, African green monkey (*Cercopithecus aethiops*); *9*, rhesus monkey (*Macaca mulatta*); *10*, olive baboon (*Papio anubis*); *11*, gorilla (*Gorilla gorilla*); *12*, chimpanzee (*Pan troglodytes*); *13*, human. While no correlation was observed between SOD and lifespan, correlations with lifespan were reported when SOD was expressed per unit specific metabolic rate. Redrawn from Tolmasoff et al., 1980.

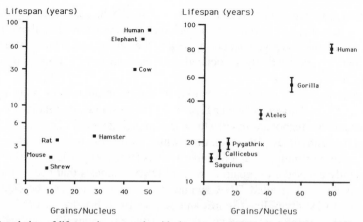

Fig. 5.21. Correlation of lifespan in mammals with the extent of UV-induced DNA repair by primary skin fibroblast cultures *in vitro*. Autoradiographic measurements (grains per nucleus of the unscheduled "repair") incorporation of [³H]-thymidine into cell nuclei after UV-irradiation. Regression lines not shown. *Left,* a general sample; *right,* a sample of primates. Redrawn from Hart and Setlow, 1974, and Hart and Daniel, 1980.

of the lifespan in other mammalian radiations. While cell lines from very long lived Carolina box turtle *Terrapene carolina* (Chapter 4.2.2) had less UV-induced DNA repair than those from mice (Woodhead et al., 1980), such comparisons between poikilothermic reptiles and homeothermic mammals may be unsound without detailed optimization of growth media and temperature for each cell type. There are also uncertainties about the equivalence across species in the varieties of UV-induced damage, as well as in the size of nucleotide pools that could influence [³H]-thymidine incorporation (Tice and Setlow, 1985). Further evaluation of the free radical hypothesis of aging will require definitive assays for the presence of macromolecular damage and for mechanisms to remove free radicals or repair their damage.

Mechanisms protecting neurons from various types of damage are of particular importance because of the irreplaceable nature of nearly all neurons in adult mammals. Calpain, a protease implicated in synaptic remodelling (Section 5.2.5), varied inversely with lifespan in the cerebral cortex, as shown in comparisons of six species (Baudry et al., 1986). The brains of small bats (*Antrozous pallidus* and *Tadarida brasiliensus*) have 80% less activity per gram than mice with the same body size (Baudry et al., 1986). *A. pallidus* can live at least 9 years (Cockrum, 1973), at least threefold longer than laboratory rodents. Furthermore, calpain is highest in the large pyramidal neurons (Siman et al., 1985) that are most at risk for age-related loss and for damage during Alzheimer's disease. One mechanism of neuronal damage could be an inappropriate proteolysis of the cytoskeleton from a sporadic, supranormal activation of calpain during fluctuations of calcium influxes (Lynch et al., 1986). In this way, species differences in neuronal calpain content could influence the risk of neuronal degeneration and, in turn, the lifespan (Lynch et al., 1986).

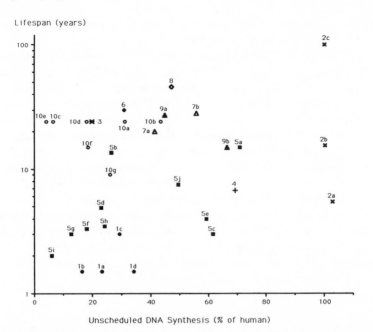

Unscheduled DNA Synthesis (% of human)

Fig. 5.22. Repair (unscheduled) DNA synthesis in early-phase lung fibroblast cultures as percentage of that in human diploid fibroblasts plotted against lifespan, using techniques like those of Figure 5.21. The points represent the following: **1** Insectivora: *1a, Sorex unguiculatus* (big-clawed shrew); *1b, S. shinto* (shinto shrew); *1c, Urotrichus talpoides* (Japanese shrew mole); *1d, Dymecodon pilirostris* (True's shrew mole); **2** Primates: *2a, Tupaia glis* (common tree shrew); *2b, Macaca irus* (crab-eating macaque); *2c,* human; **3** Edentata: *Myrmecophaga tridactyla* (giant anteater); **4** Lagomorpha: *Lepus brachyurus* (Japanese hare); **5** Rodentia: *5a, Sciurus lis* (Japanese squirrel); *5b, Petaurista leucogenys* (giant flying squirrel); *5c, Cricetulus griseus* (Chinese hamster); *5d, C. triton* (rat-like hamster); *5e, Mesocricetus auratus* (golden hamster); *5f, Rattus norvegicus* (Norway rat); *5g, R. cutchicus* (Cutch rock-rat); *5h, Mus musculus* (house mouse); *5i, Eothenomys kageus* (kage vole); *5j, Cavia porcellus* (guinea pig); **6** Cetacea: *Tursiops gilli* (bottle-nosed dolphin); **7** Carnivora: *7a, Canis familiaris* (domestic dog); *7b, Felis catus* (domestic cat); **8** Perissodactyla: *8a, Equus caballus* (domestic horse); **9** Artiodactyla: *9a, Sus scrofa* (pig); *9b, Muntiacus muntjak* (Indian muntjac); **10** Chiroptera: *10a, Rhinolophus cornutus* (little horseshoe bat); *10b, R. ferrumequinum* (greater horseshoe bat); *10c, Myotis frater* (little brown bat); *10d, M. hosonoi* (whiskered bat); *10e, M. macrodactylus* (large-footed bat); *10f, Pipistrellus abramus* (Japanese pipestrelle); *10g, Miniopterus schreibersi* (long-winged bat). In contrast to Figure 5.21, no correlation was found between unscheduled DNA synthesis and lifespan. Redrawn from Kato et al., 1980.

5.4.4. DNA Synthesis and Cell Replication in Vivo

The rates of cell turnover do not scale consistently with body size. Humans and laboratory rodents have remarkably similar cell turnover in different epithelia: cornea, uterine cervix, duodenal and rectal mucosa, seminiferous tubules (tabulated by Bertalanffy, 1967; Cameron and Thrasher, 1976). These similarities are expected from the similar genome sizes of mammals, since nuclear DNA content strongly influences the duration of the cell cycle in species comparisons (Section 5.4.1). The relation between half-life of the enucleate erythrocyte and the cell generation time of its bone marrow precursors has not been analyzed for mammals as a function of body size.

There is general agreement that the half-life of erythrocytes scales as days $\approx M^{0.18}$ (Allison, 1961), like the half-lives of albumin and other circulating proteins (Table 5.1).

The lifespan of the circulating erythrocyte is related to the binding of IgG immunoglobulins, loss of sialic acid, and other modifications of cell-surface macromolecules (Kay, 1985; Vaysse et al., 1986). The lifespans of proteins are controlled by numerous specific molecular time-tags (Dice, 1987). For example, sialic acid residues determine the lifespans of many plasma glycoproteins; the desialation ultimately signals removal from circulation and degradation (Morrell et al., 1971). Other proteins with very short lifespans (0.5–5 hours) have regions that are rich in proline, glutamic acid, serine, and threonine and that are called PEST regions, from the single-letter code for amino acids (Rogers et al., 1986; Rechsteiner, 1988; Jacobo-Molina et al., 1988. A PEST region is found in tyrosine aminotransferase, an inducible hepatic enzyme (Figure 7.8). Another time-tag revealed in yeast is the N-terminal amino acid, which determines the molecular lifespan over a 500-fold range from 2 minutes to more than 20 hours (Bachmair et al., 1986). The role of these mechanisms in scaling of protein lifespans with body size and metabolic rate is unexplored. Like the regulation of hemoglobin affinity for oxygen, the species differences in protein lifespans could be a function of genetically coded amino acid sequences in a particular protein, as well as of the modifications of the protein by other enzymes or chemical reactions, including glycation.

5.4.5. *Rates of Age-related Cell Loss* in Vivo

Adult mammals have many types of cells and cell organizations that are irreplaceable and whose stock is determined during development, including ovarian oocytes and follicles, neurons, and kidney tubules. As discussed in Chapter 3.6, ovarian oocytes are the only cells known to be predictably lost in all species throughout postnatal life. The rate of oocyte loss is clearly under some genetic influence (Figure 6.1). In contrast, the age-related loss of neurons and many other irreplaceable cell types varies widely among individuals and may be incidental; my position on this important issue is controversial (Chapter 3.10).

Comparisons of oocyte numbers and their rate of loss in mammals from twenty-two species and nine orders indicates that the number of reserve oocytes and primordial (non-growing) follicles at puberty varies with body size as $\approx M^{0.47}$ and with lifespan as $\approx L^{1.6}$ (Gosden and Telfer, 1987a). This is the first example of *hyper*allometric scaling of cell number with lifespan and suggests that larger mammals have proportionately greater ovarian reserve at the start of reproductive life. Gosden and Telfer (1987a) compare the rates of oocyte loss, which are known for only three species: mice, half-life of 0.27 years; rats, 0.75 years; humans, 7.0 years. These ten- to twentyfold differences between small rodents and humans suggest that follicular utilization scales inversely with lifespan, so that long-lived species should loose their larger stocks of oocytes more slowly. This slower depletion of oocytes in larger species with finite oocyte stocks provides a potential for extended reproductive schedules. Because the pituitary influences the rate of oocyte loss (Figure 10.23), hormones or receptors could be a factor in these species differences.

5.4.6. *Cell Replication Potential* in Vitro

In vitro cell functions indicate species differences that scale with lifespan. Conclusions are tentative, because only a few studies have directly compared cells from different species in the same experiment. One of the most rigorous studies found correlations between the proliferative capacity of diploid fibroblasts and the lifespans of eight donor species that represented intermediate and extreme lifespans (Röhme, 1981; Figure 5.23). The proliferative capacity during serial culture was defined by the number of population doublings before proliferation ceased and cultures become permanently postmitotic (phase III, or clonal senescence) (Hayflick, 1965; Chapter 7.5.3). However, another study of undefined cell types failed to find correlations of passage number with lifespan (Stanley et al., 1975). More species are needed to evaluate correlations between *in vivo* organismic and *in vitro* clonal lifespans. Nonetheless, the 100–130 population doublings observed before phase III in the Galapagos tortoise (Goldstein, 1974), which lives at least 100–150 years (Chapter 4.1.2), would appear to exceed most other vertebrates. An abstract (Simpson, 1989) described the absence of clonal senescence in fibroblast-like cells from the lizard *Anolis* (Chapter 3.4.2) during extended *in vitro* passage that exceeded 2,500 cummulative population doublings; while chromosomal abnormalities were not ruled out, the cells showed no evidence of transformation, by the absence of tumors formed after injection into immuno-suppressed hosts and the retention of density-dependent inhibition *in vitro*. There is much need for more detailed analysis of *in vitro* growth of fibroblasts and other cells of lower vertebrates.

Another technically well-done study deserves mention, although only three species were compared. The yield of smooth muscle cells that could be cloned from the aorta has not shown clear rankings by lifespan: laboratory mice (4-year lifespan) and white-footed mice, *Peromyscus leucopus* (> 7-year lifespan) gave similar yields (70 clones/mg aorta and 200 clones/mg, respectively). However, the rhesus monkey (> 35-year lifespan) gave disproportionately fewer, 1–3 clones/mg aorta (Martin et al., 1983).

The proliferative rate (cell-cycle duration or generation time) of mesenchymal cells *in vitro* ranges from 15 to 24 hours, with no marked species differences among mammals, as judged from tabulations from the literature (Altman and Katz, 1976). Again, this stability of cell-cycle duration is consistent with the close similarity of nuclear diploid DNA contents throughout mammals (Section 5.4.1).

Some of the technical problems in such studies should be considered. Although more species comparisons might be made from separate reports in the literature, it is difficult to

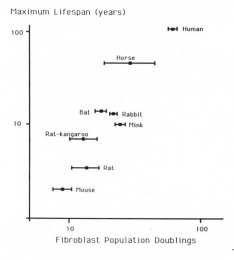

Fig. 5.23. Correlation between species mean lifespan and diploid (presumably) fibroblast lifespans *in vitro* (population doublings on serial transfer). The log-log correlation coefficient was 0.95; regression line not shown. Redrawn from Röhme, 1981.

draw general conclusions, because variations of culture conditions can alter the proliferative limit of the same cells by 25% or more even within the same laboratory. *In vitro* cell growth can also be strongly influenced by the cell type and location from which the explant was taken, for example papillary versus reticular regions of the dermis (Harper and Grove, 1979). This poses difficulties in comparing distantly related species, whose cells may differ in anatomically homologous structures. Genotypic polymorphisms may also influence the proliferative potential, as suggested by the greater differences in fibroblast population doublings observed between pairs of identical twins (21–25 doublings) than within each twin pair (only 1 doubling different; Ryan et al., 1981).

Moreover, the karyotypic instability of many transformed (immortal) rodent and human cell lines can give rise to unrecognized variants with different growth rates. Untransformed diploid fibroblasts are thus of particular value in species comparisons, but are also of interest because of their wide use as *in vitro* models of cell senescence. The well-studied human WI-38 line (diploid embryonic fibroblasts) has a cell-cycle duration or doubling time of 17 hours (Macieira-Coelho et al., 1966), which is close to the 15 hours of diploid dermal fibroblast lines from neonatal or young adult Syrian hamsters (*Mesocricetus auratus*, lifespan 2–3 years; Bruce et al., 1986). In both studies, the fibroblasts were proliferating actively as established lines (phase II of cell culture clonal lifespan). The S phase (DNA synthesis) *in vitro* does not differ among mammals in relation to the species lifespan (Altman and Katz, 1976; Cameron and Thrasher, 1976).

In conclusion, there are interesting correlations between lifespan and the numbers of *in vitro* passages before clonal senescence occurs in fibroblasts from mammals and a turtle. Much remains to be done, to bring this area to a state of conclusiveness. The important molecular genetic analyses in the mechanisms of *in vitro* clonal senescence (Chapter 7.5.3) indicate that quantitative changes in the concentrations of a small number of RNA or proteins may be crucial in the loss of proliferative capacity. If so, then the numbers of cell cycles before clonal senescence *in vitro* may be the result of a small number of changes in gene regulation during evolution.

5.4.7. Cancer, Time, and the Lifespan

Major questions arise about scaling of the risk of cancer as a function of age. Ronald Peto (1977) lucidly discusses how mice appear to have a risk that is vastly greater than humans, but still have a risk of cancer late in life that is somewhat scaled to the lifespan. In humans, the age-related incidence of spontaneous malignant disease and other abnormal growths increases exponentially as the *fourth to sixth power* of age after maturity (Armitage and Doll, 1954; Radman et al., 1982; Peto et al., 1975; Peto, 1977; Ames et al., 1985). Few analyses have calculated the exponential coefficient for cancer with age in rodents. Schach von Wittenau and Gans (1981), using data of Sheldon and Greenman (1980) on spontaneous malignant growths in BALB/c mice, found that the incidence of reticulum cell sarcomata increased as the *square* of age, while Ames et al. (1985), using data on Charles River COBS-CD outbred rats from Ross et al. (1982), found that death from cancer increased as the *sixth* power of age. Data on other abnormal growths are insufficient to evaluate if the increases are also exponential with age. Domestic and zoo animals also have increased incidence of spontaneous abnormal growths after mid-life

(Chapter 3.6.3). In hybrids of the short-lived platyfish (Chapter 3.4.1.2.2), melanoma increased as the *third* power of age (Ozato and Wakamatsu, 1983), a value closer to rodents than to humans.

An exponentially increased incidence of cancer as a function of time after exposure to specific carcinogens was also documented in a unique study by Peto and colleagues (1975). Whereas the induction of skin tumors in mice by benzo[a]pyrene increases as the *third* power of time, age *per se* was far less important than duration of exposure. The dose of carcinogen is usually scaled to smaller exponents than time, rarely more than the dose squared (Peto, 1977). One of the few corresponding analyses for humans shows that the risk of bronchial carcinoma in humans scales as the *fourth* power of time (years of smoking). Again, duration of exposure is more important than age as a risk factor (Doll and Peto, 1978). However, it is difficult to compare diverse species because the malignancies may not involve the same stem cell types and because species have different incidence of epithelial-derived carcinomas among their various malignant changes (Peto, 1979). Peto (1977, 1979) has also suggested that the risk of cancer cell per day is a million- or billion-fold greater in mice than humans. Since humans have a thousandfold more cells than a mouse and a thirtyfold greater lifespan, "exposure of two similar organisms to risk of carcinoma, one for 30 times as long as the others, would give 30^4 or 30^6 (i.e. approximating a million or billion) times the risk of carcinoma per epithelial cell" (Peto, 1977). Despite these uncertainties, these calculations point to another major unknown in the scaling of biological time. Species also differ in the transformation rate of epithelial and other cells *in vitro* (see below).

The activation of carcinogens also scales with lifespan. Fundamental links between carcinogens and mutations are suspected on many grounds (McCann et al., 1976; Saul et ·al., 1987; Radman et al., 1982). The activation of polycyclic carcinogens by enzymatic oxidation and conjugation is required for their carcinogenic effects, prior to the formation of covalent adducts to DNA bases that is considered a major step in carcinogenesis. The enzymes involved in these complex reactions include the mixed function oxidases, some of whose genes are located in the major histocompatibility complex along with other loci that influence the diseases of senescence and the lifespan in mice and humans (Chapter 6).

In a comparison of eight species, the activation of benzo[a]pyrene and dimethyl-benz[a]anthracene assayed *in vitro* with primary lung or skin fibroblast cultures (Kouri et

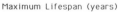

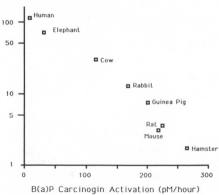

Fig. 5.24. Correlation of lifespan (log scale) and the capacity of diploid fibroblasts *in vitro* to activate the carcinogen benzo(a)pyrene (linear scale), assayed by conversion to water-soluble products that are mutagenic. Redrawn from Schwartz, 1975, and Moore and Schwartz, 1978.

al., 1974) scaled inversely with lifespan. Fibroblasts from rats made a thousandfold more mutagenic metabolites than those from elephants or humans *in vitro* (Schwartz, 1975; Moore and Schwartz, 1978; Figure 5.24). Both these carcinogens form covalent adducts with DNA (Janss and Ben, 1978; Pereira et al., 1979). The reduced *in vitro* activation of carcinogens by fibroblasts of long-lived species implies less risk of DNA damage, but this has not been documented *in vivo*.

In view of the importance of latent time to carcinogenesis, either through age or duration of exposure, a major question concerns what aspects of carcinogenesis are scaled with lifespan, body size, or metabolic rate. Are the hypothetically independent stages of carcinogenesis (initiation, promotion) correspondingly scaled? If so, do species differences result only for the different times that malignant clones need to expand to the same size relative to body size? Again, few data are available. A widely held impression is that the latency period in carcinogenesis approximates 25% of the lifespan in comparisons of humans and laboratory rodents (Cairns, 1978; Radman et al., 1982). There is no consensus that the multiple molecular steps in oncogenesis or transformation are identical in rodents and humans, or that *in vitro* comparisons of cell responses to carcinogens can be interpreted according to species differences in lifespan (McKormick and Maher, 1989). According to the hypothesis that tumor promotion is primarily associated with clonal growth of an initiated cell (Trosko and Chang, 1980; Potter, 1980), it is pertinent that the cell-cycle length in fibroblasts and mucosal epithelial cells does not differ between humans and rodents (Section 5.4.4). Thus, species differences in the latent period may be more associated with initiation than promotion.

The correlations of UV-induced repair and carcinogen activation *in vitro* to species lifespan might also be related to the frequency of spontaneous transformation by diploid

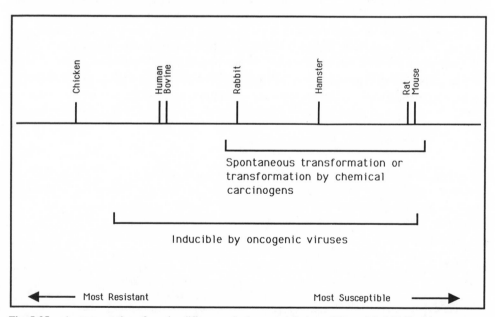

Fig. 5.25. A representation of species differences in the range of susceptibility of diploid fibroblasts to transformation (acquisition of an infinite proliferative capacity). Redrawn from Macieira-Coelho et al., 1977.

fibroblasts *in vitro*. Fibroblasts from mice particularly, but also other rodents, spontaneously transform with much higher frequency than those from humans (Figure 5.25; Ponten, 1971; Macieira-Coelho et al., 1977). Chicken cells, however, are remarkably resistant to transformation, either spontaneously or in response to viruses or chemicals that yield permanent, transformed cell lines in fibroblasts from other birds (Ponten, 1971). Human epithelial cells (the source of carcinomas) are far more resistant to transformation *in vitro* than epithelial cells of rodents. These species differences indicate the difficulty in obtaining robust generalizations about mechanisms in abnormal growths in relation to species lifespan.

5.5. Summary

Comparisons of related species for systematic variations of size and physiological characteristics have provoked important questions in gerontology, no less than for evolution and development. The root question still remains open: To what extent are variations of lifespan linked to differences in the rate of functions at various levels of the organism? The first three chapters described the wide range of different proximal pathophysiological mechanisms during senescence. But we do not yet know if initiating (distal) mechanisms are the same or different, even within related organisms. The fascinating images of metric coherence that can be conjured with some statistical support from morphological and physiological data are still far from giving a clear picture about the ultimate basis for species differences in lifespan. Until we know more about the causes of age-related dysfunctions and their species differences, little more can be said about the scaling of various parameters with lifespan. Thus, scaling of size or metabolic rate between species as young adults and scaling of various parameters with lifespan does not prove a common mechanism.

Data showing that the number of reserve oocytes in mammals scales hyperallometrically with lifespan, while the rate of oocyte loss scales inversely with lifespan (Section 5.4.5), are the first evidence concerning cell death in these size comparisons. More such information may allow judgement on whether the scaling of certain age-related changes in proportion to the lifespan of short- and long-lived mammals holds across the size range (Table 3.3). A detailed analysis of size and the mortality rate coefficients is badly needed; the limited data available do indicate good correlation of the MRDT with size (Section 5.3.1.1). According to expectations from evolutionary theory, the weakening of natural selection at later ages should permit a widely varying outcome in the accumulation of germ-line mutations during evolution that have delayed effects. If so, size correlations with the rates of specific processes of senescence should be much weaker than size correlations between species as young adults.

Several questions of scaling bear on the potential maintenance of cell functions across the lifespan. A major unknown concerns the mechanisms that regulate energy metabolism and the production of proteins by the liver and some other cells in register with body size, as revealed in the five- to tenfold differences in total body metabolism between rodents and humans (Sections 5.2.1, 5.2.5). It would be important to know if ribosome turnover and albumin synthesis, for example, were also scaled similarly in hepatocytes as a function of body size. The constant rate of protein synthesis in humans throughout postnatal development and into adult life, when expressed as $M^{0.75}$ (Section 5.2.5), implies that species differences in metabolism that are related to body size could reflect regulatory effects on gene expression. If so, species differences in metabolic rates may trivially derive from size, rather than from mutational changes like those in promoter efficiency that might intrinsically alter the rate of transcription, or in the machinery of protein synthesis that might influence translation rates. The epigenetic interpretation of size variations has been given little attention. It is much less complicated to consider systemic (physiological) influences mediated by the availability of nutrients through the circulation. It would also be of much interest to extend the findings of the few reports that suggest major species differences in the numbers of ribosomes and mitochondria per cell. The similar cell-cycle lengths in rapidly proliferating cells (Section 5.2.5) suggest that only limited subsets of cells have biosynthetic rates that are scaled with body size, basal metabolism, or lifespan.

Moreover, we do not know if rates of cell dysfunction during senescence are linked to the intensity of cell activities. In this regard, more attention is warranted on the rates of biosynthesis in liver cells that may scale allometrically, in contrast to some neurons that may not (Section 5.2.5). If neurons of humans prove to have the same or greater rates of biosynthesis than in mice (Section 5.2.4), this result would raise a new set of questions on the relationship between cell longevity *in vivo* and metabolic work over the lifespan. There is thus some reason to doubt the generalization that physiological time scales to all cell and molecular processes in an organism, whatever its lifespan (also see Prothero, 1986). Rate comparisons are also implicit in the use of short-lived animals as models for cancer in humans (Section 5.4.7). Throughout, we could be badly misled by the small number of different species that are known in detail.

Yet major new opportunities may be within reach to overcome the inherent limitations of cross-species comparisons. For example, these questions might be approached with genetically engineered animals that vary in cell number and physiology. The giant transgenic mice carrying extra copies of the rat growth hormone gene (Section 5.3.1) have already tested allometric coefficients across a wider size range than usually expressed in the same species. Transgenic animals could also be used to study scaling between cell and physiological functions. Thus, it may be possible to identify species differences that are under direct genetic control, in distinction to differences resulting from variations in gene activity during development under the control of regulators extrinsic to that particular cell type. This is, of course, a recapitulation of the adaptationist versus epigeneticist positions (Gould and Lewontin, 1979). By genetically engineering variants with different scaling, the question of epigenetic versus adaptational mechanisms may be directly addressed.

From the epigenetics perspective, many species differences in rates of senescence and longevity may be understood as direct consequences of their developmental patterns. For it is during development that the numbers of divisions in stem cells are determined, from which arise the adult size and shape. Such an approach might also epigenetically, through gene expression, adjust the levels of antioxidant enzymes or compounds in accord with the intensity of oxidative metabolism. Supraoxide dismutase, for example, can be induced by environmental oxygen (Section 5.4.3) and could be regulated by tissue-specific levels of metabolites or radicals. In contrast to these possible epigenetic manipulations, there is a mutational basis for species differences in the oxygen binding properties of hemoglobin (Section 5.2.4), which illustrates the adaptationist model but does not preclude epigenetic causes for other species variations, for example, blood levels of 2,3-diphosphoglycerate that influence oxygen dissociation from hemoglobin.

The next chapter describes genotypic variations that influence particular age-related changes and diseases within species and that also bear on the diversity of mechanisms in senescence, like those used to describe allometric relationships, in closely related species. Chapter 9 discusses the relationships between the duration of developmental stages and lifespan, addressing another aspect of time correlations not considered here.

6

Genetic Influences on Lifespan, Mortality Rates, and Age-related Diseases

6.1.   Introduction

This chapter addresses genetic influences on the adult lifespan and characteristics of senescence that have been indicated within species. Genetics has not been discussed explicitly in the previous five chapters, which described species differences in lifespan and senescence that ranged widely within taxonomic groupings such as teleostan fish, birds, and mammals, but even more widely among phyla such as the rotifers and the mollusks. Implicitly, the relatively stable species characteristics of lifespan in different environments must derive from a genetic basis, but this basis is largely mysterious. Few inbred lines in animals have major differences of lifespan. There are, however, more examples of higher plant strains with major differences in lifespan. While most of the genotypic variations of senescence are chromosomally determined, a few examples show maternal and cytoplasmic influences from viral or episomal genes, in mice and fungi.

Total lifespans are not invariant inherited characteristics in any species, in view of the many exogenous influences on timing (Chapters 2, 3, 4, 9, 10), much less so, for example, than the timing of developmental stages *in utero*. Even so, the range of maximum lifespans among species of like organisms far exceeds variations of maximum lifespans among populations or cohorts of the same species. Lifespans of closely related, senescing species can vary ten- to a hundredfold within a phylum or even a class (Chapter 2). None would dispute that the maximum lifespan of a mouse is tenfold more than a fruit fly's, or that humans live thirty times longer than mice. Genetic analyses of longevity and senescence have thus far focussed on genetic variants *within* a species that can be isolated by classical genetic approaches. Another approach was indicated in the previous chapter: species differences might be traced to variations of gene expression that did not for most parameters require different DNA sequences. In such a case, a limited number of genes might be determinants of lifespan through pleiotropic effects.

Recent achievements in artificial selection for lifespan in the nematode *Caenorhabditis* are the first to show that the age-related acceleration of mortality as represented in the MRDT (Chapter 1) is under the control of a definable genetic locus, *age–1*. Artificial selection of genetically heterogenous flies (*Drosophila*) and beetles (*Tribolium*) has yielded populations with different lifespans and reproductive schedules that show important influences from yet unidentified genetic polymorphisms. Insects and rodent genotypes with markedly different lifespans remain to be genetically resolved for differences in IMR and MRDT. While inbred mouse genotypes differ widely in age-related diseases and dysfunctions, their MRDTs are quite similar. This phenomenon recalls the different distributions of age-related diseases in human populations that nonetheless had similar MRDTs (Chapters 1, 3). The genetic influences on lifespan and senescence in mammals largely consist of alleles and genotypes with different patterns of diseases. Genetic influences on the diseases of senescence are reviewed in some detail, with emphasis on mice and humans, but not to the exclusion of scattered information on genotypes within other mammals.

The major question of what specific genes or other DNA sequences determine species differences in lifespans and the characteristics of senescence is not yet addressable. The myriad pleiotropic influences on the characteristics of senescence that can differ so widely among very similar genotypes in mice hint that a large number of genes need not be involved. If heritable delays in diseases of senescence in rodents could be combined with the severalfold increases of lifespan from slowing in the acceleration of mortality (MRDT), as in certain nematode genotypes, this would constitute an evolutionary jump or saltatory change in lifespan.

The design of studies for artificial selection for increased lifespan of flies and beetles draws from evolutionary theory on the optimization of reproductive scheduling and negative pleiotropy, in which genotypes (polymorphisms at one or more loci) that confer early advantages to reproduction may also have delayed deleterious effects (Chapter 1.5.2). A naturally occurring example of negative pleiotropy is a polymorphism of *Drosophila mercatorum,* the abnormal abdomen (*aa*) genotype which is advantageous in some microclimates, while shortening lifespan. These observations support the hypothesis that the force of natural selection declines with age and that genes causing delayed adverse effects in the postreproductive phase will not ordinarily be selected against. While this chapter presents examples of genotypes with delayed adverse effects, a more detailed discussion of selection for lifespan is postponed until the book's end (Chapters 11, 12).

Except for very short lived species with total lifespans of 1–3 months like *Drosophila* and *Caenorhabditis,* it is difficult to experimentally select for lifespan variants, since the desired information is usually not available until later in the lifespan when fertility often is greatly reduced. So far, most of the laboratory rodent strains with genotypic variations in lifespan were established for different purposes, for example, to confirm genetic differences in the incidence of tumors or other diseases or in immune functions; the differences in lifespan were established secondarily in most cases.

An important recognition is that individuals of highly inbred strains, or even genetically identical siblings, do not have synchronous or identical patterns of senescence or lifespans (Figure 1.12; Chapters 9, 10). Thus, certain aspects of senescence cannot be under rigid genetic determination and are influenced by environmental and developmental fluctuations. The nonheritable range of variations in lifespan within highly inbred mouse strains are not obviously different from those in far more genetically polymorphic human populations. Variations among individuals may vary from the probabilistic determination of the numbers of cells in some lineages during differentiation (Chapter 9). Chance traumas throughout the lifespan are bound to cause individual variations in the details of senescence and lifespan. The cumulative impact of events over time will nonetheless be constrained by genotypic factors that, for example, influence the capacity to repair molecular damage or tissue injury, or determine dangerous behaviors.

This chapter is organized taxonomically, giving examples of genotypic influences on a wide range of total lifespans that represent both rapid senescence (the soil nematode *Caenorhabditis,* several drosophilids, periodical cicadas, filamentous fungi) and gradual senescence (the flour beetle *Tribolium,* mice, rats, rabbits, humans).

6.2. The Nematode *Caenorhabditis*

Genetic influences on lifespan and in the rate of senescence were recently shown in the short-lived hermaphroditic soil nematode *Caenorhabditis elegans,* with a 20-day mean lifespan and a MRDT of about 4 days (Chapter 2.2.2.2). Somatic cell division ceases at hatching; while oocyte production continues for the first few days of the adult phase, the total numbers of sperm and eggs are limited. Males occasionally are produced by nondisjunction and loss of an X chromosome (Chapter 8.5.3.2) and may have shorter lifespans depending on the strain (Johnson and Wood, 1982). Because reproduction in nature is largely through hermaphroditic self-fertilization, *Caenorhabditis* populations in nature consist of highly inbred individuals. Although population studies have not been conducted, it is clear that a range of genotypes is available in wild populations (Thomas Johnson, pers. comm.). Many aspects of its rapid senescence can be used as markers for studying genetic manipulation of age-related changes, but the underlying causes of senescence are unknown (Chapter 2). The detailed genetic maps of the six chromosomes and conveniently small genome, which is half of that in *Drosophila* and 1% of that in mammals, favor the genetic analysis of senescence. Two different approaches have been used to identify genetic influences on senescence and lifespan: (1) the isolation of long-lived, recombinant inbred strains (RI) by self-fertilization of progenitor strain hermaphrodites; and (2) chemically induced mutants with increased lifespan.

RI strains had mean total and adult lifespans that varied from 13 to 30 days, which was mainly the result of differences in the accelerations of mortality, equivalent to twofold differences in MRDT, without significant differences in the IMR (Figure 6.1; Johnson, 1987). These inbred strains had the same durations of development, through the four molts of the free-swimming larvae (Chapter 9; Figure 9.1). The longer-lived strains had slower declines in spontaneous movement. However, all strains ceased reproducing at the same age, 10 days. This indicates a conclusion that is pertinent to population genetics theories of senescence (Chapter 1.5.2), that the duration of lifespan and the duration of adult fertility are determined by separate genes and processes. Thus, the duration of the postreproductive stage can in principal be varied independently of the reproductive schedule.

At least two factors that contribute to the duration of fertility are not obviously related to lifespan: the numbers of sperm formed during development, and the rate at which the eggs and sperm are released. There is no evidence under normal culture conditions for reproductive costs in *Caenorhabditis*. Correlations between strains in lifespans and the slowing of movement suggest a common process. While these inbred lines show strong effects of genotype on the rates of senescence, the particular loci have not been identified. Influences from a minimum of three to six genetic loci on the lifespan were estimated from Sewell Wright's formula (Johnson, 1986), which corresponds to at least one locus on each chromosome that influences the lifespan.

In another approach, mutants with lengthened adult lifespan were produced by ethyl-methanesulfonate (EMS; Klass, 1983). For example, MK542 and MK546 have 50% greater mean lifespans and twofold greater maximum lifespans than the progenitors, from which they were derived by EMS mutagenesis, the wild-type Bristol (N2) and Bergerac (BO) strains (Friedman and Johnson, 1988a, 1988b). The lengthened lifespans of the mutants are not associated with reduced food intake (Friedman and Johnson, 1988a), contrary to first impressions (Klass, 1983). This is a crucial point, in view of the increase of lifespan that so many species show with dietary restriction, including *Caenorhabditis* (Chapter 10). A genetic analysis of MK546 indicated a mutant gene that increases the lifespan, designated *age–1*, which was recessive in F_1 hybrids with the progenitor strains and segregated with the expected distribution of lifespans in backcrosses of the F_1 to long-lived mutants (Friedman and Johnson, 1988a). *Age–1* is tightly linked to a temperature-sensitive locus influencing fertility, *fer–15*, that is located in the middle of chromosome

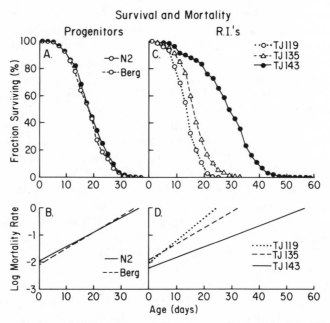

Fig. 6.1. *Top*, survivorship and, *bottom*, log mortality rates as a function of age in different recombinant inbred strains of the nematode *Caenorhabditis elegans*. These strains were made by backcrossing two progenitor wild-type strains: N2 (Bristol line) and Berg (Bergerac BO), followed by fixation through hermaphroditic selfing. The recombinant inbred strains TJ119, TJ135, and TJ143 contain different combinations of the six chromosomes from each of the progenitor strains. Mortality rates were calculated for IMR (initial mortality rate) and MRDT (mortality rate doubling time) according to equations [1.2] and [1.3], respectively (Thomas Johnson, pers. comm.). N2 (IMR, 0.15/day; MRDT, 13 days; Berg (0.12; 12); TJ119 (0.13; 12); TJ135 (0.14; 12); TJ143 (0.11; 18). TJ143 had a 50% lengthening of MRDT (slowing of the acceleration of the mortality rate), while that of TJ119 was hastened by 30%. IMRs did not differ among genotypes. Redrawn from Johnson, 1987.

(linkage group) 2. Males with *age-1* on several genetic backgrounds also have increased lifespans. Analysis of mortality rates shows that the *age-1* allele lengthens the MRDT by up to twofold (Figure 6.2; Johnson, in prep.). This is the first explicit demonstration of a gene that slows the age-related acceleration of mortality.

The absence of heterosis effects in the F_1 hybrids of these and other inbred strains (Johnson, 1986, 1987) is in contrast to the enhancement of lifespan observed in the F_1 hybrids of many strains of inbred mice (Section 6.5.1.2) and many other species (Lints, 1978). Because *Caenorhabditis* is considered to be 100% inbred in nature (Johnson, 1986), the absence of heterosis in the F_1 hybrids implies the elimination of disadvantageous alleles during reproduction by repeated self-fertilization, as may be more tolerated by inbred laboratory mice in their protected habitats. An exception is the greater lifespan of males than hermaphrodites of many different F_1 hybrids, which suggests the complete dominance of long-lived male genotypes (Johnson, 1987).

The mutant MK546 has the same duration of development as the progenitor strain. Other longevity mutants, however, extended total lifespan by forming dauer larvae, which is a state of diapause with reduced activity that extends the duration of development (Chapter 9.2.1.1; Figure 9.1); in these cases, the adult lifespan was not altered, nor was senescence slowed. Stages critical for dauer larval development were identified by temperature-sensitive mutants (Swanson and Riddle, 1981) and may lead to a genetic analysis of the transitions between development and senescence.

MK546 had greatly reduced fertility ($-$ 80%), while the duration of the reproductive phase (10 days) was unchanged. Thus, MK546 has a greatly extended postreproductive phase of its adult lifespan, as observed above for certain inbred strains. Lethargic movements are another characteristic of MK542 and MK546 (Friedman and Johnson, 1988b), which might, speaking hypothetically, spare some vital nutrient store or slow the accumulation of a toxic product of metabolism. The extrapolated decline of movements to zero predicted the life expectancy of the different genotypes, a correlate not observed in another study of a single genotype (Bolanowski et al., 1981). Complementation tests with male reisolates provisionally indicate that MK542 and MK546 contain mutant alleles at the same locus, *age–1* (Friedman and Johnson, 1988a). These striking findings on the *age–1* locus and the RI strains are the first genetic manipulations of the Gompertz rate in any animal.

To summarize, these studies show that major genetic manipulation of the lifespan and rate of mortality need not alter the usual development or the senescent changes in all physiological functions simultaneously (spontaneous movements versus reproduction). This view is also supported in other species, for example, in the diverse effects of main histocompatibility locus alleles on the age-related diseases in mice and humans, and the difficulties of finding, in any species at a single genetic locus, variants that uniformly alter

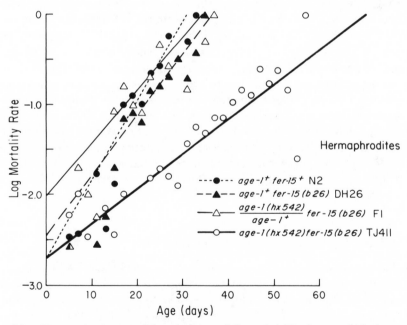

Fig. 6.2. Mortality rates in a mutant of the *age–1* gene of *Caenorhabditis elegans,* which shows a twofold lengthening of MRDT compared with a wild-type strain (N2) and the DH26 strain, which was included as a control for possible influences for a linked temperature-sensitive allele *(fer–15[b26])*. The strain TJ411 carries a mutant allele in the *age–1* gene. Representative MRDTs are N2, 3.5 days; DH26, 4.5 days; TJ411, 7.3 days; F_1, 4.9 days. The IMRs did not differ among genotypes. See Figure 6.1 legend and Chapter 1.2 for description of mortality rate coefficients. Photograph supplied by Thomas Johnson, 1990.

the rate of senescent physiological changes relative to the norm. It seems more likely that many different and independently acting genetic loci determine the rates of different senescent changes and the lifespan of most species. The studies described above should yield DNA sequences with influences on mortality rates that could be either structural (coding for specific proteins) or regulatory (influencing the transcription of other gene sequences).

6.3. Insects

Information about genetic influences on lifespan in insects includes mutant genes that were subsequently found to lengthen or shorten lifespan, and population selection experiments. The search for inbred and genetically identical nematodes with increased lifespan (Section 6.2) has as its goal the isolation of specific DNA sequences that influence the mortality rate coefficients and therefore lifespan. In contrast, some approaches to genetic influences on lifespan of insects are based on genetically heterogeneous populations, according to premises from evolutionary theory. Thus, we find powerful studies of *Drosophila* and *Tribolium* focussing on selected populations with genetic polymorphisms that influence macroscopic phenotypes such as the statistical lifespan and the reproductive schedule (the age of onset of reproduction, reproductive lifespan, and the total fecundity).

   While these studies have selected for populations with different lifespans and reproductive schedules, the combinations of alleles at numerous loci in these populations precludes identification of specific genes as determinants of lifespan. This situation is perfectly appropriate for sexually reproducing species with extensive genetic heterogeneity in natural populations, in which the genetic polymorphisms at numerous loci are adaptive for relatively rapid responses to environmental changes. However, as noted above, *Caenorhabditis,* which reproduces largely by self-fertilization, is considered to be nearly completely inbred in nature (Johnson, 1986) and therefore may have little genetic variation *within* a local population. Thus, the responses to artificial selection for increased lifespan in the laboratory may differ among species that were adapted to major differences in the extent of balanced (and perhaps even of unstable) polymorphisms in their natural populations. The next section briefly reviews the specific genetic influences on lifespan; the subject is taken up again in Chapters 11 and 12. (See Rose [1991] for detailed treatment of this and related material.)

6.3.1. Genetic Influences on Lifespan and Senescence

   First, I note some examples of mutants that influence the lifespan, which are interesting historically as well as for future research directions. These mutants are useful models for specific questions but may not be informative about more general mechanisms of senescence in the wild type. Heritable variations of lifespan of the fruit fly *Drosophila melanogaster* have long been known for inbred lines (e.g. Pearl et al., 1923; Maynard Smith, 1959; Clark and Rubin, 1964; Lints, 1978; Mayer and Baker, 1985). The lifespans of F_1 hybrids from inbred strains usually are much increased (Maynard Smith, 1959), as are hybrids from wild types captured at diverse places in the U.S. (Gowen, 1952).

   The role of superoxide dismutase in lifespan potential of flies is being investigated with newly isolated mutants. A null mutant, in which adults have no detectible activity of the copper/zinc-containing superoxide dismutase, was recently isolated after ethylmethane-sulfonate mutagenesis of *Drosophila* (Phillips et al., 1989). Homozygotes develop normally but have short adult life phases (10-day versus 60-day means). Nonetheless, the deficits of superoxide dismutase in homozygotes are associated with greatly increased larval sensitivity to paraquat and transition metals that generate free radicals. The fact that heterozygotes have normal survival curves and lifespans predicts that the maximum lifespan of *Drosophila* would not be increased by further elevations of the enzyme. Adult homozygotes lack motile sperm, which may indicate a crucial role of superoxide dismu-tases in protecting against DNA damage during gametogenesis. Phillips et al. (1989) noted that the adult tissues that arise from imaginal discs at metamorphosis may have been damaged in homozygotes even before adulthood.

Two interesting short-lived neurological mutants are mentioned, drop-dead and dunce. An interesting short-lived genotype causes rapid degeneration of brain cells and death within 1 week after hatching, the drop-dead (*drd*) mutant (Hotta and Benzer, 1972). It is unclear if the neurodegeneration in *drd* is an acceleration of a normal process, which appears to involve some neuron loss and atrophy (Chapter 2.2.2.1). The *dunce* gene gives a model for studying effects of frequent mating, which was implicated as a factor that shortens lifespan through mechanical trauma (Chapter 2.2.2.1). In contrast to wild type, *dunce*-containing genotypes mated again more than threefold more frequently within 2 days after initial mating (Bellen and Kiger, 1987). While the lifespan was also shortened by reproduction, *dunce* genotypes had mean lifespans of only 2 weeks in this study; *dunce* is a mutant in the phosphodiesterase form II and is associated with fivefold elevations of cAMP in whole flies. That *dunce* has any viability is startling. Numerous other mutations are known to shorten the lifespan; see reviews of Arking and Dudas (1989), Rose and Graves (1990), and Mayer and Baker (1985).

Another concern is the possible influences of ploidy, which are pertinent to hypotheses that senescence results from cumulative damage to crucial genes (Szilard, 1959). Several studies showed no influence of ploidy on lifespan in diploid and triploid females of *Drosophila melanogaster* (Gowen, 1931) or in haploid versus diploid males of the wasp *Habrobracon serinopae* (Clark and Rubin, 1964). Because the extent of gene dosage compensation is not known, we cannot conclude that the lifespans of these species is uninflu-enced by the number of gene copies.

Several clever experimental attempts have been made to genetically alter rates of senescence. While most approaches have not been particularly successful, it is important to recognize these efforts in the design of future strategies. Roberts and Iredale (1985) mu-tagenized with EMS or X rays and examined flies at 75% of survivorship for their mating latency and climbing ability; by this age, most flies show marked impairments in these parameters. Screening of 18,000 flies failed to identify autosomal dominants that altered the senescent decline in these motor abilities, from which it can be concluded that such dominant mutations are rare, or that the actions of several dominant mutations must be coordinated for significant effect. Mosaic genetic mapping was used in a different ap-proach that might be useful in identifying particular cells as pacemakers of senescence. Two neurological mutants, Hyperkinetic[1] (*Hk*[1]) and Shaker[3] (*Sh*[3]) shorten the lifespan by about 20% (Trout and Kaplan, 1981), in association with increased physical activity and metabolic rate. By placing these mutant genes on an unstable chromosome (ring-X) that breaks down with a predictable frequency during development, it is possible to form genetically mosaic embryos in which a mutant gene is present in only certain cells. Using this powerful technique, Trout and Kaplan (1981) showed that the presence of Hk^1 or Sh^3 in ventral thorax tissues of the adult caused slightly shorter lifespans than mosaic controls without the mutant genes. Although the mosaic controls had shorter lifespans than the wild type, this study shows the feasibility of detecting subtle effects on lifespan through genetic mosaics. As noted by Trout and Kaplan, this approach could be used to detect the cells containing pacemakers of senescence.

To end this section on a positive note, the remarkable opportunities for genetic manipulation in insects are being exploited in studies of senescence. A considerable extension of lifespan was achieved by introducing an extra copy of the gene coding the translation factor EF–1α by an engineered plasmid (Shepherd et al., 1989). The possible role of deficiencies in EF–1α in *Drosophila* senescence are discussed in Chapter 7.4.3.1. Such extensions of lifespan through the manipulation of a single gene are probably exerted through a complex pleiotropy. Moreover, when a complex system is failing, as may be the case with biosynthesis during senescence in flies, then a major change can be accomplished through mechanisms that are not necessarily the originating cause of senescence. Nonetheless, recombinant genetic approaches hold great promise in identifying the candidates for fundamental mechanisms in senescence in flies and other organisms.

6.3.2. Artificial Selection for Lifespan and Reproductive Scheduling in Genetically Heterogeneous Populations: Drosophila and Tribolium

Artificial genetic selection for greater lifespan has been achieved in conjunction with influences on reproduction in several laboratories. These studies draw from the theory of population genetics (Chapter 1.5) that genes with adverse effects during senescence are either silent or have favorable effects in early life; the later case is known as negative pleiotropy. Powerful studies have been carried out by Michael Rose, Brian Charlesworth, and colleagues, with independent and concurring efforts from other colleagues (Rose and Charlesworth, 1981; Rose, 1984a, 1984b; Rose et al., 1984; Luckinbill et al., 1984, 1988; Arking, 1987; Arking and Dudas, 1989; Service, 1989). The experimental design is based on selection of *Drosophila* from early- or late-reproducing females, that is, from eggs shed early or late in the lifespan. Overall, the same results were found by these independent laboratories. A critical feature of these studies is their use of samples from wild populations of flies in order to increase the genetic variation available for selection. Different outcomes occur when inbred lines are used as the basis for selection, as discussed below.

For example, flies with delayed senescence were selected by the capacity for reproduction in the few females surviving to 70 days. After fifteen generations, the mean and maximum lifespans of females were increased about 30% (Figure 6.3), while male lifespans were increased by 15% (Rose, 1984a). Controls were propagated from eggs harvested from females at the age of 2 weeks. This selection protocol also modified the patterns of fecundity, so that the late-selected and longer-lived lines produced fewer eggs as young adults, but continued to produce eggs at later ages (Rose, 1984c; Luckinbill et al., 1984), which I designate as delayed senescence (DS) in Figure 6.4; the various investigators each use different nomenclature for similar procedures that select early- and late-reproducing lines. A high larval density was crucial for successful selection (Luckinbill and Clare, 1985; Arking, 1987; Service et al., 1988), suggesting density-dependent neuroendocrine influences on larvae that might increase the range of phenotypic variance and selection. Chapter 9.3.4.1 gives an example from mice, in which the sex of the neighboring fetus influences the reproductive senescence.

The populations selected for delayed senescence had much smaller (75%) ovaries as early adults and later peak egg production (fecundity), followed by a slower decline in egg production (Figure 6.4; Rose, 1984c; Luckinbill et al., 1984). In Arking's (1987) study, all lines had the same interval of 12 days between the mean age when fertility begins to decrease and the mean lifespan, which Arking suggests could indicate a genetic influence on some function like protein synthesis that is shared in part by reproduction and somatic maintenance and that could be under neuroendocrine influence. The DS-selected flies also had greater resistance to starvation and desiccation (Service et al., 1985), implying greater metabolic reserves, possibly in the larval or adult fat body (Chapter 2.2.1.1); these were not characterized. The coselection of early reproductive vigor and shorter lifespan was also seen in *D. subobscura* (Wattiaux, 1968) and *D. pseudoobscura* (Taylor and Condra, 1980). This later study showed shifts in the frequencies of structural polymorphisms in the third chromosome.

The third chromosome is further implicated by studies from Arking's and Rose's groups

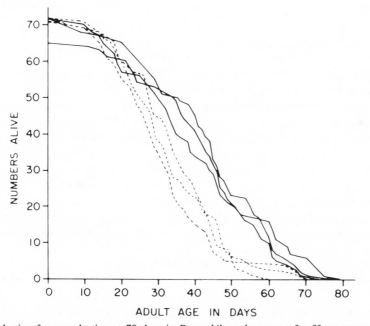

Fig. 6.3. Selection for reproduction at 70 days in *Drosophila melanogaster* for fifteen generations of an outbred stock increased the mean lifespan of females by 30% (from 33.3 days to 42.8 days) in three independent experiments. *Dotted lines,* control populations; *solid lines,* populations selected for late reproduction as an index of delayed senescence. Selection increased male lifespans less, by 15% (data not shown) (Rose, 1984c). Photograph supplied by Michael Rose.

(Arking and Dudas, 1989; Michael Rose, pers. comm.). In particular, the long-lived strains have 70% more superoxide dismutase activity, an enzyme coded by the *sod* gene in the same region of chromosome 3 that may also contain determinants for extended lifespan (Arking and Dudas, 1989). Higher superoxide dismutase could protect against free radical damage (Chapters 7.5.6, 10.3.2.5). Moreover, the long-lived lines were more resistant to paraquat. A null mutation ($cSOD^{-108}$) that obliterates the activity of Cu/Zn-superoxide dismutase greatly shortened the mean lifespan (Phillips et al., 1989).

The selection for late reproduction was interpreted as an example of negative pleiotropy (Chapter 1.5), in which genotypes that influence the efficiency of reproduction have a reciprocal influence on the lifespan. A major open question concerns the extent of changes in the mortality rate coefficients, particularly whether the acceleration of mortality is slowed in the flies selected for late reproduction, or whether the overall mortality rate was reduced. Mortality rates have not yet been calculated.

Nonetheless, the slower decline of egg production in the unselected flies implies more delayed physiological senescence. The reduced egg production and increased lifespan of late-selected flies recalls the increased lifespan of virgin flies (Partridge et al., 1986; Chapter 2.2.2.1). Possibly, delayed onset of maximum egg production also delays mechanical senescence through the efforts of mating or the utilization of nutrient stores. However, both sexes in early- and late-selected lines maintained their differences in lifespan, whether mated or not (Service, 1989). The condition of the larval and adult fat bodies and total body composition might be informative in future studies. The delayed schedule of reproduction and the much smaller ovarian weight in early adults of the flies selected for delayed senescence (Figure 6.4) suggest neuroendocrine involvements that are also implicated in the control of anautogeny, or the need for a blood meal for the first batch of eggs in biting flies (Section 6.3.3).

The failure of other efforts to select for genetic effects on lifespan of *Drosophila* (Lints et al., 1979) may be explained by the use of stocks with limited variation, which were subject to inbreeding depression (Rose, 1984b, 1991). Environmental factors also influence the outcome of selection for reproductive schedules, including temperature (Scheiner et al., 1989) and larval density (Clare and Luckinbill, 1985; Rose, 1984a; Service et al., 1988). Attempts to find conditions favoring the accumulation of late-acting deleterious genes showed that the fecundity of *D. melanogaster* females was decreased during later life by low larval density during development (Mueller, 1987); fecundity of crowded and uncrowded groups was similar in young adult flies. That is, an uncrowded environment during development appeared to favor the accumulation of harmful alleles with delayed expression, whereas crowded environments with high larval mortality favored genotypes with greater lifelong fecundity. In view of the continued *de novo* gametogenesis in both sexes throughout the lifespan (Chapter 2.2.2), it seems unlikely that these genetic variants altered germ cell reserves. The covariations of reproductive scheduling and ovarian size in the population selection studies suggest genetic influences on neuroendocrine mechanisms, as are indicated for natural populations of mosquitos (Section 6.3.3).

As a final example of selection in *Drosophila,* reverse selection was applied to the lines with delayed senescence (Service et al., 1988). That is, eggs were harvested from flies aged 2 weeks, the same interval used to propagate the control lines (Rose, 1984a). During twenty-two generations in crowded larval density, egg production of young (5-day-old)

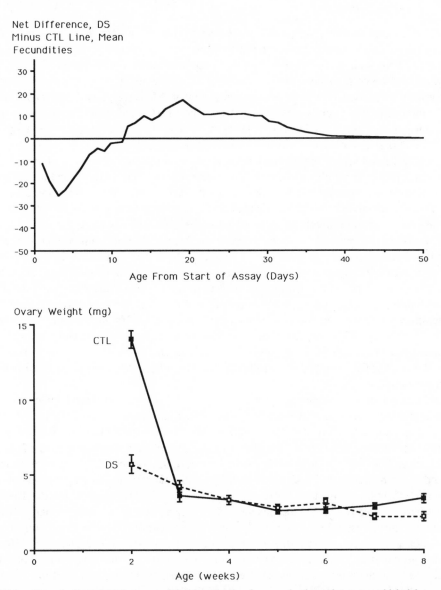

Fig. 6.4. Reproductive schedules are modified by selection for reproduction at late ages, which delays reproductive senescence and increases lifespan by 10 days in female *Drosophila melanogaster*. See Figure 6.3 legend and text for details of selection. *DS*, delayed senescence; *CTL*, unselected controls. *Top*, differences in daily egg production (CTL minus DS), showing the delayed onset of reproduction in DS flies. *Bottom*, ovary dry weight, showing that DS flies had lower ovarian weights only in early adult life. Egg weights did not differ between groups (CTL, DS) or with age. Redrawn from Rose, 1984c, and Rose et al., 1984.

adults increased progressively, while resistance to starvation decreased. This reciprocity again implies an important role for nutrient reserves. Importantly, no effects of reverse selection were seen when larval density was kept low. Under both extremes of larval density, the resistance to desiccation was unchanged. These results suggest that resistance to starvation and desiccation is under polygenic control. Moreover, larval density influences the penetrance of particular alleles that determine fitness. The rapid restoration of early fecundity during reverse selection supports the reciprocity of early reproduction and longevity that is such a recurrent phenomenon in life histories.

Thus, artificial selection for populations with different adult lifespans and reproductive schedules proceeds rapidly and effectively in *Drosophila*. These studies indicate that natural populations have genetic polymorphisms that can cause substantial differences (10–35%) in the mean and maximum lifespan and in the reproductive schedule. New spontaneous mutations are unlikely to be involved in these studies. However, while highly significant by statistical criteria, these shifts in life history do not compare with the ten- to fiftyfold longer lifespans of some social insects. Recall from Chapter 2.2.3.1 that insect queens, with lifespans that may be 5–10 years, arise from the same genome as short-lived castes like workers, whose lifespan of several months is about the same as *Drosophila*. Nonetheless, the basic body construction of queen bees is not *that* different from flies. Alterations in flying and fighting behavior might considerably increase the lifespan. A more important difference, however, may be in the capacity of gut epithelial cells for replacement, indicated by the presence of mitotic figures in the intestines of worker bees (Snodgrass, 1956). Intestinal cell proliferation has not been described in adult *Drosophila* (Chapter 2.2.2.1).

Indirect selection for delayed influences on mean lifespan and reproductive schedules was also successful with the flour beetle *Tribolium castaneum*. The life history of this species is more extended than *Drosophila* (Chapter 3.3.5): development takes 1 month and the mean lifespan is about 7 months; a maximum lifespan varies among studies, from 10 to more than 20 months (Sokal, 1970; Mertz, 1975; Boyer, 1978). Beetles were selected as young adults (< 1 month) to identify genotypic variants that might have delayed, adverse effects on lifespan, as postulated by Haldane, Medawar, and Williams. The lines that emerged after 10–50 generations had about 25% shorter mean lifespans and enhanced fecundity during early life. Importantly, the total egg production and maximum lifespan were not altered. Sokal's (1970) survivorship graph suggests a larger effect on IMR than on the acceleration of mortality. Another study demonstrated that high density of young beetles altered the reproductive schedules at later ages, so that the total lifetime fecundity was unaltered (Boyer, 1978).

6.3.3. *Natural Genetic Variations of Insect Lifespans*

This section considers natural genetic variations in the life history, which act on both the adult phase and the duration of development. The first example, and the best understood, concerns natural populations of *Drosophila mercatorum* in Hawaii that have the polymorphism abnormal abdomen (*aa*; Johnston and Templeton, 1982; reviewed in Templeton et al., 1989, and Templeton et al., 1990). The phenotype of *aa* results from retention of a juvenilized cuticle by adults, as well as abnormalities in segmentation, bristles, and color; *aa* also advances the age of maturity relative to eclosion by 2 days. Most importantly, *aa* increases egg production of young adults, giving 70% more early fecundity over the population mean (Templeton et al., 1989). Laboratory stocks of *aa* have 50% shorter median lifespans, but the same maximum lifespans (Templeton and Rankin, 1978; Templeton et al., 1990).

Aridity strongly selects for *aa,* as reflected in clines of this genotype across local gradients of humidity. The basis for selection is the impact of aridity on the age-group structure of the adult population. Under transiently arid conditions, nearly all adults die within a week after eclosion (Johnston and Templeton, 1982). The *aa* flies are not more resistant to aridity but are selected by their early fecundity. A series of field studies during meteorological fluctuations has revealed much about the dynamics of selection. When a drought in 1981 eliminated the humidity ecotone, the cline of *aa* disappeared, while the incidence of *aa* nearly doubled to 40%. Aridity greatly increased mortality (AMR) to 0.19/day, from which it is easily calculated that less than 15% could survive to 1 week, when fertility is normally greatest in *Drosophila.* When less arid years eliminated the humidity ecotone, *aa* became rarer at all locations. The effect of *aa* on the acceleration of the mortality rate (MRDT) was not analyzed, so that we do not know if the *aa* genotype has an altered rate of senescence, however shortened its lifespan. In any case, the *aa* genotype shows rapid selection for a pleiotropic trait that, while statistically shortening lifespan, is nonetheless adaptive for transient aridity.

The mechanism of *aa* involves a cascade effect from the molecular to the endocrine levels. The basis is impaired transcription of ribosomal RNA genes, which reduces the numbers of ribosomes below a crucial threshold during development, particularly in the fat body (DeSalle et al., 1986). Two genetic abnormalities are involved, one that influences the extent of replication of ribosomal genes in polytenized tissues like the fat body and another DNA sequence abnormality in the ribosomal genes themselves. *D. mercatorum* has several hundred copies of the 18S and 28S ribosomal RNA genes in tandem array on the X chromosome (Templeton et al., 1989). During larval development, there are many cycles of replication of the ribosomal RNA genes and numerous others, but without mitosis, forming giant "polytene" chromosomes.

Both the *aa* and non-*aa* flies have a large (5-kilobase) abnormal noncoding DNA sequence that is inserted within the X-linked 28S ribosomal RNA gene cluster. More than one-third of the 28S genes contain this insert, which has a transposon-like structure, with long, direct repeats and fourteen nucleotide inverted repeats at the ends of the direct repeats. The 28S ribosomal genes with the insert have impaired transcription, though various compensatory mechanisms maintain sufficient ribosomes for protein synthesis in most tissues. However, the fat body is one of the tissues with a critical and uncompensated deficiency of ribosomes.

As a consequence of fewer ribosomes, the larval fat body of *aa* flies does not step-up the synthesis of juvenile hormone esterase that usually degrades the hormone and that is normally synthesized very rapidly for a short time in development. In the absence of sufficient esterase, juvenile hormone is not degraded at the usual stage of development and consequently remains elevated into the third larval instar (Templeton and Rankin, 1978). It is likely that the persistent elevations of juvenile hormone are the basis for the precocious ovarian maturation and other anatomic abnormalities (DeSalle et al., 1986). Juvenile hormone can induce precocious ovarian development in other diptera (O'Meara and Lounibos, 1981; Wilson et al., 1983).

The most recent study indicates another gene that explains why only *aa* flies are deficient in ribosomes although both non-*aa* and *aa* have the five-kilobase insert in similar proportions (Templeton et al., 1989). Another locus on the X chromosome governs the extent of ribosomal RNA genes during polytenization, so that wild-type flies preferentially underreplicate *only* the inserted 28S genes, while *aa* underreplicate all ribosomal RNA genes across the cluster (DeSalle et al., 1986). In effect, the noninserted and functional 28S repeats are thereby *over*replicated in fat body of the non-*aa* genotype. The natural fly population appears to be polymorphic for alleles at the X-linked underreplication locus (Templeton et al., 1989).

Thus, *aa* gives a powerful and novel example of a polymorphic genetic alteration that shows negative pleiotropy in life history regulation and that is co-adapted with another gene at the molecular level (Templeton et al., 1989, 1990). There are early advantages to higher fecundity but later disadvantages with life shortening. The genetic alteration result from the dispersal of a transposable element that, like many viruses, becomes inserted into specific DNA sequences. Transposition of the insert does not appear to continue in present populations. Finally, these studies show that even noncoding sequences can have powerful influences on life history variations.

Another example of hereditary variations in life history is the mosquito, whose populations display different needs for a blood meal to complete oocyte maturation (autogeny and anautogeny, Chapter 2.2.1.1.2). Mosquitos and other biting flies that are "autogenous," that is, that do not require food during their adult phase to produce mature eggs, have shorter lifespans than races or related species that do need to feed for egg laying. A limited genetic analysis of natural populations indicates dominant genes and other genetic modifiers that determine the capacity for autogeny in the mosquitos *Culex pipiens* and *Aedes atropalpus* (Spielman, 1964, 1971; O'Meara and Krasnick, 1970; Trpis, 1978; reviewed in Anderson, 1987). Genetic influences are suspected in other biting flies (Downes, 1971). Genetic influences may account for the variations among individuals in the size of the fat body stores that are present at hatching and that will influence the need for adult feeding. Genetic influences might act through neuroendocrine functions, since juvenile hormone can induce oocyte maturation in anautogenous mosquitos even without the usually required blood meal (Lea, 1970; O'Meara and Lounibos, 1981). Larval diet also influences the penetrance of these genes. Underfeeding of larvae increased from 0 to 50% the numbers in some autogenous strains that fed as adults (Trpis, 1978; O'Meara and Krasnick, 1970). Autogeny in mosquitos, like the *aa* genotype of *Drosophila mercatorum* discussed above, reduces dependency on the environment immediately after hatching and advances the age of reproductive maturity.

Extremely cold environments with brief summers select for Diptera with other adaptations that influence the adult life history through reproductive scheduling. The next examples present the shortening as well as lengthening of the adult phase. At the shorter extreme, from the subpolar regions of both hemispheres, are species that have developed autogeny, reduced mouthparts, and parthenogenesis (Edwards, 1929; Downes, 1965; Ross and Merritt, 1987). These traits would reduce dependence of adult flies on unpredictable resources of food or weather conditions suitable for mating. At the extreme in this trend are the arctic blackflies *Prosimulium ursinum* Edwards and several species of *Gymnopais,* which lack mouthparts for biting.

Moreover, these species reproduce parthenogenetically: males are not found and the females are all triploid (Carlsson, 1962; Downes, 1965). Wood (1987) suggests that parthenogenetic autogenous reproduction favored the wide dispersal of these species; he also discusses embryological similarities and a possible phylogeny. It is also notable that such species lay fewer eggs (20–150) than temperate species (200–500) (Davies and Peterson, 1956; Downes, 1965). The adult life phase is very brief, since the eggs are already mature before the pupa hatches in these species.

Another remarkable life cycle variant is indicated for *P. ursinum* in Norway. When conditions are not favorable for hatching, eggs already ripe in the pupal shell *may* propagate the next generation, although evidence for this is indirect. Carlsson (1962, 63) noticed that ruptured pupae shed ripe eggs into streams, which he proposed to be the source of *P. ursinum* on those years following a summer that was too cold for hatching. Support for this is the frequent finding of pupal shells containing larvae that had just died, Carlsson supposed, because they could not escape the pupa. The crucial experiment remains, to observe complete development from such eggs collected in the field. Here we may have an example of a life cycle in which "aging" could not occur, since an adult phase is missing!

At the other extreme from insects with very short lifespans and sudden death are the geographic variants found in several drosophilids (*D. littoralis* and *D. lummei*) that extend their adult phase to nearly a year by entering diapause soon after hatching in the prior summer. These and certain other drosophilids overwinter in diapause as adults, when the long months of cold prevent activity except in flies that are domestic commensals (Lumme et al., 1974; Lumme and Lakovaara, 1983). In contrast, most drosophilids, including *D. melanogaster,* do not show diapause as adults. Examination of *D. littoralis* from elsewhere in Europe revealed that day length was the critical factor in whether a given geographic race delayed reproduction until the following summer by immediately entering diapause, or whether mating and reproduction occurred on the usual rapid schedule (Lumme and Oikarinen, 1977). The critical day length for inducing diapause varied linearly with longitude between 43° and 68°N (Lumme, 1982). Flies that overwintered could still lay eggs up through September of the following year (Jakko Lumme, pers. comm.). The overwintering eggs were maintained in the previtellogenic stage (Lumme et al., 1974).

Single genes are indicated as the basis for these geographic variants. F_1 hybrids show that long critical day lengths are dominant over shorter days in triggering obligate diapause upon hatching. Backcrossing indicated the segregation of a dominant autosomal allele, with no hint of polygenic control (Lumme and Oikarinen, 1977). A gene for *Cdl* (critical day length) appears to be on the fourth chromosome. The *Cdl* gene thus is distinct from the *per* locus that controls circadian rhythms (Konopka and Benzer, 1971), since *per* is located on the X chromosome in this species (Lumme, 1982; Jaako Lumme, pers. comm.). *Per* remains of interest because it also controls the onset of vitellogenesis at hatching (Handler and Postlethwait, 1977). The emerging picture is of a single dominant gene that considerably extends the adult life phase in subarctic drosophilids by controlling the photoperiodic threshold for a diapause that delays reproduction. This gene and its alleles thus may give options within the population for different reproductive schedules (Chapter 1.5) from which optima can be rapidly selected (Taylor, 1989). There is every reason to expect a neuroendocrine role in the photoperiodic control of diapause that indirectly extends the adult phase.

To close the discussion of genetic influences on dipteran life history, a major role of neuroendocrine regulation is indicated throughout the adult phase. The artificial selection for *D. melanogaster* populations with the capacity for later reproduction is associated with smaller ovarian weight and delayed onset of peak fecundity, whereas the accelerated life history of *D. mercatorum* is associated with altered regulation of juvenile hormone. The potential for experimental extension of the drosophilid lifespans can be gauged by the 5- to 15-year lifespan potentials of social insect queens, which greatly exceed those of shorter-lived castes (Chapter 2.2.3.1). In social insects, the nervous system appears to play a key role in extended longevity, through behavioral modifications that keep the long-lived queens well fed and groomed and freed from the need to forage. The capacity for renewal of enteric epithelial cells may also be important, however. Future genetic manipulations of fly adult phases may thus involve modification of behaviors as well as biochemical and cellular defenses against molecular and mechanical wear-and-tear.

The last example discussed here concerns genetic influences on the duration of development in the periodical cicadas, which differ up to fivefold in total lifespan among species (Chapter 2.2.3.3). Moreover, major lifespan differences are found between neighboring populations. Three different species of periodical cicadas (*Magicicada*) in the eastern U.S. have broods that take either 13 or 17 years to develop, terminated by synchronous emergence of adults within the same 1- to 2-day period and death within 4–6 weeks (Alexander and Moore, 1962; Lloyd et al., 1983; Lloyd, 1987). The alternate durations of development cause about 25% differences in total lifespan, without differences in the length of the short adult phase. The slow growth is attributed to the low nutritional content of their diet, the xylem fluid sucked from underground roots of trees and grasses (White and Strehl, 1978; Lloyd and White, 1987). The 17-year broods grow slowly as subterranean nymphs during their first 4 years (second instar), so that they are smaller than the 13-year brood at the same age; thereafter, they grow as fast (White and Lloyd, 1975). Both have five instars despite 4-year differences in development. The near stasis of initial growth in the 17-year cicadas is attributed to reduced feeding in early instars. The 13- and 17-year broods are often intermixed and can produce viable offspring (White, 1973), although the fertility of the hybrids is not known.

An incisive analysis of historical records suggests that the 13- and 17-year broods hybridized in 1868 when their emergence concurred and that the 17-year variant was competitively eliminated at one site that now has two 13-year broods (Lloyd et al., 1983). However, adjacent 13- and 17-year broods have different distributions of polymorphic markers (Martin and Simon, 1988). Markers included nuclear genes (three enzymes with alloenzymic variants), mitochondrial DNA (restriction fragment length polymorphisms, RFLPs), and the color of abdominal sterites. Populations of 13-year cicadas in the neighborhood of 17-year cicadas had distributions of markers that were more like the 17-year brood than like other 13-year cicadas outside of the interzone on that same year.

These different distributions of genetic polymorphisms do not decide between two possibilities: genetic versus epigenetic mechanisms (Hewitt et al., 1988). The 13- and 17-

year broods may have hybridized, with dominance of the 13-year life history (Lloyd et al., 1983). Alternatively, Martin and Simon (1988) argue for an epigenetic mechanism that shifts from 17- to 13-year development as a consequence of larval density; they note that the examples of 4-year accelerations apparently all occurred in high-density populations. However, the limited dispersal of the polymorphisms in the 13-year brood and the known viability of 13- and 17-year hybrids (White, 1973) would favor the genetic hypothesis. Moreover, because of the importance of larval density on the outcomes of artificial selection for life history variants in *Drosophila* (see above), larval density could also directly select for these different genotypes.

The cicadas are useful models for considering how life cycle variation through very extended development can influence lifespan. The facility with which development may be genetically and epigenetically varied in length in closely related species argues powerfully against cell-level limitations in the total lifespan. Genetic variation in lifespan also provides a mechanism for speciation, through reproductive isolation (Martin and Simon, 1988).

Together, these studies showing the results of artificial and natural selection demonstrate genetic influences on the schedule of reproduction and on the adult lifespan. In many examples, particularly those from dipteran species, neuroendocrine mechanisms are implicated. So far, genetic variants in soil nematodes or dipterans have increased the adult life phase by 50–100%: these results, while very encouraging for further experimental analysis, are small by comparison with the vast differences in adult lifespans between queen and worker castes that arise epigenetically in social insects (Chapter 2.2.3.1).

The ability to select for these important covariations of life history characteristics might favor their localization nearby on the same chromosome. However, transacting factors can coregulate the transcription of unlinked genes throughout the chromosomal array, as shown for numerous steroid regulatory elements in vertebrates (e.g. Yamamoto, 1985; Evans, 1988; Davidson, 1990). This type of mechanism makes virtually any combination of locations possible for genes that interact in various ways. Nonetheless, mice have a cluster of genes that influence total lifespan, in their main histocompatibility complex (*H–2*), as discussed in Section 6.5.1.4. Such linkages of genes are generally more important to the cosegregation of mutually adapted alleles at meiosis than to the regulation of transcription. The consistency of findings among independent studies with these two different species is noteworthy. The reciprocal effects of the genes that increase lifespan in decreasing early fecundity support Williams' (1957) hypothesis (Chapter 1.5.2) about negative pleiotropy and senescence.

6.4. Fungi and Plants

This section cites from a literature that is huge because of the century of genetics in fungi and plants, which led to many mutants with different growth properties and lifespans. The examples selected continue the discussion of themes from earlier sections of this book. (Although not discussed here, the senescence of yeast mother cells as a function of the number of budding cycles is influenced by genotype [Egilmez and Jazwinski, 1989].)

The lifespan of filamentous fungi can be shortened by mutations that are briefly reviewed here; there is no reason to think that the effects of these genotypes represent a widely distributed type of senescence; rather, these genotypes illustrate that mutations that have delayed harmful effects can accumulate. As described in Chapter 2.4, the rapid senescence and death of some strains of *Podospora* and *Neurospora* are caused by an endogenous, but infectious, cytoplasmic plasmid not found in other genotypes of the species. The plasmid, α-senDNA, arises from mitochondria and is transmitted maternally through the cytoplasm. Specific chromosomal alleles are required, so that without the nuclear gene mutants *i* and *viv* in *Podospora,* the lethal plasmid is not formed and rapid senescence does not occur (Tudzynski et al., 1980). Cytoplasmically transmitted senescence may be favored in filamentous fungi and other organisms whose syncytial construction allows ready passage of rapidly growing episomes (suggested to me by Thomas Johnson). Further examples might be sought on this basis.

In *Aspergillus glaucus,* Jinks (1957, 1958, 1959) selected for cytoplasmic factors that slowed the vegetative hyphal growth. Within three generations, lines that germinated from single conidia (asexual spores) were selected for slow or fast growth. The slowly growing lines had yellow-brown pigments that seemed like the lipofuscins found in many mammals. These lines differed in cytoplasmic, not nuclear, genes, presumably through their mitochondrial genomes or an extra-mitochondrial plasmid. These characteristics were reversible with back-selection. Although these well-executed studies are sometimes cited to illustrate cytoplasmic influence on senescence, Jinks never stated in these reports that the slowly growing, pigmented lines ultimately failed to grow, or died out. The implied similarity to the clonal senescence of diploid human fibroblasts (the Hayflick phenomenon, Chapter 7.6.4) thus has a very limited basis. The widely observed ability of filamentous fungi to propagate in vegetative culture over many years, even decades, suggests the continual selection against cytoplasmic heritable factors that would slow growth.

In *Neurospora crassa,* the natural death (*nd*) mutant causes vegetatively growing, haploid mycelia to degenerate and die after a week in culture (Holliday, 1969; Munkres and Minssen, 1976). During the decline of growth in *nd,* conidiophore production is greatly reduced, the hyphal wall fragments abnormally, and brownish globules accumulate, resembling in some respects the lipofuscin found in many animals (Munkres and Minssen, 1976). The *nd* allele is recessive; heterokarya with non-*nd* carrying strains grow normally. These phenomena suggest a specific metabolic defect that has not yet been identified.

The viability ("longevity") of *Neurospora* conidia during extended storage is also influenced by various mutations (Munkres et al., 1984; Munkres and Furtek, 1984a, 1984b). I doubt that conidial viability during prolonged storage is regulated by the same mechanisms that regulate longevity in metabolically active organisms. Nonetheless, it is of interest to the free radical theory of senescence that growth of vegetative cells on cyclic-GMP-supplemented media increases conidial viability in "short-lived" conidial mutants, possibly by increasing superoxide dismutase, an enzyme that degrades free radicals formed from oxygen (Munkres and Rana, 1984).

In regard to higher plants, there are innumerable examples of genotypes and clones that differ in growth properties and in their life history. As mentioned in Chapter 2.3.2, cultivated rice (*Oryza perennis*) occurs in numerous strains that differ in their perenniality. Some are obligate annuals; others are obligate perennials; yet others have both life histories depending on the conditions. All are interfertile, which should facilitate isolation of the genes responsible. Corresponding variations in life history are found within many genera.

As readers may also recall from Chapter 2.3.2, senescence in many annual plants can be prevented by removing the fruits or flowers, which indicates the importance of reproductive structures in regulating this process. The duration of the diurnal photoperiod and temperature are also crucial in the transitions from growth to senescence, depending on the species. It is therefore of much interest that two genes have been described in the pea (*Pisum sativum*) that influence the responses to bud removal. For example, the gene *Sn* delayed flower initiation and apical tissue degeneration under short-day lighting, while *Hr* enhanced this effect (Reid, 1980). The effects of *Sn* were transmissible by grafting (Murfet, 1971). These examples show that the mechanisms of plant senescence should be wide open to the techniques of transformational genetics.

6.5. Mammals

Information about genetic influence on mammals almost entirely derives from genotypes that *shorten* lifespan. While no studies have succeeded in selecting for long-lived genotypes, some genotypes have longer lifespans than in general populations. The shorter-lived genotypes are in *all* cases associated with early onset diseases that do not otherwise generally accelerate senescence. Genotypes for early onset diseases are important to the discussion of senescence, because in many cases clinical or experimental interventions like diet restriction can extend the lifespan at least to the normal range (Chapter 10.3). Comparisons of the mechanisms by which diet restriction, for example, suppresses early onset diseases in certain genotypes could more easily reveal the mechanisms of diet restriction than in longer-lived genotypes that may have a larger number of age-related changes. This literature gives a precedent for examining similar interventions for long-lived genotypes that may harbor alleles causing milder versions of these same disorders. This section is organized by species to facilitate discussion of specific genotypes.

6.5.1. Mice

Numerous inbred strains of laboratory mice have different mean lifespans and diseases of senescence. Many genotypes have mean lifespans of 22–32 months and maximum lifespans of 32–44 months (Storer, 1978; Harrison and Archer, 1983), although even greater lifespans can be achieved by special or restricted diets (Chapter 10.3). To assist discussion of these myriad genotypes with their different nomenclature and symbols, the next section gives background about some common genotypes, which the reader may wish to skip. I also briefly discuss some issues about animal husbandry that may not be generally known but that are pertinent to the difficulties in replicating work between and within laboratories, even with the same carefully standardized genetic strain.

6.5.1.1. Background Information on Mouse Genotypes and Husbandry

Many inbred mouse strains used in gerontological and genetic studies consist of several superfamilies with independent founders. Most strains were developed early in this century to obtain genotypes with different cancer incidence in early adult life. None of the genotypes used for gerontological studies were initially selected for longevity, although some strains with a low cancer incidence also proved to be long-lived. Most strains mentioned in this book are shown in Figure 6.5.

The C57BL family of inbred strains, particularly C57BL/6J and C57BL/10 mice, are widely used in gerontology because they have a low incidence of tumors and are easy to keep in general good health during the first 18–24 months of their relatively long lifespans. The C57BL/6J and C57BL/10J lines have been inbred more than one hundred generations at the Jackson Laboratory since 1936 (Bailey, 1978; Staats, 1976) and are each more than 99% identical according to the standard formula (Klein, 1975). The designation *J* for a strain means that it is produced at the Jackson Laboratory, a world center of mouse genetic research and a major mouse supplier in Bar Harbor, Maine; *Sn* refers to Snell's and *Ks* refers to Kaliss' sublines at the Jackson Laboratory.

The C57BL/6J (Finch, 1971; Dunn, 1954) and C57BL/10Sn strains (G. S. Smith et al., 1973) resemble each other in lifespan and incidence of age-related lymphoma. However, there are important differences in female reproductive characteristics (Section 6.5.1.4). While considered to have identical sets of alleles in the *H–2* gene complex that govern most histocompatibility functions, there are three known allelic differences: levulinate dehydratase (*LV*); histocompatibility–9 (*H–9*); and erythrocyte antigen–430 (*Ea–430*) (Staats, 1976; Marianne Cherry, unpublished; see Lerner et al., 1988). Another subline is the C57BL/KsJ strain, which was returned to the Jackson Laboratory after the catastrophic fire of 1947. Due to its long separation from the C57BL/6J line in other laboratories, chance breeding errors may have introduced other alleles. For example, in the *H–2* complex, C57BL/Ks mice are *H–2d*, while C57BL/6J are *H–2b* (Green, 1981). C57BL/Ks mice are considered closer to C57BL/10Sn than to C57BL/6J (David Harrison, pers. comm.).

The CBA family of strains includes CBA/J and CBA/Ca. The CBA/Ca strain (also available from Jackson Laboratory) is considered genetically identical to the CBA/HT6J (= CBA/H-T6 = CBA/HT6), which contains the T6 chromosomal translocation marker that is used in marrow transplantation studies. The CBA/J mice have slightly shorter lifespans than CBA/Ca and differ in non-*H–2* antigens that cause graft rejection (David Harrison, pers. comm.). Other CBA substrains are used in Europe (Chapter 8.5.1.1, n. 4). Bailey (1978) discusses subline divergence in CBA and other strains.

DBA/2J mice are often used for comparison with C57BL-related mice in gerontological studies, because of independent origins (Figure 6.5). C57BL/6J and DBA/2J descend from lines that were independently established at the beginning of this century in different parts of the United States (Staats, 1980; Russell, 1978). However, comparisons of several hundred alleles indicates similar extents of differences between C57BL/6J, CBA, DBA/2J, and BALB/c strains, in the range of 28–41% allelic divergence (Thomas H. Roderick, pers. comm.; data from MATRIX, a component of the genomic database GBASE at the Jackson Laboratory). The greatest divergence, 41%, was indicated for C57BL/6J and CBA mice.

The convention of calling a line of rodents as inbred after about twenty generations (Klein, 1975; Bailey, 1978) does not preclude haphazard mutations or genetic rearrangements. Assay for genetic homogeneity in inbred lines by skin grafting is a routine proce-

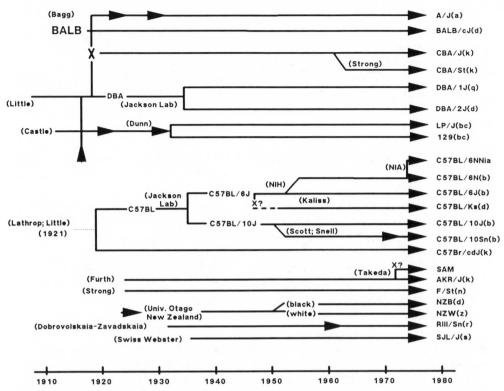

Fig. 6.5. Abbreviated genealogy of common mouse strains. Letters in parentheses represent the *H–2* haplotype (combinations of alleles at different loci in the large gene cluster of the *H–2* locus; see Figure 6.7). Origins of C57BL/6N and C57BL/6NNia are given in Section 6.5.1.1. In two instances (C57BL/Ks and SAM), outbreeding probably occurred either with rogue mice or with misidentified partners, as indicated by *X*. The SAM (senescence-accelerated mouse) differs in pathophysiology from its attributed AKR origins, as described in Section 6.5.1.7.2. Redrawn in part from Klein, 1986, Staats, 1980, and in consultation with David Harrison.

dure at the Jackson Laboratory, serving as a quality control for many loci. Nonetheless, some stray alleles might go unnoticed, unless they influenced the *H–2* haplotype or other allelically varying characteristics that are widely studied. The concern with genotype is crucial, because partly inbred mice, only a few generations apart, can differ in baseline physiologic functions (e.g. Terpstra and Raaijmaker, 1976).

Many strains have been derived from stocks at the Jackson Laboratory, like the Nia substrain of C57BL/6, which cannot be presumed genetically identical. The C57BL/6NNia, for example, traces to the C57BL/6N strain, which was an offshoot of the C57BL/6J mouse maintained since 1951 at the National Institutes of Health (NIH). Subsequently, the C57BL/6NNia substrain was established by the National Institute on Aging (NIA) under contract at the Charles River Laboratories (Boston) to provide mice for gerontological studies. Reassuringly, we found some of the same neurochemical changes during senescence in C57BL/6J and C57BL/6NNia mice, a blunted induction of dopaminergic supersensitization in response to chronic dopaminergic receptor blockade by the drug haloperidol (Randall et al., 1981).

Under good conditions, long-lived mouse strains have mean lifespans of 26–32 months, or about 2.5 years (Harrison and Archer, 1983; Finch, 1971). Thus, mean lifespans of 20 months or less are usually atypical for laboratory rodents and could result from two independent factors: unhealthy conditions in the animal colony, or genotype-specific early onset diseases (Section 6.5.1.7). An alarming number of environmental variables can compromise the replicability of studies on the lifespan and phenotypes of senescence, both between and within laboratories (Zurcher, van Zweiten, Solleveld, and Hollander, 1982; Harrison and Archer, 1983). For example, C57BL/6J male mice raised during the 1940s at the Jackson Laboratory had mean lifespans of 22 months (Storer, 1966), while later populations of this highly inbred strain had mean lifespans of 28 months (Goodrick, 1975; Harrison and Archer, 1983) and 30 months (Finch, 1971; Kunstyr and Leuenberger, 1975). Similar variations were observed for maximum lifespan for different cohorts (Hollander et al., 1984). These variations seem unlikely to reflect genotypic drift, because the strains are strictly monitored at the Jackson Laboratory. Because of improved husbandry during the last 50 years, lifespans of laboratory rodents overall have increased by up to 50% (tabulated by Goodrick, 1975, and Myers, 1978). Further increases will not be surprising. Husbandry problems are also noted for *Drosophila* (Rose, 1984c).

Numerous environmental factors must be considered. Diet can vary in minor components that are rarely recognized even by major manufacturers, for example, the phytoestrogens genisten and daidzein, which are uterotrophic and which are present in some commercial chows at 0.4 mg/gram chow (Patricia Whitten and Frederick Naftolin, pers. comm.). Infectious agents can be introduced by other nearby animals. The Sendai virus is particularly dreaded, causing a pneumonia with long-lasting effects on immune functions and life shortening (Kay, 1979; Hollander, 1979). Apparently healthy laboratory rats often carry the virus, which can be transmitted with minimal contact from nearby laboratories. Some mouse genotypes are more susceptible (Smith and Walford, 1978). Seasonal and geographic variations of climate often perturb the temperature and humidity controls of most facilities; the season of birth appears to influence the lifespan (Yunis et al., 1984). The sympathies of the caretakers in handling animals and their attention to details like leaky water bottles, which often dampen the bedding, can influence the well-being of laboratory rodents. Such variables are notoriously difficult to control, and their variations among laboratories set limits on the reliability of small (< 25%) lifespan differences between genotypes that are raised in different laboratories. Different cohorts of Jackson mice within the same laboratory, however, appear to vary less.

Reproductive experience is another important influence on senescence. The "retired breeders" often used in gerontological studies tend to have shorter lifespans than virgin male or female rodents (Suntzeff et al., 1962; Wexler, 1964b, 1976a; Ingram et al., 1983; Chapter 10.6). Likely factors are the stressful competition arising from the confined housing of several of each sex, and the rapid succession of litters. The effects of breeding on lifespan may vary by strain and sex. No differences were found in lifespan of C57BL/6J males, whether virgins or retired breeders (David Harrison, pers. comm.). The extent that the gerontological literature represents breeding stress rather than endogenous factors in senescence is unclear. Clearly worse stresses from breeding are suffered by the dasyurid marsupial mice (Chapter 2.3.1.5).

Mice tending to have long lifespans include the C57BL/6J, C57BL/10Sn, BALB/cJ,

LP/J (Figure 6.6), the CBA/HT6J (Harrison and Archer, 1983), and the congenic C57BL/ 10Sn.RIII (Section 6.5.1.4). Shorter lifespans within the normal range are reported for the CBA/CaJ and CBA/J, the DBA/2J, and the A/J strains (Harrison and Archer, 1983; G. S. Smith et al., 1973; Goodrick, 1978; Storer, 1978; Yunis et al., 1984) (Figure 6.6). The characteristic age-related pathologic lesions of select strains and estimated mortality coefficients are given in Table 6.1; the calculations were based on data extracted from the survivorship curves and thus represent averages by age group. The acceleration of mortality rates (MRDT and IMR) of different inbred strains is statistically indistinguishable, even for the NZB and SJL/J strains that die prematurely from lymphoreticular sarcomas (Storer, 1978; Figure 6.6). Future studies may resolve genotypic influences by analysis within each strain and cohort of individual mouse lifespans that were not published. The possibility may also be considered that strains differ in the age when the acceleration of adult mortality rate begins by the hypothetical Δ factor (Chapter 1.4.6; Figure 1.8). No genotype of rodents or other species that dies prematurely, relative to its normative pattern, shows a general acceleration of senescence.

6.5.1.2. F_1 Hybrid Vigor

The F_1 hybrids that are offspring from matings of different inbred strains generally have longer mean and maximum lifespans than the parental strains (G. S. Smith et al., 1973; Ingram and Reynolds, 1982; Myers, 1978; Storer, 1978; Harrison and Archer, 1983).

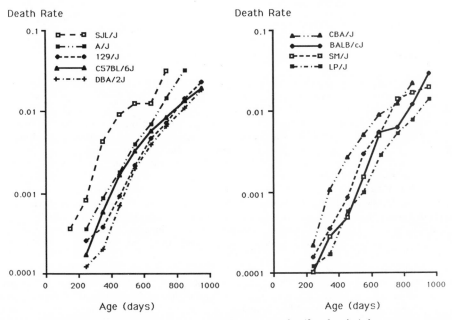

Fig. 6.6. Mortality rates by age group of some common mouse strains (female mice) that represent a range of lifespans. Mice were maintained at the Jackson Laboratory, 1961 to 1967. Mortality rate coefficients are given in Table 6.1 from this and other sources. These strains have very similar MRDTs but may differ in IMRs. Redrawn from Storer, 1978.

Table 6.1 Senescence of Inbred Mice

Genotypes	Lifespan, years (Mean/Max)	IMR/year[1]	MRDT, years[1]	Major Degenerative Changes
Inbred strains				
AKR[2]	0.7/1.5			lymphoid leukemia with ecotropic MuLV
C57BL/6J males[3]	2.2/3.2–3.5	0.024	0.21	pituitary tumors, 80%, females only
females	2.1/3.2			(lactotrophe adenomas); 50% of both genders have lymphomas (reticulum cell sarcoma of lymphoid tissues, including spleen, mesentery, and liver)
C57BL/10Sn males[4]	2.6/3.0			lymphomas, 30%; other tumors < 5%
females	2.3/2.9			
CBA/J males[5]	2.3/2.9	0.027	0.26	hepatomas (> 90% in male);
females	2.3/3.2			reticuloendothelial tumors, 10–50%
CBA/Ca males[6]	2.1/2.7			presumably as for CBA/J
females	2.4/3.5			
DBA/2J[7]	1.9/2.5	0.034	0.31	myocardial calcification common (100%); lymphomas (10%)
NZB/Lac[8]	0.8/1.5	0.015	0.19	autoimmune and kidney disease
SAM[9]	0.7/1		0.4	amyloidosis and nonthymic lymphoma
SJL/J[10]	1.1/1.6	0.073	0.27	amyloidosis
SM/J[11]	1.9/2.5	0.008	0.21	amyloidosis
Congenic strains				
B10.F[12]	0.8/1.3			graying and lymphoma with xenotropic MuLV
B10.RIII males[13]	2.7/3.5			lymphomas (35%)
females	2.7/3.3			(55%)
Hybrids (F₁)				
B6CBA males[14]	2.7/3.6–4.8			50% lymphomas like C57BL/6J
D2SM females[15]	1.7/2.7	0.2	0.04	
SMD2 females[16]	2.3/3.1	0.008	0.30	

[1]IMR and MRDT approximate, within a twofold range. The calculated values for mortality coefficients are based on survivorship curves. No significance should be attributed to apparent differences until calculations are available as based on mortality schedule data itself. IMR, the initial mortality rate, calculated at 3 months (equation [1.2]). MRDT, the mortality rate doubling time, calculated according to equation (1.3) (Gompertz model) for the age-related acceleration of mortality rate. These mortality rate coefficients were calculated by extracting the mortality rate by age group from the survivorship plots, with exceptions indicated. In all cases the regressions of ln mortality rate on age were highly significant: $r^2 > 0.99$. In the few strains for which there are several published survivorship plots, the values were averaged. The sources of the data are listed for each genotype below. Values of IMR and MRDT were not statistically different between genders (overlapping 90% confidence intervals) and were averaged as indicated. Because the mortality rate data were already consolidated by age group on the graphs, no strong conclusions should be drawn about the absence of gender differences in mortality rate coefficients. It was not possible in the time available to do a full analysis of mortality rate trends with complete lifespan data from each individual mouse, an approach that might reveal significant differences between strains and genders.

[2]References in Section 6.5.1.7.2.

[3]IMR and MRDT calculated from mortality rate plots of Storer, 1978, that had shorter mean and maximum lifespans than more recent cohorts also at the Jackson Laboratory in Bar Harbor, Maine. The lifespans shown are from Harrison and Archer (1983, 1987) at the Jackson Laboratory. The related, but not identical, C57BL/KsLwRij, a subline of C57BL/Ks mice (Figure 6.6) from the Radiobiological Institute, Rijswijk, Holland (Zurcher et al., 1982) gave similar values for IMR (males, 0.012; females, 0.06) and MRDT (males, 0.28; females, 0.22). For descriptions of the reticulum cell sarcoma of C57BL mice, see Dunn, 1954, and *Mammalian Models*, 1981.

Table 6.1 (*continued*)

[4]Lifespans and pathology from Smith and Walford, 1978.

[5]IMR and MRDT calculated from mortality rate plots of CBA/J of Storer, 1978, with genders averaged: IMR (males, 0.06; females, 0.008); MRDT (males, 0.25; females, 0.37). CBA/BrARij from Zurcher et al., 1982, gave similar values: IMR (males, 0.002; females, 0.001); MRDT (males, 0.25; females, 0.28). Mean and maximum lifespans for CBA/HT6J from Harrison and Archer, 1983; the HT6 strain has a chromosome marker used in transplantation studies.

[6]Lifespans from Harrison and Archer, 1983. CBA/HT6J mice are similar. Myocardial calcification by midlife (Nabors & Ball, 1975); age-distribution of lymphomas (Staats, 1976) is unclear.

[7]Combined values for DBA/2J males and females from mortality rate plots of Storer, 1978.

[8]Combined values for males and females, calculated from survivorship plots of Zurcher et al., 1982; IMR (males, 0.016; females, 0.013); MRDT (males, 0.21; females, 0.16). The NZB/Lac subline was obtained from the Laboratory Animals Center, Surrey, United Kingdom.

[9]SAM strain derived from AKR. From calculations of Takeda et al., 1981.

[10]Calculated from mortality rate plots of SJL/J females from Storer, 1978; amyloidosis cited by Staats, 1976.

[11]Calculated from mortality rate plots of SM/J males from Storer, 1978; amyloidosis cited by Staats, 1976.

[12]From Storer, 1978; Section 6.5.1.7.1.

[13]From Smith and Walford, 1978. Lymphomas found at natural death, ca. lifespan.

[14]From Harrison and Archer, 1987. The longest-lived mouse of 4.8 years (1,742 days) was food-restricted 1 day/week after 28 days, but given *ad libitum* food after 1,541 days. This may be the record lifespan of the species *Mus*.

[15]D2SM females, the reciprocal cross (DBA/2J females × SM/J males) gave an exception to the general rule of hybrid vigor, in the mean lifespan was markedly lower than parental strains (Storer, 1978).

[16]Calculated from mortality rate plots of (SM/J females × DBA/2J males)F_1, from Storer, 1978.

According to the traditional interpretation, this result indicates that lifespan is under polygenic control. The basis for increased lifespan is probably through reduced disease in the hybrids. For example, male C57BL/6J and CBA/HT6J and their F_1 hybrids showed consistent parallels of lifespan with glomerular filtration rate, a kidney function that is closely linked to glomerular sclerosis and renal vascular degeneration (Hackbarth and Harrison, 1982). The F_1 hybrids had the least kidney degeneration and lived 17% longer; parental strains showed sharp decreases of kidney function after 15 months. The overall order for lifespan and kidney function was F_1 > CBA > C57BL. Yet kidney failure is not a major identified cause of death in these strains (David Harrison, pers. comm.). In a large study of F_1 hybrids with parental strains, there was no correlation between the age-related reductions in the individual lifespan and the capacity to concentrate urine (Harrison and Archer, 1983). A tentative conclusion is that hybrid advantages in resisting age-related kidney dysfunctions are not the primary factor in lifespan but could nonetheless increase the physiologic reserve needed to cope with the duress from other age-related lesions.

One of the most thorough analysis of diseases found at death showed major differences between short- and long-lived strains, and even described new diseases in the F_1 hybrids (G. S. Smith et al., 1973). For example, hepatomas were rare (< 2%) in the male DBA/2J and 129/J strains, but occurred in 29% of the F_1's. A pilot analysis of Storer's (1978) age-group mortality rates did not indicate differences between the F_1 and parental SMD2 and D2SM strains in the MRDT (Table 6.1). The F_1 hybrids tend to have smaller IMRs. While these differences were not significant, a more detailed analysis is warranted, using the original lifespan data of individual mice. Provisionally, the lifespan is increased in the F_1's without altering the rate of senescence by actuarial criteria. While the general rule of increased mean lifespan in F_1 hybrids over parental strains holds up well for most genotypes (Storer, 1978; Harrison and Archer, 1983, in prep.; Harrison et al., in prep.), Storer noted that D2SM female hybrids had markedly lower mean lifespans than parental strains (Table 6.1). Further study with larger groups may be needed to show if the MRDT and IMR are under separate genetic control.

6.5.1.3.   Sex Chromosomes and Other Specific Chromosomal Determinants

Sex chromosomes are the first case of specific chromosomal determinants of lifespan to be discussed. Most inbred mouse strains have small or variable differences between sexes (G. S. Smith et al., 1973; Storer, 1978; Ingram and Reynolds, 1982; Harrison and Archer, 1983). Even in the strain with the greatest sex difference, C57BL/6J, males had slightly greater mean lifespans, 1 month longer than in females, while the maximum lifespans were the same in both sexes (Harrison and Archer, 1983). These observations do not conform to the general trend for female mammals to live slightly (5–15%) longer than males (Comfort, 1979). In mice, effects of the Y chromosome on lifespan may be influenced by autosomal genes. For example, in laboratory confinement male BALB/cJ mice fight viciously and must be isolated, while C57BL/6J and CBA males fight less. While the effects of castration on male-mouse lifespan have not been systematically studied, the reductions of fighting and other dangerous behaviors in most castrated mammals (discussed in Section 6.5.2) indicates that lifespan should increase from a reduction of age-independent mortality coefficients (A_o and M_o; equation [1.1]), represented in the IMR as calculated at puberty.

A different example of Y-chromosome influences that depend on the autosomal background is found in autoimmune-prone strains, which are described in Section 6.5.1.7.3. Briefly, the Y chromosome from BXSB mice accelerates autoimmune and related pathologic lesions only in certain short-lived strains that have a higher basal manifestation of these early onset diseases (Hudgins et al., 1985). These strain differences in behavior and autoimmune disorders illustrate how testicular hormones and other factors that are influenced by Y-chromosome alleles have effects on pathogenesis that depends on the genetic background. The strength of selection on a given allele that influences postmaturational mortality will depend on many other genetic loci and need not be the same in each sex. Consequently, there are numerous examples of diseases with strong sex bias during senescence.

Other chromosomes containing genes that influence lifespan have been identified by backcross strategies. Backcrossing of F_1 hybrids also tests hypotheses that a small number of genes are major determinants of the mammalian lifespan (Cutler, 1975; G. S. Smith et al., 1973). Backcrosses between the F_1 hybrid females of C57BL/6J and DBA/2J parental strains with DBA/2J males had differences in median lifespan of up to 25%, which could be in part associated with chromosomes 4, 17, and the sex chromosomes (XX or XY; Yunis et al., 1984). Chromosome 17 had more effect on male lifespans, while chromosome 4 had more on females. These sex differences suggest that some autosomal influences on lifespan are dependent on sex hormones. However, other of the eighteen chromosomes may be involved. Another study showed that the albino coat color allele of A/J and BALB/cJ mice was not associated with lifespan in F_2 hybrid backcrosses of C57BL/6J and DBA/2J mice (Goodrick, 1975, 1978).

6.5.1.4.   H–2 Influences

Genotypic variants that alter the patterns of senescence in different organs and the life-span are associated with the major histocompatibility complex (MHC). In large part, these studies derive from Roy Walford's (1969, 1987) hypotheses about the importance of the immune system to lifespan regulation, in particular that genes with strong genetic determinants of adult viability and lifespan occur in the MHC. Informed readers may wish to skip the following short background on the MHC.

The MHC of mice is called the *H–2* complex on chromosome 17, while humans have a corresponding region on chromosome 6, the *HLA* complex (human lymphocyte antigen; Figure 6.7). These arrays of genes encompass one to ten centimorgans in mice (Klein et al., 1982), or at least 1 million DNA base pairs. This extensive cluster of often-related genes is considered to have many pleiotropic influences throughout the lifespan. In work of great importance begun by George Snell at the Jackson Laboratory, mouse strains were inbred specifically to resolve the different genetic loci within the *H–2* complex (Klein, 1970, 1986). The set of alleles of an inbred strains is collectively called the haplotype, as indicated in the lowercase letters in parentheses next to the strains listed in Figure 6.5. The Class I, II, and III gene clusters are outlined in Figure 6.7.

The congenic strains were a key approach established by Snell. Chromosomal segments containing *H–2* regions from different inbred strains were introduced into several strains (C57BL/10Sn, C3H/HeDiSn, A/WySn) by backcrosses and were classified for different histocompatibility alleles by serology for immunoglobulins and by skin-grafting tests. For example, the symbol B10.A designates the congenic strain on a C57BL/10 background that contains the general *H–2* region from the A strain. In view of the many examples in which the influence of particular alleles is influenced by other genes (epistatic influences), the congenics on different background strains were invaluable for resolving actions of particular alleles. A caveat is that congenic strains differ by many gene loci contained within the *H–2* haplotypes, as well as variable and unknown amounts of passenger DNA from adjacent regions of the chromosome, carried along during the backcrossing (Klein et al., 1982). Nonetheless, the congenics have revealed unexpected genetic influences on lifespan and details of senescence.

Besides their influence on numerous cell-mediated immunological responses, genes in or near to the *H–2* or MHC of other species also influence diverse aspects of development and endocrine functions of adult mammals. Other functions include resistance to viruses, since some viruses use MHC glycoproteins as receptors (Klein, 1986); destruction of tumor cells by tumor necrosis factor (Müller, Jongeneel, et al., 1987); enzymes that degrade steroids, drugs, and environmental toxins (the mixed-function oxidase system, Section 6.5.1.5); target cell responses to hormones (Ivanyi, 1978; Lafuse et al., 1979); cAMP levels (Lafuse et al., 1979); and chemosensory recognition (Yamazaki et al., 1983). Several different receptors for peptide hormones have MHC Class I glycoproteins as part of the receptor complex, for example, the insulin (Phillips et al., 1986; Due et al., 1986) and the LH receptors (Solano et al., 1988). Different MHC Class I alleles may be involved in the quantitatively different cAMP production and glucagon-binding affinities of mouse strains of different *H–2* haplotypes (Lafuse and Edidin, 1980). The transcription of Class I genes is also regulated by a nuclear hormone receptor that also binds to an estrogen-

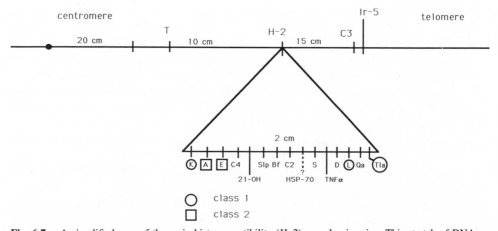

Fig. 6.7. A simplified map of the main histocompatibility (*H–2*) complex in mice. This stretch of DNA on chromosome 17 of about 1 million base pairs contains a cluster of genes that regulate many immune-related phenomena. The length of 2 cm, or 2 centimorgans, is equivalent to a recombination frequency of 2%. The largest families of related genes in the MHC (major histocompatibility complex) concern the Class I and Class II proteins. MHC Class I proteins (and genes) include the major transplantation antigens that are found in nearly all cells (*K, D, L* loci), which are involved in antigen presentation to cytotoxic T lymphocytes. Class II antigens (*A, E* loci) are implicated in T helper lymphocyte functions and are expressed in fewer cells that include macrophages and select lymphocyte subpopulations. *H–2* antigens are unusually polymorphic, with more than fifty alleles at the *H–2K* and *H–2D* loci. Inbred mice have numerous copies of each gene, with varying numbers among strains. The Class III genes code for proteins of the complement system, which are much less polymorphic and are present in one copy each. Other loci are: *Slp*, sex-linked protein; *21-OH*, steroid 21-hydroxylase (P-450) group; *TNF$_\alpha$*, tumor necrosis factor; *HSP–70*, main heat shock protein, which is located between *C2* and *TNF$_\alpha$* in humans and rats. For general reviews, see Klein, 1985, Lew et al., 1986, Flavell et al., 1986, Sargent et al., 1989; and Müller, Jongeneel, et al., 1987.

response element (Hamada et al.,1989). Thus, the MHC proteins are well integrated into physiological control mechanisms that allow many pleiotropic effects (Lerner and Finch, 1991).

Recently, we found that the length of estrous cycles is influenced by *H–2* alleles, as proven by the cosegregation of cycle lengths with the *H–2* haplotypes in $F_2 \times F_1$ backcrosses (Lerner and Finch, 1991), which could be related to the MHC through influences on the affinity of the LH receptor as observed for the glucagon receptor (Lafuse and Edidin, 1980). Some aspects of growth and development (Bonner and Slavkin, 1975; Kunz et al., 1980) are influenced by the *H–2* haplotype, through alleles that are distinct from the *t* loci (Klein, 1986). These and other examples are discussed by Walford (1987), Ivanyi (1978), and Klein (1978, 1986).

The role of *H–2* polymorphisms in age-related diseases and the lifespan is being studied by examining congenic mice (see above) that are genetically identical at autosomal loci except for segments of the *H–2* region. The lifespans and age-related diseases of the congenic *H–2* strains differed significantly, from 50 to 140 weeks (mean); most strains had mean lifespans that differed within a range of 30 weeks (Figure 6.8; Smith and Walford, 1977, 1978). The shortest lifespans within the C57BL/10Sn congenics are held by the B10.AKM and the B10.F, the later dying with lymphomas in association with a vertically transmitted leukemia virus (see below). The B10.RIII strain lived longest and was the only one of the fourteen congenics studied to outlive the background strain. The congenics on A and C3H background strains had shorter lifespans than on the B10 background, as expected from the short lifespans of these background stocks. The range of lifespans approximated the inbred, noncongenic strains that were previously studied in this colony (G. S. Smith et al., 1973). Other, but small, influences of these *H–2* haplotypes on lifespan were shown by backcrosses of B10 congenics to DBA/2J mice (Williams et al., 1981) and of A/J × C57BL/6J mice (Meyer et al., 1989). The evidence so far indicates that several *H–2* haplotypes, particularly *H–2ᵃ* and *H–2ʳ*, contain alleles that can slightly increase the lifespan of certain background mouse strains. These genetic influences should

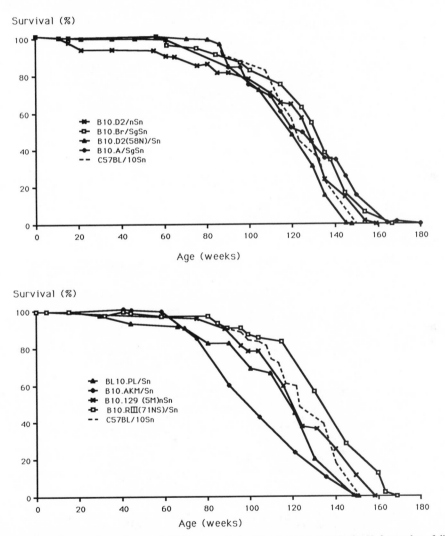

Fig. 6.8. Survivorship plots of congenic *H–2* mouse strains (females) that contain the *H–2* complex of different strains on common backgrounds. In these and subsequent cohorts, the B10.RIII is the longest-lived genotype, longer than the parental strain C57BL/10Sn. From Smith and Walford, 1977.

not be dismissed because they are small or because they vary between genders and background strains. As described next, the *H–2* haplotype has substantial influences on the schedule and extent of physiological and cellular changes throughout adult life.

The different lifespans of the *H–2* congenics raise questions about the rates of age-related changes in different cellular and physiological functions. Evidence is suggesting that rates of age-related changes are scaled with the differences in lifespan among the congenics. For example, the responses of spleen cells to T-cell-specific mitogens *in vitro* (Figure 6.9) shows slower age-related decline in the longer-lived genotypes of A and B10 backgrounds (Meredith and Walford, 1977). In C57BL/10 strains, the B10.RIII had a slower loss of T-cell mitogenic responses to phytohemagglutinin, for example. Moreover, the congenics have different rates of maturation in T-cell mitogenic responses, but B-cell mitogenic responses were not impaired over the lifespan. Major effects of the *H–2* haplotypes on spontaneous age-related tumors also generally follow the same ranking for each

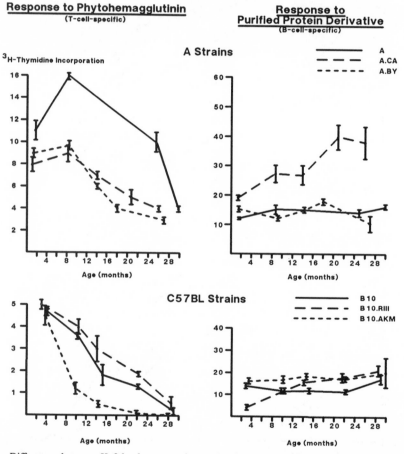

Fig. 6.9.   Differences between *H–2* haplotypes and age-related changes in spleen cell functions. The *in vitro* assays represent mitogenic responses by [³H]-thymidine incorporation into DNA with mitogens that are selective for T or B cells. Two different sets of congenic mice are used, on A and C57BL/10 backgrounds. In general, the longest-lived genotype (e.g. B10.RIII) has the most robust immune responses at later ages. Redrawn from Meredith and Walford, 1977.

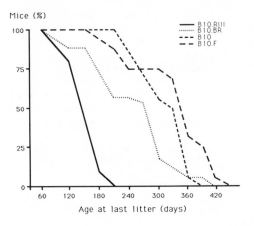

Fig. 6.10. Reproductive senescence in congenic mouse strains. In general, age at last litter corresponded to the lifespan. B10.RIII mice were the most robust and B10.F the least in both parameters. The B10.F has an unusually short lifespan in association with MuLV-caused lymphomas. From Lerner et al., 1988.

tumor type; for example, the incidence of lymphoma at death is very similar in B10.AKM and B10.RIII mice despite their 10-month difference in lifespan (Walford, 1986).

The congenics also differ in the schedule of female reproductive senescence (Lerner et al., 1988). As observed for T-cell mitogenic responses, the B10.RIII mouse had a superior reproductive performance, producing about 20% more pups during its reproductive lifespan than its nearest competitor, the B10.Br strain. The age of last litter has the order B10.RIII > B10 > B10.Br > B10.F (Figure 6.10), whereas the mean survival is ranked B10.RIII > B10.Br > B10 > B10.F (Figure 6.8; the latter strains are not shown in the figure). At 8 months, the B10.F mice had 50% fewer primary ovarian oocytes, while other strains were equivalent; 8 months was chosen as the age group to be used for the tedious job of counting total follicles because it marks the beginning of reproductive decline in most mouse strains. The relative longevity of B10.Br and B10 differs slightly by sex (Smith and Walford, 1977, 1978). Other strain differences in reproductive senescence (Finch, 1978; vom Saal and Finch, 1988) need not be related to *H–2* influences.

Besides the effects of *H–2* alleles, there are influences on reproduction from other loci, since the C57BL/10Sn mice did not show the age-related lengthening of estrous cycles that is characteristic of the virtually identical C57BL/6J strain with the same *H–2* haplotype as the C57BL/10Sn (see Section 6.5.1.1 for known allelic differences). Moreover, the C57BL/10Sn have few 4-day estrous cycles at any age, in contrast to the predominantly 4-day cycles that are the norm for young laboratory mice and rats. Other *H–2* alleles restore 4-day cycles to C57BL/10Sn mice (Steven Lerner and Caleb Finch, unpublished) and, as noted above, may influence the hormone-receptor interactions on ovarian follicles. These genetic variations in the periodicity of fertility cycles are important because they indicate potential regulation by a number of different genes that are subject to selection as balanced polymorphisms. In view of the trade-off (negative pleiotropy) of reduced early reproduction that resulted from selecting insects for reproduction at late ages (Section 6.3.2), it is of interest that, in contrast to the selected *Drosophila*, the B10.RIII mice showed no delay in achieving peak fecundity. During establishment of the strain, B10.RIII mice were selected for low cancer incidence and passively also for fecundity (David Harrison, pers. comm.). The robustness of both traits, lifespan and fecundity, in B10.RIII mice suggests that high sustained reproductive capacities in mammals is not

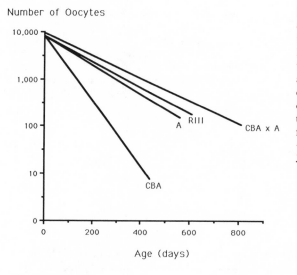

Number of Oocytes

Fig. 6.11. Ovarian oocyte loss with age differs among mouse strains because of two factors: the initial stock of primordial follicles and oocytes at birth and the rate at which oocytes are lost, which approximates an exponential decline. Most of the loss occurs from follicular involution and oocyte death during follicular maturation, through the poorly understood process of atresia. The fraction of follicles that ovulate is small, < 5% of the total at birth. Redrawn from Jones and Krohn, 1961b.

incompatible with long lifespans and that the genes involved can become co-adapted, even in inbred situations.

The rate of ovarian oocyte loss is also subject to genotype influences that have not been identified with *H–2* alleles. For example, English sublines of the A(a), CBA(a), and RIII(r) have different rates of ovarian oocyte loss with age (Figure 6.11). While each strain has a different *H–2* haplotype (denoted in parentheses), the strains are not congenic and therefore differ in numerous other alleles throughout the genome. Congenic strains are available that could resolve the contribution from *H–2* and non-*H–2* alleles to the rate of oocyte loss.

6.5.1.5. The Mixed Function Oxidase System

Mixed-function oxidases (the *Ah* and cytochrome *P–450* system) are important defenses against toxic polycyclic compounds and are associated with differences in lifespan, both between and within species. First, a brief summary about the genetics and functions. This multienzyme system is complex and includes many loci. Two genes are on chromosome 17: that of the *Ah* receptor maps at some distance from the *H–2* complex, while the gene coding the steroid 21-hydroxylase maps within the *H–2* complex in mice (White et al., 1984) and in the *HLA* in humans (White et al., 1985). In both species, the $P-450_{C21}$ gene maps close to the *C4* (a complement protein; Figure 6.7) locus and with the same 5′-to-3′ orientation. The biological significance of the close association between the $P-450_{C21}$ genes with major histocompatibility complex is unknown (Klein, 1986); structural genes for other P–450 enzymes are on different chromosomes. Other *P–450* loci also influence *Ah* expression (Koizumi et al., 1986). The major components are the cytosolic *Ah* receptor with high affinity for aromatic compounds (e.g. 3-methyl-benzanthracene) and the P–450 enzymes, which are a heterogeneous group of mixed-function oxidases that catabolize a wide range of foreign and endogenous substances. One subset of these enzymes is inducible by phenobarbital, and catabolizes steroids and other endogenous compounds, for example steroid 21-hydroxylase, while another subset activates carcinogens.

The mixed function oxidases that activate carcinogens (e.g. aryl hydrocarbon hydroxylase) showed inverse correlations with the lifespans of different species (Figure 5.22). In comparisons of the congenic mice with different *H–2* haplotypes, the inducibility of aryl hydrocarbon hydroxylase was correlated with the lifespan (A < C3H < C57BL/10Sn), but not with lifespans of congenics within a group (e.g. B10.RIII mice lived longer than C57BL/10Sn but had the smallest induction of the enzyme [Koizumi et al., 1986]). Two genotypic variants at the *Ah* locus in mice (*Ah-b* and *Ah-h*), which code for *Ah* receptors with high and low affinity for 3-methylcholanthrene respectively, were used to select RI (recombinant inbred) lines. Lines with the higher affinity *Ah-b* receptor were associated in both sexes with greater fertility and longer lifespans (Nebert et al., 1984). Because the RI lines derived in this study had shorter lifespans than the norm for the parental strains, the generality of longevity promotion by the *Ah-b* allele is not established.

6.5.1.6. Ath–1 *and Atherosclerosis Susceptibility*

As will be described in Section 6.5.3.1, numerous important hereditary influences on lipid disorders in humans cause premature atherosclerosis. Mouse strains are well known to differ in their susceptibility to atherosclerosis when given diets rich in cholesterol and fats ("atherogenic diets"), but only recently have the genetic factors been identified (reviewed in Paigen et al., 1990). For example, young adults fed atherogenic diets for 3 months developed fatty lesions in the aortic wall that contain intracellular fat and foam cells in C57BL/6J mice to a vastly greater extent than in C3H/J mice; BALB/cJ were intermediate (Paigen et al., 1985). No C57BL/6J mice survived to 10 months on this diet. These extensive arterial lesions were not observed by 12 months in mice kept on a diet of moderately high fat content, the "breeder chow" that is preferred for milk production. As in humans, the size of aortic lesions in the strains correlated inversely with blood levels of HDL cholesterol over a threefold range. However, sex differences were less consistent, being influenced by diet and strain (Paigen, Holmes, Mitchell, and Albee, 1987).

The genetic basis for these strain differences is the presence of several genes, *Ath–1* and *Ath–2*, each with alleles that influence atherosclerosis susceptibility: *r* (resistant) and *s* (susceptible). The proteins coded by *Ath–1* and *Ath–2* are unknown. *Ath–1* was mapped to chromosome 1 and is close (5.9 centimorgans) to *Alp–2*, a gene encoding apolipoprotein A-II (apoA-II), a major structural component of HDL (Paigen, Mitchell, Reue, Morrow, Lusis, and LeBoeuf, 1987). Human chromosome 1 contains a homologous region with these and other shared loci (renin, peptidase), which suggests a supergene complex that may contain mutually adapted alleles. *Ath–2* has not been mapped but does not appear to be linked to *Ath–1* or the genes coding apolipoproteins A-I, A-II, or E (Paigen et al., 1989). Discovery of these distinct genes revealed an important feature: resistant alleles at either locus confer resistance to atherogenic diets through influences on HDL cholesterol. There are intriguing similarities between the human hereditary trait for inverse correlations of atherosclerosis with plasma HDL and certain effects of *Ath–1* and *Ath–2*. Nonetheless, as noted above, *Ath–1* and *Ath–2* have little influence on normal diets yielding high HDL cholesterol. Thus, mice give a model for gene-environment interactions in atherogenesis, as is observed for humans.

6.5.1.7.   Unusually Short Lived Strains and Their Diseases

All mouse genotypes with unusually short lifespans of 12–18 months have a premature onset of organ-specific degenerative diseases, which range from malignant to autoimmune diseases (Table 6.1). So far, there is no evidence for a general acceleration of senescence in short-lived mouse strains. However, several of these genotypes show reduced extents of UV-induced DNA repair, a parameter that shows some correlation with species lifespans (recall the relationships between UV-induced DNA repair and species lifespans in Chapter 5.4.3 and Figures 5.21, 5.22).

6.5.1.7.1.   The Virology of Graying

The inbred congenic B10.F strain is of particular interest because its chromosomal genes influence a leukemia virus that causes graying of their coat and a greatly shortened lifespan. These mice originated from crosses of the C57BL/10ScSn and F/St (Stimpfling). By comparison with its background strain (C57BL/10ScSn), the short-lived B10.F has many characteristics that superficially resemble accelerated senescence, including graying of the coat that frequently begins by 2–4 months (Morse et al., 1985; Popp, 1978, 1982; Popp et al., 1986). The mice grow normally and have a normal fecundity during their reproductive lifespan (Popp, 1978). By 10 months most mice have enlarged spleens and soon develop lymphomas. Severe weight loss before death is common, unlike the C57BL/6J mouse, in which only a minority shows senile weight loss (Finch et al., 1969). The UV-induced DNA repair of splenic lymphocytes becomes greatly impaired by 10 months (Licastro and Walford, 1985). B10.F mice have mean lifespans of 15 months or less, although some survive at least 22 months (Figure 6.8; Licastro and Walford, 1985; Popp, 1978).

Age-related graying is not a general trait of senescence in mice[1] but can arise from infections of the xenotropic murine leukemia virus (MuLV). Laboratory mice are known to have a variety of MuLV sequences integrated into their germ-line DNA, typically about fifty proviruses of various host range that differ widely among strains (e.g. Stoye and Coffin, 1988). Generally, mice carry multiple germ-line MuLV sequences that are characteristically xenotropic, that is, replicating on cells from species other than mice (Stoye and Coffin, 1988). The expression of xenotropic MuLV varies among strains and is influenced by the *Cxv–1* locus, which is *H–2* linked.

---

1. A prevalent but erroneous impression is that extensive graying is a general age-related trait in mice, just as in humans, monkeys, and domestic animals. In 20 years of experience with C57BL/6J mice obtained from the Jackson Laboratory, the extent of graying is slight and usually restricted to scattered white hairs that at the most give a "salt and pepper" impression for the oldest survivors of up to 44 months (Finch, 1973c; verified for other black-coated mice by David Harrison). At least some other strains are normally nongraying, including the most long lived *H-2* congenic B10.RIII (Morse et al., 1985). The expectation of graying as a general feature of senescence in mice may have originated from a widely known photograph of mice that became gray or white from treatments with irradiation, which were used in attempts to accelerate "rates of aging" (Curtis,1966; Chapter 8.2.1). However, no mention was made about the extent of graying in nonirradiated controls. This matter was a great concern to me as a graduate student when I was first starting my work in gerontology. My first colony of C57BL/6J mice failed to show the expected graying, and I feared that somehow the ages of my mice had become confused. In a phone conversation Curtis did not at first know if his nonirradiated controls became gray with age. He soon called back in chagrin after visiting his mouse colony to see for himself that they did not. This anecdote reminds us that we are all susceptible to hidden assumptions, and its telling is not meant to defame Curtis, a strong contributor whose other observations are better founded.

B10.F mice show atypical early graying and are also high producers of xenotropic MuLV (Yetter et al., 1983). The graying phenotype was vertically transmitted by cross-foster nursing the nongraying C57BL/10ScSn or B10.BR mice on the rapidly graying B10.F strain, from which they acquired episomal xenotropic MuLV (Morse et al., 1985). Cesarean-derived B10.F mice are nearly freed from the graying trait by cross-fostering on C57BL/10 mice, and their lifespan is increased in association with a lower incidence of lymphoma (Morse et al., 1985). Besides in the B10.F, graying is also found in other congenic strains, including the B10.P, whose *H–2* donor strain P is unrelated to F/St and which also produces abundant xenotropic MuLV. Moreover, dysfunctional hair follicles in the hairless mutation (*hr/hr*) are associated with an integrated MuLV (Stoye et al., 1988).

Mice infected with MuLV *in utero* often show a patterned graying that resembles the banding pattern of allophenic mice formed from strains with different coat colors (Mintz, 1967). Electron microscopy showed large numbers of C-type virus particles in the dermis and hair bulbs of graying mice, which may impair melanocyte functions (Morse et al., 1985). Correspondingly, the thymus and spleen of graying mice produced abundant infectious xenotropic MuLV soon after weaning. Nongraying strains did not. The nongraying strains or cesarean-derived, formerly graying strains produced little MuLV initially but did express it by 5 months. In general, the nongraying long-lived strains have a sporadic age-related increase of infectious ecotropic MuLV in the thymus and spleen (Rowe et al., 1970; Morse et al., 1985; Getz, 1985; Chapter 7).

Southern blot hybridization analysis of B10.F mice with a probe specific for the *env* locus of the ecotropic MuLV virus showed that liver DNA had a single copy of the integrated provirus (Morse et al., 1985; Popp et al., 1986); this is observed for most strains (Jenkins et al., 1982). Thus, the high virus-shedding strains do not appear to have additional ecotropic MuLV sequences in their germ-line DNA. However, B10.F mice had additional ecotropic MuLV sequences in DNA from the lymph nodes and spleen, which may result from the integration of recombinant MuLV (Popp et al., 1986), as discussed in Section 6.5.1.7.2.

The short lifespans of B10.F mice are associated with at least two loci. Backcrosses to C57BL/10 mice show that the $H–2^{n/n}$ genotype is necessary but not sufficient for short lifespan. Another locus that determines the levels of serum immunoglobulin A (IgA; Popp et al., 1986) also segregates independently of the IgA phenotype in these matings (Popp, 1979). Mice with high serum IgA and the *n/n* haplotype have normal lifespans. Lifelong reduction of serum IgA can also result from exposure of neonatal C57BL/10 mice to virus by cross-fostering, by splenocyte transfers, or by cell-free spleen supernatants (Popp et al., 1986, in prep.). Conversely, mice originating from B10.F blastocysts that were transferred to C57BL/10 surrogates had normal levels of IgA.

A working hypothesis is that maternally transmitted xenotropic MuLV can recombine with the endogenous proviral ecotropic MuLV sequence to produce a *polytropic* virus that can infect lymphoid and other cells. Rare integration near a c-oncogene of a permissive cell type then would produce a malignant clone. Depigmentation from dysfunctional melanocytes might only result from MuLV infection. The spread of the infection may be limited by antiviral immune responses under *H–2* control, as well as by the IgA levels (Morse et al., 1985; Popp et al., 1986). The high xenotropic viral load transmitted vertically in B10.F mice thus is hypothesized to cause a high risk of malignant lymphoma by interacting with the endogenous ecotropic MuLV provirus.

6.5.1.7.2.   AKR Mice and Senescence-accelerated Mice (SAM)

The AKR strain is also short-lived and dies from MuLV-related lymphoid leukemias like the B10.F and B10.P strains described above. But unlike those strains, the leukemias are not eliminated by cross-fostering. AKR mice produce abundant ecotropic virus at birth but mature and reproduce normally (Richardson, 1972). The AKR/J strain carries two or more integrated ecotropic MuLV provirus sequences on different chromosomes. Variable numbers of MuLV sequences among inbred lines of AKR/J mice suggest that germ-line reintegration of MuLv is an ongoing (evolving) process (Herr and Gilbert, 1982). Nearly all develop leukemia and die by 8–12 months (Furth, 1946; Kohn and Novak, 1973). This long lag suggests that the activation of provirus gene expression is a necessary but insufficient event in leukemogenesis. The thymus appears to be a major locus for the spread of MuLV during early life, since neonatal thymectomy prevents the later formation of leukemia (Furth, 1946). Extensive individual variations were found in the locations of non-germ-line-integrated MuLV proviral sequences by DNA restriction site mapping (Quint et al., 1981). These postnatally acquired sequences are recombinant and polytropic, unlike the native ecotropic virus, which implies that the tumors had monoclonal origins (Getz, 1985). As proposed for the graying B10.F mice, the high constitutive viral load from birth onwards could favor the rare formation of recombinant viruses as a necessary step in leukemogenesis (Getz, 1985).

The lifespan of neonatally thymectomized AKR mice was increased by 50% towards the normal range, and some mice survived to 24 months. At necropsy, such mice manifested degenerative conditions that were never seen at younger ages during the usual life course in this genotype, including bone tumors (osteomata; Furth, 1946). The later occurrence of bone tumors, a rare disease of mice, illustrates the important point that there may be very late onset degenerative conditions that are not manifested during the usual lifespan in other genotypes.

Another unusual aspect of AKR/J is age-related hearing loss, which impairs high-frequency sensitivity soon after puberty, or about 6 months before impairments in other strains, including the C57BL/6J and CBA/J mice (Henry, 1982). Humans characteristically lose sensitivity to high-frequency tones at later ages (Chapter 3.6.11).

A further derivative of AKR/J mice at Kyoto University yielded sublines from brother-sister mating since 1968 that have a strikingly different set of early onset disorders (Takeda et al., 1981). These lines, designated SAM for "senescence-accelerated mice," are being thoroughly studied through the efforts of Professor Takeda and his department. After normal growth and maturation, several SAM lines die by 10 months with gross lesions associated with amyloid deposits. These amyloid deposits have a different amino acid sequence from the β-amyloid protein found in the brains of people with Alzheimer's disease and Down's syndrome, and do not, in fact, occur in the brains of SAM mice (Takeda et al., 1981).

By 6–8 months, most SAM mice have a decrepit hunched-up appearance with myriad pathologic lesions including systemic amyloid deposits, particularly in the liver, cataracts, and nonthymic lymphoma (Takeda et al., 1981; Higuchi et al., 1983, 1984; Hosokawa et al., 1984). Immune functions decline soon after maturation, including T helper cell activities (Hosokawa et al., 1987). In striking contrast to the high incidence of thymomas (thymic leukemia) and little amyloid in the original AKR/J stock (Ebbesen, 1974b; Russell, 1966), the Kyoto lines have little of this particular type of leukemia. Other changes are increased numbers of granular carbohydrate-containing depots (PAS positive) in the hippocampus (Akiyama et al., 1986) and increased monoamine oxidase-B (Nomura et al., 1989). Behavioral changes are limited, for example, to impairments in the passive avoidance test, while retention tests and shock sensitivity showed no change (Yagi et al., 1988). Control lines lived 13 months, slightly longer than the original AKR/J strain, and had a lower incidence of these same conditions (Takeda et al., 1981; Hosokawa et al., 1984). As will be discussed for SAM and other early onset lesions (Chapter 10.3.2), diet restriction substantially retards this syndrome (Kohno et al., 1985).

Another molecular lesion in SAM lines is increased levels of a high-density serum lipoprotein (apoSAS$_{SAM}$). ApoSAS$_{SAM}$ cross-reacts with antibodies to murine senile amyloid fibril protein (Higuchi et al., 1984) and is homologous to apoA-II of humans and monkeys. However, there is a codon change to glutamine (CAG) from the proline (CCA) codon found in the apoA-II coding region that occurs in senescence-resistant lines and also in a random-bred line unrelated to AKR (Kunisada et al., 1986; Yonezu et al., 1987). This double substitution should cause a major structural change and could account for the deposition of amyloid fibrils in the liver. These major, stable genotypic differences between SAM mice and their alleged source, the well-characterized and inbred AKR/J stock, could be due to many factors, including new mutations or rearrangements such as viral gene shuffling. Most likely, genes were introduced by rogue mice. While SAM mice clearly have early onset disorders with ramifying pathophysiological consequences, these amyloid-associated disorders are not usual in the senescence of other long-lived genotypes but are still useful as models of early onset diseases that can be manipulated by diet (Chapter 10.3).

6.5.1.7.3.   Autoimmune and Kidney Diseases

Several strains are widely studied for their early onset autoimmune and kidney diseases: the NZB, NZW (New Zealand black and white; Gardner et al., 1977; Gottesman and Walford, 1982), and MRL/l strains (Izui et al., 1979; Kubo et al., 1984), which develop these conditions soon after birth and have short lifespans rarely exceeding 15 months. The T-lymphocyte–mediated immune functions of NZB and NZW mice are initially normal but drop sharply by 6 months (Fernandes et al., 1976). Autoantibodies to erythrocytes cause a hemolytic anemia, whereas other autoantibodies are thymocytotoxic (Shirai and Mellors, 1971) or react with cell nuclei, DNA, and RNA (Steinberg et al., 1969). The anti–nucleic acid autoantibodies specific for both single- and double-stranded DNA make these strains useful as models for the human autoimmune disease of systemic lupus erythematosus (Dubois et al., 1966). In both the human and mouse autoimmune diseases, there is less UV-induced DNA repair *in vitro* (Paffenholz, 1978; Hall et al., 1981).

The kidneys of NZB mice degenerate progressively, with glomerular hyalinization and the deposition of antinuclear immunoglobulins (IgG and IgM) in the glomerular capillaries, in most colonies (Friend et al., 1978; Hurd et al., 1981). This "autoimmune complex–glomerulonephritis" can be monitored by proteinuria, and ultimately causes kidney failure with lethal consequences. However, early kidney failure and antinuclear antibodies are not found in all colonies of NZB and NZW mice (Knight and Adams, 1978). Similarly, lymphoid tumors in association with transmissible, murine type C leukemia oncoviruses vary widely among colonies of NZB mice (Knight and Adams, 1978; Simpson, 1976) and among wild populations of house mice (*Mus musculus;* Gardner et al., 1973). In some colonies the development of kidney disease is predictable from serum levels of the endogenous retroviral envelope protein, gp70 (Nakai et al., 1980). An obligatory link between retrovirus infections and the autoimmune disorders has been ruled out (Knight and Adams, 1978). These variable virus infections suggest environmental influences in the phenotypes of senescence but could also be due to genetic change during inbreeding, as discussed above for AKR mice.

The (NZB × NZW)F$_1$ hybrids show the most severe and early autoimmune and kidney disorders. Backcrosses suggest that the *H–2* complex may be involved, since a closely linked locus on chromosome 17 influences the severity of kidney disease (Knight and Adams, 1978) and may be an immune-response (Ir) gene (Walford, 1987). NZB and NZW mice have different *H–2* haplotypes (*d* and *z,* respectively). Genetic studies indicate influences from loci distant from the *H–2* complex on chromosome 17, a locus on chromosome 4, and yet other loci that independently influence the kidney disease, the immune complexes with gp70 (Nakai et al., 1980), and antibodies to single- and double-stranded DNA (Yoshida et al., 1981). The expression of these genetic influences on the heterozygote F$_1$ hybrid shows that the alleles involved are dominant or codominant.

Normal lifespans in NZB and NZW mice of up to 30 months can be achieved and their kidney disease can be virtually prevented by caloric dietary restriction (Gardner et al., 1977; Friend et al., 1978), by restriction of essential fatty acids (Hurd et al., 1981), or by treatment with the dehydroepiandrosterone (DHEA, an adrenal steroid; Tannen and Schwartz, 1982; Chapter 10.3.2.6). The greatest benefit occurred with diet restriction from weaning. Even so, the lifespan was still prolonged and the kidneys were still protected if diet restriction was begun at 4 months, which is after the autoantibodies are greatly elevated (Friend et al., 1978). Diet restriction, however, did not much influence the levels of antinuclear autoantibodies (Hurd et al., 1981) or the production of xenotropic MuLV (Gardner et al., 1977). With prolonged lifespans, mice over 20 months then developed lymphomas, a lesion unknown during the short lifespans of *ad libitum* fed mice (Gardner et al., 1977).

Another short-lived mutant is the (7- to 9-month) *kd/kd* genotype (Lyon and Hulse, 1971), a mutant that soon after birth develops proteinuria and a type of progressive kidney disease with glomerulosclerosis that is considered to be closer to the human condition of nephronophthisis than to the kidney diseases of most laboratory rodents. Unlike the New Zealand strains, *kd/kd* mice have no autoimmune involvement. But like them, diet restriction prevents the kidney disease and doubles the lifespan (Fernandes, Yunis, et al., 1978). Within 2 months of restoring *ad libitum* feeding, the diet-restricted *kd/kd* mice developed kidney disease and died.

The MRL/l mouse strain shows a different face of early onset autoimmune changes (Hang et al., 1982). Like the New Zealand mice, MRL/l mice develop autoantibodies and immune complex–mediated glomerulonephritis. However, they show an additional trait that makes them a model for rheumatoid arthritis. By 6 months, most females have inflamed joints and synovia, with arthritic erosions, and deposits of immune complexes including IgM/RF (rheumatoid factor). The multiple manifestations of early onset autoimmune disorders in these genotypes suggest that minor genetic changes, possibly in the MHC alleles, suffice to channel a very common genetic predisposition in different directions.

As a closing example, the long-lived C57BL/6J and BALB/cJ strains also show sporadic (0–25% of mice) thymocytotoxic autoantibodies after 12 months (Shirai and Mellors, 1971). Nuclear autoantibodies also increase with age in many mouse strains (Goidl et al., 1980; Teague et al., 1968; Teague et al., 1970; Walford, 1969; Hausman and Weksler, 1985) and in humans (Chapter 3.6.6). Moreover, some degree of glomerular hyalinization and kidney disease is very common by the average lifespan in most laboratory rodents (Cotchin and Roe, 1967; Linder et al., 1972), unless the mice are maintained on restricted or special diets (Yu et al., 1982). The NZB and NZW strains thus may be considered as models with early onset lesions, which, like the SAM mice, are useful for studying manipulations by diet and other interventions that are also effective in long-lived laboratory rodent genotypes (Chapter 10.3).

6.5.1.7.4.   Obesity, Diabetes, Breast Tumors, and DHEA

Obesity with shortened lifespan results from the mutations obese (*ob*) and diabetes (*db*) when carried on the C57BL/KsJ strain (Coleman, 1978). These mutations have milder effects on related strains, so that *ob/ob* C57BL/6J mice, for example, have only transient compensated diabetes and normal pancreatic insulin content, but are still very obese (Leiter, Coleman, Eisenstein, and Strack, 1981; Leiter, Coleman, and Hummel, 1981; Coleman et al., 1984). There are also interactions of the *H–2* haplotype and sex with the effects of *db,* an autosomal recessive on a different location, chromosome 4, which causes premature death from pancreatic β-cell necrosis at 5–8 months in some haplotypes; sex also influenced the progression of the disease (Leiter, Coleman, and Hummel, 1981). Among the aberrant aspects of gene regulation in obesity is a great reduction of adipsin, a serine protease, in fat cells of *ob/ob* and *db/db* strains (Flier et al., 1987). Transgenic studies with the adipsin gene promoter introduced to obese and nonobese genotypes show that trans-acting factors decrease transcription of a reporter gene in adipose cells of obese mice, but also in other tissues as well (Platt et al., 1989). This indicates that the numerous cellular dysfunctions from hereditary obesity can arise from aberrant gene regulation in a variety of cells, as well as those storing fat.

The short lifespan (15 months) of *ob/ob* mice can be increased to the normal 27 months by dietary restriction (Lane and Dickie, 1958). Moreover, diet restricted *ob/ob* mice stay fat and remain immune deficient, despite increased lifespan (Harrison et al., 1984; Harrison and Archer, 1987). The steroid DHEA (Section 6.5.1.7.3 and Chapter 10.3.2.6) also reduces body weight and improves the diabetes of *ob/ob* mice when given in the diet, but with little effect on food consumption (Coleman et al., 1984). The rationale for trying DHEA came from its ability to virtually suppress breast tumors in the short-lived, moderately obese C3H(A^{vy}/a) strain (Schwartz, 1979). As above, feeding of DHEA reduced body weight without major effects on influencing food intake and increased lifespan to the normal range. Low plasma DHEA is implicated as a risk factor for breast cancer and can slow other age-related diseases in rodents (Chapter 10.3.2.6).

6.5.1.7.5.   Spinal Cord Motor Neuron Disease: A Model for ALS?

There are two occurrences of highly select degeneration of motor neurons, both under genetic control, which may be models for amyotrophic lateral sclerosis (ALS). ALS, a disease of middle age in humans, causes a similar degeneration of spinal cord neurons, but with no degeneration of the brain (Hirano and Iwata, 1979). So far, there is no rodent model for spontaneous, that is, age-related Alzheimer's or Parkinson's disease that causes the same type of cellular atrophy and degeneration.

A dominant mutation causing progressive motor neuron degeneration (*Mnd*) was found in the B6.KB2/Rn substrain of C57Bl/6, a local derivative of C57Bl/6J mice (Messer and Flaherty, 1986; Messer et al., 1987). The B6.KB2 strain is congenic with C57BL/6 mice, but differs in several *t* and *H–2* loci: $+^T$ $H–2K^b$ $H–2D^b$ Q^b TL^k; the chromosomal location of the *Mnd* mutant is unknown. By 8 months, both males and females carrying *Mnd* have a clumsiness and hind-limb weakness which progresses to a spastic paralysis. Weight loss ensues, and most die by 9–14 months, while the parental stocks typically live at least 24–30 months. The basis is a progressive degeneration of anterior horn motor neurons in the spinal cord, but also in cranial nerves. Large neurons are rare, from atrophy or death. The remaining neurons in severely affected mice show loss of Nissl bodies (endoplasmic reticulum), cell body swelling, loss of nuclear staining with thionin, and lipofuscin accumulation. These degenerative changes are evidently slower than apparently similar degenerative changes in peripheral neurons that are deprived of nerve growth factor (Chapter 7.7). Neurons of *Mnd* carriers appear normal at 3 months.

A natural occurrence of similar lesions results from indigenous ecotropic MuLV in feral populations of *Mus musculus domesticus* (Gardner, 1985). Wild mice in certain areas of southern California have persistent infections with complete immunologic tolerance, but with mid-life onset of hind-limb paralysis that is caused by damage to the motor neurons of the lower spinal cord. The lesion is highly selective for these particular motor neurons and involves vacuolar (spongy) neurodegeneration with reactive gliosis but not other inflammatory cell types, unlike the frequent inflammatory or leukocytic infiltrates around nerve fibers in AKR mice. Neuron death is associated with viral replication in the perikaryon, but could also result from virally mediated indirect effects.

Susceptibility to the indigenous, vertically transmitted virus is controlled by a dominant cellular gene, *Akvr–1^R/Fv–4^R*, that is distinct from another well-known gene that controls MuLV infections (the *Fv–1* system) in laboratory mice (Gardner et al., 1980). The viral infection is blocked at the level of cell-membrane receptors, by expression of a virus-related envelope protein (gp70) that inhibits binding of the ecotropic virus to cell membranes (Rasheed and Gardner, 1983). Resistant mice have a defective endogenous retrovirus (Dandekar et al., 1987). This provirus encodes a glycoprotein, related to the MuLV envelope, that confers resistance by competing with MuLV at its cellular receptors. The relation of these lesions to those of the *Mnd* mouse mutation has not been explored. The protection against effects of viral infection by expression of related viral sequences epitomize the positive and negative effects of retroviruses, through which numerous traits may become co-adapted.

6.5.2. Rats and Other Short-lived Mammals

Genetic effects on the patterns of senescence or lifespan are not established in many mammals. Some breeds of domestic dogs (Comfort, 1960b, 1979) are short-lived, but no systematic genetic studies have been done because of inordinate time and cost. Laboratory rats (*Rattus norvegicus*) all have similar lifespans (mean of 24–32 months) if given good care: Long-Evans, Fischer 344 (F344), Sprague-Dawley, and Wistar (*Mammalian Models*, 1981). As in mice, rats commonly have glomerulosclerosis by the average lifespan.

Kidney dysfunctions are greatly modified by diet (Chapter 10.3.2.2). Although few laboratory strains are inbred to the criteria of mouse strains (F344, ACI, and brown Norway rats are exceptions), most noninbred lines and stocks have a characteristic incidence of age-related degenerative diseases. For example, testicular interstitial cell tumors are common by mid-life in F344 rats, whereas Sprague-Dawley rats have more abnormal growths of the adrenal medulla and pancreas (Anver et al., 1982; Cohen et al., 1978). Other rat lines have different susceptibility to arteriosclerosis in mesenteric vessels (Opie et al., 1970).

Although no ultra-short-lived genotypes like the B10.F or AKR mice have been established, some varieties with early onset diseases have subnormal lifespans. Lines of Sprague-Dawley rats that were selectively inbred for high blood pressure had shorter lifespans and more intense kidney degeneration than those with low blood pressure (Rapp, 1973). It is frustrating that the size of the rat, while favoring studies of physiology, leads to prohibitive costs in keeping rats in the large numbers needed for genetic studies. Nonetheless, the genetics of rats is developing, and more inbred strains should be available.

Laboratory rabbits (*Oryctolagus cuniculus*) also offer interesting models. There are tantalizing descriptions of a hereditary progeroid syndrome in rabbits (owny-rusty-dwarf, *DRW*), which developed muscular wasting, skin lesions, and infertility after 1–2 years (Pearce and Brown, 1960a, 1960b). Unfortunately, the investigators died before full pathological findings were reported, and the colony was also lost.

More recently, the WHHL strain (Watanabe heritable hyperlipidemic) rabbit was inbred to homozygosity for its faulty low-density-lipoprotein (LDL) receptors and chronically elevated blood cholesterol and lipoproteins (Brown and Goldstein, 1986; Kita et al., 1987). These abnormalities are implicated in the development of fatty streaks in the aorta soon after birth and of other precursors of severe atherosclerosis (Rosenfeld et al., 1987). The WHHL rabbit is widely used as a model for the common human disease, familial hypercholesterolemia, because of its defective LDL receptor. Moreover, WHHL rabbits respond to the drug probucol in retarding atherosclerosis, as do humans with familial hypercholesterolemia (Kita et al., 1987). Probucol is of interest to the theory that free radicals and lipid peroxides are important to atherosclerosis (Harman, 1968, 1981), since LDL from rabbits treated with this antioxidant were more protected from oxidation and since peroxidized lipids occur in atherosclerotic lesions (Kita et al., 1987). Oxidized LDL may be a molecular pathogen of atherogenesis, by favoring the formation of foam cells in developing lesions; foam cells are macrophages with accumulated LDL.

Another strong effect seen in domestic animals is the increased lifespan of male castrates (Hamilton, 1965; Hamilton et al., 1969; Comfort, 1979; Smith, 1989). In cats and many other species, sex effects are associated with testicular hormones determined by the Y chromosome, since postnatal castration of males increases their lifespans towards that of females (Hamilton et al., 1969). Effects of castration are partly associated with decreased fighting and other dangerous androgen-dependent behavior.

6.5.3. Humans

6.5.3.1. Familial Influences on Lifespan and Senescent Changes

Little is known about the genetics of the human lifespan. The maximum reported lifespan does not vary distinctively among populations, and the alleged supracentenarians of the Caucasus and the Andes could not be documented with adequate records of birth date (Finch, 1976b; Leaf, 1990). The recording of fingerprints and footprints on birth records by many countries during the last 100 years gives a future basis for validating extraordinary longevities (Smith, 1991). Meanwhile the extent of amino acid racemization in tooth dentine or eye lens proteins could be used postmortem to evaluate claims of great longevity (Chapter 7.5.4). Provisionally, it seems likely that some humans have survived beyond 110 years.

Familial trends in longevity are hard to interpret because of the huge impact of environmental variables and the many common causes of premature death. Nonetheless, the available studies give consistent indications. A study of 7,000 offspring from parents who lived for 90 years or more (Abbott et al., 1974; Abbott et al., 1978) showed that longer-lived parents had longer-lived children, while parents who lived at least 81 years had children that lived 6 years longer than children whose parents died earlier, by 60. Similarly, a prospective study of about 25,000 U.S. adults showed that mortality from coronary heart disease was about 50% less in those with very long lived parents or grandparents (Hammond and Garfinkel, 1971).

Twins confirm the trend: monozygotes died within 3 years of each other, whereas the lifespans of same-sex dizygotes differed by more than 6 years (Kallman and Sander, 1949; Jarvik et al., 1960, 1980). Monozygous twins aged 77–88 years generally had well-preserved intellects and had more similar test results than dizygous twins (Jarvik et al., 1972). In male twin veterans of U.S. military service that died early (51–61 years) from disease (i.e. not accidents or trauma), monozygotes had lifespans that were closer than dizygotes; no difference was found for deaths due to cancer alone in this group (Hrubec and Neel, 1982).

These trends do not prove a genetic basis for greater longevity but suggest that specific genes may set limits on lifespan through their influence on age-related diseases, particularly atherosclerosis and cancer. As may be recalled from Chapter 3.6.2, all human populations show marked age-related atherosclerosis, to a general extent such that only rare individuals escape degenerative changes in one or more vital vessels (Figures 3.28 and 3.29). In contrast, the incidence of cancers and other abnormal growths vary widely between populations and sexes (Figures 3.17 and 3.18). These well-established phenomena suggest that atherogenesis is a canonical or normative-phenotype of senescence in most genotypes and environments. There is some information on genetic influences in human atherogenesis.

Some genotypes may enhance longevity by reducing the risk of myocardial infarction and ischemic heart disease from atherosclerosis of coronary vessels (coronary artery disease). A possible example is the rare heritable hypo-β- or hyper-α-lipoproteinemias that are associated with lifespans that are 5 to 10 years longer than the general population or their spouses (Glueck et al., 1976; Glueck et al., 1977; Saito, 1984). Both genotypes had two- to threefold less LDL (the major carrier of cholesterol) than the general population. This is also reflected in the ratio of LDL to HDL. HDL are involved in reverse cholesterol transport, whereby cholesterol is removed from peripheral tissues and returned to the liver, where it may be recycled. Each step of the transport of cholesterol and triglycerides involves specific carrier proteins and cell receptors (Bachorik and Kwiterovich, 1988; Grundy, 1986; Sedlis et al., 1986; Brown et al., 1983). As described next, important heritable risk factors for heart disease are associated with molecular defects in these lipid transport proteins and receptors.

A formidable literature shows strong links between atherosclerotic heart disease causing premature death and increases in the ratio of LDL to HDL: loss or relative deficiencies of HDL are a major risk factor in heart disease during middle age, whereas elevated HDL protects (Castelli et al., 1975; Inkeles and Eisenburg, 1981; Dawber, 1980; Bierman, 1985; Ross, 1986; Sing et al., 1985). Familial cholesterolemia, an autosomal dominant causing a high rate of premature death from coronary artery disease, is associated with elevated total and LDL cholesterol because of defective LDL receptors on target cells, the result of mutations at various locations in the LDL gene (Brown and Goldstein, 1986). Heterozygotes with this dominant mutation are 0.2% of the general population and represent 5% of patients who had a heart attack before 60 years. Another autosomal dominant is familial combined hyperlipidemia, which increases LDL through overproduction of VLDL that transport triglycerides and are converted to LDL (Chait et al., 1980; Janus et al., 1980).

Many other genetic variants influence lipid transport and promote atherogenesis. Familial premature atherosclerosis is associated with a defective apolipoprotein *apo AI* gene containing a 6.5-kilobase DNA insert in its coding region that codes for the major protein of HDL and causes deficiencies of apo^a- (Karathanasis et al., 1983). The apo^b variants influence LDL levels through the characteristics of apolipoprotein B, the main protein in LDL through which it interacts with the LDL receptor (Gavish et al., 1989). There is also a positive association with plasma Lp(a) concentrations and premature myocardial infarction and stroke (Utermann, 1989). Lp(a) is a complex formed from LDL and apolipoprotein(a), a glycoprotein with homologies to plasminogen. The plasma concentrations of Lp(a) vary a thousandfold among individuals, but the mechanism relating Lp(a) levels to vascular accidents is obscure. The *apo(a)* gene is unusual because of its recent evolutionary origin: it occurs in Old World monkeys, but not in New World monkeys (Utermann, 1989).

Together, these different loci indicate a polygenic set of determinants. Several groups (Templeton et al., 1987; Paulweber et al., 1988) are investigating the possibilities of haplotypes with different risk factors in loci on chromosome 11, which contains a tandem cluster of genes for apolipoproteins *apo AI, CIII,* and *AIV* (Karathanasis, 1985). We may anticipate that most of these haplotypes should be subject to numerous influences from diet and other physiological influences.

Another type of genetic influence on lifespan through heart disease is suggested by anatomical differences in the coronary arteries. Both Finland and Israel have high-risk subpopulations, in which children who died from accidents or infections had coronary arteries with thicker walls and more musculoelastic tissue (Vlodaver et al., 1969; Pesonen et al., 1975; Pesonen et al., 1982; Neufeld and Goldbourt, 1983). Neufeld and Goldbourt give other examples of vessel anatomy that may influence the outcome of atherogenesis. These findings suggest possible genetic influences during heart organogenesis that may underlie different individual responses to diet and stress and that predispose to later onset affecting heart disease after maturation. However, environmental influences could also be transmitted transplacentally.

Alzheimer's disease is the last condition to be considered here that influences mental health in later years, but that generally does not shorten the lifespan. As described in Chapter 3.6.10, particular pathways and groups of large neurons in the forebrain degenerate during Alzheimer's disease. The boundaries between normal and pathological age-related changes are sometimes elusive, particularly at later ages (Figure 3.39; Table 3.3).

The genetics of familial Alzheimer's disease is proving to be complex, although most studies indicate an autosomal dominant in both the early and later onset disease and a high risk of first-degree relatives (Farrer, Myers, et al., 1990; Farrer, O'Sullivan, Cupples, Growdon, and Myers, 1989; Heyman et al., 1983; Breitner et al., 1988; Mohs et al., 1987; Bird et al., 1988; Bird, Schellenberg, Wijsman, and Martin, 1989; Bird, Sumi, et al., 1989). However, there is no consensus that a particular chromosome assignment can yet be made in general. The original report of St. George-Hyslop et al. (1987) showed that several early onset families have a strong association to the middle of the long arm of chromosome 21 by RFLPs (Figure 6.12). This conclusion has not been extended with these and other RFLP markers to other families, including descendants of Volga Germans in North America (Bird et al., 1988; Bird, Sumi, et al., 1989). Linkage homogeneity has not been established for other families in which a dominant Alzheimer's gene is indicated (Pericak-Vance et al., 1988; Schellenberg et al., 1988; Schellenberg, Moore, et al., 1990; Schellenberg, Pericak-Vance, et al., 1990; Schellenberg, Wijsman, et al., 1990). Further analysis of RFLPs nonetheless has strengthened the chromosome 21 assignment for the first few of early onset families (Tanzi et al., 1989). The precision of localization is hampered by the scarcity of large, multigenerational pedigrees and by a 10–20% inaccuracy of clinical diagnosis. The absence of these problems made the chromosomal assignments of cystic fibrosis and Huntington's disease relatively easy. Whatever the number of genes and their chromosomal location may be, Alzheimer's disease appears to be more heterogeneous. Environmental factors have not been ruled out that could alter the threshold for penetrance of some genotypes. The high prevalence of Alzheimer's disease, about 2 million afflicted or 0.8% in the U.S., implies that the dominant genes may be among the most common inherited disorders. By comparison, as discussed by Farrer et al. (1990), the inherited disorder with the highest frequency is familial hypercholesterolemia, with a gene frequency of 0.1% (McKusick, 1986). Linkage, gene frequency, and heterogeneity are insightfully discussed by Farrer et al. (1990) and Tanzi et al. (1989).[2]

Whatever the ultimate resolution of the genetic basis, it seems likely from progressive accumulation of neurofibrillary tangles and neuritic plaques in virtually all humans to some degree at later ages, that the phenotype encoded depends on some age-dependent brain changes. From this perspective, the dominant genes for Alzheimer's disease might be said to accelerate particular features of age-related neurodegeneration (Farrer, Myers, et al., 1990; Mann et al., 1985). The fact remains, however, that nearly adjacent neurons remain intact throughout Alzheimer's disease; for example, the granule neurons and the CA_3 pyramidal neurons of the hippocampus are generally spared neurofibrillary tangles, while the neighboring CA_1 neurons are badly affected (Chapter 3.6.10). Thus, if the Alzheimer's disease gene is expressed generally in neurons, its product is apparently not sufficient to cause damage without some further change or lesion. Multistep mechanisms may be as important in neurodegenerative diseases as they are in oncogenesis.

---

2. Research on this problem is made difficult by the late age of onset, which removes possible carriers from analysis through death by other causes, and by the scarcity of thorough postmortem analysis in earlier generations of families. Thus, other age-related dementias from vascular or viral diseases may be easily misdiagnosed as Alzheimer's disease.

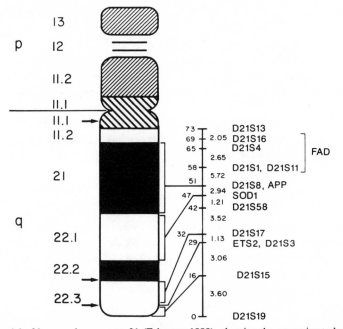

Fig. 6.12. A model of human chromosome 21 (February 1990), showing the approximate location of genes discussed in the text: *FAD*, familial Alzheimer's disease, as indicated from three family pedigrees showing an autosomal dominant pattern for early onset dementia; *APP*, amyloid precursor protein from which is derived the β-amyloid cleavage fragment that accumulates in cerebral blood vessels and neuritic plaques of Alzheimer's disease, in trisomy 21 Down's at much earlier ages, and to lesser extent than in the nondemented general population; *SOD1*, soluble superoxide dismutase; *ETS2*, homologue of avian erythroblastosis virus E26 *v-ets* homologue 2. The other markers (e.g. *D21S13*) are RFLPs (restriction-fragment-length polymorphisms) that are used to study linkage. The map distances are strongly affected in some regions by gender. See Tanzi et al., 1988, 1989. The figure, not previously published, was supplied by Rudolph Tanzi.

6.5.3.2. HLA-*Associated Diseases of Senescence*

Some age-related diseases are associated with different haplotypes or combinations of alleles in the major histocompatibility complex of humans, the *HLA* locus on chromosome 6, as found for the *H–2* complex of mice (Tiwari and Terasaki, 1985; Brown, 1979; Klein, 1986). *HLA*-linked joint diseases are of particular interest in studies of genetic influences on diseases of senescence, and include rheumatoid arthritis, juvenile arthritis, and anky- losing spondylitis; these diseases are not necessarily life-shortening in some socioeco- nomic subgroups. Rheumatoid arthritis is a chronic inflammation involving the deposition of immunoglobulin complexes in the connective tissues of joints; the immunoglobulin complexes then stimulate the infiltration of lymphocytes. A vicious cycle of inflammatory responses can cause stiffness (ankylosis) or even destruction of the joint. The precipitating causes are still unknown. Some extent of rheumatoid disease increases steadily with age and is extremely common in both sexes by 70 years (Chapter 3.6.5). Ankylosing spon- dylitis has a strong association with the *HLA-B27* haplotype (about eightyfold above con- trols), while the risk of juvenile and adult onset rheumatoid arthritis are about fourfold greater in carriers of their alleles, *HLA-DR4* and *HLA-DRw4*, respectively, as found in many different populations (Svejgaard et al., 1983; Stastny, 1978). An endocrine influence is indicated for *HLA-B15* (*Bw62*), in association with below-average blood testosterone in healthy and rheumatoid men (Ollier et al., 1989). Viral infections may be influenced by *HLA* haplotypes. For example, the Epstein-Barr virus has a coat protein (gp110) with an amino acid sequence that is shared by T-cell epitopes of protein coded by the *HLA-DRw4* allele (Roudier et al., 1989); because 70% of patients with rheumatoid arthritis are seropositive for HLA-DR4, infections with Epstein-Barr virus could trigger autoimmune reactions in those with this haplotype, leading to rheumatoid arthritis in a subset of these. No strong association with *HLA* haplotypes has been established for Alzheimer's disease or great longevity (Walford, 1987).

6.5.3.3. Genotypes with Short Lifespans and Early Onset Diseases: The Progeroid Syndromes

On the shorter extreme, a variety of heritable disorders are associated with short lifespans and premature senescence and are known as the progeroid syndromes. None of these causes a uniform acceleration or intensification of the usual spectrum of age-related degenerative changes. Each of them, as described below and summarized in Tables 6.2 and 6.3, presents a different caricature of senescence.

The *Hutchinson-Guilford syndrome* is the earliest onset progeria, with the development during childhood of balding, loss of subcutaneous fat, and wrinkling of the skin, giving facial features sometimes seen in the very elderly (Brown et al., 1984; DeBusk, 1972; Martin, 1978). This very rare condition (10^{-6} to 10^{-7} births/year) is thought to be a rare sporadic autosomal dominant, rather than recessive as first classified (Brown et al., 1984; McKusick, 1986). An intense progressive atherosclerosis during childhood usually results in cardiovascular-associated death at a median age of 12 years; only a few survive to early adult ages. Bone abnormalities include diminished growth of the lower jaw and a generalized osteoporosis that differs from the common age-related osteoporosis because it does not cause loss of bone mineral in the femur or other long bones (Reichel et al., 1971). Other differences from the usual age-related changes of humans include the low incidence of degenerative osteoarthritis, diabetes, cataracts, presbyopia, or abnormal growths. Intelligence and memory are normal; there is no loss of neurons, and no senile plaques, neurofibrillary tangles, aging pigment, or other signs for accelerated brain senescence (Spence and Herman, 1973). A biochemical lesion of possible importance to blood vessels and connective tissues is the excessive production of hyaluronic acid, as observed in urine (Takunaga et al., 1978) and in diploid fibroblast cultures (Zebrower et al., 1985); excess hyaluronic acid can inhibit vascular development (Brown et al., 1984). Diploid fibroblast cultures show rapid clonal senescence (Martin et al., 1970). The basic causes remain unknown.

Werner's syndrome begins after puberty and is considered an adult form of progeria (Brown et al., 1984; Martin, 1978, 1982; Epstein et al., 1966; Salk, 1982; Imura et al., 1985). Most die before 50 years from atherosclerosis-related heart disease. The syndrome is clearly an autosomal recessive, with an incidence of homozygotes within the range of 10^{-5} to 10^{-6} in both sexes (Epstein et al., 1985). Major degenerative changes include premature graying, hypogonadism, skeletal muscle atrophy, intense atherosclerosis, calcification of heart valves and soft tissues, and neoplasms in connective tissue and elsewhere of types that are unusual in adults. In contrast, traits of senescence that are not

Table 6.2 Genetic Progeroid Syndromes of Humans

Syndrome	Genotype	Lifespan, years	Characteristics
Ataxia telangiectasia	rare autosomal recessive	40	progressive cerebeller ataxia; nonconstitutional chromosomal aberration and deficient DNA repair
Cockayne	very rare autosomal recessive	50	demyelination; lipofuscin
Down's	trisomy 21	50–70	leukemia; mental deficiency; early onset Alzheimer-type neuropathology; by 20–40 years, amyloid β-protein in cerebral blood vessels and in neuritic plaques
Huntington's	autosomal dominant	40–70	spontaneous chorea and other movement disorders; degeneration of neostriatum; substantial diabetes mellitus before onset of neurologic symptoms; wrinkled skin and weight loss common
Hutchinson-Guilford	very rare autosomal recessive	10–20	intense atherosclerosis; cancer
Seip	very rare recessive	short	generalized lipodystrophy; diabetes mellitus
Werner's	very rare autosomal recessive	50–60	intense atherosclerosis; cancer

Sources: McKusick, 1986; Martin, 1978; Moosey, 1967; Seip, 1971. Additional sources in Section 6.5.3.3.

Table 6.3 Progerias and Age-related Changes in Humans

	Normal	Hutchinson-Guilford Syndrome	Werner's Syndrome
Lifespan (years)	75	10–20	50–60
Major degenerative conditions			
Atherosclerosis	common	severe	severe
Cancer	common	none	common, but include unusual sarcomas and connective-tissue cancers
Diabetes	variable	rare	rare
Neurofibrillary tangles and senile plaques in brain	common	none	none
Benign prostatic hypertrophy	common	none	
Hypertension	common	rare	rare
Senile cataracts	common	none	rare
Soft-tissue calcification	rare	common	rare
Brain lipofuscin	common	none	normal for age
Senile osteoarthritis	common	rare	rare
Intelligence		normal	normal

Source: Adapted from Martin, 1978.

found in Werner's include the low incidence of diabetes, of benign prostatic hypertrophy, of the senile type of cataract, of dementia, or of other neurodegenerative conditions. The basic causes are unknown, but there is intriguing evidence for selective cellular abnormalities. While diploid fibroblast cultures show rapid clonal senescence as observed in the Hutchinson-Guilford syndrome (Martin et al., 1970), glialike cells from the cerebrum and cerebellum grew well (Martin, 1978; Hoehn et al., 1975). Werner's victims produce excessive hyaluronic acid (Tajima et al., 1981; Takunaga et al., 1975; Zebrower et al., 1985; Brown et al., 1984) like the Hutchinson-Guilford syndrome. A particularly intriguing finding is that Werner's fibroblasts show a "mutator" phenotype in culture, with a tenfold increase of spontaneous mutants at the *HRPT* locus showing resistance to 6-thioguanine and a twofold increase in deletions of DNA within this gene (Fukuchi et al., 1989a, 1989b). This tendency for DNA deletions is consistent with the increase of translocation chromosomal mosaicism in lymphocytes obtained from Werner's syndrome patients (Salk, 1982). While these observations do not distinguish causes and effects, the gene for Werner's syndrome clearly has important consequences to both genomic stability and cell proliferation.

Xeroderma pigmentosum (XP) is a rare autosomal recessive that causes disease in 1 person in 250,000. In association with impaired DNA repair mechanisms, XP victims are extremely susceptible to sunlight-induced skin cancer (Robbins et al., 1974; Cleaver, 1968; Epstein et al., 1970). Besides the skin lesions, XP victims also often have neurological abnormalities that include smaller brains and deficits of large pyramidal neurons (Robbins et al., 1974). No data are available to resolve the important issue of whether the neuronal deficits occur during neuronogenesis or after neurons are formed. Because the most neurologically impaired XP subjects show more deficits of DNA repair, as assayed *in vitro* with fibroblasts (Andrews et al., 1976, 1978), it is proposed that some types of neurodegeneration could be associated with DNA damage even in cells that receive minimal UV exposure (Kidson et al., 1984; Andrews et al., 1976). Influences of the *H–2* genotype in mice on UV-induced DNA repair (Walford and Bergmann, 1979) raise the possibility that subtle genetic influences on the extent of DNA repair (unscheduled DNA synthesis) could be a factor in resistance to age-related neurodegeneration.

Down's syndrome, the result of trisomy at chromosome 21 which causes abnormal brain development and severe mental deficiencies, is of particular interest as a progeroid syndrome because of the premature occurrence of a select feature of senescence, the neuropathologic changes of Alzheimer's disease (Martin, 1982). Although other features of senescence (atherosclerosis, senile cataracts, abnormal growths) are not general findings, there is some evidence for five- to tenfold greater incidence of diabetes mellitus than in age-matched groups (e.g. Milunsky and Neurath, 1968), which would be expected to accelerate the formation of cataracts and atherosclerosis, among other diseases that are associated with adult onset diabetes (Chapter 3.6.8). Whether there is any separate aspect of accelerated senescence remains to be determined. The characteristic childhood leukemias are not clearly followed by higher rates of malignant growths in adults (Fabia and Drolette, 1970; Martin, 1982). Down's syndrome is the most common progeroid syndrome, occurring at a rate of about 1 per 700 live births (McKusick, 1982), with a very strong increase at later maternal ages (Chapter 8.5.1.1). Very short lifespans in the past may represent neglect of these mentally deficient individuals, since substantial numbers are now living beyond 40 years, when mortality rates begin to accelerate (Oster et al.,

1975). Data on mortality schedules and the incidence of different peripheral tissue diseases under favorable living conditions are not yet available.

Although initial cognitive defects make the judgement difficult without longitudinal studies, which have only recently begun, a high proportion of Down's syndrome adults express a new aspect of cognitive dysfunctions that are reminiscent of Alzheimer's disease, including loss of short-term memory and other behavioral changes (reviewed in Coyle et al., 1986; Mann, 1988). After 40 years, these moderately short lived, mentally deficient humans develop senile plaques, neurofibrillary tangles, and other cellular and biochemical changes, which strikingly resemble those of Alzheimer's disease (Wisniewski et al., 1978; Ball and Nuttall, 1980; Burger and Vogel, 1973; Reid and Maloney, 1974; Mann, 1988; Table 6.4). Alzheimer's disease is a nonvascular dementia that may be an autosomal dominant in some families, as discussed above, and that is characterized by select neuronal loss and neurochemical deficits in the frontal cortex and hippocampus (Mann, 1985; Morgan et al., 1987). A distinctive feature of Down's is that the neuropathologic changes (β-amyloid in the cerebral blood vessels, neuritic plaques, and neurofibrillary tangles in neurons) are manifested in all brains by 35–40 years, which is 10 or more years earlier than in the general population. The cerebrovascular β-amyloid protein of Down's syndrome has the same amino acid sequence as that in Alzheimer's disease (Masters, Simms, Weinman, Multhaup, McDonald, and Beyreuther, 1985; Masters, Multhaup, Simms, Pottgiesser, Martins, and Beyreuther, 1985; Glenner and Wong, 1984; Robakis et al., 1987). Hence, this feature common to both is attributable to an increased "dose" of active genes, rather than a mutation in the amyloid gene coding sequence in Alzheimer's disease. The gene for β-amyloid protein is located on chromosome 21, near the locus for early onset Alzheimer's disease that is indicated in a few families (Goldgaber et al., 1987; St. George-Hyslop et al., 1987; Blanquet et al., 1987; Tanzi et al., 1987; Figure 6.12). The two genes are distinct, however, and show segregation by RFLP analysis (Van Broeckhoven et al., 1987).

While trisomy of chromosome 21 from either parent inevitably leads to developmental abnormalities of Down's, an extra gene dose from only a small fragment may be sufficient, as judged by absence of the characteristic facial features and mental deficiencies in one individual who was trisomic except for the 21q22 band (Hagemeijer and Smit, 1977; Figure 6.12). Mosaic translocations of chromosome 21 were found in an adult without mental retardation who became demented in the early 40s in association with cerebral cortical atrophy (Schapiro et al., 1989). If future autopsy establishes the premature occurrence of vascular β-amyloid, senile plaques, and neurofibrillary tangles, then this case will clearly dissociate the genes whose overexpression through increased dose causes abnormal development from those that accelerate the Alzheimer's neuropathology. Extra copies of DNA from the proximal part of the 21q22.3 band were found in two Down's children with translocations of overlapping chromosome 21 fragments (Rahmani et al., 1989). Both had segmental trisomy for the sequence *D21S35*, which includes about 10^6 nucleotides of DNA and is located distally to the familial Alzheimer's locus of Tanzi et al. (1987), the gene for β-amyloid, and another gene of theoretical interest, coding for superoxide dismutase type 1 (Figure 6.12); these children are considered too young to show the neuropathologic changes associated with trisomy of the entire chromosome 21.

Huntington's disease may also be considered as a progeria. This autosomal dominant typically causes deterioration of neurons in the caudate and putamen, with marked uncon-

Table 6.4 Symptoms of Down's Syndrome and Alzheimer Disease

Similarities	Down's Syndrome	Alzheimer Disease
(Senile) plaques (cortex and hippocampus)	major increase	variable increase; less different from normal at later ages
Loss of pyramidal neurons in hippocampus and cortex	variable loss	variable loss
Neurofibrillary tangles	major increase	variable increase
Cortical choline acetyltransferase	loss	variable loss
Amyloid in cerebral vessels	major increase	modest to major increase
Cognitive impairments	strong trend during mid-life for memory impairments; data are preliminary	characteristic loss of short-term memory

Note: Summarized from Section 6.5.3.3, with reference to age-matched normals.

trollable jerky movements that gave rise to its historical name, Huntington's chorea (*chorea* is derived from the Greek root for dance). The memory loss and dementia associated with Huntingtonism differs from that of Alzheimer's disease (Moss et al., 1986). The disease is usually first manifested during mid-life and causes death within 15 years. In some regards, the changes of Huntingtonism resemble accelerated senescence in these regions of the brain, particularly the loss of dopaminergic (D–2) receptors (Finch, 1980; Morgan et al., 1987). The higher levels of 3-hydroxyanthranilate oxygenase in the Huntington brain (Schwarcz et al., 1988) could contribute to neuron death: the enzyme's product, quinilinate, is an excitotoxin that damages many of the same cells, when introduced to the rat brain.

In addition to brain changes in Huntington's disease, there are also peripheral signs of accelerated senescence that are often detectable before cognitive degeneration is obvious, including diabetes mellitus, brittle bones, weight loss, and pale, wrinkled skin (Bruyn, 1968; Farrer et al., 1984; Farrer, 1985a, 1985b). Family lifespans suggest an important interaction of the Huntingtonian gene with other loci that influence longevity (Farrer and Conneally, 1985). As shown in Figure 6.13, the age of onset correlates with the longevity of nonaffected (noncarrier) siblings (Farrer et al., 1984). The Huntington locus may interact with other genes that influence age-related changes and longevity, possibly through neurotransmitters, neurotrophic factors, or hormones (Finch, 1980).

Besides the classic progerias described above and in Tables 6.3 and 6.5, George Martin (1978, 1982) proposed in an elegant analysis that about 7% of the more than two thousand inherited diseases in humans (McKusick, 1975) resemble one or more pathologic conditions common to senescence. Because no one condition encompasses all of the common changes in the human senescent phenotype and because of their early onset, Martin considers these to be segmental progeroid syndromes and has classified them according to various traits that also occur in elderly human populations (Tables 6.2–6.6). The segmental progerias in humans are about equally divided between autosomal dominants and recessives, and sex-linked recessives. These observations suggest that a number of independently acting genes determine the phenotypes of senescence.

Martin (1982) identifies another set of heritable life-shortening diseases as unimodal

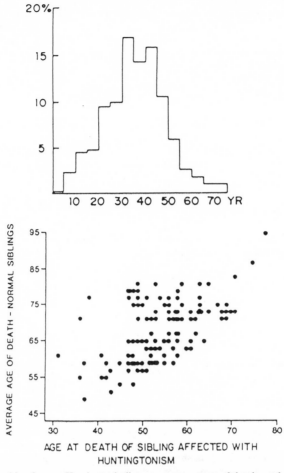

Fig. 6.13. The relationship of age to Huntington's disease, a rare autosomal dominant that causes deterioration of select neurons in the basal ganglia that are different from those affected by Parkinsonism. The symptoms include spontaneous jerky movements (chorea) and cognitive impairments (Moss et al., 1986). Besides neurological disorders, Huntingtonism is associated with peripheral signs of accelerated senescence, including diabetes mellitus, weight loss, and pale, wrinkled skin (see text). Milder cellular and neurochemical changes also occur in these same brain regions at later ages in the general population (Finch, 1980, 1988; Morgan et al., 1987). *Top,* the onset of the disease, showing its characteristic mid-life peak. Redrawn from Brackenridge, 1973. *Bottom,* the correlation between age at death of Huntingtonians and age at death of their normal siblings, while highly significant, accounts for 26% of the variance. Redrawn from Farrer et al., 1984. Because the duration of the disease does not vary with its age of onset, this correlation suggests that familial factors related to age may trigger expression of the latent phenotype in carriers through accelerated senescence according to a threshold model (Finch, 1980, 1988; Figures 3.41, 5.14). These factors could be genetic, environmental, or both.

progeroid syndromes, in which a single defective gene intensifies only one aspect of senescence. Examples are familial cataracts, lipofuscinoses, and vascular diseases (Table 6.3). Some of the examples of rapid senescence from Chapter 2 fit nicely into this concept (e.g. the corticosteroid-associated death of Pacific salmon, which does not occur in closely related species), as do the short-lived rodent genotypes with early onset diseases.

Table 6.5 Genetic Progeroid Syndromes in Humans, by Category

Type of Disorder	Number of Examples from McKusick's Catalogue
Susceptibility to age-related neoplasms	25
Graying or loss of hair	18
Senile brain-type neuropathology	50
Amyloidosis	11
Lipofuscinoses	8
Diabetes	30
Hypogonadism	75
Autoimmune disorders	6
Hypertension	12
Degenerative vascular disease	30
Osteoporosis	11
Cataracts	18
Regional fibrosis	29

Source: McKusick, 1975; excerpted from Martin, 1978.

Table 6.6 Comparisons of Common Age-related Diseases to Diseases of Short-lived Genotypes

Disease	Normal Populations	Short-lived Genotypes
Atherosclerosis	While early stages of atherosclerosis are common before mid-life in the general human population (Chapter 3.6.7), some individuals show minimal changes even at 100 years (Lie and Hammond, 1988; Chapter 3.6.2).	Accelerated in mice by the *Ath-1* and *Ath-2* genes, which increase the vulnerability to atherogenesis on high-fat diets; accelerated in humans by a number of alleles altering lipoprotein transport, but retarded by familial hypo-β- and hyper-α-lipoproteinemia.
Autoimmune reactions	Mild autoimmune reactions are common at later ages in mice and humans, particularly in the arthritides. (Chapter 3.6.5.2)	Accelerated in NZB mice; (Section 6.5.1.7.3)
Amyloid	Amyloid is not a pure substance and is not necessarily the same in all tissues or species, referring to proteinaceous fibrils with a β-configuration, which imparts birrefringence with Congo red under polarizing light (Chapter 3.6.7). Amyloids of various origins from different proteins are deposited in myocardium, kidney, and blood vessels at advanced ages in numerous species, including humans.	Accumulations of amyloid increase with age in SAM mice (Section 6.5.1.7.2) as well as in other strains (Dunn, 1944; Thung, 1957). Amyloidotic syndromes include familial amyloidotic neuropathy (Benson, 1989) and familial Alzheimer disease (Glenner and Wong, 1984; Tanzi et al., 1987).

Table 6.6 (*continued*)

Disease	Normal Populations	Short-lived Genotypes
Alzheimer neuropathology	Neuritic plaques, neurofibrillary tangles, and cerebrovascular amyloid occur nearly universally in humans at later ages, whether or not they become demented; the cerebrovascular amyloid and neuritic plaques are found in mammals of intermediate lifespan, but not in rodents.	Neuritic plaques, neurofibrillary tangles, and cerebrovascular amyloid occur precociously in Down's syndrome and in familial Alzheimer disease. Age-related accumulation of cerebrovascular amyloid occurs in certain other mammals, but has not been observed in rodents.
Other neurodegenerative trends	Neuronal atrophy with cell-body shrinkage is common in large neurons of humans at later ages through Alzheimer disease and unknown causes. (Chapter 3.6.10)	Neuronal atrophy with cell-body shrinkage occurs in spinal motor neurons of mice homozygous for *Mnd* and in wild mice that are susceptible to certain MuLV infections.
Diabetic trends in glucose regulation	Mild trends for glucose resistance during mid-life are general in humans. (Chapter 3.6.8).	Mice with the *ob/ob* and *db/db* genotypes become diabetic, with elevated blood glucose and insulin resistance, by mid-life, depending on other genes.
Obesity	A general trend for increased body fat is common by mid-life in *ad libitum*-feeding humans (Chapter 3.6.8).	Intensified in mice by the *ob/ob* and *db/db* genotypes.
Polycystic kidney disease (PKD)	PKD may be a useful model for the age-related degeneration of kidney function, although the specific cystic pathology is relatively rare.	PKD, which causes premature death during middle age, is one of the most common hereditary diseases (0.1–0.2%) and accounts for at least 10% of the end-stage renal disease in the U.S (Vollmer et al., 1983). Impairments from the multiple cysts in the kidney and other visceral organs occur in most carriers by 50 years, though some children are affected (Grantham et al., 1987). The disease is an autosomal dominant, with at least two loci, one of which is on the short arm of chromsome 16 (Kimberling et al., 1988).
Cancer and other abnormal growths		Examples that are too numerous to review (Coburn et al., 1971; Storer, 1966) show that the risk of spontaneous abnormal growths is strongly influenced by genotype (high at early ages in AKR and C3H; low in A/J mice) and gender (male C57BL/6J mice almost never develop pituitary tumors, while most females have them by 18 months) (Felicio et al., 1980; Schechter et al., 1981). Humans show familial trends for leukemia and breast cancer by mid-life.

These short-lived genotypes may be most useful as models for specific diseases that may be studied in the absence of more complex changes associated at later ages with senescence. An open question is whether the early onset diseases may also accelerate other age-related changes that would precipitate premature senescence if survival were extended.

6.5.3.4. *Gender*

Most human populations fit the general pattern in mammals that males have slightly shorter lifespans than females (Hamilton, 1948; Comfort, 1979). Differences in mean lifespan of 2 to 10 years between the sexes are common in all the world's populations (Smith, 1989, 1991; Brock et al., 1990; Smith and Warner, 1989). In the U.S. age group of 85 years and older, women are twice as numerous (National Center for Health Statistics, 1979). While the survival of women into the 90s much exceeds that of men ($> 50\%$ in the U.S., Guralnik et al., 1988), the record human lifespan is usually attributed to a man (Smith, 1989, 1991). The sex difference, however, is very sensitive to the socioeconomic ranking and may be less than 1 year in some privileged populations (Metropolitan Life Insurance Co., 1988). Thus, there are numerous complex environmental and behavioral influences on the gender-correlated phenotypes of senescence (Smith and Warner, 1989; Hazzard, 1986, 1987; Waldron, 1987).

While virtually all studies of human populations show that men have a higher mortality rate throughout most adult ages, the basis for this has not been seriously investigated in terms of age changes in the mortality rate. The influence of chromosomal sex on lifespan may be better understood on the basis of the two types of mortality coefficients, that associated with the acceleration of mortality with age (MRDT) as distinguished from the age-independent mortality rate coefficient IMR (calculated at puberty; Chapter 1.4). One Australian population analyzed by Abernethy (1979) indicates that the MRDT is identical in both sexes, while the males had a higher IMR. Further analysis could be readily done from published data in many countries.

A hormonal basis for the favored lower mortality during mid-life of females is suggested by many influences of menopause and estrogens on diseases with sex differentials. The influence of sex steroids on lipoproteins is a major factor in sex differences of atherosclerosis. Mammary cancer and female reproductive tract cancers are also strongly influenced by ovarian hormones (Neufeld and Goldbourt, 1983; Dawber, 1980). Environmental and behavioral differences are also a factor in these sex differentials, through alcohol consumption, occupational hazards, and dangerous behavior (Smith and Warner, 1989). Many parameters related to the environment, such as exposure to toxins or head injury, may have gender differentials with delayed manifestations (Chapter 3.6.11).

6.6. Summary

This chapter documented genetic influences on the lifespans of a few species. In mammals, most examples of genes that shorten the lifespan do so by inducing diseases early in adult life, but without general acceleration of senescence. Thus, premature death from atherosclerosis, from tumors, or from autoimmune diseases can be influenced by genes with varying penetrance in laboratory rodents and humans. Factors in penetrance of genes that influence the lifespan include chromosomal sex, other alleles, diet, and stress (Chapter 10). Some specific genes or genotypes, however, increase lifespan above the average for the species by lowering the incidence of specific diseases.

A major question concerns the relationships of short-lived genotypes to aspects of senescence that usually occur at later ages. How do such genotypes relate to the phenomena of senescence in outbred populations that allow such remarkable overall stability of the acceleration of mortality rate (MRDT) across human populations, which nonetheless show very different patterns of disease by age group (Chapter 1.4)? While it is premature to draw any firm conclusions until we have more data on the mortality rate coefficients, the argument can be considered that these shorter-lived genotypes represent sampling from the species gene pool that were revealed by inbreeding and recombination.

Thus, I propose that the numerous short-lived rodent genotypes, most of which were produced during inbreeding without regard to lifespan, may represent polymorphisms of genes for diseases with delayed manifestation that are merely more penetrant than other alleles in longer-lived individuals. In rodents as in humans, many of the short-lived genotypes may be viewed as segmental progerias, since numerous diseases precipitated in the short-lived genotypes are qualitatively similar to diseases that arise more or less generally at later ages. Examples summarized in Tables 6.5 and 6.6 include atherosclerosis, autoimmune reactions, accumulation of various amyloids, Alzheimer's neuropathology and other neurodegenerative trends, diabetic trends in glucose regulation, obesity, and cancer and other abnormal growths. The only class of age-related change that should *not* be accelerated in shorter-lived genotypes is the purely temperature-driven racemization of L-amino acids (Chapter 7.5.4). Continuing study of the molecular basis of pathology in short-lived genotypes promises insights about the genetic basis for lower disease risk with alternate alleles or combinations of alleles that influence penetrance and increase potential longevity. The time course of diseases in different genotypes also should reveal molecular pacemakers of general importance to pathogenesis during senescence, as in the autoimmune disorders of NZB mice that are modified by the *H–2* haplotypes and sex, or the myriad familial lipoprotein disorders.

Conversely, other alleles may protect against diseases during mid-life. An example is the mouse *Ath–1r* and *Ath–2r* alleles that confer resistance to atherogenic diets. There may also be life-extending protection against atherosclerosis in the case of the familial hypo-β- and hyper-α-liproproteinemias in humans, for which mice with the *Ath–1r* and *Ath–2r* alleles may be a model. In view of alleles and genes that predispose to cancer at various ages, could there not be genes that protect against cancer? The varying extents of mild age-related autoimmune manifestations in mouse strains and in human populations (Chapter 3.6.6) suggest that some yet unknown MHC haplotypes may protect against anti-idiotypic cascades that trigger autoimmune dysfunctions. An open question is whether the search for genetic regulators of senescence and the lifespan will ultimately be reduced to the multifarious DNA sequences in any species that control its major age-related diseases. So far in mammals, death during senescence has not been described in the absence of specific pathologic lesions (Chapter 3.6.2). Speculatively, however, homeostatic dysfunctions—for example, in blood glucose regulation—might cause death in the absence of substantive organ pathology.

In an example that may be a model for other slow infectious diseases that may interact with age-related changes, the latency of scrapie prion infections in mice is determined by alleles of the *Prn-i* gene on chromosome 2; this locus is closely linked to the gene (*PrP*) encoding the murine prion (Carlson et al., 1986). Moreover, a unique allele in the human prion gene on chromosome 20 is associated with the Gerstmann-Sträussler syndrome, a rare autosomal dominant that penetrates by mid-life to cause neurodegeneration with amyloid plaques and transmissible prions (Hsiao et al., 1989; Westaway et al., 1987). Transgene techniques now show that introduction of the hamster prion protein gene into the mouse germ line greatly increased susceptibility to hamster scrapie (Scott et al., 1989).

These examples point to many different genetic loci that can serve as pacemakers for senescence and longevity in any species. There is no evidence for a single gene or cluster of genes that uniquely determines the duration of the adult phase in any gradually senescing species with multiple-organ system pathologies. This is as predicted by population genetics theories (Chapter 1.5.2): since the force of selection generally declines after maturity, numerous variant genes scattered throughout the genome may accumulate in the population gene pool and may have deleterious consequences later in life when selection is very weak. The variety and location of these genes may nonetheless be constrained by other genes to which they are linked in function or by other functions that the same genes code for.

Much can be learned through continued pursuit of the molecular basis for pathogenesis in short-lived genotypes. As shown for the examples of naturally occurring polymorphisms that influence adult lifespans of flies and mosquitos (*aa* and autogeny, respectively), much may be learned from the natural history and microecology of the circumstances where these genotypes are found. This view is in fact familiar to population geneticists who endeavor to account for balanced polymorphisms, as in resistance to malaria that is conferred as a balanced polymorphism by the gene for sickle cell hemoglobin. Further attention to such questions may illuminate factors accounting for the presence of other alleles and combinations that shorten the adult lifespan. However, we should not yet conclude that alleles at the same loci are necessarily involved in lengthening lifespan during evolution. A major hope from transformational genetics is that we can test the possible roles of variant genes found within a species as a cause of differences in lifespan between related species. The major differences in senescence observed within genera (paired species of lampreys [Chapter 2.3.1.4.1]; *Peromyscus* [Chapter 3.6.1]) give opportunities with genetic transduction to study candidate genes that may alter the schedule of senescence.

The exciting finding of genotypes and alleles in the nematode *Caenorhabditis* that increase lifespan by slowing the acceleration of mortality also point in this direction, although the relation to disease as understood by vertebrate pathologists is not worked out. Identification of the *age–1* gene of *Caenorhabditis* and its products could bridge this gap, however. Studies on natural polymorphisms should be very informative, as well as on related genes in parasitic nematode species that live a hundredfold longer (Chapter 4.2.2). So far no variations at a single genetic locus or even a small number of favorable alleles have increased lifespan beyond 50% the normal maximum range of wild-type populations in any species. The five- to thirtyfold differences in lifespan between insect queens and short-lived worker castes (Chapter 2.2.3.1) that arise epigenetically during development from the same genome give a target for those seeking genetic manipulation of the lifespan.

Important insights are also given by studies of natural and artificial selection in various *Drosophila* species. The *aa* genotype of *D. mercatorum,* which resulted from transposible elements infecting ribosomal 28S RNA genes, has shortened adult lifespan yet has an advantage in some natural conditions because of its precocious sexual maturity (Section 6.3.3). The presence of genetic variants with different reproductive schedules appears to confer to populations advantages that are consistent with population genetics theory that reproductive schedules can rapidly evolve by drawing on existing polymorphisms (Chapter 1.5). Genetic variants in reproductive scheduling were also revealed by artificial selection for reproduction at advanced ages in *D. melanogaster,* which delayed the age of maximum fecundity (Section 6.3.2).

Mouse strains with different rates of ovarian oocyte loss and reproductive scheduling (Section 6.5.1.4) indicate similar genetic polymorphisms in mammalian populations. Among inbred mice, some strains have short lifespans but normal lifetime fecundity (B10.F), while others have early loss of fertility but with normal lifespans (CBA). Despite the consistent outcomes of selection for reproduction at later ages in fruit flies, reducing reproduction earlier in life, other gene combinations may permit both vigorous early reproduction and long life, as shown by B10.RIII mice. In the examples of fish and plants that continue to increase their production of gametes in some proportion to body size long after maturity (Chapter 4.3), there may be a physiological coupling of the genes regulating growth and those regulating gametogenesis.

Hybrids of genotypes with different lifespans do not give any general outcomes across species. While hybrids of inbred rodents generally have longer lifespans and fewer diseases without changes in the acceleration of mortality, hybrid vigor was not found in *Caenorhabditis.* The segmental progerias in humans are about equally divided between autosomal dominants and recessives, and sex-linked recessives; heterozygotes of the recessives would generally have increased lifespan. It is intriguing to consider whether hybrid vigor through reduced consanguinity might be a nonenvironmental factor in the general increase of life expectancy during the last 200 years in industrial countries (Figure 1.2). While improved hygiene, diet, and medical care most likely account for most of the increase in lifespan, hybrid genotype-environmental interactions could still be a factor in these historical trends, but would be difficult to establish.

In regard to the chromosomal distribution of loci influencing lifespan, it is pertinent that chromosomal linkage of genetic loci is not required for their coordinate control, because of hormones and other trans-acting regulators that can interact with loci scattered throughout the genome. Even so, the genes linked to the MHC may have fundamental roles in the pathogenesis during senescence in many mammals. Future studies of RI strains and congenic strains should reveal many more loci throughout the chromosomes that influence lifespan. I suspect that human populations also will be found to have similar alleles in combinations of balanced lethal or deleterious polymorphisms that would segregate with inbreeding. A DNA sequence analysis of loci that influence lifespan and patterns of senescence should soon be possible through transformational genetics.

How do these varied pleiotropic genetic influences bear on the conclusions from Chapter 3, that certain age-related changes in mammals apparently occur in proportion to the lifespan? Changes that begin before mid-life include declining T-cell responses, decreased striatal dopamine receptors, and thickening of the lens with impaired ocular accommodation (Table 3.3). As noted in Chapter 3.7, the similar scaling of these patterns across the thirtyfold range of mammalian lifespans seems inconsistent with predictions from population genetics theory that senescence and its precursors should vary widely through evolution (Chapter 1.5). In contrast, the varied genetic influences predisposing to diverse age-related diseases are consistent with evolutionary theory. Because most of these genotypes do not express their adverse traits until after puberty, it is possible that their penetrance is under neurohumoral control. As described in Chapter 10, the phenotypes of many genes causing early onset diseases are subject to manipulation through hormones or diet: this evidence implicates pleiotropic interactions between the expression of these genes and physiological changes during maturation.

I suggest then that the similar scaling of certain pre-mid-life changes that occur in mammals of widely different lifespans may reveal a developmental program under neural and endocrine control that is fundamental to the mammals represented in Table 3.3. Such a genetic system of controls might be resistant to some types of genetic variations, by compensatory responses in the expression of other genes. The resistance to genetic variations should, according to evolutionary theory, diminish at later ages so that the range of individual variations increases, as it indeed does, even within a highly inbred genotype. The relatively stable accelerations of mortality rate (MRDT) that characterize a mammal might be explained in terms of the evolutionary persistence of the hypothesized fundamental developmental program that sets the homeostatic capacity and determines the resilience to dysfunctions from organ-specific pathology.

Having now examined how specific genes influence lifespan and diseases of senescence, we next address how the functions of the genome itself may change during the adult lifespan in regard to the control of cell characteristics (Chapter 7) and the fidelity of DNA replication and gamete production (Chapter 8). The extent to which gene functions are maintained during the lifespan has great bearing on any hypotheses concerning penetrance of genes that causes age-related diseases or dysfunctions.

Section II

GENOMIC FUNCTIONS DURING SENESCENCE

That the genome has major influence on the characteristics of senescence is undeniable from the many examples of species differences in senescence and lifespan and the genotypic variations in organ-specific degenerative conditions within species (Chapters 1–6). Major differences in the length of the adult phase of closely related species recur throughout virtually all phyla (rotifers are a major exception) and point to a vast ability for genetic variations in the schedule of senescence. This plasticity is also shown by the alternate life histories that can be derived by epigenetic influences on the same set of genes, as illustrated by the five- to thirtyfold differences among the lifespans of castes of social insects that originate from the same clutch of eggs (Chapter 2.2.3.1). Other developmental influences on senescence include adult characteristics such as the capacity for regeneration, and cell and molecular replacement are predestined by selective changes in gene expression before maturity. This developmental theater of genomic influences on senescence is discussed in Chapter 9. In numerous examples of semelparous animals and plants, interventions that delay reproduction also postpone senescence. The subject of environmental influences and experimental manipulations of senescence is treated in Chapters 9 and 10.

To prepare for evaluations of epigenetic mechanisms that may act through variations in gene expression during development or after maturation, I next address evidence concerning genomic functions that may be altered during the lifespan and during senescence. This section reviews evidence on the regulation of macromolecular synthesis (Chapter 7) and the fidelity of cell replication and DNA maintenance (Chapter 8). Throughout, most of the information comes from mammals. Questions concerning evolutionary theory are taken up at the end of Chapter 11.

7

Gene Expression and Macromolecular Biosynthesis

7.1.   Introduction

Analysis of gene expression and macromolecular biosynthesis is giving extraordinary insights into the mechanisms of development, the basis for functions of differentiated cells, and many diseases, including cancer. At every turn, the variety and subtlety of controls on macromolecular synthesis continues to expand. In general, mechanisms of gene regulation do not alter the nucleotide sequence of the DNA in somatic cells. Most mechanisms of quantitative change in gene expression act by modifying the rates of RNA and protein synthesis and turnover. However, another mechanism altering the final products of gene expression involves changes in the differential processing, through which alternate mature mRNA molecules are formed by the differential splicing of exons or peptides that may be formed post-translationally by endopeptidases or transpeptidases.

A major question concerns the extent to which, across the lifespan, somatic cells retain genetic totipotency (the functional inventory of genes that were inherited from the parents).[1] Two major exceptions to the exact equivalence of somatic cell nuclear DNA to the germ-line inheritance are found in certain immune cells, where immunoglobulin genes are rearranged and mutated during differentiation, and in oncogenes at later stages of oncogenesis, where DNA sequences are changed through rearrangement and mutational substitution. DNA sequences may also be modified during differentiation by methylation, but without changing the primary sequence. Changes of gene expression occur during senescence through many mechanisms that might include alterations in DNA sequences, methylation, rates of transcription, and splicing of nascent macromolecules. Evidence for age-related changes in nuclear DNA sequence is discussed in the next chapter. First, I discuss some often neglected issues that bear on the design of studies of the molecular biology of senescence; informed readers may wish to skip this section.

7.2.   Caveats and Perspectives

Analysis of age-related changes in RNA and protein synthesis is frustrated because of difficulties in distinguishing changes that are secondary to age-related pathophysiologic changes from those that may be primary changes. Moreover, the age groups chosen sometimes confuse maturation and senescence, and there are wide species differences in the particular features of senescence, even between species that are closely related. Examples are given to guide thinking about each of these thorny issues.

7.2.1.   Confounds from Age-related Diseases

Some age-related changes in macromolecular biosynthesis are epiphenomena that can be traced to specific diseases that, more than is often recognized, are cryptic pallbearers

---

1. Nearly all somatic cells in young adult plants and animals appear to maintain nuclear totipotency. That is, they retain the complete set of nuclear genes derived from the zygote. Well-known exceptions in mammals include the rearranged immunoglobulin genes of select lymphoid clones, and the loss of cell nuclei during the differentiation of erythrocytes and mature lens fiber cells. Other differentiated somatic cell types could, in principle, provide the genomic information for any other cell type by redifferentiation. Somatic cell nuclear totipotency has been difficult to prove except in a few favored circumstances in vertebrates, such as the regeneration of

of many otherwise elegant studies. For example, albumin synthesis in the rodent liver offers valuable opportunities to study a major plasma protein, from transcription as an abundant class of mRNA through synthesis, post-translational processing, and secretion. Several labs found 50% increases in albumin synthesis by the average lifespan (Beauchene et al., 1970; Chen et al., 1973; Richardson, 1981), as well as increases of albumin mRNA (Horbach et al., 1984; Wellinger and Guigoz, 1986). There are considerable individual variations among studies, some reporting no change (Ara et al., 1983).

The age-related increase of albumin synthesis was once discussed as a possible compensation for hypothetical genomic damage from accumulated somatic mutations or other errors (Szilard, 1959; Orgel, 1963; Clarke and Maynard Smith, 1966). But a mundane explanation suffices. Laboratory rodents commonly develop kidney diseases after mid-life (Chapter 10.3.2.2; Tables 10.3, 10.4), which often cause a gross loss of plasma proteins into the urine (proteinuria) and in turn stimulate compensatory synthesis of albumin by the liver. As discussed in Section 7.4, the synthesis of other liver proteins may decrease with age. Decreases in protein synthesis in the kidney (Goldspink and Kelly, 1984) may also be confounded by gross cellular damage from age-related glomerulosclerosis (Chapter 10.3.2.2; Tables 10.3, 10.4).

This sad scenario of pathogenesis spreading from unrecognized disease into the data is probably more common than realized. The extent of age-related diseases often differs widely among individuals of gradually senescing species, as expected from the wide individual differences in life expectancy. Moreover, the same spectrum of age-related diseases may not prevail in other colonies of the same inbred genotype (recall the differing presence of leukemia oncoviruses in mouse populations, Chapter 6.5.1.7). Diet composition also strongly influences kidney and other diseases (e.g. Maeda et al., 1985; Chapter 10.3.2.2). While diets rich in soy proteins may minimize the scourge of age-related renal disease in rodents (Iwasaki et al., 1988a), such diets have unknown influences on other aspects of organismic senescence. Moreover, the role of individual differences in diseases in studies on senescence is not considered by most investigators of invertebrates; the report of Slob and Janse (1988) is a welcome exception.

Confounds from age-related diseases are natural and predictable aspects of research in this area, but only recently were taken into account by investigators. This can be judged by how few reports even mention this issue as part of the experimental design. It is often useful to distinguish changes consequent to age-related diseases as *pathogeric*, which often do not occur in all individuals of a species, in contrast to *eugeric* changes (e.g. menopause), which are more generally distributed and less a function of specific diseases, diets, or exposure to pathogens, which vary among individuals (Finch, 1972a; Chapter 3.6.2). So far, pathophysiologic confounds of senescence have not been evaluated for any invertebrate. It is likely that many older reports of major organ loss of protein or RNA were primarily due to disease: contrast the reports of gross loss of RNA from rodent liver (Detwiler and Draper, 1962) and brain (Andrew, 1959) with subsequent studies showing the overall stability of bulk RNA, poly(A)RNA, and protein content in these and other tissues (Colman et al., 1980; Barrows and Kokkonen, 1981; Birchenall-Sparks, Roberts,

---

lens from iris cells in urodeles (Chapter 9.4) or the redifferentiation of fibroblasts from macrophages (Davidson, 1976). However, somatic cells with complete germ-line gene sets are proven by the numerous plants and animals that reproduce vegetatively and asexually, by somatic cell cloning (Chapter 4.4).

Rutherford, and Richardson, 1985). As described below, such organ-level studies do not preclude changes in specific cell types or in specific gene products. In sum, each study must design protocols to identify the pertinent pathological variables most likely to produce confounds. There is no general or absolute solution to this problem.

7.2.2. *Physiological Changes and Gene Regulation*

Besides the confounds from age-related diseases, another issue bears on macromolecular synthesis in mammals and other species with integrative physiologic functions in which cell functions may be influenced by other cells elsewhere in the body. Alterations in the synthesis of a particular macromolecule may be consequent to age changes in hormones, neural factors, or other physiological regulators. The depletion of ovarian follicles and attendant loss of estrogen secretion by mid-life in mammals is a eugeric change that generally triggers a (futile) compensatory response: the increased secretion of gonadotropins by the pituitary, as observed in mice, monkeys, and humans after mid-life (Chapter 3.6.4.1). Protein synthesis in the pituitary has been little studied for age-related changes, but the increased gonadotropin secretion should be prominent. The sensitivity of the same gene to a particular hormone may vary among the cell types expressing it. For example, the chromatin protein H1° is strongly influenced by androgens in mouse prostate, but not in retina or skeletal muscle (Gjerstet et al., 1982). Accordingly, mechanisms of age-related changes in a particular gene or its encoded macromolecule may vary among cell types.

Physiological regulation of genes may also vary by genotype within a species, as it certainly varies among species. No general classification is yet available of genes according to the types of physiological regulators. The expanding literature on regulatory DNA sequences that reside both within and without the coding regions, suggests that most genes are subject to physiological regulators in adult cells. The process leading to selective gene expression during development and differentiation also requires selective regulation of gene expression that must vary among cell types. The tissue-specific patterns of gene regulation in mammals apparently arise from inductive influences between cells during embryogenesis, rather than from cytoplasmically localized tissue determinants existing in the egg before fertilization, as found in many lower organisms (Davidson, 1986; Melton, 1987).

7.2.3. *Inappropriate Comparisons of Age Groups*

Many studies until recently have compared very young and very old individuals. Some comparisons of extreme ages may be well justified to detect subtle age-group differences of some putatively linear age-related changes, for example, racemization in long-lived proteins (Section 7.5.4). However, other changes occur early in life. For example, the inducibility of liver a-glycerophosphate dehydrogenase (Schapiro and Percin, 1966) and malic enzyme (Forciea et al., 1981) decline most between weaning and maturation in rats, but change little thereafter. Other changes are multiphasic, as in basal levels of liver arginase and carbamyl-phosphate synthase, which decrease sharply during maturation, increase by mid-life, and decrease gradually thereafter (Lamers and Mooren, 1981). Similarly, female reproductive cycles progressively shorten after puberty, then stabilize during the young adult years, to again lengthen before their cessation in mice, rats, and humans (Finch and Gosden, 1986; Treloar et al., 1967). If very young rodents were compared with much older groups, misleading conclusions could easily be drawn. Most studies now include a range of intermediate age groups, but many reports cannot be evaluated for the want of this. Editorial policy in some scientific journals is beginning at last to insist on concern for this issue (Coleman et al., 1990).

7.2.4. *Species Differences in the Characteristics of Senescence*

The literature on age-related changes in biosynthesis is based on very few species, mostly a few laboratory rodents, nematodes, diptera, and plants. The diversity of species differences in the age of onset and the particular characteristics of senescence described in Chapters 2–6 frustrate those who would establish changes and their causes as universal. Since many different pathophysiological conditions can contribute to exponential (Gompertzian) increases of mortality, it also seems likely that the macromolecular changes in those same age groups may differ among species. Moreover, even if the same molecular changes did occur during senescence in some species, we could not conclude that such parallel changes had the same role in physiological dysfunctions or mortality rate increases.

7.2.5. **In Vitro** *Cell Models of Organismic Senescence*

The discovery established by Leonard Hayflick that diploid human fibroblasts sustain only about fifty population doublings during serial transfer *in vitro* provided a model system that has major influence on thinking about senescence (Hayflick and Moorhead, 1961; Hayflick, 1965, 1977; Cristofalo, 1985). A finite proliferative capacity of many types of diploid cells during serial transfer *in vitro* is also found in fibroblasts from birds and other vertebrates (Cristofalo, 1985; Ponten, 1971, 1977; Williams and Dearfield, 1985; Norwood et al., 1990). These phenomena represent the loss of replicative capacity, which is equivalent to clonal senescence. Many molecular and cell biologists hoped that, at last, they had a model system for fundamental mechanisms of senescence under well-controlled conditions *in vitro*. Standardized early passage fibroblast cultures (WI–38, IMR–90) are used by most *in vitro* researchers and have helped give a consistent set of results among laboratories. However, there are various views on how cell replication during serial transfer constitutes a model for *in vivo* cell or organismic senescence.

Here lurks, I believe, a serious conceptual issue. Some cell types become postmitotic during their differentiation, as observed for most neurons in the brains of primates and humans (Rakic, 1985). The cytoarchectonics of many cells that have morphologically complex interactions with their neighbors, as in many neurons, could be a major limitation on consummating cell division. Even certain neurons, while postmitotic, may become polyploid (Lapham, 1968; Jacobson, 1970). Yet numerous types of neurons remain fully functional for many decades, and some may never show impairments, for example, those secreting vasopressin (Chapter 3.6.9). Thus, a prolonged and healthy existence is shown by many nonreplicating and/or postmitotic cells *in vivo*.

What, then, about postmitotic diploid cells during extended *in vitro* culture? While the potential for extended survival after the loss of replication with maintenance of many cell functions is well known to *in vitro* researchers, some others acquired the impression that diploid fibroblasts degenerate as their proliferative limit is approached (phase III).[2] In fact, cultures can remain more than a year in a permanently postreplicative state if properly cared for (Matsumura et al., 1979; Bayreuther et al., 1988) and retain numerous functions (Section 7.6.4). Degenerative changes in phase III that may lead to eventual cell death appear to occur on an extremely gradual schedule, at least by comparison with cell death during development (Section 7.7). The following discussions of gene expression and macromolecular synthesis will give examples from *in vitro* studies of diploid human fibroblasts, but without assuming that senescence *in vivo* shares the same mechanisms.

7.3. Differential Gene Expression

7.3.1. The Stability of Tissue-specific Characteristics

Because RNA and protein synthesis are so closely tied to genomic mechanisms of differentiation, a point of major significance is the extent to which differentiated cell characteristics persist throughout life. For example, Cutler (1982) hypothesized that an age-related dysdifferentiation caused impairments in cell function as the result of inappropriate regulation of gene expression. As judged by histological sections with ordinary protocols, cells generally retain their specific differentiated characteristics throughout the lifespan. Cells are clearly recognizable as neurons, hepatocytes, myocardial cells, etc., throughout life. Moreover, few enzymes and proteins in liver, kidney, heart, and brain change by more than 30% from values at maturity during adult life in rodents (Finch, 1972a; Wilson, 1973). Histochemical indications for dedifferentiation of skeletal muscle fiber types as judged by shifts in ATPase and glycolytic enzymes (Bass et al., 1975; Gutmann and Hanzlikova, 1976) merit reexamination with more quantitative assays. Moreover, analysis of proteins by two-dimensional gel electrophoresis showed that all abundant protein species continue to be made throughout the lifespan in the superior cervical ganglion of rats (Wilson, Hall, and Stone, 1978). These results imply continuing cell-type–specific patterns of transcription in mammals; nematodes and flies are discussed later.

With certain exceptions, most cells appear to maintain their main functions and differentiated characteristics throughout the lifespan. Obvious exceptions include: cells that are affected by age-related wasting diseases from gross organ pathology in brain, kidney, or vascular system that do not occur in all individuals; the major changes in the thymus during its medullary involution at maturation; the atrophy of ovarian steroid-dependent cells during reproductive senescence in female mammals; and the involution of leaves and other organs of plants. By comparison with these gross involutional changes, most

---

2. For example, several reviews on biochemical and molecular changes compared senescence *in vivo* (organismic) and *in vitro* by assuming that phase III was equivalent to lifespan and calculating changes in early phase III relative to resting cells in early passage. This seems questionable in view of the potentially long duration of phase III. It is fair to say that the main emphasis has been on impairments of DNA synthesis early in phase III and that slowly evolving changes during phase III have been relatively neglected (reviewed in Section 7.6.4). There also may be considerable differences among laboratories in the stage of phase III sampled for biochemical analyses.

changes in cell functions are selective and subtle. Few differentiated cell characteristics are lost, and in those that are, cell population changes may be suspected, with loss of cells that are competent to respond to a particular stimulus, as described below for the age-related decreases of α_{2u}-globulin in hepatocytes and IL–2 in lymphocytes (Section 7.4.1.2).

Isozymes might also inform about the stability of gene expression, since their relative proportion is often tissue-specific. However, the relative proportion of isozymes for hepatic glucose–6-phosphate dehydrogenase did not change consistently with age between genders and genotypes in rats (Wang and Mays, 1978). These observations suggest that age-related changes in gene expression are not gross and that protein-DNA complexes in the cell nucleus that determine the selectivity of gene expression must therefore be relatively stable over the lifespan.

Replacement of short-lived cell types continues throughout the lifespan. This implies that the mechanisms of differentiation from progenitor cells also continue to function normally. Hemopoiesis is a good example of this (Section 7.6.1). The rates of cell replacement may, however, become slowed. Age-related changes in replaceable immune cells are more complex than for erythroid cells, because of clonal regulation through autoanti-idiotypic antibodies (Chapter 3.6.6). Each cell type must be separately evaluated in regard to these issues.

7.3.2. Nonprogrammed Derepression of Nuclear Genes

Some repressed genes may spontaneously reactivate during the lifespan with considerable variations between similar cells. X-linked genes in mammals are useful for studying this question, since one X chromosome is genetically repressed in somatic cells. Autosomal genes influencing coat color that were introduced to the X chromosome by translocation have increased expression at later ages (Cattanach et al., 1969; Cattanach, 1974). These phenomena are also called "variegated position effects." The closeness of these translocated genes to the inactive X-heterochromatin is thought to have initially prevented their expression. However, select patches of fur in some mice became progressively darker between 3 and 12 months, indicating changes in melanocyte enzyme functions, apparently through derepression of the translocated genes. The patches did not spread, suggesting that the changes are restricted to discrete melanocyte clones. Moreover, the adjacent, translocated autosomal coat color genes were observed to be reactivated in a sequence in time that was inverse to their distance from the inactivating X-heterochromatin. More detailed investigation of another gene at the cellular level has recently extended these phenomena to the liver. The X-linked gene for ornithine carbamoyl transferase, which can be histochemically discriminated, shows a progressive, apparently random increase in activation in mice, increasing fiftyfold between 2 and 17 months (Wareham et al., 1987).

The generality of this intriguing phenomenon is unclear. A subsequent study gave a counterexample by failing to detect any age-related trend for reactivation of the X-linked *HPRT* locus (hypoxanthine phosphoribosyl phosphotransferase) in fibroblasts of women who were heterozygous for HPRT deficiency (Lesch-Nyhan syndrome). Since X inactivation occurs randomly, if the *HPRT*$^+$ allele became reactivated, then an increasing fraction of fibroblasts should be positive. Fibroblast cultures established from donors of ages across the lifespan were examined for enzyme activity. Although there was a 30% increase of *HPRT*+ clones at ages between birth and maturation, there were no further changes up to 72 years (Migeon et al., 1988).

These examples also pertain to the hypothesis of nonprogrammed derepression of genes (the dysdifferentiation.hypothesis), as a mechanism in senescence (Ono and Cutler, 1978; Cutler, 1982). A search for inappropriate expression of tissue-specific genes in 2-year-old rats, however, failed to detect aberrant expression of myelin basic protein (brain-specific), hepatic albumin, skin type II keratin, atrial natriuretic factor, or spleen κ-chain immunoglobulin outside of the tissues that normally contain these mRNA types (E. L. Schneider et al., in prep.; Edward Schneider, pers. comm.). Use of polymerase chain reactions to detect novel sequences is an obvious approach to the dysdifferentiation hypothesis.

7.3.3. *Spontaneous Reversion of Genetic Defects in Somatic Cells*

Two examples have come to light that present one of the most intriguing new puzzles of molecular gerontology, in which rats that are homozygous for a defined mutation in a mapped sequence show progressive reversion of the phenotype in individual cells, as detected immunocytochemically. The sequences of the revertant proteins, RNA, or DNA have not been characterized, and the discussion of these phenotypes must be considered provisional. These examples are discussed in the context of other changes in gene expression, although the ultimate basis may prove to be mutational.

Analbuminemic rats with a genetic defect that prevents albumin synthesis display an age-related increase of albumin synthesis, which superficially resembles the breakthrough of repressed X-linked genes described above. Homozygotes carrying this autosomal recessive trait have a seven-nucleotide deletion in their albumin gene across the 5'-splicing junction of the HI intron, which blocks splicing and increases the content of HI intron sequences in liver nuclear RNA (Esumi et al., 1985). Nonetheless, homozygotes still produce trace quantities of an albumin-like protein and mRNA (Makino et al., 1982; Esumi et al., 1982). A similar low-level "leakiness" is described in the analbuminemic humans (Ruffner and Dugaiczyk, 1988). Contrary to expectations, the near absence of this major, multifunctional plasma protein has little consequence to adults.

In the analbuminemic mutants, the production of albumin is restricted to rare hepatocytes by immunohistochemical staining, $< 10^{-3}$ in neonates. Remarkably, there is a fiftyfold increase of immunoreactive hepatocytes by 22 months (Esumi et al., 1985). Moreover, the carcinogen 3'-methyl–4-dimethylaminoazobenzene further augmented the age-related increase of serum albumin (Makino et al., 1982). Several mechanisms could be involved: activation of an albumin pseudogene with open reading frames by analogy with the derepression of X-linked genes; increased efficiency of splicing; or mutations. It seems unlikely that mutations could create new splice junctions with such frequency, however. Albumin-immunopositive hepatocytes persisted after hepatectomy and regeneration. The reduction of single cells and increase of clusters in the regenerated liver suggests that the change permitting albumin synthesis is clonally heritable (Makino et al., 1986).

A similar phenomenon was just reported in the Brattleboro rat, which is homozygous for a single DNA base deletion in part of the vasopressin gene that causes a frame-shift mutation in the C-terminal glycoprotein (Schmale and Richter, 1984). Consequently, the intracellular processing and axoplasmic transport of vasopressin are blocked. Despite chronic diabetes insipidus (*di*) with polydipsia-polyuria, the homozygotes (*di/di*) can survive if carefully maintained. With age from birth onwards, solitary neurons that are immunopositive for vasopressin and the normal C-terminal peptide increase progressively (Figure 7.1; van Leeuwen et al., 1989). By immunocytochemistry, the phenotypic revertant cells also contain the mutant C-terminal peptide, and therefore are heterozygous (*di/+*). Nonetheless, vasopressin is not released into the blood circulation, and the diabetes insipidus persists.

A mechanism is hypothesized through somatic intrachromosomal gene conversion (van Leeuwen et al., 1989; Ivell, 1987; Ivell and Richter, 1984), in which the mutant gene interacts through various known mechanisms with the nearby oxytocin gene, which has normal sequences for the mutant C-terminal peptide (Figure 7.1). There is also some evidence for age-related loss of ribosomal RNA cistrons through intragenic recombination, but this is highly controversial in view of recent findings discussed in Chapter 8.2.3. An alternative mechanism in the recovery of vasopressin production by Brattleboro rats may be considered on the basis of suggestions that glucose–6-phosphate dehydrogenase in humans is a fusion protein, representing coding sequences on genes from chromosome 6 and the X chromosome (Kanno et al., 1989; Luzzatto, 1989). The mechanism might involve a single template formed by trans-splicing of hnRNAs or two templates for different peptides that are ligated during synthesis (ribosome hopping), or through post-translational transpeptidation.

Whatever mechanisms may account for the reversion of vasopressin deficits in Brattleboro rats, I predict the reciprocal process, in which neighboring genes with homologous sequences become inactivated over time through such mechanisms. Tentative evidence for

A

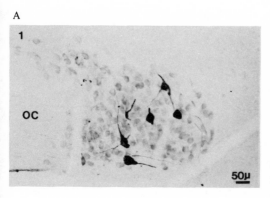

Fig. 7.1. Age-related spontaneous incidence of heterozygous phenotypes in neurons of the homozygous Brattleboro rat. The Brattleboro rat has a frame-shift mutation in a portion of the vasopressin gene that codes for a C-terminal glycoprotein (Schmale and Richter, 1984). The mutation blocks the intracellular processing and axoplasmic transport of vasopressin, thereby causing chronic diabetes insipidus in the homozygotes (*di/di*). However, solitary neurons that are immunopositive for vasopressin and the normal C-terminal peptide increase progressively from birth onwards. The phenotypic revertant cells still contain the mutant C-terminal peptide as determined immunocytochemically (not shown) and therefore are heterozygous phenotypes (*di/+*). Nonetheless, no vasopressin is released into the blood circulation, and the diabetes insipidus persists. Panel A shows neurons that are immunopositive for the C-terminal fragment of the vasopressin precursor in the magnocellular nucleus of Brattleboro homozygotes (*di/di*).

(continued)

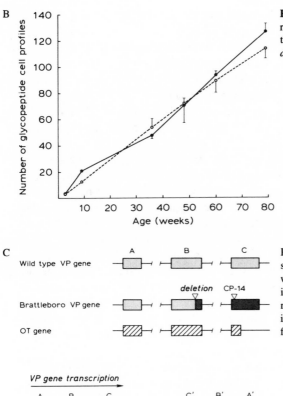

Fig. 7.1. (*continued*)   Panel B shows the linear age-related increase of neurons that are immunopositive for the vasopressin glycoprotein precursor (see Panel A) in *di/di* homozygotes.

Panel C and D: A mechanism is hypothesized through somatic intrachromosomal gene conversion (van Leeuwen et al., 1989; Ivell, 1987), in which the mutant gene interacts with the nearby oxytocin gene, which has normal sequences for the mutant C-terminal peptide and an inverted orientation (Mohr et al., 1988). Photographed from van Leeuwen et al., 1989.

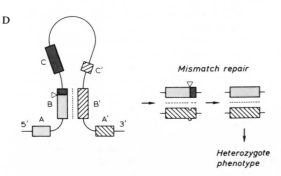

inactivation of the vasopressin gene, the opposite trend shown by Brattleboro rats, is indicated by neurons in the suprachiasmatic nucleus of the male Wistar: while the total number of suprachiasmatic neurons did not decrease in rats up through 33 months, the number of vasopressin immunoreactive neurons decreased in this region by 25% (Roozendale et al., 1987; Fliers and Swaab, 1983; also shown in Richardson, 1990, 291). The disappearance of mRNA through such mechanisms would be very rare if both the maternal and paternal gene copies were transcribed.

7.3.4. DNA, Chromatin, and Changes in DNA Methylation

At a simple level of analysis, there are few notable changes in the genomic apparatus during the adult lifespan in most cells of mammals (reviewed in Rothstein, 1982). As for DNA itself, there is no nonintegral loss or gain of nuclear diploid content of DNA (Enesco, 1967; Shima and Sugahara, 1976). The polyploidy of hepatocytes (e.g. Shima and Sugahara, 1976) and some brain cells (Jacobson, 1970) begins to be manifested soon after birth and is not associated with any known alteration in function. Moreover, no age-related changes in proportions of DNA frequency classes or in the amount of single-copy DNA have been detected by reassociation kinetics of DNA from brain and liver (Cutler, 1975; Rothstein, 1982). Most studies have failed to detect robust age-related changes in the protein composition of chromatin or in modifications such as acetylation (Rothstein, 1982).

Other evidence, however, suggests changes in chromatin structure. The gene for α-fetoprotein becomes more sensitive in 2-year-old rats to nucleases during autodigestion of liver chromatin (John Papaconstantinou, pers. comm.); during maturation, the capacity to express this gene is sharply reduced (Section 7.6.2). Increased sensitivity to nucleases frequently accompanies activation of genes and usually indicates a local change in chromatin structure (Burch and Weintraub, 1983). The mechanisms are obscure, but their probable occurrence in humans and rodents suggests an interesting new topic. Scattered reports also show age-related changes in chromatin structure that imply reduced transcription. The nucleosomal DNA segment length increases with age in human cortical neurons (Ermini et al., 1978), while the DNA of brain chromatin becomes less digestible by nucleases (Chaturvedi and Kanungo, 1985; Modak et al., 1986). Cytological evidence, mainly from the USSR, indicates increased condensation of chromatin in peripheral lymphoblasts (Lezhava, 1984), while the number of Barr bodies, the heterochromatinized X chromosome of females, progressively decreased (Voitenko, 1980); this later report would be consistent with the activation of repressed genes (Section 7.3.2). As described below for specific liver genes, there is good evidence that changes in chromatin structure are highly selective.

Age changes in DNA methylation are of interest because demethylation of CpG dinucleotides in vertebrate DNA is correlated with increased transcription of the associated gene (Cedar, 1988; Mays-Hoopes, 1989). For example, 5-azacytidine, which can block methylation, can also reactivate repressed genes on the X chromosome (Mohandas et al., 1981). In general, guanine is the methylated base, at the 5' position; the CpG dinucleotides, which contain most of the methyl groups, are relatively rare. Methylation in promoter sequences has been found in one case to influence transcription by altering binding of the transcription factor MLTF (Watt and Molloy, 1988).

Age-related DNA demethylation was first indicated by Vanyushin et al. (1973), who found marked tissue differences in rats, with greater decrease in brain than liver; most of the 15% decrease in brain occurred between 1 and 12 months. These findings have been extended and verified. Two families of repeated sequences become progressively demethylated during the mouse lifespan, as judged from the breaking of DNA by methylation-sensitive restriction nucleases: the IAP sequence family (1,000 copies; retrovirus-related; Mays-Hoopes et al., 1983) and the interspersed L1Md (or M1F1) family (10,000–100,000 copies), of which 8% became demethylated (Mays-Hoopes et al.,

1986). Another approach using high-performance liquid chromatography (HPLC) gave similar results of about a 25% total loss by the average lifespan for total liver DNA (Singhal et al., 1987) and for the major mouse satellite sequences (tandem repeats) in C57BL/6J mice (Mays-Hoopes, 1989). An interesting comparison of C57BL/6J mice with *Peromyscus leucopus* (Chapter 3.6.1) showed a slower rate of demethylation in the latter, which lives twice as long (Wilson, Smith, Ma, and Cutler, 1987). The only specific gene yet examined for demethylation is the proviral mammary tumor viral *MTV–1* locus, where demethylation may be a factor in late onset viral-associated mammary tumors (Etkind and Sarkar, 1983; Section 7.4.1.4).

Increased DNA methylation is also recently reported, in ribosomal RNA gene clusters, where CBA mice showed hypermethylation by midlife in 5' spacer sequences from brain, liver, and spleen. Consistent with the inverse relation implied between methylation and transcription, the ribosomal gene cluster on chromosome 16 appears to be selectively inactivated (Swisshelm et al., 1990). Rath and Kanungo (1989) detected increased methylation of incompletely defined repetitive sequences in rats. Taken together, the data on DNA methylation indicate multiphasic and multidirectional age-related changes with considerable sequence selectivity that implies corresponding selectivity in changes in gene expression. Evidence for this selectivity is given in the next section.

7.4. RNA and Protein Synthesis

The following view of a large, but spotty, literature illustrates the reported changes that I consider the most reliable, but cannot be comprehensive. Because of the issues of species differences, mammals, nematodes, insects, and plants are treated separately. Table 7.1 summarizes age-related changes at different steps in gene expression, from the characteristics of DNA and macromolecular processing from transcription through post-translational modifications.

7.4.1. Mammals

7.4.1.1. Bulk RNA and Protein Synthesis

Data on humans are reviewed separately from rodents since different techniques are used. Several sophisticated studies show no evidence for age changes of bulk protein synthesis or amino acid metabolism of healthy, nonobese, and nondiabetic men to ages beyond the average lifespan. As determined by the metabolism of ^{15}N-glycine, daily total body protein metabolism did not change when normalized for whole-body weight. Albumin synthesis, estimated by the fractional distribution of ^{15}N-glycine, was also not altered with age; however, albumin synthesis was less responsive during a shift to a low protein diet (Gersovitz, Munro, Udall, and Young, 1980; Gersovitz, Bier, Matthews, Udall, Munro, and Young, 1980). The stability of body protein synthesis rates at ages beyond the average lifespan was confirmed by the infusion of ^{13}C-leucine for 6 hours under conditions that stabilized the blood glucose, a glucose clamp (Fukugawa et al., 1989). Moreover, all ages showed similar increases in protein synthesis in response to elevated amino acids and insulin. On the other hand, glucose disposal decreased with age (Fukugawa et

7.1   Postmaturational Age-related Changes in Macromolecules and Gene Expression: Sundry Examples

Change	Species
Genomic apparatus	
DNA adducts	H (brain); R (kidney)
DNA rearrangements	no evidence except for that associated with immune cell differentiation
DNA demethylation*	M (liver, brain); R (liver)
Chromatin nuclease sensitivity decreases	R (brain)
Nuclear matrix, redistribution of active genes	R (liver)
Transcription and hnRNA processing	
Whole-tissue poly(A)RNA unchanged*	R (liver, brain)
Rate of transcription	
a. Unchanged, c-myc*	M (lymphocyte)
b. Slows, α_{2u}-globulin*	R (liver)
c. Increases, SMP-2	R (liver)
Rearrangements of hnRNA during processing	R (liver, brain)??
Altered ratio of alternately spliced APP-mRNA	H (brain, Alzheimer disease)
Prevalence of hnRNA increases	
Ovalbumin	A (oviduct)
Albumin (analbuminemic)	R (liver)
Slowed hnRNA processing	no example
Translation	
Rates of protein synthesis	
Unchanged, total*	H (nonmuscle, albumin)
Decreases, total*	R (liver, brain)
Decreases, preproinsulin	R (pancreas)??
Post-translational modifications	
Propeptide maturation, glycosylation	R (liver)
Passage through the endoplasmic reticulum and Golgi, transferrin	R (liver)
Damage to mature proteins	
Deamidation and oxidation, numerous examples*	R, M (liver, brain, lymphocyte)
Racemization*	H, D (lens crystallin)
Glycation and further modifications (formation of AGE)*	H (collagen, osteocalcin, lens crystallin)

Sources: Given in text.
Notes: A, avian; D, dog; H, human; M, mouse; R, rat; ??, inferred; *, independent confirmation published. These examples are collected from many sources, and only in a few cases have independent confirmations been published. This listing should not be interpreted to mean that any or all changes occur in all cells, either singly or in combination.

al., 1988; Fukugawa et al., 1989), which is consistent with numerous other studies (Chapter 3.6.8). These studies do not of course inform about the numerous individual steps of macromolecular synthesis in the major organs that might differ subtly. Most of the plasma amino acids in the postabsorptive state are thought to be taken up by the liver and other viscera served by the splanchnic vascular bed, whereas glucose is mainly metabolized by skeletal muscle and brain. These results set the stage for considering that macromolecular biosynthesis is not generally impaired throughout the human lifespan, notwithstanding the impact of specific diseases. These findings are in strong contrast to many reports of major age-related decreases in protein synthesis in rodents.

Before reviewing tracer studies on rodent protein synthesis with age, I will first consider basal levels of bulk constituents. Few enzyme activities in most tissues of rodents change substantially across the lifespan (Finch, 1972a; Wilson, 1973). These generally modest changes indicate the continuing transcription of the respective genes, since the half-lives of most mRNAs are less than 2 weeks in vertebrate cells. Similar conclusions are drawn from studies of RNA sequence complexity by my laboratory, which gives a direct measure of the number of different active genes and which did not detect age-related changes in polysomal poly(A)RNA from the whole brain of two strains of rats (Colman et al., 1980). The technical limits of about ± 5% do not exclude major changes of individual mRNA species, however (see below). Most studies agree that bulk levels and properties of rodent polysomal RNA or poly(A)RNA fractions do not change with age in liver and brain, including size, translatability, length of the poly(A) tract, or cap structure (Birchenall-Sparks, Roberts, Rutherford, and Richardson, 1985; Moudgil et al., 1979; Colman et al., 1980; Cosgrove and Rapoport, 1987).

The maintained content of bulk RNA and protein in most cells of rodents (reviewed in Barrows and Kokkonen, 1981) is consistent with stability of most enzyme activities, but does not inform about the rates of synthesis and turnover. Many conclude from studies with radiolabelled precursors that declining RNA and protein synthesis occurs generally in most cells during organismic senescence (Richardson et al., 1985; Rothstein, 1982; Webster, 1986). However, I and others (Makrides, 1983) doubt the interpretability of most tracer studies on bulk RNA and protein synthesis, especially when the differences are small (< 25%). Few penetrating analyses are available, and the absence of pool corrections in most studies severely weakens conclusions. The half-life of labelled bulk polysomal poly(A)RNA from liver (Moore et al., 1980) or bulk (ribosomal) RNA from brain (Menzies and Gold, 1971) was not different in young and old rats. However, another study that assayed precursor pools showed in old rats a slower clearance of the tracer, which, when accounted for, indicated a twofold slowing of poly(A)RNA turnover (Moore et al., 1980). More studies of turnover of specific proteins are needed to resolve these questions. In view of the steady-state levels of most cellular proteins across the lifespan, decreased protein synthesis must be coordinated by commensurate decreases in protein turnover, and there is some evidence for this (Section 7.5.1). Age changes in particular proteins and cells are described in Section 7.4.1.2.

The rate of protein synthesis peaks several months after puberty in rodents and declines progressively, but with major differences among organs. Age-related changes in cell-free systems from the rodent liver, brain, skeletal muscle, etc., vary widely, from + 20 to − 80%, though most studies report declines (Moldave et al., 1979; Ekstrom et al., 1980; Richardson et al., 1985; Sojar and Rothstein, 1986; Blazejowski and Webster, 1984). The size of the changes *in vivo* (Ingvar et al., 1985; Moore et al., 1980; Gozes et al., 1981) is comparable to short-term incubations of isolated cells from kidney (Ricketts et al., 1985) and liver (Birchenall-Sparks, Roberts, Rutherford, and Richardson, 1985).

The most definitive studies used perfusion to give direct exposure of the tissue to the same concentrations of precursors. With *in situ* perfusion of the liver to control the specific activities of valyl-tRNA, liver protein synthesis and degradation showed parallel changes, with peak at 6 months, decline by 12 months, no changes to 18 months, and decline to 40% of the 3-month value by 24 months (Ward, 1988; Figure 10.10); these studies also showed that diet restriction blunts and postpones decrements with age, as discussed in Chapter 10. Similar age-related decreases in total heart protein synthesis between 9 and 25 months were described from studies of the *in vitro* perfused mouse heart, again with corrections for pools (Geary and Florini, 1972). In contrast to these studies, C57BL/6J mice did not show age changes up through the average lifespan in leucine pool size or in the entry of radiolabelled leucine into the pool in liver and heart; nor did protein-specific activity decrease with age in liver or heart (Du et al., 1977). More studies of specific proteins are needed to resolve this question. The age-related increases of albumin mRNA and albumin synthesis in rat liver (Section 7.2.1) or the increased gonadotropin synthesis in the female pituitary (Section 7.2.2) would confound studies on the levels, synthesis, or turnover of total RNA or protein in these particular tissues.

Individual steps of polypeptide synthesis do not give consistent age-related changes; nearly all outcomes (increase, no change, decrease) are reported for the aminoacylation of tRNA in whole rodent organs (Richardson and Birchenall-Sparks, 1983; Richardson et al., 1985) and other steps. Deficiencies in elongation factor 1 (EF–1α, or EF–Tu) reported for *Drosophila* (Section 7.4.3.1) were not found in rat liver cytosols by the average lifespan (Sojar and Rothstein, 1986), while progressive decreases in EF–1 may occur in early postnatal life up through 1 year in mice and rats (Castenada et al., 1986). Elongation factors are generally thought to be in large excess and not rate limiting for protein synthesis. In some mammalian cells, their mRNA can be stored on ribonucleoprotein particles and shows major changes in the amount loaded on polysomes at different stages in the cell cycle (Rao and Slobin, 1987). Formation of 40S and 80S initiation complexes by a cell-free system from whole rat brains was not impaired up to 24 months of age, nor was translation of endogenous or exogenous (β-hemoglobin) mRNA (Cosgrove and Rapoport, 1987).

In conclusion, whatever changes occur clearly do not constitute evidence for general age-related shutdown of RNA or protein synthesis, or genomic failure in any organ. Balancing decreases in synthesis and turnover of bulk RNA would be needed to maintain the stable levels of RNA and proteins. Whatever the changes, they must be precisely regulated. Moreover, many individual proteins show no impairments in synthesis.

7.4.1.2. *Specific Genes and Proteins*

Besides albumin of rodents (Section 7.2.1) and humans (Section 7.4.1.1), a few other individual mRNA and proteins have been examined. The first examples concern unusual data showing tissue specific age changes in keratin synthesis. Then I review examples of molecular changes in cells that are capable of proliferating (liver, lymphocytes), followed by a discussion of RNA changes in brain.

Keratin, the major protein of hair and nails, shows marked age changes in production as deduced from the simplest of measurements. To no embarrassment of modern molecular techniques, it is possible to directly observe the delayed penetrance at mid-life of the gene hairy pinna (Figure 7.2). The growth of coarse hairs on the pinna of the ear is a Y-linked trait and is typically expressed after 30 years (Slatis and Apelbaum, 1963; Slatis, 1964). This example indicates increased keratin synthesis by pinnal hair follicles in this genotype. In general, the fine vellus hairs on the ear become coarser (Montagna and Giacometti, 1969). Elsewhere, hair growth shows regionally selective changes, with no slowing on the eyebrows but a general trend in both men and women for decreased growth of

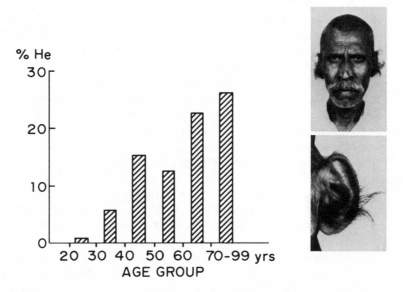

Fig. 7.2. The cumulative age-related expression of the gene "hairy pinna" (*He*) in a Middle Eastern population where the genotype is common. The phenotype causes mid-life onset of coarse hair growth on the pinna of the ear. The Y-linkage assignment has not been verified with Y-specific probes in RFLP analysis. Graph was redrawn from Slatis, 1964; picture was rephotographed from Stern et al., 1964.

scalp hair after 65 (Giacometti, 1965; Selmanowitz et al., 1977; Myers and Hamilton, 1961). In male-pattern baldness, which is determined by an androgen-regulated gene that is not Y-linked (Chapter 10.4.1), hair follicles become miniaturized with changes that are the converse of those in the ear that produce vellus hair. The temporal and regional specificity of these changes suggests exquisitely specific shifts in gene activity (Selmanowitz et al., 1977).

Nails also contain predominantly keratin. Nail growth decreases progressively across adult life, showing remarkable generalizability in humans (Orentreich and Sharp, 1967; Selmanowitz et al., 1977; Bean, 1980), rhesus monkeys (Bowden and Williams, 1985), and dogs (Selmanowitz et al., 1977). Nail growth is easy to measure through the outgrowth of a shallow nick, and could be usefully added to the ongoing studies of metabolism and age.

At a molecular level, many convincing examples show altered regulation of mRNA. One of the earliest is the decrease of mRNA sequences related to the major urinary protein (MUP) in mouse submaxillary glands, between 7 weeks (adolescence) and 6 months (pre–middle age) (Shaw et al., 1983). Like the changes in hair and nail growth described above, these early changes give precedent for considering that a mechanism in mammalian senescence could involve sequential changes in gene expression.

Other changes in the liver involve α_{2u}-globulin mRNAs that are also in the MUP group. The laboratories of Arlan Richardson and Arun Roy have reported marked decreases with age in MUP synthesis in male rodents, which are major illustrations of age-related changes at a transcriptional level with cell type selectivity. These urinary proteins are lacking in female rodents, but are induced by androgens (Motwani et al., 1984). Hepatic cytosol content and urinary excretion in F344 rats increase sharply to a pubertal peak and then gradually decline by more than 90% to below pubertal levels by the average lifespan (Roy et al., 1983; Motwani et al., 1984; Richardson, Butler, Rutherford, Semsei, Gu, Fernandes, and Chiang, 1987). Male C57BL/6J mice also showed decreases, but these were less marked than in rats of the same age (Richardson, Butler, Rutherford, Semsei, Gu, Fernandes, and Chiang, 1987). Others of the MUP gene family differentially change with age (Figure 7.3; Roy et al., 1983; Richardson, Butler, Rutherford, Semsei, Gu, Fernandes, and Chiang, 1987).

In contrast to α_{2u}-globulin, other mRNAs show marked increases, including α_1-antitrypsin, albumin mRNA (Richardson, Butler, Rutherford, Semsei, Gu, Fernandes, and

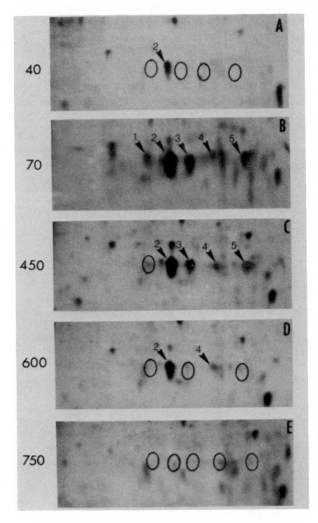

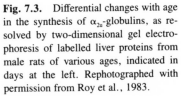

Fig. 7.3. Differential changes with age in the synthesis of α_{2u}-globulins, as resolved by two-dimensional gel electrophoresis of labelled liver proteins from male rats of various ages, indicated in days at the left. Rephotographed with permission from Roy et al., 1983.

Chiang, 1987), and a cloned RNA, SMP–2, that was not similar to other known sequences (Chatterjee et al., 1987). SMP–2 has the interesting property of being present at high levels in pubertal and senescent rats, when hepatic androgen sensitivity is least. Other liver mRNAs show a small (< 25%) increases or decreases across the lifespan in rats (α_1-glycoprotein; complement factors B, C3, C4; Rutherford et al., 1986). Thus, there is little support for the widely presumed generalized deficit in liver mRNA during organismic senescence.

The decrease of α_{2u}-globulin mRNA was traced to the transcriptional level. Both α_{2u}-gobulin mRNA and its transcription (assayed by nuclear run-on) decline in parallel after maturation (Figure 7.4; Chatterjee et al., 1981; Roy et al., 1983; Richardson, Butler, Rutherford, Semsei, Gu, Fernandes, and Chiang, 1987; Chatterjee et al., 1988). Analysis of nuclear components indicates that the α_{2u}-globulin gene dissociates from the nuclear matrix later in life when transcription decreases, while the albumin gene remains matrix-associated throughout life, as would be consistent with its continued activity (Murty et al., 1988). In contrast to α_{2u}-globulin transcription, the 65% lower levels of c-myc mRNA in lymphocytes from senescent mice occurred despite equivalent rates of transcription, as determined by nuclear run-on (Buckler et al., 1988). These studies show a selectivity that I suspect will emerge repeatedly as more genes are analyzed in other tissues.

Studies on whole liver can obscure changes in cell subpopulations that can be observed in relation to the hepatic arteries and veins. Perivenous hepatocytes have relatively more α_{2u}-globulin (Roy et al., 1986). Presumably this subpopulation in young adult male rats has one-third of hepatocytes separated by fluorescence-activated cell sorting with the highest levels of α_{2u}-globulins (Motwani et al., 1984); and this same fraction of cell is induced to form α_{2u}-globulins in females after androgen treatment. Up through middle age (18 months), this fraction of hepatocytes continues to have abundant α_{2u}-globulin, though the levels decrease progressively. But by 30 months (average lifespan), hepatocytes that synthesize α_{2u}-globulin have virtually disappeared (Figure 7.5). This major change is startling and difficult to explain. Issues of changes in cell subpopulations arise in nearly all tissues.

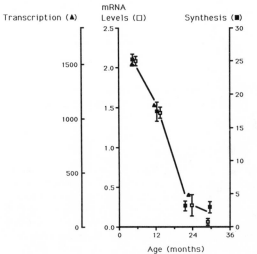

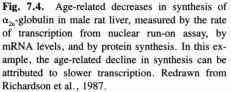

Fig. 7.4. Age-related decreases in synthesis of α_{2u}-globulin in male rat liver, measured by the rate of transcription from nuclear run-on assay, by mRNA levels, and by protein synthesis. In this example, the age-related decline in synthesis can be attributed to slower transcription. Redrawn from Richardson et al., 1987.

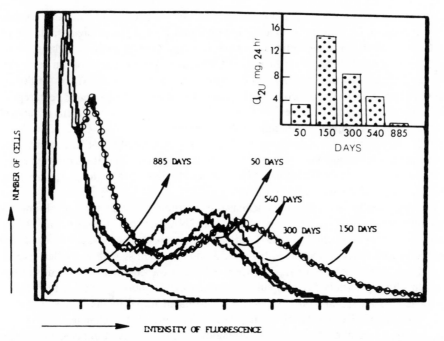

Fig. 7.5. Cell subpopulation shifts in the male rat liver as a function of age, measured by fluorescence-activated cell sorting with antibodies to α_{2u}-globulin. The cell subpopulation with high levels of α_{2u}-globulin nearly disappears by the mean lifespan. The insert shows the decrease in urinary excretion of α_{2u}-globulin. Photographed with permission from Motwani et al., 1984.

Other genes in the MUP gene cluster, among thirty different coding sequences on mouse chromosome 4, are expressed in secretory cells, including the submaxillary gland, and show decreases in mRNA after maturation of male and female mice with a time course like that of α_{2u}-globulins in the liver (Shaw et al., 1983). These changes may have several hormonal causes. Testosterone secretion typically decreases markedly by mid-life in laboratory rats, a change that is best documented by sampling across the diurnal cycle (Steger et al., 1981; Miller and Riegle, 1982). The F344 rats used in most of these studies also have a high, strain-specific incidence of benign testicular (interstitial cell) tumors by middle age (Goodman et al., 1979). Besides the age-related reductions of androgenic stimulation, the capacity of hepatocytes to respond may be reduced because of major decreases in cytosolic androgen-binding protein, but neither the cytosolic androgen-binding proteins nor α_{2u}-globulins were restored by 25 days of dihydrotestosterone treatment (Roy et al., 1974).

Calorically restricted rats did maintain transcription of hepatic α_{2u}-globulin mRNA (Richardson, Butler, Rutherford, Semsei, Gu, Fernandes, and Chiang, 1987). Other effects of diet restriction on transcription are described in Section 7.4.1.3 and in Chapter 10.3. There may be relationships between decreased testosterone stimulation, loss of androgen binding proteins, and the loss of α_{2u}-globulin synthesis. Other hormones could also be involved, since α_{2u}-globulin synthesis is stimulated by glucocorticoids, growth hormone, insulin, and thyroxine (Lynch et al., 1982), but inhibited by estrogens (Roy et al., 1975); age changes are reported for all of these hormones in male rodents.

Shifts in lymphocyte subpopulations may also account for age changes in lymphocyte gene expression. In young mammals T lymphocytes (Lyt–1⁺) that are stimulated to proliferate by antigens or plant lectins (concanavalin A) secrete the polypeptide interleukin 2 (IL–2; also called T-cell growth factor, TCGF) during late G_1 phase of the cell cycle. In turn, IL–2 stimulates the proliferation of cytotoxic T lymphocytes. Major ($\geq$ 80%) age-related decreases in the production of IL–2 and its receptor, IL–2R, have been repeatedly shown, by lymphocytes from peripheral blood of humans and from spleens of lab rodents (Weksler, 1985; Thoman and Weigle, 1981; Kennes et al., 1983; Cheung et al., 1983; Nagel et al., 1988). (The decreased IL–2 probably has ramifying consequences on immune cell responses dependent on this lymphokine; supplemental IL–2 restores proliferation of splenic B cells from old mice [Thoman and Weigle, 1981] and peripheral lymphocytes in humans [Kennes et al., 1983].)

There are parallel decreases in concanavalin A–stimulated spleen cell proliferation, total protein synthesis, IL–2 secretion, and IL–2 mRNA from rats (Cheung et al., 1983; Wu et al., 1986) and IL–2 mRNA and IL–2R mRNA from humans (Nagel et al., 1988). In rats, the largest decrease occurred between 5 and 13 months, with a smaller decrease to 29 months; the time course of IL–2 mRNA induction was identical. A recent study, however, with highly enriched T-cell subpopulations from mouse spleens, which constitute 20% or less of all spleen cells, showed no age-related impairments in the production of IL–2 or IL–2 mRNA (Holbrook et al., 1989). The major age change is most likely a decrease in the number of splenic T lymphocytes that can be stimulated to proliferate, rather than an intrinsic change in their capacity to produce IL–2. Deficiencies in the pool of pre–T cells in spleens of old mice are suspected to underlie the loss of responsive cells (Thoman and Weigle, 1981); this question remains open for peripheral T-lymphocyte populations. These results again indicate the importance of distinguishing changes that result from cell population shifts, which can easily lead to mistaken impressions that responses *per cell* are altered.

The processing of heterogeneous nuclear RNA (hnRNA) into mature mRNA in the cell nucleus has not been well studied as a function of age, but there are indications of changes. In the study of *c-myc* mRNA of lymphocytes from senescent rats mentioned above, the absence of changes in transcription and in the half-life of cytoplasmic mRNA imply that post-transcriptional events could account for the major reduction of mRNA (Buckler et al., 1988). In another model, twofold increases of high-molecular-weight RNA that contained ovalbumin sequence were found in the oviduct of nonlaying, 8-year-old hens (*Gallus gallus domesticus*; Schroeder et al., 1985); these forms could be precursors of ovalbumin mRNA and, if so, could result from decreased efficiency of hnRNA processing. Oviduct dysfunctions are further discussed in Section 7.4.1.3. The age-related increases in the synthesis of albumin mRNA in an analbuminemic strain of rat might also represent altered splicing (Section 7.3.3).

Finally, I mention indications of post-transcriptional and post-translational changes. The synthesis of the insulin precursor, preproinsulin, in response to increased glucose becomes impaired by middle age in rats, as studied in isolated pancreatic islets (Wang, Halban, and Rowe, 1988). Because there were no age differences in preproinsulin mRNA or in its responses to glucose, the locus of age-related change may be at the translational level (Wang and Rowe, 1988). A different age-related change concerns post-translational processing, as shown for transferrin. Transferrin, a glycoprotein secreted by the liver, has a 25% decreased rate of synthesis in senescent mice (Ara et al., 1983, in prep.). In contrast to typical trends in most senescent rats (Section 7.2.1), albumin synthesis was also decreased. Striking age changes occurred in the post-translational processing of transferrin, which took three times longer in 2-year-old mice to transit from the rough endoplasmic reticulum (ER; see below) through the Golgi complex before secretion. The glycosylation of transferrin was also slowed. Albumin, which is not a glycoprotein, was secreted by the liver without delay in senescent mice.

The selectivity of age-related changes in gene expression in peripheral cellular subpopulations is well supported in the brain. Brain-region selective decreases are found for total RNA (predominantly ribosomal RNA) in the striatum (Chaconas and Finch, 1973; Shaskan, 1977). Nucleolar shrinkage is much greater in neurons of the substantia nigra, which project to the striatum, than in locus ceruleus in neurologically normal humans (Figure 7.6; Mann and Yates, 1979). This differential nucleolar shrinkage is correlated with reduced cytoplasmic RNA in these cells (Mann et al., 1977). Differential age changes in bulk RNA are seen in many other types of neurons (reviewed in Chaconas and Finch, 1973; Uemura and Hartmann, 1979; Uemura, 1980). While these morphological studies do not inform about RNA turnover, the nucleolar size is a good index for the synthesis of ribosomal RNA. Recent studies from my laboratory have produced a new model for these changes, in which the substantia nigra neurons that survive 3 to 9 months after lesions by the neurotoxin 6-OHDA have smaller nucleoli and selective decreases in tyrosine hydroxylase mRNA (Pasinetti et al., 1989).

In terms of specific mRNA, no changes occurred in ID mRNA of whole mouse brain, a brain-specific transcript related to the alu-2 sequence family (Anzai and Goto, 1987), while hypothalamic content of proopiomelanocortin (POMC) mRNA decreased by 30% (Nelson, Bender, and Schachter, 1988). The first indications of changes in alternately processed mRNA transcripts are given by the increase of a mRNA subspecies for the β-amyloid precursor protein (APP). The APP–751 mRNA, which has an exon coding for a

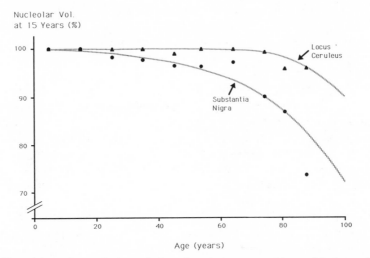

Fig. 7.6. Differential age changes in the nucleolar volume in two neuronal types of the human brain that are at risk for age-related neuron loss (McGeer et al., 1977; Vijayashankar and Brody, 1979). The subjects were neurologically normal, without Parkinson's or Alzheimer's disease. Redrawn from Mann and Yates, 1979.

Kunitz-type protease inhibitor (KPI), increases relative to APP–695 mRNA, which lacks the KPI exon. This shift only occurred in neurons of regions in the Alzheimer's brain that have a high density of amyloid-containing neuritic plaques, which are characteristic of Alzheimer's disease (Johnson et al., 1988, 1989, 1990). These results could derive from different hnRNA processing or mRNA stability in the cytoplasm. The turnover rates of these mRNA species are unknown.

Morphometric changes in ER, which could be important to polypeptide assembly or processing, are also cell-specific. The rough ER area increased in Purkinje neurons (cerebellum) but was unchanged through 30 months in mitral neurons of the olfactory bulb (Hinds and McNelly, 1978) and liver (Schmucker et al., 1977). Studies on protein synthesis in the rat brain, using a nonrecyclable precursor, also showed regional selectivity in age changes. Hypothalamic zones were among the twenty-six of thirty-nine regions that did not decrease between 6 and 23 months, while whole brain showed an average 15% decrease (Ingvar et al., 1985). Axonal flow is another assay for bulk protein synthesis. The slow kinetic compartment of axonal flow shows progressive slowing to 40% by the lifespan in motor and optic nerves of F344 rats; again, some selectivity is indicated, since transport of cytoskeletal elements was much more altered than that of membranous vesicles (McQuarrie et al., 1989).

To sum up, age-related decreases in bulk protein synthesis are consistently observed in many organs of laboratory rodents, but the generalizability to humans and other longer-lived species is not established. The analysis of whole organs may obscure selective changes in different cell types. The capacity for increased synthesis does not similarly decrease in any of the few individual proteins that have been studied. As described next, the synthesis of some mRNA and proteins can be greatly increased by hormones.

7.4.1.3. Modulation of Nuclear Gene Expression

Studies on basal RNA and protein synthesis do not inform about the capacity for dynamic modulations of gene expression. Major questions can be posed in the context of age-related declines of homeostatic capacity at a physiological level, for example in thermoregulation (Table 3.3). We now ask to what extent differentiated cells are responsive to endogenous physiological stimulae and environmental changes. Most studies concern hepatocytes of laboratory rodents.

Many liver enzymes are physiologically regulated at the level of protein or RNA synthesis, or both. The following examples of age-related changes show important differences among enzyme systems. The induction of tyrosine aminotransferase (TAT) and tryptophan oxygenase (TO) by corticosterone in the rat is understood in particular detail, and has been an important system for probing age-related changes in gene expression. Within 30 minutes after increased occupation of chromatin-bound receptors from elevations of blood corticosterone following adrenal activation or injection of this steroid, there are increased levels of cytoplasmic TAT and TO mRNA (Rowenkamp et al., 1976; Wellinger and Guigoz, 1986). Corticosterone acts directly on hepatocytes (Goldstein et al., 1962; Britton et al., 1976), where it increases the rates of transcription as shown by run-on assays (Wellinger and Guigoz, 1986; Danesch et al., 1983). The increased synthesis of TAT and TO polypeptides and increased enzyme activity is attributed to increased mRNA availability (Kenney, 1962).

In 24- to 30-month-old male rats and mice (average lifespan), the increases of TAT and TO enzyme activity are just as rapid and large as in younger adults in response to large doses of injected corticosterone (Gregerman, 1959; Finch et al., 1969; Weber et al., 1980; Wellinger and Guigoz, 1986). Hepatocytes in short-term culture also can induce TAT in response to corticosteroids without effects from age (Britton et al., 1976; Weber et al., 1980). Moreover, the increases of TAT and TO mRNA show no influence of age (Figure 7.7, left; Wellinger and Guigoz, 1986). These observations indicate that major segments in the complex mechanisms of gene expression remain unimpaired through the average lifespan of laboratory rodents for this gene.

During responses to brief cold stress, in contrast to the above, there are marked age-related delays in TAT induction (Finch et al., 1969; Wellinger and Guigoz, 1986). Correspondingly, increases of TAT mRNA were smaller (Figure 7.7, right); the absence of age-

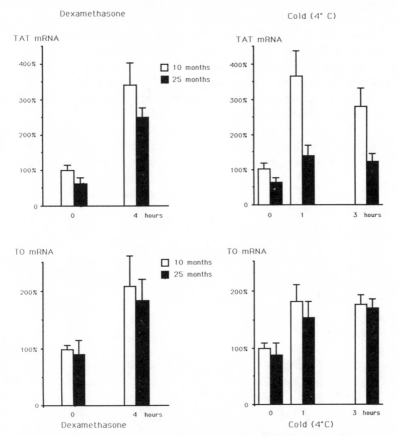

Fig. 7.7. The induction of mRNA for liver tyrosine aminotransferase (TAT) or tryptophan oxygenase (TO) in rat liver as a function of age: *open bars,* 3 months; *solid bars,* 28 months. *Left,* induction by injection of the corticosteroid dexamethasone, which acts directly on liver cells. *Right,* induction in response to cold stress of the whole rat, which causes complex neuroendocrine and autonomic responses that act on the liver indirectly. During cold stress, senescent rats show slower induction of TAT mRNA, whereas TO mRNA was induced as fast as in the young. The induction of TAT enzyme activity was also slower (see text). Redrawn from Wellinger and Guigoz, 1986.

related changes in the increase of TO mRNA emphasizes the selectivity of age changes, as discussed below (Wellinger and Guigoz, 1986). Eventually, TAT activity in old mice can reach the same levels as in the young (Finch et al., 1969). Similarly, hepatic glucokinase, which showed no effects of age on induction by directly acting hormones, also manifested age-related delays during responses to the stress of fasting; delayed increases of corticosterone are the likely proximal cause in this case (Figure 7.8; Adelman, 1970a, 1970b; Adelman et al., 1978).

It is of interest that the 40% decrease of plasma transferrin and slowed intracellular processing in senescent mice (Section 7.4.1.1) are reversed by treatment with corticosteroids (Ara et al., 1983) and may also be attributed to age-related changes in plasma corticosteroid levels. Because transferrin mRNA is increased by corticosteroids, estradiol, thyroid, and iron deficiency (McKnight et al., 1980; Hager et al., 1980), age-related changes could be secondary to a number of different endocrine and neural changes. These examples suggest an important interpretation, that some genes with no impairments in their responses to directly acting stimulae nonetheless manifest impairments during complex physiological responses.

Hepatic drug-metabolizing enzymes also show age-related changes in regulation that are of general interest because of the greater sensitivity that older humans show to a wide range of drugs, as is epitomized by the increased side effects and slower clearance of barbiturates, digitalis, and many other drugs (Greenblatt et al., 1982; Vestal and Dawson, 1985). The induction by phenobarbital of NADPH–cytochrome c reductase and cytochrome-P450 is delayed and impaired in older rats (Adelman, 1971; Schmucker and Wang, 1981). Age-related decreases in some other drug-metabolizing enzymes, which are

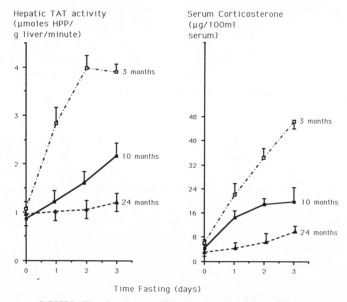

Time Fasting (days)

Fig. 7.8. Responses of C57BL/6J male mice at different ages to the stress of fasting. *Left,* the induction of liver tyrosine aminotransferase (TAT) assayed by the product HPP (hydroxyphenyl pyruvate); *right,* the increase of serum corticosterone. Similar effects of age were seen in two rat strains. Evidently, the threshold for responding to stress increases with age. Compare with Figure 7.7. Redrawn from Adelman et al., 1978.

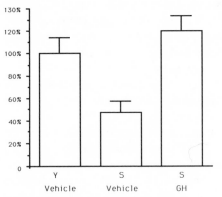

Fig. 7.9. Protein synthesis in whole rat diaphragm is decreased about 50% in 3-month-old rats (*Y*) as compared to 20-month-old male rats (*S*), judging by incorporation of ^{3}H-phenylalanine with correction for specific activity of phenylalanine. This age change is reversed by injections of growth hormone (GH) for 8 days. Redrawn from Sonntag et al., 1985.

considered to be under androgen regulation, were not reversed by testosterone treatment of 2-year-old rats: cytochrome *c* reductase decreased by 50%, while aniline hydroxylase decreased by 50% but was restored by testosterone (Rikans and Notley, 1984). Older humans retain the ability to induce some hepatic drug-metabolizing enzymes, but the rate and efficacy of response to various compounds is not well studied (Pearson and Roberts, 1984).

Reversal of age-related changes in transcription (e.g. α_{2u}-globulin) and of protein synthesis has recently been accomplished by dietary restriction, as will be discussed in Chapter 10.3.2.7. Neuroendocrine involvement is suspected, though the mechanisms are obscure. The age-related delays of blood corticosterone elevations during fasting (Figure 7.8) and in the induction of hepatic glycogen phosphorylase after hypothalamic stimulation (Shimazu et al., 1978; Shimazu, 1978) implicate neuroendocrine and autonomic changes that cascade to the liver. It is thus important in analyzing age-related changes in gene activity to resolve changes which may be *intrinsic* from those which are *extrinsic,* potentially originating elsewhere in the homeostatic network.

Another example of maintained response is the hypertrophy of myocardial cells in senescent mice in response to thyroxine, which ultimately had the same increment of protein synthesis and reached the same mass, despite a slower onset (Florini et al., 1973). The capacity for full myocardial hypertrophy is also shown by 2-year-old rats after experimental aortic narrowing to increase the cardiac workload (Hügin and Verzar, 1956). Humans suffering from arteriosclerosis characteristically have myocardial hypertrophy that can be massive and which argues against impaired protein synthesis (Van Peenen and Gerstl, 1962). Here a disease of senescence reveals the integrity of a fundamental genomic response.

Reversal of age-related changes in gene expression through endocrine manipulations also extends to skeletal muscle. Growth hormone injections restore skeletal muscle protein synthesis in senescent rats (Figure 7.9; Sonntag et al., 1985). A neuroendocrine basis is indicated for these changes, since there are marked decreases in growth hormone secretion after middle age (Chapter 3.6.9; Figure 3.31). The capacity for maintenence of muscle functions in humans is being intensively studied, especially in "master" athletes in endurance events, who maintain far better cardiovascular performance in the age range 60–75 years than do untrained individuals (Hagberg et al., 1985; Pollack et al., 1987). The long-range potential for arresting the common age-related decline in maximum aerobic capacity and cardiac stroke volume is unclear, however. Mitochondrial oxidative capacity in human skeletal muscle can be considerably enhanced by training in older men (Orlander and Aniansson, 1980). There is much to be investigated in terms of the different responses to aerobic and anaerobic conditioning at later ages.

Major impairments of responses that depend on gene regulation and hormones are unusual during organismic senescence. This and the following paragraphs describe several examples in different cell types. The induction of hepatic α_{2u}-globulin by testosterone is apparently lost in old male rats, probably as a consequence of decreased levels of androgen receptor (Roy et al., 1974; Chatterjee et al., 1988). In contrast, the failure of the estrogen-induced surge of gonadotropins in senescent female rodents (Figure 3.3.3) is not associated with deficits of cytoplasmic estrogen receptors in the hypothalamus or pituitary (Wise, 1983; Wise and Parsons, 1984; Bergman et al. 1989). The loss of this responsiveness is a mystery and contrasts with the facile reversal of atrophic changes in reproductive tract cells of postreproductive women and rodents by estrogen replacements (Finch and Gosden, 1986). At the physiological level, an important factor in the loss of hypothalamic-pituitary responsiveness to estradiol in rodents is the exposure to chronic endogenous estrogens during reproductive senescence. Most changes are prematurely induced by chronic exposure of young rodents to exogenous estradiol (Chapters 3.6.9; Table 10.7).

Age changes in the nuclear binding of the estradiol-receptor complex are not understood and could be complex. In the 2-year-old rat uterus, there is evidence for reduced maximum binding capacity of the nuclear-receptor complex, which parallels the decreased stimulation by estradiol of RNA polymerase II activity (Chuknyiska and Roth, 1985). Another study indicates diffences in the dissociation of the receptor complex from the nucleus. In 2-year-old female C57BL/6J mice, the estrogen receptor has a faster rate of release from cell nuclei of the hypothalamus, pituitary, and uterus (Bergman et al., 1989). The characteristics of the cytoplasmic estradiol receptor are not altered with age, according to the equilibrium binding constant (K_a) and stereoselectivity (Gesell and Roth, 1981; Bergman et al., 1989; Wise and Parsons, 1984) for humans (Pellika et al., 1983). Nearly all studies of steroid receptors agree on this point (Roth and Hess, 1982; Chuknyiska et al., 1985). A recent exception is the 70% age-related decrease of affinity in corticosteroid type II receptors in the rat hippocampus (Landfield and Eldridge, 1989). The lack of change in estradiol-receptor immunoprecipitability reported for the rat uterus (Gesell and Roth, 1981) bears on age-related phenomena seen in enzymes that lose activity while retaining their immunoprecipitability (Section 7.5.1).

Another example of lost steroidal responsiveness is the atrophy of oviduct observed in domesticated birds. The oviduct of middle-aged quails (*Coturnix coturnix*) atrophies, with great loss of ovalbumin and avidin (Bernd et al., 1983; Muller et al., 1985); domestic hens show similar atrophy (Okulicz et al., 1985). (Although oviduct atrophy is common in middle-aged domestic birds, it may not be like ovarian exhaustion in mammals [Chapter 3.6.4.1]. Old hens can continue to lay eggs, though at decreasing rates, at least through 10 years, while other avians remain fertile beyond 30 years [Chapter 3.5].) It is striking that the senescent oviduct of quail does not respond to diethylstilbestrol (estrogen analogue) or progesterone (Bernd et al., 1983). Oviduct atrophy in nonlaying, middle-aged hens is accompanied by 80% decreases in progesterone and estradiol receptors but no changes in plasma estradiol, which is considered to regulate the progesterone receptor (Boyd-Leinen et al., 1982; Okulicz et al., 1985). Laying hens of the same age as the nonlayers have oviduct progesterone receptors and hormone levels like those of young laying hens (Bahr and Palmer, 1989; Okulicz et al., 1985).

A possible model for these changes in hens and for the estrogen-dependent changes in the rodent hypothalamic-pituitary axis is the estrogen memory of avian liver and a human hepatoma cell line (HepG2), which retain enhanced inducibility of several genes after brief exposure to estradiol for numbers of cell replications *in vivo* and *in vitro*, respectively (Tam et al., 1986; Haché et al., 1987). In avian liver, a single injection of estradiol causes persistent DNA demethylation sites after increases in transcription of vitellogenin (Wilks et al., 1982), while HepG2 cells retain type II estrogen-binding sites in the nuclear matrix (Tam et al., 1986). The estrogen memory that gives a greater responsiveness to estradiol is a change in the opposite direction to the loss of estrogen responsiveness seen in the hen oviduct and rodent hypothalamus-pituitary. But both could share molecular mechanisms at the genomic level.

As a last example, Adler and Black (1984) described a change with much potential for further study. Explants of sympathetic (catecholaminergic) neurons from the superior cervical ganglion are observed to express substance P, a neuropeptide, in addition to the continuing expression of tyrosine hydroxylase, the enzyme of catecholamine synthesis. In 6-month-old adult rats, explants showed four- to eightfold increases above low basal values during 7 days in culture. While explants of sympathetic neurons from 2-year-old rats showed no change in the outgrowth of neurites or in tyrosine hydroxylase, there was no increase of substance P. These changes beg to be analyzed.

Taken together, many data suggest that age-related changes in RNA and protein synthesis in diverse cell types of mammals can be readily modified through physiological controls. The age-related reductions in bulk protein synthesis thus do not prevent considerable up-regulation through physiological factors. Consequently, the origins of altered gene expression in many cases may be sought in the network of hormones, neurosecretions, and other physiological controls. No broad generalization has yet been possible, and each gene apparently must be considered separately in terms of its particular set of physiological regulators and their synergies and antagonisms. For organisms like mammals, with numerous regulatory influences on most cells, the researcher must *a priori* consider how a change at any time in the lifespan may represent the regulatory factors, not only at that instant but also at preceding times during the diurnal cycle and even during development, that may have lingering effects. The smaller *in vitro* proliferative capacity of fibroblasts from diabetics (Section 7.6.3) suggests the great potential for persistence of prior environmental changes, in metabolism and hormones for example. The presumption should no longer be made that age-related changes in gene expression necessarily result from intrinsic cellular senescence in isolation from the physiological milieu.

7.4.1.4. *Delayed Expression of Latent Viral Genes*

Latent viral genes and latent viruses that are manifested with extensive delays after maturation are another important example of age-related changes in gene expression that could pertain to other somatic cell genes as well. The first set of examples draws from mice.

There are numerous examples in mice of age-related increases in the expression of sequences of C-type RNA tumor viruses (retroviruses). These chromosomally integrated ecotropic (host-limited) viruses cause murine leukemia, and are also known as MuLV (Chapter 6.5.1.7). Mouse genotypes can be classified by the timing of expression of MuLV sequences, antigens, and infectious particles in different age groups (Getz, 1985; Rowe et al., 1970). Some inbred strains already have high levels of infectious virus at birth (e.g. AKR); whereas others initially expressed little infectious virus or viral group-specific (gs) antigens, while a subpopulation manifests them at later ages (e.g. BALB/cJ, C57BL/6J, DBA/2J). Yet other strains never produce infectious virus but do make viral gs antigen (Huebner et al., 1970; Peters et al., 1972). Wild mice have shown late onset viral production (Gardner et al., 1976; Chapter 6.5.1.7.5), with the increased expression of gs antigen and lymphomas among other malignancies also seen in subgoups of strains with late onset viral production. The early onset, MuLV-producing strains differ in their ubiquitous viral production (Chapter 6.5.1.7). The delayed expression of viral genes was the basis for Huebner and Todaro's (1969) viral oncogene hypothesis, which proposed that the expression of genes from vertically transmitted, latent C-type retroviruses was a key event in oncogenesis. Relatively few tumors, however, have yielded RNA or DNA viruses that are oncogenic on subsequent transfer (Duesberg, 1987). Nonetheless, the age-related incidence of many types of cancer is closely linked to age-related changes in the structure or expression of select genes, whatever the role of mutation or recombination (Duesberg, 1987; Weiss, 1986).

C3H mice have a delayed development of mammary tumors that suggests another aspect of age-related changes in expression of chromosomally located oncogenes. The C3H mice very effectively transfer the murine mammary tumor virus (MuMTV) by nursing to their offspring, 90% of whom develop mammary tumors by 12 months; genotypes vary widely in their transmissibility of this agent (Andervont, 1940; Nandi and McGrath, 1973). In C3H mice, the dominant gene *MTV–1* on chromosome 7 regulates expression of MuMTV antigens and tumor development (van Nie and Verstraeten, 1975). By cross-fostering pups on MuMTV-free nurses, the resulting C3Hf mice have few mammary tumors until later, when they develop in up to 50% of mice by the average lifespan (Andervont, 1940; Bentvelzen, 1974; van Nie and Verstraeten, 1975). C3Hf, like all mice, carry other vertically transmitted proviruses, which appear to be an endogenous source of the MuMTV viruses causing the late onset mammary tumors. The late-arising tumors in C3Hf mice have a different virus with much lower oncogenicity. One of the three proviruses in C3Hf mice is inserted at the *MTV–1* locus (Traina et al., 1981) and shows an interesting change. After maturity, spleen DNA from tumorless C3Hf mice becomes progressively demethylated at proviral sequences in the *MTV–1* locus; this demethylation is tissue-specific and does not occur in DNA from liver or mammary glands (Etkind and Sarkar, 1983). Late-arising mammary tumors also become demethylated at this locus in their nuclear DNA. Etkind and Sarkar (1983) proposed that C3Hf mice are prone to develop late onset mammary tumors by endogenous infections of MuMTV that initially arise in spleen cells. (Readers may recall the role of infectious introns in the senescence of certain fungi [Chapter 2.4].)

The nervous system is another locus of age-related viral gene expression. The first example is not associated with any pathologic change. The brains of the late onset virus-producing C57BL/6J strain show age-related increases in MuLV-related sequences that represent 50–75% of the MuLV genome. Of even more interest is that the fraction of the

MuLV genome expressed increases with age in only nuclear, but not polysomal, RNA from brain and liver (Florine et al., 1980; Getz, 1985). There is no proof that this leakiness of the MuLV sequences results from production of infectious virus in lymphocytes.

Readers may recall from Chapter 6.5.1.7.5 the MuLV-associated degeneration of spinal cord motor neurons in some wild mouse genotypes by mid-life (Gardner et al., 1980; Rasheed and Gardner, 1983; Gardner, 1985). Similarly, selective lesions of spinal motor neurons are induced in C58 and AKR mice by infections with the lactate dehydrogenase–elevating virus (LDV), an RNA virus (Brinton, 1982), and emerge after delays of many months (Contag and Plagemann, 1988; Contag et al., 1989). Susceptible strains are homozygous for a permissive allele ($Fv-1^{n/n}$); an additional requirement for the LDV-induced disease is that the host genome has multiple copies of an N-tropic, ecotropic, AKR MuLV (Murphy et al., 1983). An age-related loss of a specific T-cell population is also implicated. Those spinal neurons that are most vulnerable to LDV-induced degeneration in C58 mice show major increases of a 3-kilobase RNA coded by the AKR MuLV between 6 and 7 months (Figure 7.10; Contag and Plagemann, 1988). In contrast, this RNA was not detected in brain. The relation between the MuLV RNA and the neurodegeneration induced by LDV is not known, but suggests further important roles for endogenous retroviruses in age-related disease of mice. Friend virus also shows age-related

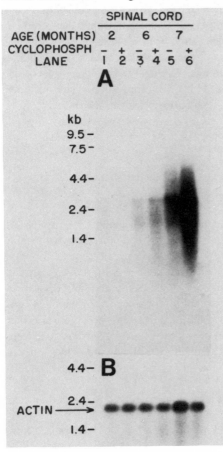

Fig. 7.10. Age-related increase in a three-kilobase MuLV RNA (murine leukemia virus) from spinal cords of C58 mice. The MuLV RNA is transcribed from endogenous MuLV provirus sequences. The increase is augmented by a 2-day pretreatment with cyclophosphamide. The increase of the three-kilobase MuLV is associated with susceptibility to neurodegeneration by middle age, subsequent to infection with LDV (lactic dehydrogenase–activating virus). From Contag and Plagemann, 1988. Photograph kindly supplied by Peter Plageman.

enhancement of replication in C57BL/6J mice after 10 months (Cinader et al., 1987), suggesting important changes in immune or cellular regulatory factors.

Genetic control of the duration of latency has been shown for the prion disease scrapie, which causes fatal neurodegeneration. Mice and other mammals have endogenous sequences identical to those found in the scrapie prion protein, which are coded at the prion gene on chromosome 20 (Westaway et al., 1987). Closely linked ($<$ 10 map units) on mouse chromosome 2 is the *Prn-i* locus, whose variants in inbred mice double the latency of scrapie infections in mice (Carlson et al., 1986; Westaway et al., 1987). Other genes that determine the latency of virus expression may be anticipated. A prion disease of humans also involves alternate prion alleles, the Gerstmann-Sträussler syndrome. This very rare autosomal dominant has a delayed onset, typically by mid-life, with neurodegeneration and the production of infectious prions in the brain. It was recently associated with a codon change in the human prion gene (Hsiao et al., 1989; Chapter 6.6). Genes that determine the latency for viral and other diseases with separate hereditary determinants could be major regulators of senescence in some genotypes.

In humans, the most general example is associated with the disease shingles (herpes zoster), which causes painful skin eruptions. Shingles is caused by reemergence of the herpes simplex (varicella zoster) DNA virus. Herpes simplex is acquired in childhood from infections of chicken pox and can remain latent for decades. During the latent phase, the virus is harbored in dorsal root ganglia and sensory neurons, where the complete genome can be detected even if infectious particles cannot (Gilden et al., 1983). In mice with latent herpes, only one of the herpes genes is transcribed (Stevens et al., 1987). The incidence of shingles increases progressively with age; ultimately, more than 10% of the elderly may be afflicted (Ragozzino et al., 1982; Hope-Simpson, 1965). The basis for delayed onset of shingles is unknown, but decreased immune functions or stress from diseases of senescence may be a factor.

7.4.1.5. Fidelity of Protein Synthesis

The error hypotheses of Medvedev (1962) and Orgel (1963) predict self-catalyzing accumulations of errors in macromolecular biosynthesis, as codons that mutate loosen the fidelity of the enzymes required for transcrition and translation. Numerous efforts have been made to detect errors in the amino acid sequence in proteins synthesized by tissues from senescent animals, but most have given negative or equivocal outcomes (Rothstein, 1982). I briefly mention a few studies.

In an effort to increase the detectability of misincorporation from putatively altered ribosomes, Mori et al. (1983) used trout protamine mRNA as a template, with ribosomes from rats of different ages and soluble factors from rabbit reticulocytes. Protamines consist mostly of arginine and only six other types of amino acids. Misincorportion errors were in the range of 10^{-4} irrespective of the age of the mouse from which ribosomes were obtained. Numerous other studies with different template RNA and sources of protein synthetic components have also failed to show effects of age on peptide synthesis in cell-free systems.

A general critique of cell-free (reconstituted) assays for protein synthesis is that their poor efficiency, relative to intact cells, could give misleading results because the suboptimal kinetics could be sensitive to factors that need not be important in intact cells. Thus, small changes in protein synthesis with components from rodent viscera of different ages (reviewed in Reff, 1985; Richardson and Birchenall-Sparks, 1983) cannot be easily interpreted under these conditions. From this perspective, it is reassuring that the *in vivo* incorporation of the leucine analogue, aminoisobutyric acid, did not indicate any age-related change in the fidelity of aminoacyl-tRNA synthetase from liver, brain, heart, or kidney of mice (Hirsch et al., 1976); any decrease in fidelity could hypothetically indicate error catastrophes.

In a different approach, Rabinovitch and Martin (1982) looked for effects of errors *in vivo*, by examining viral production in mice of different ages. The encephalmyocarditis (EMC) virus, a picornavirus, was used because it targets nondividing cells (heart, brain), which might be particularly vulnerable to accumulating erroneous proteins, since damaged cells are not competitively outgrown. However, biologically active virus grew just as well in tissues from mouse hosts aged 8 or 24 months. Since growth *in vitro* in the presence of amino acid analogues was greatly diminished, the *in vivo* result indicates that the host machinery for synthesizing viable EMC virus is fully competent. This study did not, however, give a direct estimate of the errors in viral synthesis.

7.4.2. Nematodes

This and the next section briefly review the limited information available on macromolecular biosynthesis during senescence of nematodes and a few other lower species. With few exceptions, biochemical assays are based on whole worm organisms and without regard to possible individual differences in diseases that may be in future studies distinguished from other age-related changes.

The short (2- to 3-week) lifespan of some soil nematodes, their genetic homogeneity, and their establishment in axenic culture have lead to many molecular studies of senescence (Chapter 2.5.4). Several approaches demonstrate age-related slowing in the synthesis of total body proteins. Tracer studies indicated twofold slowing in total protein synthesis and turnover during the lifespan of *Turbatrix aceti* and *Caenorhabditis elegans* (Sharma et al., 1979; Karey and Rothstein, 1986; Rothstein, 1982). The turnover of enolase and aldolase was also slowed (Sharma et al., 1979; Zeelon et al., 1973), as assayed after cycloheximide blockade of protein synthesis. (Altered forms of these enzymes are discussed in Section 7.5.1). Analysis of ^{35}S-proteins by two-dimensional gel electrophoresis, after the nematodes ingest labelled bacteria, did not detect any changes during the adult lifespan in relative labelling among the two hundred most abundant polypeptides, nor were new proteins detected (Johnson and McCaffrey, 1985). This approach was sensitive enough to detect changes in 10% of abundant proteins during nematode development (Johnson and Hirsch, 1979). Nothing is known about RNA synthesis during nematode senescence.

The absence of age-related changes in relative labelling among the most abundant proteins from the entire nematode does not preclude select changes in particular cells. It would be interesting to know whether senescent changes in biosynthesis were differentially distributed among the nematode's precisely determined cell lineages. Cell ablation during development might also have interesting effects on organismic senescence. The decreased ingestion and processing of food, especially several days before death (Chapter

2.5.4), are consistent with slowed somatic protein synthesis and could be an important cause. Hookworms (parasitic nematodes), which have multiyear lifespans (Chapter 4.2.2), would appear to have very slow declines in protein synthesis.

7.4.3. Insects

Insect species may differ as much in their patterns of senescence as in development, and far more than among homeothermic mammals (Chapters 2, 3). Age-related changes in macromolecular biosynthesis must also vary widely among very short lived aphagous insects, other short-lived species that continue to feed, and the vastly longer lived queens of social species. Most is known about age-related changes of protein synthesis in *Drosophila*. For general reviews, see Richardson, 1981, Makrides, 1983, Baker et al., 1985, Arking and Dudas, 1989, and a symposium volume edited by Collatz and Sohal (1986).

7.4.3.1. Flies, Mosquitos, and Silk Moths

Drosophila melanogaster resembles the short-lived soil nematodes in regard to the absence of somatic cell replication in the adult (Chapter 2.2.2.1), and the major 20–70% age-related reductions of total protein synthesis in whole flies, as judged by the incorporation of radiolabelled amino acids (Baumann and Chen, 1969; Levenbook, 1986). *Drosophila* apparently continue to eat throughout their lifespan (Chapter 2.2.2), but the efficiency of ingestion is not known for specific nutrients. The major loss of larval fat body proteins during the first 4 days of life (Maynard Smith et al., 1970) suggests a potential depletion of crucial nutrient stores. There are few characterizations of nucleic acids in senescent *Drosophila*. While total nuclear DNA (0.3 g/fly) did not change throughout the lifespan, striking decreases occurred in a minor density-separated fraction of mitochondrial DNA that was 3% of the total DNA in young flies (Massie and Williams, 1987). The mitochondrial DNA loss was slowed by removing the wings. This implies that wing usage causes damage to mitochondria, which are mostly in thoracic muscles. Total fly RNA content decreased slowly between 10 and 80 days by 25% in males and 40% in females (Samis et al., 1971); the greater loss in females presumably is consequent to egg laying.

Most reports show general decreases in protein synthesis during fly senescence (reviewed in Webster, 1986; Richardson et al., 1985; Makrides, 1983; Niedzwiecki and Fleming, 1990).[3] Despite the major overall decrease in protein synthesis, the most abundant proteins as resolved by two-dimensional gel electrophoresis continue to be labelled in the same proportions (Parker et al., 1981), excepting a few spots with increased labelling at middle age (Fleming et al., 1986). Among those synthesized are proteins that move slowly from the cytoplasm to the nucleolus (Maynard Smith et al., 1970), which presumably include ribosomal proteins. There is no synthesis or turnover of major muscle proteins during adult life, as judged from autoradiographic studies by the persistence of label incorporated by pupae, which, with nonreplaced proteins in other cells, constitute the bulk (80%) of the body proteins (Maynard Smith et al., 1970); the remaining 20% of proteins have half-lives of about 10 days.

Select age-related changes occur in isoenzyme components, including loss of bands in

---

3. An early report of major *increases* in protein synthesis during senescence (Clarke and Maynard Smith, 1966) is considered anomalous, probably due to lack of precursor pool corrections.

female *D. melanogaster* (Hall, 1969) but new bands in males (Dunn et al., 1969). Most enzyme activities that change after hatching do so during maturation in the first week (Baker et al., 1985). However, four mitochondrial enzymes do not change (Massie and Kogut, 1987). This report concluded that total body protein increased with age, a finding that needs further explanation. The absence of changes in mitochondrial enzymes is surprising because of another report by these investigators that 85% of mitochondrial DNA is lost by 25 days (Massie and Williams, 1987). In *D. mojavensis,* posteclosional and maturational changes of alcohol dehydrogenase isozymes reflect the histolysis of the larval fat body by 10 days (Batterham et al., 1983; Figure 2.1). Again, this change is more accurately designated as maturational than senescent. This last example demonstrates the importance of examining individual cells or particular tissues, rather than continuing the uninformative tradition of whole-fly biogerontology. Confounds from microorganisms harbored within whole flies are another often neglected concern.[4] Few studies of mammals would be welcomed by the leading journals on the basis of whole-body homogenates! The role of RNA synthesis or protein synthesis in the selective changes of enzymes is unknown.

Despite the decrease of protein synthesis, several studies show the capacity to make new proteins. One of the first indications of this was the response to brief heat shocks (38.5°C/1 hour), which transiently increased mortality in 6-day-old flies without altering the mortality rate at later ages (Strehler, 1961). This observation implies replacement of heat-denatured proteins. Subsequently, heat-shock responses that have been described in adult flies rapidly induce the set of proteins thought to repair denatured proteins. For example, a 20-minute heat shock of whole flies, followed by exposure to [^{35}S]-methionine for 2 hours to radiolabel nascent proteins, caused dramatic induction of fifty or more "new" spots on two-dimensional gels (Fleming et al., 1988).

Senescent flies (> 40 days) had a larger number of polypeptide responses to heat shock than did younger flies (Figure 7.11). Moreover, senescent flies have a large net *increase* in protein synthesis after heat, whereas young flies show a decrease (Niedzwiecki and Fleming, 1990). Presumably, some of these "new" spots represent the heat-shock proteins described elsewhere (Ashburner and Bonner, 1979). A model of these age changes is the larger number of responses shown by young flies that were fed the amino acid analogue canavine; subsequent heat shock induced most of the same spots seen in senescent, heat-shocked flies. Similarly, pretreating young flies with puromycin, which causes premature termination of nascent peptide chains, also increased the number of responses. This suggests that senescent flies have unstable or abnormal proteins, like those expected from the incorporation of canavanine or puromycin (Niedzwiecki and Fleming, 1990). In view of the decrease of catalase and peroxidase activities with age in *Drosophila* and the increased formation of peroxides (Armstrong et al., 1978; Sohal et al., 1983; Arking and Dudas, 1989), it is possible that oxidized proteins contribute to the larger number of polypeptides seen in senescent flies after heat shock.

Another demonstration of continued biosynthesis is that senescent flies can replace cat-

---

4. Because *Drosophila* are often fed on diets of fresh yeast, whole-body homogenates may contain these and other microorganisms. Furthermore, bacteria from the fly gut may be present in subcellular fractions and, even at low titres, may contribute proteins detected on gel electropherograms (Fleming and Kwak, 1986). This situation raises concerns about the biochemical studies that use whole-fly homogenates. Depending on the details, there may be no detectable effect on the qualitative patterns of proteins synthesized in flies raised on sterile or xenobiotic conditions (Spicer, 1988).

Age (days)

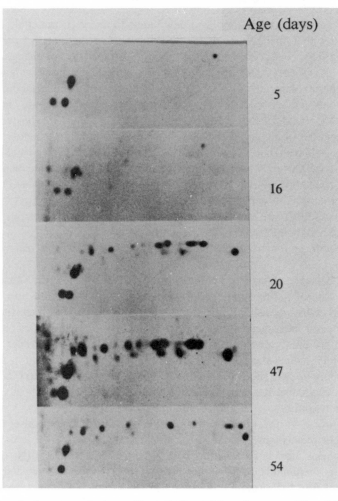

5

16

20

47

54

Fig. 7.11. Heat-shock responses increase with age in *Drosophila melanogaster*. Flies of different ages were given a 20-minute heat shock (37°C) and then exposed for 2 hours to [^{35}S]-methionine to label nascent proteins. Total body proteins were resolved by two-dimensional gel electrophoresis. While about eighteen "new" spots were detected in young adult flies (10 days), senescent flies (> 45 days) yielded at least fifty "new" spots. The present analysis does not inform whether the "new" spots correspond to the heat-shock proteins described by others, but some overlap seems likely. Rephotographed with permission from Fleming et al., 1988.

alase after its irreversible inactivation (Figure 7.12) (Ganetzky and Flanagan, 1978; Nicolosi et al., 1973). These responses show the capacity for selective induction of proteins. The role of transcription is not known in the induction of catalase and the heat-shock gene set.

A major question concerns the histospecific distribution of reductions in protein synthesis. Not all cells are equally altered. The thorax, which is mostly muscle cells, had the most drastic decrease in microsomal protein synthesis (> 90%), the abdomen decreased by 35%, but the head decreased little if at all (Webster et al., 1980). The cells involved in the major decrease of thorax synthesis probably include neurons and gut cells, which replace proteins in the adult fly, in contrast to muscle (Maynard Smith et al., 1970). Neurosecretory cells in the brain shrink markedly, with decreased free ribosomes and granular ER (Herman et al., 1971); changes of neurosecretions and related functions are unknown. In ovarian nurse cells, bulk RNA synthesis (mostly ribosomal RNA) may even increase with age (Wattiaux and Tsien, 1971). This histospecificity warns against generalizations based on whole-fly homogenates.

The rate of polypeptide chain elongation is reported to slow considerably at later ages (Webster and Webster, 1983; Webster, 1986). However, there are few changes in amino-acylation of tRNA (Webster and Webster, 1981), formation of initiation complexes (Webster et al., 1981), the rate of translation (Webster and Webster, 1982), or in the availability of ribosomes (Baker and Schmidt, 1976). Sharp decreases in EF–1α appear to precede those in protein synthesis (Webster and Webster, 1983). The supplementation of ribosomes from senescent flies with EF–1α from young flies restored the activity. Preliminary data indicate decreases of EF–1α mRNA (Webster and Webster, 1984). A recent genetic approach extends these findings, that flies transfected with a P element containing an additional copy of the gene for EF–1α lived longer (Shepherd et al., 1989).

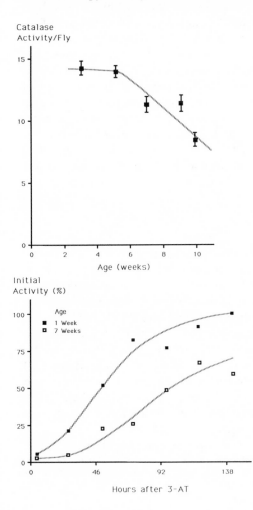

Fig. 7.12. Regulation of catalase in *Drosophila melanogaster*, as measured in whole-body homogenates. *Top*, age-related changes in basal catalase; *bottom*, recovery of catalase activity after treatment of young (1-week-old) and senescent (7-week-old) flies with an herbicide, 3-amino–1,2,4-triazole (3-AT), which irreversibly inactivates the catalase. Redrawn from Nicolosi et al., 1973.

The yellow-fever mosquito *Aedes aegypti,* another dipteran with a short 4- to 6-week lifespan like *Drosophila* (Chapter 2.2.2.1), has early age-related decreases of bulk protein synthesis in association with maturation by 10 days, but possible increases much later in life (Lang and Smith, 1972). The half-life of bulk proteins was 2–3 days, decreasing shortly after hatching; the analysis used corrections for slowed turnover of the leucine pool (Smith, 1971). Synthesis of the tripeptide glutathione (GSH) decreased progressively by about 50% up to the average lifespan, but was stable thereafter at even later ages; levels of most free amino acids did not change with age (Hazelton and Lang, 1983). The decreased GSH may be important as a chemical antioxidant; feeding GSH increased the median lifespan by 35% (Richie et al., 1987). GSH is also being considered in relation to the accumulation of altered enzymes in mammals (Section 7.5.1). Bulk RNA and protein did not change during the adult lifespan (Lang et al., 1965). Patterns of senescence seem generally similar in *Aedes* and *Drosophila. Aedes* offers a major experimental advantage, that it can be grown under axenic conditions (Lang et al., 1972), which permits biochemical and radiotracer studies of whole insects without contamination by microorganisms.[5]

Senescence in the silk moth *Bombyx mori* is of interest to nutritional hypotheses about senescence (Chapter 10.3). Silk moths, like some other Lepidoptera but unlike most Diptera, are completely aphagous as adults and must draw upon fat body reserves as the sole source of nutrients (Chapter 2.2.1). Even so, adult silk moths live as long as flies, 1–9 weeks depending on the temperature (Osani and Yonezawa, 1984). Although protein synthesis has not been studied throughout the lifespan, the free amino acids show a variety of age-related changes depending on temperature: proline, serine, and some other amino acids generally decreased at all temperatures in both sexes, while males showed increases in leucine and others when kept at 4°C (Osani and Kikuta, 1981; Osani and Yonezawa, 1984); most changes were less than 30%. Despite the drain from egg laying, most free amino acids did not change. Unless there is a remarkable recycling, the stability of free amino acids implies scant somatic cell protein synthesis during most of adult life, except possibly in the gonads.

7.4.3.2. *Endosymbiont Bacteria in Aphids*

The pea aphids (*Acyrthosiphon pisum* and *A. kondo,* Homoptera) offer an unusual experimental model for changes in protein synthesis of endosymbionts during senescence. While little-known to gerontologists, pea aphids are well-studied for the energetics of reproduction because these plant parasites are an agricultural pest. Among alternate life history characteristics of the aphid races or genotypes (Lowe and Taylor, 1964; Auclair and Aroga, 1987; MacKay, 1987), some show a photoperiod-controlled switch between parthenogenesis and sexual reproduction (MacKay, 1987). The parthenogenetic forms are viviparous, with internally developing embryos that are released as larvae. During total lifespans of at least 65 days, they begin reproducing by 10 days and show reproductive senescence by 30, with subsequent accelerations of mortality (Randolph et al., 1975; Gutierrez et al., 1987). In the laboratory, parthenogenetic aphids may spend up to 50% of

---

5. Another consideration is that adult mosquitos survive well on sucrose alone and efficiently recover nitrogen, as shown by the absence of age-related changes in amino acids and protein when mosquitos were fed on sucrose as in the studies described above. Feeding of protein through blood, which females require for egg production and laying, shortens lifespan (Clements, 1963; Chapter 2.2.1.1).

their adult life in a postreproductive state (Ishikawa, 1984a), which is unusual among invertebrates, so far as is known.

Intracellular endosymbionts live in special cells (mycetocytes) and appear to be gram-negative bacteria that are transmitted between generations. The endosymbionts synthesize cholesterol, lipid substrates, and essential amino acids; the lipids and amino acids that are synthesized differ among endosymbionts of *Acyrthosiphon* strains (Srivastava et al., 1985). While the endosymbionts have not been grown in culture, their protein synthesis continues *in vitro* at least 2 hours (Ishikawa, 1982a, 1982b).

A series of reports on parthenogenetic aphids indicate that the regulation of protein synthesis by endosymbionts in *A. kondo* and *A. pisum* changes during host senescence (Ishikawa, 1984a, 1984b, 1984c, 1984d; Ishikawa and Yamaji, 1985a, 1985b). Among proteins synthesized by young adult aphids is symbionin (63 kDa), as characterized on two-dimensional gels of whole extracts after labelling with methionine or leucine. Symbionin may be a symbiont protein since its production in young adults *in vivo* is inhibited by chloroamphenicol, but not cycloheximide (Ishikawa, 1984a). Chloramphenicol blocks prokaryotic protein synthesis, while cycloheximide blocks protein synthesis in eukaryotes. *In vitro,* endosymbionts make several hundred polypeptides not found in whole aphids, implying that young adult aphids limit the types of proteins made by their endosymbionts.

In senescent aphids (65 days old), the production of symbionin was greatly decreased in both *A. kondo* and *A. pisum* (Ishikawa, 1984a; Ishikawa and Yamaji, 1985a, 1985b). In contrast, seven new proteins appeared; the persistence of these spots after cycloheximide treatment prior to labelling, and their inhibition by chloroamphenicol prior to labelling, suggests that the new spots are made by the endosymbionts. These proteins and their mRNA have not been sequenced or their genes located on aphid or endosymbiont genomes. A provisional hypothesis is that the endosymbiont genome become derepressed during senescence. These studies raise interesting questions about the role of reproductive senescence and the redistribution of nutrients in regulating macromolecular biosynthesis in host and endosymbiont, and whether similar changes occur in aphids reproducing sexually with eggs that developed externally.

7.4.4. Plants

Senescence in higher plants is thought to be primarily under hormonal control, whether the involution is restricted to leaves and other organs or to the whole plant (Chapter 2.3.2). Because the roles of particular hormones in these processes vary widely among species, generalizations are difficult. A general trend is for the rest of the plant to resorb nutrients from a leaf or other nongerminal tissue during its involution (Noodén and Thompson, 1985; Woolhouse, 1984). Leaf senescence[6] involves a well-regulated, rapid series of events that gives the impression of a systematic disassembly proceeding to leaf yellowing, because of chlorophyll loss, through to leaf collapse. The mechanisms involve proteolysis and possibly free radical damage (Strother, 1988).

Major loss of protein and arrest of protein synthesis generally occur during leaf senes-

---

6. Here I follow the current botanical usage of *senescence,* which includes the degeneration of particular organs and need not involve degeneration of the whole plant.

cence (Woolhouse, 1983, 1984). For example, during the 6-week-long senescence of leaves in the labiate (*Perilla fructescens*), polyribosomes and the synthesis of ribosomal RNA decreased, but did so much more slowly from the cytoplasm than from chloroplasts (Callow et al., 1972; Figure 7.13). In cotyledons of the soybean (*Glycine max*), senescence was associated with numerous and select changes in nascent polypeptides, as resolved by two-dimensional gel electrophoresis (Skadsen and Cherry, 1983). The decrease of chloroplast RNA synthesis is attributed to loss of RNA polymerase in this organelle (Woolhouse, 1984). Paralleling the loss of polyribosomes, enzymes that are made entirely or partly under the control of the chloroplast genome (ribulose–1,5-bisphosphate carboxylase) are decreased much more than enzymes synthesized under nuclear control on cytoplasmic ribosomes, for example phosphoglycerokinase (Batt and Woolhouse, 1975; Thomas and Stoddart, 1980). In *Perilla* and other dicotyledons, chloroplast ribosomal RNA synthesis ceases abruptly at maximum leaf expansion, while wheat and other monocots continue synthesizing chloroplast RNA to later stages. During leaf senescence, chloroplasts are leached of many constituents, degraded chlorophylls appear, and the number of chloroplasts decreases (Maunders et al., 1983; Woolhouse, 1984). In contrast to these losses, there are select increases of chromosomally coded isozymes (Thomas and Stoddard, 1980). Mitochondria are better preserved than chloroplasts during leaf senescence. The continuing respiration of leaves, presumably fueled by mitochondrial ATP, may provide energy for nutrient transport and resorption (Woolhouse, 1984).

The degradation of chloroplasts and the yellowing of leaves during leaf senescence are actively controlled by the nucleus. Yoshida (1961) neatly showed that chloroplast degradation during excision-induced leaf senescence in the pondweed *Elodea* depended on contact with the cell nucleus; chloroplasts in enucleated cells did not degenerate and even grew. Moreover, the onset of leaf senescence in *Lolium temulentum* coincides with a burst of protein synthesis (Hedley and Stoddart, 1972). If protein synthesis is prevented at this time by inhibitors acting at cytoplasmic, but not chloroplast, ribosomes, then chlorophyll loss and leaf senescence are prevented (Thomas, 1976). Moreover, cytokinins, which delay leaf senescence, will also prevent this burst of protein synthesis (Thomas and Stoddard, 1980). The identity of the senescence-promoting proteins is unknown, as is their time of transcription. Attempts to manipulate leaf senescence by inhibitors of RNA synthesis have failed so far. It will be interesting to see if these proteins are similar to thanatins and other proteins associated with hormonally induced cell death in mammals (Section 7.7).

The potential for genetic dissection of these phenomena is demonstrated by a nonyellowing mutant (Bf993) of the grass *Festuca pratensis,* which follows Mendelian inheritance (Thomas and Stoddard, 1980). Although the mutant leaves show the normal loss of protein and RNA, their chlorophyll was lost to only a slight extent (Thomas and Stoddard, 1975). The genetic lesion appears to stabilize membranes, thereby influencing the release of linolenic acid from hydrolysis of chloroplast membrane lipids that in the wild type induces loss of chlorophyll and, in turn, promotes the degradation of chlorophyll-associated proteins (Thomas et al., 1985). It is unknown how generalizable are the above mechanisms in leaf senescence to the whole-plant senescence of annual plants. The changes that do occur differ widely among mRNA types and appear to be as selective as those observed during development.

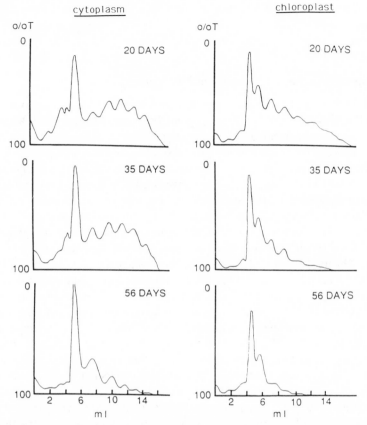

Fig. 7.13. Polyribosome profiles during leaf senescence in *Perilla fructescens,* showing polyribosomes from the cytoplasm and chloroplasts at maturity (20 days), at the onset of senescence (35 days), and later in senescence (56 days). Chloroplast polyribosome sizes decrease earlier and more severely than in the cytoplasm. Redrawn from Callow et al., 1972.

7.5. Post-translational Modifications

Numerous, but selective, post-translational changes are reported in long-lived extracellular proteins, as well as in proteins from cells of senescent organisms. Some of these changes, like the nonenzymatic addition of glucose to proteins (glycation) or the accumulation of lipofuscins, do not reflect direct or immediate regulation by genes. Yet these phenomena are of fundamental importance in diseases associated with organismic senescence that increase mortality risk and ultimately reflect organism-specific genes. In the case of nonenzymatic glycation and deamidation, specific molecular repair and removal processes have been described that one suspects are important innovations during tetrapod evolution. Specialized mechanisms may have been required to permit the slow development, extended reproductive schedules, and long lifespans of birds and mammals. These animals have sustained elevations of body temperature that should, from thermodynamic principles, accelerate nonenzymatic glycation, deamidation, and racemization of proteins. The following summary of this rapidly expanding area mostly draws from laboratory rodents.

7.5.1. *Inactive Enzymes and Impaired Protein Degradation*

Age-related increases of inactive or altered enzymes were discovered from immunotitration studies and appear to represent degradative intermediates, which accumulate because of slowed protein turnover. The immunoprecipitation of isocitrate lyase in senescent nematodes (Gershon and Gershon, 1970) and of aldolase from erythrocytes at the end of their circulating lifespans in rabbits (Mennecier, 1970; Mennecier and Dreyfus, 1974) required relatively more antisera. In more than twenty examples, enzymatically inactive forms of the same polypeptide compose from 20 to 60% of the immunoreactivity in nematodes and tissues from rodents and humans by the lifespan (Gershon, 1979; Rothstein, 1982, 1985). Beside numerous enzymes of intermediary metabolism, inactive enzymes of protein synthesis accumulate, including aminoacyl-tRNA synthetases and elongation factor 2 (EF–2; Takahashi et al., 1985a, 1985b; Takahashi and Goto, 1987a, 1987b).

Altered enzymes occur in largely postmitotic tissues such as skeletal muscle and brain, as well as in liver, which can be stimulated to proliferate. Thus, there is no obligate relationship of the formation of altered enzymes to the postmitotic status. Small amounts of such inactive forms were recently detected in young liver cells (Reznick, Rosenfelder, Shpund, and Gershon, 1985; Reznick, Dovrat, Rosenfelder, Shpund, and Gershon, 1985). Structural alterations are also indicated by the greater thermolability of some enzymes from tissues of senescent animals (Resnick and Gershon, 1977). In the rodent and human lens, glucose–6-phosphate dehydrogenase becomes so altered that inactive forms no longer cross-react with polyclonal antibodies to the native enzyme, and vice versa (Dovrat et al., 1986). The remaining enzyme activities in most cases were biochemically and kinetically normal, but diluted by inactive forms. There is general agreement that not all enzymes in these same tissues become altered (Rothstein, 1982; Reff, 1985); this diversity indicates that the cause of the changes may vary among enzymes.

The causes of lost activity are post-translational without exception and involve alterations that differ for each case, as determined on highly purified material. In some cases inactive enzymes appear to represent altered conformations and not irreversibly damaged molecules. Technically strong studies show the reversibility of certain changes; correspondingly, altered enzymes can be formed from "young" enzymes. Enolase from senescent nematodes had conformational changes without evident chemical modification that could be reversed through manipulations of structure to recover activity (Sharma and Rothstein, 1980a, 1980b). Controlled chemical oxidation of phosphoglycerate kinase yielded an inactive form that was indistinguishable from that in skeletal muscle from senescent rats (Cook and Gafni, 1988). However, controlled oxidation of glyceraldehyde–3-phosphate dehydrogenase produced alterations that differed from the senescent enzyme forms in altered cofactor affinity (Dulic and Gafni, 1987). As for other altered enzymes, the age-related changes in glyceraldehyde–3-phosphate dehydrogenase involved post-translational modification, in this case an oxidized cysteine near the active site (Gafni, 1983; Gafni and Noy, 1984).

The factors favoring the oxidation of these and other altered enzymes is unclear and could be either local, that is, in the immediate region of the cell and its circulatory bed, or systemic. The mixed-function oxidases cytochrome P450 and NADH/NADPH oxidase are implicated in the degradation of proteins, by increasing their susceptibility to degradation by specific proteases (Oliver et al., 1987; Stadtman, 1988). Oxidized proteins are then rapidly degraded to amino acids by a variety of intracellular proteolytic mechanisms (Davies, 1987; Davies et al., 1990; Wolff et al., 1986). I suggest that the accumulation of inactive proteins could be influenced by genetic variants in the cytochrome P450 system, which maps in the MHC complex (Figure 6.7). Another factor could be changes in the redox state of glutathione and other metabolites, as described next.

Among the nonenzymatic defenses against free radical damage, GSH, which is the most abundant antioxidant, is hypothesized to maintain the reduced (native) state of sulfhydryl groups (Noy et al., 1985; Flohe and Gunzler, 1976). The nearly twofold age-related increase in the ratio of reduced to oxidized GSH (GSSG/GSH) in skeletal muscles of senescent rats (Noy et al., 1985) could thus contribute, over time, to the oxidation of cysteine residues. A shift in redox status towards reduced buffers against oxidation is indicated by the parallel age-related increases in the ratios of $NAD^+/NADH$ and $NADP^+/NADPH$, 1.5- to 2-fold. Age-related trends for decreased glutathione are also reported in plasma of humans (Naryshkin et al., 1981; Schneider et al., 1982) and in tissues of mice (Abraham et al., 1978; Hazelton and Lang, 1980; Chen et al., in press) and mosquitos (Hazleton and Lang, 1983; Richie and Lang, 1988; Richie et al., 1987).

Another oxidative mechanism has a potential role in atherogenesis: the modification of apolipoprotein B (apoB) by malondialdehyde, an end product of lipid peroxidation. ApoB is a major constituent of LDL (Chapter 6.5.1.3). Haberland et al. (1988) find immunocytochemical evidence for malondialdehyde-modified apoB in atheroma of Watanabe heritable hyperlipidemic rabbits (WHHL; Chapter 6.5.2). The malondialdehyde reacts with lysine residues of apoB; when a critical amount is modified, macrophages are attracted in test systems *in vitro,* evidently through a specific receptor (Haberland et al., 1982). Macrophages may also phagocytose glycated proteins (Section 7.5.2).

A major hypothesis is that the altered enzyme forms are normal intermediates in protein degradation and accumulate because of slowed protein turnover during organismic senescence (Gershon, 1979; Reiss and Rothstein, 1975). For example, the same forms of aldolase, some of them truncated, are detected in livers of young and senescent mice, but much more so in the latter (Reznick, Dovrat, Rosenfelder, Shpund, and Gershon, 1985). There are several types of direct evidence for impaired removal of abnormal proteins. For example, puromycinyl-peptides are removed sevenfold more slowly in senescent mice (Lavie et al., 1982). In a different approach, Ishigami and Goto (1988) showed that microinjected horseradish peroxidase was inactivated 50% more slowly in primary hepatocyte culture from senescent mice. Senescent nematodes also are less able to remove abnormal proteins that incorporate amino acid analogues (Reznick and Gershon, 1979). The half-life of aldolase, an altered liver enzyme in senescent mice, was also increased by 50% (Reznick et al., 1981). Other examples of slowed protein turnover in the liver are discussed in Section 7.4.1.

Nonetheless, the altered forms can be readily manipulated. An elegant study showed that inactive enzymes can be transiently purged from the liver of senescent rodents by inducing liver regeneration, which caused two enzymes in the regenerating tissue to recover "young" immunotitrability; by 9 days, a complement of inactive forms (presumably newly formed) reappeared (Hiremath and Rothstein, 1982). In view of the increased liver protein turnover of diet-restricted rats (Section 7.4.1.3), it is important that altered enzyme forms were also decreased by diet restriction (Takahashi and Goto, 1987b). Another interesting aspect concerns possible relationships between the rate of turnover and the propensity for accumulation of altered proteins. TAT, an inducible liver enzyme (Section 7.4.1.3), has a short, 1- to 2-hour half-life (Kenney, 1976) and no altered forms (Weber et al., 1980). While Jacobus and Gershon (1980) disagree, this variability could easily reflect a recent stress in their animals, which produced a transient enzyme induction.

On the other extreme are changes in cells like erythrocytes and lens cells with no protein synthesis. As another example of protein changes in cells without protein turnover, erythrocytes accumulate a variety of altered proteins during their several-month lifespans, including altered hemoglobins (Low et al., 1985) as well as enzymes (Mennecier and Dreyfus, 1974; Turner, 1975; Gracy et al., 1985; Oliver et al., 1987). Changes during the erythrocyte lifespan in the circulation include the spontaneous denaturation of hemoglobin and the addition of tyrosine residues to an erythrocyte-membrane-spanning protein (band 3, the anion transporter). Senescent band 3 then becomes a "neoantigen" that binds circulating antibodies against senescent red cells. This molecular tagging promotes removal of the senescent cell (Waugh and Low, 1985; Low et al., 1985; Kay et al., 1986; Kay et al., 1988; Kay et al., 1989). Thus, molecular degeneration arising through protein denaturation could be a pacemaker for lifespan of some cells. Deficiencies of vitamin E, a powerful membrane-soluble antioxidant, promote breakdown of band 3 and the binding of immunoglobulins, which suggests the importance of protein oxidation in the erythrocyte lifespan (Kay et al., 1986). Altered forms of enzymes are prominent in the lens of the eye (Dovrat et al., 1984, 1986; Gracy et al., 1985).

The accumulation of abnormal fibrillar proteins (NFT, or neurofibrillary tangles) in some neurons during organismic senescence in normal humans and the greater accumulation in Alzheimer's disease (Wilcock and Esiri, 1982; Roth et al., 1967) could also be manifestations of age-related impairments in protein degradation (Finch and Morgan, 1990). NFT accumulate extremely selectively in Alzheimer's disease and in the neurologically normal elderly: for example, NFT are absent in in the granule layer of the hippocampus and the CA_1 pyramidal cells, but are present in surrounding neurons (Chapter 3.6.10); this selectivity implies localized changes in turnover or in other conditions. Ubiquitin, a small-molecular-weight heat-shock protein that is involved in the proteolysis of abnormal or short-lived proteins, is covalently linked to NFT proteins (Mori et al., 1987). Other proteins not usually found in neuronal perikarya (Nukina et al., 1987; Ksiezak-Reding et al., 1987) are attached to these very insoluble and protease-resistant proteins. The accumulation of NFT and β-amyloid in neuritic plaques could be consequent to age-related impairments in protein catabolism (Finch and Morgan, 1990).

7.5.2. Glycation

The nonenzymation addition of sugars to form glycated proteins and further chemical changes in these adducts is a burgeoning topic in molecular gerontology. These developments are surveyed briefly for their implications for genetic aspects of lifespan, including biochemical mechanisms that may permit flexible reproductive scheduling in species that are prone to accumulating these changes. Several bold theoretical proposals concerning the role of glycated adducts and their derivatives in the pathophysiology of senescence— Cerami et al., 1987; Lee and Cerami, 1990; Baynes et al., 1990; Monnier, 1989—are being pursued by these and other investigators. There are substantial differences among laboratories concerning assays and possible artifacts that are not discussed below, but that should soon be resolved; see the most recent papers cited.

The nonenzymatic addition of sugars to free amino groups of proteins occurs in model reactions and *in vivo* at rates that depend on the type of sugar and its thermodynamically determined alternate structures (conformers). Of the circulating monosaccharides, D-glucose is the least spontaneously reactive, while D-galactose reacts fivefold faster (Bunn and Higgins, 1981). This reaction, the Maillard reaction, is favored by the open form of the sugar that has a free aldehyde group that can bond covalently with a free amino group. Besides the amino groups of proteins, amino groups of DNA bases are also potential targets of glycation (Bucala et al., 1984, 1985; Baynes and Monnier, 1989). In general, aldose sugars react faster with the free amino groups of proteins than do ketoses (Bunn and Higgins, 1981). The initial adduct then undergoes an Amadori rearrangement into a fairly stable ketoamine known as the Amadori product. *In vivo* exposure of hemoglobin in erythrocytes to blood glucose results in a progressive formation of glycated hemoglobin, HbA_{1c}, during the 3-month lifespan in circulation (Bunn et al., 1974). Because circulating erythrocytes have finite lifespans with a half-life of about 4 months, HbA_{1c} does not normally accumulate beyond 3–5% of total hemoglobin.

A controversy concerns the extent to which proteins with very long lifespans may show progressive age-related increases of glycation, as observed in collagens from human skin (Schnider and Kohn, 1980, 1981; but see Garlick et al., 1988; Figure 7.14) and basement membranes (Garlick et al., 1988) and in osteocalcin (Gundberg et al., 1986). However,

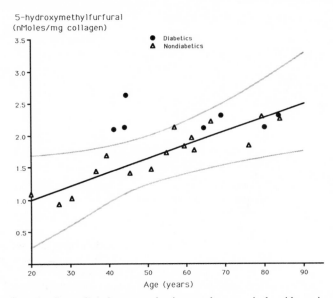

Fig. 7.14. Glycation of collagen from human tendon increased progressively with age in nondiabetics. Diabetics showed greater increases. Redrawn from Schnider and Kohn, 1980.

not all long-lived proteins show progressive age-related glycation in the absence of diabetes. Lens crystallins from normal humans, for example, show a barely perceptible increase in glycation, that is, less than 10% between 10 and 80 years (Patrick et al., 1990). Similarly, rat lens proteins show little increases in glycation from 4 to 28 months (Perry et al., 1987). The reversibility of glycation, with a half-life of about 5 days for HbA_{1c} under physiological conditions (Mortensen et al., 1984), may be one reason why glycation reaches a steady state in certain proteins with respect to the endogenous sugar concentrations, even if the protein has a very long lifespan. Another reason is that glycated proteins undergo further modifications, as described below.

There is general agreement that adult onset diabetes with increased blood glucose intensifies the extent of glycation in hemoglobin (Koenig et al., 1976; Bunn, 1981) and in collagen (Schnider and Kohn, 1980, 1981; Garlick et al., 1988; Figure 7.14). The amount of HbA_{1c} can represent the prior integrated exposure to elevated blood glucose, and in diabetics, for example, can reach 30% of hemoglobins. Besides glucose, the elevated fructose of diabetics may also be a reactant for the formation of fructosylated proteins (Suárez et al., 1988). Even diabetics, however, have less than 1–2% of glycated lysines in collagen.

A complex chemistry is emerging for glycated proteins. The initial sugar-protein adduct, that is, the Amadori product, is known from model reactions to fragment or rearrange into fluorescent and yellow-brownish products (Figure 7.15) that can act as protein cross-links. These endproducts of glycation are collectively called AGE (advanced glycation endproducts; Lee and Cerami, 1990; Brownlee et al., 1988) or advanced Maillard products. Rat aorta also accumulates a fluorophore resembling that in Figure 7.15 (Oimomi et al., 1988). Similar chromophores have been studied intensively in regard to

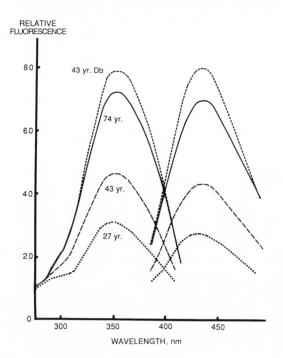

Fig. 7.15. Fluorescent spectrograms of collagen from brain dura matter that was purified from nondiabetics aged 27, 43, and 74 years and from a 43-year-old diabetic. The fluorophore is an advanced glycation endproduct (AGE) derived from glycation. Redrawn from Ulrich et al., 1985.

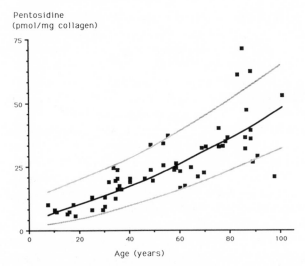

Pentosidine
(pmol/mg collagen)

Age (years)

Fig. 7.16. Pentosidine, arginine, and lysine residue cross-linkage by a pentose, increased with age in skin collagen from nondiabetic adults. Redrawn from Sell and Monnier, 1990.

browning reactions of food products (Cerami et al., 1987). The structure and chemical mechanisms of the glycation intermediates and the resulting chromophores and other end-products are in an early stage of characterization (Pongor et al., 1984; Horiuchi et al.,1988; Njoroge et al., 1988).

Tissues may have scavengers or quenchers of nascent AGE, particularly free amines like guanidino compounds (Brownlee et al., 1988; Nicholls and Mandel, 1989) and the amino acids, including taurine (Monnier et al., 1990). Even in the presence of elevated glucose, aminoguanidine blocked the formation of fluorescent membrane products in streptozotocin-induced diabetes of rats (Nicholls and Mandel, 1989). Model reactions also indicate the potential importance of multivalent metal ions in the oxidation of gly-cated proteins and the local redox state (Ahmed et al., 1986; Monnier et al., 1990). The age-related decreases in the redox state of glutathione and in plasma glutathione reported for humans, rodents, and mosquitos (Section 7.5.1) and in the sulfhydryl content of lens of rodents (Kuck et al., 1982) could be important to the formation of AGE. Cellular mechanisms involving macrophages are also indicated in the removal of glycated products (Vlassara et al., 1984, 1985a, 1985b, 1986).

Several new AGE were recently characterized and are candidates for age-related in-creases. *Pentosidine* (arginine and lysine residues cross-linked by a pentose) increased with age in healthy nondiabetic adults, sixfold over the adult lifespan in the skin (Figure 7.16; Sell and Monnier, 1990) and dura mater (Monnier et al., 1990). As for most age-related changes in mammals (Chapter 3.6), a few older individuals had no more pentosi-dine than the young adults; these deviations from the usual trend are important because they indicate the potential for environmental or genotypic factors that may minimize these chemical changes over very extended lifespans.

Diabetes and renal failure also increase skin pentosidine. *CML* (carboxymethyllysine), an oxidation product of glycated proteins, is barely detectable in the neonatal lens but increases at a constant rate after puberty to reach a fivefold increase by 80 years (Dunn et al., 1990). The consequences of modifying the positively charged lysine amino group to lens properties are unknown. Another glucose-derived AGE is *pyrroline* (5-hydroxy-methyl-alkylpyrrole–2-carbaldehyde), which increases by 160% in the blood albumin of diabetics (Hayase et al., 1989).

An important general observation about the above findings is that, *while formation of the initial glycated adduct is reversible, further products accumulate irreversibly.* These reactions may be modelled as trajectories on dose-duration surfaces (Figure 3.41). The rate of accumulated AGE may also be subject to local tissue variations and age-related changes in the tissue concentrations of sugar substrates, or metals and metabolites that influence the degradation of glycated intermediates, so that the trajectory may differ among tissues. For example, it would be interesting to look for local accumulations of glycated proteins in brain regions that accumulate iron, copper, and other metals with age (Coburn et al., 1971; Massie et al., 1979; Hallgren and Sourander, 1958). The zinc-rich mossy fiber zone of the hippocampus might also be vulnerable to the accumulation of glycated proteins. Moreover, systemic physiological conditions besides glucose could be important. Because each reaction is driven at a specific redox potential, there may be threshold levels for changes in plasma glutathione and other antioxidants, beyond which the formation of AGE is accelerated.

These findings are important molecular correlates of adult onset diabetes, particularly because many senescent changes are accelerated in adult diabetics (Chapter 3.6.8; Monnier, 1989; Vlassara et al., 1988; Baynes et al., 1990), including degenerative changes in the lens crystallins in association with cataracts (Monnier and Cerami, 1981) and altered biophysical properties of collagen in skin, blood vessels, and basement membranes, as well as in other proteins with long lifespans (Hamlin et al., 1975; Hamlin and Kohn, 1971; Kohn et al., 1984; Schnider and Kohn, 1982). Plausible roles of AGE to dysfunctions include increase of blood pressure from arterial rigidity; the loss of accommodation of the eye, which appears to proceed at a rate that scales with lifespan in rhesus monkeys and humans (Figure 3.14); impaired functions of the very long lived "memory T cells"; and atherogeneis, through the glycation of LDL, also impairs its binding to LDL receptors (Witztum et al., 1989; Sternberg et al., 1989). Another mechanism involves oxidative metabolism and the destruction of free radicals. Glycation can inactivate copper/zinc superoxide dismutase, as observed in erythrocytes from diabetics and in erythrocytes at the end of their lifespan in circulation (Arai, Iizuka, Tada, Oikawa, and Taniguchi, 1987; Arai, Maguchi, Fujii, Ishibashi, Oikawa, and Taniguchi, 1987). Protein oxidation was mentioned in Section 7.5.1 as a possible mechanism in the accumulation of inactive enzymes.

Diet restriction studies give an important approach to the specific roles of AGE in age-related dysfunctions. As will be discussed further in Chapter 10.3.2.3, restricted diets delay many age-related changes but also decrease glycated hemoglobins (Masoro et al., 1989). This important finding indicates a net reduction in the 24-hour levels of blood glucose and other aldoses. Thus, the stage is being set for detailed analysis of the relationships among blood sugars, protein turnover, and probably a host of local factors that determine various chemical changes in the initial adducts.

To close this brief overview of glycation, it is interesting to reconsider the scaling (allometric) correlations of metabolic rates and lifespan from Chapter 5.2.1. Blood glucose showed a relatively modest inverse scaling with body size ($\approx M^{-0.1}$ [Umminger, 1975]) that comprises a fourfold range from bats (200 mg/100 ml serum) to cattle (50 mg/100 ml serum). There is no obvious correlation of blood glucose to lifespan, as also noted by Monnier et al. (1990). However, there could be an unrecognized correlation of blood glucose or other adductifiable sugars to protein turnover, in comparisons between species of the same protein. Such a relationship is implied by the consistent inverse correlation of protein half-lives and oxygen consumption that scale per gram body mass as $M^{-0.25}$ (equation [5.5] and subsequent discussion in Chapter 5.2.1).

Many birds have long lifespans (> 30–40 years, with extended reproductive schedules; Chapter 3.5) that introduce yet other questions. Birds have two- to tenfold higher levels of blood glucose (Monnier et al., 1990; Hazelwood, 1986; Altman and Kirmayer, 1976) and body temperatures that are typically 2 to 4°C greater than in mammals. Both the higher temperature and blood glucose should accelerate glycation and the formation of glycation endproducts in birds. There are no data on AGE from any bird, although cataracts are occasionally reported in wild birds (Chapter 3.5). Conversely, evidence that opossums have short lifespans and early onset of cataracts (Austad, 1988; Chapter 3.6.1) suggests either diabetic-like elevations of blood glucose or deficits in scavengers of AGE.

Numerous counteracting factors must have been acquired during vertebrate evolution to allow long lifespans and high metabolic rates despite the risk of accumulating AGE to a dysfunctional limit. First, glucose is less reactive in the formation of glycated adducts than most other monosaccharides, which, as noted by Bunn and Higgins (1981), may have been the basis for its selection as a major carbohydrate fuel. Second, there are a number of scavengers that may quench formation of AGE, as noted above. Thus, the reproductive schedules that commonly extend into one or more decades in mammals and birds could be analyzed for adjustments in intermediary metabolism that influence the formation of AGE in vital long-lived proteins like lens crystallins. Other proteins may have their turnover rates adjusted inversely to the rate of AGE-related dysfunctions. No analysis has been made of nonenzymatic glycation of crystallins or collagen, for example, in relation to blood glucose, body size, and lifespan. These questions suggest a new area for study of molecular scalings with lifespan.

7.5.3. Cross-linking of Collagen

Collagen, which accounts for about one-third of the total body protein in mammals, is one of the most slowly replaced macromolecules. The negligible or very slow rate of replacement of insoluble collagens makes them particular targets of damage, as described in Section 7.5.2 for glycation in diabetes and possibly throughout the lifespan in nondiabetics. A definitive study with nonreutilized $^{18}O_2$ labelling in rats showed that about 50% of the $^{18}O_2$ in soluble collagen eventually was converted to insoluble collagens, and that there was no detectable turnover of the insoluble collagens in rats older than 30 days (Molnar et al., 1986). The comparison of cross-links derived from lysyl oxidase (unrelated to AGE) showed major differences between rat and monkey in the changes of several different cross-links that were detected by HPLC (Reiser et al., 1987), such that collagen from 2.5-year-old rat skin and lung had fewer total cross-links than at younger ages; rat and monkey differed extensively at mid-life in cross-links. These results caution against generalizations among species and tissue sources in collagen changes, and a considerable variety of cross-links is implied. In humans, a newly characterized collagen cross-link (histidinohydroxyllysinenorleucine) showed progressive increases during adult life (Yamauchi et al., 1988). The formation of cross-links in collagen after glycation is indicated from modelling studies to increase the trapping of other molecules, including immunoglobulins, lipoproteins, and other plasma proteins (Cerami et al., 1987). Such entrapment of proteins in blood vessels was hypothesized to promote atherosclerosis and the aggregation of platelets, both of which are major factors in ischemic heart disease and stroke.

Another age-related change concerns collagen tensile strength. A widely used assay for biological age in rodents is the breaking time of tail tendon, suspended in urea solutions under a constant weight, which is attributed to the tensile strength of collagen, its main constituent (Verzar, 1956; Gal and Everitt, 1970; Nimni, 1983; Harrison and Archer, 1978). The tensile strength of tail tendons increases linearly over most of the adult lifespan (Figure 10.21). The absolute rate of change was not accelerated in NZB mice (Harrison and Archer, 1978), a short-lived genotype with early onset autoimmune-related dysfunctions (Chapter 6.5.1.7.3). However, pre-morbid pathophysiological changes partially reverse the age-related increase in rat-tail-tendon tensile strength (Figure 10.26). The mechanisms in these biphasic changes are unknown and could involve covalent cross-links and noncovalent interactions with collagen and other macromolecules, as well as interactions with the cells surrounding the tendons.

Elastin may also be a target of glycation or other cross-linking agents. Like collagen elastin has a very long lifespan, with a half-life of years in mouse lung (Dubick et al., 1981) and in quail aorta (Lefevre and Rucker, 1980). Elastin is a major constituent of the arterial wall, and in the aorta represents 50% of the total proteins (Sandberg, 1975). Age-related changes in elasticity of the aorta are well known and contribute to increased blood pressure. The role of chemical modification of elastin in blood pressure changes is undetermined.

7.5.4. Racemization

This and the next section concern related examples of modifications, largely temperature-driven, of amino acids in long-lived proteins. Unlike glycation, the extent of racemization or deamidation do not initially depend on local metabolites or other molecules. The time- and temperature-driven racemization and deamidation approach pure examples of "intrinsic senescence" at a molecular level, although specialized enzymatic repair processes counteract these changes in some cells.

D-amino acids accumulate very slowly in nonreplaceable proteins over the human lifespan (Masters-Helfman and Bada, 1975; Masters et al., 1977, 1978; DeLong and Poplin, 1977; McKerrow, 1979). The occurrence of D-amino acids in newly synthesized proteins is negligible, but human lens crystallins and tooth dentine accumulate D-aspartate as a linear function of age at about 0.1% per year; over the human lifespan, more than 10% of aspartyl residues could become racemized. The rate of racemization, however, varies fourfold and may be slowest in collagen (Figure 7.17; Masters, 1983). Crystallins of brunnescent (advanced) cataracts also have nearly twofold more D-aspartate than do yellow cataracts or normal lenses of the same age group (Masters et al., 1977). The variations

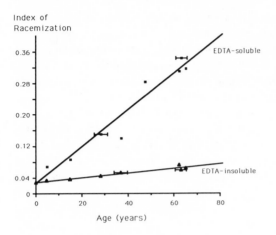

Index of Racemization

Age (years)

Fig. 7.17. The age-related racemization of L-aspartate acid in proteins, as measured by the parameter ln ([1+D/L]/[1–D/L]), where D and L are the stereoisomers of aspartate. *Square,* dentin of human teeth; *triangle,* the collagen-containing fraction (EDTA-insoluble), which racemizes more slowly. Redrawn from Masters, 1983.

may result from the accessibility of D-residues to an enzymatic repair process, as described below.

The rate of racemization is consistent with a nonenzymatic, thermodynamically driven mechanism. Amino acids racemize in polypeptides at the same relative rates as if free in solution, conforming to the Hammett-Taft relationships for effects from substituent groups on the free energy of hydrolysis (Bada, 1972) and to Arrhenius' law for temperature (DeLong and Poplin, 1977). Rates differ among amino acids; for example, aspartate racemizes sixfold faster than leucine. The most likely mechanism is a nonenzymatic abstraction of a proton from the chiral carbon of an amino acid to form a planar carbanion intermediate, which is then reprotonated to either the D or L form (Neuberger, 1948; Masters, 1982). Other chiral carbon atoms could also enantiomerize in carbohydrates or lipids and might accumulate if these had slow enough turnover times (DeLong and Poplin, 1977).

Bada and Helfman (1975) predicted this unusual phenomenon from their discovery of racemized aspartic acid in fossilized human bone, and calculated its detectability within the human lifespan. Although the slowness makes it difficult to detect changes in less than 10 years, the phenomenon also is documented in the dentine of other animals that live 20–40 years (Masters, 1982). The higher body temperature of birds predicts a twofold faster accumulation of D-aspartate than in mammals (Masters, 1982). No specific dysfunctions have been ascribed to proteins with racemized amino acids, but structural perturbations would be expected because of reoriented bond angles that might influence the amino acids' interactions with other molecules or their antigenicity.

Erythrocyte proteins also figure in this story, as shown in rigorous studies by Steven Clarke, Clare O'Connor, and others. Dense erythrocytes approaching their lifetime in circulation have an increased content of D-amino acids. D-aspartate is enzymatically converted to the methyl ester (McFadden and Clarke, 1982; O'Connor and Clarke, 1983, 1984; Galletti et al., 1983). The accumulation of carboxymethylated D-aspartate over the 3-month erythrocyte lifespan in humans is consistent with the rate of D-aspartate accumulation in lens protein (Barber and Clarke, 1983). Enzymatic carboxymethylation is proposed as a repair process to counteract racemization. Upon spontaneous hydrolysis of this labile adduct, the L configuration may be restored (McFadden and Clarke, 1982; Clarke and O'Connor, 1983). Most tissues have substantial activities of the responsible enzyme, carboxyl methyltransferase, which uses S-adenosyl-L-methionine as the methyl donor, including the brain, erythrocyte, testes, and eye lens (McFadden et al., 1983a; O'Connor et al., 1989). Cataractous lenses of humans have decreased enzyme activity, particularly in brunnescent cataracts (McFadden et al., 1983b) that have fourfold more D-aspartate (Masters, 1982). In erythrocytes, the carboxyl methyltransferase selectively acts on partially denatured hemoglobins (O'Connor and Yutzey, 1988). The increase of carboxymethylated D-aspartate in "senescent" erythrocytes could be due to either a decreased efficacy of repair or to an increased isomerization/deamidation process. Proof is still lacking that protein repair actually occurs through these mechanisms *in vivo* (Clare O'Connor, pers. comm.).

7.5.5. Deamidation

Deamidation is another slow change of long-lived proteins that also produces substrate residues for methyltransferases. Electrophoretically distinguished variants of triosephosphate isomerase and glucose-6-phosphate isomerase that accumulate in the oldest portions of the lens may arise by deamidation (Gracy et al., 1985; Cini and Gracy, 1986). Evidence for this is circumstantial but persuasive: the youngest portion of the lens contains a single basic form, while up to four acidic electrophoretic variants of glucose-6-phosphate isomerase occur in the older portions (Cini and Gracy, 1986). This observation is consistent with the observed deamidation of other proteins, which should increase the acidity of the protein. Such electrophoretic variants are not isozymes in the sense of multimeric molecules that differ in the content of peptide chains with distinct amino acid sequences.

The deamidation of glutaminyl and asparaginyl residues in proteins is a spontaneous and nonenzymatic event, which occurs *in vivo* and which is another molecular chronometer (Robinson and Rudd, 1974). The half-life of amide residues in proteins can be days to years, depending on neighboring residues. For example, the half-life for *in vitro* deamidation of gly *gln* gly is 100 days, and for gly *gln* ala gly 3,400 days (Robinson et al., 1970; Robinson and Rudd, 1974). The half-life of proteins may be in part regulated by their content of amide residues (Robinson and Rudd, 1974), as suggested by correlations of amide content and *in vivo* half-life (Figure 7.18). Long-lived proteins in birds and mammals have few amide groups.

The inverse correlation of amide content and protein turnover, while tentative, suggests an unexplored constraint on codon distribution and changes during evolution: genetic changes influencing the frequency of the two amidated amino acids would need to be compatible with the lifespan of the particular molecule or organism, particularly in proteins that are replaced slowly or not at all. Evaluation of potential codon constraints is complicated because of effects from neighboring amino acids and the three-dimensional structure. For example, a survey of proteins indicated the occurrence of aspartyl and asparaginyl residues in conformations that limit the approach of peptide-bond nitrogens to side-chain carbonyls in one mechanism of racemization and deamidation (Clarke, 1987). The L-isoaspartyl residue, a by-product of deamidation, may then be a substrate of methyltransferases. According to model reactions with peptides under physiological condi-

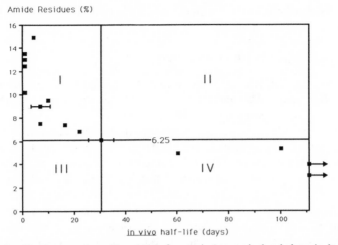

Amide Residues (%)

in vivo half-life (days)

Fig. 7.18. Relationships between the amide content of a protein (asparaginyl and glutaminyl residues) and its *in vivo* half-life. Short-lived proteins have relatively more amide residues (Sector I versus III). The distribution of amide content is nonrandom with respect to *in vivo* protein half-life for fifteen avian and mammalian proteins. The longest-lived protein represented is collagen from human bone, with an *in vivo* half-life of more than 5 years. Redrawn from Robinson and Rudd, 1974.

tions, the peptide nitrogen and side-chain carbonyls would otherwise form a succinimide derivative that, in turn, causes racemization and deamidation (Geiger and Clarke, 1987). Because particular structural domains of a protein may determine the extent of thermodynamically driven racemization and deamidation, evolutionary changes in susceptibility to these post-translational changes may be significantly constrained.

The functional consequences of deamidation are not established, but include increased susceptibility to proteolytic degradation because of structural perturbations from changes in charge that result from conversion of the amide group, which is near neutral under physiological conditions, to an acidic group. The best-studied case is triosephosphate isomerase, in which spontaneous deamidation creates new isozyme-like electrophoretic variants (Section 7.5.1). Deamidated residues also accumulate in crystallins of the eye lens in humans (Kramps et al., 1978) and cattle (Van Kleef et al., 1976; Zigler and Goosey, 1981). Deamidation should have little significance to very short lived species such as *Caenorhabditis* or *Drosophila*. Other longer-lived species of nematodes and insects (Chapters 2–4) would be interesting to study for possible species differences in molecular repair.

7.5.6. *Lipofuscins and Neuromelanin*

Progressive age-related accumulation of intracellular aging pigments, or lipofuscins, is one of the most long-standing observations in gerontology (for early history, see Strehler et al., 1959). The chemistry of these autofluorescent pigments is partly known. Those isolated from human brain contain phospholipids with unusual long-chain polyisoprenols (dolichol-like derivatives; Taubold et al., 1975; Wolfe et al., 1983; Hooghwinkel et al.,

1986). The chromophore fluoresces in the range 430–470 nanometers and is thought to include the iminopropene group

$$+$$
R-NH-CH = CH-CH = NH-R (Tappel, 1973).

The content of hydrolytic enzymes (Hendley and Strehler, 1965; Bjorkerud, 1964; Siakatos and Koppang, 1973) and their intracellular histochemistry (Barden, 1970) supports suggestions that lipofuscins are derived from lysozomes (Deduve and Wattiaux, 1966). A new approach to the genesis of lipofuscins uses leupeptin, a protease inhibitor, to rapidly induce pigment deposits in brain and retina (Ivy et al., 1984; Katz and Shanker, 1989). Partly degraded proteins may be an important constituent of lipofuscin granules, but there are also lipophilic fluorophores that are extractable with organic solvents (Eldred, 1987).

Lipofuscins accumulate at different rates in different cells. For example, for two types of large neurons, Purkinje cells accumulate lipofuscins more slowly than hippocampal pyramidal cells in mice (Reichel et al., 1968) and humans (Mann et al., 1978); Toth (1968) gives other examples of cell differences. While lipofuscins increase with age in many cells, these pigments are not unique to any age group. For example, lipoidal pigments also occur in the corpus luteum during adult ovarian cycles (Rossman, 1942). Another aspect is the presence of the β-amyloid protein fragment in the brains of patients with ceroid lipofuscin storage disease, in which lipofuscin-like pigments are accumulated precociously (Wisniewski and Maslinska, 1989).

Oxidative activity is implicated in the accumulation of lipofuscins. For example, histochemical studies show that DPN diaphorase is greatest in neurons with prominent accumulation of lipofuscin (Friede, 1962). Another hint is given by the lateral geniculate body of a 71-year-old woman who had lost one eye when 16 years old. Because the adjacent layers of the lateral geniculate receive the optic fibers from separate eyes, it was intriguing that adjacent layers showed alternately high or low amounts of lipofuscins; again, the lipofuscin levels correponded to the intensity of DPN diaphorase (Friede, 1962). These observations imply a relationship between certain functional activities and lipofuscin accumulation. Neuronal lipofuscin accumulation can be modified in rodents by diet and drugs (Chapter 10.3.2.5).

If there is a general relationship between oxidative metabolism and the accumulation of lipofuscins, then this should be reflected in species differences, because the most metabolically active tissues have oxygen consumption that scale inversely with body size, that is, negative allometry on body size. Few comparative studies have been made of the rate of lipofuscin accumulation, and it is necessary to make post hoc comparisons from separate studies. In at least one case, the myocardium, a scaling is indicated that is consistent with other findings on organ metabolic scaling. Comparisons of separate studies suggest that the rate of lipofuscin accumulation scales with the lifespan of dogs and humans and is fivefold faster in dog myocardium (Munnell and Getty, 1968) than in human (Strehler et al., 1959). This agrees with the fivefold difference in metabolic rate per gram body weight between dogs and humans, which should also pertain to the heart, since the frequency of heartbeats and the circulation of blood also show similar scaling to body weight (Chapter 5.2.1). In regard to neurons, Hofman (1983) calculated that the oxygen consumption per cortical neuron is the same within 10% for humans and rats. In large part, the differences in brain metabolism among species of different body size are attributable to different cell density (Chapter 5.2.4). On the basis of these provisional conclusions, the rate of lipofuscin accumulation need not be faster in rodents than humans. These important questions are wide open to further study.

Neuromelanins are another age-related pigment, distinct from lipofuscin in color, absence of autofluorescence, and greatly restricted occurrence. They are found in the locus ceruleus, substantia nigra, and a few other neurons of primates and carnivores; they are unknown in laboratory rodents. Neuromelanins are absent at birth and increase as a linear function of age (Brown, 1943; Barden, 1970). Both lipofuscins and neuromelanins may occur in the same neurons, substantia nigra, or locus ceruleus (Barden, 1970). Although the proximity of neuromelanin to the Golgi suggests some relation to lysozymes (Barden, 1970), others hypothesize that they arise as by-products of catecholamine metabolism (Graham, 1979); neuromelanins are usually restricted to catecholamine-containing neurons. Their absence in the locus ceruleus and in other catecholaminergic neurons of 2-year-old laboratory rodents could be accounted for by the shorter lifespan, since they are not evident until nearly the first decade in humans. It would be of interest to know if the accumulation rate of lipofuscin and neuromelanin had different scalings with lifespan in mammals. The present limited data suggest that neuromelanins accumulate more slowly.

Evidence is mixed on whether these intracellular masses are harmful. For example, despite accumulations in the neurons of the inferior olivary nucleus to the extent that the cell nucleus is displaced, this neuronal cluster shows no loss of neurons at all, even at advanced ages in humans (Monagle and Brody, 1974). Similarly, the amount of lipofuscin in the human myocardium was unrelated to heart disease (Strehler et al., 1959). On the other hand, neurons in the locus ceruleus of *Macaca nemestrina* (pig-tailed macaques)

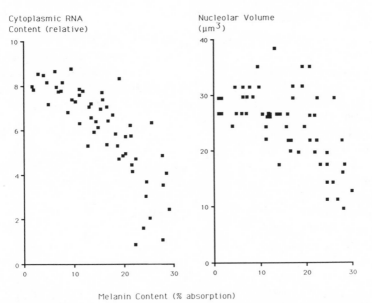

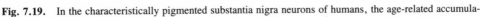

Melanin Content (% absorption)

Fig. 7.19. In the characteristically pigmented substantia nigra neurons of humans, the age-related accumulation of neuromelanin granules is accompanied by striking decreases of RNA. *Left*, cytoplasmic RNA; *right*, nucleolar size in the same neurons. The age-related shrinkage of nucleoli is also shown in Figure 7.6. Redrawn from Mann et al., 1977.

showed greater loss of the neurotransmitter norepinephrine in cell bodies with more lipo-fuscin (Sladek and Sladek, 1978). In the human substantia nigra, increased deposits of neuromelanin appear to be at the expense of cell body RNA (Figure 7.19, left; Mann et al., 1977). Quantitative histochemistry on a series of normal brains suggests reciprocal relationships between cell body RNA and nucleolar size versus lipofuscin (Mann et al., 1978). Like the decrease of norepinephrine in association with the increase of lipofuscin, the loss of RNA could merely be from displacement by the increased amount of neuro-melanin. However, the shrinkage of nucleolar volume at the highest levels of neurome-lanin (Figure 7.19, right) suggests that ribosomal gene transcription and possibly other genomic functions become impaired above some threshold (Mann et al., 1977). Finally, the neurotoxin MPP+ binds to neuromelanin (D'Amato et al., 1987). MPP+ is enzy-matically derived from a contaminant in synthetic heroin, MPTP, that kills substantia ni-gra neurons in some rodent species, primates, and humans and causes neurologic dis-orders that are very similar to Parkinsonism (Langston et al., 1983; Poirier, 1987). I suggest that age-related increases in neuromelanin could increase the risk for side effects from oxidative metabolism.

7.6. Cell Proliferation and DNA Synthesis

This section briefly reviews age-related changes in the capacity for cell proliferation and biosynthesis of DNA. Questions about fidelity of DNA synthesis and chromosomal ab-normalities are considered in Chapter 8.

7.6.1. Continuously Proliferating Cells

Many cell populations continue to proliferate actively throughout the lifespan in mam-mals, particularly the hemopoietic cells in the bone marrow and the exfoliating epithelia of the skin and gastrointestinal mucosa. As in so many aspects of gerontology, the extent of changes varies among cell types. Despite changes in cell-cycle length, some marrow and mucosal cells show no failure in continued formation from stem cell populations during the lifespan.

Hemopoiesis has been studied in the most detail. Slight decreases in the hematocrit (an estimate of the number of circulating erythrocytes) may arise beyond the average lifespan, but major decreases are associated with age-related diseases in C57BL/6J male mice (Finch and Foster, 1973) and humans (Zauber and Zauber, 1987). Hemopoietic powers across the lifespan have been investigated in elegant studies by David Harrison, among others (Ogden and Micklem, 1976; Tyan, 1980; Harrison, 1985; Harrison et al., 1988; Harrison et al., 1989). Mice aged 25 months show substantial impairments in replacing lost blood cells after bleeding; the hematocrit recovered more slowly and was never re-stored to the original levels (Figure 7.20). These impairments were not restored by trans-planting marrow cells from young mice under conditions that cured the anemic W/W^v strain, a genotype with intrinsic defects of hemopoiesis (Harrison, 1975a). This result implies alterations in the hemopoietic microenvironment of the bone marrow in which stem cells differentiate. Any depletion of hemopoietic stem cell pools is small during the lifespan, since serial transplantation of marrow through successive W/W^v hosts did not detect effects from the donor age (Harrison, 1975b, 1983). Serial transplantation was sustained for several mouse lifespans (73 months; four transplants), but ultimately failed for reasons that are attributed to transplantation trauma rather than the exhaustion of stem cell proliferative capacity (Harrison, 1985; Harrison et al., 1988). Moreover, the marrow was not influenced by donor age in its ability to reseed the hemopoietic capacity of hosts during several cycles of sublethal irradiation (Harrison, Astle, and Lerner, 1984; Harrison, Astle, and DeLaittre, 1988). Most recently, the numbers of primitive stem cells in the marrow that give rise to circulating lymphocytes and erythrocytes were assayed by a new procedure, and show a twofold *increase* in senescent C57BL/6J male mice (Harrison et al., 1989). In conclusion, present data argue that there is no intrinsic limit to the proliferation of hemopoietic stem cells, nor a loss of primitive stem cells, and that any effects of age on stem cells are dwarfed by comparison with effects from the transplantation procedures (Harrison et al., 1988; Harrison et al., 1989). Thus, the primitive marrow stem cells do not become exhausted over the lifespan, unlike the embryonically determined fixed pools of ovarian oocytes (Figure 3.19).

The proliferation of differentiated peripheral lymphocytes shows marked impairments, as illustrated by the smaller numbers of mitogen-responsive human T- and B-type lymphocytes in short-term culture (e.g. Hefton et al., 1980; Murasko et al., 1987; Figure 3.26). The study of Murasko et al. revealed major age-related decrements in mitogenic responses between individuals. Different responses to several mitogens suggest changes in lymphocyte subpopulations; gender differences were also found and the decrements were greatest before the average lifespan. Mitogenic impairments of spleen cells are seen in rodents (Meredith and Walford, 1977; Weksler, 1985). The fewer successive divisions of peripheral T cells from older humans (Hefton et al., 1980) and the limited proliferative capacity of antibody-forming clones *in vivo* (Moller, 1968; Williamson and Askonas, 1972) suggests that certain differentiated lymphocytes undergo clonal senescence *in vivo*, like the Hayflick phenomenon of diploid fibroblasts *in vitro* (Section 7.2.5). However, it

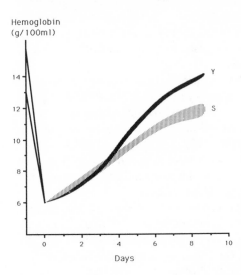

Hemoglobin (g/100ml)

Days

Fig. 7.20. The concentration of hemoglobin in mouse blood before and after bleeding, showing the faster recovery of 8-month-old mice (*Y*) than 25-month-old mice (*S*). The results were the same when 25-month-old mice were given supplements of young marrow 8 days before bleeding, implying that the hemopoietic microenvironment in the bone marrow is a larger factor in the age-related deficit in recovery from bleeding than a possible lack of hemopoietic stem cells. Redrawn from Harrison, 1975a.

is unknown whether diploid cell proliferation *in vitro* and *in vivo* is limited by the same mechanisms. Another open question is the extent to which tissues may accumulate post-replicative and clonally senescent cells after maturation.

Transplantation studies to investigate the proliferative potentials of lymphoid stem cells from bone marrow have given mixed outcomes in regard to donor age, possibly because of variable content of immune suppressor cells (Harrison, 1985). There is, however, agreement that formation of the granulocyte-macrophage cell series is not impaired during the lifespan in humans (Resnitzky et al., 1987) and mice (Sharp et al., 1989). The various outcomes of the studies on lymphocyte differentiation from bone marrow may depend on genotype as well as on these complex assays. To summarize a large and sophisticated literature, the capacity of pluripotential bone marrow cells to form the red- and the white-cell lineages seems subject to age-related shifts in differentiation, rather than to generalized cellular damage. Some issues about age-related changes in differentiation during lymphopoiesis, and the influences of genotype and species differences, should be resolved through better understanding of lymphokines, cytokines, and immunodifferentiative regulators.

The epithelial cells in the gut mucosa and in the epidermis are also subject to age changes that vary by region. These short-lived cells differentiate continuously from stem cells in the germinative zone and are continuously shed from the mucosal lining. In the duodenum, their transit from the site of formation in the crypts of Lieberkühn to the tips of the villae slows by 25% (12 hours) in 2-year-old mice (Lesher et al., 1961). It was also observed, however, that the transit time in the ileum was not slowed (Fry et al., 1962); here again, age-related changes differentially modify subpopulations of the same cell types. Age-related lengthening of the intestinal cell cycle during the mouse lifespan is mainly due to increases of G_1 phase (Thrasher and Greulich, 1965a, 1965b), while the S phase (duration of DNA synthesis) is relatively unaffected (Cameron and Thrasher, 1976); the duration of the S phase is also unaltered during clonal senescence of fibroblasts (Section 7.6.4). Because the transit time can be shortened to the same duration in young and senescent mice by chronic irradiation (Lesher, 1966), age differences in the cell cycle may not be intrinsic to intestinal epithelial cells. In any case, the normal histological appearance of intestinal crypts and villae in senescent mice (Thrasher and Greulich, 1965a, 1965b) and humans, as repeatedly told me by pathologists, suggests that that the progenitor cell pools are not exhausted during the lifespan.

Epidermal cells in humans may also have limited proliferative capacity *in vivo*. A slowed transit time through the stratum corneum after 50 years was indicated from a study of normal volunteers that used a fluorescent dye marker (Grove and Kligman, 1983). Consistent with slowed epidermal cell renewal, wound healing and reepithelialization is slower, according to numerous studies (e.g. Orentreich and Selmanowitz, 1969; Grove, 1982). In view of changes in collagen and other connective-tissue molecules that may influence epidermal progenitor cell proliferation, we do not know if these changes represent a partial depletion of a limited pool of epidermal stem cells.

In contrast to the apparently limitless production of cells in some exfoliating tissues, mammary epithelial cells are an important example of limited proliferative capacity *in vivo*. In a series of probing studies, Charles Daniel transplanted mammary ducts through a series of host mice using mammary fat pads (reviewed in Daniel, 1977). Following transplantation, the ductal epithelium outgrew and filled the fat pad. However, serial transplantation revealed an inevitable decline in the outgrowth of the ductal epithelial cells, such that a smaller fraction of the fat pad was filled after each consecutive transfer (Daniel et al., 1968; Daniel, 1973). The decline of proliferative capacity was progressive (about 15% per transplant) and not related to mouse strain or to the mammary tumor virus (MMTV). Even after outgrowth and replication had ceased completely, cholera toxin still could restimulate DNA synthesis (Daniel et al., 1984). While cholera toxin is an inducer of cAMP, the action of this external mitogen could be indirect through other cells in the fat pad. A further index of functions that are distinct from replication is that the postreplicative mammary epithelial cells remained competent for certain stimulae in the abundant production of milk by limit-transplanted mammary cells in lactating hosts (Daniel, 1977, 143). This example shows that experimentally induced exhaustion of proliferative capacity in specialized cells nonetheless did not cause loss of a separate differentiated function that can be elicited with appropriate stimulae.

7.6.2. Regeneration and Repair

Regeneration and repair of damaged or lost cells in adult mammals is limited to those organs that maintain the potential for continuous cell proliferation at high levels (exfoliating mucosa, skin) as well as at lower levels (liver, bone, endocrine glands). Regeneration of the rodent liver is discussed because, among organs with low ongoing proliferation, the most is known about its age-related changes.

In adult mammals, the liver ordinarily has a very low mitotic index in association with a slow replacement of hepatocytes, about once a year in rats (Swick et al., 1956; McDonald, 1961). Particularly between birth and mid-life, slowly continuing DNA synthesis without mitosis slowly increases the number of polyploid hepatocytes (Epstein, 1967). However, partial hepatectomy or injection of chloroform rapidly stimulates major shifts in gene activity, which are followed by DNA synthesis, cell replication, and regeneration of all cell types and the original mass in about 7 days (Bucher and Glinos, 1950; Bucher, 1963). Although regeneration becomes progressively slower (Bucher and Glinos, 1950; Bourlière and Molimard, 1957) and the onset of DNA synthesis is delayed (Bucher et al., 1964; Stöcker and Heine, 1971; Stöcker, 1976; Schapiro et al., 1982), the capacity for regeneration is retained throughout the rodent lifespan.

A progressive slowing of regeneration occurs at two distinct times, maturation and senescence, as shown by the experimental stimulation of regeneration by partial hepatectomy or chemical injury. In the maturational phase between weaning (1 month) and sexual reproductive maturity (4 months) in experimentally treated rats, the fraction of hepatocytes that incorporate thymidine decreases slightly from that in neonates ($-$ 10%; Stöcker and Heine, 1971). The onset of DNA synthesis is delayed by several hours, and increases of α-fetoprotein synthesis during regeneration become smaller and may not always occur (De Nechaud and Uriel, 1971; Sell et al., 1974). A related phenomenon may be the response to the carcinogen aflatoxin B_1, which induced more mixed tumors and higher α-fetoprotein when chronic treatment began at 6 weeks than at 26 weeks (Kroes et al., 1975). Potter (1978) suggests that these changes identify subclasses of G_0 hepatocytes which may also be important in carcinogenesis.

During senescence (by average lifespan), there are further decreases, from 90% to 70%, in the fraction of hepatocytes that are induced to synthesize DNA during regeneration

(Stöcker and Heine, 1971). Because the liver can regenerate at least twelve times in succession (Ingle and Baker, 1957) and because of the ordinarily slow turnover of hepatocytes, it is an open question about how close the subset of hepatocytes that still regenerate in senescent rats may be to a proliferative limit. It would be interesting to study the relationship of this subpopulation to the subpopulation of α_{2u}-globulin-synthesizing hepatocytes (Motwani et al., 1984; Section 7.4.1.2). The role of age-related hormone changes on changes in regeneration has not been investigated but is implied by hormonal influences on regeneration in the young (Bucher, 1963; White and Gershbein, 1987).

The ability of the gut to regenerate its mucosal surface after surgical wounds was investigated in rabbits aged 1 to 5 years (middle-aged; Thompson et al., 1990). By several criteria (morphometry, glucose uptake *in vitro*), the regenerated surface showed no age-related effects; the oldest rabbits, however, showed a slight decrease in crypt cell production rate, although the change was not statistically significant.

This and the preceding section demonstrate tissue differences in the proliferative capacity of rapidly or slowly proliferating tissue types of young adults with further changes during senescence. Other tissues of adult mammals have cell types that are already postmitotic, such as brain neurons and myocardial cells, and that will continue to function for many more decades in humans. For these permanently postmitotic cells as for the limit-transferred mammary epithelial cells, the acquisition of a postmitotic status is not accompanied by general dysfunctions. However, at the level of the organism, species with organs that have permanently postmitotic and irreplaceable cells are at risk for accumulating deficits in these cells at *any* time in life, whether through cell death or through the inability to repair or replace crucial macromolecules. The progressive decrease in the size of proliferating subpopulations in the liver and immune system may be viewed as another aspect of declining resilience or homeostatic capacity. The postmaturational decrements of proliferating cells thus are risk factors that contribute variably among individuals to later increases of morbidity and mortality.

7.6.3. In Vitro *Cell Proliferation as a Function of Donor Age*

The capacity for cell proliferation *in vitro* is another assay for age-related changes *in vivo*. The limited capacity of those clones for serial passage *in vitro* was established for diploid fibroblast-like cells by Hayflick and Moorhead (1961) and has been extended to many other mammalian cell types (Hayflick, 1977; Stanulis-Praeger, 1987; Norwood et al., 1990). Major examples include lymphocytes from human peripheral blood as mentioned above; human lens epithelial cells (Tassin et al., 1979); rabbit articular chondrocytes (Adolphe et al., 1983,; Dominice et al., 1986); satellite cells from skeletal muscle of rats (Schultz and Lipton, 1982); and endothelial cells (Levine et al., 1987). Because the cells grew as attached monolayers in most of these examples, it is important that similar results are given by liquid cultures with cells that do not normally attach, for example, human lymphocyte cultures (Perillo et al., 1989). In contrast, cancer cells and other transformed cells that are usually heteroploid can proliferate indefinitely by serial passage *in vitro* or as tumors *in vivo*. Cloned individual cells at different stages in serial passage show remarkable differences in proliferative potential (Smith and Whitney, 1980).

Many cells acquire distinctive limits to cell proliferation during differentiation, including neurons and myoblasts during prenatal life (Quinn et al., 1984; Jacobson, 1985a) and erythroid and certain immune cells throughout the adult lifespan (Holtzer, 1978; Rosendaal et al., 1976). The limited proliferative capacity of diploid cells *in vitro* is commonly described as cell senescence or aging *in vitro* (Chapter 1.2.1). Because the cessation of cell proliferation would constitute genetic death only in the case of those somatic cells in vegetatively reproducing organisms (Chapter 4.4), it seems more accurate to describe the limited proliferative capacity of somatic cells *in vitro* as clonal senescence. While clones may have lifespans defined by serial culture, the relationship of culture lifespan to cell functions during organismic senescence is unknown.

An important aspect of finite clonal lifespans is the correlation between the proliferative potential *in vitro* and the characteristic lifespan of the species. As discussed in Chapter 5.4.6 and shown in Figure 5.23, Röhme (1981) found strong correlations across a thirty-fold time range for seven mammals in a well-controlled study. Others, however, found less convincing evidence in mammals. A study of fibroblast-like cells from the lizard *Anolis* suggests a much greater proliferative capacity *in vitro* than shown by diploid mammalian cells (Simpson and Rausch, 1989), although subtle cell transformation may have occurred. Other comparisons between lifespan of the organism and clonal lifespans of fibroblasts are summarized by Hayflick (1977). While no general conclusions can be drawn about species comparisons, the approximate scaling of organismic lifespans in the same rank order as the capacity for *in vitro* proliferation has retained a major impact on thought in molecular gerontology. Other correlations with lifespan and biochemical properties are described in Chapter 5.4.

Besides the limited proliferative capacity *in vitro*, many cell types show reduced ability to proliferate *in vitro* as a function of donor age in mammals, as assayed by two different parameters: the ability to form colonies from initial tissue explants (primary cultures) and the proliferative capacity of primary cultures as judged by serial passsage. An example of the later is that fibroblasts from human neonates or human fetuses at the age of organogenesis can sustain about twice the number of serial passages *in vitro* as from adults (Hayflick, 1965).

While no cells obtained from mammals at later developmental stages have shown unlimited proliferative capacities *in vitro* under well-controlled conditions, one report suggests that early embryonic stem cells may behave differently. Based on the delayed segregation of germ cells from somatic cells until day 7–8 in mice, Suda et al. (1987) followed cultures of day 4 embryo cells. At this time, the embryo consists of a blastocyst at the egg cylinder stage, before implantation and before organogenesis. The cultures were maintained on γ-irradiated feeder layers. The frequency of stable diploid lines was low, but several lines remained diploid over prolonged serial culture and showed no diminution of proliferative power to an unprecedented ≥ 250 cumulative doublings. In contrast, numerous other studies show that mouse fibroblasts can sustain only 10 to 30 cumulative doublings *in vitro* (Todaro and Green, 1963; Röhme, 1981; Figure 5.23). Upon injection into embryos, the diploid lines had no ill effects on the resulting chimeric mice, indicating that the cells had not transformed. Moreover, the F_1 offspring of the chimeras showed markers of the donor genotype. While these early embryo cells showed extended proliferation, it is unclear what endpoint should be used: clonal senescence arises very slowly in some ciliates over a span of years and hundreds of fissions (Chapter 3.2). A major question, then, concerns the distribution in embryo cell lineages of differentiations that cause finite proliferation. Other relationships between the time of germ-line segregation and somatic cell senescence are discussed in Chapter 4.

Age effects from adult donors on the *in vitro* proliferative potential of fibroblasts are not as big as the change between birth and maturation. Several studies show an inverse age-related trend between donor age and the numbers of passages before clonal senescence (Figure 7.21; Martin et al., 1970; Goldstein et al., 1978; Schneider and Mitsui, 1976). In the F344 rat, fibroblasts from donors aged 24 months showed a 50% decrease compared with 6-month-old donors, in number of serial passages before clonal senescence (Vincent Cristofalo and Edward Masoro, pers. comm.). Increased heterogeneity of *in vitro* proliferative capacity is generally observed with donor age (Norwood et al., 1990), which makes firm conclusions difficult. One suspects that subpopulations with different systemic pathophysiology affect the donors' fibroblasts *in vivo,* as shown below.

An inverse correlation with donor age is also reported for primary cloning efficiency, or the yield of cells from primary explants. A good example is the smaller yield of arterial smooth muscle cells from several rodents (Figure 7.22; Martin et al., 1983) and humans (Bierman, 1978). The study of *Mus musculus* and *Peromyscus leucopus* (Martin et al., 1983) is important because of the twofold greater lifespan of *P. leucopus,* a valuable comparison of two rodents with similar size (Figure 3.11; Sacher and Hart, 1978). The yield of primary smooth muscle cell clones from the aorta decreased progressively with age, paralleling differences in lifespan (Figure 7.22). Secondary cloning became less efficient with older mouse donors; this study did not evaluate clonal senescence. In contrast to the smaller capacity of cells from older donors to proliferate during long-term culture, the primary cultures of aortic smooth muscle cells from 24-month-old F344 rats had a *larger* subpopulation in S phase than those from 3-month-old rats (Hariri et al., 1988). This observation is consistent with age-related increases of smooth muscle proliferation during

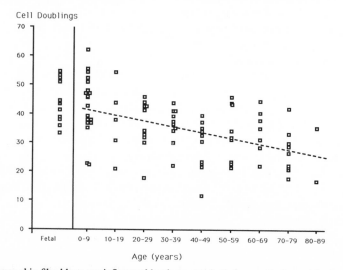

Fig. 7.21. Human skin fibroblasts are influenced by donor age in their capacity for serial passage *in vitro;* the statistical analysis indicates that about 30% of the proliferative capacity is lost during the human lifespan, though some very old donors clearly fall into much a younger range. Age accounts for 25% of the variance. Redrawn from Martin et al., 1970.

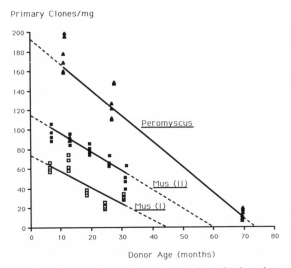

Primary Clones/mg

Donor Age (months)

Fig. 7.22. The primary cloning efficiency of smooth muscle cells from the thoracic aorta after elastase disso-ciation in *Peromyscus leucopus* (lifespan 5 years) and laboratory mice (lifespan 2.5 years; Figure 3.11). The number of primary clones is per milligram wet weight of the initial aorta fragment. The assay shows variations between two studies of laboratory mice; nonetheless, both show smaller primary cloning capacity than cells from *Peromyscus*. Both species show marked age-related loss of primary clones. Redrawn from Martin et al., 1983.

atherogenesis. There is also agreement between the *in vitro* proliferation of dermal fibro-blasts from Syrian hamsters and wound healing by the donors, at the site from which the explant was taken; the impairments did not predict the lifespan, however (Bruce and Dea-mond, 1990). For other correlations between donor age and *in vitro* proliferation, see Norwood et al., 1990.

The decreased *in vitro* proliferative capacity as a function of donor age is poorly under-stood and may be influenced by age-related changes that are distinct from intrinsic char-acteristics of fibroblasts. Diabetes and other age-related diseases may impinge, since fi-broblasts from diabetics have reduced and variable proliferative capacity *in vitro* (Goldstein et al., 1969; Goldstein et al., 1978; Vracko and McFarland, 1980; Mooradian, 1988; Section 7.8). Changes in the connective tissue matrix might also alter the efficiency of primary cloning (Norwood and Smith, 1985). Host genetic factors are indicated by the smaller differences in fibroblast proliferation of monozygotic versus dizygotic twins (Ryan et al., 1981). Other genetic-pathologic influences are indicated by the smaller *in vitro* proliferation of fibroblasts from humans with cystic fibrosis, Hutchinson-Guilford progeria (Danes, 1971), and Werner's progeria (Martin et al., 1970) and of lens epithelial cells from a cataractous strain of mice (Lipman and Muggleton-Harris, 1982). These di-verse examples of impaired *in vitro* proliferation suggest that the donor genotype has influences on various stem cell lineages, possibly through initial influences on differentia-tion during development, or through ambient metabolic influences on stem cell differentia-tion. Thus far, it cannot be concluded that earlier onset of clonal senescence *in vitro* results from the depletion of an equivalent number of cell divisions during the donor's lifespan. Moreover, the increased heterogeneity of cells from late-passage cultures makes simple conclusions difficult to defend.

The acquisition of proliferative limits during differentiation is exemplified by the fixed numbers of cell divisions of erythropoietic cells once committed to this lineage, which leads to a cell without a nucleus and finite lifespan of 2 months in mice and 4 months in humans. To pick a heuristic if extreme example, the membraneous changes in the mammalian erythrocyte that has no protein synthesis during its 2- to 4-month "lifespan" in circulation (Section 7.5.1) are not obviously related to the loss of proliferation during serial passage of fibroblasts *in vitro*. Another example of this is given by mammary epithelial cells that have a maximum proliferative capacity (clonal senescence) during serial transplantation in adult mice, yet can still produce milk in their postreplicative state (Section 7.6.1). The slow degenerative processes in certain postmitotic neurons that lead to cell atrophy and the accumulations of neurofibrillar tangles in humans may not share common mechanisms with other types of age-related cell degeneration. The next section shows how impairments in proliferation are independent from continued functioning of other gene sets.

7.6.4.   The Cessation of DNA Synthesis during Clonal Senescence and the Characteristics of Postreplicative Fibroblasts

Thirty years ago, the concept that cultured cells had a limited proliferative capacity was heresy. New data had to confront the legendary, but artifactual, long-term cultures of chick fibroblasts begun by Alexis Carrell at the Rockefeller Institute earlier in the century (Carrel, 1935, 172; Hayflick, 1977; Norwood and Smith, 1985). However, diploid mammalian fibroblasts were ultimately proven to have limited proliferative capacity during serial passage *in vitro* (Hayflick and Moorhead, 1961; Hayflick, 1965), a major finding extended to many other cell types. This recognition also sharpened focus on the mechanisms of cell immortalization. Serially passaged diploid cells go through a "crisis" of declining proliferative capacity, followed by the emergence of cells with indefinite proliferative capacity; the frequency of spontaneous transformation varies with species and is vastly lower in human and chick fibroblasts than in mouse (Hayflick, 1977; Norwood et al., 1990; Stanulis-Praeger, 1987). Experimentally, diploid fibroblasts can be transformed into cell lines with limitless proliferative capacity *in vitro* by infection with animal viruses, particularly DNA viruses like simian virus 40 (SV40). The sequences required for transformation express a DNA-binding protein, large T antigen, which has given an important tool for analyzing the process of immortalization (Wright et al., 1989).

A major approach was based on the proliferative potential of hybrid cells and heterokarya. Hybrid cells formed from diploid fibroblasts and immortalized (transformed) cells with indefinite replicative potential have unlimited proliferative capacity (Bunn and Tarant, 1980; Muggleton-Harris and Aroian, 1982; Pereira-Smith and Smith, 1983; Stanulis-Praeger, 1987). In all cases, the phenotype for clonal senescence was recessive in the hybrids. Four complementation groups were shown in the fusions of immortal HeLa cells with each other that all led to lines with finite proliferative capacity (Pereira-Smith and Smith, 1988). Human chromosome 1, in particular, when introduced by interspecies hybrids, can cause finite proliferative lifespans in immortal hamster cells (Sugawara et al., 1990). These data implicate specific genes in the proliferative limits that are not randomly distributed on the chromosomes.

The causes of *in vitro* clonal senescence and the cessation of DNA synthesis are beginning to be understood as a blockade of clonally senescent cells in the G_1 (pre–DNA synthesis phase) of the cell cycle at the G_1/S phase boundary (Yanishevsky et al., 1974; Sherwood et al., 1988; Norwood et al., 1990; Olashaw et al., 1983). A major effort maintained by Vincent Cristofalo, Sam Goldstein, Thomas Norwood, James Smith, and Olivia Pereira-Smith for 20 years shows that selective changes in the regulation of a limited number of genes can explain the loss of replicative potential during serial passage. As the clonal limit approaches, the G_1 phase lengthens (Grove and Cristofalo, 1977). As noted above, the postreplicative cells actively synthesize many mRNA and proteins, including those of importance to the cell cycle such as thymidine kinase and histone H3 (Rittling et al., 1986). Thus, there is no general failure of biosynthesis. Despite overall reductions of proliferative capacity, a few cells continue to synthesize DNA with varying cell-cycle durations at the last stages of *in vitro* clonal senescence (Burmer and Norwood, 1980; Macieira-Coelho and Taboury, 1982), but most do not initiate DNA synthesis.

The impaired initiation of DNA synthesis can be overcome by the introduction of specific DNA sequences, by acute infections with SV40 mutants that were unable to transform fibroblasts but that expressed the T antigen (Gorman and Cristofalo, 1985). This study showed that the machinery of DNA synthesis is intact at the time of clonal senescence. However, acquisition of continued replication, that is, transformation of diploid fibroblasts, may require further steps, since the T-antigen protein alone is insufficient for transformation (Wright et al., 1989). Several studies indicate a role for the cellular oncogene *c-fos*, which is transcribed in early G_1 phase. While *c-fos* mRNA increased rapidly after serum stimulation in early-passage diploid fibroblasts, late-passage fibroblasts failed to respond (Doggett et al., 1989; Seshadri and Campisi, 1990). The role of *c-fos* mRNA in passing the G_1/S boundary is being studied by transfections with plasmids that contain *c-fos* sequences under the control of an inducible promoter. With transient increases of *c-fos* mRNA, increased numbers of late-phase fibroblasts were stimulated to DNA synthesis (Phillips, Nishikura, and Cristafalo, 1990).

Evidently, numerous macromolecules can influence these complex changes. Much work remains to identify the relative roles in what may be viewed as a range of points in multiple regulatory pathways that control the cell replication cycle. Acting on these putative control points are trophic factors and hormones that influence proliferation *in vitro*, including hydrocortisone (Grove and Cristofalo, 1977), IL-2 (Fathman and Frelinger, 1983), and insulin-like growth factor (Phillips et al., 1987). These extrinsic influences are consistent with many *in vivo* cellular changes during organismic senescence that are influenced by hormones (Chapter 10).

Whichever genes are implicated, the data point to selective changes in gene regulation that become altered during clonal senescence, resulting in this specific blockade in late G_1. DNA methylation could be involved (Section 7.3.4), as suggested by the decreased 5-methylcytosine at the approach to clonal senescence (Wilson and Jones, 1983; Fairweather et al., 1987) and by the earlier onset of clonal senescence in cultures treated with 5-azacytidine, a compound that causes undermethylation during DNA synthesis (Fairweather et al., 1987). Recombinant gene technology may soon reveal the controls on transcription of mRNA with antiproliferative activities, the peptides encoded, and their presence in other cells *in vivo* that become postmitotic or possibly postreplicative during fetal or adult life.

Postreplicative diploid fibroblasts may acquire an unhealthy appearance, with the accumulation of lipofuscins and other biochemical changes suggestive of degeneration (Deamer and Gonzales, 1974; Hayflick, 1977; Matsumura et al., 1979). While there are eventually involutional changes, cell death is not prominent until about 9 months (Section 7.6.4; Bayreuther et al., 1988). Nonetheless, cell survival extends beyond at least 12 months (Matsumura et al., 1979) and phase III fibroblasts still support virus infections with the same yield as "young" (early transfer) fibroblasts (Holland et al., 1973; Tomkins et al., 1974). Many other aspects of biosynthesis are not different between quiescent, early passage fibroblasts and late passage cells (Norwood and Smith, 1985; Cristofalo and Stanulis-Praeger, 1982), including the mRNA levels of thymidine kinase, histone H3 (expressed at the G_1/S boundary), and other genes expressed during G_1 (Rittling et al., 1986). On these grounds, postreplicative phase III diploid fibroblasts, at least in the initial weeks, are senescent *only* by virtue of their inability to initiate DNA synthesis.

On the other hand, late passage adult bovine adrenocortical cells maintained the full induction of cAMP in response to prostaglandin PGE_1, while responses to ACTH and induction of steroid 17α-hydroxylase steadily declined (Hornsby and Gill, 1978; Hornsby et al., 1989; Hornsby et al., 1990). This could be a model for diminishing hormone responsiveness seen in some reproductive target cells (Section 7.4). As discussed throughout this chapter, there are few indications of general loss in differentiated cell characteristics during organismic senescence *in vivo*.

The biochemistry of clonally senescing cultures forms a huge literature, from which items were selected in relation to this discussion. Phase III cultures show progressive increases of altered enzymes, that is, enzymes in which a portion of the total immunoreactive population is catalytically inactive through post-translational modifications (Section 7.5.1; Holliday and Tarant, 1972; Holliday et al., 1974; Goldstein and Moerman, 1975; Stanulis-Praeger, 1987). Increased susceptibility of cellular proteins to proteolysis (M. O. Bradley et al., 1975) also indicates alterations in protein structure. However, as observed *in vivo*, many enzymes do not change during clonal senescence (Rothstein, 1982; Stanulis-Praeger, 1987). Fibroblasts from Werner's syndrome, an adult onset progeria (Chapter 6.5.3.3), grow poorly, cease proliferating early in serial passage (Goldstein and Moerman, 1975; Martin et al., 1970; Salk et al., 1981), and show an increase of some enzymes with altered forms at a corresponding stage in culture lifespan (Goldstein and Moerman, 1975). The mechanisms leading to altered enzymes during clonal senescence are unknown but may be related to slowed protein turnover. Direct evidence includes the slower degradation of exogenous proteins that were introduced by microinjection (Dice, 1982) or cell fusion (Dice, 1989). Thus, slowed turnover and impaired protein catabolism could be common to organismic and clonal senescence.

A portentious finding is that phase III fibroblasts produce much more procollagenase, but less of a metalloproteinase inhibitor (TIMP; West et al., 1989). For a corresponding change to occur in the fibroblasts of connective tissues *in vivo*, the local fibroblasts would need to be undergoing clonal senescence *in vivo*, which is neither proven nor disproven. The direction of these changes, whatever their occurrence *in vivo*, is consistent with increased collagenase and decreased TIMP activity in bones during osteoarthritis (Reynolds, 1986) and osteoporosis (Otsuka et al., 1984).

The macromolecular changes in phase III fibroblasts have not been studied in detail over prolonged times. A recent report on postreplicative fibroblasts indicates a complex temporal sequence of changes in the sets of polypeptides being synthesized (Bayreuther et al., 1988). Stage VII leads to cell degeneration and death and occurs after about 9 months in the postreplicative state. These are among the first studies to suggest an ultimate loss of viability in the postreplicative phase. The potential of stationary cultured diploid cells for a prolonged postreplicative existence has long been recognized (Section 7.2.5), but the extent of subsequent postreplicative changes is not known in detail and cannot be simply equated with postreplicative changes and proliferative limitations *in vitro*.[7]

In conclusion, certain rapidly proliferating cells maintain this capacity with few impairments throughout the organismic lifespan, as shown for hemopoietic stem cells and intestinal mucosal cells. However, mammary epithelial duct cells, certain lymphocytes, fibroblasts, and probably other types have a limited proliferative ability in mature adults, as shown by *in vivo* transplantation studies and by serial transfer *in vitro*. It remains to be established whether cell types with limited proliferative capacity may also show clonal senescence *in vivo*. Thus, it cannot be said that any *in vivo* cell-cycle changes during organismic senescence (e.g. extended G_1 in the mouse intestinal mucosa) represent early manifestations of clonal senescence.

7.7. Cell Death

Recent findings on mechanisms of "programmed" cell death, or apoptosis (Wyllie et al., 1980), show the active role of gene expression. Apoptosis represents cell death that is predictably triggered by a defined stimulus like the removal of a hormone, and does not include adventitious cell death, which might occur, for example, in the kidney during autoimmune degenerative conditions in the NZB mouse or through random trauma. A major search is underway for changes in gene expression that lead to cell death and for proteins that may be the active agents. A precedent for hypothetical killer proteins, prospectively named *thanatins* (Johnson, 1990), is the proteins secreted by lymphocytes and cytotoxic T cells that kill select target cells (Marx, 1986). The requirement during apoptosis for gene activity leading to the synthesis of new proteins is also indicated by the arrest of cell death through inhibitors of protein or RNA synthesis during development in insects (Lockshin, 1969) and chickens (Martin, Schmidt, DiStefano, Lowry, Carter, and Johnson, 1988). Three examples of adult cells are briefly described; all involve hormones. The likely role of calcium in many of these mechanisms will not be discussed, although there may be important, proximal genomic controls over calcium and other intracellular metabolic components during apoptosis.

Glucocorticoids cause the rapid death of thymocytes by a mechanism that leads to nucleolytic degradation of nuclear DNA (DNA ladders), which is one of the best-understood examples of apoptosis (Munck and Crabtree, 1981; Compton and Cidlowski, 1986; Wielckins and Delfs, 1986). Cell killing by glucocorticoids takes about 1–3 days and can be protected against by inhibiting transcription through actinomycin D (Mosher et al., 1971). The gene products required for cell death are unknown. Because glucocorticoids and excitatory neurotoxins synergize to kill select hippocampal neurons *in vivo* (Sapolsky,

---

7. This work comes from the laboratory of Klaus Bayreuther, no relation to Konrad Beyreuther who studies amyloid and other abnormal proteins of Alzheimer's disease.

1985) and *in vitro* (Sapolsky et al., 1988; Masters et al., 1989), the absence of DNA ladders in the latter study suggests other mechanisms.

The next examples of apoptosis result from the withdrawal of trophic hormones. Castration of adult rats causes rapid death of prostatic epithelial cells with DNA fragmentation ladders as observed in lymphocytes exposed to glucocorticoids (Kyprianou and Isaacs, 1988), and concomitant increases of a mRNA, testosterone-repressed message-2 (TRPM-2; Buttyan et al., 1989). This RNA is induced by other circumstances involving cell death, but in the absence of obvious steroid hormone involvement: in the limb buds of developing mouse embryos (Buttyan et al., 1989), in the brain during Alzheimer's disease (May et al., 1990), or in response to deafferenting lesions (Lampert-Etchells et al., 1990). TRPM-2 is identical with a protein normally present in Sertoli cells of the testis, sulfated glycoprotein–2 (SCG-2; Collard and Griswold, 1987), and with a serum protein that inhibits the complement cascade, SP-40,40 (Kirszbaum et al., 1989) or complement lysis inhibitor (CLI; Jenne and Tschopp, 1989). This multifunctional protein has roles in viable cells as well as those that are degenerating.

The last example is the death of sympathetic neurons from the superior cervical ganglion after withdrawal of their required hormone, nerve growth factor (NGF; Levi-Montalcini and Booker, 1960; Hamburger et al., 1981). As observed for glucocorticoids on lymphocytes, cell death from withdrawal of NGF in sympathetic neuron cultures depends on continuing RNA and protein synthesis (Martin, Schmidt, DiStefano, Lowry, Carter, and Johnson, 1988). Together, these examples demonstrate that death of nucleated cells is an active process involving dynamic changes in gene activity and is very different from the more passive degeneration of erythrocytes during their cell lifespan. The causes of slow neurodegeneration over the course of years should be open to study with the molecular markers described above.

7.8. Memories and Gene Expression

Besides degenerative molecular and cellular changes, some differentiated cell types show intriguing changes in response to physiological stimulae that constitute a type of memory at the level of gene expression and that might be important in limiting homeostatic responses as age progresses. Acquisition of immunological memories in lymphocytes is a familiar example. The severalfold increase during the mouse lifespan of memory T lymphocytes with the *Pgp-1* surface marker (Lerner et al., 1989; Miller, 1990) might represent a lifetime of exposure to environmental antigens. While lifetime exposures to antigens seem unlikely to deplete cells capable of primary responses, the homeostatic control of lymphocyte progenitor pools could be altered in concert with anti-idiotypic responses (Chapter 3.6.6). Evidence for an increase of marrow progenitor cells (Harrison et al., 1989) is consistent with this possibility.

Endothelial cells show a different type of memory effect that lingers after exposure to elevated blood glucose (Roy et al., 1990). Recognizing that fibronectin accumulates during clinical and experimental diabetes in basement membranes of the kidney and elsewhere, Roy et al. showed threefold increases of fibronectin mRNA in kidney of rats made diabetic by streptozotocin. Moreover, the mRNA elevations persisted after normalization of blood glucose. Primary umbilical endothelial cells grown in media with 50 mM glucose (tenfold normal) for 22 days retained elevated fibronectin mRNA and collagen IV mRNA when returned to 5 mM glucose for 9 days. During exposure to high glucose, *c-myc* mRNA was not elevated, nor did fibronectin or collagen IV mRNAs have slower turnover. Whether exposure to elevated glucose alters transcription rates is unknown.

These interesting findings suggest another mechanism in diabetes at the level of mRNA regulation that should be studied for interactions with clonal senescence *in vivo* or *in vitro* and that could be a subtle consequence of age-related glucose intolerance (Chapter 3.6.8). While the molecular mechanism is unknown, a precedent for transcriptional influences is given by other memory effects from estrogen, which persist for several weeks as DNA demethylation (Wilks et al., 1982) or several months as chromatin estrogen-binding sites (Haché et al., 1987) or estrogen receptor mRNA (Barton and Shapiro, 1988) after estrogen removal in liver cells. Similar changes may be considered as a mechanism in the cumulative and only partially reversible effects of estrogen on neuroendocrine functions in rodents that are implicated as a mechanism in rodent reproductive senescence (Chapter 10.4; Table 10.7). These diverse examples of long-lasting influences on gene expression hint at phenomena of general importance to the pathophysiology of senescence in mammals.

7.9. Summary

This chapter presented examples of changes in gene expression from a sampling of mammals, invertebrates, and plants that represent a few of the diverse types of organismic senescence described in Chapters 1–4. Most data on specific cell changes come from laboratory rodents. Besides limited data on humans, there is virtually no data on long-lived birds, invertebrates, or plants with slower senescence. Nonetheless, this limited sample of species shows recurrent characteristics.

The cell-type specificity of age-related changes is remarkable, so that closely related cell types show quantitatively and qualitatively different changes during the lifespan, in regard to the accumulation of lipofuscins (Section 7.5.6); accumulations of NFT (Section 7.5.1); altered cell-cycle kinetics (Section 7.6.1); mRNA levels that show differential increases (SMP-2) and decreases (α_{2u}-globulin) in the same tissue (liver) (Section 7.4.1.2), and enzyme inducibility (Section 7.4.1.3). An impressive range of cell functions in most organs remain unimpaired throughout the lifespan. Exceptions in which impairments are general may be sought in ovarian steroid target cells after ovarian secretions cease (Chapter 3.6.4.1) and in other cells that may become deprived of trophic support. This selectivity shows that subtle aspects of cell differentiation strongly influence cell functions during organismic senescence.

Besides the selectivity at the level of which mRNA and proteins are increased or decreased, there is also temporal selectivity as to when the changes occur. While evidence is sketchy, select changes of macromolecular biosynthesis occur in the transition between puberty and young adulthood, for example, MUP mRNA in the submaxillary glands of mice, and male-pattern baldness. Many further changes occur at later ages. The hypothesis can be considered that some events in senescence represent endpoints in an ontogenic cascade of gene regulation (Finch, 1976a, 1988). Whether or not slowly evolving cascades of gene regulation prove to be important in senescence, the demonstration of any postmaturational changes in gene expression pertain to questions about genes with de-layed adverse effects. The negative pleiotropy hypothesis from population genetics (Chap-ter 1.5.2) requires genotypes with delayed adverse effects that were advantageous during development or during the major reproductive phase. Thus, changes in gene expression during the transition to mid-life are of keen interest, whatever their immediate impact on health. The many changes in hormones and other regulatory factors that begin shortly after maturation (DHEA, secretion of pituitary gonadotropins; Chapter 3.6) might be a cause of altered gene activity.

At a molecular level, an important age-related change is a widely observed trend for the accumulation of "altered enzymes," which show post-translational modifications and loss of catalytic activity (Section 7.5.1). No changes in primary amino acid sequence are known in these altered proteins in senescent nematodes, flies, mice, and humans. A work-ing hypothesis is that altered forms accumulate as a consequence of slowed protein turnover; for example, the accumulation of abnormal proteins in the brain (NFT and β-amyloid), particularly during Alzheimer's disease, could be explained by slowed pro-tein turnover. While there is good evidence that protein turnover slows in rat liver, the evidence for other organs is less certain. The maintenance of protein synthesis rates as judged from total body metabolism in healthy elderly humans (Section 7.4.1.1) suggests that only certain cells may be compromised. The cell-type specificity would be consistent with the rarity of altered proteins in some brain regions at advanced ages.

Progressive accumulation of racemized amino acids is another age-related change of proteins very slowly turning over that arises from thermodynamic (nonenzymatic) causes (Section 7.5.4). Lens crystallins and dentine proteins become progressivley racemized as a function of time and temperature. Unlike many other age-related changes, the accumu-lation of racemized amino acids does not appear to scale with the lifespan in species comparisons. The accumulation of glycated adducts on proteins is also nonenzymatically induced, though in this case the local glucose or fructose concentrations allow physiolog-ical variations that may prove to scale with species lifespans. There are indications of enzymatic mechanisms to repair racemized amino acids, as in the carboxyl methyltrans-ferase of erythrocytes. Other cellular mechanisms may remove glycated proteins. These repair and removal mechanisms, while provisionally shown, may have evolved to meet a need for maintaining long-lived macromolecules and cells during extended reproductive schedules. The accumulation of abnormal proteins from a variety of mechanisms may be considered provisionally as a generalizable or canonical trait of mammalian senescence.

Another major theme is the manipulability of age-related changes by diet or hormones (Section 7.4.1.3; examples are given in more detail in Chapters 9 and 10). The manipu-lation of age-related changes by such physiological means points to the importance of extrinsic factors in cell degeneration during senescence. Two working hypotheses may be stated: (1) cell degenerative changes during senescence are generally controlled by factors that are extrinsic to the cell or exogenous to the organism; (2) extrinsic factors act by modulating the expression of specific genes through influences on transcription and trans-lation efficiency. The maintenance of at least some mechanisms of genomic expression throughout the lifespan in mammals bears on how genes influence the lifespan, particu-larly in regard to how the lifespan is shortened by genes that cause early onset diseases. Both circumstances would depend on the dynamic regulation of genes that may be related not only to a specific senescent phenotype like Alzheimer's disease, kidney degeneration, or arteriosclerosis, but also to the compensatory and repair processes that compensate for local or systemic injuries.

The absence of an intrinsic genetic program for degeneration in some cells would be consistent with the ability of many plants and animals to reproduce asexually from somatic cell lineages that show no clonal senescence (Chapter 4). This capacity for vegetative reproduction demonstrates the maintenance of a complete and functional set of nuclear genes as inherited from the parental genome(s), that is, somatic cell nuclear gene totipotency. The numerous genomic responses that are maintained in mammalian somatic cells throughout adult life show that these gene sets are functionally intact; the extent to which nuclear gene totipotency is maintained in adult somatic cells is a major open question. As described next, the evidence is mixed for genomic changes in somatic cells. The extent of genomic totipotency in somatic cells will be an important factor in experimental and clinical interventions into cell degeneration.

The next chapter takes up the fidelity of genomic replication, which has been a major focus of molecular theories in gerontology. The extent of normal gametogenesis bears on another basic issue, the weakening of selection against genotypes with adverse effects during adult life, as well as the question of the genetic load of gametes formed during later life.

8

Fidelity in the Maintenance of Genomic DNA and Its Replication during Senescence

8.1. Introduction

The previous chapter discussed evidence for age-related changes in macromolecular biosynthesis that, depending on the species and cell type, gave examples of global changes as well as changes that were specific to the expression of particular genes in particular cells. Global changes are represented by decreased synthesis of bulk RNA and protein in several rodent organs. Specific changes at later adult ages are represented by changes in the levels of particular mRNAs or proteins, some of which increase, while others decrease. Changes in gene expression are not restricted to later stages in the lifespan and in some cases begin during mid-life. The overall impression from studies of mammals, nematodes, and flies is that changes in macromolecular biosynthesis are specific according to cell type and schedule. It is unclear how many of the changes represent genomically programmed responses that are part of homeostasis and compensation for diseases as distinct from more "intrinsic" genomically programmed events. Yet another class of "nonprogrammed" change is represented in the spontaneous and sporadic reactivation of genes and the spontaneous curing of mutant phenotypes for specific defective proteins. These examples (Chapter 7.3.3) are poorly understood and could involve changes in the DNA of somatic cells as well as epigenetic changes in gene activation and splicing of defective RNA or proteins.

This chapter considers evidence on possible age-related changes in the structure of genomic and organellar DNA and the fidelity of its replication. This continues to be one of the most controversial topics in molecular gerontology. A well-publicized, but still scantly supported, hypothesis concerns age-related accumulations of mutations in somatic cells. The *somatic mutation hypothesis* has been discussed for nearly 40 years as a major mechanism in somatic cell degeneration during senescence (see reviews by Comfort, 1979, 210; Gensler and Bernstein, 1981; Strehler, 1986; Kirkwood, 1988). Other developments of this concept include proposals for the accumulation of mutations in genes that regulate the fidelity of DNA replication, transcription, or translation. The *Orgel hypothesis* (1963) that errors in protein synthesis would be self-amplifying and lead to an error catastrophe with many alterations in proteins through mistaken substitution of amino acid residues was considered in Chapter 7.3.1.1 from the viewpoint of errors in protein synthesis. The genomic side of the Orgel hypothesis will be considered here. Changes in DNA sequence organization are represented in Strehler's proposal for age-related loss of contiguous repetitive sequences, such as ribosomal RNA cistrons (Section 8.2.3).

I also consider evidence on age-related changes in the fidelity of DNA and chromosomal replication. Abnormal karyotypes and structural changes in chromosomes increase with age in some cell types. In particular, abnormal karyotypes are a major factor in Down's syndrome and other chromosomal abnormalities of increased maternal age of mammals. It is unclear if chromosomal abnormalities arise from causes related to mutations or abnormalities in DNA synthesis. Alternatively, there is evidence for age-related changes in the mitotic spindle. The relationships are also unclear between these changes in somatic cells and changes during gametogenesis and during abnormal growth. Again, most information is available for mammals.

8.2. Somatic Cells and Age-related Modifications of Genetic Structure

8.2.1. *Evidence for and against Accumulated Point Mutations*

The somatic mutation theory of organismic senescence was one of the earliest attempts to understand senescence at a molecular level and continues to have much influence on the field. This theory arose from observations that sublethal radiation usually shortens the lifespan of mammals. These "late effects" of radiation are a great concern because many humans have been chronically exposed to low-level radiation (Storer and Graham, 1960). As has so often prevailed among gerontologists, early observations on radiation in relation to organismic senescence were prejudiced by implicit assumptions about the nature of senescence. Influential early observations that implied a relationship between senescence and mutations came from several studies of mice that were exposed to sublethal doses of radiation. Irradiated mice predictably showed "generalized atrophy (premature aging)" and graying, and had shortened lifespans (Henshaw et al., 1947; Curtis, 1963, 1966; Russ and Scott, 1937). In fact, the graying of coat hair is not a usual outcome of senescence in nonirradiated mice (Finch, 1973c; Chapter 6, n. 1). For further discussions about the similarities and differences of radiation-induced life shortening and accelerated senescence see Walburg (1975) and Casarett (1964).

Single exposure of mice to sublethal ionizing irradiation shortens lifespan at some doses by increasing the IMR, but without further accelerating the age-dependent mortality rate, or MRDT (Lindop and Rotblat, 1961a; Sacher, 1956, 1977; Figure 8.1). The Gompertz mortality rate equation is discussed in Chapter 1 (equation [1.3]). The general increase of IMR without increase in the acceleration of mortality suggests that the rate of organismic senescence is not accelerated by life-shortening radiation. The delayed effects of radiation include increased neoplastic diseases and kidney degeneration (glomerulosclerosis), to degrees that vary with the dose and tissues (e.g. Lindop and Rotblat, 1961b).

The role of gene dose in responses to X-ray irradiation was evaluated with respect to haploidy in the wasp (*Habrobracon juglandis*). In their classic study, Clark and Rubin (1961) showed that haploid and diploid male wasps have identical median lifespans (62 days); however, X-ray treatment decreased the haploid lifespan more than the diploid. This outcome argues that spontaneous mutations and chromosomal breakage are not critical factors in the lifespan under ordinary conditions. This outcome also argues for the continuing importance of gene activity in the short adult lifespan of wasps, since radiation damage is usually restricted to one chromosomal locus in diploid cells, leaving the other gene copy intact. Because radiation causes point mutations in DNA as well as·deletions and chromosomal breaks and rearrangements, a popular hypothesis was advanced by many, that radiation shortened lifespan by intensifying an underlying process of accumulating somatic mutations across the lifespan (Failla, 1958; Curtis and Healy, 1958; Upton, 1957).[1] There are many variations of this hypothesis. Among the most historically prominent is Szilard's (1959) proposal that genes on chromosomes of somatic cells are inactivated by a random "aging hit"; the accumulation of aging hits leads to dysfunctional cells, and age-related death is postulated to occur when a threshold number of cells become dysfunctional.

---

1. Also, see Strehler et al., 1960, and Strehler, 1986, for accounts of a seminal meeting at Gatlinburg, Tennessee, in 1957, which stimulated much further work on this topic.

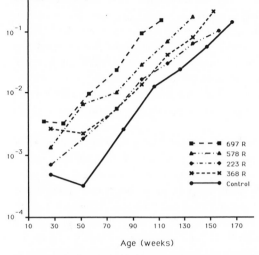

Fig. 8.1. Effects of gamma radiation from an experimental nuclear detonation ("Operation Greenhouse") on the mortality rates of LAF$_1$ female mice. Life shortening results from increased IMR, without changes in the Gompertz exponential mortality rate (slope, or MRDT). Redrawn from Upton et al., 1960.

The life-shortening effects of sublethal irradiation have stimulated a major effort over the years to identify age-related changes in DNA. Only recently have the molecular tools become sharp enough to address this issue. Data come from several sources: the occurrence of spontaneous mutations as indirectly assayed by amino acid substitution, errors in DNA synthesis, and resistance to cytotoxic purines.

Two different conclusions are indicated by searches for errors that increase with age and that may be related to somatic cell mutations. The first examples concern attempts to detect mutations in erythrocyte proteins. Human hemoglobin A lacks isoleucine and thereby provides a favorable circumstance for chromatographically detecting minute increases of isoleucine against a very low background in hydrolysates of pure hemoglobin. The frequency of basal substitution, 3×10^{-5} per amino acid residue (Popp et al., 1976), was much higher than indicated by an exhaustive analysis of electrophoretic variants, 10^{-8} mutations per residue per generation (Neel et al., 1986). A single brief report indicated that Marshall Island natives who were exposed to nuclear fallout 20 years before had a twofold higher frequency of isoleucine above nonirradiated controls, which would be consistent with mutations in hemopoietic stem cells. The population also showed a slight, but not significant, age-related increase in the trace amounts of isoleucine from 20 to 51 years (Popp et al., 1976). These data do not indicate if the radiation-related increase of isoleucine is due to single base substitutions (e.g. in one of the four valine codons, where *GUU* can be mutated at a single step to *AUU* for isoleucine), or if translational errors are involved that cannot be attributed to single base substitutions. A caveat is that the frequency of misincorporation of isoleucine from mistakes in synthesis may be larger than that from mutations, for example at the N terminus of sheep globin (Hirsch et al., 1980). Another study of erythrocytes looked for age-related changes in the loss of blood-group antigens in families in which both parents were AB, but found no age-related incidence of AO and BO cells (Atwood and Pepper, 1961).

In a novel test of the Orgel hypothesis, errors made by mammalian DNA polymerase were assayed by reversion of amber mutants in bacteriophage ΦX174. With DNA polymerase β that was isolated from purified neurons (6 versus 40 months) or DNA polymerases α and β isolated from regenerating liver (6 versus 28 months), the age of the mouse did not influence the enzymatic fidelity of copying the bacteriophage template (Silber et al., 1985; Subba Rao et al., 1985). However, several caveats must be considered. A trend for relative increases in livers of older mice of the DNA polymerase β that has lower fidelity could allow creeping infidelity of replication at later ages. Moreover, it is unknown if the efficiency of neuronal nuclear isolation was the same across ages. Nonetheless, these studies give the most direct evidence to date against somatic mutations in the enzymes of replication or protein synthesis that were predicted by the error catastrophe hypothesis (Orgel, 1963; Holliday and Tarrant, 1972).

Another approach to testing the error hypothesis is to promote misincorporation. Edelmann and Gallant (1977), for example, examined misincorporation of cysteine into flagellin, which normally lacks this amino acid. When bacteria (*Escherichia coli*) were grown in the presence of the error-promoting drug streptomycin, the rate of cysteine misincorporation into flagellin increased to a maximum of about twentyfold after six generations. In the new steady state, there was no loss of cell viability, though growth was slower; after removal of streptomycin, the error frequency rapidly returned to baseline. This study shows that major increases in errors are not incompatible with continued cell proliferation in bacteria, under conditions where there is little concern that selection for resistance occurred. This study shows that major increases of errors need not precipitate error catastrophes. To test for the possible effects of missubstitutions in human diploid fibroblast cultures, the RNA base analogue 5-fluorouracil was given at concentrations that did not inhibit cell proliferation (Holliday and Tarrant, 1972). The treated cells stopped replicating after ten doublings (interpreted as premature cell senescence) and increased amounts of heat-labile glucose–6-phosphate dehydrogenase, as observed at late passage in control cultures (Chapter 7.6.4). The premature cessation of growth need not involve an error catastrophe, because of evidence that age-related enzyme alterations represent reversible changes in structure rather than in amino acid sequence (Chapter 7.5.1). This and other conceptually attractive experiments from an earlier era (Holliday, 1969) should be repeated with new techniques that can detect single base changes.

Resistance to metabolic blockers gives a direct approach to detecting somatic mutations. The cytotoxic purine analogue 6-thioguanine (TG) provides an assay for mutations in somatic cells that have lost activity of the enzyme hypoxanthine-guanine phosphoribosyl transferase (HPRT), an X-linked gene. TG, when phosphorylated by HPRT and incorporated into DNA, is normally cytotoxic. As assayed by the ability for DNA synthesis with [^{3}H]-thymidine, only rare cells (10^{-5} in lymphocytes or fibroblasts) are normally TG resistant. Cells selected for TG resistance are increased manyfold by X-rays or mutagens (e.g. Evans and Vijayalaxmi, 1981). In mouse cells, stable mutants are associated with large deletions and other major structural alterations in their DNA (Nicklas et al., 1986; Bradley et al., 1987). Not all TG-resistant cell lines are stable, however. In primary clones from human and dog kidneys, there was a surprisingly high (80%) reversion to TG sensitivity (Turker et al., 1988), a result that is incompatible with the spontaneous reversion of mutations, particularly those arising from DNA deletions. The high reversion frequency implies epigenetic mechanisms of TG resistance or unsuspected wild-type contaminants. Conclusions cannot yet be drawn from these studies.

In primary lymphocyte cultures from humans, the incidence of TG-resistant cells increased severalfold across the lifespan to a similar extent in normal men and women (Figure 8.2, top; Evans and Vijayalaxmi, 1981; Vijayalaxmi and Evans, 1984; Trainor et al., 1984; Morley et al., 1982). In contrast, Strauss and Albertini (1979) found no change. The studies of human lymphocytes did not examine the stability of TG resistance. However, studies of mice showed a lack of reversion that is consistent with true mutations. In female BALB/c mice, TG resistance increased slightly with donor age between 3 and 32 months in splenic T-cell cultures (Inamizu et al., 1986), but did not increase in fibroblastoid cultures from males of other hybrid genotypes up to the very advanced age of 40 months (Figure 8.2, bottom; Horn et al., 1984). Inamizu et al. found large differences

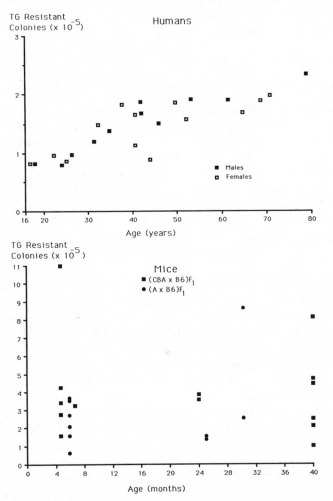

Fig. 8.2. Resistance to thioguanine in short-term cell cultures as a function of donor age—an indirect assay for mutations in the enzyme hypoxanthine phosphoribosyl transferase. *Top,* humans, peripheral blood lymphocytes. Redrawn from Vijayalaxmi and Evans, 1984. *Bottom,* mice, fibroblastoid cells from kidney (*squares*) or skeletal muscle (*circles*) of (CBA × B6)F$_1$ and (A × B6)F$_1$ male mice, where B6 is C57BL/6J. Redrawn from Horn et al., 1984.

among individual mice, such that the ability of spleen cells to produce the lymphokine IL–2 at the time of sacrifice was inversely related to TG resistance in the clones. This suggests that chronological age is less important than "physiological age" and weakens the case for strict age-related increase of mutations in the donor mice. Firm conclusions from TG resistance cannot yet be drawn about age and somatic mutations.

Techniques being developed should give more definitive estimates of the somatic mutation frequency *in vivo*. For example, a shuttle vector system was developed for efficient rescue of lambda vector DNA sequences from transgenic mice that then allows their cloning in bacterial hosts (Vijg and Uitterlinden, 1987; Gossen et al., 1989). For example, DNA from the brains of mice exposed to ethyl nitrosourea (a mutagen and DNA alkylating agent) showed dose-dependent increases of point mutations (Gossen et al., 1989). Age-related changes are being characterized.

8.2.2. Modified DNA Bases Accumulate with Age in Some Tissues

Modified DNA bases appear to increase with age in mammals, according to new data. Using an extraordinarily sensitive assay, Randerath and colleagues detected severalfold increases during the lifespan in modified bases of DNA from rat liver and kidney (Figure 8.3, bottom; Randerath et al., 1986) and human brain (Figure 8.3, top; Kurt Randerath and Caleb Finch, unpublished). The modified bases accumulated progressively (1–9 months) in purified mitochondrial DNA as well as in that of nuclei from rat liver (Gupta et al., 1990). The assay involves 32P labelling of enzymatically digested DNA, followed by chromatography under conditions that resolve as few as 10^{-9} modified nucleotides (Reddy and Randerath, 1986). The identity of the modified bases is unknown, but their chromatographic profiles resemble the covalent DNA adducts formed in tissues after exposure to heterocyclic carcinogens, for example in the placental DNA of women who smoked during pregnancy (Everson et al., 1986) or in the DNA of estrogen-induced tumors of mice (Liehr et al., 1986). While detectable by these exquisite assays, the number of changes per cell nucleus is minute, < 1 per 10^7 bases in nuclear DNA (Figure 8.3, top). Another lab examined abnormal compounds that they found in acid hydrolysates of liver nuclear DNA from rabbits (Yamamoto et al., 1988). Age-related increases were detected in unknown fluorescent compounds that are several times heavier than DNA bases by mass spectrometry and that might be modified DNA bases, cross-linked to each other or to peptides; the amount of these peculiar compounds increased sharply after 6 months and was variable. In proliferating cells, covalent modifications of DNA are thought to increase the risk of mutations; the effect of the adducts on transcription is unknown. These modifications are the first direct indication for age-related changes in DNA structure, though different than proposed in early somatic mutation theories. The extremely low incidence makes it unlikely that these are functional consequences to gene expression in nonreplicating cells.

Another approach to assaying damage to DNA is based on urinary excretion of thymine glycol, a modified nucleotide produced by hydroxyl radical attack on thymidine which is greatly increased by ionizing radiation (Leadon and Hanawalt, 1983). Thymine glycol can be excised by a DNA glycosylase in a repair step and can be recovered in the urine (Cathcart et al., 1984; Saul et al., 1987). The daily production of thymine glycol in the

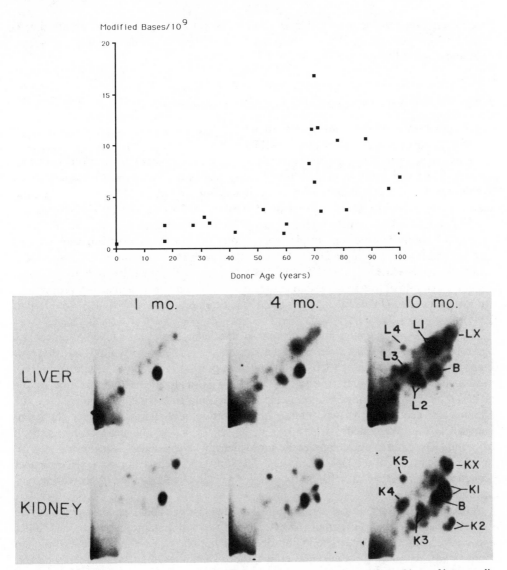

Fig. 8.3. Age-related increases in modified DNA bases, which resemble the covalent adducts of heterocyclic carcinogens to DNA (see text). *Top*, human brain: age-related increases of total modified bases in DNA from neurologically normal humans as determined chromatographically. Redrawn from C. Finch and K. Randerath, unpublished. *Bottom*, rat tissues: chromatograms from liver (*L*) and kidney (*K*) DNA of untreated male Sprague-Dawley rats, showing an age-related increase of six or more modified bases up through 10 months (early middle age); *B* identifies a spot that did not change with age. Rephotographed from Randerath et al., 1986.

absence of radiation thus may be related to repair of endogenous oxidative damage to DNA; however, the cellular sources of thymine glycol are not known. Species comparisons (mice, rats, humans) suggest that the daily thymine glycol production per gram correlates with metabolic rate (Chapter 5.4.3; Saul et al., 1987). No age-related trends were detected in adults aged 20–85 years, which implies that oxidative damage to DNA and its repair do not increase across the human lifespan. Another report from this group, that 7-methylguanine adducts in DNA increased with age (Park and Ames, 1988), was subsequently retracted (Ames, 1988).

8.2.3. Ribosomal Cistrons and the Nucleolus Organizer

The possibility of deletions in ribosomal DNA cistrons was raised on theoretical grounds (Johnson and Strehler, 1972; Strehler, 1986). A classic observation from *Drosophila* cytogenetics is that the *Bar* locus, which contains a tandem duplication, is genetically unstable and tends to pair asymmetrically with a high aberrant frequency of 5% during meiosis to form progeny with three (hyper-*Bar*) or one (wild-type) copy. Strehler then proposed that the tandemly repeated ribosomal RNA cistrons, 50–500 copies per haploid genome in mammals, might be at similar risk for deletions over sufficient time, even in the absence of meiosis. Exploratory studies of DNA from heart and brain regions of dogs and humans, using solution hybridization with rat ribosomal RNA, indicated a progressive age-related loss of ribosomal DNA cistrons that reached 10–40% by the mean lifespan. The rate of loss was fivefold slower in humans than dogs, apparently proportionate to the lifespan (Johnson and Strehler, 1972; Strehler et al., 1979a, 1979b; Strehler, 1986).

Unfortunately, subsequent studies from two other laboratories have failed to detect any age-related change in DNA from mouse myocardial cells (Peterson et al., 1984), liver, or brain up through 33 months (Swisshelm et al., in prep.). The study of Swisshelm et al. used extensively deproteinized nuclear DNA and normalized the hybridization of the cloned mouse rRNA probes (nontranscribed spacer and transcribed sequences from the 18S and 28S sequences) to two different single-copy probes. Further rigorous studies of human DNA are warranted. At least in rodents, the evidence indicates that decreases in ribosomal RNA and in nucleolar size are more plausibly attributed to changes in transcriptional rate than loss of the DNA templates. The increased methylation of 5′ rRNA spacer sequences (Swisshelm et al., 1990; Chapter 7.3.4) could be a correlate of decreased transcription.

A decreased transcription of some or all ribosomal cistrons could be consistent with the loss of nucleolus organizer regions (NOR). Detected by silver stains, the number of NORs appears to decline with donor age in fibroblasts and peripheral blood lymphocytes, though more slowly (Buys et al., 1979; Das et al., 1986; Denton et al., 1981). Taken together, these reports indicate more metaphase plates with low numbers of stained NOR in cells from older adults. Silver staining of NOR can be related to the number of ribosomal cistrons that differ between individual chromosomes (Warburton and Henderson, 1979). Because silver staining of NOR also indicates transcriptional activity of rRNA cistrons (Miller et al., 1976), fewer NOR would be consistent with age-related decreases of total RNA in select neurons and other cells (see Chapter 7.5.6). Thus, changes in silver staining of NOR need not be caused by loss of ribosomal genes.

8.2.4. Evidence for the Stability of DNA Sequences

Two other approaches indicate that most DNA sequences are stable across the lifespan in nondividing or slowly dividing cells. A novel study exploited the small but reliable differences in the ^{13}C to ^{12}C ratio from natural isotopes in DNA from the cerebellum that are caused by slight differences in the diets of Europeans and Americans (Slatkin et al., 1985). Cerebellar DNA from emigrants to America retained the European characteristic ^{13}C to ^{12}C ratio for many decades, as measured in postmortem tissues. The same result was given by the complete retention of [^{3}H]-thymidine label in the retinas of rats injected as neonates, up through 2 years (Ishikawa et al., 1983; Ishikawi et al., 1984). Taken together, these reports imply that there is little turnover of DNA in neural tissues, which include permanently postmitotic neurons as well as glia and other cells that can be stimulated to proliferate.

Another approach analyzed the restriction nuclease digestion patterns of DNA from brain, liver, and spleens of mice up to 28 months old for the MMTV provirus and another eight sequences that show rearrangement or amplification under special circumstances as detected by DNA blot hybridization (Southern blots; Ono et al., 1985). Within the limited sensitivity of the technique, no changes were detected. These studies do not rule out subtle changes in DNA structure or sequence organization, but set useful limits by excluding gross DNA lability. A different view comes from studies of chromosome structure.

8.2.5. Chromosomal Abnormalities in Somatic Cells

Chromosomal abnormalities increase with donor age in cells sampled from peripheral leukocytes and other tissues of humans and rodents, according to an extensive literature. Subpopulations of cells with two types of chromosomal abnormalities are seen during mitosis: aneuploidy (extra or missing chromosomes) and structural abnormalities (amplifications, breaks). The causes of age-related increases are not known in either case. Although the structural abnormalities are hypothesized by some to be equivalent to somatic mutations, no data are available to evaluate the extent of changes in DNA sequences.

8.2.5.1. Aneuploidy

In peripheral blood lymphocytes of women, aneuploidy and hypodiploidy increase severalfold by the average lifespan, as assayed after stimulation by mitogens during short term *in vitro* culture. The incidence approaches 5–10% of total cells after 65 years, as particularly shown for women (Figure 8.4). Men do not show as strong age-related increases in most studies (Galloway and Buckton, 1978; Jacobs et al., 1963; Dutkowski et al., 1985; fourteen other studies are reviewed by Schneider, 1985). (However, even cultured lymphocytes from newborns show a substantial incidence of aneuploidy [ca. 2%]. An extensive analysis [one thousand metaphases of 2-day culture of peripheral lymphocytes] showed that the types of chromosomal abnormalities of adults were more random than those of neonates, and increased twofold with age [Prieur et al., 1988].) While most studies are cross-sectional and may be influenced by cohort effects for exposure to drugs or environmental factors, similar trends were reported in the only longitudinal study published to date (Jarvik et al., 1976). Age-related aneuploidy is established only for peripheral lymphocytes in humans and was not obvious in studies of bone marrow of women (Pierre and Hoagland, 1972). Moreover, primary cultures of ovarian granulosa cells showed marked age-related *decreases* in aneuploidy after 35 years (Wertheim et al., 1986). Age-related aneuploidy is not well established in rodent somatic cells *in vivo*, although structural abnormalities are found (Section 8.2.5.2). The most common abnor-

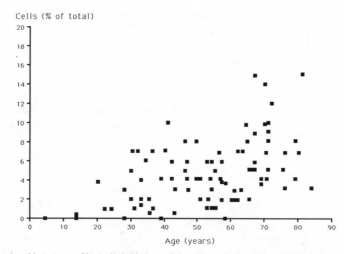

Cells (% of total)

Fig. 8.4. Age-related increases of hypodiploidy in peripheral lymphocytes from 102 women selected randomly from a general practitioner's patient roster in Edinburgh. Metaphases were observed after stimulation of mitosis by phytohemagglutinin *in vitro*. Redrawn from Galloway and Buckton, 1978.

mality seen in cultured lymphocytes is a loss of the X chromosome, which is generally the inactive or heterochromatinized X (Abruzzo et al., 1985). Other chromosomes are lost half as frequently as the X.

Mechanical factors related to chromosome segregation and cytokinesis could be factors in the age-related increase of aneuploidy. One indication of this comes from the age-related increase of micronuclei. Micronuclei are small DNA-containing bodies in the cytoplasm that often arise from chromosomes that lagged at metaphase or from acentric fragments; they are increased by exposure to X-rays or mutagens that cause chromosomal damage (Heddle et al., 1978). Micronuclei account for a minority, less than 10%, of overall aneuploidy in human lymphocytes. A fourfold age-related increase in micronuclei is reported in cultured lymphocytes from humans (Norman et al., 1984; Fenech and Morley, 1985). Bone marrow cells from middle-aged mice showed smaller increases of micronuclei, which were enhanced by exposure to mutagens (Singh et al., 1985). Micronuclei may result from impairments of centromere or spindle functions, as well as from structural damage to chromosomes (Section 8.2.5.2).

Alterations in the centromere are suggested by a threefold increase to 1% of lymphocytes from older women, in which the centromere was histochemically undetectable by the "Cd-banding" technique (Nakagome et al., 1984). The trend for premature separation of the centromere in X chromosomes of short-term lymphocyte cultures (Mukherjee and Weinstein, 1988) is consistent with the increase of X-chromosome aneuploidy. Little is known about age-related changes in the centromere or mitotic spindle of mammals or lower species; abnormalities of the spindle during meiosis in oocytes are described in Section 8.5.1.1.

The age-related increase of aneuploidy in certain cell types could involve increased production of aneuploidy or reduced selection against it. The replication of sex chromosomes late in the S phase may predispose them to loss, especially in rapidly proliferating cells where the G_2 phase is shortened (Schneider, 1985). In view of the weaker age effects of aneuploidy in men, endocrine changes at menopause may have a role in these changes (Chapter 3.6.4). It would be interesting to know if hypodiploidy was related to blood estrogens in women of different body weight, since fat cells produce estrogens from adrenal steroid precursors, or with estrogen replacement therapy.

8.2.5.2. Structural Abnormalities in Chromosomes and DNA Size

Structural abnormalities of chromosomes were historically the strongest source of support for the somatic mutation hypothesis of aging, because they increase with age spontaneously in several species and with life-shortening treatments of radiation (Curtis, 1963). Early studies induced mitosis in the liver by injections of CCl_4, a hepatotoxin that stimulates regeneration. A progressive age-related increase of chromosome bridges and fragments was seen in regenerating liver after CCl_4 in two mouse strains (Crowley and Curtis, 1963), in guinea pigs (Curtis and Miller, 1971), and in dogs, using surgical hepatectomy to induce regeneration (Curtis et al., 1966).[2]

Chromosomal structural abnormalities also increased with donor age in the *in vitro* mitosis of cells that grew out from collagenase-digested kidneys (Martin et al., 1985). Of the eight types of structural abnormalities seen in chromosomes from 8-month-old donors (Figure 8.5), all increased twofold or more by 40 months; the largest was a fiftyfold increase in "pulverized" chromosomes (*pvz* in Figure 8.5). Bone marrow cells obtained 2 hours after treatment with colcemid also showed increased chromosomal gaps and fragments; the greatest increase (tenfold) occurred between 2 and 6 months, with smaller changes up to 25 months (Fabricant and Parkening, 1982). Cultured ovarian granulosa cells from donors over 35 years also had severalfold increases of fragments and gaps (Wertheim et al., 1986). Species differences revealed in these assays are puzzling. Translocations occur in human and rat cells at 0.01–0.3/mitosis, while hamsters have far fewer, 10^{-4}/mitosis (Duesberg, 1987).

In a novel approach to age effects on chromosomal instability, cell hybrid cotransfer selected for hybrid hamster cells that contained the human X chromosome from donors of different ages (Esposito et al., 1989). Human T lymphocytes were fused with hamster cells and then selected for the presence of the human X chromosome. The basis for selection is the X-linked enzyme HPRT, which allows growth in "HAT" medium, that is, the converse of TG resistance (Section 8.2.1). In addition to the presence of HPRT, the hybrid cells were screened for the retention of two genetic markers that are located at opposite ends of the human X chromosome (*G6PDH* and *MIC2*) and that are also lacking in the hamster CHO-YH21 cells. Nearly all cells had the *G6PDH* marker, which is located near the *XPRT* gene. However, twice as many hybrids from male human donors aged 75 years had lost the *MIC2* marker as from 35-year-olds. DNA blot hybridization directly proved the loss of the loci. Moreover, the regions of the X chromosome that were preferentially lost are also prone to meiotic rearrangements, hereditary fragility, and viral integration. These important results add much to the evidence for increased chromosomal instability with age.

Changes of DNA size might be related to the age-related increases of chromosome breakage. Sedimentation profiles on alkaline sucrose gradients indicate that the size of

---

2. These studies might be confounded by age differences in clearance of CCl_4 as shown for many drugs (Vestal and Dawson, 1985), or in the susceptibility to chromosomal abnormalities induced by CCl_4 (Section 8.2.6).

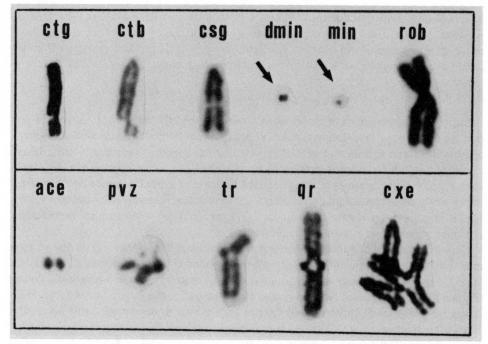

Fig. 8.5. Structural abnormalities in chromosomes seen at metaphase from primary cultures of fibroblastoid kidney cells of (CBA × B6)F₁ hybrid mice. The structural abnormalities, and their extent of increase between 8 and 40 months, are *ctg,* chromatid gap, a nonstaining region in a single chromatid (fourfold); *ctb,* chromatid break, a wider nonstaining region in a single chromatid (fivefold); *csg,* chromosome gap, a nonstaining region involving both chromosomes (> fivefold); *dmin,* double minute, paired acentric fragments (fourfold); *min,* a minute fragment lacking a centromere (acentric fragment); *rob,* Robertsonian translocations that fuse two centromeres and convert two of the usual telocentric chromosomes of mice into a single metacentric chromosome (twofold); *ace,* paired acentric chromatids (sixfold); *pvz,* pulverization of chromosomes (> fiftyfold); *tr,* triradial chromosomes with three arms; *qr,* quadriradial chromosomes with four arms (fivefold); *cxe,* complex exchanges involving at least three chromosomes (threefold). From Martin et al., 1985. Photograph was supplied by George Martin.

DNA (single-strand) in some tissues becomes smaller with age, but data are insufficient to indicate consistency. While DNA size in a few samples of neurons from the cerebellar granule layer of dogs decreased progressively up to 10 years (Wheeler and Lett, 1974; the mean canine lifespan), another study found no size changes in whole-tissue DNA from mouse cerebellum, liver, spleen, or thymus (Ono et al., 1976). In another approach, the labelling *in situ* of cell nuclei with terminal nucleotidyl transferase and dGTP showed increased incorporation into mouse brain neurons (cortex layer 3) after 25 months (Modak et al., 1986); this technique detects increased 3'-OH termini of DNA after γ-irradiation or terminal differentiation of lens cells and keratinocytes.

Breaks in DNA may also be assayed by specific DNAases, such as S1 nuclease, which, under appropriate conditions, preferentially digests single-stranded DNA, including unpaired regions in nicked double-stranded DNA. In what must be one of the largest samples of human tissues ever assembled for molecular analysis (470 donors of known health status), Zahn and colleagues measured the contour length of double-stranded DNA (S1 treated) from muscle biopsies, a prolonged labor (Zahn, 1983; Zahn et al., 1987). The size of the DNA pieces (mean of 75,000 base pairs; range from ca. 35,000 to 90,000) decreased slightly ($< 10\%$) across the lifespan in the whole group. The extensive scatter of DNA size tended to increase with age. Because single-strand breaks are induced in cultured cells by components of cigarette smoke (Nakayama et al., 1985), it is interesting that a subgroup that did not smoke and used few drugs had similar-sized DNA to a subgroup that smoked at least ten cigarettes per day (Zahn et al., 1987); the absence of effects of smoking indicates effective repair of single-strand breaks throughout the lifespan in a nonproliferating tissue. In rodents, variable results were reported, with age-related increases in S1 sensitivity of semipurified liver DNA observed by some (Chetsanga et al., 1975; Chetsanga et al., 1977; Nakanishi et al., 1979; Mullaart et al., 1988) but not by others (Finch, 1979; Dean and Cutler, 1978). Cerebellar neuron DNA from rats showed no change in susceptibility to micrococcal nuclease up through 2 years (Jaberaboansari et al., 1989).

Taken together, these studies consistently suggest that chromosome structure becomes increasingly at risk for disruption in several cell types. Although radiation also causes similar damage, there is no proof that age-related abnormalities in chromosomal structure have the same origin as those induced by radiation, or if there are primary changes in DNA.

8.2.6. Age and Repair of DNA Damage: UDS, Strand Breaks, and SCE

Processes leading to the repair of damage to DNA in eukaryotes are complex and their mechanisms are generally not well understood. A substantial number of enzymes encoded on different chromosomes are being implicated for different repair mechanisms (Tice and Setlow, 1985; Friedberg, 1985), some of which may be shared, for example the RecA-LexA functions of SOS repair in *E. coli* (Little and Mount, 1982). Some commonly discussed eukaryotic repair systems are summarized in Table 8.1. Because age changes in DNA repair were recently well reviewed (Hanawalt, 1987; Schneider, 1985; Tice and Setlow, 1985; Kidson et al., 1984), the present section summarizes their conclusions with a few examples.

Most data on repair come from responses to UV irradiation *in vitro*, as assayed by the incorporation of [³H]-thymidine into DNA of cells after cell-cycle arrest, usually by hydroxyurea or other inhibitors of "scheduled" DNA synthesis. The extent of this "unscheduled" DNA synthesis (UDS) is believed to indicate a long patch of repair, through which adjacent covalently linked pyrimidine dimers are excised, while the undamaged opposite strand is used as a template. Before discussing data on age, here are some caveats about the technique. Figures 5.21 and 5.22 show intriguing species trends for greater thymidine incorporation for the same dose of UV irradiation in fibroblasts and other cells from short-lived mammals. However, the relationship of UDS to cell or organismic senescence is not clearly shown. Because the amount of repair in response to UV and chemical agents is greatest on actively transcribed DNA strands (Bohr et al., 1986: Hanawalt, 1987), species comparisons might be confounded by *in vitro* influences on gene activity or by subtle differences in the state of differentiation. Another concern is the absence of a relationship between UV-induced cell death and UDS (Hanawalt, 1987; Tice and Setlow, 1985): mouse and human cells are equally susceptible to killing, yet differ fivefold in UDS.

Table 8.1 DNA Repair Processes and Effects of Donor Age on Mammalian Cells

Repair System	Characteristics	Age-related Change
Single-strand breaks	endogenous breaks are rare in rodents; induced by ionizing radiation, largely at phosphate backbone	basal age changes in rodents are controversial; notably slowed in cells from older animals
Double-strand breaks	rarer than single-strand breaks; induced by ionizing radiation; rapidly joined	
Nucleotide excision; also called unscheduled DNA synthesis (UDS)	UV causes adjacent pyrimidines in same strand to become covalently linked; the dimer is excised by glycosylases in conjunction with apurinic/apyrimidinic endonucleases; uses the undamaged strand as template	age changes are controversial
Long-patch repair	patch of repair involves up to several hundred nucleotides; known in most detail	
Short-patch repair	patch of repair, 5–10 nucleotides	
Photoreactivation	UV interacts with enzyme that binds to and monomerizes a pyrimidine dimer	
Sister chromatid exchange (SCE)	increased by mutagens and carcinogens; represents recombination	decreased with age in several cell types

Source: Based on Friedberg, 1985, in consultation with Myron Goodman.

Human skin is the tissue most extensively studied for influences of donor age because of easy access; several of its cell types vigorously repair UV damage, principally the removal of pyrimidine dimers (D'Ambrosio et al., 1981). An excellent series of studies comes from the laboratories of Philip Hanawalt and Barbara Gilchrest. Primary cultures of keratinocytes showed no effects of age on the dose response or time course of UDS, from neonates up to 90 years (Liu et al., 1982), or after periodic priming doses, followed by a larger challenge (Liu et al., 1983). Keratinocytes from sun-exposed regions also showed no donor age effects on UDS. Similar results were obtained by Gilchrest (Barbara Gilchrest, unpublished, pers. comm.). However, keratinocytes from older donors or from sun-exposed regions *at all ages* grew and survived less well *in vitro* (Liu et al., 1985; Gilchrest et al., 1979; Gilchrest et al., 1983; Stanulis-Praeger and Gilchrest, 1989; Chapters 3.6.11, 7.6.3). Other studies of primary cell cultures report divergent trends: age-related decreases of UV-induced UDS from epidermal cell cultures (Nette et al., 1984) and peripheral leukocytes (Lambert et al., 1979), no age changes (Kovacs et al., 1984; Hall et al., 1984), or increased UDS (Pero and Ostlund, 1980). These differences might result from different responses to culture or from sampling of different phases of repair. Pero and Ostlund (1980) propose two phases of response to UV, early and late repair, which are independently influenced by donor age.

A direct assay showed no age difference between 6- and 36-month-old rats in the rates of removal of pyrimidine dimers after UV irradiation of skin fibroblasts and keratinocytes (Mullaart et al., 1989; Figure 8.6). The assay was based on the single-strand breaks next to pyrimidine dimers that are made by *Micrococcus luteus* endonuclease. Antibodies to thymine dimers (see legend to Figure 8.6) may be useful in immunocytochemical detection of DNA lesions in human tissues as a function of age. UDS is also induced by methyl nitrosourea (a mutagen and alkylating agent); a comparison of six mouse strains at 2, 5, and 16 months showed decreases with age that differed among strains: UDS was least in C3H and largest in 129/ReJ males (Bond and Singh, 1987; other strain differences in DNA repair are discussed in Chapter 6).

In one of the few studies of UDS in fish, *Oryzias latipes* (Chapter 3.4.3) showed incorporation of [³H]-thymidine into adult brain neurons (Ishikawa et al., 1978). Labelling was stimulated by the carcinogens ethyl nitrosourea and nitroquinoline oxide. No impairments in basal or nitroquinoline-stimulated thymidine incorporation were found, up to 3 years (the maximum lifespan is 6 years). There was no relation of UDS to the presence of neuronal lipofuscins, which is the first time this question has been addressed.

Several cell types show age-related susceptibility to chromosomal damage or killing by radiation or chemical mutagens. For example, X-rays, which are vastly more energetic photons than those in the UV range, had more lethal effects on short-term cultures of lymphocytes from older humans (Figure 8.7; Kutlaca et al., 1982). Similarly increased radiosensitivity is seen for the increase of chromosomal aberrations at metaphase in lymphocytes from older donors, as a function of the amount (μCi/mM) of [³H]-thymidine in the medium (Dutkowski et al., 1985). The age-related increase of radiosensitivity could reflect changes in many different functions, besides deficiencies in repair. In one of the few direct approaches to this question which followed the decrease of DNA size after irradiation by gamma-rays, cerebellar granule layer neurons from old dogs showed no loss of repair that removed DNA strand breaks (Wheeler and Lett, 1974). In contrast, the fragmentation of intestinal mucosal cell DNA caused by dimethylhydrazine (alkylating agent) was less rapidly repaired in rats at 15 months (mid-life) than at 2 months (Kanagalingam and Balis, 1975), a change that may pertain more to maturation than to senescence.

Sister chromatid exchanges (SCE) also probe the lability of chromosome structural

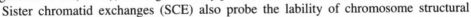

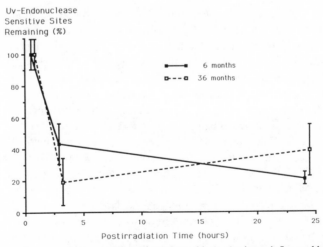

Fig. 8.6. The enzymatic removal of pyrimidine dimers from skin *in vivo* is not influenced by age in rats 6 and 36 months old. Shaved rats were irradiated with UV (4,000 joules/m²); a timed series of biopsies were taken for assay of thymidine dimers. The dimers mainly formed in keratinocytes, as shown immunocytochemically with an antibody against thymine dimers. Redrawn from Mullaart et al., 1989.

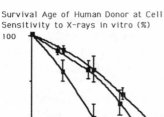

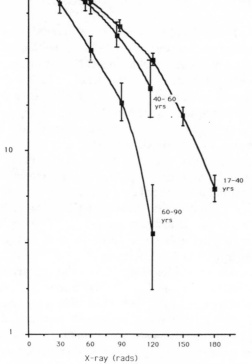

Fig. 8.7. Age-related increases in the cell-killing effects of X-rays *in vitro*. Redrawn from Kutlaca et al., 1982.

changes with age. SCE of homologous chromosomal segments occur at low frequency in somatic cells and strikingly resemble the results of synapsis between homologous parental chromosomes at meiosis. Because SCE are increased by mutagens and X-rays (e.g. Evans and Vijayalaxmi, 1981), their presence may indicate DNA repair processes. Bloom's syndrome is a rare autosomal recessive in which there are unusual large increases of SCE as well as spontaneous chromosomal breakage and rearrangements (German, 1964; German et al., 1977). A relationship between SCE and recombination is suggested by the numerous quadriradial chromosomes in Bloom's syndrome (German, 1965).

In contrast to age-related increases of radiosensitivity, the induction of SCE by mitomycin C decreases with age in spleen and bone marrow cells from laboratory rodents (Kram et al., 1978; Mann et al., 1981; Schneider, 1985; Figure 8.8). (Mitomycin C is a mutagen that alkylates and cross-links DNA.) Similar age-related decreases in the induction of SCE are seen in response to the mutagens adriamycin and cyclophosphamide (Nakanishi et al., 1979). A study of eight mouse strains reported that marrow cells from 20-month-old compared to 3-month-old mice had more SCE in some strains, while other strains showed opposite changes (Singh et al., 1985). The basal frequency of SCE is not clearly changed with the donor age, but this question is difficult to assess because SCE are induced by the bromodeoxyuridine that is used to detect their presence. The incidence of

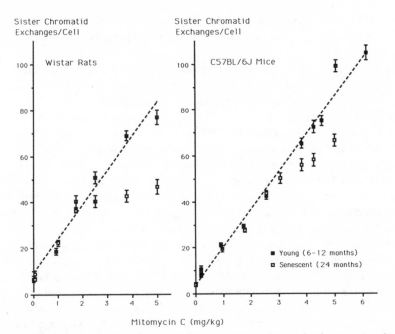

Fig. 8.8. Age-related changes in the induction of sister chromatid exchanges by the mutagen mitomycin C. Bone marrow cells from young (6- to 12-month) and senescent (24-month) rodents were exposed *in vitro* to mitomycin C at different concentrations and in the presence of bromodeoxyuridine, for the identification of chromatid exchanges in subsequent cell cycles by fluorescent dyes. Sister chromatid exchanges increased with mitomycin C, but cells from older rodents had smaller responses at high levels of mitomycin C. Redrawn from Kram et al., 1978.

quadriradial chromosome figures at mitosis is not known as a function of age, but may be predicted to decrease with age in parallel with the incidence of SCE by analogy with Bloom's syndrome cells (see above).

In summary, the evidence is mixed concerning the accumulation of mutations or damage to somatic cell DNA throughout the lifespan. Whatever the extent of DNA modification, whether random or not, the capacity for repair is not markedly altered. The incidence of gross chromosomal structural abnormalities is much higher than modifications of DNA. The relationship of chromosome structural abnormalities to DNA damage is unclear. Overall, the evidence on somatic DNA and chromosome damage with age is far from resolved.

8.3. Cancer and Age-related Genetic Changes

Cancer and other abnormal growths have distinct incidence as a function of age, which could be related to a wide variety of genomic changes that include accumulated mutations, gene amplification, and structural changes in chromosomes. As described in Section 8.2.5.2, chromosome structural abnormalities increase with age in some cell types. Many adult onset malignancies in humans have an age-related incidence that closely parallels the age-related mortality rate, that is, the fourth to sixth power of age (Burch, 1984;

Chapter 5.4.7). This parallelism and the demonstration of genetic alterations in many malignancies have suggested a fundamental relationship between the causes of cancer and organismic senescence.

In a few examples, the origins of cancer strongly imply particular chromosomal structural changes in an original clone. Various types of leukemia each have one or a few characteristic chromosomal translocations that relocate a protooncogene, that is, a normal gene that has regions of strong homology to a viral oncogene. Oncogenes of animal viruses are identified by their ability to transform susceptible cell types *in vitro* or *in vivo* into cell lines with unlimited growth potential, which are models for spontaneous malignant transformation. According to the oncogene hypothesis, endogenous protooncogenes become activated through various genetic changes that are postulated to transform the involved cell (Weiss, 1986). Protooncogene activation might occur through a chromosomal translocation that inserts the protooncogene into a new regulatory environment where transcription is activated through the influence of heterologous promoters or enhancers. Alternative routes to activation include point mutations, gene amplification, and inactivation of suppressors (Duesberg, 1987; Weiss, 1986). For example, chronic myeloid leukemia, a disease that increases exponentially during middle age in humans, shows a characteristic translocation, in which most of the long arm of chromosome 22 is translocated to chromosome 9. The 22q⁻ product, the "Philadelphia chromosome," also has a small amount of DNA from chromosome 9, including *abl,* a protooncogene related to the transforming sequences of Abelson leukemia virus, a murine retrovirus (De Klein et al., 1982).

Other adult onset leukemias and carcinomas are characterized by specific translocations that are thought to place protooncogenes in a new regulatory environment (Yunis, 1983; Sandberg, 1984; Bennington, 1986). An important early example of such transformations was given by Burkitt's lymphoma, a childhood disease associated with infections of Epstein-Barr virus and translocations bringing the proto-*myc* gene near to transcriptional enhancers for the heavy chain γ-globulin gene (Leder et al., 1983).

The oncogene hypothesis in its simplest form is open to question because no protooncogene from human tumors causes transformation of diploid cells and because the chromosomal rearrangements and mutations of protooncogenes are not universal in any given type of tumor (Duesberg, 1986). Nonetheless, it would be of much interest to analyze age-related chromosomal structural abnormalities for possible correspondences with chromosomal abnormalities in spontaneous malignancies of that age group. A related question concerns possible age changes in the location or incidence of "fragile sites" seen in some conditions of cell culture, which coincide in some cases with translocations of malignant cells (Yunis, 1983).

However convincing may be the evidence for specific chromosomal changes in malignant cells and for the ability of sequences related to protooncogenes to transform some normal cell types, the *originating* events may not have been these genomic changes (Duesberg, 1986). Intense selection during latent phases often lasts many years after initial exposure to viruses, X-rays, carcinogens, etc., and permits the accumulation of many genomic changes in the surviving clones. The originating events remain elusive and are unknown for any spontaneous malignancy of humans. It is unknown for most malignancies whether the crucial factor is the passing of time during which additional genetic changes may accrue, or age-related physiological changes. For example, the slope of the

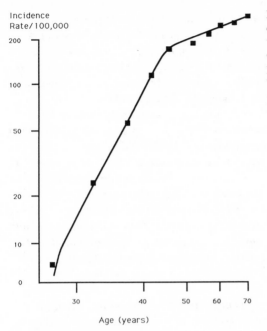

Fig. 8.9. The age-related incidence of breast cancer in U.S. Caucasian females slows after 50 years, when there is a major decrease in blood levels of ovarian steroids (Chapter 3.6.4.1). The decrease of progesterone is implicated as a cause, since progesterone is mitogenic for breast epithelial cells and since epithelial mitoses decrease after menopause. Redrawn from Pike et al., 1983.

exponential increase of breast cancer decreases to a much smaller rate of increase after menopause (Figure 8.9). This deceleration may be related to decreased proliferation of breast epithelial cells (Haagensen, 1971), consequent to menopause-related decreases in the blood levels of ovarian steroids, particularly progesterone (Pike et al., 1983).

Several clever experiments to resolve time and age in rodent tumorigenesis show that both are factors. For example, the induction of carcinomas by dimethylbenzanthrocene on mouse skin transplants from 2- and 14-month-old donors was increased threefold in the older transplants to young hosts (Ebbesen, 1974a). However, UV-induced carcinomas showed opposite effects of age with skin from the ear versus dorsum (Ebbesen and Kripke, 1982). Varying the duration of exposure to benzpyrene was found to induce skin cancer independently of age in mice (Peto et al., 1975), while older mice showed reduced response to delayed promotion by phorbol esters (Stenback et al., 1981). These complex results argue against a simple, time-dependent accumulation of genetic damage as the cause of age-related increases of cancer.

8.4.   Plasmids and Episomes in Somatic Cells

8.4.1.   *Mitochondria*

Mitochondrial DNA has a much higher mutation frequency than nuclear DNA, according to the incidence of RFLP in human populations (Giles et al., 1980) and the formation of mitochondrial mutants in yeast (Linnane and Nagley, 1978). Much evidence indicates that mitochondrial DNA mutations cause deficiencies in respiration and ATP synthase complexes. A new version of the somatic mutation hypothesis is that somatic mitochondrial DNA mutations will accumulate with age, leading to deficiencies in oxidative phosphorylation (Linnane et al., 1989). I review first the evidence for alterations in mitochondrial DNA with age in mammalian tissues and then the evidence for altered respiration.

Pikó and colleagues report from extensive electron-microscopic analyses that mitochondrial DNA of mammals develops age-related structural alterations of circular dimers, concatenated circles, and replicative intermediates that vary between tissues of individuals. Firstly, mitochondrial DNA from blood leukocytes changed little across the human lifespan; circular dimers were elevated in a few older men, which may indicate a tissue-specific disorder. This material mainly represents DNA from short-lived granulocytes (Pikó et al., 1978). Circular dimers are normally rare (< 0.1%) and are commonly associated with leukemia; their presence does not indicate dysfunction, since the vigorous LD line of cultured mouse L cells has only circular dimers (Clayton and Smith, 1975). Circular dimers also increased manyfold over the lifespan in different tissues of BALB/c mice and F344 rats, particularly the brain (Pikó et al., 1984; Pikó and Matsumoto, 1977). Among brain regions, the medulla had fifteenfold more dimers than the striatum (Bulpitt and Pikó, 1984). Concatanes (interlinked circles) also generally increased with age. The vast majority of mitochondrial circles were normal throughout the mouse lifespan, including the frequencies of replicative (D-loop) forms. Replication is obviously not grossly impaired in view of the minority of mitochondrial DNA variants.

In contrast to the appearance of native mitochondrial DNA, however, deletions and insertions were recently shown to increase by fivefold in liver mitochondria during the mouse lifespan. The deletions were revealed by the single-strand loops that formed after thermal denaturation and reannealing of mitochondrial DNA (Pikó et al., 1988). The deletions/insertions may be nonrandomly distributed, with a possible cluster between the *ND2* and *COI* mitochondrial genes, and another in the D-loop region. Readers may recall from Section 8.2.2 that there is also an accumulation of modified bases in mitochondrial DNA, at low levels such that about 1/15 mitochondria has a modified base molecule (Gupta et al., 1990). The presence of dimers, concatanes, and deletions in mitochondrial DNA of some tissues of older mammals may indicate local pathologic changes, rather than a general dysfunction of mitochondrial replication. These results are possibly consistent with considerable evidence for decrements in oxidative phosphorylation, particularly state 3, which represents the maximum capacity of the respiratory chain (Trounce et al., 1989; Sanadi, 1977). There are, however, considerable individual variations. In the absence of information on the mitochondrial DNA deletions or polymorphisms, we cannot yet conclude that the respiratory deficiencies arose from mitochondrial DNA damage. Other questions concern tissue differences in cell replacement, which might remove failing cells, and in the differential survival and clonal expansion of mitochondrial mutants.

8.4.2. Episomes

Except for episome-linked organismic senescence in specific genotypes of filamentous fungi, there is no general indication of age changes in episome replication. Some strains of *Podospora* and *Neurospora* show a rapid senescent death from an infectious plasmid that integrates into mitochondrial DNA and that is maternally inherited. Other filamentous fungi show no evidence of such unstable sequences (Chapter 2.4).

8.5. Parental Age and Abnormalities in Gametogenesis, Fertility, and Development

Parental age has many influences on gametogenesis, fertility, and development. These parental age effects have important bearing on the potential to select for changes in reproductive schedule that also indirectly influence the adult lifespan. For example, the limited stock of ovarian oocytes and follicles of mammals is implicated in aneuploidy and birth defects by an experiment in which one ovary was removed (hemi-ovariectomy; Brook et al., 1984). Young CBA/Ca mice that were hemi-ovariectomized within several months suffered a precocious increase of fetal aneuploidy. One interpretation is that the dwindling follicular pool caused disturbances in hormone secretion, involving either ovarian follicular secretions or neuroendocrine responses to reduced follicular pools. Whatever the mechanism, this important result draws attention to other species with fixed stocks of gametes, for example soil nematodes and rotifers.

A major question concerns the generality among species of age-related effects on gamete viability and chromosomal stability. Should instability in quality of gametes arise at later ages in any species, the contribution of survivors in that age group to the population gene pool will be further reduced. This compounded jeopardy is an additional factor that could favor the accumulation of late-manifesting genes with adverse effects, according to an evolutionary theory of senescence (Chapter 1.5). Gender differences in age effects on gamete quality are also important to different systems of mating strategy, which is another major question in population genetics.

8.5.1. *Mammals*

Loss of fertility during mid-life in female mammals is one of the most general age-related changes, and contrasts with the more gradual and haphazard reproductive impairments of males (Chapter 3.6.4). Apart from mutations arising from chromosomal abnormalities, which become common at later maternal ages, or from the much rarer effects of paternal age on specific hereditary diseases, parental age has little effect on postnatal development in mammals.

8.5.1.1. *Maternal Age Effects*

The major maternal age-related changes in mammals are increases in fetal aneuploidy later in reproductive life and the exhaustion of ovarian oocyte stores. The production of mature ova is ultimately limited by the finite ovarian stock of primary follicles and oocytes. These irreplaceable cells are produced only before birth in most mammals and are progressively lost throughout postnatal life (Figure 6.11). Of the vast numbers present at birth, only a tiny fraction survive to become mature ova: in humans, less than 0.1% ($500/10^6$) of the oocytes present at birth are shed as mature ova during the reproductive lifespan.

The vast majority of follicles degenerate through the poorly understood process of atresia. A recurrent speculation is that this huge-scale follicular degeneration may be a mechanism for removing defective oocytes, but there is no direct evidence for this. Readers may recall that oocyte resorption can be used by some biting flies as a source of yolk for maturation of the remaining oocytes, and a similar phenomenon is suspected in brook lampreys (Chapter 2.3.1.4.1). At menopause in humans the stock of oocytes is depleted, while laboratory rodents can have extensive numbers of remaining oocytes after the end of fertility cycles (more discussion in Chapter 3.6.4).

The oocytes of humans and laboratory rodents are postmitotic since their formation in the fetus. In humans, about twenty-two cell divisions separate the ovum from its primordial germ cell progenitors, whether the ovum is shed soon after puberty or just before menopause (Vogel and Motulsky, 1986). In mammals, meiosis is arrested at the dictyate stage of meiotic prophase I,[3] which is reached by virtually all germ cells of humans in the twenty-fourth week of gestation (Speed, 1985). Meiosis is resumed in the oocytes of mature (Graafian) follicles under the control of gonadotropins just before ovulation, which may be more than 40 years later in humans and elephants. The remarkably prolonged prophase of meiosis may require unique mechanisms for repair of DNA damage and replacement of proteins in the formed chromosomes. Because actively transcribed genes have more rapid DNA repair mechanisms in somatic cells (Bohr et al., 1986), perhaps the extent of transcription in the extended meiotic prophase may influence the accumulation of DNA damage at various loci. There is no counterpart of extended meiotic prophase in nondividing somatic cells, which generally are in the G_1 phase.

Ovarian oocytes manifest strong age-related increases of aneuploidy from chromosomal nondisjunction, or the failure of chromosome pairs to separate. A consequence of nondisjunction is a marked age-related increase of fetal aneuploidy at greater maternal ages in women and laboratory mice, rats, and rabbits (Fujimoto et al., 1978; Gosden, 1985; Fabricant and Schneider, 1978; Harman and Talbert, 1985; Finch and Gosden, 1986). Maternal age increase of aneuploidy, however, has not been shown in some rodent strains (Beerman et al., 1987). Aneuploidy is considered to arise through nondisjunction, specifically during meiotic anaphase I (Jacobs and Hassold, 1980; Mikkelsen et al., 1980; Sankaranarayanan, 1979; De Boer and van der Hoeven, 1980).

In humans at birth, the most common maternal age-related karyotypic abnormalities are the hyperdiploidies: in order of frequency, Down's syndrome (trisomy of chromosome 21), Klinefelter's syndrome (XXY), Edward's syndrome (trisomy 18), and Patau's syndrome (trisomy 13) (Figure 8.10; Hook, 1986; Carothers et al., 1978, 1980). The rate of increase with maternal age is exponential, with an acceleration that is slightly faster than the 8-year MRDT in the general human population (Figure 1.5) (Larry Butcher and Caleb Finch, unpublished). Besides by maternal age, the risk of Down's syndrome is also increased by environmental insults, which include maternal exposure to X-rays (Alberman et al., 1972) and by cigarette smoking (Kline et al., 1983). Although by the maternal age of 50 years about 10% of the viable term fetuses would be aneuploid according to Figure 8.10, the initial incidence may be even higher if spontaneous abortions are included. Several researchers have suggested that a factor in the increased aneuploidy with maternal age could be decreased efficiency of uterine rejection mechanisms (pros and cons discussed by Erickson, 1978; Sved and Sandler, 1981). The basis for uterine rejection of abnormal fetuses is poorly understood.

The incidence of aneuploidy is astonishingly high in spontaneous abortuses, about 50% (Boué et al., 1975a, 1975b; Hassold, Chen, et al., 1980; Schneider, 1985; Gosden, 1985;

---

3. The dictyate stage is also called dictyotene or diplotene.

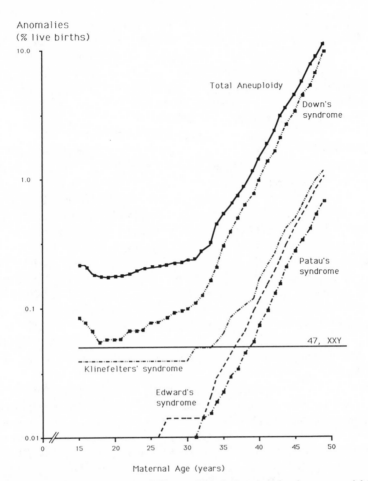

Fig. 8.10. Maternal age-related increases in fetal aneuploidy in humans, showing exponential increases of Down's syndrome (trisomy 21), Klinefelter's syndrome (XXY), Edward's syndrome (trisomy 18), and Patau's syndrome (trisomy 13). The majority of children with Down's and Klinefelter's syndromes can survive to adulthood, while Edward's and Patau's syndromes are usually lethal within a year of birth. The doubling time of incidence is about 6 years, that is, close to the MRDT. Based on data of Hook, 1981; redrawn from Gosden, 1985.

Wilmut et al., 1986). Among these gross abnormalities, trisomies of the smallest chromosomes tend to preferentially increase with maternal age (Hassold, Jacobs, Kline, Stein, and Warburton, 1980); this effect is thought to result from an age-related loss of chiasmata, which could particularly affect smaller chromosomes, as discussed below. Fetal aneuploidy also increases with maternal age in laboratory rodents, though relatively more abnormal fetuses may be resorbed than in humans (Fabricant and Schneider, 1978). The literature does not resolve whether there are species differences in the initial incidence of aneuploidy that might scale with lifespan or the duration of the dictyate stage of meiosis.

An important unknown concerns chromosomal abnormalities in oocytes before ovulation. A study of surgical specimens without ovarian pathology detected aneuploidy in oocytes from women 35 and older, but not before (Wertheim et al., 1986). The small number of aneuploid oocytes (5%; 5/99 in metaphase) available do not either prove or disprove the age-related increase expected from fetal aneuploidy. Whatever the case for oocytes, granulosa cells from the ovaries of these same women showed an age-related *decrease* in aneuploidy. The extent of relationships between aneuploidy in somatic cells and oocytes in the same individuals will be of great interest to elucidate. More data should be forthcoming from *in vitro* fertilization studies.

The inverse relationship between chromosomal crossing over (chiasma formation) and nondisjunction has long been noticed (Dobzhansky, 1933; Mather, 1938). As described in the classic study of Henderson and Edwards (1968) and verified in most subsequent studies (Speed, 1977; Sugawara and Mikamo, 1983; Luthardt et al., 1973; Polani and Jagiello, 1976; Harman and Talbert, 1985), several mouse strains show strong age-related decreases of chiasmata in their oocytes but not spermatocytes (Figure 8.11, top). Limited observations of human oocytes also suggest age-related decreases of chiasmata (Luthardt, 1977). Analysis of genetic markers also indicates an age trend for decreased crossovers in mice, for example between the loci for pallid and fidget (Bodmer, 1961) and leaden and fuzzy (Reid and Parsons, 1963). A mechanism in these phenomena could be greater chromosomal contraction at the time of crossing over. Speed (1977) observed that chromosomes from oocytes of older C57BL mice were more condensed than those from young during short-term culture. This indication of altered kinetics of chromosome condensation during meiosis recalls the trend in some somatic chromosomes for earlier separation of centromeres in the X chromosome (Mukherjee and Weinstein, 1988; Section 8.2.5.1). The time and technology are ripe for detailed analysis of chromosome structure and mechanics.

Univalents, which are unpaired chromosomes that failed to form chiasmata with their homologues during meiotic prophase I, may increase with age. Univalents vary inversely with the number of chiasmata in CBA and C57BL mice (Figure 8.11, bottom) according to one study (Henderson and Edwards, 1968), but no changes were found in the same strains by Speed (1977).[4] Overall, univalents are tenfold more common in human ova (> 10%) than in mice. This species difference is also consistent with the tenfold higher incidence of aneuploid conceptions in man (Speed, 1985; Eichenlaub-Ritter et al., 1988).

The age-related increase of univalents was proposed as a major source of aneuploidy (Henderson and Edwards, 1968). The univalents, being unpaired, should randomly distribute into daughter cells at anaphase I instead of remaining with their homologous chromosome. The greater incidence of trisomies involving smaller chromosomes in spontaneous abortions of humans is consistent with these predictions, since the smaller chromosomes have fewer chiasmata and should be relatively more influenced by age-related decreases in chiasma formation (Hassold, Jacobs, Kline, Stein, and Warburton, 1980). However, hamsters give contrary data; their small chromosome number (11 pairs versus 23 in humans or 20 in mice) favors resolution of univalent distributions in chromosomes of different sizes. While most univalents were among the smallest chromosomes, there was no size bias in chromosomes of hamsters in regard to nondisjunction (Suguwara and Mikamo, 1983), which argues against the ultimate importance of univalents in aneuploidy of the oocytes that become mature ova. This observation suggests

---

4. The particular CBA substrains are not identified in some of these reports from the U.K., but probably include CBA/H and CBA/Ca. See Chapter 6.5.1.1.

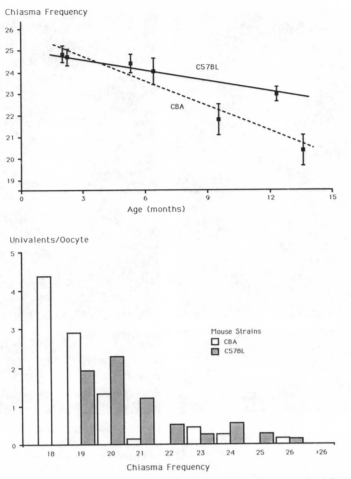

Fig. 8.11. Age-related changes of oocyte chromosomes in inbred mice. *Top,* age-related decrease of chiasma frequency; *bottom,* inverse relationship between chiasma frequency and univalent chromosomes (chromosomes that failed to pair and form chiasma during by diplotene of meiotic prophase I). The age-related increase of univalents is proposed as a major source of aneuploidy. The unpaired univalents are randomly distributed into daughter cells at anaphase I, instead of remaining with their homologous chromosome. Redrawn from Henderson and Edwards, 1968.

defective spindles, a mechanism that could equally apply to aneuploidy during somatic cell mitosis.

Age-related alterations in spindle function during oogenesis are indicated by a careful study of spontaneously ovulated oocytes in CBA/Ca mice. By 9 months, oocytes showed sixfold increases of spindles, with irregular chromosome alignments on the equatorial plate increasing to 65% (Figure 8.12a and b). The age-related incidence of disorganized meiotic metaphase plates in mice is accelerated by hemi-ovariectomy (Eichenlaub-Ritter et al., 1988). As noted above, hemi-ovariectomized CBA/Ca mice have precocious increases of fetal aneuploidy (Brook et al., 1984). CBA/Ca mice are a favored genotype for these manipulations because of their faster loss of oocytes and earlier reproductive failure

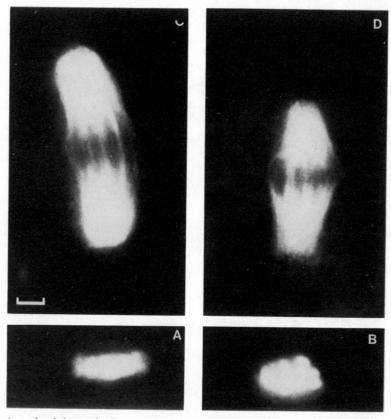

Fig. 8.12. Age-related changes in chromosomes and spindles of spontaneously ovulated oocytes at metaphase II (CBA/Ca mice). Equatorial plates in oocytes from 4-month-old (*A*) and 9-month-old (*B*), showing chromosomal misalignment and spreading, which are fivefold more frequent in older mice; stained by the fluorescent dye, DAPI. Spindles were visualized with indirect antitubulin immunofluorescence and were 15% shorter in older mice. *Bar* in *C,* 10 μm. From Eichenlaub-Ritter et al., 1988. Photographs were kindly supplied by Ursula Eichenlaub-Ritter and Roger Gosden.

(Finch and Gosden, 1986; Figure 6.11). In a possible human counterpart of this study, Chandley (1985) noted that in the rare Turner's syndrome (45, X karyotype), women who are able to conceive despite almost complete gonadal dysgenesis have a very high incidence of Down's offspring, possibly in association with their greatly reduced ovarian follicular pool, which also causes premature menopause (King et al., 1978; Reyes et al., 1976).

Returning to CBA/Ca mice, the length of the oocyte spindle is shortened by about 15% between 6 and 9 months, a change that was not influenced by hemi-ovariectomy (Figure 8.12c and d); otherwise, the spindle was normal in oocytes (Eichenlaub-Ritter et al., 1988). These researchers suggest that disorders in chromosomal arrangements at metaphase II could predispose to aneuploidy. The acceleration of aneuploidy and abnormalities at metaphase I by hemi-ovariectomy argues against these changes as a consequence of chronological age, and favors the role of endocrine abnormalities consequent to the imminent exhaustion of ovarian follicles. The greater susceptibility of oocytes from 12-

month-old mice to arrest at metaphase I by a low dose of colchicine (Tease and Fisher, 1986) also suggests unstable spindle functions. Another mechanism proposed by Peluso, England-Charlesworth, and Hutz (1980) invokes a decreased tendency of oocytes from middle-aged mice to form a polar body, which would result in triploidy upon fertilization.

The evidence for age-related changes in spindle function suggests that mechanisms leading to genetic change through aneuploidy need not originate from primary changes in DNA itself. Nicklas (1988) discusses how the dynamics of the spindle give many opportunities for error, since microtubules of the spindle are continually undergoing cycles of assembly-disassembly with chance attachments of chromosomes to the spindle. Unlike error correction in DNA, there does not seem to be any process based on molecular templates for detecting and correcting mistakes in spindle attachment that may cause nondisjunction and aneuploidy.

At a physiological level, age changes in the timing of ovulation are implicated in increased aneuploidy, for example through a lengthened follicular phase (Hertig, 1967). This proposal is supported by studies showing that delayed ovulation, by impairing the preovulatory gonadotropin surge, leads to subsequent increases of fetal aneuploidy in laboratory rodents (Butcher and Fugo, 1967; Butcher, 1975; Laing et al., 1984). Lapolt et al. (1986) also delayed reproductive senescence by extending fertility 2–4 months in middle-aged rats through progesterone treatment. This result implies the potential for endocrine influences on fetal aneuploidy, since aneuploid fetuses in middle-aged rodents are generally resorbed. Progesterone might have several different actions here, including correction of hormonal imbalances that are proposed to slow the rate of meiosis and increase the loss of chiasmata by delaying their attachment to the spindle (Crowley et al., 1979).

Another hypothesis about timing is that the decreased frequency of intercourse during middle age in men predisposes to aneuploidy by the persistence of "outdated" sperm in the reproductive tract, which still may retain sufficient viability for fertilization despite other abnormalities (German, 1968). This mechanism may have less importance because most nondisjunction occurs during meiosis I, that is, *before* ovulation and fertilization. Like the onset of fetal aneuploidy as mentioned above (Brook et al., 1984), the onset of cycle regularities and cessation can be accelerated in early middle-aged mice by hemiovariectomy (Sopelak and Butcher, 1982). These manipulations argue the case against an intrinsic age-related degeneration of ovarian oocytes that increases the risk of aneuploidy, and shift the focus to the depletion of steroid-secreting ovarian follicles and the effects of altered steroid secretion on hypothalamic-pituitary functions that determine the timing of ovulation (Chapters 3.6.9, 10.4.2).

Prolonged pregnancy from delayed parturition becomes a significant factor in the mid-life fertility decline of laboratory rodents. For example, we found that C57BL/6J mice by 12 months have prolonged pregnancy such that parturition is typically delayed by 1 to 3 days (Holinka et al., 1978). Concomitantly, there is a major increase in stillborn pups (about 25%), whose number increases with the delay of parturition. Thus, litter size of older rodents decreases from two factors: the loss of abnormal fetuses through resorption and the delay of parturition. Smaller litter sizes alone do not account for the delay in parturition. The mechanisms include abnormalities in the preparturitional changes of ovarian steroids: the drop of progesterone and the elevations of estrogens are retarded (Holinka et al., 1979a, 1979b). Similar changes occur at middle age in hamsters (Soderwall et al., 1960). Evidence for delayed parturition through prolonged pregnancy at later maternal ages is equivocal in domestic animals (Clegg, 1959), while this phenomenon clearly does not occur at later maternal ages in humans (Beischer et al., 1969).

In summary, some maternal age effects on fetal aneuploidy in mammals may be associated with spindle abnormalities, which could lead to the elucidation of molecular mechanisms. It is unknown whether the massive loss of oocytes through atresia from birth onwards wastes damaged as well as potentially viable oocytes. The role of age-related change in the uterine environment that may permit survival of more aneuploid embryos is unknown. Whatever the basis for these numerous changes at a proximal level, there is strong evidence for a role of hormones, through the increase of fetal aneuploidy by hemi-ovariectomy in mice and the rescue of fetal loss by progesterone treatment before fertilization. These findings argue that maternal age effects on fetal abnormalities are not due to intrinsic changes in the oocytes. It would be timely to extend these studies to other vertebrates, including reptiles and birds, for which there is little information on maternal reproductive age effects, besides limited observations on alligators (Chapter 3.4.2).

8.5.1.2. Paternal Age Effects

Mammalian males maintain some capacity for spermatogenesis in their seminiferous tubules throughout adult lifespan, although the extent of spermatogenesis declines on the average in population samples (Chapter 3.6.4.2). Many pathologic epiphenomena of senescence can impair testicular functions, leading to reduced numbers and viability of sperm, and androgen secretion. Major causes include arteriosclerosis and effects of diabetes. Male mice of several strains showed age-related decreases in urinary pheromones that may also be a factor in declining female reproduction (Wilson and Harrison, 1983). The male's capacity for intercourse also decreases in all species examined, although individuals vary widely in humans (Kinsey, 1948) and mice (Bronson and Desjardins, 1977; Huber et al., 1980). Male dysfunctions are not analogous to menopause, however, because only exceptionally do testicular functions cease completely (Figure 3.21). A qualitative impression is that the cellular aspects of spermatogenesis remain normal throughout adult life.

Concerning age changes in chromosomal mechanics at spermatogenesis, few data exist for species besides mice. Chiasma frequency did not change with age up to 15 months in C57BL and CBA mice (Speed, 1977), while C57BL/6J mice showed 1.5- to 2-fold increases of morphologically abnormal sperm by 28 months (Fabricant and Parkening, 1982; Parkening et al., 1988); Several analyses of fetal chromosomes show that paternal age contributes much less than maternal age to fetal aneuploidy in humans (Penrose, 1961; Hook, 1986). About one-third of all trisomies are attributed to nondisjunction in men (Vogel, 1983).

Turner's syndrome does not show any effect from paternal age. This is particularly striking and contrasts with the age-related increase of the corresponding somatic cell aneuploidy (45, X) that is observed in subpopulations of lymphocytes and fibroblasts from middle-aged women during short-term culture (Section 8.5.1.2). Turner's syndrome arises in about 50% of the cases from loss of either the paternal X or Y by nondisjunction during spermatogenesis; the remaining cases are associated with partial deletions, chromosomal rearrangements, or loss of an X during development that causes mosaicism (Hook, 1986). Several studies agree that there is no maternal or paternal age effect on the incidence of Turner's syndrome at term (Carothers, Frackiewicz, et al., 1980; Carothers,

Collyer, De Mey, and Frackiewicz, 1980) or from abortuses (Boué et al., 1975a, 1975b; Kajii and Ohama, 1979). Similarly, Down's syndrome does not indicate a paternal age contribution (Erickson, 1978). The absence of paternal age effects in Turner's and Down's syndromes is consistent with the absence of age-related decline in chiasma formation during spermatogenesis (Hamerton, 1971).

Paternal age effects on fetal abnormalities are most strongly indicated for some sporadic hereditary conditions, particularly autosomal dominant diseases that arise from new mutations rather than from abnormal distributions of chromosomes. The risk of achondroplasia, myositis ossificans, Apert's syndrome (acrocephalosyndactyly), and Marfan's syndrome all increase by fourfold between paternal ages of 20 and 50 years (Vogel, 1983; Hook, 1986). The overall risk of a paternal age contribution to fresh mutations, while increasing sharply, still is vastly less, about 10^{-3}, than the maternal age contribution to fetal aneuploidy.

Next we consider an interesting, but unfulfilled, theoretical argument. Penrose (1955) noted that ongoing proliferation of spermatogonia and undifferentiated spermatogonia implies that the sperm produced at later ages resulted from more cell divisions. Calculations by Vogel and Motulsky (1986) estimated that sperm present at puberty are derived from an estimated thirty divisions from the male primordial germ cells; by 35 years, sperm result from more than five hundred divisions. Because mutations are primarily thought to occur during DNA replication, the cumulative number of spermatogonial divisions could be an important factor in paternal age effects. However, the evidence argues against this theoretical prediction, since *males and females have about the same incidence of mutation rates per generation* (Vogel and Motulsky, 1986). The numbers of divisions in such calculations can be greatly influenced by the stem cell model chosen. Further inconsistencies are that many loci do not show paternal age effects for autosomal dominant diseases, including bilateral retinoblastoma and neurofibromatosis; the latter has the highest-known mutation rate at a specific locus, 10^{-4} per gamete per generation. In reading this literature, one must keep in mind an unknown, the incidence of mutations in all the sperm produced; only those that yield viable adults are usually scored. The causes of locus specificity in mutation rate are probably multiple and are just being elucidated; for example, the efficacy of DNA repair may vary with the transcription at a particular locus and the cell-cycle stage when the lesion occurred (Hanawalt, 1987).

I raise another controversial point. Does not the maintained production of viable sperm throughout the lifespan of most men imply powerful mechanisms for repairing or eliminating DNA or chromosomal abnormalities that may arise? If the genes for the same hypothetical repair functions are present and expressed in somatic cells, this would explain why rapidly proliferating somatic cells change little with age in production of normal cells, for example, erythropoiesis in the bone marrow and in the gut mucosa and epidermis (Chapter 7.6.1). A future topic is whether the genes encoding DNA repair processes that are needed for maintenance of gametogenesis in long-lived species might also be differentially expressed in somatic cells that show the least damage at later ages.

8.5.2. Other Vertebrates

Little information on vertebrate species other than mammals is available. A variety of different age-related patterns should be expected because of species differences in the extent of gametogenesis during adult life. For example, *de novo* oogenesis appears to continue beyond sexual maturity in at least some teleostan fish, amphibia, and probably reptiles (Franchi et al., 1962; Tokarz, 1978; Zuckerman and Baker, 1977). As described in Chapter 4.2.2, long-lived and slowly growing teleosts continue to increase the production of ova consistent with size allometry, with no indication of senile decline. Except for some reports on maternal age-related increases in abnormal embryos in alligators (Chapter 3.4.2), little is known about age-related changes in gametogenesis and abnormalities in other long-lived lower vertebrates. The possibility that there is a continuing age-related increase in gamete production would give these species a selective advantage against accumulating deleterious genes with late onset effects, if indeed there is no decline in gamete quality. A few short-lived fish (*Gambusia affinis* and *Betta splendens*) have postreproductive stages in females or males, respectively (Chapter 3.4.3). Abnormal gametogenesis might be predicted during the transition to infertility.

Certain species of female rays store sperm up to at least 2 years before fertilization may occur (Dodd, 1983). A specialized organ, the shell gland, holds sperm for successive fertilizations, which are universally internal in the rays and other elasmobranchs. Nothing is known about the effects of prolonged sperm storage on embryo abnormalities.

In birds, the decline of egg production among domestic chickens and quail is not matched by many species in the wild (Chapter 3.5). Although there are indications of declining egg volume in the herring gull, others like the fulmar do not show this trend in field populations. Like long-lived fish, the long-lived birds with no decline in reproduction up through ages when only a minority survive may permit a different dynamic of selection against genes with delayed adverse effects. The incidence of age-related abnormalities gametes in these long-lived birds is not known.

8.5.3. Invertebrates

Except for a few insects and nematodes there are few data.

8.5.3.1. Insects

In *Drosophila* most age effects on reproduction occur before mid-life. Flies become fertile within a few hours after hatching (eclosion), although egg production may not be maximum until a week later, an age that is early in their typical lifespans of 2–3 months (Chapter 2.2.1.1). *De novo* oogenesis continues throughout life, and the number of functional ovarioles is unchanged (Wattiaux, 1967). However, the daily production of eggs decreases by the average lifespan, with differences among strains (reviewed in Mayer and Baker, 1985). Although spermatogenesis also continues throughout adult life, the number of spermatogonia decreases at advanced ages. Morphological abnormalities include the occasional loss of the axial filament complex in spermatids (Miquel, 1971; Miquel, Bensch, Philpott, and Atlan, 1972).

In flies, chiasmata form in the pro-oocytes of pupa just before hatching and thus do not have the extended dictyate stage observed in mammals; oogenesis takes about 10 days from oogonia to maturity (King et al., 1968). Although male *D. melanogaster* do not have chiasmata or crossing over, males of other drosophalines do form chiasmata, for example, *D. ananassae*. Age-related decreases in recombination varied amoung chromosomes as well as among different regions of the same chromosome (Stern, 1926; Bridges, 1927; Swanson, 1957, 235; Parsons, 1962) and are of continuing concern because they can differentially influence map distances between loci. For example, in chromosome 3, crossovers in region *st-p*^{p} decreased by 90% between 3 and 11 days (early middle age), while in *ru-h* they decreased by only 15% (Bridges, 1927). At later ages, crossovers may increase (Bridges, 1927; Valentin, 1972; Lake, 1984), to give a U- or W-shaped, multiphasic time function of age that varies among chromosome regions. The absence of maternal age effects on crossovers around the vermilion locus of the X chromosome is attributed to counterbalancing effects of age on crossing-over flanking regions (Valentin, 1972). In male *D. ananassae,* crossovers at three loci decreased sharply by 2 weeks in one strain, but not in another (Tobari et al., 1983). The incidence of nondisjunction in X chromosomes increases by the second week (Tokunaga, 1970), reaching a frequency of about 10^{-3} of flies that hatched. Although similar to the incidence of aneuploidy in live births of humans (Figure 8.10), these figures cannot be compared because the extent of embryo death is unknown. The increased mortality of larvae and pupae with maternal age in some genotypes (David, 1959b; Baker et al., 1985) has not been analyzed for aneuploidy.

There is some evidence for changes of macromolecular biosynthesis within *Drosophila* ovaries. The duration of oocyte maturation appears to shorten and the incorporation of [^{3}H]-uridine into ovarian nurse cells may increase, particularly in the lower ploidy class cells (Wattiaux and Tsien, 1971). Considered with the acceleration of oogenesis, the formation of ribosomal RNA, which is the major oocyte RNA produced by the enveloping nurse cells (Davidson, 1986), may also be accelerated. The RNA content of the eggs decreases by 30% after 12 days (Tsien and Wattiaux, 1971). The incorporation of [^{3}H]-leucine into total ovarian proteins, however, decreases (Wattiaux et al., 1971); this is consistent with the overall decline in egg production at later ages. Up through at least 5 weeks, ovarian protein synthesis is rapidly inhibited by actinomycin D, indicating the continuing importance of transcription in the ovary.

A few other examples are mentioned. Mosquitos lay fewer eggs at later ages (Woke et al., 1956; Chapter 2.2.1.1) but I am unaware of any studies on abnormal gametes or embryos. A little-cited report describes interesting maternal age effects on developments in a wasp. In a partly inbred line of *Habrobracon juglandis* characterized by developmental abnormalities, reduced genitalia occurred mostly from eggs laid in the first 2 weeks; age-related changes in recombination were suggested as a cause of this variation (Whiting, 1926).

A subject for future research is the extent, if any, of age-related abnormalities in the gametes of social insect queens. As discussed in Chapter 2.2.3.1, many species mate only when young and draw on sperm stores throughout reproductive lifespans that may last 5 years or more. While fecundity is thought to persist into later ages, the offspring generated by these long-stored sperm have not been studied for developmental abnormalities or viability. Oogenesis appears to continue *de novo*.

8.5.3.2. Soil Nematodes, Rotifers, and Other Species

The soil nematode *Caenorhabditis* (Chapters 2.5.4, 6.2) has both self-fertilizing hermaphroditic forms and occasional males, which arise by nondisjunction leading to loss of an X chromosome. The karyotype is 5AA + XX in hermaphrodites and 5AA + X0 in males (Brenner, 1974). Both the hermaphrodites and the males have determined numbers of somatic cells, although the numbers of germ cells may vary severalfold. Some parasitic nematodes evidently produce massive numbers of viable eggs during lifespans of many years. Nothing is known about the extent of *de novo* gametogenesis in the long-lived parasitic nematodes, or the causes of the ultimate decline in fecundity after many years (Chapter 4.2.2). Hermaphroditic *Caenorhabditis* show sharp declines in the numbers of progeny early in the second week of their typical 3-week lifespan, and few were fertile after 10 days (Johnson, 1987). Lines selected for different lifespans showed no correlations between the onset of sterility, numbers of progeny, and lifespan. The sharp reduction of fertility largely results from exhaustion of the fixed numbers of sperm. The viability of progeny does not change markedly during reproductive decline, although this has not been systematically studied (Thomas Johnson, pers. comm.).

In oocytes of *Caenorhabditis*, pachytene cannot be seen after day 9; other cytological changes include condensation of chromatin and nucleolar shrinkage (Goldstein and Curis, 1987). Similar senescent changes occur in vinegar eelworms (*Turbatrix aceti*; Pai, 1928). Recombination decreases progressively with age in two chromosomes (Rose and Baillie, 1979). Nondisjunction of the X chromosome increases with age in several strains and causes severalfold or more increases in the number of males, but the number is always small and less than 1% (Rose and Baillie, 1979). Correspondingly, there are decreases of "disjunction regulator regions" in the synaptinemal complex of oocytes early in the second week (Goldstein and Curis, 1987). These changes in chromosomal dynamics during oogenesis in the short-lived nematodes are remarkably similar to changes in flies and mammals.

Rotifers demonstrate several unusual features of maternal age. Like soil nematodes, this phylum has fixed numbers of somatic cells and gametes (Chapter 2.2.2.2). Typically, rotifers lay less than fifty eggs; eggs are produced one at a time and are huge relative to these tiny organisms. Despite their very short lifespans, females have a definite post-reproductive phase in most species; *Floscularia conifera* may not (Edmondson, 1945). Male lifespans are particularly short, a few days to a week, because they lack a digestive tract (Chapter 2.2.1.2). In the order Monogononta, two types of females occur: *amictic* females lay diploid eggs, which become females, and *mictic* females lay haploid eggs. Amictic females have shorter lifespans and lay more and larger eggs than mictic females (Miller, 1931). In some monogononts, males are mostly produced parthenogenetically from eggs of mictic females laid after the initial group, and these later eggs cannot be fertilized, for example *Asplanchna sieboldi* (Thane, 1974; Buchner and Kiechle, 1967; Gilbert, 1983a) and *Euchlanis triqueta* (Lehmensick, 1926). In *Lecane inermis* the eggs that yield males are relatively smaller (Miller, 1931).

At later ages, the eggs become extremely varied in their size and shape, for example *Proales sordida* (monogonont; Jennings and Lynch, 1928a), and change in color and density, for example *Philodina citrina* (bdelloid) and *Euchlanis triqueta* (monogonont; Lansing, 1947). The process of oogenesis, involving secretion of a large mass of yolk through the oviduct, is observed to show loss of coordinate with the formation of the egg shell by glandular secretions from the rotifer foot (Jennings and Lynch, 1928a). Reduced yolk production is consistent with ultrastructural changes in the yolk gland (vitellarium) of *Philodina*, which include decreases of endoplasmic reticulum and loss of roundness in nuclei (Herold and Meadow, 1970). The viability of the eggs from older rotifers decreases to nearly 50% (Jennings and Lynch, 1928a), a loss that still is less than in eggs from very old *Drosophila*. Perhaps the most remarkable aspect of maternal age effects on oogenesis is the reduced fecundity and shorter lifespans of females that hatch from the eggs of older mothers, the "Lansing effect." Because these maternal age effects don't cause abnormalities in the genetic material, they are treated as an aspect of developmental influences on senescence (Chapter 9.3.2.2).

Information on age-related changes in gametogenesis for other species is very limited and mainly documents decreased egg production with no information on mutations or aneuploidy. For example, the flatworm *Dugesia lugubris* shows modest decreases in egg production and larger decreases of fertility (Figure 3.1), whereas spermatogenesis continues throughout life (Chapter 3.3.2). Other examples are described in Chapter 2, Comfort, 1979, and Rockstein and Miquel, 1973.

8.5.4. Apical Meristems in Plants

Vascular plants share an important trait of potential importance to age changes in reproduction. Unlike most animals, plants segregate their germ cell lineages from somatic cells just before reproduction, when the meiocytes arise from somatic cells in various meristems (Raven et al., 1986; Klekowski, 1988a, 1988b; Chapter 4.2). Thus, the germ cells of plants are proposed to be more subject to somatic mutations that may have arisen during the vegetative growth than are the oogonia of species like mammals where germ cells are segregated from somatic cells early in development and, in females, may be held in extended meiotic arrest.

Meristems usually form gametogenic tissue at the apex of growing shoots (apical meristems), although gametogenesis occurs in other species in the root tips or cambium (Klekowski, 1988a, 85). Several types of meristems are distinguished by whether daughter cells remain in the rapidly dividing cells or are randomly sampled to divide or remain quiescent for extended time, the "meristem d'attente" (Esau, 1977; Klekowski, 1988a, 1988b; Buvat, 1952; Sussex and Rosenthal, 1973). The extent of cell turnover has not been measured directly, however. These different compartmentalizations have implications for the accumulation of somatic mutations by the different numbers of cell divisions that a gametic cell line may have experienced since the original zygote (Klekowski, 1988a, 1988b). The consequences of these anatomically defined meristem types are largely theoretical, lacking definitive evidence that meristematic cell compartmentalization influences the mutation rates in subsequently produced gametes.

To test the proposal that prolonged vegetative growth leads to the accumulation of deleterious mutations in somatic cells, Klekowski (1984, 1988a, 1988b) compared two fern species that had different reproductive patterns. In nature, *Onoclea sensibilis* (the sensitive fern) reproduces both vegetatively and sexually; *Matteuccia struthiopteris* (the ostrich fern) reproduces sexually, but rarely vegetatively. These species share habitats in the Connecticut River Valley and have the same number of chromosomes. While both species had similar mutation rates, they differed greatly in genetic load, as judged by a fourfold greater incidence of lethal mutations per spore in *Matteuccia,* the species that rarely reproduces sexually in nature. Because the somatic cells are diploid, so goes the argument, recessive mutations should tend to be accumulated and then may be manifested after formation of the haploid gametes. It is thought that accumulated mutations have a major impact on the population through inbreeding depression (Klekowski, 1988a, 1988b).

If somatic mutations are accumulated during meristematic expansion, then mutation rates should be greater in plants with long generation cycles. Consistent with this prediction, the frequency of mutations in the red mangrove (*Rhizophora mangle*), a viviparous plant with a 20-year generation cycle, is about 25-fold greater (7×10^{-3} mutations/ genome/generation) than in several annuals from the literature (0.3×10^{-3} mutations/ genome/generation) (Klekowski and Godfrey, 1989). The mutation frequency was estimated from the incidence of albinism, a somatic cell phenotype that is attributed to deficiencies of chlorophyll production through mutations of any of about three hundred genes. By a simple calculation, the mutation rates per year are similar for both species.

8.6. Summary

The most convincing evidence for age-related disorders in genome maintenance and replication is at the chromosomal level, where the incidence of chromosomal breakage and other structural abnormalities and aneuploidy increases with adult age in some mammalian somatic cells and in oocytes. A paternal age contribution to fetal aneuploidy and new mutations can be demonstrated in humans but is much less than that from women at all ages, by a factor of 10^{-3}. This major gender differential in the contribution to germ-line abnormalities may be related to the continuing proliferation of spermatogonia, in contrast to the extended arrested prophase of oocytes since midgestation.

Aneuploidy of embryos is also indicated for *Drosophila* and *Caenorhabditis* with about the same frequency as in mammals when scaled to the lifespan. Nothing is known about the age pattern of birth defects in long-lived species of fish and birds that do not demonstrate an upper age limit to reproduction (Chapters 3, 4). This question should be pressed for species like sharks and sturgeon that may have a limited oocyte store, as in mammals, versus those that may have continuing *de novo* oogenesis during adult life, as in some teleosts and tetrapod radiations other than mammals. Neither the extent of *de novo* oogenesis and the extent of abnormal gametes have been characterized as in different adult age groups in any detail for lower species.

Besides increased aneuploidy, structural chromosomal abnormalities and chromosomal fragility increase with donor age in rodent and human cells. The relationship of these changes to the fidelity of DNA replication is unknown. Chromosome breaks and aneuploidy need not have the same cause or originate from changes in DNA sequence or organization. The extent of accumulated mutations in somatic cells is unclear, with some evidence for no change in some cells and increases in others. The evidence for altered repair of induced DNA damage is similarly unclear as a function of donor age. Taken together, the evidence indicates that DNA replication in somatic and germ cells is less subject to age errors than is the movement of chromosomes.

The extent of mutational or structural changes in the nuclear genome of somatic cells remains controversial and without a major accepted example. The remarkable age-related spontaneous curing of genetic defects for albumin and vasopressin in individual cells of rodents (Chapter 7.3.3) could prove to involve changes in DNA structure. If so, then there may soon be examples of structural changes in nuclear DNA with age that inactivate genes. A promising example of organellar DNA change is indicated for mitochondrial DNA; the generality of deletions in mtDNA is unknown.

The provisional conclusions about differences in somatic cell mutation rates that scale with the generation length in plants raise questions about the viability of spores and seeds produced by the very long lived hardwoods and conifers, but also about animals like ascidians that do not have early segregation of their germ line (Table 4.4). However, the persistence of both asexual and sexual reproductive modes in plants and animals since the Paleozoic (Chapter 11.2) implies that mechanisms were evolved to cope with somatic mutations before late onset germ-line segregation.

Many findings discussed on this and Chapter 7 bear on theories about the evolution of senescence (Chapter 1.5). The hypothesis that the strength of natural selection must progressively weaken after maturity because mortality removes potential gene carriers and because fecundity decreases with age in some species leads to predictions that species will accumulate a wide range of mutations with delayed onset that should be expected to have deleterious effects on virtually all functions. This prediction is not fulfilled for some aspects of cell replication (e.g. erythropoiesis) that show few if any impairments at later ages and for some nondividing cells that appear to maintain biosynthetic responses (myocardium, vasopressin-secreting neurons).

My overall impression is that organismic senescence, when it occurs, does not result from fundamental and general impairments of macromolecular genesis that were initiated after maturation. In some species, the capacity for macromolecular synthesis and cell replacement may become sharply limited during development or before maturation, for example, the nematodes and flies, which lack somatic cell proliferation. If so, then differential changes in gene regulation during development or in the adult phase may be looked to for changes during organismic senescence. As noted throughout Chapter 7, many examples of altered gene expression show a high degree of selectivity with respect to the cell and gene product that is altered and the age when that change occurs. This evidence, while still provisional, supports the hypothesis that the major age-related changes in gene expression are selective and mediated by particular axes of physiological regulation. While there is some level of random genomic damage through various types of point mutations, deletions, and adducts, there is no evidence that these have general importance to cellular changes across the lifespan, with the important exception of malignancy. Because some of the mechanisms that repair DNA are linked to transcription of those sequences, or are subject to other aspects of gene activity, age-related damage to DNA of somatic or germ cells may also be traced to selective changes in gene activity.

The last two chapters have analyzed evidence concerning macromolecular biosynthesis and genomic propagation through somatic and gametogenic cell lineages throughout the lifespan. The next two chapters examine variations in the senescent phenotype that arise during development or in adult life, that are attributable to extrinsic influences in the natural environment, or that arise from experimental manipulations.

9

Developmental Influences on Lifespan and Senescence

9.1. Introduction

This and the next chapter review influences on senescence and longevity that are demonstrated by a range of experimental and environmental manipulations of metabolism and of cell functions. The chosen examples show extensive phenotypic variations in temporal organization during development and adult life that are revealed by environmental influences and experimental interventions. This temporal plasticity pertains to the role of selective regulation of gene activities in the changes of senescence (Chapter 7). In many cases, these phenotypic variations in the scheduling of life history phases are paralleled by species differences that imply close genetic control. The phenotypic variations thus may give important clues to the mechanisms of genotypic variations in life history scheduling.

Three major ways are considered by which the phenotypes of senescence are determined during development, with emphasis on epigenetic and environmental influences. (1) The duration of developmental stages can vary total lifespan in some species. (2) A variety of epigenetic influences on senescence arise during oogenesis or development, including the effects of maternal age and alternative developmental pathways. (3) The developmental determinants of the capacity for cell renewal and regeneration can influence potential longevity (a difficult subject). The wide species differences in longevity and patterns of senescence among multicellular organisms (Chapters 1–4) can be ascribed in many cases to heritable differences that are consequent to specific differentiated cell characteristics, such as the capacity for regeneration, the presence of vital organs, and adult body size.

A general question is traced through this and the next chapter: Do species differences in senescence arise from differences in the *regulation* of gene activities or from differences in enzymatic activities and other protein *functions* that have been hypothesized to be accumulated during the lifespan in somatic cells from mutations and other damage to the genome? Genetic elements crucial for the control of senescence could involve either, or both, coding and noncoding DNA sequences. Trans-acting regulators of transcription, by which the products of one gene influence the activities of distant genes, are being studied intensively and will soon provide powerful new approaches to investigating mechanisms of senescence through manipulating development. For the present, we can only get glimpses of these possibilities, through examples of environmental influences on lifespan and senescence that act before maturation, and in some cases must penetrate to the level of gene regulation. Even at this early stage of understanding how the patterns of senescence are predestined, the evidence clearly shows extensive plasticity in the developmental determinants of senescence. The manipulation of senescence during prematurational stages also gives insight into pacemakers for senescence.

9.2. Influences on Total Lifespan through Variable Durations of Developmental Stages

Some of the most spectacular variations of total lifespan occur through variable durations of developmental (prematurational) stages. By a convention that is generally unstated and unscrutinized, the organismic lifespan is often calculated from birth onwards, to the exclusion of earlier stages. This convention is particularly questionable for the many invertebrates whose larval or juvenile stages last longer than the adult. Among examples from Chapter 2 of prolonged developmental or juvenile phases that last for a decade or more are the periodical cicadas, tarantulas, and bamboos. There is no basis for excluding prematurational stages from discussions of time-dependent processes that could influence senescence or total lifespan. To do so would presume that time-dependent processes in differentiated cells and in mature macromolecules are qualitatively different during earlier than during later stages of the lifespan, for which there is no evidence. Although lifespan can be lengthened from low temperatures or diet restriction during development, many studies continued this regime into adult life, as discussed in Chapter 10.

9.2.1. Diapause and Dormancy

9.2.1.1. Plants and Invertebrates

Diapause and dormancy are phases of slowed growth or metabolism and can occur during development or in adult life in nearly all types of organisms (Andrewartha, 1952; Grossowicz et al., 1961; Krishnakumaran, 1983; Aitken, 1977; de Wilde, 1983). Though both diapause and dormancy can be induced by adverse environments (e.g. developmental arrest in some fish by desiccation, or cold-induced lethargy in some mammals), they differ because dormancy is immediately relieved by restoration of growth-favoring conditions, while diapause may continue beyond. This distinction is sometimes not made in the literature but is not a crucial issue here. The capacity to arrest development or delay maturation is broadly interpreted in life history theory as allowing flexibility in reproductive scheduling, pending more optimum conditions. For many organisms with short life cycles, this means, in effect, that the embryos from one generation can be present and contribute to the gene pool of future generations.

Diapause and dormancy have a bearing on this discussion of longevity and senescence, because organisms usually remain at temperatures where they are subject to thermodynamic and biochemical time-dependent changes, even if total oxidative metabolism is greatly reduced. The efficacy of molecular repair mechanisms during diapause or dormancy is little studied, and would be interesting to consider in the many examples of closely related species with major differences in the presence of diapause or its duration. There are often no obvious constitutional or evolutionary reasons for these species differences.

Diapause has been demonstrated at virtually all stages of development in one or another species. Few species can enter diapause at more than one stage, while others, like humans, do not show the phenomenon. The molecular mechanisms that determine the potential for diapause or dormancy are not known, and may include diverse hormonal and nutritional controls. For perspective, the primary triggers for sexual maturation are also little understood, and might share some control points with diapause. Genetic polymorphisms that influence diapause are indicated in *Drosophila* (Chapter 6.3.2), but their relation to other control points in development is not indicated.

The following present the range of these phenomena, using as examples, where possible, organisms discussed elsewhere in this book. Two general responses are documented: no effects of prolonging the duration of development on total lifespan; and an inverse relation of dormancy duration to subsequent lifespan, which could be a finite metabolic resource of some type.

First we note the vast capacity of some species for total metabolic arrest. The most extreme are seeds of plants of seasonally flowering plants containing an embryo that can remain dormant for decades or centuries. Few of the legendary claims are proven, as Osborne (1980) thoughtfully discusses. The best example with radiocarbon dating is the successful germination of *Canna compacta* after more than 500 years (Lerman and Cigliano, 1971), although initial growth was slow and showed abnormal geotropism (Sivori et al., 1968). In an extensive study of stored seeds of twenty-one local plants, begun by W. J. Beal at Michigan State University in 1879 (Beal, 1905), periodic sampling of *Verbascum blatteria* still germinated normally though with increased slowness up through 90 years, while the others failed to germinate after the fiftieth year (Kivilaan and Bandurski, 1973). In general, seed viability deteriorates during prolonged storage, even under optimum conditions, and is associated with increased chromosomal aberrations (Sax, 1962; Osborne, 1980).

Nonetheless, many plants are seasonally induced to produce fertilized seed containing a dormant embryo that can survive for 10 or more years. Seeds with the greatest capacity for dormancy during prolonged unfavorable conditions of cold, heat, or aridity are thick, are impervious to water and gases, and have well-developed embryos. A great variety of seed structures appear to have evolved to cope with ecological fluctuations. At the other extreme, some seeds have no capacity for dormancy, particularly aquatic species like the red mangrove (*Rhizophora mangle*), which forms seedlings on the mother plant even before the seed is shed, like vivipary in mammals (Osborne, 1980); the red mangrove was mentioned in Chapter 8.5.4 in regard to somatic mutations. Unfortunately, there are no data on whether arrest influences the lifespan in these examples.

Many invertebrates also survive extended dormancy in a complete metabolic arrest, which may be better described as quiescence rather than diapause or dormancy (Andrewartha, 1952). For example, resting eggs of monogonont rotifers can remain encysted without losing viability for at least 30 years (Jennings and Lynch, 1928b), like the seeds of plants as described above. Nematode phytoparasites also may survive decades of desiccation (Hyman, 1951, 405). The tardigradia (water bears) may hold the record (100 years) for revival from the "tuns," encystments that are induced by desiccation. Tardigradia also have extraordinary resistance to temperature extremes ($-270°$ to $150°C$) and high doses of X-rays ($> 500,000$ roentgens; Margulis and Schwartz, 1988), which is pertinent to mechanisms in longevity because it implies great capacity to prevent or repair macromolecular damage.

Next are examples of dormancy or diapause with several influences on the adult phase duration. The best-studied is the soil nematode *Caenorhabditis*. Considerable extensions of the total lifespan (hatching to senescent death) are induced in *C. elegans* by starving free-living larvae at either of three stages (Figure 9.1). Up to 15 days are added to the total lifespan if embryos are hatched into medium lacking food (Johnson et al., 1984). The stage-1 larvae continue to swim actively and maintain pharyngeal pumping, through which food could be ingested. Upon refeeding, the stage-1 larvae resume normal development without changing the adult lifespan. However, there are limits to the duration of the quiescent state. Starvation beyond 2 to 4 weeks causes a progressive increase of mor-

Stages of Caenorhabditis Life Cycle
free–living larvae
embryo   larval stage   adult

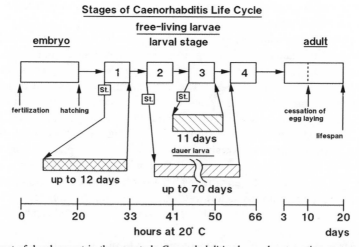

Fig. 9.1. Arrest of development in the nematode *Caenorhabditis elegans* by starvation at any of three larval stages extends the total lifespan by this duration of the development arrest. If eggs are hatched into a medium lacking food (starved, *St*), development is arrested; the starved larvae remain healthy and swim vigorously for up to 10 days. If refed after up to 12 days of starvation, development resumes normally and the adult lifespan is not altered; the total lifespan, counting hatching, is increased by the duration of developmental arrest (Johnson et al., 1984). If starvation is imposed during the next (second) larval stage and before the second molt, an alternative larval state is induced, the dauer larva, which is characterized by formation of an impermeable cuticle and much reduced movement. Dauer larvae are reactivated by exposure to food after starvation for at least up to 70 days (Klass and Hirsh, 1976). Again, the duration of the adult lifespan is unaffected by arrest as dauer larvae, so that the total lifespan of the worm from hatching can be increased at least threefold. Starvation during larval stage 3 also induces dormancy (Johnson et al., 1984). Figure drawn by author in consultation with Thomas Johnson.

phological abnormalities in the adults. Premature death within 2 days of maturity results from starvation of stage-4 larvae, because the self-fertilized eggs continue to develop within and then hatch by rupturing the body wall. For examples of matricidal hatching in other nematodes and invertebrates, see Chapter 2.3.3. The induction of matricidal hatching by starvation in *Caenorhabditis* demonstrates the potential in this species for alternate pathways of death through epigenetic events.

Starvation can also arrest development at the next (second) larval stage to produce the *dauer* (German for "enduring") larva, which is an alternative to the third larval stage (Figure 9.1). High population density also favors dauer formation (Albert et al., 1981). Dauer larvae are semiquiescent, with reduced swimming and oxygen consumption, and cease pharyngeal pumping (note contrast to starvation at stage 1). Dauer larvae also form a tough cuticle that resists detergents (Cassada and Russell, 1975). Contrary to a previous report (Yeargers, 1981), dauer larvae are not particularly resistant to radiation damage and show about the same sensitivity to life-shortening effects of ionizing radiation as the stage-3 larvae they replace (Johnson and Hartman, 1988). Refeeding after up to 70 days relieves the developmental arrest without altering the adult lifespan. In effect, the total lifespan from hatching to senescence can be varied threefold through the formation of dauer larvae. The mortality of refed dauer larvae increases with the duration of starvation, with $\leq 70\%$ surviving dauer stages of 60 days (Cassada and Russell, 1975; Johnson et al., 1984; Anderson, 1978). Nor is the egg production or lifespan of the progeny affected by experiencing the dauer stage (Klass and Hirsh, 1976).

In general, the dauer stage postpones senescence in adults by extending the duration of juvenile stages. This result suggests that major lifespan extension in *Caenorhabditis* occurs without complete suspension of metabolism. Moreover, this example shows that senescence is not linked to *total* lapsed chronological time, but is coupled to the completion of a developmental program. Dauer larvae also are found in *C. briggsae* and other soil nematodes (Bird, 1971; Yarwood and Hansen, 1969). Although less studied, dormancy is induced by starvation during stage 3 (Johnson et al., 1984).

More than twenty mapped mutant genes are known to influence dauer larva (Swanson and Riddle, 1981; Cassada and Russell, 1975; Riddle et al., 1981). Several mutants that did not form dauer larvae proved to have abnormalities in sensory neural ultrastructure and neural functions that impaired chemotaxis and mating (Albert et al., 1981). These results suggest that the triggering of dauer larval formation by starvation or crowding involves chemosensory and/or integrative neuronal mechanisms. This example implicates a neural role in regulating the total lifespan.

Many examples of variation in lifespan through development can be found in insects. Readers may recall the overwintering of worker bees, which extends total lifespan by 6 or more months (Chapter 2.2.3.1). There are numerous other examples. The aphagous moth *Ephestia elutella* (Chapter 2.2.1.1.1) shows geographic variations of larval stages through diapause, which influence total lifespan. In England, most larvae have a prolonged diapause of up to 10 months during the fall, winter, and spring. During diapause, the larvae lose up to 50% of fat and dry weight at the expense of reserves in the fat body, which were acquired by the larvae through their feeding before diapause (Waloff, 1949). Although larvae usually metamorphose after May, precocious metamorphosis can be induced by warming as early as February. Consistent with their adult aphagy, the early hatchers are heavier and have slightly longer (25%) adult life phases (Waloff et al., 1948). The larval diet (maize versus wheat) also influences the adult lifespan (by 50%) and the number of eggs (by 300%). Elsewhere, diapause may be much shorter, permitting up to three generations per year (Waloff, 1949). These striking variations identify nutrition during development as important to adult lifespan. Some insects that are aphagous as adults might have increased adult lifespans, through prolongation of larval stages with active feeding, which could augment nutrients stored for the adult phase.

Crickets and grasshoppers give another major set of examples. Some cricket species, particularly tropical or domestic ones, breed continuously and have rapid development of 2 to 3 months, while other crickets have life cycles that span 2 or more years with extended diapause at the egg, larval, or adult stages (Alexander, 1968). Some species exhibit two annual generations with diapause during the winter only, for example the common field cricket *Gryllus rubens*. These variations account for tenfold differences in total lifespan. This species and other crickets have alternate adult forms, with long or short wings in either sex, which might prove to influence mortality risk and lifespan according to factors of mechanical senescence and exposure to predation. Both genetic and environmental factors are indicated (Walker, 1987).

Other species show geographic differences in the incidence of diapause that are attributed to local variations of temperature and light. Andrewartha (1945, 1952) described races of Australian grasshoppers (e.g. *Austroicetes cruciata*) in which diapause is differentially induced by temperature and which may regulate their population age structuress according to seasonal differences in food availability. The characteristics of diapause vary widely among species, including the fraction of eggs that become diapausal, the temperature that determines if diapause will occur, and the duration of diapause. In place of diapause, *Chortoicetes terminifera* can become dormant in early development under aridity, but rapidly resumes growth upon moistening. There is no information on whether these variations arise from genetic polymorphisms or from variable epigenetic influences during oogenesis or development. Effects on the adult life phase from extending development through dormancy or diapause are not reported. Maternal age can also influence the incidence of diapause, as discussed in Section 9.3.2.1.

Variations in total lifespan through hormonal influences on diapause and dormancy are implied in *Chelonus annulipes*, a braconid wasp that completes its entire development as a parasite of other insects (Bradley and Arbuthnott, 1938). *Chelonus* oviposits into the eggs of the cornborer *Pyraustra nubilis*, which overwinters in diapause in certain strains of its host. Different host strains produce either one (univoltine) or two generations in one summer (bivoltine). In either case, the development of parasite and host are remarkably synchronized, so that the parasite remains in a dormant first larval stage until the host is ready to metamorphose. In diapausing (univoltine) hosts, *Chelonus* has a 12-month life cycle, whereas development takes only 2 months in nondiapausing strains. However, the life cycle can be even shorter, 6 weeks, if *Chelonus* is grown in the laboratory on the moth *Ephestia kühniella* (cited by Andrewartha, 1952). Other examples of synchronized development in parasitic insects suggest control of dormancy by host hormones (Bradley and Arbuthnott, 1938; Andrewartha, 1952).

Diapause in some insects is coordinated with the senescence of the deciduous leaves on which they feed. The caterpillar *Euproctis*, which lives on apple trees, could be induced to develop without diapause if fed on young leaves. However, a diet of older leaves caused diapause in the larvae from these insects that usually spend the winter in diapause (Grison, 1947). This example suggests how seasonally induced senescence of plants and their leaves can regulate lifespan in ecologically associated organisms.

9.2.1.2. *Vertebrates*

Diapause and dormancy also occur in vertebrates. The most detailed studies have been done on the cyprinodont "annual" fish (Chapter 3.4.1.2.2). In a number of these species, seasonal desiccation with erratic rains during the yearly cycle can arrest early development at one or more stages (Figure 9.2; Wourms, 1972, 1973). Diapause can last 18 months (Wourms, 1973), although the norm is about 8 months (Myers, 1952). Under permissive conditions without desiccation, "annual" fish can live for 2 to 3 years. South American species (e.g. *Austrofundulus myersi*, *Cynolebias bellottii*, *Rachovia brevis*) and African species (e.g. *Nothobranchius guentheri*) can arrest development for variable times (as little as 1 day to many months) at early cleavage when the ameboid blastocysts remain dispersed at epiboly just before axiation and organogenesis (diapause I). The duration of diapause I depends on the environment, and ends when favorable conditions permit blastocyst reaggregation. Some species escape diapause I altogether, while subpopulations of species with obligate diapause I may also escape. Diapause II arrests differentiation after

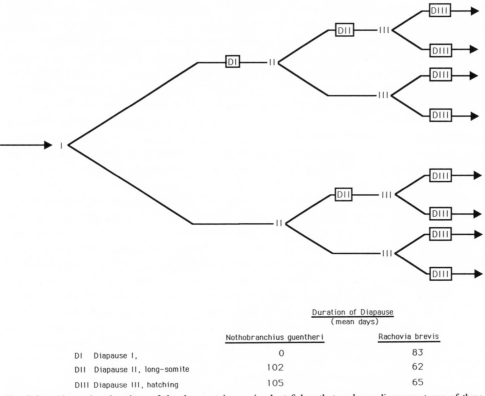

	Duration of Diapause (mean days)	
	Nothobranchius guentheri	Rachovia brevis
DI Diapause I,	0	83
DII Diapause II, long-somite	102	62
DIII Diapause III, hatching	105	65

Fig. 9.2. Alternative durations of development in cyprinodont fishes that undergo diapause at one of three stages before hatching (DI through DIII, see text). The choice points in the South American annuals (*Nothobranchius guentheri* and *Rachovia brevis*) create a wide range of development times from little more than a month to more than 18 months. Redrawn from Wourms, 1973.

formation of the body axis and simple tubular heart, but before the start of blood circulation. Diapause III occurs just before hatching, when organogenesis is nearly complete; metabolism must be nearly completely arrested, since the heart beats only sporadically and yolk reserves are not depleted. Depending on the availability of water, development can take as little as 40 days to more than 500 days. This amounts to tenfold variations of total lifespan.

In nature, *Austrofundulus* and related fish display individual variation in the duration of each diapause (Figure 9.2). Some in the population even escape diapause and develop nearly as fast as nonannual *Fundulus* species, though their blastocysts disperse transiently. Wourms (1973) proposed that this heterogeneity in developmental timing represents bet-hedging to optimize survival of the population during sensitive developmental stages in the face of widely variable rainfall. In contrast, perennial cyprinodonts that inhabit more permanent pools do not have developmental arrest at stages corresponding to diapause I through III. A characteristic of annuals during diapause I is that they have fewer blastocysts at epiboly, as little as 1% of the number in nonannual species. In effect, cleavage is shortened, and epiboly is precocious. How these developmental variations may influence the lifespan or the lengths of diapause in the next generation is unknown.

Prolonged embryonic diapause in mammals is important to theories of senescence through macromolecular damage. The temperatures typical of mammals (37°C) imply mechanisms that combat oxidative and other damage that is time- and temperature-dependent. Diapause at preimplantation stages is reported in eight mammalian orders and is associated with arrested cell division and greatly slowed metabolism (Aitken, 1977; Saidler, 1969). The mechanisms controlling the onset and duration of preimplantation diapause in mammals are poorly understood but are likely to involve neuroendocrine influences transduced from sensory stimulae through nursing or other sociobehavioral cues, as well as through the photoperiod. The following examples again bring into view the issues of how to calculate the lifespan and when during development there may be manifestations of mechanisms that permit extensive plasticity of developmental schedules that are also pertinent to the duration of the adult phase.

The record duration for preimplantation diapause is a startling 2 years in the long-nosed armadillo *Dasypus novemcintus* (Storrs et al., 1988, 1989). Ordinarily, diapause lasts 3 to 4 months in this species, but at this extreme, diapause would add 50% to the reported postnatal lifespan of 4 years. However, 4 years probably underestimates the potential lifespan, since other armadillos live for up to 16 years (Nowak and Paradiso, 1983). Other species with prolonged diapause (8–10 months) are the badger *Meles* (Canivenc and Bonnin-Laffargue, 1963; Canivenc, 1965) and the red kangaroo *Macropus rufus* (Russell, 1974). In these, diapause accounts for a small fraction (ca. 5%) of their maximum lifespans (Nowak and Paradiso, 1983). In addition to preimplantation diapause, the striped skunk *Mephitis* may also have a lethargic winter phase (Nowak and Paradiso, 1983), with unknown effects on the lifespan (which is about 13 years). As discussed in Chapter 10.2.2, hibernation in hamsters adds to their total lifespan. These examples show that the total mammalian lifespan is subject to extensive and variable durations of hypometabolism during early development and the adult phase.

The species distribution of diapause is puzzling and does not always fit the conventional wisdom that embryonic diapause is an adaptation to optimize seasonally the interval between mating and parturition. Diapause is unknown in humans and primates. Embryonic diapause never occurs in a subspecies of the spotted skunk *Spilogale putorius* from eastern North America, while the western subspecies always has a diapause lasting 6 months (Mead, 1968); effects of diapause on the adult lifespan are unknown. In badgers and many other species with obligatory diapause of relatively fixed duration, diapause is thought to be controlled by photoperiod, probably via the pineal gland.

In other species, diapause can vary extensively according to the time of mating or the presence of nurslings. In laboratory mice and rats, for example, mating and fertilization can occur within hours of the birth of a previous litter. If the pups are nursed, diapause will be maintained at least to weaning at about 3 weeks (Kirkham, 1916), though the upper limit is not described. At least in laboratory rodents, delayed implantation development would not add much to the total lifespan. Although there are no recognized ill effects of delayed implantation on development, this question merits specific study, especially in relation to possible interactions with maternal age that might potentiate conditions for fetal aneuploidy. As discussed in Chapter 8.5.1.1, delayed ovulation may be a factor in the maternal age contribution to fetal abnormalities in rodents. Another approach would be to compare lifespans and fetal defects in closely related genotypes that differ in the occurrence of diapause. These phenomena may be important in populations with multigenerational age structures.

9.2.2. Delayed Maturation through Prolonged Development

Total lifespan can also be increased in some species by delaying maturation, independently of any extensions from diapause or other hypometabolic states. The relationships among growth rate, age at maturity, distribution of reproduction during the lifespan, and longevity are a major topic in life history theory. As discussed in Chapter 11, it is hard to support the tenet that earlier maturation is strongly or necessarily linked to more intense reproduction and shorter lifespans (Pianka, 1970). Unfortunately data bearing on these issues are spotty, particularly for mortality statistics. This complex, important subject is thoughtfully reviewed by Stearns (1977, 1984).

A relationship between the cessation of growth and the onset of senescence was hypothesized by Bidder (1932). As discussed in Chapter 4.5.3, Bidder's hypothesis that species with continued growth lack senescence is not universally demonstrated. The converse proposal that cessation of growth leads to senescence is shown by numerous species that die at maturation in association with rapid onset senescence (Chapter 2). Examples include the many annual flowering plants that can survive their usual lifespans after removal of flowers or buds, for example soybeans (Figure 2.16). Triggers for flowering, rapid senescence, and death in other species are photoperiod and climate, for example the century plant, whose lifespan can vary tenfold depending on geography. Examples for animals also include manipulation of lifespan by preventing maturation, for example in some species of octopus and salmon. These are special cases, however, since the pathophysiology that causes senescence is almost certainly sufficient to interrupt growth. Most data indicate that rapid senescence does not arise *because* of failure of cellular capacity for continued growth, but is rather a dysorganizational consequence of hormonal changes associated with maturation. The relation between growth cessation and senescence in iteroparous species, however, is unclear. As discussed in Chapter 3.4.1.2.2, some trout populations that continue to grow because of access to particular food organisms live much longer than those that reached smaller maximum sizes. Thus, continued growth delays or prevents senescence in some species.

9.2.2.1. Invertebrates

The relation of the duration of development to total lifespan has long been noted. One of the first neuroendocrine manipulations of total lifespan was the demonstration by Wigglesworth (1934) that decapitation of the larvae of the bloodsucking bug *Rhodnius prolixus* arrested development; one such nymph survived more than a year. Environmentally caused variations in the duration of development (independent of diapause or dormancy) were studied in insects many decades ago to quantify relationships between lifespan and metabolism. These studies also pertain to the "rate of living" hypothesis (Chapter 5). First we consider diet, then temperature, and finally chemical signals that influence the duration of development.

A classic study by Northrop (1917) on *Drosophila* showed that the larval stages were prolonged twofold by diet restriction, while the adult (imaginal) phase was unchanged. This was the first indication of the independent regulation of the length of larval and adult phases. However, the reported adult lifespans (10–12 days) were much shorter than the 2- to 3-month lifespans of today's laboratory flies. Recent studies show that the environmental temperature during development in *Drosophila* does not influence the adult lifespan, although the growth rate varied twofold and the adult body weight varied by 30% (Economos and Lints, 1986). These studies of food and temperature variations during development both agree on the autonomy of the schedule of development and senescence in flies. Similar conclusions were drawn for the crustacean *Daphnia longispina,* such that starvation during larval stages slowed development without altering the duration of adult lifespan (Ingle et al., 1937).

The duration of development may also have an impact on genetic variations in the population. According to one report (Bergner, 1928), diets prolonging the larval stage increased crossing over in adults. This observation raises an interesting possibility that environmental factors during development may influence genetic variations by altering the amount of recombination. Mechanisms may have evolved to modulate population-level genetic diversity through recombination or mutation, in accord with developmental exigencies.

The periodical cicadas also show variable lengths of development, causing major differences in total lifespan without corresponding influences on the adult phase. In brief, *Magicicada* races have extremely prolonged development, with interfertile broods that take either 13 or 17 years to develop (Chapter 6.3.3). Synchronous emergence of adults within the same 1- to 2-day period for each brood is followed by death within 4 to 6 weeks. In cicadas, the adult phase lasts less than 1% of the total lifespan and is the same in 13- and 17-year broods. The slow growth of *Magicicada* is attributed by Lloyd and White (1987) to the low nutritional content of their diet, which is xylem fluid from underground roots.

Aphagous adult insects present different questions, since the adult metabolic reserves are acquired prior to eclosion (Chapter 2.2.1.1). Contrary to intuition, alternate-day starvation of caterpillars of the moth *Lymantria dispar* did not change the adult lifespan (ca. 1 week), though the larval stage lasted 50% longer (Kopec, 1924).[1] Perhaps the smaller size of the aphagous adults that arose from starved caterpillars compensated for net decrease in nutrient reserves. The fat body itself was smaller, but in proportion to body size.

Some species can arrest development pending a chemical signal, but without hypometabolism from starvation or cold. The nudibranch mollusk *Phestilla sibogae* has a variable larval stage, also without effect on the duration of the adult phase (Miller and Hadfield, 1990). This short-lived laboratory species completes its developmental cycle from egg to egg in as little as 33 days. The adult phase of 45 days is associated with decreased daily egg production preceding death. Metamorphosis of the veliger larvae can be triggered by exposure to a chemical signal from its coral prey at varying times. During this variable hiatus, the larvae cease growth, despite free access to their phytoplankton diet. Metamorphosis can be delayed fivefold longer than its minimum (8–42 days; Kempf and Hadfield,

---

1. This and other papers by Kopec, who worked at the Government Institute for Agriculture Research in Pulawy, Poland, are worth reading because of their careful scholarship, experimental design, and insights on the neuroendocrine controls of insect development, which were to be further established a decade later. In fact, Wigglesworth, in his classic 1934 paper, acknowledged the precedence of Kopec's discoveries on the importance of the brain for molting in *Lymantria.*

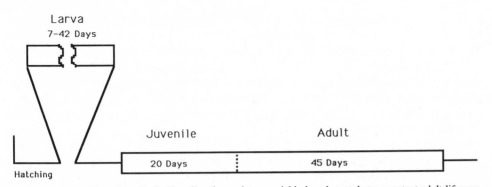

Larva
7–42 Days

Juvenile

Adult

Hatching

20 Days

45 Days

Fig. 9.3. The nudibranch mollusk *Phestilla sibogae* has a variable larval stage but a constant adult lifespan. This short-lived species can complete its life cycle from egg to egg in as little as 33 days in the laboratory, with total lifespans of 77–112 days (Kempf and Hadfield, 1985; Miller and Hadfield, 1990). Hatching of veliger larvae occurs 5 days after fertilization. Depending on exposure to chemoattractant signals from its coral prey, the larvae undergoes metamorphosis at any time from 8 to 42 days after hatching. Metamorphosis can be delayed more than fivefold without impact on length of the adult phase or egg production. Drawn in consultation with M. G. Hadfield.

1985) but without impact on length of the adult phase (Figure 9.3) or egg production. The developmental hiatus of *Phestilla* is thought to allow flexibility in larval duration pending settlement in a favorable habitat that contains the coral prey.

In the snail *Physella virgata virgata,* a common freshwater pulmonate, waterborne chemical cues from a crayfish predator delayed the onset of egg laying until a much larger size was reached; total lifespan was more than doubled, from about 4 months to about 12 months (Crowl and Covich, 1990). Larger snails suffered less mortality from predation by crayfish, which is a size-specific mortality risk in the natural environment. The report gives no information on the mortality rates or on the extent of pathologic lesions or physiological dysfunction in the older snails. Thus, we cannot assess the contribution of senescence to these major differences in lifespan. *Physella*'s short lifespan should favor further laboratory studies that would be valuable for comparisons in another pulmonate, *Lymnaea stagnalis,* that shows marked senescence (Chapter 3.3.6).

9.2.2.2.   Vertebrates

Vertebrates also show many variations in the age of reproduction within a species, but there are few examples in which environmental influences are proven to increase lifespan through slowed development and delayed maturation. The following examples are generally more suggestive than definitive; few vertebrates show the major link of delayed maturation and greater total lifespan that is demonstrable in lower forms.

9.2.2.2.1.   Fish and Frogs

Rigorous studies on the age of maturation in relation to lifespan are being done with the platyfish *Xiphophorus maculatus* (Chapter 3.4.1.2.2). Strong genetic influences on the age of maturation are linked to the sex-linked gene *P.* The alleles *P1–P5* control sexual maturation over a twelvefold range, from 2 to 24 months. Yet the earliest maturing fish with the *P1* allele live slightly longer (Schreibman and Margolis-Nunno, 1989). In contrast, males of *Nothobranchius guentheri* (Section 9.2.1.2) lived 25% longer if they matured later (Markofsky and Perlmutter, 1973); one suspects genetic factors, as for platy-fish. The longer lifespans of later maturing fish may be related to the large metabolic demand and stress associated with reproduction in fish, reproductive drain. For example, Orton (1929) hypothesized that the reproductive drain is greater on larger and older fish. Since ovarian weight and fecundity increase at a disproportionately greater rate than body weight in some species (Gerking, 1959), older fish might have increased mortality at spawning. Strong correlations were found between the age at maturation and subsequent lifespan (maximum reported) in comparisons of twenty-six species (Garrod and Hor-wood, 1984; Roff, 1981). Outliers in this trend were the turbot (*Scophthalmus maximus*) and plaice (*Pleuronectes platessa*), with disproportionately long lifespans by comparison with other species that mature at 5 years (Roff, 1981). Because the maximum observed lifespan may be subject to much variation among cohorts, even in inbred laboratory mice (Hollander et al., 1984; Chapter 1.4), post hoc analyses of literature data should not be pressed too far.

The mosquito fish (*Gambusia affinis*), a viviparous poeciliid, varies in growth rate and lifespan among populations in Hawaii (Chapter 3.4.1.2.2) that are discussed by Stearns (1983a, 1983b) as an example of natural selection for different reproductive schedules. Under standard conditions in the laboratory, six F_1 stocks differed nearly twofold in fecundity, with smaller differences in age at maturity, size at maturity, and the interval between broods. Although the mean lifespan of about 9 months did not differ among stocks, there was a strong correlation between body length at maturity and lifespan.

These results are consistent with Krumholz's (1948) observations on mosquito fish from the midwestern U.S., which showed two patterns of life history, early- and late-reproducing according to whether they overwinter before becoming gravid. In this study, maturation was evaluated by the presence of fertilized eggs, which may be more of an index of hormonal and behavioral receptivity than of anatomical maturation; males are considered promiscuous and limited only by female receptivity. Fish that became gravid during the first summer of life (ca. 1 month) were smaller and had fewer and smaller broods; these disappeared soon after their last clutch was born. Those born later in the summer delayed their reproduction until the second summer; these fish were larger and had twofold more clutches. In mosquito fish the larger females produce more eggs and offspring. Irrespective of the reproductive schedule, all mosquito fish have a sterile postre-productive phase before death, in which the ovary is histologically "senile" and devoid of ova; further studies are needed to evaluate the relative roles of ovarian oocyte depletion (which cannot be assumed from the brief account given) and possible neuroendocrine influences.

The coexistence of fish of the same size and apparent maturity that were gravid or were not reproductive in late summer populations suggests neuroendocrine variables. Although feral populations of *Gambusia* rarely survive more than a year in the temperate zones, they may live 4 to 5 years in aquaria, which Krumholz (1948) attributed to effects of "semistarvation"; the prevention of reproduction may also be important. These studies show that the total lifespan of female mosquito fish is more related to body size than to the age at maturation. The main factor appears to be the smaller stress of reproduction experienced by larger females. These studies also challenge the early view that "reproductive drain" is greater on larger and older fish (Orton, 1929). There are other examples of late-maturing members of a population that have greater reproductive lifespans (Garrod and Horwood, 1984; Roff, 1981).

As in most vertebrates, gender is often associated with differences in the lifespan of fish. A novel study used sex steroids to cause phenotypic sex reversal in rice fish, or medaka (*Oryzias latipes*), that also caused reversal of lifespan differences (Fineman et al., 1974). Just after hatching, genotypic males were fed estrone; these grew to become phenotypic females with fertile ova and had shorter lifespans that were characteristic of untreated females. Converse effects were induced in genotypic females that had male phenotypes after being fed methyltestosterone. Most of the effects on lifespan are attributed to the IMR (equation [1.2]), which is represented by Fineman et al. as a "constant hazard" factor. Because the life expectancies of untreated control females (5 months) and males (11 months) were much less than the several years reported elsewhere (Chapter 3.4.1.2.2), the conditions of this study might not have allowed enough fish to reach ages when the mortality rate acceleration was robust enough to compare experimental groups. In any case, the risk factor of producing fertile ova was clearly shown.

Variations in the duration of development of salmonids merit future consideration as possible lifespan alternates. Atlantic salmon (*Salmo salar*) have an immature river phase of 1 to 3 years that is followed by transformation into smolts in preparation for migration to the sea. The sea phase, when 99% growth occurs, varies 1 to 2 years. Subsequent return to the home stream for spawning is associated with much (not universal!) postspawning mortality (Thorpe et al., 1982; Thorpe et al., 1984; Hane and Robertson, 1959). Local populations have distinct growth and mortality rates that are influenced by parental stocks in hatchery-bred juveniles (Thorpe and Morgan, 1978). Genetic isolation of these home stream populations is indicated by transferrin polymorphisms and chromosomal number.

In further contrast to Pacific salmon that nearly always die after spawning in association with elevated corticosteroids (Chapter 2.3.1.4.2), Atlantic salmon have *decreased* plasma corticosteroids during spawning (Leloup-Hatey, 1964; Hane and Robertson, 1959). The reduction of corticosteroids at a stressful time could very well favor greater survival after spawning. Thorpe and Mitchell (1981) showed an inverse relationship between the age at migration as smolts and the time at sea before spawning; this implies that the duration of development to spawning may be relatively constant. Nonetheless, the duration of the river phase influences adult fecundity and hence could be an aspect of natural history that is subject to strong selection. Fish that spent 3 years in the river before smoltification produced fewer eggs than after 1 or 2 river years (Thorpe et al., 1984). Egg size increased with adult size in all groups, as did the size of the fry at hatching.

Coho salmon (semelparous *Oncorhynchus kisutch*) have variants, jack males that mature 1 year early at 2 years. While coho jacks have the same universal postspawning death as the normative adult Pacific salmon, jacks of *O. tschawytscha* and *O. masou* can survive spawning (Table 2.3). Despite their smaller size and smaller teeth, the precociously mature coho jacks are reproductively competitive, because of 50% lower mortality to maturation (Gross, 1985). Jacks are also "specialized at sneaking" access for females, whereas the larger normative adult males must fight their way past competing males. Breeding studies of the two variants indicate genetic differences between them that influence growth rate (Childs and Law, 1972; Iwamoto et al., 1983). Precocious maturation also occurs in males of other Pacific salmon (Table 2.4). Precocious maturation is rarer in female salmonids; an occurrence in Atlantic salmon (*Salmo salar*) was recently reported (Hindar and Nordland, 1989). Gross (1985) interpreted these alternate growth forms to represent a mixed evolutionary strategy that allowed a single population to produce adults with maturity at different ages; these variations in developmental schedules may level the effects of environmental fluctuations, including population size. Social dominance hierarchies also influence growth rates in salmon. In rainbow trout (*O. mykiss*, conditionally semelparous; Table 2.4), subordinate fish grew more rapidly (Yamagishi, 1962). This is probably an epigenetic influence.

Other salmonid species give evidence of alternative schedules of development and adult lifespan in which both environmental and hereditary factors may interact. The subspecies of brown trout (*Salmo trutta*) include the ferox, which grows to unusual size and may be a sympatric but noninterbreeding variant in the same waters as much shorter lived and smaller browns; these subpopulations are distinguished by very different frequencies of *LDH* polymorphism (Chapter 3.4.1.2.2). Other examples are from arctic char (*Salvelinus alpinus*), a polytypic species with controversial taxonomy (Behnke, 1990a). Over their broad geographic range, coexisting pairs of forms are often found that have different life histories. Smaller freshwater benthic forms grow more slowly, mature earlier, and have shorter lifespans. Coresident and interfertile with the small benthic forms are larger anadromous pelagic char that grow more rapidly and live longer. The variants differ by 2 to 8 years in the age at maturity, twofold in the size (length) at maturity, and severalfold in adult lifespan. Arctic char are typically iteroparous. While the benthic chars are dwarfs in the southern parts of the range and may live only a few years, pelagic and anadromous char mature later, grow larger, and may live 20 to 30 years (Sprules, 1952; Robert Behnke, pers. comm.). These paired forms are suspected to differ in a few polymorphic loci that influence the schedules of development and reproduction (Nordeng, 1983; Behnke, 1990a).

Temperature and food availability probably have ubiquitous influences on the age of maturation in poikilothermic vertebrates. Another example from tetrapod radiations is shown by the age at metamorphosis of bullfrogs (*Rana catesbeiana*). There are tenfold effects of climate, ranging from 3 months after fertilization in Arizona up to 3 years in the Northeast (Collins, 1979). During the winter, growth is continuous but slowed. Nothing is known about the lifespans of bullfrogs that mature at different ages. Because other anurans live at least 13 to 15 years (Chapter 4.2.2), delayed maturation may amount to 20% of the bullfrog lifespan.

9.2.2.2.2.   Mammals

Puberty can be delayed severalfold through restriction of diet and light (Frisch, 1985; Ojeda et al., 1980; Magee et al., 1970; McCay, Ellis, Barnes, Smith, and Sperling, 1939; McCay, Maynard, Sperling, and Barnes, 1939). A smaller variation of menarche is widely observed among different human populations (Tanner, 1962). Because diet restriction can interrupt fertility cycles of mature adults (Frisch, 1985), the absence of fertility cycles does not necessarily prove retarded sex organ development. That is, brain regulatory mechanisms for gonadotropin regulation might mature, albeit slowly, and be masked by the effects of low body fat, which are the strongest body compositional predictor of cyclicity (Frisch, 1985). Diet restriction in adults is discussed in Chapter 10.

No general relationship is established between rate of growth and adult lifespan in mammals. Two approaches have been used, apart from diet restriction: increasing the growth rate by making more food available and by genetic variations in growth rates. A classic study by Widdowson and Kennedy (1962) varied the number of pups being nursed, to form two experimental groups with three pups and more than fifteen pups (small versus large litters), which is equivalent to very early diet restriction. After weaning at 21 days, all rats were given the same diet *ad libitum*. Those from large litters were smaller and had relatively more fat throughout life, although they achieved some catch-up growth. While the mean lifespans were only slightly different, being 15% more for the pups from large litters, the groups had a different pathophysiology of senescence. The larger adults from small litters had a prominent weight loss before death and a higher incidence of kidney disease, particularly in males. Thus, early diet can have a major impact on later adult diseases. The role of *ad libitum* early feeding in adult fat depots is discussed in Section 9.3.3.

Several studies of genetic influences have also failed to establish consistent relationships between growth rates and lifespan. A thorough analysis of growth rates in both sexes of nine inbred Jackson Laboratory mouse strains and six F_1 hybrids showed that the adult body weight, growth rate, and lifespans were positively correlated to lifespan in males, but inversely correlated in females (Ingram and Reynolds, 1982). I also mention a study of mice that were selected for rapid growth (Eklund and Bradford, 1977). The large adults of the rapidly growing strain had short lifespans (12–18 months) but also an earlier incidence of tumors (type not specified); the unselected controls had shorter lifespans than typical of most Jackson Laboratory strains.

Further insight might come from analysis of the age of puberty and the lifespan in humans or other mammals from individual records. However, the historical trends in many populations for earlier puberty (Tanner, 1962) and for greater average lifespan (Figures 1.2, 1.3), both continuing into this century, argue against a fixed duration of adult human (postmaturational) lifespan. Thus, it seems unlikely that several or more years earlier exposure to adult levels of sex steroids has adverse effects on the potential lifespan. The greater rates of growth during childhood are attributed to the reduced impact of childhood infectious disease and improved year-round nutrition (Tanner, 1962). These better conditions could also enhance the immune responses throughout the lifespan.

In summary, the examples discussed indicate a huge plasticity in the temporal scheduling during development that has major influences on the total lifespan. The capacity for extended arrest through diapause at various developmental stages is found scattered throughout most phyla and can be, perhaps too easily, rationalized as adaptive for environmental fluctuations according to standard arguments about reproductive strategies (Chapter 1.5.1). Little is known about the ecological distributions of genotypes that allow environmental variations in the duration of development. One of the best examples is the geographic clines in the inducibility of photoperiodic controlled diapause in *Drosophila lummei* and *D. littoralis*, which were attributed by Lumme and colleagues to polymorphisms of a single gene, *Cdl* (Chapter 6.3.3). In *Caenorhabditis*, experimental mutagenesis of numerous loci can influence diapause, as described in Section 9.2.1.1. These studies have thus identified genes that regulate specific phases in the duration of the lifespan that are under environmental control.

9.3. Developmental Influences on Patterns of Senescence

9.3.1. *Alternate Pathways of Morphogenesis*

Alternate patterns of morphogenesis can have major influences on adult lifespan and patterns of senescence, through gene regulation during development. The present examples of variable developmental phenotypes are germane to the important concept that alternate phenotypes are important in speciation (West-Eberhard, 1986). I suspect that alternate phenotypes in many other species mentioned by West-Eberhard—for example, aphids, salamanders, stickleback fish—will also differ in mortality rates and longevity. The role of lifespan in speciation from alternate phenotypes will be considered in Chapter 11.

The best-understood examples of alternate developmental pathways that influence the adult lifespan are given by social insects. As readers may recall from Chapter 2.2.3.1, social insects have remarkable differences in lifespan among castes. The same nuclear genome can be programmed to alternate developmental pathways through the diet. These examples indicate a general phenomenon among the many thousand species of ants, bees, termites, and wasps (Wilson, 1971). Parasitic wasps show another type of phenotypic polymorphism in morphology and lifespan, which are nutritionally determined during oogenesis. These are clearly different phenomena from the effects of maternal age on the incidence of diapause in the progeny (Section 9.3.2.1), though in both cases the presence of diapause considerably lengthens the life cycle. Another example is the alternate life histories of the flatworm *Polystoma integerrimum* (Chapter 2.2.3.3), in which major morphological differences and total lifespans may arise from alternate developmental pathways of the same genome.

Male rotifers represent another variation through anatomic differences. Their degenerate alimentary tracts probably are the major cause of their 80% shorter adult phases and accelerated mortality (shorter MRDTs; Table 1.1). Males are rare or never seen in many species and are generally thought to be haploid (Gilbert, 1983b; Bell, 1982). The gender differences in lifespan can be attributed to different patterns of gene expression in the same set of genes that, when diploid, yield females. In *Euchlanis*, males are produced mainly by young rotifers and pediaclones (Lansing, 1947), while crowding favors the production of males in other species (Bell, 1982). Males develop from smaller eggs than those yielding females; the stomach begins to degenerate later in embryogenesis (Gilbert, 1983b). Thus, the greater production of males by young *Euchlanis* cannot be attributed to insufficient yolk production by the vitellarium. As described in Section 9.3.2.2, young rotifers generally have larger eggs than do older ones.

Mammals commonly manifest major gender differences in age-related diseases and smaller but important differences in lifespan (Chapter 6.5). These gender differences may also be considered as alternate pathways of morphogenesis, because of the strong evidence that gender phenotypes in mammals are interconvertible through alternate programming of somatic cells by exposure to testicular hormones during development, irrespective of the sex chromosome karyotype (Jost, 1953; George and Wilson, 1988). In placental mammals, the testis-determining genes on the Y chromosome determine the gender (Eicher and Washburn, 1986; George and Wilson, 1988). In the absence of these genes or in the absence of two hormones, Müllerian inhibiting factor and testosterone, which are secreted by the testes at critical stages in development, the embryo acquires a female phenotype. Besides the testis-determining genes, nineteen other loci are known to influence sex differentiation in humans (Wilson and Goldstein, 1975).

It seems reasonable to hypothesize that many gender differences in patterns of senescence could also be altered by exposure to sex steroids during development. The sensitivity of bone remodelling in women to estrogen deficits which accelerate osteoporosis (Chapter 3.6.5), for example, contrast with the lesser role of estrogens in bone metabolism of men. This gender difference has been little noted, and it is unknown whether it arises during development or later. The gender differences in the distribution of body fat on the hips versus waist are also notable and are tied to different risks of vascular disease (Chapter 3.6.7; Shimokata et al., 1989). There are gender differences in age-related neuroendocrine functions in mice, however, that arise during development (Section 9.3.4.1). Yet other gender differences in age-related changes may be evoked in adults by hormonal manipulations, for example the induction of male-pattern baldness in women by androgens (Chapter 10.4.2).

9.3.2. Influences from Maternal Age

This section discusses influences of maternal age on the phenotypes of the next generation: transgenerational effects of maternal age that have been defined only for the next generation, and transgenerational effects that are cumulative across generations, the famous "Lansing effect."

9.3.2.1. Effects on the Next Generation

Most information on how maternal age influences the development of invertebrates comes from flies and wasps. Adult flies that hatched from the eggs of older mothers had different total numbers of chaetae (sternopleural bristles; Durrant, 1955) and ratios of chaetae (asymmetry) between the left and right sides (Parsons, 1962). The wing size is also a complex (cyclic) function of maternal age, which may indicate maternal age effects on the rate of cell proliferation in the wing anlage (Delcour, 1968; Delcour and Heuts, 1968). Offspring from older flies were more fecund (Lints and Hoste, 1977). These authors cite reports of other insects with parental age effects on the fecundity of their offspring.

There is also an intriguing report of maternal age effects on the penetrance of the recessive homeotic mutation bicaudal. Within a week of birth, the incidence of the bicaudal phenotype decreases by 70%, while normal embryos increase tenfold (calculated from the total egg production; Nusslein-Volhard, 1977). Temperature is also crucial, so that bicaudal is expressed only at $\geq 25°C$. Because bicaudal is transiently observed for just 1 day in older females shifted to higher temperature, the temperature-sensitive period appears to end during oogenesis before eggs are shed. Starvation of older females at 28°C, followed by refeeding, also transiently reinduces bicaudal. These transient changes again show the plasticity of maternal age-related phenomena. In view of the successful curing of some other homeotic mutants by injections of poly(A)RNA (Anderson and Nusslein-Volhard, 1984), the molecules responsible for these differential effects on the embryo genome could be identified.

Several wasps show maternal age influences on the incidence of diapause and the duration of larval stages. The best-studied is the chalcid *Melittobia chalybii*, which parasitizes larvae of flies and other wasps and which produces alternative morphologic forms, with or without a diapause (Schmieder, 1933, 1939). Adult females have immobile, dwarfed wings and abnormalities in the differentiation of segments, which suggest variations of homeotic gene expression. The alternate forms have an accelerated development (2 weeks versus 13) and shortened lifespan (1 week versus 10) that permit an additional generation concurrent with the longer life cycle of the typical morphology (type form) in the same individual host. The alternate forms are produced by the first batch of eggs laid by the parasite, whereas later eggs produce diapausing, longer-lived adults of the type form. In contrast to the normative female type, alternate-form adult females usually do not feed and have an exhausted appearance within a week. However, occasional alternate-form females can feed and continue to oviposit up to 4 weeks, showing that the aphagy is not obligatory.

Schmieder (1933) showed that the production of alternate forms is unrelated to maternal age at oviposition and can be extensively manipulated. If type-form females were transferred before oviposition to used hosts that had already yielded a generation of the alternate form, all offspring were the longer-lived type form. Conversely, transfer of type-form females after initial oviposition to a fresh host induced formation of the shorter-lived variant. Diapause apparently did not occur if alternate-form larvae were transferred to a used host, though larval transfers produced morphologic intergrades with intermediate durations of development; the lifespans were not stated. Access to hosts' blood by the first parasites to feed was proposed by Schmieder (1939) as a determinant of the alternative form, which would therefore have programmed these differences during oogenesis.

Diapausal larvae of the parasitic wasps *Spalangia drosophila* and *Cryptus inortus* become severalfold more frequent at later maternal ages (4 weeks), when there are also sharp decreases in egg production (Simmonds, 1948). Another parasitic wasp, *Nasonia vitripennis,* showed exponential increases with maternal age in the incidence of diapause in offspring from some outbred individuals but not from others (Saunders, 1962b). This heterogeneity helps explain the failure of other researchers to find maternal age effects in this species (Schneiderman and Horwitz, 1958). Other species in which a variable incidence of diapause is determined during oogenesis include the ichneumonid wasp *Sphecophaga burra* (Schmieder, 1939), the dipteran *Lucilia sericata* (Cragg and Cole, 1952), and the silkworm *Bombyx mori* (Fukuda, 1951; Andrewartha, 1952). These maternal age effects might be adaptive, since diapause of late-born larvae and adults is common at the end of the growing season in temperate zones, as shown in bees and moths (Chapter 2). Other maternal age effects on the duration of development are distinct from diapause. In the mealworm *Tenebrio molitor,* older parents (10 weeks of the 12- to 14-week adult stage) had offspring with 30% shorter larval periods and 10% fewer molts; these larvae also grew markedly faster, whereas the number of eggs and their viability decreased sharply (Ludwig and Fiore, 1960, 1961; Tracey, 1958).

A few observations are available for the laboratory nematode. Offspring from senescent *Caenorhabditis* produced slightly fewer (15%) sperm, while the fecundity of the offspring decreased more sharply (Beguet and Brun, 1972). Although no direct observations of egg characteristics were reported, the reduced fecundity of eggs from old worms appears due to their eggs rather than sperm, because crossing senescent worms with young males of the original strain did not restore viability of the later eggs; under these conditions, spermatozoa from senescent F_1 parents are thought to compete with sperm from the younger worm. Offspring from senescent worms reached sexual maturity slightly earlier (6 hours). Together, these findings imply that changes in the composition of the eggs from older parents could be investigated with molecular techniques.

In mice, there is suggestive but limited evidence for maternal age effects on lifespan. A 2-decade-long study of litter seriation effects was made of the tumor incidence in offspring from mice of different ages (Strong, 1951, 1954, 1968). At each generation, the litters were inbred to produce age-group-specific lines. These reports give some data about life-spans of mice born from old versus young mothers. The lifespans of offspring from inbred lines of young (< 100-day) or older (501–600-day) parents had slightly shorter mean lifespans than from intermediate ages (Strong and Johnson, 1962; Strong, 1968). In the < 100-day line, males died earlier by 24 days and females by 65 days. Offspring from mothers in the age group 201–300 days lived longest. The shorter lifespan of the females in the < 100-day line was associated with more mammary tumors, while leukemias and lung tumors were more common in later parental age lines. Unfortunately, the data are not given in much detail. The different lines according to maternal age maintained the same age at first litter during fifty-four generations of inbreeding (Strong and Johnson, 1965). These phenomena could involve age effects on the expression of MMTV and other retro-viral genes implicated in abnormal growth (Chapter 7.4.1.4). The age-related pattern of retroviral gene expression in the lines from this hybrid are unknown. Maternal age could influence viral transmissibility, with effects on age-related diseases and longevity.

After marching through this scattered literature, we now reach firmer ground with the well-documented maternal age effects on chromosomal abnormalities in humans, partic-ularly the accelerated increase at later maternal ages of Down's and Turner's syndromes (details in Chapters 6.5.3, 8.5.1). These developmental abnormalities arise through so-matic cell aneuploidy and are thus different, so far as has been studied, from the above examples.[2] The shortened lifespan of Down's syndrome offspring of older mothers is as-sociated with an accelerated onset of Alzheimer's disease–like neuropathology in associa-tion with gene dose effects from the whole or segments of chromosome 21 (Chapter 6.5.3.3). It is unknown whether physical features of Down's vary with maternal age. Here we see a concurrence of the two syndromes with accelerated reproductive senescence. As described in Chapter 8.5.1.1, offspring from the occasionally fertile women with Turner's syndrome show an increased incidence of Down's syndrome, presumably through their reduced ovarian oocyte pool.

An area for future study is the possible enhancement of adult health and longevity from the greater experience acquired with increasing maternal age, as shown by field studies. In gulls and other long-lived birds with extended reproductive schedules (Chapter 3.5), younger parents are less successful in rearing clutches to maturity. Similar trends are re-ported for primiparous monkeys, whose offspring have higher mortality rates up through maturity (Robert Sapolsky, pers. comm.). Data to pursue questions on the effects from maternal age on characteristics of the offspring in later adult life may be found in the archives of field studies that have been ongoing for several decades.

9.3.2.2. *Transmissible and Cumulative Effects (Lansing Effect)*

Transmissible maternal age effects on senescence that accumulate with successive gen-erations are indicated in rotifers and a few other species. These effects are variable, pos-sibly because of subtle nutritional deficits, as indicated by studies on the water flea. Trans-

---

2. Maternal age effects on fetal aneuploidy are presently, and are likely to remain, important in industrialized nations, which provide extensive health care to infants with congenital abnormalities. Two other factors may contribute to an increase of aneuploid subgroups: the delay of childbearing until later maternal ages in some socioeconomic groups and the politics of the antiabortion movements. Taken together, these factors could pro-gressively increase the size of the adult subpopulation with Down's syndrome that suffer further with an inevi-table development of Alzheimer's disease (Chapter 6.5.3.3).

missible influences of parental age on the characteristics of the offspring were discovered in rotifers by Albert Lansing, who established clones from the offspring of mothers at different ages, such that each line was derived in successive generations from mothers of the same age. Geriaclones, derived from successive generations of older offspring soon died out, while pediaclones from successive generations of younger mothers had greater viability. This procedure is most straightforward with parthenogenetic-like rotifers, but is also used with sexual reproducers.

In the rotifers *Philodina citrina* and *Euchlanis dilatata* obtained from mothers of different ages, geriaclones had progressively shorter lifespans and became extinct (Figure 9.4; Lansing 1947, 1948; reviews in Lansing, 1964, 1956). Lints and Hoste (1974, 1977) provide valuable discussion of Lansing's three major experiments and point out the peculiarity that none of Lansing's age-selected clones had the unlimited viability of the parental rotifer stocks, but those from the extreme ages died out earlier. However, varying the maternal age in successive generations protected against clonal extinction. In all lines, fecundity decreases sharply by the final generation. A crucial age-related transition occurred after day 5, when *P. citrina* reaches maximum body size (Lansing, 1947). By the fifteenth generation of clones obtained from 7-day-old mothers, the adults laid no eggs and had lifespans of 8 days, versus 23 days in the unselected line. In another striking finding, the reduced fecundity and lifespan of geriaclones could be reversed by starting a pediaclone (Figure 9.5). In extensive experiments with different diets, the Lansing effect was confirmed for *E. dilatata* (King, 1967) but not for *Philodina acutocornis odiosa* (Meadow and Barrows, 1971a).

The composition of the egg probably has a major role in this fascinating phenomenon, as pointed out by Meadow and Barrows (1971b). The eggs produced by senescent rotifers vary widely in size and viability (Jennings and Lynch, 1928b). Moreover, Lansing (1947, 1954) noted that eggs of pediaclones were paler and less dense than those of geriaclones,

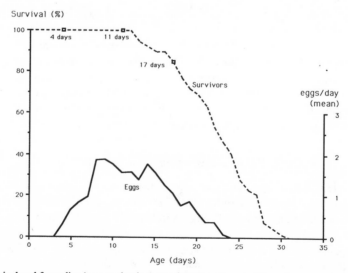

Fig. 9.4. Survival and fecundity (egg production) as a function of age in the parthenogenetic rotifer *Philodina citrina*. The ages indicated on the survival curve were used for selecting mothers of the same age. Redrawn from Lansing, 1947.

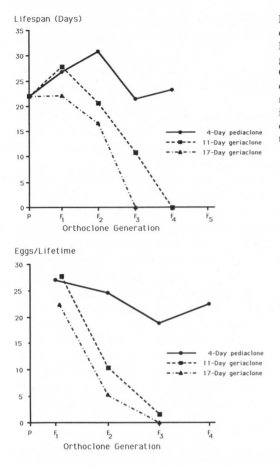

Lifespan (Days)

Orthoclone Generation

- 4-Day pediaclone
- 11-Day geriaclone
- 17-Day geriaclone

Eggs/Lifetime

Orthoclone Generation

- 4-Day pediaclone
- 11-Day geriaclone
- 17-Day geriaclone

Fig. 9.5. *Top,* selection for phenotypes of senescence in rotifers on the basis of maternal age effects. Mean lifespan in clones derived from successive generations (F_1–F_4) of mothers at the same age. Clones from young (4-day) mothers at peak fecundity are pediaclones; those from 11- or 17-day-old mothers at mid-life or later are geriaclones that rapidly loose viability. *Bottom,* total lifetime egg production in clones derived from successive generations. Data graphed from tables in Lansing, 1947.

which were a dense yellow. The eggs of old *Proales sordida* are also darker as well as larger, though no major effect on lifespan was seen (Jennings and Lynch, 1928a). These observations suggest that specific cytoplasmic determinants could be found for senescence in rotifers, which might be ultimately traced to the vitellarium, a syncytium of nurse cells that form and extrude yolk into the developing oocytes. Prominent age-related structural and histochemical changes are reported in the vitellarium of *Philodina acuticornis odiosa,* including reduction of rough endoplasmic reticulum (Chapter 8.5.3.2; Herold and Meadow, 1970). Ribosomes also decrease in stomach cells (Lansing, 1964). Rotifers give a model for the relationship between specific cytoplasmic determinants during oogenesis and the epigenetic control of senescence.

Lansing effects are also seen in the water flea *Moina macrocopa,* a small (2–3 mm) self-fertilizing crustacean with an unusually short, 10- to 18-day lifespan, which can be maintained in monoxenic culture (Murphy, 1970). About one hundred eggs are produced parthenogenetically in five clutches. The fourth and fifth geriaclones derived at the approach to the lifespan rapidly died out; shortened lifespans were most clearly seen in the fourth geriaclone (Murphy and Davidoff, 1972). Besides age, nutritional factors can be crucial in the survival of orthoclones, since the addition of liver extracts permitted the survival of fourth and fifth geriaclones and increased the number of offspring. Lifespan, however, was not altered. These studies suggest that nutritional deficiencies could underlie the "transmissibility" of lifespan shortening in the rotifer geriaclones. These nutritional effects must be subtle because of the vigorous reproduction of young organisms; nonetheless, the liver supplements in *Moina* caused reproduction to start even earlier than with conventional media.[3]

In flies, the influence of parental age on the lifespan of progeny has not been conclusively shown. Divergence among studies may be subject to the conditions of husbandry, the species and genotypes, and the particular age used to select the pediaclones and geriaclones (Mayer and Baker, 1985; Lints and Hoste, 1974, 1977; Comfort, 1979). In *Drosophila pseudoobscura* (Comfort, 1953; Wattiaux, 1968) and *Musca domestica* (Rockstein, 1959) several attempts failed to found geriaclones; there was no progressive shortening of lifespans with late offspring. Lints and Hoste (1974, 1977) showed progressive parental age effects on lifespan in several hybrid lines of *D. melanogaster*, although progressive lifespan shortening occurred in lines selected from both young and old parents. Yet lifespan spontaneously lengthened in later generations in parents from either age. This reversibility may arise through selection of genetic polymorphisms during development. The lines were derived from hybrid stocks started a few generations before the studies, which may have disrupted balanced polymorphisms (Rose, 1983).

Nonetheless, several aspects of Lansing's observations on rotifers were corroborated in flies, including the greater fecundity, viability, and lifespans in pediaclones and the reversibility of these changes. The reduced RNA content of eggs from older flies (Chapter 8.5.3.1) could contribute to these phenomena. Another result is of interest for later discussion of the evolutionary changes in scheduling (Chapter 11.4.2): the age of maximum egg production became progressively earlier in pediaclones and later in geriaclones (Lints and Hoste, 1977).

Nematodes are attractive as models for studies of Lansing effects. Because they can reproduce by self-fertilization as well as by mating, one can study these effects in eggs or sperm from different ages. So far, there is no evidence for Lansing effect in regard to fecundity. Pediaclones from 5-day-old *Caenorhabditis elegans* initially had greater fecundity than geriaclones from 8-day-old worms, but these differences were unstable and vanished at later generations (Beguet, 1972). Unpublished studies by Thomas Johnson (pers. comm.) did not demonstrate Lansing effect in *C. elegans*.

9.3.3.   Variation in Organ Size and Cell Populations

Developmental variations in the size of coronary vessels could be important factors in the impact of coronary arteriosclerosis. In subpopulations with high adult incidence of cardiovascular-associated mortality in Finland and Israel, autopsies of children indicated

---

3. These painstakingly constructed clones were destroyed and the study tragically terminated by exposure to traces of Dursban being used to exterminate cockroaches at the Rockefeller University (James S. Murphy, pers. comm.). The sensitivity of *Moina* cultures to insecticides was known (Murphy, 1970) but could not in practicality be absolutely protected against. Similar accidents that in principle are preventable by better maintenance have destroyed years of work in other animal populations. Our gerontologic mouse colony at the University of Southern California was destroyed on August 29, 1980, from exposure to > 45°C for 3 hours, due to a malfunctioning valve in the vivarium temperature controls.

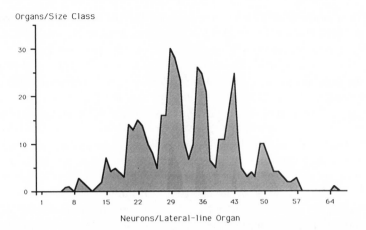

Fig. 9.6. Neuron number varies tenfold among lateral-line organs within an individual in the frog *Xenopus*. This variation is ascribed to statistical fluctuations during clonal descent, such that there is a fixed probability at each division that a neuron will remain in a lineage. Redrawn from Winklbauer and Hausen, 1983.

coronary arteries with thicker walls and more musculoelastic tissue (Chapter 6.5.3.1). These variations could reflect environmental as well as genetic factors and raise the possibility of other developmental variations that could alter responses to disease in adults.

Developmental variation in the size of cell populations is another potential influence on the outcome of senescence. Elegant studies of neuronogenesis in frogs show that the variations in the number of differentiated neurons in Rohon-Beard and lateral-line cells are consistent with a random process during clonal descent, in which there is a fixed probability of a given cell remaining in the lineage (Jacobson, 1985a, 1985b; Jacobson and Moody, 1984; Winklbauer and Hausen, 1983). The final number of cells varies according to the binomial distribution, as shown in different portions of the lateral-line system in the same frog, which can vary tenfold in the number of neurons (Figure 9.6).

The fivefold range of ovarian oocytes within any of several mouse strains at the same age (Jones and Krohn, 1961b; Gosden et al., 1983) is astonishing at first consideration, given their inbred status. Similar variations are seen in oocytes of adult lampreys (Hardisty and Cosh, 1966). These variations could arise from stochastic variation during clonal descent. Another mechanism could be imprecision of primordial germ cell migration. Ectopic germ cells are seen in the adrenal and elsewhere, but apparently disappear before puberty (Upadhyay and Zamboni, 1982). Imprecisions of migration could also be important in variations among adult brains.

The variation of neuron number among individuals is unknown in mammals. Data would be hard to obtain, even in inbred rodents, since the boundaries of neuronal groups often differ among individuals. For example, neurobiologists widely recognize that the stereotaxic atlases used for location of electrodes or lesions only give an approximation of boundaries. Moreover, an analysis of the brains of adult twins by a new two-dimensional mapping procedure shows considerable differences in the surface areas of cortical gyri. Identical twins differed in gyral areas and in the asymmetry between the left and right sides, in the range of 5 to 40% (Jouandet et al., 1989). There is as yet no analysis of neuron density or neuron number per gyrus in the brains of identical twins. The extensive differences in gyral surface areas could arise during development from different neuron or glial numbers and/or different elaboration of dendrites and other neuronal processes. Information on variations in neuron number would be valuable as a means to estimate the factors of safety in neuronal populations (Figure 5.14; Meltzer, 1906).

There is, however, information on developmental variations in neuronal branching patterns. The optic ganglion of the water flea *Daphnia magna* has developmentally determinate numbers of neurons (unlike the above examples). Nonetheless, the branching patterns of dendrites from the corresponding cells of genetically identical individuals from the same clone vary widely (Levinthal et al., 1975); the stability of these patterns over time is unknown, however. These variations in cell number and dendritic branching could be crucial in determining the impact of neuronal dysfunction or neuronal loss on brain functions. Another type of variation could arise through random aspects of neuronal connectivity. The extent of variations in immune stem cell numbers could also be important in the outcome of age-related changes, especially for autoregulatory functions.

Adult adipose depots are also subject to early postnatal influences from food availability. Varying access to milk before weaning influenced the numbers of adult adipose cells, so that milk-restricted rats had fewer fat cells as adults (Knittle and Hirsch, 1968). These effects can be initiated at ages after weaning, moreover. Similarly, rats from very large litters have less adult fat (Widdowson and Kennedy, 1962; Section 9.2.2.2.2). Moreover, diet restriction of rats between 6 weeks (early puberty) and 6 months followed by *ad libitum* feeding markedly decreased the numbers of adipose cells by 30% and their size by 20% (Bertrand and Masoro, 1977). These studies demonstrate that diet during early postnatal development can program adults for varying extents of adiposity.

9.3.4. *Influences on Neuroendocrine Functions in Adult Mammals*

9.3.4.1. *Intrauterine Influences on Reproductive Senescence in Mice*

The uterine environment has major influences on reproductive senescence in female mice, among other aspects of adult reproductive functions. The sex of the neighboring fetus determines gradations in hypothalamic regional and reproductive organ size, fertility, and behavior (vom Saal and Bronson, 1980; vom Saal et al., 1984; Rines and vom Saal, 1984; Kinsley, Miele, Konen, Ghiraldi, Broida, and Svare, 1986; Kinsley, Miele, Konen, Ghiraldi, and Svare, 1986; vom Saal and Finch, 1988). These shadings of gender occur in both males and females and arise from variations of testosterone and perhaps other hormones. For example, in CF–1 mice from an outbred line, females flanked *in utero* by males (2M females, MFM)) are more aggressive as young adults than females flanked by females (0M females, FFF) (vom Saal and Bronson, 1980). The 0M females become infertile sooner and have smaller last litters than the 2M females (Figure 9.7; vom Saal and Moyer, 1985). The mechanisms of this earlier cessation of reproduction in 0M females appear to involve delayed parturition rather than a failure of ovulation (Fred vom Saal, pers. comm.). We showed that the incidence of stillborn pups increases strikingly with delayed parturition in middle-aged female mice (Holinka et al., 1979a, 1979b; Chapter 8.5.1.1).

Fetal neighbor effects extend to adult age changes in behavioral responses to steroids. The male behaviors of aggression and mounting, as induced in female rodents by testosterone treatment, become more intense during middle age in 0M CF–1 mice, but not in

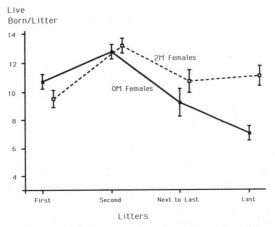

Fig. 9.7. Uterine position effects on adult phenotypes in rodents; the number of live-born pups as a function of litter number in mice classified according to their intrauterine neighbors. *In utero,* 2M females were flanked by males (MFM), while 0M females were flanked by females (FFF). The 0M females had earlier loss of fertility after fewer litters and had fewer pups that survived to birth in their last litters. Redrawn from vom Saal and Moyer, 1985.

2M females (Rines and vom Saal, 1984). This suggests that 0M females become progressively more masculinized from endogenous steroids during adult life and that the prenatal exposure to steroids determines a different outcome to subsequent steroid exposure. Effects of steroid exposure during adult life on the senescent phenotype are discussed in Chapter 10.4, 10.8.

Two other findings from vom Saal's group have potential impact on adult aspects of senescence. If pregnant females are subjected to brief restraint stress, the anatomical and behavioral differences of all females resembled the 2M (vom Saal et al., 1990). It is not yet known if the later onset differences in reproduction and hormonal sensitivity according to uterine position are also suppressed. Another recent finding concerns the short-lived NZB/NZY hybrid that develops early onset autoimmune disease (Chapter 6.5.1.7.3). Implants of testosterone at day 12 during pregnancy increased the lifespan by 25% beyond the 9.5-month mean lifespans of controls (Fred vom Saal et al., unpublished). This result implicates developmental effects of gonadal steroids on immune dysfunctions that have a postmaturational onset.

9.3.4.2. Neonatal Influences on Hippocampal Senescence in Rats

The handling of neonatal laboratory rats strongly influences adult responses to stress (Levine and Mullins, 1966; Meaney et al., 1987). Neonates that were merely put in an unfamiliar cage for 30 minutes each day for 20 days had modified stress responses as adults, as shown by smaller blood elevations of corticosterone and quicker return of corticosteroids to baseline after release from restraint stress (Meaney et al., 1988). A likely cause of this reduced response is the larger number of receptors for corticosterone in the large pyramidal neurons of the hippocampus (Meaney and Aitken, 1985), a brain region that influences the sensitivity of negative feedback to corticosteroids (Keller-Wood and Dallman, 1984). Damage to the pyramidal neurons or interruption of their output to the

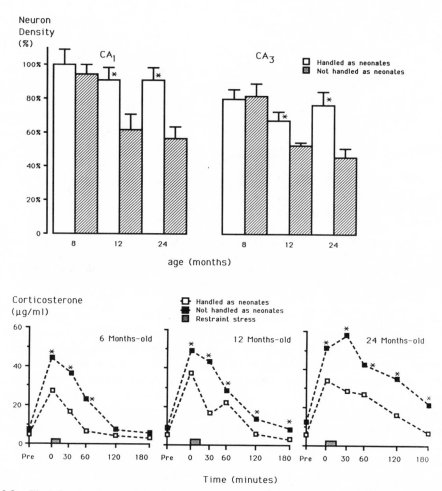

Fig. 9.8. The influence of neonatal handling in male rats on age-related changes in hippocampal-related functions. *Top,* age-related changes in the density of pyramidal neurons in the CA_1 and CA_3 regions of the hippocampus, showing that handled rats had less neuron loss in both regions. These neurons contain very high concentrations of corticosterone receptors, which may be a factor in the smaller age-related increases of plasma corticosterone. *Bottom,* plasma corticosterone before (*pre*), during, and after 15-minute restraint stress in rats at three adult ages, corresponding to maturity (6 months), middle age (12 months), and early senescence (24 months). The maximum elevations of plasma corticosterone are smaller and the return to basal levels is faster in neonatally handled rats at all ages. The handled rats show smaller responses to stress, earlier recovery, and fewer effects of age, which is consistent with their smaller age-related loss of pyramidal neurons. *, significant difference between handled and control rats of the same age. Redrawn from Meany et al., 1988.

hypothalamus through the fornix increases the secretion of corticosteroids and reduces the negative feedback sensitivity. The additional corticosteroid receptors of handled rats is thus consistent with increased sensitivity, since the basal levels of corticosteroids should then have partially activated the negative feedback inhibition on ACTH and therefore should reduce maximum elevations. Similarly, after stress, the greater sensitivity to negative feedback by corticosteroids would lead to faster shutoff of ACTH.

The consequences of neonatal handling extend to senescence, since handled rats suffer fewer age-related decrements in hippocampal-related functions, including less loss of pyramidal neurons, milder age changes in stress responses with earlier return to basal corticosterone, and better preservation of spatial learning tasks (Figure 9.8; Meaney et al., 1988; Meaney et al., 1990). One cause of these age changes is exposure to endogenous corticosteroids, since the changes are accelerated in young rats by exposure to chronic elevations of corticosteroids and are retarded by adrenalectomy (see Chapter 10.5). This phenomenon and the fetal neighbor effects described in Section 9.3.4.1 imply that many features of brain senescence may be modified by perinatal influences. So far rodents are the only species for which these phenomena are described.

9.3.4.3.   Hot Flushes and Estrogen

Hot flushes, a transient vasodilation in skin of the upper body, are a common consequence of diminished estrogen secretions at menopause (Chapter 3.6.4.1). The capacity for hot flushes is acquired in women as an aspect of maturation, and thus can be discussed as an age-related change that is developmentally determined. The basis for hot flushes in adult women is the response of hypothalamic thermoregulatory centers to deficits of estrogens, and flushes soon disappear with estrogen-replacement therapy. Their absence in adult men despite menopause-equivalent estrogen levels can be ascribed to insufficient exposure to estrogens during puberty, since adult men treated with estrogens for prostatic cancer will develop hot flushes on withdrawal from estrogen (Ginsberg and O'Reilly, 1983). Similarly, adult Turner's syndrome patients with severe gonadal dysgenesis who were treated with estrogens have hot flushes on estrogen withdrawal (Yen, 1977). These reports suggest that the human nervous system has loci, presumably in the hypothalamus, that can acquire an estrogen requirement at some time during maturation or perhaps before and that this susceptibility to estrogen dependency is maintained throughout most of adult life, irrespective of chromosomal gender. Thus, adult men can acquire a characteristic change of menopause, which itself is normally acquired during maturation in girls.

9.3.5.   Childhood Disease and Adult Mortality in Humans

Hardin Jones' pioneering analyses of diseases and mortality rates in various human populations (Figures 1.2, 1.5) also identified a relationship between childhood mortality rates and adult mortality rates (Jones, 1956, 1959). Cohorts born between 1850 and 1903 (in the U.S. and four northern European countries) showed linear relations between mortality rates at ages 5 and 50 (Figure 9.9). Even though childhood mortality rate had decreased typically by 75%, middle-aged men and women in the same environment with the lowered childhood mortality retained much higher adult mortality rates. Moreover, there is no evidence that the high childhood mortality selected out a subgroup of individuals who would have been weak as adults, since the entire surviving cohort appears to be weaker throughout its adult life. Jones (1956, 298) proposed the important hypothesis that childhood diseases inflict an increment of "physiological aging" that places an adult at greater risk for a wide range of adult age-related diseases, including cardiovascular diseases and cancer. That is, "every disease episode does some damage to physiological function." The reductions of childhood diseases would then reduce the risk of these diseases and postpone their contribution to mortality rates. Although the relationships of

Death Rate, Age 5 (per 1000)

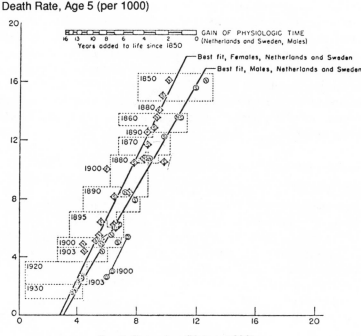

Death Rate, Age 50 (per 1000)

Fig. 9.9. Relationships between mortality in children aged 5 years and in the same cohort as adults aged 50. The cohorts are displayed by country. Jones (1956) hypothesized that childhood infectious diseases accelerated senescence (presumably through cryptic damage rather than lingering infectious agents; see Figure 3.41), and that populations experiencing less childhood disease are physiologically "younger" as adults. Redrawn from Jones, 1956.

childhood and adult mortality rates are statistically complex and difficult to analyze, the data analysis argues convincingly that mortality curves have shifted to later ages during the last 100 years, with relatively minor changes in slope in association with reduced earlier mortality (Figure 1.2).

The mechanisms causing such global effects of childhood diseases on adult mortality are obscure for most conditions. One link is through the long-term consequences of rheumatic fever from streptococcus infections, which are associated with a high incidence of cardiovascular-related mortality in adults (Jones, 1959). Lesions in the heart and valves of children persist into adulthood and are associated with heart murmurs (aortic regurgitation). The focal myocardial damage that is also a common consequence of streptococcus infections might also increase the consequences of transient ischemia and arteriosclerosis. Streptococcus and other infectious scourges of childhood have been greatly reduced through antibiotics and public hygiene in most populations. Jones also noted a nearly twofold higher risk of cancer in adults with heart murmurs, according to a life insurance study (Society of Actuaries, 1951).

Childhood viral infections may also compromise other vital organ functions. Kidney damage is common during early infections and could potentiate further damage by common chronic adult diseases that are also associated with kidney damage, such as diabetes and hypertension. An example of this in the laboratory is the multiorgan damage that persisted from a single epizootic infection of a mouse colony by Sendai virus (parainfluenza type 1). Mice infected prepubertally (1 month) had deposits of immunoglobulins on the kidney glomerular basement membranes and in the mesangium when examined 12 months later (Kay, 1979). Moreover these, as well as mice infected at later ages, had apparently permanent depression of T-cell functions and autoimmune manifestations, despite a healthy appearance. The Sendai survivors also had increased sensitivity to radiation (Kay, 1978b). Thus, childhood disease might also cause adult immune deficits with far-reaching consequences. Another example, for which there is persistent, albeit circumstantial, evidence, concerns the relationship between Epstein-Barr virus infections and Hodgkin's disease, a malignant disease of young and middle-aged adults (Mueller et al., 1989). The relationship of immunodeficiencies to most types of adult malignancy remains controversial.

Several viral diseases of early life also have later manifestations. Epstein-Barr infections are implicated in the susceptibility of particular *HLA* haplotypes to later onset rheumatoid arthritis (Chapter 6.5.3.2). Another example is shingles, a viral disease that increases after middle age in association with activation of latent herpes simplex virus from childhood infections of chicken pox (Chapter 7.4.1.4). I also suggest that neuronal impairments might also be an outcome of the sustained high fevers that were once common during childhood and could compromise many brain-directed homeostatic functions. Few quantitative data are available, for example, on neuronal loss in children who suffered sustained fevers or most infectious diseases. On the other hand, early exposure to some microorganisms may reduce the risk of adult onset neurologic conditions. For example, seroepidemiologic studies indicate an inverse relationship between the incidence of multiple sclerosis in adults and childhood infections (Alter et al., 1986; Sullivan et al., 1984).

The age at menarche in girls shows an inverse correlation with the risk of breast cancer, and that can account for twofold or greater differences within and between populations (Pike et al., 1983). An important risk factor in breast cancer appears to be the duration of exposure to progesterone and other reproductive hormones, which have mitogenic effects on breast epithelial cells and which become elevated at puberty. The earlier menarche in subpopulations is usually attributed to improved nutrition and reduced growth-stunting childhood disease (Tanner, 1962).

Exposure to toxins during development may also interact with senescence. Although there are as yet no examples relating to age-related mortality or disease, it will be of interest in the future to examine the adult neurologic diseases of children whose mother smoked during pregnancy. The smaller head sizes of children from heavy-smoking mothers were retained at least to 5 years of age (Elwood et al., 1987); the persistence of these influences into adulthood is not known. In conclusion, a host of mechanisms could relate childhood and adult mortality, which could be investigated with controlled lesions to neonatal rodents.

9.4. Developmental Determinants of Regeneration and Cell Renewal

Many features of the senescent phenotype of any species must be consequences of the developmental determinants of organogenesis and differentiation. This is clearly demonstrated by species variations in the capacity for regeneration and cell replacement and in defects of adult organs. Compare, for example, the irreplaceable wings of adult flies (Chapter 2.2.2) and the limited regenerative capacity of most organs in adult mammals versus the capacity of adult salamanders for limb regeneration (Schmidt, 1968; Sicard, 1985; Anton, 1988); the ability to regenerate a whole functional lens from pigmented retinal iris cells is retained in adult newts of certain species in the family Salamandridae versus its absence in other amphibia (Stone, 1960; Yamada, 1968); the irreplaceable loss of ovarian oocytes in adult mammals versus the possibly unlimited capacity for *de novo* oogenesis in long-lived fish (Chapters 3.4, 4.2). Such differences will strongly influence the incidence of age-related lesions and changes, which it seems reasonable to presume are reflected in different mortality rate constants.

Proliferative potential varies widely among cell types in different species. At one extreme are somatic cell lineages with no manifested limit to proliferation, as is observed in the many species of plants and animals that reproduce asexually through budding, fragmentation, and other types of somatic cell cloning (Chapter 4.4). At the other extreme are mammalian erythrocytes whose differentiation proceeds through a fixed number of divisions and whose potential for further proliferation is terminated by loss of the nucleus. Species comparisons reveal many examples of similar cell types with widely differing proliferation. In teleostan ovaries, new oogonia and primordial follicles continue to be formed during adult life, so that the production of eggs increases as a function of body size and age, whereas mammals have an embryonically determined number of oocytes and follicles (Chapter 3.6.4.1). Lampreys (Chapter 2.3.1.4.1) differ fifteenfold among species in the number of larval oogonia, despite similar numbers of primordial germ cell precursors of the oogonia (Hardisty and Cosh, 1966; Davidson, 1968). Because the number of oogonial divisions determines the maximum size of the adult oocyte stocks, the mechanisms governing this cell characteristic during differentiation have a direct bearing on reproductive senescence. The basis for species differences in the number of oogonial divisions is unknown, but quantitative changes in the activities of only a small number of genes may be involved (Chapter 7.6.4).

The evidence that oogonia vary enormously among species in their proliferative potentials extends to many other cell types. The intestinal epithelia continue to proliferate throughout the lifespan in mammals (Chapter 7.6.2). While intestinal mitoses are not seen in adult flies, adult worker bees show them (Snodgrass, 1956, 189; Chapter 2.2.4.1). The capacity for regenerating exfoliating epithelial cells in the gut and skin is clearly essential for long-lived and slowly maturing mammals, and much less so for short-lived insects. In contrast to the negligible proliferation of cells in short-lived adult flies, the imaginal disc precursors of the adult organs have an extraordinary potential for proliferation. Imaginal discs have been propagated by 160 serial transfers in adult females during 6 years, which is equivalent to at least 320 population doublings (Hadorn, 1966). Exposure to ecdysone, the metamorphosis-inducing hormone, at any point (e.g. Schweitzer and Bodenstein, 1975) induces the imaginal disc cells to differentiate into wings or other adult structures, as predetermined for imaginal discs from particular locations. Thus, the absence of proliferation in cells of the adult wing and elsewhere contrasts with the apparently unlimited proliferative capacity of precursor cells in the imaginal disc. The distal cause of the mechanical senescence of wings and other irreplaceable appendages thus can be traced to the hormonally controlled mechanisms of cell differentiation.

This perspective also applies to many other cell types that do not replicate in adult mammals and other vertebrates and that are therefore at risk for accumulating damage and suffering irreversible cell loss during the lifespan. Irreplaceable cells and cell organizations in adult mammals include brain, myocardium, adipocytes, sweat glands, and hair follicles (Schmidt, 1968). While the liver may be unique in its ability for complete regeneration after two-thirds is removed (Chapter 7.6.2), the pituitary and adrenal cortex also retain considerable capacity to replace lost cells. Other organs can respond to loss or increased demand by compensatory hypertrophy of existing functional units, such as the kidney and most skeletal muscle bundles, in which few new cells are added. Many vertebrates show a progressive impairment in the capacity for regeneration during development and maturation, for example, in reinnervation after spinal cord transection of fish (Keil, 1940; Bernstein, 1964) and amphibia (Piatt and Piatt, 1958) and in the ability to replace adipocytes in rats (Faust et al., 1974); regeneration is absent or greatly reduced in adults. This remarkable neural regenerative capacity is not shown by other adult tetrapods. If a functional complete set of germ-line genes is retained in vertebrate somatic cells (genomic totipotency), then manipulation of gene expression could, in principle, reactivate this capacity in other species and other cell types.

9.5. Summary

This chapter considered two ways that development influences the lifespan and the patterns of senescence: delayed development and characteristics of the adult that are determined during development. First, I reviewed examples of variations in the duration of development, either through diapause or dormancy, or through delayed maturation. Some organisms can multiply their total lifespan manyfold by hypometabolic states of unlimited duration, while others have less plasticity. For example, the starvation-induced dauer larval state of *Caenorhabditis* extends the total lifespan by its duration but cannot be extended indefinitely without eventually increasing mortality. In this and several other examples, the adult-phase duration was not altered despite major increases in developmental duration.

In the preimplantation diapause of mammals, the blastocyst remains in the uterus for months, or even up to 2 years in the unusual case of the armadillo. While little is known about biosynthesis in arrested blastocysts, their maintenance at body temperature without loss of viability implies effective molecular repair mechanisms. The onset of sexual maturation is also subject to major changes in scheduling. Many years may elapse in mammals between the time in development when the reproductive tract first acquires the competence to respond to hormones and the time when maturation is induced by rising hormones. This flexibility in reproductive scheduling is widely observed among animals and plants, whether semelparous or iteroparous, and gives another statement about the chronological plasticity of some genomic mechanisms.

Many types of events during development can alter the patterns of senescence. The potential during development for programming the same genome for drastically different alternative phenotypes of senescence is shown by queens and workers in social insects. There are numerous and scattered other examples, but no generalizations seem possible at this time.

When cell replication is programmed to cease is an important developmental influence. In the case of adult Diptera, the embryonic cells from which the adult organs are formed cease proliferating and are transformed by exposure to ecdysone and other hormones before hatching. In such species, the distal cause of mechanical senescence in wings and other irreplaceable parts must be attributed to the hormonal regulation of select genes during development. Other regulatory mechanisms must also determine whether a given type of cell will continue to proliferate in adults. Species differences in the continuation of *de novo* oogenesis and the capacity for regeneration of appendages and tissues during adult life must be major determinants of the patterns of senescence and the potential lifespan, because of their influence on reproduction and their influence on mortality rates through the ability to repair injuries. Thus, the specifics of gene regulation during development have the most fundamental impact by determining the body-plan of the adult, which, in turn, sets the statistics of adult mortality.

The relationships between development and the adult lifespan seem unconstrained, upon consideration of the range of experimental and environmental variations of the same genotype. Parallel variations that can be observed between species suggest that the mechanisms by which variations in temporal organization arise need not depend on species differences in structural gene sequence. Together with the evidence for selective changes in gene regulation during the adult life phase in mammals, the present examples support the working hypothesis that the timing of events throughout life is determined by quantitative variations in gene expression.

10

Postmaturational Influences on the Phenotypes of Senescence

10.1. Introduction

This chapter has selected examples from a huge literature on how the phenotypes of senescence are subject to environmental and experimental influences on adult organisms. Some species, particularly rodents and humans, have demonstrated extensive plasticity in the schedule and phenotype of senescence throughout adult life. Although most manipulations have heterochronic effects that alter the timing of age-related changes during development or in adult life, others are *isochronic* and modify the intensity of age-related changes without altering the time of onset. Externally induced changes in the patterns of senescence and in lifespan, such as those accomplished through diet or hormones, give approaches to identifying subsets of epigenetic mechanisms through which selective alterations in gene regulation may influence the phenotype of senescence. A goal of this subject is to identify the different levels of gene regulation that are accessible to variations at different times in the life cycle. In some cases, genotypic variations have important influences on the outcome of these manipulations. Manipulatable age-related changes are also superposed on developmental determinants (Chapter 9).

Age-related traits that are altered in register with external manipulations of lifespan are candidates for root causes of age-related changes. Examples include age-related diseases that are influenced by nutrition, such as the complexes of vascular disease in humans or kidney disease in rodents. Except for the effects of diet restriction in slowing the age-related increase of mortality in rodents, there are few other examples in which variations in longevity have been resolved according to the two mortality rate coefficients. In considering the increased longevity from dietary restriction or from lower temperature on poikilotherms, I argue that the mechanisms may not be the same throughout the phyla. These findings lead to a reevaluation of the rate-of-living hypothesis. Finally, I review the perplexing problem of finding "biomarkers of aging," parameters that hypothetically predict individual lifespans or future mortality risk and that are sought after in evaluating the effects of experimental manipulations on the schedule of changes in gene expression and other functions.

10.2. Ambient Temperature and Lifespan

Ambient temperature variations have been used to test the classic, but oversimplifying, "rate of living" hypotheses, for example Rubner's calculation (now discounted) that the amount of metabolic energy expended during the lifespan is a relatively constant function of mass in mammals, about 200 kcal per gram body weight over the lifespan (Chapter 5.3.4). However, environmental temperature affects metabolic rate quite differently in homeothermic mammals than in flies, fish, and other poikilotherms. Responses vary even among mammals, so that some species become hypometabolic for an extended time.

10.2.1. Temperature and Metabolic Rate

Chronic exposure of mammals to cold environments would be predicted by the rate-of-living hypothesis to shorten lifespan commensurately with increasing metabolism. The increased oxygen consumption needed to support increased metabolism during cold exposure also bears on the free radical theory of senescence, since faster oxidative metabolism might also increase the production of potentially damaging superoxide radicals and their derivatives (Harman, 1981; Pryor, 1987; Saul et al., 1987; Ames, 1983). Only recently have interpretable results been obtained for the effects of cold on longevity of mammals: Rodents adapt remarkably well even to subzero temperatures, given adequate nesting material and food (Barnett, 1965). Exposure of laboratory rodents to cool temperatures for most of their adult lifespans caused marked (ca. 30%) shortening of average lifespan in several studies (Johnson et al., 1963; Heroux and Campbell, 1960). However, these studies were often confounded by respiratory infections and other endemic pathogens that can be common in laboratory rodent colonies even today.

A new challenge to the concepts of limited lifetime energy comes from the well-controlled study of Holloszy and Smith (1986). These investigators used specific-pathogen-free rats and increased their metabolism by standing them in a shallow bath of slightly cool (23°C) water for 4 hours, five times a week. During the periods of immersion, heat production increased about threefold. This increase of metabolism was fueled by 44% increases of food intake per day, but had no effect on lifespan. The survival plots showed negligible mortality before 24 months (mean lifespan, 2.7 years; maximum, 3.5 years). Another interesting feature is that the exposure to cold altered the spectrum of age-related degenerative diseases. Whereas nearly all rats had some type of kidney disease, the cold exposure reduced the incidence of sarcomas and carcinomas by 50% or more. However, cardiovascular lesions increased from 65% in controls to 100% in the cold-exposed. The similar lifespans and mortality statistics despite these different disease patterns recall the difficulties of assigning proximal causes of death (Chapters 1, 3, 6). In sum, this study argues that any limits in the potential caloric expenditure over the lifespan are $\geq$ 40% in excess of the usual caloric consumption. The increased food intake associated with this mild cold stress, which had no effect on longevity, argues against simple interpretations of diet or caloric restriction studies (Section 10.3).

In fish the rate of development and the lifespan are readily manipulated by environmental temperatures, much more than in homeothermic mammals. Readers may recall the example of the Bunny Lake brook trout, which grew very slowly and lived far beyond their usual lifespan without signs of senescence because of the cold temperature and restricted food (Reimers, 1979; Chapter 3.4.1.2.2). Beverton's (1987) analysis of temperature, growth, and lifespan in the walleye (*Stizostedion vitreum*) shows that age at maturity, maximum size, and longevity are inverse monotonic functions of temperature in thirteen different locations (Figure 10.1). The estimated lifetime egg production was also quite constant, but fell off in the warmest locations. Although genetic polymorphisms have been described in muscle and plasma protein isozymes (reviewed in Colby and Nepszy, 1981), these population differences in growth and longevity are most likely related to temperature. The lifespans in Figure 10.1 that range below 19 years probably underestimate the maximum, in view of survival to 28 years in northerly lakes (Colby and Nepszy, 1981; Colby et al., 1979). Although age of walleyes at maturity varies fourfold and maximum lifespan varies sixfold, the size at maturity was invariant in these and most other teleosts (literature cited by Beverton and Holt, 1959, and Beverton, 1987).

Several laboratory studies defined these phenomena convincingly. *Cynolebias adolffi* (a South American annual, Chapter 3.3.1) has a doubling of mean lifespan and a decreased mortality rate at 16°C versus 22°C; contrary to the walleye, growth is faster at 22°C (Liu

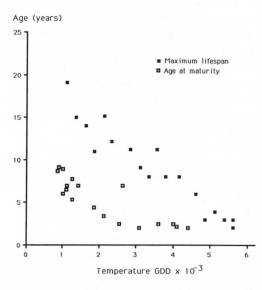

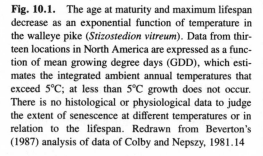

Fig. 10.1. The age at maturity and maximum lifespan decrease as an exponential function of temperature in the walleye pike *(Stizostedion vitreum)*. Data from thirteen locations in North America are expressed as a function of mean growing degree days (GDD), which estimates the integrated ambient annual temperatures that exceed 5°C; at less than 5°C growth does not occur. There is no histological or physiological data to judge the extent of senescence at different temperatures or in relation to the lifespan. Redrawn from Beverton's (1987) analysis of data of Colby and Nepszy, 1981.[14]

and Walford, 1966). Relationships of ambient temperature to the intensity of age-related changes are little studied and may be divergent. Age-related decreases of collagen solubility are retarded at cooler temperatures (Walford et al., 1969), whereas the strength of graft rejection, a measure of cellular immunity, increases with temperature up through 20°C (Hildemann, 1957). The lifespan was longer if temperature was shifted at 8 months from 20°C to 15°C (maximum 38 months, mean 22.5 months) rather than held at a constant 15°C (maximum 32 months, mean 16.5 months) (Liu and Walford, 1975). This result suggests a little-considered possibility that relatively constant laboratory conditions may not be optimal for longevity in poikilotherms, which ordinarily experience major seasonal fluctuations in nature. For example, walleyes fail to mature unless exposed seasonally to temperatures less than 12°C, despite continuing growth (Colby and Nepszy, 1981). Another factor is that the temperature optimum for maximum growth can change with age; for example, the optimum temperature decreases about 5°C by 6 months in the desert pupfish *Cyprinodon macularis* (Kinne, 1960). Effects of ambient temperature on fish of other species are thoughtfully compared by Liu and Walford (1972).

The lifespans of many other invertebrates also lengthen at lower temperature (Pearl, 1928; Comfort, 1979; Sacher, 1977; Strehler, 1977; Liu and Walford, 1972). Jacques Loeb and John Northrop (1916, 1917) began this influential line of investigation by showing that lifespan was inverse to ambient temperature in *Drosophila melanogaster,* decreasing nearly tenfold between 10°C and 30°C. Temperature variations had proportionately the same effects on the duration of the larval, pupal, and adult stages. These investigators proposed a common mechanism, "production or destruction of a substance" and proposed a similarity of these temperature responses to the kinetics of typical chemical reactions (Loeb, 1908; Loeb and Northrop, 1917). These early efforts to quantitatively apply chemical principles to gerontology drew from the popular discussions of that time, about how biological temperature responses resembled temperature effects on chemical kinetics (Arrhenius, 1907; Hollingsworth, 1969b).

Temperature and longevity relationships can also be represented in the Arrhenius plot from chemical kinetics, which estimates the activation energy (E_a) of chemical reactions from the dependence of reaction rate on temperature (Figure 10.2). E_a for the "rate of aging" is about 18 kcal/mole (Strehler, 1961, 1977), which is within the range of activation energy for enzyme reactions (< 30 kcal/mole), but is much less than for protein denaturation (60 kcal/mole) (Sacher, 1967, 1977; Strehler, 1961). In a penetrating biophysical analysis of how environmental temperature influenced maturation, adult metabolism, and mortality in flies, Sacher (1967) argued against regulation by a single biochemical process. Though perhaps more obvious today than then, this conclusion helped lay the foundation for discussing multiple mechanisms in senescence. The molecules responsible have not been specified, nor is the age-independent mortality factor identified that may increase with temperature. Although temperature coefficients are similar for developmental and adult stages in flies and for total lifespan in many species, we should not presume that identical molecular mechanisms are involved. For example, similar mortality rate coefficients occur in organisms with very different pathophysiologic changes during senescence (Chapter 2).

More sophisticated experiments followed these early studies, particularly temperature-shift studies that indicated distinct phases of temperature sensitivity within the adult phase. This approach was also used, for example, to identify stages in early amphibian development that are critical for organogenesis (Gilchrist, 1933), possibly through influences on transcription (Davidson, 1968, 32). A series of elegant studies examined temperature shifts with adult *D. subobscura* (Maynard Smith, 1958; Clarke and Maynard Smith, 1961a, 1961b). When shifted from 30°C to 20°C at different ages up to mid-life (6-day mean lifespan at 30°C), the lifespan remained unchanged with reference to the lower temperature (83-day lifespan at 20°C). This result could mean that the "rate of aging" was temperature-independent in this range. However, the converse experiment gave a different outcome. If flies were up-shifted from 20°C to higher temperatures before mid-life, lifespan decreased nearly linearly for every day at 20°C, though after mid-life the decrease of lifespan was much less.

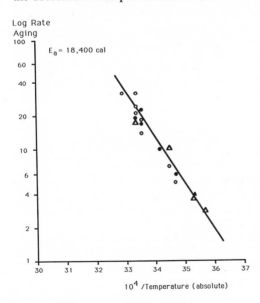

Log Rate
Aging

$E_a = 18,400$ cal

10^4 /Temperature (absolute)

Fig. 10.2. Arrhenius plot of the relationship between mean lifespan and ambient temperature for the flies *Drosophila melanogaster* (*open circles*) and *D. subobscura* (*closed circles*), the crustacean *Daphnia magna* (*open triangles*), and the bark beetle *Pinus tectus* (*closed triangle*). By analogy to the Arrhenius equation (the log of the rate constant is proportional to $- E_a/RT$, where E_a is the activation energy of a chemical reaction), an activation energy for the effect of temperature on lifespan can be calculated at about 18 kcal/mole in this set of invertebrates. An equivalent rate constant for aging was estimated as the reciprocal of mean longevity as normalized to 20°C (Bernard Strehler, pers. comm.). Redrawn from Strehler, 1977, who used data from seven previous reports by other authors.

These data imply phases of the adult lifespan with different temperature sensitivities, such that the first half of lifespan is roughly temperature-independent from 15°C to 30°C ("aging phase"), while a subsequent "dying phase" is extended at lower temperatures. Influences of genotype and sex on exposure to higher temperatures (30°C; Hollingsworth, 1969a, 1969b) imply polymorphisms in adaptability to high temperatures. For example, female flies showed no ill effects from being shifted from 25°C to 30°C for 3 hours a day, while male lifespans were 25% shorter (Hollingsworth, 1969a). Clarke and Maynard Smith (1961b) hypothesized several molecular processes with different temperature coefficients; for example, protein catabolism during the dying phase could have a higher temperature coefficient than protein synthesis.

An indication that temperature may qualitatively change the characteristics of senescence is that, at 18°C, fruit flies (*Drosophila melanogaster;* Arking et al., 1988) and milkweed bugs (*Oncopeltus fasciatus;* McArthur and Sohal, 1982) did not show the prominent age-related decrease in hourly oxygen use that occurred at higher temperatures. These results argue against the view that lifespan is determined by a single molecular process. Histological data are regrettably absent in all of these studies, which could help evaluate the importance of temperature to mechanical aspects of senescence and internal organ pathology. The impact of temperature on age-related changes in macromolecular biosynthesis also needs study.

Using oxygen consumption to directly assay metabolic rate, Arking et al. (1988) showed that lifetime oxygen consumption varied inversely with temperature in two random-bred stocks of *D. melanogaster* and in a long-lived inbred strain. The latter, at all temperatures and without sparing its metabolism, lived longer than the shorter-lived genotype (Figure 10.3). Similarly, mean respiration rate (Lints and Lints, 1968; Arking et al., 1988) and spontaneous activity (Le Bourg and Lints, 1984) were not correlated with lifespan in other genotypes. These findings show that total metabolic expenditure is not a strong lifespan determinant in *Drosophila*.

In contrast to *Drosophila*, milkweed bugs have an essentially constant lifetime oxygen consumption at temperatures that caused threefold variations in lifespan and twofold variations in the acceleration of mortality rate (MRDT; McArthur and Sohal, 1982). Moreover, accumulation of lipofuscins scaled similarly with temperature, indicating a stronger

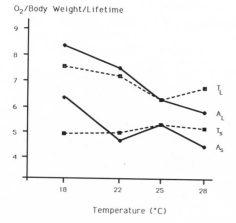

O₂/Body Weight/Lifetime

Fig. 10.3. Lifespan oxygen consumption is not a fixed constant. The lifetime oxygen consumption (ml O₂ per mg body weight per lifespan) is shown in two genotypes of *Drosophila melanogaster* (*S*, random bred; *L*, inbred for longevity) kept at different temperatures (*T*, from development throughout life; *A*, adult phase only). Lifespans were about threefold longer at 18°C than at 28°C. Redrawn from Arking et al., 1988.

linkage to metabolism within a species than is known in mammals (Chapters 5.4.3, 7.5.6). Relationships among metabolism, lifespan, and rates of senescence in other species invite further inquiry.

10.2.2. Temperature and Hypometabolic States

The species described above appear to have smooth metabolic responses to temperature variations. Other species, however, become hypometabolic at some temperatures. Well-known examples include the hibernation of many mammals during winter and aestivation during summer, or the daily torpor of temperate-zone bats. These metabolic variations bear on Rubner's and Pearl's "rate of living" hypotheses (Chapter 5.3.4; Lints, 1989), but few species have been systematically studied for effects on the total lifespan. A remarkable variety of metabolic adaptations have been evolved by different vertebrates, for example, that permit hypometabolic states in association with hibernation or oxygen deficits. As reviewed in a recent monograph (Hochachka and Guppy, 1987), both the duration of hypometabolism and species differences are factors in the extent of suppression of oxidative metabolism and macromolecular synthesis. Nearly universally, hypometabolic states adjust the energy costly ion-specific pumps at cell membranes. The extent to which hypometabolic states arrest senescence is unclear.

In Turkish hamsters (*Mesocricetus brandti*) the lifespan was increased in general proportion to the degree of hibernation induced by exposure to 5°C during a simulated winter season (23 weeks; Figure 10.4; Lyman et al., 1981). The most extreme hibernators lived about 1 year longer than the longest-lived control, while the average lifespan was about 3 months longer in the cold-exposed. The duration of hibernation varied from almost 0% to 35% of the lifespan. The metabolic rate was not directly measured, but there is little doubt that the hibernating hamsters were hypometabolic. Because hibernating hamsters spontaneously awakened to feed, the best hibernators may have metabolized more. The influence of torpor on the lifespan of bats does not seem to be as obvious as in these hamsters, because species without torpor apparently live at least as long (Chapter 5.3.4). Nonetheless, further studies are needed, especially in species with very different mortality rates, as in bats, which have high IMR but long lifespans (Chapter 3.6.1; Table 3.1). Because torpor is induced by cool temperatures for several weeks or more at any time of the year in some bats (Racey, 1969), experiments could be designed to test different models for the impact of cumulative time in hibernation on mortality rates.

In summary, the lifespan of many invertebrates varies inversely with ambient temperature within a range that permits normal functions. The limited studies of mammals suggest that the adult life phase may also be lengthened in accord with the length of hibernation, at least in the Turkish hamster. The interpretation of this fairly general phenomenon is not straightforward. One factor could involve mechanical senescence, such as erosion of wings and other nonreplaceable parts, since physical activities have a complicated temperature dependence. The "rate of living" hypothesis that lifespan is limited by the expenditure of a fixed quota of energy was disproved for *Drosophila* (Arking et al., 1988; Arking and Dudas, 1989) and is also discounted by the many species that have prolonged juvenile phases (Chapter 2.2.3.3) or that continue growth as adults, especially plants (Chapter 4.2.1).

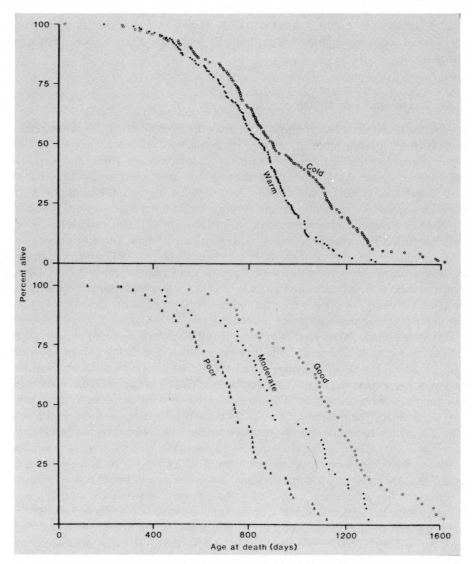

10.3. Diet and Caloric Restriction

Dietary influences on lifespan are shown throughout the animal kingdom, where 25% or larger increases of mean and maximum longevity are widely observed from decreased intake of bulk nutrients. In general, to be effective, diets are restricted in energy sources, particularly protein, carbohydrate, and fat, but without deficiencies of vitamins or minerals. Most investigators conclude that diet-restriction effects on lifespan really represent *caloric restriction* (Weindruch and Walford, 1988; Masoro, 1988). Readers should be aware that the terminology of diet restriction in the literature may represent quite different diets and conditions among studies. Recently, most investigators supplement the diet with trace nutrients and minerals to avoid malnutrition, but diets may still vary importantly in the types and amounts of fat and protein. There is as yet no equivalent in the diets of

Age at Death (days)

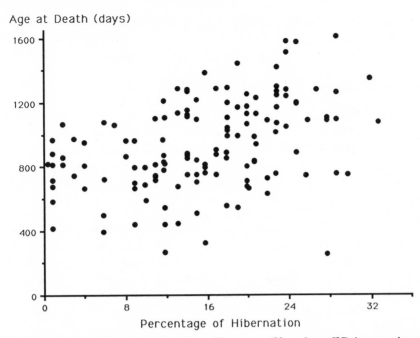

Percentage of Hibernation

Fig. 10.4. *Opposite page*, hibernation induced by cold exposure (23 weeks at 5°C) increases longevity in Turkish hamsters. Redrawn from Lyman et al., 1981. Comparison of controls (at warm temperatures) and hibernators (cold). Each group had 144 animals. *Above*, cold-exposed hamsters subdivided into three groups that had various durations of hibernation. Individual lifespans were correlated with the duration of hibernation, which accounted for about 25% of the variance ($r^2 = 0.52$; $p < 0.001$).

laboratory animals to the standards of chemically defined microbial nutrition. An early view that lifespan increased because of slowing of growth and reduced metabolism has been refuted, at least for mammals. The mechanisms in these phenomena are basically unknown and probably diverse, and include effects of diet on hormones, cell loss, transcription, and the glycation of long-lived proteins. There is no evidence for a single mechanism.

Two issues are important here: (1) Diet has major influences on the incidence of specific life-threatening diseases, for example cancer, vascular disease, and kidney diseases. (2) The impact of diet is also strongly influenced by physical activities and metabolic needs that are imposed by temperature and other environmental conditions. As discussed above, rats can eat 44% more than their usual *ad libitum* intake with no effects on mortality rates or lifespan whatsoever, if they are placed under increased metabolic demand by mild cold stress (Section 10.2). Thus, physiological adjustments to ambient temperature, the behavioral state, and other environmental factors can strongly influence the metabolism independently of the bulk calories ingested. This issue is important in evaluating the effects of temperature on lifespan in both poikilotherms and mammals. There are substantial limitations in comparing studies in the literature, because metabolism is subject to unknowable variations in the behavioral and physical environment. That is, few reports describe the environment, even in laboratory studies, in enough detail to evaluate differences in metabolic demands. Moreover, the local colony conditions are not necessarily within the tolerated range; investigators generally are unaware of local fluctuations short of disasters like those described in Chapter 9, note 3. The following is a selection of examples from a huge literature on manipulations of senescence.

10.3.1. Diet, Lifespans, and Mortality Rates

10.3.1.1. Mammals

Rodents are the best-studied examples relating caloric restriction to lifespan, an approach that is still the most robust phenotypic manipulation of senescence in mammals. Early demonstrations of increased longevity from caloric restriction of pubertal or young adult rats include Osborne et al., 1917; McCay et al., 1935; McCay, Ellis, Barnes, Smith, and Sperling, 1939; McCay, Maynard, Sperling, and Barnes, 1939; McCay et al., 1943; McCay, 1947; and Ross, 1972. Although rodents in early studies were often relatively short lived due to infectious disease, most subsequent studies with healthier and longer-lived controls give the same result of increased mean and maximum lifespan, typically by 10 to 30% (reviewed in Comfort, 1979; Yu, 1987; Weindruch and Walford, 1988; Figures 10.5, 10.6), but occasionally up to 50% (Weindruch et al., 1986). Recent protocols supplemented the required nutrients to the same daily intake as *ad libitum* controls. On nutritionally balanced and restricted diets, rodents appear outstandingly healthy throughout their lifespans.

Alternate-day *ad libitum* feeding is another mode of diet restriction, favored by its con-

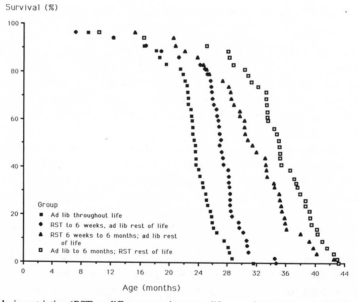

Fig. 10.5. Caloric restriction (*RST*) at different ages increases lifespan of rats. Male rats (Fischer 344) in good health were restricted to 60% of caloric intake supplemented with trace nutrients (vitamins, minerals, and choline) that gave the same daily intake as controls. Body weights were about 50% less in rats restricted throughout life or after 6 months, yet the caloric consumption per gram body weight was the same. Analysis of mortality rate coefficients (Table 10.1) shows that the MRDT was lengthened by about 50% by continual restriction during the adult phase. Redrawn from Yu et al., 1985.

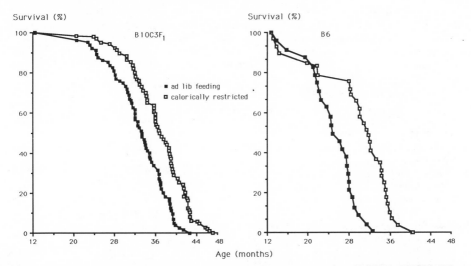

Fig. 10.6. Caloric restriction in mice. Males of two long-lived mouse strains: B10C3F₁ ([C57BL/10Sn × C3H/HeDiSn]F₁) and B6 (C57BL/6J) were subjected at 12 months to caloric restriction with trace nutrient supplements. Lifespans were increased by 10 to 20% and lymphomas of B10C3F₁ were reduced by 35%. Redrawn from Weindruch and Walford, 1982.

venience. In effect, alternate-day *ad libitum* feeding is a schedule for alternate-day fasting that reduces average food intake and body weight by 10 to 30%. One of the few studies to directly compare these regimens showed that C57BL/6J male mice had increased lifespans on both, but alternate-day feeding was slightly better than 50% daily restriction (Ingram and Reynolds, 1987a, 1987b). Possible differences between daily restriction and alternate-day feeding have not been investigated and could be important for blood sugar and other aspects of metabolism. Because diet restriction may alter circadian feeding and activity cycles, detailed studies are needed across the 24-hour cycle.

Caloric restriction is the only known environmental means to alter the acceleration of mortality rate (MRDT) in any mammal. As discussed in Chapter 1.4, the MRDT is a fundamental measure of the rate of senescence at the population level. Sacher (1977) first showed that diet-restricted rats from several early studies had longer MRDT. A confirmation of this important result (Yu et al., 1985; Figure 10.5), showed MRDTs that were 30 to 40% longer on restricted diets.[1] Table 10.1 presents an analysis of the mortality rate coefficients from studies of rats that presented mortality schedules throughout the lifespan from which the Gompertz slopes could be reliably calculated. This analysis shows a robust phenomenon in fourteen comparisons of control versus various caloric restriction

1. The IMR has not shown consistent changes in response to caloric restriction, however. Two extensive studies on long-lived inbred mice demonstrated similar increases in lifespan, but showed no effect on MRDT (Cheney et al., 1980; Cheney et al., 1983); the main effect of diet restriction was to decrease the IMR by 30 to 70%. An important difference between these analyses and those of rats (Table 10.1) is that Cheney et al. calculated mortality rates after 120 weeks, when mortality rates had just begun to increase sharply; this different approach might account for the ten- to one hundredfold higher values of IMR than when IMR is calculated across the lifespan (Tables 1.1, 10.1). Despite the lack of change in MRDT, diet restriction in these studies clearly reduced the incidence of tumors and other diseases that are a hallmark of senescence in these genotypes (Cheney et al., 1980).

Table 10.1 Diet Restriction Slows Mortality Rate Accelerations in Long-lived Strains of Laboratory Rats

Study	IMR/year	MRDT, years[1]	*tmax*, years
Ross, 1959 (Sprague-Dawley males)[2]			
Ad libitum	0.001	0.17	2.4
Restricted	0.003	0.38 (2.23)	3.0
Berg and Simms, 1960 (Sprague-Dawley males)[2]			
Ad libitum	0.001	0.17	2.6
Restricted	0.003	0.27 (1.59)	3.1
Ross, 1961 (Sprague-Dawley males)[3]			
Ad libitum	0.01	0.29	3.0
Restricted			
High casein–high sucrose	0.013	0.40 (1.58)	3.8
High casein–low sucrose	0.017	0.47 (1.62)	4.1
Low casein–high sucrose	0.021	0.42 (1.45)	3.6
Low casein–low sucrose	0.03	0.60 (2.0)	4.8
Ross and Bras, 1973 (Sprague-Dawley males)[3]			
Ad libitum, 10% protein	0.026	0.30	2.8
Restricted, 10% protein	0.020	0.42 (1.4)	3.5
Ad libitum, 22% protein	0.018	0.28	2.8
Restricted, 22% protein	0.063	0.65 (2.32)	4.1
Ad libitum, 51% protein	0.027	0.33	3.1
Restricted, 51% protein	0.029	0.57 (1.72)	4.3
Goodrick et al., 1983 (Wistar males)[3,4]			
Ad libitum	0.02	0.25	2.5
EOD, 10 months	0.06	0.44 (1.76)	2.9
Ad libitum	0.05	0.31	2.4
EOD, 18 months	0.11	0.59 (1.9)	3.1
Yu et al., 1982 (SPF Fischer 344 males)[3]			
Ad libitum	0.020	0.30	2.6
Restricted	0.029	0.53 (1.77)	4.0
Yu et al., 1985 (Fischer 344 males)[5]			
Ad libitum	0.002	0.19	
Restricted			
After 6 weeks[6]	0.0007	0.27 (1.42)	
6 weeks to 6 months[7]	0.0007	0.19	
After 6 months	0.006	0.32 (1.63)	

Grand mean: MRDT of restricted ÷ *ad libitum*[8] 1.74 ± 0.27

Note: The calculated values for IMR (initial mortality rate) and MRDT (mortality rate doubling time) assume equation (1.2) and puberty at 0.25 year.

[1] Numbers in parentheses are the ratio of restricted to *ad libitum* for each experimental group.

[2] Calculated from Sacher, 1977, by Antonius van Hagen.

[3] Calculated from mortality schedule data by Antonius van Hagen.

[4] 24% protein; EOD, every-other-day fasting alternating with *ad libitum* feeding. A voluntary-use exercise wheel was in cage, unlike other studies.

[5] Calculated from mortality schedule data by Matthew Witten.

[6] Average daily consumption was reduced by about 60%. Portions of food were made available each day. Trace nutrients were supplemented to the same intake as unrestricted rats. Body weights changed little after 14 months (ca. 250–300 gm), whereas unrestricted rats were twofold heavier and continued to increase until just before death.

[7] Body weight decreased sharply in the first 3 months after restriction and remained slightly higher than rats restricted from 6 weeks onwards (table entry above).

[8] $n = 14$. The ratio was calculated for each category of restricted to its repective control, with the omission of Yu et al., 1985, restricted 6 weeks to 6 months.

protocols from seven reports; the grand average indicates that caloric restriction causes a normative 1.7-fold lengthening of MRDT.

Short-lived genotypes that succumb to unusual early onset degenerative diseases also have increased lifespans from caloric restriction (Table 10.2); mortality rate coefficients remain to be calculated. Because the type and onset of age-related degenerative diseases are also altered by caloric restriction (Section 10.3.2.1), a major, unresolved question concerns whether these life extensions are mainly a result of reducing and delaying diseases.

Another major question appears to be answered, the possibility of a linkage between life prolongation and retarded growth, which was not resolved by earlier studies in which diet was restricted from weaning onwards (McCay et al., 1943; Chapter 9.2.2.2.2). It might be hypothesized, for example, that the activation of mechanisms in senescence could be postponed by delaying maturation, as described in Chapter 2.3 for species that show rapid senescence in association with mating and reproduction. However, such a mechanism is not shown in laboratory rodents. For example, rats restricted by alternate-day feeding beginning at 1 week after weaning showed no correlation between rates of weight gain and longevity; in *ad libitum*–fed controls, smaller weights at weaning but faster growth were positively correlated with lifespan, a trend that is opposite to delay of senescence by delay of growth (Goodrick et al., 1982; Ingram and Reynolds, 1982).

Moreover, long-lived mouse and rat strains clearly showed increased longevity even when caloric restriction is initiated long after maturity and the cessation of growth. Restriction beginning at 6 or 12 months increased the mean and maximum lifespan by 10 to 50% (Figures 10.5, 10.6; Yu et al., 1985; Beauchene et al., 1986; Cheney et al., 1983; Ingram and Reynolds, 1987a, 1987b).

Diet composition influences lifespan independently of caloric consumption. A study of

Table 10.2 Dietary Influences on Short-lived Rodent Genotypes

	Mean Lifespan, Months	
	Ad libitum	Diet-Restricted
Spontaneous hypertensive rat (SHR)[1]	18	30
systolic blood pressure elevations, kidney and myocardial lesions		
Senescence-accelerated mouse (SAM-P/1)[2]	10	15
diffuse amyloid deposits		
NZB mouse[3]	12	16
autoantibodies to cyrthrocytes, DNA, and nuclei; hemolytic anemia;		
kidney disease		
kd/kd mouse[3]	8	18
chronic interstitial nephritis		
MRL/1 mouse[3]	6	> 15
massive lymphoadenopathy; autoimmune and kidney disease		
Obese mouse (*ob/ob*-C57BL/6J)[4]	14	26
adult onset diabetes and obesity		

[1]Lloyd, 1984.
[2]Chapter 6.5.1.7.2.
[3]Chapter 6.5.1.7.3.
[4]Section 10.3.2.1

isocaloric diets varying in protein content (casein) from 10 to 51% showed greater effects on lifespan in restricted than *ad libitum*–fed groups; lifespans on both the *ad libitum* and the restricted diets increased progressively with more protein (Ross and Bras, 1971, 1973). Until recently, high-fat diets ($\geq$ 10% fat; breeder chows), which favor milk production, were widely used in long-term rodent studies; diets with lower fat, more typically 5%, may have been important in the circa 25% increase of mean lifespans of laboratory rodents after 1970, although improved husbandry procedures that minimize pneumonia and other infectious diseases also contributed. In humans, high fat in diets is widely thought to promote vascular disease and shorten lifespan, though other factors are involved. Genotypic susceptibility to atherogenesis on fatty diets is discussed in Section 10.3.2.1.

In regard to humans, data on lifespan and diet are largely anecdotal (Weindruch and Walford, 1988) and cannot yet be analyzed for their effects on mortality rates. Nonetheless, one analysis suggests that the MRDT may resist dietary influences in humans. The mortality rates of two different civilian groups under dire circumstances during World War II, which imposed severe diet insufficiencies among many other adversities, indicate no change in the MRDT across the adult age range; yet the IMR increased severalfold (Figure 1.5; Appendix 1). Future studies of human diet and lifespan could use the case-control approach of epidemiologists to resolve the impact of habitual differences in the amount and composition of diet and in the frequency of voluntary fasting.

Thus, there is no evidence for a critical period in regard to caloric restriction that is comparable to critical periods during development, for example, for effects of steroids on the sexual differentiation of the hypothalamus in rodents or for the hemispheric transference of language in humans. The increase of lifespan generally scales with the duration of restriction, whether restriction was begun at 6 weeks or 6 months in rats (Figure 10.5; Yu et al., 1985) or in mice (Cheney et al., 1983). Although caloric restriction might not be effective once severe degenerative diseases develop, the progression of severe kidney disease (glomerulosclerosis) in humans can be retarded by reducing dietary protein (Brenner et al., 1982).

10.3.1.2. Fish

Some fish respond to diet restriction as do mammals, by slowing growth and delaying maturation, while others show quite opposite effects. First, the conventional outcomes. In Comfort's (1960a, 1963) pioneering studies of female guppies (*Lebistes* [= *Poecilia*] *reticulatus;* Chapter 3.4.1.2.2), diet was varied to manipulate growth, by providing the same ration of live worms once per week, or every other week; in effect this protocol alternates a week of feeding and of fasting. In both groups, growth continued throughout the lifespan, but was slower in diet-restricted fish. Mortality patterns were compared in fish that were given alternate-week feeding until 600 days with those given *ad libitum* food throughout their lifespans. Those kept up to 600 days on 50% restricted diets had 35% longer mean lifespans. Inspection of the mortality rate graphs shows no change in the MRDT. Even when growth was severely retarded by food restriction up to the age of 1,200 days, the guppies were able, with full feeding, to catch up to the size normal for their age. Woodhead (1978) also discusses catch-up growth in fish.

However, greater lifespan need not be linked to smaller size or limited growth in fish any more than in mammals. Field observations of brown trout and ferox salmon show increased lifespan and growth to unusual sizes when other fish and larger prey are eaten after maturity (Behnke, 1989; Campbell, 1979; Ferguson and Mason, 1981; Chapters 3.4.1.2.2, 4.5.2). Ordinarily, the drift food and small organisms otherwise available may be equivalent to caloric restriction, and under these conditions lifespan is much shorter. There are suggestions that this response to increased food after maturity is influenced by genotype (Ferguson and Mason, 1981; Hamilton et al., 1989), but the phenomenon is poorly understood. These observations show the complexity of interactions among diet, growth, and lifespan and the desirability of examining other species with different life histories.

10.3.1.3. *Invertebrates*

The impact of diet restriction on lifespan should vary widely among species according to whether adults are completely or partially aphagous. As described in Chapter 2.2.1, many species of butterflies, moths, and other invertebrates either cannot eat at all or cannot ingest a complete diet. Their potential adult life phases are ultimately limited by the nutrient reserves stored in the fat body or elsewhere. According to the species, access to carbohydrate foods can *extend*, rather than decrease, lifespan, as in the examples of several *Ephestia* moths (30% increase) and the dog tick *Rhipicephalus* ($\geq$ 500%). Adult *Drosophila* are sensitive to partial starvation, as shown in Kopec's (1928) detailed study: removal of food for 6 or 12 hours per day had little effect, while starvation for 18 of 24, 48 of 72, and 72 of 96 hours reduced lifespan with increasing severity that approached the drastic effects of total starvation. Adults given water but no food lived an average of 3 days.

Mosquitos are a special case. Female *Aedes aegypti* var. *queenslandiensis* (Pena and Lavoipierre, 1960a, 1960b; Putnam and Shannon, 1934) lived longer when fed on sucrose only rather than on the usual complete diet of vertebrate blood; the maximum lifespan, however, was not much influenced by diets of sugar versus blood (Pena and Lavoipierre, 1960a, 1960b). Other examples are reviewed by Downes (1958). Apparently this improvement from a restricted and incomplete diet stems from the adverse effects of egg laying. The blood meal required by many mosquitos and other biting flies to complete oogenesis (Downes, 1958a; Lavoipierre, 1961) thus shortens lifespan because reproduction is stimulated. Moreover, frequent blood meals and mating commonly increase mortality (Lavoipierre, 1961; Clements, 1963). With few exceptions, male biting flies feed only on sugar-rich nectar (Downes, 1958a; Lang, 1986).

In regard to evidence for age-related nutrient deficiencies in mosquitos, there are deficits of cysteine and glutathione as assayed in whole mosquitos; the deficits increased progressively throughout the lifespan in female *Aedes aegypti*, reaching $-$ 70% (Richie and Lang, 1988). The age-related deficit of glutathione was corrected by supplemental feeding with a cysteine precursor (magnesium thiazolidine–4-carboxylic acid) that increased the average lifespan by 35%, with smaller effects on the maximum lifespan (Richie et al., 1987). Similar decreases of glutathione in human and rodent blood (Section 10.3.2.5) are proposed by Lang as a possible general biochemical change that may have profound impact on lipid peroxidation and detoxification mechanisms (Lang, 1986).

Another special example is the more than 50% extension of lifespan in two flatworms by providing "superabundant food," as briefly described for *Polycelis tenuis* (Planariidae; iteroparous; Reynoldson, 1960) and *Dendrocoelum lacteum* (Dendrocoelidae; normally semelparous; Jennings and Calow, 1975; Calow and Woollhead, 1977). These manipulations resemble the ferox salmon and brook trout, whose lifespans were prolonged by increasing their food intake. Together, these various examples show that effects of diet restriction cannot be generalized among species with different food requirements as adults and different growth patterns in association with alternative natural history phenotypes.

10.3.2. *Mechanisms of Dietary Influences on Senescence*

This section reviews age-related changes, from molecules through behavior, that are modified by diet restriction.

10.3.2.1. *Diseases and Genotypes*

Laboratory rodents dominate this discussion, since little is known about the effects of genotypes on diet restriction in the senescence of other species. I discuss in detail particular mouse genotypes, which may be considered as models for unrecognized genotypic polymorphisms in humans and other mammals. Caloric restriction and diet composition have striking effects on age-related diseases that appear to parallel the decrease of mortality rate in many studies. Numerous studies have established that caloric restriction of rodents yields major delays and reductions in the genotype-specific disease patterns that generally parallel the increase of longevity (Yu, 1987; Weindruch and Walford, 1988). A remarkable diversity of age-related lesions is altered by caloric restriction, including kidney diseases, autoimmune conditions, musculoskeletal degeneration, cardiomyopathy, spinal neuron degeneration, and tumors (Table 10.3). Tumor incidence also depends on *when* mice were restricted; for example, hepatomas were threefold *more* frequent in mice kept on restricted diets for up to 6 months versus mice restricted from weaning onwards, or in *ad libitum*–fed controls (Cheney et al., 1983). Despite the often coordinate shifts in disease patterns and in longevity, it is unknown why so many types of diseases are diminished and why lifespan is increased. As discussed below, some diseases *increase* with diets that reduce the incidence of others.

One of the most extensive investigations of diet restriction is ongoing at the University of Texas in San Antonio with the Fischer 344 (F344) rat. This well-characterized inbred strain has an early onset of testicular tumors (Table 10.4; mortality patterns are shown in Figure 10.5; Yu et al., 1985; Maeda et al., 1985). Testicular tumors, chronic progressive kidney disease, and a host of other disorders are delayed by caloric restriction, begun at weaning or later ages (Table 10.4). However, not all age-related disorders were altered by caloric restriction, for example prostatic inflammations. Kidney disease (Maeda et al., 1985; Berg and Simms, 1960; Bras and Ross, 1964) is usually more suppressed by caloric restriction than abnormal growths. Tumors may even be increased on diets that reduce kidney disease (Iwasaki et al., 1988a, 1988b). A high incidence of kidney disease was observed consistently by the San Antonio group and was greater than in another colony of F344 males fed a similar diet (Coleman et al., 1977). One factor in this difference between rat populations may be greater body weights in the San Antonio study, implying greater food intake (Edward Masoro, pers. comm.).

Table 10.3 Diet-Restriction Delays of Age-related Pathologic Changes in Laboratory Rodents

Organ	Species	References
Bone (femur)		
Senile calcium loss	rat	Kalu, Hardin, Cockerham, and Yu, 1984;
		Kalu, Hardin, Cockerham, Yu, Norling, and Egan, 1984
Heart		
Myocardial degeneration	rat	Maeda et al., 1985
Aortic atherosclerosis	mouse	Paigen et al., 1987a,b
Kidney		
Glomerulosclerosis	rat	Maeda et al., 1985
		Berg and Simms, 1960
		Bras and Ross, 1964
		Tucker et al., 1976
Liver		
Bile-duct hyperplasia	rat	Maeda et al., 1985
Fatty degeneration	rat	Maeda et al., 1985
Hepatomas (increase)	mouse	Cheney et al., 1983
Lymphoreticular system		
Lymphomas	mouse	Cheney et al., 1983
	rat	Maeda et al., 1985
Thymomas	rat	Maeda et al., 1985
Nervous system		
Spinal neuron degeneration	rat	Everitt et al., 1980
Muscle (skeletal)		
Atrophy and degeneration	rat	Maeda et al., 1985
		Berg et al., 1962
Calcification	rat	Maeda et al., 1985
Parathyroid		
Hyperplasia	rat	Maeda et al., 1985
Pituitary		
Prolactin cell tumors	rat	Ross and Bras, 1971
Testes		
Interstitial cell tumors	rat	Maeda et al., 1985
Atrophy of semeniferous tubules	rat	Maeda et al., 1985

Note: Selected examples from rodents with usual 2- to 3-year lifespans; genotypes within each species may respond quite differently (see text).

The standardization of diet restriction protocols has just begun, and there are frustrating variations among studies in the impact of caloric restriction on disease and mortality rates, also seen for C57BL/6J mice within (Cheney et al., 1980) and among laboratories (Harrison and Archer, 1987). A new program of the National Institute on Aging to supply investigators with rodents from colonies with standardized caloric restrictions and diets should improve the generalizability of results. The bulk components in diets prepared in the laboratory or in diets from most commercial sources generally fall far short of being reagent grade, and may vary importantly in trace toxins or nutrients among batches. Ambient temperature could also be important, since diet-restricted rats have wider diurnal oscillations of body temperature and a lower mean 24-hour core temperature (Nakamura et al., 1989; Duffy et al., 1989).

Gender influences the responses of different genotypes to diet. In general, male rodents

Table 10.4 Effect of Diet Restriction on Spontaneous Degenerative Diseases in Fischer 344 Rats

Age, Months	Nephropathy[1] Ad lib[4] (%)	Restricted[5] (%)	Testicular tumors[2] Ad lib (%)	Restricted (%)	Bile-Duct Hyperplasia[3] Ad lib (%)	Restricted (%)
6	0	0	0	0	0	0
12	60	0	0	0	0	0
18	100	10	90	10	60	0
24	100	0	90	60	60	20
> 24	100	100	95	95	80	30

Source: Excerpted from Maeda et al., 1985. Three other diets were used.

[1]Moderate to severe kidney degeneration,, with proteinaceous casts within the tubules, sclerosis of the glomeruli, and interstitial fibrosis; the end stage with imminent kidney failure as indicated by fivefold elevations of blood urea was threefold more prevalent in the *ad libitum*–fed rats.

[2]Neoplasia of testicular interstitial cells is characteristic of the F344 rat, but uncommon in other lab rodents (Coleman et al., 1977). These cells, which make testosterone, became hyperplastic on all diets by 18 months.

[3]Besides hyperplasia of bile-duct epithelia, basement membranes become thickened and sclerotic and portal blood vessels may be blocked (Coleman et al., 1977). These impairments of liver function were rare in diet-restricted rats.

[4]The *ad libitum* diet contained 21% protein (casein) and 10% fat (corn oil) with supplements of trace nutrients (vitamins, minerals, methionine, choline) (Yu et al., 1985).

[5]The restricted group from 6 weeks onwards was given 60% fewer calories than the *ad libitum* group, supplemented with trace nutrients to the same daily intake.

are thought to be more susceptible to kidney disease than females (Dunn, 1967). On the other hand, females of the short-lived NZB mice respond less than males to low-fat diets, which increase lifespan and reduce formation of autoimmune-related disease (Figure 10.7; Fernandes et al., 1972). Other effects of sex are mentioned in Section 10.4. Although long-lived inbred mouse strains differ substantially in kidney disease and age-related dysfunctions (Hackbarth and Harrison, 1982; Tables 6.1, 6.6), it is unclear how genotype influences responses of kidney disease to diet.

The strain background influences the effects of caloric restriction on expression of disease-promoting alleles. For example, the recessive mutants of mice, *ob* and *db*, have very different impact on the C57BL/6J and C57BL/Ks strains. (Refer to Chapter 6.5.1.1 for information on these different genotypes.) When homozygously present in C57BL/KsJ mice, the *db* and *ob* mutations yield hyperphagia, obesity, hyperglycemia, insulin deficits from pancreatic β-cell necrosis, and lifespans of ≤ 10 months (e.g. Leiter, Coleman, Eisenstein, and Strack, 1981; Leiter, Coleman, and Hummel, 1981; Herberg and Coleman, 1977; Coleman, 1982). Although thermoregulation is deficient at normal temperatures, these mice adapt nearly as well as normal strains to life at 4°C, which indicates the integrity of metabolic up-regulation required for thermogenesis despite persistent pancreatic abnormalities (Coleman, 1982); at 4°C, food intake was the same in mutants and normals, about 25% of body weight per day, which may approach the maximum.

In contrast to their expression on the C57BL/KsJ background genotype, *ob* or *db* homozygotes of C57BL/6J mice, β cells are capable of compensatory hyperplasia so that the diabetes is much milder and lifespan reaches 20 months; these mice, while very obese, were not diabetic in terms of blood sugar and insulin (Herberg and Coleman, 1977).

Nonetheless, *db* homozygotes on both C57BL backgrounds and sexes showed severe kidney degeneration by 6 months (Leiter, Coleman, and Hummel, 1981). Restriction of calories by 50% failed to suppress the obesity and diabetes of *ob/ob* C57BL/Ks mice (Coleman, 1982), whereas obesity was reduced and lifespan was increased in *ob/ob* C57BL/6J mice (Lane and Dickie, 1958; Harrison, Archer, and Astle, 1984; Harrison and Archer, 1987). The important conclusion can be drawn that diet restriction in this mouse strain acted to increase lifespan *independently* of high body fat content (Harrison, Archer, and Astle, 1984). Analysis of *H–2* haplotypes in these and other strains showed that β-cell atrophy also occurred with the *db* mutation on the DBA/2J as on the C57BL/Ks (both *H–2d*), while resistance to diabetes occurred on other haplotypes as well as on *H–2b*; females were less susceptible to diabetes in nearly all strains (Leiter, Coleman, and Hummel, 1981). These genotypes should help to identify the molecular basis for genotypic influences on the susceptibility to age-related lesions in the kidney and pancreas. Glycation is only one candidate: it would valuable to know how genotype influences the relationships between plasma glucose and the glycation of kidney basement membranes.

Diet fat content has important effects on the penetrance of genes associated with atherosclerosis in mice, for example *Ath–1* (Chapter 6.5.1.6). *Ath–1r* may be homologous to the unmapped human gene for familial hyper-α-lipoproteinemia, which is implicated in reduced heart disease and increased longevity (Chapter 6.5.3.1). The inverse correlation between HDL and atherogenesis, and the sex differences, imply that the *Ath–1* alleles and other loci adjust the threshold levels of HDL-lipids below which atherogenesis is favored. Here as elsewhere, the penetrance of genotypes that favor pathogenesis is modifiable through diet and hormones.

Decreased diet fat suppressed mammary tumors in C3H mice. This genotype usually develops mammary tumors by 9 to 16 months, in association with vertically transmitted MMTV proviral genomes (Chapter 7.4.1.4). However, reducing fat content to below 5% on a series of diets with the same caloric content nearly eliminated visible mammary tumors during mid-life (Fernandes et al., 1979). It is not possible to extrapolate from these findings to the breast cancer risk in humans. Although epidemiological studies indicate associations between body weight and breast cancer in women over 60, the role of fat is

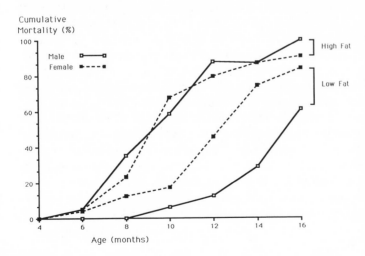

Cumulative Mortality (%)

Male □——□
Female ■----■

High Fat
Low Fat

Age (months)

Fig. 10.7. Gender and diet interaction on age-related mortality in the short-lived NZB mice that were maintained on diets with different fat and protein content. The characteristic autoimmune-related disorders of NZB mice (see text) were suppressed more and longevity was increased more in males than in females on a low-fat diet (4.5% fat and 23% protein, standard breeder chow) than a high-fat diet (11.5% fat and 17% protein, standard lab chow). Redrawn from Fernandes et al., 1972.

unclear (Henderson et al., 1984). Returning again to the age-related kidney disease of rodents, low fat content also reduces its severity in comparisons of isocaloric diets (Iwasaki et al., 1988b). The effects of decreasing dietary fat probably involve different mechanisms than does caloric restriction.

It is hard to imagine how such widely different pathological changes (Tables 10.2–10.4) can be caused by a single common mechanism through diet restriction. I propose two general categories of age-related disorders that are indicated by responses to dietary manipulation: (1) disorders that are related as aspects of a syndrome, for example changes linked to kidney disorders or to glycation, and that are more accurately *time-dependent*, since their progress depends on the duration of exposure; and (2) disorders that are consequent to general changes that ultimately occur on all diets and in all environments in *organisms* of that age group, hence are *age-related*. An example would be consequences of ovarian follicle depletion. This classification is also pertinent to other manipulations of senescence by hormones and other external factors discussed in Sections 10.4–10.9.

10.3.2.2. Diet and Kidney Disease of Laboratory Rodents

The kidney diseases of laboratory rodents are among the most labile and manipulatable of senescent disorders. Kidney lesions have an important role in the pathogenesis of many conditions that are often described as age-related (Chapters 3.6, 7.2). Chronic nephropathy to varying extents is common and perhaps ubiquitous in conventionally fed laboratory rodent colonies, where it is associated with degeneration of the glomerulus (Chapter 3.6). Maeda et al. (1985) describe this complex of lesions as "renal-hyperparathyroid disease" and hypothesize that its root cause is kidney disease. In rats fed *ad libitum* with casein as the protein source, circulating parathyroid hormone increased progressively from middle age onwards, reaching massive 25-fold elevations when kidney disease reached its "end stage" (Figure 10.8). Diet restriction greatly reduces the progression of kidney degeneration, and also blunts the elevations of parathyroid hormone and the decrease of serum 1,25-dihydroxyvitamin D (Armbrecht et al., 1988). These results have counterparts in human kidney disease, where progression to the end stage can be delayed by reducing protein intake (Brenner et al., 1982).

A striking recent finding is that substituting soy for casein as the protein source in *ad libitum*–fed rats prevented the age-related progression of chronic nephropathy and increases of plasma parathyroid hormone at least as effectively as diet restriction. Consequently, there was a major amelioration of the abnormal (metastatic) deposition of calcium in soft tissues and the loss of bone calcium, which are strongly linked to the parathyroid gland hyperactivity that often accompanies major kidney impairments (Table 10.5). However, the usual age-related increase of plasma calcitonin also occurred on the soy-substituted diet (Figure 10.8; Kalu et al., 1988). The mechanism by which different diets have such major influences on kidney and parathyroid functions is unknown, and Kalu et al. suggest two factors: The ratio of lysine to arginine, which is implicated in atherogenesis and hypercholesterolemia (Kritchevsky et al., 1982) and which differs between soy and casein, could influence degeneration in kidney blood vessels. Alternatively, casein may be more antigenic because of its complex glycoprotein content and might promote kidney degeneration through the formation of immune complexes in kidney tubules. Another factor could be blood glucose (Section 10.3.2.3). The dramatic influences of substi-

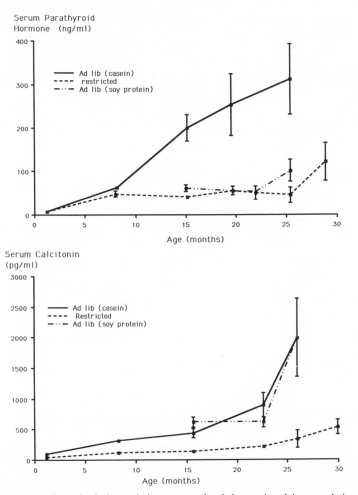

Fig. 10.8. Effects of soy diet and caloric restriction on age-related changes in calcium-regulating hormones in male Fischer 344 rats. *Top,* serum parathyroid hormone; *bottom,* calcitonin. The radioimmunoassay detected the N-terminal bioactive form and minimized interference by degraded forms of the hormones. Redrawn from Kalu et al., 1988.

tuting soy for casein has major implications for many published studies that will take some time to evaluate.

The kidney is thus a likely origin of many changes in Table 10.5, in view of the overall correlation between histopathological changes in the kidney and elevations of parathyroid hormone. However, the early increases in parathyroid hormone precede kidney impairments (Kalu et al., 1988). Age-related declines in calcium uptake from the gut are not likely to be a factor, since parathyroid hormone increases also precede changes in plasma calcium and its uptake from the gut (Armbrecht et al., 1984). Serum calcium itself is not notably altered during the adult lifespan on *ad libitum* or restricted diets (Kalu, Hardin, Cockerham, and Yu, 1984; Kalu et al., 1988). Whatever their origin, elevations of parathyroid hormone are implicated in the mobilization of bone calcium and soft tissue calci-

Table 10.5 Dietary Influences on the Age-related Renal-Parathyroid Disease Complex of
Male Fischer 344 Rats

	Casein Diet, Caloric Restriction	*Ad libitum* Diet, Soy Protein
Kidney lesions		
Glomerulosclerosis	↓	↓
Parathyroid and thyroid gland lesions		
Elevated serum parathyroid hormone[1]	↓	↓
Elevated serum calcitonin	↓	↓
Parathyroid glandular hyperplasia	↓	0[2]
Calcium deposit in soft tissues		
Heart muscle	↓	↓
Kidney	↓	↓
Skeletal muscle	↓	↓
Osteodystrophy		
Gradual loss of bone mineral	↓	↓
Loss of bone mineral just before death	↓	0
Replacement of bone mineral by fibrous connective tissue	↓	↓
Gastric ulcers	↓	↓
Liver lesions[3]		
Fatty degeneration	↓	↓
Bile-duct hyperplasia	↓	↓
Periductal fibrosis	↓	↓

Sources: Compiled from Kalu, Hardin, Cockerham, and Yu, 1984; Kalu, Hardin, Cockerham, Yu, Norling, and Egan, 1984; Kalu et al., 1988; Iwasaki et al., 1988a,b; Maeda et al., 1985.

Note: ↓ , decrease of age-related trend; ↑ , increase; 0, no change.

[1]Although the age-related elevations of serum parathyroid hormone on *ad libitum* casein diets are delayed and greatly blunted, modest elevations occur by 30 months (Kalu et al., 1988).

[2]Edward Masoro, pers. comm.

[3]An overall relationship between liver and kidney lesions on restricted and soy diets is indicated from the reports of Maeda et al., 1985, and Iwasaki et al., 1988a,b. However, the inclusion of liver disorders in the renal-hyperparathyroid disease complex is speculation by C. E. Finch and was not revealed by correlations of the extent of liver and kidney lesions in individual rats in studies with casein diets in Maeda et al., 1985; possible histopathologic correlations were not examined for the soy diet (Edward Masoro, pers. comm.).

fication. Gastric ulcers are also associated with severe kidney disease, possibly through uremia, which predisposes mammals to ulcers (discussed in Maeda et al., 1985).

The renal-hyperparathyroid complex cannot account for all the features of senescence, as shown by two examples: the incidence of testicular tumors was unaltered by the soy diet; and there was a marked loss of bone calcium just before death on soy diets, yet without marked elevations of parathyroid hormone or calcitonin (Kalu et al., 1988). Moreover, the incidence of lymphomas and mesenchymal tumors increased severalfold, while mean and maximum lifespan improved by only 10% (Iwasaki et al., 1988a). The fact that kidney disease and its concomitant can be suppressed on some diets with but little increase of lifespan argues that age-related mortality must have other causes. These extremely common chronic degenerative kidney conditions must be taken into account in evaluating effects of diet on age changes in other systems and functions, for example alterations of protein synthesis in liver on casein diet, which could be linked to fatty changes and bile-duct pathology (Table 10.5).

In essence, the kidney is a vulnerable organ that, by exposure to the demands of metabolism imposed by certain diets or in association with diseases like hypertension and diabetes, can suffer cryptic damage that accumulates irreversibly, whether the damage is sporadic or regular. Alone or in combination with other lesions, kidney lesions lurk as a potential cause of mortality risk that increases with age, but at ages that are generally beyond the major phase of reproduction. According to the premises of Haldane and others (Chapter 1.5.2), the design features of the mammalian kidney are sufficient for propagation of the species because sufficient functions are maintained into mid-life. The modulations of kidney disease by diet demonstrate the conditionality of these design features, which are determined during development by the regulation of genes that can be considered equivalent to conditional lethals. As discussed in Section 10.3.3, the protective effect of diet restriction on kidney and other diseases can be considered in the context of evolutionary strategies for reproduction.

Although more optimum diets may minimize these kidney conditions, we should not invalidate previous studies of laboratory rodents on this account. Analogous situations doubtless occur throughout human populations (Brenner et al., 1982). Even though the incidence of specific diseases varies widely among nations and subpopulations of older age groups (Jones, 1956; Figure 8.9), the acceleration rate of mortality is remarkably stable (Figure 1.2). Whatever the prevalence of particular diseases, if a strong age-related incidence is manifested, then the mechanisms of this age-related incidence can be sought and are of basic interest. In this regard, some studies of humans detect an age-related trend for elevations of circulating parathyroid hormone (Jowsey and Offord, 1978; Wiske et al., 1979; Marcus et al., 1984), which could be a response to decreased calcium absorption and which also could facilitate bone resorption (Riggs and Melton, 1986). The common but not universal occurrence of age-related kidney impairments (Chapter 3.6) thus could represent subgroups showing aspects of this same disease complex seen in rodents.

10.3.2.3. Glycation

Another mechanism in diet restriction may proceed through the nonenzymatic addition of glucose to long-lived proteins and its spontaneous conversion to AGE (advanced glycation or Amadori endproducts; Chapter 7.5.2). Ongoing studies of F344 rats show that diet restriction reduces blood glucose by about 15% (Masoro et al., 1989). Correspondingly, there is a decrease of glycated hemoglobin (HbA_{1c}), as observed during the correction of hyperglycemia in human diabetics. The glycation of collagen and the formation of AGE could be the molecular basis for thickening of the basement membrane in the kidney mesangial matrix, a change that is associated with kidney failure in diabetes and that may be a factor in age-related reductions of kidney function, even in the absence of clinically defined diabetes. Another possibility concerns insulin transport. Because insulin transport from plasma across capillary endothelia to the interstitial fluid is the rate-limiting step by which insulin stimulates glucose uptake (Yang et al., 1989), glycation of capillary membranes could be a factor in age-related insulin resistance (Marilyn Ader, pers. comm.).

Demyelinating disorders, which are common during diabetic neuropathy and which occur to a varying but usually lesser extent at later ages in nondiabetics, could involve formation of AGE in myelin (Vlassara et al., 1984). The reduction of an age-related peripheral neuropathy in rats by diet restriction (Table 10.3; Everitt et al., 1980) could be mediated by reducing myelin AGE. Because experimental diabetes in rats induces abnormalities in hypothalamic LHRH neurons (Bestetti et al., 1985) and slows axonal transport in the sciatic nerve (Medori et al., 1985), the impact of glycation on brain functions could be profound, especially in brain regions that influence metabolism. For example, elevations of blood glucose that may impair neuronal functions in the ventromedial nucleus and other hypothalamic regions could provoke a vicious feedback cycle, by the formation of AGE that could impair neural control of pituitary secretions of growth hormone or autonomic influences on glucose metabolism (Mobbs, 1990).

Cataracts, an important age-related disorder, may also involve AGE in lens crystallins. Cataracts and loss of soluble eye proteins in the Emory mouse strain, which has a high spontaneous incidence by 12 months, occurred less frequently during moderate caloric restriction (Taylor et al., 1989). Caloric restriction also strikingly retarded the loss of soluble γ-crystallins in normal hybrid mice (Leveille et al., 1984). Loss of γ-crystallins is associated with cataract development in humans (Bloemendal, 1981) and could be linked to the loss of free sulfhydryl (–SH) groups observed in total lens proteins (Kuck et al., 1982; Chapter 5.4.2). An imposing array of age-related degenerative conditions could be linked to effects of the diet on blood glucose levels (Tables 10.2–10.5). These and many other changes should be modifiable by favoring conditions that reduce the formation or enhance the removal of AGE. As noted in Chapter 7.5.2, the initial glycation event is reversible, with a half-life of about 5 days.

10.3.2.4. *Metabolic Rate and Oxidative Metabolism*

A major new approach was recently made to Sacher's hypothesis (1977, 628) that caloric restriction increases longevity by reducing metabolic rate. Several studies showed that daily caloric intake per gram body weight is not altered by caloric restriction of F344 male rats (Masoro et al., 1989; Yu et al., 1985) or of a long-lived hybrid, (C3H × B10.R, i.e. C3H.SW/Sn × C57BL1O.RIII/Sn)F$_1$, female mice (Weindruch et al., 1986). Moreover, daily oxygen consumption per gram lean body weight was not influenced by caloric restriction at 6, 12, or 18 months (McCarter et al., 1985; Yu, 1987). The absence of effect on maximum lifespan by exercise under most conditions (Holloszy et al., 1985; Goodrick et al., 1983; Ingram and Reynolds, 1983) or cool temperatures (Section 10.2) also shows that modest variations (ca. 50% or less) in metabolic flux have little influence on lifespan in laboratory rats. However, the lower mean core temperature of diet-restricted rodents that has been well documented for mice (Weindruch and Walford, 1988, 209) but only recently established in rats (Nakamura et al., 1989; Duffy et al., 1989) raises the possibility that temperature-driven processes could be retarded from this cause alone. While the boundary values within which lifespan is independent of metabolic flux are unknown, their upper limit could be estimated from thermodynamics as a function of body size and ambient temperature.

The unknowns concern just how diet restriction, ambient temperature, and exercise influence cell metabolism. There is every reason to expect differential responses of particular organs and cells to varying metabolic fluxes. Although total body metabolism per gram lean body mass might not be altered by caloric restriction, metabolism in specific organs and cells could be differentially altered. It is becoming clear that diet-age interactions differ in their effect on oxidative metabolism between liver and skeletal muscle. In homogenates of rat liver at ages across the lifespan, the capacity for oxidation of different substrates representing the different pathways for amino acids, carbohydrates, and lipids did not decrease with age up to 30 months, whether rats were fed *ad libitum,* calorically restricted by alternate-day feeding, or allowed regular exercise (Figure 10.9; Rumsey et al., 1987).

Diet-restricted rats had elevated respiratory capacities at all ages, which could be an adaptive response to the phase of fasting imposed by alternate-day feeding (Rumsey et al., 1987). Similarly, liver mitochondria from restricted mice showed increased oxidative efficiency (Weindruch et al., 1980). A study of liver enzymes of intermediary metabolism found evidence that diet restriction decreased glycolytic pathways and lipid synthesis pathways, while increasing the efficiency of glucose synthesis (Feuers et al., 1989). For example, pyruvate kinase, which is a key enzyme in glycolysis, had 50% lower activity, while glycerol kinase and several amino acid dehydrogenases and transaminases had increased activity, consistent with enhanced gluconeogenesis.

Skeletal muscle respiratory capacity shows age-related decreases, irrespective of access to exercise in rodents (Figure 10.9); the mechanisms involved could be atrophy through denervation or through mitochondrial DNA damage (Chapter 8.4.1). The content of cytochrome *c* in skeletal muscle decreased after 18 months, indicating a net loss of mitochondria. The levels were maintained at later ages in liver by diet restriction and in muscle by exercise (Figure 10.9). The turnover number of cytochrome *c* oxidase was not altered by age or treatment in liver or muscle. Because of the decreased density of mitochondria at advanced ages, as determined morphometrically from electron micrographs in liver and myocardium of C57BL/6J mice after 30 months (Herbener, 1976), decreased respiratory capacity could be expected. These studies extend earlier conclusions that the capacity for oxidative metabolism and the generation of ATP is not compromised at later ages in liver, heart, brain, or most other tissues in healthy humans or rodents (Masoro, 1985; Sanadi, 1977; Meyer et al., 1985; Duara et al., 1985; Weindruch et al., 1980). In a thorough review, Hansford (1983) also concluded that mitochondrial functions were intact in senescent mammals. Though many report reduced activities for some substrates, most agree that the maximum respiratory capacity is intact throughout the lifespan.

The enhancement of metabolic capacity by diet restriction or exercise in liver and muscle further weakens the hypothesis that longevity is inversely related to metabolic rate, either at the level of the whole organism or of major organs. As indicated from the review of macromolecular biosynthesis (Chapter 7), age-related changes in the respiratory capacity vary widely among tissues. The general trend for age-related increase of fat depots at the expense of skeletal muscle will probably account for much of the reduction in whole-body metabolism of mammals (Masoro, 1985; Rumsey et al., 1987). As described in Chapter 7.2 for age-related changes in macromolecular biosynthesis, the literature on respiratory capacity is controversial. Some reports that inferred major mitochondrial deficits (e.g. Weinbach and Garbus, 1959) may have been biased by comparisons of very

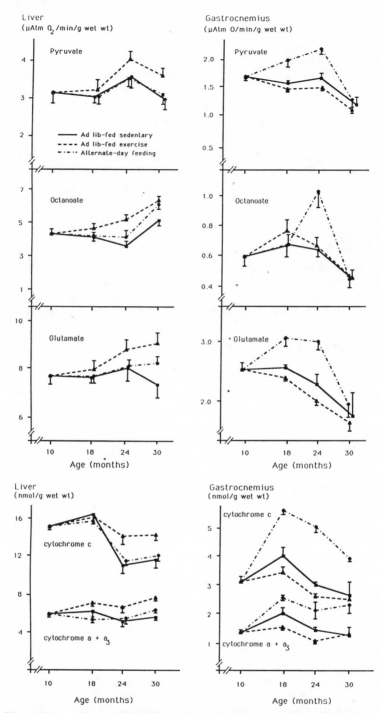

Fig. 10.9. Diet restriction and exercise modify age-related changes in oxidative metabolism in liver and skeletal muscle of Fischer 344 male rats. Rats were fed on alternate days to about 60% of the average daily food intake of *ad libitum*–fed sedentary rats, or were given access to exercise wheels while fed *ad libitum. Top,* the maximum oxidative (respiratory) capacity of homogenates with substrates that represent carbohydrates (pyruvate), lipids (octanoate), and amino acids (glutamate). *Bottom,* cytochromes. Redrawn from Rumsey et al., 1987.

young and senescent rats to the omission of intermediate ages, or to artifacts arising from increased fragility of mitochondria from senescent rats (Murfitt and Sanadi, 1978) that could cause differential loss during isolation (Rumsey et al., 1987).

10.3.2.5. Free Radicals and Lipid Peroxidation

Diet restriction may also influence free radical formation. A decreased synthesis of lipids was proposed by Feuers et al. (1989) as a factor in the decreased liver lipid peroxidation observed during diet restriction (Ruggeri et al., 1987; Koizumi et al., 1987), which could, in turn, decrease the formation of fatty acid epoxides: these later compounds are sources of free radicals and are highly mutagenic (Frankel, 1987). Diet restriction also increases free radical scavengers in rodent liver, for example catalase (Koizumi et al., 1987) and free glutathione (Lang et al., 1990); other neutralizing agents of free radicals are reviewed by Weindruch and Walford (1988, 256).

A potent free radical scavenger, diethylhydroxylamine, increased the lifespan of Swiss mice by about 10% without influencing body weight (Heicklen and Brown, 1987). A mechanism could involve the slowing of kidney disease, since the treated mice had smaller increase of daily water intake at advanced ages; increased water intake is often associated with kidney dysfunction. The effect on kidney diseases of other free radical scavengers that increase lifespan is unknown. Similar increases of lifespan are reported for 2-mercaptoethylamine (Harman, 1968), ethoxyquin (Santoquin; Comfort et al., 1971), and 2-ethyl–6-methyl–3-hydroxypyridine (Emmanuel, 1976). Unfortunately, evaluation was not done of the effects of these treatments on pathology in the kidneys or other organs.

The role of free radical damage remains speculative in most age-related pathology, though the case is strongest for cancers as the result of cumulative effects from free radical–inducing endogenous and dietary carcinogens (Pryor, 1987; Harman, 1981; Ames, 1983). Possible influences of diet restriction on damage to DNA were not shown in urinary excretion of thymine glycol, a modified DNA base that can be produced by free radicals and then excised by a DNA glycosylase in a repair step (Chapter 8.2.2; Cathcart et al., 1984). Diet restriction may, in fact, *increase* the production of thymine glycol, according to preliminary analysis with normalization to active metabolic mass (Saul et al., 1987). Much work is needed to establish the relationship between urinary thymine glycol and oxidative damage to DNA. While diet-restricted rats had enhanced DNA repair, as assayed in UV-irradiated skin cells (Lipman et al., 1989) and splenocytes (Licastro et al., 1988), these observations do not prove that the extent of endogenous DNA damage was altered by diet.

A new topic concerns the relationships between diet and lipid peroxidation. Because membrane lipid peroxides are also formed by free radical attack, it is of interest that diet restriction reduced lipid peroxides by about 50% during the lifespan in mitochondria and microsomes from F344 rat livers, but nonetheless did not alter the progressive increase with age (Laganiere and Yu, 1987). Lipid peroxidation is strongly implicated as a factor in atherogenesis (e.g. Steinberg et al., 1989; Morel et al., 1984). The uptake of LDL particles by vascular endothelial cells is enhanced by lipid peroxidation, as is the chemotactic activity; both changes are involved in the complex, multifactorial influences on the accumulation of blood lipids in vessel walls. The manipulations of diet to reduce LDL can thus be further refined to influence lipid peroxidation. Ongoing studies show that antioxidants like probucol reduce atherogenesis in rabbits independently of influences on cholesterol levels (Carew et al., 1987; Kita et al., 1987). Thus, genetic influences on the constitutive levels of antioxidants or on changes in endogenous antioxidants with age could be mechanisms through which genes influence atherogenesis.

Lipid peroxides are also implicated in the formation of lipofuscins (Chapter 7.5.6), and diet restriction decreased the amount of lipofuscin in rat myocardium and brain (Chipalkatti et al., 1983; Enesco and Kruk, 1981). Other chronic treatments that inhibit the formation of lipofuscins have used vitamin E (Tappel et al., 1973; Freund, 1979), a well-known antioxidant, and in regard to the brain two mild drugs with minor effects on behavior in the doses used, centrophenoxine (Nandy and Bourne, 1966) and Hydergine, an ergot mixture (Amenta et al., 1988). These drug manipulations may involve indirect mechanisms but are pertinent because they demonstrate the manipulability of even such a general phenomenon as lipofuscin accumulation in permanently nondividing cells.

In regard to the general rationale for free radical theories, at least we now know that free radicals can transiently increase in the heart during reperfusion with oxygen after ischemia (Baker et al., 1988). An open question is the linkage of diet to the risk of damage by free radicals during transients of tissue hypoxia and reoxygenation, which may be expected to increase with age, possibly during sleep apneas that occur with increasing frequency in older age groups (Chapter 3.6.9).

10.3.2.6. DHEA

The weak adrenal androgen DHEA (Chapter 3.6) is receiving much attention because of its large and early decreases during mid-life in both men and women (Figure 3.36) and because it has remarkable effects in inhibiting diseases of rodents that include cancer, kidney, and vascular disease. Experimental studies generally administer DHEA in the diet, in the proportion of about 0.5% by weight of the total food intake. For general reviews, see Schwartz, Whitcomb, Nyce, Lewbart, and Pashko, 1988, and Gordon et al., 1987. In human populations, prospective studies indicate that DHEA levels are inversely correlated with breast cancer incidence (Bulbrook et al., 1971) and with death from cardiovascular disease after 50 (Barrett-Conner et al., 1986). These broad effects may be related to metabolism and diet and are thus considered now, rather than in later sections on other steroid hormones. This thriving area of investigation is resurgent with serendipity.

Mammary tumors are greatly inhibited by chronic treatment with DHEA in female C3H(A^{vy}/a) mice (Schwartz, 1979), a strain that has a 50% tumor incidence by 10 months. DHEA inhibits lipogenesis in VY(A^{vy}/a) mice and blunts obesity in this strain (Yen et al., 1977; Schwartz, 1979).[2] These results are consistent with the prospective study of Bulbrook et al. (1971), which showed that risk of mammary cancers was inversely correlated with urinary excretion of steroids derived from DHEA. The chemical induction of tumors in other organs is also blocked by DHEA (Schwartz and Tannen,

---

2. Some readers may be tantalized that obesity in adult mice (Schwartz, 1979) was prevented without altering *ad libitum* food intake, but should also be aware that DHEA impairs postpubertal growth in rats (Garcea et al., 1988).

1981; Nyce et al., 1984; Pashko et al., 1984; Moore et al., 1986). These effects are consistently shown by rodents.

The molecular mechanisms of these effects involve distinct stages in experimental carcinogenesis. DHEA blocks carcinogen activation by mixed-function oxidases and blocks tumor promotion by phorbol esters as assayed by formation of superoxide radicals in neutrophils and by the stimulation of DNA synthesis in carcinogen epidermal target cells (Hastings et al., 1988). A commonality of DHEA and diet restriction is suggested by their similar inhibitions of the binding of the carcinogen dimethylbenz(a)anthracene to mouse skin DNA and the stimulation of epidermal DNA synthesis by phorbols (Schwartz and Pashko, 1986). DHEA also inhibits the differentiation of 3T3 fibroblasts to adipocytes at confluence or in response to insulin (Gordon et al., 1987; Shantz et al., 1989).

DHEA also has several links to vascular disease that are consistent with its inverse relation to coronary disease risk in middle-aged humans (Barrett-Conner et al., 1986). Feeding DHEA to rabbits reduced the size of atherogenic plaques after aortic endothelial injury (Gordon et al., 1988) and the size of fatty aortic streaks on cholesterol-rich diets without prior injury (Arad et al., 1989). No clear relationships of the induced vascular lesions to blood lipid components were found by these studies. Scattered reports on possible relationships of blood cholesterol and lipids to DHEA in various human groups are reviewed by Arad et al. (1989). In view of the implied role of lipid peroxidation of LDL particles in atherogenesis (Steinberg et al., 1989), the impact of DHEA on lipid peroxidation may be pertinent to atherogenesis.

Returning to systemic pathophysiology, the feeding of DHEA for 1 year but not 1 month, markedly improved glucose tolerance, measured by plasma glucose after a glucose load, in 2-year-old C57BL/6J male mice (Coleman et al., 1984). Readers may recall the pancreatic β-cell degeneration associated with the *db* mutation in mice. Treatment with DHEA (see above) from weaning through 4 months suppressed most effects of the *db* mutation on both C57BL/Ks and C57BL/6J mice, including necrosis of pancreatic β-cells; these effects were largely independent of food intake and not related to the inhibition of glucose-6-phosphate dehydrogenase in several tissues (Coleman et al., 1984). DHEA added to a low-fat diet at 10 months reduced urinary protein excretion (a measure of kidney degeneration) by 80% at 17 months in male Sprague-Dawley rats and C57BL/6J mice; body weight and food intake were slightly less (Pashko et al., 1986). In contrast to these findings, another genotype, the (C3H × B10.RIII)F$_1$ hybrid, did not eat all DHEA-containing food provided but also gained less weight per gram of food eaten; enhanced phytohemagglutinin-induced splenocyte proliferation at 18 weeks was probably due to the effects of reduced caloric intake (Weindruch et al., 1984). The interactions of genotype and DHEA have not been studied in much detail.

A common mechanism in these diverse actions appears to involve intermediary metabolism, particularly through the cofactor NADPH and ribose-5-phosphate. DHEA can directly and uncompetitively inhibit glucose-6-phosphate dehydrogenase (GPDHase), which is a major source for NADPH and a contributor to lipogenesis (Oertel and Rebelein, 1969; Marks and Banks, 1960; Ranieri and Levy, 1970; Lopes and Rene, 1973). By inhibiting lipogenesis, it is possible that DHEA mimics an aspect of diet restriction. A recent report showed clear histochemical evidence of decreased hepatic GPDHase in DHEA-treated rats; moreover, there were 50% decreases in the flux of glucose through the hexose monophosphate shunt or pentose phosphate pathway and in the levels of ribulose-5-phosphate (Garcea et al., 1988), which is produced only by this pathway and is the immediate precursor of D-ribose. On the basis of this finding and evidence from several laboratories that exogenous deoxy- and ribonucleosides reverse the inhibition of proliferation from DHEA, Garcea et al. hypothesize that DHEA causes a critical limitation of nucleosides.

The link to GPDHase is controversial, however, because DHEA treatment greatly reduced diabetes and obesity without influencing GPDHase in several tissues in some mouse strains (Coleman et al., 1984). Moreover, 16-α-bromo-epiandrosterone, a DHEA analogue that gives sixtyfold better inhibitor of GPDHase, had no effect on blood sugar or weight gain in C57BL/Ks mice. Another mechanism could involve sex steroids, since DHEA can be converted to estrogen or testosterone and since estrogens reduce fat mass, among other metabolic effects (Leiter et al., 1984). However, a new fluoridated analogue of DHEA (16α-fluoro–5-androsten–17-one) that lacks the estrogenic and androgenic activities of DHEA is more effective in suppressing weight gain (Schwartz, Lewbart, and Pashko, 1988) and in inhibiting skin-tumor initiation and promotion than DHEA itself (Schwartz et al., 1989). The estrogenic and androgenic actions of DHEA thus can be distinguished from its antiobesity and anticancer actions in rodents. The influence of DHEA on the glycation of proteins is unknown. The effects of DHEA on glycolytic metabolism point to the need for more comparisons with diet restriction.

10.3.2.7. Diet Restriction and Macromolecular Biosynthesis

Age-related decreases in transcription and protein synthesis in the liver (Chapter 7.4) can be extensively manipulated by diet restriction. Changes in bulk protein synthesis also show extensive amelioration during diet restriction, though decrements are not ultimately prevented. Ward (1988), from the San Antonio group, showed that diet-restricted F344 rats maintained 35% greater rates of bulk hepatic protein synthesis throughout their longer lifespans, although decreases occurred irrespective of diet during mid-life (Figure 10.10). This result differs from most other effects of diet restriction, which delay as well as reduce histopathologic changes. Thus, diet restriction can have both *heterochronic* and *isochronic* effects on age-related changes (Section 10.1). The major decrease between 6 and 12 months is difficult to attribute to kidney or liver histopathology, although up to 50% of restricted rats have a low-grade hyperplasia of their bile ducts by 12 months (Maeda et al., 1985), a cellular change that could conceivably alter protein synthesis.

Few specific genes and their products have been characterized for effects of diet restriction. One example is the liver proteins, α_{2u}-globulin and cytosol androgen-binding protein, which decrease strikingly by 2 years in rats (Figures 7.3–7.5; Richardson et al., 1987; Chatterjee et al., 1988). Another protein, SMP-2, changes reciprocally, being high at early and then at later ages. These studies used the F344 male rat, which has mid-life onset of benign testicular tumors and larger decreases of plasma testosterone (Amador et al., 1985; Chatterjee et al., 1988) than in some mouse strains (Nelson et al., 1975). Diet restriction after 6 weeks blunted the changes in protein and mRNA. In particular, the alterations of α_{2u}-globulin and SMP-2 by diet restriction were at the transcriptional level, as determined by nuclear run-on assays (Richardson et al., 1987; Chatterjee et al., 1988).[3] The transcription of catalase and superoxide dismutase by liver nuclei showed similar effects of diet restriction in retarding age changes (Richardson et al., 1987).

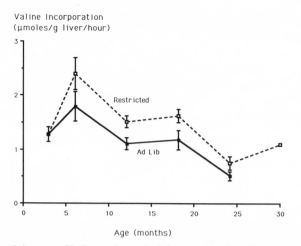

Valine Incorporation
(μmoles/g liver/hour)

Age (months)

Fig. 10.10. Age-related decrease of bulk protein synthesis in *in situ*–perfused livers of Fischer 344 male rats. Valine was used as a precursor that rapidly equilibrated with valyl-tRNA (Khairallah and Mortimore, 1976). The restricted rats on 60% *ad libitum* daily food from 6 weeks onwards had 35% greater rates of protein synthesis at all ages after 6 months. An older age group could be assayed in the restricted rats because of their greater lifespans. Redrawn from Ward, 1988.

These data are the first to show that diet restriction influences transcription and are also consistent with the elevated hepatic respiration described above. In summary, diet restriction has extensive influences on the extent of age-related changes in macromolecular biosynthesis in the few tissues of rodents studied, but much work is needed to establish the generality of these changes. In view of the influence of diet on metabolic hormones, hormones are an obvious class of candidates for the mechanisms by which diet restriction might influence transcription.

10.3.2.8. Homeostasis, Hormones, and Defense Mechanisms

Diet restriction has important influences on the regulation of immune functions and homeostasis for which the specific relation to increased lifespan is unclear. Examples from this intriguing new area also show the potential for major manipulations of the senescent phenotype.

10.3.2.8.1. Immunological Functions

Diet influences numerous aspects of immune changes with age in rodents. Among the early changes during maturation are thymic involution and decreased blood levels of

---

3. The retention of the cytosol androgen-binding protein is consistent with maintained transcription of the testosterone-responsive α_{2u}-globulin and SMP-2 genes, which are regulated positively and negatively, respectively, by testosterone. However, diet restriction did not alter the age-related decrease of plasma testosterone (Chatterjee et al., 1988); this is puzzling in view of the reduction of testicular tumors in diet-restricted F344 rats (Table 10.4).

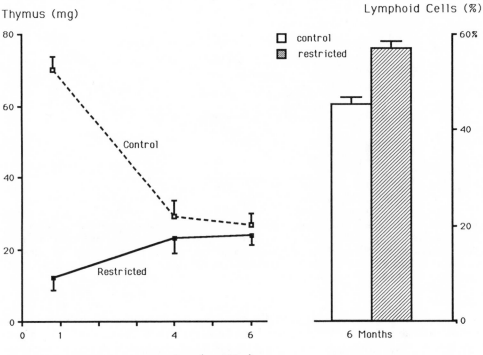

Fig. 10.11. Thymus-gland involution during maturation is prevented by diet restriction. Alternate-day feeding was begun at weaning of (C57BL/10Sn × C3H/HeDiSn)F₁ female mice. The figure represents restriction on the 35% casein diet, which yielded 60% of the average caloric intake on an *ad libitum* diet with 20% casein. The fraction of the thymic cortex occupied by large and small lymphoidal cells was determined morphometrically at 6 months; the fractional volume of lymphoid cells in the medulla was not altered by diet restriction. Redrawn from the data of Weindruch and Suffin, 1980.

thymic hormones (Chapter 3.6.6), which show quite different impacts of diet on mice. Caloric restriction begun at weaning counteracted the usual thymic involution ($> 50\%$ weight loss) and reduction of thymic cortex lymphoid cells as measured at 6 months (Figure 10.11; Weindruch and Suffin, 1980). Because the thymic cortex is a major site for the differentiation of T lymphocytes, the greater cortex lymphocyte pool in restricted mice predicts a greater pool of circulating T cells. Responses of spleen cells to T-cell mitogens were also improved by diet restriction (Weindruch et al., 1979; Weindruch, Kristie, Naeim, Mullen, and Walford, 1982; Cheney et al., 1983). Diet restriction may delay the age of maximum responses by T and B cells (Cheney et al., 1983). Moreover, mice restricted from weaning had threefold more T helper and cytotoxic T cells when nearly 3 years old, indicating a delayed loss of key T-cell precursors (Miller and Harrison, 1985). By contrast, diet restriction did not prevent the major decrease of serum thymosin$_{\alpha 1}$ between weaning and 2 months or the more modest and variable decrease by 26 months (Weindruch et al., 1988). Thymosin$_{\alpha 1}$ is a systemic immunomodulator that acts on T helper cells and stimulates IL-1 among other lymphokines. The consequences of the decrease of thymosin$_{\alpha 1}$ at puberty are unknown, but seem less important than alterations of T-cell precursors.

The mitogenic response of spleen T cells continues to decrease throughout the life of long-lived strains, yet this decrease can be postponed (a heterochronic shift) if diet restriction is begun at 12 months. Some amelioration occurs with later onset restriction, though the impact is less (Figure 10.12; Weindruch and Walford, 1982). Two effects of adult onset restriction are salient: that diet influences the kinetics of the plaque-forming spleen cell responses to sheep erythrocytes more than it does the peak response; and that spleens of restricted mice were smaller, yet had nearly twofold more T cells. It is also of interest to theories relating DNA repair to senescence that splenic lymphocyte DNA repair capacity (Chapters 5, 8.2.6) declines slightly more slowly in restricted mice (Licastro et al., 1988). Diet restriction also blunted the early onset autoimmune disease of short-lived NZB and MRL/l mice (Chapter 6.5.1.7.3; Table 10.2) and reduced the incidence of brain reactive autoantibodies in senescent mice (Nandy, 1982).

Finally, a different aspect of diet may be pertinent to cellular immunity changes that have been associated with decreased membrane fluidity (Rivnay et al., 1983; Naeim and Walford, 1985). Lipids obtained from egg yolk ("active lipid" AL 721) reportedly restore mitogenic responses of lymphocytes from senescent mice (Rivnay et al., 1983) and of older volunteers (Rabinowich et al., 1987). It is not surprising that various lipids could alter membrane fluidity and mitogenic responses, but more detailed cellular studies are needed.

Together these data suggest a complex picture of dietary influences on T-cell differen-

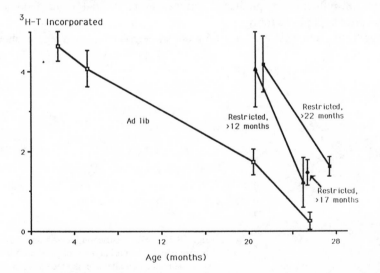

Fig. 10.12. Age-related decreases in the mitogenic response of spleen T cells is slowed by caloric restriction in (C57BL/10Sn × C3H/HeDiSn)F$_1$ male mice. Splenocytes were stimulated *in vitro* by phytohemagglutinin and mitogenic responses were assayed by the incorporation of [³H]-thymidine. Age-related declines of mitogenic responses are postponed by caloric restriction if begun at 12 or 17 months; values at 21 months were identical to young *ad libitum*–fed controls. By 28 months, the effects were nearly gone. *Open square, ad libitum*–fed; *solid triangle,* restricted after 12 months; *solid circle,* restricted after 17 months; *solid square,* restricted after 22 months. Redrawn from Weindruch, Gottesman, et al., 1982.

tiation with diverse effects on T-cell populations in various reticuloendothelial compartments. But the picture is far from clear. Lymphocyte subtypes in various locations need to be characterized to establish the generality of effects from diet restriction. The most likely interpretation of restored mitogenic responses by various lymphocyte populations is that diet restriction increased the number of responding cells. Although improved immune functions are correlated with increased lifespan, this is not likely to be the only cause, in view of the reduction of kidney diseases (Section 10.3.2.2). There is little evidence that infectious disease is a major cause of death or morbidity in most contemporary laboratory colonies. However, in times when infectious lung diseases were endemic (up through the 1960s), diet restriction greatly reduced the chronic pneumonia that was considered to be the major cause of death in senescent rodents (Saxton et al., 1946). No specific mechanism is known that explains the reduction of spontaneous tumors and enhancement of immune functions during diet restriction. Longitudinal studies might show how lifespan, tumor incidence, and immune functions are interrelated.

10.3.2.8.2. Female Reproductive Senescence and Oocyte Loss

Loss of ovarian follicles and oocytes is retarded by alternate-day feeding of adult mice, beginning at 3 months (Figure 10.13; Nelson et al., 1985); the oocytes and primordial follicles remaining at 12 months corresponded to the numbers in *ad libitum*–fed mice at 10 months. These influences on a major age-related cell-loss phenomenon show that the death of irreplaceable cells in adult mammals is subject to extrinsic factors; as discussed in Section 10.8, influences by pituitary hormones are implicated by the major slowing of oocyte loss after hypophysectomy.

When diet restriction was ended at 10.5 months, the mice regained regular short estrous

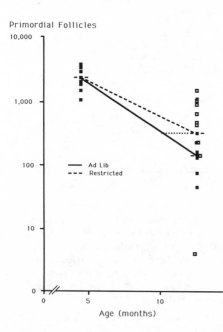

Fig. 10.13. Diet restriction slowed the age-related loss of ovarian oocytes and primordial follicles in C57BL/6J mice that were maintained on alternate-day feeding from 3 months to 10 months. Diet-restricted mice initially had longer estrous cycles and ceased cycling by 7 months. After restoration of *ad libitum* feeding at 10 months, the formerly restricted mice rapidly regained cycles of a length typical of younger mice, while 80% of controls showed the usual age-related acyclicity. The number of primordial follicles at 12.5 months was equivalent to *ad libitum*–fed controls at 10 months. Redrawn from Nelson et al., 1985.

cycles, which resembled those of mice 2 to 5 months younger (Nelson et al., 1985). These results are extended by other studies that showed that diet restriction of varying severity retarded aspects of female reproductive senescence (Huseby et al., 1945; Ball et al., 1947; Merry and Holehan, 1981; Holehan and Merry, 1985; also reviewed by Weindruch and Walford, 1988, 138). Severe tryptophan deficiencies (very different from the usual caloric restriction!), when corrected at middle age, were shown to have retarded reproductive senescence sufficiently for a few rats to bear offspring as late as 28 months, an age 12 months later than in controls (Segall and Timiras, 1976, 1983).

Although the mechanism is not established, altered hormone regulation through diet-restriction influences on hypothalamic-pituitary functions is implicated. For example, diet restriction of rats caused the preovulatory surge of LH to occur 5–6 hours earlier and diminished the serum levels of FSH and progesterone throughout the cycle (Holehan and Merry, 1985). Other studies indicate altered feedback sensitivity of estrogen on LH from diet restriction (Piacsek, 1985; Howland and Ibrahim, 1973). The early concept that malnutrition causes "nutritional hypophysectomy" (Mulinos and Pomerantz, 1940) clearly does not pertain here, since enough LH and FSH were secreted to induce preovulatory surges and to maintain estrogen-secreting follicles. Extreme deficits of LH and FSH from hypophysectomy also retard ovarian follicle loss (Jones and Krohn, 1961a; Section 10.8). Besides LH levels, nothing is known about effects of diet restriction on pulsatile LH secretion, for which particular frequencies are critical to follicular growth. These humoral influences on a major age-related cell-loss phenomenon are important as the leading example in which loss of irreplaceable cells can be modified by hormonal interventions.

10.3.2.8.3. Metabolic Hormones

Diet has many influences on the regulation of insulin and other hormones that could be pertinent to the salutary aspects of diet restriction, especially in regard to the regulation of blood glucose as a substrate for protein glycation (Section 10.3.2.3; Chapter 7.5.2). With middle age, humans generally tend to have slight increases in fasting blood sugar, and larger and longer-lasting elevations of blood glucose after a glucose challenge (Chapter 3.6.8), which occurs without obesity (Chen et al., 1988). These phenomena are described as the glucose intolerance of aging and are presently attributed to several factors that include pancreatic β-cell dysfunctions and/or to insulin resistance. Insulin resistance is the reduced uptake of glucose by cells in response to an increment of insulin, as measured at constant glucose levels.

A recent study indicates that the glucose intolerance may be rapidly improved by shifting to a high carbohydrate diet for 3 to 5 days, as determined in select nonobese and healthy elderly men (Chen et al., 1988). Comprehensive studies are needed to evaluate the effects of diet and metabolic demand (exercise, ambient temperature) on age-related changes in the regulation of pancreatic and pituitary hormones, which should be sampled throughout the diurnal cycle. The age-related impairments of growth hormone secretion in humans and rodents (Figure 3.31) could be important in view of the direct stimulation of insulin secretion by growth hormone (Altszuler et al., 1968; Fineberg et al., 1970) and the insulin resistance following infusions of growth hormone (Ader et al., 1987; McGorman et al., 1981).

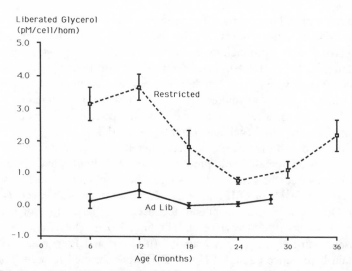

Fig. 10.14. Caloric restriction (60% *ad libitum* after 6 weeks) delayed age-related impairments in the lipolytic response of epididymal adipocytes from male Fischer 344 rats to 1 μg/ml glucagon *in vitro*. Similar results were obtained for perirenal adipocytes. Redrawn from Bertrand et al., 1980.

Age-related changes in the accessibility of metabolic reserves are also subject to influence by diet restriction. One of the earliest age-related impairments in metabolism regulation is the lipolytic action of glucagon on epididymal adipocytes, which usually disappears by 6 months in rats fed *ad libitum* diets. This change is one of the few nongonadal hormonal responses that disappear postmaturationally in rats. Diet restriction, however, preserves this lipolytic response until after 12 months (Figure 10.14; Bertrand et al., 1980). There is no obvious relation between this phenomenon and the reversal of age-related glucose intolerance by short-term high-carbohydrate diet.

10.3.2.8.4.   Responses to Cold Stress

Impairments of thermoregulation during cold stress are a hallmark of senescence (Table 3.3) and are subject to effects of long-term diets. Thermoregulation during senescence as tested by responses to short-term cold exposure was enhanced in mice by chronic alternate-day *ad libitum* feeding (Figure 10.15; Talan and Ingram, 1985) or in rats made severely tryptophan deficient, with subsequent restoration of a balanced diet at mid-life (Segall and Timiras, 1976). Another study of rats found the opposite result, however (Campbell and Richardson, 1988). The mechanisms could involve lesions in the hypothalamic thermoregulatory centers as well as in the peripheral capacity to generate energy through shivering and the readily released energy reserves (glycogen, fatty acids) that fuel nonshivering thermogenesis, for example the enhanced lipolysis of diet-restricted rats (Figure 10.14). It would also be of interest to learn how diet influences control of liver glycogen metabolism by the hypothalamus. Major age-related impairments were observed in *ad libitum*–fed 2-year-old rats by direct electrical stimulation of the ventromedial nucleus, which rapidly induced liver glycogen phosphorylase in the young (Shimazu et al., 1978).

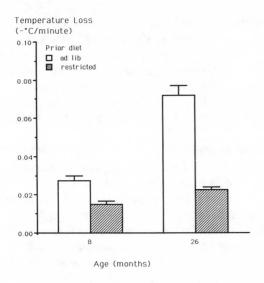

Temperature Loss
(-°C/minute)

Fig. 10.15. Alternate-day fasting eliminated age-related impairments of thermoregulation during acute responses to cold stress. C57BL/6J male mice were kept on alternate-day *ad libitum* feeding after 6 months; different groups were tested at 8 months or 26 months. Thermoregulation was measured with a rectal thermometer during a 3-hour exposure to 10°C on the morning of the day food was removed. The rate of temperature loss was threefold greater in the older daily *ad libitum*–fed mice than in either young group; diet scheduling did not significantly alter responses of young mice. Redrawn from the data of Talan and Ingram, 1985.

10.3.2.8.5. The Nervous System and Behavioral Changes

Caloric restriction influences a variety of age-related changes in the central and peripheral nervous system. In the rat brain, the well-documented loss of D–2 receptors in the striatum is slowed by alternate-day *ad libitum* feeding from 6 weeks onwards (Figure 10.16). Alternate-day feeding also lessened decreases of cerebellar and cortical choline acetyltransferase of rats from this same colony (London et al., 1985). Continuous diet restriction (60% *ad libitum*) lowered striatal dopamine and monoamine levels elsewhere (Kolta et al., 1989). In view of the importance of the cerebellum and striatum to motor control but also to cognitive functions, it is of interest that diet restriction not only maintained the capacity of senescent female mice to balance on a rotating rod, but also improved maze learning (Ingram et al., 1987).

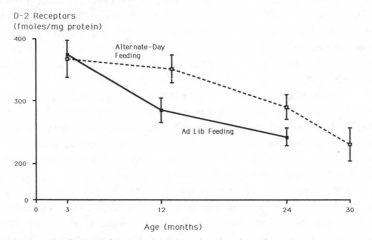

Fig. 10.16. Alternate-day feeding after 6 weeks delays the age-related loss of striatal D–2 receptors in male Wistar rats (spiperone binding to membranes). Diet and age did not influence the dissociation constant for spiperone. Redrawn from Roth et al., 1984.

In view of the positive relation between corticosteroid receptors in the hippocampus and hippocampal neuron loss (Chapter 9.3.4.2; Figure 9.8; Meany et al., 1988), it is pertinent that diet restriction blunted the age-related decline of hippocampal corticosteroid receptors (Robert Sapolsky, pers. comm.). This may mean that hippocampal neuron integrity at later ages is enhanced by diet restriction, which would be consistent with the enhanced maze learning cited above. The effects of diet on age-related ovarian follicle loss and on thermoregulation could be mediated through effects on brain loci, though this remains to be proven.

Several structural changes are also retarded by diet restriction. The loss of dendritic spines in the parietal cortex of male Wistar rats (35% by 24 months) was substantially lessened by alternate-day feeding when begun at 24 months (Moroi-Fetters et al., 1989). Cortical neuron atrophy, however, was not ameliorated and eventually occurred (Peters et al., 1986). The accumulation of brain lipofuscins (Section 10.3.2.5; Chapter 7.5.6) and of brain reactive autoantibodies that Nandy has described in serum of several species at later ages (Nandy, 1982) was also retarded by diet restriction. Autoantibodies could be a factor in the peripheral demyelinating lesions of senescent rats, which are also attenuated by diet restriction (Everitt et al., 1980).

Diet restriction also has important influences on the peripheral nervous system that were first considered as a mechanism in hypertension and that may be extremely pertinent to slowing of senescence. An extensive series of studies on the relationship between obesity and hypertension in humans and various rodent models (reviewed in Landsberg and Krieger, 1989) shows that fasting or caloric restriction suppresses sympathetic activity, as measured by norepinephrine turnover in myocardium and brown fat. Conversely, diets rich in carbohydrate and fat that are associated with insulin elevations enhance norepinephrine turnover. The modulations are thought to be regulated through the glucose-responsive ventromedial nucleus of the hypothalamus.

These observations, while based on relatively shorter duration manipulations of diet, suggest important links to metabolic changes, for example to the age-related trend for elevated plasma norepinephrine and reduced glucose tolerance (Chapter 3.6.8). Thus, the age-related trend for hypertension that is a factor in vascular diseases could have a basis in glucose and insulin regulation. The relationships between blood glucose and sympathetic activities should be investigated during the chronic course of diet restriction. Other dietary influences that may be mediated through blood glucose on hormonal and neuronal age changes can be considered: they are too numerous to list here.

To summarize this lengthy section, the diet composition and the caloric intake can profoundly influence the pathogenesis of a wide range of disorders that increase with age in mammals. These effects of diet are particularly clear in laboratory rodents, where not only is lifespan increased by slowing the acceleration of mortality as shown in five studies (Table 10.1), but numerous specific pathophysiological changes are blunted. Because many age-related diseases and specific biochemical or physiological changes are also modified, a major hypothesis is being discussed, that *diet restriction slows the fundamental rate of senescence.* Moreover, diet restriction selectively changes the alterations with age in the rates of transcription in the liver. Thus, the elements of a causal framework are in place for analyzing the role of changes in genomic expression in relation to manipulations of senescent phenotypes.

Many different mechanisms may be involved in these complex processes. A current topic is glycation, which could be a common factor. Yet other factors could operate through other aspects of metabolism, such as through the influence of glucose on sympathetic activity, or growth hormone on protein synthesis in the liver and muscle. The lower body temperature of diet-restricted rats (Section 10.3.2.4) suggests that temperature-dependent changes like racemization of amino acids in long-lived proteins (Chapter 7.5.4) should also proceed more slowly. The independence that many changes show from each other suggests that diet restriction has a mosaic effect on senescence (Weindruch and Walford, 1988). At present there is no clear example of an age-related change that is not influenced by diet restriction.

10.3.3. Evolutionary Questions

The general influences of diet restriction in lengthening the lifespan are paralleled by delayed reproduction in some species. Several examples have already been discussed: laboratory rodents, in which estrous cycles are arrested during restriction, and the spider *Frontinella*, in which oviposition depends on prior feeding (Austad, 1989; Chapter 2.3.1.2). Depending on when diet restriction is imposed, sexual maturation in mammals may be extensively delayed; when food is freely available, fecundity appears (Chapter 9.2.2.2.2). Other examples of diet restriction causing delayed maturation are the supernumerary molts of the beetle *Trogoderma* during starvation (Chapter 2.2.3.3) and the dauer larvae of *Caenorhabditis* (Chapter 9.2.1.1).

Several recent discussions take account of these phenomena in rodents as an evolutionary strategy to cope with transience in food supply. Caloric restriction may improve fitness by preventing breeding in a "semi-starved animal . . . which would necessarily have limited resources for prenatal development and the subsequent feeding of the offspring" (Holliday, 1989, 126). Similarly, Harrison and Archer (1989, 3) argued that "the beneficial effect of food restriction on mammalian lifespan evolved due to the selective advantages for females whose reproductive lifespans were extended by food restriction." Moreover, Harrison and Archer (1988b) predicted that diet restriction should have more effect on the lifespan of species with condensed reproductive schedules, because these animals would be under the greatest selective pressure to cope with seasonal exigencies and be least able to delay reproduction until more food was available. Comparisons of *Mus* and the twice longer lived *Peromyscus leucopus* would be very interesting, as Harrison and Archer note. The retardation of oocyte loss and extension of the duration of fertility in laboratory mice and rats by diet restriction (Section 10.3.2.8.2) are consistent with these arguments. Holliday (1989) also considers that longer-lived animals should invest more resources in somatic repair, consistent with their need to reproduce at later ages. The argument is thus made that the retardation of senescence and the inhibition of reproduction, pending food availability, can be considered an evolutionary adaptation that favors fitness by increasing the potential reproductive lifespan.

This argument must be considered in a more general context, however. Phelan and Austad (1989) believe that diet restriction may extend lifespan mainly by delaying maturation. The ergonomic postulate that vital resources are partitioned between somatic maintenance and reproduction clearly applies to many organisms that require extra food to complete egg maturation, as in the spider *Frontinella* (see above) and in the anautogenous mosquitos (Chapter 2.2.1.1.2). It would be interesting to look for species differences in somatic repair and regeneration mechanisms in various flies and mosquitos, according to whether they were autogenous or anautogenous. The range of adult lifespans in Diptera is severalfold, exclusive of diapause (Chapter 2.2.1.1). There is also a concern about the scarcity of data on the effects of diet restriction on the lifespans of mammals that have slow reproductive rates even under *ad libitum* feeding (Weindruch and Walford, 1988). McNab (1980) made an important observation that high metabolic rates in placental mammals are associated with greater reproductive capacity and population growth rates, the Malthusian growth rate *r* in equation (1.11), as shown by a series of species comparisons. Marsupials and egg-laying mammals have lower metabolic rates per gram body size than placental mammals (Figure 5.17). Thus, the evolutionary success of placental mammals with more extended lifespans than shown by marsupials in ecologically overlapping species may be linked to their capacity for higher metabolism.

The brain has a general role in regulating both appetite and reproduction. Since the *ad libitum* food intake is determined by the mammalian ventromedial nucleus and other integrative centers in the hypothalamus and limbic system, I propose that we can also view the set point of appetite to maximize reproduction in young adults. Recall from Chapter 1.5 that young adults are statistically the major progenitors in nearly all populations that have high overall mortality rates. The setting of appetite is a species characteristic, adapted according to the reproductive schedule required by the environment, and must be genetically specified. We have already encountered mouse strains that are hereditarily obese (Section 10.3.2.1). The obese strains are thought to have a hypothalamic defect that underlies their greater food consumption in association with altered regulation of the adipsin gene (Platt et al., 1989; Chapter 6.5.1.7.4). Thus, the effects of diet restriction in increasing lifespan and delaying reproduction implicate a complex of gene activities in the nervous system as genomic influences on senescence. The balance of metabolism among glycolysis, gluconeogenesis, and lipogenesis, which is influenced by diet restriction or by DHEA, is under the control of brain regions that, from this perspective, are pacemakers for senescence and longevity.

10.3.4. Conclusions

Some tentative conclusions are offered about the multifarious effects of diet restriction on increasing lifespan by retarding the progression and reducing the intensity of senescence. There is no reason to expect a single mechanism of diet retriction that accounts for increased lifespan in mammals. The data show multiple outcomes from shifts in carbohydrate metabolism that reduce lipid synthesis. Moreover, feeding the weak androgen DHEA mimics certain aspects of diet restriction, particularly the diminution of lipid synthesis. The favorable consequences of diet restriction and DHEA both include reductions in atherogenesis and tumor promotion, and possibly reduced nucleoside pools. The data available for humans show major decreases of plasma DHEA in both sexes after 30 years (Figure 3.36). While further studies are needed, two prospective studies indicate inverse relationships between plasma DHEA sulfate and the risks of breast cancer in women (Bulbrook et al., 1971) and heart disease in both sexes (Barrett-Conner et al., 1986). Studies with DHEA have yet to show increases of lifespan in long-lived rodent strains, as demonstrated by diet restriction.

Several types of invertebrates also show life-extending effects of diet restriction, but the mechanisms may be entirely different. Several invertebrates require food as adults to complete the maturation of eggs or for reproduction, for example, in the spider *Frontinella*. Thus, the life-threatening event in these cases is, paradoxically, reproduction, for which food is required. Although mammals certainly must have adequate food for maturation and fecundity, reproduction is far less life threatening than in these insects. The example of the ciliate *Tokophrya infusionum* is also unique, in that this unicellular species lacks an excretory organelle (Chapter 2.2.1.2). Together, the responses of diverse species to diet restriction seem to me mainly a statement that energy is required for living and propagating, and that where lifespan is increased, it does so because diet restriction limits passage of the organism through its normal life history, however that may be regulated.

10.4. Sex Hormones

While numerous hormonal modifications of the senescent phenotype are described in a huge literature from basic and clinical studies, fewer effects are documented on lifespan. Two major categories are considered: age-related changes that are consequent to hormonal deficits (e.g. estrogen deficiencies of menopause) and changes that result from effects of exposure to hormones. Because of the important influences of steroids and other hormones on gene expression in target cells, many changes should prove to be mediated by selective effects on gene regulation. At present, however, this is mostly speculation.

10.4.1. Male Hormones and Lifespan

The two genders may differ modestly, up to 20%, in mean and maximum lifespan in many mammals. Usually, but not always, males have shorter lifespans (Chapter 6.5.3.4; Smith, 1981). But for a few studies on humans, sex differences of the two mortality coefficients remain to be analyzed. As described in Chapter 6.5.3.4, the IMR may be slightly lower for women than for men, while no sex differences were obvious in the acceleration rate of mortality in the few populations where this was analyzed.

Hormonal factors in the greater statistical lifespans of women are indicated by data that eunuchs live slightly longer than normal men (Hamilton and Mestler, 1969). Data that are appalling to contemplate indicate that postpubertal castration of men increased lifespan by reducing the IMR (Figure 10.17). This study used a case control approach: institutionalized eunuchs were matched for testing by year of birth and intelligence with other institutionalized men; neither inmate group lived as long as normal males. Two mechanisms could reduce the IMR of the eunuchs: inhibitory effects of androgens on immune functions (Grossman, 1984), and altered behaviors that must include consequences of the smaller size and strength of the eunuchs.

In cats, males usually have shorter lifespans. Castration increased the mean and maximum lifespan of males for at least two reasons: fewer deaths from trauma and a later onset of urinary tract stones (Hamilton, 1965; Hamilton et al., 1969); analysis of mortality rates has not been done. In contrast, no effects was found for DBA$_f$ mice (Mühlbock, 1959), which might be expected because of the slightly greater lifespan of males in most strains

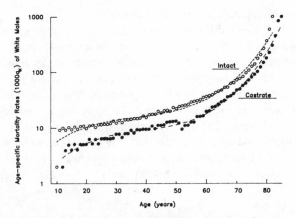

Fig. 10.17. Age-related mortality of mentally retarded eunuchs and a group that was matched for age. The practice of castrating the mentally deficient even as youths was a judicially supported offshoot of the popular eugenics movement that continued in the United States after World War II (Kevles, 1985). Some hospital administrators required "voluntary" sterilization before a person judged mentally deficient would be released. The acceleration of mortality (MRDT; Gompertz slope) was nearly identical, while the IMR was much smaller in the eunuchs. These mortality rate curves differ from normative larger populations (Figure 1.1) and may reflect subgroups with differential mortality risk. Redrawn from Hamilton and Mestler, 1969.

(Chapter 6.5.1). It remains to be determined when during development the sex-determining genes begin to influence lifespan potential. Although some insight is obtained by examining removal of sex steroids by castration at various stages or by administering sex steroids, such studies must also consider the prior developmental history, when sex-determining genes may establish chromatin conformations that set up future patterns of genomic responses.

10.4.2. Hormonal Manipulations of Gender Differences in Senescence

This section gives a sampling of gender differences in age-related changes, most of which can be manipulated in adults. A fundamental gender difference that leads to divergent outcomes in the phenotypes of senescence is the exhaustion of germ cells and the virtual cessation of sex steroid production in females (Chapter 3.6.4.1). Consequently, female mammals manifest two major aspects of reproductive senescence that can be extensively manipulated by hormonal treatment in adults. *Ovariprival changes* that emerge during the depletion of the ovarian oocyte pool are attributed to loss of ovarian estradiol and can be blunted or reversed by estrogen replacements (Table 10.6). In addition, laboratory rodents show *estrogen-induced changes* in neuroendocrine functions and other cells, which are attributed to chronic exposure to endogenous estradiol, since these changes are blunted by removing estrogens through ovariectomy. Both types of changes are prematurely inducible in the young by chronic estrogen treatments. Analysis at the level of gene regulation has been achieved only for uterine atrophy.

Cell atrophy is reversed in many parts of the uterus and other cells of the reproductive tract by estrogen replacement in women (reviewed in Gosden, 1985) and mice (Papadaki et al., 1979). These findings imply that decreases in the estradiol-receptor dependent mechanisms in uterine cell nuclei, as described in rats (Chuknyiska et al., 1985; Gesell and Roth, 1981), are also reversible by restoring sufficient hormonal stimulation. In contrast, other cell responses to estradiol are lost. The cornification of vaginal epithelial cells (keratinization through terminal differentiation) is highly sensitive to estrogen levels and is greatly diminished in 2-year-old C57BL/6J mice (Adler and Nelson, 1988). Although estrogen implants stimulated mitosis in the vaginal epithelia, epithelial cornification responses remained greatly impaired. Because the same impairments of cornification were induced by exposure of young mice to estradiol for 3 or more months without lost of proliferative capacity, there is not likely to be a general defect in the estrogen-receptor interactions with chromatin, implying select impairments of gene regulation in vaginal epithelia cells. Analysis of cell-cycle choice points for proliferation versus terminal differentiation could be rewarding. It is pertinent to recall from Chapter 7.4.1.3 that the physical properties of the estrogen receptor do not change notably with age.

Although there is no male equivalent to the near total deficits of gonadal steroids at menopause in rodents or humans (Chapter 3.6.4), some rodent strains have deficits of testosterone that are large enough for atrophy of testosterone-dependent cell populations. For example, the poly(A)RNA content of the ventral lobe of the prostate gland in AXC/SSh rats decreased progressively after 6 months to an 80% deficit by 26 months; after 4 days of testosterone injections, the poly(A)RNA was increased fourfold in 26-month-old rats, but remained 50% less than young injected rats (Schultz and Shain, 1988). At all ages and treatments, the mRNA yield for S-adenosyl-L-methionine decarboxylase and for L-ornithine decarboxylase remained unchanged. Enzyme activities were similar to the mRNA levels (Shain et al., 1986).

Similar results were found for a brain neuropeptide, vasopressin. Male rats have a higher density of vasopressin fibers than do females, which is eliminated in the hippocampus and locus ceruleus, for example, by castration. Consistent with this testosterone dependence in young rats and decreased plasma testosterone at later ages, 3-year-old brown Norway rats had decreased vasopressinergic fiber density in the hippocampus and locus ceruleus, which was restored by testosterone implants (Goudsmit et al., 1988). These partly reversible changes occur in the absence of cell loss and differ from the dorsal prostate lobe where atro-

Table 10.6 Changes in Females that Are Secondary to Decreased Estrogen (Ovariprival Syndromes)

	Mice	Humans
Decrease of plasma estradiol to castrate levels	Gee et al., 1983	Sherman et al., 1976
Elevation of LH to castrate levels	Gee et al., 1983	Sherman et al., 1976
Hot flushes	not observed	Judd, 1983
Uterine epithelial atrophy	Papadaki et al., 1979	Lieblum et al., 1983, and Lin et al., 1973
Acceleration of osteoporosis	castration induces osteoporosis in young rodents (Aitken et al., 1972; Wronski et al., 1985; osteoporosis occurs in 2-year-old female mice (Shah et al., 1967; Silberberg and Silberberg, 1941; Silberberg and Silberberg, 1962)	Albright et al., 1941, and Avioli, 1982

phy does not occur (see above). These molecular markers provide excellent laboratory models for further studies on intrinsic and extrinsic factors in senescence.

Age-related osteoporosis is often accelerated after menopause to varying degrees in humans in association with deficits of ovarian steroids, of which the most important is thought to be estradiol (Chapter 3.6.4). Men also suffer from osteoporosis, but the changes are more gradual and not linked to estrogen. Castration of adult men caused bone loss (Stepan et al., 1989), although the loss appears to be less severe than after ovariectomy. There must be major sex differences in the influence of estradiol on bone cells, yet little is known about the stage in development when these differences are acquired, or their possible conversion between sexes by hormone treatment. The efficacy of postmenopausal estradiol treatments in reducing the risk of several types of fractures and bone loss (Horsman et al., 1983; Riggs and Melton, 1986) is one of the major examples of intervention into the phenotype of senescence. Osteoporosis may be intensified in female rodents during ovarian senescence (Chapter 3.6.4). Although the relationship of age-related osteoporosis to ovarian senescence is less understood for rats than for humans, osteoporotic changes are induced in young rats by ovariectomy (Aitkin et al., 1972; Wronski et al., 1985). Correspondingly, estradiol treatment increased bone mass in the femur of middle-aged mice (Papadaki et al., 1979). While osteoblasts have high-affinity estradiol receptors (Eriksen et al., 1988), estradiol may also act indirectly.

Besides estradiol deficiencies, clinical studies of osteoporotic postmenopausal women have shown often decreased blood levels of the physiologically active vitamin D metabolite 1,25-dihydroxyvitamin D_3; this hormone, which regulates calcium uptake from the intestine, could be a major factor in their decreased calcium absorption from the intestine (Gallagher et al., 1979; Tsai et al., 1984). Treatment of postmenopausal osteoporosis by vitamin D_3 has not given clear results. However, in ovariectomized dogs, vitamin D reversed bone loss by stimulating osteoblasts (Malluche et al., 1988). Calcitonin, which inhibits bone resorption by inhibiting the osteoclasts, is another approach. A 2-year study of postmenopausal women indicated that calcitonin was as effective as estradiol in preventing bone loss (MacIntyre et al., 1988).

Strenuous regular exercise without estrogen replacement showed intriguing effects on hormones and bone (Nelson, Meredith, Dawson-Hughes, and Evans, 1988). Postmenopausal Caucasians who ran 10 miles or more per week exceeded sedentary women of the same age and height in bone mineral of the spinal column, but also in an arm bone, the radius, that presumably was less directly stimulated by the exercise. Of particular interest was plasma growth hormone, which was three- to sevenfold higher at rest and just after exercise in the trained women, who also had higher plasma 1,25-dihydroxyvitamin D_3 and somatomedin C. Parathyroid hormone and estrone were lower in the exercisers, while calcitonin and estradiol were unchanged. The elevated vitamin D_3 should favor increased retention of calcium. The elevated growth hormone and somatomedin C, a growth factor that growth hormone stimulates liver cells to produce, might be among the systemic factors in the exercisers that promoted bone growth or mineral retention. Although estrogens markedly increase growth hormone secretions (Ho et al., 1987), this mechanism is not involved here. Systemic factors are indicated by the increased bone in the radius, since running would not be expected to promote growth of this bone through mechanical stress. In any case, the striking increase of circulating growth hormone is the first evidence that exercise can reverse this age-related decrease, a finding of much significance since it occurred without estrogen replacements or elevations of endogenous estrogens.

These studies give hope that the tragic and frequent bone fractures in elderly men and women can be greatly reduced by supplemental hormones, calcium, fluoride, or exercise. Smoking or severe dieting can have cumulative effects on bone mineral stores over the lifespan, which can alter individual responses to estrogen deficits at menopause (Riggs and Melton, 1986). Age-related osteoporosis may soon be understood as an example of an age-related disease involving selective changes in gene regulation through hormonal influences that are reversible.

The relatively smaller incidence of coronary artery disease in women than men tends to diminish in many populations after menopause, for example as shown by the Framingham Study during a 26-year follow-up (Wilson, Castelli, and Kannel, 1987; Kannel, 1985b). Diet and sex interact, and diabetes (glucose intolerance) increases coronary disease more in older women than in men (Kannel, 1985b, 1986). Recall from Section 10.3.2.1 and Chapter 6.5.1.7.4 examples of sex differences in response to atherogenic diets and diabetes. In regard to menopause, several studies indicate that estrogen replacement reduces the risk of coronary disease (Sullivan et al., 1988; Stampfer et al., 1985) and stroke (Paganini-Hill et al., 1988) by about 50%. These provisional results are consistent with the recognized relation of estrogen levels to plasma levels of HDL (Bush et al., 1987; Wallentin and Larsson-Cohn, 1977), which epidemiological studies indicate is protective (e.g. Gordon et al., 1977).

Another sex difference in humans concerns male-pattern baldness or alopecia, which depends on testosterone in conjunction with a susceptible genotype. Male-pattern baldness results from the miniaturization of hair follicles, by which the type of hair is modified from long, pigmented, and coarse hairs (terminal hairs) to short, nonpigmented, and fine hairs (vellus hairs) (Selmanowitz et al., 1977; Kligman et al., 1985). A reciprocal change in the type of hair growth may occur on the ear (Figure 7.2). Consequently, the hairline recedes on the temples and forehead, although the number of active hair follicles is unchanged. The change in follicular cell activities progress for years or decades. While poorly understood at cellular and genetic levels, male-pattern baldness is clearly under endocrine control. Baldness does not occur in adult men castrated prepubertally, and varies in men castrated after puberty but before mid-life (Hamilton, 1942; Hamilton et al., 1969). Upon treatment of eunuchs with testosterone, baldness emerged in a subgroup. Moreover, women sometimes show the trait when exposed to androgens secreted by androgen secreting (virilizing) tumors of the adrenal or ovary (Hamilton, 1942; Mintz and Geist, 1941; Ludwig, 1964). Although the genetics of alopecia is poorly understood (Rook and Dowber, 1982), at least some determinants of male-pattern baldness are not Y-linked. The delayed onset and slow progression of male-pattern baldness after puberty suggests more complex mechanisms than the classical direct model of steroid-gene interactions.

Alterations in neuronal functions could also have important roles in age changes of macromolecular biosynthesis in bone and other tissues. In 2-year-old male rats, L-dopa restored the amplitude of growth hormone pulses (Sonntag et al., 1982). Age-related increases in hypothalamic somatostatin, a neuronally secreted antagonist of growth hormone releasing factor at the pituitary level, are implicated in the decreased growth hormone secretion (Sonntag and Gough, 1988). The effects of exercise and body fat on somatostatin are unknown. The decrease of growth hormone could have general significance, since supplements nearly restored the incorporation of amino acids into muscle proteins of 2-year-old rats (Figure 7.9; Sonntag et al., 1985). Returning to humans, the vigorous secretion of growth hormone provoked by the α-adrenergic agonist clonidine decreases with age in normal humans, and sooner in women than men (Gil-ad et al., 1984). This suggests a declining central responsiveness that is compounded in women by ovarian steroid deficits at menopause.

A clear example of a brain-based postmenopausal change is the hot flush, a transient vasodilation in skin of the upper body, which is a common consequence of diminished estrogen secretions at menopause and which is thought to be triggered by hypothalamic thermoregulatory centers (Judd, 1983). Flushes soon disappear with estrogen replacement therapy. Their absence in adult men despite menopause-equivalent estrogen levels can be ascribed to insufficient exposure to estrogens during puberty, since adult men treated with estrogens for prostatic cancer will develop hot flushes on withdrawal from estrogen (Ginsberg and O'Reilly, 1983; Steinfeld and Rheinhardt, 1980). Similarly, adult Turner's syndrome patients with severe gonadal dysgenesis who were treated with estrogens have hot flushes on estrogen withdrawal (Yen, 1977).

These reports suggest that the human nervous system has loci, presumably in the hypothalamus, that can acquire an estrogen requirement at some time during maturation or perhaps before. This susceptibility to estrogen dependency is maintained throughout most of adult life, irrespective of chromosomal sex. Thus can adult men acquire a characteristic change of menopause, which itself is normally acquired during maturation in girls. We do not know either the neuronal mechanisms or the minimum time and dose requirements for estradiol to induce this specific effect of chronic estradiol exposure.

A potentially related phenomenon in laboratory rats and mice is the ovary-induced neuroendocrine syndrome (Table 10.7). During mid-life, female rodents show marked impairments of the estradiol-driven gonadotropin surges that precede ovulation.[4] There is also an early proliferation of prolactin cells in the pituitary, which may be related to the later formation of lactotrope adenomas. The impairments of gonadotropin regulation and many other aspects of reproductive senescence in female rodents can be greatly attenuated by removing the ovary in rodents when they are young (Figure 10.18; Finch et al., 1984). Conversely, nearly all of these phenomena are induced in young rodents by chronic exposure to physiological levels of estradiol (Table 10.7).

We are characterizing the time-dose requirements for exogenous estradiol damage to neuroendocrine functions. Daily exposure for 9 to 12 weeks (about 2,000 pg-days/ml plasma) suffices to cause irreversible impairments in the mechanisms of cycling (Kohama et al., 1989a, 1989b). After exposure to 6 weeks of oral estradiol, mice regained cycling, but then soon ceased cycling prematurely (Kohama et al., 1989b). This result fits a model of cryptic damage from the exogenous oral estrogens that then summate with endogeneous estrogens to a threshold at which acyclicity ensues (Figure 3.41). The estrogen-mediated damage was at a neuroendocrine level, since replacement of ovaries in the acyclic mice did not restore cyclicity. The age-related loss of cyclicity shown by intact

---

4. Humans show no parallel to this, since robust surges can be hormonally induced long after menopause (Odell and Swerdloff, 1968).

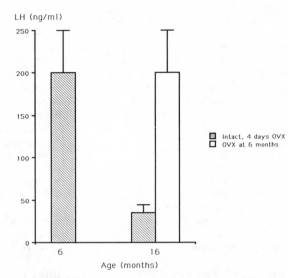

Fig. 10.18. Age-related impairments of the LH surge in mice are prevented by ovariectomy as young adults. C57BL/6J mice were ovariectomized (OVX) at 6 months and then tested by comparison with 6- and 16-month-old mice that were ovariectomized 4 days previously. Estradiol implants were given to simulate the preovulatory elevations of estradiol during the normal estrous cycle. Robust surges of LH, appropriately timed in relation to daily light-dark cycles, occurred in young and long-term ovariectomized 16-month-old mice but not in 16-month-old mice that were exposed to their endogenous ovarian secretions until just before testing. Redrawn from Mobbs et al., 1984a.

mice may represent aspects of these phenomena. Replacement of ovaries in middle-aged mice just after the loss of cyclicity did not restore cycles in the majority (Felicio et al., 1983; Mobbs, Gee, and Finch, 1984), whereas ovarian replacement just before the loss of cyclicity nearly doubled the duration of cycling (Felicio et al., 1986). Provisionally, ovary-induced neuroendocrine changes that cause acyclicity during reproductive senescence do not integrate cumulatively across all cycles; the damage may occur when cycles lengthen, which is associated with increased plasma estrogen:progesterone ratios (Nelson et al., 1981; Finch et al., 1984). Recall from above, that progesterone can antagonize the effects of exogenous estrogens in causing cyclicity (Kohama et al., 1989a).

The neuronal targets of estradiol in these changes are unknown and are probably at multiple loci. We ruled out major loss of LHRH-containing neurons, which are retained in normal numbers during reproductive senescence up to 21 months (Hoffman and Finch, 1986; Don Carlos et al., 1986), although there is a subsequent loss by the average lifespan (Miller et al., 1990). These phenomena, like the ovariprival aspects of osteoporosis discussed above, can be extensively manipulated by exposure to estradiol but, unlike osteoporosis, are driven by the continuing presence of estradiol, rather than its absence.

The reversibility of these phenomena varies. For example, the increase of pituitary dopamine by middle age is readily reversed by ovariectomy (Telford et al., 1986), while the transition from 4- to 5-day cycles is not recovered by ovariectomy for 2 months (Mobbs, Gee, and Finch, 1984a; Felicio et al., 1986). The extent of reversibility should be useful in analyzing the causes. Irreversible changes are candidates for a mechanism involving neuron death, a possibility that so far is speculative, at least in the rodent hypothalamus.

Table 10.7   Ovary-induced Neuroendocrine Syndrome in Rodents

Markers of Reproductive Senescence	Delayed, by Chronic Ovariectomy in Aging Rodents	Acceleration, by Chronic Estrogen Treatment in Young Rodents
Estrous cycles lengthen	Felicio et al., 1986	not demonstrated
Estrous cycles cease	Aschheim, 1965, 1976	Brawer et al., 1983
	Felicio et al., 1983	Kawashima, 1960
		Mobbs, Flurkey, et al., 1984
		Kohama et al., 1989a
Induced LH surge impairment	Blake et al., 1983	Mobbs, Flurkey, et al., 1984
	Mobbs et al., 1984	
Pulsatile release of LH	Blake et al., 1983	not known
Postovariectomy LH elevations decrease	Blake et al., 1983	Mobbs, Flurkey, et al., 1984
	Gee et al., 1983	
Glial hyperactivity in arcuate nucleus	Schipper et al., 1981*	Brawer et al., 1978, 1983
		Brawer and Sonnenschein, 1975
Reduced pituitary stalk blood dopamine	not studied	Sarker et al., 1984
Pituitary dopamine increases	Telford et al., 1986	Telford et al., 1986
Pituitary glucose-6-phosphate dehydrogenase increases	Gordon et al., 1988	Gordon et al., 1988
Lactotrope adenomas	Mobbs et al., 1984	Brawer and Sonnenschein, 1975
	Nelson et al., 1980	Giok, 1961
	Takahashi and Kawashima, 1983	Wiklund and Gorski, 1982
Lacrotropes increase	Takahashi and Kawashima, 1983	Brawer and Finch, 1983
Prolactinemia	Nelson et al., 1980	Casanueva et al., 1982
	Takahashi and Kawashima, 1983	

*Schipper, H., Brawer, J. R., Nelson, J. F., Felicio, L. S., Finch, C. E. (1981) Role of the gonads in histological aging of the hypothalamic arcuate nucleus. *Biol. Reprod.* 25:413–419.

Closely related mouse genotypes have different patterns of reproductive senescence. For example, the C57BL/6J and C57BL/10Sn strains, which were separated in 1936 from an already inbred line, differ extensively: the /6J shows a marked cycle lengthening from 4- to 5-days by 8 months, whereas the /10Sn have few 4-day cycles when young and maintain their 5-day modal cycle distribution until they become acyclic about 3 months after the C57BL/6J (Lerner et al., 1988). This and ovarian grafting studies with /6J mice (Felicio et al., 1983, 1986) show that cycle lengthening and cessation can be regulated independently and may involve separate mechanisms. Because both strains had similar numbers of ovarian oocytes at 8 months (Lerner et al., 1988), it seems unlikely that the different ages of acyclicity obtain from different rates of oocyte exhaustion. We are investigating the hypothesis that the C57BL/6J strain is more susceptible to estradiol-induced neuroendocrine damage than the C57BL/10Sn strain. Because they share the *H–2b* haplotype, the differences appear to involve non-*H–2* alleles, of which three are known (Chapter 6.5.1.1). Another example of strain background influences was discussed in Section 10.3.2.1 for the *db* and *ob* mutations. Within the C57BL/10Sn *H–2* congenics that have different *H–2* haplotypes on the C57BL/10Sn background, the age when fertility is lost followed the same rank order (Lerner et al., 1988) as the lifespan determined in a prior study (Smith and Walford, 1977). Other changes did not, however. These substantial effects of genotype on reproductive senescence, which arise from a limited number of genetic loci, open the possibility for a fine-structure genetic analysis of reproductive senescence.

Future studies may reveal roles for changes of gene regulation in responses to estradiol, in view of the influence of estradiol in regulating brain enzymes, neurotransmitter receptors, and nucleolar morphology. Other mechanisms could also involve the modification of DNA itself. For example, estradiol increases the formation of DNA adducts in trace quantities within 1 month in the kidneys of Syrian hamsters, which also develop malignant tumors during more prolonged treatments (Liehr et al., 1986). Although the chemical nature of these adducts is unknown, they chromatographically resemble covalent adducts to nucleic acid bases that are formed with dimethylbenz(a)anthracene and other carcinogens (Reddy and Randerath, 1986). Exploratory studies indicate that such adducts increase progressively with age in brain and other tissues (Chapter 8.2.2; Figure 8.3).

10.5. Stress and Corticosteroids

This section treats two topics from a huge literature of chronic stress effects that cause multifarious damage and dysfunction throughout the body, including the nervous and immune systems (Munck et al., 1984; Selye, 1976; Sapolsky et al., 1985; Baxter and Forsham, 1972). My focus here is on irreversible changes in two targets of senescent change that were discussed in Section 10.3.2.8.5: the hippocampus and the myocardium.

10.5.1. Hippocampal Damage

Impairments of memory and other cognitive functions show age-related trends in humans, with often subtle deficits that appear progressive after 60 years. Numerous aspects of cognition can be distinguished. Some features of learning and memory show more age-related risk of deficit than others, and there are major individual differences in the trajectories of change, just as found, for example, in kidney function (Figure 3.12). This complex literature is reviewed in Albert and Moss, 1988, Salthouse and Prill, 1987, Craik, 1984, and Schaie and Herzog, 1985. The hippocampus is implicated in memory functions, particularly those involving retrieval of recent information (Squire, 1986; Rawlins, 1985). For example, stroke-related lesions in the large CA_1 pyramidal neurons, which are a main output from the hippocampus to the cortex, are associated with recent memory loss but with no obvious effects on intelligence (the "WAIS" test) (Zola-Morgan et al., 1986; Squire, 1986). There are many individual variations, in the rates of change and in the extent of neuropathology, that range fivefold or more within older age groups in the extent of neuron loss (Figure 3.37). It is possible that some individuals do not lose any hippocampal neurons, though this is difficult to ascertain in view of the wide range of hippocampal cell numbers even in young individuals (Chapter 3.6.10).

Several factors may account for these individual variations. Older human groups have an increasing prevalence of Alzheimer's disease, which may not have been recognized in its early stages. Until very recently, collections of postmortem specimens frequently included this condition; diagnosis is still controversial in the oldest age groups where age-related trends for neuropathology may overlap extensively with those of Alzheimer's disease (Mann, 1985). In Alzheimer's disease, the hippocampus is nearly always damaged, particularly in the CA_1 zone, which has the highest density of intraneuronal neurofibrillary tangles and senile plaques with degenerating nerve terminals, which are hallmarks of Alzheimer's disease; in contrast, neurofibrillary tangles are much rarer in the CA_3 zone (Ball, 1977, 1978; Hyman et al., 1984; Hyman et al., 1986). Strokes and transient ischemia are another source of variability. The hippocampus is well known for the susceptibility of its pyramidal neurons to damage from oxygen deficits, particularly those in the CA_1 zone (e.g. discussed in Zola-Morgan et al., 1986).

Cumulative damage from corticosteroids is being considered as another mechanism of neuronal damage during senescence. This story began with observations that male F344 rats aged 25 months showed loss of pyramidal cells and astrocytic hypertrophy (Landfield et al., 1977) and that the astrocytic hypertrophy in rats aged 13 or 25 months correlated with plasma corticosterone (Landfield et al., 1978). Because astrocytic hypertrophy often occurs during degeneration of neurons, the correlation with corticosterone levels implies that neuronal degeneration is also a graded function of exposure to corticosterone. Subsequent studies confirm the age-related loss of hippocampal pyramidal neurons, including decreased numbers of CA_3 neurons with high-affinity receptors for corticosterone (Sapolsky et al., 1983; Sapolsky, 1990; Meany et al., 1988). Other studies indicate that these CA_3 neurons are particularly susceptible to damage from chronic elevations of corticosteroids in young rats (Sapolsky et al., 1985, 1986). Primates are also susceptible to these phenomena, as indicated by effects of chronic stress on vervet monkeys (*Cercopithecus aethiops;* Uno et al., 1989). The hippocampus of vervet monkeys also shows damage by stereotaxically implanted cortisol pellets, which is consistent with a local neurotoxic effect (Sapolsky et al., 1990).

A possible mechanism involving stress and elevated corticosteroids involves feedback sensitivity with brain loci. Because hippocampal damage disinhibits ACTH secretion, it is possible that cumulative effects from corticosterone that damage the hippocampus cause a vicious cycle with glucocorticoid cascade (Landfield, 1978; Sapolsky et al., 1986,; Sapolsky, 1990). As discussed in Chapter 9.3.4.2., neonatal handling diminished the age-related loss of CA_3 neurons in rats, possibly as the result of additional corticosteroid receptors in the handled rats (Figure 9.8). This effect suggests that the set point of negative feedback for ACTH regulation by corticosterone has a major influence on the loss of pyramidal cells throughout the lifespan (Meany et al., 1988). Thus, handled rats had faster recovery from stress to baseline corticosterone levels, which would be predicted over the lifespan to give less cumulative exposure to corticosterone and less potential for neuronal damage from corticosteroids.

If lowering of exposure to endogenous corticosterone had a long-term consequence of reducing hippocampal damage, then adrenalectomy should also have a salutory effect. Adrenalectomy does diminish several morphological signs of age-related hippocampal neuron damage (Landfield et al., 1979), which supports the concept of cumulative effects from endogenous corticosteroids. Intraneuronal levels of calcium are a possible transducer of these steroid effects, as suggested by the opposite effects of age and adrenalectomy on the calcium currents of CA_1 pyramidal neurons (Kerr et al., 1989). Greater calcium transients could cause damage through activation of proteases, like calpain (Lynch et al., 1986).

In many regards, these phenomena parallel the influences of estradiol on reproductive senescence in female rodents as described in Section 10.4.2 and demonstrate at least in two major brain centers, the hippocampus and the hypothalamus, that the phenotypes of senescence can be extensively manipulated in either direction, by chronic increases or decreases of circulating steroids. But the generalization to humans is far from established (Sapolsky, 1990). Two studies of healthy elderly men have different outcomes: A slight trend for elevated cortisol was reported for a sample from the Baltimore Longitudinal Study (Pavlov et al., 1986), while a lowered basal cortisol was seen in a specially selected sample of "successfully aging men" from Boston (Greenspan et al., 1990). Tests of cortisol feedback sensitivity through the dexamethasone suppression test generally show no age-related trend, in the absence of specific diseases like depression or Alzheimer's disease (De Leon et al., 1988; Sapolsky, 1990).

10.5.2. Myocardial Damage

The heart muscle (myocardium) may also accumulate damage from stress, but through effects of elevated catecholamines rather than steroids. A substantial portion of sudden deaths from heart attacks occur without significant narrowing of the crucial coronary arteries (e.g. Baroldi et al., 1979). However, these individuals show a histopathological change, coagulative myocytoloysis, which is an acute myocardial lesion associated with tetanic contraction and hypercontracted sarcomeres. Because similar lesions are induced in dog myocardium by perfusion with isoproterenol (β-adrenergic agonist) and because these effects are protected by propranolol (β-blocker), it is plausible that endogenous catecholamines can cause cumulative myocardial damage during successive bouts of stress (Eliot et al., 1978; Melville et al., 1973). The 50 to 100% age-related trend for elevations of basal plasma norepinephrine in normal humans (Chapter 3.6.9) might have a role in these phenomena.

10.6. Costs of Reproduction and Accelerated Senescence

This section discusses how reproduction may increase mortality rates or accelerate senescence, a topic known as the "cost of reproduction" (Sibly and Calow, 1986; Clutton-Brock et al., 1982; Clutton-Brock et al., 1983; Clutton-Brock et al., 1988; Reznick, 1985; Partridge and Harvey, 1985; Bell and Koufopanou, 1986; Rickleffs, 1977). There is abundant evidence in natural and laboratory populations that reproduction is associated with shortened lifespan. This relationship is powerfully shown in numerous semelparous species whose usual rapid senescence or sudden death after one bout of reproduction can be delayed by preventing maturation (Chapter 2). The first examples concern repeatedly bred laboratory rodents and flies that continue the preceding discussion on the adverse consequences of stress. None of these examples has been analyzed by the Gompertz mortality model, for the IMR or MRDT.

Under certain conditions that are still to be precisely defined, the continuous breeding of rodents is highly stressful and causes severe degeneration of numerous organs in association with elevated corticosteroids and diabetes. Bernard Wexler, who first described these changes in detail, was alerted to such phenomena from his earlier work with Robertson on corticosteroid elevations of Pacific salmon during spawning (Chapter 2.3.1.4.2). During repeated breeding by single pairs under normal caging conditions, young Sprague-

Dawley rats develop the following lesions: adrenal cortex hypertrophy; pancreatic β-cell hyperplasia, degranulation, and glucose intolerance; fatty degeneration of the liver; kidney stones; gastric ulcers; and hyperlipidemia and atherosclerosis in coronary and other arteries throughout the body (Wexler 1964a, 1964b, 1964c, 1964d; Wexler and Fischer, 1963a, 1963b; Wexler et al., 1964; Lewis and Wexler, 1974; Wexler and Kittinger, 1965; Wexler and True, 1963; reviewed in Wexler, 1976a). Both sexes showed these changes by 7 months, though the sexes differed in locations of the most severe vascular lesions. The repeatedly bred rats died prematurely, often with myocardial infarctions. These lesions arose on a low-fat (4%) diet and were intensified on fattier chows. The progression of these changes with each cycle of breeding in males and females implicates a stress-related syndrome, which is hypothesized to be driven by adrenocortical hypersecretion. The eventual development of most of these changes in virgin rats after 2 years in Wexler's laboratory (Wexler, 1976b) identified this syndrome as accelerated senescence. The importance of the adrenal to these phenomena is shown by adrenalectomy, which abolished the vascular, pancreatic, and liver degeneration (Iams and Wexler, 1977). The adrenalectomized females were still fertile and had four or more litters of reduced size.

The generalizability of Wexler's extensive findings is unclear. There is good agreement between Wexler's pathophysiologic findings and age-related changes of virgin male F344 rats in Table 10.3. Similarities include myocardial lesions, fatty liver, and kidney stones; specific tests of kidney function were apparently not done by Wexler under nonbreeding conditions. Retired breeder DBA/2J male mice at 22 months, however, showed symptoms of kidney disease (more water intake) and motor dysfunction (less spontaneous movement and poorer balance on a rotorod) than virgins of the same age (Ingram et al., 1983). Whatever one may conclude about repeatedly bred rodents, repeated copulation cannot be intrinsically adverse, as shown in a study of male Norway (outbred) rats that grouped with hysterectomized females and that lived 10 to 20% longer than unmated males, also in group caging (Drori and Folman, 1969).

Besides these laboratory studies, field studies show major increases in mortality risk from reproduction-related activities. Marsupial mice (*Antechinus*) show seasonal die-off in association with pathophysiological stress effects that are more extreme than observed in repeatedly bred laboratory rodents; isolation of males before breeding prevents this stress syndrome and allows survival to at least threefold the usual lifespan (Chapter 2.3.1.5). Large mammals also show these phenomena, as exemplified by the increased mortality of red deer (*Cervus elaphus*) that bore and nursed a calf (Clutton-Brock et al., 1982; Clutton-Brock et al., 1983; Clutton-Brock, 1988). The fat accumulated during summer grazing is thought to be an important factor, since females nursing calves have much smaller fat reserves than those without. The annual mortality rate increase from reproduction accelerated at later ages (Figure 10.19). In view of the progressive tooth wear suffered by ungulates (Chapter 3.6.11.4), accelerated tooth erosion could be a factor that increases mortality rates as a consequence of the increased foraging associated with reproduction.

Flies also have adverse consequences from mating (Partridge, 1988; Partridge et al., 1986; Partridge and Harvey, 1985; Partridge and Andrews, 1985; Partridge and Farquhar, 1981), which can be attributed to mechanical damage to wings (Ragland and Sohal, 1973). For example, mortality of adult male flies increased strikingly in the presence of females and was sharply reduced by their removal. In the presence of males, survival appeared to decrease exponentially, but without any evident increase in senescence (Fig-

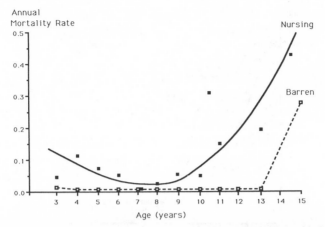

Fig. 10.19. The influence of bearing and nursing a calf on mortality of feral red deer hinds (*Cervus elaphus;* Clutton-Brock et al., 1982; Clutton-Brock, 1988); the fat accumulated during summer grazing is thought to be an important factor, since the females that are nursing a calf (hinds) have much smaller fat depots than those without (barren, or yeld). The annual mortality rate increase from reproduction accelerated at later ages. A free-drawn curve is shown to assist discrimination of the data points. Redrawn from Clutton-Brock et al., 1983.

ure 10.20; Partridge and Andrews, 1985). The causes of mortality could include mechanical senescence of wings through the flight required for courtship and mating, fighting between males, and expenditure of accessory ("seminal") fluid, which is rich in amino acids (Partridge, 1988). Reproduction also shortens lifespan in females, as indicated by the increased lifespan of flies that were sterilized by brief heat shock, permanently reducing egg laying, or that had a genetic defect "ovariless" (Maynard Smith, 1958). Readers may also recall from Section 10.3.2.3 that mosquitos live longer when fed on a sucrose diet that prevents egg maturation. A counterbalancing factor is that females in some *Drosophila* species (Partridge, 1988) and some other insects (Chapter 2.2.1.1) receive nutrients in the seminal fluid, which would be a further cost to males. An extreme of sex-specific costs is the beetle *Micromalthus,* which is cannibalized by the female after mating (Chapter 2.3.3).

Also recall two invertebrates described earlier: the spider *Frontinella,* where the inhibition of reproduction through limitation of food increased the adult lifespan (Section 10.3.3; Austad, 1989) and the rotifer *Philodina,* where mortality rates increased rapidly after reproduction ceased (Meadow and Barrows, 1971b; Chapter 2.2.3.2). The metabolic costs of reproduction are particularly large in small invertebrates whose eggs are large relative to their small size, like *Philodina,* which produces its own dry-weight equivalent in eggs per day. In animals that must care for multiple offspring, the risk may increase with the numbers that must be fed and tended at one time. A good example is the blue tit (*Parus caeruleus*) in which the brood size was manipulated. Maternal weight loss increased with the brood size, as did mortality: 40% of females with small broods of three survived to the next year, while only 20% did with broods larger than twelve (Nur, 1984a, 1984b).

Taken together, the evidence shows that, depending on the species and gender, reproduction can be stressful, with adverse consequences that interact with slower ongoing senescent processes, such as atherogenesis in rodents and mechanical senescence in in-

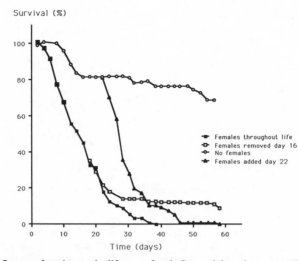

Fig. 10.20. The influence of mating on the lifespan of male *Drosophila melanogaster*. The removal of females after day 16 sharply reduced mortality to the rate of celibate males. Mating increases mortality in males, by a variety of factors that may include metabolic drains of flying, by mechanical damage to the wings, and by the loss of proteins in the seminal fluid. Partridge and Andrews (1985) concluded that reproductive activity caused major loss of life expectancy in male fruit flies that was unrelated to senescence; no analysis was done to evaluate whether mortality rates were accelerated by mating. Redrawn from Partridge and Andrews, 1985.

sects. In each species, the intensity of reproductive-related mortality may be considered an adaptive trade-off among the natural history parameters of generation time, rate of fecundity, and age-independent mortality (Sibly and Calow, 1986; Reznick, 1985).

Despite such widely discussed examples, there are numerous exceptions in which reproduction is not associated with increased mortality. For example, eleven out of thirty-three studies reviewed by Reznick (1985) concluded that there was no increased mortality in association with reproduction. For example, song sparrows (*Melospiza melodia*) that survived to breed the next year showed slightly better egg production and success in rearing their hatchlings to independence (Smith, 1981). Similar conclusions were drawn for a range of species by Bell and Koufopanou (1986). Under optimal conditions of nutrition and hygiene, reproduction may not cause increased risk of morbidity or mortality. However, variable nutrition, infections and parasites, and behavioral stresses that are common in nature must all contribute to the widespread occurrence of reproductive costs in shortening lifespan. These external influences thus can be considered as not genetically programmed in most iteroparous and long-lived organisms.

10.7. Immunological Changes and Hormones

Hormonal manipulations influence age-related changes in select immune functions as seen at two levels, described mostly in rodents: effects of growth hormone and hypophysectomy and effects of secreted hormones from the thymus gland. These phenomena put age-related changes of immune functions into a neuroendocrine regulatory context, so that alterations in specific lymphocyte functions may be linked to nonimmune functions. A rich emerging literature demonstrates many neuroimmune correspondences, for example the inverse relationship of estradiol to thymosin$_{\alpha 1}$ (Michael et al., 1981), the ability of IL–1 to stimulate secretion of hypothalamic corticotropin-releasing factors (Sapolsky et al., 1987), or the presence of neuropeptides in immune tissues (Geenen et al., 1986). The reach of these interactions extends from early development through puberty to late ages, and supports speculation that these interactions could determine or modulate many age-related changes outside of the immune system (Fabris, 1982; Besedovsky et al., 1985), including the loss of ovarian oocytes (Michael, 1979; Rebar, 1980). Moreover, blood monocytes of postmenopausal women release increased amounts of IL–1, a lymphokine that also may stimulate bone resorption; estrogen replacements eliminated the IL–1 elevations (Pacifici et al., 1989).

During sexual maturation, the cortex of the thymus gland undergoes a major involution, which precedes the maximum cellular immune response and decline of serum thymic hormones by 5 to 15% of the lifespan in rodents and humans (Chapter 3.6.6). Correspondingly, the capacity of thymic transplants to promote T-cell maturation declines sharply as studied in thymectomized, irradiated young host mice with syngeneic bone marrow transplants (Hirokawa and Makinodan, 1975). However, thymectomy of adult mice has less obvious effects on cellular immunity than in neonates, and there is extensive self-renewal of peripheral immune cells, so that the role of the thymus in adult immune function is considered to be unclear (e.g. Miller, 1989). In any case, the pubertal thymic changes are not comparable to ovarian involution at menopause because there is no corresponding precipitant decline of functions.

Thymic involution is strongly influenced by steroids. Castration before puberty delays reductions of thymus weight, while postpubertal castration increases thymic weight (reviews in Grossman, 1984; Wise, 1988). Estrogen treatments, conversely, cause involution. The strength of many cell-mediated immune functions follows similar patterns. Adrenalectomy also retards pubertal changes, while glucocorticoids cause involution (Ritter, 1977). The mechanism is not resolved.

The trend of changes begun at puberty continues, so that mice at late mid-life (18–24 months) had major loss of cortical and medullary tissue. A remarkable rejuvenation of thymus structure was brought about by implanted GH$_3$ cells which secrete growth hormone and prolactin (Kelley et al., 1986). This and a subsequent study from this group (Davila et al., 1987; Kelley, Davila, Brief, Simon, and Arkins, 1988) showed that many cell immune functions were also greatly improved, nearly to levels of young mice (mitogenic responses of splenic lymphocyte to concanavalin A and phytohemagglutinin), while IL–2 production by splenic lymphocytes was only partially restored. The reconstituted thymus nearly regained its weight at 3 months and had T-cell subsets in near normal proportions for this age (Thy 1.1, T helper, and Ia). The presence of these cell types indicates a major reversal of the age-related decline of T-lymphocyte formation from bone marrow stem cells. Because a major secretion of GH$_3$ cells is growth hormone, it is puzzling that others failed to show effects of growth hormone on age-related immune changes (Flurkey and Harrison, 1990). Prolactin and insulin-like growth factor were also elevated in rats grafted with the GH$_3$ cells and could have a role (Kelley et al., 1986), as could the host reaction to antigens on the GH$_3$ cells (Kevin Flurkey, pers. comm.).

In senescent male mice and rats, castration is reported in an abstract (Chiodi, 1976) to increase thymus weight. While no data on immune functions were given, this result is consistent with the inhibitory effects of testosterone on immune functions (Grossman,

1984, 1985) and the continuing secretion of testosterone at some level throughout the lifespan (Figure 3.21). The age-related loss of ovarian steroids during female reproductive senescence might interact with age-related immune changes, in view of the inverse relationship between estradiol and thymosin$_{\alpha 1}$ (see above) among other inhibitory effects of estrogen on the thymus. On the other side, estradiol positively regulates the fraction of splenic lymphocytes bearing the B-cell marker Ig, with the most responsiveness at low plasma estradiol (Flurkey et al., 1986).

Immune hormones themselves have some capacity to reverse changes. For example, thymosin$_{\alpha 1}$ injection for 5 days increased helper T cell activity by threefold in 15- to 24-month-old mice, but had less effect on *in vitro* proliferation (Frasca et al., 1982; Frasca et al., 1987). Others report positive responses to thymosin$_{\alpha 1}$ in mice (Effros et al., 1988; Weksler et al., 1978) and humans (Meroni et al., 1987), though not all attempts succeeded (Meroni et al., 1987; reviewed in Miller, 1989). Another example is that exogenous IL–2 restored the mitogenic responses of spleen cells from senescent mice (Chang et al., 1982). It would be important to learn if immune hormone treatments could reverse age-related impairments of c-myc mRNA accumulation in lectin-stimulated lymphocytes (Chapter 7.3.1).

10.8. Hypophysectomy

A general observation is that removal of the anterior pituitary (hypophysectomy) superficially maintains health in rodents at ages when controls appear senescent. The impact on mean and maximum lifespan, however, is unclear. In several mouse genotypes hypophysectomized at 8 months, only life shortening was observed (Harrison et al., 1982), while hypophysectomy of younger rats may have increased lifespan (Everitt et al., 1980; Donner Denckla, unpublished, cited in Harrison et al., 1982).[5] That age-related changes could be blunted by hypophysectomy may seem paradoxical in view of the many examples where changes are reversed or prevented by hormone supplementation, since in the absence of the pituitary trophic hormones, the levels of gonadal and adrenal steroids and thyroid hormones should be very low. However, the adverse effects that can occur under certain conditions from chronic exposure to steroids and other hormones (Section 10.4.2, Table 10.7) and the improvement of some age-related changes by removal of hormone-producing endocrine glands (Table 10.8) are consistent with positive results of hypophysectomy. Some effects of hypophysectomy may involve reduced food intake and, in this respect, could be similar to diet restriction. Most studies of chronic hypophysectomy in rodents address physiological functions and gross pathology, with few molecular analyses.

Early support for the concept that hypophysectomy could retard senescence came from the slower age-related increases of tail-tendon tensile strength (breaking time under isometric tension) (Figure 10.21; Olsen and Everitt, 1965; Verzar and Spichtin, 1966), a

---

5. This literature, however, is often difficult to interpret because the completeness of hypophysectomy is rarely verified by measurements of steroids and other hormones at later times when limited regeneration could occur from residual pituitary cells. Another difficulty is that studies vary considerably in the replacements of glucocorticoids, mineralocorticoids, and thyroxine, which are found necessary to some extent to prevent extensive mortality. Glucocorticoids could have adverse effects on their own, as discussed in Section 10.5.

Table 10.8 Summary of Age-related Changes which are Blunted by Removal of Hormones

1. Ovary-dependent hypothalmic-pituitary syndrome in female rodents (Table 10.7)
2. Adrenal-dependent hippocampal syndrome in male rodents (Chapter 10.5.1)
3. Kidney dysfunctions associated with exposure of young male rats to 28 daily injections of vasopressin, as assayed *in vitro* by cAMP production (Miller, 1985, 1987)
4. Generalized effects of hypophysectomy (Table 10.9)

Table 10.9 Hypophysectomy and the Retardation of Senescence

	Improved by Hypophysectomy	Worsened by Hypophysectomy
Collagen	tail-tendon tensile strength[1,2]	
Immune functions	thymus weight;[3] delayed hypersensitivity;[3] graft rejection;[4] *in vivo* phagocytosis;[4] primary immune responses[5]	lymphocyte responses to mitogens;[3] primary immune responses[3]
Kidney disease	proteinurea;[1] thickened basement membrane[6]	
Metabolism	responsiveness of oxygen consumption to thyroxine;[7] oxygen consumption *in vivo* (MOC);[7] liver RNA synthesis[7]	
Neuropathy	hind-limb paralysis through spinal neuron degeneration[1]	
Vascular (aorta)	β-adrenergic–mediated relaxation of smooth muscle;[8] aortic thickness[1]	
Tumors	reduced incidence in endocrine glands;[1] other organs[1,9]	
Wounds	healing improved[10]	

[1]Everitt et al., 1980.
[2]Harrison and Archer, 1978; Harrison et al., 1978; Figure 10.21.
[3]Harrison et al., 1982.
[4]Bilder and Denckla, 1980.
[5]Scott et al., 1979.
[6]Everitt, 1976.
[7]Denckla, 1974; Bolla and Denckla, 1979.
[8]Parker et al., 1978.
[9]Heston, 1963.
[10]Harrison et al., 1982.

measure thought to be related to collagen cross-linking (Chapter 7.5.3). Hypophysectomy at middle age also had a small but definite effect on tail-tendon tensile strength in rats (Everitt et al., 1980), but a less significant effect in mice (David Harrison, pers. comm.). These different outcomes of hypophysectomy in rats and mice may be secondary to food intake (see below). Arthur Everitt, working quite alone at the University of Sydney for several decades, established a diverse array of age-related changes that are modified by hypophysectomy (Table 10.9) and that brought general attention to adverse interactions of hormones with senescence (Everitt, 1980).

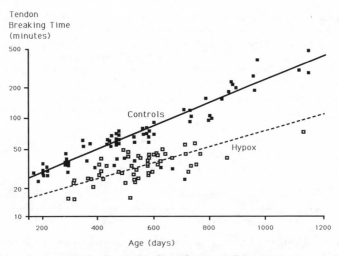

Fig. 10.21. The tensile strength of collagen-rich fibers from rat tail increases progressively with age, but increases more slowly in rats that were hypophysectomized at 50 days. This assay is based on the breakage time of fibers suspended with a weight in 7 M urea at 40°C and is thought to measure covalent cross-links in collagen, though the molecular bonds have not been specified. Redrawn from Everitt et al., 1968.

Hypophysectomy also decreased kidney disease, the decreased extent of proteinuria (Everitt and Duval, 1965) and slowed thickening of basement membranes in the proximal tubules found with electron microscopy (Everitt, 1976b). Thickened renal basement membranes are often associated with diabetes and could involve glycation (Section 10.3.2.3, Chapter 7.5.2). Amelioration of these changes by hypophysectomy may, however, be related to the 70% reduction of *ad libitum* food intake, which is maintained throughout the lifespan of male Wistar rats hypophysectomized at 70 or 400 days (Everitt et al., 1980); body weights were lower, with the greatest effects on the younger rats. Because diet restriction greatly reduces the intensity of kidney diseases (Section 10.3.2.2), the decreased kidney disease could be merely consequent to reduced protein intake. On the other hand, C57BL/6J male mice hypophysectomized at mid-life and fed *ad libitum* have slightly *increased* food intake, despite slightly lower body weight (Harrison et al., 1982). The influences of genotypes and species on responses to hypophysectomy are not well characterized, but may be expected in view of allelic influences on responses to estrogens (Section 10.4.2).

The mechanism by which hypophysectomy reduces kidney disease could involve the protein component of diet or blood glucose, as discussed with effects of food restriction in Section 10.3.2.3. The influence of hypophysectomy on body weight has also been compared with diet restriction (Denckla, 1973). The physiological consequences of hypophysectomy and diet restriction cannot be simply equated, although retarded glycation and formation of advanced glycation or Amadori endproducts could be a common outcome.

The immune system has complex responses to hypophysectomy. There is general agreement that short-term hypophysectomy markedly impairs primary immune responses in young adult rats (e.g. Nagy and Berczi, 1978; Harrison et al., 1982). The partial restoration of thymus weight and cellular structure (increased ratio of cortical to medullary cells) in middle-aged mice (hypophysectomized at 8 months, tested at 15 months) implies regeneration at an age when considerable postpubertal atrophy had occurred (Harrison et al., 1982). This response is opposite to the results predicted from the thymus regeneration induced by pituitary (GH₃) cells (Section 10.7; Kelley et al., 1986). While *in vitro* mitogenic responses to phytohemagglutinin and delayed-type hypersensitivity responses were much improved by hypophysectomy, a primary immune response was much *worsened* (Figure 10.22). A study of xenografts indicated faster rejection in hypophysectomized middle-aged rats than in intact controls (Bilder and Denckla, 1977). Such treatments improved clearance of colloidal carbon, an assay for phagocytic capacity (Bilder and Denckla, 1977), and, in contrast to results on mice, also improved primary immune responses (Scott et al., 1979). Pursuit of these complex responses to deficits of hormones could be fruitful but will require detailed characterization of thymic and other hormones.

In view of Section 10.5.1, showing that corticosteroids can adversely interact with

A. Mitogenic Responses

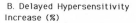

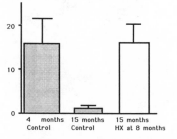

B. Delayed Hypersensitivity
Increase (%)

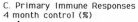

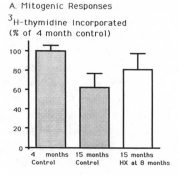

C. Primary Immune Responses
4 month control (%)

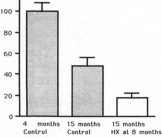

Fig. 10.22. Long-term hypophysectomy prevents age-related declines in certain immunologic responses. Male C57BL/6J mice were hypophysectomized (HX) at 8 months, maintained on replacements of corticosteroids and thyroxine, and examined at 15 months. *Top,* mitogenic responses of spleen cells to phytohemagglutinin were assayed by [³H]-thymidine incorporation *in vitro. Middle,* delayed hypersensitivity was assayed by footpad weight increases 1 day after injection of sheep erythrocytes in mice given priming injections 5 days before, into the contralateral footpad. *Bottom,* primary immune responses to sheep erythrocytes were measured 5 days after immunization by the Jerne plaque assay with cells from the same spleens used in top. Redrawn from Harrison et al., 1982.

hippocampal neuronal senescence, hypophysectomy should be as effective as adrenalectomy in minimizing neuronal damage and loss. Although there are no data on brain cell loss, hypophysectomy prevents the hind-limb paralysis in association with degeneration of skeletal muscle spinal motor nerves, as also occurs from diet restriction (Berg et al., 1962; Everitt et al., 1980). The origin of these changes is obscure and might involve autoimmune phenomena, which are suppressed by diet restriction.

Hypophysectomy also has major impact in retarding the age-related loss of another irreplaceable cell type, ovarian oocytes (Figure 10.23; Jones and Krohn, 1961a). This is the only example known so far in mammals, in which a universal and irreversible age-related loss of a postreplicative cell type is delayed by manipulations of a defined hormonal axis. Conversely, there is evidence for accelerated loss of the remaining oocytes as menopause approaches, possibly as a consequence to elevated FSH (S. J. Richardson et al., 1987).

Tumors are also greatly suppressed by hypophysectomy, as expected for growths in endocrine glands such as testis, adrenal, and thyroid (Everitt et al., 1980), which become atrophic without pituitary support. These effects also extend to the suppression of hepatomas by hypophysectomy in (C3H × YBR)F$_1$ mice (Heston, 1963); there are many influences on hepatocytes from all endocrine glands, any of which could be involved.

Several cardiovascular changes are modified by hypophysectomy. An early change concerns β-adrenergic regulation of contraction in aortic smooth muscle, which is studied *in vitro* in aortic strips pretreated with the adrenergic antagonist phentolamine to cause contraction. Under these conditions, the β-agonist isoproterenol causes smooth muscle relaxation with well-behaved dose responses that are markedly impaired by mid-life in rats and rabbits (Fleisch et al., 1970; Cohen and Berkowitz, 1974). Although the β-mediated relaxation is too rapid to involve modulation of ongoing gene activity, it is of interest because the change is considered to involve alterations in receptor coupling with ionic

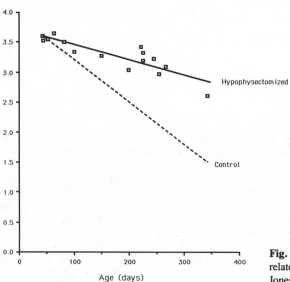

Fig. 10.23. Hypophysectomy slows the age-related oocyte loss in CBA mice. Redrawn from Jones and Krohn, 1961a.

fluxes, which could be important to subsequent steps that influence gene regulation in other mechanisms; nitroglycerine, which induces relaxation independently of the adreno-receptor, eliminates the age differences. The extensive recovery of this response in hypophysectomized rats (Parker et al., 1978) suggests its endocrine basis.

Aortic wall thickening, usually attributed to smooth muscle proliferation, was also markedly reduced by hypophysectomy (Everitt, 1976a; Everitt et al., 1980); the latter reference cites unpublished histological findings that show retarded senescent changes in the aortic medial layer after hypophysectomy. It would be interesting to know if spontaneously hypertensive rats (SHR strains) had similar vascular responses to hypophysectomy in view of the increased lifespans with diet restriction (Lloyd et al., 1984). Hypopituitarism in humans is associated with low incidence of heart disease (Everitt, 1976b). Several hormones under pituitary influence also enhance *in vitro* proliferation of arterial smooth muscle (Weinstein et al., 1981). It would be valuable to know the influence of hypophysectomy on platelet-derived growth factor (PDGF), a mitogen for arterial smooth muscle cell proliferation. In aortic smooth muscle there could also be an adrenoreceptor-dependent mitogenesis that is influenced by hypophysectomy, though the conventional interpretation concerns mitogenic stimulation of vascular smooth muscle through elevated blood pressure.

Analysis of how hypophysectomy influences age-related changes in metabolism is controversial. An assay developed by Donner Denckla in the most extensive studies to date (Denckla, 1974) defined a new parameter, minimal oxygen consumption (MOC), at thermoneutrality in eviscerated rats anesthetized with pentobarbital (Denckla, 1970, 1973). Masoro (1985) estimated that MOC is 25% less than basal metabolic rate (BMR) and 50% less than average daily metabolism. Denckla was the only user of this approach. Of all endocrine glands, only the removal of the pituitary or thyroid influenced MOC (Denckla, 1973). Under his conditions, MOC decreased progressively from puberty onwards, in general similarity to the decrease of BMR or average daily metabolism (Masoro, 1985). In part, the decline in MOC could be attributed to decreased responsiveness to thyroxine (Denckla, 1974, 1975). Moreover, it was adduced from studies with hypophysectomized rats that endogenous thyroxine at about the time of puberty induced secretion of a new pituitary factor, declining consumption of oxygen (DECO), that reduced sensitivity to thyroxine itself.

Hypophysectomy of adult rats slowly restored responsiveness to thyroxine over 6 months. The restoration was incomplete and never reached the thyroxine responsiveness of rats hypophysectomized before puberty (3 weeks; Figure 10.24). Chronic treatment of hypophysectomized rats with thyroxine appeared to enhance the effects of hypophysectomy in the responses to thyroxine.

The basis for these intriguing changes is unknown. Possible mechanisms are changes in subpopulations of differentially metabolizing cell populations and altered gene expression. The later is suggested by age-related decreases in the incorporation of [^{32}P]-ATP and -GTP by isolated liver nuclei, which are reversed by hypophysectomy (J. K. Miller et al., 1980). Uncertainties about these assays make further studies highly desirable, for example, on the transcription rate of specific mRNA by nuclear run-on assays. DECO is occasionally described as a death hormone, though this misquotes Denckla (1975): there is no evidence for cell killing. DECO's main effects, as inferred from its counteractions of hypophysectomy, are to modify cellular responses to hormones and neuronal signals. The

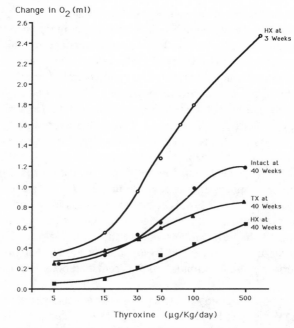

Fig. 10.24. Hypophysectomy (HX) before puberty maintains whole-body metabolic responses to thyroxine (T_4). Metabolic responses to T_4 were assayed, after injection for 3 weeks at various doses, by measurement of oxygen consumption (MOC). Adult female rats hypophysectomized at 9 months and examined 6 months later had enhanced responses to T_4 by comparison with those thyroidectomized (TX) at 9 months and assayed 6 months later, or in intact controls that were assayed at 9 months. By comparison with rats hypophysectomized at 3 weeks and tested immediately, the middle-aged hypophysectomized rats responded nearly as well at lower doses of T_4, but did not achieve the responses at the highest doses. Redrawn from Denckla, 1974.

hypothesized DECO has not been isolated and Denckla, who is barely 50, has not published for a decade.

In conclusion, hypophysectomy has numerous important effects on early and later age-related changes that are documented at physiological and cellular levels and that merit further study at the level of gene expression.

10.9.   The Biomarker Problem: How to Measure Influences on Senescence

Despite the manipulability of the many age-related changes described above, its has proved difficult to link any of them definitively to lifespan, though the argument can be made that the suppression or reduction of age-related diseases by manipulation of diet or hormones should reduce mortality risk. Here we encounter an important difference between senescence as encountered at the individual level and at the population level. The exponential increase of mortality at the *population level* of species manifesting senescence has lead to searches for easily measured parameters, the "biomarkers of aging" that change with age and that predict changes in *individual* mortality risk with advancing age (Reff and Schneider, 1982; Baker and Sprott, 1988).

In analyses of "biomarkers of aging," it stands out that simple chronological age remains the best general predictor of mortality in adults of most species. As discussed in

Chapters 2, 3, 8, and 9, age-related changes are rarely linear functions of time in individuals. Many changes of potential relevance to mortality risk require postmortem tissues or invasive biopsy procedures that preclude repeated measurements on the same individual. These include enzyme induction, intracellular lipofuscins, neuron number, and lymphocyte subpopulations. Other parameters may change but require expensive technologies that preclude longitudinal studies of large samples, for example studies of circadian patterns of hormone levels or tissue volume using imaging procedures such as PET, CAT, and NMR scans. The most accessible biomarkers are behavior patterns and routine physical data, but these have not yielded predictors of mortality risk for adults that are any better than age itself.

The following examples show the problems that are encountered in organisms of very different natural histories. Similar approaches are being developed as biomarkers for senescence in *Drosophila* (Baker and Sprott, 1988; Arking and Dudas, 1989).

10.9.1. Nematodes

The short-lived *Caenorhabditis elegans* (Chapter 2.2.2.2) illustrates the difficulties in determining biomarkers that predict lifespan. Some readers may be startled by the fact that even genetically homozygous cohorts of this self-fertilizing species have a wide range of lifespans that vary up to 5 days on either side of the 12-day mean lifespan (Figure 10.25). These individual variations of lifespan have been used to analyze phenotypic differences among individuals. The frequency of wavelike swimming movements and the frequency of defecation showed linear decreases with age from daily measurements over the lifetime when calculated for either the entire population or for individuals (Figure 2.20). However, there was no correlation between the rate of swimming, or its extrapolated age of cessation, and the individual lifespan. Individuals differed by severalfold in their rates of movement decrease with age; moreover, even the young showed remarkable differences in swimming movement frequency. Thus, even the genetically identical individuals of a simple invertebrate differ widely in their rates of senescence.

Physiological age may be estimated by a combination of variables. For example, a variable P is compared between ages to calculate the changes as a function of time since birth, t:

$$P = f(t). \tag{10.1}$$

Hence, P is almost always less than the postmaturational lifespan. Since mortality rate increases with age, there is an implied relationship between any age-related (or time-dependent) change to mortality risk or remaining life expectancy. But, as shown below, no parameter is better than age.

Another function, f^*, can be considered,

$$T = f^*(P), \tag{10.2}$$

which represents the chronological age of the organism in terms of the biological variable P. In this case, age or time might become an independent variable determined by the function f^*. Calculated in this way, the chronological age, T, could be considered to estimate the "functional" or the "physiological" age.

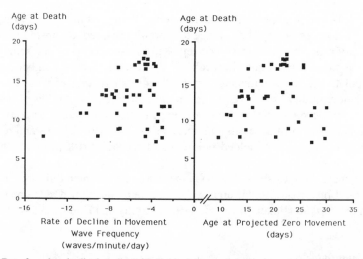

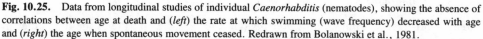

Fig. 10.25.   Data from longitudinal studies of individual *Caenorhabditis* (nematodes), showing the absence of correlations between age at death and (*left*) the rate at which swimming (wave frequency) decreased with age and (*right*) the age when spontaneous movement ceased. Redrawn from Bolanowski et al., 1981.

Following this approach for the nematode data, a multiparametric index of physiologic age from individual behaviors gave a regression on age with lower variance than the swimming movement frequency alone (Figure 10.25). Such multiparametric indexes treat each variable separately and presume independence, except as a function of age. However, the integrative basis for physiological functions makes it likely that most age-related changes have *some* specific relationship to each other; hence the presumption of independence is generally unjustifiable. Multivariate terms to represent statistical interactions should be useful in furthering this approach. In the above case, swimming and defecation rates must be linked because they both depend on metabolism.

10.9.2.   Mice

Mice are important models for developing biomarkers to predict lifespan because of their inbred status and because of the imminent achievement of genetic characterization at the nucleotide sequence level. In the future it should be possible to prove the influence of particular DNA sequences on the lifespan and on specific changes of senescence by recombinant genetic manipulations of the embryo and germ line, such as the insertion of specific DNA sequences or site-specific mutagenesis. Meanwhile, few biomarkers are good predictors of individual lifespans.

Although many parameters show some age-related change in humans, few tests have been adapted to mice. Even fewer meet the requirement that serial measurements are reliable and have no effect on lifespan. The limits of this approach on inbred mice were tested with more than ten behavioral, physiological, and biochemical parameters that had no adverse effects (Harrison and Archer, 1983, 1987, 1988a). For example, the tensile strength of tail tendons increases progressively during most of the lifespan in laboratory rodents (Figure 10.21). The age-related changes in tail tendon and other parameters, however, were not correlated with individual longevity (Harrison and Archer, 1986).

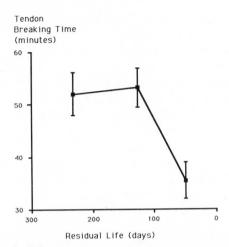

Fig. 10.26. The tensile strength of rat tail tendons decreases sharply about 2 months before death in rats. Age at death ranged from 874 to 897 days: 12–14 rats per time point. This change sharply departs from the progressive age-related trend for increased tensile strength at earlier ages (e.g. Figure 10.21) and indicates general involution before death that alters even the structure of extracellular collagen in tendons. The mechanisms have not been investigated. Redrawn from Everitt, 1969.

Moreover, in the 2 to 3 months before death, the tensile strength of tail tendons actually *decreased,* so that mice at 30 months with longer lifespans had weaker tail tendons (Figure 10.26), as also observed by Everitt (1969). This reversal of the prior age-related trend is probably associated with a pre-morbid involutionary phase that typically lasts 2 to 3 months. Rats (Everitt, 1957) more often than mice (Finch et al., 1969) suffer big decreases of body weight before death in senescence. Adding to the confusion, the strength of tail tendons, tested at 16 or 23 months, showed *no* correlation with individual lifespan in C57BL/6J males, females, or in five other genotypes (Harrison and Archer, 1983; Harrison, Ingram, and Archer, in prep.).

In any case, no single test given at 16 and 22 months was a good predictor of subsequent individual lifespan in all sexes and genotypes. Even combinations of parameters from the different tests (all were done twice on each mouse, at 16 and 22 months) as multiple regressions accounted for only 32% of the variance of subsequent lifespan at 22 months (Harrison, Ingram, and Archer, in prep., 1986). Moreover, the best combinations of parameters that correlated with individual lifespan (movements in open field, number of fecal boluses deposited, paw strength) nonetheless were *not* correlated with chronological age throughout the entire lifespan in the population. These rigorously done studies suggest that no single pacemaker determines age-related changes throughout the body, at least in mice, and that in mice as in nematodes, the determinants of mortality risk vary widely even among individuals with minimal genetic differences.

10.9.3. *Humans*

Humans represent yet another level of complexity in this analysis, in view of their genetic heterogeneity and the huge individual differences during their long lifespans from the impact of diet, infectious disease, and other physical and mental trauma, as well as the salutary aspects of sociobehavioral experiences. At present, we poorly understand how postnatal environmental factors modify the phenotype of senescence, its time course, and the risk of mortality with advancing age (Chapters 2, 9). Nonetheless, there has been some encouraging, if limited, success in defining physiological predictors of mortality risk with age. The most attractive data come from the Framingham Study, which considered the risk factors of cardiovascular disease in the community of Framingham, Massachusetts. This outstanding longitudinal analysis of 5,209 men and women, aged 30 to 62 years at entry, began in 1954, with testing every 2 years. The mortality risk factors identified include age, hypertension, body weight, and hypertrophy of the left cardiac ventricle (Kannel et al., 1980). Hypertension and other parameters linked to the risk of cardiac disease were correlated with mortality risk in the 65- to 74-year age group (Kannel, 1985a). The cardiovascular diseases show weaker age correlations than does hypertension, for which the risk increases progressively with age, as in most human populations (Kannel, 1985a).

A key factor in the Framingham Study is its longitudinal design, in which the same individuals are tested at regular intervals during several decades, as distinct from the cross-sectional paradigm, in which cohorts from different age groups are tested at roughly the same time.[6] Most studies of age-related changes by necessity are cross-sectional. However, cross-sectional studies are susceptible to cohort effects from the different histories of the older and younger cohorts. It is difficult to resolve effects of age from the far-ranging differences between cohorts born in different historical periods and exposed to different diets, infections, and use of alcohol, tobacco, drugs, etc., and other sociobehavioral variables. Cohort effects are known to influence the incidence of major diseases such as hypertension and diabetes, as well as performance on physiological tests (Shock et al., 1984; Shock, 1985), intelligence, and other behaviors (Schaie and Herzog, 1985; Nesselroade et al., 1979; Sørensen et al., 1986). Different outcomes of longitudinal and cross-sectional measurements of blood pressure are illustrated in Figure 3.16.

The mortality predictor in this particular Framingham sample was the lung volume, forced vital capacity, which is the largest air volume that can be voluntarily expelled from the lungs (Kannel et al., 1980). The lung volume decreased progressively with age in nonsmokers as well as smokers (Figure 10.27; Ashley et al., 1975; Kannel and Hubert, 1982), whether measured from cross-sectional or longitudinal data. The cause of the decrease is unknown but is thought to derive from muscle weakness and other changes in the chest wall, rather than from changes in the lung parenchyma (Ashley et al., 1975; Kannel and Hubert, 1982). The lung volume was correlated with risk of death from cardiovascular disease as well as other causes at ages 45 to 74 years, in smokers as well as nonsmokers (Figure 10.28). However, these correlations were *still* slightly weaker than chronological age, which remains the best predictor of mortality overall risk (Kannel et al., 1980).

A strong correlation of lung volume with hand-grip strength over a 30-year age range suggests lung volume is a general measure of musculoskeletal function and of health in general (Kannel and Hubert, 1982). The possible relationship of lung volume to distur-

6. The Baltimore Longitudinal Study (BLS), the Framingham Study, and other longitudinal studies spanning middle and older age in humans are inventoried by Migdal et al. (1981). The BLS is discussed in Shock, 1985; a collection of its reports was assembled in Shock et al., 1984. Longitudinal studies of rodents include those of Richter (1922, perhaps the first behavioral study over the lifespan of any laboratory species), Nelson et al. (1982), Talan et al. (1985), and LeFevre and McClintock (1988).

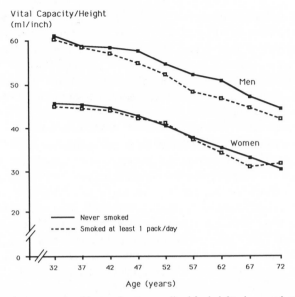

Fig. 10.27. Vital capacity, a measure of lung volume normalized for height, decreased progressively with age, as determined from longitudinal measurements in the Framingham Study, years 1948–1968, in smokers and nonsmokers. Redrawn from Kannel and Hubert, 1982.

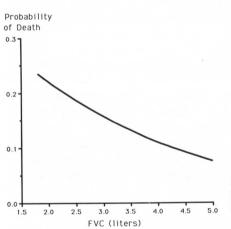

Fig. 10.28. Mortality risk is inversely correlated with forced vital capacity, a measure of lung volume, in men of the Framingham Study, aged 50 to 59 years. Redrawn from Ashley et al., 1975.

bances of breathing during sleep (sleep apnea) could be important because of the linkage of sleep apnea, stroke, and mortality that is indicated by a small number of subjects examined in another study (Dement et al., 1985). The elderly are given to irregular and disturbed breathing during sleep, about tenfold more at 75 years than at 50, which may cause transient deoxygenation (desaturation) of blood hemoglobin (Block et al., 1979).

Another attempt to find biomarkers of senescence evaluated the survivors of irradiation from the atomic bombing of Hiroshima, and included survivors who had shown acute radiation sickness in 1945 as well as nonexposed comparison groups (Hollingsworth et al., 1965). Measurements of nine parameters were made in 1960, in an effort to find evidence for the acceleration of senescence predicted by the somatic mutation hypothesis (Szilard, 1959; Strehler and Mildvan, 1959; Curtis, 1963), because the irradiation should, by this hypothesis, have added significantly to the mutations that were accumulated endogenously. It is remarkable that *no* effects of irradiation were detected in most age-related changes after several decades, for example hair graying and skin elasticity (Hollingsworth et al., 1961) and cellular immunological responses (Bloom et al., 1988). Solid tumors, however, occur more frequently (Preston et al., 1987).

Based on the 1960 results, Hollingsworth et al. (1965) attempted a mutivariate analysis with nine measures of physiological function. A regression was fitted by least squares to a third-degree polynomial for each of the nine parameters, and a composite estimate of functional age, T^*, was determined by assuming a linear relationship. The individual parameters all changed with age, but not necessarily in a regular or monotonic way; in general the variance increased considerably with age. There were no correlations of change before 40 years. Both sexes showed very similar correlations between age and the polynomial estimate. The strongest correlations were in the ranking of skin elasticity, followed closely by forced vital capacity and then sensitivity of hearing and touch. The aggregate estimate of functional age, T^*, correlated better with age than did the individual parameters, accounting for 80% of the variance in age.

Why is it so difficult to obtain biomarkers that predict longevity better than mere chronological age? This vexing question involves uncertainty about causes of death at ages beyond the average lifespan. The causes of death before the average lifespan, that is, in middle age, 45–65 years in humans and 18–24 months in laboratory rodents, are usually readily determined because there are few other concurrent pathological changes besides an occluded blood vessel in the brain or heart, or a metastasized cancer. Death before the average lifespan with one major pathologic condition would usually be considered premature and is the category most clearly correlated with mid-life or premature onset of disease. At the mean lifespan or later, however, most individual mammals have more than one type of degenerative condition (Figure 1.10), so that a unique cause of death is hard to prove; this is particularly so for victims of chronic diseases like Parkinson's or Alzheimer's disease, in which death may result from unrelated conditions like pneumonia. The biomarkers-for-longevity problem may be equivalent to establishing the relationship of the range of late onset disorders of senescence to preceding changes at mid-life or before. Many of the best-documented age-related changes (e.g. rat tail-tendon tensile strength, graying of hair in humans) may have little or no linkage to the degenerative changes that accelerate mortality risk.

10.10.   Summary

The capacity for experimental and environmental variations in the phenotypes of senescence is well demonstrated in different phyla. Particularly dramatic examples are given by the semelparous plants and animals, whose lifespan can be extended severalfold or more through interventions that block reproduction-related senescence (Chapter 2.3). In these species, preparations for reproduction are a trigger for senescence. Although little is known about mortality rates in these species, the slowing of rapid senescence is associated with a great reduction of mortality rates, as clearly occurs in Pacific salmon or monocarpic plants that are prevented from maturing. Some interventions convert annuals into perennials with multiple seasons or bouts of reproduction. Most such interventions involve neural and humoral functions related to reproduction, and imply mechanisms of organismic senescence that are based on selective gene regulation of responses of various target cells to changes in hormones and other regulatory factors.

Manipulations of senescence by interventions during adult life have also been shown for iteroparous species throughout animal phyla. We see here major shifts in the phenotypes of senescence. Fifty years of research has established numerous interventions through diet and hormones that modify or attenuate many age-related changes and diseases in laboratory rodents and, to an increasing extent, clinically important conditions of humans. The impact on total lifespan so far is much more modest than achieved for most semelparous species, however. Thus, the most effective diet-restriction study has increased rat maximum lifespan by less than 50% beyond controls (Table 10.1), that is, far less than achieved for Pacific salmon. Caloric restriction remains of high interest as the only example that slows the acceleration of mortality in mammals.

Contrary to these familiar examples, other species show increased lifespan when food is available in superabundance at a certain time in their adult phase. Examples include several salmoniforms and flatworms. These exceptions do not contradict many studies showing salutary effects of diet restriction, but point to alternative phenotypes of senescence that are also nutritionally determined but in opposite directions to diet restriction. The genotypic basis is not known, but is suspected for ferox salmon (Ferguson and Mason, 1981; Hamilton et al., 1989; Chapter 3.4.1.2.2).

The mechanisms of diet restriction are unclear but probably are multiple. The possibilities in vertebrates include modulation of various hormones that influence metabolism (insulin, growth hormone, glucagon, glucocorticoids) and protein glycation, which are not mutually exclusive. There may be an important nexus of biochemical mechanisms shared by the antisenescent effects of diet restriction, DHEA, and reduced stress. The lowered body temperature of diet-restricted mice and rats could also be an important factor, possibly through slowing the rate of protein glycation or of oxidative damage. Rubner's old hypothesis that lifespan is set by a maximum expenditure of caloric energy (Chapter 5.1) is clearly inconsistent with the studies on diet restriction in rats and on environmental temperature in flies that showed the independence of caloric expenditure per gram body weight with extended lifespans. The major manipulations of delaying or reversing age-related diseases, resulting in the physiological, cellular, and molecular changes described above, also point away from theories that require essentially irreversible molecular changes such as accumulations of mutations. Taken together, the manipulatable aspects of senescence argue for epigenetic mechanisms, in contrast to mechanisms of accumulated and irreversible modification of DNA sequence or structure.

Subtle shifts in the balance of metabolic pathways contributing to peroxidizable lipids in various cells and to blood glucose and lipids may thus promote insidious damage to slowly replaced molecule and cells. The rate at which such damage accumulates, according to population genetics–based evolutionary theory (Chapter 1.5), should be scheduled by forces of selection to delay substantive dysfunctions until ages when individuals in the population contribute negligibly to the gene pool. It thus appears that many mammals are adapted for surviving seasonal and reproduction-related metabolic transients earlier rather than later in adult life. As argued in Chapter 6.6, this explanation more easily accounts for the diversity of later onset age-related diseases within and among species than it does the overall scaling of many changes that begin before mid-life in proportion to the life-span.

The ergonomic aspects of reproductive schedule may be very different for other verte-brates, like the fish that continuously increase in reproductive capacity, without limitation on the numbers of gametes produced. Other species with age-related reproductive declines have reproductive schedules that are adapted to more intense senescence. Extreme ex-amples include aphagous (autogenous) insects, while mammals and birds, which gener-ally require parental care, occupy less intense zones in the spectrum of senescence. The possible role of caloric limitations in delaying reproduction to more favorable times, nec-essarily at later ages, was hypothesized as adaptive. I further suggested that these phe-nomena revealed the important role of the brain in setting the *ad libitum* food intake, through yet unknown genes whose polymorphisms are adapted to various ergonomic schedules as needed for reproduction. Thus, the life-prolonging effects of diet restriction have revealed the brain as an important pacemaker of senescence, through its role in co-ordinating the balance of metabolism with food availability and the reproductive schedule.

Caloric restriction, DHEA, and hypophysectomy alike also delay or slow early onset diseases in numerous short-lived and in long-lived rodent genotypes alike. This observa-tion helps restore shorter-lived genotypes to the mainstream of discussions on senescence. Even though senescence may not be generally accelerated in the short-lived genotypes, the fact that the same external manipulations retard their lesions as in long-lived genotypes suggests common denominators in pathogenesis.

In view of the inhibition of reproduction caused by caloric restriction and hypophysec-tomy in iteroparous species, it is necessary to reconsider the possible role of reproduction to senescence. This relationship is fundamental in most semelparous species, in which the prevention of reproduction delays senescence. For the aphagous mayflies and other orga-nisms that loose the capacity to ingest a complete diet during development through ac-quired anatomical defects, the adult stage itself is doomed to a finite lifespan. Iteroparous species also present a varied picture of reproductive costs. On one hand are the Diptera, for which reproduction is only indirectly debilitating through the demands that flying and fighting make on mechanically vulnerable wings. In a similar mode, some ungulates have increased mortality in association with reproduction, through wounds incurred during fighting in the case of males (bighorn sheep, Chapter 1.4.3) and from the nutritional drain of nursing (red deer, Section 10.6). While many such examples show that various degen-erative changes are accelerated by reproduction-related stress, there is no evidence that other features of senescence are also accelerated. The relatively small impact of castration on the lifespan of adult mammals also supports this view.

The next chapter discusses phylogenetic and evolutionary aspects of senescence and makes frequent reference to the phenotypic plasticity of senescence as demonstrated by interventions during development or in the adult stage, as described in this and the previ-ous chapters. In some cases, evolutionary changes in life history patterns appear to draw from the same fundamental plasticity shown by environmental and experimental manipu-lations of senescent phenotypes.

11

Phylogeny and Evolutionary Changes in Longevity and Senescence

11.1.   Introduction

This chapter scouts largely unexplored areas of the evolutionary record that pertain to longevity and senescence. Questions of how the potential lifespan and the patterns of senescence have evolved seem at first to be even less accessible than those relating to the evolutionary record of development (Jefferies, 1986; Buss, 1987). We may never enjoy certain knowledge of when in evolution most patterns of senescence arose, because morphologic or even biochemical relicts rarely disclose enough to reconstruct much about life history. Fossils of salmon, for example, cannot show which species were destined to die at first spawning. Also elusive are the mortality rate parameters that determined the lifespan of extinct organisms. Even so, I hope to convince you that the fossil record reveals patterns of adult life history and senescence which emerged numerous times during evolution. Furthermore, something can be said about evolutionary trends in lifespan among morphologically similar ancestors of present species, since some lineages show evidence for both shortening and lengthening of adult phases.

Life history theory has focussed on the age and rate of reproduction in relation to lifespan but has not dealt much with taxon-specific patterns of senescence. A major controversy concerns the extent to which variations in natural history are constrained according to the anatomical and physiological organization (body-plan)[1] of a particular evolutionary lineage. As an approach to this difficult question, Section 11.3 considers phylogenetic variations in life histories that relate to major themes discussed earlier in the book: vegetative (nongametogenic) reproduction in corals and vascular plants; taxonomic clustering of semelparity and of limited oogenesis in fish; clustering of aphagy in insects; the evolution of long-lived vascular plants and the independent origins of finite growth and leaf abscission; the divergent lifespans and rates of mortality acceleration in different birds and mammals; tooth replacement in elephants and other mammals, which is implicated in setting the potential lifespan; and Alzheimer's disease. Intraspecific variants found in present species that were discussed earlier are also recounted in relation to the question of variant genotypes and phenotypes from which different life histories may evolve. These discussions are necessarily focussed on certain species to complement the foregoing descriptions of senescence and cannot be comprehensive. Full discussion of the paleontologic evidence, for example on insect mouthparts, is impractical and would justify another volume at least as long as this already lengthy book.

Evolutionary trends in senescence are then surveyed according to the body-plan, with a discussion of how derived anatomic deficiencies, size, and the capacity for regeneration and cell replacement may influence adult life history (Section 11.4). The extensive variations in the age and patterns of reproduction are then discussed in the general context of heterochrony, or variations in the timing of developmental events, that depends on plasticity in the scheduling of changes in gene expression. I will argue that heterochronic changes in the scheduling of senescence, like those of development, require the maintenance of genomic competence. The plasticity in the phenotypes of senescence described in Chapters 9 and 10 is another manifestation of this potential.

---

1. *Body-plan* has the same meaning as *Bauplan* in Gould, 1977.

11.2.   An Outline of Evolution and a Discussion of Early Biochemical and Molecular Changes

An overview of animal and plant evolution is given to set the stage for discussions of particular issues and species. My sources are contemporary (Gifford and Foster, 1988; Boucot and Gray, 1983; Cloud and Glaessner, 1982; Margulis and Schwartz, 1988; Jefferies, 1986; Romer, 1966; Carroll, 1988; Grzimek, 1976; Seilacher, 1984; Stewart, 1983; Tiffney and Niklas, 1985; Whittington, 1985; Gould, 1989). Figure 11.1 shows major geologic stages and the first occurrence of organisms discussed below.

Precambrian animals were highly diverse by 700 Myr (million years) ago. For example, the Ediacarian fossils include forms resembling present soft-bodied annelids, arthropods, jellyfish and other cnidarians, and echinoderms. Most multicellular Precambrian animals were aquatic. None had mineralized shells or solid skeletons, though some may have used chitin for strengthening. Some reached large sizes, about one meter in diameter, that imply growth for months or years.

Precambrian plants were simple, alga-like forms with filamentous or spheroidal aggregates (probable chlorophyte ancestors) and may have arisen long before 1,000 Myr ago. Although other forms with branching but nonseptate filaments suggest fungi, unequivocal fossils of filamentous fungi occur much later (Devonian). Precambrian multicellular plant life was entirely aquatic and appears to have been less diverse than animal life.

Asexual reproduction is suspected as a major reproductive mode long before the Cambrian, at least for those organisms whose morphologically similar descendants reproduce vegetatively as well as sexually (cnidarians, annelids, etc.; Chapter 4.3.3; Table 4.3). The mechanisms would most likely have included budding or fragmentation, with redifferentiation of somatic cells as found in extant cnidarians and annelids. Sexually dimorphic characteristics and sexual reproduction, however, are also found in Precambrian organisms. Evidence on mosaic or partial senescence in early fossils appears to be lacking. Readers may recall from Chapters 2 and 4 that some species of cnidarians, colonial tunicates, and rhizomatous plants show a mosaic senescence that causes regional cell death.

A change that is important to the potential for longevity is the evolution of the present oxygen-rich atmosphere. According to the Berkner and Marshall hypothesis (1965), the evolution of an oxidizing atmosphere was a critical step in the evolution of complex metazoa that allowed increased rates of metabolism. Atmospheric oxygen is generally agreed to have increased greatly during the Precambrian (Cloud, 1980; Holland, 1984). There is, however, debate on the sequence of changes and when the present levels were reached, and it cannot be supposed that the increases were monotonic. Most evidence indicates that the transition to an oxidizing atmosphere had occurred by 1 billion years ago or earlier. Sharp increases of atmospheric oxygen are indicated during the Paleozoic, particularly in the Devonian (McLean, 1978) as the land became populated by photosynthesizing plants. Land plants presently produce more than 50% of biogenic oxygen (Holland, 1984, 518).

Because oxygen is highly toxic from the formation of free radicals that damage nucleic acids, proteins, and lipids (Chapter 7.5.1), we may suppose that enzymes had been evolved long before the Cambrian to counteract free radicals. In present aerobic organisms, superoxide dismutases and catalase are ubiquitous quenchers of oxygen-derived free radicals. Other antioxidant free radical quenchers include small molecules like glutathione and vitamin E. Besides these protectants, virtually all cells have proteolytic mech-

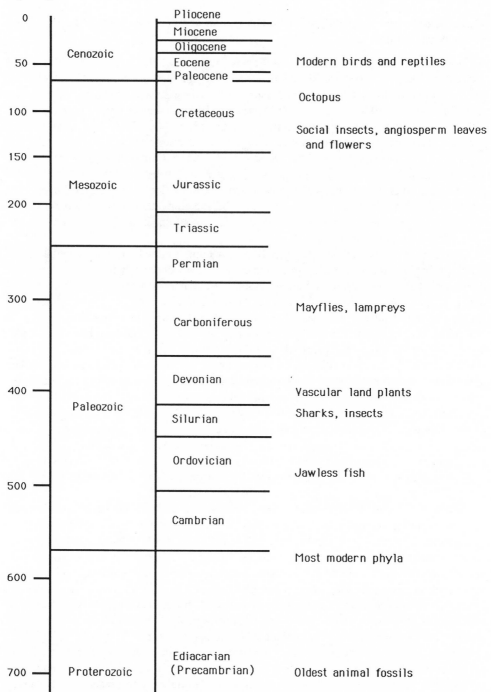

Fig. 11.1. Major geologic stages, a chart based on standard geochronology (Geological Society of America), showing the first record of select organisms discussed in the text.

anisms for rapidly degrading proteins that are damaged by oxygen radicals and enzyme systems for repairing oxidative damage to DNA. The generality of this biochemical and enzymatic machinery throughout prokaryotic and eukaryotic cells suggests its existence before multicellular organizations and sharply increased atmospheric oxygen. Exposure of terrestrial life to UV-irradiation was also decreased through formation of the ozone layer, when atmospheric oxygen increased to a critical level. Protection from UV damage may have enhanced population growth in exposed habitats.

The evolution of enzymatic and metabolic mechanisms for free radical detoxication and macromolecular repair may be considered crucial in the evolution organisms with long-lived DNA and proteins, and slowly replaced membrane lipids such as found in myelin sheaths and neurons. Without effective detoxication and repair, major energy would have to be continuously diverted to biosynthesis for molecular replacement in somatic cells. I propose that the evolution of antioxidant mechanisms was a fundamental early step in the evolution of extended durations of development or extended reproductive schedules that resulted in prolonged lifespans. The capacity of mammals and birds to sustain constant high body temperatures (Section 11.3.5.3) may then be traced to the efficacy of antioxidant mechanisms that were established before or by the time the modern atmosphere had evolved.

Another biochemical attribute that must have been tested evolutionarily in the Paleozoic or before is the prevalent carbohydrate fuel, which in most vertebrates is glucose. Bunn and Higgins (1981) note that D-glucose is one of the least reactive of the circulating monosaccharides in the nonenzymatic formation of glycated adducts with proteins. Glycation and further modifications to advanced Maillard products (AGE) can cause functional damage, particularly to proteins with long lifespans like collagen, elastin, and lens proteins (Chapter 7.5.2). The extended reproductive schedules in species that segregate germ-line cells early in development may have depended on minimization of molecular damage to the long-lived oocytes, but also to organs like the eye lens that, presently, have little protein turnover. The evolution of metabolic pathways that yield monosaccharides active in glycation should be further considered as important in the evolution of different life history patterns.

Among fossils from the Cambrian period (590–505 Myr ago, the beginning of the Paleozoic era, Figure 11.1), most animal phyla are represented. The remarkable fossils of the Burgess shale (530 Myr ago) include arthropods and bivalve and cephalopod mollusks, as well as other protostomate-like creatures with body-plans that are unlike any subsequent phyla. We may suppose that their life histories were as diverse as their anatomy! Cambrian fossils reveal the first sponges, which reproduce asexually by fragmentation and lack senescence (Chapter 4.4). Many phyla whose descendants show senescence also appeared by the Cambrian, including cephalopods, chordates, crustaceans, snails, and primitive vascular plants. In contrast, Cambrian plants had not achieved these higher grades of organization and were restricted to bryophytes and primitive vascular types with little phyletic variety. Vascular land plants of any size do not appear until there are land-living vertebrates and flying insects, about 400 Myr ago, in the Silurian. Transitions to the Cambrian are marked by a great increase of animals that constructed complex burrows and had mineralization of soft tissues (including teeth). These innovations imply specializations such as tunneling for protection and maceration of food. The importance of predators and the higher trophic level of carnivores in the Burgess ecological community (Conway Morris, 1986) suggests that some of these early invertebrates had evolved nervous systems capable of complex behaviors.

There is also clear evidence for deuterostomes and even primitive chordates in the Burgess lode. The fishlike *Pikaia gracilens* had a notochord and myotomes (Jefferies, 1986; Gould, 1989). Like an extant primitive chordate, *Amphioxus, Pikaia* lacked skeleton and brain. Predecessors of the vertebrate immune and endocrine systems may have existed in Precambrian chordates or soon thereafter. Whenever the immune and nervous system acquired definitive roles in physiological organization, their very long time of existence may have promoted the establishment of specialized gene sets.

An important example of an ancient gene set that influences adult life history through its influence on various diseases is the main histocompatibility complex (MHC), which is clustered on a single chromosome in present vertebrates (Chapter 6.5.1.4; Figure 6.7). The mammalian MHC shows an unusual extent of allelic polymorphisms in certain loci that influence immune functions, but also hormonal responses and reproduction. Moreover, the mitochondrial form of superoxide dismutase and the P_{450} enzymes that are important in detoxification are also linked to the MHC. An MHC may also occur in invertebrate chordates, for example tunicates, and manifestly has a very long evolutionary history (Klein, 1986; Scofield et al., 1982) that Walford (1987) suggests included many antisenescent functions, such as protection against free radical damage and environmental toxins. Another example that differs by being dispersed throughout the genome is that of hormone receptor families and their associated DNA regulatory elements that allow transacting factors to control multiple genes on different chromosomes (Britten and Davidson, 1971; O'Malley, 1989; Davidson, 1990; Evans, 1988; Beato, 1989; Yamamoto, 1985; Miesfeld, 1989). Physiological networks that are built up from enmeshed regulatory systems may be relatively stabilized against some types of mutational change and genomic rearrangements. Further studies of the types of functions influenced by MHC polymorphisms might reveal genomic constraints on life history patterns. Suggestions (Gould, 1989) that the Cambrian organisms explosively invaded underpopulated environments raise intriguing questions about the evolution of life histories and potential longevity when molecular mechanisms to minimize free radical damage almost surely had evolved.

In the next geological period, the Ordovician (505–438 Myr ago), many of the Burgess wonders had disappeared, leaving behind the present phyla and body-plans. Jawless fish were represented by several groups that are supposed to include the ancestors of hagfish and lampreys; each is thought to be a distinct line. The jawless ostracaderms from the middle Ordovician were protected by bony plates but did not have ossified (calcified) internal skeletons. Soon after in the Silurian period (438–408 Myr ago), jawed fish first appear, including bony fish and sharks. However, no fossil agnathan is established as their ancestors. Jawless fish radiated extensively into the Devonian. Vascular land plants, the slender leafless *Cooksonia,* are found by the mid-Silurian.

In the Devonian (408–360 Myr ago), the terrestrial vertebrates (Amphibia) and lungfish (a lobe-finned fish, or sarcopterygian) are represented. It is not clear whether the lungfish are in the same lineage or from some other stem as a sister group to tetrapods (Carroll, 1988; Nelson, 1984). Primitive fish of the Upper Devonian are considered the stem of modern elasmobranchs (sharks, skates, and rays). Ray-finned fish (actinopterygians) also emerged by the Devonian and had well-ossified skeletons. The rainy Devonian climates are thought to have favored the evolution of tall, forest-forming vascular plants. Coal deposits of the Carboniferous swamps (360–286 Myr ago) contain abundant, finely detailed fossils of many plants and insects, including mayflies.

As the Paleozoic closed in the Permian (286–248 Myr ago), the land was inhabited by numerous reptiles and other vertebrates, flying insects, and early conifers. Climates became cooler and drier, no longer supporting tall swamp plants. There was also a mysterious and massive extinction of more than 90% of marine and land species. The land differed vastly from the present. The continental plates had become joined in the supercontinent, Pangaea, which was again tectonically fragmented.

At the beginning of the Mesozoic era (the Triassic period, 248–213 Myr ago), the record shows fossils of the first flying vertebrates and early mammal-like reptiles (cynodonts). Climates continued to become drier but were generally mild throughout the Mesozoic. Dinosaurs were most abundant in the Jurassic (213–144 Myr ago). The earliest sturgeonlike fish (*Chondrosteus*) is from the Lower Jurassic; the reduced extent of ossification is a derived characteristic that evolved during an unknown sequence since the derivation of sturgeon from Paleozoic ray-finned fish (actinopterygians). The major radiations leading to modern teleosts occurred during the Mesozoic, from ancestors that had diverged from a lineage of sturgeonlike fish in the Paleozoic. Eels (anguilliforms) were established by the Upper Cretaceous, while salmonids are not found until after the Cretaceous.

By the end of the Cretaceous (144–66 Myr ago) which closed the Mesozoic, most of the present types of higher plants and animals are in the fossil record. The breakup of Pangaea into the northerly continent of Laurasia and the southerly Gondwana began in the Jurassic. Gondwana fragmented further during the Cretaceous into the present southern continents. Another mysterious and even more massive extinction that occurred late in the Cretaceous eliminated the dinosaurs and is thought to have allowed the radiation of the then much smaller mammals. Although marsupials survived, they never regained their diversity and abundance. After the Cretaceous extinctions (Clemens, 1986), there was a delay before radiations during the Cretaceous and into the Cenozoic yielded numerous new orders of placental mammals. Extant species include about 230,000 flowering plants and 500 conifers; 1,000,000 insects, probably a gross underestimate; 22,000 bony fish; 9,000 birds; and 4,000 mammals.

11.3. Examples of Phylogenetic Variations in Senescence and Longevity

This section gives select aspects of the evolutionary record that pertain particularly to the occurrence of vegetative (asexual) reproduction, semelparity, and growth patterns (determinate versus indeterminate size). I also discuss anatomic features associated with permanent aphagy and tooth replacement; the latter is an example of mechanical senescence in mammals. These topics represent various levels in which senescence may occur, from molecules to complex cell organizations.

11.3.1. Corals

Animals and plants that reproduce vegetatively by somatic cell budding or fission contain potentially immortal somatic cell lineages. In view of the emphasis given to the clonal senescence of cultured mammalian fibroblasts as a model for organismic senescence (Chapter 7), it is valuable to consider the evolutionary record on vegetatively reproducing species. Readers may recall from Chapter 4 that vegetative reproduction is found in numerous, but not all, phyla of animals, while it occurs in all plant phyla. Vegetatively reproducing species lack segregation of the germ line from somatic cells during development, although not all somatic cells need be genomically totipotent and capable of originating another complete adult. The immortality of some somatic cell lineages through vegetative reproduction does not preclude senescence of the whole organism, or senescence of its parts (Chapter 4.4).

A further feature found in some vegetatively reproducing organisms is that the reproductive capacity of the individual increases with size without signs of a limit; this is particularly shown for perennial, iteroparous plants that show continued growth and increasing fecundity. This phenomenon demonstrates the crucial point that some organisms are exceptions to theoretical conclusions (Chapter 1) about the inevitability of senescence in two vital regards: that somatic cell lineages are necessarily destined to senescence and that reproduction is necessarily truncated with advancing age. While Watkinson and White (1985) among others note that such organisms challenge Hamilton's (1966) theoretical arguments about the inevitability of senescence (Chapters 1.5.2, 4.5.3), it is pertinent that Hamilton argued from the assumption that mutations enhancing early fecundity do so at the expense of an equal decrement later. This assumption, which generally prevails in population genetics thinking about costs of reproduction and senescence, may not hold for all organisms.

We now look to the evolutionary record of the persistence of vegetative reproduction, first in corals and then in plants. The corals (Cnidaria) demonstrate the evolutionary persistence of asexual (vegetative) reproduction side by side with sexual reproduction. Corals have a rich fossil record of 450 Myr, extending from the Ordovician to the present, that shows fluctuations in the predominance of species with the two major reproductive modes: sexual (through gametic fusion) versus asexual and clonal (vegetative reproduction from somatic cells; Chapter 4.4.3). The clonal genera may be presumed to reproduce sexually also, whereas aclonal genera lacked the capacity for vegetative reproduction as judged by extant corals. Paleozoic corals formed massive reefs, while subsequent species had several cycles of size reduction during the mid-Mesozoic and early Cretaceous.

In present corals and bryozoans, the aclonally reproducing (sexual) corals have larger modules (zooids or polyps; multicellular individuals) than clonal species that reproduce by budding (Coates and Jackson, 1985). Coates and Jackson hypothesize that larger solitary individuals of these marine species have reduced mortality, a size correlation also observed in many vertebrates (Chapter 5.3.1). The prediction is that larger corals should reproduce more offspring than smaller species; the greater lifetime fecundity should thus select for larger sizes. However, actual mortality data are lacking. On the other hand, selection for module size should be more relaxed in clonally reproducing colonial corals, since survival depends less on the size of a module than on the size and organization of the colony. As shown in Figure 11.2, clonal and aclonal genera of rugosan, scleractinian, and tabulate corals were present early in the Paleozoic. Clonal coral genera increased relatively during the Paleozoic, but became less common later on when the massive reefs began to disappear in the Upper Devonian (Frasnian stage); based on present corals, clonal genera also reproduced sexually. Throughout these shifts the aclonal genera had larger anatomical modules than the clonal ones.

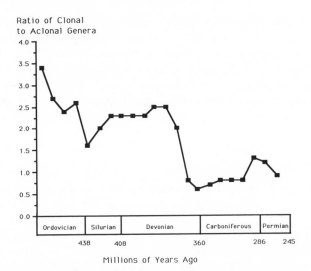

Fig. 11.2. Reproductive modes in corals (anthozoan cnidarians) from the fossil record, as deduced from rugosan, scleractinian, and tabulate forms. Aclonal genera are characterized by solitary individuals, whereas clonal genera show a range of sizes and sometimes indeterminate growth. There is no reason to doubt that clonal genera also reproduced sexually, as observed in extant clonal corals. Clonal genera increased relative to aclonal genera during the Paleozoic, but became less common later on when reefs began to disappear in the later Devonian (Frasnian stage). Throughout these shifts the aclonal genera had larger modules than the aclonal ones. Redrawn from Coates and Jackson, 1985.

Analysis of the species duration in the fossil record for hundreds of scleractinian corals with different growth forms shows an important feature: the species capable of clonal reproduction (i.e. vegetative reproduction, Chapter 4.4, but also presumably sexual) persisted during evolution at least as well as the species that reproduced aclonally (i.e. gametically; 9.06 versus 9.57 Myr, respectively; Jackson and Coates, 1986). The influences of clonal reproduction on the evolutionary persistence of a species have less impact on evolutionary persistence than growth form. This conclusion is very important to the argument that evolutionary parsimony selects for a disposable soma (Section 11.4.5).

Why there are slow shifts in proportions between clonal and aclonal species is obscure but may be related to the ubiquitous extinctions at the Upper Paleozoic. Jackson (1985) comments that the individual organismic lifespan in many colonial marine species (corals, bryozoans, sponges, etc.) is much longer than in solitary aclonal species and that, except for the ascidians, most examples of senescence are found in solitary aclonal species (also discussed in Chapter 4.4.3). By this premise, the increase of aclonal corals implies a statistical shortening of average individual lifespans across the different genera.

In sum, both clonal and sexually reproducing cnidarians have persisted since the Paleozoic, as in numerous other phyla (Chapter 4). However, the fossil record has not yet been analyzed for influences by the mode of reproduction (sexual, parthenogenetic, vegetative, etc.) on the evolutionary lifespan of plants or animals, in terms of the body-plan or genomic complexity. This information bears on whether recombination is essential for evolutionary persistence, and whether mutational changes are accumulated in the somatic cells that gave rise to new progeny. Somatic mutation may be used as a source of genetic variation (Watkinson and White, 1985). As discussed in Chapter 8.5, the gametogenic tissues of plants are formed from somatic cells, often after extended vegetative growth. Klekowski and colleagues (Klekowski, 1988a, 1988b; Klekowski and Godfrey, 1989) argue that species with long generation cycles and asexually reproducing species accumulate more mutations. It is possible that vegetative reproducers are ultimately less responsive to selection in a changing environment, as conventional theory predicts, and that only sexual species survive, from which new asexual species could be derived. Nonetheless, as described next, clonal propagation in seed plants is long-standing and also evolutionarily robust.

11.3.2. Plants

The evolution of plants gives further insights about clonal reproduction, but also about determinate growth. Vascular land plants appeared by the Silurian period (495 Myr ago) and are thought to have propagated clonally for the most part (Tiffney and Niklas, 1985). The fossil record is detailed enough to indicate the absence of advanced sexual organs like cones or flowers in the first vascular land plants. Tracheophytes of the late Silurian and early Devonian had very limited vascular systems and proliferated vegetatively through outgrowing horizontal rhizomes. The only known late Silurian tracheophyte, *Cooksonia,* was a leafless surface plant three cm or less tall (Gifford and Foster, 1989; Chaloner and Sheerin, 1979). Its successors, including those with like forms, grew much larger.

A characteristic of rhizomatous growth is the formation of apical meristematic growth zones, which can separate to form a new plant. This growth pattern, which is nearly universal in present vascular plants (Chapter 4.2.1), probably is the primitive condition and, if not, occurred very early in vascular land plants. Early evidence of vigorous clonal growth is found in the rock deposits that contain a single species of vascular plant, implying the dominance of one clone over others in that neighborhood (Tiffney and Niklas, 1985; Edwards and Fanning, 1985). The lack of roots in *Cooksonia* and other early vascular land plants may have favored clonal growth, which allowed sequestration of nutrients through space occupation; this is less efficient than nutrient sequestration through roots (Knoll et al., 1985). Since, according to conventional wisdom, the formation of sporangia requires considerable energy, the cost of sexual reproduction may have been high for these early plants.[2] This argument may also apply to the maintenance of asexual reproduction as an option in certain animals.

Bryophytes, specifically the Hepaticae (liverworts), are another plant phylum that became established on land in the Devonian (ca. 380 Myr ago; Schuster, 1981). Their lack of vascular tissue requires growth close to the ground in moist environments, with reproduction mainly by clonal offshoots. It is noteworthy that sexual cycles are lacking in some present bryophytes (Chapter 4.4.2).

The capacity for extensive vertical, aerial growth (arborescence) arose in various phylogenetic lineages of vascular plants during the Devonian, including the progymnos-

---

2. The literature of plant life history is perfused with the assumption that nutrients and energy resources are nearly always limiting in growth and that reproduction is nearly always costly. These complex questions deserve renewed scrutiny, since the regulation of life history events like the choices of reproductive mode may be influenced by chemical and physical signals that have little impact on energy partitioning or on mortality risk. As noted in Chapter 10.6, the concept of reproductive costs, found in the animal literature, has recently come under scrutiny, and many of the same issues pertain as well to plants.

perms, lycopods, and ferns. Seeds did not occur until the late Devonian. The middle to late Devonian flora included large treelike progymnosperms and lycopods, which evolved the capacity to produce secondary support tissues: wood in progymnosperms; bark in lycopods. These changes were necessarily accompanied by increased diameters of the tracheids, which are fluid-conducting, elongate dead cells in the xylem (Niklas, 1985; Knoll and Niklas, 1987). According to the physics of capillary action, the longitudinal volume flow through a tube increases as the fourth power of the radius (r^4, Hagen-Poiseuille equation). Thus, minor increases in tracheid diameter will greatly increase the conductance of fluids to aerial structures above (Tomlinson, 1983; Zimmerman, 1983).

Through apparently minor quantitative variations in tracheid morphogenesis, land plants attained the capacity to support taller vertical axes and thus formed forests with high canopies. This greater aerial development also created the potential for complex microenvironments at different levels above ground. Microenvironments may have favored the evolution of sexual cycles that became independent of water for fertilization through aerial and wind-driven dispersion of spores and seeds. Diverse gametic vehicles are widely thought to have arisen independently and polyphyletically during the Devonian (Chaloner and Sheerin, 1979). Concurrently with the evolution of these forms in the late Devonian, clonal vascular plants became proportionately less important in the landscape during the progressive colonization of new environments (Figures 11.3, 11.4).

Progymnosperms and lycopods are characterized by deciduous leaves and determinate growth. Some apparently had short and others long lifespans. Progymnosperms are thought to be the ancestors of modern seed plants, and some species grew quite tall (C. B. Beck, 1971; Stewart, 1983). The largest was *Archeopteris* from the Upper Devonian; reconstructions suggest heights of thirty meters and stumps up to two meters across (Banks, 1970; Beck, 1962; Stewart, 1983; Gensel and Andrews, 1984). Other arborescent Devonian progymnosperms were shrubby, for example, *Tetraxylopteris* and *Aneurophyton,* in which the arbors had up to fourth-order branching, with planar leaves of determi-

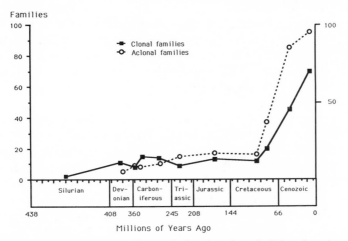

Fig. 11.3. Clonally (vegetatively) and sexually reproducing plants from the Paleozoic to the present, showing the absolute numbers of families of each type. There is no reason to doubt that clonal forms also reproduced sexually. Radiations of angiosperms in the Cretaceous and Paleocene (Figure 11.6) account for most of the increase in sexual forms. Redrawn and adapted from Tiffney and Niklas, 1985.

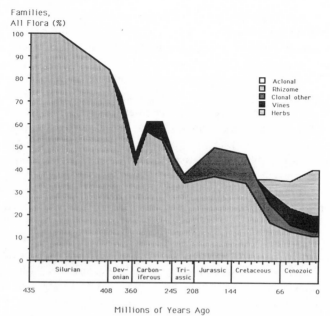

Fig. 11.4. Relative numbers of vascular land plant families, according to different modes of clonal (vegetative) growth versus sexual reproduction. *Aclonal,* all families where sexual reproduction predominates; *rhizome,* families with predominantly rhizomatous growth; *clonal other,* families dominated by basal sprouting, root suckering, gemmae, etc.; *vines,* families that form lianes (vines or aerial extensions), but with only adventitious formation of roots; *herbs,* dicotyledonous families of herbs with rhizomatous or stoliniferous growth. Redrawn and adapted from Tiffney and Niklas, 1985.

nate size and webbed divisions like present leaves. Moreover, like present redwoods, the leafed branches showed seasonal cycles of abscission and regrowth, which implies hormonal mechanisms of senescence.

A characteristic of progymnosperms was a cambium that produced a secondary xylem with radially aligned tracheids that resembled the water-conducting cells of modern conifers. Some fossils also show a few growth rings (ten or fewer) in the secondary xylem (Arnold, 1929, 1931; Banks, 1970; Figure 11.5). These growth rings could represent periodic fluctuations of moisture or temperature, but probably do not reflect changes in the photoperiod in view of the then equatorial location of these forests. Growth to such large sizes might have taken 50 years or more, as estimated from present growth rates.[3]

An important characteristic of the progymnosperms is determinate growth. A size maximum resulted from the cessation of growth, because the apical meristem became progressively smaller at the top branches (Scheckler, 1978; Stewart, 1983, 224). The diminution of the apical meristem implies a limit to cell proliferation. The cellular basis for determinate aerial growth by limited meristematic proliferation is poorly understood but could be analogous to the limited replicative potential of many mammalian cell types that is ac-

---

3. The estimate of 50 years for a 1.5-meter-diameter stump is based on tree growth rates in present tropical rain forests, in which the base (stump) diameter can increase up to thirty millieters a year (P. S. Savill and T. Whitmore, Oxford Forestry Institute, U.K., pers. comm.). Since stump expansion ranges down to one millimeter a year, the age could be more than tenfold greater.

Fig. 11.5. Fossil of *Archeopteris* from the Devonian. This transverse section of the trunk shows growth rings, which suggest seasonal fluctuations of temperature, moisture, or both. From Banks, 1970. The photograph was supplied by Harlan Banks.

quired during differentiation, for example the finite number of primordial germ cells in the mammalian ovary (Chapter 3.6.4.1) or the finite replicative capacity of fibroblasts as assayed *in vitro* (Chapter 7.6.3). It is important to recall that growth-arrested apical meristems occasionally can be reactivated, as shown by the formation of new shoots from the needles of the Monterey pine (*Pinus radiata*), which can then regenerate a whole plant (Chapter 4.4; Raven et al., 1986, 338). Such examples argue against a rigid proliferative limit in plant meristems, even in those with determinate growth.

Lycopods are of interest here because they also showed determinate growth. The lycopods are a division of seedless vascular land plants that was distinct from the rest of the vascular plants by the Silurian, but also contributed to Devonian forests and evolved from rhizomatous ancestors. Present lycopods include the club mosses (*Lycopodium*). The Upper Devonian *Cyclostigma* and *Lepidosigillaria* were five to eight meters tall, with trunks that had a constant diameter, diminishing only at the terminal portion. This growth pattern suggests that the meristematic growth was relatively constant until just before the maximum height was reached. As argued for the Archeopteridales, some arborescent lycopods are thought to have had a maximum aerial size and determinate growth that resulted from winnowing of the apical meristem at the top branches to a size insufficent for further growth (Andrews and Murdy, 1958; Eggert, 1961; Delevoryas, 1964).[4] The presence of fruiting bodies at the terminals of the branches suggests to some paleobotanists that arborescent lycopods were monocarpic and underwent whole plant senescence, or "top senescence." By the Upper Devonian, several aclonal lycopod lineages became established and reached their greatest morphological complexity and diversity in the Carboniferous (Tiffney and Niklas, 1985). Most Devonian lycopods had determinate growth and were spore-forming (DiMichele and Phillips, 1985).

A question of major interest is the evolution of organ senescence. Because mechanisms in the senescence of leaves or of whole plants appear similar (Chapter 2.3.2), the early occurrence of leaf abscission could be taken as evidence for hormonally dependent processes. Leaf abscission is indicated in *Cyclostigma*, an Upper Devonian lycopod (Chaloner, 1968; Chaloner and Sheerin, 1979; Stewart, 1983, 109; Addicot, 1982, 268). Abscission of branches is indicated for *Lepidodendron* and for some members of the

---

4. Stewart (1983) goes so far as to say that the "apical meristems producing branches were 'used up.'"

Cordaitales of the Upper Carboniferous (Addicot, 1982, 271); *Lepidodendron* grew over thirty meters tall, forming branches above heights of ten to twenty meters (Gifford and Foster, 1989, 145). These occurrences suggest hormonal mechanisms such as described in the abscission of leaves and branches in present seed plants (Chapter 2.3.2). Because the roles of abscisic acid and cytokinins are still unclear, little can be inferred about the role of such hormones in the Devonian plants, whose phylogenetic linkage to modern angiosperms and gymnosperms is also uncertain.

On the other hand, these leaves might have more simply withered ("die-back") from mechanical factors after growing beyond a more or less fixed breakage point, since there is no sign of abscission layers, at least in leaves of arborescent lycopods (William Di-Michele, pers. comm.). As mentioned above, *Tetraxylopteris* and other Devonian pro-gymnosperms appear to have had seasonal cycles of abscission and regrowth of individual shoots, which implies hormonal mechanisms of senescence. In regard to the long-standing debate about growth cessation as a trigger for senescence in animals (Chapter 4.5.2), an important conclusion by paleobotanists is that determinate growth and leaf or branch abscission (organ senescence) arose independently during evolution in plants. For example, the primitive fern *Pseudosporochnus* (middle Devonian) had determinate growth but no abscission of its fronds (Stewart, 1983). Moreover, growth in present angiosperms can cease without triggering flowering or whole-plant senescence, as extensively studied in soybeans (Chapter 9; Noodén, 1988c, 401).

Most lycopods of the Devonian and subsequent periods were small and herbaceous and are considered the primitive stock of present lycopods (Stewart, 1983; Gifford and Foster, 1989, 124). Several authorities argue that major size reduction from large treelike forms began in the later Devonian (Stewart, 1983, 125–127). Arborescent aclonal lycopods may have disappeared as their swampy habitats dried out, beginning in the Upper Pennsylvanian and continuing through the Permian-Triassic transition (DiMichele et al., 1985). Surviving lycopods are clonal, herbaceous, perennial, and lack leaf abscission. Examples include the club moss *Lycopodium lucidulum,* which strikingly resembles the Devonian *Baragwanathia longifolia* (Gensel and Andrews, 1984). The aerial vertical axes of modern *Lycopodium* have determinate growth like its ancestors (Banks, 1970, 105). *Lycopodium* also reproduces clonally from underground shoots with indeterminate growth that form ever expanding circles ("fairy rings") whose growth continues a century or more (Gifford and Foster, 1989, 107).

By the Lower Carboniferous, other large trees in conifer lineages formed forests with high canopies, for example, *Pitys primaeva* (with stumps up to 2.5 meters in diameter; Long, 1979; Gensel and Andrews, 1984), *Protopitys* (1.5 meters), and *Cordaites* (≥ 1 meter) (Stewart, 1983). Besides the large trees, most others were small and herbaceous; none of these species has survived. The first conifers appeared in the Upper Carboniferous (310 Myr ago), but large numbers of growth rings are not found until much later (Scott and Chaloner, 1983). The paucity of growth rings in large trees prevailed into the Mesozoic, a time of tropical climates, for example the one-meter-diameter logs of *Araucarioxylon arizonicum* (225 Myr ago) in the petrified forest of Arizona. Trees like the present long-lived conifers with numerous rings are common in the Upper Cretaceous (65 Myr ago), for example the swamp cyprus (*Taxodium distichum*) and sequoia (*Sequoia langsdorfi*), which appear to have indefinite growth and lifespans (Chapter 4.2.1). With the possible exception of the Pinaceae, all present families of conifers originated earlier in the

Mesozoic. While most Mesozoic conifers were aclonal, Tiffney and Niklas (1985) think that some may have propagated vegetatively through their roots (root suckering), as found in conifers today.

Phenomenal numbers of short-lived plants and associated insects have evolved since the end of the Mesozoic. Flowering plants (angiosperms) are found by the Lower Cretaceous (100–140 Myr ago). Angiosperms diversified extensively, while conifers (gymnosperms) had little radiation at this time (Knoll and Niklas, 1987; Lidgard and Crane, 1988). Figure 11.6 shows the expansion of angiosperms on a massive scale in temperate zones worldwide. Most angiosperms are perennial. During the extensive angiosperm radiations, monocarpy is thought to have arisen independently on numerous occasions. Since the Cretaceous, the number of conifers generally remained stable, while the number of cycad genera dwindled by 50%. Surviving conifers and cycads are polycarpic perennials and do not show whole-plant senescence.

Angiosperms appear to have had a monophyletic origin (Doyle and Donoghue, 1986). Clonal growth is much less common in dicots (ca. 10% of families). Those dicots that are aquatic or semi-aquatic are thought to have evolved from aclonal ancestors since the Pliocene (5–2 Myr ago) and generally have short life cycles (Tiffney and Niklas, 1985). Monocots, however, have maintained the capacity for rhizomatous reproduction from the outset, as found in most families. For example, grasses, a huge family (500 genera, 8,000 species) propagate both sexually and rhizomatously.

While Mesozoic gymnosperms may have depended on wind for pollination, flower-loving (anthophilous) insects apparently promoted angiosperm diversification to an

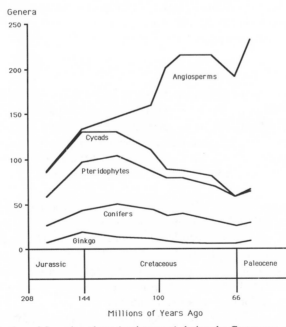

Fig. 11.6. Diversification of flowering plants (angiosperms) during the Cretaceous, showing a threefold increase in the number of genera. Also see Knoll and Niklas, 1987. Redrawn and adapted from Lidgard and Crane, 1988.

astounding 230,000 species since the Cretaceous by cross-pollination (Regal, 1977a; Crepet, 1979; Knoll and Niklas, 1987). Pollination by insects may extend back at least to the Carboniferous (Dilcher, 1979) but has acquired greater significance since the Mesozoic. In particular, the great diversification of bees (20,000 species) since the Cretaceous is hypothesized to have resulted from co-evolution with flowering plants. However, the emphasis given to social insects in this regard may neglect the importance of seed dispersal by dinosaurs, mammals, and birds (Regal, 1977a; Wing and Tiffney, 1987). The case for early associations of social insects with flowers is deduced from flower and pollen morphology but is lacking direct fossil evidence. Flowers like those pollinated by extant bees certainly occurred by the middle Eocene (45 Myr ago) (Crepet, 1979; Stewart, 1983). It would be interesting if wind-pollinated angiosperms (Amentiferae) have different patterns of senescence than insect-pollinated species.

To summarize, as tree architecture evolved in the late Devonian and early Carboniferous, clonal propagation declined in relative importance, but probably not in absolute abundance. Sexual mechanisms were present throughout, however. The absolute number of clonal plant families did not decline over this great span of time; some representatives from those times may have survived because of their capacity for clonal propagation (Tiffney and Nicklas, 1985), despite associations of clonal reproduction with greater genetic load (Chapter 8.5). Moreover, clonal (vegetative) reproduction is found in many species whose ancestors appear to have lacked it at some point in their lineage, for example pines, dicotyledonous angiosperms, and cycads. Whatever genetic changes occurred during evolution did not compromise the capacity for indefinite somatic cell replication and for the redifferentiation that requires genomic totipotency. Sexually propagating plants have far greater capacities for opportunistic aerial dispersal and, of course, for the more rapid production of genetic variation through recombination.

11.3.3. Insects

This section discusses the evolution of insects, with a focus on mouthparts. Insect phylogeny is summarized in Figure 11.7. As discussed in Chapter 2.2.2, reductions of oral anatomy that limit capacity for food intake are found in many short-lived adult insects. While mayflies epitomize the phenomena, aphagy occurs sporadically in other insects. The evolution of these striking limitations on adult lifespan is of much interest. Social insects are also briefly considered. Further information on the distribution of adult aphagy in modern insects is in Chapter 2.2.1 and Table 2.2.

Primitive insects tended to be small (< 1 cm) and wingless and resembled the bristletails (order Archeognatha). The earliest fossil, from the Lower Devonian (Emsian, 390 Myr ago), had a bristletail-like head less than one millimeter long and mouthparts that were better adapted for macerating soft foods than for tearing (Labandeira et al., 1988). Other Devonian insects, however, had mouthparts suitable for piercing. Labandeira suggests that the divergence of lineages equipped for piercing versus chewing occurred in the Silurian during early radiations of vascular land plants.

Extensive radiations followed after a major gap (60 Myr) in the fossil record of insects. At least twelve insect orders occurred by the Pennsylvanian. Aphagous insects like the mayflies (Ephemeroptera) stimulate questions about evolutionary changes in the potential

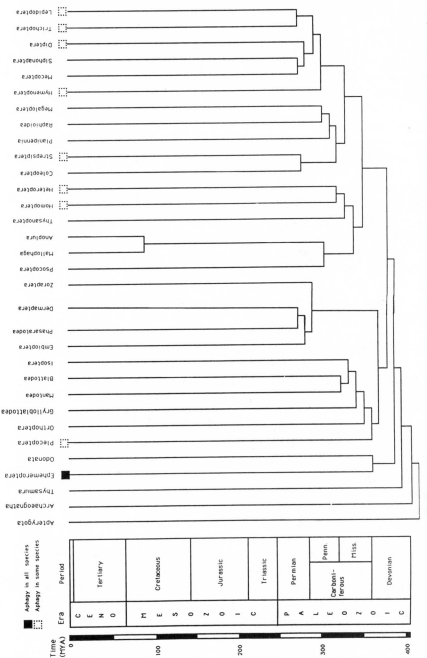

Fig. 11.7. Insect phylogeny, showing the first occurrence of species discussed in the text and the present distribution of adult aphagy through characteristic reductions of oral anatomy. In mayflies (Paleoptera), adult aphagy appears to prevail throughout the order; no other insect order manifests adult aphagy so consistently. Paleozoic mayflies had well-formed mandibles and beaks that represent the primitive state. Aphagy has not been reported in the other extant paleopteran order, the Odonata (dragonflies). In the Neoptera (all orders to the right of Odonata), adult aphagy is sporadic. Examples of adult aphagy can be anticipated to turn up in more orders than shown, as the vast literature of insect morphology is further combed. The phylogeny is based on cladistic relationships and on the earliest occurrence of fossils in their respective orders. For documentation of aphagy, see Table 2.2. From Conrad Labandeira, 1990, pers. comm.

for long-lived adults. All present mayflies have degenerate mouthparts as adults and extremely short adult phases, typically lasting a few days or less (Chapter 2.2.1.1.1). This
consistent life history and anatomic pattern suggest radiations under the influence of selection for very rapid reproduction from one ancestral lineage with aphagous adults. The
earliest mayflies in the middle Upper Carboniferous (ca. 300 Myr ago) were very different
from their small and fragile descendants. Not only were some of the Carboniferous mayflies gigantic—for example *Bojophlebia* had wingspans up to forty-five centimeters—the
adults also had well-formed chewing mandibles and beaks (Kukalová-Peck, 1985; Carpenter and Burnham, 1985).

While their total lifespans are unknown, a short adult phase with rapid senescence
seems unlikely in such large and apparently rugged creatures. The scaling of mortality
rates with size (Chapter 5) would also predict long adult phases. Even if adult aphagy
through mouthpart reduction had occurred when mayflies were still large, a substantial
adult lifespan might have been possible, as judged from *Pleocoma,* a large aphagous beetle that lives up to a year (Chapter 2.2.1.3). The time needed to pass through the many
instars (Kukalová-Peck, 1985) and reach the large sizes of adult Paleozoic mayflies could
have been years. Present mayflies, which also have numerous instars, take up to 5 years
before hatching. Mayflies show two ranges of adult phase lengths, from minutes to hours,
and from days to a few weeks (Chapter 2.2.1.1.1).

The sketchy fossil record does not tell when reduction of mouthparts began in mayflies.
Reduced mouthparts were, however, represented by the Jurassic in various Coleoptera,
Diptera, and Hymenoptera that with other insects compose a "reduced trophic mouthparts
class" (Labandeira, 1990). By the Eocene (57–36 Myr ago), mayflies resembled those in
modern families (Edmunds, 1972). With regard to the large size that often accompanies
longevity, it is interesting that most Carboniferous insects were larger than modern insects, which predicts considerable adult lifespans. The record is *Meganeuropsis per-
miana,* a dragonfly-like insect (Protodonata) with a seventy-one-centimeter wingspan.
The presence of sturdy mouthparts and long, spiny legs that were able to reach forward
suggests that the Protodonata, like present dragonflies, caught prey while flying (Carpenter and Burnham, 1985). The small extant dragonflies have adult phases of several weeks
to months, after development that takes up to 3 years. The longest adult phase is in *Sym-
pecma,* a small dragonfly with a winter diapause that extends the adult life phase for up to
6 months (Corbet et al., 1960).

Aphagy because of reduced mouthparts occurs in the adult stages of at least nine insect
orders besides the mayflies (Table 2.2) and evidently arose independently numerous times
during radiations after the Carboniferous (Labandeira, 1990). The multiple independent
(polyphyletic) occurrence of adult aphagy is shown by its occurrence in different Neoptera
(Figure 11.7). The Paleoptera (mayflies and dragonflies) and Neoptera are sister groups
that evolved from a common ancestor, during or after the Devonian. All surviving Ephemeroptera have reduced mouthparts as adults, while the other surviving paleopteran order,
the dragonflies (Odonata) have well-developed mouthparts and prey efficiently on other
adult flying insects. Thus, the reduction of mouthparts and the ultrashort adult lifespans
of modern mayflies is a clearly derived condition and a major departure from ancestral
mayflies.

Numerous examples of adult aphagy from reduced mouthparts are found scattered throughout the Neoptera (Figure 11.7). The exopterygote stone flies (Plecoptera) include species with degenerated mouthparts, although not all species are aphagous as adults (Porsch, 1958).

Among endopterygotes, at least three modern orders have some species that are completely or partially aphagous as adults. Butterflies (Lepidoptera), caddis flies (Trichoptera), Diptera, and beetles (Coleoptera) show striking intrataxonomic variations in oral anatomy that cover the range of oral anatomic deficiencies. For example, among the Diptera, all botflies (Oestridae) but only some hoverflies (Syrphidae) have degenerate mouths as adults, while many other families retain powerful mandibles, for example adult female mosquitos (Culicidae). The phylogeny of these is discussed in Wood and Borkent, 1989, Woodley, 1989, and McAlpine, 1989. Many related dipteran genera show major differences in oral anatomy. Among the few archaic Tanyderidae, the Australian *Radinoderus* has well-developed mandibles, while the North American *Protoplasa* has nonfunctional vestiges (Downes and Colless, 1967).

The evolution of aphagy is not well understood in endopterygotes, many of which are known as body fossils since the early Permian (Carpenter and Burnham, 1985). Downes (1971) makes a good case that Diptera evolved adult oral deficiencies numerous times since the Triassic period, throughout both hemispheres. Diptera first appeared in the late Triassic (Rohdendorf, 1974) and had features of predators with robust mouthparts (Kalugina and Kovalev, 1985), like the early chewing mayflies and dragonflies. Insectivorous habits, as retained by dragonflies, are also found in Diptera, for example Ceratopogonidae, but also in many other insects.

Selection for reduced mouthparts appears to be a continuing phenomenon in modern Diptera, Hymenoptera, and Lepidoptera. Aphagy may be adaptive for some flies and mosquitos by eliminating the need for a delay to feed after hatching in environments where nutrition and temperature may fluctuate widely (Chapter 6.3.3). Here as in mayflies (Edmunds and McAfferty, 1988), precocious oocyte maturation by the time of hatching is a common adaptation in species with partial or complete adult aphagy. The pitcher-plant mosquito (*Wyeomyia smithii*) has races that vary in their need for blood meals after hatching before eggs are shed (Chapter 2.2.1.1.2). These and other examples show that the need for feeding in adult Diptera is a highly labile trait of life history and can rapidly respond to selection. In species where autogeny has not become fixed through the reduction of mouthparts, neuroendocrine mechanisms appear to be important, as illustrated by Lea's (1970) demonstration that supplemental neuroendocrine grafts induce egg maturation even without the blood meal that an anautogenous mosquito *Aedes taeniorhynchus* usually requires (Chapter 2.2.1.1.2). This adjustment of the timing of gametogenesis is discussed below (Section 11.4.2) as an example of heterochrony under neuroendocrine influences.

In many present mosquitos and other biting flies, there is a common trend for males to hatch and breed without depending on a complete diet (Chapter 2.2.1.1.2). The gender differences in reduced oral anatomy in some Diptera imply that the mutant genes are regulated by other genes for sex determination, or are not proximal to the controls for morphogenesis of oral structures. These large changes might derive from mutations. For example, mutations in receptors could alter the binding of hormones (juvenile hormone or ecdysone) or binding of the receptor-hormone complex to the DNA sequences that regulate the proliferation and differentiation of the labial imaginal disc, which in turn gives rise to the proboscis. It seems feasible to examine paired species such as *Cnephia eremites* and *C. mutata* for differences in genes that regulate formation of the labial imaginal disc.

Beetles also show scattered aphagy (Table 2.2), but with less apparent consequence to the length of the adult phase than in mayflies or Diptera. For example, the rain beetle (*Pleocoma*), which is aphagous through degenerate mouthparts, has an adult phase lasting up to 1 year (Chapter 2.2.1.3). As noted for Diptera, reduced mouthparts and adult aphagy appear to be more common in males, who may live just a few days as adults (Chapters 2.2.1.1.1, 2.2.1.3, 3.3.5). These scattered occurrences suggest independent origins of adult aphagy and short lifespans in at least three distinct lineages from predecessors in the Triassic (Ademosynidae; Crowson, 1981, 661).

However, most beetles eat well as adults and live long. Beetles that feed as adults have lifespans ranging from the very short (days to weeks, like the Diptera) to 10 years (Chapter 3.3.5). This broad range of lifespans is matched only by the social insects (Hymenoptera and Isoptera); the next longest may be crickets and other Orthoptera with adult phases of up to a year (Masaki and Walker, 1987). Many beetles continue to reproduce in several consecutive years. It seems plausible that these long adult phases have been retained since the Permian (Figure 11.7). Some Triassic insects, which Crowson (1981) believes were beetles, had mouths equipped for boring into large trees. A possible record of this is the tunneling found in Mesozoic trace fossils in tree trunks from the petrified forest of Arizona (Walker, 1938) and from deposits in Germany (Linck, 1949).

It would be interesting if abnormal oral development were ever reversed by subsequent mutations during evolution. The gender differences in mouthpart reduction suggest that this could be so, since the genes permitting normal oral development are manifestly intact in females. Unless separate gene sets are used for each sex, these gender dimorphisms imply that such anatomic defects can arise from regulatory genes that are controlled themselves by genes on sex chromosomes. Reduction of mouthparts, however, does not preclude alternate evolutionary replacement of this function. For example, tsetse flies and some muscid flies, while lacking mandibles, nonetheless can use their prestomal teeth for making wounds from which they sop up nutrients (Downes, 1971; Paysinger et al., 1978; Elzinga and Broce, 1986). Another question is whether insects with reduced oral anatomy in *both* sexes accumulate mutational changes in loci coding for the anatomically defective parts, since in this case there would be no selective pressure to maintain the genes for a single sex.

Social insects arose polyphyletically from various neopteran lines (Burnham, 1978; Figure 11.7). As described in Chapter 2.2.3.1, either a queen or a worker can develop from the same egg, depending on the diet given it during development. The termites (Isoptera) are well represented by the early Cretaceous (ca. 138 Myr ago; Whalley and Jarzembowski, 1985) and may have arisen as early as the Triassic from ancestral cockroaches (Blattodea; exopterygotes); cockroaches extend at least to the Upper Carboniferous period (300 Myr ago; Burnham, 1978; Carpenter and Burnham, 1985). In contrast to the monophyletic origin of termites, sociality evolved at least ten times independently in hymenopteran lineages of endopterygotes (Wilson, 1987).

The earliest signs of castes are from the late Cretaceous period (80–100 Myr ago), consisting of a single bee (*Trigona;* Michener and Grimaldi, 1988) and ants (*Sphecomyrma;* Wilson, 1987; Grimaldi et al., 1989). No fossils of queens have been found from this period. Vespid wasps appeared shortly before, in the late Cretaceous (Rasnitsyn, 1980a, 1980b; Carpenter and Burnham, 1985) and apparently radiated during the late Cretaceous. Social insects have changed little since the middle Eocene to early Oligocene, as indicated by many extant genera in Baltic amber (50–32 Myr ago) (Larsson, 1978; Keilbach, 1982). Masaki and Walker (1987) suggest that the potential for long adult life-spans was prerequisite to the evolution of sociality in insects. However, some social insects are annual, for example, the bumblebee (*Bombus terrestrias*) is annual in severe climates but may form perennial colonies during better conditions (Cumber, 1949). This example shows that the first social insects could have been either annual or perennial.

A major hypothesis is that the evolution of insect sociality was closely coordinated with that of angiosperms, which show explosive speciation during the Cretaceous (Figure 11.6; Lidgard and Crane, 1988). Since angiosperms were present by the earliest Cretaceous (130 Myr ago; Lidgard and Crane, 1988; Knoll and Niklas, 1987), it was argued by Michener and Grimaldi (1988) that sociality could have existed even earlier. The association of plants and insects was surely established by the Upper Paleozoic (ca. 300 Myr ago; Dilcher, 1979; Labandeira, 1990). The evolution of mouthparts suitable for chewing and of wings for rapid aerial locomotion was a foundation for the crucial roles of insects as pollinators in the subsequent evolution of short-lived flowering plants and castes in social insects with different lifespans. Characteristically, social insects have overlapping of two or more generations, which enables the sharing of work by offspring and parents (Wilson, 1975). The overlap of generations is particularly dramatic for the queens, which are generally perennial and show negligible senescence (Chapter 2.2.3.1). Short generation cycles and rapid development could be factors in evolving the specialized interactions of social insects and flowers involved in pollination, for example.

It is interesting to consider how the control of flowering and senescence in angiosperms coordinates rapid senescence on a global scale, enmeshing the life histories of countless vertebrates, insects, and angiosperms. The scale of this coordinated rapid senescence was probably less before social insects and annual angiosperms became dominant in the Upper Mesozoic. It would be interesting to examine social insects that pollinated annual versus perennial plants. An intriguing feature of social insects is the capacity for queens to store viable sperm, certainly for a year but possibly beyond a decade after their usually single mating (nuptial) flight (Chapter 2.2.3.1). This capacity to maintain viable sperm for prolonged durations is not unique to insects. For example, sperm is stored for at least 2 years by elasmobranchs (Dodd, 1983). In contrast, mammalian sperm rapidly deteriorates at ambient temperature. Analyses of why eusociality is rarely found outside of the Hymenoptera have stressed the haplodiploid mode of sex determination, in which haploid males develop from unfertilized eggs, while the fertilized eggs become diploid females; few other arthropods use haplodiploidy for sex determination (Wilson, 1975; Richards, 1965; Hamilton, 1964).

11.3.4. Fish

We now pick up the trace of early fish by returning to the Paleozoic. Of particular interest are the lineages that have different patterns of oogenesis and tooth replacement, and the distribution of fish that die at first spawning. Whether or not *de novo* oogenesis continues beyond maturation bears on the theoretically important question of the relation of oocyte exhaustion to senescence (Chapters 1.5.2, 3.6.4, 8.5). The capacity to replace teeth relates to a different issue, the extent of wear and tear on this vital organ. Finally, the distribution of fish that die at first spawning will reveal an unexpected clustering. A brief review of fish taxonomy based on Nelson, 1984, is given first.

Fish represent about half of the existing forty to fifty thousand species of vertebrates and are divided into two major branches, the Agnatha (jawless fish) with just over seventy species, and the Gnathostomata (jawed fish), which includes all others (Figure 11.8). The Agnatha (hagfish and lampreys) are the most primitive fish, based on retention of the jawless condition since the Paleozoic period. Jawed fish are divided into the cartilaginous Chondrichthyes and the Osteichthyes, most of which have well-ossified skeletons. The Chondrichthyes include numerous Elasmobranchii (sharks and rays) and a few surviving Holocephali (Chimaeriformes). Osteichthyes include the Actinopterygii (ray-finned fish, a huge and complex subclass) and several subclasses with few extant species (bichirs, coelocanths, and lungfish). Ray-finned fish are subdivided into the Chondrostei (sturgeon and paddlefish) and the Neopterygii; the latter represent more than 99% of the present twenty thousand species of fish, of which most are the familiar teleosts. While the jawless hagfish and most of the recently evolved teleosts appear to have continuing oogenesis, the lampreys and at least some sharks and sturgeon have fixed oocyte numbers. In contrast to lampreys, the sharks and sturgeon are iteroparous, and their spawning is not associated with increased mortality (Chapters 3.4.1).

The Ordovician witnessed the first vertebrates, the Agnatha, but it is not until much later that fossils are found that resemble modern lampreys and hagfish. Except for these, most other Agnatha became extinct in the Devonian. Bony fish with jaws occurred by the Silurian, but the stem line of sharks and other cartilaginous elasmobranchs is recorded later, in the Devonian (Figure 11.8). These early fish continuously replaced their teeth during adult life (polyphyodonty; Peyer, 1968; Carroll, 1988), a trait retained by most teleosts, amphibia, and reptiles. Tooth replacement would be important to predators because of the risk of tooth breakage and the tooth abrasions encountered by mammals in ways that limit their lifespans, as discussed in Section 11.3.5.2. Modern selachian sharks have a continuous, rapid replacement of true (dentine-containing) teeth that are not fused to their jaw bones and that move into position in a row; the new teeth are slightly larger, in register with continuing body growth (Peyer, 1968; Reif, 1976, 1984). Teeth of *Cladoselache* from the late Devonian show considerable wear, to an extent rarely found in modern selachians (Williams, 1990); tooth replacement was continuous but probably slower. The larger size of ingrowing teeth in the innermost row implies anticipation of overall body growth. Other Paleozoic sharks also apparently had slow tooth replacement (copedonts and psammodonts; Carroll, 1988; Nelson, 1984). The first sharks to have more rapid tooth replacement, as evidenced by the absence of wear, occur in the late Paleozoic (the Symmoriidae, Stethocanthidae, and Xenocanthidae). By the Mesozoic, shark teeth show the modern growth pattern, but whenever it evolved, the fossil record indicates its "derived" nature. The acquisition of rapid tooth replacement during evolution of selachians and some other chondrichthyans diverges from the general trend of higher vertebrates to have diminished regeneration as adults.

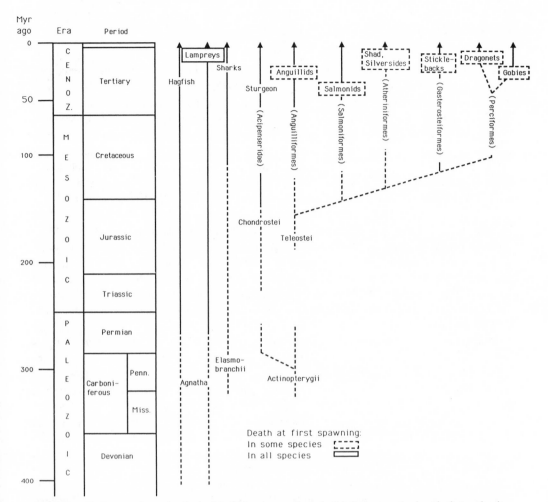

Fig. 11.8. Phylogeny of semelparity in fish. Lampreys are the only fish that appear to show both semelparity and determinate oocyte numbers. The occurrence of semelparity is a minor and sporadic phenomenon in most families and orders except for lampreys, anguillids, and certain salmonids. Phylogeny adapted from Carrol, 1988, and Nelson, 1984.

11.3.4.1. Lampreys

Lampreys are of particular interest in the evolution of senescence, because all species that have been studied die after first spawning, in association with rapid senescence and aphagy. Moreover, in all species studied, oogenesis is limited and the numbers of oocytes are predetermined (Chapter 2.3.1.4.1). Several lampreys occur by the Carboniferous period, for example *Mayomyzon pieckoensis* (Bardack and Zangerl, 1968, 1971; Janvier and Lund, 1983), which resembled modern lampreys. The fossil lamprey assemblage is considered monophyletic (Janvier and Lund, 1983; Maisey, 1986). The small size and absence of teeth (the latter may be artifactual due to poor fossilization) might represent a juvenile phase, while the large eyes and gill chambers of *Mayomyzon* are consistent with adult development. Other small Pennsylvanian agnathans had teeth (Bardack and Richardson, 1977). Lamprey teeth differ fundamentally from those of higher vertebrates by their formation from the palatal epidermis and by the absence of dentine (Peyer, 1968). *Mayomazon* probably lived in marine or brackish waters (David Bardack, pers. comm.), which would be consistent with a juvenile oceanic phase as in some extant lampreys. Migratory life histories are often found in semelparous fish, as discussed in Section 11.3.4.4. Three lamprey subfamilies diverged long before radiations of modern teleosts: the Petromyzontinae in the northern hemisphere, and the Geotriinae and Mordaciinae, both in the far southern hemisphere (Nelson, 1984).

Modern genera of lampreys commonly have species with two different life history patterns, both of which are, without known exception, semelparous. Lampreys may be anadromous (migrating from the ocean where they spend most of their lives to freshwater where they spawn) or confined to freshwater. Adult anadromous lampreys are all external parasites of other fish, from which they obtain blood and other tissues (Chapter 2.3.1.4.1). Feeding ceases just before spawning and extensive visceral atrophy begins; the adult (postlarval) phase lasts about 2 years. In contrast, freshwater lampreys in the northern hemisphere may or may not be parasitic; freshwater lampreys from the southern hemisphere are nonparasitic, however. Nonparasitic freshwater lampreys are generally smaller at maturity; they are aphagous throughout their short adult lifespan, and their guts usually collapse by metamorphosis, unlike the parasitic species in which the adults feed in estuaries or the open sea before returning to spawn upstream. The parasitic and nonparasitic forms within a genus are referred to as "paired species" and are often difficult to distinguish before the adult stage.

Nonparasitic freshwater lampreys have a shortened adult phase of 6 to 9 months, which is about 25% of the duration of the adult phase in the parasitic forms, and complete maturation while completely aphagic. Nonparasitic lampreys do not grow larger than the premetamorphic ammocoete stage and differ from anadromous species in reproductive scheduling: oogenesis and sexual differentiation occur earlier, while vitellogenesis is later (Figure 2.9). They produce fewer mature ova, possibly because fewer protein reserves are available for vitellogenesis. Because nonparasitic lampreys take 2 years longer to mature, the total lifespan is about the same in both types of life history.

The fossil record does not indicate when parasitism evolved in Agnatha. In general, parasitic lampreys are more widely distributed than the dwarf nonparasitic species. For example, the anadromous northern Pacific *Lampetra tridentata* spawns from California to Alaska and is related to several freshwater nonparasitic species with narrower distribution:

L. minima (recently extinct), *L. lethophaga,* and *L. hubbsi.* Similar paired species are described for other genera in both hemispheres. Occasional nonparasitic forms occur without a parasitic counterpart; for example, *L. zanandreai* occurs in the Po basin without its pair, *L. fluviatilis* (Zanandrea, 1954; Hardisty and Potter, 1971; Hardisty, 1979). Furthermore, no freshwater parasitic lampreys are reported in the southern hemisphere (Nelson, 1984).

The occurrence of parasitic and nonparasitic species in most genera implies that the nonparasitic species arose from the parasitic forms, by haphazard isolation from their host fish. About half (twenty-three out of forty-one) of the species are nonparasitic dwarfs (Nelson, 1984). Speciation of brook lampreys may have occurred during the last glacial age (10,000 years ago) when access to the sea was blocked (Hardisty and Potter, 1971). Recent derivations of nonparasitic forms are suspected in *L. tridentata,* which has intergrade forms (Hubbs, 1971). The adaptability of sea lampreys to freshwater is shown by the landlocked populations of *Petromyzon marinus* that first entered the upper Great Lakes after 1839, when the Welland Canal allowed them to bypass Niagara Falls. Landlocked *P. marinus* continue to be parasitic.

The trend for elimination of parasitic feeding during the adult phase may be favored by natural variations in the seasonal timing of maturation. A capacity for earlier maturation is suggested by the wide range of sizes or intergrades described in spawning river lampreys; Hardisty (1963) gives examples at several locales. Moreover, precocious maturation of ammocoetes with ripe eggs is described, but without signs of impending metamorphosis or the full "nuptial attributes" of darker coloration and enlargement of dorsal fins (Zanandrea, 1957, 1958; Hubbs, 1971; Walsh and Burr, 1981).

The life cycle of the other surviving jawless fish, Myxinoidae (hagfish) is not known in much detail. Like lampreys, hagfish lack jaws, ossification, and scales. Unlike lampreys, hagfish are exclusively marine, develop directly without a larval stage, and with continuing growth and *de novo* oogenesis (Chapter 4.2.2). Moreover, there is evidence that hagfish spawn repeatedly and ovulate continuously, as judged from corpora lutea that represent previous ovulations and the presence of various stages of gametogenesis, from the formation of primordial germ cells to mature oocytes. In contrast to adult lampreys, adult hagfish are broad-range scavengers, feeding on invertebrates and on other fish (Hardisty, 1978, 64). Although similar feeding mechanisms in lampreys and hagfish were used to argue for a monophyletic origin (Yalden, 1985; Carroll, 1988), other evidence indicates that hagfish preceded lampreys, possibly as early as the Cambrian (Nelson, 1984; Bardack, 1985). The continuing *de novo* oogenesis of present teleosts and hagfish thus would be consistent with this as a primitive trait in early fish from which the surviving lamprey lineage departed.

11.3.4.2. Sturgeon

An early Mesozoic fish *Chondrosteus* resembled sturgeon, whose predatory descendants (Acipenseridae, Chondrostei) are extremely long lived and iteroparous. It is predicted that sturgeon eventually show gradual senescence because of evidence for a finite oocyte stock (Chapter 3.4.1.1.2). The scant ossification of *Chondrosteus* is considered a derived characteristic, unlike their ancestors in the Paleozoic, which had bony skeletons. The limited skeletal ossification and the limited oogenesis as indicated in two species of *Aci-*

penser are characteristics that may be shared with at least some modern sharks. Sharks and sturgeon are considered to have long been separated in the Paleozoic, however (Maisey, 1986; Carroll, 1988); the chondrosteans are not closely related to the Chondrichthyes, a similarly named taxon that includes sharks. In view of the general trend for continuing oogenesis in teleosts (also actinopterygians), the limited oogenesis of sturgeons may be provisionally considered a derived characteristic. The distribution of limited oogenesis is not known in sturgeons.

Despite the absence of feeding before spawning in some migratory species, mortality does not increase during spawning, giving another example of iteroparous species with no reproductive costs (Chapter 10.6). The partially anadromous Russian sturgeon (*Acipenser güldenstädti*) and the sevrjuga (*A. stellatus*) show some signs of stress (Krivobok and Tarkovskaya, 1970) that are evidently not comparable to the severe consequences of spawning in salmon (Chapter 2.3.1.4.2). About half of the sturgeon species (sixteen out of twenty-seven) do not migrate. It would be interesting to know the patterns of oogenesis in paddlefish (family Polyodontidae, another Chondrostei) and in gars (family Lepisosteidae), neopterygians that are a sister group to modern teleosts.

11.3.4.3. Teleosts

The evolutionary history of teleosts (bony ray-finned fish) poses interesting questions about the geographic distribution of different life history characteristics. For example, comparisons of life histories in teleosts from Africa and South America can be considered in the context that these land masses became separated during the breakup of Gondwana in the late Cretaceous. The annual cyprinodonts of seasonal freshwater pools of Africa and South America have alternative stages for diapause during development (Chapter 2.3.1.4) that are adaptive for surviving climatic fluctuations. The widely scattered distribution of diapause during teleostan development (Chapter 9.2.2.2.1) is consistent with independent origins of this trait. In contrast, certain semelparous species in the family Salmoniformes may have preserved semelparity in association with an assumed ancient characteristic of anadromous migrations, as described next. First I review aspects about semelparity in fish from Chapter 2.3.1.4.

Only a small fraction ($< 1\%$) of the twenty thousand teleosts are semelparous, and even fewer may be obligately semelparous in all populations (Figure 11.8). My survey of this phenomenon indicates that the majority of semelparous fish share another unusual trait with lampreys: they or their likely ancestors migrate between freshwater and the ocean for reproduction. Moreover, migration in either direction is by itself unusual, occurring in less than 10% of the 445 families of fish. McDowall (1988) emphasizes that migration is common in only the most primitive teleost families, being sporadic elsewhere. Anadromy (migration from oceans to freshwater for spawning) is far more common than the reverse route. Catadromy (maturation in fresh water, followed by migration to the ocean) is mainly represented in the European eels and other Anguillidae, and is rare in other families. Again, it should be stressed that not all anadromous fish show semelparity.

Among eels (nineteen families in the order Anguilliformes), certain species in three families die off completely or nearly so at first spawning (Chapter 2.3.1.4.2). Semelparity has been described in detail for a few of the sixteen species of Anguillidae and is also reported for at least one species each in snipe eels (Nemichthyidae) and conger eels (Congridae). The details have apparently not been characterized. Searches should be made for semelparity in the other eels, which include about six hundred species.

The best known examples of semelparity in eels are in the catadromous American eel *Anguilla rostrata* and the European eel *Anguilla anguilla*. Death occurs at spawning in association with an aphagous adult phase after its return from fresh water to the Sargasso Sea. Nearly all species (fourteen out of sixteen *Anguilla*) occur in the Sargasso Sea (Moriarity, 1987; Post and Tesch, 1982). The other two species are offshore of southeastern Africa and in the North and South Pacific. All Anguillidae have the same life cycle phases, but the durations vary much more than in lampreys. Nutrient reserves are accumulated during slow maturation in fresh water, followed by migration for thousands of kilometers to spawn and die. The immature leptocephalus larvae and the metamorphosed glass eels then return to fresh water (Nelson, 1984, 408; Bruton, Bok, and Davies, 1987; Jellyman, 1987). The age at migration varies widely, and maturation is not tightly linked to body size, unlike most other vertebrates. One specimen of *Anguilla dieffenbachii* showed sixty "annual" growth rings from its freshwater phase (Todd, 1980), while others of unknown age grew to huge lengths of nearly two meters (McDowall, 1978).

This enormous plasticity in the duration of the growth phase before maturation and the cessation of feeding is entirely consistent with decades-long lifespans of other teleosts (Chapter 4.2.2), and shows that the 60- to 88-year lifespans of aquarium eels is not merely an artifact of a protected environment. These extended lifespans contrast strikingly with the short-lived lampreys and are the best vertebrate equivalent of the very long lived, but ultimately semelparous and rapidly senescing, plants like the bamboo (Chapter 2.3.2). The evidence for universal death at spawning in sixteen species of freshwater eels, though indirect, suggests a monophyletic origin of semelparity. Eels can be traced through fossil *Anguilla* in the Eocene that resemble the modern genus, back to ancestral elopomorphs in the Mesozoic (Nelson, 1984; Carroll, 1988; Figure 11.8).

Semelparity may occur in other Anguilliformes, according to scattered reports on a snipe eel (*Nemichthys scolopaceus;* one of nine species in the family Nemichthyidae) and on a conger eel (*Conger vulgaris;* one of 109 species in the family Congridae). Both conger and snipe eels live entirely in the ocean (Chapter 2.3.1.4.2). At maturity, both sexes of these species lose their teeth, although only males lose them all. Fossils of snipe eels are not yet reported. Nelson (1984) places the Anguillidae and Nemichthyidae as families in the same infraorder (Anguilloidea), which contains six other families (155 species) for which I have found no information about spawning death. Tooth loss could be secondary to the cessation of feeding in *Nemichthys* and *Conger,* through the resorption of metabolically labile tissues that may hold the teeth in place. Both *Nemichthys* and *Conger* mature in deep water, but their leptocephalic immature stages are found in shallow water, near the shore (Nielsen and Smith, 1978; Wheeler, 1969, 229), implying a history of migration. I have not found indications of death at spawning in the other seventeen families of eels, which are entirely marine. Because larvae of all eel families are pelagic, whether or not they migrate to shallow seas or to fresh water, it is equally possible that the Anguillidae and Nemichthyidae shared a common ancestor that spawned once, or that semelparity evolved independently in both families. We do not know if all conger and snipe eels are semelparous.

Semelparity is also found in ten out of ninety genera of the order Salmoniformes. The best known is the Pacific salmon, genus *Oncorhynchus* (Table 2.4; the common names of

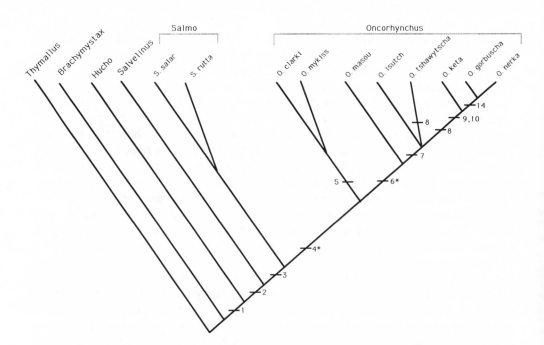

Fig. 11.9. Phylogenetic relationships of salmonid fishes. The distributions of biochemical, ecological, and morphological characteristics indicate a primitive group, the trouts (*Salmo*) with anadromous forms that are salmonlike, and an advanced group, salmon (*Oncorhynchus*), in which the earliest members were troutlike. *Oncorhynchus* presumably diverged from the Atlantic *Salmo* after isolation in the Pacific, about 1 Myr ago. The universal death after spawning in all species from lineages "to the right" of *O. masou* suggests that this trait occurred in a common ancestor after divergence from Pacific *Salmo*, probably 1 million years or more ago (Hoar, 1976). The numbers represent the earliest inferred occurrence of the following life history traits: *1*, egg diameter > 4.5 mm; dig redds; *2*, fall spawning season; *3*, capable of long migrations at sea; *4*, often irreversible damage from hyperadrenocorticism after spawning; *5*, spring spawning season; *6*, all anadromous forms die after spawning; *7*, stream-resident residual fish do not reproduce; *8*, most smolt at age 0 +; some go to sea as fry; *9*, juveniles show strong schooling behavior; *10*, reduced freshwater phase. Traits 4 and 6 are asterisked because of their importance to semelparous life histories. This recent tree is consistent with the two stem lines of salmon proposed by Hikita (1962) on the basis of ecological and morphological traits. This phylogeny is also supported by systematic species differences in resistance to salinity and temperature (Hoar, 1976), and differences between *O. keta* and *O. kisutch* in growth-hormone immunoreactivity, composition, and bioactivity (Wagner and McKeown, 1985). *O. kisutch* appears to have diverged more than *O. keta* from their common ancestor, as judged by comparisons with the biochemistry and bioactivity of growth hormone from sturgeon (*Acipenser güldenstädti*) and tilapia (*Tilapia mombassa*) (Wagner and McKeown, 1985). Mitochondrial DNA restriction fragment analysis indicates little divergence between *O. kisutch, O. mykiss,* and *O. tschawytscha* (Thomas et al., 1986); mitochondrial DNA of the latter two also showed less distance than to *Salmo trutta* (Berg and Ferris, 1984). The present Pacific *Salmo* are closer to *Oncorhynchus* on the basis of neurocranial morphology (Sanford, 1987), among other traits, and were recently renamed *O. mykiss* (Behnke, 1989; Smith and Stearly, 1989); they retain iteroparity like their ancestors, the Atlantic *Salmo*. Redrawn from Smith and Stearly, 1989, Figures 2 and 3.

these fish are given there). Of the seven species, six show obligate semelparity with a dramatic senescence at first spawning, whether they are anadromous or freshwater. The famous pathophysiology of death from rapid senescence in association with hyperadrenocorticism at spawning is documented for *O. nerka* and *O. tschawytscha* (Chapter 2.2.1.4.2). A similar rapid senescence is indicated for the Asiatic *O. masou*. The perennially spawning freshwater rainbow trout (*O. mykiss*) is an exception to the 100% mortality of the other anadromous Pacific salmon. Stem lines of *Oncorhynchus* are distinguished by life history, morphology, immunology, and mitochondrial DNA (legend to Figure 11.9). The Pacific *Salmo gairdneri* as studied by Robertson and colleagues (Robertson, Krupp, Thomas, Favour, Hane, and Wexler, 1961) was recently renamed *O. mykiss* because of its closer affinities to this genus (Smith and Stearly, 1989; Behnke, 1990b).

In contrast to the Pacific salmon, all other anadromous and freshwater salmonid species are perennial spawners like the rainbow trout (*O. mykiss*), arctic char (*Salvelinus alpinus*), or lake trout (*Salvelinus namaycush;* Chapter 4.2.2). Covariational analysis of twenty-two salmonids showed that *Oncorhynchus* produces fewer but larger eggs, and matures earlier and at larger sizes than members of other salmonid subfamilies (Hutchings and Morris, 1985). Statistical clusters of life history traits such as semelparity, overwintering of eggs with slowed development, and size at maturity follow taxonomic lines, but do not support life history theories that predict covariation of semelparity with larger numbers of eggs. Iteroparity and perennial spawning was probably the primitive characteristic in ancestral oncorhynchids (Section 11.3.4.4).

Contrary to the general rule of obligate (100%) postspawning mortality in the Pacific salmon, there is some postspawning survival of precociously mature parr (juvenile males) in *O. tschawytscha* and *O. masou* (Table 2.4). Iteroparous salmon also show precocious maturation as parr (Chapter 9.2.2.2.1). Thus, an escape from mortality in usually semelparous *Oncorhynchus* species is consistent with life history variants in closely related species. These examples again show preservation of the plasticity and the potential for rapid evolutionary changes in life history through variants of reproductive scheduling.

The migratory anadromous pattern itself must be a major factor in the (near) universal death at spawning. For example, in migratory steelheads (*O. mykiss*), while only a small fraction of anadromous fish survive, these can spawn again. In contrast, the nonmigratory rainbows have much greater survival to second spawning (Robertson, Krupp, Thomas, Favour, Hane, and Wexler, 1961). The extent of death at spawning increases with the length of the upstream run in steelheads (Robert Behnke, pers. comm.). Freshwater *O. masou* often occur in the same streams as the anadromes (Tanaka, 1965), which suggests a facile transition; some freshwater populations of *O. masou* are iteroparous (Table 2.4). A landlocked stock of *O. gorbuscha* was recently established in Lake Superior (Kwain, 1987). A notable change in these Lake Superior *O. gorbuscha* is the delay of maturity up to 3 years in some fish, which departs from the invariant 2-year life cycles of anadromous populations; this change in entire populations is so rapid as to represent an alternative phenotype, rather than selection of genetic variants.

Of particular interest to the genetic basis for semelparity are phenotypes of conditional semelparity in at least two teleostan orders. Conditional semelparity varies with geographic latitude in *O. mykiss* and the American shad (*Alosa sapidissima;* Figure 2.14; Table 11.1). Smelt (*Hypomesus olidus*), also salmoniforms, are also conditionally semelparous (Table 11.1; Chapter 2.3.1.4.2). These examples of conditional semelparity contrast with the obligate semelparity of certain freshwater eels, lampreys, and Pacific salmon. The conditionality of semelparity in most other species in Table 11.1 is unknown. Conditional semelparity could be an intermediate stage leading to obligate semelparity, in conjunction with genotype changes that may be easily reversed (genetic polymorphisms) or irreversible (mutations).

Coexistent populations of brown trout (*Salmo trutta*) with different lifespans are described in Scotland and Ireland (Chapter 3.4.1.2.2). Their different LDH isozymes suggest reproductive isolation (Hamilton et al., 1989; Ferguson and Mason, 1981). The short-lived, smaller trout are close to semelparous, while the long-lived and larger ferox trout are clearly iteroparous (Campbell, 1979) and may represent an ancestral genotype (Ferguson and Mason, 1981; Hamilton et al., 1989; Fahy and Warren, 1984). Thus, coexisting but noninterbreeding salmonid subspecies may contain genetic variants that account for major differences in growth patterns and response to larger food, allowing some to become giants with continued growth and lifespans fivefold longer than others in the same lake. Numerous glacial phases in northern Europe during the last 2 million years are thought to have been a major factor in reproductive isolation and subtype evolution in salmonids (Svardson, 1979; Ferguson and Mason, 1981; Fahy and Warren, 1984; Hamilton et al., 1989). Further molecular studies could show the extent of recent demographic isolation.

Semelparity is also found in other types of salmoniforms from both the southern and northern hemispheres. Examples are scattered throughout the superfamilies Galaxioidea and Osmeroidea (the latter includes the smelts; Table 11.1). It seems important that many of these fish are also diadromous (migrate in one or the other direction) and inhabit cold water. Unfortunately, no histological or physiological data indicate whether semelparity is associated with glucocorticoid pathophysiology. This background enables consideration of possible evolutionary relationships.

The evolutionary radiations of the order Salmoniformes is known only in general outline, and there are many gaps in the fossil record (McDowall, 1988; Behnke, 1990a). Two characteristics of extant salmoniforms, the restriction to cold water and reproduction in fresh water, are probably ancient in these radiations and imply freshwater origins. Nonetheless, the widespread capacity of extant salmoniforms for osmoregulation in the sea, exhibited at particular stages of life history and requiring the retention of water and the excretion of salts, suggests also that ancestral salmonids were anadromous. Certain examples of freshwater and anadromous forms in the same species (*O. mykiss* and *O. nerka*, Table 2.4; Figure 11.9) demonstrate the phenotypic plasticity during certain phases of their life history that would have allowed facile expansions of range. All three subfamilies of extant Salmonidae, the Coregoninae (whitefishes), Thymallinae (graylings), and Salmoninae, are tetraploid relative to other salmoniform orders by nuclear DNA content, suggesting that polyploidization occurred before the Salmonidae diverged into subfamilies, at least 25 Myr ago (Allendorf and Thorgaard, 1984; Ohno, 1970; Ohno et al., 1969).

Oncorhynchus probably stemmed from the ancestral Atlantic basin trout (*Salmo*) in mid-Miocene, about 10 to 20 Myr ago, which gave rise to Pacific trout (Behnke, 1990a). By the Upper Miocene, three lines are indicated from fossils of western North America: a side branch to the ancestral rainbow-cutthroat line, *Rhabdofario;* ancestors of the present rainbow-cutthroat trout (classified as *Salmo* in the older literature); and *Oncorhynchus,* the ancestors of the present Pacific salmon and trout. The streambed location of these

Table 11.1 Migration and Semelparity in Fish

Classification	Type of Migration
Class Agnatha (jawless fish)	
Order Petromyzontiformes (lampreys)	
Subfamilies **Petromyzontinae, Geotriinae, Mordaciinae**	most species anadromous;
Nelson (1984) designates these as subfamilies, but others accord family status. 41 species; obligate semelparity; lifespans 6 to ≥ 20 years.	landlocked "dwarf" lampreys clearly derived from anadromous species.
Class Osteichthyes (bony fish)	
Order Anguilliformes	
Infraorder Anguilloidea	
Family **Anguillidae** (freshwater eels)	100% catadromous
16 species with obligate semelparity; widely varying lifespans 8 to ≥ 88 years.	
Family **Nemichthyidae** (snipe eels)	exclusively marine; leptocephalus
9 species; semelparity indicated but not rigorously shown for *Nemichthys scolopaceus* (Nielsen and Smith, 1978); status unknown for other species.	stage found in shallow waters (Nielsen and Smith, 1978)
Infraorder Congroidea	
Family **Congridae** (conger eels)	exclusively marine; leptocephalus
109 species. *Conger vulgaris* (Cunningham, 1891; Wheeler, 1969, 229) is semelparous; status unknown for other species.	stage found in shallow waters (Wheeler, 1969, 229)
Order Clupeiformes	
Family **Clupeiformidae** (herring)	< 20% anadromous
190 species.	
Alosa pseudoharengus (alewives) cease feeding at migration from seas to rivers; mortality at spawning is 39–57%; a few spawn again next year (Durbin et al., 1979).	anadromous
Alosa sapidissima (American shad), death at spawning is complete below 32°N, but decreases greatly at northerly latitudes (Leggett and Carscadden, 1978; Figure 2.14. Most shad species are iteroparous (McDowall, 1988).	anadromous
Order Cypriniformes	
Family **Cyprinidae** (minnows and carps)	freshwater only
2,070 species. In the genus *Pimephales,* at least 3 of the 4 species are semelparous: *P. promelas* (fathead minnow), *P. notatus* (bluntnose minnow); *P. vigilax* (bullhead minnow); some survive spawning and overwinter but die before the next breeding season; repeat spawning is atypical (Markus, 1934: Larry Page, Illinois Natural History Survey, pers. comm.); lifespan, 1 year. There are no historical or hormonal data to indicate the adrenal status or the cause of death. The life history of most minnows native to North America is incompletely known.	
Order Salmoniformes	
Suborder Salmonoidei	
Superfamily Galaxioidea	
Family **Galaxiidae**	
Subfamily **Aplochitoninae** (Tasmanian whitebait)	some populations diadromous;
4 species, but only 1 with obligate semelparity, *Lovettia sealii;* may cease feeding at migration from sea; dies at spawning in rivers (McDowall, 1988; Blackburn, 1950); *tmax,* 1 year.	others landlocked (McDowall, 1971a)

Table 11.1 *(continued)*

Classification	Type of Migration
Subfamily **Galaxiinae**	most marginally catadromous
5 species show varying degrees of semelparity.	(McDowall, 1988, 1971b)
Brachygalaxias bullocki, apparently annual (McDowall, 1988).	
Galaxias divergens (dwarf galaxias), mature at 1–2 years; semelparity is implied by the *tmax* of 2–3 years (Hopkins, 1971; McDowall, 1978, 77).	landlocked only
Galaxias maculatus (inanga), famous for spawning on tidal banks; nearly all die as the high tide recedes (McDowall, 1978, 72; McDowall, 1971b); semelparity here would appear to be consequent to behavior rather than pathophysiology.	spawns in brackish water
Galaxiella Pusilla (dwarf galaxiid; Backhouse and Vanner, 1978). R. M. McDowall (pers. comm.) considers that all species in the genus are semelparous.	freshwater
Most of the other 45 galaxiids are perennial and iteropaarous.	freshwater rivers
Family **Retropinnidae** (southern smelts)	
4 species.	
Retropinna retropinna and *R. tasmanica*, semelparous; *tmax*, 2 years (R. M. McDowall, pers. comm.).	anadromous or landlocked
Stokellia anisodon is considered semelparous; some may mature 1–2 years later than the norm of 1 (*Retropinna*) or 2 years (*Stokellia*) (McDowall, 1978, 43; McDowall, 1979; McDowall, pers. comm.). The cause of death is unknown; *Stokellia* is eaten by gulls at spawning (McDowall, 1978, 46). Lifespan, 1–2 years (McDowall; pers. comm.).	anadromous; spawns in estuaries
Superfamily Osmeroidea	
Family **Osmeridae** (smelts)	anadromous
Hypomesus olidus (= *transpacificus*), conditionally semelparous in Japan (Hamada, 1961); a-type, all spring migrants die at spawning; natural history like *Oncorhynchus masou;* b-type, autumn anadromous migrants and landlocked populations are iteroparous.	
Thaleichthys pacificus (eulachon), most die at first spawning, but a few survive to spawn 2–3 times.	anadromous
Mallotus vellosus (capelin), most die at first spawning, by 4–5 years (Leggett and Frank, 1990; Whitehead and Carscadden, 1985).	spawns on beaches; entirely marine, maturation in deep waters (Whitehead and Carscadden, 1985)
Family **Plecoglossidae** (Japanese ayu)	anadromous
1 species.	
Plecoglossus altivelis, semelparous with adrenocortical hypertrophy during gonadal maturation (Honma, 1960); *tmax,* 1 year (Honma, 1959).	
The landlocked ko-ayu (dwarf ayu; *P. altivelis*) has higher postspawning survival than the ayu. Also shows adrenocortical hypertrophy during gonadal maturation (Honma and Tamura, 1963).	landlocked
Family **Salangidae** (ice-noodle fish)	anadromous
Salangichthys ishikawae and *S. microdon* are considered to die at first spawning (McDowall, 1988); lifespan, 1–2 years.	
Family **Sundasalangidae** (smallest salmonid < 2 cm)	landlocked
2 species; *Sundasalanx* (Roberts, 1981; Nelson, 1984); one of the few samoniforms to invade tropical waters (Borneo).	

Table 11.1 *(continued)*

Classification	Type of Migration
Superfamily Salmonoidea Family **Salmonidae** 　Most of the 68 species are long-lived and iteroparous. 　Subfamily **Salmoninae** 　　*Oncorhynchus* is the only genus with semelparity, nearly universal 　　in 5 species: *O. gorbuscha, O. keta, O. nerka, O. kisutch, O.* 　　*tschawytscha* (see Table 2.4; Figure 11.9); lifespans 3–8 years. 　　*O. mykiss* (steelhead) tends to semelparity at more northerly 　　latitudes (Withler, 1966; McDowall, 1988, 113).	most anadromous, but some 　landlocked anadromous and landlocked
Order Atheriniformes Family **Atherinidae** (silversides) 　160 species; *Labidesthes sicculus* is semelparous; *tmax,* 1 year 　(Hubbs, 1921; Nelson, 1984).	freshwater; most other atherinids 　marine (Nelson, 1984)
Order Gasterosteiformes Family **Gasterosteidae** (sticklebacks) 　7 species with complex taxonomy. *Spinachia spinachia,* 　semelparous; *tmax,* 1 year (Johnsen, 1944).	coastal oceanic; 2 other species 　besides *Spinachia* anadromous 　(McDowall, 1988)
Order Perciformes Family **Callionymidae** (dragonets) 　130 species; semelparity described only for *Callionymus lyra* 　males, which mature at 1–5 years and show rapid weight loss 　suggesting aphagy; *tmax,* 5 years. Females may spawn perennially; 　*tmax,* ≥ 7 years (Chang, 1951).	coastal oceanic
Family **Gobiidae** (gobies) 　2,000 species; 6 species with proven or suspected semelparity.	mostly oceanic; two species enter 　rivers (Nelson, 1984; McDowall, 　1988)
Leucopsarion petersi (Japanese ice goby), aphagic during 　maturation; obligate semelparity (Tamura and Honma, 1969, 1970; 　McDowall, 1988). The interrenal gland does not hypertrophy 　during maturation (Tamura and Honma, 1970a); there is extensive 　thymus atrophy (Tamura and Honma, 1970b); *tmax,* 1 year.	anadromous
Pomatoschistus microps (common goby), semelparous; growth 　resumes after breeding; but mortality increases for unknown causes; 　*tmax,* 2 years (Healey, 1972; Miller, 1975).	marginally catadromous
Pseudaphya ferreri, semelparous (Miller, 1973); empty gut at 　spawning (Fage, 1910): *tmax,* 1 year. 　*Aphia minuta; Crystallogobius linearis, Pomatoschistus minutus,* 　and *Pseudaphya pelagica* are stated to be semelparous; *tmax,* 1 　year (Miller, 1973, 1975, 1979; Swedmark; 1957).	coastal oceanic; spawns on sandy 　beaches
Family **Mugilidae** (mullets) 　95 species; most are iteroparous.	20% migratory
Myxus capensis (South African), aphagic during maturation and 　semelparous (Bok, 1979; McDowall, 1988, 88); lifespan, 2–5 years 　(Bruton, Jackson, and Merron, 1987).	anadromous

Notes: Anadromous: most of life spent in ocean; migration to spawn in fresh water; *catadromous:* most of life spent in fresh water; migration to spawn in ocean; *tmax,* maximum lifespan, where reported; otherwise lifespans are typical. Families and subfamilies are indicated in bold letters; the sequence is based on the cladistic order of Nelson, 1984, and is entirely deduced from morphology. The left column duplicates entries from Table 2.3 that are semelparous.

fossils suggests anadromy. The earliest fossil, *Eosalmo,* was a troutlike fish found in British Columbia about 50 Myr ago (Eocene). The order Salmoniformes may have originated by the Upper Mesozoic, at least 100 Myr ago. All salmoniforms share certain characteristics that are regarded as primitive: the absence of oviducts, the formation of the upper jaw from the maxillary bone, and smooth scales (Nikolsky, 1963). The rapid senescence of most *Oncorhynchys* is almost certainly a recent trait (Figure 11.9). Since no North Atlantic salmon or trout is semelparous and since the Pacific *Salmo* tend to be perennial, obligate semelparity would appear to have arisen in an ancestor common to both stem lines after isolation from the North Atlantic in the Miocene.

The origins of the southern hemisphere salmoniforms are unclear, but could have involved anadromous migration with dispersal in their marine phases, and tectonic shifts (McDowall, 1988). Nearly all southern salmoniforms are anadromous, like their northern relatives. It is, of course, arguable whether the anadromy is a primitive trait or one that evolved in parallel. The breakup of Pangaea into the northerly Laurasia and the southerly Gondwana began in the Jurassic, that is, 50–100 Myr *before* the most recent date proposed for the establishment of the salmoniforms. Thus, the northward-moving Laurasia could have conveyed ancestors of salmonids, while the southward-moving Gondwana conveyed ancestral aplochitonids, galaxids, and retropinnids. Fragmentations of Gondwana that produced South America and Australia in the Upper Cretaceous (70 Myr ago) are consistent with broad distributions of southern hemispheric salmoniforms in the cold waters of Australia, New Zealand, Tasmania, and southern South America (Table 11.1). According to McDowall (1987), families with consistent habits of migration also were established in the Mesozoic before Pangaea broke up: the ancestors of freshwater eels, lampreys, salmon, and sturgeon. Where landlocked populations were established, the entire life cycle could occur in environments like those used for reproduction by migratory species.

11.3.4.4. Semelparity and Migration

Why is there a disproportionate incidence of semelparity in fish species that migrate in one or the other direction (diadromy), or whose ancestors were diadromous? It is striking that nearly all of the eleven orders with at least one semelparous species (Table 11.1) are also diadromous or had diadromous ancestors. Most of these belong to two major orders, the Anguilliformes and the Salmoniformes, which are found in temperate zones of both hemispheres. The taxonomic relationships among Salmoniformes shows striking clustering in the suborder Salmonoidei (Figure 11.10). This cluster includes about ten species from the South and North Pacific. The genera *Galaxias* and *Retropinna* show semelparity in most species, while other genera may have a more scattered incidence. Other orders (Atheriniformes, Clupeiformes, Cypriniformes, Gasterosteiformes, and Perciformes) each have at least one semelparous species. With the exception of the North American minnow *Pimephales* and the capelin (*Mallotus villosus*), most others show migration between fresh and salt water. An unusual example is the gender-related semelparity of *Callionymus lyra;* males disappear from natural populations after first spawning, while females spawn for several years (Chang, 1951). The associations of migration and semelparity, while provisional, may give clues to finding semelparity in species with poorly known natural history, as well as to elucidating possible associations of semelparity and ancestral diadromy.

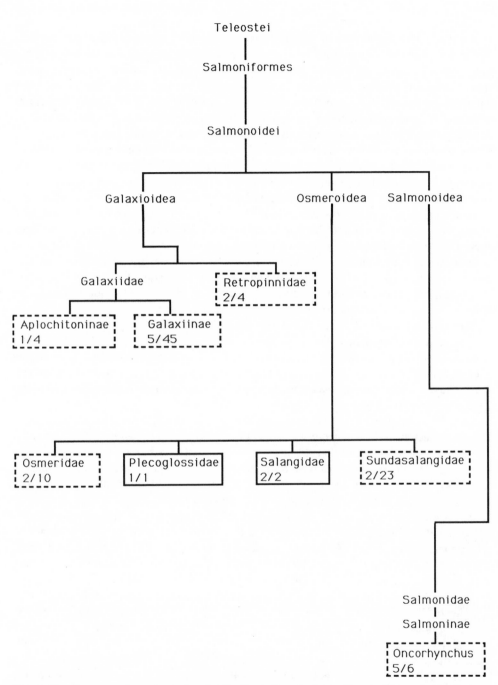

Fig. 11.10. Taxonomy of semelparity in select families in the order Salmoniformes. The *numbers in the boxes* indicate the fractional number of species that show obligate or predominant semelparity. See Tables 11.1, 2.4. Based on information in Nelson, 1984, and McDowall, 1988.

Another common attribute of semelparous and migratory species is that most are temperate zone and cold-water fish (Table 11.1). An important exception is the catadromous eels, which occur in tropical as well as temperate zones (McDowall, 1988). It is also notable that the smallest salmonid, *Sundasalanx* (family Sundasalangidae), is the only salmoniform known to live in tropical waters; it is freshwater (Table 11.1). Temperate climates are characterized by major seasonal shifts in the abundance of nutrients. Here may be a clue about the co-occurrence of semelparity and migration between fresh water and the ocean. A life history based on migration for spawning requires special adaptations that engender an increased risk of mortality from some or all of three major factors: (1) Migration sends fish through environments with different and possibly unpredictable food sources and predators. Indeed, eels, lampreys, and salmon, as well as others noted above, cease feeding before spawning, which in itself may weaken resistance to disease or predation. As discussed in Chapter 2.3.1.4.2, spawning Pacific salmon have immune dysfunctions and increased fungal infections. (2) The physiological stresses arise from changes in environmental osmolality and in hormones, particularly the glucocorticoids that also cause organ damage. (3) The mating-related behaviors, for example, territorial fighting, nest (redd) building, are highly demanding at the end of a long foodless journey and must also be considered a metabolic stress.

These conditions, as far as they have been studied, are met by elevated adrenal steroids, as typified in *Oncorhynchus*. The anadromous and semelparous ice goby (Table 11.1; not a salmoniform) also shows adrenal enlargement and splenic degeneration at spawning (Tamura and Honma, 1970). Evidently most diadromous fish can survive these exigencies. Thus, semelparous species present a special phenomenon that may be much less likely to evolve in the absence of adaptations to diadromy.

Spawning-associated death is associated with two factors that may give advantages to the next generation. The aphagic adults will not consume invertebrate larvae and other food in the spawning beds that are needed by the next generation. In two species the aphagic fish and their carcasses have been shown to import large amounts of nutrients to the spawning and nursery areas, contributing importantly to the nutrient resources available to young fish of the next generation. Alewives and Pacific salmon spawn in nutrient-sparse streams and rivers and stimulate growths of plankton that are important to the growth of young fish (Table 11.1; Richey et al., 1975; Durbin et al., 1979). Alewives, with a mortality of 39 to 57%, survive in sufficient numbers to spawn in subsequent years. Spawning itself causes considerable exhaustion of alewives, with a 35% loss of dry weight and carbon and a 17% loss of phosphorus and nitrogen. The nutrients contributed by alewives to the small streams and lakes where they spawn is a hundredfold greater than calculated for sockeyes (*O. nerka;* Durbin et al., 1979). This efflux stimulates transient growth of microorganisms (planktonic bloom), which in turn enhances the decomposition of leaf detritus, releasing further nutrients into the nursery areas. Besides nutrient effluxes, the parental carcasses in Siberian sockeye populations are often preserved by the cold after fall spawning and are fed upon by the young chum salmon in the following spring (Nikolsky, 1963, 190). As a further example, blowfly larvae that grew in the parental carcasses of coho, chinooks, and steelheads are eaten by their young the next spring (Johnson and Ringler, 1979). The importance of the parental carcasses to food chains needed by the progeny may extend beyond these examples of semelparous fish, and suggests a constraint on life history evolution, since decreased postspawning death could in turn reduce nutrients available to the progeny.

Despite the coincidence of semelparity and diadromy in these examples, we can be sure that there is no *obligate* link between migration and semelparity, as shown in numerous anadromous *Salmo* that are repeat spawners. Similarly, sturgeon, primitive ray-finned fishes, generally survive to spawn repeatedly despite the absence of feeding for many months during migration and spawning. Most species of these large fish are partially anadromous, preferring brackish water, while a minority are landlocked; a few species are completely anadromous (Doroshov, 1985). Sturgeon may better survive the rigors of migration and spawning, which include elevations of corticosteroids and aphagy (Chapter 3.4.1.2.1), because their large size and nutrient reserves protect them better than smaller fish. Other migratory fish cease feeding before spawning, for example several species of herring (Table 11.1), but the relation to mortality is not generally known. The distribution of aphagy during reproduction in nonmigratory fish is indicated in scattered references and, if better known, might lead to further discoveries of semelparity. The scattered semelparity in eight out of ninety of the salmoniform genera and the absence of semelparity in the Atlantic *Salmo* from which *Oncorhynchus* is derived suggest that semelparity in salmoniform fish had polyphyletic origins.

To close the discussion of fish, I note that certain other short-lived species are iteroparous. These include Siamese fighting fish (*Betta splendens,* order Perciformes); anchovies (*Engraulis,* Clupeiformes); mosquito fish (*Gambusia affinis*) and platyfish (*Xiphophorus maculatus*), both Cyprinidontiformes (Chapter 3.4.1.2.2). Of these, *B. splendens* and *G. affinis* show reproductive senescence. Because other species in these orders live longer with no known age-related loss of fertility, their reproductive senescence arose independently. More examples will surely be discovered. The present information about reproductive senescence in these species does not preclude exhaustion of a determinate oocyte stock; firm data are lacking.

Manifestations of senescence in present teleostan fish are more scattered and less uniform than in mammals, which radiated at about the same time in the Upper Mesozoic. There are no examples so far of long-lived mammals lacking an obvious pattern of senescence. Rockfish, among others discussed in Chapter 4.2.2, show no evidence for senescence, as judged by the constancy of mortality rates and continuing egg production at least beyond 100 years. Except for the few species in which senescence was studied, often fortuitously, little is known about the natural history of the vast majority of fish. Surprises surely await.

11.3.5. Tetrapod Vertebrates

Arguments about evolutionary changes in the patterns of senescence in tetrapods can be made from the comparisons of extant species. Moreover, fossil evidence, while fragmentary, is informative about a few questions. Major areas of interest are largely open, particularly in regard to the evolutionary traces of pathophysiological disorders of senescence involving the brain and proliferating tissues. Quite a lot can be said about teeth, a favorite material of paleontologists. Some things can also be said about changes in possible lifespan during evolution by comparisons of age-related mortality rate accelerations in birds and mammals. The discussion also is threaded by a synopsis of tetrapod evolution and mentions various fragments of information about the evolutionary record of senescence that I have chanced on.

11.3.5.1.   The Distribution of Mortality Rates: Evidence That Slow Acceleration of Mortality Is a Primitive Trait

As described in Chapter 3, very long lifespans are documented in three vertebrate classes: mammals, birds, and reptiles (turtles). Turtles represent the most extreme tetrapod lifespan potentials, probably more than several decades beyond the longest-lived human, and appear to have slow or negligible senescence (Chapters 3.4.2, 4.2.2). Fossilized shells that closely resemble extant species are found in the Upper Triassic in association with radiations of the Protorothyridae, a line that had long since diverged from the synapsid lineage that lead to mammals. By the Triassic, Testudinata were distinct from the diapsid lineage that led to dinosaurs, crocodiles, and birds (Carroll, 1988). Many extant turtle genera were established in the Cretaceous. The evidence that senescence is slow in turtles as well as in most birds and reptiles implies that early tetrapods had slow senescence.

Early avian fossils are rarer than those of mammals, though recent efforts are filling in many gaps (Carroll, 1988; Olson, 1985). The stem of the radiation is traced to diapsids in the Pennsylvanian. Feathers, which are rarely preserved, are found in strata of the Upper Jurassic, 150 Myr ago. Unlike amphibia and reptiles, whose Mesozoic predecessors resembled present species, modern avian forms are found only late in the Cretaceous, 70 Myr ago. The mortality rate coefficients for birds (Figure 11.11; Table 3.1) suggest that some orders differ in the clustering of slow and fast accelerations of the age-related mortality rates, as compared by the rate at which the mortality rate doubles with adult age (MRDT; Chapter 1.4). Most gulls, ibises, and terns (Charadriiformes) appear to be long-lived with slow MRDTs and few signs of senescence. In contrast, quail, pheasant, peafowl, and turkeys (all Galliformes) are shorter-lived and show fast MRDTs. The best documented examples of senescence, particularly in reproduction of both genders, are in the galliform birds, although short-lived species occur in other orders (Chapter 3.5).

The absence of birds that show rapid senescence or semelparity is striking, in view of the scattered occurrence of semelparity in fish and in some small mammals. It would be important if senescence proved restricted to the Galliformes, since this would imply that senescence arose relatively recently in this order, within 50 million years. Most galliforms are ground-living and nonmigratory, which could be a factor selecting against extended reproductive schedules. The weakly developed flight of most galliforms, however, cannot be the only factor in their trend for shorter lifespans and more rapid senescence, since there is no indication of these traits in the flightless ostriches and other ratites. The ability to forage outside of the nesting area could also be important as a factor permitting a greater range of food, and merits further inquiry in relation to extended reproductive schedules and longevity. Migratory birds differ in an important trait from migratory fish: continued feeding is essential during egg laying and nurture of the young. Examples of more rapid senescence in birds might be sought in those species that are selected for rapid and early reproduction. Perhaps the extended nurture that is characteristic of most long-

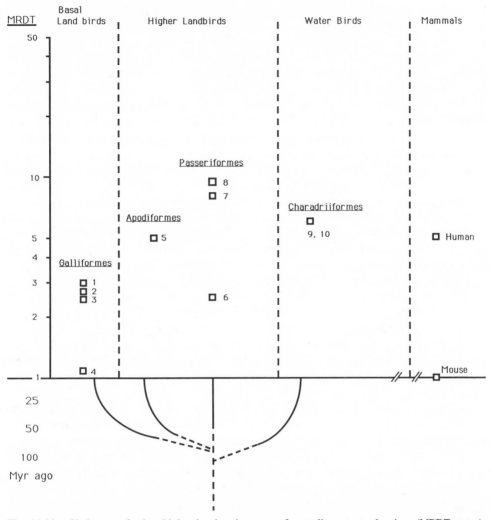

Fig. 11.11. Phylogeny of select birds, showing the range of mortality rate accelerations (MRDT, years). Mortality rates are from Table 3.1. Galliformes [*1, Pavo cristatus* (peafowl); *2, Syrmaticus reevesi* (Reeves pheasant); *3, Alectura lathami* (brush turkey) *4, Coturnix coturnix japonica* (Japanese quail)]; Apodiformes [*5, Apus apus* (common swift)]; Passeriformes [*6, Lonchura striata* (Bengal finch); *7, Erithacus rubecula* (European robin); *8, Sternus vulgaris* (starling)]; Charadriiformes [*9, Vanellus vanellus* (lapwing); *10, Larus argentatus* (herring gull)]. MRDT for mouse (0.30 yr) is not drawn to scale. Phylogeny adapted from Olson, 1985.

lived birds became well enough established to select against mutations with early onset deleterious effects. In view of the indications that senescence is slow in most avian orders, as well as in turtles and other reptiles, I speculate that the stem lines of tetrapods in the Upper Paleozoic and during the Mesozoic also had slow senescence as a primitive trait.

Several marsupial families have species with short lifespans and varying extent of rapid senescence. At one extreme are the marsupial mice (Dasyuridae) with rapid death and sudden senescence within 1 year in association with reproductive-related stress, as best known in the genus *Antechinus* (Chapter 2.3.1.5); other dasyurids are much larger and longer-lived, for example the marsupial cats and dogs. The American opossums include short-lived species like *Didelphis marsupialis* and *D. virginianus* that may survive to breed a second season, but rarely live beyond 30 months (Chapter 3.6.1).

Monotremes (egg-laying mammals) separated during the Triassic from other mammals that were not much advanced beyond species like *Morganucodon* (a mammal-like reptile). However, the sparse fossil record does not sharply date the divergence of marsupial and placental mammals. Ancestors in the early Cretaceous resembled *Peramus*, a genus known from the Upper Jurassic, which had already diverged from the Morganucodontidae. Marsupials are found in North America by the Upper Cretaceous (84–65 Myr ago), forming the superfamily Didelphoidea, from which opossums survive. Australian marsupials date from the Oligocene but may have arrived even before from Asia or South America (Benton, 1985). Miocene (24–5 Myr ago) fossils include dasyurids. The phenomenon of stress-related death after first mating probably arose more than once in the small dasyurids, since the sex of *Antechinus* that dies seasonally varies among species and since some very similar species are perennials. Larger Australian marsupials are longer-lived, but not as long as similar-sized placental herbivores (Table 3.1). The lifespan of large kangaroos (Macropodidae; found by the later Oligocene [Carroll, 1988]) may be limited by wear on irreplaceable teeth (Chapter 3.6.11.4). One macropod, the Australian rock wallaby (*Petrogale*), has continuous tooth replacement, an extremely unusual trait that gives a potential escape from age-related edentulism; the reported lifespans are not distinctively longer than in other macropods (Nowak and Paradiso, 1983).

Placental mammals display a thirtyfold range of maximum lifespan and MRDT, from laboratory rodents to humans (Table 3.1). Although no mortality rate data are available for monotremes, the 50-year lifespan of an Australian spiny anteater (*Tachyglossus aculeatus*; Jones, 1982) is consistent with a slow MRDT.[5] Mesozoic mammal-like reptiles (cynodonts) were relatively small, ten centimeters long, weighing thirty grams or less, as represented by *Morganucodon* from the Upper Triassic (ca. 210 Myr ago; Carroll, 1988, 408; Jenkins and Crompton, 1979). The much larger fossils of mid-Triassic cynodonts (Jenkins, 1984) suggest a size reduction, which also implies a statistically shorter lifespan, on the basis of the general correlations of body size and lifespan shown by birds and mammals (Chapter 5). Because all lineages of marsupial and placental mammals show progressive increases in size during the Tertiary, it is likely that increases of maximum potential lifespan occurred independently, and for different reasons.

The earliest placental mammals are from the Upper Cretaceous period, 80 Myr ago (Figure 11.12). Modern mammals are traced to small, twenty- to thirty-centimeter-long, condylarth ancestors (an extinct, archaic order of herbivorous placental mammals) in the Paleocene, although the ancestry of sea cows and other Sirenia is less certain. These animals had primitive dentition and brains; for example, *Asiorcytes'* skull was three centimeters long with contours indicating a brain with large olfactory bulbs and little neocortex (Kielan-Jaworowska, 1984). Primates, bats, and carnivores may have shared an earlier ancestor with *Asiorcytes*. The extinction of dinosaurs by the end of the Cretaceous may have opened ecological niches that could be filled by large terrestrial carnivores. The

---

5. In regard to the following discussion of tooth wear and replacement, it is notable that spiny anteaters have no teeth, while the platypus loses all teeth by maturation. In contrast, ancestral monotremes retained molars as adults (Carroll, 1988).

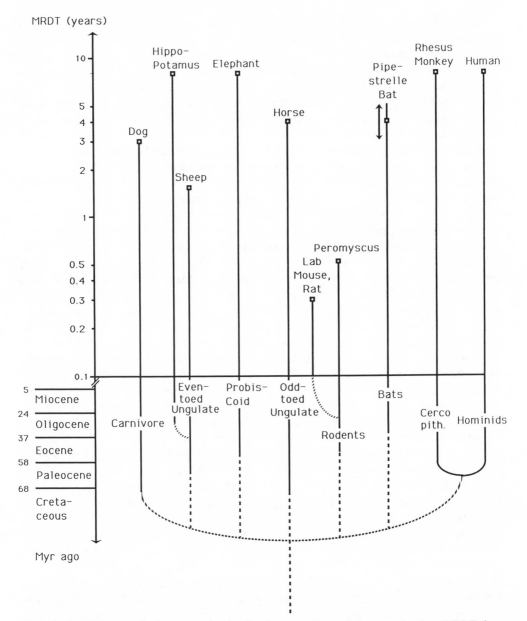

Fig. 11.12. Phylogeny of select mammals, showing the range of mortality rate accelerations (MRDT) from Table 3.1. The uncertainty of MRDT for the pipestrelle bat is indicated by the arrow. Phylogeny adapted from Carroll, 1988.

presence of birds in the Eocene that were large and probably predatory, however, indicates that early mammals were not the only contenders (Gould, 1989, 297). Whatever the extent of that contest, most radiations of terrestrial mammals had begun to increase body size by the Paleocene, which would generally require longer development. Perhaps the statistical lifespan, in turn, expanded concurrently.

Late Cretaceous fossils represent an ancestral primate and various condylarths. Bats may have become established as a lineage by the Upper Cretaceous (75 Myr ago). Their skeletons have changed little since the early Eocene; the vespertilionids are known from the mid-Eocene (Jepsen, 1970; Carroll, 1988). Based on the slow accelerations of mortality rate (long MRDT) indicated for a bat, a primate, and humans (Chapter 3.6.1; Table 3.1), I argue that the common ancestors of these species in the *Asiorcytes* lineage could also have had long potential lifespans. This suggestion opposes the arguments for a progressive increase of potential maximum lifespan of extinct primates, which are based on predictions from body and brain size correlates with lifespan, and on the apparently short maximum lifespans of lower primates (Cutler, 1976a; Hofman, 1984). The mean and maximum lifespans can be misleading indexes and do not necessarily correspond to the rate of pathophysiological senescence or the acceleration of mortality (Chapters 1.4, 3.6.1). Moreover, evidence argues against a constraint on lifespan potential from small size (Promislow and Harvey, 1990; Austad and Fischer, in press; Chapter 5.3.1.1).

Laboratory rodents (hamsters, mice, rats) have the shortest lifespans and fastest MRDT (Chapter 3.6.1). The much longer lifespans of some large rodents suggests long MRDT, for example in the paca (*Agouti paca*, 6–10 kg, zoo lifespans > 16 years) and capybara (*Hydrochaeris*, 50–70 kg, > 12 years) (Nowak and Paradiso, 1983). Although laboratory rabbits are short-lived (< 10 years), there is insufficient information to estimate their MRDT. The common ancestor of rodents and lagomorphs (rabbits) was probably a condylarth in the early Paleocene (65 Myr ago; Carroll, 1988). Both rodents and lagomorphs have continuous growth of lower and upper incisors, though other aspects of dentition differ widely between them (Peyer, 1968; Carroll, 1988). Differences in dentition and in the muscular control of the jaw may have limited lagomorphs to fewer niches than the rodents, and are thought to have favored the much greater (fortyfold) radiation of rodents, which include seventeen hundred living species versus just forty-six in the lagomorphs.

Skeletons of *Paramys*, a primitive rodent from the Lower Eocene, resemble squirrels (Wood, 1962). The Cricetidae (hamsters and voles) and the Muridae (mice and rats) diverged in the Oligocene (ca. 30 Myr ago; Carroll, 1988). Explosive radiations occurred several times, even recently in the case of the two hundred subspecies and species of *Microtus* (voles), which arose during the past 2 million years in North America (Chaline, 1977). Although mortality rate data are available only for a few species each of rats, mice, hamsters, and voles (Figure 11.12; Table 3.1), there is general consistency within a two-fold range for the MRDTs, which are more than 90% faster than for humans. Most data indicate that lagomorphs and rodents live less than 20 years, and more typically 3 to 6 years, which would be consistent with generally rapid accelerations of mortality rates. Provisionally, the lagomorph-rodent lineages appear to be inherently short-lived, with more rapid senescence than in the bat-primate lineages, and represent a departure from most other placental mammals. Although seasonal mortality may be very high in small rodents (Chapter 2.3.1.6), no placental mammal is characteristically semelparous as in the dasyurids.

Several ungulates have values for maximum lifespans and MRDTs that are intermediate between humans and laboratory rodents, such as domestic horses (Figure 11.12; Table 3.1). Most ungulates have short maximum lifespans of 20 years or less, for example sheep (Artiodactyla), but some achieve 40 to 50 years, for example horse and rhinoceros (Perissodactyla) and sea cows (Sirenia). Ungulates are particularly vulnerable to erosion of their molars through mechanical abrasion, and few live more than 50 years. The best documented are the elephants, which have evolved a unique mechanism that permits extended reproductive schedules despite severe dental wear, as discussed next.

11.3.5.2. Teeth

The extent of tooth replacement is a factor in senescence and in the age-related increase of mortality risks of placental mammals. Several examples (Chapters 3.6.1, 3.6.11.4) show how tooth erosion contributes to age-related disability and constitutes an example of mechanical senescence. The irreversible erosion of irreplaceable adult teeth in certain mammals is developmentally programmed. For elephants and many herbivores, but also carnivores, eventual disabilities from tooth erosion are a major limiting factor on the potential longevity. However, senescence in these species is the result not only of tooth erosion (Chapter 3.6.11.4). The fossil record indicates that a major transition in tooth formation occurred during the Mesozoic, through which present mammals acquired what proved to be, for some, a factor limiting the potential lifespan.

A distinction should be made here. Other than humans, few species that rely on teeth during their adult life can adapt to tooth loss later in life by changing their diets. However, numerous species are well adapted to the absence of teeth throughout the adult life. Birds, for example, have lacked teeth since the Cretaceous (Carroll, 1988). Edentate egg-laying mammals also are long-lived. Clearly, the lack of teeth in some adult vertebrates does not preclude extended reproductive schedules and long lifespans.

The capacity for polyphyodonty, or continuous tooth replacement, is the rule in reptiles. Mammals, however, generally are diphyodont, with only two complete sets of teeth (Peyer, 1964; Kemp, 1982; Carroll, 1988). Polyphyodonty is attributed to the reptilian need for new sets of slightly larger teeth during the continuing growth of their jaw and other bones throughout their lifespans (Osborn, 1973). Mammals have cheek teeth (molars and premolars) that "wear in" to form efficient grinding and shearing surfaces and whose function would be disrupted if frequently replaced. The teeth of mammals are exquisitely differentiated by shape and size (heterodonty). In contrast to the masticatory functions of mammalian dentition, reptiles mostly use their teeth for holding food prior to swallowing. A mutually adapted suite of characteristics in jaw bones and muscles was evolved by mammals in association with these differences in dentition (Osborn, 1973; Kemp, 1982; Carroll, 1988). Depending on the species, the incisors may grow continuously, as do the incisors of rodents and the tusks of elephants, while horses have continual growth of their molars. However, the single adult set of teeth in most mammals have a major consequence to potential longevity, since the inevitable erosion of particular teeth will impair feeding at later ages. This is an example of mechanical senescence in mammals, in which the wearing out of an irreplaceable part contributes to systemic dysfunctions that must increase the risk of mortality.

The teeth and skulls of the Mesozoic mammal-like reptiles are more anatomically diverse than any other tetrapod except their mammalian descendants (Kemp, 1982; Jenkins, 1984). Polyphyodonty persisted in certain radiations of Mesozoic mammals (Carroll, 1988; Kemp, 1982); however, by the Upper Triassic (225 Myr ago) *Morganucodon* and other cynodonts had evolved mammal-like jaws and teeth with characteristic incisors, canines, premolars, and molars. Of particular importance, Parrington (1971, 1978) showed that the Morganucodontidae did not replace their adult molars, and the molars showed signs of wear (also see Kemp, 1982; Carroll, 1988). These small animals may have had short individual lifespans (Parrington, 1971). *Morganucodon* also had a relatively large brain, threefold bigger than in other therapsids of the same size (Crompton and Jenkins, 1979). During these transitions, other cynodont lines retained some extent of adult tooth replacement, for example *Gobiconodon* (Jenkins, 1984). Carroll (1988, 416) suggests that *Gobiconodon's* longevity as a genus in the fossil record was related to tooth replacement.

Tooth development may have interacted with the evolution of mammary glands and prolonged maternal care. Hopson (1973) proposed that diphyodonty implies the capacity for lactation and extended maternal care. Mammals at birth do not have complete sets of teeth that are as developed as those of reptiles. Nursing thus would favor survival of the young during postnatal growth of the mammalian jaw to a stage where permanent teeth could wear in as grinding surfaces. In contrast, reptiles generally have fully functional dentition at hatching and do not need to wear in their teeth as grinding surfaces. Successive generations of teeth can be lost and replaced during continuing jaw enlargement, with no loss of function. Throughout the Jurassic and Cretaceous, mammals remained small and apparently had limited or no replacement of adult molars. A major question is whether the evolutionary stage may have been set for an intrinsic limitation on lifespan for certain mammalian radiations with species that depended on extended reproductive schedules and dentally abrasive diets.

Recent evolutionary changes in elephants give a unique example of how lifespan potential can be rapidly influenced by modulating a developmental process, the scheduling of tooth eruption or emergence, which postpones the age when tooth wear becomes a vital deficiency. It is curious that Africa was the immediate source of the two longest lived mammals, human and elephant; their lifespans of 70 or more years appear to result from very different traits, however. Elderly (> 50 years) elephants of both extant species (*Loxodonta africana,* African; *Elephas maximus,* Asian) typically wear out their molars during their long lifespans, in association with the huge daily consumption of fodder that often contains abrasive phytoliths. Tooth erosion eventually prevents them from feeding adequately and is considered a major cause of senescent death, even in zoos (Chapter 3.6.11.4). Severe tooth erosion is common in other herbivores and is a significant factor in mortality at later ages. Elephants also can delay the first pregnancy until after 20 years under exceptionally harsh conditions, an age when few herbivores are fertile, or even alive.

Lifespans exceeding 50 years are unusual in field populations of mammals. While the following comments are controversial, it is interesting to consider if long lifespans of elephants and some whales (Chapter 3.6.1) could be adaptive through information accumulated over lifespans of many decades. Strategic information could be transmitted transgenerationally as a "living tradition" (Eisenberg, 1981, 183). For elephants, this information might include the locations of water and forage during seasonal fluctuations, as well as behaviors to defend against specific predators, which must have included early humans throughout the Pleistocene. But, being anecdotal and necessarily so, these notions must be considered speculative. The leadership roles of senior females in their herds is well known (Moss, 1988; Douglas-Hamilton, 1973; Laws and Parker, 1968) and is consistent with the generally matrifocal organization of large ungulates (Wilson, 1975; Eisenberg, 1981).

The exceptional lifespan of elephants among terrestrial herbivores is almost certainly due to a feature that is unlike nearly all other mammals: African and Asian elephants have up to five replacements of their adult molars (Chapter 3.6.11.4). Their specialized dentition is also adapted for the horizontal shearing of food against the elevated (hypsodont) enamel ridges of these unusually elongate teeth (Turnbull, 1970; Figure 3.44). The enamel of these ridges, which is more durable than dentine or cementum, thus can be considered as a pacemaker for tooth erosion. In present African and Asian elephants, the cheek teeth on each side of the jaw (in the anatomic position of molars) are replaced sequentially during adult life; the total number, however, does not exceed the three molars per jaw quadrant that are characteristic of placental mammals (Nowak and Paradiso, 1983; Laws, 1966). Tooth replacement results from sequentially delayed tooth eruption, so that just one of these unusually elongate molars is fully used at any time in adult elephants (Chapter 3.6.11.4). Worn teeth are replaced as the lengthening molar moves forward (Figure 3.45), which is unusual; in nearly all mammals, teeth are generally replaced vertically, rather than horizontally. Other herbivore adaptations that reduce the impact of tooth wear include increased tooth height (hypsodonty) and molarization of premolars into a single functional tooth battery. Eventual tooth erosion in older elephants draws them to moist locations that provide softer stems, which Sikes, 1971, 222) suggests can explain the frequent location of elephant skeletons, both modern and prehistoric, in river beds and swamps.

The phylogeny of elephants is deduced from the types of teeth, most of which are molars, and their patterns of wear (Beden, 1983; Maglio, 1972, 1973; Harris, 1975). There is general agreement that the large proboscideans (e.g. *Paleomastodon*) in the Eocene to Oligocene did not have delayed tooth eruption. The first proboscideans showing it were the deinotherians, which lived in Africa from about 25 to 1.5 Myr ago. Deinotheres had a different pattern of odontogenesis than present elephants, in that more than one tooth was active in each jaw and new teeth were replaced vertically, rather than horizontally. Despite extensive wear that is suggestive of advanced age, the contact (occlusal) surfaces apparently were continually resharpened by abrasion from chewing (Harris, 1975). By the Eocene deinotheres had diverged from the gomphotheriid line leading to present elephants (Figure 11.13).

The ancestors of elephants are represented in the Miocene by *Gomphotherium angustidens,* which had three pairs of elongate molars differing from those of modern elephants by elevated cones and the simultaneous presence of two molars on each jaw quadrant in young adults. "Very old" individuals retained only one molar (Maglio, 1973, 87). These short-legged proboscideans had both upper and lower tusks, four in all, and may have been forest browsers (Maglio, 1972). Anticipating the direction of tooth formation in their descendants, the gomphotherian milk teeth were replaced horizontally (Carroll, 1988; Peyer, 1968). Some specimens show marked wear of the enamel edges, which Maglio deduces to have caused gradual shifts in the mechanics of chewing, from a pattern of grinding motions to more horizontal shearing in "old age." Maglio also notes evidence for abscesses in jaws that would have impaired chewing. Descendants of *Gomphotherium* in the late Miocene include the extinct *Stegotetrabelodon* and *Primelephas,* which show

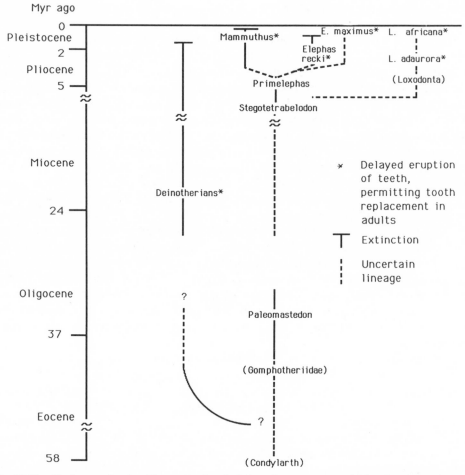

Fig. 11.13. Phylogeny of proboscideans, showing the independent occurrence of delayed tooth eruption and adult tooth replacement in deinotherians and in three elephant genera, of which only African elephants (*Loxodonta africana*) and Asian elephants (*Elephas maximus*) have survived. Delayed tooth eruption permits modern adult elephants to replace up to six sets of worn molars. The erosion of the last set at advanced ages (50 to 70 years; Figure 3.44) causes gradual weakening (morbidity), which is considered a specific geriatric cause of death (Chapter 3.6.11.4; see text). *Elephas, Mammuthus,* and *Loxodonta* evolved specialized, elongate molars during the last 5 million years. Their immediate ancestors in the Miocene, in contrast, had smaller molars and premolars, which developed as a single set of adult teeth. Radiations from the gomphotherians *Stegotetrabelodon* and *Primelephas* at the Miocene-Pliocene boundary are based on fragmentary evidence that is subject to revision. Redrawn from Beden, 1983, Harris, 1975, and Maglio, 1973, accepting Beden's conclusion that *Loxodonta* separated before *Elephas* and *Mammuthus.*

transitions to the teeth of modern elephants, including further lengthening, increased plate number, and the disappearance of the cones. The adults of these genera had less than two full molars in use at any one time and may have already developed delayed eruption. Together, these changes are consistent with a shift in the mechanics of chewing, from grinding to shearing. Fossil grasses from the Miocene of East Africa had siliceous and highly abrasive phytoliths at least by 14 Myr ago; these grassy woodlands show evidence of heavy grazing (Retallack et al., 1990).

Maglio (1972) attributes the success of elephants, indicated by their rapid radiation and geographic dispersal during the last 5 Myr, to the adaptive shifts in chewing with horizontal shearing motions, which is, perhaps not so coincidentally, the very shift indicated during the gomphotherian life cycle. Increasingly cooler and drier climates in the Pliocene of Africa were the presumed cause of dwindling forests and woodlands, forcing many proboscideans into savannah grasslands. A similar shift from forest dwelling during these times is thought to have influenced hominid dental evolution (Howell, 1979; Szalay and Delson, 1979). The habitat shift is hypothesized to have selected elephants for elongate molars with increased shearing edges (increased numbers of lamellas and convolutions of enamel folding), which enhanced horizontal shearing at the expense of vertical grinding, as is better suited for masticating grasses (Maglio, 1973). As noted above, their immediate gomphothere ancestors had already shown a phylogenetic trend for tooth lengthening. These evolutionary changes are associated with numerous modifications in the development of the teeth, jaws, and oral musculature (Maglio, 1970, 1972, 1973; Beden, 1983).

In regard to the origins of variations in teeth that could be subject to selection, there are several examples. Genetic polymorphisms of tooth development are implied by the tendency of the Ceylon race of Asian elephants to lack tusks (incisors; Peyer, 1968). It is unknown if the occasional seventh set of molars in modern elephants is a genetic polymorphism or a sporadic occurrence. Supranumerary molars (M^4) are also occasionally found in extinct elephants, for example in an early subspecies of *Elephas recki* (2 Myr ago; Beden, 1983, 74). As discussed in Section 11.4.2, the variations of tooth eruption schedules in proboscideans represent heterochronic changes in development.

Loxodonta was the first lineage to achieve the modern tooth construction by the mid-Pliocene (*L. adaurora*, 4.5 Myr ago) and had the same number of lamellar plates in corresponding teeth as its surviving descendent, *L. africana*, which appeared in the Pleistocene (Figure 11.13). The numbers and shapes of the adult teeth indicate the establishment of delayed tooth eruption in the earliest known (Pliocene) representatives of *Elephas*, *Loxodonta*, and *Mammuthus*. It seems more likely that a common ancestor had acquired this trait than that each genus did so independently. The fossil record is too sparse to say. The lines in the genus *Mammuthus* leading to *Elephas* and the recently extinct mammoth diverged from *Loxodonta* about 5 Myr ago. Of the more than twenty species identified by teeth and jaws, some existed for 0.5 Myr or less, a very brief record (Maglio, 1973).

Tooth structures in these genera changed extremely rapidly, successively increasing the number of lamellar plates to more than twenty (Figure 11.14). All three lineages, however, also evolved the reorientation of the plates and broadening of the teeth, which, Maglio (1972) proposed, might have maintained the efficiency of shearing for longer duration, possibly "into old age." The longevity of extinct elephants is estimated by comparison of tooth wear with present elephants. By this criterion and with the further assumption of a similar tooth succession, *Loxodonta adaurora* is predicted to have lived 50 years (Beden, 1983, 52). Maglio (1973, 10) also argues that the greater height of molars in earlier forms allowed a similar total tooth and organismic lifespan. While this is plausible, no firm conclusions are possible.

Manatees or sea cows (Sirenia) have another unusual dental adaptation, molar progres-

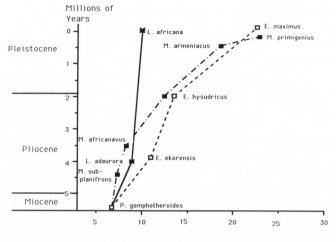

Fig. 11.14. Evolutionary changes in elephant teeth, showing the progressive increase in the number of lamellar plates in species from the three genera *Elephas, Mammuthus,* and *Loxodonta. Loxodonta* teeth have changed little since the early Pliocene. Phylogeny taken from Maglio 1972, 1973, which differs from that of Beden, 1983.

sion, whose effects resemble the delayed tooth eruption of elephants but which arises through a different mechanism. All three species of *Trichechus* (*T. inunguis* of the Amazon basin; *T. manatus* of Florida and the Caribbean; *T. senegalensis* of West Africa) continuously produce extra molars (Chapter 3.6.11.4). Tooth replacement is horizontal, with new teeth formed at the back of the jaw as worn teeth are shed from the front of the mouth as in elephants. Unlike elephants, however, manatees make unlimited extra molars, at least thirty per jaw quadrant over their lifespans, and never become toothless (Domning, 1983). The rate of replacement is about one per year, varying with diet, apparently because of the mechanical stimulation from chewing fibrous plants. This manatee trait is traced back to *Ribodon,* 5 Myr ago in the late Miocene (Domning, 1982, 1987). No sea cow besides *Trichechus* has unlimited tooth replacement.

Little is known about the cellular mechanisms of the delayed tooth eruption of proboscideans or the apparently limitless polyphyodonty in *Trichechus* and *Petrogale.* Peyer (1968) suggests that the dental laminas of *Trichechus* continue to proliferate and sustain the histotypic interactions that produce "germs" of new teeth; an unusual system of fibers connects posteriorly emergent teeth to the in-wear teeth, conveying them forward (Miller, Sanson, and Odell, 1980). It might now be possible to reconstruct the genetic changes that terminate odontogenesis during development. The neural crest is implicated in odontogenesis and contributes cells of the dental mesenchyme in lower vertebrates and possibly in mammals (Ledouarin et al., 1981; Lumsden, 1988). Inductive interactions with the stomatodeal ectoderm (mesenchymal-epithelial interactions) determine the characteristics of different teeth, such as their height and the folding of enamel. Genes expressed in neuroectodermal cells that determine these characteristics of teeth thus could also indirectly serve as pacemakers for the lifespan of numerous mammals by scheduling tooth eruption and by determining tooth durability through variations in shape and in the amount of enamel.[6]

11.3.5.3. Molecular and Pathologic Traces

The tetrapod lines leading to birds and mammals branched from terrestrial reptiles at about 300 Myr ago in the Upper Carboniferous period. An interesting set of questions concerns evolutionary changes in thermoregulation as it may relate to the acceleration of damage to proteins through oxidation and glycation. While all vertebrates have some capacity for thermoregulation through metabolic variations, a very high set point of body temperature is found in placental mammals and birds. Monotremes have lower body temperatures than most placental mammals, about 32°C, whereas birds are hotter by several degrees, about 40°C (Taylor, 1980; Bligh, 1973; Whittow, 1970, 1971, 1973). The origins of homeothermy are speculative, some arguing for an independent origin during the Mesozoic avian and mammalian lines (Crompton et al., 1978; Taylor, 1980). The small size of Mesozoic mammals also implies endothermy (Hopson, 1973; Taylor, 1980). The evolution of consistently high body temperatures may have required adaptive changes in the antioxidant systems to protect against the greater production of free radicals that is chemically predicted at higher temperatures. The glycation of proteins and its further chemical transformations into various cross-links (Chapter 7.5.2) merit attention during the evolutionary transitions to sustained elevations of body temperature in tetrapod radiations. The elevated temperatures would be predicted to accelerate glycation, but the formation of advanced glycation or Amadori products also depends on blood glucose levels that tend to be manyfold higher in birds than in placental mammals. Glucose may have already been selected as a major carbohydrate fuel in ancestral vertebrates because it is one of the least reactive monosaccharides in forming glycated adducts with proteins (Bunn and Higgins, 1981). The evolution of long total lifespans in hot-blooded tetrapods could therefore have required mechanisms to remove or repair glycated proteins. Little is presently known about the phylogenetic distribution of these mechanisms. Evolutionarily new molecular pathologic changes that interact with senescence may thus have arisen during the acquisition of homeothermy in the Mesozoic. Another biochemical attribute of birds and reptiles is their high blood levels of uric acid, which is one of the most powerful antioxidants (Ames et al., 1981; Chapter 5.4.3). A high level of circulating uric acid is a consequence of the characteristic uricotelic metabolism of birds.

Dinosaurs, which prevailed during the Mesozoic era, ranged in size from *Staurikosaurus* (two meters long, middle Triassic) to *Diplodocus* and other giant sauropods (thirty meters long, Upper Jurassic) (Carroll, 1988). Lifespans of decades or more would be consistent with present reptiles and amphibia (Chapters 3.4.2, 4.2.2). The time required to reach large sizes need not, however, have been long if food were not limiting, based on the age of puberty in large reptiles (8–10 years in alligators) and in large mammals (3–5 years for horses and the giant baleen whales). There is some evidence for degenerative joint lesions that are an age-related characteristic in present mammals. For example, six out of ten specimens of the sauropod *Diplodocus* display a coalescence of caudal vertebrae that suggests traumatic osteoarthritis (Sokoloff, 1969; Blumberg and Sokoloff, 1961). Cu-

---

6. Dentine is an evolutionary ancient protein traceable to antigens in hagfish (Slavkin et al., 1983).

riously, the weight-bearing joints of these huge creatures did not show degenerative changes.

If the early placental mammals had slow senescence, as argued in Section 11.3.5.1, then how can we explain the shorter lifespan potential and the faster onset of many features of senescence seen in laboratory rodents? I suggest that these phenomena have resulted from unremitting selection for short generation cycles and high fecundity, that is, *r*-selection that calibrated fitness by the most rapid and fecund reproductive schedule (Chapter 1.5.1). The small rodents, for whatever cause, have the largest litters and shortest generation times found in the mammals. The vulnerability to kidney disease that is associated with the ingestion of certain diets that are high in animal proteins (Chapter 7.2) need not increase mortality until after two or three litters are produced; such ages are attained by a small fraction in natural populations (Chapter 1.5.2). This explanation then requires that further genetic changes accumulate, for example, altering the sensitivity of retroviral gene expression to age-related changes in endocrine regulations, which then yield abnormal growths.

The adverse impact of ovarian estrogens on the neuroendocrine axis of female mice and rats (Chapter 3.6; Table 10.7) would be another example, not selected against because its later manifestations should affect such a minority of the population. There is no evidence for such phenomena in humans and primates, where ovarian oocyte depletion occurs without evidence for neuroendocrine dysfunctions in the regulation of the ovary. In view of the data, while limited, suggesting that humans do not show the corresponding decrease of protein metabolism at later ages as described in rodents (Chapter 7.4.1.1), we may suspect that laboratory rodents also have some relatively unusual neuroendocrine changes that down-regulate metabolism at later ages.

The search for metabolic and biochemical correlates of lifespan in mammals may be recalled here from Chapter 5.4. The first group of these correlations concerns mechanisms that protect against damage. There are good correlations of primate lifespan with the plasma urate levels. This major antioxidant is present at particularly high values in humans, about twofold higher than in chimpanzees or gorillas (Figure 5.19). Superoxide dismutase, however, shows weaker correlations with lifespan and varies less between species (Figure 5.20). The extent of unscheduled DNA synthesis in fibroblasts after irradiation with UV (an index of DNA repair capacity) varies in the strength of correlations between studies (Figures 5.21, 5.22). The capacity to activate heterocyclic carcinogens varies inversely with lifespan, being least in humans (Figure 5.25). The proliferative capacity of cultured diploid fibroblasts showed good correlation with lifespan in a limited sample of species (Figure 5.24).

While arguments can be made pro and con the relevance of these measurements, the comparative approach does not allow strong conclusions, pending direct evidence for species differences in free radical damage to DNA or proteins that *cause* age-related pathology and the accelerations of mortality rate that ultimately define the process of senescence. More efforts are needed to acquire tissue specimens under standardized conditions from species that represent the range of rates of senescence in each order. For example, it is hard to interpret comparisons of fibroblast clonal lifespans from laboratory rodents, bats, horses, and humans (Figure 5.23) from the four orders represented. There is a need for data from other species in each order that represent different ranges of the mortality rate coefficients, particularly the MRDT.

In the absence of data on the mortality coefficients of primates other than rhesus monkeys, little can be said about evolutionary changes in the pace of senescence during anthropoid and human evolution. The similar MRDT of rhesus monkeys and humans (Chapter 3.6.1; Table 3.1) suggests that slow actuarial senescence is not recent in primates. The limited data show that the MRDTs of bats, elephants, and hippopotamus are as about long as in humans. Tentatively, the greater human lifespan may be ascribed to smaller IMR in humans than in the healthiest colonies of rhesus monkeys. As noted in Chapter 3.6.1, these data on mortality are puzzling in view of evidence for many aspects of senescence at early ages in rhesus monkeys (*Macaca mulatta*) versus humans. Like human males, primates show male-pattern baldness, for example the stump-tailed macaque (*Macaca nemestrina*, Old World primate; Uno et al., 1967) and the red uacari (*Carajao rubicundus*, New World; Uno et al., 1969). The phylogenetic distribution of androgen-dependent and postpubertal male-pattern baldness in primates is barely known.

The growth of prehistoric human populations was most likely limited by high mortality rates across all ages. Judged from a review of data on present hunter-gatherers and late Paleolithic groups (Hassan, 1980), prehistoric populations would have an IMR of about 0.03/year, approximating that in rhesus monkey populations in captivity (IMR, 0.02/year, Table 3.1). Demographic reconstructions indicate that prehistoric *Homo sapiens* had life expectancies at birth in the range of 15 to 20 years, approximating that of present hunter-gatherers (Weiss, 1973; Hassan, 1980). About 50% would die before 15 years (nominal age of puberty), which is about tenfold more than the current figure in the U.S of less than 6%. Those surviving to the age of 15 years would have had lifespans of about 35 years, compared to 75 years in the U.S. today. Based on trends for earlier menarche and earlier end of adolescent sterility since the eighteenth century in Europe and North America (Ashley-Montague, 1957; Tanner, 1962), Hassan argues that nubility (onset of childbearing) occurred in prehistoric populations at about 18 years (Hassan, 1978, 1980). However, reproductive schedules must have varied among populations, no less then than now, according to fluctuations in diet, infectious disease, and stress. Nonetheless, the statistics suggest that survival to menopausal ages, with some multigenerational age structure, is a long-standing feature of human populations. While extensive survival to multigenerational ages is documented only in recent times, the biological potential is not new and, moreover, could have allowed populations at any time in human evolution to sporadically achieve much greater life expectancy after puberty than is generally thought.

Several biochemical features appear unique to humans and would appear to have evolved recently. Humans lack uricase, which results in plasma urate concentrations that are a hundredfold higher than in rats or prosimians (Ames et al., 1981). Because urate is a powerful antioxidant, an interesting speculation is that the loss in the capacity to make vitamin C during human evolution may have been compensated by the increased plasma urate (Ames et al., 1981; Hochstein et al., 1986). Among primates, plasma urate levels correlate positively with lifespan (Figure 5.19). Such correlations of biochemical parameters, as measured in young adults, with lifespans of different species do not inform about the rate of senescence. While urate may protect against free oxidative damage, high blood urate may also increase the risk of gout, or deposits of urate salts, which are apparently more common in humans than in primates (indicated by informal discussions with comparative pathologists).

Age-related increases in the numbers of neurofibrillary tangles (NFT) are a crucial feature of senescence that also may have been acquired recently during anthropoid evolution. These intraneuronal filamentous protein aggregates are found in select brain regions associated with Alzheimer's disease. NFT also accumulate to a lesser extent at later ages in healthy elderly (Chapter 3.6.10; Figure 3.39). NFT have only been described in humans, where they are most prominent in recently evolved neocortical neuronal pathways (Rapoport, 1988). The proteins of the NFT have been only partially characterized. Besides covalently linked ubiquitin and microtubule-associated proteins (MAPs), the bulk of NFT remains a mystery.

In Alzheimer's disease, but also to a considerable extent in the nondemented elderly, β-amyloid protein accumulates in cerebral blood vessels (amyloid angiopathy) and in the extracellular space around degenerating nerve terminals (neuritic plaques). Accumulation of β-amyloid protein after mid-life occurs in few other species; those from only two orders have been examined so far, carnivores and herbivores (Selkoe et al., 1987). No rodent accumulates β-amyloid protein during senescence (Chapter 3.6.10). The gene for β-amyloid protein has been sequenced and located (Figure 6.12). Other species should be studied for amyloid accumulations at later ages, for example, bats and large rodents like the capybara that live more than 10 years.

Brain β-amyloid protein is indistinguishable in amino acid sequence and biophysical properties from that deposited during Alzheimer's disease and Down's syndrome and in neurologically normal elderly (Chapter 3.6.10). However, Alzheimer's amyloid deposits of senile plaques are found in association with tau and neurofilament proteins, whereas those from nondemented controls of the same age group do not contain these epitopes (Arai et al., 1990). A provisional conclusion is that amyloid accumulation in hereditary Alzheimer's disease is not caused by a mutation in the coding sequences of the amyloid protein complex. Based on these similarities of brain amyloid, genes acquired during evolution that predispose to Alzheimer's disease may act pleiotropically by influencing the expression of other genes. Although the phylogeny of the gene encoding β-amyloid protein is not known in detail, portions of the protein may be ancient. In *Drosophila*, RNA coding for a related protein occurs in the adult head, presumably representing a brain mRNA; amino acid sequences in three regions show about 40% identity to corresponding extracellular and cytoplasmic domains of the human β-amyloid protein APP–695 (Rosen et al., 1989).

A calcium-binding protein (calbindin-D_{28}) shows species differences in its neuronal distribution that bear on age-related loss of neurons through mechanisms involving calcium influx. Neurons lacking calbindin are the most vulnerable to epileptic damage or recovery from oxygen deficits, for example human CA_1 pyramidal neurons. In contrast, the CA_3 neurons of rats and baboons have abundant calbindin immunoreactivity, and suffer little lasting damage during experimental seizures (Sloviter, 1989; Sloviter et al., 1989). Other cells in rats and baboons, nonetheless, have less calbindin and are vulnerable to seizures.

Calbindin and other calcium-binding proteins may act as a buffer during sustained calcium influxes that are thought to be important in neuron swelling and death after seizures, possibly through the activation of calcium-dependent proteases like calpain. Lynch et al. (1986) hypothesize that sporadic fluctuations of calpain activation in association with intracellular calcium transients could damage the cytoskeleton. Calpain occurs at high concentrations in pyramidal neurons (Siman et al., 1985), of which certain subsets in the frontal cortex and hippocampus are at high risk for degeneration as a function of age and Alzheimer's disease. The phylogenetic distribution of calcium-binding proteins in different brain regions may tell approximately when during mammalian evolution the hippocampus acquired or lost protection against calcium-influx-dependent neurodegeneration. Thus, the presence of calcium-binding proteins is a potentially crucial factor in the interactions between age and corticosteroids on calcium currents of hippocampal pyramidal neurons (Kerr et al., 1989; Chapter 10.5.1). The phylogeny of calpain is incompletely known. In limited sampling of different mammals, cerebral cortical calpain content of young animals varied inversely with the mean lifespan (Baudry et al., 1986; Chapter 5.4.3).

To close this discussion on mammals, I point out a major issue that arises from the evidently similar changes of senescence across the thirtyfold range of lifespans. The comparisons shown in Table 3.3 suggest a canonical pattern of senescence that is remarkably general across the species that are well characterized. Many changes are shared, particularly those associated with female reproductive senescence at mid-life. Both sexes, however, show osteoporosis and osteoarthritis, vascular lesions, glucose intolerance, cataracts, various abnormal growths, etc., in the last third of the potential lifespan. Nonetheless, there are clearly species-specific disorders, like two that are characteristic of humans: the accumulation of NFT (Chapter 3.6.10) and benign prostatic hyperplasia (Chapter 3.6.3). Overall, mammals share remarkably similar pathophysiological changes during mid-life and later. These similarities imply two major alternatives: this canonical pattern is a primitive trait of mammals for which genetically determined characteristics have persisted for 50 to 100 Myr. Alternatively, these multisystem age changes might represent convergent evolution, through unknown selection pressures. The concept of convergent evolution of characteristics that arise late in life when the force of selection is expected to be least is not predicted by contemporary population genetics models (Chapter 1.5). Another part of this puzzle is the controversial analysis suggesting that mice have a faster germ-line mutation rate per Myr than in humans (Easteal, 1985; Wu and Li, 1985; see commentary of Fitch, 1987). If so, mouse senescence should be less canonical because of greater genetic drift.

Some insight to these complex issues might come from considering embryonic determinants of limited stem cell pools. Because the numbers of ovarian oocytes is fixed during development in mammals, it seems likely that the gene controls on primordial germ cell proliferation in females were established early in mammalian evolution. We may then ask, Could not other changes shared in common across short- and long-lived species also represent consequences of limited stem cell numbers that were embryonically determined? This possibility may apply to certain cells in the immune system that, in turn, might influence many other functions such as vascular lesions, defenses against malignancy, and bone and joint disorders. However, the numbers of stem cells do not obviously pertain to the reduced dendritic arbor in brain neurons during senescence in rodents and humans, or to the loss of ocular accommodation in rhesus monkeys and humans. The similarity of each of these changes in animals of ages rarely found in nature challenges population genetics theories of senescence that predict extensive diversity in the patterns of senescence due to the weakness of selection at later ages. There is a real puzzle here.

11.3.6. Conclusion: Senescence Has Polyphyletic Origins

Senescence is a very common cause of maximum potential lifespans in higher plants and animals. Relatively few individuals in iteroparous species achieve ages when senescence would be manifested, however, because of a high level of mortality at all ages in most natural populations (Chapters 1, 3). The clearest case concerns rapid senescence in association with reproduction in semelparous life histories, in which, by definition, senescence is ubiquitous. Semelparity and rapid senescence is distributed unevenly across the taxonomic levels (Table 4.4). Most phyla have a broad range of lifespans and degrees of senescence, Rotifera being the major exception, since all species in this phylum have short adult lifespans with rapid senescence. Moreover, in only a few phyla (Arthropoda, Chordata, and Mollusca) are entire orders semelparous. There is no example from plants in which all species in an order are monocarpic. Even in these phyla, however, numerous orders have few or no species that are semelparous or show rapid senescence. The other part of this spectrum, gradual to negligible senescence with adult phases lasting months to years, occurs in almost all phyla without obvious restriction. Similarly, the capacity for asexual (clonal) vegetative reproduction is scattered throughout eukaryotes. Besides the Coelenterata and Porifera, few other phyla show this capacity in all species (Table 4.3).

A major inference is that the total lifespan and the length of the adult phase have expanded and contracted throughout the evolutionary history in most taxa on innumerable occasions. The capacity of dipteran populations for reversible alterations in lifespan was demonstrated by artificial selection on the basis of age at reproduction that modulated lifespan by about 30% in either direction within about fifteen generations (Chapter 6.3.2). Over the much greater span of generations in evolution, a much greater range of variations is implied. Most phyla show a tenfold or greater range of lifespans and corresponding differences in the intensity of senescence. Up to a hundredfold differences in lifespan arise from the same genome in the different castes of social insects, showing that this plasticity of biological time can be regulated at the level of gene expression.

Despite the major variations in lifespan and the rate of senescence that are manifested in social insects for example, the adult anatomical features also indicate canonical patterns of senescence. In insects and other invertebrates that do not regenerate their exoskeleton and its appendages, mechanical senescence is a canonical pattern (Chapter 2.2.2), as is the ultrashort lifespan of mayflies in association with anatomic deficiencies in mouthparts (Chapter 2.2.1). Similarly, the elevated glucocorticoids in migratory salmon cause characteristic vascular degeneration from endothelial proliferation in the heart and elsewhere, not only in semelparous Pacific salmon but also in the iteroparous Atlantic salmon (Chapter 2.3.1.4.2). In mammals, the ovariprival syndromes that result from ovarian follicle exhaustion also show considerable stability across widely varying species lifespans (Chapter 3.6.4.1). Because of the vast potential for environmental influences on the phenotypes of senescence during developmental and adult phases (Chapters 9, 10), the concept of canonical senescence represents a strong species tendency that may be overwhelmed on occasion by external forces.

In searching for further examples of semelparity in fish, I found evidence that nearly all examples are associated with migration. This is well known for the Pacific salmon but also occurs in other salmoniforms, including those in the southern hemisphere. These preliminary findings may give clues to the identification of other migratory fish with obligate or conditional semelparity. The association between semelparity and migration is not general, however, being absent in sturgeon, which also cease feeding before migration and spawning. Unlike the Pacific and other semelparous salmon, sturgeon generally survive to spawn perennially. The cessation of feeding before spawning could be adaptive by reducing competition for resources in the spawning beds that will be used by the young of the next generation.

The reduction of mouthparts in many adult insects may be seen in a different light, as the result of mutational changes that alter the morphogenesis of the imaginal discs. That this can derive from differential gene regulation is suggested by gender differences in oral anatomical difficiencies that are often more severe in males than females, particularly in the Nematocera. Mayflies are unique in their universally ultrashort adult phases and inability to feed as adults, which suggests a monophyletic derivation of this trait from the Paleozoic ancestors that had fully developed chewing mouthparts.

Elephants illustrate the recent rescheduling of tooth eruption, which, together with their elongate teeth, provides both the African and Asian elephants with fresh grinding surfaces up through the sixth and seventh decades. Few other herbivores can delay first reproduction to ages later than 20 years, which occasionally is found in elephants, and few continue to reproduce at such late ages. This adaptation compensates for a major aspect of mechanical senescence that is a later factor in senescence and that limits lifespan of these creatures that must spend nearly all their waking hours chewing.

The taxon-specific distribution of senescence strongly indicates that the different patterns of senescence had independent origins and should be considered polyphyletic. Present evidence allows the argument that throughout evolution eukaryotes have not been destined at any grade of organization to particular types of senescence. Moreover, the representation of vegetative reproduction by somatic cell cloning in numerous phyla (Chapter 4.4) indicates that somatic cells need not be committed to finite replication. While controversial, there is evidence that hematopoietic stem cells in mammals are also capable of renewal to vast extents that certainly exceed the organismic lifespan and may be unlimited (Chapter 7.6).

Conversely, we can consider the proposition of frequent evolutionary escapes from clonal and organismic senescence that, for this argument, would be primordial and monophyletic in Precambrian eukaryotes. The persistence of clonal (vegetative) reproduction in numerous cnidarians, ascidians, and vascular plants would require frequent and independent evolutionary derivation from gametogenically reproducing species whose somatic cells all showed clonal senescence. This situation, moreover, requires that Precambrian annelids and cnidarians could not reproduce vegetatively, unlike their descendants.

Furthermore, Precambrian organisms do not support the case for primordial senescence, because the pertinent anatomy and physiology had not yet evolved, for example, for senescence from aphagy and from mechanical or hormonal causes. Demonstrably, the numerous types of rapid senescence arose independently. It remains possible, but unknowable, that some differentiated cells of Precambrian animals had slow degeneration that increased organismic mortality rates after maturation and that could be a diffuse underlying constraint on life history patterns down to the present. The ubiquity of lipofuscin accumulation could be an example of slow changes that might ultimately cause dysfunction in nondividing cells. So far, lipofuscins have not been associated with cellular dysfunctions (Chapter 7.5.6).

Finally, I note that several novel pathophysiological aspects of senescence that emerged during the evolution of tetrapod vertebrates that have not been documented elsewhere. While humans are the only species to show the accumulation of NFT during Alzheimer's disease, long-lived primates and species of two other orders (carnivores and ungulates) accumulate amyloid proteins in their cerebral blood vessels at later ages. Data are not yet available for vascular β-amyloid protein in other long-lived mammals or birds. We also lack data on the accumulation of glycated proteins and the extent of kidney lesions at later ages in lower vertebrates. New molecular defenses may have been required against oxidative damage at the higher temperatures that evolved in birds and mammals. Further comparative studies may indicate when specific pathophysiological aspects of senescence appeared during tetrapod evolution.

11.4. Factors in the Evolution of Senescence and Longevity

Section 11.3 showed the taxon specificity of life history patterns and the recurrence of lifespan shortening and lengthening during evolution. We now ask more generally about limits to the evolutionary modifiability of senescence and adult life history. For example, what constraints may account for the absence of any long-lived adults in rotifers or mayflies, two taxa with little variation among species in the canonical form of senescence? The following discussion emphasizes structural characteristics that are established during development. These include anatomical features and physiological organization, but also the capacity for repair and cell replacement, collectively known as the body-plan. Section 11.4.3 addresses a general issue, the variations in reproductive modes. Many of these questions are "old-fashioned" and are presently difficult to pose in the terms of modern molecular, cellular, and physiological mechanisms. Even so, their discussion points to the important conclusion that there is a great plasticity in the scheduling of life history, whatever the body-plan characteristics.

11.4.1. The Body-plan

Body-plan constraints are invoked to account for why flying insects never evolved to as large sizes as have birds (Gould and Lewontin, 1979), and may also constrain the options of adult life history or the rates of evolutionary change in life history variables. Similar questions have been raised about speciation through developmental variations and the genetically interlocking processes that may limit the scope of viable developmental alternatives. This line of thought has often been influenced by Waddington's (1940) "epigenetic landscape," wherein the steepness of the walls represents the barriers to modification of the developmental path. A related concept is Levinton's (1986) "evolutionary rachet," whereby accumulated genetic constraints during evolution become progressively more difficult to disrupt because of interlocking developmental processes.

Systems of trans-acting regulators of gene activity may be a major substrate for evolutionary innovation (Britten and Davidson, 1971). Now familiar examples include the steroid response elements in promoter regions that activate or repress transcription of numerous genes (Yamamoto, 1985; Evans, 1988; Beato, 1989; Miesfeld, 1989). However, the extent of hierarchic constraints on genetic changes is by no means clear. For example, the nineteenth-century doctrine of Haeckel that "ontogeny recapitulates phylogeny" is in serious question because of evidence for major intertaxa variations in cell migratory patterns during early development (Davidson, 1990). Moreover, the importance of cross-regulatory interactions between homeotic genes during *Drosophila* development may be less than supposed (Gonzales-Reyes et al., 1990).

Genetic analysis of development will eventually identify the specific sets of genes that produce adult structures predisposing to particular phenotypes of senescence. Four constraints are discussed, in which the potential for modifying adult life histories differs according to the body-plan: anatomic deficiencies, such as loss of mouthparts; body size; capacity for regeneration of appendages; and capacity for cell replacement.

11.4.1.1. Anatomic Deficiencies

Anatomic features that influence the adult lifespan and mortality rate coefficients can be found deficient only by comparison with other species. A clear example is the deficienct oral anatomy in aphagous insects. Species comparisons show, however, that even reduction of mouthparts does not preclude recovery of function through the evolution of alternate anatomy. For example, tsetse flies lack the mandibles of most biting flies, but use prestomal serations for bleeding their hosts (Downes, 1971). Biting flies with gender differences in deficiencies of their oral anatomy (Chapter 2.2.1; Section 11.3.3) might be able to recover lost oral anatomy, since the gender differences imply that anatomic defects are regulated by genes on sex chromosomes.

It is unknown whether mayflies and dipterans with defective oral anatomy in both genders accumulate mutational changes in loci regulating oral morphogenesis. Unless these gene sets served other functions, there would be no selective pressure to preserve them. Examples of multiple usage include lens crystallins that are identical to enzymes of intermediary metabolism (Piatagorski and Wistow, 1989).[7] The accumulation of mutational changes that might alter the protein structure or impair translation in genes that control oral morphogenesis in mayflies would indicate that these genes had organ-specific functions. Another approach was evolved by parasitic Crustacea lacking a digestive tract, which absorb nutrients directly from their hosts and which can have long lifespans, for example *Sacculina* (Chapter 2.2.1.2). Major body-plan deviations like the elimination of adult organs thus do not preclude adaptations for recovering lost functions that would increase lifespan potential.

11.4.1.2. Size

Large size is often but not universally associated with lower overall mortality rates and longer lifespans in mammals (Chapter 5.3.1.1). Several taxonomic analyses of mammals by family and order (Stearns 1983b, 1984; Read and Harvey, 1989; Promislow and Harvey, 1990) showed covariations in life history parameters that distinguished different evolutionary radiations. The phylogenetic lineage influenced the covariations with lifespan of body size, age at maturity, gestation length, and mass and number of offspring.

---

7. An obvious prediction is for the preservation of genes that encode the lens crystallins and other eye proteins in blind cave fish that have impaired eye development (Avise and Selander, 1972; Wilkens, 1971; Sadoglu, 1967).

Clusters were according to small size, early maturation, and short lifespan, as distinct from large size, late maturing, few offspring, and long lifespan. After statistically controlling for influences of size, lifespans still differed among orders (Stearns, 1983b). Nonetheless, there was no indication that body size is a limiting factor in lifespan potential. These analyses give a foundation for future correlates of life history variables with the age-dependent acceleration of mortality (e.g. MRDT) and the age-independent mortality rate coefficients (Chapter 1) according to taxonomy and phylogeny.

Birds and fish do not show as strong a correlation between body size and lifespan. Some very long lived birds weighing less than one kilogram show no signs of reproductive senescence at least to 30 years and have MRDTs at least as short as in humans (Chapter 3.5; Table 3.1). Hence phylogenetic trends for the differential survival of smaller or larger species need not constrain the lifespan. Since there is general agreement that Mesozoic ancestors of placental mammals were small (Section 11.3.5), the mammalian radiation was characterized in all orders by increased size of the largest species (Carroll, 1988; Stanley, 1973). The trend for increasing size during evolution, or "Cope's rule" (Cope, 1896), recurs in most phyla. However, the size of the smallest extant species did not necessarily increase by comparison with evolutionary ancestors (Stanley, 1973; Brown and Maurer, 1986).

A major unknown is the impact of size changes during evolution on mortality rates. The MRDT is similar in small bats, rhesus monkeys, and humans despite manyfold differences in reported maximum lifespan (Chapter 3.6.1). The IMR may influence both mean and maximum lifespan as much as any slowing of senescence during mammalian evolution (Chapter 1.4.8). Correlations between IMR and body size in mammals (Figure 5.8) and fish (Figure 5.10) could account for much of the increase in maximum lifespan with body size. While the mean lifespan in mammals seems likely to have increased during evolution (Cutler, 1976a, 1979), the limited data do not inform about changes in IMR or MRDT.

Body-size increases during mammalian evolution since the Cretaceous imply later maturation throughout the different radiations (Chapter 5; Figure 5.11). However, body size can be statistically partitioned out of the strong correlations between life expectancy at birth and the age at first reproduction in mammals (Figure 1.13; Harvey and Zammuto, 1985). The relation between lifespan and onset of reproduction in mammals is consistent with the studies of *Drosophila* showing that artificial selection of increased lifespan also delayed reproduction (Chapter 6.3). Although larger mammals generally have smaller litters and longer generation times, the potential lifetime numbers of births were independent of body size, so that most females can produce ten to twenty offspring in their lifespan (Figure 5.12). However, species differences in those surviving to reproduce at least once (not equivalent to maturity!) could alter the impression that lifetime reproductivity is independent of body size. The highest producers are small, short-lived, and effectively semelparous because few survive to a second breeding season. In general, large mammals (of more than one hundred kilograms) have one or two offspring per pregnancy, an exception being pigs. Relationships between size and the duration of the fertile phase of adult life shown by mammals (Chapters 3.6, 5.3.2) are not apparent in birds. Besides two galliforms (chicken and Japanese quail), few other species have yet shown female reproductive senescence up through the oldest ages documented (Chapter 3.5).

Larger animals may have a later exhaustion of ovarian oocytes because the stock of oocytes scales hypoallometrically with body size, but also because the rate of oocyte loss through atresia varies inversely with lifespan and with the number of oocytes present at puberty in the few species for which data are available (Chapter 5.4.5). It would be of much interest to know the age of oocyte depletion in small bats. Pipestrelle bats are mouse-sized and have about the same numbers of oocytes as laboratory mice (Gosden and Telfer, 1987a), but live at least four times as long with an MRDT that is relatively long among mammals (Appendix 1; Chapter 3.6.1). Shrews may represent the lower body size that can contain a sufficient ovarian mass. A trade-off may be necessitated by the largest ovarian mass compatible with body size and the thermal economy of these hyperactive creatures, but that would still give enough oocytes to reach puberty. An example of limiting endowments of oocytes is the W alleles in mice, the ([C57BL/6J $\times$ C3He/J]F$_1$ $- W^x/$ W^v) strain with fifty oocytes or less, or less than 1% of the parental mice (Murphy, 1972). In this genotype, the number of primordial oocytes at birth is so low that few adults are fertile. Even so, this situation might be escaped by slowing follicle loss through reduced atresia.

Correlations between body size and the duration of fertility (puberty to menopause) in mammals are less evident in teleosts or other species with continuing *de novo* oogenesis. For example, the small (seven to nine centimeters) Fries' goby (*Lesueurigobius friesii*) lives 11 years or more and breeds perennially; an asymptotic limit to growth is likely by 5 years after maturity. In contrast, the same-sized sand goby (*Pomatoschistus minutus*) lives only 2 years and usually breeds once (Chapter 3.4.1.2.2). Fries' gobies apparently make more eggs than other small gobies and yield multiple generations of oocytes in each breeding season (Gibson and Ezzi, 1978). Similar examples might be found in amphibia and reptiles. The semelparity of large Pacific salmon and tiny gobies (Table 11.1) indicate that semelparity has no size constraints in teleosts. Moreover, unlike mammals, egg production in iteroparous teleosts is found to scale with body size and shows no hint of decreasing during continuing postmaturational growth (Figure 4.8).

Does small size ultimately constrain life history patterns? The lower size of teleosts (seven millimeters or less) might be limited by the required size of mature oocytes (Miller, 1979), which must contain sufficient stores of maternally produced molecules to enable autonomous development up to the onset of feeding (Davidson, 1986). Some Coleoptera may have reached this lower size limit. The minute (smaller than one millimeter) featherwing beetles (*Ptinellodes*) produce only one egg at a time (Dybas, 1978), which is relatively huge, about 10% of the body volume. During short adult phases of 1 to 2 weeks, they produce thirty to sixty eggs (Utida, 1972, cited by Dybas); there is no information on whether new oogonia are formed during oocyte maturation. Many *Ptinellodes* are parthenogenetic, a rare trait in Coleoptera which is hypothesized to be adaptive in species with low fecundity and short lifespans (Dybas, 1978; Williams, 1975). Beetles of other genera live much longer and lay eggs during several successive years (Chapter 3.3.5). Aphagous species that depend on stored nutrients to complete their adult phases might also have a lower size limitation, but this seems unlikely. The aphagous insects have a wide size range, from the very small autogenous biting flies to the more than hundredfold larger aphagous beetles (Chapter 2.2.1); this range does not reveal any obvious body-size limitation in the nutrient stores needed to complete an aphagous semelparous life cycle. In summary, most evidence shows that body size does not dictate whether an animal is semelparous or iteroparous, or its rate of senescence.

11.4.1.3. Regeneration

The capacity for regeneration of appendages like limbs and wings as well as internal organs bears on the potential longevity through the ability to repair damage from environmental hazards. Free-living organisms are at risk for incurring external damage that would increase their mortality risk. The brittle and erodible appendages of insects, for example, become increasingly worn as a function of age (Chapter 2.2.2.1) and cannot be replaced or repaired. The absence of regeneration in insects must contribute to the shortness of adult phases. In contrast to the short adult phases of most insects, however, queens in social insects of both the Isoptera and Hymenoptera can live more than twentyfold longer than workers and soldiers, yet all have the same body-plan with fragile irreplaceable appendages (Chapter 2.2.3.1). The capacity of queens for multiyear lifespans may derive from two factors: the few flights made outside of the hive, thereby saving their wings from abrasion; and the grooming and cleansing of their bodies by attendant workers. Theoretically, long adult life phases in insects could also evolve through precocious gametogenesis (progenesis) in a larval phase that retains the capacity for cell proliferation; such examples may yet turn up.

The phylogenetic distribution of organ regenerative capacity does not clearly correlate with the adult lifespan, or with the level of anatomical or physiological complexity. Thus, we find examples of very short lived and rather long lived insects that are not known to differ in their capacity for regeneration of appendages. On the other hand, the regeneration of legs observed during successive molts in spiders (Chapter 2.3.1.2), in a different subphylum of the Arthropoda than the insects, could be a factor in longevity, for example in tarantulas that begin reproduction when 10 or more years old (Chapter 2.3.1.2). The indeterminately growing Crustacea can also regenerate lost appendages during the successive molts of their adult phases, for example lobsters (Aiken, 1980). Major genetic changes would be required to enable the largely postmitotic adult insects to regenerate their appendages that form from imaginal disc cells during pupation.

Regeneration through cell replacement in mammals is limited, but includes regeneration of the liver and intestinal mucosa (Chapter 7.6.2). Compensatory neuronal outgrowths in response to injury in the brain and elsewhere involve remodelling of neuronal projections and dendritic fields. With a few exceptions, there is no capacity in mammals for neuronal replacement. Wound healing at epithelial surfaces (skin, gut) and healing of bone fractures require to varying degrees cell proliferation and differentiation that are equivalent to regeneration; the normal turnover of erythrocytes and of the epithelial cells of the skin and gut (Chapter 7.6.1) is equivalent to limited regeneration.

The capacity for tooth replacement and tooth elongation in adults is rarely discussed as an influence on lifespan, although tooth wear is well known as a geriatric trait in humans and other mammals (Chapter 3.11.4). Resistance to tooth wear or loss is pertinent to reproductive scheduling, particularly for large herbivores like elephants, which spend the bulk of their waking time in chewing and which show debilitating tooth erosion with advancing age. The rate of tooth wear thus must be adaptively balanced by selection for initial tooth height (hypsodonty), the thickness of tooth enamel, continual tooth growth, or tooth replacement to attain the age for bearing and rearing the young (Van Valen, 1960). For example, in elephants under suboptimal conditions, puberty can be delayed until after 20 years and intervals between calving extended up to 8.5 years (Moss, 1988; Laws and

Parker, 1968). Few other vertebrates manage such an extensive flexibility in reproductive scheduling without diapause or other hypometabolic phases. In comparison, African lions rarely reproduce after 20 years and show extensive tooth wear at ages when elephants are just raising their first offspring (Chapter 3.6.11.4). Whether or not lions reach an age when they could die from specific effects of wear on their single set of adult teeth, there is no obvious way that lions could extend their reproductive schedules to the upper range observed in elephants. Variable numbers of adult teeth in elephants (Section 11.3.5.2) could be a basis for selection; whether these are determined by genetic polymorphisms is unknown.

11.4.1.4. Cell Proliferation

Somatic cell proliferation is prerequisite for wound healing or regeneration, as well as for vegetative reproduction. The determinants of fixed cell number in soil nematodes, rotifers, and other animals with postmitotic adult cell populations are established during development. Somatic cell proliferation in adult insects is considered very limited in most short-lived species, which are essentially postmitotic as adults (Chapter 2.2.2). While DNA synthesis may continue, leading to polyploidization, mitotic figures are rare in somatic cells. Two exceptions seem important. The gut cells of adult worker bees show mitotic figures (Snodgrass, 1956, 189; Chapter 2.2.3.1). There is no report on cell turnover in queens, but it is reasonable to consider that gut cell replacement may be crucial to absorption of the nutrients needed for oocyte maturation during reproductive lifespans of many years. Grasshoppers also have continuing cell proliferation in the cecum and intestine (John and Hewitt, 1966; John and Freeman, 1976; Chapter 2.2.2.1). The importance of cell proliferation to lifespans might be tested with mitotic inhibitors, as in the studies with FUdR, which even slightly increased the 3-week lifespan of *Caenorhabditis* (Ghandi et al., 1980; Chapter 2.6.4).

What evidence is there for variations in adult somatic cell proliferation that could be used by selection? Several phyla show major differences among *species* in the extent of somatic cell replication. In contrast to the short-lived *Caenorhabditis*, whose somatic cells are entirely postmitotic, the long-lived parasitic ascarids evidently acquire additional gut and muscle cells during their fiftyfold elongation after hatching (Chapter 4.2.2). Postmaturational addition of neurons to the brain is documented in fish and birds (Chapter 3). There is only one example of differences within a species in the extent of adult somatic cell proliferation, however. Neurons continue to be added to the granule layer in the dentate gyrus in Sprague-Dawley rats beyond 9 months, but not in other laboratory-rat genotypes (Boss et al., 1985; Kaplan and Bell, 1984). While there are no reports of intraspecific variations in the age when oogenesis ceases, mouse strains differ in the numbers of oocytes at birth and in their rate of loss (Figure 6.11). A search for more examples is needed because the maintenance of intraspecific variations in the persistence of proliferative potential into adult life would allow rapid adaptive response of a population to changes in trade-offs between energy-expensive repair processes and expedient reproduction.

These examples of continuing cell replacement or regeneration depend on the preservation of genomic totipotency in somatic cells, to which there are rare exceptions. The nematode *Ascaris megalocephala*, the gall midge *Wachtliella persicariae*, and other scattered insects and crustaceans lose considerable amounts of somatic cell nuclear DNA at specific stages of development in a process historically known as chromatin elimination (Davidson 1986; Wilson, 1925). A constraint on DNA elimination in different somatic cells would be the sharing of tissue-specific genes and of genes that code for more than one function, either by alternate splicing of nuclear RNA or by multifunctional polypeptides like lens crystallins (Piatigorsky and Wistow, 1989). Recovery of the capacity for somatic regeneration or vegetative reproduction might be more difficult in species with somatic cell chromatin elimination, which may lose genomic totipotency by discarding subsets of genes in their somatic cells. It is unclear how closely the capacity for regeneration in adults follows that for vegetative reproduction, since some lower vertebrates can regenerate eye lenses, limbs or fins, and tails (Goss, 1968; Hay, 1966; Singer, 1973; Yamada, 1967). No chordates besides the tunicates reproduce vegetatively (Chapter 4.4.3). Species lacking somatic cell proliferation must also lack vegetative reproduction.

11.4.2. Heterochrony

Variations in life history can influence longevity by the dissociation of age-related biological changes. The concept of heterochrony, or species variations in the timing of developmental events, arose from comparisons of developmental schedules in related taxa by Haeckel, von Baer, and others early in this century, as beautifully summarized by Gould (1977). Many examples indicate that differences in developmental timing are a major substrate in evolution (Alberch et al., 1979; Gould, 1966, 1975, 1977; de Beer, 1958; Fleagle, 1985). These dissociations of developmental processes can alter the scheduling (timing) of differentiation and the size of organs and body at different stages of development, as well as in adults. The breakdown of phylogenetic recapitulation during development as a general rule (Davidson, 1990) further emphasizes the plasticity of schedules in gene activity throughout the life history.

A special lexicon describes the varieties of heterochrony (Gould, 1977; de Beer, 1958), but only a few of these arcane terms are used here. In *paedogenesis,* sexual maturity occurs before the organism reaches the full size or relative proportions (shape) of closely related species. Paedogenesis can result from either (1) precocious sexual maturation with other somatic tissues at their usual state of immature development for that age (*progenesis*); or (2) sexual maturation at the usual age, but with retarded development of the other somatic tissues (*neoteny*).

The many phenotypic variations of senescence (Chapters 9, 10) demonstrate the crux of heterochrony: the dissociable scheduling of different changes (Gould, 1977, 234). This book has considered two heterochronic characteristics of age-related changes: (1) The timing of events during postnatal development as well as later during the adult life phase is subject to dissociability, as seen in variations among individuals and among species. (2) Proximal controls for changes are often sex hormones that regulate somatic changes during puberty and menopause.

The precocious maturation of oocytes before hatching in mayflies and in autogenous aphagous dipterans illustrates progenesis, while the sexual maturation of the nonparasitic dwarf lampreys at the same age as the much larger parasitic forms of the paired species is neoteny. Neoteny can also occur to varying extent within a species. Lamprey ammocoetes sometimes have precociously mature ovaries, which might permit life cycles with a short-ened adult phase, for example the nonparasitic dwarf lampreys (Section 11.3.4.1). These species variations in the timing of maturation can have major impact on adaptiveness. The acceleration of oogenesis permits extremely short adult phases without dependency on unpredictable food sources.

The other extreme in heterochrony is delayed maturation and continued growth, which permits increased body size at maturity, even though sex hormone target cells can respond at much earlier ages. Humans reach puberty at an age (12–14 years) that is 75-fold later than in mice and 5-fold their lifespan. While the mechanisms governing the different rates of maturation in mammals are still obscure, elevated sex steroids have indisputable impor-tance (Ojeda et al., 1986; Reiter and Grumbach, 1982). The concept of competence from classical embryology is used to describe the potential for responding to morphogenetic stimulae like the induction of neurulation. Correspondingly, sex-steroid target tissues ac-quire competence for hormonal responses early during differentiation in the fetus, so that pubic hair, breast development, and other pubertal changes can occur precociously in very young children if sex steroids are abnormally elevated (Money and Ehrhardt, 1972; Ro-senthal and Weiss, 1980; Tanner, 1969). Conversely, pubertal phenotypes of humans can be delayed by gonadal insuffiency, as in Turner's syndrome (XO, gonadal dysgenesis), at least beyond 30 years when the secondary sexual changes can be induced by steroid ther-apy. Another source of delayed puberty is through emotional (psychogenic) factors that impair hypothalamic-pituitary functions (Money and Ehrhardt, 1972; Tanner, 1969). The age limits of response to delayed induction of puberty have not been tested beyond middle age.

Provisionally, age alone has little effect on the competence of peripheral reproductive-system target cells to respond to sex steroids in mammalian species that have extreme differences of lifespan. The species differences in timing of puberty in short- and long-lived mammals depend on the absence of time constraints from age alone on at least some peripheral hormone target cells. The hypothalamic pacemaker controlling the frequency of secretory bursts of the neuropeptide gonadotropin-releasing hormone (GnRH) is impli-cated in the onset of ovulatory cycles in primates (Terasawa et al., 1984; Terasawa, 1985). Ultimately, the development of this pacemaker may be determined by genes regulating synaptogenesis, in which changes in the regulation of a few genes could cause major differences among species.

The capacity for temporal independence of some hormone target cells may be regarded as a subcase of genomic totipotency. As reviewed in Chapter 5, many aspects of cell function do not correlate with body size, rate of maturation, or lifespan; this suggests extensive potential for changes in the scheduling of developmental events. However, spe-cies differences in the time required for development could result from autonomous cell-level controls or from different timing in the secretion of hormonal and other signals for differentiation. Other than the short adult phases reported for rotifers of 2 months or less (Comfort, 1979), species in nearly all other phyla have adult phases that range to a decade or more. The duration of development further adds to the potential for extended total lifespan. On the other hand, some taxa may have time constraints for the durations of development or differentiation, which would restrict heterochronic options.

A genetic basis for heterochronic variations is being studied in *Caenorhabditis*. Muta-tions in mapped genes influence the larval molt when cell proliferation permanently ceases; other mutations cause supernumerary molts (Ambros, 1989) by retarding or accelerating development. A nuclear protein made by the *lin–14* gene dwindles at later larval stages (Ruvkun and Guisto, 1989); the amount of this protein could be crucial to these heterochronic variations. In semelparous species with obligatory rapid senescence at spawning (certain octopods, lampreys, bamboo, etc.), at least some of these same gene sets *must* also be the direct determinants of the particular form of senescence and the particular lifespan, for example those that determine the hormonal or neural changes for the first and only reproduction. For iteroparous species with gradual senescence, one can speculate about influences from genes scheduling developmental events in the later time course of senescence. One example would be the genes determining the numbers of oocytes. The general proportionality of life history phases in mammals across their thirty-fold range of lifespans (Table 3.3) hints at common and stable genetic controls. Clearly, certain characteristics programmed during development, such as the initial populations of oocytes and the capacity for regeneration or repair, will have many *indirect* consequences for the timing of senescence.

11.4.3. Reproductive Scheduling and Modes in Relation to Longevity and Senescence

This section discusses issues in reproduction that are related to evolutionary changes in longevity and senescence. As described above, heterochronic variations in age at maturation are a major source of variations in total lifespan of different species. Furthermore, individuals may differ manyfold within a species in age at maturation or first reproduction, depending on temperature, light, nutrition, or chemical signals. Examples from Chapter 9 include the flowering plants, which may be perennial or annual, depending on light and temperature; the mollusk *Phestilla sibogae,* which can delay metamorphosis over a five-fold range, depending on chemical signals from its coral prey; and vertebrates whose puberty is delayed by stress, limitations of food, or cold temperatures. The catadromous eels throughout the world also vary widely, up to 50 years, in age at maturity, before returning to the sea for spawning that ends their adult life (Chapter 2.3.1.4.2). Another phenotypic variation in lifespan is the development of long-lived queen bees from the same genome that yield short-lived workers (Chapter 2.2.3.1).

Few species show restricted plasticity in reproductive scheduling. The relatively invariant 2- to 3-year lifespan in *Oncorhynchus gorbuscha* contrasts with variations from 3 to 8 years in its close relative *O. keta* (Table 2.4; Figure 11.9). The narrow range of lifespan in *O. gorbuscha* is an exception to the generally large variations in reproductive scheduling shown within many fish species. One explanation could be the fixation of alleles that increase the susceptibility of target cells to toxic effects from elevated corticosteroids, possibly by slightly increasing the tightness of steroid binding to the receptor complex, which would prolong residency in the cell nucleus. Genetic determinants of death after spawning might be subject to the age and size at spawning. For example, there is considerable survival of precociously mature male *O. tschawytscha* after spawning, in contrast to the usual complete mortality of larger-sized and older fish after spawning (Table 2.4). Semelparity varies geographically in *O. mykiss* and American shad (*Alosa sapidissima;* Figure 2.14). The semelparity found in only a small zone of their range could provide a substrate for fixation of semelparity in these populations. These examples support West-Eberhard's (1986) concept that evolutionary variations in life history patterns arise from the genetic fixation of epigenetic alternatives.

The potential, during evolution, for reversing semelparity or a high mortality at spawning may hinge on another issue, that of transgenerational ecological coordination of life histories. The carcasses of alewives and salmon contribute nutrients or food organisms to nursery areas used by the next generation (Section 11.3.4.4). Thus, evolutionary selection for reduction of mortality at spawning implies an adjustment of nutrient flux that might require a reorganization of ecologically coordinated life histories of numerous species. Flowering plants and insects also have closely coordinated life cycles that are indicated throughout their long evolutionary record of association. For example, weather changes can trigger the onset of seasonal senescence in annual plants, thereby increasing the mortality of butterfly larvae that feed on them (Ehrlich et al., 1975). The interlocking of life histories, in nutrient flux for example, is well recognized by ecologists. However, it is unclear whether transgenerational ecological interactions constrain the plasticity of life history and the extent of its genetic fixation.

Ecological coordination may influence the relative distributions of genetic and epigenetic determinations of life history plasticity. The epigenetic determination of life history alternatives could be favored by climate fluctuations in temperate zones, which might select *against* fixation of genes for obligate semelparity. This trend is seen in many flowering plants that have reproductive options on conditional semelparity or vegetative propagation. The Pacific salmon and other teleosts that show obligate semelparity thus have only narrowly departed from the majority of fish that preserve options for iteroparity, even though certain populations may be virtually semelparous. The transgenerational aspect of ecological coordination clearly does not require complete semelparity, as shown by alewives, some of which survive to the next season, and by geographic variations in semelparity of shad.

Evolution of reproductive scheduling and reproductive modes may have initially reversible transitions between iteroparity and semelparity, and occasional emergence of parthenogenesis. Some species have the potential for either iteroparity or semelparity. Examples include flowering plants (Chapter 2.3.2), marsupial mice (Chapter 2.3.1.5), and anadromous rainbow trout (Robertson, Krupp, Thomas, Favour, Hane, and Wexler, 1961); and the geographic variations in semelparity of salmon and shad noted above. Other species show conditional parthenogenesis, as in the pea aphids in which parthenogenesis is under photoperiodic control (Chapter 7.4.3.2; MacKay, 1987).

An open question is whether parthenogenesis favors the shortening or elimination of the adult phase. Parthenogenetic species that are not dependent on sexual reproduction could mature precociously (neoteny) and escape their adult phase more readily than sexual reproducers because they would not be dependent on adult mating behavior. Several parthenogenetic species of arctic blackflies (*Prosimulium* and *Gymnopais*) also are aphagic and have reduced mouthparts, which gives them virtual independence of their environment for propagation (Chapter 2.2.1; Section 11.3.3). Other examples of parthenogenesis with short lifespans occur in species that kill their mother by hatching or digesting her from within (endotokic matricide; Chapter 2.3.3). It would be helpful to have a more complete comparison between parthenogenetic and bisexual species' lifespans over a range of different life histories and durations of reproductive schedules.

The loss of sexual reproduction and its replacement by various types of parthenogenesis has arisen numerous times from sexually reproducing lineages throughout animal phyla (Bell, 1982). Numerous examples show parthenogenetic populations and species in *Drosophila,* among other insects (MacKay, 1987; Templeton, 1983; Downes, 1965; Suomalainen et al., 1976; Bell, 1982). The nematode *Caenorhabditis* is almost exclusively a self-fertilizing hermaphrodite, with rare males. The absence of parthenogenesis in birds and mammals suggests a change in gender-related gene regulation during the Mesozoic or earlier, since scattered species of amphibia, lizards, and snakes are exclusively female (Bell, 1982). Barriers to parthenogenesis through DNA methylation patterns present molecular imprinting of parental genomes as observed in mice (McGrath and Solter, 1984, reviewed in Mays-Hoopes, 1989; Mays-Hoopes et al., 1990; Reik et al., 1987); the generality of this mechanism in inhibiting parthenogenesis is not known. Parthenogenesis can arise irrespective of whether the stem-line species was iteroparous (trematodes and lizards) or effectively semelparous because of their short season (arctic blackflies). Although parthenogenesis is not known to evolve from species that reproduced only vegetatively, this might happen in a vegetative reproducer that had lost its sexual phase through mutations impairing spermatogenesis, but without compromising genes for oogenesis.

The reversibility of parthenogenesis leading to the recovery of bisexual reproduction is shown by the occasional production of fertile males of parthenogenetic strains of *Drosophila* (Templeton, 1983). Moreover, reversion to bisexuality is indicated for natural populations of rock lizards (*Lacerta*), where rare fertile males can hybridize with sympatric unisexual species (Darevski et al., 1977). Another route might be taken by species whose sex determination is not based on a heterogametic sex chromosome, as in pythons and other Boidae, where an autosomal pair is thought to carry sex determinants (Jones and Singh, 1981). Although Boidae reproduce sexually, parthenogenesis is suspected in a few other snakes, for example, *Typhelina bramina* (Bell, 1982).

The scheduling of reproduction has important influences on the selection for or against senescence. Without doubt, the patterns of reproduction and mortality across the lifespan must be mutually adapted, for both sexual and vegetative (clonal, nongametic) species. Among those dependent on gametogenic reproduction (whether sexual or parthenogenetic), it is tautological that the onset of senescence with increased mortality is adapted to the duration of the reproductive phase. Were this not so at the population level, senescence would soon extinguish the species. Similarly, numerous examples describe how fecundity and generation time are trade-offs against the mortality rates of developing and young adults, to statistically determine the lifespan needed for population maintenance (Sibly and Calow, 1986; Reznick, 1985; Calow, 1979; Schaffer, 1974a, 1974b; Horn, 1978; Stearns, 1976).

As discussed in Chapter 1.5, selection against adverse effects would generally be weaker at later than at younger ages, because the reproductive contribution of older individuals to the population *must* diminish with age when their number is reduced by mortality, whether or not reproduction has declined by that age. Organisms that *increase* their reproductive capacity with age, however, are an opposite case, as represented in Williams' (1957) hypothesis that senescence should be slower and more strongly selected against in species that show increasing reproduction during their lifespan. This prediction is realized in the long-lived teleosts and trees whose fecundity increases with continued growth (Chapter 4.2) and in those birds that continue to increase efficacy as parents (Chapter 3.5; Table 3.2). This important issue is discussed further in Section 11.4.4.

The extent of senescence in sharks may become an important test of Williams' hypothesis. In contrast to large teleosts, sharks ovulate a fixed number of eggs at all body sizes. Histological studies of adult sharks in a few species indicate a finite stock of oocytes, as in mammals and lampreys; all germ cells enter meiotic prophase I and become primary oocytes before puberty (Chapter 3.4.1.1). Limited data suggest age-related decline in the numbers of maturing ova in ovaries of bat rays and could indicate exhaustion of a finite oocyte pool, since elasmobranchs are thought to cease *de novo* oogenesis. (Readers should recall that *de novo* oogenesis leading to the pool of primary follicles with oocytes is distinct from maturation of oocytes and follicular growth.) Although some sharks may live many decades (dogfish, 70 years), the maximum age achieved by even very large and late-maturing sharks appears to be less than by teleosts. Other specific changes of senescence may be evaluated in ongoing studies of sharks. Early data from ongoing studies also suggest that mortality rates of sharks may increase at later ages, which would be consistent with a gradual senescence.

The capacity for continuing *de novo* gametogenesis has important consequences to the adult reproductive schedule. The available evidence suggests that organisms with finite gametogenesis show more pronounced age declines of fertility than those with continuing gametogenesis in either gender (Chapters 3, 4). The different outcomes of finite versus continuing gametogenesis have not been explicitly addressed in population genetics approaches to the evolution of senescence, but are pertinent to the age-related changes in fertility that are widely presumed to underlie the declining force of natural selection against deleterious genes.

There is no example yet of genetic variations within a species that alter the timing of oogenesis outside of its characteristic stage(s). In *Drosophila,* experiments show that lines selected for reproduction at later ages have later onset of reproduction and longer lifespans (Chapter 6). The effectiveness of this selection may depend on the continuation of *de novo* oogenesis and spermatogenesis for several months in adult flies (Chapter 2.2.2.1). The limit of such selection in flies remains to be established and might ultimately involve mechanical senescence of irreplaceable parts.

Oogenesis during the adult phase varies widely among species. The many animals that acquire a limited lifelong stock (determinate number) of primordial germ cells and cease *de novo* oogenesis before maturation present extremes of lifespan, for example, humans versus rotifers. Female humans and elephants have the longest documented duration of fertility (up to 30–40 years after puberty) among species with a determinate number of oocytes and rotifers the shortest (< 2 weeks). Fish show both characteristics, from the determinate oocyte stocks of lampreys to the continuing *de novo* oogenesis of different orders, such as some short-lived teleosts like the medaka and apparently the long-lived flounder (Section 11.3.4). While teleosts produce numerous eggs that are externally fertilized with huge losses, elasmobranchs produce many fewer eggs that are fertilized internally. Most amphibia and reptiles continue *de novo* oogenesis throughout adult life, and thus do not have a determinate stock. However, birds are thought to be like mammals in their fixed oocyte stocks, though few species have been studied besides the galliforms that generally have short lifespans (Chapter 3.5). Chickens show major decreases in egg production, consistent with oocyte exhaustion.

A limited capacity for spermatogenesis is much less common than limited oogenesis. Species with a limited number of spermatids include the very short lived soil nematodes (e.g. *Caenorhabditis*), rotifers (Chapter 2), and probably others with fixed cell numbers.

These finite stocks will fundamentally limit the reproductive strategies possible. The capacity to store viable sperm is very great in some species, such as the social insects (Chapter 2.2.3.1), in contrast to mammals, which evolved during the same time but apparently never required this adaptation. Spermatogenesis is not determinate in any long-lived species that I know of, and most vertebrates maintain *de novo* spermatogenesis throughout adult life (Chapters 3.6.4.2, 8.5).

When gamete stocks are finite and irreplaceable, the extent of gamete wastage before release or fertilization may have major impact on the plasticity of reproductive schedules. The best example is follicular atresia, which causes loss of more than 50% of the original oocyte stock before puberty in laboratory rodents and humans (Chapter 3.6.4); lampreys and elasmobranchs show similar follicular atresia. The concurrence of determinate oocyte stocks and follicular atresia could limit the delay of reproduction that would still permit reproduction. In turn, the increased ovarian stock predicts a later age limit for the exhaustion of oocytes. For example, female elephants continue to bear, though at reduced rates, into the late fifties (Chapter 3.6.11.4), implying that their ovarian oocyte stocks may exhaust later than in humans. A time limit to the adult reproductive phase is predicted by the allometric relationships of oocyte stock and body size, since small mammals would be predicted to exhaust their fixed oocyte stocks sooner than larger species. Because atresia is modified by hormones and diet (Chapters 10.3, 10.8), the plasticity of reproductive scheduling, despite fixed oocyte stocks, is probably greater than it might seem.

The timing and extent of continuing gametogenesis in relation to life history could influence the rate of evolution. Environmental mutagens could have different impact on germ-line mutations in species where gametogenesis was complete during development than in those with continuing *de novo* gametogenesis during adult life. The germ-line mutation rate is generally thought to depend on the frequency of DNA replication (Chapter 8.5). Social insects with a high rate of oviposition might be an interesting case if *de novo* oogenesis persists into adult life. A highly speculative suggestion is that exposure to heterocyclic floral compounds brought back to the hive by worker bees could be mutagenic and promote natural variations.

The selectability of reproductive scheduling and the impact of its natural polymorphisms on lifespan were experimentally shown for *Drosophila melanogaster* by several laboratories (Chapter 6.3.2). Responses to selection were relatively rapid. Within fifteen generations of selection for fertility late in life, the peak fecundity was delayed and the lifespan extended by 15 to 30% (Chapter 6.3.2). While these changes are substantial, the time extensions (≤ 1 month) are small on an absolute time scale, and are minor by comparison with the longevity of social insect queens. Nonetheless, the success of selection studies in influencing mortality rates and lifespan supports the view that natural history traits can change relatively rapidly through selection of polymorphisms (e.g. Hairston and Walton, 1986; West-Eberhard, 1986). While the distribution of these variants or polymorphisms has been characterized in few natural populations (Chapter 6), this source of genetic variation is probably quite general in higher organisms, particularly the many plants and animals that show seasonal and annual options in their age at first reproduction (Chapter 9).

Genetic variants of neuroendocrine functions appear to be a major target of evolutionary change in life history. The success of the artificial selection studies of *D. melanogaster* depended on polymorphisms that altered the reproductive schedule and that most likely, therefore, act on neuroendocrine functions (data on hormones, etc., are not yet available for these populations). In the case of *D. mercatorum* (Chapter 6.3.3), the *aa* genotype that alters reproductive schedules appears to affect the levels of juvenile hormone esterase. The extent of polymorphisms that influence neuroendocrine functions in natural populations of other species is not known. The influence of *H–2* loci on reproduction and peptide hormone receptor responses in inbred laboratory mice (Chapter 6.5.1.4), however, suggests important effects from such polymorphisms. The major influence of diet restriction in delaying the onset of reproduction involves neuroendocrine mechanisms at the level of the hypothalamus and pituitary that may also be subject to genetic influence (this has not been investigated). Besides polymorphisms that influence the levels of enzymes or receptor responses, there may be others that modulate hormone responsiveness at the transcriptional level. Androgen regulation, for example, was apparently introduced to a homologue of the *C4*-complement gene in mice by the insertion of a provirus with a hormone-sensitive element that functions as an enhancer in the *Slp* (sex-limited protein) of mice (Stavenhagen and Robins, 1988).

Work is needed on the extent to which any postnatal differences among individuals are heritable, as distinct from traits that arise from stochastic epigenetic events during development. For example, individuals from the same litter of inbred mice show major differences in individual reproductive schedules that are attributed to hormonal influences from neighboring fetuses on neuroendocrine development (Chapter 9.3.4.1). An important example, not yet analyzed for genetic versus epigenetic differences, is the individualization of optimal clutch size in birds. Great tits (*Parus major*) have clutch sizes ranging from five to thirteen eggs; remarkably, individuals had the same clutch size each year (Pettifor et al., 1988). This 9-year study showed that, despite experimental interventions of adding or taking eggs, individual birds maintained their characteristic clutch number. The implied individual optimization of clutch size does not support the generality of the "cost of reproduction" hypothesis (Chapter 10.6), but does not discount its validity for species of other construction, like insects, as discussed in Section 11.4.5.

The known genetic variants of reproductive scheduling within a species are very modest compared to species differences, which show a vast plasticity of timing. Some species can complete their life cycles from fertilization to senescent death in a few weeks (rapidly senescing rotifers, flies), while others span decades (slowly senescing large mammals) and even exceed 100 years (bamboo species that senesce rapidly after many decades of vegetative growth). Moreover, small rodents with high mortality rates and short maximum lifespans are iteroparous. These huge differences must to a large extent be hereditary because they are characteristic of the species under such a wide range of environments. In any case, there does not seem to be a necessary link between the *duration* of the adult phase or the total lifespan and whether an organism is semelparous or iteroparous. Species that reproduce vegetatively by somatic cell cloning may be considered nonsenescent clones, although senescence may limit the individual's lifespan (Chapters 2, 4).

Age-related changes of germ-line mutations should differ among species that reproduce vegetatively as well as sexually (e.g. tunicates and vascular plants; Chapter 4). In such cases, somatic cell mutations could be introduced into future germ-line cells during vegetative reproduction. While Buss (1987, 143) argues that asexual propagules have a "genetic" age equivalent to the last sexually produced generation, the accumulation of somatic mutations could influence the mutations seen at the next production of gametes. As discussed in Chapter 8.5.4, plants show evidence for increased chromosomal abnormalities and lethal mutations during prolonged vegetative growth. So far as I know, this phenomenon has not been explored in animals that are capable of vegetative reproduction. A major question concerns whether species that segregate their germ lines early in development have different mutation rates or error correcting mechanisms than species that differentiate the germ line from somatic cells after vegetative replication in adults.

I note here an issue affecting hypotheses that link senescence and reproduction (Chapter 1.5). Vegetatively reproducing colonial species like tunicates (Chapter 2.3.5) often show mosaic somatic degeneration. Thus, the presence of somatic degeneration is not restricted to a particular mode of reproduction, whether sexual (mictic), parthenogenetic (amictic), or vegetative (amictic) (Table 4.4). The coexistent potential in numerous vegetatively reproducing organisms for somatic cell immortality through vegetative reproduction and for somatic senescence epitomizes the importance of differential gene expression in the genetic basis for the evolution of senescence.

11.4.4. Genetic Influences on Senescence

A major conclusion of population genetics theories is that the force of natural selection weakens with advancing age, simply because of the high overall mortality demonstrated for individuals in virtually all natural populations (Chapter 1.5). Even without age-related increases of mortality, diminishingly fewer individuals will survive to each next adult age group. Consequently, the genes of older individuals are diluted by the proportionately more numerous young of subsequent cohorts. This perspective suggests that genetic determinants of senescence are largely haphazard, being subject to selection only in regard to their interference with fitness as defined reproductively. That is, the adult life phase should last long enough to yield sufficient progeny to balance mortality, from birth through adult ages. Fluctuations in mortality rate are also considered in these models (e.g. Charlesworth, 1980). Fitness must be recognized as a highly complex aggregate of individual characters that are related to innumerable environmental parameters, including environmental fluctuations.

Genes that influence senescence have historically been discussed in terms of whether they are initially *neutral*, with delayed adverse effects that arise later. Alternatively, genes may accumulate in the germ line of populations because they are initially advantageous, with subsequent adverse effects manifested later in adult life, that is, *negative* or *antagonistic pleiotropy*.

Neutralism in genes, according to a common population genetics definition, requires that selection is not effective, such that a given character is not correlated with fitness, as defined by reproductive efficacy. For macroscopic functional or morphological characters that arise from alternate alleles, the extent to which selection is ineffective may depend on local conditions such as population size and on fluctuations in selection intensity (Reeve et al., 1990). Assessment of genetic neutralism in laboratory conditions is also problematic because the range of environmental fluctuations is necessarily curtailed.

Neutral alleles with delayed adverse effects hypothetically accumulate in a population without specific selection because they may not alter fitness except at later adult ages. Most discussions of selection for genes by their contributions to fecundity emphasize younger adults, which are proportionately the largest contributors to reproduction in most populations (Chapter 1.5; Rose, 1991). Observing that many genotype-specific diseases of inbred mice emerged at middle age, when reproduction was diminishing or had ceased, Haldane (1941) hypothesized that genotypes with delayed manifestations of harmful effects would not be strongly selected against. This type of genetic influence on senescence might arise from random mutations that had no effect on fitness until middle age.

Examples of neutral genes with delayed adverse effects in humans might include dominant genes for familial Alzheimer's disease that have no known early benefits to or adverse effect on their carriers up through middle age or later (Chapter 6.5.3.3). However, the dominant gene for Huntington's disease, a life-shortening neurodegenerative condition of mid-life, is associated with a slightly higher ($< 10\%$) fecundity (Albin, 1988), which suggests nonneutrality. Life-shortening hereditary disorders with late ages of onset are relatively rare (Rose, 1991), although familial Alzheimer's disease may prove to be the exception. Nonetheless, the lifespans and mortality acceleration rates that are characteristic of mammals, flies, and many other species with perceptible senescence suggest a strong hereditary basis that has not been defined for individual alleles.

Medawar (1952) suggested that genetic variations in fitness that are manifested *before* maturation affect the whole breeding population, whereas variations in fitness at later ages with few survivors have little influence. Senescence thus could evolve from the selection of traits that are favorable early in life, but become harmful later on. Moreover, a late-acting gene with favorable effects early in life would establish itself in a population, but much more slowly. Extending concepts from Haldane and Medawar, Williams (1957, 1966b) proposed the important negative or antagonistic pleiotropy hypothesis, that genes might nonetheless be selected for advantageous early effects that had harmful later characteristics.

As a hypothetical example of antagonistic pleiotropy, Williams (1957) suggested a mutation that was selected because it promoted calcification of bone during development (fitness enhancing), but also caused arterial calcification later in adult life (mortality promoting). Another example might be genotypes favoring the deposition of fat stores in children, which might be adaptive for fluctuating food supply but which would predispose to diabetes and other metabolic disorders in adults. The adaptive value of stored nutrients as a hedge for uncertainties in food availability is basic in allowing reproduction during fluctuations of food (Chapter 10.6). Long-delayed adverse consequences from blood lipids and other nutrients in mammals include the promotion of atherosclerosis. Genes that set the appetite level through the hypothalamic centers that control metabolism could be pacemakers of senescence. In certain mammals, appetite is evidently optimized for growth and reproduction in early adult life, rather than for long survival at later ages.

Evidence for antagonistic pleiotropy was obtained in artificial selection for increased lifespan on the basis of reproduction at late ages in *Drosophila melanogaster* (Chapter 6.3.2). Forward and reverse selection showed that fecundity at early adult ages is associated with shorter lifespans, apparently through segregating alleles. Other studies on *D. mercatorum* also indicate antagonistic pleiotropy for the genotype *aa*, which is associated with early maturation that favors survival during arid conditions, yet which shortens the lifespan (Chapter 6.3.3). The *aa* genotype is an interesting example because molecular genetic analysis shows involvement of a noncoding sequence inserted into 28S ribosomal RNA genes.

None of these studies on *Drosophila* characterized the acceleration of mortality, which is a fundamental measure of senescence (Chapter 1.4). Thus, the reciprocity between fecundity and statistical lifespan may be independent of effects on biological senescence and could be heavily influenced by mortality risks that are independent of age or senescence. Whether genes have delayed adverse effects on the statistical lifespan through overall mortality or through acceleration of senescence does not invalidate population genetics models about selection for total lifespan *per se*. Here the distinction between increased overall mortality rates and age-related mortality accelerations becomes important if the influence on senescence by natural selection is the objective of study.

Further studies are needed to examine negative pleiotropy in perennial species with extended reproductive schedules and low or negligible reproductive costs. Particularly for semelparous species that show the ultimate extreme in reproductive costs, delayed reproduction, by definition, increases longevity by delaying senescence. Modelling studies show that the impact of mutations on the age distribution in populations is sensitive to assumptions about the effects of mutations on mortality in different age groups (Williams and Taylor, 1987). For example, mutations causing *proportionate* increases in mortality across the lifespan have a different effect than mutations that cause the same *absolute* increment of mortality. The latter is predicted to yield semelparity.

Although senescence is generally thought to be an unselected trait, there may be circumstances when senescence might be actively selected for. The case for this possibility can be made for fish like the Pacific salmon, shad, and alewife that have high incidence of death at first spawning (Chapter 2.3.1.4.2). Their carcasses provide nutrients vital to the development of the next generation in the food chains of their oligotrophic spawning and nursery beds (Section 11.3.4.4). The examples of geographic variations in semelparity at spawning in shad and steelhead could be evaluated for the need to add nutrients to the spawning beds. The establishment of this pattern would appear to favor the accumulation of further germ-line mutations that shift these geographic variants towards obligate semelparity. Other examples might be sought in plants, which produce vast amounts of gametes that might enhance local nutrient availability for the growth of young plants. Specialized situations that select for senescence by eutrophication are probably uncommon in semelparous species and even rarer in iteroparous species, particularly those with perennial reproductive schedules.

The seasonal florescence of angiosperms also coordinates the life cycles not only of innumerable short-lived insects but also of animals in these food chains. This suggests another aspect of selection for lifespan with ecological consequences, since alterations in the life history of "pacemaker" organisms in trophic chains would impact adversely on trophically dependent species. While it has long been recognized that benthic invertebrate larval density is associated with plankton blooms, a mechanism was only recently found in certain sea urchins and mussels in which spawning is triggered by phytoplanktonic metabolites; the larval urchins and mussels thus are produced when phytoplanktonic food is available (Starr et al., 1989). The mollusk *Phestila* delays metamorphosis and extends total lifespan in accord with a chemoattractant produced by its prey, a coral (Chapter 9.2.2.1). Ecological interdependency thus might widely constrain the genetic variations in life history beyond those apparently unusual examples like the fish in which semelparity directly benefits the next generation.

To close this section, I review certain animals that show progressive age-related increases of fecundity and so far have shown no indication of senescence. Examples that continually grow with scaled increases in gamete production are lobsters, bivalves, and rockfish (Chapter 4.2). Hamilton (1966) presented a widely believed mathematical argument that senescence will "eventually creep in" in such life histories (Chapter 1.5.2). However, there is no evidence from the real world that this always happens. The counterevidence is given by the long-lived rockfish *Sebastes* and the clam *Mercenaria,* both of which maintain gamete production at advanced ages beyond the human lifespan (Chapter 4.2). Major efforts are warranted to characterize the incidence of tumors and other diseases often found during senescence, the extent of abnormalities in gametes, and the mortality rates. Even if senescence does eventually creep in, it will be important to population genetic models to learn what factors favor the slowest forms of senescence.

11.4.5. *Parsimony and "Leftover Lifespan"*

An open question is the stringency of selection for lifespan potential, that is, the degree of parsimony in selecting for the lifespan potential needed to satisfy the reproductive schedule in the face of general and age-group mortality (Chapter 1.5.1). It is useful to discuss this question in terms of the common assumption that useless somatic features must soon be genetically discarded because of limited energy that could be otherwise devoted to features of fitness that serve more immediate purposes. For example, in the disposable soma theory (Kirkwood and Cremer, 1982; Kirkwood, 1985), the energy allocated to somatic repair processes is postulated to be optimized according to the reproductive schedule. Species with statistically short lifespans and high fecundity are predicted to have less investment in somatic repair than those with low fecundity and extended reproductive schedules. Examples include mayflies and many other insects with short adult phases and little or no capacity for organ repair. Other examples interpreted by this concept in general body-plan, irrespective of life history, are the loss of eyes and pigments in cave fish (Barr, 1968; Wilkens, 1971; Sadoglu, 1967); the loss of flight in birds from different taxa, for example, in habitats with reduced predators (Diamond, 1986; Kosswig, 1948); and the trend for reduction of the number of bones in the vertebrate skull (Prout, 1964; Regal, 1977b). These and many other examples are viewed as "streamlining" or noise reduction during evolution (Regal, 1977b), or efforts to conserve wasted energy (Diamond, 1986; Lwoff, 1944).

However, the causes of specific deficits relative to other species may not conform to the rule of parsimony. Diamond (1986) reviewed studies of mutants in the bacterium *Escherichia coli* that failed to support the wasted energy hypothesis; these studies showed growth rates on media in which certain enzymes were not required for viability. Dykhuizen and Hartl (1983) compared auxotrophs requiring an amino acid with the wild type in complete media, while Andrews and Hegeman (1976) compared mutants that gratuitously made *lac*-operon proteins in the absence of lactose as a substrate. Contrary to expectations, growth of these genotypes was at least as good as the wild type. There was no sign that growth or viability was limited by "wasteful" production of proteins that served no known need. Another example is the presence of a tooth type (the second lower molar, M_2) in the lynx (*Felis lynx*) not found since Miocene felids; M_2 may also occur at low frequency in some other modern cats (Kurtén, 1963). This finding indicates the preservation of genes controlling odontogenesis that were not expressed for at least 5 Myr; all teeth evidently shared the same proteins.

The maintenance of allegedly unused or unneeded genes may also depend on multifunctional usage of proteins, as in the lens crystallins that have identical amino acid sequences to enzymes used in other tissues (Piatagorski and Wistow, 1989). Many multimeric proteins share common subunits, for example pituitary hormones, hormone and neurotransmitter receptors, and nucleotide cyclase complexes. These examples caution that many traits are interwoven at a molecular level such that an apparently useless characteristic may be an essential part of some other mechanism. Thus, the principle of parsimony may need more careful scrutiny than usually given.

What can be said about genotypes that allow potentially greater lifespans than usually observed in natural populations where so few survive to a second or third generation? Would not the principle of parsimony or minimizing wasted energy select against genotypes that were "over built," leading to unused adult lifespan? Supporting examples are the mayflies and other aphagous insects that have very short adult lifespans because of developmentally programmed defects of oral anatomy. Similarly, the body-plan of flies and their flying and fighting behavior limit adult lifespans through mechanical senescence (Chapter 2.2.2). In these species, little unused lifespan may be expected. Nonetheless, the example of similar body-plans in the far greater lifespans of social insect queens (Chapter 2.2.3.1) shows that behavioral modifications can overcome many apparent limitations and greatly extend the lifespan potential.

Moreover, species with other body-plans may have much greater potential lifespans than in nature. According to the limited data available, *Peromyscus leucopus* has a statistically short lifespan in nature, about the same as *Mus musculus*, such that few may survive to 1 year, or three generations (reviewed by Phelan and Austad, 1989); the data from these field studies, however, are not highly reliable and, for example, do not evaluate the extent of migration. In the laboratory, however, far greater lifespans are achieved. In particular, *P. leucopus* lives up to 6–7 years, or twice as long as *Mus*, and with correspondingly slower acceleration of mortality (Figures 3.11, 11.12; Appendix 1). There is a great need to have more comparisons of lifespan of species in the field and under protected conditions.

While there does not appear to be any requirement for the extra lifespan potential of *P. leucopus* in nature, the potential lifespan might be advantageous when conditions required increased duration of reproduction. Thus, the ordinarily latent potential for longevity might be selectively advantageous during ecological fluctuations from climate or infectious disease that opened new ecological niches. The periodic extension of reproductive schedules might thus purge some populations of genotypes for early senescence. Genotypes allowing rapid prolongations of life history may arise against a background of short adult lifespans, as in social insects, as well as in many vertebrates where adult attrition is high but evidently not at the expense of unused lifespan potential.

It is unclear whether extended reproductive schedules generally occur at the expense of capacity for enhanced early reproduction. Although this result was obtained by selection for *Drosophila* genotypes (Chapter 6.3.2), there are mouse strains with genotypes constructed for other purposes that showed the combination of greater fecundity throughout their lifespan and longer lifespan (B10.RIII, Lerner et al., 1988; Chapter 6.5.1.4). Some iteroparous species like mice show few reproductive costs under favorable conditions that would limit the reproductive lifespan by the number of progeny. A provisional conclusion is that, despite short lifespans in nature, it does not always follow that the population must accumulate genotypes that predispose to senescence in ages immediately after the most probable ages of reproduction.

An open question concerns the occurrence of senescence in natural populations (Chapter 4.5.4). Many examples may turn up for iteroparous species and for species that are conditionally semelparous. Menopause or its equivalent is reported in at least four mammalian orders in natural populations (Chapter 3.6.4.1). It will also be important if there are more examples of species showing obligate rapid senescence with a postreproductive phase in nature. Sightings of senescent *Aplysia juliana* in nature (Hadfield and Switzer-Dunlap, 1990) indicate this possibility, although this species does not live long in the laboratory (Chapter 2.3.1.3). Soil nematodes could yield examples of this in species with small numbers of ova that might be shed early. Also, it is worth inquiring whether long-lived parasitic nematodes (Chapter 4.2.2) remain alive in their hosts after egg production ceases. Further examples of postreproductive individuals may require rethinking the premise that nonreproductive individuals are strongly selected against because of their competition for resources. The possible social role of menopausal females in populations of pilot whales and elephants needs further study. On the other hand, social insects give a clear counterexample, since queens are killed quickly after they become infertile, it is thought, upon exhausting their sperm supply (Chapter 2.2.3.1). A strategy for finding postreproductive individuals might consider the role of older individuals in various nurturing or leadership roles, particularly in iteroparous species that reproduce perennially.

In some cases, the characteristics needed for survival to maturation may also confer long lifespan potential. Powerful enzymatic and metabolic antioxidant mechanisms to combat free radical damage to macromolecules and lipids must be essential in the evolution of prolonged development and prematurational phases lasting 5 to 10 years, and sometimes much more. Examples from earlier chapters include cicadas, tarantulas, rockfish, alligators, fulmars, and humans. Most data show that the efficacy of these mechanisms remains quite constant after maturation in long-lived species (Chapters 7, 8). I suggest that genes for molecular repair will be expressed at high levels in radiations that have frequently experienced selection for lengthened time to maturation or extensions of the adult reproductive schedule. Such genes are important in preserving options in reproductive strategies, whether the germ line is segregated during early development or just before reproduction.

In organisms with early segregation of the germ line, the gamete stem cells *must* be protected against damage to the nuclear and organellar genomes. This protection particularly applies to oocytes that remain in meiotic arrest for decades in many mammals. In late segregation of germ cells, particularly in long-lived plants that may delay reproduction for more than 100 years (Chapter 2.3.2), protection mechanisms for somatic cell genes must be important, although somatic cell selection might eliminate some abnormalities. In both early and late segregation of germ line of long-lived species, molecular repair is thus necessary. There is no evidence that these repair genes are repressed in other somatic cells, for example, those in erythropoiesis and lymphopoiesis whose stem cells show no impairments at late ages in rodents (Chapters 7.6.1, 7.6.2). If, as seems generally true, the genome of somatic cells contains the same genes for molecular repair processes as are found in dividing germ cells (proliferating oogonia and spermatogonia) and in resting oocytes, then age-related damage to somatic cells would arise only if the repair genes were not expressed.

Moreover, the evolutionary record for corals and vascular plants clearly shows numerous genera with the capacity for vegetative reproduction from somatic cells that survived as well as those that reproduced sexually (Figures 11.2–11.4). This documentation strongly indicates that the maintenance of somatic cell repair at levels consistent with preserving genomic totipotency does not reduce fitness. These considerations form a new argument that eukaryotic somatic cells have been constitutively protected against time-dependent molecular damage since early in the evolution of multicellular organisms.

In conclusion, parsimony need not eliminate unused lifespan because deeper evolutionary strategies may favor the maintenance of temporal flexibility and because allegedly unneeded features may be nested integrally within other vital functions.

11.5. Summary

This chapter has assembled information that leads to a new set of questions about evolutionary trends and constraints on scheduling of life cycle phases. In contrast to the major stabilization in body-plans since the Cambrian, there seems to be no limit to the varieties of temporal organization in life histories, either among genotypes in the same or different species, or even within a genotype, as represented by social insects. Thus, evolution seems free to adjust the rates of senescence, whenever its onset during life history.

The extensive manipulations of the phenotypes of senescence through diet, temperature, and hormones that were described in Chapters 9 and 10 may be viewed as revealing potential variations in senescent phenotypes that give a basis for the selection of particular genotypes from which adaptive changes could be selected. Although selection is based on genotypes, the range of phenotypic variations in response to different environments throughout life must also be under genetic influence. The rapidity with which dipteran reproductive scheduling and lifespan may be selected in the laboratory and in field populations (Chapter 6.3) suggests that much of senescence is superficial and open to genetic manipulation. In general, the evolution of life history phases shows no constraint by calendar time.

The evolutionary record and taxonomic distribution of vegetative reproduction (Table 4.3) indicates its persistence since the Precambrian. Moreover, the capacity for vegetative reproduction, as well as presumably sexual modes, does not adaptively penalize the scleractinian corals, since the aclonal species have a slightly shorter average evolutionary persistence. Certain somatic cell lineages in these organisms may have long been exempted from clonal senescence; the most secure examples are in very long lived plant clones (Chapter 4.4.2; Table 4.2). The fossil record for corals (Figure 11.2) and angiosperms (Figure 11.6) indicates dynamic shifts in the balance between sexual and clonal species that might be looked for in other taxa. The evolutionary persistence of vegetative (clonal) reproduction from somatic cells in plants and animals at all grades of organization cap the argument that differentiation does not inevitably lead to clonal senescence. Thus, many lineages of eukaryotic somatic cells are no less immortal than the master germ lines of the Precambrian.

Several examples in this chapter bear on earlier discussions of evolutionary theories of senescence that are based on population genetics models (Chapters 1.5, 4.5, 6.6, 10.10). The rodent strains with different organ-specific diseases during senescence could be considered illustrations of the concept that age-related dysfunctions are weakly selected for, and that senescence in any species or taxonomic group should therefore present a great variety of dysfunctions. However, many age-related changes in mammals are found in corresponding fractions of the lifespan, across a thirtyfold range (Table 3.3). The sharing of certain features of senescence in mammals despite major evolutionary changes during their 50- to 100-Myr radiations raises issues for population genetics models for senescence (Chapter 1.5). These common features might have evolved independently and convergently from some yet unrecognized selection mechanisms that operate independently of absolute time in the reproductive schedule. Alternatively, the genomic determinants may have been primitive traits that were present from the beginning of the radiations and were protected from extensive divergence, possibly because of their role in genetic networks. Either possibility merits much further thought.

There are, however, numerous examples in which evolutionary changes in one characteristic appear to be selected for their relationship to other evolving characters. The example of delayed tooth eruption in elephants discussed above is the epitome of this process, in which a suite of characters (e.g. the shape and developmental schedule of teeth; the jaw and skull shape) evolved rapidly as the result, it is thought, of an adaptation for changes in the type of food that was available. The acquisition of delayed eruption of molars that became established more than 5 Myr ago in ancestors of the African and Asian elephant thus coherently extended the lifespan to about double that of other large ungulates. It will be interesting to learn if other features of senescence are also postponed by several decades in elephants.

Another exception to the generality of decreased coherence during senescence is the remarkable concentration of semelparity in fish that are migratory, or whose ancestors were (Section 11.3.4.4). Because semelparity outside of the migratory fish is rare, I suggest that the suite of characteristics associated with migration (elevated corticosteroids; cessation of feeding; spawning in small streams where the next generation must grow) favored death at spawning and in some of these cases became genetically fixed. Another factor may be the contribution that dead Pacific salmon and alewives make to the nutrition of the next generation (Section 11.3.4.4). Whether this was important in fixation of the genotypes that lead to obligate semelparity in other migratory fish is unknown. Many fish that are characteristically semelparous can also slide towards iteroparity, as shown by the geographic gradients of semelparity in American shad and in the postspawning survival of precociously mature parr of several Pacific salmon (Chapter 2.3.1.4.2; Figure 2.14).

The molecular basis for semelparity or iteroparity has not been identified in any example. Only minor mutational changes might be required, for example, that might alter corticosteroid-receptor interactions such that cell death programs are activated to varying extent. Altered balance of semilethal genes in polymorphic combinations could also be easily imagined. The migratory habit itself need not be lethal, as in the examples of the unexceptionably iteroparous sturgeon that show elevated corticosteroids and also cease feeding (Chapter 3.4.1.2.1). The distribution of semelparity in fish, from the agnathan lampreys to the salmon, suggests that this trait arose numerous times during fish evolution, and recently in *Oncorhynchus*.

Mayflies give one of the few examples of semelparity through rapid senescence and sudden death that extends to all in their taxonomic order (Ephemeroptera). The Mesozoic ancestors of mayflies were well equipped as adults with mouthparts and digestive tracts, unlike their descendants. The geological age is unknown when developmental changes occurred that led to anatomical deficiencies in adults. Adult aphagy without major anatomic deficiencies is found in many Diptera, some of which show geographic clines. Thus, adult aphagy that causes rapid senescence and sudden death has arisen many times and for many reasons during evolution, both in the past and recently. A complex question is the role of rapid senescence in ecological coordination, which I suspect will reveal a greater coherence of adaptive mechanisms than apparent in the senescence of individual organisms.

In closing this long chapter, I suggest two general hypotheses: (1) The aspects of senescence that are most open to experimental manipulation are clues to the most evolutionarily labile changes in adult phase life history in a given radiation. (2) The limit to the plasticity of senescence is determined by the body-plan, and is least for body parts that consist of irreplaceable molecules and greatest for parts in which the molecules undergo continuous turnover, that is, molecules that are under active gene regulation.

12

Genomic Mechanisms in the Biology of Extended Time

12.1. Introduction

This closing chapter recapitulates discussions of the book's root questions: what is the nature of mechanisms that regulate the potential lifespan and what is the genomic basis of these mechanisms. The vast reach of this subject and the scarcity of crucial information on many issues preclude a global synthesis. Even so, certain general features of genomic mechanisms in longevity and senescence can be discerned, while some mechanisms can be discounted.

After reviewing the agenda from Chapter 1, I summarize material from all chapters that concerns the relative roles of endogenous versus exogenous factors and various mechanisms involving feedback and cascades. Then I consider genes that influence the lifespan, either through their role in specifying characteristics that constrain the lifespan or that more directly set upper limits to the potential lifespan. A wide range of organisms have been discussed to identify phenomena that are not apparent from studies of the most popular models, which are largely short-lived rodents, flies, and nematodes.

The evidence assembled challenges the assumption that the processes of senescence require pervasive deteriorative changes that progress inexorably throughout all parts and functions of the organism, from molecules to behavior. On the contrary, most species show relatively few proximal causes of death during senescence, whether senescence occurs rapidly or very slowly. In many life species, senescence involves dysorganizational consequences that emanate from the physiological or organ function level, rather than through diffuse cellular or molecular degenerative changes that spread "upwards."

Senescence in association with relatively few causes is also consistent with the possibility that other organisms have negligible senescence. Species and even populations can vary enormously in regard to the duration of development and the adult lifespan and phenotypic variations. This extensive plasticity of age-related phenotypes is shown by environmental and experimental manipulability, and by variant developmental schedules between populations and between species. There is little evidence that the lifespan is generally set by molecular or cellular mechanisms that are intrinsically time-dependent. The book closes with a discussion that the plasticity of age-related changes across the lifespan can be considered as the *biology of extended time*. I thus hope to have extended gerontological thought beyond its usual focus on biological, medical, and social phenomena of late-in-life involution to basic issues that are central to developmental, evolutionary, and reproductive biology.

12.2. The Agenda Revisited

The opening agenda (Chapter 1.1) posed issues about genomic influences on the phenotypes of senescence and the potential lifespan. These matters pertain to gerontological theory and experiment, but also to specific physiological and cellular mechanisms in natural selection for life history schedules.

(a) *How valid is the lifespan as a measure of the rate of senescence?* The *total* lifespan, from egg to the end of the adult phase, is meaningless as a measure of the rate of senescence, because many life histories include prolonged developmental or juvenile phases that dwarf the duration of the adult (postmaturational) phase. Moreover, the mean or maximum duration of the adult phase in natural or captive populations cannot resolve contributions from environmental dangers, as distinct from age-related accelerations of mortality rates that, provisionally, are a more fundamental assay of senescence. The maximum lifespan is especially vulnerable to artifacts from sampling strategies and migration. As an alternative to lifespan in characterizing senescence, I calculated the mortality rate doubling time (MRDT) for species that represented various modes and rates of senescence (summarized in Appendix 1). As may be calculated by either the Gompertz or Weibull models (Chapter 1.4), the MRDT estimates of the acceleration of mortality that accompanies senescence in many populations facilitates comparisons, because the MRDT changes in the same direction as the length of the adult phase.

The generality of simple power function models for mortality rate changes with adult age is not yet clear, although examples of good fit to the Gompertz or Weibull models can be found throughout the animal kingdom, from the most short lived aphagous rotifers and insects (Chapter 2.2) to very long lived birds and mammals (Chapters 3.5, 3.6). Very long MRDT in the range observed for human populations are indicated for marine birds that have not shown reproductive declines during observations over 20 to 30 years. An empirical basis for mortality models with delayed onset of mortality accelerations is the delayed mortality rate increases in plants and animals that delay maturation for one or more decade and then die after their first and only reproductive season (Figure 1.8).

Other natural populations give no indication of senescent deterioration according to the absence of age changes in mortality rates (Chapter 4.2). It is an open question as to whether populations under more protected circumstances would eventually show mortality rate accelerations. The apparent stability or mortality rates in some populations of rockfish (*Sebastes*) for at least 100 years (Chapter 4.2.1) suggests a very gradual or even negligible process of senescence. Data are insufficient to evaluate the mortality rates of some very long lived bivalves and turtles. The progressive increases of gamete numbers with body size in long-lived lobsters, bivalves, and certain teleosts is strong provisional evidence for the negligible senescence. However, we do not know if the gametes produced are as viable as those made by younger and smaller individuals. Major age-related increases in abnormal gametes have been shown only for mammals, which have finite oocyte stocks.

The relationships of age-related changes in biochemical and cellular functions to mortality rates is in general unclear. The case for relationships of biochemical-cellular changes to lifespan may be strongest in connection with particular age-related diseases. Attempts to find correlates between biochemical and cellular characteristics of young adults and the species lifespan in cross-species comparisons have so far been frustrating. Similarly, the formerly well-regarded correlations between metabolic rate, body size, and lifespan have been recently reevaluated and found lacking in comprehensive analyses of mammals from different orders (Chapter 5). Further work is needed to evaluate the two types of mortality rate coefficients: the initial or age-independent term (IMR) or separable coefficients which it aggregates (Equation [1.1]); and the accelerations of mortality (MRDT). The relationship between mortality rate coefficients and specific molecular, cellular, and physiological characteristics is obscure. Meanwhile, the MRDT seems to give more information about the rate of senescence than does the total lifespan.

(b) *When does senescence begin?* This question leads us to deal with the great duration of larval or juvenile stages in numerous animals and plants. Examples of extended juvenile phases are the 17-year life cycle of slowly growing cicadas with adult phases of 6 weeks that are < 1% of the total lifespan (Chapters 2.2.3.3, 6.3.3) and perennial mono-carps, like certain bamboo and *Puya raimondii* that grow vegetatively for more than 100 years before flowering and rapid senescence (Chapter 2.3.2). Other species do not repro-duce for a decade or more after developing into free-living forms, as in tarantulas (Chapter 2.3.1.2). Extended prematurational phases occur in both semelparous or iteroparous life histories (Chapter 9.2.2). By comparison with related species of the same body-plan, the prematurational phases can last many times longer than the adult, reproductive phase. The data do not show general limits in the delay of reproduction by delaying maturation, in-dependently of diapause or hypometabolic states (Chapter 9.2.1).

The plasticity in the duration of prematurational phases gives insight about genes that may cause senescence. Organisms with very extended development *must* have powerful mechanisms that protect them from damage before maturation. These genetically coded mechanisms could include biochemical machinery for removing toxins or free radicals; repairing damaged cells or molecules; or, the capacity for cell proliferation to replace damaged cells or regenerate damaged body parts. In the face of evidence against general-ized loss of somatic cell genes throughout the lifespan (Chapter 8.2), somatic cell degen-eration during senescence in some species may be attributed to genetically controlled lim-itations on repair or regeneration. The extensive flexibility in reproductive scheduling that is shown by species comparisons in most taxa is further evidence for potentially very long lifespans, whatever the body-plan.

In the case of organisms with fixed numbers of gametes (e.g. the fixed store of ovarian oocytes (Chapter 3.6.4), reproductive senescence might be said to begin during develop-ment. Similarly, senescence in mayflies and other insects that are aphagous can be as-cribed to developmentally programmed defects in oral anatomy (Chapter 2.2.1). In these examples, selective changes in gene regulation during development are explicit determi-nants of senescence.

(c) *At what levels of organization do genes influence senescence?* Age-related dysfunc-tions can occur across the entire range of biological organization, in macromolecules, cells, organs, behavior, etc., all of which are specified by the genome at some time during the life cycle. The codon distribution in some long-lived proteins can, theoretically, influ-ence the extent of slow racemization, which, for example, varies among amino acids (Chapter 7.4.4). At the behavioral level, the regulation of the appetite by satiety centers in the brain is revealed by diet restriction studies in rodents to influence the progression of numerous age- related diseases (Chapter 10.3). Thus, age-related deteriorative changes can be under genetic influence at many and possibly any level of organization.

Because of links between reproduction and the onset of senescence in many species, certain neural and hormonal functions have key roles in the regulation of senescence throughout plants and animals. Besides the triggering of reproductive-related senescence in semelparous species, considerable evidence implicates neural and endocrine functions in numerous mechanisms of senescence in iteroparous mammals. This is discussed further in Section 12.3.

The evidence assembled here challenges assumptions by many gerontologists and pop-ulation geneticists that the processes of senescence require pervasive deteriorative changes that progress inexorably throughout all compartments and functions of the organism, from molecules to behavior. On the contrary, there may be few proximal causes of death during senescence, whether senescence occurs rapidly or very slowly. While it is premature to judge generally, senescence and morbidity in a large number of species appear to involve dysorganizational consequences that emanate from the physiological or organ function level, rather than through diffuse cellular or molecular degenerative changes that spread "upwards." Types of senescence that are associated with relatively few causes give further support to the possibility that some organisms have negligible senescence.

(d) *What is the relative contribution of selective (nonrandom) versus random changes in gene expression and other cell functions during senescence?* This question can now be refocussed on the structure of DNA and the types of mRNA that are present. Many changes in mammalian mRNA levels are selective, so that the expression of some genes may become more active, while other have decreased activity or remain unchanged in the same tissue. Some evidence for demethylation of DNA is found in humans and rodents, including demethylation of specific sequences that might be a factor in altered transcription rates (Chapter 7.2.4). The present data, though limited, do not show many shared changes in macromolecular biosynthesis in different tissues of mammals during senescence. In general, the few genes that have been studied in depth show selective changes in the extent and direction of change, some showing no change in transcription rates or mRNA abundance (Chapter 7.3). It is unknown whether this selectivity will be found in comparisons of transcription rates of the same gene in different tissues. Altered episomes during senescence in certain fungal genotypes (*Podospora* and *Neurospora*) also show selectivity for the DNA sequences that are involved (Chapter 2.4). The importance of selective gene regulation throughout development and cell differentiation is clear, but the role of dynamic changes of gene regulation in senescence has been described for a very small number of cell types in a few organisms. The general impression prevails that most differentiated cell functions persist throughout the adult life phase. For example, the oldest human tissues display cells that are morphologically indistinguishable from those in the brain, liver, or muscles of young adults.

What then about changes in the structure of the genomic apparatus of somatic cells, which might include mutations and other forms of DNA damage and which two generations ago radiation biologists hypothesized to be a mechanism in senescence (Chapter 8.2.1). Numerous attempts to document the accumulation of somatic mutations have failed to reach conclusions. The nucleotide adducts detected by Randerath that increase with age in nuclear and mitochondrial DNA occur at a very low level per cell over the lifespan, ca. < 1 per million nucleotides in humans and rodents (Chapters 8.2.2, 8.4.1). Another abnormality was found in muscle mitochondrial DNA structure, nonrandom deletions and insertions (Chapter 8.4.1). The impact of such trace changes is hypothetical, but might propagate errors as hypothesized by Orgel (Chapter 8.1).

On the other hand, there is evidence for spontaneous reversion of mutant genes in cells of genotypes with germ-line mutations that cause the absence of albumin (analbuminemic rat) or of vasopressin (Brattleboro rat) (Chapter 7.2.2; Figure 7.1). The mechanisms of reversion are unknown, and could involve recombinant events within nuclear DNA, or post-transcriptional or translational splicing. The mechanisms here will be extremely interesting to resolve, and could conceivably extend to the *deletion* of gene functions that have not been searched for on a cell by cell basis. The frequency of reversion appears to be much higher in the analbuminemic and Brattleboro rats than accounted for by the DNA changes mentioned above.

Proceeding to the near macroscopic level, chromosomal structural instability and abnormal metaphase distributions increase with age in several somatic cell types of mammals (Chapter 8.2.5). The incidence of abnormal karyotypes in short-term cultures of somatic cells increases severalfold or more with donor age, with substantial variations between species and gender. Germ cell karyotypes show characteristic and striking increases of abnormalities in female rodents and women, in which hyperdiploidies are the most common. The abnormal karyotypes suggest changes in the stability of the spindle apparatus (Figure 8.12). While a variety of random and of selective changes in genomic structure and functions are demonstrated during the mammalian lifespan, the relative importance of random versus nonrandom changes in DNA and chromosomal structure remains an open question.

(e) *How directly do genes operating during development or adult life specify senescence?* Genes may be *direct determinants* or may *set indirect constraints* on the lifespan. Direct effects are epitomized by the dominant gene for Huntington's disease that causes premature death in humans at middle age and by the genotypes that cause universal death after reproduction in most species of Pacific salmon (Chapter 2.3.1.4.2) and marsupial mice (Chapter 2.3.1.5). Genes that indirectly constrain lifespan include those determining the replacement of vital cells (e.g. neurons, at risk for a variety of degenerative changes; Chapter 3.6.10) or vital organs (insect wings, at risk for mechanical senescence; Chapter 2.2.2). The failure of complete mouthparts to develope in mayflies and certain species of other orders (Table 2.2) is another example of genetic constraints. Genetic constraints and controls on senescence are discussed further in Section 12.4.

(f) *How do species vary in reproductive senescence, including the total production of gametes and incidence of developmental abnormalities?* Reproductive age changes are important to the selection of genes that influence senescence. Continuing gametogenesis in adult life is found in some iteroparous fish and in invertebrates that show no signs of reproductive senescence (Chapter 4.1.2). Moreover, there is a positive scaling of gamete production with body size in some plants (Figure 4.3) and animals (Figures 4.6, 4.8) that have continued growth after maturation. It is unknown if the expanding production of ova with continued growth in these animals results from *de novo* oogenesis. Continuing gametogenesis is characteristic in vascular plants, which do not segregate their germ lines (Chapter 4.1.1). Mammals generally show age-related increases of birth defects with maternal age in association with imminent exhaustion of a fixed oocyte pool; paternal age contributions and the diminution of spermatogenesis are much less in mammals. Limited gamete production, when it occurs as a species characteristic in either gender, is considered in population genetics models to be an important factor in the accumulation of mutations with delayed adverse effects that contribute to senescence (Chapter 1.5).

(g) *How much plasticity in senescence is there from environmental (i.e. nongenetic) sources?* Extensive variations are seen in the patterns of senescence between populations of the same genotype, as well as in outbred populations. Many examples show modulations in rates of changes, e.g. the slowed rate of ovarian oocyte loss during caloric restriction (Chapters 10.3.2.8.2, 10.8). The relative rarity of gross coronary vessel occlusions in the hearts of nonagenarians from a Mayo Clinic study (Chapter 3.6.2) cannot now be resolved as heredity or environment, but underlines the great individuality of age-related pathology. These individual variations indicate that there are few limitations on the lifespan and schedule of senescence. Plasticity in senescence is seen virtually everywhere and is a strong argument against the intrinsicality of any specific phenotypes of senescence. The limits of plasticity in any phenotype of senescence, however, are unclear, and there is no basis at present for concluding that somatic senescence could generally be eliminated if all adverse environmental factors were eliminated. Ultimately, somatic senescence could be minimized or eliminated, however, by the regeneration of damaged molecules, cells, or organs.

Nonetheless, some species may have lifespans that result entirely from exogenous factors, exemplified by long-lived hardwood tree stands that become at increasing risk to damage from storms as they grow larger (Chapter 4.3). Examples with short lifespans attributable to exogenous (ecological) causes are the gastropods *Lottia* and *Nucella*. The plasticity of lifespan and senescence in the gastropods can be readily studied under protected conditions.

(h) *How universal are age-related degenerative changes at the organismic, cellular, and molecular levels as may be judged by species comparisons?* The universality of senescence is commonly presumed by many clans of biologists. A body of irrefutable evidence shows the prevalence of limited potential lifespan in innumerable species throughout eukaryotes that manifest organismic senescence and accelerations of mortality rate. Patterns of senescence, while subject to numerous environmental influences, nonetheless persist in flies, soil nematodes, and mice under the best environments that can be contrived. However, while lifespans of individual organisms are statistically indeterminate because of environmental hazards even if mortality did not accelerate, some species do not show a perceptible degree of senescence. Provisionally, such species are characterized as showing negligible senescence (Chapter 4), as judged by the criteria of no age-related increases in mortality rate nor manifestations of age-related dysfunction from endogenous causes. Whether some species altogether escape senescence is an open question.

Chapters 2–4 described the great diversity of somatic changes during senescence that varied widely between species, and in some cases between populations. Rapid senescence and sudden death from many different causes terminates the adult phase of many species, across a 10,000-fold range of total lifespans. Rapid senescence and sudden death usually results from highly select changes in particular systems of molecules or cells that do not compromise most other cells (Chapter 2). Intermediate in this spectrum of senescence are mammals which show gradual senescence, with genotype-specific pathophysiological changes in particular organs (Chapters 3.6, 6.5). The general occurrence of intracellular lipofuscins or aging pigments during the lifespan of many invertebrates as well as mammals (Chapter 7.4.6) suggests that nonetheless there may be generalized cell changes in organisms with slower senescence, that in this example of lipofuscins probably result from free radical–mediated lipid peroxidation. Another universal is the exponential acceleration of mortality rate in organisms with demonstrable pathophysiological senescence.

Certain evolutionary lineages show strong trends for similar mechanisms in senescence that suggest phylogenetic trends. Semelparous fish, particularly Pacific salmon and other salmoniforms, are migratory or had an evolutionary background of migration (Table 11.1). Other close relatives, however, are perennial spawners. Other taxa showing major variations of life history are the anguillids (Chapter 2.3.1.4.3) and cephalopods (Chapter 2.3.1.3), each having semelparous and iteroparous representatives. Aphagy through reduced mouthparts is characteristic of present mayflies, but not their ancestors, and is sporadic in other insects (Table 2.2). The reduced mouthparts of all mayflies and the lampreys with terminal aphagy (Chapter 2.2.1.1) suggest monophyletic descents from ancestors with these traits. There is no general explanation for monophyletic versus polyphyletic distributions of semelparity in radiations of similar body-plans. The aphagous mermithid nematodes with long adult phases (Chapter 2.2.1.3) suggest that even gross anatomic deficiencies present no barrier to the evolution of extended lifespans. Other comparisons also show that the body-plan does not limit the rate of senescence or the pathophysiological mechanisms of senescence. The years-long lifespans of insect queens show that modest changes of neural and endocrine mechanisms (Chapter 2.2.3.1) can extend the lifespan far beyond the norm for other species of the same body-plan.

Mammals present several puzzles. First, how can age-related diseases vary so widely between populations in humans and between inbred strains in mice (Table 6.1) without altering the species-characteristic MRDT (acceleration of mortality with age). Another puzzle concerns the major discrepancy that is emerging from studies on age changes in protein synthesis: the evidence abundantly indicates slowed protein synthesis in rodents, from studies of several organs, but several studies have failed to find decreased whole body protein synthesis in healthy elderly men (Chapter 7.3.1.1). Studies of biosynthesis of particular proteins in several cell types are needed in a range of mammals and birds with different lifespans.

Many population genetics arguments *a priori* exclude species that reproduce vegetatively (Chapter 1.5). A theoretical concern may be that delayed differentiation of germ cells from somatic cells will permit a fundamentally different type of senescence than in those organisms (the bulk of species) in which the germ line is segregated early in development (Weismann's classic model of the senescent soma versus the immortal germ line). However, senescence with death of the individual clearly occurs in many plants and animals with delayed segregation of the germ line from somatic cells in the adult (Chapter 4.3.3). Arguments that a disposable soma is favored by ergonomic parsimony in evolution (Chapter 1.5.2) are not supported by evidence of the persistence of clonal reproduction for several hundred million years in corals (Chapter 11.2) and vascular plants (Chapter 11.3). These species show no penalty for expressing the genes that maintain somatic cell potentials for immortality by differentiating into germ cells.

Whether any organism escapes senescence has great pertinence to population genetics theories of senescence, for example, predictions of an inexorable accumulation of germline mutations in the population that have delayed adverse effects (Chapters 1.5, 4.4.3, 11.3.4). Organisms with very gradual or negligible senescence thus challenge the generality of these conclusions from population genetics theory. Time will tell if these prospective examples of animals and plants do indeed escape somatic senescence.

12.3. Exogenous Factors, Endogenous Factors, and Cascades

Mechanisms in senescence can be categorized according to the role of *exogenous* (i.e. environmental and ecological) versus *endogenous* (i.e. intrinsic) factors. Collectively, these mechanisms represent the physiology of senescence, from molecular to behavioral levels of organization. Within an organism, endogenous factors can have influences at the different levels of organization. This simple typology gives a useful framework for discussing diverse age-related phenomena, especially the hierarchical and cascading physiological changes that often underlie age-related changes in mammals (Finch, 1976a). The best-known examples of neurally or hormonally dependent rapid senescence are given by Pacific salmon, marsupial mice, and flowering plants (Chapter 2) and by many senescent changes in mammals, particularly those associated with ovarian follicle depletion. There is every reason to expect many further examples of neural and humoral influences on senescence in other taxa.

At the *organismic* level, numerous examples show how seasonal variations in energy (light and temperature) trigger senescence (monocarpy) of whole plants. Some semelparous angiosperms and some teleosts acquire extended iteroparous (polycarpic) reproductive schedules under slightly different ecologic conditions. For example, conditional semelparity in American shad varies systematically with longitude (Figure 2.14). Marsupial mice exemplify a type of senescence that is indirectly coordinated by seasonal light and temperature cycles. Social isolation of individuals prevents their death during the seasonal die-offs (Chapter 2.3.1.5.), showing that in this case, seasonal triggers for rapid senescence are transduced to the behavioral environment. The potential for iteroparity in marsupial mice is less clear than for angiosperms. Yet both examples show strong exogenous causes of rapid senescence. Seasonal cycles of solar energy and climate are clearly a grand exogenous force that has selected for and continues to regulate the seasonal senescence and other life history traits of flowering plants and their associated insects and soil nematodes, to mention but a few. Throughout the larval, juvenile, and adult phases of species from most phyla, the nervous system has a crucial role in regulating the life history events, so as to be compatible with the physical and population (social) environment. However, there is no evidence that the nervous system contains explicit pacemakers for senescence in any species that operate independently of the external environment.

Carcinogens are another exogenous influence. Many abnormal growths of humans have been attributed to carcinogens like those from tobacco. Numerous other environmental toxins that are by-products of civilization may be suspected to have insidious impact on senescence, but these are much less known. Another more pervasive danger is skin cancer through cumulative exposure to UV from sunlight over extended time. Besides cancer, other age-correlated changes in the human skin are associated with cumulative effects of sunlight (senile keratosis, Chapter 3.6.11.3). In other species besides humans, the role of environmental factors is unclear in abnormal growths of internal tissues during senescence. However, some tumors occur in laboratory rodents from endogenous causes entirely, for example, the pituitary tumors of female rodents during reproductive senescence (Table 10.7).

Nutrition also has a major role in the phenotypes of senescence as shown by the effects of caloric restriction in extending the adult lifespan and in delaying numerous age-related changes and diseases, including slowed transcription of several genes in the liver, delayed onset of kidney dysfunctions, and reduced abnormal growths in rodents (Chapter 10.3). Life-prolonging effects of caloric restriction are a robust manipulation of senescent phenotypes in numerous animals. The mechanisms are not well understood and could differ fundamentally between taxa as physiologically divergent as mammals, nematodes, and insects. Mechanisms range from the influence of diet on protein glycation via blood glucose to influences on neuroendocrine functions. Insects with different degrees of aphagy as adults are a special case. In contrast, some species may have lengthened lifespans with access to extra nutrients, as indicated for ferox salmon (Chapter 3.4.3), and certain ticks and flatworms (Chapter 3.3.1.3).

Mechanisms in physiological functions show age-related changes that can be described in terms of hormones or other regulators that influence functions hierarchically. Major examples include the role of ovarian follicle exhaustion and ovarian estrogen loss in neuroendocrine responses (altered secretion of LH, FSH, and GH during reproductive senescence; hot flushes; Chapters 3.6.4.1, 10.4.2); the role of kidney disease in altered mineral metabolism of senescent rodents (increased parathyroid hormone and ectopic calcium deposits (Chapters 3.6.5, 10.3.2.2); and the role of blood sugar through protein glycation in vascular lesions and neurological lesions of diabetics (Chapters 7.4.2, 10.3.2.3). The interventions into the rapid senescence of flowering plants, Pacific salmon, marsupial mice, and octopus (Chapter 2) all demonstrate cascading effects through physiological networks that represent extrinsic influences on target cell senescence.

The role of local intrinsic mechanisms in age-related changes of physiological hierarchies is not generally established. Hypothetically, changes in a regulatory system could arise from physiological circuit properties that did not require deleterious changes in any subcomponent, though the outcome may be deleterious. Progressive auto-antiidiotypic responses might lead to age-correlated autoimmune pathology (Chapter 3.6.6) and could represent deleterious outcomes from regulatory changes. The long-lasting effects of glucose and steroids on transcription and chromatin structure of certain target cells of adults (Chapter 7.7) constitutes a memory effect that might be an important component in senescence.

At the cellular level, many age-related changes are attributed to altered levels or time patterns of secretion in hormones, neurotransmitters, or other physiological regulators. The numerous cases of this include the atrophy of estrogen-dependent target cells as a consequence of low blood estrogens after menopause; the delayed induction of tyrosine aminotransferase during fasting in rodents; and the expanded dendritic arbor of granule neurones in the hippocampus during Alzheimer's disease which are associated with degeneration of neuronal afferents from the entorhinal cortex (Chapter 3.6.10).

While these examples concern frank deficits, other age-related changes may involve subtle alterations in the frequency of episodically secreted neurohormones. Such examples are predicted by the influence of the frequency of secretions of hormones on target cell output, for example, the gender differences in liver prolactin receptors that derive from gender differences in growth hormone secretory patterns (Norstedt and Palmiter, 1984; other examples in Chapter 3.6.9). The age-related prolongation of blood lipid and glucose elevations after a meal (Chapter 3.6.8) might influence other functions through variant timing, as well as providing increased substrates for atherogenesis and glycation of proteins.

Local intrinsic causes of degenerative changes in cells *in vivo* are speculative. Consider the evidently unimpaired capacity of hypothalamic neurons to secrete GnRH and the absence of loss in these cells, except perhaps at advanced ages (Chapter 3.6.10). Thus, differentiation into a nondividing neuron does not lead to predictable or programmed dysfunction during the mammalian lifespan. On the other hand, numerous other types of neurons show age-related impairments. In several examples, the cell body RNA varies inversely with the amount of neuromelanin and aging pigments (lipofuscins) in some neurons (Chapter 7.4.6; Figure 7.18); it is unknown if this change merely represents the displacement of cytoplasm by these pigment granules, or if pigment accumulation impairs transcription of ribosomal or other genes. The consequences of pigment accumulation are unknown and may differ for neuromelanin which occurs in fewer cell types than lipofuscin. Since these different extents of pigment accumulation *directly* result from differentiation, the intrinsicality of different extent of pigment accumulation is ultimately an outcome of selective gene expression.

The age-related loss of ovarian oocytes, while irreversible in the adult, was acquired during gonadal differentiation. Ovarian oocytes are more vulnerable to spontaneous death (atresia) than neurons and myocardial cells, among other postmitotic cells. Removal of the pituitary gonadotropins through hypophysectomy greatly slows, but does not eliminate spontaneous death of oocytes (Figure 10.23). These examples lead to the conclusion that the extent of intrinsicality in cell senescence should be viewed as the outcome of cell differentiation, no less so than the choice between finite and infinite proliferative capacity of somatic cells in lower species that reproduce vegetatively (Chapter 4.3). This viewpoint focuses on the mechanisms of selective gene expression which create cell phenotypes with varying susceptibilities to senescence.

The molecular level shows other presentations of intrinsic and extrinsic senescence. The spontaneous age-related racemization of L-aspartic acid and other amino acid residues is a thermodynamic (non-enzymatic) phenomenon that is driven solely by local temperature in dentine, lens crystallin, and other long-lived proteins (Chapter 7.4.4). Since the amount of each L-amino acid is determined directly by genetic coding in nearly all proteins upon their biosynthesis, this phenomenon is the purest example of an intrinsic age-related change that I know. Even so, the manifestation still depends on a parameter extrinsic to these molecules, namely the body temperature.

Another cumulative molecular change is glycation, in which collagen and many other proteins with long lifespans spontaneously form adducts with blood glucose in mammals (Chapter 7.4.2). Glycated proteins have altered structures that are implicated in atherosclerosis and kidney dysfunction. The exposure to glucose drives the reaction, but there may also be influences from the redox state in the local intra- or extra-cellular environment, for example, through glutathione concentration, that might, theoretically, alter thresholds for further changes in the Amadori (AGE) products. The mammalian age-related trends for blood glucose elevations (Chapter 3.6.8) and for altered oxidative metabolism (Chapters 10.3.2.3, 10.3.2.4) are thus locally extrinsic factors in protein glycation. The striking premorbid reversal of the trend for increased tail tendon tensile strength in rats (Figure 10.26) shows the continuing importance of the local environment, even to extracellular molecules like collagen.

An intangible is the extent of repair or removal of damaged macromolecules. While DNA repair is well demonstrated (Table 8.1), little is known about the repair of proteins and other macromolecules. Other changes that have unknown mechanisms include age-related DNA demethylation (Chapter 7.2.4), sporadic derepression of X-linked genes (Chapter 7.2.2), and chromosomal structural abnormalities (Chapter 8.2). Ovarian follicular mass influences on fetal aneuploidy (Chapter 8.5.1.1) suggests the importance of hormones to the fidelity of chromosomal mechanics during oocyte maturation.

General mechanisms in senescence are represented by cascading mechanisms of two major types: (1) cascades caused by wear-and-tear damage, or loss of irreplaceable organs, cells, or molecules; and (2) cascades caused by regulatory shifts involving various levels of feedback and feed-forward that may not be initiated by damage to, or loss of cells or macromolecules. Collectively, these phenomena may be called *dysorganizational se-*

nescence (Chapter 2.6). Numerous examples of both types of cascades occur across the phyla and evidently are widely distributed with few species barriers. This diversity arises from both the genetically determined anatomic endowment at maturity, as well as numerous individual experiences in the vicissitudes of living that may cause lasting deficits. Thus, age-related increases of mortality rates can have multiple causes, even within a species, as a function of haphazard environmental factors that may damage irreplaceable cells or organs.

The potential for reversing these cascades appears to vary widely. When cell degeneration is limited and when loss of irreplaceable cells has not occurred, then cascades should be easily arrested. Some examples from clinical medicine include estrogen replacement therapy for menopausal symptoms and the control of age changes in glucose-insulin tolerance through diet (Chapter 10.3). However, once postmenopausal bone loss has occurred, estrogen does not restore it. Similarly, stress-induced hippocampal pyramidal neuron loss (Chapter 10.5.1) cannot be restored by endogenous mechanisms in mammals, though this might occur in birds, which retain limited postnatal neuronogenesis (Paton and Nottebohm, 1984). The manipulability of immune system changes is considerable (Chapter 10.7), but the prematurational complement of immune cells has yet to be fully restored. Further studies may lead to the recognition of limitations within a taxonomic group in the plasticity of senescent phenotypes.

It is hard to generalize about these diverse phenomena of senescence. Nonetheless, the evidence supports the larger principle that the timing of age-related changes and species differences are often mediated by regulatory systems of which the neural and hormonal controls are representative. Because humoral, neural, and hormonal factors are such general regulators of the developmental and maturational schedules in plants and animals, it is not surprising that postmaturational changes are also often regulated by these same systems. These types of controls are important, too, in the life cycles of invertebrates whose senescence is ultimately caused by mechanical erosion or by deficiencies of oral anatomy. Humoral and neural mechanisms in senescence can be superposed as subsidiary pacemakers on cascades that arise from separate causes.

12.4. Types of Genetic Influences on the Lifespan

Without question, specific genes and combinations of genes ultimately determine the characteristics of senescence and the potential lifespan. This synopsis of genetic influences on the length of the adult life phase and on the phenotypes of senescence considers genes that may be described as *constraints* and *controls*. The concept that some genes indirectly constrain the lifespan is consistent with the view that not all aspects of the phenotype are directly selected for (Gould and Lewontin, 1979). More can be said about genetic constraints on lifespan than on genetic controls. Except for the unusual fungal genotypes that cause senescence in association with episomal proliferation in the filamentous fungi *Neurospora* and *Podospora* (Chapter 2.4), no other species has a cellular involution during senescence that causes cell death throughout the entire organism. That is, "death" genes have not been demonstrated for any other organism.

There are various types of genetic constraints on lifespan. Vast differences in potential lifespans arise through genes that determine dangerous behaviors (flying) and anatomical features (wings) that are vulnerable to irreparable damage and mechanical senescence (Chapter 2.2). Flies do not live as long as humans for a clear reason: the initial mortality rates at maturation (IMR) are more than 1,000 times greater in flies than humans (Table 1.1; Appendix 1). Consequently, even in the absence of senescence, the statistical lifespan of flies must be correspondingly shorter in all real (finite) populations (Chapter 1.4.8). Innate behaviors like flying are adult characteristics that, like the anatomy of wings, are direct consequences of the body-plan laid down under genetic control during development. The genes that determine the body-plan thus represent indirect constraints on adult life history and constitute a major class of genes that influence the lifespan and characteristics of senescence. Other examples include the numbers of cells or anatomical units that provide factors of safety in the amount of cumulative damage that can be tolerated. A model for different brain regions (Figure 5.14) may also represent the factors of safety for loss of nephrons and a critical threshold for kidney damage.

Yet undescribed genes determine the numbers of ovarian oocytes. It is unknown whether some of the same genes that set the proliferative limit to the postmitotic cells of the adult also set the proliferative limit of cultured diploid fibroblasts and other cells (Chapter 7.6.3). The number of ovarian oocytes in mammals are a major determinant of the age at menopause and thence the mortality risk associated with estrogen deficits through coronary disease and osteoporosis (Chapter 3.6.4.1).

The reduction of anatomic features like mouthparts in mayflies that allows the absence of feeding (autogeny) in adult insects also permits them to reproduce immediately on hatching in environments with variable food supplies. While compatible with reproduction immediately after hatching, the loss of mouthparts indirectly constrains their adult phase.

Vulnerabilities to "wear and tear" through normal usage may be traced to genes that determine the capacity for cell replacement. Examples include replacement of exfoliated epithelia in the gut and of circulating blood cells, which is characteristic of mammals and absent in adult flies (Chapters 2, 3, 7, 8). The vulnerability of blood vessels to damage from hypertension, particularly at points of turbulent flow in humans (Chapter 3.6.11.1) can be considered as mechanical senescence. Tooth erosion is another example of mechanical senescence. Loss of teeth from abrasion, gingival disease, or trauma was very common at later ages in humans until recently (Chapter 3.6.11.4). Edentularity must have increased mortality risk independently of age for many causes, including difficulties in ingesting an adequate diet at all seasons. The example of tooth replacement in elephants and manatees represents genotypes that increased potential longevity by relaxing a constraint.

Vulnerability to environmental insults are influenced by the capacity for wound healing and organ regeneration, and the repair of damaged molecules such as UV-induced DNA repair. The *H–2* complex in mice contains alleles in some strains that influence the efficacy of DNA repair (Chapter 6.5.1.4). The distribution of codons that determine the amino acid composition of proteins and specific sequences like PEST that regulate protein lifespan, in turn, predestine the extent of L-amino acid racemization and deamidation (Chapter 5.4.4). Greater amounts and more rapid accumulation of these thermodynamically driven changes would be tolerated in the lens crystallins of short-lived species, for example. None of these body-plan characteristics is attributable to a single gene or could be said to directly control the lifespan. Their influence on senescence and longevity is indirect, through various risk factors, and largely pertain to iteroparous species with adult phases of nominally 1 year or more.

Vulnerability to injury from dangerous behaviors (e.g. fighting) and dangerous modes of locomotion (e.g. jumping) can increase mortality, despite their evolution as adaptations that, for example, enhanced the ability to defend against predators. Genetic influences on vulnerability to predators, parasites, and microbial infections may set the initial mortality rate (IMR) through efficacy of avoiding predation. Effects on the mortality rate acceleration (MRDT) could be mediated through senescent changes in the immune system. We may also include the genes that influence appetite in rodents, through which *ad libitum* feeding of certain diets causes a range of pathological changes with delayed onset (Chapter 10.3.3). The genes programming the neural circuitry of the brain satiety centers would thus be considered as setting a pacemaker for senescence.

In contrast to genetic constraints are another class of genes that act more immediately and constitute direct controls. Examples of genetic control on the total lifespan include genes that regulate the length of development or diapause; genes encoding triggers for the onset of reproduction in semelparous species; and genes that determine specific diseases and causes of morbidity. Other genes have a direct role in senescence and longevity of semelparous species whose reproduction is linked to mortal injury and rapid senescence. Genetic determinants of semelparity could include the numbers of hormone receptors or other molecules which influence the toxic effects of corticosteroids on various target cells, for example, and might be single genes in some cases. There is no reason to expect that these genes will be the same between taxa which have different humoral mechanisms of rapid senescence, for example, octopus versus salmon.

In iteroparous species, genes causing fatal diseases that arise postmaturationally— whether early during adult life or later—are perhaps the clearest cases of genes that explicitly control senescence and longevity. Examples of these rare genes in humans include the progeroid syndromes, with short lifespans and organ-specific degeneration, such as Werner's and Huntington's (Chapter 6.5.3.3). Nonetheless, the lesions in most progeroid syndromes are atypical of senescence in normal humans. Life-shortening genotypes are found in inbred lab animals (Chapters 6.5.1, 6.5.2), but again cause specific diseases with early onset. The genes predisposing to Alzheimer disease may be widely distributed in view of the neuropathologic changes (amyloid in blood vessels; neuritic plaques; neurofibrillary tangles) that occur to some degree in every human by 70–80 years (Chapter 3.6.10). A major question is whether some alleles influence the onset of Alzheimer disease, as suggested by genetic polymorphisms that influence the latency of prion diseases (Chapter 6.5.1).

Other fatal diseases with hereditary control include a variety of malignant growths and kidney diseases (Chapter 6.5). Benign prostatic hypertrophy must also have a genetic basis, since this ubiquitous condition of older men is rarely reported in mammals besides humans and domestic dogs (Chapter 3.6.3).

In sum, there is little data on genes that constrain or control longevity and senescence, and we have only limited genetic explanations for differences in longevity and senescence between species and differences between higher taxonomic levels. We cannot yet explain why age-related changes of mammals are generally so similar in their specific details and in the fraction of the lifespan in which they occur, despite thirtyfold absolute differences in lifespan (Chapters 3.6.1, 5.3). The generality of age-related changes that prevails as a backdrop to specific diseases would seem to differ in its genetic origin from heritable disorders with delayed onset, for example, Huntington's and Alzheimer's disease, that are less ubiquitous. The conservation of these gene sequences must be high, in view of the basically similar patterns of senescence in short- and long-lived placental mammals that descended from common ancestors more than 100 million years ago in the Mesozoic (Chapter 11.2.5). As an index of this conservation, vast changes have taken place in germ-line DNA through base-pair substitutions and in repetitive DNA sequence frequency and location.

While it is easy to explain the generality of pathophysiological changes from ovarian senescence in female mammals, it is harder to explain the generality of dendritic regression shown by brain neurons of both sexes at later ages in rodents and humans (Chapter 3.6.10). Pleiotropic genetic influences could account for the proportionate variations in senescence among mammals. The overall scaling of so many physiological events of postnatal development and senescence with body size in mammals (Chapters 3.6.1, 5.3) suggests that the pacemakers of senescence are in some way linked to those that regulate the duration of growth or the timing of maturation. Experimental approaches to these questions can be imagined; perhaps using transgenic mice with human genes that have altered activity at middle age or later. Candidates for such senescence-reporter genes are those coding for brain diseases with delayed onset, for example, Huntington's and Alzheimer's disease.

Genes influencing senescence and longevity appear to include those classified by either *housekeeping* or *specialized functions*. Housekeeping genes code for enzymes and proteins that maintain generalized cell functions such as macromolecular biosynthesis, intermediary metabolism, or repair (e.g. ribosomal RNA cistrons; enolase; heat shock proteins). Genes like those for Huntington's and Alzheimer's disease are also specialized, since their adverse effects are restricted to particular neural pathways (Chapters 3.6.10, 6.5.3). Examples of housekeeping genes that influence senescence might be found in the genes coding for the complex enzymatic machinery for protein degradation and turnover. Senescent rodents accumulate altered (inactive) enzymes, for example, aldolase or superoxide dismutase that differ by enzyme and cell type (Chapter 7.4.1). Slowed protein turnover is implicated in the accumulation of these inactive enzymes. In turn, protein turnover might be slowed through neuroendocrine or autonomic changes, which themselves are controlled by specialized genes with systemic effects on metabolism. The distinction between housekeeping and special function genes may not ultimately be very useful, because the same polypeptide sequences may serve both types of functions, as illustrated by the lens crystallins that have the same amino acid sequence as enzymes of intermediary metabolism in liver (Piatigorsky and Wistow, 1989).

The statistical lifespan results from two mortality rate parameters, the initial mortality rate (IMR), which is calculated here at puberty when its value is usually lowest, and the rate at which mortality accelerates after puberty, the mortality rate doubling time (MRDT) (Chapter 1). The ranges of both parameters, particularly the MRDT, are species characteristics and must be under genetic control. Genetic variants in *Caenorhabditis* influence the MRDT (Figure 6.2). The gender differences that inbred mice show in IMR indicate genetic influences on this parameter. The genetic constraints and controls will determine the range of IMR for populations in different environments. These is no obvious way to yet predict from the polymorphisms of a particular gene the quantitative variations in responses to environmental hazards that set the range of IMR. Examples might be found in genes that determine the efficacy of specific repair or detoxifying enzymes through variations in the specific activities, the basal concentration of these proteins, or their inducibility. Genes that influence repair may also influence the MRDT by the accumulation of various types of internal and external damage. Thus, the efficacy of enzymes that repair UV-induced DNA damage in skin cells will influence the risk of skin cancer from exposure to sunlight, which increases exponentially with age (Chapter 3.6.11.3).

In closing, I note major issues that remain obscure, particularly for slowly senescing iteroparous species: (1) Why the MRDT varies so little between different genotypes in some species like laboratory mice which have major differences in pathologic lesions; and (2) Why the MRDT of humans is also so similar in populations that are in widely differing environments, eat different diets, are exposed to different pathogens, and have different frequencies of organ-specific lesions (Chapter 3.6.2).

12.5. Aging in the Biology of Extended Time

Aging has a legitimate usage in discussing the biology of time-dependent phenomena, of which organismic senescence is only one phase. I return *aging* to this lexicon because the term quite reasonably represents the entire manifold of age-related changes, whether good, bad, or indifferent with respect to the functions and mortality risks of the individual. The passage of time does not necessarily lead to generalized organismic dysfunctions, as clearly shown by species with prolonged development followed by adult phases lasting less than 1/100 of their lifespan (bamboo; 17-year cicadas), or by highly fecund adult phases that are over 100-fold longer than the duration of development (insect queens). Moreover, aging in adults may bring increased reproductive success and efficacy of parenting, as judged by improved survival of the second or third clutches raised by some bird species (Chapter 3.5). The examples of animals and plants that produce increasing amounts of gametes as they grow larger (Chapter 4.1) also suggest that aging can occur without senescent dysfunction. I suggest that organismic aging and senescence be considered as aspects of a nascent subject, *the biology of extended time*.

Foremost, the biology of extended time describes the enormous plasticity of life history scheduling and the wide range of phenotypes that are possible during lifespans across the milliontold range of adult life phases. While lacking a scale for comparing the ranges of temporal and morphological varieties, the evidence suggests that temporal schedules vary with no fewer limits than do morphological variants among the body-plans of most radiations. Evidence from experimental and field studies (Chapter 6.3) indicates that the duration of life history phases may be as readily lengthened as shortened during evolution.

The biology of extended time differs from approaches to biology that seek to avoid individual differences in the pursuit of general conclusions. It is a canon of aging that individual phenotypes become increasingly different. Humans, for example, show individual patterns of physiological change in kidney function (Figure 3.14), in which these differences may represent genotypic interactions with life-style characteristics such as diet and smoking. Moreover, short-lived inbred laboratory flies, mice, and worms develop similar individual variations in age-related change. Even identical genotypes and standardized environments do not prevent the individualization of aging during extended time.

Some differences between aging changes of individuals may be traced back to statistical fluctuations in cell numbers that arose during development (Chapter 9.3). Other differences in life history trajectories represent individual experiences during adult life, and the superposition of these experiences on individual differences in cell number and other nuances of body-plan that were acquired during development. Thus, studies in the biology of extended time must address the expanding range of individual variations in many parameters that occurs with advancing age and experience. The individual trajectories, nonetheless, are constrained by various taxon-specific canonical changes, like menopause and presbyopia that are inevitable, sooner or later, in all present environments.

The mechanisms of senescence as viewed across the huge range of lifespans have not indicated any new biological, chemical, or physical principles that are not already familiar from development or that occur during cancer and other chronic adult diseases. Not all of the changes, however, have known functions elsewhere in the lifespan, for example, accumulations of racemized amino acids and Amadori (AGE) products of glycation in lens crystallins (Chapter 7.4). Unless unknown mechanisms turn up, we may already know the causes of senescence that limit lifespan in many species.

Mechanisms in the biology of extended time involve many of the same mechanisms that also schedule daily and seasonal cycles of reproduction. In cases where senescence is coupled to reproduction (Chapters 2, 10), the mechanisms regulating biorhythmicity must overlap or interact with those causing senescence, because of the crucial role of light, temperature, and nutrition in scheduling reproduction. Yet other mechanisms may come into play over extended time for repair and regeneration, which differ from those regulating short-term rhythms.

Besides the familiar changes of aging, other phenomena may not have been recognized because extended time may be needed for their observation. For example, transgenerational effects of diet were found in the F_2 generation derived from rats whose dams were malnourished during pregnancy (Zamenhof et al., 1972). Despite cross-fostering on *ad libitum* fed nurses, the brain DNA of the F_2 generation was slightly below normal. Another example is indicated by the oocytes that are formed before birth, as in mammals. These ova will give rise to the F_2 generation and might transmit a range of environmental influences through the grandmother. Possible transgenerational influences of nutrition, toxins, or stress on longevity or senescence have not been investigated.

It is interesting to consider whether functions of memory also develop different properties over extended time. While there is no evidence that the various memory systems in the brain become saturated during the present human lifespan, there must be some physical limit. A little studied question is whether information processing and retrieval acquire different dynamics with aging beyond some threshold amount of information stored. It is unclear to what extent changes in information processing result from age-related alterations in brain structure (the prevalent view), or from the amount and type of information that is stored. Morgan (in press), for example, proposes that benign senescent forgetfulness in the absence of Alzheimer's disease may result from saturation of available "synaptic space." Either memory saturation or organic changes could be factors in the age-related changes in cognitive functions that are described as transitions from fluid to crystallized intelligence (Horn, 1982).

In the immune system, cumulative exposure to environmental and endogenous antigens on immune function may alter functions, as indicated by the increase of memory T-cells with the Pgp–1 marker (Chapter 3.6.6). The consequences of increasing the auto-antiidiotypic repertoire are unknown, and might be involved in the selective changes of immune functions during aging, some of which may be specifically harmful.

There may be other types of emergent changes in these and other integrative functions that record events of various types, either through specialized memory and retrieval processes, as in the nervous and immune systems, or through other processes, like those through which glucose and steroid hormones alter genomic functions (Chapter 7.8). The entire organism, extracellular matrix molecules to neurons, may be considered as an engram for aging with beginnings in the previous generation or before when the ovum was formed. To whatever extent experiences are transduced and recorded on the engram, the duration and intensity of impact will vary according to cellular and molecular replacement and regeneration in a given tissue.

The biology of extended time transcends aging and senescence in the usual sense, to include changes that are good, bad, or indifferent to mortality risk, and those that are ephemeral or may extend transgenerationally. The mechanisms in senescent and nonsenescent changes over extended time may be studied experimentally in familiar models, as documented extensively throughout the book. Besides these established lines of study, it may be fruitful to consider species with different life histories that have been given little attention by experimental gerontologists, especially those species with extended reproductive schedules. There are many new experimental opportunities and a good basis of knowledge to plan aggressive attacks on mechanisms of aging. Motivations for pursuing these promising issues are the very human desire to enhance or at least maintain functions throughout the lifespan and the curiosity of the natural philosopher.

Appendix 1

Species Differences in Mortality Rate Constants and Lifespans

	IMR/year	MRDT, Years	*tmax*, Years
Phylum:	Animalia		
Annelida			
Lumbricus terrestris[T3.1]	0.1		> 6
Arthropoda			
CRUSTACEA			
Daphnia longispina[T2.1]	0.7	0.02	0.1
Homarus americanus[C4.2.2]			> 50
INSECTA			
Coleoptera			
Carabus coerulescens, C. glabratus[T3.1]			3–4
Macronychus glabratus[T3.1]			> 9
Microcylloepus pussilus[T3.1]			> 9
Tribolium castaneum[T3.1]	0.12		2–3
Diptera			
Calliphora erythrocephala[T2.1]	0.4	0.07	0.3
Drosophila melanogaster[T2.1]	0.01–4	0.02–0.04	0.3
Musca domestica[T2.1]	4–12	0.02–0.04	0.3
Hymenoptera			
Apis mellifera[T2.1]			
Worker bees (June)	0.2	0.02	0.2
Worker bees (winter, >240 days)	< 0.001	0.03	0.9
Queens	< 0.1		> 5
Lepidoptera (aphagous)			
Automeris boucardi[T2.1]	10	0.005	0.03
Automeris junonia[T2.1]	20	0.005	0.03
Dirphia eumedide[T2.1]	7	0.005	0.05
Dirphia hircia[T2.1]	20	0.01	0.05
Lonomia cynira[T2.1]	45	0.007	0.02
Chordata			
TELEOSTA			
Anguilla[C2.3.1.4.2]			88
Acipenser fulvescens[T3.1]	0.013	10	> 150

Species Differences in Mortality Rate Constants and Lifespans (*continued*)

	IMR/year	MRDT, Years	*tmax*, Years
	Animalia		
Betta splendens[T3.1]			1.5–3
Cynolebias adolffi[T3.1]	0.07	0.1	1–2
Lebistes reticulatus[T3.1]	0.07	0.8	5
Oryzias latipes[T3.1]			3–6
Perca fluviatalis[T3.1]	0.07	27	25
Sebastes aleutianus[C4.2.2]	0.05		> 140
Xiphophorus maculatus[T3.1]			2–3
AMPHIBIA			
Ambystoma maculata[C4.2.2]			30
Megalobatrachus japonicus[C4.2.2]			> 55
Xenopus laevis[C4.2.2]			> 15
REPTILIA			
Alligator mississipiensis[C3.4.2]			> 50–60
Calotes versicolor[C3.4.2]			> 4
Geochelone gigantea[C4.2.2]			> 100
Terrapene carolina[C4.2.2]			> 50
AVES			
Apodiformes			
Apus apus[T3.1]	0.1	5	21
Selasphorus platycerus[T3.1]	0.25		> 12
Charadriiformes			
Larus argentatus[T3.1, A2]	0.004	6	49
Vanellus vanellus[A2]	0.2	6	16
Galliformes			
Alectura lathami[T3.1]	0.05	3.3	12.5
Coturnix coturnix japonica[T3.1]	0.7	1.2	5
Pavo cristatus[T3.1]	0.06	2.2	9.2
Syrmaticus reevesi[T3.1]	0.02	1.6	9.2
Passeriformes			
Erithacus rubecula[T3.1, A2]	0.5	8	12
Lonchura striata[T3.1, A2]	0.1	2.5	10
Sternus vulgaris[T3.1, A2]	0.5	> 8	20
MAMMALIA			
Artiodactyla			
Hippopotamous amphibius[T3.1]	0.01	7	> 45
Ovis dalli[T3.1]	0.05	1.5	15–20
Chiroptera			
Pipistrellus pipistrellus[T3.1, A2]	0.36	3–8	> 11
Myotis lucifugis[T3.1]			> 32
Carnivora			
Canis familiaris[T3.1]	0.02	3	20

Species Differences in Mortality Rate Constants and Lifespans (*continued*)

	IMR/year	MRDT, Years	*tmax*, Years
	Animalia		
Perissodactyla			
Equus caballus[T3.1]	0.0002	4	> 45
Rhinosceros unicornis[T3.1]			> 50
Primates			
Homo sapiens			
U.S. female, 1980[T3.1]	0.0002	8.9	> 110
Australian, 1944–1945[F1.5]	0.0013	8.2	
POW[F1.5]	0.0070	7.7	
Dutch civilian, 1945[F1.5]	0.0014	7.8	
Dutch civilian, 1946[F1.5]	0.0008	7.6	
Hunter-gatherers[C11.2.5.3]	0.03		
Macaca mulatta[T3.1]	0.02	8	> 35
Proboscidea			
Loxodonta africana[T3.1]	0.002	8	> 70
Rodentia			
Mus musculus[T3.1]	0.01	0.3	4–5
Peromyscus leucopus[T3.1]	0.001	0.5	7–8
Rattus norvegicus[T3.1]	0.002	0.3	5–6
Mollusca			
GASTROPODA			
Aplysia juliana[T2.1]	0.007	0.13	0.9
Lottia digitalis[F4.10]	0.06	ND	> 1
Lottia insessa[F4.10]	2.5	ND	< 1
Lymnaea stagnalis[T3.1]	0.15	0.25	2
Nucella emarginata[C4.3]	0.9	ND	> 5
Nucella lamellosa[C4.3]	0.5	ND	
PELYCOPODA			
Arctica islandica[C4.2.2]			220
Margeritifera[C4.2.2]			120
Mercenaria[C4.2.2]			46
Panope generosa[C4.2.2]			120
POLYPLACOPHORA			
Chiton tuberculatus[C4.3]	0.01	1.5	> 12
Nematoda			
Caenorhabditis elegans[T2.1]	1–7	0.02–0.04	0.16
Rotifera			
Lecane inermis[T2.1]			
Female, amictic	6	0.005	0.10
Female, mictic	20	0.03	0.08
Male, aphagic	0.4	0.002	0.02

Species Differences in Mortality Rate Constants and Lifespans (*continued*)

	IMR/year	MRDT, Years	*tmax*, Years
Eumycota			
Saccharomyces cerevisiae[T2.1]		0.004	
Angiosperm			
Cereus giganteus[C4.3]	0.001	18	> 200

Notes: This table consolidates information from appendix tables, text, or figures, as indicated in superscripts (A, T, C, F). The IMR (initial mortality rate, calculated at sexual maturity) and MRDT (mortality rate doubling time) should be regarded as first approximations within a twofold range. ND, not detectable; blank space, data judged insufficient for estimations.

Appendix 2

Estimation of Mortality Rates in the Absence of Mortality Schedules by Age Group

	Rav/Year	*tmax*, Years	N	IMR/Year	MRDT, Years
Pipestrelle bat	0.36	15*			
			10^3	0.32	14.9
			10^4	0.28	7.5
			10^5	0.26	5.7
			10^6	0.25	4.7
		11*	$10^{3}**$	0.25	4.7
			10^4	0.22	3.4
			10^5	0.20	2.8
			10^6	0.19	2.5
European robin	0.62	12			
			$10^{3}**$		
			10^4	0.58	15.3
			10^5	0.54	7.9
			10^6	0.52	5.8
Lapwing	0.34	16			
			10^3	0.30	16.4
			10^4	0.27	8.2
			10^5	0.25	6.0
			10^6	0.24	5.1
Starling	0.52	20			
			$10^{3-4}**$		
			10^5	0.51	56.6
			10^6	0.49	21.2
			10^6	0.49	21.2
Common swift	0.18	21			
			10^3	0.12	8.2
			10^4	0.10	6.0
			10^5	0.094	5.1
			10^6	0.088	4.5
Herring gull	0.04	49			
			10^3	0.0060	7.2
			10^4	0.0046	6.3
			10^5	0.0037	5.7
			10^6	0.0032	5.4

Notes: The table is based on equations (1.4)–(1.7) in Chapter 1.4.5 and Finch et al., 1990. The best-fit values were determined for a range of test population sizes as indicated. The algorithm is described below.

Estimation of Mortality Rates in the Absence of Mortality Schedules by Age Group (*continued*)

AMR, average population mortality rate, as determined by banding and recapture. IMR, initial mortality rate; the value of A in equation (1.2), calulated at puberty. MRDT, mortality rate doubling time; a measure of the acceleration of mortality rate as a function of adult age as determined from the Gompertz coefficient G (equation [1.2]): MRDT $= (\ln 2)/G$. *tmax*, maximum observed lifespan.

Calculations were based on equation (1.6) according to a procedure devised by Malcolm Pike and Matthew Witten (Finch et al., 1990) that used AMR from banding and recapture studies and *tmax*, either from banding and recapture or from captive specimens (see Table 3.1 notes). A computer program fitted values of A and G for equation (1.4) so as to be consistent with AMR and *tmax* for populations of different sizes. A must always be less than AMR, since the latter represents contributions from both A and G. In a related approach, Calder (1984) derived equation (1.6) for calculations of life expectancy as a function of body size. The present calculations require no assumptions about the relationship of body size to the mortality coefficients.

The program integrated $s(t)$ (equation [1.4]) from zero to twice *tmax*. The algorithm used a series of values for A (20 steps from 0.001 to 0.999 of the value of Rav) to estimate G for each trial A_0, so as to minimize the difference between the entered A and the A calculated for each trial A_0 and the estimated G. Before proceeding from one trial A to the next, the program calculated *tmax* using the current trial A_0 and G, and compared its value to the entered *tmax*. If the difference did not exceed a specified tolerance percentage, the current trial A and G were accepted. However, if the best trial did not reach the accepted tolerance for *tmax* at that population size, then a new series of trial A's was defined to fall between the best two values for the current trial A. Trial A's and G's were iteratively fitted in this manner for successively larger populations, $10^{N=3-6}$.

*Calculations were based on two values of *tmax:* 11 years as reported for *Pipistrellus pipistrellus* (Thompson, 1987; which is probably an underestimate) and 15 years as reported for *P. subflavus* (Nowak and Paradiso, 1983: which is closer to the generally reported 20-year-long *tmax* of vespertilionids; Austad and Fischer, in prep.). See Table 3.1, note 26.

**Could not reach *tmax*.

Appendix 3

Alfred R. Wallace's Notes on Natural Selection and Senescence

Quoted from E. B. Poulton, who translated this edition of Weismann (1889a,b, 23).

> Dr. AR Wallace kindly sent me [EBP] an unpublished note upon the production of death by means of natural selection, written by him some time between 1865 and 1870. The note contains some ideas on the subject, which were jotted down for further elaboration, and were then forgotten until recalled by the argument of this Essay. . . . with Dr. Wallace's permission I print it in full.":[1]

"The Action of Natural Selection in Producing Old Age, Decay, and Death.

"Supposing organisms ever existed that had not the power of natural reproduction, then since the absorptive surface would only increase as the square of the dimensions while the bulk to be nourished and renewed would increase as the cube, there must soon arrive a limit of growth. Now if such an organism did not produce its like, accidental destruction would put an end to the species. Any organism therefore that, by accidental or spontaneous fission, could become two organisms, and thus multiply itself indefinitely without increasing in size beyond the limits most favourable for nourishment and existence, could not be thus exterminated: since the individual only could be accidentally destroyed,—the race would survive. But if individuals did not die, they would soon multiply inordinately and would interfere with each other's healthy existence. Food would become scarce, and hence the larger individuals did not die they would decompose or dimnish in size. The deficiency of nourishment would lead to parts of the organism not being renewed; they would become fixed, and liable to more or less slow decomposition as dead parts within a living body. The smaller organisms would have a better chance of finding food, the larger ones less chance. That one which gave off several small portions to form each a new organism would have a better chance of leaving descendants like itself than one which divided equally or gave off a large part of itself. Hence it would happen that those which gave off very small portions would probably soon after cease to maintain their own existence while they would leave a numerous offspring. This state of things would be in any case for the advantage of the race, and would therefore, by natural selection, soon become

---

1. Author's Note: It is historically pertinent that this view was not discussed in the otherwise comprehensive monograph of Lankester (1870). That Lankester and Wallace knew each other is shown by a personal communication from Wallace, concerning the evolution of lifespan, that orangutans lived as long as the human inhabitants of Borneo (Lankester, 1870, 72).

established as the regular course of things, and thus we have the origin of *old age, decay, and death;* for it is evident that when one or more individuals have provided a sufficient number of successors they themselves, as consumers of nourishment in a constantly increasing degree, are an injury to those successors. Natural selection therefore weeds them out, and in many cases favours such races as die almost immediately after they have left successors. Many moths and other insects are in this condition, living only to propagate their kind and then immediately dying, some not even taking any food in the perfect and reproductive state."

Glossary

Advanced glycation endproducts (AGE): Glucose, among other sugars with free aldehyde groups, can nonenzymatically form covalent bonds to free amino groups (glycation) through the Maillard reaction (Chapter 7.5.2). The initial glycation adduct then undergoes an Amadori rearrangement to form ketoamines that are known as Amadori products. Further spontaneous modifications of Amadori products can yield covalently crossed and colored endproducts that are alternatively described as advanced glycation endproducts (AGE) or advanced Maillard products.

Advanced Maillard products: *See* Advanced glycation endproducts

Agametic reproduction: Reproduction without the formation of gametes by modes that include budding, fission, and other forms of vegetative reproduction from somatic cells, but not parthenogenetic reproduction from an egg.

Age-independent mortality: A mortality rate coefficient with a numerical value that is constant across all age groups. The Gompertz-Makeham model has two age-independent mortality coefficients (Equation [1.1]): an additive constant M_0 (the Makeham term) and a constant multiplier of the mortality rate exponential term at any age, A_0. These coefficients are aggregated as the initial mortality rate, IMR (Equation [1.2]), which is calculated at sexual maturity.

Age-related: A biological difference that is statistically significant between two age groups and that does not imply dysfunction, specific cause, or a progressive trend continuing to later ages.

Aging: A nondescript colloquialism that can mean any change over time, whether during development, young adult life, or senescence. Aging changes may be good (acquisition of wisdom); of no consequence to vitality or mortality risk (male pattern baldness); or adverse (arteriosclerosis). Although avoided during most of the book, the word aging has a legitimate use once the context is made clear.

Allometric: Differences between species or changes during development that fit exponential functions of body size or weight where the exponential coefficient is greater than one (hyperallometric) or less than one (hypoallometric) (Equation [5.1]). Differences that scale in proportion to size are isometric.

Amadori product: *See* Advanced glycation endproducts

Amictic: A term from population genetics that means reproduction by parthenogenesis and without the combination of chromosomes from separate parents.

Ammocoete: A larval stage of lampreys that may last 3.5 to 7 years, during which they reside in mud or silt of freshwater streams; followed by metamorphosis into subadults.

Amyloid: A general term for fibrillar protein aggregates that are classically identified by their capacity to bind the dye Congo red and show green birefringence under polarized light. X-ray diffraction patterns indicate a β-pleated sheet structure that can be formed from many different polypeptides, including the β-amyloid of Alzheimer's disease and the prion proteins of scrapie infections (Chapter 6). Amyloids can occur in many tissues.

Anadromy: Migration of fish from the ocean to fresh water.

Anautogenous/anautogeny: Adults, particularly insects, that must ingest some nutrients to complete maturation of the first clutch of oocytes (Chapter 2.2.1.1.2).

Angina pectoris: Sudden chest pain due to ischemia that is characteristic of occluded coronary arteries in the heart, usually from advanced coronary arteriosclerosis. Angina usually is transient during exercise or emotional stress and can be treated with vasodilators such as nitroglycerine.

Annuals: Plants particularly, but also animals, that have an annual life cycle terminated by rapid senescence and death after their first and only season of reproduction. Under different conditions, many annuals can survive to reproduce for one or more years, that is, they can become perennials.

Aphagy: The absence of feeding on solid or liquid nutrients which is characteristic of mayflies and many other invertebrates during their adult phase; the failure to feed is a consequent to the absence of mouthparts or other organs concerned with ingestion that fail to develop in these species (Chapter 2.2.1). Depending on the species, aphagy can be complete (solids and liquids) or partial (allowing some ingestion of nutrients through liquids). Some fish stop feeding during migration and spawning (Chapter 2.3.1.4), but for reasons unrelated to anatomical deficiencies.

Apoptosis: A popular descriptor of cell death, in which the cell fragments and dies because of specific physiological or developmental triggers that cause a regulated or programmed sequence of events in macromolecular synthesis and structure (Wyllie et al., 1981). The triggers may include loss of trophic support or exposure to steroids (Chapter 7.7). Apoptosis is held to be distinct from cell death due to mechanical injury or toxins, and may be different from cell death from ischemia.

Atresia: Spontaneous death of ovarian oocytes and their surrounding follicles that can occur at any stage of oocyte maturation.

Autogenous/autogeny: Animals, particularly insects, that can complete maturation of oocytes without requiring the ingestion of food after hatching (Chapter 2.2.1.1.2).

Average mortality rate (AMR): The mortality rate in a population as averaged across age groups. The AMR as used here is calculated on the basis of adult ages.

Biological chronometers: Molecular or cellular changes that measure the passage of time or age, as in the accumulation of racemized amino acids in dentine (Chapter 7.5.4).

Biomarkers of aging: Age-related biological parameters, the absolute values or rates of change of which might estimate the subsequent life expectancy. Few single measures are better than chronological age, even in highly inbred laboratory species (Chapters 1.4.10, 10.9).

Body-plan: The characteristic details of body construction that may be useful in comparisons of species, for example, the presence of an exoskeleton in insects or the presence of continuously replaced teeth in sharks. Same meaning as *Bauplan* as used by Gould (1977).

Browning reactions: Further modifications of glycation adducts in proteins which lead to yellow or brown pigments. *See also* Advanced glycation endproducts

Canonical: A change during development (Davidson, 1990), extended here to age-related changes in adult phases, that is characteristic of a number of species in a taxon and occurs ubiquitously in populations. Canonical sequences in proteins and nucleic acids are shared by a number of species.

Catadromy: Migration of fish from fresh water to the ocean.

Cell death: Cell death occurs during development and continues throughout life in circulating and exfoliating cells. Cell death through normal turnover is distinct from organismic senescence in most species.

Chondrocalcinosis: Calcification of cartilage, with calcium deposits that are distinct from the ureate crystals of gout.

Clonal senescence: Approach to a limiting number of cell divisions that in diploid fibroblasts *in vitro* is commonly referred to as phase III (Chapter 7.2.5). Clonally senescent cells need not die, as represented in postmitotic neurons.

Cohort senescence: A group born from the same generation (cohort) that dies out due to senescence, whether of exogenous or endogenous causes.

Competence: A term from developmental biology that indicates the emergence during differentiation of the ability to respond to stimulae like embryonic inductors, trophic factors, or hormones.

Compression of morbidity: James Fries' (1980, 1988) concept that the lengthening of human life in the face of an assumed rigid maximum lifespan requires "in payment" an accelerated phase of disability before death at later years.

Conditional changes: Age-related changes that are dependent on some other exogenous or endogenous factors and that otherwise may not occur.

Congenic strains: Inbred genotypes that contain a segment of a chromosome from one strain that was introduced to other strains by back-crossing. Congenic strains in principle allow study of one set of alleles on the background of other sets of alleles elsewhere in that or other chromosomes. Congenic strains, however, may also differ in passenger genes that flank the donor chromosome segment and that were carried in by back-crossing. Congenic mice are used to define the *H-2* complex (Chapter 6.5.1.4).

Coronary artery disease: Occlusive narrowing of the coronary arteries in the heart through atherogenesis. Even young healthy adults often show early stages of atherogenesis, with fatty streaks that could form occlusive vascular plaques.

Correlation effects: A term from plant physiology to represent how hormones emanating from one part of a plant then cause "correlated" degeneration of other parts, for example, the induction of whole plant senescence by flowering (Chapter 2.3.2).

Dependent variable: A parameter that is observed to depend for its values on another, the independent variable. By identifying the specific and extrinsic causes of age-related changes such as steroid-dependent neuron death, noise-dependent hearing loss, and turbulence-dependent arterial lesions, one may consider biological time in phenomena of aging as equivalent to a dependent variable that can be manipulated within certain limits as determined experimentally for each system (Finch, 1988).

Diadromy: Migration of fish in either direction between fresh water and the ocean. McDowall (1988) discusses a variety of migratory life histories.

Differential gene expression: Expression or activity of particular genes in different cell types. The mechanisms of differential gene expression involve differential transcription of select genes, but may also involve differential processing of primary transcripts to assemble mRNA with different arrays of exons (alternative splicing).

Diphyodont: Animals that produce two sets of teeth; for example, mammals are typically born with deciduous milk teeth, which are replaced by permanent adult teeth.

Disposable soma hypothesis: A hypothesis articulated by Thomas Kirkwood that ascribes the existence of senescence to the insufficiency of repair according to the concept that organisms must optimize the allocation of resources for reproduction versus somatic repair. Excessive allocation of metabolic resources for somatic repair that might limit somatic senescence would be selected against because, according to this view, energy diverted to repair would be from that amount required to meet the minimum reproductive schedule necessary for population stability (Chapter 1.5.2).

Down's syndrome: Developmental abnormalities in humans caused by three copies of chromosome 21 that include abnormalities of the brain and mental retardation (Chapter 6.5.3). Some individuals are mosaics for trisomy 21; very rarely, translocations of a small part of chromosome 21 are found to cause Down's. Down's syndrome also leads to precocious onset of Alzheimer disease–like neuropathology.

Dysdifferentiation hypothesis: A hypothesis articulated by Richard G. Cutler that ascribes senescence to the loss of selective gene regulation and the inappropriate expression of genes (Chapter 7.3.1; Cutler, 1982).

Dysorganizational senescence: A description of senescence as a decrease of physiological integration, which in many examples arises from cascading effects from lesions at a small number of anatomical loci or in a small number of physiological functions that precipitate major impairments (Chapters 1.5.2, 2.6, 3.6.4, 3.6.5.2). Dysorganizational senescence may be most severe for co-adapted traits in a "suite of characters." The pathophysiological changes associated with rapid senescence and sudden death vary widely and in most cases represent supracellular and dysorganizational disturbances, rather than insidious cellular or molecular defects (Table 2.7).

Ecdysis: Molting, or shedding of the skin, usually in reference to transitions between developmental stages in insects.

Ecological hazards: The mortality risk factors due to the natural environment, from predators, weather, infectious disease, etc.

Ecotropic: A virus that replicates only or mainly in a single species. *See also* Xenotropic

Error catastrophe: A hypothesis articulated by Leslie Orgel (1963) that ascribes senescence to a self-catalyzing propagation of errors in macromolecular biosynthesis. The enzymatic machinery of transcription or translation might, for example, accumulate changes in amino acids that reduce the fidelity of macromolecular biosynthesis and consequently increase the likelihood of further errors (Chapters 7.4.1.5, 8.2).

Eugeric: An age-related change that is general (canonical) and not related to specific pathological changes, for example, menopause (Finch, 1972b).

Event-dependent: A change that is identified with a specific cause or event, as distinguished from a change that might be independent of the physical environment, such as radioactive decay.

Exon: A segment of DNA corresponding to sequences of RNA that are spliced together after transcription to produce mature mRNA. Not all exons contain coding sequences.

Familial cholesterolemia: A dominant hereditary disorder with elevations of blood cholesterol bound to low-density lipoprotein (LDL) and high risk of coronary artery disease. In most populations, heterozygotes are at least 0.2%; heterozygous familial cholesterolemia is the most common Mendelian disorder (McKusick, 1986, 391).

Fatty streaks: Deposits of lipids in the intimal layer of arteries that give a white or yellow appearance and that may be found in the aortas of very young children, without disturbing vessel morphology. Some fatty streaks may be precursors to elevated fibrous vascular plaques which increase the risk of death from vascular blockade.

Forced vital capacity: Lung capacity as measured by the largest volume of air that a person can exhale after taking as deep a breath as possible. Forced vital capacity shows one of the largest and most predictable declines with age (Figures 10.27, 10.28).

Genet: In species that reproduce clonally, the entire group of clones with the same set of genes or genotype is called a genet. A ramet is any of these clones that may be formed from adults by budding or fragmentation (Chapter 4.4).

Genome: The total inventory of heritable nucleic acids, usually DNA, including chromosomal DNA in cell nuclei (nuclear genome), but also mitochondrial and chloroplast DNA. One may argue that the genomic inventory of an organism should also include DNA in symbionts, RNA or DNA in non-integrated viruses, and the informational macromolecules in prions.

Genomic totipotency: The concept that somatic cells retain a complete set of germ-line genes that in principle could give rise to another complete adult. Nuclear totipotency is documented by the many animals and plants that reproduce "vegetatively" from somatic cells (Chapter 4.4) and by nuclear transplantation from somatic cells to enucleated ova in limited species (Gurdon, 1962; Davidson, 1986). The capacity for regeneration of limbs and other complex tissues (Chapter 9.4) is consistent with genomic totipotency. The generality of genomic totipotency in somatic cells is not established and may vary among species and cell types. Genomic reorganization in immunoglobulin genes of certain lymphopoietic cells is a clear exception to totipotency.

Genotype: The genes or alleles that distinguish individuals in a species.

Glucose intolerance in aging: The age-related slowing in the clearance of glucose, as assayed in the oral-glucose-tolerance test. The slowed removal of glucose from circulation is mostly due to a decreasing sensitivity of peripheral tissues to utilize glucose in response to insulin (insulin resistance).

Gompertz-Makeham equation: A mortality rate model for age-related increases in mortality that assumes a monotonic exponential acceleration of mortality rates during adult ages (Equation [1.1]).

Haplotype: A set of alleles at one or more genetic loci; used here in reference to antigens of the major histocompatibility locus.

Heterochronic transplantation: Transplantation of cells or tissues between individuals of different ages.

Huntington's disease: A rare dominant hereditary neurodegenerative disease that affects movement and cognition, with typical onset in middle-age (Chapter 6.5.3.3; Table 6.2).

Hutchinson-Guilford syndrome: An extremely rare hereditary progeria with onset in childhood and atherosclerosis, but not involving any mental impairments (Chapter 6.5.3.3; Table 6.2)

Imago: The adult stage of insects.

Incidental cell death: Cell death that occurs at random rather than by any regular process, such as neuron loss after stroke.

Independent variable: A parameter that does not depend on another variable in an experiment or study and is used as a basis for comparisons, for example, age.

Indeterminate growth: Growth without upper limit, as shown by many trees and mollusks, in contrast to growth in mammals which, with rare exceptions, cease long bone growth early in adult life. Nonetheless, indeterminate growth may asymptotically approach a limit.

Initial mortality rate (IMR): The age-independent mortality rate calculated here at the age of maturation when neonatal mortality rate components approach zero (Equation [1.2]). The IMR may be viewed as an aggregate value of several age-independent coefficients.

Insulin resistance: Insensitivity of peripheral tissues to insulin as measured by glucose uptake. *See also* Glucose intolerance in aging

Iteroparous: The capacity of an organism to survive the first round or season of reproduction to reproduce in a subsequent season; in plants, also called perennial.

Karyotype: The set of chromosomes.

K-selection: Selection for life histories with slow development and low rates of reproduction that are usually associated with large size, iteroparity, and long lifespans (Chapter 1.5.1).

Lifespan: A vague term that may refer to maximum observed lifespan or mean lifespan.

Lifespan reproductive capacity: The total potential reproductive yield over the lifespan, however the latter is defined.

Mechanical senescence: Changes in an organ that are consequent to mechanical usage, as in the abrasions of insect wings and mammalian joints and teeth.

Mictic: Reproductive modes that allow exchange of genetic information between individuals in the formation of the egg.

Modular: Organisms which contain repeating anatomical units that are essentially self-sufficient physiological units and often show continuous regeneration of their structures, for example, colonial ascidians, but also sponges and vascular plants (Chapter 4.4). To be distinguished from unitary organisms.

Monocarps: Plants that die after maturation and flowering; equivalent to semelparous animals.

Mortality rate doubling time (MRDT): The time required for the mortality rate to double, as calculated for populations that show age-related accelerations of mortality rate. The MRDT can be calculated readily by Equation (1.3) from the exponential coefficient G of the Gompertz model which has a monotonic acceleration of mortality. MRDTs may also be calculated from the Weibull mortality rate model (Chapter 1.4.1), but because mortality decelerates at later ages, the MRDT would be an average and somewhat arbitrary.

Negative or antagonistic pleiotropy hypothesis: The hypothesis proposed by George Williams (1957) that genes might be selected for on the basis of advantages to fitness early in life, but that might also have harmful characteristics (Chapter 1.5.2).

Neoteny: Sexual maturation at the usual age, but with retarded development of other tissues (Chapter 11.4.2).

Obligate changes of senescence: Changes that are inescapable in any environment of the normative life history, as in mechanical erosion of insect wings or ovarian follicular atresia in mammals.

Oncogene: Genes that can induce tumor formation, for example, as found in retroviruses and other transforming viruses.

Oogenesis: The production of eggs. *De novo* oogenesis is the formation of new oocytes from primordial germ cells, which in mammals generally ceases by birth. Oocyte maturation is a distinct process that may be extensively delayed after *de novo* oogenesis.

Organismic senescence: Senescence of the organism that leads to greatly increased risk of death, as distinguished from senescence of cells that may have no impact on organismic viability.

Osteoarthritis (osteoarthrosis): Degenerative joint disease (Chapter 3.6.5.2).

Ovariprival syndromes of mammalian senescence: The consequences of loss of estrogen and progesterone that commonly ensue after ovarian follicles are depleted, usually during mid-life. Individuals in all species show wide variations in the physiological and pathological sequelae (Table 10.6).

Paedogenesis: Sexual maturation before the organism reaches the full size or proportions of closely related species (Chapter 11.4.2).

Paired species: The comparison of species in the same genus that have different life history patterns, for example, landlocked versus anadromous lampreys (Chapter 2.3.1.4.1).

Pathogeric: An age-related change that is secondary to specific pathological lesions, for example, increased albumin mRNA in old rats that are producing increased amounts of albumin to compensate for loss of protein in the urine from kidney disease. *See also* Eugeric

Perennials: Plants that survive successive years. Some may be monocarps (semelparous) and flower once; others may be polycarps (iteroparous) and flower in successive seasons.

Phenotype: Characteristics of an organism, usually in reference to specific genes.

Polycarps: Plants that flower or produce gametes in successive years.

Polyphyodonty: The capacity for continued production of new teeth throughout the lifespan, for example, as found in sharks.

Prion: A small proteinaceous particle with infectivity that resists treatments that usually destroy RNA and DNA, and that has a corresponding form (isoform) of a protein in host cells. Scrapie, kuru, Creutzfeldt-Jacob disease, and Gerstmann-Sträussler syndrome are associated with prions. This term was introduced by Stanley Prusiner (Prusiner, 1990; Westaway et al., 1987). The existence of infectious particles without nucleic acids continues to be an important controversy.

Progenesis: Precocious sexual maturation, but with other somatic tissues remaining at the immature state of development that is typical for that age (Chapter 11.4.2).

Programmed cell death: *See* Apoptosis

Protooncogene: A gene that may be converted into an oncogene, for example, by incorporation into a retrovirus with subsequent changes in base sequence.

Racemization: The interconversion of optical isomers, as in the racemization of the levorotatory amino acids into a mixture of dextro- and levorotatory forms (Chapter 7.5.4).

Ramet: *See* Genet

Rate-of-living hypothesis: A longstanding hypothesis that the duration of potential lifespan varies inversely with metabolic intensity, such that there is a relatively fixed amount of energy that can be used before senescence sets in. This view is not supported by recent data (Chapters 5.3.4, 10.3)

Renal-osteodystrophic syndrome: A disorder of senescence in laboratory rats on certain diets that appears to originate in kidney damage, with cascading consequences to mineral metabolism that cause osteoporosis (Chapter 10.3.2.5). Similar changes are described in humans (Chapter 3.6.5.1).

Reproductive costs: Increased mortality risk from reproductive activities, for example, through metabolic drain from nursing or from dangerous behaviors associated with mating.

Reproductive schedule: The schedule of reproductive activities, including the onset of reproduction and the intervals between reproduction throughout the lifespan.

Retrovirus: A virus that uses RNA rather than DNA to transmit its genome to the next generation; once in the host cell, retroviruses replicate by forming double-stranded DNA.

Rheumatoid arthritis: Degenerative joint diseases that are usually associated with inflammation and immune components.

***r*-selection:** Selection for life histories with rapid development and high rates of reproduction that are usually associated with small size, semelparity, and short lifespans; generally the converse of *K*-selection (Chapter 1.5.1).

Segmental progeroid syndromes: A concept introduced by George M. Martin to represent degenerative syndromes that show limited features of senescence. Thus Hutchinson-Guilford syndrome is associated with intense arteriosclerosis, but without neurodegeneration or dementia.

Semelparous: Organisms that reproduce only once, or in only one season, usually followed by death.

Senescence: A deteriorative change that causes increased mortality. Senescence is mostly discussed in this book in regard to organisms, but also applies to clones of cells that cease replicating (clonal senescence).

Senility: A colloquialism that encompasses all ills of aging, and that can represent the end stage of senescence when mortality risk approaches 100%.

Somatic mutation hypothesis: The hypothesis that senescence results from the accumulation of mutations in the genome of somatic cells (Chapter 8.2).

Subimago: In mayflies, a winged and very brief stage after hatching that is followed by another molt yielding the adult, often with further reduction of body parts.

Thromboses: Blood clots in blood vessels that may be attached to the wall without blocking flow or may occlude blood flow.

Transcription: The production of RNA through an enzymatic process that uses a DNA template. In retroviruses, the genome is RNA, which is then used as a template for the enzyme reverse transcriptase to produce a DNA strand.

Translation: The enzymatic production of proteins on ribosomes by *de novo* synthesis using a messenger RNA (mRNA) template.

Unitary: Organisms, such as flies and fish, with a nonrepeating internal structure that is physiologically independent. Unitary organisms typically do not reproduce vegetatively, and often show senescence. As distinct from clonal or colonial organisms that consist of many repeating modules (Chapter 4.4; Grosberg and Patterson, 1989).

Vascular-ischemic syndromes: Disorders that result from insufficient blood supply, usually as the consequence of arteriosclerosis or stroke.

Vegetative reproduction: Reproduction in plants and animals from somatic cells without formation of an egg or of gametes.

Wear-and-tear syndromes: Disorders from the mechanical or biochemical effects of repeated usage, as typified by the mechanical senescence of insect wings or the overuse of joints that causes arthritis.

Werner's syndrome: A very rare progeroid syndrome with adult onset (Chapter 6.5.3.3).

Xenotropic: A virus that replicates in cells from other species as well as in those of the usual host. *See also* Ecotropic

Xeroderma pigmentosum: A skin disorder with hereditary forms associated with extreme sensitivity to UV-radiation that causes pigmented spots which frequently become malignant. Hereditary forms are caused by enzymatic deficiencies in the excision of dimerized bases.

Bibliography

Abbott, M. H., Abbey, H., Bolling, D. R., and Murphy, E. A. (1978) The familial component in longevity. A study of the offspring of nonagenarians. 3. Intrafamilial studies. *Am. J. Med. Genet.* 2:105–120.

Abbott, M. H., Murphy, E. A., Bolling, D. R., and Abbey, H. (1974) The familial component in longevity. A study of the offspring of nonagenarians. 2. Preliminary analyses of the completed study. *Johns Hopkins Med. J.* 134:1–16.

Abernethy, J. D. (1979) The exponential increase in mortality rate with age attributed to wearing-out of biological components. *J. Theoret. Biol.* 80:333–354.

———. (1981) Erratum. The exponential increase in mortality rate with age attributed to wearing-out of biological components. *J. Theoret. Biol.* 90:159.

Abraham, E. C., Taylor, J. F., and Lang, C. A. (1978) Influence of mouse age and erythrocyte age on glutathione metabolism. *Biochem. J.* 178:819–825.

Abruzzo, M. A., Mayer, M., and Jacobs, P. A. (1985) Aging and aneuploidy. Evidence for the preferential involvement of the inactive X chromosome. *Cytogenet. Cell. Genet.* 39:275–278.

Achan, P. D. (1961) On prolonging the longevity of the dog-tick *Rhipicephalus sanguineus latr.* after oviposition. *Curr. Sci.* 30:265–266.

Adams, M. R., and Bond, M. G. (1979) Benign prostatic hyperplasia in a squirrel monkey (*Saimiri sciureus*). *Lab. Anim. Sci.* 29:674–676.

Addicott, F. T. (1982) *Abscission*. Berkeley and Los Angeles: Univ. of California Press.

Adelman, R. C. (1970a) An age-dependent modification of enzyme regulation. *J. Biol. Chem.* 245:1032–1035.

———. (1970b) Reappraisal of biological aging. *Nature* 228:1095–1096.

———. (1971) Age-dependent effects in enzyme induction. A biochemical expression of aging. *Exp. Gerontol.* 6:75–87.

Adelman, R. C., Britton, G. W., Rotenberg, S., Ceci, L., and Karoly, K. (1978) Endocrine regulation of gene activity in aging animals of different genotypes. In *Genetic Effects on Aging*, D. Bergsma and D. E. Harrison (eds.), 355–364. New York: Liss.

Ader, M., Agajanian, T., Finegood, D. T., and Bergman, R. (1987) Recombinant deoxyribonucleic acid–derived 22k—and 20k human growth hormone generate equivalent diabetogenic effects during chronic infusion in dogs. *Endocrinology* 120:725–731.

Adler, A. J., and Nelson, J. F. (1988) Aging and chronic estradiol exposure impair estradiol-induced cornification but not proliferation of vaginal epithelium in C57BL/6J mice. *Biol. Reprod.* 38:175–182.

Adler, J. E., and Black, I. B. (1984) Plasticity of substance P in mature and aged sympathetic neurons in culture. *Science* 225:1499–1500.

Adolph, E. D. (1949) Quantitative relations in the physiological constitution of mammals. *Science* 109:579–585.

Adolphe, M., Ronot, X., Jaffray, P., Hecquet, C., Fontagne, J., and Lechat, P. (1983) Effects of donor's age on growth kinetics of rabbit articular chondrocytes in culture. *Mech. Age. Dev.* 23:191–198.

Agid, Y., Blin, J., Bonnet, A. M., Dubois, B., Javoy-Agid, F., Ruberg, M., and Sherman, D. (1989) Does aging contribute to aggravation of Parkinson's disease? In *Parkisonism and Aging*, D. B. Calne (ed.), 115–124. New York: Raven Press.

Ahmed, M. U., Thorpe, S. R., and Baynes, J. W. (1986) Identification of N-carboxymethyllysine, a degradation product of fructoselysine in glycated protein. *J.Biol.Chem.* 261:4889–4894.

Aiken, D. E. (1980) Molting and growth. In *The Biology and Management of Lobsters*, vol. 1, J. S. Cobb and B. F. Phillips (eds.), 147–161. New York: Academic Press.

Aiken, D. E., and Waddy, S. L. (1980) Reproductive biology. In *The Biology and Management of Lobsters*, vol.1, J. S. Cobb and B. F. Phillips (eds.), 215–276. New York: Academic Press.

Aikten, R. J. (1977) Embryonic diapause. In *Development in Mammals*, vol. 1, M. H. Johnson (ed.), 307–359. New York: North-Holland.

Aitken, J. M., Armstrong, E., and Anderson, J. R. (1972) Osteoporosis after oophorectomy in the mature female rat and the effect of oestrogen and/or progesterone replacement therapy in its prevention. *J. Endocrinol.* 55:79–87.

Akiyama, H., Yonezu, T., Tsunasawa, S., Sakiyama, F., and Takeda, T. (1986) Periodic acid-Schiff (PAS)-positive, granular structures increase in the brain of senescence-accelerated mouse (SAM). *Acta Neuropathol.* (Berlin) 72:124–129.

Alberch, P., Gould, S. J., Oster, G. F., Wake, D. B. (1979) Size and shape in ontogeny and phylogeny. *Paleobiology* 5:296–317. Alberman, E., Polani, P. E., Fraser Roberts, J. A., Spicer, C. C., Elliot, M., and Armstrong, E. (1972) Parental exposure to X-irradiation and Down's syndrome. *Ann. Human Genet.* (London) 36:195–208.

Albert, M. S., and Moss, M. B. (eds.) (1984) *Geriatric Neuropsychology*. New York: Guilford Press.

Albert, P. S., Brown, S. J., and Riddle, D. L. (1981) Sensory control of dauer larva formation in *Caenorhabditus elegans*. *J. Comp. Neurol.* 198:435–451.

Albin, R. L. (1988) The pleiotropic gene theory of senescence: Supportive evidence from human genetic disease. *Eth. Sociobiol.* 9:371–382.

Albright, F., Smith, P. H., and Richardson, A. M. (1941) Post-menopausal osteoporosis: Its clinical features. *J.A.M.A.* 116:2465–2474.

Alevizaki, C. C., Ikkos, D. G., and Singhelakis, P. (1973) Progressive decrease of true intestinal calcium absorption with age in normal man. *J. Nucl. Med.* 14:760–762. [not read]

Alexander, P. (1966) Is there a relationship between aging, the shortening of life-span by radiation, and the induction of somatic mutations? In *Perspectives in Experimental Gerontology*, N. W. Shock (ed.), 266–279. Springfield, IL: C. C. Thomas.

Alexander, R. D. (1968) Life cycle origins, speciation, and related phenomena in crickets. *Q. Rev. Biol.* 43:1–41.

Alexander, R. D., and Moore, T. E. (1962) The evolutionary relationships of 17–year and 13–year cicadas, and three new species (Homoptera, Cicadae, Magicicada). *Misc. Publs. Mus. Zool. Univ. Mich.*, no. 121:5–59.

Allen, G. R., and Berra, T. M. (1989) Life history aspects of the West Australian salamanderfish, *Lepidogalaxia salamandroides* Mees. *Rec. West. Austral. Mus.* 14:253–267.

Allendorf, F. A., and Thorgaard, G. H. (1984) Tetraploidy and evolution of salmonid fishes. In *Evolutionary Genetics of Fishes*, B. J. Turner (ed.), 1–53. New York: Plenum.

Allison, A. C. (1961) Turnover of erythrocytes and plasma proteins in mammals. *Nature* 188:37–40.

Allison, J. C., and Weinmann, H. C. (1970) Effect of the absence of the developing grain on carbohydrate content and senescence of maize leaves. *Plant Physiol.* 46:435–436. [not read]

Allsopp, P. G. (1979) Determination of age and mated state of adult *Pterohelaeus darlingensis* Carter (Coleoptera: Tenebrionidae). *J. Austral. Entomol. Soc.* 18:235–239.

Alm, G. (1952) Year class fluctuations and span of life of perch. *Rep. Inst. Freshwater Res. Drottningham* 33:17–38.

Alsum, P., Richardson, R., Houser, W. D., and

Uno, H. (1990) Geriatric diseases in a captive colony of rhesus macaques (*Macaca mulatta*). Proceedings of the American Society of Primatologists, Thirteenth Annual Meeting (abstract). In press.

Alter, M., Zhen-xin, Z., Davanipour, Z., Sobel, E., Zibulewski, J., Schwartz, G., and Friday, G. (1986) Multiple sclerosis and childhood infections. *Neurology* 36:1386–1389.

Altland, P. D. (1951) Observations on the structure of the reproductive organs of the box turtle. *J. Morphol.* 89:599–621.

Altman, P. L., and Katz, D. D. (1976) *Cell Biology.* Biological Handbooks. Bethesda, MD: Federation of the American Society of Experimental Biology.

Altman, R. B., and Kirmayer, A. H. (1976) Diabetes mellitus in the avian species. *J. Am. Anim. Hosp. Assoc.* 12:531.

Altszuler, N., Rathzeb, I., Winkler, B., de Bodo, R. C., and Steele, R. (1968) The effects of growth hormone on carbohydrate and lipid metabolism in the dog. *Ann. N.Y. Acad. Sci.* 148:441–458.

Amador, A., Steger, R. W., Bartke, A., Johns, A., Silker-Khodr, T. M., Parker, C. R., and Shepherd, A. M. (1985) Testicular LH receptors during aging in Fischer 344 rats. *J. Androl.* 6:61–64. [not read]

Ambros, V. (1989) A hierarchy of regulatory genes controls a larva-to-adult switch in *C. elegans. Cell* 57:49–57.

Amenta, D., Ferrante, F., Franch, F., and Amenta, F. (1988) Effects of long-term hydergine administration on lipofuscin accumulation in senescent rat brain. *Gerontology* 34:250–256.

Ames, B. N. (1983) Dietary carcinogens and anticarcinogens (oxygen radicals and degenerative diseases). *Science* 221:1256–1264.

———. (1988) Correction. *Proc. Natl. Acad. Sci.* 85:9508.

Ames, B. N., Cathcart, R., Schwiers, E., and Hochstein, P. (1981) Uric acid provides an antioxidant defense in humans against oxidant- and radical-caused aging and cancer: A hypothesis. *Proc. Natl. Acad. Sci.* 78:6858–6862.

Ames, B. N., Saul, R. L., Schwiers, E., Adelman, R., and Cathcart, R. (1985) Oxidative DNA damage as related to cancer and aging.

The assay of thymine glycol, thymidine glycol, and hydroxymethyluracil in human and rat urine. In *Molecular Biology of Aging: Gene Stability and Gene Expression,* R. S. Sohal, L. S. Birnbaum, and R. G. Cutler (eds.), 137–144. New York: Raven Press.

Amin, A., Chai, C. K., and Reineke, E. P. (1957) Differences in thyroid activity of several strains of mice and hybrids. *Am. J. Physiol.* 191:34–36.

Anders, F., Schartl, M., and Scholl, E. (1981) Evaluation of environmental and hereditary factors in carcinogenesis, based on studies in *Xiphophorus.* In *Phyletic Approaches to Cancer,* C. J. Dawe et al. (ed.), 289–309. Tokyo: Japan Scientific Society Press.

Andersen, F. S. (1951) Contributions to the biology of the ruff (*Philomacus pugnax*). *Dansk orn Foren. Tidsskr.* 45:145–173.

Anderson, A. C. (ed.) (1970) *The Beagle as an Experimental Dog.* Ames: Iowa State Univ. Press.

Anderson, A. D., Barrett, S. F., and Robbins, J. H. (1976) Relation of DNA repair processes to pathological ageing of the nervous system in xeroderma pigmentosum. *Lancet* 1:1318–1320.

Anderson, C. L. (1978) Responses of dauer larvae of *Caenorhabditis elegans* to thermal stress and oxygen deprivation. *Can. J. Zool.* 56:1786–1791.

Anderson, J. R. (1987) Reproductive strategies and gonotrophic cycles of black flies. In *Black Flies. Ecology, Population Management, and Annotated World List,* K. C. Kim and R. W. Merritt (eds.), 276–293. University Park: Pennsylvania State Univ. Press.

Anderson, K. V., and Nusslein-Volhard, C. (1984) Information for the dorsal-ventral pattern of the *Drosophila* embryo is stored as maternal mRNA. *Nature* 311:223–227.

Anderson, O. R. (1988) Ecology, physiology, life history. In *Comparative Protozoology.* New York: Springer-Verlag.

Anderson, R. M., and May, R. M. (1985) Helminth infections of humans: Mathematical models, population dynamics, and control. *Adv. Parasitol.* 24:1–102.

Andervont, H. B. (1940) The influence of foster nursing upon the incidence of spontaneous mammary cancer in resistant and susceptible mice. *J. Natl. Canc. Inst.* 1:147–153.

Andres, R., and Tobin, J. D. (1977) Endocrine systems. In *Handbook of the Biology of Aging*, 1st ed., C. E. Finch and L. Hayflick (eds.), 357–378. New York: Van Nostrand.

Andrew, W. (1959) The reality of age differences in nervous tissue. *J. Gerontol.* 14:259–267.

Andrewartha, H. G. (1945) Some differences in the physiology and ecology of locusts and grasshoppers. *Bull. Entomol. Res.* 35:379–389.

———. (1952) Diapause in relation to the ecology of insects. *Biol. Rev.* 27:50–107.

Andrews, A. D., Barrett, S. F., and Robbins, J. H. (1976) Relation of DNA repair processes to pathological ageing of the nervous system in xeroderma pigmentosum. *Lancet* 1:1318–1320.

———. (1978) Xeroderma pigmentosum neurological abnormalities correlate with colony-forming ability after ultraviolet radiation. *Proc. Natl. Acad. Sci.* 75:1984–1988.

Andrews, H. N., and Murdy, W. H. (1958) *Lepidophloios* and ontogeny in arborescent lycopods. *Am. J. Bot.* 45:552–560.

Andrews, K. J., and Hegeman, G. D. (1976) Selective disadvantage of non-functional protein synthesis in *Escherichia coli. J. Mol. Evol.* 8:317–328.

Anton, H. J. (ed.) (1988) *Control of Cell Proliferation and Differentiation during Regeneration.* Monographs in Developmental Biology, vol. 21. Basel: Karger.

Anver, M. R., Cohen, B. J., Lattuada, C. P., and Foster, S. J. (1982) Age-associated lesions in barrier-reared male Sprague-Dawley rats: A comparison between Hap:(SD) and Crl:COBS[R]CD[R] (SD) stocks. *Exp. Aging Res.* 8:3–24.

Anzai, K., and Goto, S. (1987) Brain-specific small RNA during development and ageing of mice. *Mech. Age. Dev.* 39:129–135.

Aoki, K. (1963) Experimental studies on the relationship between endocrine organs and hypertension in spontaneously hypertensive rats. 1. Effects of hypophysectomy, adrenalectomy, thyroidectomy, nephrectomy, and sympathectomy on blood pressure. *Jpn. Heart J.* 14:443.

Ara, G., Ritchie, D. G., and Papaconstantinou, J. (1983) Reversal of age-related alteration in transferrin secretion. *J. Cell Biol.* 97:1307a.

———. (In prep.) Effect of aging on post-transcriptional processing of transferrin in mouse liver. University of Texas, Galveston, Dept. of Biochemistry.

Arad, Y., Badimon, J. J., Badimon, L., Hembree, W. C., and Ginsberg, H. N. (1989) Dehydroepiandrosterone feeding prevents aortic fatty streak formation and cholesterol accumulation in cholesterol-fed rabbit. *Arteriosclerosis* 9:159–166.

Arai, H., Lee, V. M.-Y., Otvos, L., Greenberg, B. D., Lowery, D. E., Sharma, S. K., Schmidt, M. L., and Trojanowski, J. Q. (1990) Defined neurofilament, T, and β-amyloid precursor protein epitopes distinguish Alzheimer from non-Alzheimer senile plaques. *Proc. Natl. Acad. Sci.* 87:2249–2253.

Arai, K., Iizuka, S., Tada, Y., Oikawa, K., and Taniguchi, N. (1987) Increase in the glucosylated form of erythrocyte Cu-Zn-superoxide dismutase in diabetes and close association of the nonenzymatic glucosylation with the enzyme activity. *Biochim. Biophys. Acta* 924:292–296.

Arai, K., Maguchi, S., Fujii, S., Ishibashi, H., Oikawa, K., and Taniguchi, N. (1987) Glycation and inactivation of human Cu-Zn-superoxide dismutase. Identification of the *in vitro* glycated sites. *J. Biol. Chem.* 262:16,969–16,972.

Arking, R. (1987) Successful selection for increased longevity in *Drosophila:* Analysis of the survival data and presentation of a hypothesis on the genetic regulation of longevity. *Exp. Gerontol.* 22:199–220.

Arking, R., Buck, S., Wells, R. A., and Pretzlaff, R. (1988) Metabolic rates in genetically based long-lived strains of *Drosophila. Exp. Gerontol.* 23:59–76.

Arking, R., and Dudas, S. P. (1989) Review of genetic investigations into the aging processes of *Drosophila. J. Am. Geriatr. Soc.* 37:757–773.

Armbrecht, H. J. (1984) Changes in calcium and vitamin D metabolism with age. In *Nutritional Intervention in the Aging Process,* H. J. Armbrecht, J. M. Prendergast, and

R. M. Coe (eds.), 69–83. New York: Springer-Verlag.

———. (1986) Age-related changes in calcium and phosphorus uptake by rat small intestine. *Biochim. Biophys. Acta* 882:281–286.

Armbrecht, H. J., Forte, L. R., and Halloran, B. P. (1984) Effect of age and dietary calcium on renal 25–OH-D metabolism, serum 1,25(OH)₂D, and PTH. *Am. J. Physiol.* 246:E266–E270.

Armbrecht, H. J., Strong, R., Boltz, M., Rocco, D., Wood, W. G., and Richardson, A. (1988) Modulation of age-related changes in serum 1,25–dihydroxyvitamin D and parathyroid hormone by dietary restriction of Fischer 344 rats. *J. Nutrit.* 118:1360–1365.

Armitage, P., and Doll, R. (1954) The age distribution of cancer and a multistage theory of carcinogenesis. *Brit. J. Canc.* 8:1–12.

Armstrong, B. K., Brown, J. B., Clarke, H. T., Crooke, D. K., Hahnel, R., Maserei, J. R., and Ratajczak, T. (1981) Diet and reproductive hormones: A study of vegetarian and nonvegetarian postmenopausal women. *J. Natl. Canc. Inst.* 67:761–767.

Armstrong, D., Rinehart, R., Dixon, L., and Reigh, D. (1978) Changes of peroxidase with age in *Drosophila. Age* 1:8–12.

Armstrong, E. (1989) Comparative review of primate motor systems. *J. Motor Behav.* 21:493–517.

Armstrong, J. A. (1989) Biotic pollination mechanisms in the Australian flora: A review. *New Zealand J. Bot.* 17:467–508.

Arnal, M., Obled, C., and Attaix, D. (1983) Renouvellement des proteines et flux d'acides amines au cours du développement. In *IVe Symp. Int. Métab. et Nutrit. Azotés,* 117–136. Clermont-Ferrand, France.

Arndt, W. (1928) Lebensdauer, Altern, and Tod der Schwammer. *Ges. Naturf. Fr.* (Berlin) 23:23–44.

Arnold, C. A. [1929] (1939) *The Genus Callixylon from the Upper Devonian of Central and Western New York.* Reprinted from *Papers of the Michigan Acadamy of Science, Arts, and Letters,* vol. 11, 1–50.

———. (1931) On *Callixylon Newberryi* (Dawson) Elkins et Wieland. *Contr. Museum of Paleontol.* 3 (12): 207–232.

Arnold, J. M., and Carlson, B. A. (1986) Living *Nautilus* embryos: Preliminary observations. *Science* 232:73–76.

Arnold, J. W. (1959) Observations on living hemocytes in wing veins of the cockroach *Blaberus giganteus* (L.) (Orthoptera: Blattidae). *Ann. Entomol. Soc. Am.* 52:229–236.

———. (1961) Further observations on amoeboid haemocytes in *Blaberus giganteus. Can. J. Zool.* 39:755–766.

———. (1964) *Blood Circulation in Insect Wings.* Memoires Entomol. Soc. Canada, no. 38.

Arrhenius, S. (1907) *Immunochemistry.* London: Macmillan. [not read]

Asayama, K., Janco, R. L., and Burr, I. M. (1985) Selective induction of manganous superoxide dismutase in human monocytes. *Am. J. Physiol.* 249 (Cell Physiology 18): C393–397.

Aschheim, P. (1965) Resultats fournis par la greffe heterochrone des ovarie dan's l'etude de la regulation hypothalamus-hypophyso-ovarienne de la ratte senile. *Gerontologia* 10:65–75.

———. (1976) Aging in the hypothalamic-hypophysical-ovarian axis in the rat. In *Hypothalamus, Pituitary, and Aging,* A. V. Everitt and J. A. Burgess (eds.), 376–418. Springfield, IL: C. C. Thomas.

Aschoff, J., Gunther, B., and Kramer, K. (1971) *Energiehaushalt und Temperaturregulation.* Munich: Urban and Schwarzenberg. [not read]

Asdell, A. A. (1946) Comparative chronologic age in man and other mammals. *J. Gerontol.* 1:224–236.

Asencot, M., and Lensky, Y. (1984) Juvenile hormone induction of "queenliness" on female honey bee (*Apis mellifera L.*) larvae reared on worker jelly and on stored royal jelly. *Comp. Biochem. Physiol.* 78B:109–117.

Ashburner, M., and Bonner, J. J. (1979) The induction of gene activity in *Drosophila* by heat shock. *Cell* 17:241–254.

Ashby, E. (1950) Studies in the morphogenesis of leaves. 6. Some effects of length of day upon leaf shape in *Ipomoea caerulea. New Phytol.* 49:375–387.

Ashley, F., Kannel, W. B., Sorlie, P. D., and Masson, R. (1975) Pulmonary function: Relation to aging, cigarette habit, and mortality. *Ann. Intern. Med.* 82:739–745.

Ashley-Montague, M. F. (1957) *The Reproductive Development of the Female with Special Reference to the Period of Adolescent Sterility.* New York: Julian Press.

Atwood, K. C., and Pepper, F. J. (1961) Erythrocyte automosaicism in some persons of known genotype. *Science* 134:2100–2101.

Atz, J. W. (1957) The relation of the pituitary to reproduction in fishes. In *The Physiology of the Pituitary Gland of Fishes,* G. E. Pickford and J. W. Atz (eds.), 178–270. New York: New York Zoological Society.

Auclair, J.-L., and Aroga, R. (1987) Influence de la température sur la croissance et la reproduction de quatre biotypes du puceron du pois, *Acyrthosiphon pisum* (Homoptera). *Ann. Soc. Entomol. Fr.,* n.s. 23:279–286.

Aufderheide, K. J. (1984) Clonal aging in *Paramecium tetraurelia.* Absence of evidence for a cytoplasmic factor. *Mech. Age. Dev.* 28:57–66.

———. (1987) Clonal aging in *Paramecium tetraurelia.* 2. Evidence of functional changes in the macronucleus with age. *Mech. Age. Dev.* 37:265–279.

Aufderheide, K. J., and Schneller, M. V. (1985) Phenotypes associated with early clonal death in *Paramecium tetraurelia. Mech. Age. Dev.* 32:299–309.

Austad, S. N. (1988) The adaptable opossum. *Sci. Am.,* Feb., 98–104.

———. (1989) Life extension by dietary restriction in the bowl and doily spider, *Frontinella pyramitela. Exp. Gerontol.* 24:83–92.

Austad, S. N., and Fischer, K. E. (In press) Mammalian aging, metabolism, and ecology: Evidence from the bats and marsupials. *J. Geront.*

Austin, C. R. (1961) *The Mammalian Egg.* Oxford: Blackwell.

Avioli, L. V. (1982) Aging, bone, osteoporosis. In *Endocrine Aspects of Ageing,* S. G. Korenman (ed.), 199–230. London: Elsevier.

Avise, J. C., and Selander, R. K. (1972) Evolutionary genetics of cave-dwelling fishes of the genus *Astyanax. Evolution* 26:1–19.

Bachmair, A., Finley, D., and Varshavsky, A. (1986) In vivo half-life of a protein is a function of its amino-terminal residue. *Science* 234:170–186.

Bachmann, K. (1972) Genome size in mammals. *Chromosoma* (Berlin) 37:85–93.

Bachmann, K., Goin, O. B., and Goin, C. J. (1972) Nuclear DNA amounts in vertebrates. *Brookhaven Symp. Biol.* 23:419–447.

Bachorik, P. S., and Kwiterovich, P. O. (1988) Apolipoprotein measurements in clinical biochemistry and their utility vis-à-vis conventional assays. *Clin. Chem. Acta* 178:1–34.

Backhouse, G. N., and Vanner, R. W. (1978) Observations on the biology of the dwarf galaxiid *Galaxiella pusilla* (Mack) (Pisces: Galaxiidae). *Vic. Nat.* 95:128–133.

Bada, J. L. (1972) Kinetics of racemization of amino acids as a function of pH. *J. Am. Chem. Soc.* 95:1371–1373.

Bada, J. L., Brown, S., and Masters, P. M. (1980) Age determination of marine mammals based on aspartic acid racemization in the teeth and lens nucleus. *Rep. Int. Whal. Commn.* (special issue 3): 113–118.

Bada, J. L., and Helfman, P. M. (1975) Amino acid racemization of fossil bones. *World Arch.* 7:160–173.

Baerg, W. J. (1920) The life cycle and mating habits of the male tarantula. *Q. Rev. Biol.* 3:109–116.

———. (1958) *The Tarantula.* Lawrence: Univ. of Kansas Press.

———. (1963) Tarantula life history records. *J. N.Y. Entomol. Soc.* 71:233–238.

Baerg, W. J., and Peck, W. B. (1970) A note on the longevity and molt cycle of two tropical theraphosids. *Bull. Brit. Arachnol. Soc.* 7:107.

Bahr, J. M., and Palmer, S. S. (1989) The influence of aging on ovarian function. *Poultry Biol.* 2:103–110.

Bailey, C. H., Castellucci, V. F., Koester, J., and Chen, M. (1983) Behavioral changes in aging *Aplysia.* A model system for studying the cellular basis of age-impaired learning, memory, and arousal. *Behav. Neurol. Biol.* 38:70–81.

Bailey, D. W. (1978) Sources of subline divergence and their relative importance for sublines of six major inbred strains of mice. In *Origins of Inbred Mice,* H. C. Morse, III (ed.), 197–216. New York: Academic Press.

Baker, G. H. B. (1982) Life events before the onset of rheumatoid arthritis. *Psychother. Psychosom.* 38:173–177.

Baker, G. T. (1975) Age-dependent arginine phosphokinase activity changes in male vestigial and wild-type *Drosophila melanogaster. Gerontologia* (Basel) 21:203–210.

Baker, G. T., Jacobson, M., and Mokrynski, G. (1985) Aging in *Drosophila.* In *CRC Handbook of Cell Biology of Aging*, V. J. Cristofalo, R. C. Adelman, and G. S. Roth (eds.), 511–578. Boca Raton, FL: CRC Press.

Baker, G. T., and Schmidt, T. (1976) Changes in 80S ribosomes from *Drosophila melanogaster* with age. *Experientia* 32:1505–1506.

Baker, G. T., and Sprott, R. L. (1988) Biomarkers of aging. *Exp. Gerontol.* 23:223–239.

Baker, J. E., Felix, C. C., Olinger, G. N., and Kalyanaraman, B. (1988) Myocardial ischemia and reperfusion. Direct evidence for free radical generation by electron spin resonance spectroscopy. *Proc. Natl. Acad. Sci.* 85:2786–2789.

Balázs, A. (1966) Experimental investigations with the regenerative ability of planarians in old age. *Seventh International Congress on Gerontology,* A. Balázs (ed.), 111–117. Budapest.

Balázs, A., and Burg, M. (1962) Span of life and senescence of *Dugesia lugubris. Gerontologia* (Basel) 6:227–236.

Balázs, A., and Burg, M. (1963) Influence of copulation on the longevity of the great wax moth (*Galleria mellonella*). *Gerontologia* (Basel) 7:233–244.

Balázs, A., Kovats, A., and Burg, M. (1962) Biochemical analysis of premortal involution processes in aphagous imagines. *Acta Biol. Hung.* 13:169–176.

Ball, M. J. (1977) Neuronal loss, neurofibrillary tangles, and granulovacuolar degeneration in the hippocampus with ageing and dementia. *Acta Neuropathol.* (Berlin) 37:111–118.

———. (1978) Topographic distribution of neurofibrillary tangles and granulovacuolar degeneration in hippocampal cortex of aging and demented patients. A quantitative study. *Acta Neuropathol.* (Berlin) 42:73–80.

Ball, M. J., and Nuttall, K. (1980) Neurofibrillary tangles, granulovacuolar degeneration, and neuron loss in Down syndrome. Quantitative comparison with Alzheimer dementia. *Ann. Neurol.* 7:462–465.

Ball, Z., Barnes, R., and Visscher, M. (1947) The effects of dietary caloric restriction on maturity and senescence with particular reference to fertility and longevity. *Am. J. Physiol.* 150:511–519.

Ballard, F. J., Hanson, R. W., and Kronfeld, D. S. (1969) Gluconeogenesis and lipogenesis in tissue from ruminant and nonruminant animals. *Fed. Proc.* 28:218–227.

Balthazart, J., Turek, R., and Ottinger, M. A. (1984) Altered brain metabolism of testosterone is correlated with reproductive decline in aging quail. *Horm. Behav.* 18:330–345.

Bancroft, F. W. (1903) Variation and fusion in colonies of compound ascidians. *Proc. Calif. Acad. Sci.* (3d ser.) 3:137–186.

Banks, H. P. (1970) *Evolution of Plants of the Past.* Berkeley: Wadsworth.

Bantle, J. A., and Hahn, W. E. (1976) Complexity and characterization of polyadenylated RNA in the mouse brain. *Cell* 8:139–150.

Barber, J. R., and Clarke, S. (1983) Membrane protein carboxyl methylation increases with human erythrocyte age. Evidence for an increase in the number of methylatable sites. *J. Biol. Chem.* 258:1189–1196.

Bardack, D. (1985) Les premiers fossiles de hagfish (Myxiniformes) et entéropneusta (Hemichordata) dépôts de la faune (Pennsylvanienne) du Mazon Creek dans l'Illinois, U.S.A. *Bull. Soc. Hist. Nat. Autun.* 116:97–99 (abstract).

Bardack, D., and Richardson, E. S. (1977) New agnathous fishes from the Pennsylvanian of Illinois. *Fieldiana Geol.* 33:489–510.

Bardack, D., and Zangerl, R. (1968) First fossil lamprey. A record from the Pennsylvanian of Illinois. *Science* 162:1265–1267.

———. (1971) Lampreys in the fossil record. In *The Biology of Lampreys,* vol. 1, M. W. Hardisty and I. C. Potter (eds.), 67–84. New York: Academic Press.

Barden, H. (1970) Relationship of Golgi thiaminepyrophosphatase and lysosomal acid phosphatase to neuromelanin and lipofuscin in cerebral neurons of the aging rhesus monkey. *J. Neuropathol. Exp. Neurol.* 29:225–240.

Barker, K., Beveridge, I., Bradley, A. J., and Lee, A. K. (1978) Observations on spontaneous stress-related mortality among males of the dasyurid marsupial *Antechinus stuartii* (MacLeay). *Austral. J. Zool.* 26:435–447.

Barker, W. G. (1953) Proliferative capacity of the medullary sheath region in the stem of *Tilla americani. Am. J. Bot.* 40:773–778.

Barnett, J. L. (1973) A stress response in *Antechinus stuartii* (MacLeay). *Austral. J. Zool.* 21:501–513.

Barnett, S. A. (1965) Adaptation of mice to cold. *Biol. Rev.* 40:5–51.

Barney, J. L., and Neukom, J. E. (1979) Use of arthritis care by the elderly. *Gerontologist* 19:548–554.

Baroldi, G., Falzi, G., and Mariani, F. (1979) Sudden coronary death. A postmortem study in 208 selected cases compared to 97 "control" subjects. *Am. Heart J.* 98:20–31.

Barr, T. C., Jr. (1968) Cave ecology and the evolution of troglobites. In *Evolutionary Biology,* vol. 2, T. H. Dobzhansky, M. R. Hecht, and W. C. Steer (eds.), 35–102. Amsterdam: North-Holland.

Barrett-Conner, E., Khaw, K.-T., Yen, S. S. C. (1986) A prospective study of dehydroepiandrosterone sulfate, mortality, and cardiovascular disease. *N. Engl. J. Med.* 315:1519–1524.

Barrows, C. H., Jr., and Kokkonen, G. C. (1981) Effect of age on the DNA, RNA, and protein content of tissues. In *CRC Handbook of Biochemistry of Aging,* J. R. Florini (ed.), 115–134. Boca Raton, FL: CRC Press.

———. (1987) The effect of age and diet on the cellular protein synthesis of liver of male mice. *Age* 10:54–57.

Bartels, H. (1964) Comparative physiology of oxygen transport in mammals. *Lancet* 594–604.

———. (1982) Metabolic rate of mammals equals the 0.75 power of their body weight. *Exp. Biol. Med.* 7:1–11.

Barth, L. G., and Barth, L. J. (1967) A study of regression and budding in *Perophora vividis. J. Morphol.* 118:451–459.

Bartholomew, J. W., and Mittwer, T. (1953) Demonstration of yeast bud scars with the electron microscope. *J. Bacteriol.* 65:272–275.

Barton, A. A. (1950) Some aspects of cell division in *Saccharomyces cerevisiae. J. Gen. Microbiol.* 4:84–86.

Barton, M. C., and Shapiro, D. J. (1988) Transient administration of estradiol-17β establishes an autoregulatory loop permanently inducing estrogen receptor mRNA. *Proc. Natl. Acad. Sci.* 85:7119–7123.

Bass, A., Gutmann, E., and Hanzlikova, V. (1975) Biochemical and histochemical changes in energy supply–enzyme pattern of muscles of the rat during old age. *Gerontologia* (Basel) 21:31–45.

Batt, T., and Woolhouse, H. W. (1975) Changing activities during senescence and sites of synthesis of photosynthetic enzymes in leaves of the labiate *Perilla fructescens* (L.) Britt. *J. Exp. Bot.* 26:569–579.

Batterham, P., Lovett, J. A., Starmer, W. T., and Sullivan, D. T. (1983) Differential regulation of duplicate alcohol dehydrogenase genes in *Drosophila mojavensis. Dev. Biol.* 96:346–354.

Baudry, M., Simonson, L., Dubrin, R.. and Lynch, G. (1986) A comparative study of soluble calcium-dependent proteolytic activity in brain. *J. Neurobiol.* 17:15–28. Bauer, K. A., Weiss, L. M., Sparrow, D., Vokonas, P. S., and Rosenberg, R. D. (1987) Aging-associated changes in indices of thrombin generation and protein C activation in humans. *J. Clin. Invest.* 80:1527–1534.

Baumann, P., and Chen, P. S. (1969) Alterung und Proteinsynthesese bei *Drosophila melanogaster. Rev. Suisse Zool.* 75:1051–1055.

Baumann, R. W. (1982) Plecoptera. In *Synopsis and Classification of Living Organisms,* vol. 2, S. P. Parker (ed.), 389–393. New York: McGraw-Hill.

Baxter, J. D., and Forsham, P. H. (1972) Tissue effects of glucocorticoids. *Am. J. Med.* 53:573–589.

Baylis, C., Fredericks, M., Leypoldt, J., Frigon, R., Wilson, C., and Henderson, L. (1988) The mechanisms of proteinuria in aging rats. *Mech. Age. Dev.* 45:111–126.

Baynes, J. W., Dunn, J. A., Dyer, D. G., Knecht, K. J., Ahmed, M. U., and Thorpe, S. R. (1990) Role of glycation in development of pathophysiology in diabetes and aging. In *Glycated Proteins in Diabetes Mellitus,* Proceedings of Satellite Symposium of International Diabetes Federation, 1989–1990.

Baynes, J. W., and Monnier, V. M. (eds.) (1989) *The Maillard Reaction in Aging, Diabetes, and Nutrition.* New York: Liss.

Bayreuther, K., Rodemann, H. P., Hommel, R., Dittman, K., Albiez, M., and Francz, P. I. (1988) Human skin fibroblasts *in vitro* differentiate along a terminal cell lineage. *Proc. Natl. Acad. Sci.* 85:5112–5116.

Beach, F. A. (1950) The snark was a boojum. *Am. Psychol.* 5:115–124.

Beal, W. J. (1905) The vitality of seeds. *Bot. Gaz.* 40:140–143.

Beale, G. H. (1958) The role of cytoplasm in antigen determination in *Paramecium aurelia. Proc. Roy. Soc. B.* 148:308–314.

Beamish, R. J., and McFarlane, G. A. (1985) Annulus development on the second dorsal spine of the spiny dogfish (*Squalus acanthias*) and its validity for age determination. *Can. J. Fish. Aquat. Sci.* 42:1799–1805.

Bean, W. B. (1980) Nail growth. Thirty-five years of observation. *Arch. Intern. Med.* 140:73–76.

Beato, M. (1989) Gene regulation by steroid hormones. *Cell* 56:335–344.

Beauchamp, P. de. (1956) Le développement de *Ploesoma hudsoni* (Imhof) et l'origine des feuillets chez les rotifers. *Bull. Soc. Zool. Fr.* 81:374–383.

Beauchene, R. E., Bales, C. W., Bragg, C. S., Hawkins, S. T., and Mason, R. L. (1986) Effect of age initiation of food restriction on growth, body composition, and longevity of rats. *J. Gerontol.* 41:13–19.

Beauchene, R. E., Roeder, L. M. and Barrows, C. H., Jr. (1970) The inter-relationships of age, tissue protein synthesis, and proteinuria. *J. Gerontol.* 25:359–363.

Beauregard, S., and Gilchrest, B. A. (1987) A survey of skin problems and skin care regimens in the elderly. *Arch. Dermatol.* 123:1638–1643.

Beck, C. B. (1962) Reconstruction of *Archeopteris* and further consideration of its phylogenetic position. *Am. J. Bot.* 49:373–382.

———. (1971) On the anatomy and morphology of the lateral branch systems of *Archeopteris. Am. J. Bot.* 58:758–784.

Beck, S. D. (1971a) Growth and retrogression in larvae of *Trogoderma glabrum* (Coleoptera: Dermestidae). 1. Characteristics under feeding and starvation conditions. *Entomol. Soc. Am. Ann.* 64:149–155.

———. (1971b) Growth and retrogression in larvae of *Trogoderma glabrum* (Coleoptera: Dermestidae). 2. Factors influencing pupation. *Entomol. Soc. Am. Ann.* 64:946–949.

Beden, M. (1983) Family Elephantidae. In *Koobi Fora Research Project,* J. M. Harris (ed.). Vol. 2, *The fossil ungulates: Proboscidea, Perrissodactyla and Suidae,* 40–129. Oxford: Clarendon Press.

Beerman, F., Bartels, I., Franke, U., and Hansmann, I. (1987) Chromosome segregation at meiosis I in female T (2;4)1Go/+ mice. No evidence for a decreased crossover frequency with maternal age. *Chromosoma* (Berlin) 95:1–7.

Begg, R. J. (1981) The small mammals of little Nourlangie rock N.T.2. Ecology of *Antechinus bilarni.* The sandstone *Antechinus* (Marsupalia: Dasuridae). *Austral. Wildlife Res.* 8:57–72.

Beguet, B. (1972) The persistence of processes regulating the level of reproduction in the hermaphrodite nematode *Caenorhabditis elegans,* despite the influence of parental aging, over several consecutive generations. *Exp. Gerontol.* 7:207–218.

Beguet, B., and Brun, J. L. (1972) Influence of parental aging on the reproduction of the F_1 generation in a hemaphrodite nematode *Caenorhabditis elegans. Exp. Gerontol.* 7:195–206.

Behnke, R. J. (1989) We're putting them back alive. *Trout,* Autumn.

———. (1990a) Interpreting the phylogeny of *Salvelinus. Physiol. Ecol. Jpn.,* special issue, 1.

————. (1990b) Still a rainbow by any other name. Why are rainbow trout now *Oncorhynchus mykiss?* Does this make them salmon? *Trout,* 42–45.

Beischer, N. A., Evans, J. H., and Townsend, L. (1969) Studies in prolonged pregnancy. *Am. J. Obstet. Gynecol.* 103:476–482.

Belcour, L., and Begel, O. (1978) Lethal mitochondrial genotypes in *Podospora anserina.* A model for senescence. *Mol. Gen. Genet.* 163:113–123.

Belcour, L., and Vierny, C. (1986) Variable DNA splicing sites of a mitochondrial intron: Relationship to the senescence process in *Podospora. EMBO J.* 5:609–614.

Bell, E., Marek, L. F., Levinstone, D. S., Merrill, C., Sher, S., Young, I. T., and Eden, M. (1978) Loss of division potential in vitro. Aging or differentiation. *Science* 202:1158–1163.

Bell, E., Marek, L. F., Merrill, C., Levinstone, D. S., Young, I. T., Eden, M., and Sher, S. (1980) Loss of division in culture. Aging or differentiation. *Science* 208:1483.

Bell, G. (1980) The costs of reproduction and their consequences. *Am. Nat.* 116:45–76.

————. (1982) *The Masterpiece of Nature: The Evolution and Genetics of Sexuality.* London: Croom Helm.

————. (1984) Measuring the cost of reproduction. 2. The correlation structure of the life tables of five freshwater invertebrates. *Evolution* 38:314–326.

————. (1985) Evolutionary and nonevolutionary theories of senescence. *Am. Nat.* 124:600–603.

————. (1988) *Sex and Death in Protozoa. The History of an Obsession.* Cambridge: Cambridge Univ. Press.

Bell, G., and Koufopanou, V. (1986) The cost of reproduction. *Oxford Surveys Evol. Biol.* 3:83–131.

Bell, W. J., and Bohm, M. K. (1975) Oosorption in insects. *Biol. Rev.* 50:373–396.

Bellen, H. J., and Kiger, J. A., Jr. (1987) Sexual hyperactivity and reduced longevity of *dunce* females of *Drosophila melanogaster. Genetics* 119:153–160.

Benditt, E. P. (1977) The origin of atherosclerosis. *Sci. Am.* 236:74–85.

Benditt, E. P., and Benditt, J. M. (1973) Evidence for a monoclonal origin of human atherosclerotic plaques. *Proc. Natl. Acad. Sci.* 70:1753–1756.

Benedict, F. G. (1938) *Vital Energetics: A Study in Comparative Basal Metabolism.* Washington, DC: Carnegie Institute.

Benfey, T., and Sutterlin, A. (1984) Triploidy induced by heat shock and hydrostatic pressure in landlocked Atlantic salmon (*Salmo salar L.*). *Aquaculture* 36:359–367.

Bennett, J. T., Boehlert, G. W., and Turekian, K. K. (1982) Confirmation of longevity in *Sebastes diploproa* (Pisces: Scorpaenidae), from Pb/Ra measurements in otoliths. *Mar. Biol.* 71:209–215.

Bennett, K. L., and Truman, J. W. (1985) Steroid-dependent survival of identifiable neurons in cultured ganglia of the moth *Manduca sexta. Science* 229:58–59.

Bennett, M. D. (1972) Nuclear DNA content and minimum generation time in herbaceous plants. *Proc. Roy. Soc. Lond.,* ser. B, 181:109–135.

————. (1977) The time and duration of meiosis. *Phil. Trans. Roy. Soc.* (London), ser. B, 277:201–226.

Bennett, M. D., and Smith, J. B. (1976) Nuclear DNA amounts in angiosperms. *Phil. Trans. Roy. Soc.* (London), ser. B, 274:227–274.

Bennington, J. L. (1986) Cancer and aging: Pathology. *Front. Radiat. Ther. Oncol.* 20:45–51.

Benson, M. D. (1989) Familial amyloidotic polyneuropathy. *Trends in Neurosciences* 12:88–92.

Bentley, P. J., and Follett, B. K. (1965) Fat and carbohydrate reserves in the river lamprey during spawning migration. *Life Sci.* 4:2003–2007.

Benton, M. J. (1985) First marsupial fossil from Asia. *Nature* 318:313.

Bentvelzen, P. (1974) Host-virus interactions in murine mammary carcinogenesis. *Biochim. Biophys. Acta* 355:236–259.

Berg, B. N., and Simms, H. S. (1960) Nutrition and longevity in the rat. 3. Longevity and onset of disease with different levels of food intake. *J. Nutrit.* 71:255–263.

Berg, B. N., Wolf, A., and Simms, H. S. (1962) Nutrition and longevity in the rat. 4. Food restriction and the radiculoneuropathy of aging rats. *J. Nutrit.* 77:439–442.

Berg, W. J., and Ferris, S. D. (1984) Restriction endonuclease analysis of salmonid mitochondrial DNA. *Can. J. Fish. Aquat. Sci.* 41:1041–1047.

Berger, S., and Schweiger, H. G. (1975) Cytoplasmic induction of changes in the ultrastructure of the *Acetabularia* nucleus and perinuclear cytoplasm. *J. Cell Sci.* 17:517–529.

———. (1986) Perinuclear dense bodies: Characterization as DNA-containing structures using enzyme-linked gold granules. *J. Cell Sci.* 80:1–11.

Bergman, M. D., Karelus, K., Felicio, L. S., and Nelson, J. F. (1989) Differential effects of aging on estrogen receptor dynamics in hypothalamus, pituitary, and uterus of the C57BL/6J mouse. *J. Steroid Biochem.* 33:1027–1033.

Bergman, R. A. (1948) Who is old? Death rate in a Japanese concentration camp. *J. Gerontol.* 3:14–20.

Bergman, R. N. (1990) Quantitative approaches to the pathogenesis of age-related metabolic conditions. In, *Biomedical Advances in Aging*, A. L. Goldstein (ed.), 229–245. New York: Plenum.

Bergner, A. D. (1928) The effect of prolongation of each stage of the life-cycle on crossing-over in the second and third chromosomes of *Drosophila melanogaster*. *J. Exp. Zool.* 50:107–163.

Berkman, L. F., and Breslow, L. (1983) *Health and Ways of Living. The Alameda County Study.* Oxford: Oxford Univ. Press.

Berkner, L. V., and Marshall, L. C. (1965) On the origin and rise of oxygen concentration in the earth's atmosphere. *J. Atm. Sci.* 22:225–261.

Bernardi, G., Olofsson, B., Filipski, J., Zerial, M., Salinas, J., Cuny, G., Meunier-Rotival, M., Rodier, F. (1985) The mosaic genome of warm-blooded vertebrates. *Science* 228:953–958.

Bernd, A., Schroder, H. C., Leyhausen, G., Zahn, R. K., and Muller, W. E. G. (1983) Alternation of activity of nuclear envelope nucleoside triphosphatase in quail oviduct and liver in dependence on physiological factors. *Gerontology* 29:394–398.

Bernheimer, H., Birkmayer, W., Hornykiewicz, O., Jellinger, K., and Seitelberger, F. (1973) Brain dopamine and the syndromes of Parkinson and Huntington: Clinical, morphological, and neurochemical correlations. *J. Neurol. Sci.* 20:415–455.

Bernstein, J. J. (1964) Relation of spinal cord regeneration to age in adult goldfish. *Exp. Neurol.* 9:161–174.

Berra, T. M., and Allen, G. R. (1989) Burrowing, emergence, behavior, and functional morphology of the Australian salamanderfish, *Lepidogalaxias salamandroides*. *Fisheries* 14:2–10.

Berrill, N. J. (1935) Studies on tunicate development. 4. Asexual reproduction. *Phil. Trans. Roy. Soc.* (London), ser. B, 225:327–379.

———. (1951) Regeneration and budding in tunicates. *Biol. Rev.* 26:456–475.

Berrill, N. J., and Karp, G. (1976) *Development.* New York: McGraw-Hill.

Bertalanffy, L. von. (1938) A quantitative theory of organic growth (inquiries on growth laws). 2. *Human Biol.* 10:181–213.

Bertalanffy, F. D. (1967) The basis of cell exfoliation and cytochemical cancer diagnosis. *Canc. Cyt.* 7:13–19.

Bertin, L. (1956) *Eels: A Biological Study.* London: Cleaver-Hume.

Bertrand, H., Chan, B. S., and Griffiths, A. J. F. (1985) Insertion of a foreign nucleotide sequence into mitochondrial DNA causes senescence in Neurospora intermedia. *Cell* 41:877–884.

Bertrand, H., Griffiths, A. J. F., Court, D. A., and Cheng, C. K. (1986) An extrachromosomal plasmid is the etiological precursor of kalDNA insertion sequences in the mitochrondrial chromosome of senescent neurospora. *Cell* 47:829–837.

Bertrand, H. A., and Masoro, E. J. (1977) Post-weaning food restriction reduces adipose cellularity. *Nature* 266:62–63.

Bertrand, H. A., Masoro, E. J., and Yu, B. P. (1980) Maintenance of glucagon-promoted lipidosis in adipocytes by food restriction. *Endocrinology* 107:591–595.

Beschel, R. (1955) Individuum und Alter bei Flechten. *Phyton* (Horn, Austria) 6:60–68.

Besedovsky, H. O., del Rey, A. E., and Sorkin, E. (1985) Immune-neuroendocrine interactions. *J. Immunol.* 135:750s–754s.

Bestetti, G., Locatelli, V., Tirone, F., Rossi, G. L., and Muller, E. E. (1985) One month of streptozotocin-diabetes induces different neuroendocrine and morphological alterations in the hypothalami-pituitary axis of male and female rats. *Endocrinology* 117:208–216.

Beverton, R. J. (1987) Longevity in fish: Some ecological and evolutionary considerations. In *Evolution of Senescence. A Comparative Approach,* A. D. Woodhead and K. H. Thompson (eds.), 145–160. New York: Plenum.

Beverton, R. J., and Holt, S. J. (1959) A review of the lifespans and mortality rates of fish in nature and their relation to growth and other physical characteristics. In *The Lifespan of Animals, CIBA Foundation Colloqium on Aging,* vol. 5, 142–180.

Bickel, D. J. (1982) Diptera. In *Synopsis and Classification of Living Organisms,* vol. 2, S. P. Parker (ed.), 563–599. New York: McGraw-Hill.

Bidder, G. P. (1932) Senescence. *Brit. Med. J.* 2:583–585.

Bierbaum, T. J., Mueller, L. D., and Ayala, F. J. (1989) Density-dependent selection of life-history traits in *Drosophila melanogaster. Evolution* 43:382–392.

Bierman, E. L. (1978) The effect of donor age on the *in vitro* lifespan of cultured human arterial smooth muscle cells. *In Vitro* 14:951–955.

———. (1985) Arteriosclerosis and aging. In *Handbook of the Biology of Aging,* 2d ed., C. E. Finch and E. L. Schneider (eds.), 842–858. New York: Van Nostrand.

Bilder, G. E., and Denckla, W. D. (1977) Restoration of ability to reject xenografts and clear carbon after hypophysectomy of adult rats. *Mech. Age. Dev.* 6:153–163.

Birchenall-Sparks, M. C., Roberts, M. S., Rutherford, M. S., and Richardson, A. (1985) The effect of aging on the structure and function of liver messenger RNA. *Mech. Age. Dev.* 32:99–111.

Birchenall-Sparks, M. C., Roberts, M. S., Staecker, J., Hardwick, J. P., and Richardson, A. (1985). Effect of dietary restriction on liver protein synthesis in rats. *J. Nutrit.* 115:944–950. Bird, A. F. (1971) *The Structure of Nematodes.* New York: Academic Press.

Bird, T. D., Lampe, T. H., Nemens, E. J., Miner, G. W., Sumi, S. M., and Schellenberg, G. D. (1988) Familial Alzheimer's disease in American descendants of the Volga Germans: Probable genetic founder effect. *Ann. Neurol.* 23:25–31.

Bird, T. D., Schellenberg, G. D., Wijsman, E. M., and Martin, G. M. (1989) Evidence for etiologic heterogeneity in Alzheimer's disease. *Neurobiol. Aging* 10:432–434.

Bird, T. D., Sumi, S. M., Nemens, E. J., Nochlin, D., Schellenberg, G. D., Lampe, T. H., Sadovnick, A., Chui, H., Miner, G. W., and Tinklenberg, J. (1989) Phenotypic heterogeneity in familial Alzheimer's disease: A study of 24 kindreds: *Ann. Neurol.* 25:12–25.

Birkeland, C. (1981) Mortality of didemnid ascidian colonies. *Bull. Mar. Sci.* 31:170–173.

Birren, J. E. (1964) *Psychology of Aging.* Summit, NJ: Prentice-Hall.

Bishop, M. H. W. (1970) Aging and reproduction in the male. *J. Reprod. Fertil.,* suppl. 12:65–87.

Bito, L. Z., DeRousseau, C. J., Kaufman, P. L., and Bito, J. W. (1982) Age-dependent loss of accommodative amplitude in rhesus monkeys: An animal model for presbyopia. *Invest. Ophthalmol. Vis. Sci.* 23:23–31.

Bjorkerud, S. (1964) Isolated lipofuscin granules: A survey of a new field. *Adv. Gerontol. Res.* 1:257–288.

Blackburn, M. (1950) The Tasmanian whitebait, *Lovettia sealii* (Johnston) and the whitebait fishery. *Austral. J. Mar. Freshwater Res.* 1:155–198.

Blaha, G. C. (1964a) Effect of age of the donor and recipient on the development of transferred golden hamster ova. *Anat. Rec.* 150:413–416.

————. (1964b) Reproductive senescence in the female golden hamster. *Anat. Rec.* 150:405–412.

Blake, C. A., Elias, K. A., and Huffman, L. J. (1983) Ovariectomy of young adult rats has a sparing effect on the suppressed ability of aged rats to release luteinizing hormone. *Biol. Reprod.* 28:575–585.

Blanquet, V., Goldgaber, D., Turleau, C., Creau-Goldberg, N., Delabar, J., Sinet, P. M., Roudier, M., and de Grouchy, J. (1987) The β amyloid protein (AD-AP) cDNA hybridizes in normal and Alzheimer individuals near the interface of 21q21 and q22.1. *Ann. Genet.* 30:68.

Blazejowski, C. A., and Webster, G. C. (1984) Effect of age on peptide elongation in preparations from brain, liver, kidney, and skeletal muscle of the C57BL/6J mouse. *Mech. Age. Dev.* 25:323–333.

Blest, A. D. (1963) Longevity, palatability, and natural selection in five species of New World saturniid moth. *Nature* 197:1183–1186.

Blichert-Toft, M. (1975) Secretion of corticotrophin and somatotrophin by the senescent adenohypophysis in man. *Acta Endocrinol.* 78 (suppl. 195): 1–57.

Blichert-Toft, M., and Hummer, L. (1977) Serum immunoreactive corticotrophin and response to metyrapone in old age in man. *Gerontology* 23:236–243.

Bligh, J. (1973) *Temperature Regulation in Mammals and Other Vertebrates*, A. Neuberger and E. L. Tatum (eds.). New York: North-Holland.

Block, E. (1952) Quantitative morphological investigations of the follicular system in women. *Acta Anat.* 14:108–123.

Block, J., Boysen, P., Wynne, J., and Hunt, L. (1979) Sleep apnea, hypopnea, and oxygen desaturation in normal subjects. *N. Engl. J. Med.* 300:513–517.

Bloemendal, H. (ed.) (1981) *Molecular and Cellular Biology of the Eye Lens*. New York: Wiley.

Bloom, E. T., Akiyama, M., Korn, E. L., Kusunoki, Y., and Makinodan, T. (1988) Immunological responses of aging Japanese A-bomb survivors. *Rad. Res.* 116:343–355.

Blumberg, B. S., and Sokoloff, L. (1961) Coa-lescence of caudal vertebrae in the giant dinosaur *Diplodocus*. *Arthrit. Rheum.* 4:592–601.

Blumenthal, H. T., Handler, F. P., and Blache, J. O. (1954) The histogenesis of arteriosclerosis of the larger cerebral arteries, with an analysis of the importance of mechanical factors. *Am. J. Med.* 17:337–347.

Bochenek, Z., and Jachowska, A. (1969) Atherosclerosis, accelerated presbyacusis, and acoustic trauma. *Audiology* 8:312–316. Boddington, M. J. (1978) An absolute metabolic scope for activity. *J. Theoret. Biol.* 75:443–449.

Bodmer, W. F. (1961) Effects of maternal age on the incidence of congenital abnormalities in mouse and man. *Nature* 190:1134–1135.

Boëtius, I., and Boëtius, J. (1967) Studies in the European eel *Anguilla anguilla (L)*. Experimental induction of the male sexual cycle, its relation to temperature and other factors. *Medd. Danm. Fiskeri-og Havunders*, n.s. 4:339–405.

————. (1980) Experimental maturation of female silver eels, *Anguilla anguilla*. Estimates of fecundity and energy reserves for migration and spawning. *Dana* 1:1–28.

Boggs, C. L., and Watt, W. B. (1981) Population structure of pierid butterflies. 4. Genetic and physiological investment in offspring by male *Colias*. *Oecologia* 50:320–324.

Bohnet, H. G., Dahlen, H. G., Wuttek, W., and Schneider, H. P. (1976) Hyperprolactinaemic anovulatory syndrome. *J. Clin. Endocrinol. Metab.* 42:132–143.

Bohr, V. A., Okumoto, D. S., and Hanawalt, P. C. (1986) Survival of UV-irradiated mammalian cells correlates with efficient DNA repair in an essential gene. *Proc. Natl. Acad. Sci.* 83:3830–3833.

Bok, A. H. (1979) The distribution and ecology of two mullet species in some freshwater rivers in eastern cape, South Africa. *J. Limnol. Soc. S. Africa* 5:97–102. [not read]

Bok, S. T. (1959) *Histonomy of the Cerebral Cortex*. Princeton, NJ: Van Nostrand Reinhold.

Bolanowski, M. A., Russell, R. L., and Jacobson, L. A. (1981) Quantitative measures of aging in the nematode *Caenorhabditis ele-*

gans. 1. Population and longitudinal studies of two behavioral parameters. *Mech. Age. Dev.* 15:279–295.

Bolla, R. I., and Denckla, W. D. (1979) Effect of hypophysectomy on liver nuclear ribonucleic acid synthesis in aging rats. *Biochem. J.* 184:669.

Bolla, R. I., and Roberts, L. S. (1971) Developmental physiology of cestodes. 9. Cytological characteristics of the germinative region of *Hymenolepis diminuta. J. Parisitol.* 57:967–977.

Bolli, R., Jeroudi, M. O., Patel, B. S., DuBose, C. M., Lai, E. K., Roberts, R., and McCay, P. B. (1989) Direct evidence that oxygen-derived free radicals contribute to postischemic myocardial dysfunction in the intact dog. *Proc. Natl. Acad. Sci.* 86:4695–4699.

Bond, S. L., and Singh, S. M. (1987) Methyl nitrosourea induced unscheduled DNA synthesis *in vivo* in mice. Effects of background genotype on excision repair during aging. *Mech. Age. Dev.* 41:177–187.

Bonner, J. J., and Slavkin, H. C. (1975) M523, a new end mutant. Study of *H–2* mutations in mice. *Immunogenetics* 2:291–295.

Bonner, J. T. (1965) *Size and Cycle: An Essay on the Structure of Biology.* Princeton, NJ: Princeton Univ. Press.

Boorman, L. A., and Fuller, R. M. (1984) The comparative ecology of two sand dune biennials: *Lactuca virosa L.* and *Cynoglossum officinale L. New Phytol.* 69:609–629.

Borovsky, D., Thomas, B. R., Carlson, D. A., Whisenton, L. R., and Fuchs, M. S. (1985) Juvenile hormone and 20–hydroxyecdysone as primary and secondary stimuli of vitellogenesis in *Aedes aegypti. Arch. Insect Biochem. Physiol.* 2:75–90.

Borror, D. J., and Delong, D. M. (1954) *Introduction to the Study of Insects.* New York: Rhinehart.

Bosch, C. A. (1971) Redwoods: A population model. *Science* 172:345–349.

Boss, B. D., Peterson, G. M., and Cowan, W. M. (1985) On the number of neurons in the dentate gyrus of the rat. *Brain Res.* 338:144–150.

Botkin, D. B., and Miller, R. S. (1974) Mortal-

ity rates and survival of birds. *Am. Nat.* 108:181–192.

Boucot, A. J., and Gray, J. (1983) A Paleozoic Pangaea. *Science* 222:571–581.

Boué, J., Boué, A., and Lazar, P. (1975a) The epidemiology of human spontaneous abortions with chromosomal anomalies. In *Aging Gametes: Their Biology and Pathology,* R. J. Blandau (ed.), 330–348. Basel: Karger.

———. (1975b) Retrospective and prospective epidemiological studies of 1,500 karyotyped spontaneous human abortions. *Teratology* 12:11–26.

Bounoure, L. (1940) *Continuité germinale et reproduction agame.* Paris: Gauthier-Villais.

Bourlière, F. (1958) The comparative biology of aging. *J. Gerontol.* 13:16–24.

———. (1959) Lifespans of mammalian and bird population in nature. *CIBA Found. Colloq. Aging* 5:90–105.

Bourlière, F., and Molimard, R. (1957) L'action de l'age sur la régénération du foie chez le rat. *Compt. Rend. Soc. Biol.* (Paris) 151:1345–1348.

Bowden, D. M., and Williams, D. D. (1985) Aging. *Adv. Vet. Sci. Comp. Med.* 28:305–341.

Bowen, B. J., Codd, C. G., and Gwynne, D. T. (1984) The katydid spermatophore (*Orthoptera tettigoniidae*): Male nutritional investment and its fate in the mated female. *Austral. J. Zool.* 32:23–31.

Bowler, J. K. (1977) Longevity of reptiles and amphibians in N. American collections as of Nov. 1, 1975. *Soc. Amphib. Reptiles Herpetol.* 6:1–32.

Boyd-Leinen, P. A., Fournier, D., and Spelsberg, T. C. (1982) Nonfunctioning progesterone receptors in the developed oviducts from estrogen-withdrawn immature chicks and in aged nonlaying hens. *Endocrinology* 111:30–36.

Boyer, J. F. (1978) Reproductive compensation in *Tribolium castaneum. Evolution* 32:519–528.

Bozcuk, A. N. (1972) DNA synthesis in the absence of somatic cell division associated with ageing in *Drosophila subobscura. Exp. Gerontol.* 7:147–156.

Bozina, K. D. (1961) How long does the queen

live? (in Russian). *Pchelovodstvo* 38:13. Cited in Winston, 1987.

Brackenridge, C. J. (1973) The relation of sex of affected parent to the age at onset of Huntington's disease. *J. Med. Genet.* 10:333–336.

Bradley, A. J. (1980) Stress and mortality in a small marsupial (*Antechinus stuartii* (Macleay). *Gen. Comp. Endocrinol.* 40:188–200.

———. (1982) Steroid binding proteins in the plasma of dasyurid marsupials. In *Carnivorous Marsupials*, M. Archer (ed.), 651–657. Sydney: Royal Zoological Society of New South Wales.

Bradley, A. J., McDonald, I. R., and Lee, A. K. (1975) Effect of exogenous cortisol on mortality of a dasyurid marsupial. *J. Endocrinol.* 66:281–282.

———. (1976) Corticosteroid binding globulin and mortality in a dasyurid marsupial. *J. Endocrinol.* 70:323–324.

———. (1980) Stress and mortality in a small marsupial (*Antechinus stuartii*, Macleay). *Gen. Comp. Endocrinol.* 40:188– 200.

Bradley, M. O., Dice, J. F., Hayflick, L., and Schimke, R. T. (1975) Protein alterations in aging WI–38 cells as determined by proteolytic susceptibility. *Exp. Cell Res.* 96:103–112.

Bradley, W. E. C., Gareau, J. L., Seifert, A. M., and Messing, K. (1987) Molecular characterization of 15 rearrangements among 90 human in vivo somatic mutants shows that deletions predominate. *Mol. Cell. Biol.* 7:956–960.

Bradley, W. G., and Arbuthnott, K. D. (1938) The relation of the host physiology to the development of the braconid parasite *Chelonus annulipes*. *Ann. Entomol. Soc. Am.* 31:359–365.

Bradshaw, W. E., and Lounibos, L. P. (1977) Evolution of dormancy and its photo-periodic control in pitcher-plant mosquitos. *Evolution* 31:546–567.

Braithwaite, R. W. (1974) Behavioral changes associated with the population cycle of *Antechinus stuartii* (Marsupialia). *Austral. J. Zool.* 22:45–62.

Braithwaite, R. W., and Lee, A. K. (1977) A mammalian example of semelparity. *Am. Nat.* 113:151–155.

Bras, G., and Ross, M. H. (1964) Kidney disease and nutrition in rats. *Toxicol. Pharmacol.* 6:246–262.

Braun, A. C. (1959) A demonstration of the recovery of the crown-gall tumor cell with the use of complex tumors of single-cell origin. *Proc. Natl. Acad. Sci.* 45:932–938.

Braun, E. (1942) One-year- and two-year-old queens. *Am. Bee J.* 82:356–357.

Braverman, I. M., and Fonferko, E. (1982) Studies in cutaneous aging. 1. The elastic fiber network. *J. Invest. Dermatol.* 78:434–443.

Brawer, J. R., and Finch, C. E. (1983) Normal and experimentally altered aging processes in the rodent hypothalamus and pituitary. In *Experimental and Clinical Intervention in Aging*, R. F. Walker and R. L. Cooper (eds.), 45–66. New York: Dekker.

Brawer, J. R., Naftolin, F., Martin, J., and Sonnenschein, C. (1978) Effects of a single injection of estradiol valerate on the hypothalamic arcuate nucleus and on reproductive function in the female rat. *Endocrinology* 103:501–512.

Brawer, J. R., Schipper, H., and Robaire, B. (1983) Effects of long-term androgen and estradiol exposure on the hypothalamus. *Endocrinology* 112:194–199.

Brawer, J. R., and Sonnenschein, C. (1975) Cytopathological effects of estradiol on the arcuate nucleus of the female rat. A possible mechanism for pituitary tumorigenesis. *Am. J. Anat.* 144:57–87.

Breeden, S. (1967) *The Life of the Kangaroo*. New York: Taplinger.

Breitner, J. C. S., Silverman, J. M., Mohs, R. C., and Davis, K. L. (1988) Familial aggregation in Alzheimer's disease: Comparison of risk among relatives of early- and late-onset cases, and among male and female relatives in successive generations. *Neurology* 38:207– 212.

Bremner, W. J., Vijtello, M. V., and Prinz, P. N. (1983) Loss of circadian rhythmicity in blood testosterone levels with aging in normal men. *J. Clin. Endocrinol. Metab.* 56:1278–1281.

Brenner, B. M., Meyer, T. W., and Hostetter,

T. H. (1982) Dietary protein intake and the progressive nature of kidney disease. *N. Engl. J. Med.* 307:652–659.

Brenner, S. (1974) The genetics of *Caenorhabditis elegans*. *Genetics* 77:71–94.

Brian, M. V. (1976) Endocrine control over caste differentiation in a myrmicine ant. In *Phase and Caste Determination in Insects*, M. Luscher (ed.), 63–70. New York: Pergamon Press.

Bridges, C. B. (1927) The relation of the age of the female to crossing over in the third chromosome of *Drosophila melanogaster*. *J. Gen. Physiol.* 81:689–701.

Brien, P. (1948) Embranchement des tuniciers morphologie et reproduction. In *Traité de Zoologie Masson*, vol. 2, P. P. Grasse (ed.), 553–894. Paris.

———. (1953) La pérennité somatique. *Biol. Rev.* 28:308.

Brinton, M. A. (1982) Lactate dehydrogenase-elevating virus. In *The Mouse in Biomedical Research*, vol. 2, H. L. Foster and J. D. Small (eds.), 194–208. New York: Academic Press.

Bristowe, W. J. (1958) *The World of Spiders*. London: Collins.

Brittain, J. E. (1982) Biology of mayflies. *Ann. Rev. Entomol.* 27:119–147.

Britten, R. J. (1986) Rates of DNA sequence evolution differ between taxonomic groups. *Science* 231:1394–1398.

Britten, R. J., and Davidson, E. H. (1971) Repetitive and non-repetitive DNA sequences and a speculation on the origins of evolutionary novelty. *Q. Rev. Biol.* 46:111–138.

Britton, G. W., Britton, V. J., Gold, G., and Adelman, R. C. (1976) The capability for hormone-stimulated enzyme adaptation in liver cells isolated from aging rats. *Exp. Gerontol.* 11:1–4.

Brocas, J., and Verzar, F. (1961) The aging of *Xenopus laevis*, a South African frog. *Gerontologia* (Basel) 5:228–240.

Brock, D. B., Guralnik, J. M., and Brody, J. A. (1990) Demography and epidemiology of aging in the United States. In *The Handbook of the Biology of Aging*, 3d ed., E. L. Schneider and J. W. Rowe (eds.), 3–23. San Diego: Academic Press.

Brock, M. A. (1970) Ultrastructural studies on the life cycle of a short-lived metazoan *Campanularia flexuosa*. 2. Structure of the old adult. *J. Ultrastruct. Res.* 32:118–141.

———. (1984) Senescence in *Campanularia flexuosa* and other cnidarians. In *Invertebrate Models in Aging Research*, D. H. Mitchell and T. E. Johnson (eds.), 16–44. Boca Raton, FL: CRC Press.

Brock, M. A., and Strehler, B. L. (1963) Studies on the comparative physiology of aging. 4. Age and mortality of some marine Cnidaria in the laboratory. *J. Gerontol.* 18:23–28.

Brock, M. A., Strehler, B. L., and Brandes, D. (1968) Ultrastructural studies on the life cycle of a short-lived metazoan, *Campanularia flexuosa*. 1. Structure of a young adult. *J. Ultrastruct. Res.* 21:281–312.

Brodal, A., and Fänge, R. (eds.) (1963) *The Biology of Myxine*. Olso: Universitetsforlaget.

Brody, H. (1955) Organization of cerebral cortex. 3. A study of aging in the human cerebral cortex. *J. Comp. Neurol.* 102:511–556.

Brody, J. A. (1983) Limited importance of cancer and competing risk theories in aging. *J. Clin. Exp. Gerontol.* 5:141–154.

Brody, S. (1924) The kinetics of senescence. *J. Gen. Physiol.* 6: 245–257.

———. (1945) *Bioenergetics and Growth*. New York: Reinhold.

Brody, S., Henderson, E. W., and Kempster, H. L. (1923) The rate of senescence of the domestic fowl as measured by the decline in egg production with age. *J. Gen. Physiol.* 6:41–45.

Bronson, F. H., and Desjardins, C. (1977) Reproductive failure in aged CBF male mice: Interrelationships between pituitary gonadotrophic hormones, testicular function, and mating success. *Endocrinology* 101:931–945.

———. (1982) Reproductive aging in male mice. In *Biological Markers of Aging*, M. E. Reff and E. L. Schneider (eds.), 87–93. NIH Publication 82–2221. Washington, DC: USDHHS.

Brook, J. D., Gosden, R. G., and Chandley, A. C. (1984) Maternal aging and aneuploid embryos: Evidence from the mouse that biological and not chronological age is the important influence. *Human Genet.* 66:41–45.

Brown, G. W., and Flood, M. M. (1947) Tumbler mortality. *J. Am. Statist. Assn.* 42:562–574.

Brown, H. P. (1973) Survival records for elmid beetles. *Entomol. News* 84:278–284.

Brown, J. H., and Maurer, B. A. (1986) Body size, ecological dominance, and Cope's rule. *Nature* 324:248–250.

Brown, J. O. (1943) Pigmentation of substantia nigra and locus in certain carnivores. *J. Comp. Neurol.* 79:393–404.

Brown, M. S., and Goldstein, J. L. (1986) A receptor-mediated pathway for cholesterol homeostasis. *Science* 232:34–47.

Brown, M. S., Goldstein, J. L., and Fredrickson, D. S. (1983) Familial type 3 hyperlipoproteinemia (dysbetalipoproteinemia). In *The Metabolic Basis of Inherited Disease*, 5th ed., J. B. Stanbury, J. B. Wyngaarden, D. S. Fredrickson, J. L. Goldstein, and M. S. Brown (eds.), 655–671. New York: McGraw-Hill.

Brown, W. E. (1979) *HLA and Disease: A Comprehensive Review.* Boca Raton, FL: CRC Press.

Brown, W. T., Zebrower, M., and Kieras, F. J. (1984) Progeria, a model disease for the study of accelerated aging. In *Molecular Biology of Aging*, Basic Life Sciences, vol. 35, A. V. Woodhead, A. D. Blackett, and A. Hollaender (eds.), 375–396. New York: Plenum.

Brownlee, M., Cerami, A., and Vlassara, H. (1988) Advanced glycosylation end products in tissue and the biochemical basis of diabetic complications. *N. Engl. J. Med.* 318:1315–1321.

Bruce, M. E., Dickinson, A. G., and Fraser H. (1976) Cerebral amyloidosis in scrapie in the mouse: Effect of agent strain and mouse genotype. *Neuropath. Appl. Neurobiol.* 2:471–478.

Bruce, S. A. (1990) Ultrastructure of dermal fibroblasts during development and aging: Relationship to in vitro senescence of dermal fibroblasts. *Exp. Gerontol.*, in press.

Bruce, S. A., and Deamond, S. F. (1990) Longitudinal study of in vivo wound repair and in vitro cellular senescence of dermal fibroblasts. In prep.

Bruce, S. A., Deamond, S. F., Ts'O, P. O. (1986) In vitro senescence of Syrian hamster mesenchymal cells of fetal to aged adult origin. Inverse relationship between in vivo donor age and in vitro proliferative capacity. *Mech. Age. Dev.* 34:151–173.

Brues, C. T. (1946) *Insect Dietary. An Account of the Food Habits of Insects.* Cambridge, MA: Harvard Univ. Press.

Bruley-Rosset, M., and Vergnon, I. (1984) Interleukin–1 synthesis and activity in aged mice. *Mech. Age. Dev.* 24:247–264.

Brunetti, R. (1974) Observations on the life cycle of *Botryllus schlosseri* (Pallas) (Ascidiacea) in the Venetian Lagoon. *Bull. Zool.* 41:225–251.

Bruton, M. N., Bok, A. H., and Davies, M. T. T. (1987) Life history styles of diadromous fishes in inland waters of southern Africa. *Am. Fish. Soc. Symp.* 1:104–121.

Bruton, M. N., Jackson, P., and Merron, E. (1987) Fishes and wetlands. *African Wildlife* 41:261–263. [not read]

Bruun, A. F. (1940) A study of the fish Schindleria from South Pacific waters. *Dana Rep.*, no. 12:1–12.

Bruyn, G. W. (1968) Huntington's chorea: Historical, clinical, and laboratory synopsis. In *Handbook of Clinical Neurology*, vol. 6, G. W. Bruyn and P. R. Vinken (eds.), 298–378. New York: Wiley.

Bucala, R., Model, P., and Cerami, A. (1984) Modification of DNA by reducing sugars: A possible mechanism for nucleic acid aging and age-related dysfunction in gene expression. *Proc. Natl. Acad. Sci.* 81:105–109.

Bucala, R., Model, P., Russell, M., and Cerami, A. (1985) Modification of DNA by glucose–6–phosphate induces DNA rearrangements in an *Escherichia coli* plasmid. *Proc. Natl. Acad. Sci.* 82:8439–8442.

Bucher, N. L. (1963) Regeneration of mammalian liver. *Int. Rev. Cytol.* 15:245–300.

Bucher, N. L., and Glinos, A. D. (1950) The effect of age on the regeneration of rat liver. *Canc. Res.* 10:324–332.

Bucher, N. L., Swaffield, M. N., and Di Troia, J. F. (1964) The influence of age upon the incorporation of thymidine–2–^{14}C into the DNA of regenerating rat liver. *Canc. Res.* 24:509–512.

Buchner, H., and Kiechle, H. (1967) Die Determination der Mannchen und Dariereiproduktion bei *Asplanchna sieboldi. Biol. Zentralt.* 86:599–621.

Buckland, S. T. (1982) A mark-recapture survival analysis. *J. Anim. Ecol.* 51:833–847.

Buckler, A. J., Vie, H., Sonenshein, G. E., and Miller, R. A. (1988) Diminished production of mature c-myc RNA after mitogen exposure not attributable to alterations in transcription or RNA stability. *J. Immunol.* 140:2442–2446.

Buell, S. J., and Coleman, P. D. (1981) Quantitative evidence for selective dendritic growth in normal human aging but not in senile dementia. *Brain Res.* 214:23–41.

Bulbrook, R. D., Hayward, J. L., and Spicer, C. C. (1971) Relation between urinary androgen and corticoid excretion and subsequent breast cancer. *Lancet* 2:395–398.

Bulkley, B. H., and Hutchins, G. M. (1977) Accelerated "atherosclerosis." A morphologic study of 97 saphenous vein coronary artery bypass grafts. *Circulation* 55:163–169.

Bull, J. J., and Shine, R. (1979) Iteroparous animals that skip opportunities for reproduction. *Am. Nat.* 114:296–303.

Bulpitt, K. J., and Piko, L. (1984) Variation in the frequency of complex forms of mitochondrial DNA in different brain regions of senescent mice. *Brain Res.* 300:41–48.

Bunn, C. L., and Tarrant, G. M. (1980) Limited lifespan in somatic cell hybrids and cybrids. *Exp. Cell Res.* 127:385–396.

Bunn, F., and Higgins, P. J. (1981) Reaction of monosaccharides with proteins: Possible evolutionary significance. *Science* 213:222–224.

Bunn, H. F. (1981) Nonenzymatic glycosylation of protein: Relevance to diabetes. *Am. J. Med.* 70:325–330.

Bunn, H. F., Seal, U. S., and Scott, A. F. (1974) The role of 2,3–diphosphoglycerate in mediating hemoglobin function of mammalian red cells. *Ann. N.Y. Acad. Sci.* 241:498–512.

Burch, J. B., and Weintraub, H. (1983) Temporal order of chromatin structural changes associated with activation of the major chicken vitellogenin gene. *Cell* 33:65–76.

Burch, P. R. (1984) Cancer and senescence: Is there a biological link? *Acta Genet. Med. Gemellol.* 33:457–465.

Burch, P. R., Jackson, D., Fairpo, C. G., and Murray, J. J. (1973) Gingival recession ("getting long in the tooth"), colorectal cancer, degenerative and malignant changes as errors of growth control. *Mech. Age. Dev.* 2:251–273.

Burger, P. C., and Vogel, S. F. (1973) The development of the pathologic changes of Alzheimer's disease and senile dementia in patients with Down's syndrome. *Am. J. Pathol.* 73:457–476.

Burgoon, C. F., Jr., Burgoon, J. S., and Baldridge, G. D. (1957) The natural history of herpes zoster. *J.A.M.A.* 164:265–269.

Burke, J. D. (1966) Vertebrate blood oxygen capacity and body weight. *Nature* 212:46–48.

Burmer, G. C., and Norwood, T. H. (1980) Selective elimination of proliferating cells in human diploid cell cultures by treatment with BrdU, 33258 Hoechst, and visible light. *Mech. Age. Dev.* 12:151–159.

Burnet, F. M. (1974) *Intrinsic Mutagenesis: A Genetic Approach.* New York: Wiley.

Burnett, A. L., and Diehl, N. A. (1964) The initiation of sexuality in *Hydra. J. Exp. Zool.* 157:237–249.

Burnham, L. (1978) Survey of social insects in the fossil record. *Psyche* 85:85–134.

Burns, B. D. (1958) *The Mammalian Cerebral Cortex.* E. Arnold (ed.), Monograph of the Physiological Society 5. London.

Bush, T. L., Barrett-Conner, E., Cowan, L. D., Criqui, M. H., Wallace, R. B., Suchindron, C. M., Tyroler, H. A., and Rifkind, B. M. (1987) Cardiovascular mortality and noncontraceptive use of estrogen in women: Results from the Lipid Research Clinics Program Follow-up Study. *Circulation* 75:1102–1109.

Buss, L. W. (1987) *The Evolution of Individuality.* New York: Princeton Univ. Press.

Butcher, R. L. (1975) The role of intrauterine environment and intrafollicular aging of the oocyte on implantation rates and development. *Aging Gametes: Their Biology and Pa-*

thology, R. J. Blandau (ed.), 201–218. Basel: Karger.

Butcher, R. L., and Fugo, M. W. (1967) Overripeness and the mammalian ova. 2. Delayed ovulation and chromosome anomalies. *Fertil. Steril.* 18:297–302.

Butler, C. G. (1957) The process of queen supersedure in colonies of honeybees (*Apis mellifera L.*). *Insectes Sociaux* 4:211–223.

———. (1960) The significance of queen substance in swarming and supersedure in honeybee (*Apis mellifera L.*) colonies. *Proc. Roy. Entomol. Soc.* (London), ser. A, 35:129–132.

Butler, H., and Juma, M. B. (1970) Oogenesis in an adult prosimian. *Nature* 226:552–553.

Butschli, O. (1876) Studien über die Entwicklungsvorgange der Eizelle, die Zellteilung, und die Conjugation der Infusorien. *Abh. Senchkenberg Naturf. Ges.* 10:213–452. [not read]

Buttyan, R., Olsson, C. A., Pintar, J., Chang, C., Bandyk, M., Ng, P.-Y., and Sawczuk, I. S. (1989) Induction of the TRPM–2 gene in cells undergoing programmed death. *Mol. Cell. Biol.* 9:3473–3481.

Buus, O., and Larsen, L. O. (1975) Absence of known corticosteroids in blood of river lampreys (*Lampetra fluviatilis*) after treatment with mammalian corticotropin. *Gen. Comp. Endocrinol.* 26:96–99.

Buvat, R. (1952) Structure, évolution, et fonctionnement du meristeme apical de quelques dicotyledones. *Ann. Sci. Nat. Bot. Ser.* 13:199–300.

Buys, C. C. M., Osinga, J., and Anders, G. J. P. A. (1979) Age-dependent variability of ribosomal RNA-gene activity in man as determined from frequencies of silver staining nucleolus organizing regions on metaphase chromosomes of lymphocytes and fibroblasts. *Mech. Age. Dev.* 11:55–75.

Cabib, E., Ulane, R., and Bowers, B. (1974) A molecular model for morphogenesis: The primary septum of yeast. *Curr. Top. Cell. Regul.* 8:1–32.

Cailliet, G. M. (1985) Sharks. An inquiry into biology, behavior, fisheries, and use. Proceedings of the Conference. Portland, Ore-

gon, October 13–15. Oregon State University Extension Service.

Cailliet, G. M., and Bedford, D. W. (1983) The biology of three pelagic sharks from California waters and their emerging fisheries: A review. *Calif. Curr. Calif. Coop. Oceanic Fish Invest. Rep.* 24:57–69.

Cailliet, G. M., Martin, L. K., Harvey, J. T., Kusher, D., and Welden, B. A. (1983) Preliminary studies on the age and growth of blue, *Prionace glauca,* common thresher, *Alopias vulpinus,* and shortfin mako, *Isurus oxyrinchus,* sharks from California waters. In Proceedings, International Workshop on Age Determination in Ocean Pelagic Fishes, Tunas, Bullfishes, Sharks, E. D. Prince and L. M. Pulos (eds.). U.S. Department of Commerce, *NOAA Tech. Rep. NMFS* 8:179–188.

Cailliet, G. M., Martin, L. K., Kusher, D., Wolf, P., and Welden, B. A. (1983) Techniques for enhancing vertebral bands in age estimation of California elasmobranchs. In Proceedings, International Workshop on Age Determination in Ocean Pelagic Fishes, Tunas, Bullfishes, Sharks, E. D. Prince and L. M. Pulos (eds.). U.S. Department of Commerce, *NOAA Tech. Rep. NMFS* 8:157–165.

Cailliet, G. M., Natanson, L. J., Welden, B. A., and Ebert D. A. (1985) Preliminary studies on the age and growth of the white shark, *Carcharodon carcharias,* using vertebral bands. *Memoirs S. Calif. Acad. Sci.* 9:49–60.

Cailliet, G. M., Radtke, R. L., and Welden, B. A. (1986) Elasmobranch age determination and verification: A review. *Indo-Pacific Fish Biology: Proceedings of the Second International Conference of Indo-Pacific Fishes,* T. Uyeno, R. Arai, R. Taniuchi, and K. Matsuura, (eds.), 345–360. Tokyo: Ichthyological Society of Japan.

Cairns, J. (1978) *Cancer, Science, and Society.* San Francisco: W. H. Freeman.

Calaby, J. H., and Taylor, J. M. (1981) Reproduction in two marsupial mice, *Antechinus bellus* and *A. bilarni* (Dasyuridae), of tropical Australia. *J. Mammal.* 62:329–341.

Calder, W. A., III. (1968) Respiratory and heart rates of birds at rest. *Condor* 70:358–365.

————. (1976) Aging in vertebrates: Allometric considerations of spleen size and lifespan. *Fed. Proc.* 35:96–97.

————. (1982) The relationship of the Gompertz constant and maximum potential lifespan to body mass. *Exp. Gerontol.* 17:383–385.

————. (1983). Body size, mortality, and longevity. *J. Theoret. Biol.* 102:135–144.

————. (1984) *Size, Function, and Life History.* Cambridge, MA: Harvard Univ. Press.

————. (1985) The comparative biology of longevity and lifetime energetics. *Exp. Gerontol.* 20:161–170.

Calkins, E., and Challa, H. R. (1985) Disorders of the joints and connective tissue. In *Principles of Geriatric Medicine,* W. R. Hazzard, R. Andres, E. L. Bierman, and J. P. Bass (eds.), 813–843. New York: McGraw-Hill.

Calkins, E., and Wright, J. R. (1989) Amyloid. In *Principles of Geriatric Medicine and Gerontology,* 2d ed., W. R. Hazzard, R. Andres, E. L. Bierman, and J. P. Bass (eds.), 897–903. New York: McGraw-Hill.

Callow, J. A., Callow, M. E., and Woolhouse, H. W. (1972) *In vivo* protein synthesis, RNA synthesis, and polyribosomes in senescing leaves of *Perilla. Cell Diff.* 1:79–90.

Calow, P. (1973) The relationship between fecundity, phenology, and longevity: A systems approach. *Am. Nat.* 107:559–574.

————. (1978a) The evolution of life-cycle strategies in fresh-water gastropods. *Malacologia* 17:351–364.

————. (1978b) Bidder's hypothesis revisited. Solution to some key problems associated with general molecular theory of ageing. *Gerontology* 24:448–458.

————. (1979) The cost of reproduction—A physiological approach. *Biol. Rev.* 54:23–40.

————. (1983) Life cycle patterns and evolution. In *The Mollusca,* vol. 6, *Ecology,* W. D. Russell-Hunter (ed.), 649–678. New York: Academic Press.

Calow, P., Beveridge, M., and Sibley, R. (1979) Heads and tails: Adaptational aspects of asexual reproduction in freshwater triclads. *Am. Zool.* 19:715–727.

Calow, P., and Woollhead, A. S. (1977) The relationship between ration, reproductive effort, and age-specific mortality in the evolution of life-history strategies—Some observations on freshwater triclads. *J. Anim. Ecol.* 46:765–781.

Camboué, P. (1926) Prolongation de la vie chez les papillons decapités. *Compt. Rend. Soc. Biol.* (Paris) 183:372.

Cameron, I. L., and Thrasher, J. D. (1976) Cell renewal and cell loss in the tissues of aging mammals. In *Interdisciplinary Topics in Gerontology,* vol. 10, R. G. Cutler (ed.), 83–133. Basel: Karger.

Campbell, B. A., Krauter, E. E., and Wallace, J. E. (1980) Animal models of aging: Sensorimotor and cognitive function in the aged rat. In *Psychobiology of Aging: Problems and Perspectives,* D. G. Stein (ed.), 201–226. Amsterdam: Elsevier–North-Holland.

Campbell, B. A., and Richardson, R. (1988) Effect of chronic undernutrition on susceptibility to cold stress in young adult and aged rats. *Mech. Age. Dev.* 44:193–202.

Campbell, R. N. (1979) Ferox trout, *Salmo trutta* L., and charr, *Salvelinus alpinus* (L.), in Scottish lochs. *J. Fish. Biol.* 14:1–29.

Campos Carcera, H. (1969) Reproducción del *Aplochiton taeniatus Jenyns. Bol. Museo Nacional de Hist. Nat.* (Santiago, Chile) 29:207–222.

Canivenc, R. (1965) A study of progestation in the European badger (*Meles meles* L.). In *Comparative Biology of Reproduction in Mammals,* R. W. Rowlands (ed.), 15–26. New York: Academic Press.

Canivenc, R., and Bonnin-Laffargue, M. (1963) Inventory of problems raised by the delayed ova implantation in the European badger (*Meles meles* L.). In *Delayed Implantation,* A. C. Enders (ed.), 115–128. Chicago: Univ. of Chicago Press.

Carew, T. E., Schwenke, D. C., and Steinberg, D. (1987) Antiatherogenic effort of probucol unrelated to its hypocholesterolemic effect: Evidence that antioxidants *in vivo* can selectively inhibit low density lipoprotein degradation in macrophage-rich fatty streaks and slow the progression of atherosclerosis in the Watanabe heritable hyperlipidemic rabbit. *Proc. Natl. Acad. Sci.* 84:7725–7729.

Carlson, G. A., Kingsbury, D. T., Goodman, P. A., Coleman, S., Marshall, S. T., De-Armond, S., Westaway, D., and Prusiner, S. B. (1986) Linkage of prion protein and scrapie incubation time genes. *Cell* 46:503–511.

Carlsson, G. (1962) Studies on Scandinavian black flies (fam. Simuliidae Latr.). *Opuscula Entomol.*, suppl., 21:1–280.

Carothers, A. D., Collyer, S., De Mey, R., and Frackiewicz, A. (1978) Parental age and birth order in the aeticology of some sex chromosome aneuploidies. *Ann. Human Genet.* 41:227–287.

———. (1980) Parental age and birth order in the aetiology of some sex chromosome aneuploidies. *Ann. Human Genet.* (London) 41:277–287.

Carothers, A. D., Frackiewicz, A., De Mey, R., Collyer, S., Polani, P. E., Osztovics, M., Horvath, K., Papp, Z., May, H. M., and Ferguson-Smith, M. A. (1980) A collaborative study of the aetiology of Turner syndrome. *Ann. Human Genet.* (London) 43:355–375.

Carpenter, F. M., and Burnham, L. (1985) The geological record of insects. *Ann. Rev. Earth Planet. Sci.* 13:297–314.

Carrel, A. (1935) *Man the Unknown*. New York: Halcyon House. [not read].

Carroll, R. L. (1988) *Vertebrate Paleontology and Evolution*. New York: Freeman.

Carskadon, M. A., and Dement, W. C. (1981) Respiration during sleep in the aging human. *J. Gerontol.* 36:420–423.

Casanueva, F., Cocchi, D., Locatelli, V., Flauto, C., Zambotti, F., Bestetti, G., Rossi, G. L., and Muller, E. (1982) Defective central nervous system dopaminergic function in rats with estrogen-induced pituitary tumors, as assessed by plasma prolactin concentrations. *Endocrinology* 110:590–599.

Casarett, G. W. (1963) Concept and criteria of radiologic ageing. In *Cellular Basis and Aetiology of Late Somatic Effects of Ionizing Radiation*, R. J. Harris (ed.), 189–205. New York: Academic Press.

———. (1964) Similarities and contrasts between radiation and time pathology. *Adv. Gerontol. Res.* 1:109–163.

Cassada, R. C., and Russell, R. L. (1975) The dauer larva, a post-embryonic developmental variant of the nematode *Caenorhabditis elegans*. *Dev. Biol.* 46:326–342.

Castanet, J., and Naulleau, G. (1985) Skeletochronology in reptiles. 2. Experimental data about the signification of the skeletal growth marks used among snakes. Comments on growth and longevity of *Asp viper*. *Ann. Sci. Nat. Zool.* (Paris) 7:41–62.

Castelli, W. P., Doyle, J. T., Gordon, T., Hames, C., Hulley, S. B., Kagan, A., McGee, D., Vicic, W. J., Zukel, W. J. (1975) HDL cholesterol levels (HDLC) in coronary heart disease (CHD): A cooperative lipoprotein phenotyping study. *Circulation* 52 (suppl. 2): II–97.

Castenada, M., Vargas, R., and Galvan, S. C. (1986) Stagewise decline in the activity of brain protein synthesis factors and relationship between this decline and longevity in two rodent species. *Mech. Age. Dev.* 36:197–210.

Caswell, H. (1986) The evolutionary demography of clonal reproduction. In *The Growth and Form of Modular Organisms*, J. L. Harper, B. R. Rosen, and J. White (eds.), 187–224. *Phil. Trans. Roy. Soc.* (London), special issue, ser. B, 313:1–250.

Cathcart, R., Schwiers, E., Saul, R. L., and Ames, B. N. (1984) Thymine glycol and thymidine glycol in human and rat urine: A possible assay for oxidative DNA damage. *Proc. Natl. Acad. Sci.* 81:5633–5637.

Cattanach, B. M., Pollard, C. F., and Perez, J. N. (1969) Controlling elements in the mouse X-chromosome. *Z. Vererbungslehre* 96:313–323.

Cattanach, B. M. (1974) Position effect variegation in the mouse. *Genet. Res.* 23:291–306.

Caughley, G. (1977) *Analysis of Vertebrate Populations*. New York: Wiley and Sons.

Caullery, M. (1952) *Parasitism and Symbiosis*. London: Sidgwick and Jackson.

Cedar, H. (1988) DNA methylation and gene activity. *Cell* 53:3–4.

Center for Disease Control. (1981) Morbidity and mortality report. DHHS Publication (CDC) 81–8017 30. Atlanta, GA.

Cerami, A., Vlassara, H., and Brownlee, M.

(1987) Glucose and aging. *Sci. Am.* 256:90–96

Chaconas, G., and Finch, C. E. (1973) The effect of aging on RNA/DNA ratios in brain regions of the C57BL/6J mouse. *J. Neurochem.* 21:1466–1473.

Chait, A., Albers, J. J., and Brunzell, J. D. (1980) Very low density lipoprotein overproduction in genetic forms of hypertriglyceridemia. *Eur. J. Clin. Invest.* 10:17–22.

Chaline, J. (1977) Rodents, evolution, and prehistory. *Endeavour,* n.s. 1:41–51.

Chaloner, W. G. (1968) The cone of *Cyclostigma kiltorkense* Haughton from the upper Devonian of Ireland. *J. Linn. Soc.* (London), Botany, 61:25–36.

———. (1983) Leaf and stem growth in the lepidodendrales. *Bot. J. Linnean Soc.* 86:135–148.

Chaloner, W. G., and Boureau, E. (1967) Lycophyta. In *Traité de Paléobotanique,* vol. 2, E. Boureau (ed.), 437–802. Paris: Masson.

Chaloner, W. G., and Sheerin, A. (1979) Devonian microfloras. In *The Devonian System,* Special Papers in Paleontology, vol. 23, M. R. House, C. T. Scrutton, and M. G. Bassett (eds.), 145–161. London: The Paleontological Association.

Chance, B., Sies, H., and Boveris, A. (1979) Hydroperoxide metabolism in mammalian organs. *Physiol. Rev.* 59:527–603.

Chandler, A. C., and Read, C. P. (1961). *Introduction to Parasitology.* 10th ed. New York: Wiley.

Chandley, A. C. (1985) Maternal aging as the important etiological factor in human aneuploidy. *Basic Life Sci.* 36:409–416.

Chang, H. S. (1951) Age and growth of *Callionymus lyra. J. Mar. Biol. Assoc.* (U.K.) 30:281–288.

Chang, M.-P., Makinodan, T., Peterson, W. J., and Strehler, B. L. (1982) Role of T cells and adherent cells in age-related decline in murine interleukin 2 production. *J. Immunol.* 129:2426–2430.

Chapman, A., Gonzales, G., Burrows, W. R., Assanah, P., Iannone, B., Leung, M. K., and Stefano, G. B. (1984) Alterations in high-affinity binding characteristics and levels of opioids in invertebrate ganglia during aging:

Evidence for an opioid compensatory mechanism. *Cell. Mol. Neurobiol.* 4:143–155.

Charlesworth, B. (1980) *Evolution in Age-Structured Populations.* Cambridge: Cambridge Univ. Press.

———. (1990a) Natural selection and life-history patterns. In *Genetic Effects on Aging II,* D. E. Harrison (ed.). Caldwell, NJ: Telford Press, in press.

———. (1990b) Optimization models, quantitative genetics, and mutation. University of Chicago, Dept. of Evolution.

Chase, M., Babb, M., Stone, M., and Rich, S. (1976) Active sleep patterns in the old cat. *Sleep Res.* 5:20.

Chatterjee, B., Fernandes, G., Yu, B. P., Song, C., Kim, J. M., Demyan, W., and Roy, A. K. (1988) Caloric restriction delays age-dependent loss in androgen responsiveness of the rat liver. *Fed. Am. Soc. Exp. Biol.* 3:169–173.

Chatterjee, B., Majumdar, D., Ozbilen, O., Murty, C. V. R., and Roy, A. K. (1987) Molecular cloning and characterization of cDNA for androgen-repressible rat liver protein, SMP-2. *J. Biol. Chem.* 262:822–825.

Chatterjee, B., Nath, T. S., and Roy, A. K. (1981) Differential regulation of the messenger RNA for three major senescence marker proteins in male rat liver. *J. Biol. Chem.* 256:5939–5941.

Chaturvedi, M. M., and Kanungo, M. S. (1985) Analysis of conformation and function of the chromatin of the brain of young and old rats. *Mol. Biol. Rep.* 10:215–219.

Chaudhuri, M., Sartin, J. L., and Adelman, R. C. (1983) A role for somatostatin in the impaired insulin secretory response to glucose by islets from aging rats. *J. Gerontol.* 38:431–435.

Cheal, P. D., Lee, A. K., and Barnett, J. L. (1976) Changes in the haematology of *Antechinus stuartii* (Marsupialia) and their association with male mortality. *Austral. J. Zool.* 24:299–311.

Chen, J. C., Ove, P., and Lansing, A. I. (1973) *In vitro* synthesis of microsomal protein and albumin in young and old rats. *Biochim. Biophys. Acta* 312:598–607.

Chen, M., Bergman, R. N., and Porte, D., Jr.

(1988) Insulin resistance and β-cell dysfunction in aging: The importance of dietary carbohydrate. *J. Clin. Endocrinol. Metab.* 67:951–957.

Chen, T. S., Richie, J. P., and Lang, C. A. Lifespan production of glutathione and acetaminophen detoxification. (In press) *Drugs Metab. Disposition.*

Cheney, K. E., Liu, R. K., Smith, G. S., Leung, R. E., Mickey, M. R., and Walford, R. L. (1980) Survival and disease patterns in C57BL/6J mice subjected to undernutrition. *Exp. Gerontol.* 15:237.

Cheney, K. E., Liu, R. K., Smith, G. S., Meredith, P. J., Mickey, M. R., and Walford, R. L. (1983) The effect of dietary restriction of varying duration on survival, tumor patterns, immune function, and body temperature in B10C3F₁ female mice. *J. Gerontol.* 38:420–430.

Cherkin, A., and Eckhardt, M. J. (1977) Effects of dimethylaminoethanol upon life-span and behavior of aged Japanese quail. *J. Gerontol.* 32:38–45.

Chetsanga, C. J., Boyd, V., Peterson, L., and Rusholow, K. (1975) Single-stranded regions in DNA of old mice. *Nature* 253:130–131.

Chetsanga, C. J., Tuttle, M., Jacobini, A., and Johnson, C. (1977) Age-associated structural alterations in senescent mouse brain DNA. *Biochim. Biophys. Acta* 474:180–187.

Cheung, H. T., Twu, J.-S., and Richardson, A. (1983) Mechanism of the age-related decline in lymphocyte proliferation: Role of IL-2 production and protein synthesis. *Exp. Gerontol.* 18:451–460.

Cheung, W. W. K., and Marshall, A. T. (1973) Water and ion regulation in cicadas in relation to xylem feeding. *J. Insect Physiol.* 19:1801–1816.

Chia, F. S. (1976) Sea anemone reproduction patterns and adaptive radiations. In *Coelenterate Behavior and Ecology,* G. O. Mackie (ed.), 261–270. New York: Plenum.

Child, C. M. (1913) The asexual cycle of *Planaria velata* in relation to senescence or rejuvenescence. *Biol. Bull.* 25:181–203.

———. (1914) Asexual breeding and prevention of senescence in *Planaria velata. Biol. Bull.* 26:286–293.

———. (1915) *Senescence and Rejuvenescence.* Chicago: Univ. of Chicago Press.

Childs, E. A., and Law, D. K. (1972) Growth characteristics of progeny of salmon with different maximum life spans. *Exp. Gerontol.* 7:405–407.

Chilton, D. E., and Beamish, R. J. (1982) Age determination methods for fishes studied by the groundfish program at the Pacific biological station. *Can. Special Pub. Fish. Aquat. Sci.* 60.

Chiodi, H. (1976) Thymus hypertrophy induced by castration in old male rats and mice. *Fed. Proc.* 35:277 (abstract 395).

Chipalkatti, S., De, A. K., and Aiyar, A. S. (1983) Effect of diet restriction on some biochemical parameters related to aging. *Mech. Age. Dev.* 21:37–48.

Choat, J. H., and Black, R. (1979) Life histories of limpets and the limpet-laminarian relationship. *J. Exp. Mar. Biol. Ecol.* 41:25–50.

Choongkittaworn, N., Hosick, H. L., and Jones, W. (1987) *In vitro* replication potential of serially passaged mammary parenchyma from mice with different reproductive histories. *Mech. Age. Dev.* 39:147–175.

Christian, J. J. (1950) The adreno-pituitary system and population cycles in mammals. *J. Mammal.* 31:247–259.

Christie, J. R. (1929) Some observations on sex in the Mermithidae. *J. Exp. Zool.* 53:59–77.

Chu, Y. E., Morishima, H., and Oka, I. (1969) Reproductive barriers distributed in cultivated rice species and their wild relatives. *Jpn. J. Genet.* 44:207–223.

Chuknyiska, R. S., Haji, M., Foote, R. H., and Roth, G. S. (1985) Age-associated changes in nuclear binding of rat uterine estradiol receptor complexes. *Endocrinology* 116:547–551.

Chuknyiska, R. S., and Roth, G. S. (1985) Decreased estrogenic stimulation of RNA polymerase II in aged rat uteri is apparently due to reduced nuclear binding of receptor-estradiol complexes. *J. Biol. Chem.* 260:8661–8663.

Cinader, B., Van Der Gaag, H. C., Koh, S.-Y. W., and Axelrad, A. A. (1987) Friend virus replication as a function of age. *Mech. Age. Dev.* 40:181–191.

Cini, J. K., and Gracy, R. W. (1986) Molecular basis of the isozymes of bovine glucose–6–phosphate isomerase. *Arch. Biochem. Biophys.* 249:500–505.

Clancy, D. W. (1946) The insect parasites of the Chrysopidae (Neuroptera). *Univ. Calif. Publ. Entomol.* 7:403–496.

Clapp, R. B., Klimkiewicz, M. K., and Kennard, J. H. (1982) Longevity records of North American birds: Gavidae through Alcidae. *J. Field Orthnithol.* 53:81–124.

Clare, M. J., Luckinbill, L. S. (1985) The effects of gene-environment interaction on the expression of longevity. *Heredity* 55:19–29.

Clark, A. M. (1964) Genetic factors associated with aging. *Adv. Gerontol. Res.* 1:107–255.

Clark, A. M., and Rubin, M. A. (1961) The modification by X-irradiation of the life span of haploids and diploids of the wasp, *Habrobracon* sp. *Rad. Res.* 15:244–253.

Clark, E. (1963) The maintenance of sharks in captivity, with a report on their instrumental conditioning. In *Sharks and Survival*, P. W. Gilbert (ed.), 115–149. Boston: D. C. Heath.

Clark, T. B. (1940) The relation of production and egg weight to age—White Leghorn fowl. *Poultry Sci.* 14:54.

Clarke, J. M., and Maynard-Smith, J. (1961a) Independence of temperature of the rate of ageing in *Drosophila subobscura*. *Nature* (London) 190:1027–1028.

———. (1961b) Two phases of aging in the *Drosophila subobscura*. *J. Exp. Biol.* 38:679–684.

———. (1966) Increase in the rate of protein synthesis with age in *Drosophila subobscura*. *Nature* 209:627–629.

Clarke, S. (1987) Propensity for spontaneous succinimide formation from aspartyl and asparaginyl residues in cellular proteins. *Int. J. Peptide Res.* 30:808–821.

Clarke, S., and O'Connor, C. M. (1983) Do eukaryotic carboxyl methyltransferases regulate protein function? *Trends in Biochem. Sci.* 8:391–394.

Clayton, D. A., and Smith, C. A. (1975) Complex mitochondrial DNA. *Int. Rev. Exp. Pathol.* 14:1–67.

Cleaver, J. E. (1968) Defective repair replication of DNA in xeroderma pigmentosum. *Nature* 218:652–656.

Clegg, M. T. (1959) Factors affecting gestation length and parturition. In *Reproduction in Domestic Animals*, H. H. Cole and P. T. Cuppo (eds.), 509–538. New York: Academic Press.

Clemens, W. A. (1986) Evolution of the terrestrial vertebrate fauna during the Cretaceous-Tertiary boundary. In *Dynamics of Extinction*, D. K. Elliott (ed.), 63–85. New York: Wiley.

Clements, A. N. (1963) *The Physiology of Mosquitos*. New York: Pergamon Press.

Cleveland, I. R. (1947a) The origin and evolution of meiosis. *Science* 105:207–289.

———. (1947b) Sex produced in the protozoa of *Cryptocercus* by mating. *Science* 105:16–17.

Clokey, G. V., and Jacobsen, L. A. (1986) The autofluorescent "lipofuscin granules" in the intestinal cells of *Caenorhabditis elegans* are secondary lysosomes. *Mech. Age. Dev.* 35:79–94.

Cloud, P. (1980) Early biogeochemical systems. In *Biogeochemistry of Ancient and Modern Environments*, P. A. Trudinger, M. R. Walter, and B. J. Ralph (eds.), 7–27. New York: Springer-Verlag.

Cloud, P., and Glaessner, M. F. (1982) The Ediacarian period and system: Metazoa inherit the earth. *Science* 217:783–792.

Clutton-Brock, T. H., Albon, S. D., and Guinness, F. E. (1988) Reproductive success in male and female red deer. In *Reproductive Success: Studies of Individual Variation in Contrasting Breeding Systems*, T. H. Clutton-Brock (ed.), 325–343. Chicago: Univ. of Chicago Press.

Clutton-Brock, T. H., Guinness, F. E., and Albon, S. D. (1982) *Red Deer: Behavior and Ecology of Two Sexes*. Chicago: Univ. of Chicago Press.

———. (1983) The costs of reproduction to red deer hinds. *J. Anim. Ecol.* 52:367–383.

Coates, A. G., and Jackson, J. B. (1985) Morphological themes in the evolution of clonal and aclonal marine invertebrates. In *Population Biology and Evolution of Clonal Organisms*, J. B. Jackson, L. W. Buss, and R. E. Cook (eds.), 67–106. New Haven: Yale Univ. Press.

Coburn, H. G., Grey, R. M., and Rivera,

S. M. (1971) Observations of the alteration of heart rate, life span, weight, and mineralization in the diagoxin-treated A/J mouse. *Johns Hopkins Med. J.* 128:169–193.

Cockrum, E. L. (1973) Additional longevity records for American bats. *J. Ariz. Acad. Sci.* 8:108–110.

Cockburn, A., Lee, A. K., and Martin, R. W. (1983) Macrogeographic variation in litter size in *Antechinus* (Marsupialia: Dasyuridae). *Evolution* 37:86–95.

Cohen, A. M., Aberdroth, R. E., and Hochstein, P. (1984) Inhibition of free radical–induced DNA damage by uric acid. *FEBS Lett.* 174:147–150.

Cohen, B. J., Anver, M. R., Ringler, D. H., and Adelman, R. C. (1978) Age-associated pathological changes in male rats. *Fed. Proc.* 37:2848–2850.

Cohen, B. J., Danon, D., and Roth, G. S. (1987) Wound repair in mice as influenced by age and antimacrophage serum. *J. Gerontol.* 42:295–301.

Cohen, M. L., and Berkowitz, B. A. (1974) Age-related changes in vascular responsiveness to cyclic nucleotides and contractile agonists. *J. Pharmacol. Exp. Ther.* 191:147–155.

Cohen, M. N., Malpass, R. S., and Klein, H. G. (eds.) (1980) *Biosocial Mechanisms of Population Regulation.* New Haven: Yale Univ. Press.

Cohn, S. H., Abesamis, C., Yasumura, S., Aloia, J. F., and Ellis, K. J. (1977) Comparative skeletal mass and radial bone mineral content in black and white women. *Metabolism* 26:171–178.

Colby, P. J., McNicol, R. E., and Ryder, R. A. (1979) Synopsis of biological data on the walleye, *Stizostedion v. vitreum. Food Agric. Org. Fish. Synopsis,* no. 119:1–139.

Colby, P. J., and Nepszy, S. J. (1981) Variation among stocks of walleye (*Stizostedion vitreum vitreum*). Management implications. *Can. J. Fish. Aquat. Sci.* 38:1814–1831.

Cole, L. C. (1954) The population consequences of life history phenomena. *Q. Rev. Biol.* 29:103–137.

Coleman, D. L. (1978) Obese and diabetes: Two mutant genes causing diabetes-obesity syndromes in mice. *Diabetologia* 14:41–48.

———. (1982) Thermogenesis in diabetes-obesity syndromes in mice. *Diabetologia* 22:205–11.

Coleman, D. L., Schwizer, R. W., and Leiter, E. H. (1984) Effect of genetic background on the therapeutic effects of dehydroepiandrosterone (DHEA) in diabetes-obesity mutants and in aged normal mice. *Diabetes* 33:26–32.

Coleman, G. L., Barthold, S. W., Osbaldiston, G. W., Foster, S. J., and Jonas, A. M. (1977) Pathological changes during aging in barrier-reared Fischer 344 male rats. *J. Gerontol.* 32:258–278.

Coleman, J. W. (1976) Hair cell loss as a function of age in the normal cochlea of the guinea pig. *Acta Otolaryngol.* 82:33–40.

Coleman, P., Finch, C. E., and Joseph, J. (1990) On two time point aging studies. *Neurobiol. Aging* 11:1–2.

Coleman, P. D., and Flood, D. G. (1987) Neuron numbers and dendritic extent in normal aging and Alzheimer's disease. *Neurobiol. Aging* 8:521–545.

Colgin-Bukovsan, L. A. (1979) Life cycles and conditions for conjugation in the Suctoria *Tokophrya lemnarum. Arch. Protistenk* 121:223–237.

Collard, M. W., and Griswold, M. D. (1987) Biosynthesis and molecular cloning of sulfated glycoprotein 2 secreted by rat Sertoli cells. *Biochemistry* 26:3297–3303.

Collatz, K.-G., and Sohal, R. S. (eds.) (1986) *Insect Aging: Strategies and Mechanisms.* Heidelberg: Springer-Verlag.

Collatz, K.-G., and Wilps, H. (1986) Aging of flight mechanism. In *Insect Aging: Strategies and Mechanisms,* K.-G. Collatz and R. S. Sohal (eds.), 54–72. Heidelberg: Springer-Verlag.

Collins, J. P. (1979) Intrapopulation variation in the body size at metamorphosis and timing of metamorphosis in the bullfrog, *Rana catesbeiana. Ecology* 60:738–749.

Collins, K. J., Dore, C., Exton-Smith, A. N., Fox, R. H., MacDonald, I. C., and Woodward, P. M. (1977) Accidental hypothermia and impaired temperature homeostasis in the elderly. *Brit. Med. J.* 1:353–356.

Colman, P. C., Kaplan, B. B., Osterburg, H. H., and Finch, C. E. (1980) Brain poly(A)RNA during aging: Stability of yield and sequence complexity in two rat strains. *J. Neurochem.* 34:335–345.

Comfort, A. (1953) Absence of a Lansing effect in *Drosophila subobscura. Nature* 172:83–84.

———. (1957) The duration of life in molluscs. *Proc. Malacol. Soc.* 32:219–249.

———. (1959) The longevity and mortality of thoroughbred stallions. *J. Gerontol.* 14:9–10.

———. (1960a) The effect of age on growth-resumption in fish (*Lebistes*) checked by food restriction. *Gerontologia* (Basel) 4:177–186.

———. (1960b) Longevity and mortality in dogs of four breeds. *J. Gerontol.* 15:126–129.

———. (1961) The longevity and mortality of a fish (*Lebistes reticulatus* Peters) in captivity. *Gerontologia* (Basel) 5:209–222.

———. (1962) Survival curves of some birds in the London Zoo. *Ibis* 104:115–117.

———. (1963) Effect of delayed and resumed growth on the longevity of a fish. (*Lebistes reticulatus* Peters) in captivity. *Gerontologia* (Basel) 8:150–155.

———. (1979) *The Biology of Senescence.* 3d ed. Edinburgh and London: Churchill Livingstone.

———. (1980). Sexuality in later life. In *Handbook of Mental Health and Aging,* J. E. Birren and R. B. Sloane (eds.), 885–982. Englewood Cliffs, NJ: Prentice Hall.

Comfort, A., and Doljanski, F. (1958) The relation of size and age to rate of tail regeneration in *Lebistes reticulatus. Gerontologia* (Basel) 2:266–283.

Comfort, A., Youhotsky-Gore, I., and Pathmanathan, K. (1971) Effect of ethoxyquin on the longevity of C_3H mice. *Nature* (London) 229:254–255.

Commoner, B. (1964) Roles of deoxyribonucleic acid in inheritance. *Nature* 202:960–968.

Compton, M. M., and Cidlowski, J. A. (1986) Rapid *in vivo* effects of glucocorticoids on the integrity of rat lymphocyte genomic deoxyribonucleic acid. *Endocrinology* 118:38–45.

Comstock, J. H. (1950) *Introduction to Entomology,* 9th ed. Ithaca, NY: Comstock.

Connell, J. H. (1970) A predator-prey system in the marine intertidal region. 1. *Balanus glan-*

dula and several predatory species of *Thais. Ecol. Monographs* 40:49–78.

———. (1973) Population ecology of reef-building corals. In *Biology and Geology of Coral Reefs,* vol. 2, *Biology 1,* O. A. Jones and R. Endean (eds.), 205–245. New York: Academic Press.

Conner-Johnson, B., Gaijar, A., Kubo, C., and Good, R. A. (1986) Calories versus protein in onset of renal disease in NZB × NZW mice. *Proc. Natl. Acad. Sci.* 83:5659–5662.

Contag, C. H., Harty, J. T., and Plagemann, G. W. (1989) Dual virus etiology of age-dependent poliomyelitis of mice. A potential model for human motor neuron diseases. *Microbiol. Pathogen.* 6:391–401.

Contag, C. H., and Plagemann, G. W. (1988) Susceptibility of C58 mice to paralytic disease induced by lactate dehydrogenase-elevating virus correlates with increased expression of endogenous retrovirus in motor neurons. *Microbiol. Pathogen.* 5:287–296.

Conway Morris, S. (1986) The community structure of the middle Cambrian phyllopod bed (Burgess shale). *Paleontology* 29:423–467.

Cook, L. L., and Gafni, A. (1988) Protection of phosphoglycerate kinase against *in vitro* aging by selective cysteine methylation. *J. Biol. Chem.* 263:13991–13993.

Cook, R. E. (1983) Clonal plant populations. *Am. Sci.* 71:244–253.

———. (1985) Growth and development in clonal plant populations. In *Population Biology and Evolution of Clonal Organisms,* J. B. C. Jackson, L. W. Buss, and R. E. Cook (eds.), 259–296. New Haven: Yale Univ. Press.

Cooke, J., and Smith, J. C. (1990) Measurement of developmental time by cells of early embryos. *Cell* 60:891–894.

Cooper, G. J. S., Leighton, B., Dimitriadis, G. D., Parry-Billings, M., Kowalchuk, J. M., Howland, K., Rothbard, J. B., Willis, A. C., and Reid, K. B. M. (1988) Amylin found in amyloid deposits in human type 2 diabetes mellitus may be a hormone that regulates glycogen metabolism in skeletal muscle. *Proc. Natl. Acad. Sci.* 85:7763–7766.

Cooper, R. A., and Uzmann, J. R. (1977) Ecology of juvenile and adult clawed lobsters, *Homarus americanus, Homarus gammarus,* and *Nephrops norvegicus. Circ.-CSIRO, Div. Fish. Oceanog. (Austral.)* no. 7:187–208. [not read]

Cope, E. D. (1896) *The Primary Factors of Organic Evolution.* Chicago: Open Court. [not read]

Coquelin, A., and Desjardins, C. (1982) Luteinizing hormone and testosterone secretion in young and old male mice. *Am. J. Physiol.* 243:257–263.

Corbet, P. S. (1967) Facultative autogeny in arctic mosquitos. *Nature* 215:662–663.

Corbet, P. S., Longfield, C., and Moore, N. W. (1960) *Dragonflies.* London: N. N. Collins.

Corkin, S., Rosen, T. J., Sullivan, E. V., and Clegg, R. A. (1989) Penetrating head injury in young adulthood exacerbates cognitive decline in later years. *J. Neurosci.* 9:3876–3883.

Corliss, J. O. (1953) Comparative studies in holotrichous ciliates in the colpidium-glaucoma-leukophrys-tetrahymena group. 2. Morphology, life cycles, and systematic status of strains in pure culture. *Parasitology* 43:49–87.

Cosgrove, J. W., and Rapoport, S. (1987) Absence of age differences in protein synthesis by rat brain, measured with an initiating cell-free system. *Neurobiol. Aging* 8:27–34.

Cotchin, E., and Roe, F. J. (1967) *Pathology of Laboratory Rats and Mice.* Philadelphia: F. A. Davis.

Cotman, C. (1990) Synaptic plasticity, neurotrophic factors, and transplantation in the aged brain. In *Handbook of the Biology of Aging,* 3d ed., E. L. Schneider and J. W. Rowe (eds.), 255–274. San Diego: Academic Press.

Cott, H. B. (1961) Scientific results of an enquiry into the ecology and economic status of the Nile crocodile (*Crocodylus niloticus*) in Uganda and Northern Rhodesia. *Trans. Zool. Soc. Lond.* 29:211–356.

Cottam, W. P. (1954) Prevernal leafing of aspen in Utah mountains. *J. Arnold Arboretum* 35:239–245.

Coulson, J. C., and Horobin, J. (1976) The influence of age on the breeding biology and survival of the arctic tern *Sterna paradisaea. J. Zool.* (London) 178:247–260.

Coulson, R. A. (1986) Metabolic rate and the flow theory: A study in chemical engineering. *Comp. Biochem. Physiol.* 84A:217–229.

———. (1987) Aerobic and anaerobic glycolysis in mammals and reptiles in vivo. *Comp. Biochem. Physiol.* 87B:207–216.

Coulson, R. A., Hernandez, T., and Herbert, J. D. (1977) Metabolic rate, enzyme kinetics in vivo. *Comp. Biochem. Physiol.* 56:251–262.

Coyle, J. T., Oster-Granite, M. L., and Gearhart, J. D. (1986) The neurobiologic consequences of Down syndrome. *Brain Res. Bull.* 16:773–787.

Cragg, J. B., and Cole, P. (1952) Diapause in *Lucilia sericata* (Mg.) Diptera. *J. Exp. Biol.* 29:600–604.

Craig, J. F. (1985) Aging in fish. *Can. J. Zool.* 63:1–8.

Craik, F. I. (1984) Age differences in remembering. In *Neuropsychology of Memory,* L. R. Squire and N. Butters (eds.), 3–12. New York: Guilford.

Crepet, W. L. (1979) Some aspects of the pollination biology of middle Eocene angiosperms. *Rev. Paleobot. Palynol.* 27:213–238.

Crichton, M. I. (1957) The structure and function of the mouth parts of adult caddis flies (Trichoptera). *Phil. Trans. Roy. Soc.* (London), ser. B, 241:45–91.

Cristofalo, V. J. (1985) The destiny of cells: Mechanisms and implications of senescence. The 1983 Robert W. Kleemeier Award Lecture. *Gerontologist* 25:577–583.

Cristofalo, V. J., and Stanulis-Praeger, B. M. (1982) Cellular senescence *in vitro.* In *Advances in Tissue Culture,* vol. 2, K. Maramorosch (ed.), 1–68. New York: Academic Press.

Crocker, W. (1942) Aging in plants. In *Problems of Ageing,* 2d ed., E. V. Cowdry (ed.), 1–28. New York: Plenum.

Crompton, A. W., and Jenkins, F. A. (1979) Origins of mammals. In *Mesozoic Mammals: The First Two-thirds of Mammalian History,* J. A. Lillegraven, Z. Keilan-Jaworowska, and W. A. Clemens (eds.), 59–73. Berkeley: Univ. of California Press.

Crompton, A. W., Taylor, C. R., and Jagger,

J. A. (1978) Evolution of homeothermy in mammals. *Nature* 272:333–336.

Crowl, T. A., and Covich, A. P. (1990) Predator-induced life-history shifts in a freshwater snail. *Science* 247:949–951.

Crowley, C., and Curtis, H. J. (1963) The development of somatic mutations in mice with age. *Proc. Natl. Acad. Sci.* 49:626–628.

Crowley, P. H., Gulati, D. K., Hayden, T. L., Lopez, P., and Dyer, R. (1979) A chiasma-hormonal hypothesis relating Down's syndrome and maternal age. *Nature* 280:417–419.

Crowson, R. A. (1981) *The Biology of the Coleoptera.* New York: Academic Press.

Crozier, W. J. (1918) Growth and duration of life in *Chiton tuberculatus. Proc. Natl. Acad. Sci.* 4:322–325.

Cuellar, O. (1977) Animal parthenogenesis. *Science* 197:837–843.

Cumber, R. A. (1949) An overwintering nest of the bumble-bee *Bombus terrestris* (L.) (Hymenoptera, Apidae). *New Zealand Sci. Rev.* 7:96–97.

———. (1952) Notes on the biology of *Melampsalta cruentata fabricius* (Hemiptera-Homoptera: Cicadidae), with special reference to the nymphal stages. *Trans. Roy. Entomol. Soc.* (London) 103:219–237.

Cummings, D. J., Turker, M. S., and Domenico, J. M. (1986) Mitochondrial excision-amplification plasmids in senescent and long-lived cultures of *Podospora anserina.* In *Extrachromosomal Elements in Lower Eukaryotes,* R. B. Wickner, A. Hinnebusch, A. M. Lambowitz, I. C. Gonsalus, and A. Hollaender (eds.), 129–146. New York: Plenum.

Cummings, D. J., and Wright, R. M. (1983) DNA sequence of the excision sites of a mitochondrial plasmid from senescent *Podospora anserina. Nucleic Acids Res.* 11:2111–2119.

Cummings, S. R., Kelsey, J. L., Nevitt, M. C., O'Dowd, K. J. (1985) Epidemiology of osteoporosis and osteoporotic fractures. *Epidemiol. Rev.* 7:178–208.

Cunningham, J. T. (1891) On the reproduction and development of the conger. *J. Mar. Biol.* 2:16–42.

Curcio, C. A., Buell, S. J., and Coleman, P. D. (1982) Morphology of the aging central nervous system. In *Aging Motor Systems: Advances in Neurogerontology,* vol. 3, F. J. Pirozzolo and J. G. Maletta (eds.), 7–35. New York: Praeger.

Curtis, H. J. (1963) Biological mechanisms underlying the aging process. *Science* 141:686–694.

———. (1966) *Biological Mechanisms of Aging.* Springfield, IL: C. C. Thomas.

Curtis, H. J., and Healey, R. (1958) Effects of radiation on aging. In *Advances in Radiobiology,* G. C. Hevesy, A. G. Fossberg and J. D. Abbott (eds.), 261–265. London: Oliver and Boyd.

Curtis, H. J., Leith, J., and Tilley, J. (1966) Chromosome aberrations in liver cells of dogs of different ages. *J. Gerontol.* 21:268–270.

Curtis, H. J., and Miller, K. (1971) Chromosome aberrations in liver cells of guinea pigs. *J. Gerontol.* 26:292–293.

Cutler, R. G. (1975) Transcription of unique and reiterated DNA sequences in mouse liver and brain tissues as a function of age. *Exp. Gerontol.* 10:37–60.

———. (1976a) Evolution of longevity in primates. *J. Human Evol.* 5:169–202.

———. (1976b) Nature of aging and life maintenance processes. In *Interdisciplinary Topics in Gerontology,* vol. 9, R. G. Cutler (ed.), 83–133. Basel: Karger.

———. (1979) Evolution of longevity in ungulates and carnivores. *Gerontology* 25:69–86.

———. (1982) The dysdifferentiative hypothesis of mammalian aging and longevity. In *The Aging Brain: Cellular and Molecular Mechanisms of Aging in the Nervous System,* E. Giacobini, G. Filogamo, G. Giacobini, and A. Vernadakis. (eds.), 1–19. New York: Raven Press.

———. (1983) Superoxide dismutase, longevity, and specific metabolic rate. *Gerontology* 29:113–120.

———. (1984) Evolutionary biology of aging and longevity in mammalian species. In *Aging and Cell Function,* J. E. Johnson, Jr. (ed.), 1–148. New York: Plenum.

———. (1985) Antioxidants and longevity of mammalian species. *Basic Life Sci.* 35:15–73.

Cutler, S. J., Scotto, J., Devesa, S. S., and

Connelly, R. R. (1974) Third National Cancer Survey: An overview of available information. *J. Natl. Canc. Inst.* 53:1565–1575.

Dadswell, M. J., Klauda, R. J., Moffitt, C. M., Saunders, R. L., Rulifson, R. A., and Cooper, J. E. (1987) Common strategies of anadromous and catadromous fishes. *Am. Fish. Soc. Symp.* 1. Bethesda, MD: American Fisheries Society.

D'Amato, R. J., Alexander, G. M., Schwartzman, R. J., Kitt, C. A., Price, D. L., and Snyder, S. H. (1987) Evidence for neuromelanin in MPTP-induced neurotoxicity. *Nature* 327:324–326.

D'Ambrosio, S. M., Slazinski, L., Whetstone, J. W., and Lowney, E. (1981) Excision repair of UV-induced pyrimidine dimers in human skin in vivo. *J. Invest. Dermatol.* 77:311–313.

Dandekar, S., Rossitto, P., Pickett, S., Mockli, G., Bradshaw, H., Cardiff, R., and Gardner, M. (1987) Molecular characterization of the *Akvr–1* restriction gene: A defective endogenous retrovirus-borne gene identical to *Fy–4*. *J. Virol.* 61:308–314.

Danes, B. S. (1971) Progeria: A cell culture study on aging. *J. Clin. Invest.* 50:2000–2003.

Danesch, U., Hashimoto, R., Renkawitz, R., and Schutz, G. (1983) Transcriptional regulation of the tryptophan oxygenase gene in rat liver by glucocorticoids. *J. Biol. Chem.* 258:4750–4753.

Daniel, C. W. (1973) Finite growth span of mouse mammary gland serially progated in vitro. *Experientia* 29:1422–1424.

———. (1977) Cell longevity in vivo. In *Handbook of the Biology of Aging*, 1st ed., C. E. Finch and L. Hayflick (eds.), 122–158. New York: Van Nostrand Reinhold.

Daniel, C. W., DeOme, K. B., Young, J. T., Blair, P. B., and Faulken, L. J. (1968) The in vivo life span of normal and preneoplastic mouse mammary glands: A serial transplantation study. *Proc. Natl. Acad. Sci.* 61:52–60.

Daniel, C. W., Silberstein, G. B., and Strickland, P. (1984) Reinitiation of growth in senescent mouse mammary epithelium in response to cholera toxin. *Science* 224:1245–1247.

Daniel, C. W., and Young, L. J. T, (1971) Influence of cell division on an aging process. Life span of mouse mammary epithelium during serial propagation in vivo. *Exp. Cell Res.* 65:27–32.

Daniel, C. W., Young, L. J., Medina, D., and DeOme, K. B. (1971) The influence of hormones on serially transplanted mouse mammary gland. *Exp. Gerontol.* 6:95–101.

Darevski, I. S., Kupriyanova, L. A., and Bakradze, M. A. (1977) Residual bisexuality in parthenogenetic species of rock lizards for the genus *Lacerta*. *Zh. Obschch. Biol.* 38:772–780. [not read]

Das, B. C., Rani, R., Mitra, A. B., and Luthra, U. K. (1986) The activity of silver-stained rDNA (NORs) in newborn lymphocytes and its relation with in vivo aging in humans. *Mech. Age. Dev.* 36:117–123.

Davey, J. E., and Van Staden, J. E. (1978) Cytokinin activity in *Lupinus albus*. 2. Distribution in fruiting plants. *Physiol. Plant.* 43:82–86.

David, H. (1977) *Quantitative Ultrastructural Data of Animal and Human Cells*. New York: Gustav Fischer.

David, J. (1959a) Étude quantitative du développement de la Drosophile élevée en milieu axénique. *Bull. Biol.* 93:472–505.

———. (1959b) Influence de l'âge de la femelle sur les dimensions des oeufs de *Drosophila melanogaster*. *Compt. Rend. Hebd. Séanc. Acad. Sci. Paris* 249:1145.

———. (1962) Influence de l'âge de la mère sur les dimensions des oeufs dans une souche vestigiale de *Drosophila melanogaster* Meig. Étude expérimentale du déterminisme physiologique de ces variations. *Bull. Biol. Fr. Belg.* 96:505.

Davidson, E. H. (1968) *Gene Activity in Early Development*. 1st ed. New York: Academic Press.

———. (1976) *Gene Activity in Early Development*. 2d ed. New York: Academic Press.

———. (1986) *Gene Activity in Early Development*. 3d ed. New York: Academic Press.

———. (1990) How embryos work: A comparative view of diverse modes of cell fate specification. *Development* 108:365–389.

Davidson, J., Ross, R. K., Paganini-Hill, A., Hammond, C. D., Siiteri, P. K., and Judd, H. L. (1982) Total and free estrogen and androgens in postmenopausal women with hip fractures. *J. Clin. Endocrinol. Metab.* 54: 115–120.

Davies, D. M. (1978) Ecology and behaviour of adult black flies (Simuliidae): A review. *Quaestiones Entomol.* 14:3–12.

Davies, D. M., and Peterson, B. V. (1956) Observations on the mating, feeding, ovarian development, and oviposition of adult black flies (Simuliidae, Diptera). *Can. J. Zool.* 34:615–655.

Davies, K. J. A. (1987) Protein damage and degradation by oxygen radicals. 1. General aspects. *J. Biol. Chem.* 262:9895.

———. (1988) Proteolytic systems as secondary antioxidant defenses. In *Cellular Antioxidant Defense Mechanisms,* vol. 2, C. K. Chow (ed.), 25–67. Boca Raton, FL: CRC Press.

Davies, K. J. A., Wiese, A. G., Sevanian, A., and Kim, E. H. (1990) Repair systems in oxidative stress. In *Molecular Biology of Aging,* C. E. Finch and T. E. Johnson (eds.), 123–141. UCLA Symposia on Molecular and Cellular Biology, n.s., vol. 123.

Davies, M. (1961) On body size and tissue respiration. *J. Cell Comp. Physiol.* 57:135–147.

Davila, D. R., Brief, S., Simon, J., Hammer, R. E., Brinster, R. L., and Kelley, K. W. (1987) Role of growth hormone in regulating T-dependent immune events in aged, nude, and transgenic rodents. *J. Neurosci. Res.* 18:108–116.

Davis, D. R. (1978) A revision of the North American moths of the superfamily Edriocranioidea with the proposal of a new family, Acanthopteroctetidae (Lepidoptera). *Smithsonian Contrib. to Zool.* 251:11–31.

Davis, J. W. (1975) Age egg-size and breeding success in the herring gull *Larus argentatus.* *Ibis* 117:460–473.

Dawber, T. R. (1980) *The Framingham Study.* Cambridge, MA: Harvard Univ. Press.

Dawson, T. J. (1973) Primitive mammals. In *Comparative Physiology of Thermoregulation. Special Aspects of Thermoregulation,* vol. 3, G. C. Whittow (ed.), 1–46. New York: Academic Press.

Dawson, T. J., Fanning, D., and Bergin, T. J. (1978) Metabolism and temperature regulation in the New Guinea monotreme *Zaglossus bruijni. Austral. J. Zool.* 20:99–103.

Dawson, T. J., Grant, T. R., and Fanning, D. (1979) Standard metabolism of monotremes and the evolution of homeothermy. *Austral. J. Zool.* 27:511–515.

Deamer, D. W., and Gonzales, J. (1974) Autofluorescent structures in cultured WI–38 cells. *Arch. Biochem. Biophys.* 165:421–426.

Deamond, S. F., Portnoy, L. G., Strandberg, J. D., and Bruce, S. A. (In press) Longevity and age-related pathology of LVG outbred golden Syrian hamsters (*Mesocricetus auratus*). *Exp. Gerontol.*

Dean, R. G., and Cutler, R. G. (1978) Absence of significant age-dependent increase of single-stranded DNA extracted from mouse liver nuclei. *Exp. Gerontol.* 13:287–292.

Deansley, R. (1927) The structure and development of the thymus in fish, with special reference to *Salmo fario. Q. J. Microscop. Sci.* 71:113–145.

De Backer, G., Rosseneu, M., and Deslypere, J. P. (1982) Discriminative value of lipids and apoproteins in coronary heart disease. *Atherosclerosis* 42:197–203.

de Beer, G. (1958) *Embryos and Ancestors.* 3d ed. Oxford: Clarendon Press.

De Boer, P., and van der Hoeven, F. A. (1980) The use of translocation-derived "marker-bivalents" for studying the origin of meiotic instability in female mice. *Cytogenet. Cell. Genet.* 26:49–58.

DeBusk, F. L. (1972) The Hutchinson-Gilford progeria syndrome. *J. Pediatr.* 80:697–724.

Dedrick, R. L., Bischoff, K. B., and Zaharko, D. S. (1970) Interspecies correlation of plasma concentration history of methotrexate (NSC-740). *Canc. Chemo. Rep.* 54:95–101.

Deduve, C., and Wattiaux, R. (1966) Functions of lysozymes. *Ann. Rev. Physiol.* 23:435.

Deevey, E. S., Jr. (1947) Life tables for natural populations of animals. *Q. Rev. Biol.* 22: 283–314.

DeFronzo, R. A. (1979) Glucose intolerance in aging. Evidence for tissue insensitivity to glucose. *Diabetes* 28:1095–1101.

De Klein, A., Van Kessel, A. G., and Grosveld, G. (1982) A cellular oncogene is translocated to the Philadelphia chromosome in chronic myelocytic leukaemia. *Nature* 300:765–767.

DeKosky, S., Scheff, S., and Cotman, C. (1984) Elevated corticosterone levels. A mechanism for impaired sprouting in the aged hippocampus. *Neuroendocrinology* 38:33–40.

Delcour, J. (1968) Cell size and cell number in the wing of *Drosophila melanogaster* as related to parental aging. *Exp. Gerontol.* 3:247–255.

———. (1969) Influence de l'âge parental sur la dimension des oeufs, la durée de développement, et la taille thoracique des descendants chez *Drosophila melanogaster*. *J. Insect Physiol.* 15:1999.

Delcour, J., and Heuts, M. J. (1968) Cyclic variations in wing size related to parental ageing in *Drosophila melanogaster*. *Exp. Gerontol.* 3:45–53.

De Leon, M., McRae, T., Tsai, J., George, A., Marcus, D., Freedman, M., Wolf, A., and McEwen, B. (1988) Abnormal cortisol response in Alzheimer's disease linked to hippocampal atrophy. *Lancet* 2:391–392.

Delevoryas, T. (1964) Ontogenic studies of fossil plants. *Phytomorphology* 14:299–314.

DeLong, R., and Poplin, L. (1977) On the etiology of aging. *J. Theoret. Biol.* 67:111–120.

Dement, W., Richardson, G., Prinz, P., Carskadon, M., Kripke, D., and Czeisler, C. (1985) Changes of sleep and wakefulness with age. In *Handbook of the Biology of Aging,* 2d ed., C. E. Finch and E. L. Schneider (eds.), 692–717. New York: Van Nostrand.

Denckla, W. D. (1970) Minimal oxygen consumption in the female rat: Some new definitions and measurements. *J. Appl. Physiol.* 29:263–274.

———. (1973) Minimal O_2 consumption as an index of thyroid status: Standardization of method. *Endocrinology* 93:61–73.

———. (1974) Role of pituitary and thyroid glands in the decline of minimal O_2 consumption with age. *J. Clin. Invest.* 53:572–581.

———. (1975) A time to die. *Life Sci.* 16:31–44.

De Nechaud, B., and Uriel, J. (1971) Transitory cell antigens of rat liver. 1. The secretion and synthesis of fetospecific serum proteins during hepatic development and regeneration. *Int. J. Canc.* 8:71–80.

Denko, C. W., and Gabriel, P. (1981) Age and sex related levels of albumin, ceruloplasmin, alpha-$_1$ acid glycoprotein, and transferrin. *Ann. Clin. Lab. Sci.* 11:63–68.

Denton, T. E., Liem, S. L., Cheng, K. M., and Barrett, J. V. (1981) The relationship between aging and ribosomal gene activity in humans as evidenced by silver staining. *Mech. Age. Dev.* 15:1–7.

DeSalle, R., Slightom, J., and Zimmer, E. (1986) The molecular through ecological genetics of abnormal abdomen. 2. Ribosomal DNA polymorphism is associated with the abnormal abdomen syndrome in *Drosophila mercatorum*. *Genetics* 112:861–875.

Deslypere, J. P., Kaufman, J. M., Vermeulen, T., Vogelaers, D., Vandalem, J. L., and Vermeulen, A. (1987) Influence of age on pulsatile luteinizing hormone release and responsiveness of the gonadotrophs to sex hormone feedback in men. *J. Clin. Endocrinol. Metab.* 64:68–73.

Detinova, T. C. (1955) Fertility of the ordinary malaria mosquito *Anopheles maculopensis* (in Russian). *Meditsinskaia Parazitologia 1. Parazit. Bolezni* 24:6–11. [not read]

Detwiler, T. C., and Draper, H. H. (1962) Physiological aspects of aging. 4. Senescent changes in the metabolism and composition of nucleic acids of the liver and muscle of the rat. *J. Gerontol.* 17:138–143.

de Wilde, J. (1983) Endocrine aspects of diapause in the adult stage. In *Endocrinology of Insects,* R. G. Downer and H. Laufa (eds.), 357–367. New York: Liss.

Dhindsa, D. S., Hoversland, A. S., and Metcalfe, J. (1971) Comparative studies of the respiratory functions of mammalian blood. 7. Armadillo (*Dasypus novemcintus*). *Resp. Physiol.* 13:198–208.

Dhindsa, D. S., Metcalfe, J., and Hoversland, A. S. (1972) Comparative studies of the respiratory functions of mammalian blood. 9. Ring-tailed lemur (*Lemur catta*) and black lemur (*Lemur macaca*). *Resp. Physiol.* 15:331–342.

Dhindsa, D. S., Metcalfe, J., Hoversland, A. S., and Hartman, R. A. (1974) Comparative studies of the respiratory functions of mammalian blood. 10. Killer whale (*Orcinus orca* Linnaeus) and beluga whale (*Delphin apterus* Leucas). *Resp. Physiol.* 20:93–103.

Diamond, J. M. (1982) Big-bang reproduction and aging in male marsupial mice. *Nature* 298:115–116.

———. (1986) Why do disused proteins become genetically lost or repressed? *Nature* 321:565–567.

Dice, J. F. (1982) Altered degradation of proteins microinjected into senescent human fibroblasts. *J. Biol. Chem.* 257:14624–14627.

———. (1987) Molecular determinants of protein half-lives in eukaryotic cells. *FASEB J.* 1:349–357.

———. (1989) Altered intracellular protein degradation in aging: A possible cause of proliferative arrest arrest. *Exp. Gerontol.* 24:451–459.

Dierschke, D. J. (1985) Temperature changes suggestive of hot flashes in rhesus monkeys: Preliminary observations. *J. Med. Primatol.* 14:271–280.

Dietz, A., Hermann, H. R., and Blum, M. S. (1979) The role of exogenous JH I, JH III, and anti-JH (precocene II) on queen induction of 4.5–day-old worker honey bee larvae. *J. Insect Physiol.* 25:503–512.

Dilcher, D. L. (1979) Early angiosperm reproduction: An introductory report. *Rev. Paleobot. Palynol.* 27:291–328.

Diller, W. F. (1936) Nuclear reorganization processes in *Paramecium aurelia*, with descriptions of autogamy and "hemixis." *J. Morphol.* 59:11–67.

Dilman, V. M. (1971) Age-associated elevation of hypothalamic threshold to feedback control and its role in development, ageing, and disease. *Lancet* 2:1211–1219.

———. (1981) *The Law of Deviation of Homeostasis and Disease of Aging.* John Wright (ed.). Boston: PSG Biomedical.

———. (1984) Three models of medicine. *Med. Hypothesis* 15:185–208.

DiMichele, W. A., and Phillips, T. L. (1985) Arborescent lycopod reproduction and paleoecology. A coal-swamp environment of late middle Pennsylvanian age (Herrin coal, Ill., USA). *Rev. Paleobot. Palynol.* 44:1–26.

DiMichele, W. A., Phillips, T. L., and Peppers, R. A. (1985) The influence of climate and depositional environmental factors in the distribution and evolution of Pennsylvanian coal swamp plants. In *Geological Factors and the Evolution of Plants*, B. H. Tiffney (ed.), 223–256. New Haven: Yale Univ. Press.

Dische, Z., and Zil, H. (1951) Studies on the oxidation of cysteine to cystine in lens proteins during cataract formation. *Am. J. Opthalmol.* 34:104–113.

Dix, D. (1989) The role of aging in cancer incidence: An epidemiological study. *J. Gerontol.* 44:10–18.

Dix, D., Cohen, P., and Flannery, J. (1980) On the role of aging in cancer incidence. *J. Theoret. Biol.* 83:163–173.

Dmi'el, R. (1967) Studies on reproduction, growth and feeding in the snake *Spalerosophis cliffordi* (Colubridae). *Copeia* 2:332–346.

Dobzhansky, T. H. (1933) Studies on chromosome conjugation. 2. *Z. indukt Abstamm.-u. Vererb.-lehre* 65:269–309.

———. (1941) *Genetics and the Origin of Species.* New York: Columbia Univ. Press.

———. (1950) Evolution in the tropics. *Am. Sci.* 38:209–221.

Dodd, J. M. (1977) The ovary of nonmammalian vertebrates. In *The Ovary*, vol. 1, 2d ed., S. Zuckerman and B. J. Weir (eds.), 219–263. New York: Academic Press.

———. (1983) Reproduction in cartilaginous fishes (Chondrichthyes). In *Fish Physiology*, vol. 9A, W. S. Hoar, D. J. Randall, and E.

M. Donaldson (eds.), 31–96. New York: Academic Press.

Doebler, J. A., Markesbery, W. R., Anthony, A., and Rhoads, R. E. (1987) Neuronal RNA in relation to neuronal loss and neurofibrillary pathology in the hippocampus in Alzheimer's disease. *J. Neuropathol. Exp. Neurol.* 46:28–39.

Doggett, D. L., Phillips, P. B., Keogh, B. P., Baharloo, S., and Cristofalo, V. J. (1989) A search for differentially expressed genes in early and late population-doubling-level WI-38 cells. *Cell Biol.* 109:25a.

Doll, R. (1970) Cander and aging: The epidemiologic evidence. In *Oncology*, R. L. Clark, R. W. Cumley, J. E. McCay, and M. M. Copeland (eds.), 1–15. Chicago: Year Book Medical Publishers.

Doll, R., and Peto, R. (1978) Cigarette smoking and bronchial carcinoma: Dose and time relationships among regular smokers and lifelong nonsmokers. *J. Epidemiol. Commun. Health* 32:303–313.

Dominice, J., Levasseur, C., Larno, S., Ronot, X., and Adolphe, M. (1986) Age-related changes in rabbit articular chondrocytes. *Mech. Age. Dev.* 37:231–240.

Domning, D. P. (1982) Evolution of manatees: A speculative history. *J. Paleontol.* 56:599–619.

———. (1983) Marching teeth of the manatee. *Nat. Hist.* 92:8–10.

———. (1987) Sea cow family reunion. *Nat. Hist.* 96:64–71.

Domning, D. P., and Hayek, L. A. (1984) Horizontal tooth replacement in the Amazonian manatee (*Trichechus inunguis*). *Mammalia* 48:105–127.

Domontay, J. S. (1931) Autotomy in holothurians. *Nat. Appl. Sci. Bull.* 1:389–404.

Don Carlos, L. L., Hoffman, G. E., and Finch, C. E. (1986) Quantitative assessment and comparison of LHRH immunoreactive neurons in young, middle-aged, and old female mice. *Soc. Neurosci. Abstr.* 12:1466.

Donisthorpe, H. (1936) The oldest insect on record. *Entomol. Rec. J. Variation* 48:1–2.

Doroshov, S. I. (1985) Biology and culture of sturgeon Acipenseriformes. In *Recent Advances in Aquaculture*, J. F. Muir and R. J. Roberts (eds.), 251–274. Boulder, CO: Croom Helm/Westview Press.

Douglas-Hamilton, I. (1973) On the ecology and behaviour of the Lake Manyara elephants. *E. African Wildlife J.* 11:401–403.

Dovrat, A., Scharf, J., Eisenbach, L., and Gershon, D. (1984) Glyceraldehyde 3–phosphate dehydrogenase activity in rat and human lenses and the fate of enzyme molecules in the aging lens. *Mech. Age. Dev.* 28:187–191.

———. (1986) G6PD molecules devoid of catalytic activity are present in the nucleus of the rat lens. *Exp. Eye Res.* 42:489–496.

Downes, J. A. (1958a) The feeding habits of biting flies and their significance in classification. *Ann. Rev. Entomol.* 3:249–266.

———. (1958b) The genus *Culicoides* (Diptera: Ceratopogonidae) in Canada: An introductory review. In *Proceedings of the 10th International Congress of Entomology*, 801–808.

———. (1965) Adaptations of insects in the arctic. *Ann. Rev. Entomol.* 10:257–274.

———. (1971) The ecology of blood-sucking Diptera: An evolutionary perspective. In *Ecology and Physiology of Parasites*, A. M. Fallis (ed.), 232–258. Toronto: Univ. of Toronto Press.

Downes, J. A., and Colless, D. H. (1967) Mouthparts of the biting and blood-sucking type in Tanyderidae and Chironomidae (Diptera). *Nature* 214:1355–1356.

Doyle, J. A., and Donoghue, M. J. (1986) Seed plant phylogeny and the origin of angiosperms: An experimental cladistic approach. *Bot. Rev.* 52:321–431.

Drabkin, D. L. (1950) The distribution of the chromoproteins, hemoglobin, myoglobin, and cytochrome c, in the tissues of different species, and the relationship of the total content of each chromoprotein to body mass. *J. Biol. Chem.* 182:317–333.

Drori, D., and Folman, Y. (1969) The effect of mating on the longevity of male rats. *Exp. Gerontol.* 4:263–266.

Du, J. T., Beyer, T. A., and Lang, C. A. (1977) Protein biosynthesis in aging mouse tissues. *Exp. Gerontol.* 12:181–191.

Duara, R., Grady, C., Haxby, J., Ingvar, D., Sokoloff, L., Margolin, R. A., Manning, R. G., Cutler, N. R., and Rapoport, S. I. (1984) Human brain glucose utilization and cognitive function in relation to age. *Ann. Neurol.* 16:702–713.

Duara, R., London, E. D., and Rapoport, S. I. (1985) Changes in the structure and energy metabolism of the aging brain. In *Handbook of the Biology of Aging,* 2d ed., C. E. Finch and E. L. Schneider (eds.), 595–616. New York: Van Nostrand.

Dubick, M. A., Rucker, R. B., Cross, C. E., and Last, J. A. (1981) Elastin metabolism in rodent lung. *Biochim. Biophys. Acta* 672:303–306.

Dubois, E. L., Horowitz, R. E., Demopoulos, H. B., and Teplitz, R. (1966) NZB/NZW mice as a model of systemic lupus erythematosus. *J.A.M.A.* 195:285–289.

Ducharme, L. J. (1969) Atlantic salmon returning for their fifth and sixth consecutive spawning trips. *J. Fish. Res. Bd. Can.* 26:1661–1664.

Due, C., Simonsen, M., and Lennart, O. (1986) The major histocompatibility complex class I heavy chain as a structural subunit of the human cell membrance insulin receptor: Implications for the range of biological functions of histocompatibility antigens. *Proc. Natl. Acad. Sci.* 83:6007–6011.

Duesberg, P. H. (1987) Cancer genes: Rare recombinants instead of activated oncogenes (a review). *Proc. Natl. Acad. Sci.* 84:2117–2124.

Duffy, H. (1970) Relationship of Golgi thiaminepyrophosphatase and lysosomal acid phosphatase to neuromelanin and lipofuscin in cerebral neurons of the aging rhesus monkey. *J. Neuropath. Exp. Neurol.* 29:225–240.

Duffy, P. H., Feuers, R. J., Leakey, J. A., Nakamura, K. D., Turturro, A., and Hart, R. W. (1989) Effect of chronic caloric restriction on physiological variables related to energy metabolism in the male Fischer 344 rat. *Mech. Age. Dev.* 48:117–133.

Dulic, V., and Gafni, A. (1987) Mechanism of aging of rat muscle glyceraldehyde–3–phosphate dehydrogenase studied by selective enzyme-oxidation. *Mech. Age. Dev.* 40:289–306.

Dunlap-Pianka, H., Boggs, C. L., and Gilbert, L. E. (1977) Ovarian dynamics in heliconiine butterflies: Programmed senescence versus eternal youth. *Science* 197:487–490.

Dunn, G. R., Wilson, T. G., and Jacobson, K. G. (1969) Age-dependent changes in alcohol dehydrogenase in *Drosophila. J. Exp. Zool.* 171:185–190.

Dunn, J. A., Patrick, J. S., Thorpe, S. R., and Baynes, J. W. (1990) Oxidation of glycated proteins: Age-dependent accumulation of carboxymethyllysine in lens proteins. *Biochemistry.*

Dunn, T. B. (1944) Relationships of amyloid infiltration and renal disease in mice. *J. Natl. Canc. Inst.* 5:17–27.

———. (1954) Normal and pathologic anatomy of the reticular tissue in laboratory mice, with a classification and discussion of neoplasms. *J. Natl. Canc. Inst.* 14:1281–1390.

———. (1967) Amyloidosis in mice. In *Pathology of Laboratory Rats and Mice,* E. Cotchin and F. J. Roe (eds.), 181–212. Oxford: Blackwell Scientific Publications.

Dunnet, G. M., and Ollason, J. C. (1978) The estimation of survival rate in the fulmar, *Fulmarus glacialis. J. Anim. Ecol.* 47:507–520.

———. (1979) The fulmar. *Biologist* 26:117–122.

Durbin, A. G., Nixon, S. W., and Oviatt, C. A. (1979) Effects of the spawning migration of the alewife, *Alosa pseudoharengus,* on freshwater ecosystems. *Ecology* 60:8–17.

Durrant, A. (1955) Effect of time of embryo formation on quantitative characters in *Drosophila. Nature* 175:560.

Dutkowski, R. T., Lesh, R., Staiano-Coico, L., Thaler, H., Darlington, G. J., and Weksler, M. E. (1985) Increased chromosomal instability in lymphocytes from elderly humans. *Mutat. Res.* 149:505–512.

Dybas, H. S. (1978) Polymorphism in featherwing beetles, with a revision of the genus *Ptinellodes* (Coleoptera: Ptiliidae). *Ann. Entomol. Soc. Am.* 71:695–714.

Dyke, B., Gage, T. B., Mamelka, P. M., Goy, R. W., and Stone, W. H. (1986) A demographic analysis of the Wisconsin Regional Primate Center rhesus colony, 1962–1982. *Am. J. Primatol.* 10:257–269.

Dykhuizen, D. E., and Hartl, D. L. (1983) Selection in chemostats. *Microbiol. Rev.* 47:150–168.

Easteal, S. (1985) Generation time and the rate of molecular evolution. *Mol. Biol. Evol.* 2:450–453.

Eaton, S. B., Konner, M., and Shostak, M. (1988) Stone agers in the fast lane: Chronic degenerative diseases in evolutionary perspective. *Am. J. Med.* 84:739–749.

Ebbesen, P. (1974a) Aging increases susceptibility of mouse skin to DMBA carcinogenesis independent of general immune status. *Science* 183:217–218.

———. (1974b) Mutually exclusive occurrence of amyloids and thymic leukemia in casein treated AKR mice. *Brit. J. Canc.* 29:76–79.

Ebbesen, P., and Kripke, M. L. (1982) Influences of age and anatomical site on ultraviolet carcinogenesis in BALB/c mice. *J. Natl. Canc. Inst.* 68:691–694.

Ebbesson, S. O. E. (1968) Quantitative studies of superior cervical sympathetic ganglia in a variety of primates including man. 2. Neuronal packing density. *J. Morphol.* 124:181–186.

Ebert, T. A. (1975) Growth and mortality of post-larval echinoids. *Am. Zool.* 15:755–775.

———. (1982) Longevity, life history, and relative body wall size in sea urchins. *Ecol. Monographs* 52:353–394.

———. (1983) Recruitment in echinoderms. In *Echinoderm Studies*, vol. 1, M. Jangoux and J. M. Lawrence (eds.), 169–203. Rotterdam: A. A. Balkema.

———. (1985) Sensitivity of fitness to macroparameter changes: Analysis of survivorship and individual growth sea urchin life histories. *Oecologia* (Berlin) 65:461–467.

Economos, A. C. (1979) A non-Gompertzian paradigm for mortality kinetics of metazoan animals and failure kinetics of manufactured products. *Age* 2:74–76.

———. (1980a) Brain-lifespan conjecture: A reevaluation of the evidence. *Gerontology* 26:82–89.

———. (1980b) Taxonomic differences in the mammalian lifespan–body weight relationships and the problem of brain weight. *Gerontology* 20:90–98.

———. (1982) Rate of aging, rate of dying, and the mechanism of mortality. *Arch. Gerontol. Geriatr.* 1:3–27.

Economos, A. C., and Lints, F. A. (1986) Developmental temperature and life span in *Drosophila melanogaster*. *Gerontology* 32:18–27.

Edelmann, P., and Gallant, J. (1977) On the translational error theory of aging. *Proc. Natl. Acad. Sci.* 74:3396–3398.

Edgar, W. D. (1971) The life cycle, abundance, and seasonal movement of the wolf spider *Lycosa lugubris* in central Scotland. *J. Anim. Ecol.* 40:303–321.

Edmondson, W. T. (1945) Ecological studies of sessile Rotatoria. 2. Dynamics of populations and social structures. *Ecol. Monographs* 15:141–172.

Edmunds, G. F. (1965) May fly. *Encyclopedia Britannica* 15:11.

———. (1972) Biogeography and evolution of Ephemeroptera. *Ann. Rev. Entomol.* 17:21–42.

Edmunds, G. F., and McAfferty, W. P. (1988) The mayfly subimago. *Ann. Rev. Entomol.* 33:509–529.

Edney, E. B., and Gill, R. M. (1968) Evolution of senescence and specific longevity. *Nature* 220:281–282.

Edwards, C. A., and Lofty, J. R. (1972) *Biology of Earthworms*. London: Chapman and Hill.

Edwards, D., and Fanning. V. (1985) Evolution and environment in the late Silurian early Devonian—the rise of the Pteridophytes. *Phil. Trans. Roy. Soc.* (London), ser. B, 309:147–165.

Edwards, F. W. (1929) Blepharoceridae. In *Diptera of Patagonia and South Chile*, part 2, *Nematocera (excluding Crane-Flies and Mycetophilidae)*, 33–76. London: British Museum.

Edwards, N. A. (1975) Scaling of renal functions in mammals. *Comp. Biochem. Physiol.* 52A:63–66.

Effros, R. B., Casillas, A., and Walford, R. L. (1988) The effect of thymosin-α_1 on immunity to influenza in aged mice. *Aging: Immunol. Infect. Dis.* 1:31–40.

Egami, N. (1971) Further notes on the lifespan of the teleost, *Oryzias latipes. Exp. Gerontol.* 6:379–382.

Eggen, D. A., and Solberg, L. A. (1968) Variation of atherosclerosis with age. *Lab. Invest.* 18:111–119.

Eggert, D. A. (1961) The ontogeny of Carboniferous arborescent Lycopsida. *Palaeontog. Abt. B, Paläophytol.* 128:1–47.

Egilmez, N. K., Chen, J. B., and Jazwinski, S. M. (1989) Specific alterations in transcript prevalence during the yeast life span. *J. Biol. Chem.* 264:14312–14317.

———. (1990) Preparation and partial characterization of old yeast cells. *J. Gerontol.* 44:B9–B17.

Egilmez, N. K., and Jazwinski, S. M. (1989) Evidence for the involvement of a cytoplasmic factor in the aging of the yeast *Saccharomyces cerevisiae. J. Bacteriol.* 171:37–42.

Egoscue, H. J., Bittmenn, J. G., and Petrovich, J. A. (1970) Some fecundity and longevity records for captive small mammals. *J. Mammal.* 51:622–623.

Ehrlich, P. R., White, R. R., Singer, M. C., McKechnie, S. W., and Gilbert, L. E. (1975) Checkerspot butterflies: A historical perspective. *Science* 188:221–228.

Eichenlaub-Ritter, U., Chandley, A. C., and Gosden, R. G. (1988) The CBA mouse as a model for age-related aneuploidy in man: Studies of oocyte maturation, spindle formation, and chromosome alignment during meiosis. *Chromosoma* (Berlin) 96:220–226.

Eicher, E. M., and Washburn, L. L. (1986) Genetic control of primary sex determination in mice. *Ann. Rev. Genet.* 20:327–360.

Eisenberg, J. F. (1978) Evolution of arboreal herbivores in the class Mammalia. In *Ecology of Arboreal Folivores,* G. G. Montgomery (ed.), 135–152. Washington, DC: Smithsonian Institution Press.

———. (1981) *The Mammalian Radiations: An Analysis of Trends in Evolution, Adaptation, and Behavior.* Chicago: Univ. of Chicago Press.

Eisenberg, J. F., and Wilson, D. E. (1978) Relative brain size and feeding strategies in the Chiroptera. *Evolution* 32:740–751.

Eisner, E. (1967) Actuarial data for the Bengalese finch (*Lonchura striata:* Estrildidae) in captivity. *Exp. Gerontol.* 2:187–189.

Eisner, N., and Etoh, H. (1967) Actuarial data for the Bengalese finch (*Lonchura striata*) in captivity. *Exp. Gerontol.* 2:187–189.

Eklund, J., and Bradford, G. E. (1977) Longevity and lifetime body weight in mice selected for rapid growth. *Nature* 265:48–49.

Ekstrom, R., Liu, D. S., and Richardson, A. (1980) Changes in brain protein synthesis during the lifespan of male Fischer rats. *Gerontology* 26:121–128.

Elahi, D. (1982) Effect of age and obesity on fasting levels of glucose, insulin, glucagon, and growth hormone in man. *J. Gerontol.* 37:385–391.

———. (1983) Glucose and insulin metabolism. *Rev. Biol. Res. Aging* 1:343–356.

Eldred, G. E. (1987) Questioning the nature of the fluorophores in age pigments. In *Advances in Age Pigments Research,* E. A. Totaro, P. Glees, and F. A. Pisanti (eds.), 23–36. Oxford: Pergamon Press.

Eliot, R. S., Todd, G. L., Clayton, F. C., and Pieper, G. M. (1978) Experimental catecholamine-induced acute myocardial necrosis. *Adv. Cardiol.* 25:107–118.

Ellertson, F. E., and Ritcher, P. O. (1959) Biology of rain beetles, *Pleocoma* spp., associated with fruit trees in Wasco and Hood River counties. *Tech. Bull.* (Agricultural Experimental Station, Oregon State College, Corvallis) 44:1–42.

Elliott, A. M., and Hayes, R. E. (1955) Tetrahymena from Mexico, Panama, and Colombia, with special reference to sexuality. *J. Protozool.* 2:75–80.

Elliott, K. A. C. (1948) Metabolism of brain tissue slices and suspensions from various mammals. *J. Neurophysiol.* 11:473–484.

Ellis, F. P., Exton-Smith, A. N., Foster, K. G., and Weiner, J. S. (1976) Eccrine sweating and mortality during heat waves in very young and very old persons. *Israel J. Med. Sci.* 12:815–817.

Ellis, G. B., and Desjardins, C. (1982) Male rats secrete luteinizing hormone and testosterone episodically. *Endocrinology* 110:1618–1627.

Elwood, P. C., Sweetnam, P. M., Gray, O. P., Davies, D. P., and Wood, P. D. (1987) Growth of children from 0–5 years: With special reference to mother's smoking in pregnancy. *Ann. Human Biol.* 14:543–557.

Elzinga, R. J., and Broce, A. B. (1986) Labellar modifications of Muscomorpha flies (Diptera). *Ann. Entomol. Soc. Am.* 797:150–209.

Emmanuel, N. M. (1976) Free radicals and the action of inhibitors of radical processes under pathological states and aging in living organisms and in man. *Q. Rev. Biophys.* 9:283–308.

Emmett, B., and Hochachka, P. W. (1981) Scaling of oxidative and glycolytic enzymes in mammals. *Resp. Physiol.* 45:261–272.

Emson, R. H., and Wilkie, I. C. (1980) Fission and autotomy in echinoderms. *Oceanog. Mar. Biol. Ann. Rev.* 18:155–250.

Enesco, H. E. (1967) A cytophotometric analysis of DNA content of rat nuclei in aging. *J. Gerontol.* 22:445–448.

Enesco, H. E., Bozovic, V., and Anderson, P. D. (1989) The relationship between lifespan and reproduction in the rotifer *Asplanchna brightwelli*. *Mech. Age. Dev.* 48:281–289.

Enesco, H. E., and Kruk, P. (1981) Influence of dieting restriction and accumulation of fluorescent age pigment. *Gerontologist,* special issue, 21:87–88.

Enwonwu, C. O. (1987) Potential health hazard of the use of mercury in dentistry: Critical review of the literature. *Environ. Res.* 42:257–274.

Epstein, C. J. (1967) Cell size, nuclear content, and the development of polyploidy in the mammalian liver. *Proc. Natl. Acad. Sci.* 57:327–334.

Epstein, C. J., Martin, G. M., Schultz, A. L., and Motulsky, A. G. (1966) Werner's syndrome. A review of its symptomatology, natural history, pathologic features, genetics,

and relationship to the natural aging process. *Medicine* 45:177–221.

———. (1985) Werner's syndrome: A review of its symptomatology, natural history, pathologic features, genetics, and relationships to the natural aging process. In *Werner's Syndrome and Human Aging,* D. Salk, Y. Fujiwara, and G. M. Martin (eds.). *Adv. Exp. Biol. Med.* 190:57–120.

Epstein, J. H., Fukuyama, K., Reed, W. B., and Epstein, W. L. (1970) Defect in DNA synthesis in skin of patients with xeroderma pigmentosum demonstrated in vivo. *Science* 168:1477–1478.

Epstein, M. H., and O'Connor, J. S. (1965) Respiration of single cortical neurons and of surrounding neuropile. *J. Neurochem.* 12:389–395.

Erickson, J. D. (1978) Down syndrome, paternal age, maternal age, and birth order. *Ann. Human Genet.* (London) 41:289–298.

Erickson, J. G. (1967) Social hierarchy, territoriality, and stress reactions in sunfish. *Physiol. Zool.* 40:40–48.

Eriksen, E. F., Colvard, D. S., Berg, N. J., Graham, M. L., Mann, K. G., Spelsberg, T. C., and Riggs, B. L. (1988) Evidence of estrogen receptors in normal human osteoblast-like cells. *Science* 241:84–86.

Ermini, M., Moret, M. L., Reichlmeier, K., and Dunne, T. (1978) Age-dependent structural changes in human neuronal chromatin. *Aktuel. Gerontol.* 7:675–680.

Esau, K. (1977) *Anatomy of Seed Plants.* 2d ed. New York: Wiley.

Esposito, D., Fassina, G., Szabo, P., De Angelis, P., Rodgers, L., Weksler, M., and Siniscalco, M. (1989) Chromosomes of older humans are more prone to aminopterine-induced breakage. *Proc. Natl. Acad. Sci.* 86:1302–1306.

Esser, K., and Tudzynski, P. (1977) The prevention of senescence in the ascomycete in the filamentous fungus *Podospora anserina* by the action of plasmid-like DNA. *Nature* 265:454.

Essig, E. O. (1947) *College Entomology.* New York: Macmillan.

Estes, K. S., and Simpkins, J. S. (1982) Resumption of pulsatile leutinizing hormone release after α-adrenergic stimulation in aging constant estrous rats. *Endocrinology* 111:1778–1784.

Esumi, H., Takahashi, Y., Makino, R., Sato, S., and Sugimura, T. (1985) Appearance of albumin-producing cells in the liver of analbuminemic rats on aging and administration of carcinogens. In *Werner's Syndrome and Human Aging*, D. Salk, Y. Fujiwara, and G. M. Martin (eds.). *Adv. Exp. Biol. Med.* 190:637–650.

Esumi, H., Takahashi, Y., Sekiya, T., Sato, S., Nagase, S., and Sugimura, T. (1982) Presence of albumin mRNA precursors in nuclei of analbuminemic rat liver lacking cytoplasmic albumin mRNA. *Proc. Natl. Acad. Sci.* 79:734–738.

Etkind, P. R., and Sarkar, N. H. (1983) Integration of new endogenous mouse mammary tumor virus proviral DNA at common sites in the DNA of mammary tumors of C3Hf mice and hypomethylation of the endogenous mouse mammary tumor virus proviral DNA in C3Hf mammary tumors and spleens. *J. Virol.* 45:114–123.

Evans, H. J., and Vijayalaxmi. (1981) Induction of 8–azaguanine resistance and sister chromatid exchange in human lymphocytes exposed to mitomycin C and X-rays in vitro. *Nature* 292:601–605.

———. (1984) Measurement of spontaneous and X-irradiation-induced 6–thioguanine-resistant human blood lymphocytes using a T-cell cloning technique. *Mutat. Res.* 125:87–94.

Evans, R. M. (1988) The steroid and thyroid hormone receptor superfamily. *Science* 240:889–895.

Everitt, A. V. (1957) The senescent loss of weight in male rats. *J. Gerontol.* 12:382.

———. (1969) The effect of chronic lung disease and the process of dying on the aging of collagen fibres in the male rat. *Gerontologia* 15:24–30.

———. (1973) The hypothalamic-pituitary control of ageing and age-related pathology. *Exp. Gerontol.* 8:265–277.

———. (1976a) Cardiovascular aging and the pituitary. In *Hypothalamus, Pituitary, and Aging*, A. V. Everitt and J. A. Burgess

(eds.), 262–281. Springfield, IL: C. C. Thomas.

———. (1976b) Hypophysectomy and aging in the rat. In *Hypothalamus, Pituitary and Aging*, A. V. Everitt and J. A. Burgess (eds.), 68–93. Springfield, IL: C. C. Thomas.

———. (1980) The neuroendocrine system and aging. *Gerontology* 20:109–119.

Everitt, A. V., and Cavanagh, L. M. (1965) The aging process in the hypophysectomized rat. *Gerontologia* (Basel) 11:198.

Everitt, A. V., and Duvall, L. K. (1965) The delayed onset of proteinuria in ageing hypophysectomised rats. *Nature* 205:1015.

Everitt, A. V., Olsen, G. G., and Burrows, G. R. (1968) The effect of hypophysectomy on the aging collagen fibers in the tail tendon of the rat. *J. Gerontol.* 23:333.

Everitt, A. V., Seedsman, N. J., and Jones, F. (1980) The effects of hypophysectomy and continuous food restriction, begun at ages of 70 and 400 days on collagen aging, proteinuria, incidence of pathology, and longevity in the male rat. *Mech. Age. Dev.* 12:161–172.

Everson, R. B., Randerath, E., Santella, R. M., Cefalo, R. C., Avitts, T. A., and Randerath, K. (1986) Detection of smoking-related covalent DNA adducts in human placenta. *Science* 231:54–57.

Fabens, A. J. (1965) Properties and fitting of the von Bertalanffy growth curve. *Growth* 29:265–289.

Faber, E. (1897) Unsere Baumreisen. *Rec. Mem. Trav. Soc. Bot.* (Luxembourg) 13:51–90. [not read]

Fabia, J., and Drolette, M. (1970) Malformations and leukemia in children with Down's syndrome. *Pediatrics* 45:60–70.

Fabricant, J. D., Dunn, G., and Schneider, E. L. (1978) Maternal age-related pre- and postimplantation fetal mortality: A strain survey. *Mech. Age. Dev.* 8:227–231.

Fabricant, J. D., and Parkening, T. A. (1982) Sperm morphology and cytogenetic studies in ageing C57BL/6 mice. *J. Reprod. Fertil.* 66:485–489.

Fabricant, J. D., and Schneider, E. L. (1978) Studies of the genetic and immunologic component of the maternal age defect. *Dev. Biol.* 66:337–343.

Fabris, N. (1982) Neuroendocrine-immune network in aging. In *Developmental Immunology: Clinical Problems and Aging,* E. L. Cooper and M. P. Brazier (eds.), 291–301. UCLA Forum in Medical Sciences, vol. 25. New York: Academic Press.

Fage, L. (1910) Récherches sur les stades pélagiques de quelques teleosteens de la mer Nice (parages de Monaco) et du golfe du Lion. *Ann. Inst. Oceanog. Monaco* 1, fasc. 7: 53. [not read]

Fagerlund, U. H. M. (1967) Plasma cortisol concentration in relation to stress in adult sockeye salmon during the freshwater stage of their life cycle. *Gen. Comp. Endocrinol.* 8:197–207.

Faggiotto, A., and Ross, R. (1984) Studies of hypercholesterolemia in the nonhuman primate. 2. Fatty streak conversion to fibrous plaque. *Arteriosclerosis* 4:341–356.

Fahy, E., and Warren, W. P. (1984) Long-lived sea trout, sea-run "ferox"? *Salmon Trout Mag.* 227:72–75.

Failla, G. (1958) The aging process and carcinogenesis. *Ann. N.Y. Acad. Sci.* 71:1124–1135.

Fairbairn, D. J. (1977) The spring decline in deer mice: Death or dispersal? *Can. J. Zool.* 55:84–92.

Fairweather, D. S., Fox, M., and Margison, G. P. (1987) The in vitro lifespan of MRC–5 cells is shortened by 5–azacytidine-induced demethylation. *Exp. Cell Res.* 168:153–159.

Farrer, L. A. (1985a) Diabetes mellitus in Huntington disease. *Clin. Genet.* 27:62–67.

———. (1985b) Genetic and anthropometric studies of aging in Huntington disease. Ph.D. thesis, Department of Medical Genetics, Indiana University, Indianapolis.

Farrer, L. A., Connally, P. M., and Yu, P. I. (1984) The natural history of Huntington disease: Possible role of "aging genes." *Am. J. Med. Genet.* 18:115–123.

Farrer, L. A., and Conneally, M. (1985) A genetic model for age at onset in Huntington disease. *Am. J. Human Genet.* 37:350–357.

Farrer, L. A., Myers, R. H., Cupples, L. A.,

St. George-Hyslop, P., Bird, T. D., Rossor, M. N., Mullan, M. J., Polinsky, R., Nee, L., Heston, L., Van Broeckhoven, C., Martin, J.-J., Crapper-McLachlan, D., and Growdon, J. H. (1990) Transmission and age at onset patterns in familial Alzheimer's disease: Evidence for heterogeneity. *Neurology* 40:395–403.

Farrer, L. A., O'Sullivan, D. M., Cupples, L. A., Growdon, J. H., and Myers, R. H. (1989) Assessment of genetic risk for Alzheimer's disease among first-degree relatives. *Ann. Neurol.* 25:485–493.

Fathman, C. G., and Frelinger, J. G. (1983) T-lymphocyte clones. *Ann. Rev. Immunol.* 1:633–655.

Faust, E. C., and Russell, P. F. (1964) *Craig and Faust's Clinical Parasitology.* 7th ed. Philadelphia: Lea and Fibiger.

Faust, I. M., Johnson, P. R., and Hirsch, J. (1974) Adipose tissue regeneration following lipectomy. *Science* 197:391–393.

Feldman, H. A., and McMahon, T. A. (1983) The ¾ mass exponent for energy metabolism is not a statistical artifact. *Resp. Physiol.* 52:149–163.

Feldman-Muhsam, B., and Muhsam, H. V. (1946) Life tables for *Musca vicina* and *Calliphora erythrocephala. Proc. Zool. Soc. Lond.* 115:296–305.

Felicio, L. S., Nelson, J. F., and Finch, C. E. (1980) Spontaneous pituitary tumorigenesis and plasma oestradiol in ageing female C57BL/6J mice. *Exp. Gerontol.* 15:139–143.

———. (1986) Prolongation and cessation of estrous cycles in aging C57BL/6J female mice is caused by ovarian steroids, not intrinsic pituitary aging. *Biol. Reprod.* 34:849–858.

Felicio, L. S., Nelson, J. F., Gosden, R. G., and Finch, C. E. (1983) Restoration of ovulatory cycles by young ovarian grafts in aging mice: Potentiation by long-term ovariectomy decreases with age. *Proc. Natl. Acad. Sci.* 80:6076–6080.

Feller, B. A. (1981) Prevalence of selected impairments, U.S. 1977. *Vital Health Stat.* 10:134. Washington, DC: U.S. Government Printing Office.

Fellin, D. G. (1981) *Pleocoma* in western Oregon coniferous forests: Observations on adult flight habits and on egg and larval biology (Coleoptera: Scarabaeoidae). *Pan Pacific Entomol.* 57:461–484.

Fenech, M., and Morley, A. (1985) The effects of donor age on spontaneous and induced micronuclei. *Mutat. Res.* 148:99–105.

Ferguson, A., and Mason, F. M. (1981) Allozyme evidence for reproductively isolated sympatric populations of brown trout *Salmo trutta L.* in Lough Melvin, Ireland. *J. Fish. Biol.* 18:629–642.

Ferguson, M. W. J. (1984) Craniofacial development in *Alligator mississippiensis. Symp. Zool. Soc. Lond.* 52:223–273.

———. (1985) Reproductive biology and embryology of the crocodilians. In *Biology of the Reptilia,* vol. 14, *Development A,* C. Gans, F. Billett, and P. F. Maderson (eds.), 330–491. New York: Wiley and Sons.

Fergusson, I. L. C., Taylor, R. W., and Watson, J. M. (1982) *Records and Curiosities in Obstetrics and Gynecology.* London: Balliere.

Fernandes, G., Friend, P., Yunis, E. J., and Good, R. A. (1978) Influence of dietary restriction on immunologic function and renal disease in (NZB × NZW)F₁ mice. *Proc. Natl. Acad. Sci.* 75:1500–1504.

Fernandes, G., West, A., and Good, R. A. (1979) Nutrition, immunity, and cancer: A review. 3. Effects of diet on the diseases of aging. *Canc. Bull.* 9:91–106.

Fernandes, G., Yunis, E. J., and Good, R. A. (1976) Influence of protein restriction on immune function in NZB mice. *J. Immunol.* 116:782–790.

Fernandes, G., Yunis, E. J., Smith, J., and Good, R. A. (1972) Dietary influence on breeding behavior, hemolytic anemia, and longevity in NZB mice. *Proc. Soc. Exp. Biol. Med.* 139:1189–1196.

Fernandes, G., Yunis, E. J., Miranda, M., Smith, J., and Good, R. A. (1978) Nutritional inhibition of genetically determined renal disease and autoimmunity with prolongation of life in *kdkd* mice. *Proc. Natl. Acad. Sci.* 74:2888–2892.

Fernholm, B. (1974) Diurnal variations in the behaviour of the hagfish *Eptatretus burgeri. Mar. Biol.* 27:351–356.

Feuers, R. J., Duffy, P. H., Leakey, J. A., Turturro, A., Mittelstaedt, R. A., and Hart, R. W. (1989) Effect of chronic caloric restriction on hepatic enzymes of intermediary metabolism in the male Fischer 344 rat. *Mech. Age. Dev.* 48:179–189.

Fielde, A. M. (1904) Tenacity of life in ants. *Biol. Bull.* 7:300–309.

Finch, C. E. (1969) Cellular activities during aging in mammals. Ph.D. diss., Rockefeller University, New York. New York MSS Information Corp.

———. (1971) Comparative biology of senescence: Evolutionary and developmental considerations. *Animal Models for Biomedical Research,* vol. 4, 47–67. Washington, DC: National Academy of Science.

———. (1972a) Cellular pacemakers of ageing in mammals. In *Proceedings of First Conference on Cell Differentiation,* R. Harris and D. Viza (eds.), 259–262. Munksgaard, Copenhagen, 1971.

———. (1972b) Enzyme activities, gene function, and ageing in mammals (review). *Exp. Gerontol.* 7:53–67.

———. (1973a) Catacholamine metabolism in the brains of ageing male mice. *Brain Res.* 52:261–276.

———. (1973b) Monoamine metabolism in the aging male mouse. In *Development and Aging in the Nervous System,* M. Rockstein (ed.), 192–218. New York: Academic Press.

———. (1973c) Retardation of hair growth, a phenomenon of senescence in C57BL/6J male mice. *J. Gerontol.* 28:13–17.

———. (1976a) The regulation of physiological changes during mammalian aging. *Q. Rev. Biol.* 51:49–83.

———. (1976b) Supracentenarians. Review of *The Centenarians of the Andes,* by D. Davies. *BioScience* 27:54.

———. (1978) Genotypic influences in female reproductive senescence in rodents. In *Genetic Effects on Aging,* D. Bergsma and D. E. Harrison (eds.), 335–354. Birth Defects: Original Article Series, 14. New York: Alan R. Liss.

———. (1979) Susceptibility of mouse liver DNA to digestion by S1 nuclease: Absence of age-related change. *Age* 2:45–46.

———. (1980) The relationships of aging

changes in the basal ganglia to manifestations of Huntington's chorea. *Ann. Neurol.* 7:406–411.

———. (1982) Rodent models for aging processes in the human brain. In *Alzheimer's Disease: A Report of Progress,* S. Corkin, K. L. Davis, J. H. Growden, E. Usdin, and R. J. Wurtman (eds.), 249–256. *Aging* 19. New York: Raven Press.

———. (1987) Neural and endocrine determinants of senescence: Investigation of causality and reversibility by laboratory and clinical interventions. In *Aging,* vol. 31, *Modern Biological Theories of Aging,* H. R. Warner, R. N. Butler, R. L. Sprott, and E. L. Schneider (eds.), 261–306. New York: Raven Press.

———. (1988) Neural and endocrine approaches to the resolution of time as a dependent variable in the aging processes of mammals. 1985 Kleemeier Award Lecture. *Gerontologist* 28:29–42.

Finch, C. E., Felicio, L. S., Mobbs, C. V., and Nelson, J. F. (1984) Ovarian and steroidal influences on neuroendocrine aging processes in female rodents. *Endocrine Rev.* 5:467–497.

Finch, C. E., and Foster, J. R. (1973) Hematologic and serum electrolyte values of the C57BL/6J male mouse in maturity and senescence. *Lab. Anim. Sci.* 23:339–349.

Finch, C. E., Foster, J. R., and Mirsky, A. E. (1969) Ageing and the regulation of cell activities during exposure to cold. *J. Gen. Physiol.* 54:690–712.

Finch, C. E., and Girgis, F. G. (1974) Enlarged seminal vesicles of senescent C57BL/6J mice. *J. Gerontol.* 29:134–138.

Finch, C. E., and Gosden, R. G. (1986) Animal models for the human menopause. In *Aging, Reproduction, and the Climacteric,* L. Mastroianni and C. A. Paulsen (eds.), 3–34. New York: Plenum.

Finch, C. E., and Hayflick, L. (eds.) (1978) *Handbook of the Biology of Aging.* 1st ed. New York: Van Nostrand Reinhold.

Finch, C. E., and Landfield, P. W. (1985) Neuroendocrine and autonomic function in aging. In *Handbook of the Biology of Aging,* 2d ed., C. E. Finch and E. L. Schneider (eds.), 79–90. New York: Van Nostrand.

Finch, C. E., Marshall, F. J., and Randall, P. K. (1981) Aging and basal ganglia functions. *Ann. Rev. Gerontol. Geriatr.* 2:49–87.

Finch, C. E., and Morgan, D. G. (1987) Aging and schizophrenia: A hypothesis relating asynchrony in neural aging processes to the manifestations of schizophrenia and other neurologic diseases with age. In *Schizophrenia and Aging,* N. E. Miller and G. D. Cohen (eds.), 98–108. New York: Guilford Publications.

———. (1990) RNA and protein metabolism in the aging brain. *Ann. Rev. Neurosci.* 13:75–87.

Finch, C. E., Pike, M. C., Witten, M. (1990) Slow increases of the Gompertz mortality rate during aging in certain animals approximate that of humans. *Science* 249:902–905.

Fineberg, S. E., Merimee, T. J., Rabinowitz, D., and Edgar, P. Js-4. (1970) Insulin secretions in acromegaly. *J. Clin. Endocrinol. Metab.* 30:288–292.

Fineman, R., Hamilton, J., and Silen, W. (1974) Duration of life and mortality rates in male and female phenotypes in three sex chromosomal genotypes (XX, XY, YY) in the killifish *Oryzias latipes. J. Exp. Zool.* 188:35–40.

Fisch, U., Dobozi, M., and Greig, D. (1972) Degenerative changes of the arterial vessels of the internal auditory meatus during the process of aging. *Acta Otolaryng.* 73:259–266.

Fisher, R. A. [1930] (1958) *The Genetical Theory of Natural Selection.* Oxford: Clarendon Press. Reprint (revised). New York: Dover.

Fitch, J. E., and Lavenburg, R. J. (eds.) (1971) *Marine and Game Fish of California.* Berkeley and Los Angeles: Univ. of California Press.

Fitch, W. M. (1987) Commentary on the Li and Wu, Easteal letters. *Mol. Biol. Evol.* 4:81–82.

Fitzpatrick, J. W., and Woolfenden, G. E. (1988) Components of lifetime reproductive success in the Florida scrub jay. In *Reproductive Success. Studies of Individual Variation in Contrasting Breeding Systems,* T. H. Clutton-Brock (ed.), 305–324. Chicago: Univ. of Chicago Press.

Flanders, S. E. (1957) Ovigenic-ovisorptive

cycle in the economy of the honey bee. *Sci. Monthly* 85:176–178.

Flavell, R. A., Allen, H., Burkly, L. C., Sherman, D. H., Waneck, G. L., and Widera, G. (1986) Molecular biology of the H–2 histocompatibility complex. *Science* 233:437–443.

Fleagle, J. G. (1985) Size and adaptation in primates. In *Size and Scaling in Primate Biology*, W. L. Jungers (ed.), 1–20. Advances in Primatology. New York: Plenum.

Fleg, J. L., Tzankoff, S. P., and Lakatta, E. G. (1985) Age-related augmentation of plasma catecholamines during dynamic exercise in healthy males. *J. Appl. Physiol.* 59:1033–1039.

Fleisch, J. H., Maling, H. M., and Brodie, B. B. (1970) Beta-receptor activity in the aorta. *Circ. Res.* 26:151–162.

Fleming, J. E. (1986) Role of mitochondria in *Drosophila* aging. In *Insect Aging: Strategies and Mechanisms*, K.-G. Collatz and R. S. Sohal (eds.), 131–141. Heidelberg: Springer-Verlag.

Fleming, J. E., and Kwak, E. L. (1986) *In vitro* translation products of *Drosophila* mitochondria are contaminated with newly synthesized bacterial proteins. *Biochem. Biophys. Res. Commun.* 136:797–801.

Fleming, J. E., Quattrocki, E., Latter, G., Miquel, J., Marcuson, R., Zuckerkandl, E., and Bensch, K. G. (1986) Age-dependent changes in proteins of *Drosophila melanogaster*. *Nature* 231:1157–1159.

Fleming, J. E., Walton, J. K., Dubitsky, R., and Bensch, K. G. (1988) Aging results in an unusual expression of *Drosophila* heat shock proteins. *Proc. Natl. Acad. Sci.* 85:4099–4103.

Flier, J. S., Cook, K. S., Usher, P., and Spiegelman, B. M. (1987) Severely impaired adipsin expression in genetic and acquired obesity. *Science* 237:405–408.

Fliers, E., and Swaab, D. F. (1983) Activation of vasopressinergic and oxytocinergic neurons during aging in the Wistar rat. *Peptides* 4:165–170.

Fliers, E., Swaab, D. F., Pool, C. W., and Verwer, R. W. (1985) The vasopressin and oxytocin neurons in the human supraoptic and paraventricular nucleus: Changes with aging and in senile dementia. *Brain Res.* 342:45–53.

Flohe, L., and Gunzler, W. A. (1976) Glutathione peroxidase. In *Glutathione: Metabolism and Function*, I. M. Arias and W. B. Jacoby (eds.), 17–34. New York: Raven Press.

Flood, D. G. (1990) Region specific stability of dendritic extent in normal human aging and regression in Alzheimer's disease. 2. Subiculum. *Brain Res.*, in press.

Flood, D. G., Buell, S. J., Defiore, C. H., Horwitz, G. J., and Coleman, P. D. (1985) Age-related dendritic growth in dentate gyrus of human brain is followed by regression in the "oldest old." *Brain Res.* 345:366–368.

Flood, D. G., Buell, S. J., Horwitz, G. J., and Coleman, P. D. (1987) Dendritic extent in human dentate gyrus granule cells in normal aging and senile dementia. *Brain Res.* 402:205–216.

Flood, D. G., and Coleman, P. D. (1990) Hippocampal plasticity in normal aging and decreased plasticity in Alzheimer's disease. *Prog. Brain Res.* In press.

Flora, G. C., Baker, A. B., Loewenson, R. B., and Klassen, A. C. (1968). A comparative study of cerebral atherosclerosis in males and females. *Circulation* 38:859–869.

Florine, D. L., Ono, T., Cutler, R. G., and Getz, M. J. (1980) Regulation of endogenous murine-leukemia virus-related nuclear and cytoplasmic RNA complexity in C57BL/6J mice of increasing age. *Canc. Res.* 40:519–523.

Florini, J. R., and Regan, J. F. (1985) Age-related changes in hormone secretion and action. *Rev. Biol. Res. Aging* 2:227–250.

Florini, J. R., Saito, Y., and Manowitz, E. J. (1973) Effect of age on thyroxine-induced cardiac hypertrophy in mice. *J. Gerontol.* 28:293–297.

Flower, S. S. (1925a) Contributions to our knowledge of the duration of life in vertebrate animals. 1. Fishes. *Proc. Zool. Soc. Lond.* 247–268.

———. (1925b) Contributions to our knowledge of the duration of life in vertebrate animals. 2. Batrachians. *Proc. Zool. Soc. Lond.* 269–289.

——. (1925c) Contributions to our knowledge of the duration of life in vertebrate animals. 3. Reptiles. *Proc. Zool. Soc. Lond.* 60:911–981.

——. (1936) Further notes on the duration of life in animals. 2. Amphibians. *Proc. Zool. Soc. Lond.* 1:369–394.

——. (1937) Further notes on the duration of life in animals. 3. Reptiles. *Proc. Zool. Soc. Lond.* ser. A, 107:1–39.

——. (1938) Further notes on the duration of life in animals. 4. Birds. *Proc. Zool. Soc. Lond.* ser. A, 108:195–235.

Fluri, P., Luscher, M., Wille, H., and Gerig, L. (1982) Changes in weight of the pharyngeal gland and haemolymph titres of juvenile hormone, protein, and vitellogenin in worker honey bees. *J. Insect Physiol.* 28:61–68.

Flurkey, K., Eskanazi, D. P., and Finch, C. E. (1986) Estradiol regulates cell surface immunoglobulin in mice. *Dev. Comp. Immunol.* 10:85–91.

Flurkey, K., and Harrison, D. E. (1990) Use of genetic models to investigate the hypophyseal regulation of senescence. In *Genetic Effects on Aging II,* D. E. Harrison (ed.). Caldwell, NJ: Telford Press. In press.

Foelix, R. A. (1982) *Biology of Spiders.* Cambridge, MA: Harvard Univ. Press.

Foley, J. M., and Baxter, D. (1958) On the nature of pigment granules in the cells of the locus ceruleus and substantia nigra. *J. Neuropath. Exp. Neurol.* 17:586–598.

Fonds, M. (1973) Sand gobies in the Dutch Waden Sea (*Pomatoschistus,* Gobbiidae, Pisces). *Neth. J. Sea Res.* 6:417–478.

Foote, C. S., Chang, Y. C., and Denny, R. W. (1970) Chemistry of singlet oxygen. 10. Carotenoid quenching parallels biological protection. *J. Am. Chem. Soc.* 92:5216.

Forbes, P. D., Davies, R. E., and Urbach, F. (1979) Aging, environmental influences, and photocarcinogenesis. *J. Invest. Dermatol.* 73:131–134.

Forbes, T. R. (1947) The crowing hen: Early observations on spontaneous sex reversal in birds. *Yale J. Biol. Med.* 19:955–970.

Forciea, M. A., Schwartz, H., Towle, H. C., Mariash, C. N., Kaiser, F. E., and Oppenheimer, J. H. (1981) Thyroid hormone-carbohydrate interaction in the rat. Correlation between age-related reductions in the inducibility of hepatic malic enzyme by triiodo-L-thyronine and a high carbohydrate fat-free diet. *J. Clin. Invest.* 67:1739–1747.

Foss, G. (1963) *Myxine* in its natural surroundings. In *The Biology of Myxine,* A. Brodal and R. Fränge (eds.), 42–45. Oslo: Universitetsforlaget Oslo.

Foster, R. B. (1977) *Tachigalia versicolor* is a suicidal neotropical tree. *Nature* (London) 268:624–626.

Fox, H. (1938) Chronic arthritis in wild mammals. Being a description of lesions found in the collections of several museums and from a pathological service. *Trans. Am. Phil. Soc.* 31:73–148.

Fozard, J. L., Wolf, E., Bell, B., McFarland, R. A., and Podolsky, S. (1985) Visual perception and communication. In *Handbook of the Psychology of Aging,* 2d ed., J. E. Birren and K. W. Schaie (eds.), 497–534. New York: Van Nostrand.

Franchi, L. L., Mandl, A. M., and Zuckerman, S. (1962) The development of the ovary and the process of oogenesis. In *The Ovary,* vol. 1, S. Zuckerman (ed.), 1–88. New York: Academic Press.

Francis, A. A., Lee, W. H., and Regan, J. D. (1981) The relationship of DNA excision repair of ultraviolet induced lesions to the maximum lifespan of mammals. *Mech. Age. Dev.* 16:181–189.

Francis, D., Kidd, A. D., and Bennett, M. D. (1985) DNA replication in relation to DNA C values. In *The Cell Division Cycle in Plants,* J. A. Bryant and D. Francis (eds.), 61–82. Society for Experimental Biology, Seminar Series 26. Cambridge: Cambridge Univ. Press.

Frank, K. T., and Carscadden, J. E. (1984) Meteorological and hydrographic regulation of year-class strength in capilan (*Mallotus villosus*). *Can. J. Fisheries Aquat. Sci.* 41:1193–1201.

Frank, P. W. (1969) Growth rates and longevity of some gastropod mollusks on the coral reef at Heron Island. *Oecologia* (Berlin) 2:232–250.

Franke, W., Berger, S., Falk, H., Spring, H.,

Scheer, U., Furth, W., Trendelenburg, M. F., and Schweiger, H. G. (1974) Morphology of the nucleo-cytoplasmic interactions during the development of *Acetabularia* cells. 1. The vegetative phase. *Protoplasma* 82:249– 282.

Frankel, E. N. (1987) Secondary products of lipid oxidation. *Chem. Phys. Lipids* 44:73– 85.

Frasca, D., Adorini, L., and Doria, G. (1987) Enhanced frequency of mitogen-responsive T-cell precursors in old mice injected with thymosin α1. *Eur. J. Immunol.* 17:727–730.

Frasca, D., Garavini, M., and Doria, G. (1982) Recovery of T-cell functions in aged mice injected with synthetic thymosin$_{\alpha 1}$. *Cell. Immunol.* 72:384–391.

Freeman, B. A., Mason, R. J., Williams, M. C., and Crapo, J. D. (1986) Antioxidant enzyme activity in alveolar type II cells after exposure of rats to hyperoxia. *Exp. Lung Res.* 10:203–222.

Freeman, G. (1964) The role of blood cells in the process of asexual reproduction in the tunicate *Perophora viridis*. *J. Exp. Zool.* 156:157–184.

Freeman, R. S. (1962) Studies on the biology of *Taenia crassiceps* (Cestoda) *Can. J. Zool.* 40:969–990.

Freund, G. (1979) The effects of chronic alcohol and vitamin E consumption of aging pigments and learning performance in mice. *Life Sci.* 24:145–152.

Freytag, G. E. (1975) Die Lebensdauer einiger Querzahnmolche (Ambystomatidae) in Gefangenschaft. *Salamandra* 11:105–106.

Friedberg, E. C. (1985) *DNA Repair*. New York: Freeman.

Friede, R. L. (1962) The relation of the formation of lipofuscin to the distribution of oxidative enzymes in the human brain. *Acta Neuropathol.* (Berlin) 2:113–125.

———. (1963) The relationship of body size, nerve cell size, axon length, and glial density in the cerebellum. *Proc. Natl. Acad. Sci.* 49:187–193.

Friedel, T., and Gillott, C. (1977) Contribution of male-produced proteins to vitellogenesis in *Melanoplus sanguinipes*. *J. Insect Physiol.* 23:145–151.

Friedenthal, H. (1910) Über die Gültigkeit der Massenwirkung für der Energieumsatz der le-

bendigen Substanz. *Zentralbl. Physiologie* 24:321–327. [not read]

Friedman, D. B., and Johnson, T. E. (1988a) A mutation in the *age–1* gene in *Caenorhabditis elegans* lengthens life and reduces hermaphrodite fertility. *Genetics* 118:75–86.

———. (1988b) Three mutants that extend both mean and maximum life span of the nematode, *Caenorhabditis elegans,* define the *age–1* gene. *J. Gerontol.* 43:B102–B109.

Friend, P. S., Fernandes, G., Good, R. A., Michael, A. F., and Yunis, E. J. (1978) Dietary restrictions early and late. Effects on the nephropathy of the NZB × NZW mouse. *Lab. Invest.* 38:629–632.

Fries, J. F. (1980) Aging, natural death, and the compression of morbidity. *N. Engl. J. Med.* 303:130–135.

———. (1988) Aging, illness, and health policy: Implications of the compression of morbidity. *Persp. Biol. Med.* 31:407–428.

Fries, J. F., and Crapo, L. M. (1981) *Vitality and Aging*. San Francisco: Freeman.

Frisch, R. W. (1985) Fatness, menarche, and female fertility. *Persp. Biol. Med.* 28:611– 633. Frisch, R. W., and Revelle, R. (1971) Height and weight at menarche and a hypothesis of menarche. *Arch. Dis. Child.* 46:695– 701.

Fritz, E. (1929) Some popular fallacies concerning the California redwood. *Madroño* 1:221– 224.

Frolkis, V. V., Golovchenko, S. F., Medved, V. I., and Frolkis, R. A. (1982) Vasopressin and cardiovascular system in aging. *Gerontology* 28:290–302.

Frolkis, V. V., Martynenko, O. A., and Timchenko, A. N. (1989) Age-related changes in the function of somatic membrane potassium channels of neurons in the mollusc *Lymnaea stagnalis*. *Mech. Age. Dev.* 47:47–54.

Frolkis, V. V., Stupina, A. S., Martinenko, O. A., Toth, S., and Timchenko, A. I. (1984) Aging of neurons in the mollusc *Lymnaea stagnalis*. Structure, function, and sensitivity to neurotransmitters. *Mech. Age. Dev.* 25:91–102.

Frost, H. B. (1938) Nuclear embryony and juvenile characters in clonal varieties of citrus. *J. Heredity* 29:423–432.

Fry, R. J., Lesher, S., and Kohn, H. I. (1962)

Influence of age on the transit time of cells of the mouse intestinal epithelium. 3. Ileum. *Lab. Invest.* 11:289–293.

Fugo, N. W., and Butcher, R. L. (1971) Effects of prolonged estrous cycling on reproduction in aged rats. *Fertil. Steril.* 22:98–101.

Fujimoto, S., Sato, C., Yamagami, Y., and Arai, E. (1978) Chromosomal anomalies of preimplantation rabbit blastocysts and human artificial abortuses in relation to maternal age. *Int. J. Fertil.* 23:207–212.

Fukagawa, N. K., Minaker, K. L., Rowe, J. W., Matthews, D. E., Bier, D. M., and Young, V. R. (1988) Glucose and amino acid metabolism in aging man: Differential effects of insulin. *Metabolism* 37:371–377.

Fukagawa, N. K., Minaker, K. L., Young, V. R., Matthews, D. E., Bier, D. M., and Rowe, J. W. (1989) Leucine metabolism in aging humans: Effect of insulin and substrate availability. *Am. J. Physiol.* 256:E288–E294.

Fukuchi, K. I., Martin, G. M., and Monnat, R. J. (1989a) Mutator phenotype of Werner syndrome is characterized by extensive deletions. *Proc. Natl. Acad. Sci.* 86:5893–5897.

———. (1989b) Correction: Mutator phenotype of Werner syndrome is characterized by extensive deletions. *Proc. Natl. Acad. Sci.* 86:7994.

Fukuda, H., and Hayashi, I. (1982) Ecology of dominant plant species of early stages in secondary succession: On *Chenopodium album* L. *Jpn. J. Ecol.* 32:517–526.

Fukuda, S. (1951) Production of the diapause eggs by transplanting the subesophageal ganglion in the silkworm. *Proc. Imp. Acad. Jpn.* 27:672–677.

Fuller, J. H., Shipley, M. J., Rose, G., Jarrett, R. J., and Keem, H. (1980) Coronary heart disease risk and the oral GTT: The Whitehead study. *Lancet* 1:1373–1376.

Furth, J. (1946) Prolongation of life with prevention of leukemia by thymectomy of mice. *J. Gerontol.* 1:46–53.

Gad, A. M. (1951) The head-capsule and mouth-parts in the Ceratopogonidae. *Bull. Soc. Fouad 1ᵉʳ Entomol.* 35:17–75.

Gadgil, M., and Bossert, W. H. (1970) Life historical consequences of natural selection. *Am. Nat.* 104:1–24.

Gafni, A. (1981) Location of age-related modifications in rat muscle glycerldehyde–3–phosphate dehydrogenase. *J. Biol. Chem.* 256:8875–8877.

———. (1983) Molecular origin of the aging effects in glyceraldehyde–3–phosphate dehydrogenase. *Biochem. Biophys. Acta* 742:91–99.

Gafni, A., and Noy, N. (1984) Age-related effects in enzyme catalysis. *Mol. Cell. Biochem.* 59:113–129.

Gal, A., and Everitt, A. V. (1970) Age changes in the polymer composition of acid soluble collagen prepared from rat tail tendon. *Exp. Gerontol.* 5:1–5.

Gallagher, J. C., Riggs, B. L., Eisman, J., Hamstra, A., Arnaud, S. B., and DeLuca, H. F. (1979) Intestinal calcium absorption and serum vitamin D metabolites in normal subjects and osteoporotic patients. *J. Clin. Invest.* 64:729–736.

Gallagher, J. C., Riggs, B. L., Jerpbak, C. M., and Arnaud, C. D. (1980) The effect of age on serum immunoreactive parathyroid hormone in normal and osteoporotic women. *J. Lab. Clin. Med.* 95:373–385.

Galletti, P., Ingrosso, D., Nappi, A., Gragnaniello, V., Iolascon, A., and Pinto, L. (1983) Increased methyl esterification of membrane proteins in aged red-blood cells. Preferential esterification of ankyrin and band 4.1 cytoskeletal proteins. *Eur. J. Biochem.* 135:25–31.

Gallien, L. (1935) Récherches expérimentales sur le dimorphisme évolutif et le biologie de *Polystomum integerrimum* Frohl. *Trav. Sta. Zool. Wimereaux* 12:181–201.

Galloway, S. M., and Buckton, K. E. (1978) Aneuploidy and ageing: Chromosome studies on a random sample of the population using G-banding. *Cytogenetics* 20:78–95.

Gamlin, L. (1987) Mammals with a social conscience. *New Scientist* 1571:39–47.

Gandolfi-Hornyold, M. A. (1935) La longévité de l'anguille en liberté et en captivité. *Bull. Soc. Natl. Acclimat. Fr.* 76:98–107.

Ganetzky, B., and Flanagan, J. R. (1978) On the relationship between senescence and age-related changes in two wild-type strains of *Drosophila melanogaster. Exp. Gerontol.* 13:189–196.

Garcea, R., Daino, L., Frassetto, S., Cozzo-lino, P., Ruggiu, M. E., Vannini, M. G., Pascale, R., Lenzerini, L., Simile, M. M., Puddu, M., and Feo, F. (1988) Reversal by ribo- and deoxyribonucleosides of dehydroepiandrosterone-induced inhibition of enzyme altered foci in the liver of rats subjected to the initiation-selection process of experimental carcinogenesis. *Carcinogenesis* 9: 931–938.

Gardner, M. B. (1985) Retroviral spongiform polioencephalomyelopathy. *Rev. Infect. Dis.* 7:99–110.

Gardner, M. B., Henderson, B. E., Estes, J. D., Menck, H., Parker, J. C., and Huebner, R. J. (1973) Unusually high incidence of spontaneous lymphomas in wild house mice. *J. Natl. Canc. Inst.* 50:1571–1579.

Gardner, M. B., Henderson, B. E., Estes, J. D., Rongey, R. W., Casagrande, J., Pike, M., and Huebner, R. J. (1976) The epidemiology and virology of C-type virus-associated hematologic cancers and related diseases in wild mice. *Canc. Res.* 36:574–581.

Gardner, M. B., Ihle, J. N., Pillarisetty, R. J., and Talal, N. (1977) Type C virus expression and host response in diet-cured NZB/W mice. *Nature* 268:341–344.

Gardner, M. B., Rasheed, S., Pal, B. K., Estes, J. D., and O'Brien, S. J. (1980) *Akvr–1,* a dominant murine leukemia virus restriction gene, is polymorphic in leukemia-prone wild mice. *Proc. Natl. Acad. Sci.* 77:531–535.

Garlick, P. J., Clugston, G. A., and Waterlow, J. C. (1980) Influence of low-energy diets on whole-body protein turnover in obese subjects. *Am. J. Physiol.* 238:E235–E244.

Garlick, R. L., Bunn, H. F., and Spiro, R. G. (1988) Nonenzymatic glycation of basement membranes from human glomeruli and bovine sources: Effect of diabetes and age. *Diabetes* 37:1144–1155.

Garlick, R. L., Mazer, J. S., Chylack, L. T., Jr., Tung, W. H., and Bunn, H. F. (1984) Nonenzymatic glycation of human lens crystallin: Effect of aging and diabetes mellitus. *J. Clin. Invest.* 74:1742–1749.

Garn, S. M., Rohmann, C. G., and Wagner, B. (1967) Bone loss as a general phenomenon in man. *Fed. Proc.* 26:1729–1736.

Garrod, D. J., and Horwood, J. W. (1984) Re-productive strategies and the response to exploitation. In *Fish Reproduction: Strategies and Tactics,* G. W. Potts and R. J. Wootton (eds.), 367–384. New York: Academic Press.

Gauthier, G. F., and Padykula, H. A. (1966) Cytological studies of fiber types in skeletal muscle. A comparative study of the mammalian diaphragm. *J. Cell Biol.* 28:333–354.

Gavish, D., Brinton, E. A., and Breslow, J. L. (1989) Heritable allele-specific differences in amounts of apoB and low-density lipoproteins in plasma. *Science* 244:72–76.

Geary, S., and Florini, J. R. (1972) Effect of age on rate of protein synthesis in isolated perfused mouse hearts. *J. Gerontol.* 27:325–332.

Geddes, J. W., Monaghan, D. T., Cotman, C. W., Lott, I. T., Kim, R. C., and Chui, H. C. (1985) Plasticity of hippocampal circuitry in Alzheimer's disease. *Science* 230:1179–1181.

Gee, D. M., Flurkey, K., and Finch, C. E. (1983) Aging and the regulation of luteinizing hormone in C57BL/6J mice: Impaired elevations after ovariectomy and spontaneous elevations at advanced ages. *Biol. Reprod.* 28:598–607.

Geenen, V., Legros, J. J., Franchimont, P., Baudrihaye, M., Defresne, M. P., and Boniver, J. (1986) The neuroscience thymus: Coexistence of oxytocin and neurophysin in the human thymus. *Science* 232:508–511.

Geiger, T., and Clarke, S. (1987) Deamidation, isomerization, and racemization at asparagenyl and aspartyl residues in peptides. Succinimide-linked reactions that contribute to protein-degradation. *J. Biol. Chem.* 262:785–794.

Geist, V. (1971) *Mountain Sheep. A Study in Behavior and Evolution.* Chicago: Univ. of Chicago Press.

Gensel, P. G., and Andrews, H. N. (1984) *Plant Life in the Devonian.* New York: Praeger.

Gensler, H. L., and Bernstein, H. (1981) DNA damage as the primary cause of aging. *Q. Rev. Biol.* 56:279–303.

George, F. W., and Wilson, J. D. (1988) Sex determination and differentiation. In *The Physiology of Reproduction,* vol. 1, E. Knobil, J. Neill, L. L. Ewing, G. S. Greenwald,

C. L. Markert, and D. W. Pfaff (eds.), 3–26. New York: Raven Press.

Gepstein, S. (1988) Photosynthesis. In *Senescence and Aging in Plants,* L. D. Noodén and A. C. Leopold (eds.), 85–109. Boca Raton, FL: CRC Press.

Gere, G. (1978) Über den Wasser- und Fetthaushalt der Imagines von Rhopalocera-Schmetterlingen. *Opuscula Zool. Budapest* 15:83– 91.

Gerking, S. D. (1959) Physiological changes accompanying aging in fishes. *CIBA Found. Colloq. Aging* 5:181–211.

German, J. (1964) Cytological evidence for crossing-over in vitro in human lymphoid cells. *Science* 144:298–301.

———. (1965) Chromosomal breakage in a rare and probably genetically determined syndrome of man. *Science* 148:506–507.

———. (1968) Mongolism, delayed fertilization, and human sexual behavior. *Nature* 217:516–518.

German, J., Schonberg, S., Louis, E., and Chaganti, R. S. (1977) Bloom's syndrome. 4. Sister-chromatid exchanges in lymphocytes. *Am. J. Human Genet.* 29:248.

Gerrity, R. G., Naito, H. K., Richardson, M., and Schwartz, C. J. (1979) Dietary induced atherogenesis in swine: Morphology of the intima in prelesion stages. *Am. J. Pathol.* 95:775–792.

Gershon, D. (1979) Current status of age altered enzymes: Alternative mechanisms. *Mech. Age. Dev.* 9:189–196.

Gershon, H., and Gershon, D. (1970) Detection of inactive enzyme molecules in aging organisms. *Nature* 227:1214–1217.

Gersovitz, M., Bier, D., Matthews, D., Udall, J., Munro, H. N., and Young, V. R. (1980) Dynamic aspects of whole body glycine metabolism: Influence of protein intake in young adult and elderly males. *Metabolism* 29:1087–1094.

Gersovitz, M., Munro, H. N., Udall, J., and Young, V. R. (1980) Albumin synthesis in young and elderly subjects using a new stable isotope methodology: Response to level of protein intake. *Metabolism* 29:1075–1086.

Gertsch, W. J. (1949) *American Spiders.* New York: Van Nostrand.

Gesell, M. S., and Roth, G. S. (1981) Decrease

in rat uterine estrogen receptors during aging: Physio- and immuno-chemical properties. *Endocrinology* 109:1502–1508.

Getz, M. J. (1985) Molecular mechanisms for age-related virus expression. In *Handbook of the Biology of Aging,* 2d ed., C. E. Finch and E. L. Schneider (eds.), 255–271. New York: Van Nostrand.

Ghandi, S., Santelli, J., Mitchell, D. H., Stiles, J. W., and Sanadi, D. R. (1980) A simple method for maintaining large aging populations of *Caenorhabditis elegans. Mech. Age. Dev.* 12:137–150.

Ghoneum, M. M. H., and Egami, N. (1982) Age-related changes in morphology of the thymus of the fish *Oryzias latipes. Exp. Gerontol.* 17:33–40.

Giacometti, L. (1965) Hair growth and aging. In *Advances in Biology of the Skin,* W. Montagna (ed.), 97–118. Oxford: Pergamon Press.

Gibbons, J. W. (1976) Aging phenomena in reptiles. In special review issue, M. F. Elia, B. E. Eleftheriou, and P. K. Elias (eds.). *Exp. Aging Res.,* 454–475.

———. (1987) Why do turtles live so long? *BioScience* 37:262–269.

Gibson, C. W. D., and Hamilton, J. (1984) Population processes in a large herbivorous reptile: The giant tortoise of Aldabra atoll. *Oecologia* (Berlin) 61:230–240.

Gibson, R. N., and Ezzi, J. A. (1978) The biology of a Scottish population of Fries goby, *Lesueurigobius friesii. J. Fish. Biol.* 12:371–389.

Giese, A. C., and Pearse, J. S. (1975) *Reproduction of Marine Invertebrates,* vol. 3, *Annelids and Echiurans.* New York: Academic Press.

Gifford, E. M., and Foster, A. S. (1988) *Morphology and Evolution of Vascular Plants.* 3d ed. New York: W. H. Freeman.

Gil-ad, I., Gurewitz, R., Marcovici, O., and Rosenfeld, J. (1984) Effect of aging on human plasma growth hormone response to clonidine. *Mech. Age. Dev.* 27:97–100.

Gilbert, E. (1972) Pollen feeding and reproductive biology of *Heliconius* butterflies. *Proc. Natl. Acad. Sci.* 69:1403.

Gilbert, F. S. (1981) Foraging ecology of hoverflies: Morphology of the mouthparts in relation to feeding on nectar and pollen in some common urban species. *Ecol. Entomol.* 6:245–262.

Gilbert, J. J. (1968) Dietary control of sexuality in the rotifer *Asplanchna brightwelli* (Gosse). *Physiol. Zool.* 41:14–40.

———. (1980a) Developmental polymorphism in the rotifer *Asplanchna sieboldi.* Three distinct female morphotypes, controlled by the level of dietary vitamin E, or tocopherol, allow this small aquatic organism to adapt rapidly to environmental changes. *Am. Sci.* 68:636–646.

———. (1980b) Female polymorphism and sexual reproduction in the rotifer *Asplanchna:* Evolution of their relationship and control by dietary tocopherol. *Am. Nat.* 116:409–431.

———. (1983a) Rotifera. In *Reproductive Biology of Invertebrates* vol. 1, *Oogenesis, Oviposition, and Oosorption.* K. G. Adiyodi and R. G. Adiyodi (eds.), 181–209. New York: Wiley and Sons.

———. (1983b) Rotifera. In *Reproductive Biology of Invertebrates,* vol. 2, *Spermatogenesis and Sperm Function,* K. G. Adiyodi and R. G. Adiyodi (eds.), 181–193. New York: Wiley and Sons.

———. (1988a) Rotifera. In *Reproductive Biology of Invertebrates,* vol. 3, *Accessory Sex Glands,* K. G. Adiyodi and R. G. Adiyodi (eds.), 73–80. Oxford and New Delhi: Oxford Univ. Press.

———. (1988b) Rotifera. In *Reproductive Biology of Invertebrates,* vol. 4, part A, *Fertilization, Development, and Parental Care,* K. G. Adiyodi and R. G. Adiyodi (eds.), 179–199. Oxford and New Delhi: Oxford Univ. Press.

Gilbert, J. J., and Thompson, G. A. (1968) Alpha tocopherol control of sexuality and polymorphism in the rotifer *Asplanchna. Science* 159:734–736.

Gilchrest, B. A. (1980) Prior chronic sun exposure decreases the lifespan of human skin fibroblasts in vitro. *J. Gerontol.* 35:537–541.

———. (1990) Physiology and pathophysiology of aging skin. In *Biochemistry and Physiology of the Skin,* 2d ed., L. A. Goldsmith (ed.). In press.

Gilchrest, B. A., Blog, F. B., and Szabo, G. (1979) Effects of aging and chronic sun exposure on melanocytes in human skin. *J. Invest. Dermatol.* 73:141–143.

Gilchrest, B. A., Szabo, G., Glynn, E., and Godwyn, R. (1983) Chronologic and actinically induced aging in human facial skin. *J. Invest. Dermatol.* 80:815–855.

Gilchrist, F. G. (1933) The time relations of determination in early amphibian development. *J. Exp. Zool.* 66:15–51.

Gilden, D. H., Vafai, A., Shtram, Y., Becker, Y., Devlin, M., and Wellish, M. (1983) Varicella-zoster virus DNA in human sensory ganglia. *Nature* 306:478–480.

Giles, R. E., Blanc, H., Cann, H. M., and Wallace, D. C. (1980) Maternal inheritance of human mitochondrial DNA. *Proc. Natl. Acad. Sci.* 77:6715–6719.

Gillies, M. T. (1964) The study of longevity in biting insects. *Int. Rev. Gen. Exp. Zool.* 1:47–76.

Ginsberg, J., and O'Reilly, B. (1983) Climacteric flushing in a man. *Brit. Med. J.* 287:262.

Giok, K. H. (1961) *An Experimental Study of Pituitary Tumors: Genesis, Cytology, and Hormone Content.* Berlin: Springer-Verlag.

Gjerstet, R., Gorka, C., Hasthrope, S., Lawrence, J. J., and Eisen, H. (1982) Developmental and hormonal regulation of protein H1° in rodents. *Proc. Natl. Acad. Sci.* 79:2333–2337.

Glass, A. G., and Hoover, R. N. (1989) The emerging epidemic of melanoma and squamous cell skin cancer. *J.A.M.A.* 262:2097–2100.

Glenner, G. G. (1980) Amyloid deposits and amyloidosis. The beta-fibrilloses. *N. Engl. J. Med.* 302:1283–1292.

Glenner, G. G., and Wong, C. W. (1984) Alzheimer's disease: Initial report on the purification and characterization of a novel cerebrovascular amyloid protein. *Biochem. Biophys. Res. Commun.* 120:885–890.

Gloyna, R. E., and Wilson, J. D. (1969) A comparative study of conversion of testosterone to 17β-hydroxy–5α-androstane–3–one (dihydrotestosterone) by prostate and epididymis. *J. Clin. Endocrinol. Metab.* 29:970–977.

Glueck, C. J., Garside, P. S., Fallat, R. W., Sielski, J., and Steiner, P. M. (1976) Longevity syndromes: Familial hypobeta and familial hyperalpha lipoproteinemia. *J. Lab. Clin. Med.* 88:941–957.

Glueck, C. J., Garside, P. S., Mellies, M. J., and Steiner, P. M. (1977) Familial hypobetalipoproteinemia. Studies in 13 kindreds. *Clin. Res.* 25:517.

Goidl, E. A. (ed.) (1987) *Aging and the Immune Response: Cellular and Humoral Aspects.* New York: Marcel Dekker.

Goidl, E. A., Choy, J. W., Gibbons, J. J., Weksler, M. E., Thorbecke, G. J., and Siskind, G. W. (1983) Production of autoantiidiotype antibody during the normal immune response. 7. Analysis of the cellular basis for the increased auto-antiidiotype antibody production by aged mice. *J. Exp. Med.* 157:1635–1645.

Goidl, E. A., Thorbecke, G. J., Weksler, M. E., and Siskind, G. W. (1980) Production of auto-antiidiotypic antibody during the normal immune response: Changes in the auto-antiidiotypic antibody response and the idiotype repertoire associated with aging. *Proc. Natl. Acad. Sci.* 77:6788–6792.

Goldberg, A. P., and Coon, P. J. (1987) Non-insulin-dependent diabetes mellitus in the elderly. Influence of obesity and physical inactivity. *Endocrinol. Metab. Clin.* 16:843–865.

Goldberg, A. P., and Hagberg, J. M. (1990) Physical exercise in the elderly. In *Handbook of the Biology of Aging,* 3d ed., E. L. Schneider and J. W. Rowe (eds.), 407–428. San Diego: Academic Press.

Goldberg, F. M., L. D., and Oakley, G. P. (1979) Reducing birth defect risk in advanced middle age. *J.A.M.A.* 242:2292–2294.

Goldberg, M. A. (1976) Histocompatibility antigens in systemic Lupus Erythematosus. *Arthrit. Rheum.* 19:129–132.

Golden, J. W., and Riddle, D. L. (1984) The *Caenorhabditis elegans* dauer larva: Developmental effects of pheromone, food, and temperature. *Dev. Biol.* 102:368–378.

Goldgaber, D., Lerman, M. I., McBride, W., Saffiotti, U., and Gajdusek, D. C. (1987) Characterization and chromosomal localization of cDNA encoding brain amyloid of Alzheimer's disease. *Science* 235:877–880.

Goldspink, D. F., and Kelly, F. J. (1984) Protein turnover and growth in the whole body, liver, and kidney of the rat from the foetus to senility. *Biochem. J.* 217:507–516.

Goldstein, A. L., Hooper, J. A., Schulof, R. S., Cohen, G. H., McDaniel, M. C., White, A., and Dardenne, M. (1974) Thymosin and the immunopathology of aging. *Fed. Proc.* 33:2053–2056.

Goldstein, J. L., and Brown, M. S. (1982) The LDL defect in familial hypercholesterolemia. Implication for pathogenesis and therapy. *Med. Clin. N. Am.* 66:335–362.

Goldstein, J. L., Schrott, H. G., Hazzard, W. R., Gierman, E. L., and Motulsky, A. G. (1973) Hyperlipidemia in coronary heart disease. 2. Genetic analysis of lipid levels in 176 families and delineation of a new inherited disorder, combined hyperlipidemia. *J. Clin. Invest.* 52:1544–1568.

Goldstein, L., Steller, E. J., and Knox, W. E. (1962) The effect of hydrocortisone on tyrosine-alphaketoglutarate transaminase and tryptophan pyrolase activities in the isolated, perfused liver. *J. Biol. Chem.* 237:1723–1726.

Goldstein, P., and Curis, M. (1987) Age-related changes in the meiotic chromosomes of the nematode *Caenorhabditis elegans. Mech. Age. Dev.* 40:115–130.

Goldstein, S. (1974) Aging in vitro. Growth of cultured cells from the Galapagos tortoise. *Exp. Cell Res.* 83:297–302.

Goldstein, S., Littlefield, J. W., and Soeldner, J. S. (1969) Diabetes mellitus and aging: Diminished plating efficiency of cultured human fibroblasts. *Proc. Natl. Acad. Sci.* 64:155–160.

Goldstein, S., and Moerman, E. J. (1975) Heat-labile enzymes in skin fibroblasts from subjects with progeria. *N. Engl. J. Med.* 292:1306–1309.

Goldstein, S., Moerman, E. J., Soeldner, J. S., Gleason, R. E., and Barnett, D. M. (1978) Chronologic and physiologic age affect replicative lifespan of fibroblasts from diabetic, prediabetic, and normal donors. *Science* 199:781–782.

Goldstein, S., and Singal, D. P. (1974) Senescence of cultured human fibroblasts: Mitotic versus metabolic time. *Exp. Cell Res.* 88:359–364.

Gompertz, B. (1825) On the nature of the function expressive of the law of human mortality, and on a new mode of determining the value of life contingencies. *Phil. Trans. Roy. Soc.* (London) 115:513–585.

Gonzalez-Reyes, A., Urquia, N., Gehring, W. H., Struhl, G., and Morata, G. (1990) Are cross-regulatory interactions between homoeotic genes functionally significant? *Nature* 334:78–80.

Good, N. E. (1936) The flour beetles of the genus *Tribolium. Tech. Bull.* 498:1–54.

Goodall, J. (1979) Life and death at Gombe. *Natl. Geog.* 155:592– 621.

Goodman, D. (1974) Natural selection and a cost ceiling on reproductive effort. *Am. Nat.* 108:247–268.

Goodman, D. G., Ward, J. M., Squire, R. A., Chu, K. C., and Linhart, M. S. (1979) Neoplastic and nonneoplastic lesions in aging F344 rats. *Toxicol. Appl. Pharmacol.* 48:237–248.

Goodrick, C. L. (1975) Life-span and the inheritance of longevity of inbred mice. *J. Gerontol.* 30:257–263.

Goodrick, C. L., Ingram, D. K., Reynolds, M. A., Freeman, J. R., and Cider, N. L. (1982) Effects of intermittent feeding upon growth and life span in rats. *Gerontology* 28:233–241

———. (1983) Differential effects of intermittent feeding and voluntary exercise on body weight and lifespan in adult rats. *J. Gerontol.* 38:36–45.

Gorbman, A. (1983) Reproduction in cyclostome fishes and its regulation. In *Fish Physiology,* vol. 9A, W. S. Hoar, D. J. Randall, and E. M. Donaldson (eds.), 1–29. New York: Academic Press.

Gordon, G. B., Bush, D. E., and Weisman, H. F. (1988) Reduction of atherosclerosis by administration of dehydroepiandrosterone. A study in the hypercholesterolemic New Zealand white rabbit with aortic intimal injury. *J. Clin. Invest.* 82:712–720.

Gordon, G. B., Shantz, L. M., and Talalay, P.

(1987) Modulation of growth, differentiation, and carcinogenesis by dehydroepiandrosterone. *Adv. Enzyme Reg.* 26:355–382.

Gordon, H. A., Bruckner-Kardoss, E., and Wostman, B. S. (1966) Ageing in germ-free mice: Life tables and lesions observed at natural death. *J. Gerontol.* 21:380–387.

Gordon, T., Castelli, W. P., Hjortland, N. C., Kannel, W. B., and Dawber, T. R. (1977) High density lipoprotein as a protective factor against coronary heart disease: The Framingham Study. *Am. J. Med.* 62:707–714.

Gorham, S. L., and Ottinger, M. A. (1986) Sertoli cell tumors in Japanese quail. *Avian Dis.* 30:337–339.

Gorman, S. D., and Cristofalo, V. J. (1985) Reinitiation of cellular DNA synthesis in BrdU-selected nondividing senescent WI-38 cells by Simian virus 40 infection. *J. Cell. Physiol.* 125:122–126.

Gosden, R. G. (1973) Chromosomal anomalies of preimplantation mouse embryos in relation to maternal age. *J. Reprod. Fertil.* 35:351–354.

———. (1985) *The Biology of Menopause: The Causes and Consequences of Ovarian Aging.* New York: Academic Press.

Gosden, R. G., Laing, S. C., Felicio, L. S., Nelson, J. F., and Finch, C. E. (1983) Imminent oocyte exhaustion and reduced follicular recruitment mark the transition to acyclicity in aging C57BL/6J mice. *Biol. Reprod.* 28:255–260.

Gosden, R. G., and Telfer, E. (1987a) Numbers of follicles and oocytes in mammalian ovaries and their allometric relationships. *J. Zool.* (London) 211:169–175.

———. (1987b) Scaling of follicular sizes in mammalian ovaries. *J. Zool.* (London) 211:157–168.

Goss, R. J. (1968) *Principles of Regeneration.* New York: Academic Press.

Gossen, J. A., de Leeuw, W. J. F., Tan, C. H. T., Zwarthoff, E. C., Berends, F., Lohman, P. H., Knook, D. L., and Vijg, J. (1989) Efficient rescue of integrated shuttle vectors from transgenic mice: A new model for studying mutations in vivo. *Proc. Natl. Acad. Sci.* 86:7971–7975.

Gottesman, S. R. C., and Walford, R. L. (1982) Autoimmunity and aging. In *Testing the Theories of Aging*, R. C. Adelman and G. S. Roth (eds.), 233–279. Boca Raton, FL: CRC Press.

Goudsmit, E., Fliers, E., and Swaab, D. F. (1988) Testosterone supplementation restores vasopressin innervation in the senescent rat brain. *Brain Res.* 473:306–313.

Gould, S. J. (1966) Allometry and size in ontogeny and phylogeny. *Biol. Rev.* 41:587–640.

———. (1975) Allometry in primates, with emphasis on scaling and the evolution of the brain. In *Approaches to Primate Biology*, F. Szalay (ed.). Basel: Karger. *Contrib. Primatol.* 5:244–292.

———. (1977) *Ontogeny and Phylogeny*. Cambridge, MA: Harvard Univ. Press.

———. (1989) *Wonderful Life. The Burgess Shale and the Nature of History*. New York: W. W. Norton.

Gould, S. J., and Lewontin, R. C. (1979) The spandrels of San Marco and the Panglossian paradigm: A critique of the adaptationist programme. *Proc. Roy. Soc.* (London), ser. B, 205:581–598.

Gowen, J. W. (1931) On chromosome balance as a factor in duration of life. *J. Gen. Physiol.* 14:447–461.

———. (1952) Hybrid vigor in *Drosophila*. In *Heterosis*, J. W. Gowan (ed.), 474–493. Ames: Iowa State Univ. Press.

Goya, R. G. (1988) Neuroendocrine system, programmed cell death, and aging. *Interdiscipl. Top. Gerontol.* 24:1–7.

Goyette, D., Guenette, S., Fournier, N., Leclerc, J., Roy, G., Fortin, R., and Dumont, P. (1988) Maturité sexuelle et périodicité de la reproduction chez la femelle de l'esturgeon jaune (*Acipenser fulvescens*) du fleuve Saint-Laurent. In *Rapport de Travaux* 06–02. Montreal: Service Aménagement Exploitation.

Gozes, I., Cronin, B. L., and Moskowitz, M. A. (1981) Protein synthesis in rat microvessels decreases with aging. *J. Neurochem.* 36:1311–1315.

Gozes, Y., Umiel, T., Meshorer, A., and Trainin, N. (1978) Syngeneic GvH induced in popliteal lymph nodes by spleen cells of old C57BL/6J mice. *J. Immunol.* 121:2199–2204.

Gracy, R. W., Yuksel, K. U., Chapman, M. L., Cini, J. K., Jahani, M., Lu, B., Oray, B., and Talent, J. M. (1985) Impaired degradation may account for the accumulation of "abnormal" proteins in aging cells. In *Modifications of Proteins during Aging*, R. C. Adelman (ed.), 1–18. New York: Liss.

Graff-Radford, N. R., and Godersky, J. C. (1989) Symptomatic congenital hydrocephalus in the elderly may be mistaken for NPH. *Neurology* (suppl. 1) 39:169.

Graham, A. (1968) *The Lake Rudolf crocodile (Crocodylus niloticus Laurenti) population*. Report to the Kenya Game Department of Wildlife Services Limited, Nairobi, Kenya.

Graham, D. G. (1979) On the origin and significance of neuromelanin. *Arch. Pathol. Lab. Med.* 103:359–362.

Graham, T. E., and Hutchinson, V. H. (1969) Centenarian box turtles. *Int. Turtle Tortoise Soc. J.* 3:25–30. [not read]

Grandi, G. (1936) Morfologia ed etologia comparate di insetti a regime specializzado. 12. *Macrosiagon ferrugineum flabellatum* F. *Boll. Entomol. Bologna* 9:33–64.

Grandison, L. J., Hodson, C. A., Chen, H. T., Advis, J., Simpkins, J., and Meites, J. (1977) Inhibition by prolactin of postcastration rise in LH. *Neuroendocrinology* 23:312–322.

Grantham, J. J., and Gabow, P. A. (1987) Polycystic kidney disease. In *Diseases of the Kidney*, 4th ed., R. W. Schrier and C. W. Gottschalk (eds.), 583–615. Boston: Little, Brown.

Grasso, M. (1974) Some aspects of sexuality and agamy in planarians. *Boll. Zool.* 41:379–393.

Green, M. C. (ed.) (1981) *Genetic Variants and Strains of the Laboratory Mouse*. New York: Gustav Fischer.

Greenblatt, D. J., Sellers, E. M., and Shader, R. I. (1982) Drug disposition in old age. *N. Engl. J. Med.* 306:1081–1088.

Greenough, W. T. (1985) The possible role of experience-dependent synaptogenesis, or synapses on demand, in memory processes. In *Memory Systems of the Brain*, N. M.

Weinberger, J. L. McGaugh, and G. Lynch (eds.), 77–103. New York: Guilford.

Greenspan, S. L., Maitland, L. A., Rowe, J. W., and Elahi, D. (1990) The pituitary-adrenal stress response in successful aging. *Clin. Res.* 38:528a (abstract).

Gregerman, R. I. (1959) Adaptive enzyme responses in the senescent rat: Tryptophan peroxidase and tyrosine transaminase. *Am. J. Physiol.* 1197:63–65.

Griffond, B., and Bride, J. (1987) Germinal and non-germinal lines in the ovotestis of *Helix aspersa:* A survey. *Roux's Arch. Dev. Biol.* 196:113–118.

Grimaldi, D., Beck, C. W., and Boon, J. J. (1989) Occurrence, chemical characteristics, and paleontology of the fossil resins from New Jersey. *Am. Mus. Novit.* 2948:1–28.

Grimby, G., and Saltin, B. (1983) The ageing muscle. *Clin. Physiol.* 3:209–218.

Grison, P. (1947) Développement sans diapause des chenilles de *Euproctis phaeorrhaea. Compt. Rend. Acad. Sci.* (Paris) 225:1089–1090.

Grosberg, R. K. (1982) Ecological, genetical, and developmental factors regulating life history variation within a population of the colonial ascidian *Botryllus schlosseri* (Pallas) savigny. Ph.D. diss., Yale University, New Haven, CT.

———. (1988a) The evolution of allorecognition specificity in clonal invertebrates. *Q. Rev. Biol.* 63:377–412.

———. (1988b) Life-history variation within a population of the colonial ascidian *Botryllus schlosseri*. 1. The genetic and environmental control of seasonal variation. *Evolution* 42:900–920.

Grosberg, R. K., and Patterson, M. R. (1989) Iterated ontogenies reiterated. A review. *Paleobiology* 15:67–73.

Gross, M. R. (1985) Disruptive selection for alternative life histories in salmon. *Nature* 313:47–48.

Grossman, C. J. (1984) Regulation of the immune system by sex steroids. *Endocrine Rev.* 5:435–455.

———. (1985) Interactions between the gonadal steroids and the immune system. *Science* 227:257–261.

Grossowicz, N., Hestrin, S., and Keynan, A. (eds.) (1961) *Cryptobiotic Stages in Biological Systems.* Proceedings of a Symposium. Amsterdam: Elsevier.

Grove, G. L. (1982) Age-related differences in healing of superficial skin wounds in humans. *Arch. Dermatol. Res.* 272:381–385.

Grove, G. L., and Cristofalo, V. J. (1977) Characterization of the cell cycle of cultured human diploid cells: Effects of aging and hydrocortisone. *J. Cell Physiol.* 90:415–422.

Grove, G. L., and Kligman, A. M. (1983) Age-associated changes in human epidermal cell renewal. *J. Gerontol.* 38:137–142.

Grundy, S. M. (1986) Cholesterol and coronary heart disease. A new era. *J.A.M.A.* 256:2849–2858.

Grzimek, H. C. (1976) *Grzimek's Encyclopedia of Evolution.* New York: Van Nostrand.

Guinness Book of World Records. (1987) 15th ed., A. Russell (ed.). New York: Bantam Books.

Guiroy, D. C., Miyazaki, M., Multhaup, G., Fischer, P., Garruto, R. M., Beyreuther, K., Masters, C. L., Simms, G., Gibbs, C. J., Jr., and Gajdusek, D. C. (1987) Amyloid of neurofibrillary tangles of Guamanian Parkinsonism dementia and Alzheimer disease share identical amino acid sequence. *Proc. Natl. Acad. Sci.* 84:2073–2077.

Gumbel, E. J. (1958) *Statistics of Extremes.* New York: Columbia Univ. Press.

Gumbiner, B., Polonsky, K. S., Beltz, W. F., Wallace, P., Brechtel, G., and Fink, R. I. (1989) Effects of aging on insulin secretion. *Diabetes* 38:1549–1556.

Gundberg, C. M., Anderson, M., Dickson, I., and Gallop, P. M. (1986) Glycated osteocalcin in human and bovine bone. *J. Biol. Chem.* 261:14557–14561.

Gupta, K. P., van Golen, K. L., Randerath, E., and Randerath, K. (1990) Age-dependent covalent DNA alterations (I-compounds) in rat liver mitochondrial DNA. *Mutat. Res.* 237:17–27.

Gupta, S. K., and Rothstein, M. (1976) Triosephosphate isomerase from young and old *Turbatrix aceti. Arch. Biochem.* 174:333–338.

Guralnik, J. M., Yanagishita, M., and Schneider, E. L. (1988) Projecting from the older population of the United States from the past and prospects for the future. *Millbank Q.* 66:283–308.

Gurdon, J. B. (1962) Adult frogs derived from the nuclei of single somatic cells. *Dev. Biol.* 4:256–263.

Gutierrez, A. P., Schulthess, F., Wilson, L. T., Villacorta, A. M., Ellis, C. K., and Baumgaertner, J. U. (1987) Energy acquisition and allocation in plants and insects: A hypothesis for the possible role of hormones in insect feeding patterns. *Can. Entomol.* 119:109–129.

Gutmann, E. (1977) Muscle. In *Handbook of the Biology of Aging*, 1st ed., C. E. Finch and L. Hayflick (eds.), 445–469. New York: Van Nostrand Reinhold.

Gutmann, E., and Hanzlikova, V. (1976) Fast and slow motor units in ageing. *Gerontology* 22:280–300.

Guyton, A. (1963) *Circulatory Physiology: Cardiac Output and Its Regulation.* Philadelphia: Saunders.

Haagensen, C. D. (1971) *Diseases of the Breast,* 2d ed. Philadelphia: Saunders. [not read]

Haberland, M. E., Fogelman, A. M., and Edwards, P. A. (1982) Specificity of receptor-mediated recognition of malondialdehyde-modified low density lipoproteins. *Proc. Natl. Acad. Sci.* 79:1712–1716.

Haberland, M. E., Fong, D., and Cheng, L. (1988) Malondialdehyde-altered protein occurs in atheroma of Watanabe heritable hyperlipidemic rabbits. *Science* 241:215–218.

Haché, R. J., Tam S.-P., Cochrane, A., Nesheim, M., and Deeley, R. G. (1987) Long-term effects of estrogen on avian liver: Estrogen-inducible switch in expression of nuclear, hormone-binding proteins. *Mol. Cell. Biol.* 7:3538–3547.

Hackbarth, H., and Harrison, D. E. (1982) Changes with age in renal function and morphology in C57BL/6, CBA/HT6, and B6CBAF mice. *J. Gerontol.* 37:540–547.

Hadfield, M. G. (1986) Extinction in Hawaiian achatinelline snails. *Malacologia* 27:67–81.

Hadfield, M. G., and Miller, S. E. (1989) Demographic studies on Hawaii's endangered

tree snails: *Partulina proxima. Pacific Sci.* 43:1–16.

Hadfield, M. G., and Mountain, B. S. (1980) A field study of a vanishing species, *Achatinella mustelina* (Gastropoda, Pulmonata), in the Waianae mountains of Oahu. *Pacific Sci.* 34:345–358.

Hadfield, M. G., and Switzer-Dunlap, M. F. (1990) Environmental regulation of lifespan and reproduction in *Aplysia juliana. Advances in Invertebrate Reproduction,* vol. 5. Amsterdam: Elsevier.

Hadorn, E. (1966) Konstanz, Wechsel, und Typus der Determination in Zellen aus mannlichen Genitalanlagen von *Drosophila melanogaster* durch Dauerkulter in vivo. *Dev. Biol.* 13:424–509.

———. (1968) Transdetermination in cells. *Sci. Am.* 219:110–114.

Hafez, M., and El-Said, L. (1969) On the bionomics of *Orygia dubia judaea* Stgr. *Bull. Soc. Entomol. Egypte* 53:161–183.

Haga, N., and Karino, S. (1986). Microinjection of immaturin rejuvenates sexual activity of old *Paramecium. J. Cell Sci.* 86:263–271.

Hagberg, J. M., Allen, W. K., Seals, D. R., Hurley, B. F., Ehsani, A. A., and Holloszy, J. O. (1985) A hemodynamic comparison of younger and older endurance athletes during excercise. *J. Appl. Physiol.* 58:2041–2046.

Hagelin, L. O., and Steffner, N. (1958) Notes on the spawning habits of the river lamprey (*Petromyzon fluviatilis*). *Oikos* 9:221–238.

Hagemeijer, A., and Smit, E. M. E. (1977) Partial trisomy 21: Further evidence that trisomy of band 21q22 is essential for Down's phenotype. *Human Genet.* 38:15–23.

Hagen, G., and Kochert, G. D. (1980) Protein synthesis in a new system for the study of senescence. *Exp. Cell Res.* 127:457–475.

Hager, L., McKnight, G. S., and Palmiter, R. D. (1980) Glucocorticoid induction of egg-white messenger-RNAs in chick oviduct. *J. Biol. Chem.* 255:7796–7800.

Hahn, L. J., Kloiber, R., Vimy, M. J., Takahashi, Y., and Lorscheider, F. L. (1989) Dental "silver" tooth fillings: A source of mercury exposure revealed by whole-body image scan and tissue analysis. *FASEB J.* 3:2641–2646.

Hailey, A., and Davies, P. M. (1987) Maturity,

mating, and age-specific reproductive effort of snake *Natrix maura. J. Zool.* (London) 211:573–587.

Hairston, N. G., and Walton, W. E. (1986) Rapid evolution of a life-history trait. *Proc. Natl. Acad. Sci.* 83:4831–4833.

Haldane, J. B. (1941) *New Paths in Genetics.* London: Allen and Unwin.

Hall, A. D. (1929) *The Book of the Tulip.* London: Martin Hopkinson. [not read]

Hall, G. O., and Marble, D. R. (1931) The relationship between the first year egg production and the egg production of later years. *Poultry Sci.* 10:194–205.

Hall, J. C. (1969) Age-dependent enzyme changes in *Drosophila melanogaster. Exp. Gerontol.* 4:207–222.

Hall, K. Y., Bergmann, K., and Walford, R. L. (1981) DNA-repair, H–2, and aging in NZB and CBA mice. *Tiss. Antigens* 17:104–110.

Hall, K. Y., Hart, R. W., Benirschke, A. K., and Walford, R. L. (1984) Correlation between ultraviolet-induced DNA repair in primate lymphocytes and fibroblasts and species maximum achievable life span. *Mech. Age. Dev.* 24:163–174.

Hallé, F., Oldeman, R. A., and Tomlinson, P. B. (1978) *Tropical trees and forests. An architectural analysis.* Berlin: Springer-Verlag.

Hallen, H. (1966) Discrete gamma globulin (M)–components in serum. Clinical study of 150 subjects without myelomastosis. *Acta Med. Scand.,* suppl. 462.

Hallgren, B., and Sourander, P. (1958) The effect of age on the non-haemin iron in the human brain. *J. Neurochem.* 3:41–51.

Hallgren, H. M., Buckley, C. E., Gilbertsen, V. A., and Yunis, E. J. (1973) Lymphocyte phytohemagglutinin responsiveness, immunoglobulins and autoantibodies in aging humans. *J. Immunol.* 111:1101–1107.

Halpern-Sebold, L. R., Schreibman, M. P., and Margolis-Nunno, H. (1986) Differences between early- and late-maturing genotypes of the platyfish (*Xiphophorus maculatus*) in the morphometry of their immunoreactive luteinizing hormone releasing hormone-containing cells: A developmental study. *J. Exp. Zool.* 240:245–257.

Hamada, K. (1961) Taxonomic and ecological studies of the genus *Hypomesus* in Japan. *Memoirs Faculty Fish. Hokkaido Univ.* 9:1–56.

Hamada, K., Gleason, S. L., Levi, B.-Z., Hirschfeld, S., Appella, E., and Ozato, K. (1989) H–2RIIBP, a member of the nuclear hormone receptor superfamily that binds to both the regulatory element of major histocompatibility class I genes and the estrogen response element. *Proc. Natl. Acad. Sci.* 86:8289–8293.

Hamburger, V., Brunso-Bechtold, J. K., and Yip, J. W. (1981) Neuronal death in the spinal ganglia of the chick embryo and its reduction by nerve growth factor. *J. Neurosci.* 1:60–71.

Hamerton, J. L. (1971) Human cytogenetics. In *General Cytogenetics,* vol. 1, 102–104. New York: Academic Press.

Hamilton, J. B. (1942) Male hormone stimulation is prerequisite and an incitant in common baldness. *Am. J. Anat.* 71:451–480.

———. (1948) The role of testicular secretions as indicated by the effects of castration in man and by studies of pathological conditions and the short lifespan associated with maleness. *Rec. Adv. Horm. Res.* 3:257–324.

———. (1965) Relationship of castration, spaying, and sex to survival and duration of life in domestic cats. *J. Gerontol.* 20:96–104.

Hamilton, J. B., and Mestler, G. E. (1969) Mortality and survival: A comparison of eunuchs with intact men and women in a mentally retarded population. *J. Gerontol.* 24:395–411.

Hamilton, J. B., Terada, H., Mestler, G. E., and Tirman, W. (1969) 1. Coarse sternal hairs, a male secondary sex character that can be measured quantitatively: The influence of sex, age and genetic factors. 2. Other sex-differing characters: Relationship to age, to one another, and to values for coarse sternal hairs. In *Advances in Biology of Skin Hair Growth,* vol. 9, 129–151. Proceedings of a symposium held at the University of Oregon Medical School, 1967. Oxford and New York: Pergamon Press.

Hamilton, K. E., Ferguson, A., Taggart, J. B., Tomasson, T., Walker, A., and Fahy, E. (1989) Post-glacial colonization of brown trout, *Salmo trutta L.*: *Ldh-5* as a phylogeographic marker locus. *J. Fish. Biol.* 35:651–664.

Hamilton, W. D. (1964) The genetical evolution of social behavior. *J. Theoret. Biol.* 7:1–52.

———. (1966) The moulding of senescence by natural selection. *J. Theoret. Biol.* 12:12–45.

Hamlin, C. R., and Kohn, R. R. (1971) Evidence for progressive age-related structural changes in post-mature human collagen. *Biochim. Biophys. Acta* 236:458–467.

Hamlin, C. R., Kohn, R. R., and Luschin, J. H. (1975) Apparent accelerated aging of human collagen in diabetes mellitus. *Diabetes* 24:902–904.

Hammerling, J. (1924) Die ungeschlechtliche Fortpflanzung und Regeneration bei *Aelosoma hemprichii. Zool. Jb.* 41:581–655. [not read]

———. (1963) Nucleo-cytoplasmic interactions in *Acetabularia* and other cells. *Ann. Rev. Plant Physiol.* 14:65–92.

Hammond, E. C., and Garfinkel, L. (1971) Longevity of parents and grandparents in relation to coronary heart disease and associated variables. *Circulation* 43:31–44.

Hamrick, J. L. (1979) Genetic variation and longevity. In *Topics in Plant Population Biology*, O. T. Solbrig, S. Jain, G. B. Johnson, and P. H. Raven (eds.), 84–113. New York: Columbia Univ. Press.

Hanamura, T. (1966) Salmon of the North Pacific Ocean. 3. Review of the life history of North Pacific salmon. Sockeye salmon in the Far East. International North Pacific Fisheries Commission, Vancouver, B.C., bulletin 18:1–28.

Hanawalt, P. C. (1987) On the role of DNA damage and repair processes in aging: Evidence for and against. In *Modern Biological Theories of Aging*, H. R. Warner, R. N. Butler, R. L. Sprott, and E. L. Schneider (eds.), 183–198. New York: Raven Press.

Handler, A. M., and Postlethwait, J. H. (1977) Endocrine control of vitellogenesis in *Drosophila melanogaster:* Effects of the brain and corpus allatum. *J. Exp. Zool.* 202:389–402.

Hane, S., and Robertson, O. H. (1959) Changes in plasma of 17-hydroxycortico-

steroids accompanying sexual maturation and spawning of the Pacific salmon (*Oncorhynchus tschawytscha*) and rainbow trout (*Salmo gairdnerii*). *Proc. Natl. Acad. Sci.* 45:886–893.

Hang, L., Theofilopoulos, A. N., and Dixon, F. J. (1982) A spontaneous rheumatoid arthritis-like disease in MRL/1 mice. *J. Exp. Med.* 155:1690–1701.

Hanks, S. D., and Flood, D. G. (1990) Region specific stability of dendritic extent in normal human aging and regression in Alzheimer's disease. 1. CA$_1$ of hippocampus. *Brain Res.* In press.

Hansen, P. M. (1963) Tagging experiments with the Greenland shark (*Somniosus microcephalus* (Bloch and Schneider) in subarea L. *Int. Comm. NW Atl. Fish. Spec. Publ.* 4:172–175.

Hansford, R. G. (1983) Bioenergetics in aging. *Biochim. Biophys. Acta* 726:41–80.

Haranghy, L., and Balázs, A. (1964) Ageing and rejuvenation in planarians. *Exp. Gerontol.* 1:77–84.

Haranghy, L., Balázs, A., and Burg, M. (1964) Phenomenon of ageing in Unionidae, as example of ageing in animals of telometric growth. *Acta Biol. Hung.* 14:311–318.

Harberd, D. J. (1961a) Observations on population structure and longevity of *Festuca rubra L. New Phytol.* 60:184–206.

———. (1961b) Some observations on natural clones in *Festuca ovina. New Phytol.* 61:85–100.

———. (1967) Observations on natural clones of *Holcus mollis. New Phytol.* 66:401–408.

Hardisty, M. W. (1963) Fecundity and speciation in lampreys. *Evolution* 17:17–22.

———. (1970) The relationship of gonadal development to the life cycles of the paired species of lamprey, *Lampetra fluviatilis (L).* and *Lampetra planeri* (Bloch). *J. Fish. Biol.* 2:173–181.

———. (1979) *Biology of the Cyclostomes.* London: Chapman and Hall.

Hardisty, M. W., and Cosh, J. (1966) Primordial germ cells and fecundity. *Nature* 210:1370.

Hardisty, M. W., and Potter, I. C. (1971) Paired species. In *The Biology of Lampreys*, vol. 1,

M. W. Hardisty and I. C. Potter (eds.), 249–277. New York: Academic Press.

Hardwick, R. C. (1986) Functions of leaf fall. *Nature* 324:517.

Hariri, R. J., Hajjar, D. P., Coletti, D., Alonso, D. R., Weksler, M. E., and Rabellino, E. (1988) Aging and arteriosclerosis. Cell cycle kinetics of young and old arterial smooth muscle cells. *Am. J. Pathol.* 131:132–136.

Harman, D. (1962) Role of free radicals in mutation, cancer, aging and maintenance of life. *Rad. Res.* 16:752–763.

———. (1968) Free radical theory of aging: Effect of free radical reaction inhibitors on the mortality rate of male LAF mice. *J. Gerontol.* 23:476–482.

———. (1981) The aging process. *Proc. Natl. Acad. Sci.* 78:7124–7128.

Harman, S. M., and Talbert, G. B. (1985) Reproductive aging. In *Handbook of the Biology of Aging,* 2d ed., C. E. Finch and E. L. Schneider (eds.), 457–510. New York: Van Nostrand.

Harman, S. M., and Tsitouras, P. D. (1980) Reproductive hormones in aging men. 1. Measurement of sex steroids, basal LH, and Leydig cell response to hCG. *J. Clin. Endocrinol. Metab.* 51:35–40.

Harp, J. A., Tsuchida, C. B., Weissman, I. L., and Scofield, V. L. (1988) Autoreactive blood cells and programmed cell death in growth and development of protochordates. *J. Exp. Zool.* 257:257–262.

Harper, J. L. (1967) A Darwinian approach to plant ecology. *J. Ecol.* 55:247–270.

———. (1977) *Population Biology of Plants.* New York: Academic Press.

Harper, J. L., and Bell, A. D. (1979) The population dynamics of growth form in organisms with modular construction. In *Population Dynamics,* R. D. Anderson, B. D. Turner, and L. R. Taylor (eds.), 29–52. Oxford: Blackwell.

Harper, J. L., and White, J. (1974) The demography of plants. *Ann. Rev. Ecol. Sys.* 5:419–463.

Harper, R. A., and Grove, G. (1979) Human skin fibroblasts derived from papillary and reticular dermis: Differences in growth potential in vitro. *Science* 204:526–527.

Harris, C. C., Vahakangas, K., Newman, M. J., Trivers, G. E., Shamsuddin, A., Sinopoli, N., Mann, D. L., and Wright, W. E. (1985) Detection of benzo(a)pyrene diol epoxide-DNA adducts in peripheral blood lymphocytes and antibodies to the adducts in serum from coke oven workers. *Proc. Natl. Acad. Sci.* 82:6672–6676.

Harris, J. M. (1975) Evolution of feeding mechanisms in the family Deinotheriidae (Mammalia: Probiscidea). *J. Linn. Soc. Zool.* 56:331–362.

Harris, K., Walker, P. M., Mickle, D. A., Harding, R., Gatley, R., Wilson, G. J., Kuzon, B., McKee, N., and Romaschin, A. D. (1986) Metabolic response of skeletal muscle to ischemia. *Am. J. Physiol.* 250:H213-H220.

Harrison, D. E. (1975a) Defective erythropoietic responses of aged mice not improved by young marrow. *J. Gerontol.* 30:286–288.

———. (1975b) Normal function of transplanted marrow cell lines from aged mice. *J. Gerontol.* 30:279–285.

———. (1983) Long-term erythropoietic repopulating ability of old, young, and fetal stem cells. *J. Exp. Med.* 157:1496–1504.

———. (1985) Cell and tissue transplantation: A means of studying the aging process. In *Handbook of the Biology of Aging,* 2d ed., C. E. Finch and E. L. Schneider (eds.), 322–356. New York: Van Nostrand.

Harrison, D. E., and Archer, J. R. (1978) Measurement of changes in mouse tail collagen with age: Temperature dependence and procedural details. *Exp. Gerontol.* 13:75–82.

———. (1983) Physiological assays for biological age in mice: Relationship of collagen, renal function, and longevity. *Exp. Aging Res.* 9:245–251.

———. (1987) Genetic differences in effects of food restriction on aging in mice. *J. Nutrit.* 117:376–382.

———. (1988a) Biomarkers of aging: Tissue markers. Future research needs, strategies, directions, and priorities. *Exp. Gerontol.* 23:309–321.

———. (1988b) Natural selection for extended longevity from food restriction. *Growth Dev. Aging* 52:65.

———. (1989) Natural selection for extended longevity from food restriction. *Growth Dev. Aging* 53:3–6.

———. (In prep.) Non-lethal tests to measure changes with age in 11 different biological systems: Longevity and results in 6–12 mouse strains.

Harrison, D. E., Archer, J. R., and Astle, C. M. (1982) The effect of hypophysectomy on thymic aging in mice. *J. Immunol.* 129:2673–2677.

———. (1984) Effects of food restriction on aging: Separation of food intake and adiposity. *Proc. Natl. Acad. Sci.* 81:1835–1838.

Harrison, D. E., Archer, J. R., Sacher, G. A., and Boyce, F. M. (1978) Tail collagen aging in mice of thirteen different genotypes and two species: Relationship to biological age. *Exp. Gerontol.* 13:63–73.

Harrison, D. E., Astle, C. M., and DeLaittre, J. (1988) Effects of transplantation and age on immunohemopoietic cell growth in the splenic microenvironment. *Exp. Hematol.* 16:213–216.

Harrison, D. E., Astle, C. M., and Lerner, C. (1984) Ultimate erythropoietic repopulating abilities of fetal, young, and old adult cells compared using repeated irradiation. *J. Exp. Med.* 160:759–771.

Harrison, D. E., Astle, C. M., and Stone, M. (1989) Numbers and functions of transplantable primitive immunohematopoietic stem cells. *J. Immunol.* 142:3833–3840.

Harrison, D. E., Ingram, D. K., and Archer, J. R. (In prep.) Longevity predictions by 13 physiological, behavioral, and growth parameters in 6 mouse strains.

Hart, B. L. (1970) Reproductive system: A male. In *The Beagle as an Experimental Dog,* A. C. Anderson (ed.), 296–312. Ames: Iowa State Univ. Press.

Hart, R. W., and Daniel, F. B. (1980) Genetic stability in vitro and in vivo. *Adv. Pathobiol.* 7:123–141.

Hart, R. W., and Setlow, R. B. (1974) Correlation between deoxyribonucleic acid excision-repair and lifespan in a number of mammalian species. *Proc. Natl. Acad. Sci.* 71:2169–2173.

Hartmann, H. T., and Kester, D. E. (1968) *Plant Propagation. Principles and Practices.* 2d ed. Prentice Hall. [not read]

Hartshorn, G. S. (1975) A matrix model of tree population dynamics. In *Tropical Ecological Systems: Trends in Terrestrial and Aquatic Research,* F. B. Golley and E. Medina (eds.), 41–51. New York: Springer-Verlag. [not read]

Harvell, C. D., and Grosberg, R. K. (1988) The timing of sexual maturity in clonal animals. *Ecology* 69:1855–1864.

Harvey, P. H., and Zammuto, R. M. (1985) Patterns of mortality and age at first reproduction in natural populations of mammals. *Nature* 315:319–320.

Hascall, G. K., and Rudzinska, M. A. (1970) Metamorphosis in *Tokophrya infusionum*: An electron-microscope study. *J. Protozool.* 17:311–323.

Haskins, C. P. (1960) Note on the natural longevity of fertile females of *Aphaenogaster picea. N.Y. Entomol. Soc.* 58:66–67.

Hass, M. A., Frank, L., and Massaro, D. (1982) The effect of bacterial endotoxin on synthesis of (Cu, Zn) superoxide dismutase in lungs of oxygen-exposed rats. *J. Biol. Chem.* 257:9379–9382.

Hassan, F. A. (1978) Prehistoric demography. In *Advances in Archaeological Method and Theory,* vol. 1, M. Schiffer (ed.), 49–103. New York: Academic Press. [not read]

———. (1980) The growth and regulation of human population in prehistoric times. In *Biosocial Mechanisms of Population Regulation,* M. N. Cohen, R. L. Malpass, and H. G. Klein (eds.), 305–311. New Haven, CT: Yale Univ. Press.

Hassold, T., Chen, N., Funkhouser, J., Jooss, T., Manuel, B., Matsuura, J., Matsuyama, A., Wilson, C., Yamane, J. A., and Jacobs, P. A. (1980) A cytogenetic study of 1,000 spontaneous abortions. *Ann. Human Genet.* (London) 44:151–178.

Hassold, T., Jacobs, P., Kline, J., Stein, Z., and Warburton, D. (1980) Effect of maternal age on autosomal trisomies. *Ann. Human Genet.* (London) 44:29.

Hastings, L. A., Pashko, L. L., Lewbart, M. L., and Schwartz, A. G. (1988) Dehydroepiandrosterone and two structural analogs inhibit 12-*O*-tetradecanoylphorbol–13-acetate stimulation of prostaglandin E_2 content in mouse skin. *Carcinogenesis* 9(6): 1099–1102.

Haug, H. (1984) Macroscopic and microscopic morphometry of the human brain and cortex. A survey in the light of new results. *Brain Pathol.* 1:123–149.

Hausman, P. B., and Weksler, M. E. (1985) Changes in the immune response with age. In *Handbook of the Biology of Aging,* 2d ed., C. E. Finch and E. L. Schneider (eds.), 414–432. New York: Van Nostrand.

Haviland, M. D. (1921) On the bionomics and development of *Lygocerus testaceimanus,* Kieffer, and *Lygocerus cameroni,* Kieffer (Proctotrypoidea: Ceraphronidae), parasites of *Aphidius* (Braconidae). *Q. J. Microscop. Sci.,* n.s. 65:101–127.

Haviland, R. P. (1964) *Engineering Reliability and Long Life Design.* Princeton, NJ: Van Nostrand.

Hay, E. (1966) *Regeneration.* Holt, Rinehart and Winston.

Hayase, F., Nagaraj, R. H., Miyata, S., Njoroge, F. G., and Monnier, V. M. (1989) Aging of proteins: Immunological detection of a glucose-derived pyrrole formed during Maillard reaction *in vivo. J. Biol. Chem.* 264:3758–3764.

Hayashi, I. (1984) Secondary succession of herbaceous communities in Japan: Quantitative features of the growth in the form of successional dominants. *Jpn. J. Ecol.* 34:375–382. [not read]

Hayashi, I., and Numata, M. (1968) Ecology of pioneer species of early stages in secondary succession. 2. The seed production. *Bot. Mag.* (Tokyo) 81:55–66. [not read]

Haydak, M. H. (1957) Changes with age in the appearance of some internal organs of the honey bee. *Bee World* 38:197–203.

Hayflick, L. (1965) The limited *in vitro* lifespan of human diploid cell strains. *Exp. Cell Res.* 37:614–636.

———. (1977) The cellular basis for biological aging. In *Handbook of the Biology of Aging,* 1st ed., C. E. Finch and L. Hayflick (eds.), 159–185. New York: Van Nostrand Reinhold.

Hayflick, L., and Moorhead, P. (1961) The serial cultivation of human diploid cell strains. *Exp. Cell Res.* 25:585–621.

Hazelton, G. A., and Lang, C. A. (1980) Glutathione contents of tissues in the aging mouse. *Biochem. J.* 188:25–30.

———. (1983) Glutathione biosynthesis in the aging adult yellow-fever mosquito (*Aedes aegypti*). *Biochem. J.* 210:289–295.

Hazelwood, R. L. (1986) Carbohydrate metabolism. In *Avian Physiology,* 4th ed., P. D. Storkie (ed.), 303–325. New York: Springer-Verlag.

Hazzard, W. R. (1986) Biological basis of the sex differential in longevity. *J. Am. Gerontol. Soc.* 34:455–471.

Healey, M. C. (1972) On the population ecology of the common goby in the Ythan estuary. *J. Nat. Hist.* 6:133–145.

Heath, G. W., Hagberg, J. M., Eshani, A. A., and Holloszy, J. O. (1981) A physiological comparison of young and older endurance athletes. *J. Appl. Physiol.* 51:634–640.

Heddle, J. A., Lue, C. B., Saunders, E. F., and Benz, R. (1978) Sensitivity to five mutagens in Ganconi's anaemia as measured by the micronucleus method. *Canc. Res.* 38: 2983–2988.

Hedley, C. L., and Stoddart, J. L. (1972) Patterns of protein synthesis in *Lolium temulentum* L. 1. Changes occuring during leaf development. *J. Exp. Bot.* 23:490–501.

Hefti, F., Hartikka, J., and Knusel, B. (1990) Authors' response to commentaries. *Neurobiol. Aging* 10:515–535.

Hefton, J. M., Darlington, G. J., Casazza, B. A., and Weksler, M. E. (1980) Immunologic studies of aging. 5. Impaired proliferation of PHA responsive human lymphocytes in culture. *J. Immunol.* 125:1007–1010.

Hegler, S. (1981) The sexual function of 1,161 elderly Danish males. *Fifth World Congress of Sexology,* 21–26 June 1981, Jerusalem. In *Sexology: Sexual Biology, Behavior, and Therapy,* Z. Hoch and H. I. Lief (eds.). Amsterdam: Excerpta Medica.

Heicklen, J., and Brown, E. (1987) Increase in life expectancy for mice fed diethylhydroxylamine (DHEA). *J. Gerontol.* 42:674–680.

Helderman, J. H., Vestal, R. E., Rowe, J. W., Tobin, J. D., Andres, R., and Robertson, G. L. (1978) The response of arginine vasopressin to intravenous ethanol and hypertonic saline in man: The impact of aging. *J. Gerontol.* 33:38–47.

Helfman, P. M., and Bada, J. L. (1975) Aspartatic acid racemisation in tooth enamel from living humans. *Proc. Natl. Acad. Sci.* 79:2891–2894.

———. (1976) Amino acid racemisation in dentine as a measure of aging. *Nature* 262:279–281.

Hendelberg, J. (1960) The fresh-water pearl mussel, *Margaritifera Margaritifera. Instit. Freshwater Reports* 41:149–171.

Henderson, B. E., Pike, M. C., and Ross, R. K. (1984) Epidemiology and risk factors. In *Breast Cancer: Diagnosis and Management*, G. Bonadonna (ed.), 15–33. New York: Wiley and Sons.

Henderson, D., Hamernik, R. P., Dosanjh, D. S., and Mills, J. H. (1976) *Effects of Noise on Hearing.* New York: Raven Press.

Henderson, I. W., Jackson, B. A., and Hargreaves, G. (1975) Actions and metabolism of corticosteroids in teleost fishes. *Biochem. Soc. Trans.* 3:1168–1171.

Henderson, I. W., Sa'Di, M. N., and Hargreaves, G. (1974) Studies on the production and metabolic clearance rates of cortisol in the European eel, *Anguilla anguilla* L. *J. Steroid Biochem.* 5:701–707.

Henderson, S. A., and Edwards, R. G. (1968) Chiasma frequency and maternal age in mammals. *Nature* 218:22–28.

Hendley, D. D., and Strehler, B. L. (1965) Enzymatic activities of lipofuscin age pigments: Comparative histochemical and biochemical studies. *Biochim. Biophys. Acta* 99:406–417

Hendricks, L. C., and Heidrick, M. L. (1988) Susceptibility to lipid peroxidation and accumulation of fluorescent products with age is greater in T-cells than B-cells. *Free Rad. Biol. Med.* 5:145–154.

Hennig, W. (1981) *Insect Phylogeny.* New York: Wiley and Sons.

Henry, J. P., Ely, D. L., Stephens, H. L., Ratcliffe, G. A., Santisteban, G. A., and Shapiro, A. P. (1971) The role of psychosocial factors in the development of arteriosclerosis in CBA mice. *Atherosclerosis* 14:203–218.

Henry, K. R. (1982) Age-related auditory loss and genetics: An electrocochleographic comparison of six inbred strains of mice. *J. Gerontol.* 37:275–282.

Henshaw, P. S., Riley, E. R., and Stapleton, G. E. (1947) The biological effects of pile radiation. *Radiology* 49:349–364.

Herbener, G. H. (1976) A morphometric study of age-dependent changes in mitochondrial populations of mouse liver and heart. *J. Gerontol.* 31:8–12.

Herberg, L., and Coleman, D. L. (1977) Laboratory animals exhibiting obesity and diabetes syndromes. *Metabolism* 26:59–99.

Herman, M. E., Miquel, J., and Johnson, M. (1971) Insect brain as a model for the study of aging. *Acta Neuropathol.* (Berlin) 19:167–183.

Herold, R. C., and Meadow, N. D. (1970) Age-related changes in ultrastructure and histochemistry of rotiferan organs. *Ultrastructure Res.* 33:203–218.

Heroux, O., and Campbell, J. S. (1960) A study of the pathology and life span of 6C and 30C acclimated rats. *Lab. Invest.* 9:305–315.

Herr, W., and Gilbert, W. (1982) Germ-line MuLV reintegrations in AKR/J mice. *Nature* 296:865–868.

Herreid, C. F. (1964) Bat longevity and metabolic rate. *Exp. Gerontol.* 1:1–9.

Herrick, F. H. (1895) The American lobster: 1. A study of its habits and development. *Bull. U.S. Fish. Commn.* 15:1–252.

———. (1909) Natural history of the American lobster. *Bull. Bureau Fish.* 29:149–408.

Hertig, A. T. (1967) The overall problem in man. In *Comparative Aspects of Reproductive Failure*, K. Benirschke (ed.), 11–41. Berlin and New York: Springer-Verlag.

Hertz, L. (1966) Neuroglial localization of potassium and sodium effects on respiration in brain. *J. Neurochem.* 13:1373–1387.

Hess, H. H. (1961) The rates of respiration of neurons and neuroglia in human cerebrum. In *Regional Neurochemistry,* S. S. Kety and J.

Elkes (eds.), 200–212. London: Pergamon Press.

Heston, L. L., Mastri, A. R., Anderson, V. E., and White, J. (1981) Dementia of the Alzheimer type. *Arch. Gen. Psychiatry* 38:1085–1090.

Heston, W. E. (1963) Complete inhibition of occurrence of spontaneous hepatomas in highly susceptible (C3H × YBR)F$_1$ male mice by hypophysectomy. *J. Natl. Canc. Inst.* 31:467–474.

Heusner, A. A. (1985) Body size and energy metabolism. *Ann. Rev. Nutrit.* 5:267–293.

Hewitt, G. M., Nichols, R. A., and Ritchie, M. G. (1988) 1868 and all that for *Magicicada. Nature* 336:206–207.

Heyman, A., Wilkinson, W. E., Hurwitz, B. J., Schmechel, D., Sigmon, A. H., Weinberg, T., Helms, M. J., and Swift, M. (1983) Alzheimer's disease: Genetic aspects and associated clinical disorders. *Ann. Neurol.* 14:507–515.

Hibbs, A. R., and Walford, R. L. (1989) A mathematical model of physiological processes and its application to the study of aging. *Mech. Age. Dev.* 50:193–214.

Hickling, C. F. (1936) Seasonal changes in the ovary of the immature hake, *Merluccius merluccius L. J. Mar. Biol. Assoc.* (U.K.) 20:443–461.

Highsmith, R. C. (1979) Coral growth rates and environmental control of density banding. *J. Exp. Mar. Biol. Ecol.* 37:105–125.

Higuchi, K., Matsumura, A., Hashimoto, K., Honma, A., Takeshita, S., Hosokawa, M., Yasuhira, K., and Takeda, T. (1983) Isolation and characterization of senile amyloid-related antigenic substance (SAS$_{SAM}$) from mouse serum. Apo SAS$_{SAM}$ is a low molecular weight apoprotein of high density lipoprotein. *J. Exp. Med.* 158:1600–1614.

Higuchi, K., Matsumura, A., Honma, A., Toda, K., Takeshita, S., Matsushita, M., Yonezu, T., Hosokawa, M., and Takeda, T. (1984) Age-related changes of serum apoprotein SAS$_{SAM}$, apoprotein A–1 and low-density lipoprotein levels in senescence accelerated mouse (SAM). *Mech. Age. Dev.* 26:311–326.

Hijmans, W., Radl, J., Bottazzo, G. F., and Doniach, D. (1984) Autoantibodies in highly aged humans. *Mech. Age. Dev.* 26:83–89.

Hikita, T. (1962) Ecological and morphological studies of the genus *Oncorhynchus* (Salmonidae) with particular consideration on phylogeny. *Sci. Rep. Hokkaido Salmon Hatchery* 17:1–97.

Hildemann, W. H. (1957) Scale homotransplantation in goldfish (*Carassius auratus*). *Ann. N.Y. Acad. Sci.* 64:775–790.

———. (1978) Phylogenetic and immunogenetic aspects of aging. *Birth Defects* 14:97–107.

———. (1979) Immunocompetence and allogeneic polymorphism among invertebrates. *Transplantation* 27:1–3.

Hildemann, W. H., and Reddy, A. L. (1973) Phylogeny of immune responsiveness: Marine invertebrates. *Fed. Proc.* 32:2188–2194.

Hildemann, W. H., and Walford, R. L. (1963) Annual fishes—Promising species as biological control agents. *J. Trop. Med. Hygiene* 66:163–166.

Hill, A. V. (1950) The dimensions of animals and their muscular dynamics. *Sci. Prog.* 38:209–230.

Hill, C. C., and Emery, W. T. (1937) The biology of *Platygaster herrickii,* a parasite of the Hessian fly. *J. Agric. Res.* 55:199–213.

Hindar, K., and Nordland, J. (1989) A female Atlantic salmon, *Salmo salar L.,* maturing sexually in the parr stage. *J. Fish. Biol.* 35:461–463.

Hinds, J. W., and McNelly, N. A. (1978) Dispersion of cisternae of rough endoplasmic reticulum in aging CNS neurons: A linear trend. *Am. J. Anat.* 52:433–439.

Hines, A. H. (1979) The comparative reproduction ecology of three species of intertidal barnacles. In *Reproductive Ecology of Marine Invertebrates,* S. E. Stancyk (ed.), 213–234. Belle W. Baruch Library in Marine Science.

Hinton, S. (1962) Longevity of fishes in captivity as of September 1956. *Zoologica* 47:105–116.

Hiramoto, R. N., Ghanta, V. K., and Soong, S. J. (1987) Effect of thymic hormones on immunity and life span. In *Aging and the Immune Response: Cellular and Humoral Aspects,* E. A. Goidl (ed.), 177–198. New York: Marcel Dekker.

Hirano, A., and Iwata, M. (1979) Pathology of motor neurons with special reference to amyotrophic lateral sclerosis and related diseases. In *Amyotrophic Lateral Sclerosis,* T. Tsubaki and Y. Toyokura (eds.), 107–133. University Park.

Hiremath, L. S., and Rothstein, M. (1982) Regenerating liver in aged rats produces unaltered phosphoglycerate kinase. *J. Gerontol.* 37:680–683.

Hirokawa, K., and Hayashi, Y. (1980) Effect of adult thymectomy on immune potentials, endocrine organs, and tumor incidence in long-lived mice. *Adv. Exp. Med. Biol.* 129:243–247.

Hirokawa, K., and Makinodan, T. (1975) Thymic involution: Effect on T cell differentiation. *J. Immunol.* 114:1659–1664.

Hirokawa, K., and Utsuyama, M. (1984) The effect of sequential multiple grafting of syngeneic newborn thymus on the immune functions and life expectancy of aging mice. *Mech. Age. Dev.* 28:111–121.

Hirsch, G. P., Grunder, P., and Popp, R. A. (1976) Error analysis by amino acid analog incorporatioon in tissues of aging mice. *Interdiscipl. Top. Gerontol.* 10:1–10.

Hirsch, G. P., Popp, R. A., Francis, M. C., Bradshaw, B. S., and Bailiff, E. G. (1980) Species comparison of protein synthesis activity. *Adv. Pathobiol.* 7:142–159.

Hirsch, H. R., and Peretz, B. (1984) Survival and aging of a small laboratory population of a marine mollusc, *Aplysia californica. Mech. Age. Dev.* 27:43–62.

Hirst, S., and Maulik, S. (1926) On some arthropod remains from Rhynie chert (Old Red Sandstone). *Geol. Mag.* 63:69–71.

Ho, K. Y., Evans, W. S., Blizzard, R. M., Veldhuis, J. D., Merriam, G. R., Samojlik, E., Furlanetto, R., Rogol, A. D., Kaiser, D. L., and Thorner, M. O. (1987) Effects of sex and age on the 24-hour profile of growth hormone secretion in man: Importance of endogenous estradiol concentrations. *J. Clin. Endocrinol. Metab.* 64:51–58.

Hoar, W. S. (1976) Smolt transformation: Evolution, behaviour, and physiology. *J. Fish. Res. Bd. Can.* 33:1234–1252.

Hochachka, P. W., and Guppy, M. (1987) *Metabolic Arrest and the Control of Biological Time.* Cambridge, MA: Harvard Univ. Press.

Hochstein, P., Sevanian, A., and Davies, K. J. A. (1986) The stabilization of ascorbic acid by uric acid. In *Purine Pyrimidine Metabolism in Man,* W. F. Myhan, L. F. Thompson, and R. W. Watts (eds.). *Adv. Exp. Med. Biol.* 195:325–328.

Hodge, C. F. (1894) Changes in ganglion cells from birth to senile death: Observations on man and honey-bee. *J. Physiol.* (London) 17:129–134.

Hodgson, J. L., and Buskirk, E. R. (1977) Physical fitness and age, with emphasis on cardiovascular function in the elderly. *J. Am. Geriatr. Soc.* 25:385–392.

Hoehn, H., Bryant, E. M., Au, K., Norwood, T. H., Boman, H., Martin, G. M. (1975) Variegated translocation mosaicism in human skin fibroblast cultures. *Cytogenet. Cell. Genet.* 15:282–298.

Hoffman, G. E., and Finch, C. E. (1986) LHRH neurons in female C57BL/6J mice: No loss up to middle-age. *Neurobiol. Aging* 7:45–48.

Hofman, M. A. (1982) Encephalization in mammals in relation to the size of the cerebral cortex. *Brain Behav. Evol.* 20:84–96.

———. (1983) Energy metabolism, brain size, and longevity in mammals. *Q. Rev. Biol.* 58:495–512.

———. (1984) On the presumed coevolution of brain size and longevity in hominids. *J. Human Evol.* 13:371–376.

Hogg, N. A. S., and Wijdenes, J. S. (1979) A study of gonadal organogenesis, and the factors influencing regeneration following surgical castration in *Deroceras reticulatum* (Pulmonata: Limacidae). *Cell Tiss. Res.* 198:295–307.

Holbrook, N. J., Chopra, R. K., McCoy, M. T., Nagel, J. E., Adler, W. H., and Schneider, E. L. (1989) Expression of interleukin 2 and the interleukin 2 receptor in aging rats. *Cell. Immunol.* 120:1–9.

Holden, M. J. (1977) Elasmobranchs. In *Fish Population Dynamics,* J. A. Gulland (ed.), 187–215. New York: Wiley and Sons.

Holehan, A. M., and Merry, B. J. (1985) Modification of the oestrous cycle hormonal profile by dietary restriction. *Mech. Age. Dev.* 32:63–76.

Holinka, C. F., Tseng, Y.-C., and Finch, C. E.

(1978) Prolonged gestation, elevated preparturitional plasma progesterone, and reproductive aging in C57BL/6J mice. *Biol. Reprod.* 19:807–816.

———. (1979a) Impaired preparturitional rise of plasma estradiol in aging C67BL/6J mice. *Biol. Reprod.* 21:1009–1013.

———. (1979b) Reproductive aging in C57BL/6J mice: Plasma progesterone, viable embryos, and resorption frequency throughout pregnancy. *Biol. Reprod.* 20:1201–1211.

Holland, H. D. (1984) *The Chemical Evolution of the Atmosphere and Oceans.* Princeton, NJ: Princeton Univ. Press.

Holland, J. J., Kohne, D., and Doyle, M. V. (1973) Analysis of virus replication in aging human fibroblasts. *Nature* 245:316–318.

Hollander, C. F. (1979) Proper use of laboratory rats and mice in gerontological research. In *Physiology and Cell Biology of Aging,* vol. 8, *Aging,* A. Cherkin, C. E. Finch, N. Kharasch, T. Makinodan, L. Scott, and L. Strehler (eds.), 223–227. New York: Raven Press.

Hollander, C. F., Solleveld, H. A., Zurcher, C., Noteboom, A. L., and Van Zwieten, M. J. (1984) Biological and clinical consequences of longitudinal studies in rodents: Their possibilities and limitations. An overview. *Mech. Age. Dev.* 28:249–260.

Holliday, M. A., Potter, D., Jarrah, A., and Baerg, S. (1967) The relation of metabolic rate to body weight and organ size. *Pediatr. Res.* 1:185–195.

Holliday, R. (1969) Errors in protein synthesis and clonal senescence in fungi. *Nature* 221:1224–1228.

———. (1989) Food, reproduction, and longevity: Is the extended lifespan of calorie-restricted animals an evolutionary adaptation? *Bioessays* 10(4): 125–127.

Holliday, R., Porterfield, J. S., and Gibbs, D. D. (1974) Premature aging and the occurrence of altered enzyme in Werner's syndrome fibroblasts. *Nature* 248:762–763.

Holliday, R., and Tarrant, G. M. (1972) Altered enzymes in aging human fibroblasts. *Nature* 238:26–30.

Hollingsworth, J. W., Hashizume, A., and Jablon, S. (1965) Correlations between tests of aging in Hiroshima subjects-an attempt to de-fine "physiologic age." *Yale J. Biol. Med.* 38:11–26.

Hollingsworth, J. W., Ishii, G., and Conard, R. A. (1961) Skin aging and hair graying in Hiroshima. *Geriatrics* 16:27–36.

Hollingsworth, M. J. (1969a) The effect of fluctuating environmental temperatures on the length of life of adult *Drosophila. Exp. Gerontol.* 4:159–167.

———. (1969b) Temperature and length of life in *Drosophila. Exp. Gerontol.* 4:49–55.

Holloszy, J. O., and Smith, E. K. (1986) Longevity of cold-exposed rats: A reevaluation of the "rate-of-living theory." *J. Appl. Physiol.* 61:1656–1660.

Holloszy, J. O., Smith, E. K., Vining, M., and Adam, S. (1985) Effect of voluntary exercise on longevity of rats. *J. Appl. Physiol.* 59:826–831.

Holloway, B. A. (1976) Pollen-feeding in hover-flies (Diptera: Syrphidae). *New Zealand J. Zool.* 3:339–350.

Holt, B. F. (1972) Effect of arrival time on recruitment, mortality, and reproduction in successional plant populations. *Ecology* 53:668–673.

Holtzer, H. (1978) Cell lineages, stem cells, and the "quantal" cell cycle concept. In *Stem Cells and Tissue Homeostasis,* B. I. Lord, C. S. Potten, and R. J. Cole (eds.), 1–27. Cambridge: Cambridge Univ. Press.

Honacki, J. H., Kinman, K. E., and Koeppl, J. W. (eds.) (1982) *Mammal Species of the World. A Taxonomic and Geographic Reference.* Lawrence, Kans.: Allen Press and Association of Systematics Collections.

Honma, A., Matsumura, A., Yasuhira, K., and Takeda, T. (1984) Cataract and other ophthalmic lesions in senescence accelerated mouse (SAM). Morphology and incidence of senescence associated ophthalmic changes in mice. *Exp. Eye Res.* 38:105–114.

Honma, Y. (1959) Studies on the endocrine glands of a salmonoid fish, ayu. 1. Seasonal variation in the endocrines of the annual fish. *J. Fac. Sci.,* ser. 2, 2:225–233..

———. (1960) Studies on the endocrine glands of the salmonoid fish, the ayu, *Plecoglossus altivelis* Temminck and Schlegel. 3. Changes in the adrenal cortical tissue during the lifespan of the fish. *Annot. Zool. Jpn.* 33:234–240.

Honma, Y., and Tamura, E. (1963) Studies on the endocrine glands of the salmonoid fish, the ayu, *Plecoglossus altivelis* Temminck and Schlegel. 5. Seasonal changes in the endocrines of the land-stocked form, the koayu. *Zool. N.Y.* 48:25–32.

Hooghwinkel, G. J. M., Blaauboer, A. J., Novak, L., and Trippelvitz, L. A. (1986) On the composition of autofluorescent accumulation products: Ceroid and lipofuscin. In *Enzymes of Lipid Metabolism,* vol. 2, L. Freysz and H. Dreyfus (eds.), 827–831. NATO Series A.

Hook, E. B. (1981) Rates of chromosome abnormalities at different maternal ages. *Obstet. Gynecol.* 58:282–285.

———. (1986) Paternal age and effects on chromosomal and specific locus mutations and on other genetic outcomes in offspring. In *Aging, Reproduction, and the Climacteric,* L. Mastroianni and C. A. Paulsen (eds.), 117–146. New York: Plenum.

Hooper, A. C. B. (1981) Length, diameter, and number of ageing skeletal muscle fibres. *Gerontology* 27:121–126.

Hope-Simpson, R. E. (1965) Nature of herpes zoster—A long-term study and a new hypothesis. *Proc. Roy. Soc. Med.* 58:9–20.

Hopkins, C. L. (1971) Life history of *Galaxias divergens* (Salmonoidea: Galaxiidae). *New Zealand J. Mar. Freshwater Res.* 5:41–57.

———. (1979a) Age-growth characteristics of *Galaxias fasciatus* Gray (Salmoniformes: Galaxiidae). *New Zealand J. Mar. Freshwater Res.* 13:39–46.

———. (1979b) Reproduction in *Galaxias fasciatus* Gray (Salmoniformes: Galaxiidae). *New Zealand J. Mar. Freshwater Res.* 13:225–230.

Hopkins, H. S. (1930) Age differences and the respiration in muscle tissues of mollusks. *J. Exp. Zool.* 56:209–239.

Hopson, J. A. (1973) Endothermy, small size, and the origin of mammalian reproduction. *Am. Nat.* 107:446–452.

Horbach, G. J. M. J., Princen, M., van der Kroef, M., van Bezooijen, C. F. A., and Yap, S. H. (1984) Changes in the sequence content of albumin mRNA and its translational activity in the rat liver with age. *Biochim. Biophys. Acta* 783:60–66.

Horiuchi, S., Shiga, M., Araki, N., Takata, K., Saitoh, M., and Morino, Y. (1988) Evidence against *in vivo* presence of 2-(2-furoyl)–4(5)-(2-furanyl)–1*H*-imidazole, a major fluorescent advanced end product generated by nonenzymatic glycosylation. *J. Biol. Chem.* 263:18821–18826.

Horn, H. S. (1978) Optimal tactics of reproduction and life-history. In *Behavioural Ecology: An Evolutionary Approach,* J. R. Krebs and N. B. Davies (eds.), 411–429. Oxford: Blackwell.

Horn, J. L. (1982) The theory of fluid and crystallized intelligence in relation to concepts of cognitive psychology and aging in adulthood. In *Aging and Cognitive Processes,* F. I. Craik and S. Trehub (eds.), 237–278. New York: Plenum.

Horn, P. L., Turker, M. S., Ogburn, C. E., Disteche, C. M., and Martin, G. M. (1984) A cloning assay for 6-thioguanine resistance provide evidence against certain somatic mutational theories of aging. *J. Cell Physiol.* 121:309–315.

Hornsby, P. J., Cheng, C. Y., Ryan, R. F., and Yang, L. (1990) Stochastic changes in gene expression in adrenal cell senescence. In *Molecular Biology of Aging,* vol. 123, C. E. Finch and T. E. Johnson (eds.), 249–263. UCLA Symposia on Molecular and Cellular Biology. New York: Wiley-Liss.

Hornsby, P. J., and Gill, G. N. (1978) Characterization of adult bovine adrenocortical cells throughout their life span in tissue culture. *Endocrinology* 102:926–936.

Hornsby, P. J., Ryan, R. F., and Cheng, C. Y. (1989) Replicative senescence and differentiated gene expression in cultured adrenocortical cells. *Exp. Gerontol.* 24:539–558.

Horsman, A., Jones, M., Francis, R., and Nordia, C. (1983) The effect of estrogen dose on post-menopausal bone loss. *N. Engl. J. Med.* 309:1405–1407.

Horst, R. L., DeLuca, H. F., and Jorgenson, J. A. (1978) The effect of age on calcium absorption and accumulation of 1,25-dihydroxyvitamin D in intestinal mucosa of rats. *Metab. Bone Dis. Rel. Res.* 1:29–33.

Hosking, G. P., and Kershaw, D. J. (1985) Red beech death in the Maruia valley, South Island, New Zealand. *New Zealand J. Bot.* 23:201–211.

Hosokawa, M., Kasai, R., Higuchi, K., Takeshita, S., Shimizu, K., Hamamoto, M., Takeshita, S., Higuchi, K., Shimizu, K., Irino, M., Toda, K., Hosokawa, T., Hosono, M., Hanada, K., Aoike, A., Kawai, K., and Takeda, T. (1987) Immune responses in newly developed short-lived SAM mice. 2. Selectively impaired T helper cell activity in *in vitro* antibody response. *Immunology* 62:425–429.

Hosokawa, M., Takeshita, S., Higuchi, K., Shimizu, K., Irino, M., Toda, K., Honma, A., Matsumura, A., Yashita, K., and Takeda, T. (1984) Cataract and other ophthalmic lesions in senescence accelerated mouse (SAM). Morphology and incidence of senescence associated ophthalmic changes in mice. *Exp. Eye Res.* 38:105–114.

Hotta, Y., and Benzer, S. (1972) Mapping of behaviour in *Drosophila* mosaics. *Nature* 240:527–535.

Houston, W. W. K. (1981) The life cycles and age of *Carabus glabratus* Paykull and *C. problematicus* Herbst (Col.: Carabidae) on moorland in northern England. *Ecol. Entomol.* 6:263–271.

Hovore, F. T. (1979) Rain beetles. Small things wet and wonderful. *Terra* 17:10–14.

Howard, R. D. (1978) The evolution of mating strategies in bullfrogs, *Rana catesbeiana*. *Evolution* 32:850–871.

———. (1983) Sexual selection and variation in reproductive success in a long-lived organism. *Am. Nat.* 122:301–325.

Howell, F. C. (1979) Hominidae. In *Evolution of African Mammals*, V. J. Maglio and H. S. B. Cooke (eds.), 154–248. Cambridge, MA: Harvard Univ. Press.

Howell, T. H. (1963) Multiple pathology in nonegenarians. *Geriatrics* 18:899–902.

Howie, J. B., and Simpson, L. O. (1976) Autoimmune hemolytic disease in NZB mice. *Ser. Haematol.* 7:386.

Howland, B. E., and Ibrahim, E. A. (1973) Increased LH-suppressing effect of oestrogen in ovariectomized rats as a result of underfeeding. *J. Reprod. Fertil.* 35:545–548.

Hrdy, S. B. (1981) *The Woman That Never Evolved*. Cambridge, MA: Harvard Univ. Press.

Hrubec, Z., and Neel, J. V. (1981) Familial factors in early deaths: Twins followed 30 years to ages 51–61 in 1978. *Human Genet.* 59:39–46.

———. (1982) Contribution of familial factors to the occurrence of cancer before old age in twin veterans. *Am. J. Human Genet.* 34:658–671.

Hsiao, K., Baker, H. F., Crow, T. J., Poulter, M., Owen, F., Terwilliger, J. D., Westaway, D., Ott, J., and Prusiner, S. B. (1989) Linkage of a prion protein missense variant to Gerstmann-Straussler syndrome. *Nature* 338:342–345.

Hsie, A. W., Recio, L., Katz, D. S., Lee, C. Q., and Schneley, R. L. (1986) Evidence for reactive oxygen species inducing mutations in mammalian cells. *Proc. Natl. Acad. Sci.* 83:9616–9620.

Huang, H. H., Marshall, S., and Meites, J. (1976) Capacity of old versus young female rats to secrete LH, FSH, and prolactin. *Biol. Reprod.* 14:538–543.

Hubbell, S. P. (1980) Seed predation and the coexistence of tree species in tropical forests. *Oikos* 35:214–229.

Hubbs, C. L. (1921) An ecological study of the life-history of the fresh-water fish *Labidesthes sicculus*. *Ecology* 2:262–276.

———. (1971) *Lampetra (Entosphenus) lethophaga*, new species. The nonparasitic derivative of the Pacific lamprey. *San Diego Soc. Nat. Hist. Trans.* 16:125–164.

Hubbs, C. L., and Trautman, M. B. (1937) A revision of the lamprey genus *Ichthyomyzon*. *Misc. Publs. Mus. Zool. Univ. Mich.*, no. 35:1–109.

Huber, M. H. R., Bronson, F. H., and Desjardins, C. (1980) Sexual activity of aged male mice: Correlation with level of arousal, physical endurance, pathological status, and ejaculatory capacity. *Biol. Reprod.* 23:305–316.

Hudgins, C. C., Steinberg, R. T., Klinman, D. M., Patton Reeves, M. J., and Steinberg, A.

D. (1985) Studies of consomic mice bearing the Y chromosome of the bxsb mouse. *J. Immunol.* 134:3849–3854.

Huebner, R. J., Kelloff, G. J., Sarma, P. S., Lane, W. T., Gilden, R. T., Oroszlan, S., Meier, H., Myers, D. D., and Peters, R. L. (1970) Group specific antigen expression during embryogenesis of the genome of C-type RNA tumor virus: Implication for ontogenesis and oncogenesis. *Proc. Natl. Acad. Sci.* 67:366–376.

Huebner, R. J., and Todaro, G. J. (1969) Oncogenes of RNA tumor viruses as determinants of cancer. *Proc. Natl. Acad. Sci.* 64:1087–1094.

Huggins, C. (1943) Endocrine control of prostatic cancer. *Science* 97:541–544.

Huggins, C., and Clark, P. J. (1940) Quantitative studies of prostatic secretion. 2. The effect of castration and of estrogen injection on the normal and on the hyperplastic prostate glands of dogs. *J. Exp. Med.* 72:747–761.

Hughes, R. N. (1989) *Functional Biology of Clonal Animals.* Chapman and Hall. [not read]

Hughes, T. P. (1990) Recruitment limitation, mortality, and population regulation in open systems: A case study. *Ecology* 71:12–20.

Hughes, T. P., and Cancino, J. M. (1986) An ecological overview of cloning in Metazoa. In *The Growth and Form of Modular Organisms,* J. L. Harper, R. Rosen, and J. White (eds.), 153–186. *Phil. Trans. Roy. Soc.,* (London), ser. B, 313:1–250.

Hughes, T. P., and Connell, J. H. (1987) Population dynamics based on size or age? A reef-coral analysis. *Am. Nat.* 129:818–829.

Hughes, T. P., and Jackson, J. B. (1980) Do corals lie about their age? Some demographic consequences of partial mortality, fission, and fusion. *Science* 209:713–715.

———. (1985) Population dynamics and life histories of foliaceous corals. *Ecol. Monographs* 55:141–166.

Hugin, F., and Verzar, F. (1956) Untersuchungen über die Arbeitshypertropie des Herzens bei jungen und alten Ratten. *Pflug. Arch. ges. Physiol.* 262:181–186.

Hunsaker, D. (1977) Ecology of New World marsupials. In *The Biology of Marsupials,* D.

Hunsaker (ed.), 95–156. New York: Academic Press.

Hunter, W. R. (1961) Life cycles of four freshwater snails in limited populations in Loch Lomond, with a discussion of infraspecific variation. *Proc. Zool. Soc. Lond.* 137:135–171.

Hurd, E. R., Johnston, J. M., Okita, J. R., MacDonald, P. C., Ziff, M., and Gilliam, J. N. (1981) Prevention of glomerulonephritis and prolonged survival in New Zealand black/New Zealand white F_1 hybrid mice fed an essential fatty acid-deficient diet. *J. Clin. Invest.* 67:476–485.

Husby, G., and Sletten, K. (1986) Editorial review: Chemical and clinical classification of amyloidosis 1985. *Scand. J. Immunol.* 23:253–265.

Huseby, R. A., Ball, Z. B., and Visscher, M. B. (1945) Further observations on the influence of simple caloric restricyion on mammary cancer incidence and related phenomena in C3H mice. *Canc. Res.* 5:40–46.

Hutchings, J. A., and Morris, D. W. (1985) The influence of phylogeny, size, and behavior on patterns of covariation in salmonid life histories. *Oikos* 45:118–124.

Huxley, J. S. (1921) Studies in dedifferentiation. 2. Dedifferentiation and resorption in *Perophora. Q. J. Microscop. Sci.* 65:643–697.

———. (1932) *Problems of Relative Growth.* London: Methuen.

Huxley, J. S., and de Beer, G. R. (1923) Studies in dedifferentiation. 4. Resorption and differential inhibition in *Obelia* and *Campanularia. Q. J. Microscop. Sci.* 67:473–495.

Hyman, B. T., Damasio, A. R., Van Hoesen, G. W., and Barnes, C. L. (1984) Alzheimer's disease: Cell-specific pathology isolates the hippocampal formation. *Science* 225:1168–1170.

Hyman, B. T., Van Hoesen, G. W., Kromer, L. J., and Damasio, A. R. (1986) Perforant pathway changes and the memory impairment of Alzheimer's disease. *Ann. Neurol.* 20:472–481.

Hyman, L. H. (1940) *The Invertebrates: Protozoa through Ctenophora.* New York: McGraw-Hill.

————. (1951). *The Invertebrates.* Vol. 3, *Acanthocephala, Aschelminthes, and Ento-procta. The Pseudocoelomate Bilateria.* New York: McGraw-Hill.

Iams, S. G., and Wexler, B. C. (1977) Inhibition of spontaneously developing arteriosclerosis in female breeder rats by adrenalectomy. *Atherosclerosis* 27:311–323

Iberall, A. S. (1972) Blood flow and oxygen uptake in mammals. *Ann. Biomed. Eng.* 1:1–8.

Idler, D. R., Ronald, A. P., and Schmidt, P. J. (1959) Biochemical studies on sockeye salmon during spawning migration. 7. Steroid hormones in plasma. *Can. J. Biochem. Physiol.* 37:1227–1238.

Idler, D. R., and Sangalang, G. B. (1970) Steroids of a chondrostean: In vitro steroidogenesis in yellow bodies isolated from kidneys and along the posterior cardinal veins of the American Atlantic sturgeon, *Acipenser oxyrhynchus* Mitchell. *J. Endocrinol.* 48:627–637.

Imayama, S., and Braverman, I. M. (1989) A hypothetical explanation for the aging of skin. *Am. J. Pathol.* 134:1019–1025.

Imura, H., Nakao, Y., Kuzuya, H., Okamoto, M., Okamoto, M., and Yamada, K. (1985) Clinical, endocrine, and metabolic aspects of the Werner syndrome compared with those of normal aging. In *Werner's Syndrome and Human Aging,* D. Salk, Y. Fujiwara, and G. M. Martin (eds.). *Adv. Exp. Biol. Med.* 190:171–185.

Inagaki, H., and Berreur-Bonnenfant, J. (1970) Croissance et sénescence chez un crustace isopode *Ligia oceanica* (L.) *Compt. Rend. Acad. Sci. Paris* 271:207–210.

Inamizu, T., Kinohara, N., Chang, M. P., and Makinodan, T. (1986) Frequency of 6–thioguanine-resistant T cells is inversely related to the declining T-cell activities in aging mice. *Proc. Natl. Acad. Sci.* 83:2488–2491.

Ingle, D. W., and Baker, B. L. (1957) Histology and regenerative capacity of liver following multiple partial hepatectomies. *Proc. Soc. Exp. Biol. Med.* 95:813–815.

Ingle, L., Wood, T. R., and Banta, A. M. (1937) A study of the longevity, growth, reproduction and heart rate in *Daphnia longis-*

pina as influenced by limitations in quantity of food. *J. Exp. Zool.* 76:325.

Ingram, D. K., and Reynolds, M. A. (1982) The relationship of genotype, sex, body weight, and growth parameters to lifespan in inbred and hybrid mice. *Mech. Age. Dev.* 20:253–266.

————. (1987a) Effects of protein, dietary restriction, and exercise on survival in adult rats: A re-analysis of McCay, Maynard, Sperling, and Osgood (1941). *Exp. Aging Res.* 9:41–42.

————. (1987b) The relationship of body weight to longevity within laboratory rodent species. In *Evolution of Longevity of Animals. A Comparative Approach,* A. D. Woodhead and K. H. Thompson (eds.), 247–282. New York: Plenum.

Ingram, D. K., Spangler, E. L., and Vincent, G. P. (1983) Behavioral comparison of aged virgin and retired breeder mice. *Exp. Aging Res.* 9:111–113.

Ingram, D. K., Weindruch, R., Spangler, E. L., Freeman, J. R., and Walford, R. L. (1987) Dietary restriction benefits learning and motor performance of aged mice. *J. Gerontol.* 42:78–81.

Ingvar, M. C., Maeder, P., Sokoloff, L., and Smith, C. B. (1985) Effects of ageing on local rates of cerebral protein synthesis in Sprague-Dawley rats. *Brain* 108:155–170.

Inkeles, S., and Eisenberg, D. (1981) Hyperlipidemia and coronary atherosclerosis: A review. *Medicine* 60:110–123.

Inns, R. W. (1976) Some seasonal changes in *Antechinus flavipes* (Marsupialia, Dasyuridae). *Austral. J. Zool.* 24:523–531.

Ishigami, A., and Goto, S. (1988) Inactivation kinetics of horseradish peroxidase microinjected into hepatocytes from mice of various ages. *Mech. Age. Dev.* 46:125–133.

Ishikawa, H. (1982a) DNA, RNA, and protein synthesis in the isolated symbionts from the pea aphid, *Acyrthosiphon pisum. Insect Biochem.* 12:605–612.

————. (1982b) Isolation of the intracellular symbionts and partial characterizations of their RNA species of the elder aphid, *Acyrthosiphon magnoliae. Comp. Biochem. Physiol.* 72B:239–247.

————. (1984a) Age-dependent regulations of

protein synthesis in an aphid endosymbiont by the host insect. *Insect Biochem.* 14:427–433.

———. (1984b) Alterations with age of symbiosis of gene expression in aphid endosymbionts. *BioSystems* 17:127–134.

———. (1984c) Characterization of the protein species synthesized *in vivo* and *in vitro* by an aphid endosymbiont. *Insect Biochem.* 14:417–425.

———. (1984d) Molecular aspects of intracellular symbiosis in the aphid mycetocyte. *Zool. Sci.* 1:503–522.

Ishikawa, H., and Yamaji, M. (1985a) Protein synthesis by intracellular symbionts in two closely interrelated aphid species. *BioSystems* 17:327–335.

———. (1985b) Symbionin, an aphid endosymbiont-specific protein. 1. Production of insects deficient in symbiont. *Insect Biochem.* 15:155–163.

Ishikawa, T., Kuwabara, N., and Takayama, S. (1976) Spontaneous ovarian tumors in domestic carp (*Cyprinus carpio*): Light and electron microscopy. *J. Natl. Canc. Inst.* 57:579–584.

Ishikawa, T., Nakajima, H., Kodama, K., and Takayama, S. (1983) DNA turnover: Long-term labeling study in ganglion cells. *Trans. Soc. Pathol. Jpn.* 72:66.

Ishikawa, T., Sakurai, J., and Takayama, S. (1984) *In vivo* studies on DNA repair and turnover with age. In *Molecular Biology of Aging,* Basic Life Sciences, vol. 35, A. D. Woodhead, and A. D. Blackett (eds.), 297–313. New York: Plenum.

Ishikawa, T., Takayama, S., and Kitagawa, T. (1978) Autoradiographic demonstration of DNA repair synthesis in ganglion cells of aquarium fish at various age in vivo. *Virchow's Arch. Cell Pathol.* 28:235–242.

Ivanyi, P. (1978) Some aspects of the H–2 system, the major histocompatibility system in the mouse. *Proc. Roy. Soc. Lond.,* ser. B, 202:117–158.

Ivell, R. (1987) Vasopressinergic and oxytocinergic cells: Models in neuropeptide gene expression. In *Neuropeptides and Their Peptidases,* A. J. Turner (ed.), 31–64. Chichester, U.K.: Ellis-Horwood.

Ivell, R., and Richter, D. (1984) Structure and

comparison of the oxytocin and vasopressin genes from rat. *Proc. Natl. Acad. Sci.* 81:2006–2010.

Ivy, G. O., Schottler, F., Wenzel, J., Baudry, M., and Lynch, G. (1984) Inhibitors of lysosomal enzymes: Accumulation of lipofuscinlike dense bodies in the brain. *Science* 226:985–987.

Iwamoto, R. N., Alexander, B. A., and Hershberger, W. K. (1983) *Salmonid Reproduction.* Vol. 8. Seattle: School of Fisheries, Univ. of Washington.

Iwasaki, K., Gleiser, C. A., Masoro, E. J., McMahan, C. A., Seo, E. J., and Yu, B. P. (1988a) The influence of dietary protein source on longevity and age-related disease processes of Fischer rats. *J. Gerontol.* 43:B5–B12.

———. (1988b) Influence of the restriction of individual dietary components on longevity and age-related disease of Fischer rats: The fat component and the mineral component. *J. Gerontol.* 43:B13–B21.

Izui, S., McConahey, P. J., Theofilopoulos, A. N., and Dixon, F. J. (1979) Association of circulating retroviral gp70–anti-gp70 immune complexes with murine systemic lupus erythematosus. *J. Exp. Med.* 149:1099–1116.

Jaberaboansari, A., Fletcher, C., Wallen, C. A., and Wheeler, K. T. (1989) Organization of DNA in cerebellar neurons of ageing unirradiated and irradiated rats. *Mech. Age. Dev.* 50:257–276.

Jackson, D. J. (1961) Observations on the biology of *Caraphractus cinctus* Walker (Hymenoptera: Mymaridae), a parasitoid of the eggs of Dytiscidae (Coleoptera). *Parasitology* 51:269–294.

Jackson, J. B. (1985) Distribution and ecology of clonal and aclonal benthic invertebrates. In *Population Biology and Evolution of Clonal Organisms,* J. B. Jackson, L. W. Buss, and R. E. Cook (eds.), 297–355. New Haven: Yale Univ. Press.

Jackson, J. B., Buss, L. W., and Cook, R. E. (eds.) (1985) *Population Biology and Evolution of Clonal Organisms.* New Haven: Yale Univ. Press.

Jackson, J. B., and Coates, A. G. (1986) Life cycles and evolution of clonal (modular) animals. *Phil. Trans. Roy. Soc.* (London), ser. B, 313:7–22.

Jacobo-Molina, A., Villa-Garcia, M., Chen, H. C., and Yang, D. C. (1988) Proteolytic signal sequences (PEST) in the mammalian aminoacyl-tRNA synthetase complex. *FEBS Lett.* 232:65–68.

Jacobs, P. A., Brunton, M., Court Brown, W. M., and Doll, R. (1963) Change of human chromosome count distribution with age: Evidence for a sex difference. *Nature* 197:1080–1081.

Jacobs, P. A., and Hassold, T. J. (1980) The origin of chromosome abnormalities in spontaneous abortion. In *Human Embryonic and Fetal Death,* I. H. Porter and B. B. Hook (eds.), 289–298. New York: Academic Press.

Jacobs, S., and Ostfeld, A. (1977) An epidemiological review of the mortality of bereavement. *Psychosom. Med.* 39:344–357.

Jacobs, W. P. (1979) *Plant Hormones and Plant Development.* London and New York: Cambridge Univ. Press.

Jacobson, E. R. (1980) Reptile neoplasms. In *Reproductive Biology and Diseases of Captive Reptiles,* J. B. Murphy and J. T. Collins (eds.), 255–260. SSAR Contributions to Herpetology 1. Society for the Study of Amphibians and Reptiles.

Jacobson, M. (1970) *Developmental Neurobiology.* New York: Holt, Rinehart and Winston.

———. (1985a) Clonal analysis and cell lineages of the vertebrate central nervous system. *Ann. Rev. Neurosci.* 8:71–102.

———. (1985b) Clonal analysis of the vertebrae CNS. *Trends in Neuroscience* 8:151–155.

Jacobson, M., and Moody, S. A. (1984) Quantitative lineage analysis of the frog's nervous system. Lineages of Rohon-Beard neurons and primary motoneurons. *J. Neurosci.* 4:1361–1369.

Jacobus, S., and Gershon, D. (1980) Age-related changes in inducible mouse liver enzymes: Ornithine decarboxylase and tyrosine aminotransferase. *Mech. Age. Dev.* 12:311–322.

James, H. C. (1928) On the life-histories and economic status of certain cynipid parasites of dipterous larvae, with descriptions of some

new larval forms. *Ann. Appl. Biol.* 15:287–316.

Jamet-Vierny, C., Begel, A.-M., and Belcour, L. (1980) Senescence in *Podospora anserina:* Amplification of a mitochondrial DNA sequence. *Cell* 21:189–194.

Janse, C., Beek, A., Van Oorschot, I., and van der Roest, M. (1986) Recovery of damage in a molluscan nervous system is impaired with age. *Mech. Age. Dev.* 35:179–183.

Janse, C., and Joosse, J. (1989) Aging in mollusca nervous and neuroendocrine systems. In *Development, Maturation, and Senescence of Neuroendocrine Systems: A Comparative Approach,* M. P. Schreibman and C. G. Scanes (eds.), 43–61. San Diego: Academic Press.

Janse, C., Slob, W., Popelier, C. M., and Vogelaar, J. W. (1988) Survival characteristics of the mollusc *Lymnaea stagnalis* under constant culture conditions: Effects of aging and disease. *Mech. Age. Dev.* 42:263–274.

Janse, C., ter Maat, A., and Pieneman, A. W. (1990) Molluscan ovulation hormone containing neurons and age-related reproductive decline. *Mech. Age. Dev.* 11:457–463.

Janse, C., van der Roest, M., Bedaux, J. J., and Slob, W. (1986) Age-related decrease in electrical coupling of two identified neurons in the mollusc *Lymnaea stagnalis. Brain Res.* 376:208–212.

Janse, C., Wildering, W. C., and Popelier, C. M. (1989) Age-related changes in female reproductive activity and growth in the mollusc *Lymnaea stagnalis. J. Gerontol.* 44:B148–B155.

Jansky, L. (1961) Total cytochrome oxidase activity and its relation to basal and maximal metabolism. *Nature* 189:921–922.

Janss, D. H., and Ben, T. L. (1978) Age-related modification of 7,12–dimethylbenz-(a)anthracene binding to rat mammary gland DNA. *J. Natl. Canc. Inst.* 60:173–177.

Janus, E. D., Nicoll, A. M., Turner, P. R., Magill, P., and Lewis, B. (1980) Kinetic bases of the primary hyperlipidemia: Studies of apolipoprotein B turnover in genetically defined subjects. *Eur. J. Clin. Invest.* 10:161–172.

Janvier, P., and Lund, R. (1983) *Hardistilea montanensis:* New genera petromyzon from the Lower Carboniferous. *J. Vert. Paleontol.* 2:407–413.

Janzen, D. H. (1976) Why bamboos wait so long to flower. *Ann. Rev. Ecol. Sys.* 7:347–391.

Jarvik, L. F., Blum, J. E., and Varma, A. D. (1972) Genetic components and intellectual functioning during senescence: A 20-year study of aging twins. *Behav. Genet.* 2:159–171.

Jarvik, L. F., Falek, A., Kallman, F. J., and Lorge, I. (1960) Survival trends in a senescent twin population. *Am. J. Human Genet.* 12:170–179.

Jarvik, L. F., Ruth, V., and Matsuyama, S. S. (1980) Organic brain syndrome and aging. *Arch. Gen. Psychiatry* 37:280–286.

Jarvik, L. F., Yen, F. S., Fu, T. K., and Matsuyama, S. S. (1976) Chromosomes in old age: A six year longtitudinal study. *Human Genet.* 33:17–22.

Jarvis, J. U. (1981) Eusociality in a mammal: Cooperative breeding in naked mole-rat colonies. *Science* 212:571–573.

Jaycox, E. R., Skowronek, W., and Guynn, G. (1974) Behavioral changes in worker honey bees induced by injections of a juvenile hormone mimic. *Ann. Entomol. Soc. Am.* 67:529–534.

Jazwinski, S. M., Chen, J. B., and Jeansonne, N. E. (1990) Replication control and differential gene expression in aging yeast. In *The Molecular Biology of Aging,* C. E. Finch and T. E. Johnson (eds.), 189–204. UCLA Symposia on Molecular and Cellular Biology, n.s., vol. 123. New York: Liss.

Jazwinski, S. M., Egilmez, N. K., and Chen, J. B. (1989) Replication control and cellular lifespan. *Exp. Gerontol.* 24:423–436.

Jefferies, R. P. (1986) *The Ancestry of the Vertebrates.* London: British Museum (Natural History).

Jelinek, J., Kappen, A., Schonbaum, E., and Lomax, P. (1984) A primate model of human postmenopausal flushes. *J. Clin. Endocrinol. Metab.* 59:1224–1228.

Jelkmann, W., and Bauer, C. (1980) 2,3–DPG levels in relation to red cell enzyme activities in rat fetuses and hypoxic newborns. *Pflugers Arch.* 389:61–69.

Jellyman, D. J. (1987) Review of the marine life history of Australasia temperate species of *Anguilla. Am. Fish. Soc. Symp.* 1:276–285.

Jenkins, F. A., Jr. (1984) A survey of mammalian origins. In *Mammals. Notes for a Short Course,* D. Gingerich and C. E. Badgeley (eds.), 32–47. University of Tennessee, Studies in Geology 8.

Jenkins, F. A., Jr., and Crompton, A. W. (1979) Triconodonta. In *Mesozoic Mammals. The First Two-Thirds of Mammalian History,* J. A. Lillegraven, Z. Kielan-Jaworowska, and W. A. Clemens (eds.), 74–90. Berkeley: Univ. of California Press.

Jenkins, N. A., Copeland, N. G., Taylor, B. A., and Lee, B. K. (1982) The organization, distribution, and stability of endogenous ecotropic murine leukemia virus DNA sequences in chromosomes of *Mus musculus. J. Virol.* 43:26–36.

Jenne, D. E., and Tschopp, J. (1989) Molecular structure and functional characterization of a human complement cytolysis inhibitor found in blood and seminal plasma: Identity to sulfated glycoprotein 2, a constituent of rat testis fluid. *Proc. Natl. Acad. Sci.* 86:7123–7127.

Jennings, H. S. (1944a) Paramecium bursaria: Life history. 2. Age and death of clones in relation to the results of conjugation. *J. Exp. Zool.* 96:17–52.

———. (1944b) 3. Repeated conjugation in the same stock at different ages with and without inbreeding in relation to mortality at conjugation. *J. Exp. Zool.* 96:243–273.

———. (1944c) 4. Relation of inbreeding to mortality of exconjugant clones. *J. Exp. Zool.* 97:165–197.

———. (1945) Some relations of external conditions, past or present, to aging and to mortality of exconjugants with summary of conclusions on age at death. *J. Exp. Zool.* 99:15–31.

Jennings, H. S., and Lynch, R. S. (1928a) Age, mortality, fertility, and individual diversity in the rotifer *Proales sordida* Gosse. 1. Effect of age of the parent on characteristics of the offspring. *J. Exp. Zool.* 50:345–407.

———. (1928b). Life history in relation to

mortality and fecundity. *J. Exp. Zool.* 51:339–381.

Jennings, J. B., and Calow, P. (1975) The relationship between high fecundity and the evolution of endoparasitism. *Oecologia* 27:109–115.

Jepsen, G. L. (1970) *Biology of Bats.* Vol. 1, *Bat Origins and Evolution.* New York: Academic Press.

Jerison, H. J. (1973) *Evolution of the Brain and Intelligence.* New York: Academic Press.

Jerne, N. K. (1975) The immune system: A web of V-domains. *Harvey Lectures,* ser. 70:93–110.

Jinks, J. L. (1957) Selection for cytoplasmic differences. *Proc. Roy. Soc.,* ser. B, 146:527–540.

———. (1958) Cytoplasmic differentiation in fungi. *Proc. Roy. Soc.,* ser. B, 148:314–321.

———. (1959) Lethal suppressive cytoplasms in aged clones of *Aspergillus glaucus. J. Gen. Microbiol.* 21:397–409.

———. (1987) Selection for cytoplasmic differences. *Proc. Roy. Soc.,* ser. B, 146:527–540.

Joanen, T., and McNease, L. (1980) Reproductive biology of the American alligator in southwest Louisiana. In *Reproductive Biology and Diseases of Captive Reptiles,* J. B. Murphy and J. T. Collins (eds.), 153–159. SSAR Contributions to Herpetology 1. Society for the Study of Amphibians and Reptiles.

John, B., and Freeman, M. (1976) The cytogenetic systems of grasshoppers and locusts. 3. The genus *Tolgadia* (Oxyinae: Acrididae). *Chromosoma* (Berlin) 55:105–119.

John, B., and Hewitt, G. M. (1966) Karyotype stability and DNA variability in the Acrididae. *Chromosoma* (Berlin) 20:155–172.

Johnsen, S. (1944) *Studies on Variation in Fish in North-European Waters.* Vol. 1, *Variation in Size.* Bergens Museums Arbok 1944. Naturvitenskapelig rekke Nr. 4.

Johnson, B. (1983) Flight muscle breakdown in an aphid *Megoura viciae. Tiss. Cell* 15:529–539.

Johnson, H. A. (1985) Is aging physiological or pathological? In *Relations between Normal Aging and Disease,* H. A. Johnson (ed.), 239–247. Aging 28. New York: Plenum.

Johnson, H. D., Kintner, L. D., and Kibler, H. H. (1963) Effects of 48F (8.9C) and 83F (28.4C) on longevity and pathology of male rats. *J. Gerontol.* 18:29–36.

Johnson, J. H., and Ringler, N. H. (1979) Occurrence of blow fly larvae (*Diptera celiphordiae*) on salmon carcinogens and utilization as food by juvenile salmon and trout. *Great Lakes Entomol.* 12:137–140.

Johnson, P. A., Dickerman, R. W., and Bahr, J. M. (1986) Decreased granulosa cell luteinizing sensitivity and altered thecal estradiol concentration in the aged hen, *Gallus domesticus. Biol. Reprod.* 35:641–646.

Johnson, R., and Strehler, B. L. (1972) Loss of genes coding for ribosomal RNA in aging brain cells. *Nature* 240:512–516.

Johnson, S. A., McNeill, T., Cordell, B., and Finch, C. E. (1990) Neuronal APP–751/APP–695 mRNA ratio correlates with neuritic plaque density in Alzheimer's disease. *Science* 248:854–857.

Johnson, S. A., Pasinetti, G. M., May, P. C., Ponte, P. A., Cordell, B., and Finch, C. E. (1988) Selective reduction of mRNA for the β-amyloid precursor protein that lacks a Kuntiz-type protease inhibitor motif in cortex from AD. *Exp. Neurol.* 102:264–268.

Johnson, S. A., Rogers, J., and Finch, C. E. (1989) APP–695 transcript prevalence is selectively reduced during Alzheimer's disease in cortex and hippocampus but not in cerebellum. *Neurobiol. Aging* 10:267–272.

Johnson, T. E. (1986) Molecular and genetic analyses of a multivariate system specifying behavior and life span. *Behav. Genet.* 16:221–235.

———. (1987) Aging can be genetically dissected into component processes using long-lived lines of *Caenorhabditis elegans. Proc. Natl. Acad. Sci.* 84:3777–3781.

———. (1989) The increased lifespan of *age–1* mutants of *Caenorhabditis elegans* results from lowering the Gompertz rate of aging.

———. (1990) *Age–1* mutants of *Caenorhabditis elegans* prolong life by modifying the Gompertz rate of aging. *Science* 249:908–912.

Johnson, T. E., and Hartman, P. S. (1988) Radiation effects on life-span in *Caenorhabditis elegans. J. Gerontol.* 43:B137–B141.

Johnson, T. E., and Hirsch, D. (1979) Patterns of proteins synthesized during development of *Caenorhabditis elegans. Dev. Biol.* 70:241–248.

Johnson, T. E., and McCaffrey, G. (1985) Programmed aging or error catastrophe: An examination by two-dimensional polyacrylamide gel electrophoresis. *Mech. Age. Dev.* 30:285–287.

Johnson, T. E., Mitchell, D. H., Kline, S., Kemal, R., and Foy, J. (1984) Arresting development arrests aging in the nematode *Caenorhabditis elegans. Mech. Age. Dev.* 28:23–40.

Johnson, T. E., and Wood, W. B. (1982) Genetic analysis of lifespan in *Caenorhabditis elegans. Proc. Natl. Acad. Sci.* 79:6603–6607.

Johnsson, L., and Hawkins, J. E., Jr. (1972) Sensory and neuronal degeneration with aging, as seen in microdissections of the human inner ear. *Ann. Otol.* 81:178–183.

Johnston, J. R. (1966) Reproductive capacity and mode of death of yeast cells. *Antonie v. Leeuwenhoek* 32:94–98.

Johnston, J. S., and Ellison, J. R. (1982) Exact age determination in laboratory and field-caught *Drosophila. J. Insect Physiol.* 28:773–779.

Johnston, J. S., and Templeton, A. R. (1982) Dispersal and clines in *Opuntia* breeding *Drosophila mercatorum* and *D. hydei* at Kamuela, Hawaii. In *Ecological Genetics and Evolution: The Cactus-Yeast-Drosophila Model System,* J. S. Barker and W. T. Starmer (eds.), 241–256. New York: Academic Press.

Joly, L. (1965) Fonctionnement des corpora allata chez imago de *Locusta migratoria. Compt. Rend. Acad. Sci.* 260:7006–7009.

Jones, D. S. (1983) Sclerochronology: Reading the record of molluscan shell. *Am. Sci.* 71:384–391.

Jones, E. C., and Krohn, P. L. (1961a) The effect of hypophysectomy on age changes in the ovaries of mice. *J. Endocrinol.* 21:497–508.

———. (1961b) The relationships between age, numbers of oocytes, and fertility in virgin and multiparous mice. *J. Endocrinol.* 21:469–496.

Jones, H. B. (1956) A special consideration of the aging process, disease, and life expectancy. *Adv. Biol. Med. Phys.* 4:281–337.

———. (1959) The relation of human health to age, place, and time. In *Handbook of Aging and the Individual,* J. E. Birren (ed.), 336–363. Chicago: Univ. of Chicago Press.

Jones, K. W., and Singh, L. (1981) Conserved repeated DNA sequences in vertebrate sex chromosomes. *Human Genet.* 58:46–53.

Jones, M. L. (1982) Longevity of captive mammals. *Zool. Gart.* (Jena), N. F. 52:113–128.

Jones, P. A., and Gilbert, J. J. (1977) Polymorphism and polyploidy in the rotifer *Asplanchna sieboldi:* Relative nuclear DNA contents in tissues of saccate and campanulate females. *J. Exp. Zool.* 201:163–168.

Jordan, D. S., and Evermann, B. W. (1896) The fishes of North and Middle America: A descriptive catalogue of the species of fish-like vertebrates found in the waters of North America, north of the isthmus of Panama. *Bull. Natl. Mus.* no. 47. [not read]

Jost, A. (1953) Problems of fetal endocrinology: The gonadal and hypophyseal hormones. *Rec. Prog. Horm. Res.* 8:379–418.

Jouandet, M. L., Tramp, M. J., Herron, D. M., Hermann, A., Loftus, W. C., Bazell, J., and Gazzaniga, M. S. (1989) Brain prints: Computer-generated cerebral cortex *in vivo. J. Cog. Neurosci.* 1:88–117.

Jowsey, J. O. M., and Offord, K. P. (1978) Mechanisms of localized bone loss. In *Information Retrieval,* vol. 345, E. J. Horton, J. M. Tarpley, and W. F. Davis (eds.). Arlington, VA.

Juberthie, C. (1967) *Siro rubeus. Rev. Ecol. Biol. Sol.* 4:155–171.

Juckett, D. A., and Rosenberg, B. (1988) Integral differences among human survival distributions as a function of disease. *Mech. Age. Dev.* 43:239–257.

Judd, H. L. (1983) Pathophysiology of menopausal hot flashes. In *Neuroendocrinology of Aging,* J. Meites (ed.), 173–202. New York: Plenum.

Jürgens, K. D., and Prothero, J. (1987) Scaling of maximal lifespan in bats. *Comp. Biochem. Physiol.* 88A:361–367.

Kaestner, A. (1968) *Invertebrate Zoology.* Vol. 2. Interscience Publishers.

Kajii, T., and Ohama, K. (1979) Inverse maternal age effect in monosomy X. *Human Genet.* 51:147–151.

Kaler, L. W., Gliessman, P., Craven, J., Hill, J., and Critchlow, V. (1986) Loss of enhanced nocturnal growth hormone secretion in aging rhesus males. *Endocrinology* 119:1281–1284.

Kaler, L. W., and Neaves, W. B. (1978) Attrition of the human Leydig cell population with advancing age. *Anat. Rec.* 192:513–518.

Kallman, F. J., and Sander, G. (1949) Twin studies on senescence. *Am. J. Psychiatry* 106:29–36.

Kalu, D. N., Cockerham, R., Yu, B. P., and Roos, B. A. (1983) Lifelong dietary modulation of calcitonin levels in rats. *Endocrinology* 113:2010–2016.

Kalu, D. N., Hardin, R. R., Cockerham, R., and Yu, B. P. (1984) Aging and dietary modulation of rat skeleton parathyroid hormone. *Endocrinology* 115:1239–1247.

Kalu, D. N., Hardin, R. R., Cockerham, R., Yu, B. P., Norling, B. K., and Egan, J. W. (1984) Lifelong food restriction prevents senile osteopenia and hyperparathyroidism in F344 rats. *Mech. Age. Dev.* 26:103–112.

Kalu, D. N., Masoro, E. J., Yu, B. P., Hardin, R. R., and Hollis, B. W. (1988) Modulation of age-related hyperparathyroidism and senile bone loss in Fischer rats by soy protein and food restriction. *Endocrinology* 122:1847–1854.

Kalugina, N. S., and Kovalev, V. G. (1985) *Dipterous Insects from the Jurassic of Siberia* (in Russian). Moscow: Akedemia Nauk. [not read]

Kanagalingam, K., and Balis, M. E. (1975) In vivo repair of rat intestinal DNA damage by alkylating agents. *Cancer* 36:2364–2372.

Kanazawa, Y. (1982) Some analyses of the reproduction process of a *Quercus crispula* Blume population in Nikko. 1. A record of acorn dispersal and seedling establishment for several years at three natural stands. *Jpn. J. Ecol.* 32:325–331.

Kannel, W. B. (1985a) Hypertension and aging. In *Handbook of the Biology of Aging,* 2d ed., C. E. Finch and E. L. Schneider (eds.), 859–877. New York: Van Nostrand.

———. (1985b) Lipids, diabetes, and coronary heart disease: Insights from the Framingham Study. *Am. Heart J.* 110:1100–1106.

Kannel, W. B., Dawber, T. R., Kagon, A. Revoltskie, N., and Stokes, J. (1961) Factors of risk in the development of coronary heart disease—Six year follow-up experience. *Ann. Int. Med.* 55:33–50.

Kannel, W. B., and Hubert, H. H. (1982) Vital capacity as a biomarker of aging. In *Biological Markers of Aging,* M. E. Reff and E. L. Schneider (eds.), 145–160. NIH Publication no. 82–2221. Bethesda, MD: National Institutes of Health.

Kannel, W. B., Lew, E. A., Hubert, H. B., and Castelli, W. P. (1980) The value of measuring vital capacity for prognostic purposes. *Trans. Assoc. Life Ins. Med. Dir. Am.* 64:66–81.

Kannel, W. B., Wolf, P. A., McGee, D. L., Dawber, T. R., McNamara, P. M., and Castelli, W. P. (1981) Systolic blood pressure, arterial rigidity, and risk of stroke: The Framingham Study. *J.A.M.A.* 245:1225–1229.

Kanno, H., Huang, I.-Y., Kan, Y. W., and Yoshida, A. (1989) Two structural genes on different chromosomes are required for encoding the major subunit of human red cell glucose-6–phosphate dehydrogenase. *Cell* 58:595–606.

Kaplan, B. B., and Finch, C. E. (1982). The sequence complexity of brain ribonucleic acids. In *Molecular Approaches to Neurobiology,* I. R. Brown (ed.), 71–98. New York: Academic Press.

Kaplan, J. R., Manuck, S. B., Clarkson, T. B., Lusso, F. M., Taub, D. M., and Miller, E. W. (1983) Social stress and atherosclerosis in normocholesterolemic monkeys. *Science* 220:733–735.

Kaplan, M. S., and Bell, D. H. (1984) Mitotic neuroblasts in the 9-day-old and 11-month-old rodent hippocampus. *J. Neurosci.* 4:1429–1441.

Kara, T. C., and Patnaik, B. K. (1981) Age-related changes in endogenous respiration, respiratory control ratio, and malonate inhibition of oxygen consumption of liver homogenate of male garden lizards. *Exp. Gerontol.* 16:31–33.

———. (1985) Age-related differences in the

response of hepatic oxygen consumption to thermal stress in the male garden lizard. *Arch. Gerontol. Geriatr.* 4:29–35.

Karakashian, S. J., Lanners, H. N., and Rudzinska, M. A. (1984) Cellular and clonal aging in the suctorian protozoan *Tokophrya infusionum. Mech. Age. Dev.* 26:217–229.

Karathanasis, S. K. (1985) Apolipoprotein multigene family. Tandem organization of human apolipoprotein A–1. C–111 and A-IV genes. *Proc. Natl. Acad. Sci.* 82:6374.

Karathanasis, S. K., Zannis, V. I., and Breslow, J. L. (1983) A DNA insertion in the apolipoprotein A-I gene of patients with premature atherosclerosis. *Nature* 305:823–825.

Karey, K. P., and Rothstein, M. (1986) Evidence for the lack of lysosomal involvement in the age-related slowing of protein breakdown in *Turbatrix aceti. Mech. Age. Dev.* 35:169–178.

Karpas, A. E., Bermner, W. J., Clifton, D. K., Steiner, R. A., and Dorsa, D. M. (1983) Diminshed luteinizing hormone pulse frequency and amplitude with aging in the male rat. *Endocrinology* 112:788–792.

Karr, J. P., Kim, U., Resko, J. A., Schneider, S., Chai, L. S., Murphy, G. P., and Sandberg, A. A. (1984) Induction of benign prostatic hypertrophy in baboons. *Urology* 23:276–289.

Kasuya, T., and Marsh, H. (1984) Life-history and reproductive biology of the short-finned pilot whale, *Globicephala macrorhynchus,* off the Pacific coast of Japan. *Rep. Int. Whal. Commn.,* special issue 6, W. F. Perris, R. L. Brownell, and D. P. DeMastan (eds.), 260–309.

Kathirithamby, J. (1989) Review of the order Strepsiptera. *Sys. Entomol.* 14:41–92.

Kato, H., Harada, M., Tsuchiya, K., and Moriwaka, K. (1980) Absence of correlation between DNA repair in ultraviolet irradiated mammalian cells and life span of the donor species. *Jpn. J. Genet.* 55:99–108.

Katz, M., and Shanker, M. J. (1989) Development of lipofuscin-like fluorescence in the retinal pigment epithelium in response to protease inhibitor treatment. *Mech. Age. Dev.* 49:23–40.

Kawamura, S. (1927) On the periodical flowering of the bamboo. *Jpn. J. Bot.* 3:335–349.

Kawashima, S. (1960) Influence of continued injections of sex steroids on the estrous cycle in the adult rat. *Annot. Zool. Jpn.* 33:226–232.

Kay, M. M. B. (1978a) Effect of age on T cell differentiation. *Fed. Proc.* 37:1241–1244.

———. (1978b) Immunologic aging patterns: Effect of parainfluenza type 1 virus infection on aging mice of eight strains and hybrids. In *Genetic Effects on Aging,* D. Bergsma and D. E. Harrison (eds.), 213–240. Birth Defects: Original Article Series, no. 14. New York: Alan R. Liss.

———. (1979) Parainfluenza infection of aged mice results in autoimmune disease. *Clin. Immunol. Immunopathol.* 12:301–315.

———. (1985) Immune systems: Expression and regulation of cellular aging. In *Thresholds in Aging,* M. Bergener, M. Ermini, and H. B. Stähelin (eds.), 59–82. New York: Academic Press.

Kay, M. M., Bosman, G. J., and Lawrence, C. (1988) Functional topography of band 3: Specific structural alteration linked to functional aberrations in human erythrocytes. *Proc. Natl. Acad. Sci.* 85:492–496.

Kay, M. M., Bosman, G. J., Shapiro, S. S., Bendich, A., and Bassel, P. S. (1986) Oxidation as a possible mechanism of cellular aging: Vitamin E deficiency causes premature aging and IgG binding to erythrocytes. *Proc. Natl. Acad. Sci.* 83:2463–2467.

Kay, M. M., Flowers, N., Goodman, J., and Bosman, G. (1989) Alteration in membrane protein band 3 associated with accelerated erythrocyte aging. *Proc. Natl. Acad. Sci.* 86:5834–5838.

Kay, M. M., Mendoza, J., Diven, J., Denton, T., Union, N., and Lajiness, M. (1979) Age-related changes in the immune system of mice of eight medium and long-lived strains and hybrids. 1. Organ, cellular, and activity changes. *Mech. Age. Dev.* 11:295– 346.

Keil, J. H. (1940) Functional spinal cord regeneration in adult rainbow fish. *Proc. Soc. Exp. Biol. Med.* 43:175–177.

Keilbach, R. (1982) Bibliographie und Liste der Arten tierischer Einschlusse in fossilen Harzen sowie ihrer Aufbewahrungsorte. *Deut. Entomol. Z.*, N.F. 29:129–286.

Keilin, D. (1925) Parasitic autotomy of the host as a mode of liberation of coelomic parasites from the body of the earthworm. *Parasitology* 17:170–172.

Keller-Wood, M., and Dallman, M. F. (1984) Corticosteroid inhibition of ACTH secretion. *Endocrine Rev.* 5:1–24.

Kelley, K. W., Brief, S., Westly, H. J., Novakofski, J., Bechtel, P. J., Simon, J., and Walker, E. B. (1986) GH₃ pituitary adenoma cells can reverse thymic aging in rats. *Proc. Natl. Acad. Sci.* 83:5663–5667.

Kelley, K. W., Davila, D. R., Brief, S., Simon, J., and Arkins, S. (1988) A pituitary-thymus connection during aging. *Ann. N.Y. Acad. Sci.* 521:88–98.

Kelley, S. E., Antonovics, J., and Schmitt, J. (1988) A test of the short-term advantage of sexual reproduction. *Nature* 331:714–715.

Kellgren, J. H., Lawrence, J. S., and Bier, F. (1963) Genetic factors in generalized osteoarthrosis. *Ann. Rheum. Dis.* 22:237–255.

Kellogg, E. W., III, and Fridovich, I. (1976) Superoxide dismutase in the rat and mouse as a function of age and longevity. *J. Gerontol.* 31:405–408.

Kelsall, P. (1963) Non-disjunction and maternal age in *Drosophila melanogaster. Genet. Res. Cambridge* 4:284–289.

Kemnitz, J. W. (1984) Obesity in macaques: Spontaneous and induced. *Adv. Vet. Sci. Comp. Med.* 28:81–114.

Kemnitz, J. W., Goy, R. W., Flitsch, T. J., Lohmiller, J. J., and Robinson, J. A. (1989) Obesity in male and female rhesus monkeys: Fat distribution, glucoregulation, and serum androgen levels. *J. Clin. Endocrinol. Metab.* 69:287–293.

Kemp, T. S. (1982) *Mammal-like Reptiles and the Origin of Mammals.* New York: Academic Press.

Kemper, T. (1984) Neuroanatomical and neuropathological changes in normal aging and in dementia. In *Clinical Neurology of Aging,* M. L. Albert (ed.), 9–25. New York: Oxford Univ. Press.

Kemperman, J. A., and Barnes, B. V. (1976)

Clone size in American aspens. *Can. J. Bot.* 54:2603–2607.

Kempf, S. C., and Hadfield, M. G. (1985) Planktotrophy by the lecithotrophic larvae of a nudibranch, *Phestilla sibogae* (Gastropoda). *Biol. Bull.* 169:119–130.

Kennedy, G. C., and Mitra, J. (1963) Body weight and food intake as initiation factors for puberty in the rat. *J. Physiol.* (London) 166:408–418.

Kennes, B., Brohee, D., and Neve, P. (1983) Lymphocyte activation in human aging. 5. Acquisition of response to T cell growth factor and production of growth factors by mitogen stimulated lymphocytes. *Mech. Age. Dev.* 23:103–111.

Kenney, F. T. (1962) Induction of tyrosine-α-ketoglutarate transaminase in rat liver. 4. Evidence for an increase in rats of enzyme synthesis. *J. Biol. Chem.* 237:3495–3498.

———. (1976) Turnover of rat liver tyrosine transaminase: Stabilization after inhibition of protein synthesis. *Science* 156:525–528.

Kern, M. J. (1986) Brain aging in insects. In *Insect Aging: Strategies and Mechanisms,* K.-G. Collatz and R. S. Sohal (eds.), 90–105. Heidelberg: Springer-Verlag.

Kerr, D. S., Campbell, L. W., Hao, S.-Y., and Landfield, P. W. (1989) Corticosteroid modulation of hippocampal potentials: Increased effect with aging. *Science* 245:1505–1509.

Ketchen, K. S. (1975) Age and growth of dogfish, *(Squalus acanthias)* in British Columbia waters. *J. Fish. Res. Bd. Can.* 32:43–59.

Kevles, D. J. (1985) *In the Name of Eugenics: Genetics and the Uses of Human Heredity.* New York: Knopf.

Khairallah, E. A., and Mortimore, G. E. (1976) Assessment of protein turnover in perfused rat liver. Evidence for amino acid compartmentalization from differential labeling of free and tRNA-bound valine. *J. Biol. Chem.* 251:1375–1384.

Kidson, C., Chen, P., and Imray, F. P. (1984) DNA manipulating genes and the aging brain. In *Molecular Biology of Aging,* A. D. Woodhead, A. D. Blackett, and A. Hollaender (eds.), 285–296. Basic Life Sciences, vol. 35. New York: Plenum.

Kielan-Jaworowska, Z. (1984) Evolution of the therian mammals in the late Cretaceous of Asia. 6. Endocranial casts of eutherian mammals. In *Results of the Polish-Mongolian Paleontological Expedition*. Part 10, *Paleontologia Polonica*, Z. Kielan-Jaworowska (ed.), 46:157–171. [not read]

Kim, Y. T., Goidl, E. A., Samarut, C., Weksler, M. E., Thorbecke, G. J., and Siskind, G. W. (1985) Bone marrow function. 1. Peripheral T cells are responsible for the increased auto-antiidiotype response of older mice. *J. Exp. Med.* 161:1237–1242.

Kimberling, W. J., Fain, P. R., Kenyon, J. B., Goldgar, D., Sujansky, E., and Gabow, P. A. (1988) Linkage heterogeneity of autosomal dominant polycystic kidney disease. *N. Engl. J. Med.* 319:913–918.

Kimble, J., and Ward, S. (1988) Germ-line development and fertilization. In *The Nematode Caenorhabditis Elegans*, W. B. Wood (ed.), 191–213. New York: Cold Spring Harbor Laboratory.

Kime, D. E., and Larsen, L. O. (1987) Effect of gonadectomy and hypophysectomy on plasma steroid levels in male and female lampreys (*Lampetra fluviatilis*. L.). *Gen. Comp. Endocrinol.* 68:189–196.

Kime, D. E., and Rafter, J. J. (1981) Biosynthesis of 15-hydroxylated steroids by gonads of the river lamprey, *Lampetra fluviatilis*, in vivo. *Gen. Comp. Endocrinol.* 44:69–76.

Kimura, M. (1983) *The Neutral Theory of Molecular Evolution*. Cambridge: Cambridge Univ. Press.

King, C. E. (1967) Food, age, and the dynamics of a laboratory population of rotifers. *Ecology* 48:111–128.

———. (1970) Comparative survivorship of mictic and amictic rotifers. *Physiol. Zool.* 43:206–212.

King, C. R., Magenis, E., and Bennett, S. (1978) Pregnancy and the Turner syndrome. *Obstet. Gynecol.* 52:617–624.

King, R. C., Aggarwal, S. K., and Aggarwal, U. (1968) The development of the female *Drosophila* reproductive system. *J. Morphol.* 124:143–166.

Kingdon, J. E. (1979) *East African Mammals. An Atlas of Evolution in Mammals*. Vol. 3B, 18–20. New York: Academic Press.

Kinne, O. (1960) Growth, food intake, and food conversion in a euryplastic fish exposed to different temperatures and salinities. *Physiol. Zool.* 33:288.

Kinsey, A. C., Pomeroy, W. E., and Martin, C. E. (1948) *Sexual Behavior in the Human Male*. Philadelphia: Saunders.

Kinsley, C., Miele, J., Konen, C., Ghiraldi, L., Broida, J., and Svare, B. (1986) Prior intrauterine position influences body weight in male and female mice. *Horm. Behav.* 20:201–211.

Kinsley, C., Miele, J., Konen, C., Ghiraldi, L., and Svare, B. (1986) Intrauterine contiguity influences regulatory activity in adult female and male mice. *Horm. Behav.* 20:7–19.

Kinzie, R. A., and Sarmiento, T. (1986) Linear extension rate is independent of colony size in the coral *Pocillopora damicornis*. *Coral Reefs* 4:177–181.

Kirkham, W. B. (1916) The prolonged gestation period in suckling mice. *Anat. Rec.* 11:31–40.

Kirkwood, T. B. (1985) Comparative evolutionary aspects of longevity. In *Handbook of the Biology of Aging*, 2d ed., C. E. Finch and E. L. Schneider (eds.), 27–45. New York: Van Nostrand.

———. (1987) Immortality of the germ-line versus disposability of the soma. In *Evolution of Longevity of Animals. A Comparative Approach*, A. D. Woodhead and K. H. Thompson (eds.), 209–218. New York: Plenum.

———. (1988) DNA, mutations, and aging. Review. *Mutat. Res.*, suppl. 1:7–13.

Kirkwood, T. B., and Cremer, T. (1982) Cytogerontology since 1881: A reappraisal of August Weismann and a review of modern progress. *Human Genet.* 60:101–212.

Kirkwood, T. B., and Holliday, R. (1986) Selection for optimal accuracy and the evolution of ageing. In *Accuracy in Molecular Processes: Its Control and Relevance to Living Systems*, T. B. Kirkwood, R. F. Rosenberger, and D. J. Galas (eds.), 363–379. London: Chapman and Hill.

Kirszbaum, L., Sharpe, J. A., Murphy, B.,

d'Apice, A. J. F., Classon, B., Hudson, P., and Walker, I. D. (1989) Molecular cloning and characterization of the novel, human complement-associated protein, SP-40,40: A link between the complement and reproductive systems. *EMBO J.* 8:711–718.

Kishimoto, S., Tomino, S., Inomata, K., Kotegawa, S., Salto, T., Kuroki, M., Mitsuya, H., and Hisamitsu, S. (1978) Age-related changes in the subsets and functions of human T lymphocytes. *J. Immunol.* 121: 1773–1780.

Kita, T., Nagano, Y., Yokode, M., Ishii, K., Kume, N., Ooshima, A., Yoshida, H., and Kawai, C. (1987) Probucol prevents the progression of atherosclerosis in Watanabe heritable hyperlipidemic rabbit, an animal model for familial hypercholesterolemia. *Proc. Natl. Acad. Sci.* 84:5928–5931.

Kitaguchi, N., Takahashi, Y., Tokushima, Y., Shiojiri, S., and Ito, H. (1988) Novel precursor of Alzheimer's disease amyloid protein shows protease inhibitory activity. *Nature* 331:530–532.

Kivilaan, A., and Bandurski, R. S. (1973) The 90 year period for Dr. Beal's seed viability experiment. *Am. J. Bot.* 60:140–145.

Klass, M., and Hirsh, D. (1976) Non-ageing development variant of *Caenorhabditis elegans*. *Nature* 260:523–525.

Klass, M., and Smith-Sonneborn, J. (1976) Studies in DNA content, RNA synthesis, and DNA template activity in aging cell of *Paramecium aurelia*. *Exp. Cell Res.* 98:63–72.

Klass, M. R. (1983) A method for the isolation of longevity mutants in the nematode *Caenorhabditis elegans* and initial results. *Mech. Age. Dev.* 22:279–286.

Kleiber, M. (1932) Body and size and metabolism. *Hilgardia* 6:315–353.

———. (1947) Body size and metabolic rate. *Physiol. Rev.* 27:511–541.

———. (1975) Metabolic turnover rate: A physiological meaning of the metabolic rate per unit body weight. *J. Theoret. Biol.* 53:199–204.

Klein, D., Twearson, S., Figueroa, F., and Klein, J. (1982) The nominal length of the differential segment in H–2 congenic lines. *Immunogenetics* 16:319–328.

Klein, J. (1975) *Biology of the Mouse Histocompatibility-2 Complex.* New York: Springer-Verlag.

———. (1978) H–2 mutations: Their genetics and effect on immune functions. *Adv. Immunol.* 26:56–146.

———. (1986) *Natural History of the Major Histocompatibility Complex.* New York: Wiley and Sons.

Klekowski, E. J. (1984) Mutational load in clonal plants: A study of two fern species. *Evolution* 38:417–426.

———. (1988a) *Mutation, Development Selection, and Plant Evolution.* New York: Columbia Univ. Press.

———. (1988b) Progressive cross- and self-sterility associated with aging in fern clones and perhaps other plants. *Heredity* 61:247–253.

Klekowski, E. J., and Godfrey, P. J. (1989) Ageing and mutation in plants. *Nature* 340:389–391.

Kligman, A. M. (1969) Early destructive effects of sunlight on human skin. *J.A.M.A.* 210:2377–2380.

Kligman, A. M., Grove, G. L., and Balin, A. K. (1985) Aging of human skin. In *Handbook of the Biology of Aging,* 2d ed., C. E. Finch and E. L. Schneider (eds.), 820–841. New York: Van Nostrand.

Klimkiewicz, M. K., and Futcher, A. G. (1987) Longevity records of North American birds: Coerebinae through Estrilidae. *J. Field Ornithol.* 58:318–333.

Kline, J., Levin, B., Shrout, P., Stein, Z., Susser, M., and Warberton, D. (1983) Maternal smoking and trisomy among spontaneous aborted conceptions. *Am. J. Human Genet.* 35:421–431.

Knight, H., Millman, R. P., Gur, R. C., Saykin, A. J., Doherty, J. U., and Pack, A. I. (1987) Clinical significance of sleep apnea in the elderly. *Am. Rev. Resp. Dis.* 136:845–850.

Knight, J. G., and Adams, D. D. (1978) Three genes for lupus nephritis in NZB × NZW mice. *J. Exp. Med.* 7:1653–1660.

Knittle, J. L., and Hirsch, J. (1968). Effect of early nutrition on development of rat epididymal fat pads-cellularity and metabolism. *J. Clin. Invest.* 47:2091–2098.

Knoll, A. H., and Niklas, K. J. (1985) Adaptation and the fossil record of plants. *Am. J. Bot.* 72:886–887.

———. (1987) Adaptation, plant evolution, and the fossil record. *Rev. Paleobot. Palynol.* 50:127–149.

Koenig, R. J., Araujo, D. C., and Cerami, A. (1976) Increased hemoglobin A_{1c} in diabetic mice. *Diabetes* 25:1–5.

Koenig, R. J., and Cerami, A. (1975) Synthesis of hemoglobin A_{1c} in normal and diabetic mice: Potential model of basement membrane thickening. *Proc. Natl. Acad. Sci.* 72:3687–3691.

Koenig, R. J., Peterson, C. M., Jones, R. L., Saudek, C., Lehrman, M., and Cerami, A. (1976) Correlation of glucose regulation and hemoglobin A_{1c} in diabetes mellitus. *N. Engl. J. Med.* 295:417–420.

Kohama, S. G., Anderson, C. P., and Finch, C. E. (1989a) Progesterone implants extend the capacity for 4–day estrous cycles in aging C57BL/6J mice and protect against acyclicity induced by estradiol. *Biol. Reprod.* 41:233–240.

Kohama, S. G., Anderson, C. P., Osterburg, H. H., May, P. C., and Finch, C. E. (1989b) Oral administration of estradiol to young C57BL/6J mice induces age-like neuroendocrine dysfunctions in the regulation of estrous cycles. *Biol. Reprod.* 41:227–232.

Kohn, R. R. (1963) Human ageing and disease. *J. Chronic Dis.* 16:5–21.

———. (1971) *Principals of Mammalian Aging.* Englewood Cliffs, NJ: Prentice Hall.

———. (1982) Causes of death in very old people. *J.A.M.A.* 247:2793–2797.

Kohn, R. R., Cerami, A., and Monnier, V. M. (1984) Collagen aging in vitro by nonenzymatic glycosylation and browning. *Diabetes* 33:57–59.

Kohn, R. R., and Novak, D. (1973) Variability in AKR mouse leukemia mortality. *J. Natl. Canc. Inst.* 51:683–685.

Kohno, A., Yonezu, T., Matsushita, M., Irino, M., Higuchi, K., Takeshita, S., Hosokawa, M., and Takeda, T. (1985) Chronic food restriction modulates the advance of senescence

in senescence accelerated mouse (SAM). *J. Nutrit.* 115:1259–1266.

Koizumi, A., Hasegawa, L., Walford, R. L., and Imamura, T. (1986) H–2, *ah,* and aging: The immune response and the inducibility of P_{450} mediated monooxygenase activities, xanthine oxidase, and lipid peroxidation in H–2 congenic mice on C57BL/10, C3H, and A strain backgrounds. *Mech. Age. Dev.* 37:119–136.

Koizumi, A., Weindruch, R., and Walford, R. L. (1987) Influences of dietary restriction and age on liver enzyme activities and lipid peroxidation in mice. *J. Nutrit.* 117:361–367.

Koll, F. (1986) Does nuclear integration of mitochondrial sequences occur during senescence in *Podospora? Nature* 324:597–599.

Koll, F., Begel, O., Keller, A. M., Vierny, C., and Belcour, L. (1984) Ethidium bromide rejuvenation of senescent cultures of *Podospora anserina:* Loss of senescence: Specific DNA and recovery of normal mitochondrial DNA. *Curr. Genet.* 8:127–134.

Kolmes, S. A., Winston, M. L., and Fergusson, L. A. (1989) The division of labor worker honey bees (Hymenoptera: Apidae): The effects of multiple patrilines. *J. Kans. Entomol. Soc.* 62:80–95.

Kolta, M. G., Holson, R., Duffy, P., and Hart, R. W. (1989) Effect of long-term caloric restriction on brain monoamines in aging male and female Fischer 344 rats. *Mech. Age. Dev.* 48:191–198.

Konopka, R. J., and Benzer, S. (1971) Clock mutants of *Drosophila melanogaster. Proc. Natl. Acad. Sci.* 68:2112–2116.

Koo, E. H., Sisodia, S. S., Cork, L. C., Unterbeck, A., Bayney, R. M., and Price, D. L. (1990) Differential expression of amyloid precursor protein mRNAs in cases of Alzheimer's disease and in aged nonhuman primates. *Neuron* 4. In press.

Kopec, S. (1924) Studies on the influence of inanition on the development and the duration of life in insects. *Biol. Bull.* 46:1–21.

———. (1928) On the influence of intermittent starvation on the longevity of the imaginal stage of *Drosophila melanogaster. Brit. J. Exp. Biol.* 5:204–211.

Kornienko, G. G., Lozhichevskay, T. B., and Rekoa, G. I. (1988) Sexual maturation in sturgeon stock of Azov Sea. *Rybnoe Khozyestro* 3:38–40.

Korschelt, E. von. (1914) Über Transplantationsversuche, Ruheszustande, und Lebensdauer der Lumbriciden. *Zool. Anz.* 43:537–555.

———. (1924) *Lebensdauer, Altern, and Tod.* 3d ed. Jena: Fischer.

———. (1942) Weiteres über die Dauer der ungeschlechtlichen hortpflanzung des *Ctenodrilus monostylos. Zool. Anz.* 137:162–166.

Kosswig, C. (1948) Genetische Beitrage zur Preadaptationstheorie. *Rev. Fac. Sci. Univ. Istanbul,* ser. B, 13:176–209. [not read]

Kosztarab, M. (1982) Homoptera. In *Synopsis and Classification of Living Organisms,* vol. 2, S. P. Parker (ed.), 447–470. New York: McGraw-Hill.

Kouri, R. E., Keifer, R., and Zimmerman, E. M. (1974) Hydrocarbon-metabolizing activity of various mammalian cells in culture. *In Vitro* 10:18–25.

Kovacs, E., Weber, W., and Muller, H. (1984) Age-related variation in the DNA-repair synthesis after UV-C irradiation in unstimulated lymphocytes of healthy blood donors. *Mutat. Res.* 131:231–237.

Kovesdi, I., Reichel, R., and Nevins, J. R. (1987) Role of an adenovirus *E2* promoter binding factor in E1A-mediated coordinate gene control. *Proc. Natl. Acad. Sci.* 84:2180–2184.

Kram, D., Schneider, E. L., Tice, R. R., and Gianas, P. (1978) Aging and sister chromatid exchange. 1. The effect of aging on mitomycin-C induced sister chromatid frequencies in mouse and rat bone marrow cells in vivo. *Exp. Cell Res.* 114:471–475.

Kramps, J. A., De Jong, W. W., Wollensak, J., and Hoenders, H. (1978) The polypeptide chains of α-crystallin from old human lenses. *Biochim. Biophys. Acta* 533:487–495.

Krebs, C. J., and Boonstra, R. (1978) Demography of the spring decline in populations of the vole, *Microtus townsendii. J. Anim. Ecol.* 47:1007–1015.

Krebs, C. J., and Myers, J. H. (1974) Population cycles in small mammals. *Adv. Ecol. Res.* 8:267–399.

Krebs, H. A. (1950) Body size and tissue respiration. *Biochim. Biophys. Acta* 4:249–269.

Krishna Rao, C. V. C., and Draper, H. H. (1969) Age-related changes in the bones of adult mice. *J. Gerontol.* 24:149–151.

Krishnakumaran, A. K. (1983) Introduction: Evolution of regulatory controls in insect life cycles. In *Endocrinology of Insects,* R. G. Downer and H. Laufa (eds.), 333–336. New York: Liss.

Kritchevsky, D., Tepper, S. A., Czarnecki, S. K., and Klurfeld, D. M. (1982) Artherogenicity of animal and vegetable protein: Influence of the lysine to arginine ratio. *Atherosclerosis* 41:429–431.

Krivobok, M. N., and Tarkovskaya, O. I. (1970) Lipid and protein metabolism during the sexual maturation in the Russian sturgeon and sevrjuga from the Volga River and Caspian Sea. *Vses. Nauchno-issled. Inst, Morsk. Rybn. Khoz. Okeanogr., Trudy,* 69:109–132. [not read; citation provided by Serge Doroshov]

Krizek, D. T., McIlrath, W. J., and Vergara, B. S. (1966) Photoperiodic induction of senescence in xanthium plants. *Science* 151:95–96.

Kroes, R., Sontag, J. M., Sell, S., Williams, G. M., and Weisburger, J. H. (1975) Elevated concentrations of serum α-fetoprotein in rats with chemically induced liver tumors. *Canc. Res.* 35:1214–1217.

Krumbiegel, I. (1929) Untersuchungen über die Einwirkung der fortpflanzung auf Altern und Lebensdauer der Insekten: *Carabas* und *Drosophila. Zool. Jb. Anat. Ontog.* 51:111–162.

Krumholz, L. A. (1948) Reproduction in the western mosquitofish, *Gambusia affinis affinis* (Baird and Girard), and its use in mosquito control. *Ecol. Monographs* 18:1–43.

Ksiezak-Reding, H., Dickson, D. W., Davies, P., and Yen, S.-H. (1987) Recognition of tau epitopes by anti-neurofilament antibodies that bind to Alzheimer neurofibrillary tangles. *Proc. Natl. Acad. Sci.* 84:3410–3414.

Ku, D. N., and Giddens, D. P. (1983) Pulsatile flow in a model carotid bifurcation. *Arteriosclerosis* 3:31–39.

Kubo, C., Day, N. K., and Good, R. A. (1984) Influence of early or late dietary restriction on life span and in pathological parameters in MRL/Mp-lpr/lpr mice. *Proc. Natl. Acad. Sci.* 81:5831–5835.

Kuck, J. F., Yu, N.-T., and Askren, C. C. (1982) Total sulfhydryl by Raman spectroscopy in the intact lens of several species: Variation in the nucleus and along optical axis during aging. *Exp. Eye Res.* 34:23–37.

Kuck, U., Osiewacz, H. D., Schmidt, U., Kappelhoff, B., Schulte, E., Stahl, U., and Esser, K. (1985) The onset of senescence is affected by DNA rearrangement of a discontinuous mitochondrial gene in *Podospora anserina*. *Curr. Genet.* 9:373–382.

Kuck, U., Stahl, U., and Esser, K. (1981) Plasmid-like DNA is part of mitochondrial DNA in *Podospora anserina*. *Curr. Genet.* 3:151–156.

Kukalová-Peck, J. (1985) Ephemeroid wing veination based upon new gigantic Carboniferous mayflies and basic morphology, phylogeny, and metamorphosis of pterygote insects (Insecta, Ephemerida). *Can. J. Zool.* 63:933–955.

Kulincevic, J. M., and Rothenbuhler, W. C. (1982) Selection for length of life in the honeybee (*Apis mellifera*). *Apidologie* 13:347–352.

Kunisada, T., Higuchi, K., Aota, S.-I., Takeda, T., and Yamagishi, H. (1986) Molecular cloning and nucleotide sequence of cDNA for murine senile amyloid protein: Nucleotide substitutions found in apolipoprotein A-II cDNA of senescence accelerated mouse (SAM). *Nucleic Acids Res.* 14:5729–5739.

Kunkel, H. O., and Campbell, J. E. (1952) Tissue cytochrome oxidase activity and body weight. *J. Biol. Chem.* 198:229–236.

Kunkel, H. O., Spalding, F. J., de Franciscis, G., and Futrell, M. F. (1956) Cytochrome oxidase activity and body weight in rats and in three species of large animals. *Am. J. Physiol.* 186:203–206.

Kunstyr, I., and Leuenberger, H. G. (1975) Gerontological data of C57BL/6J mice. 1. Sex differences in survival curves. *J. Gerontol.* 30:157–162.

Kunz, H. W., Gill, T. J., III, Dixon, B. D., Taylor, F. H., and Greiner, D. L. (1980) Growth and reproduction complex in the rat. Genes linked to the major histocompatibility complex that affect development. *J. Exp. Med.* 152:1506–1518.

Kurten, B. (1963) Return of a lost structure in the evolution of the felid dentition. *Comment. Biol.* 26(4):2–12.

Kutlaca, R., Seshadri, R., and Morley, A. A. (1982) Effect of age on sensitivity of human lymphocytes to radiation: A brief note. *Mech. Age. Dev.* 19:97–101.

Kwain, W. (1987) Biology of pink salmon in the North American Great Lakes. *Am. Fish. Soc. Symp.* 1:57–65.

Kyprianou, N., and Isaacs, J. T. (1988) Activation of programmed cell death in the rat ventral prostate after castration. *Endocrinology* 122:552–562.

Labandeira, C. C. (1990) Use of phenetic analysis of recent hexapod mouthparts for the distribution of hexapod food resource guilds in the fossil record. Ph.D. diss., University of Chicago.

Labandeira, C. C., Beall, B. S., and Hueber, F. M. (1988) Early insect diversification: Evidence for a Lower Devonian bristletail from Québec. *Science* 242:913–916.

Lack, D. (1943a) The age of some more British birds. *Brit. Birds* 36:193–197.

———. (1943b) *The Life of the Robin*. London: Witherby.

———. (1954) *The Natural Regulation of Animal Numbers*. Oxford: Clarendon Press. [not read]

———. (1966) *Population Studies of Birds*. Oxford: Clarendon Press. [not read]

Lafuse, W., and Edidin, M. (1980) Influence of the mouse major histocompatibility complex, H-2, on liver adenylate cyclase activity and on glucagon binding to liver cell membranes. *Biochemistry* 19:49–54.

Lafuse, W., Meruelo, D., and Edidin, M. (1979) The genetic control of liver cAMP levels in mice. *Immunogenetics* 9:57–65.

Laganiere, S., and Yu, B. P. (1987) Antilipoperoxidation action of food restriction.

Biochem. Biophys. Res. Commun. 145:1185–1191.

Laing, S. C., Godsen, R. G., and Fraser, H. M. (1984) Cytogentic analysis of mouse oocytes after experimental induction of follicular overripening. *J. Reprod. Fertil.* 70:387–393.

Lakatta, E. (1990) Heart and circulation. In *Handbook of the Biology of Aging,* 3d ed., E. L. Schneider and J. W. Rowe (eds.), 181–218. San Diego: Academic Press.

Lake, S. (1984) Variation in the recombination frequency and the relationship to maternal age in brood analysis of the distal and centromeric regions of the X-chromosome in temperature shocked reciprocal hybrids of inbred lines of *Drosophila melanogaster. Hereditas* 100:121–129.

LaMarche, V. C. (1969) Environment in relation to age of bristlecone pines. *Ecology* 50:53–59.

Lambert, B., Ringborg, U., and Skoog, L. (1979) Age-related decrease of ultraviolet light-induced DNA repair synthesis in human peripheral leukocytes. *Canc. Res.* 39:2792–2795.

Lambowitz, A. M. (1989) Infectious introns. *Cell* 56:323–326.

Lamers, W. H., and Mooren, P. G. (1981) Changes in the control of enzyme clusters in the liver of adult and senescent rats. *Mech. Age. Dev.* 15:119–128.

Lampert-Etchells, M., May, N. J., Schrieber, S. K., and Finch, C. E. (1990) The protein encoded by pADHC9/SGP2 is expressed in Alzheimer's disease hippocampus and is elevated in the rat hippocampus following entorhinal cortex lesion. *Soc. Neurosci. Abstr.* In press.

Lamy, R. (1947) Observed spontaneous mutation rates to experimental techniques. *J. Genet.* 43:212.

Lance, V. A., Joanen, T., and McNease, L. (1983) Selenium, vitamin E, and trace elements in the plasma of wild and farm-reared alligators during the reproductive cycle. *Can. J. Zool.* 61:1744–1751.

Lance, V. A., and Elsey, R. M. (1986) Stress-induced suppression of testosterone secretion

in male alligators. *J. Exp. Zool.* 239:241–246.

Landfield, P. W. (1978) An endocrine hypothesis of brain aging and studies on brain-endocrine correlations and monosynaptic neurophysiology during aging. In *Parkinson's Disease,* vol. 2, *Aging and Neuroendocrine Relationships,* C. E. Finch (ed.), 179–199. New York: Plenum.

Landfield, P. W., and Eldridge, J. C. (1989) Increased affinity of type II corticosteroid binding in aged rat hippocampus. *Exp. Neurol.* 106:110–113.

Landfield, P. W., Rose, G., Sandles, L., Wohlstadter, T. C., and Lynch, G. (1977) Patterns of astroglial hypertrophy and neuronal degeneration in the hippocampus of aged, memory-deficient rats. *J. Gerontol.* 32:3–12.

Landfield, P. W., Waymire, J. L., and Lynch, G. (1978) Hippocampal aging and adrenocorticoids: Quantitative correlations. *Science* 202:1098–1102.

Landfield, P. W., Wurtz, G., Lindsey, J. D., and Lynch, G. (1979) Long-term adrenalectomy reduces some morphological correlates of brain aging. *Soc. Neurosci. Abstr.* 5:20.

Landsberg, L., and Krieger, D. R. (1989) Obesity, metabolism, and the sympathetic nervous system. *Am. J. Hypertension* 2:125S-132S.

Lane, P. W., and Dickie, M. M. (1958) The effect of restricted food intake on the life span of genetically obese mice. *J. Nutrit.* 64:549–554.

Lang, C. A. (1986) Research strategies for the study of nutrition and aging. In *Nutritional Aspects of Aging,* L. H. Chen (ed.), 1–18. Boca Raton, FL: CRC Press.

Lang, C. A., Basch, K. J., and Storey, R. S. (1972) Growth, composition, and longevity of the axenic mosquito. *J. Nutrit.* 102:1057–1066.

Lang, C. A., Chen, T. S., and Mills, B. J. (1990) Glutathione status and longevity are enhanced by dietary restriction. *FASEB J.* 4, abstract 601.

Lang, C. A., Lau, H. Y., and Jefferson, D. J. (1965) Protein and nucleic acid changes during growth and aging in the mosquito. *Biochem. J.* 95:372–377.

Lang, C. A., and Smith, E. R. (1972) Protein metabolism in the aging mosquito. *Fed. Proc.* 31:878 (abstract).

Langston, J. W., Ballard, P., Tetrud, J. W., Irwin, I. (1983) Chronic Parkinsonism in humans due to a product of meperidine-analog synthesis. *Science* 219:979–980.

Lankester, R. E. (1870) *On Comparative Longevity in Man and the Lower Animals.* London: Macmillan.

Lansing, A. I. (1947) A transmissable cumulative and reversible factor in ageing. *J. Gerontol.* 2:228–239.

———. (1948) Evidence for aging as a consequence of growth cessation. *Proc. Natl. Acad. Sci.* 34:304–310.

———. (1956) Comparative physiology of aging. *FASEB J.* 15:960–964.

———. (1964) Age variations in cortical membranes of rotifers. *J. Cell Biol.* 23:403–419.

Lapham, L. W. (1968) Tetraploid DNA content of Purkinje neurons of human cerebellar cortex. *Science* 159:310–312.

Lapolt, P. S., Matt, D. W., Judd, H. L., Lu, J. K. (1986) The relation of ovarian steroid levels in young female rats to subsequent estrous cyclicity and reproductive function during aging. *Biol. Reprod.* 35:1131–1139.

Larsen, L. O. (1965) Effects of hypophysectomy in the cyclostome, *Lampetra fluviatilis (L.)* Gray. *Gen. Comp. Endocrinol.* 5:16–30.

———. (1969) Effects of hypophysectomy before and during sexual maturation in the cyclostome, *Lampetra fluviatilis* L. Gray. *Gen. Comp. Endocrinol.* 12:200–208.

———. (1974) Effects of testosterone and oestradiol on gonadectomized and intact male and female river lampreys (*Lampetra fluviatilis* L. Gray). *Gen. Comp. Endocrinol.* 24:305–313.

———. (1980) Physiology of adult lampreys, with special regard to natural starvation, reproduction, and death after spawning. *Can. J. Fish. Aquat. Sci.* 37:1762–1779.

———. (1985) The role of hormones in reproduction and death in lampreys and other species which reproduce once and die. In *Current Trends in Comparative Endocrinology,* B. Lofts and W. N. Holmes (eds.), 613–616. Hong Kong: Hong Kong Univ. Press.

———. (1987) The role of hormones in initiation of sexual maturation in male river lampreys (*Lampetra fluviatilis* L.): Gonadotropin and testosterone. *Gen. Comp. Endocrinol.* 68:197–201.

Larsson, S. G. (1978) Baltic amber: A palaeobiological study. *Entomonograph* 1:1–192.

Larsson, T., Sjogren, T., and Jacobson, G. (1963) Senile dementia. *Acta Psychiatr. Scand.* 39, supp. 167:3–259.

Lassere, P. (1975) Clitellata. In *Reproduction of Marine Invertebrates,* vol. 3, *Annelids and Echiurans,* A. C. Pearse and A. Giese (eds.), 215–275. New York: Academic Press.

Latham, K., and Finch, C. E. (1976) Hepatic glucocorticoid binders in mature and senescent C57BL/6J male mice. *Endocrinology* 98:1434–1443.

Lavie, L., Reznick, A. Z., and Gershon, D. (1982) Decreased protein and puromycinylpeptide degradation in livers of senescent mice. *Biochem. J.* 202:47–51.

Lavoipierre, M. M. J. (1961) Blood-feeding, fecundity, and ageing in *Aedes aegypti* var. *queenslandensis. Nature* (London) 191:575–576.

Lawrence, J. F. (1982) Coleoptera. In *Synopsis and Classification of Living Organisms,* vol. 2, S. P. Parker (ed.), 482–552. New York: McGraw-Hill.

Laws, R. M. (1966) Age criteria for the African elephant, *Loxodonta a. africana. E. African Wildlife J.* 4:1–37.

———. (1968) Dentition and ageing of the hippopotamus. *E. African Wildlife J.* 6:19–52.

Laws, R. M., and Parker, I. S. C. (1968) Recent studies on elephant populations in East Africa. *Symp. Zool. Soc. Lond.,* no. 21:319–359.

Lea, A. O. (1970) Endocrinology of egg maturation in autogenous and anautogenous *Aedes taeniorhynchus. J. Insect Physiol.* 16:1689–1696.

———. (1982) Artifactual stimulation of vitellogenesis in *Aedes aegypti* by 20–hydroxyecdysone. *J. Insect Physiol.* 28:173–176.

Leadon, S. A., and Hanawalt, P. C. (1983) Monoclonal antibody to DNA containing thymine glycol. *Mutat. Res.* 112:191–200.

Leaf, A. (1990) Long-lived populations (extreme old age). In *Principles of Geriatric Medicine and Gerontology,* 2d ed., W. R. Hazzard, R. Andres, E. L. Bierman, and J. P. Blass (eds.), 142–145. New York: McGraw-Hill.

Leaman, B. M. (1988) Reproductive and population biology of Pacific ocean perch (*Sebastes aleutus* [Gilbert]). Ph.D. diss., University of British Columbia, Vancouver.

————. (1990) Reproductive strategies and life history theory relative to exploitation and management of *Sebastes* stocks. *Environ. Biol. Fishes.* In press.

Leaman, B. M., and Beamish, R. J. (1984) Ecological and management implications of longevity in some northeast Pacific groundfishes. *Bull. Int. N. Pac. Fish. Commn.* 42:85–97.

Le Bourg, E., and Lints, F. A. (1984) A longitudinal study of the effects of age on spontaneous locomotor activity in *Drosophila melanogaster. Gerontology* 30:79–96.

Leder, P., Battey, J., Lenoir, G., Moulding, C., Murphy, W., Potter, M., Stewart, T., and Taub, R. (1983) Translocations among antibody genes in human cancer. *Science* 227:765–771.

Ledouarin, N. M., Smith, J., and Lelievre, C. S. (1981) From the neural crest to the ganglia of the peripheral nervous system. *Ann. Rev. Physiol.* 43:53–671.

Lee, A. K., and Cockburn, A. (1985a) *Evolutionary Ecology of Marsupials.* New York: Cambridge Univ. Press.

————. (1985b) Spring declines in small mammal populations. *Acta Zool. Fennica* 173:75–76.

Lee, A. T., and Cerami, A. (1990) Modifications of proteins and nucleic acids by reducing sugars: Possible role in aging. In *Handbook of the Biology of Aging,* 3d ed., E. L. Schneider and J. W. Rowe (eds.), 116–130. San Diego: Academic Press.

Lee, J. A. (1982) Melanoma and exposure to sunlight. *Epidemiol. Rev.* 4:110–130.

Lee, P., Rooney, P. J., Sturrock, R. D., Kennedy, A. C., and Dick, W. C. (1974) The etiology and pathogenesis of osteoarthrosis: A review. *Seminars in Arthritis and Rheumatism* 3:189–218.

Lee, P. M., Rothwell, K., and Whitehead, J. K. (1977) Fractionation of mouse skin carcinogens in cigarette smoke condensate. *Brit. J. Canc.* 35:730.

Lee, R. B. (1980) Lactation, ovulation, infanticide, and women's work: A study of hunter-gatherer population regulation. In *Biosocial Mechanisms of Population Regulation,* M. N. Cohen, R. S. Malpass, and H. G. Klein (eds.), 321–348. New Haven: Yale Univ. Press.

Lee, S. C. (1974) Biology of the sand goby *Pomatoschistus minutus* (Pallas) (Teleostei: Gobioidei) in the Plymouth area. Ph.D. thesis, University of Bristol, U.K.

LeFevre, J., and McClintock, M. K. (1988) Reproductive senescence in female rats: A longitudinal study of individual differences in estrous cycles and behavior. *Biol. Reprod.* 38:780–789.

Lefevre, M., and Rucker, R. B. (1980) Aorta elastin turnover in normal and hypercholesterolemic Japanese quail. *Biochim. Biophys. Acta* 630:519–529.

Leggett, W. C., and Carscadden, J. E. (1978) Latitudinal variation in reproduction characteristics of American shad (*Alosa sapidissima*): Evidence for population-specific life history strategies in fish. *J. Fish. Res. Bd. Can.* 35:1469–1478.

Leggett, W. C., and Frank, K. T. (1990) The spawning of the capelin. *Sci. Am.* 262:102–107.

Lehmensick, R. (1926) Zur Biologie, Anatomie, und Eireifung der Radertiere. *Z. Wiss. Zool.* 128:37.

Leiter, E. H., Coleman, D. L., Eisenstein, A. B., and Strack, I. (1981a) Dietary control of pathogenesis in C57BL/KsJ *db/db* diabetes mice. *Metabolism* 30:544–562.

Leiter, E. H., Coleman, D. L., and Hummel, K. P. (1981) The influence of genetic background on the expression of mutations at the diabetes locus in the mouse. 3. Effect of H-2 haplotype and sex. *Diabetes* 30:1029–1034.

Leiter, E. H., Premdas, F., Harrison, D. E., and Lipson, L. G. (1988) Aging and glucose homeostasis in C57BL/6J male mice. *FASEB J.* 2:2807–2811.

Lekholm, G. C. (1939) *En Ålderstigen Ål.* Halsingborgs, Sweden: Halsingborgs Museum Arsskrift.

Leloup-Hatey, J. (1964) Fonctionnement de l'interrénal anterior: de deux téléostéens: le saumon atlantique et l'anguille européenne. *Ann. Inst. Oceanog.* 42:224–337.

Lenhoff, H. M., and Loomis, W. F., eds. (1961) *Biology of Hydra and Some Other Coelenterates.* Coral Gables, FL: Univ. of Miami Press.

Leopold, A. C. (1961) Senescence in plant development. *Science* 134:1727–1732.

———. (1978) The biological significance of death in plants. In *The Biology of Aging,* J. A. Behnke, C. E. Finch, and G. B. Moment (eds.), 101–114. New York: Plenum.

———. (1981) Aging and senescence in plant development. In *Senescence in Plants,* K. V. Thimann (ed.), 1–12. Boca Raton, FL: CRC Press.

Leopold, A. C., and Kreidmann, P. E. (1975) *Plant Growth and Development,* 2d ed. New York: McGraw-Hill.

Leopold, A. C., Niedergang-Kamien, E., and Janick, J. (1959) Experimental modification of plant senescence. *Plant Physiol.* 34:570–573.

Lerman, J. C., and Cigliano, E. M. (1971) New carbon-14 evidence for six hundred years old *Canna compacta* seed. *Nature* 252:568–570.

Lerner, A., Yamada, T., and Miller, R. A. (1989) Pgp-1[hi] T lymphocytes accumulate with age in mice and respond poorly to concanavalin A. *Eur. J. Immunol.* 19:977–982.

Lerner, S. P., Anderson, C. P., Walford, R. L., and Finch, C. E. (1988) Genotypic influences on reproductive aging of inbred female mice: Effects of H-2 and non-H-2 alleles. *Biol. Reprod.* 38:1035–1043.

Lerner, S. P., and Finch, C. E. (1991) The major histocompatibility complex and reproductive functions. *Endocrine Rev.,* in press.

Lesher, S. (1966) Chronic irradiation and aging in mice and rats. In *Radiation and Ageing,* P. J. Lindop and G. A. Sacher (eds.), 183–205. London: Taylor and Francis.

Lesher, S., Fry, R. J., and Kohn, H. I. (1961) Influence of age on transit time of the cells of mouse intestinal epithelium. *Lab. Invest.* 10:291–300.

Leutenegger, W. (1979) Evolution of litter size in primates. *Am. Nat.* 114:525–531.

Levav, I., Friedlander, Y., Kark, J. D., and Peritz, E. (1988) An epidemiologic study of mortality among bereaved parents. *N. Engl. J. Med.* 319:457–461.

Leveille, P. J., Weindruch, R., Walford, R. L., Bok, D., and Horwitz, J. (1984) Dietary restriction retards age-related loss of gamma crystallins in the mouse lens. *Science* 224:1247–1249.

Levenbook, L. (1986) Protein synthesis in relation to insect aging: An overview. In *Insect Aging. Strategies and Mechanisms,* K.-G. Collatz and R. S. Sohal (eds.), 200–206. New York: Springer-Verlag.

Levi-Montalcini, R., and Booker, B. (1960) Destruction of the sympathetic ganglia in mammals by an antiserum to the nerve-growth promoting factor. *Proc. Natl. Acad. Sci.* 42:384–391.

Levin, P., Janda, J. K., Joseph, J. A., Ingram, D. K., and Roth, G. S. (1981) Dietary restriction retards the age-associated loss of rat striatal dopaminergic receptors. *Science* 214:561–562.

Levine, E. M., Mueller, S. N., and Noveral, J. P. (1987) Cultured endothelial cells: Development of a new model system for cellular aging. *Rev. Biol. Res. Aging* 3:417–427.

Levine, J. D., Coderre, T. J., Helms, C., and Basbaum, A. I. (1988) α_2-adrenergic mechanisms in experimental arthritis. *Proc. Natl. Acad. Sci.* 85:4553–4556.

Levine, N. D. (1980) *Nematode Parasites of Domestic Animals and of Man,* 2d ed. Minneapolis, MN: Burgess.

Levine, S., and Mullins, R. (1966) Hormonal influences on brain organization in infant rats. *Science* 152:1585–1592.

Levinthal, F., Macagno, E., and Levinthal, C. (1975) Anatomy and development of identified cells in isogenic organisms. *Cold Spring Harbor Symp. Quant. Biol.* 40:321–332.

Levinton, J. S. (1986) Developmental constraints and evolutionary saltations: A discussion and critique. In *Genetics, Development, and Evolution,* 17th Stadler Genetics Symposium, J. P. Gustafson, G. L. Stebbins, and F. J. Ayala (eds.), 253–288. New York: Plenum.

Lew, A. M., Lillehoj, E. P., Cowan, E. P., Maloy, W. L., Van Schravendijk, M. R., and Coligan, J. E. (1986) Class I genes and molecules: An update. *Immunology* 57:3–18.

Lewis, B. K., and Wexler, B. C. (1974) Serum insulin changes in male rats associated with age and reproductive activity. *J. Gerontol.* 29:139–144.

Lewis, V. M., Twomey, J. J., Bealmear, P., Goldstein, S., and Good, R. A. (1978) Age, thymic involution, and circulating thymic hormone activity. *J. Clin. Endocrinol. Metab.* 47:145–150.

Lezhava, T. A. (1984) Heterochromatinization as a key factor in aging. *Mech. Age. Dev.* 28:279–287.

Licastro, F., and Walford, R. L. (1985) Proliferative potential and DNA repair in lymphocytes from short-lived and long-lived strains of mice, relation to aging. *Mech. Age. Dev.* 31:171–186.

Licastro, F., Weindruch, R., Davis, L. J., and Walford, R. L. (1988) Effect of dietary restriction upon the age-associated decline of lymphocyte DNA repair activity mice. *Age* 11:48–52.

Lidgard, S., and Crane, P. R. (1988) Quantative analyses of the early angiosperm radiation. *Nature* 331:344–346.

Lie, J. T., and Hammond, P. I. (1988) Pathology of the senescent heart: Anatomic observations on 237 autopsy studies of patients 90 to 105 years old. *Mayo Clin. Proc.* 63:552–564.

Lieblum, S., Bachmann, G., Kemmann, B., Colburn, D., and Swartzman, L. (1983) Vaginal atrophy in menopausal women. *J.A.M.A.* 249:2190–2195.

Liehr, J. G., Avitts, T. A., Randerath, E., and Randerath, K. (1986) Estrogen-induced endogenous DNA adduction: Possible mechanism of hormonal cancer. *Proc. Natl. Acad. Sci.* 83:5301–5305.

Lillegraven, J. A., Kielan-Jaworowska, Z., and Clemens, W. A. (1979) *Mesozoic Mammals. The First Two-Thirds of Mammalian History.* Berkeley: Univ. of California Press.

Lin, T. J., So-Bosita, J. L., Brar, H. K., and Roblete, B. V. (1973) Clinical and cytologic responses of postmenopausal women to estrogen. *Obstet. Gynecol.* 41:97–107.

Linck, O. (1949) Fossile Bohrgange (*Anobichnium simile* n.g.n.sp.) an einem Keuperholz. *Neues Jb. Mineral. Geol. Palaont.* 1949B:180–185. [not read]

Lindeman, R. D., Tobin, J. D., and Shock, N. W. (1984) Association between blood pressure and the rate of decline in renal function with age. *Kidney Int.* 26:861–868.

———. (1985) Longitudinal studies on the rate of decline in renal function with age. *J. Am. Geriatr. Soc.* 33:278–285.

Linder, E., Pasternack, A., and Edgington, T. S. (1972) Pathology and immunology of age-associated disease of mice and evidence for an autologous immune complex pathogenesis of renal disease. *Clin. Immunol. Immunopathol.* 1:104–121.

Lindh, N. O. (1957) The mitotic activity during the early regeneration in *Euplanaria polychroa. Arkiv. Zool.* 10:497–509.

Lindop, P. J., and Rotblat, J. (1961a) Long-term effect of a single whole-body exposure of mice to ionizing radiations. 1. Life-shortening. *Proc. Roy. Soc.* 154:332–349.

———. (1961b) Long-term effect of a single whole-body exposure of mice to ionizing radiations. 2. Causes of death. *Proc. Roy. Soc.* 154:350–368.

Lindstedt, S. L., and Calder, W. A. (1976) Body size and longevity in birds. *Condor* 78:91–145.

———. (1981) Body size, physiological time, and longevity of homeothermic mammals. *Q. Rev. Biol.* 56:1–16.

Linnane, A. W., Marzuki, S., Ozawa, T., and Tanaka, M. (1989) Mitochondrial DNA mutations as an important contributor to ageing and degenerative diseases. *Lancet* 1:642–645.

Linnane, A. W., and Nagley, P. (1978) Mitochondrial genetics in perspective: The derivative of a genetic and physical map of the yeast mitochondrial genome. *Plasmid* 1:324–345.

Lints, F. A. (1978) *Genetics and Aging.* Interdisciplinary Topics in Gerontology, vol. 14. Basel: Karger.

———. (1985) Insects. In *Handbook of the Biology of Aging,* 2d ed., C. E. Finch and E. L. Schneider (eds.), 146–172. New York: Van Nostrand.

———. (1989) The rate of living theory revisited. *Gerontology* 35:36–57.

Lints, F. A., and Hoste, C. (1974) The Lansing effect revisited. 1. Life-span. *Exp. Gerontol.* 9:51–69.

———. (1977) The Lansing effect revisited. 2. Cumulative and spontaneously reversible parental age effects on fecundity in *Drosophila melanogaster. Evolution* 31:387–404.

Lints, F. A., and Lints, C. V. (1968) Respiration in *Drosophila.* 2. Respiration in relation to age by wild, inbred, and hybrid *Drosophila melanogaster* images. *Exp. Gerontol.* 3:341–349.

Lints, F. A., and Soliman, M. H. (1977) Growth rate and longevity in *Drosophila melanogaster* and *Tribolium casteneum. Nature* 266:624–625.

Lints, F. A., Stoll, J., Gruwez, G., and Lints, C. V. (1979) An attempt to select for increased longevity in *Drosophila melanogaster. Gerontology* 25:192–204.

Lipman, J. M., Turturro, A., and Hart, R. W. (1989) The influence of dietary restriction on DNA repair in rodents: A preliminary study. *Mech. Age. Dev.* 48:135–143.

Lipman, R. D., and Muggleton-Harris, A. L. (1982) Modification of the cataractous phenotype by somatic cell hybridization. *Somatic Cell Genet.* 8:791–800.

Lipson, L. G. (1984a) Special problems in treatment of hypertension in the patient with diabetes mellitus. *Arch. Intern. Med.* 144:1829–1831.

———. (1984b) Treatment of hypertension in diabetic men: Problems with sexual dysfunction. *Am. J. Cardiol.* 53:46A-50A.

Little, J. W., and Mount, D. W. (1982) The SOS regulatory system of *Escherichia coli. Cell* 29:11–22.

Liu, R. K., and Walford, R. L. (1966) Increased growth and life-span with lowered ambient temperature in the annual fish *Cynolebias adolffi. Nature* 121:1277–1278.

———. (1969) Laboratory studies on life span, growth, aging, and pathology of the annual fish, *Cynolebias bellottii* Steindachner. *Zoologica* 54:1–16.

———. (1970) Observations on the lifespan of several species of annual fishes and of the world's smallest fishes. *Exp. Gerontol.* 5:241–246.

———. (1972) The effect of lowered body temperature on lifespan and immune and nonimmune processes. *Gerontologia* (Basel) 18:363–388.

———. (1975) Mid-life temperature-transfer effects on life-span of annual fish. *J. Gerontol.* 30:129–131.

Liu, S. C., Meagher, K., and Hanawalt, P. C. (1985) Role of solar conditioning in DNA repair response and survival of human epidermal keratinocytes following UV irradiation. *J. Invest. Dermatol.* 85:93–97.

Liu, S. C., Parson, C. S., and Hanawalt, P. C. (1982) DNA repair response in human epidermal keratinocytes from donors of different age. *J. Invest. Dermatol.* 79:330–335.

———. (1983) DNA repair in cultured keratinocytes. *J. Invest. Dermatol.* 81:179–183.

Lloyd, M. (1984) Periodical cicadas. *Antenna* 8:79–91.

———. (1987) A successful rearing of 13–year periodical cicadas beyond their present range and beyond that of 17–year cicadas. *Am. Midl. Nat.* 117:363–368.

Lloyd, M., Kritsky, C., and Simon, C. (1983) A simple Mendelian model for 13– and 17–year life cycles of periodical cicadas, with historical evidence of hybridization between them. *Evolution* 37:1162–1180.

Lloyd, M., and White, J. (1987) Xylem feeding by periodical cicada nymphs on pine and grass roots, with new suggestions for pest control in conifer plantations and orchards. *Ohio J. Sci.* 87:50–54.

Lloyd, T. (1984) Food restriction increases life span of hypertensive animals. *Life Sci.* 34:401–407.

Lockshin, R. A. (1969) Programmed cell death. Activation of lysis by a mechanism involving the synthesis of protein. *J. Insect Physiol.* 15:1505–1516.

Lockshin, R. A., and Zimmerman, J. A. (1983) Insects: Endocrinology and aging. In *Endocrinology of Insects,* R. G. Downer and H. Laufer (eds.), 395–406. New York: Liss.

Loeb, J. (1908) Über den Temperaturkoeffizienten für die Lebensdauer kaltblütiger Tiere usw. *Pflugers Arch.* 124:411–426.

Loeb, J., and Northrop, J. H. (1916) Is there a temperature coefficient for the duration of life? *Proc. Natl. Acad. Sci.* 2:456–457.

———. (1917) On the influence of food and temperature upon the duration of life. *J. Biol. Chem.* 32:102–121.

London, E. D., Waller, S. B., Ellis, A. T., and Ingram, D. K. (1985) Effects of intermittent feeding on neurochemical markers in aging rat brain. *Neurobiol. Aging* 6:199–204.

Long, A. G. (1979) Observations on the Lower Carboniferous genus *Pitus* Witham. *Trans. Roy. Soc. Edinburgh* 70:111–127.

Lopes, S. A., and Rene, A. (1973) Effect of 17–ketosteroids on glucose-6–phosphate dehydrogenase activity (G6PD) and on G6PD isoenzymes. *Proc. Soc. Exp. Biol. Med.* 142:258–261.

Loranger, A. W. (1984) Sex differences in age at onset of schizophrenia. *Arch. Gen. Psychiatr.* 41:157–161.

Loschiavo, S. R. (1968) Effect of oviposition sites on egg production and longevity of *Trogoderma parabile* (Coleoptera: Dermistidae). *Can. Entomol.* 100:86–89.

Lotz, M., Carson, D. A., and Vaughan, J. H. (1987) Substance P activation of rheumatoid synoviocytes: Neural pathway in pathogenesis of arthritis. *Science* 235:893–895.

Love, R. M. (1970) *The Chemical Biology of Fishes.* New York: Academic Press.

Low, P. S., Waugh, S. M., Zinke, K., and Drenckhahn, D. (1985) The role of hemoglobin denaturation and band 3 clustering in red blood cell aging. *Science* 227:531–533.

Lowe, H. J., and Taylor, L. R. (1964) Population parameters, wing production, and behaviour in red and green *Acyrthosiphon pisum* (Harris) (Homoptera: Aphididae). *Entomol. Exp. Appl.* 7:287–295.

Lu, J. K., Hopper, B. R., Vargo, T. M., and Yen, S. S. C. (1979) Chronological changes in sex steroid, gonadotropin, and prolactin secretion in aging female rats displaying different reproductive states. *Biol. Reprod.* 21:193–203.

Lubbock, J. (1885) Longevity. *J. Linn. Soc. Zool.* (London) 20:133.

———. (1888) Phytobiological observation: On the forms of seedlings and the causes to which they are due. *J. Linn. Soc. Zool.* (London) 23:62–87.

Luckinbill, L. S., Arking, R., Clare, M. J., Cirocco, W. C., and Buck, S. A. (1984) Selection for delayed senescence in *Drosophila melanogaster. Evolution* 38:996–1003.

Luckinbill, L. S., and Clare, M. J. (1985) Selection for life span in *Drosophila melanogaster. Heredity* 55:9–18.

Luckinbill, L. S., Graves, J. L., Tomkiw, A., and Sowirka, O. (1988) A qualitative analysis of life history characters in *Drosophila melanogaster. Evol. Ecol.* 2:85–94.

Ludwig, D., and Fiore, C. (1960) Further studies on the relationship between parental age and the life cycle of the mealworm *Tenebrio molitor. Ann. Entomol. Soc. Am.* 53:595–600.

———. (1961) Effects of parental age on offspring from isolated pairs of the mealworm *Tenebrio molitor. Ann. Entomol. Soc. Am.* 54:463–464.

Ludwig, E. (1964) Diffuse alopecia in women: Its clinical forms and probable causes. *J. Soc. Cosmetic Chem.* 15:437–446.

Lumme, J. (1981) Localization of the genetic unit controlling the photoperiodic adult diapause in *Drosophila littoralis. Hereditas* 94:241–244.

———. (1982) The genetic basis of the photoperiodic timing of the onset of winter dormancy in *Drosophila littoralis. Acta Univ. Ouluenisis,* ser. A, Scientiae Rerum Naturalium 129, Biologica 16.

Lumme, J., and Keranen, L. (1978) Photoperiodic diapause in *Drosophila lummei* Hackman is controlled by an X-chromosomal factor. *Hereditas* 89:261–262.

Lumme, J., and Lakovaara, S. (1983) Seasonality and diapause in *Drosophila*. In *The Genetics and Biology of the Drosophila,* vol. 3C, M. Ashburner, H. L. Carson, and J. N. Thompson, 171–220. London: Academic Press.

Lumme, J., and Oikarinen, A. (1977) The genetic basis of the geographically variable photoperiodic diapause in *Drosophila littoralis. Hereditas* 86:129–142.

Lumme, J. A., Oikarinen, S., Lakovaara, S., and Alatalo, R. (1974) The environmental regulation of adult diapause in *Drosophila littoralis. J. Insect Physiol.* 20:2023–2033.

Lumpkin, C. K., McClung, J. K., Pereira-Smith, O. M., and Smith, J. R. (1987) Existence of high abundance antiproliferative mRNA's in senescent human diploid fibroblasts. *Science* 232:393–395.

Lumsden, A. G. S. (1988) Spatial organization of the epithelium and the role of neural crest cells in the initiation of the mammalian tooth germ. *Development* 103:155–169.

Luo, S. W., and Hultin, H. (1986) Effect of age of winter flounder on some properties of the sarcoplasmic reticulum. *Mech. Age. Dev.* 35:275–289.

Lutenegger, W. (1976) Allometry of neonatal size in eutherian mammals. *Nature* 263:229–230.

Lutes, P. B., Doroshov, S. I., Chapman, F., Harrah, J., Fitzgerald R., and Fitzpatrick, M. (1987) Morpho-physiological predictors of ovulatory success in white sturgeon, *Acipenser transmontanus* Richardson. *Aquaculture* 66:43–52.

Luthardt, F. W. (1977) Cytogenetic analyses of human oocytes. *Am. J. Human Genet.* 29:71A.

Luthardt, F. W., Palmer, C. G., and Yu, P. L. (1973) Chiasma and univalent frequencies in aging female mice. *Cytogenetics* 12:68–79.

Luzzatto, L. (1989) One enzyme from two genes? *Nature* 341:286–287.

Lwoff, A. (1944) *L'Evolution Physiologique.* Paris: Herman.

Lyman, C. P., O'Brien, R. C., Greene, G. C., and Papagrangos, E. D. (1981) Hibernational

longevity in the Turkish hamster *Mesocricetus brandti. Science* 212:668–670.

Lynch, G., Larson, J., and Baudry, M. (1986) Proteases, neuronal stability, and brain aging: An hypothesis. In *Treatment and Development of Strategies for Alzheimer's Disease,* T. Crook, R. T. Bartus, S. Ferris, and S. Gershon (eds.), 119–149. Madison, CT: Mark Powley Associates.

Lynch, K. R., Dolan, K. P, Nakhasi, H. L., Unterman, R., and Feigelson, P. (1982) The role of growth hormone in alpha$_{2u}$-globulin synthesis: A reexamination. *Cell* 28:185–189.

Lyon, M. F., and Hulse, E. V. (1971) Inherited kidney disease of mice resembling human nephronophthsis. *J. Med. Genet.* 8:41–48.

McAlpine, J. F. (1989) Phylogeny and classification of the Muscomorpha. In *Manual of Nearctic Diptera,* vol. 3, J. F. McAlpine and D. M. Wood (eds.). *Res. Br. Agric. Can.* 32:1397–1518.

McAlpine, J. F., and Wood, D. M. (eds.) (1989) *Manual of Nearctic Diptera.* Vol. 3. *Res. Br. Agric. Can.* 32:1333–1518.

MacArthur, J. W., and Baillie, W. H. T. (1929) Metabolic activity and duration of life. 2. Metabolic rates and their relation to longevity in *Daphnia magna. J. Exp. Zool.* 53:243–269.

McArthur, M. C., and Sohal, R. S. (1982) Relationship between metabolic rate aging, lipid peroxidation, and fluorescent age pigment in milkweed bug, *Oncopeltus fasciatus* (Hemiptera). *J. Gerontol.* 37:268–274.

MacArthur, R. H., and Wilson, E. O. (1967) *The Theory of Island Biogeography.* Princeton, NJ: Princeton Univ. Press.

McBride, J. R., Fagerlund, U. H. M., Smith, M., and Tomlinson, N. (1965) Post-spawning death of Pacific salmon: Sockeye salmon (*Oncorhynchus nerka*) maturing and spawning in captivity. *J. Fish. Res. Bd. Can.* 22:775–782.

McBride, J. R., and van Overbeeke, A. P. (1969) Hypertrophy of the interrenal tissue in sexually maturing sockeye salmon (*Oncorhynchus nerka*) and the effect of gonadectomy. *J. Fish. Res. Bd. Can.* 26:2975–2985.

———. (1971) Histological effects of 11-keto-testosterone, 17α-methyltestosterone, estradiol, estradiol cypionate, and cortisol on the interrenal tissue, thyroid gland, and pituitary gland of gonadectomized sockeye salmon (*Oncorhynchus nerka*). *J. Fish. Res. Bd. Can.* 28:477–484.

McCafferty, W. P., and Edmunds, G. F., Jr. (1979) The higher classification of the Ephemeroptera and its evolutionary basis. *Ann. Entomol. Soc. Am.* 72:5–12.

McCann, J., Choi, E., Yamasaki, E., and Ames, B. N. (1976) Detection of carcinogens as mutagens in the *Salmonella* microsome test: Assay of 300 chemicals. *Proc. Natl. Acad. Sci.* 72:5135–5139.

McCarter, M., Masoro, E. J., and Yu, B. P. (1985) Does food restriction retard aging by reducing the metabolic rate? *Am. J. Physiol.* 248:E488–E490.

McCay, C. M. (1947) Effect of restricted feeding upon aging and chronic diseases in rats and dogs. *Am. J. Publ. Hlth.* 37:521–528.

McCay, C. M., Crowell, M. F., and Maynard, L. A. (1935) The effect of retarded growth upon the length of the life span and upon the ultimate body size. *J. Nutrit.* 10:63.

McCay, C. M., Ellis, G. H., Barnes, L. L., Smith, C. A. H., and Sperling, G. (1939) Chemical and pathological changes in aging and after retarded growth. *J. Nutrit.* 18:15–25.

McCay, C. M., Maynard, L. A., Sperling, M. G., and Barnes, L. L. (1939) Retarded growth, life span, ultimate body size, and age changes in the albino rat after feeding diets restricted in calories. *J. Nutrit.* 18:1–13.

McCay, C. M., Sperling, G., and Barnes, L. L. (1943) Growth, ageing, chronic diseases, and life span in rats. *Arch. Biochem.* 2:469–479.

McCleery, R. H., and Perrins, C. M. (1988) Lifetime reproductive success of the great tit, *Parus major*. In *Reproductive Success. Studies of Individual Variation in Contrasting Breeding Systems*, T. H. Clutton-Brock (ed.)., 136–172. Chicago: Univ. of Chicago Press.

McClure, H. M. (1980) Bacterial diseases in nonhuman primates: Literature review and observations. In *The Comparative Pathology of Zoo Animals*, R. J. Montali and G. Migaki (eds.), 197–218. Washington, DC: Smithsonian Institution Press.

McCormick, J. J., and Maher, V. M. (1989) Malignant transformation of mammalian cells in culture, including human cells. *Environ. Mol. Mutagen.* 14, supp. 16:105–113.

McCullagh, K. G. (1972) Arteriosclerosis in the African elephant. *Atherosclerosis* 16:307–335.

McDonald, R., Hegenauer, J., and Saltman, P. (1986) Age-related differences in the bone mineralization pattern of rats following excercise. *J. Gerontol.* 41:445–452.

McDonald, R. A. (1961) "Lifespan" of liver cells. Autoradiographic study using tritiated thymidine in normal, cirrhotic, and partially hepatectomized rats. *Arch. Intern. Med.* 107:335–343.

MacDougal, D. T., and Long, F. L. (1927) Characters of cells attaining great age. *Am. Nat.* 61:385–406.

McDowall, R. M. (1971a) Fishes of the family Aplochitonidae. *J. Roy. Soc. New Zealand* 1:31–52.

———. (1971b) The galaxiid fishes of South America. *J. Linn. Soc. Zool.* (London) 50:33–73.

———. (1978) *New Zealand Freshwater Fishes. A Guide and Natural History.* Auckland, New Zealand: Heineman.

———. (1979) Fishes of the family Retropinnidae (Pisces: Salmoniformes). A taxonomic revision and synopsis. *J. Roy. Soc. New Zealand* 9:85–121.

———. (1987) The occurrence and distribution of diadromy among fishes. *Am. Fish. Soc. Symp.* 1:1–13.

———. (1988) *Diadromy in Fishes. Migrations between Freshwater and Marine Environments.* Portland, OR: Timber Press.

McFadden, P. N., and Clarke, S. (1982) Methylation at D-aspartyl residues in erythrocytes: Possible steps in the repair of aged membrane proteins. *Proc. Natl. Acad. Sci.* 79:2460–2464.

McFadden, P. N., Horwitz, J., and Clarke, S. (1983a) Protein carboxyl methylation in mammalian eye lens. *Fed. Proc.* 42:1912 (abstract).

————. (1983b) Protein methyltransferase from cow eye lens. *Biochem. Biophys. Res. Commun.* 113:418–424.

MacFadyen, D. (1990) International demographic trends. In *Improving the Health of Older People: A World View,* R. L. Kane, J. G. Evans, and D. Macfadyen (eds.), 19–29. Oxford: Oxford Univ. Press.

McGeer, P. L., Itagaki, S., Akiyama, H., and McGeer, E. G. (1989) Comparison of neuronal loss. In *Parkinson's Disease and Aging,* D. B. Calne (ed.), 25–34. *Aging* 30. New York: Raven Press.

McGeer, P. L., McGeer, E. G., and Suzuki, J. S. (1977) Aging and extrapyramidal function. *Arch. Neurol. Chicago* 34:33–35.

McGill, H. C. (1984) Persistent problems in the pathogenesis of atherosclerosis. *Arteriosclerosis* 4:443–451.

MacGinitie, G. E. (1939) The natural history of the blind goby, *Typhlogobius californiensus steindachner. Am. Midl. Nat.* 21:489–505.

MacGorman, L. R., Rizza, R. A., and Gerich, J. E. (1981) Physiological concentrations of growth hormone exert insulin-like and insulin antagonistic effects on both hepatic and extrahepatic tissues in man. *J. Clin. Endocrinol. Metab.* 53:556–559.

McGrath, J., and Solter, D. (1984) Completion of mouse embryogenesis requires both the maternal and paternal genomes. *Cell* 37:179–183.

Machidori, S., and Kato, F. (1984) Spawning populations and marine life of masu salmon (*Oncorhynchus masou*). International North Pacific Fisheries Commission, Vancouver, B.C., bulletin 43:1–138.

Macieira-Coelho, A., Diatloff, C., and Malaise, E. (1977) Concept of fibroblast aging in vitro: Implications for cell biology. *Gerontology* 23:290–305.

Macieira-Coelho, A., and Taboury, F. (1982) A reevaluation of the changes in cell proliferation in human fibroblasts during aging *in vitro. Cell Tiss. Kinet.* 15:213–224.

MacIntyre, I., Stevenson, J. C., Whitehead, M. I., Malawansa, S. J., Banks, L. M., and

Healy, M. J. (1988) Calcitonin for prevention of postmenopausal bone loss. *Lancet* 1:900–902.

MacKay, P. A. (1987) Production of sexual and asexual morphs and changes in reproductive sequences associated with photoperiod in the pea aphid, *Acyrthosiphon pisum* (Harris). *Can. J. Zool.* 65:2602–2606.

McKerrow, J. H. (1979) Non-enzymatic posttranslational amino acid modifications in aging. *Mech. Age. Dev.* 10:371–377.

McKnight, G. S., Lee, D. C., and Palmiter, R. D. (1980) Transferrin gene expression. Regulation of mRNA transcription in chick liver by steroid hormones and iron deficiency. *J. Biol. Chem.* 255:148–153.

McKusick, V. A. (1975) *Mendelian Inheritance in Man: Catalogues of Autosomal Dominant, Autosomal Recessive, and X-Linked Phenotypes.* 4th ed. Baltimore: Johns Hopkins Univ. Press.

————. (1982) *Mendelian Inheritance in Man: Catalogs of Autosomal Dominant, Autosomal Recessive, and X-Linked Phenotypes.* 6th ed. Baltimore: Johns Hopkins Univ. Press.

————. (1986) *Mendelian Inheritance in Man: Catalogs of Autosomal Dominant, Autosomal Recessive, and X-linked Phenotypes.* 7th ed. Baltimore: Johns Hopkins Univ. Press.

McLaren, A. (1958) The biology of the ringed seal (*Phoca hispida* Schreber) in the eastern Canadian arctic. Bulletin 118. Ottawa: Fisheries Research Board of Canada.

McLean, D. M. (1978) Land floras: The major late Proterozoic atmospheric carbon dioxide/oxygen control. *Science* 200:1060–1062.

McMahon, T. A. (1973) Size and shape in biology. Elastic criteria impose limits on biological proportions and consequently on metabolic rates. *Science* 179:1201–1204.

————. (1975) Using body size to understand the structural design of animals: Quadrupedal locomotion. *J. Appl. Physiol.* 39:619–627.

McNab, B. K. (1980) Food habits, energetics, and the population biology of mammals. *Am. Nat.* 116:106.

McNamara, J. J., and Malot, M. A. (1971) Coronary artery disease in combat casualties in Vietnam. *J.A.M.A.* 216:1185–1187.

McQuarrie, I. G., Brady, S. T., and Lasek, R. J. (1989) Retardation in the slow axonal transport of cytoskeletal elements during maturation and aging. *Neurobiol. Aging* 10:359–365.

Maddocks, I. (1961a) The influence of standard of living on blood pressure in Fiji. *Circulation* 24:1220–1223.

———. (1961b) Possible absence of essential hypertension in two complete Pacific Island populations. *Lancet* 11:396–399.

Maeda, H., Gleiser, C. A., Masoro, E. J., Murata, I., McMahan, C. A., and Yu, B. P. (1985) Nutritional influences on aging Fischer 344 rats. 2. Pathology. *J. Gerontol.* 40:671–688.

Magee, K., Basinska, J., Quarrington, B., and Stancer, H. C. (1970) Blindness and menarche. *Life Sci.* 9:7–12.

Maggenti, A. (1981) *General Nematology.* New York: Springer-Verlag.

———. (1982) Nemata. In *Synopsis and Classification of Living Organisms,* vol. 1, S. P. Parker (ed.), 879–928. New York: McGraw-Hill.

Maglio, V. J. (1970) Early Elephantidae of Africa and a tentative correlation of African Plio-Pleistocene deposits. *Nature* 225:328–332.

———. (1972) The evolution of mastication in the Elephantidae. *Evolution* 26:638–658.

———. (1973) Origin and evolution of the Elephantidae. *Trans. Am. Phil. Soc.,* n.s. 63:1–149.

Maisey, J. G. (1986) Heads and tails: A chordate phylogeny. *Cladistics* 2:201–256.

Makeham, W. H. (1867) On the law of mortality. *J. Inst. Actuaries* 13:325–358.

Makino, R., Esumi, H., Sato, S., Takashi, Y., Nagase, S., and Sugimura, T. (1982) Elevation of serum albumin concentration in analbuminemic rats by administration of 3′-methyl-4–dimethylaminobenzene. *Biochem. Biophys. Res. Commun.* 106:863–870.

Makino, R., Sato, S., Esumi, H., Negishi, C., Takano, M., Sugimura, T., Nagase, S., and Tanaka, H. (1986) Presence of albumin-positive cells in the liver of analbuminemic rats and their increase on treatment with hepatocarcinogens. *Jpn. J. Canc.* 77:153–159.

Makinodan, T., and Adler, W. H. (1975) The effects of aging on the differentiation and proliferation potentials of cells of the immune system. *Fed. Proc.* 34:153–158.

Makinodan, T., and Hirokawa, K. (1985) Normal aging of the immune system. In *Relations between Normal Aging and Disease,* H. A. Johnson (ed.), 117–132. *Aging* 28. New York: Raven Press.

Makinodan, T., Lubinski, J., and Fong, T. C. (1987) Cellular, biochemical, and molecular basis of T-cell senescence. *Arch. Pathol. Lab. Med.* 111:910–914.

Makinodan, T., and Peterson, W. J. (1964) Growth and senescence of the primary antibody-forming potential of the spleen. *J. Immunol.* 93:886–896.

Makman, M. H., and Stefano, G. B. (1984) Marine mussels and cephalopods as models for study of neuronal aging. In *Invertebrate Models in Aging Research,* D. H. Mitchell and T. E. Johnson (eds.), 166–189. Boca Raton, FL: CRC Press.

Makrides, S. C. (1983) Protein synthesis and degradation during aging and senescence. *Biol. Rev.* 58:343–422.

Mallouk, R. S. (1975) Longevity in vertebrates is proportional to relative brain weight. *Fed. Proc.* 34:2102–2103.

Malluche, H. H., Faugere, M-C., Ruch, M., and Friedler, R. (1988) Osteoblastic insufficiency is responsible for maintenance of osteopenia after loss of ovarian function in experimental Beagle dogs. *Endocrinology* 119:2649–2654.

Mammalian Models for Research on Aging. (1981) Washington, DC: National Academy Press.

Manion, P. J., and Smith, B. R. (1978) Biology of larval and metamorphosing sea lampreys, *Petromyzon marinus,* of the 1960 year class in the Big Garlic River, Michigan: Part 2, 1966–1972. Technical Report 30, Great Lakes Fisheries Commission.

Mann, C. V., Shaffer, R. D., Anderson, R. S., and Sandstead, H. H. (1964) Cardiovascular disease in the Masai. *J. Atheroscl. Res.* 4:289–312.

Mann, D. M. (1985) The neuropathology of Alzheimer's disease: A review with pathogenetic, aetiological, and therapeutic considerations. *Mech. Age. Dev.* 31:213–255.

———. (1988) The pathological association between Down syndrome and Alzheimer disease. *Mech. Age. Dev.* 43:99–136.

Mann, D. M., and Yates, P. O. (1979) The effects of ageing on the pigmented nerve cells of the human locus ceruleus and substantia nigra. *Acta Neuropathol.* (Berlin) 47:93–97.

Mann, D. M., Yates, P. O., and Barton, C. M. (1977) Neuromelanin and RNA in cells of the substantia nigra. *J. Neuropath. Exp. Neurol.* 36:379–383.

Mann, D. M., Yates, P. O., and Marcyniuk, B. (1984) Alzheimer's presenile dementia, senile dementia of Alzheimer type, and Down's syndrome in middle age form an age related continuum of pathological changes. *Neuropath. Appl. Neurobiol.* 10:185–207.

———. (1985) Some morphometric observations on the cerebral cortex and hippocampus in presenile Alzheimer's disease, senile dementia of Alzheimer type, and Down's syndrome in middle age. *J. Neurol. Sci.* 69:139–159.

Mann, D. M., Yates, P. O., and Stamp, J. E. (1978) The relationship between lipofuscin pigment and ageing in the human nervous system. *J. Neurol. Sci.* 37:83–93.

Mann, P. L., Kern, D. E., Kram, D., and Schneider, E. L. (1981) Relationship between in vivo mitomycin C exposure, sister chromatid exchange induction, and in vitro mitogenic proliferation. 2. Effect of aging on spleen cell mitogenesis and sister chromatid exchange induction. *Mech. Age. Dev.* 17:203–209.

Marcou, D. (1961) Notion de longévité et nature cytoplasmique du déterminent de la sénescence chez quelques champignons. *Ann. Sci. Nat. Bot.* 12: 663–764.

Marcus, R., Madvig, P., and Young, G. (1984) Age-related changes in parathyroid hormone and parathyroid hormone action in normal humans. *J. Clin. Endocrinol. Metab.* 58:223–230.

Marcusson, J. O., Morgan, D. G., Winblad, B., and Finch, C. E. (1984) Serotonin-2 binding sites in human frontal cortex and hippocampus: Selective loss of S-2A sites with age. *Brain Res.* 311–51–56.

Marechal, M. M. R., Lion, Y., and Duchesne, J. (1973) Radicaux libres organiques et longévité maximale chez les maniferes et les oiseaux. *Biochim. Comp.* 277:1085–1087.

Margolis-Nunno, H., Halpern-Sebold, L., and Schreibman, M. P. (1986) Immunocytochemical changes in serotonin in the forebrain and pituitary of aging fish. *Neurobiol. Aging* 7:17–21.

Margolis-Nunno, H., Schreibman, M. P., and Halpern-Sebold, L. (1987) Sexually dimorphic age-related differences in the immunocytochemical distribution of somatostatin in the platyfish. *Mech. Age. Dev.* 41:139–148.

Margulis, L., and Schwartz, K. V. (1988) *Five Kingdoms*. 2d ed. New York: W. H. Freeman.

Markofsky, J., and Perlmutter, A. (1972) Age at sexual maturity and its relationship to longevity in the male annual cyprinodont fish *Nothobranchius guenther. Exp. Gerontol.* 7:131–135.

———. (1973) Growth differences in subgroups of varying longevities in a laboratory population of the male annual cyprinodont fish, *Nothobranchius guentheri* (Peters). *Exp. Gerontol.* 8:65–73.

Marks, P. H., and Banks, J. (1960) Inhibition of mammalian glucose-6–phosphate dehydrogenase by steroids. *Proc. Natl. Acad. Sci.* 46:447–452.

Marks, R., Rennie, G., and Selwood, T. (1988) The relationship of basal cell carcinomas and squamous cell carcinomas to solar keratoses. *Arch. Dermatol.* 124:1039–1042.

Markus, H. C. (1934) The life history of the black-headed minnow, *Pimephales promelas. Copeia* 70:116–122.

Marlow, B. J. (1961) Reproductive behaviour of the marsupial mouse, *Antechinus flavipes* (Marsupialia) and the development of the pouch young. *Austral. J. Zool.* 9:203–218.

Marsh, H. (1980) Age determination of the dugong (*Dugong dugong*) (Muller) in northern Australia and its biological implication. In *Age Determination of Toothed Whales and*

Sirenians. Rep. Int. Whaling Commn., special issue, 3:181–201.

Marsh, H., and Kasuya, T. (1986) Evidence for reproductive senescence in female cetaceans. In *Behaviour of Whales in Relation to Management*, G. P. Donovan (ed.). *Rep. Int. Whaling Commn.*, special issue, 8:57–74.

Martin, A. P., and Simon, C. (1988) Anomalous distribution of nuclear and mitochondrial DNA markers in periodical cicadas. *Nature* 336:237–239.

Martin, A. W., and Fuhrman, F. A. (1955) The relationship between summated tissue respiration and metabolic rate in the mouse and dog. *Physiol. Zool.* 28:18–34.

Martin, D. P., Schmidt, R. E., DiStefano, P. S., Lowry, O. H., Carter, J. G., and Johnson, E. M. (1988) Inhibitors of protein synthesis and RNA synthesis prevent neuronal death caused by nerve growth factor deprivation. *J. Cell Biol.* 106:829–844.

Martin, G., Sprague, C., and Epstein, C. (1970) Replicative life-span of cultivated human cells: Effects of donor's age, tissue, and genotype. *Lab. Invest.* 23:86–91.

Martin, G. M. (1978) Genetic syndromes in man with potential relevance to the pathobiology of aging. In *Genetic Effects on Aging*, D. Bergsma and D. E. Harrison (eds.), 5–39. Birth Defects: Original Article Series, no. 14. New York: Alan R. Liss.

———. (1982) Syndromes of accelerated aging. *Natl. Canc. Inst. Monograph* 60:241–247.

———. (1987) Interactions of aging and environmental agents: The gerontological perspective. In *Environmental Toxicity and the Aging Processes*, 25–80. New York: Liss.

Martin, G. M., Obgurn, C. E., and Wight, T. N., (1983) Comparative rates of decline in the primary cloning efficiencies of smooth muscle cells from the aging thoracic aorta of two murine species of contrasting maximum life span potentials. *Am. J. Pathol.* 110:236–245.

Martin, G. M., Smith, A. C., Ketter, D. J., Ogburn, C. E., and Distech, C. M. (1985) Increased chromosomal aberrations in first metaphases of cells isolated from the kidneys of aged mice. *Israel J. Med. Sci.* 21:296–301.

Martin, L. J., Cork, L. C., Koo, E. H., Sisodia, S. S., Weidemann, A., Beyreuther, K., Masters, C., and Price, D. L. (1989) Localization of amyloid precursor protein (APP) in brains of young and aged monkeys. *Soc. Neurosci. Abstr.* 15:23.

Martin, L. K., and Cailliet, G. M. (1988a) Age and growth determination of the bat ray, *Myliobatis californica* Gill, in central California. *Copeia*, no. 3: 762–773.

———. (1988b) Aspects of the reproduction of the bat ray, *Myliobatis californica*, in central California. *Copeia*, no. 3: 754–762.

Martin, R. D. (1981) Relative brain size and basal metabolic rate in terrestrial vertebrates. *Nature* 293:57–60.

———. (1984) Scaling effects and adaptive strategies in mammalian lactation. *Symp. Zool. Soc. Lond.* 51:87–117.

Marx, J. (1986) How killer cells kill their targets. *Science* 231:1367–1369.

Masaki, S., and Walker, T. J. (1987) Cricket life cycles. In *Evolutionary Biology*, vol. 21, M. K. Hecht, B. Wallace, and G. T. Prance (eds.), 349–423. New York: Plenum.

Masoro, E. J. (1985) Metabolism. In *Handbook of the Biology of Aging*, 2d ed., C. E. Finch and E. L. Schneider (eds.), 540–566. New York: Van Nostrand.

———. (1988) Food restriction in rodents: An evaluation of its role in the study of aging. *J. Gerontol.* 43:B59–B64.

Masoro, E. J., Katz, M. S., and McMahan, C. A. (1989) Evidence for the glycation hypothesis of aging from the food-restricted rodent model. *J. Gerontol.* 44:B20–B22.

Massie, H. R., Aiello, V. R., and Iodice, J. (1979) Changes with age in copper and superoxide dismutase levels in brains of C57BL/6J mice. *Mech. Age. Dev.* 10:93–99.

Massie, H. R., and Kogut, K. A. (1987) Influence of age on mitochondrial enzyme levels in *Drosophila. Mech. Age. Dev.* 38:119–126.

Massie, H. R., and Williams, T. R. (1987) Mitochondrial DNA and life span changes in normal and dewinged *Drosophila* at different temperatures. *Exp. Gerontol.* 22:139–153.

Masters, C. L., Multhaup, G., Simms, G., Pottgiesser, J., Martins, R. N., and Beyreuther, K. (1985a) Neuronal origin of a cerebral amyloid: Neurofibrillary tangles of Alzheimer's disease contain the same protein as the amyloid of plaque cores and blood vessels. *EMBO J.* 4:2757–2763.

Masters, C. L., Simms, G., Weinman, N. A., Multhaup, G., McDonald, B. L., and Beyreuther, K. (1985) Amyloid plaque core protein in Alzheimer disease and Down syndrome. *Proc. Natl. Acad. Sci.* 82:4245–4249.

Masters, J. N., Finch, C. E., and Sapolsky, R. M. (1989) Glucocorticoid endangerment of hippocampal neurons does not involve deoxyribonucleic acid cleavage. *Endocrinology* 124:3083–3088.

Masters, P. M. (1982) Amino acid racemization in structural proteins. In *Biological Markers of Aging*, M. E. Reff and E. L. Schneider (eds.), 120–137. NIH Publication 82–2221. Washington, DC: USDHHS.

———. (1983) Stereochemically altered noncollagenous protein from human dentin. *Calc. Tiss. Int.* 35:43–47.

Masters, P. M., Bada, J. L., and Zigler, J. S., Jr. (1977) Aspartatic acid racemisation in the human lens during ageing and in cataract formation. *Nature* 268:71–73.

———. (1978) Aspartic acid racemization in heavy molecular weight crystallins and water-soluble protein from normal human lens and cataracts. *Proc. Natl. Acad. Sci.* 75:1204–1208.

Masters-Helfman, P., and Bada, J. L. (1975) Aspartic acid racemization in tooth enamel from living humans. *Proc. Natl. Acad. Sci.* 72:2891–2894.

Masuda, H., Amaoka, K., Araga, C., Uyeno, T., and Yoshino, T. (1984) *The Fishes of the Japanese Archipelago*. Tokyo: Tokai Univ. Press. [not read]

Mather, K. (1938) Crossing-over. *Biol. Rev.* 13:252–292.

Matsumura, T., Zerrudo, Z., and Hayflick, L. (1979) Senescent human diploid cells in culture: Survival, DNA synthesis, and morphology. *J. Gerontol.* 34:328–334.

Matsushita, M., Tsuboyama, T., Kasai, R., Higuchi, H., Yamamuro, T., Higuchi, K., Kohno, A., Yonezu, T., Utani, A., Umezawa, M., and Takeda, T. (1986) Age-related changes in bone mass in the senescence-accelerated mouse (SAM) *Am. J. Pathol.* 125:276–283.

Matthews, L. H. (1950) Reproduction in the basking shark, *Cetorhinus maximus* (Gunner) *Phil. Trans. Roy. Soc.* (London), ser. B, 234:247–316.

Matthieu, O., Krauer, R., Hoppeler, H., Gehr, P., Lindstedt, S. L., Alexander, R. M., Taylor, C. R., and Weibel, E. R. (1981) Design of the mammalian respiratory system. 7. Scaling mitochondrial volume in skeletal muscle to body mass. *Resp. Physiol.* 44:113–128.

Maule, A. G., and Schreck, C. B. (1987) Changes in the immune system of coho salmon (*Oncorhynchus kisutch*) during the parr-to-smolt transformation and after implantation of cortisol. *Can. J. Fish. Aquat. Sci.* 44:161–166.

Maunders, M. J., Brown, S. B., and Woolhouse, H. W. (1983) The appearance of chlorophyll derivatives in senescing tissue. *Phytochemistry* 22:2443–2446.

Maupas, E. (1888) La multiplication des infusoires cilies. *Arch. Zool. Exp. Gen.*, ser. 2, 6:165–277. Discussed in Wilson, 1925, Bell, 1988.

———. (1900) Medes et formes de reproduction des nematodes. *Arch. Zool. Exp. Gen.*, ser. 13, 8:463–623.

Maurizio, A. (1946) Beobachtungen über die Lebensdauer und den Futterverbrauch gefangen gehaltener Bienen. *Beh. Schweiz. Beinen-Zeit.* 2:2–44.

———. (1954) Pollenernahrung und Lebensvorgange bei der Honigbiene. *Landw. Jb. Schweiz.* 68:115–182.

———. (1959) Factors influencing the lifespan of bees. *CIBA Found. Colloq. Aging* 5:231–234.

———. (1961) Lebensdauer and Altern bei der Honigbiene (*Apis mellifica L.*). *Gerontologia* (Basel) 5:110–128.

May, P. C., Johnson, S. A., Poirier, J., Lampert-Etchells, M., and Finch, C. E. (1989) Altered gene expression in Alzheimer's disease brain tissue. *Can. J. Neurol. Sci.* 16:473–476.

May, P. C., Lampert-Etchells, M., Johnson,

S. A., Poirier, J., Masters, J. N., and Finch, C. E. (1990) Dynamics of gene expression for a hippocampal glycoprotein elevated in Alzheimer's disease and in response to experimental lesions in rat. *Neuron.,* in press.

Mayer, P. J., and Baker, G. T. (1985) Genetic aspects of *Drosophila* as a model system of eukaryotic aging. *Int. Rev. Cytol.* 95:61–102.

Maynard Smith, J. (1958) The effects of temperature and of egg laying on the longevity of *Drosophila subobscura. J. Exp. Biol.* 35: 832–842.

———. (1959) A theory of ageing. *Nature* (London) 184:956.

———. (1962) The causes of ageing. *Proc. Roy. Soc.* (London), ser. B, 157:115–127.

Maynard Smith, J., Bozcuk, A. N., and Tebbutt, S. (1970) Protein turnover in adult *Drosophila. J. Insect Physiol.* 16:601–613.

Mayr, E. (1961) Cause and effect in biology. *Science* 134:1501– 1506.

Mays, A. W. (1967) Fecundity of Atlantic cod. *J. Fish Res. Bd. Can.* 24:1531–1551.

Mays-Hoopes, L. L. (1989) Development, aging, and DNA methylation. *Int. Rev. Cytol.* 114:118–220.

Mays-Hoopes, L. L., Brown, A., and Huang, R. C. (1983) Methylation and rearrangement of mouse intracisternal A particle genes in development, aging, and myeloma. *Mol. Cell. Biol.* 3:1371–1380.

Mays-Hoopes, L. L., Chao, W., Butcher, H. C., and Huang, R. C. (1986) Decreased methylation of the major mouse long interspersed repeated DNA during aging and in myeloma cells. *Dev. Genet.* 7:65– 73.

Mays-Hoopes, L. L., Jennings, M., Spuck, L., Chao, W., and Nelson, J. F. (1990) DNA methylation in aging mouse liver. In *Molecular Biology of Aging,* vol. 123, C. E. Finch and T. E. Johnson (eds.), 341–350. UCLA Symposia on Molecular and Cellular Biology. New York: Wiley-Liss.

Mead, J. F. (1976) Free radical mechanisms of lipid damage and consequences for cellular membranes. In *Free Radicals in Biology,* vol. 1, W. A. Pryor (ed.), 51–68. New York: Academic Press.

Mead, R. A. (1968) Reproduction in western forms of the spotted skunk (genus *Spilogale*). *J. Mammal.* 49:373–390.

———. (1971) Effects of light and blinding upon delayed implantation in the spotted skunk. *Biol. Reprod.* 5:214–220.

Meadow, N. D., and Barrows, C. H., Jr. (1971a) Studies on aging in a bdelloid rotifer. 1. The effect of various culture systems on longevity and fecundity. *J. Exp. Zool.* 176:303–314.

———. (1971b) Studies on aging in a bdelloid rotifer. 2. The effects of various environmental conditions and maternal age on longevity and fecundity. *J. Gerontol.* 26:302–309.

Meaney, M. J., and Aitken, D. H. (1985) The effects of early postnatal handling on the development of hippocampal glucocorticoid receptors: Temporal parameters. *Dev. Brain Res.* 22:301–304.

Meaney, M. J., Aitken, D. H., Bhatnagar, S., and Sapolsky, R. M. (1990) Postnatal handling attenuates certain neuroendocrine, anatomical and cognitive dysfunctions associated with aging in female rats. *Neurobiol. Aging.* In press.

Meaney, M. J., Aitken, D. H., Bhatnagar, S., Van Berkel, C., and Sapolsky, R. M. (1988) Postnatal handling attenuates neuroendocrine, anatomical, and cognitive impairments related to the aged hippocampus. *Science* 238:766–768.

Meaney, M. J., Aitken, D. H., and Sapolsky, R. M. (1987) Thyroid hormones influence the development of hippocampal glucocorticoid receptors in the rat: A mechanism for the effects of postnatal handling on the development of the adrenocortical stress response. *Neuroendocrinology* 45:278–283.

Medawar, P. B. (1946) Old age and natural death. *Mod. Q.* 1:30– 56.

———. (1952) *An Unsolved Problem of Biology.* London: H. K. Lewis.

———. (1957) *The Uniqueness of the Individual.* London: Methuen.

Medland, T. E., and Beamish, F. W. (1987) Age validation for the mountain brook lamprey, *Ichthyomyzon greeleyi. Can. J. Fish. Aquat. Sci.* 44:901–904.

Medley, G. F., Anderson, R. M., Cox, D. R., and Billard, L. (1987) Incubation period of AIDS in patients infected via blood transfusion. *Nature* 328:719–721.

Medori, R., Autilio-Gambetti, L., Monaco, S.,

and Gambetti, P. (1985) Experimental diabetic neuropathy: Impairment of slow transport with change in axon cross-sectional area. *Proc. Natl. Acad. Sci.* 82:7716–7720.

Medori, R., Jenich, H., Autilio-Gambetti, L., and Gambetti, P. (1988) Experimental diabetic neuropathy: Similar changes of slow axonal transport and axonal size in different animal models. *J. Neurosci.* 8:1814–1821.

Medvedev, Z. H. A. (1962) Ageing at the molecular level and some speculations concerning maintaining the functioning of systems for replicating specific macromolecules. In *Biological Aspects of Ageing*, N. W. Shock (ed.), 255–266. New York: Columbia Univ. Press.

———. (1966) *Protein Biosynthesis and Problems of Heredity*. New York: Plenum.

Meinecke, R. O. (1974) Retention of one-trial learning in neonate, young adult, and aged Japanese quail. *J. Gerontol.* 29:172–176.

Meites, J. (1990) Effects of aging on the hypothalamic-pituitary axis. *Rev. Biol. Res. Aging* 4:253–261.

Melton, D. A. (1987) Translocation of a localized maternal mRNA to the vegetal pole of *Xenopus* oocytes. *Nature* 328:80–82.

Melville, K. I., Garvey, H. L., and Gillis, R. A. (1973) Neurogenic lesions of heart muscle. In *Recent Advances in Studies on Cardiac Structure and Metabolism*, vol. 2, E. Bajusz and G. Rona (eds.), 433–447. Baltimore: Univeristy Park Press.

Meltzer, S. J. (1906) The factors of safety in animal structure and animal economy. *Harvey Lect.* (1906–1907):139–169.

Mennecier, F. (1970) Mise en évidence d'un matériel à reaction immunologique croisée (CRM) de l'aldolase dans les hémolysates de globules rouges et de reticulocytes de lapin. *Compt. Rend. Acad. Sci. Paris,* ser. D, 270:742–745.

Mennecier, F., and Dreyfus, J. C. (1974) Molecular ageing of fructose-diphosphates aldolase in tissues of rabbit and man. *Biochim. Biophys. Acta* 364:320–326.

Menzies, R. A., and Gold, P. H. (1971) The turnover of mitochondria in a variety of tissues of young adult and aged rats. *J. Biol. Chem.* 246:2425–2429.

———. (1972) The apparent turnover of mitochondria, ribosomes, and sRNA of the brain in young adult and aged rats. *J. Neurochem.* 19:1671–1683.

Meredith, P. J., and Walford, R. L. (1977) Effect of age on response to T- and B-cell mitogens in mice congenic at the H-2 locus. *Immunogenetics* 5:109.

Meroni, P. L., Barcellini, W., Frasca, D., Sguotti, C., Borghl, M. O., DeBartolo, G., Doria, G., and Zanussi, C. (1987) In vivo immunopotentiating activity of thymopentin in aging humans: Increase of IL-2 production. *Clin. Immunol. Immunopathol.* 42:151–159.

Merry, B. J., and Holehan, A. M. (1981) Serum profiles of LH, FSH, testosterone, and 5–alpha-DHT from 21 to 1,000 days of age in *ad libitum* fed and dietary restricted rats. *Exp. Gerontol.* 16:431–444.

Merry, B. J., Holehan, A. M., Lewis, S. E. M., and Goldspink, D. F. (1987) The effects of ageing and chronic dietary restriction on *in vivo* hepatic protein synthesis in the rat. *Mech. Age. Dev.* 39:189–199.

Mertens, R. (1970a) In memoriam Carl Koch. *Salamander* 6:1–2.

———. (1970b) Über die Lebensdauer einiger Amphibien und Reptilien in Gefangenschaft. *Zool. Gart.,* N.I. 39:193–209.

Mertz, D. B. (1971) The mathematical demography of the California condor population. *Am. Nat.* 105:437–453.

———. (1975) Senescent decline in flour beetle strains selected for early adult fitness. *Physiol. Zool.* 48:1–23.

Meruelo, D., and Edidin, M. (1975) Association of mouse liver adenosine 3'-5'-cyclic monophosphate (cyclic AMP) levels with histocompatibility-2 genotype. *Proc. Natl. Acad. Sci.* 72:2644–2648.

Messer, A., and Flaherty, L. (1986) Autosomal dominance in a late-onset motor neuron disease in the mouse. *J. Neurogenet.* 3:345–355.

Messer, A., Strominger, N. L., and Mazurkiewicz, J. E. (1987) Histopathology of the late-onset motor neuron degeneration (*Mnd*) mutant in the mouse. *J. Neurogenet.* 4:201–213.

Metchnikoff, E. (1915) La mort du papillon du Murier. Un chapitre de thanotologie. *Ann. Inst. Pasteur* 10:477–497.

Metropolitan Life Insurance Co. (1988) Women's longevity advantage declines. *Stat. Bull.* 69:18–23.

Meyer, R. A., Brown, T. R., and Dudley, G. A. (1985) PCr recovery rates in muscle of old vs adult rats. *Med. Sci. Sports Exer.* 17:257.

Meyer, T. E., Armstrong, M. J., and Warner, C. M. (1989) Effects of H-2 haplotype and gender on the lifespan of A and C57BL/6J mice and their F_1, F_2, and backcross offspring. *Growth Dev. Aging* 53:175–183.

Michael, S. D. (1979) The role of the endocrine thymus in female reproduction. *Arthrit. Rheum.* 22:1241–1245.

Michael, S. D., Taguchi, O., Nishizaka, Y., McClure, J. E., Goldstein, A. L., and Barkley, M. S. (1981) The effect of neonatal thymectomy on early follicular loss and circulating levels of corticosterone, progesterone, estradiol, and thymosin. In *Dynamics of Ovarian Function,* N. B. Schwartz and M. Hunzicker-Dunn (eds.), 279–284. New York: Raven Press.

Michel, F., and Lang, B. F. (1985) Mitochondrial class II introns encode proteins related to the reverse transcriptase of retrovirus. *Nature* (London) 316:641–643.

Michener, C. D. (1974) *The Social Behavior of the Bees. A Comparative Study.* Cambridge, MA: Belknap Press.

Michener, C. D., and Grimaldi, D. A. (1988) The oldest fossil bee: Apoid history, evolutionary stasis, and antiquity of social behavior. *Proc. Natl. Acad. Sci.* 85:6424–6426.

Michod, R. E. (1979) Evolution of life histories in response to age-specific mortality factors. *Am. Nat.* 113:531–550.

Michod, R. E., and Levin, B. R. (eds.) (1987) In *The Evolution of Sex: An Examination of Current Ideas.* Sunderland, MA: Sinauer.

Michon, J. (1953) Les phases du développement postembryonnaire chez les Lumbricidae à diapause. Un cas de réversibilité. *Compt. Rend. Acad. Sci. Paris* 236:2545–2547.

———. (1954) Contribution expérimentale à l'étude de la biologie des Lumbricidae. Les variations ponderales au cours des différentes modalités du développement postembryonnaire. Ph.D. thesis, Universitaire de Poitiers, France.

Miesfeld, R. (1989) The structure and function of steroid receptor proteins. *CRC Crit. Rev. Biochem. Mol. Biol.* 24:101–117.

Migdal, S., Abeles, R. P., and Sherrod, L. R. (1981) *An Inventory of Longitudinal Studies of Middle and Old Age.* New York: Social Science Research Council.

Migeon, B. R., Axelman, J., and Beggs, A. H. (1988) Effect of ageing on reactivation of the human X-linked HPRT locus. *Nature* 335: 93–96.

Mikkelsen, M., Poulsen, H., Grinsted, J., and Lange, A. (1980) Nondisjunction in trisomy-21: Study of chromosomal heteromorphisms in 110 families. *Ann. Human Genet.* 44:17–28.

Miklos, G. L. G. (1974) Sex-chromosome pairing and male fertility. *Cytogenet. Cell. Genet.* 13:558–577.

Milkman, R. (1967) Genetic and developmental studies of *Botryllus schlosseri. Biol. Bull.* 132:229–243.

Millar, J. S., and Zammuto, R. M. (1983) Life histories of mammals. An analysis of life tables. *Ecology* 64:631–635.

Millecchia, L. L., and Rudzinska, M. A. (1970) The ultrastructure of brood pouch formation in *Tokophrya infusionum. J. Protozool.* 17:574–583.

———. (1971) The ultrastructure of nuclear division in a suctorian, *Tokophrya infusionum. Z. Zellforsch.* 115:149–164.

Miller, A. E., and Riegle, G. D. (1982) Temporal patterns of serum luteinizing hormone and testosterone and endocrine response to luteinizing hormone releasing hormone in aging male rats. *J. Gerontol.* 37:522–528.

Miller, A. R. (1988) A set of life tables for theoretical gerontology. *J. Gerontol.* 43:B43–B49.

Miller, C. N. (1976) Early evolution in the Pinaceae. *Rev. Paleobot. Palynol.* 21:101–117.

Miller, H. M. (1931) Alteration of generations in the rotifer *Lecane inermis* Bryce. 1. Life histories of the sexual and non-sexual generations. *Biol. Bull.* (Woods Hole) 60:345–381.

Miller, J. K., Bolla, R., and Denckla, W. D. (1980) Age associated changes in initiation of ribonucleic acid synthesis in isolated rat liver nuclei. *Biochem. J.* 188:55–60.

Miller, M. (1984) Chronic exposure to vasopressin in aging rats impairs renal response to vasopressin. Presented at Seventh International Congress of Endocrinology, Montreal. *Excerpta Medica* 652:853 (abstract).

———. (1985) Beneficial effect of partial vasopressin deficiency on water regulation in aging rats. Proceedings of the 67th Endocrine Society Meeting, Baltimore, 71 (abstract).

———. (1987) Increased vasopressin secretion: An early manifestation of aging in the rat. *J. Gerontol.* 42:249–258.

Miller, M. M., Joshi, D., Billiar, R. B., and Nelson, J. F. (1990) Loss of LH-RH neurons in the rostral forebrain of old female C57BL/6J mice. *Neurobiol. Aging* 11:217–222.

Miller, O. J., Miller, D. A., Dev, V. G., Tantravahi, R., and Croce, M. L. (1976) Expression of human and suppression of mouse nucleolus organizer activity in mouse-human somatic cell hybrids. *Proc. Natl. Acad. Sci.* 73:4531–4535.

Miller, P. J. (1973) The species of *Pseudaphya* (Teleostei: Gobiidae) and the evolution of aphyiine gobies. *J. Fish. Biol.* 5:353–365.

———. (1975) Age-structure and life-span in the common goby, *Pomatoschistus microps.* *J. Zool.* (London) 177:425–448.

———. (1979) Adaptiveness and implications of small size in teleosts. *Symp. Zool. Soc. Lond.* 44:263–306.

Miller, R. A. (1989) The cell biology of aging: Immunological models. *J. Gerontol.* 44:B4–B8.

———. (1990) Aging and the immune response. In *Handbook of the Biology of Aging,* 3d ed., E. L. Schneider and J. W. Rowe (eds.), 157–180. San Diego: Academic Press.

Miller, R. A., and Harrison, D. E. (1985) Delayed reduction in T cell precursor frequencies accompanies diet-induced lifespan extension. *J. Immunol.* 134:1426–1429.

Miller, R. B., and Kennedy, W. A. (1948) Observations on the lake trout of Great Bear Lake. *J. Fish. Res. Bd. Can.* 7:176.

Miller, S. E., and Hadfield, M. G. (1990) Developmental arrest life span extension in a marine mollusc. *Science* 248:356–358.

Miller, W. A., Sanson, G. D., and Odell, D. K. (1980) Molar progression in the mana-

tee (*Trichechus manatus*). *Anat. Rec.* 196:128A (abstract).

Millington, W. F., and Chaney, W. R. (1973) Shedding of shoots and branches. In *Shedding of Plant Parts,* T. T. Kozlowski (ed.), 149–204. New York: Academic Press.

Mills, C. A., and Mann, R. H. (1983) The bullhead, *Cottus gobio,* a versatile and successful fish. *Rep. Freshwater Biol. Assoc.* 51:76–88.

Millward, D. J., and Waterlow, J. C. (1978) Effect of nutrition on protein turnover in skeletal muscle. *Fed. Proc.* 37:283–290.

Milunsky, A., and Neurath, P. (1968) Diabetes mellitus in Down's syndrome. *Arch. Environ. Health* 17:372–376.

Minaker, K. L., Meneilly, G. S., and Rowe, J. W. (1985) Endocrine systems. In *Handbook of the Biology of Aging,* 2d ed., C. E. Finch and E. L. Schneider (eds.), 423–457. New York: Van Nostrand.

Mink, J. W., Blumenschine, R. J., and Adams, D. B. (1981) Ratio of central of nervous system to body metabolism in vertebrates: Its constancy and functional basis. *Am. J. Physiol.* 241:203–212.

Mintz, B. (1974) Gene control of mammalian differentiation. *Ann. Rev. Genet.* 8:411–470.

Mintz, N., and Geist, S. (1941) The adrenal cortex in its relation to virilism. *J. Clin. Endocrinol.* 1:316–326.

Miquel, J. (1971) Aging of male *Drosophila melanogaster.* Histology, histochemistry, and ultrastructural observation. *Adv. Gerontol. Res.* 3:39–71.

Miquel, J., Bensch, K. G., Philpott, D. E., and Atlan, H. (1972) Natural aging and radiation-induced life shortening in *Drosophila melanogaster. Mech. Age. Dev.* 1:71–97.

Miquel, J., Lundgren, P. R., Bensch, K., and Atlan, H. (1976) Effects of temperature on the lifespan, vitality, and fine structure *Drosophila melanogaster. Mech. Age. Dev.* 5:347.

Miquel, J., Lundgren, P. R., and Binnard, R. (1972) Negative geotaxis and mating behavior in control and gamma-irradiated *Drosophila. Drosophila Info. Service* 48:60–61.

Miquel, J., and Philpott, D. E. (1986) Structural correlates of aging in *Drosophila*: Relevance to the cell differentiation, rate-of-living, and free radical theories of aging. In *Insect Aging: Strategies and Mechanisms,* K.-G. Collatz and R. S. Sohal (eds.), 116–129. Heidelberg: Springer-Verlag.

Mirsky, A. E., and Ris, H. (1950) Quantitative cytochemical determination of deoxyribonucleic acid: The Feulgen nucleal reaction. *J. Gen. Physiol.* 33:125–146.

Miyakawa, T., Shimoji, A., Kuramoto, R., and Higuchi, Y. (1982) The relationship between senile plaques and cerebral blood vessles in Alzheimer's disease and senile dementia. Morphological mechanism of senile plaque formation. *Virchow's Arch. B. Cell Pathol.* 40:121–129.

Mobbs, C. V. (1990). Neurotoxic effects of estrogen, glucose, and glucocorticoids: Neurohumoral hysteresis and its pathological consequences during aging. *Rev. Biol. Res. Aging* 4:201–228.

Mobbs, C. V., Flurkey, K., Gee, D. M., Yamamoto, K., Sinha, Y. N., and Finch, C. E. (1984) Estradiol-induced adult anovulatory syndrome in female C57BL/6J mice: Age-like neuroendocrine, but not ovarian, impairments. *Biol. Reprod.* 30:556–563.

Mobbs, C. V., Gee, D. M., and Finch, C. E. (1984) Reproductive senescence in female C57BL/6J mice: Ovarian impairments and neuroendocrine impairments that are partially reversible and delayable by ovariectomy. *Endocrinology* 115:1653–1662.

Mobbs, C. V., Kannegieter, L. S., and Finch, C. E. (1985) Delayed anovulatory syndrome induced by estradiol in female C57BL/6J mice: Age-like neuroendocrine, but not ovarian, impairments. *Biol. Reprod.* 32:1010–1017.

Modak, S. P., Deobagkar, D. D., Leuba-Gfeller, G., Conet, C., and Basu-Modak, S. (1986) Genetic information in aging cells. In *The Biology of Human Aging,* A. H. Bittles and K. J. Collins (eds.), 17–32. London: Cambridge Univ. Press.

Mohandas, T., Sparkes, R. S., and Shapiro, L. J. (1981) Reactivation of an inactive X chromosome: Evidence for X-chromosome inactivation by DNA methylation. *Science* 211:393–396.

Mohr, E., Schmitz, E., and Richter, D. (1988) A single-rat genomic DNA fragment encodes both the oxytocin and vasopressin genes separated by 11 kilobases and oriented in opposite transcriptional directions. *Biochemie* 70:649–654.

Mohs, R. C., Breitner, J. C., Silverman, J. M., and Davis, K. L. (1987) Alzheimer's disease: Morbid risk among first degree relatives approximates 50% by age 90. *Arch. Gen. Psychiatry* 44:405–408.

Moldave, K., Harris, J., Sabo, W., and Sadnik, I. (1979) Protein synthesis and aging: Studies with cell-free mammalian systems. *Fed. Proc.* 38:1979–1983.

Molisch, H. (1938) *The Longevity of Plants.* Translated by H. Fulling. Lancaster, PA: Science Press.

Moller, G. (1968) Regulation of cellular antibody synthesis. Cellular 7S production and longevity of antigen-sensitive cells in the absence of antibody feedback. *J. Exp. Med.* 127:291–306.

Molnar, J. A., Alpert, N., Burke, J. F., and Young, V. R. (1986) Synthesis and degradation rates of collagens *in vivo* in whole skin of rats, studied with $^{18}O_2$ labelling. *Biochem. J.* 240:431–435.

Moltoni, E. (1947) Fringuello vissuto in schiavitù per ben 29 anni. *Riv. Lat. Orn.* 17:139.

Monagle, R. D., and Brody, H. (1974) The effects of age upon the main nucleus of the inferior olive in the human. *J. Comp. Neurol.* 155:61–66.

Money, J., and Ehrhardt, A. A. (1972) *Man and Woman, Boy and Girl. The Differentiation and Dimorphism of Gender Identity from Conception to Maturity.* Baltimore and London: Johns Hopkins Univ. Press.

Monnier, V. M. (1989) Toward a Maillard reaction theory of aging. In *The Maillard Reaction in Aging, Diabetes, and Nutrition,* J. W. Barnes and V. M. Monnier (eds.). *Prog. Clin. Biol. Res.* 304:1–22.

Monnier, V. M., and Cerami, A. (1981) Nonenzymatic browning in vivo: Possible process for aging of long-lived proteins. *Science* 211:491–493.

———. (1983) Detection of nonenzymatic browning products in the human lens. *Biochim. Biophys. Acta* 760:97–103.

Monnier, V. M., Kohn, R. R., and Cerami, A. (1984) Accelerated age-related browning of human collagen in diabetes mellitus. *Proc. Natl. Acad. Sci.* 81:583–587.

Monnier, V. M., Sell, D. R., Miyata, S., and Nagara, R. H. (1990) The Maillard reaction as a basis for a theory of aging. In *Proceedings of the 4th International Symposium on the Maillard Reaction,* P. A. Finot (ed.), 393–414. *Adv. Life Sci.* Basel: Birkhausert-Verlag.

Montagna, W., and Carlisle, K. (1979) Structural change in aging and human skin. *J. Invest. Dermatol.* 73:47–53.

Montagna, W., and Giacometti, L. (1969) Histology and cytochemistry of human skin. 32. The external ear. *Arch. Dermatol.* 99:757–767.

Mooradian, A. D. (1988) Tissue specificity of premature aging in diabetes mellitus. The role of cellular replicative capacity. *J. Am. Geriatr. Soc.* 36:831–839.

Moore, C. J., and Schwartz, A. G. (1978) Inverse correlation between species lifespan and the capacity to form 7,12–dimethylbenezene(a)anthracene to a form mutagenic to a mammalian cell. *Exp. Cell Res.* 116:359–364.

Moore, G. H. (1974) Aetiology of the die-off of male *Antechinus stuartii.* Ph.D. diss., Australian National University, Canberra.

Moore, J. (1981) Asexual reproduction and environmental predictability in cestodes (Cyclopdyllidea: Taeniidae). *Evolution* 35:723–741.

Moore, M. A., Thamavit, W., Tsuda, H., Sato, K., Ichihara, A., and Ito, N. (1986) Modifying influence of dehydroepiandrosterone on the development of dihydroxy-di-*n*-propylnitrosamine-iniated lesions in the thyroid, lung, and liver of F344 rats. *Carcinogenesis* 7:311–316.

Moore, R. E., Goldsworthy, T. L., and Pitot, H. C. (1980) Turnover of 3′-polyadenylate-containing RNA in livers from aged, partially hepatectomized, neonatal, and Morris 5123C hepatoma-bearing rats. *Canc. Res.* 40:1449–1457.

Moosey, J. (1967) The neuropathology of Cockayne's syndrome. *J. Neuropathol. Exp. Neurol.* 26:654–660.

Morel, D. E., DiCorleto, P. E., and Chisolm, G. M. (1984) Endothelial and smooth muscle cells alter low density lipoprotein in vitro by free radical oxidation. *Arteriosclerosis* 4:357–364.

Morgan, D. G. (In press) Neurochemical changes with aging. Predispositions towards age-related mental disorders. In *Handbook of Mental Health and Aging,* 2d ed., J. E. Birren, R. B. Sloane, and G. D. Cohen (eds.). San Diego: Academic Press.

Morgan, D. G., and May, P. C. (1990) Age-related changes in synaptic neurochemistry. In *Handbook of the Biology of Aging,* 3d ed., E. L. Schneider and J. W. Rowe (eds.), 219–254. San Diego: Academic Press.

Morgan, D. G., May, P. C., and Finch, C. E. (1987) Dopamine and serotonin systems in human and rodent brain. Effects of age and neuro-degenerative disease. *J. Am. Geriatr. Soc.* 35:334–345.

Mori, H., Kondo, J., and Ihara, Y. (1987) Ubiquitin is a component of paired helical filaments in Alzheimer's disease. *Science* 235:1641–1644.

Mori, N., Hiruta, K., Funatsu, Y., and Goto, S. (1983) Codon recognition fidelity of ribosomes at the first and second positions does not decrease during aging. *Mech. Age. Dev.* 22:1–10.

Moriarity, C. (1987) Factors influencing recruitment of the Atlantic species of anguillid eels. *Am. Fish. Soc. Symp.* 1:483–491.

Morishima, H., Oka, H. I., and Chang, W. T. (1961) Directions of differentiation in populations of wild rice, *Oryza perennis* and *Oryza sativa f. spontanea. Evolution* 15:326–339.

Morley, A. A., Cox, S., and Holliday, R. (1982) Human lymphocytes resistant to 6–thioguanine increase with age. *Mech. Age. Dev.* 19:21–26.

Mormon, R. H. (1987) Relationship of density to growth and metamorphosis of caged larval sea lampreys, *Petromyzon marinus* Linnaeus, in Michigan streams. *J. Fish. Biol.* 30:173–182.

Moroi-Fetters, S. E., Mervis, R. F., London, E. D., and Ingram, D. K. (1989) Dietary restriction suppresses age-related changes in dendritic spines. *Neurobiol. Aging* 10:317–322.

Morrell, A. G., Gregoriadis, G., and Scheinberg, I. H. (1971) The role of siliac acid in determining the survival of glycoproteins in the circulation. *J. Biol. Chem.* 246:1461–1467.

Morrison, P. R., and Ryser, F. A. (1952) Weight and body temperature in mammals. *Science* 116:231–232.

Morse, H. C., Yetter, R. A., Stimpfling, J. H., Pitts, O. M., Frederickson, T. N., and Hartley, J. W. (1985) Greying with age in mice: Relation to expression of leukemia viruses. *Cell* 41:439–448.

Mortensen, H. B., Volund, A., and Christopherson, C. (1984) Glucosylation of human haemoglobin A. Dynamic variation in HbA$_{1c}$ described by a biokinetic model. *Clin. Chem. Acta* 136:75–81.

Mortimer, J. A., Schuman, L. M., and French, L. R. (1981) Epidemiology of dementing illness. In *The Epidemiology of Dementia*, J. A. Mortimer and L. M. Schuman (eds.), 3–23. New York: Oxford Univ. Press.

Mortimer, R. K., and Johnston, J. R. (1959) Life span of individual yeast cells. *Nature* 183:1751–1752.

Mosher, K. M., Young, D. A., and Munck, A. (1971) Evidence for irreversible, actinomycin D–sensitive, and temperature-sensitive steps following binding of cortisol to glucocorticoid receptors and preceding effects on glucose metabolism in rat thymus cells. *J. Biol. Chem.* 246:654–659.

Moss, C. (1988) *Elephant Memories. Thirteen Years in the Life of an Elephant Family.* New York: William Morrow.

Moss, M., Albert, M., Butters, N., and Payne, M. (1986) Differential patterns of memory loss among patients with Alzheimer's disease, Huntington's disease, and alcoholic Korskakoff's syndrome. *Arch. Neurol.* 43:239–246.

Motwani, N. M., Caron, D., Demyan, W. F., Chatterjee, B., Hunter, S., Poulik, M. D., and Roy, A. K. (1984) Monoclonal antibodies to alpha$_{2u}$-globulin synthesizing hepatocytes during androgenic induction and aging. *J. Biol. Chem.* 259:3653–3657.

Moudgil, P. G., Cook, J. R., and Buetow, D. E. (1979) The proportion of ribosomes active in protein synthesis and the content of polyribosomal poly(A)-containing RNA in adult and senescent rat livers. *Gerontology* 25:322–326.

Moulton, C. R. (1923) Age and chemical development in mammals. *J. Biol. Chem.* 57:79–97.

Mouritizen Dam, A. M. (1979) The density of neurons in the human hippocampus. *Neuropath. Appl. Neurobiol.* 5:249–264.

Mueller, L. D. (1987) Evolution of accelerated senescence in laboratory populations of *Drosophila. Proc. Natl. Acad. Sci.* 84:1974–1977.

Mueller, N., Evans, A., Harris, N. L., Comstock, G. W., Jellum, E., Magnus, K., Orentreich, N., Polk, F., and Vogelman, J. (1989) Hodgkin's disease and Epstein-Barr virus. *N. Engl. J. Med.* 320:689–695.

Mueller-Dombois, D. (1983) Canopy dieback and successional processes in Pacific forests. *Pacific Sci.* 37:317–325.

Muggleton-Harris, A. L., and Aroian, M. A. (1982) Replicative potential of individual cell hybrids derived from young and old donor human skin fibroblasts. *Somatic Cell Genet.* 8:41–50.

Muggleton-Harris, A. L., Hardy, K., and Higbee, N. (1987) Rescue of developmental lens abnormalities in chimaeras of noncataractous and congenital cataractous mice. *Development* 99:473–480.

Mühlbock, O. (1959) Factors influencing the life-span of inbred mice. *Gerontologia* 3:177–183.

Mukherjee, A. B., and Weinstein, M. E. (1988) Sequence of centromere separation of mitotic chromosomes during human cellular aging. *Mech. Age. Dev.* 45:59–64.

Mulinos, M. G., and Pomerantz, L. (1940) Pseudohypophysectomy. A condition resembling hypophysectomy produced by malnutrition. *J. Nutrit.* 19:493–504.

Mullaart, E., Boerrigter, M. E. T. I., Brouwer, A., Berends, F., and Vijg, J. (1988) Age-dependent accumulation of alkali-labile sites in DNA of post-mitotic but not in that of mitotic rat liver cells. *Mech. Age. Dev.* 45:41–49.

Mullaart, E., Roza, L., Lohman, P. H., and Vijg, J. (1989) The removal of UV-induced pyrimidine dimers from DNA of rat skin cell *in vitro* and *in vivo* in relation to aging. *Mech. Age. Dev.* 47:253–264.

Muller, H. J. (1964) The relation of recombination to mutational advance. *Mutat. Res.* 1:2–9.

Muller, I. (1971) Experiments on ageing in single cells of *Saccharomyces cerevisiae. Arch. Microbiol.* 77:20–25.

———. (1985) Parental age and the life-span of zygotes of *Saccharomyces cerevisiae. Antonie van Leeuwenhoek* 51:1–10.

Muller, I., Zimmerman, M., Becker, D., and Flomer, M. (1980) Calendar life span versus budding life span of *Saccharomyces cerevisiae. Mech. Age. Dev.* 12:47–52.

Müller, U., Jongeneel, C. V., Nedospasov, S. A., Lindahl, K. F., and Steinmetz, M. (1987) Tumour necrosis factor and lymphotoxin genes map close to H-2D in the mouse major histocompatibility complex. *Nature* 325:265–267.

Müller, U., Stephan, D., Philippsen, P., and Steinmetz, M. (1987) Orientation and molecular map position of the complement genes in the mouse MHC. *EMBO J.* 6:369–373.

Muller, W. E. G., Agutter, P. S., Bernd, A., Bachmann, M., and Schroder, W. E. G. (1985) Role of post-transcriptional events in aging: Consequences for gene expression in eukaryotic cells. In *Thresholds in Aging,* M. Bergener, M. Ermini, and H. B. Stähelin (eds.), 21–57. New York: Academic Press.

Mulligan, T. J., and Leaman, B. M. (In prep.) Length-at-age analysis: Can you get what you see?

Munck, A., and Crabtree, G. R. (1981) Glucocorticoid-induced lymphocyte death. In *Cell Death in Biology and Pathology,* I. D. Bowen and R. A. Lockshin (eds.), 329–359. New York: Chapman and Hall.

Munck, A., Guyre, P., and Holbrook, N. (1984) Physiological functions of glucocorticoids during stress and their relation to pharmacological actions. *Endocrine Rev.* 5:25–49.

Munkres, K. D., and Furtek, C. A. (1984a) Linkage of conidial longevity determinant genes in *Neurospora crassa. Mech. Age. Dev.* 25:63–77.

———. (1984b) Selection of conidial longevity mutants of *Neurospora crassa. Mech. Age. Dev.* 25:47–62.

Munkres, K. D., and Minssen, M. (1976) Ageing of *Neurospora crassa.* 1. Evidence for the free radical theory of ageing from studies of a natural death mutant. *Mech. Age. Dev.* 5:79–98.

Munkres, K. D., and Rana, R. S. (1984) Genetic control of cellular longevity in *Neurospora crassa*: A relationship between cyclic nucleotides, antioxidants, and antioxygenic enzymes. *Age* 7:30–35.

Munkres, K. D., Rana, R. S., and Goldstein, E. (1984) Genetically determined conidial longevity is positively correlated with superoxide dismutase, catalase, glutathione peroxidase, cytochrome c peroxidase, and ascorbate free radical reductase activities in *Neurospora crassa. Mech. Age. Dev.* 24:83–100.

Munnell, J. F., and Getty, R. (1968) Rate of accumulation of cardiac lipofuscin in the aging canine. *J. Gerontol.* 23:154–158.

Munro, H. N. (1969) Evolution of protein metabolism in mammals. In *Mammalian Protein Metabolism,* vol. 3, H. N. Munro and J. B. Allison (eds.), 133–182. New York: Academic Press.

Munro, H. N., and Downie, E. D. (1964) Relationship of liver composition to intensity of protein metabolism in different mammals. *Nature* 203:603–605.

Munro, H. N., and Gray, J. A. (1969) The nucleic acid content of skeletal muscle and liver in mammals of different body size. *Comp. Biochem. Physiol.* 28:897–905.

Murakami, A., Kuroda, Y., and Fukase, Y. (1985) Sex and strain differences in the lifespan of the adult silkmoth, *Bombyx mori. Natl. Inst. Genet. Jpn. Ann. Rep.* 36:49–50.

Murashige, T., and Nakano, R. (1965) Morphogenetic behavior of tobacco tissue cultures and implications of plant senescence. *Am. J. Bot.* 52:819–827.

Murasako, D. M., Weiner, P., and Kaye, D. (1987) Decline in mitogen induced proliferation of lymphocytes with increasing age. *Clin. Exp. Immunol.* 70:440–448.

———. (1988) Association of lack of mitogen-induced lymphocyte proliferation with increased mortality in the elderly. *Aging: Immunol. Infect. Dis.* 1:1–6.

Murchie, W. R. (1960) Biology of the oligochaete *Bimastos zeteki* Smith and Gittins (Lumbricidae) in northern Michigan. *Am. Midl. Nat.* 64:194–215.

Murfet, I. C. (1971) Flowering in *Pisum:* Reciprocal grafts between known genotypes. *Austral. J. Biol. Sci.* 24:1089–1101.

Murfitt, R. R., and Sanadi, D. R. (1978) Evidence for increased degeneration of mitochondria of old rats. *Mech. Age. Dev.* 8:197–201.

Murphy, E. D. (1972) Hyperplastic and neoplastic changes in the ovaries of mice after genetic deletion of germ cells. *J. Natl. Canc. Inst.* 48:1283–1295.

Murphy, J. S. (1970) A general method for the monoxenic cultivation of the Daphnidae. *Biol. Bull.* 139:321–332.

Murphy, J. S., and Davidoff, M. (1972) The result of improved nutrition on the Lansing effect in *Moina macrocopa. Biol. Bull.* 142:302–309.

Murphy, W. H., Nawrocki, J. F., and Pease, L. R. (1983) Age-dependent paralytic viral infection in C58 mice: Possible implications in human neurologic disease. *Prog. Brain Res.* 59:291–303.

Murray, B. G. (1979) Population dynamics: Alternative models. In *Physiological Ecology. A Series of Monographs, Texts, and Treatises,* T. T. Kozlowski (ed.). New York: Academic Press.

Murty, C. V. R., Mancini, M. A., Chatterjee, B., and Roy, A. K. (1988) Changes in transcriptional activity and matrix association of α_{2u}-globulin gene family in the rat liver during maturation and aging. *Biochim. Biophys. Acta* 949:27–34.

Mutch, W. J., Dingwall-Fordyce, I., Downie, A. W., Paterson, J. G., and Roy, S. K. (1986) Parkinson's disease in a Scottish city. *Brit. Med. J.* 292:534–536.

Myers, D. D. (1978) Review of disease patterns and life span in aging mice: Genetic and environmental interactions. In *Genetic Effects on Aging,* D. Bergsma and D. E. Harrison (eds.), 41–53. Birth Defects: Original Article Series, 14. New York: Alan R. Liss.

Myers, G. S. (1952) Annual fishes. *Aquarium J.* 23:125–141.

Myers, R. J., and Hamilton, J. B. (1961) Regeneration and rate of growth of hairs in man. *Ann. N. Y. Acad. Sci.* 53:562–568.

Nabors, C. E., and Ball, C. R. (1970) Spontaneous calcification in hearts of DBA mice. *Anat. Rec.* 164:153–162.

Nadgauda, R. S., Parasharami, V. A., and Mascarenhas, A. F. (1990) Precocious flowering and seeding behaviour in tissue-cultured bamboos. *Nature* 344:335–336.

Naeim, F., and Walford, R. L. (1985) Aging and cell membrane complexes: The lipid bilayer, integral proteins, and cytoskeleton. In *Handbook of the Biology of Aging,* 2d ed., C. E. Finch and E. L. Schneider (eds.), 272–289. New York: Van Nostrand.

Nagel, J. E., Chopra, R. K., Chrest, F. J., McCoy, M. T., Schneider, E. L., Holbrook, N. J., and Adler, W. H. (1988) Decreased proliferation, interleukin 2 synthesis, and interleukin 2 receptor expression are accompanied by decreased mRNA expression in phytohemagglutinin-stimulated cells from elderly donors. *J. Clin. Invest.* 81:1096–1102.

Nagy, E., and Berczi, I. (1978) Immunodeficiency in hypophysectomized rats. *Acta Endocrinol.* 89:530–537.

Naiki, H., Higuchi, K., Yonezu, T., Hosokawa, M., and Takeda, T. (1988) Metabolism of senile amyloid precursor and amyloidogenesis: Age related acceleration of apolipoprotein A-II clearance in senescence accelerated mouse. *Am. J. Pathol.* 130:579–587.

Nakagome, Y., Abe, T., Misawa, S., Takeshita, T., and Iinuma, K. (1984) The "loss" of centromeres from chromosomes of aged women. *Am. J. Human Genet.* 36:398–404.

Nakai, Y., Maruyama, N., Ohta, K., Yoshida, H., Hirose, S., and Shirai, T. (1980) Genetic studies of autoimmunity in New Zealand mice: Association of circulating retroviral gp70 immune complex with proteinuria. *Immunol. Lett.* 2:53.

Nakamura, K. D., Duffy, P. H., Lu, M. H., Turturro, A., and Hart, R. W. (1989) The effect of dietary restriction on *myc* protooncogene expression in mice: A preliminary study. *Mech. Age. Dev.* 48:199–205.

Nakanishi, Y., Kram, D., and Schneider, E. L. (1979) Aging and sister chromatid exchange. 4. Reduced frequencies of mutagen-induced sister chromatid exchanges in vivo in mouse bone marrow cells with aging. *Cytogenet. Cell Genet.* 24:61–67.

Nakashima, M., Noda, H., Hasegaea, M., and Ikai, A. (1985) The oxygen affinity of mammalian hemoglobins in the absence of 2,3–diphosphoglycerate in relation to body weight. *Comp. Biochem. Physiol.* 82A:583–589.

Nakayama, T., Kaneko, M., Kodama, M., and Nagata, C. (1985) Cigarette smoke induces DNA single-strand breaks in human cells. *Nature* 314:462–464.

Nandi, S., and McGrath, C. M. (1973) Mammary neoplasia in mice. *Adv. Canc. Res.* 17:353–414.

Nandy, K. (1982) Effects of controlled dietary restriction on brain-reactive antibodies in sera of aging mice. *Mech. Age. Dev.* 18:97–102.

Nandy, K., and Bourne, G. H. (1966) Effect of centrophenoxine on the lipofuscin pigment in neurons of senile guinea pig. *Nature* 210:313–314.

Nanney, D. L. (1957) Inbreeding degeneration in *Tetrahymena*. *Genetics* 42:137–146.

———. (1959) Vegetative mutants and clonal senility in *Tetrahymema*. *J. Protozool.* 6:171–177.

———. (1974) Aging and long term temporal regulation in ciliated protozoa. A critical review. *Mech. Age. Dev.* 3:81–105.

Naryshkin, S., Miller, L., Lindeman, R., and Lang, C. A. (1981) Blood glutathione: A biochemical index of human aging. *Fed. Proc.* 40:789.

Nathan, D. M., Singer, D. E., Hurxthal, K., and Goodson, J. D. (1984) Clinical information value of the glycosylated hemoglobin assay. *N. Engl. Med. J.* 310:341–346.

National Center for Health Statistics. (1980) Monthly vital statistics report: Annual summary for the United States, 1979. DHHS Publication PHS 81-1120 28(13).

Neave, F. (1958) The origin and speciation of *Oncorhynchus*. *Trans. Roy. Soc. Can.*, sect. 5, 52:25–39.

Nebert, D. W., Brown, D. D., Towne, D. W., and Eisen, H. J. (1984) Association of fertility, fitness, and longevity with the murine Ah locus among (C57BL/6N) × (C3H/HeN) recombinant inbred lines. *Biol. Reprod.* 30:363–373.

Needham, A. E. (1950) Growth and regeneration rates in relation to age in the Crustacea. With special reference to the isopod, *Asellus aquaticus*. *J. Gerontol.* 5:15–16.

Neel, J. V., Satoh, C., Goriki, K., Fujita, M., Takahashi, N., Asakawa, J. I., and Hazama, R. (1986) The rate with which spontaneous mutation alters the electrophoretic mobility of polypeptides. *Proc. Natl. Acad. Sci.* 83:389–393.

Nelson, J. F., Bender, M., and Schachter, B. S. (1988) Age-related changes in proopiomelanocortin messenger ribonucleic acid levels in hypothalamus and pituitary of female C57BL/6J mice. *Endocrinology* 123:340–344.

Nelson, J. F., and Felicio, L. S. (1985) Reproductive aging in the female: An etiological perspective. *Rev. Biol. Res. Aging* 2:251–314.

Nelson, J. F., Felicio, L. S., Osterburg, H. H., and Finch, C. E. (1981) Altered profiles of estradiol and progesterone associated with prolonged estrous cycles and persistent vaginal cornification in aging C57BL/6J mice. *Biol. Reprod.* 25:413–419.

Nelson, J. F., Felicio, L. S., Randall, P. K., Simms, C., and Finch, C. E. (1982) A longitudinal study of estrous cyclicity in aging C57BL/6J mice. 1. Cycle frequency, length, and vaginal cytology. *Biol. Reprod.* 27:327–339.

Nelson, J. F., Felicio, L. S., Sintra, Y. N., and Finch, C. E. (1980) An ovarian role in the spontaneous pituitary tumorigenesis and hyperprolactinemia of aging female mice. *Gerontologist* 20:171 (abstract).

Nelson, J. F., Goodrick, G., Karelus, K., and Felicio, L. S. (1989) Longitudinal studies of estrous cyclicity in C57BL/6J mice. 3. Dietary modulation declines during aging. *Mech. Age. Dev.* 48:73–84.

Nelson, J. F., Gosden, R. G., and Felicio, L. S. (1985) Effect of dietary restriction on estrous cyclicity and follicular reserves in aging C57BL/6J mice. *Biol. Reprod.* 32:512–522.

Nelson, J. F., Karelus, K., Felicio, L. S., and Johnson, T. E. (1990) Genetic influences on the timing of puberty in mice. *Biol. Reprod.*

Nelson, J. F., Latham, K. R., and Finch, C. E. (1975) Plasma testosterone levels in C57BL/6J male mice. Effects of age and disease. *Acta Endocrinol.* 80:744–752.

Nelson, J. S. (1968) Life history of the brook silverside *Labidesthes sicculus* in Crooked Lake, Indiana. *Trans. Am. Fish. Soc.* 97:293–296.

———. (1984) *Fishes of the World.* 2d ed. New York: Wiley.

Nelson, M. E., Meredith, C. N., Dawson-Hughes, B., and Evans, W. J. (1988) Hormone and bone mineral status in endurance-trained and sedentary postmenopausal women. *J. Clin. Endocrinol. Metab.* 66:927–935.

Nesse, R. M. (1988) Life table tests of evolutionary theories of senescence. *Exp. Gerontol.* 23:445–453.

Nesselroade, J. R., Stigler, S. M., and Baltes, P. B. (eds.) (1979) *Longitudinal Research in the Study of Behavior and Development.* New York: Academic Press.

Nette, E. G., Xi, Y. P., Sun, Y. K., Andrews, A. D., and King, D. W. (1984) A correlation between aging and DNA repair in human epidermal cells. *Mech. Age. Dev.* 24:283–292.

Neuberger, A. (1948) Stereochemistry of amino acids. *Adv. Protein Chem.* 4:297–383.

Neufeld, H. N., and Goldbourt, U. (1983) Coronary heart disease: Genetic aspects. *Circulation* 67:943–954.

Neukirch, A. (1982) Dependence of the life span of the honeybee (*Apis mellifica*) upon flight performance and energy consumption. *J. Comp. Physiol.* 146:35–40.

Newbold, R. F., Overell, R. W., and Connell, J. R. (1982) Induction of immortality is an early event in malignant transformation of mammalian cells by carcinogens. *Nature* 299:633–635.

Newton, I. (1988) Age and reproduction in the sparrowhawk. In *Reproductive Success. Studies of Individual Variation in Contrasting Breeding Systems,* T. H. Clutton-Brock (ed.), 201–219. Chicago: Univ. of Chicago Press.

Newton-John, H. F., and Morgan, D. B. (1970) The loss of bone with age, osteoporosis, and fractures. *Clin. Orthop. Related Res.* 71:229–252.

Nicholls, K., and Mandel, T. E. (1989) Advanced glycosylation end-products in experimental murine diabetic nephropathy: Effect of islet isografting and of aminoguanidine. *Lab. Invest.* 60:486–491.

Nicklas, J. A., O'Neill, J. P., and Albertini, R. J. (1986) Use of T-cell receptor gene probes to quantify the in vivo *hprt* mutations in human T-lymphocytes. *Mutat. Res.* 173:7–72.

Nicklas, R. B. (1988) Chance encounter and precision in mitosis. *J. Cell Sci.* 89:283–285.

Nicolosi, R. J., Baird, M. B., Massie, H. R., and Samis, H. V. (1973) Senescence in *Drosophila.* 2. Renewal of catalase activity in flies of different ages. *Exp. Gerontol.* 8:101–110.

Niedzwiecki, A., and Fleming, J. E. (1990) Changes in protein turnover after heat shock are related to accumulation of abnormal proteins in aging *Drosophila melanogaster. Mech. Age. Dev.* 52:295–304.

Nielsen, E. S., and Scoble, M. J. (1986) *Afrotheora,* a new genus of primitive Hepialidae from Africa (Lepidoptera: Hepialoidea). *Entomol. Scand.* 17:29–54.

Nielsen, J. G., and Smith, D. G. (1978) The eel family Nemichthyidae (Pisces, Anguilliformes). *Dana Rep.,* no. 88:1–71.

Nieuwkoop, P. D., and Sutasurya, L. A. (1979) *Primordial Germ Cells in the Chordates. Embryogenesis and Phylogenesis.* Cambridge: Cambridge Univ. Press.

Niklas, K. J. (1985) The evolution of tracheid diameter in early vascular land plants and its implications on the hydraulic conductance of the primary xylem strand. *Evolution* 39:1110–1122.

Nikolsky, G. V. (1963) *The Ecology of Fishes.* Translated from Russian by L. Birkett. New York: Academic Press.

Nimni, M. E. (1983) Collagen: Structure, function, and metabolism in normal and fibrotic tissues. *Seminars Arthr. Rheum.* 13:1–88.

Njoroge, F. G., Fernandes, A. A., and Monnier, V. M. (1988) Mechanism of formation of the putative advanced glycosylation end product and protein cross-link 2–(2–furoyl)-4(5)-(2–furanyl)-1*H*-imidazole. *J. Biol. Chem.* 263:10646–10652.

Noble, E. R., and Noble, G. A. (1971) *Parasitology: The Biology of Animal Parasites.* New York: Lee and Fibiger.

Noble, G. K. [1931] (1954) *The Biology of the Amphibia.* McGraw-Hill. Reprint. New York: Dover.

Nomura, Y., Wang, B.-X., Qi, S. B., Namba, T., and Kaneko, S. (1989) Biochemical changes related to aging in the senescence-accelerated mouse. *Exp. Gerontol.* 24:49–55.

Noodén, L. D. (1980a) Regulation of senescence. In *World Soybean Research Conference 2 Proceedings,* F. T. Corbin (ed.), 139–152. Boulder, CO: Westview Press.

———. (1980b) Senescence in the whole plant. In *Senescence in Plants,* K. V. Thimann (ed.), 219–258. Boca Raton, FL: CRC Press.

———. (1988a) Abscissic acid, auxin, or other regulators of senescence. In *Senescence and Aging in Plants,* L. D. Noodén and A. C. Leopold (eds.), 329–367. San Diego: Academic Press.

———. (1988b) The phenomena of senescence and aging. In *Senescence and Aging in Plants,* L. D. Noodén and A. C. Leopold (eds.), 1–51. San Diego: Academic Press.

———. (1988c) Whole plant senescence. In *Senescence and Aging in Plants,* L. D. Noodén and A. C. Leopold (eds.), 391–442. San Diego: Academic Press.

Noodén, L. D., and Thompson, J. E. (1985) Aging and senescence in plants. In *Handbook of the Biology of Aging,* 2d ed., C. E. Finch and E. L. Schneider (eds.), 105–127. New York: Van Nostrand.

Noodén, L. D., Van Staden, J., and Cook, E. L. (1988) Cytokinins and senescence. In *Senescence and Aging in Plants,* L. D.

Noodén and A. C. Leopold (eds.), 282–329. San Diego: Academic Press.

Nordeng, H. (1983) Solution to the "char problem": based on arctic char (*Salvelinus alpinus*) in Norway. *Can. J. Fish. Aquat. Sci.* 40:1372–1387.

Norman, A., Cochran, S., Bass, D., and Roe, D. (1984) Effects of age, sex, and diagnostic X-rays on chromosome damage. *Int. J. Radiat. Biol.* 46:317–321.

Norris, D. M., Chu, H. M., and Rao, K. D. P. (1983) Changes in ovarian ultrastructure and ecdysteroid titer during the aging process of female *Xyleborus ferrugineus* (Coleoptera: Scolytidae). *J. Morphol.* 177:245–254.

Norris, D. M., and Moore, C. L. (1980) Lack of dietary Δ^7-sterol markedly shorten periods of locomotor vigor, reproduction, and longevity of adult *Xyleborus ferrugineus. Exp. Gerontol.* 15:359–364.

Norris, D. M., Rao, K. D. P., and Chu, H. M. (1986) Role of steroids in aging. In *Insect Aging: Strategies and Mechanisms,* K.-G. Collatz and R. S. Sohal (eds.), 182–199. Heidelberg: Springer-Verlag.

Norris, M. J. (1934) Contributions towards the study of insect fertility. 3. Adult nutrition, fecundity, and longevity in the genus *Ephestia:* Epidoptera, Phycitidea. *Proc. Zool. Soc. Lond.* 1:333–366.

Norstedt, G., and Palmiter, R. (1984) Secretary rhythm of growth hormone regulates sexual differentiation of mouse liver. *Cell* 36:805–812.

Northrop, J. H. (1917) The effect of prolongation of the period of growth on the total duration of life. *J. Biol. Chem.* 32:123–126.

Norwood, T. H., and Smith, J. R. (1985) The cultured fibroblast-like cell as a model for the study of aging. In *Handbook of the Biology of Aging,* 2d ed., C. E. Finch and E. L. Schneider (eds.), 291–321. New York: Van Nostrand.

Norwood, T. H., Smith, J. R., and Stein, G. H. (1990) Aging at the cellular level: The human fibroblastlike cell model. In *Handbook of the Biology of Aging,* 3d ed., E. L. Schneider and J. W. Rowe (eds.) 131–156. San Diego: Academic Press.

Nowak, R. M., and Paradiso, J. L. (1983)

Walker's Mammals of the World. 2 vols. Baltimore: Johns Hopkins Univ. Press.

Noy, N., Schwartz, H., and Gafni, A. (1985) Age-related changes in the redox status of rat muscle cells and their role in enzyme-aging. *Mech. Age. Dev.* 29:63–69.

Nukina, N., Kosik, K., and Selkoe, D. J. (1987) Recognition of Alzheimer paired helical filaments by monoclonal neurofilament antibodies is due to cross-reaction with tau protein. *Proc. Natl. Acad. Sci.* 84:3415–3419.

Nur, N. (1984a) The consequences of brood size for breeding blue tits. 1. Adult survival, weight change, and the cost of reproduction. *J. Anim. Ecol.* 53:479–496.

———. (1984b) The consequences of brood size for breeding blue tits. 2. Nestling weight, offspring survival, and optimal brood size. *J. Anim. Ecol.* 53:497–517.

Nusslein-Volhard, C. (1977) Genetic analysis of pattern-formation in the embryo of *Drosophila melanogaster. Roux's Arch. Dev. Biol.* 183:249–268.

Nyce, J. W., Magee, P. N., Hard, G. C., and Schwartz, A. G. (1984) Inhibition of 1,2-dimethylhydrazine-induced colon tumorigenesis in BALB/c mice by dehydroepiandrosterone. *Carcinogenesis* 5:57–62.

O'Connor, C. M., and Clarke, S. (1983) Methylation of erythrocyte membrane proteins at extracellular and intracellular D-aspartate sites *in vitro. J. Biol. Chem.* 258:8485–8492.

———. (1984) Carboxyl methylation of cytosolic proteins in intact human erythrocytes. Identification of numerous methyl-accepting proteins including hemoglobin and carbonic anhydrase. *J. Biol. Chem.* 259:2570–2578.

O'Connor, C. M., Germain, B. J., Guthrie, K. M., Aswad, D. W., and Millette, C. F. (1989) Protein carboxyl specific for age-modified aspartyl residues in mouse testes and ovaries: Evidence for translation during spermiogenesis. *Gamete Res.* 22:307–319.

O'Connor, C. M., and Yutzey, K. E. (1988) Enhanced carboxyl methylation of membrane-associated hemoglobin in human erythrocytes. *J. Biol. Chem.* 263:1386–1390.

Odell, W. D. and Swerdloff, R. S. (1968) Progesterone-induced luteinizing and follicle-stimulating surge in postmenopausal women: A simulating ovulatory peak. *Proc. Natl. Acad. Sci.* 61:529–536.

Oertel, G. W., and Benes, P. (1972) The effects of steriods on glucose-6–phosphate dehydrogenase. *J. Steroid Biochem.* 3:493– 496.

Oertel, G. W., and Rebelein, I. (1969) Effects of dehydroepiandrosterone and its conjugates upon the activity of glucose-6–phosphate dehydrogenase in human erythrocytes. *Biochim. Biophys. Acta.* 184:459–460.

Ogden, D. A., and Micklem, H. S. (1976) The fate of serially transplanted bone marrow cell populations from young and old donors. *Transplantation* 22:287–293.

Ogden, J. (1978) On the dendrochronological potential of Australian trees. *Austral. J. Ecol.* 3:339–356.

———. (1985a) An introduction to plant demography with special reference to New Zealand trees. *New Zealand J. Bot.* 23:751–772.

———. (1985b) Past, present, and future: Studies on the population dynamics of some long-lived trees. In *Studies on Plant Demography: A Festschrift for John L. Harper,* 3–16. New York: Academic Press.

———. (1988) Forest dynamics and stand-level dieback in New Zealand's *Nothofagus* forests. *Geojournal* 17:225–230.

Ogomori, K., Kitamoto, T., Tateishi, J., Sato, Y., and Tashima, T. (1988) Aging and cerebral amyloid: Early detection of amyloid in the human brain using biochemical extraction and immunostain. *J. Gerontol.* 43:B157–B162.

Oguri, M. (1960) Studies on the adrenal glands of teleosts. 6. On the interrenal tissue of chum salmon, *Oncorhynchus keta* (Walbaum), migrating up river to spawn. *Bull. Jpn. Soc. Sci. Fish.* 26:981–984. [not read]

Ohno, S. (1970) *Evolution by Gene Duplication.* Heidelberg: Springer-Verlag.

Ohno, S., Muramoto, J., Klein, J., and Aitkin, N. B. (1969) Diploid-tetraploid relationship in clupeoid and salmonid fish. In *Chromosomes Today,* vol. 21, C. D. Darlington and K. R. Lewis (eds.), 139–147. Edinburgh: Oliver and Boyd. [not read]

Ohsumi, S. (1979) Interspecies relationships among some biological parameters in cetaceans and estimation of the natural mortality coefficient of the Southern Hemisphere minke whale. *Rep. Int. Whal. Commn.* 29:397–405.

Oimomi, M., Maeda, Y., Hata, F., Kitamura, Y., Matsumoto, S., Hatanaka, H., and Baba, S. (1988) A study of the age-related acceleration of glycation of tissue proteins in rats. *J. Gerontol.* 43:B98–B101.

Oinonen, E. (1967a) The correlation between the size of Finnish bracken (*Pteridium aquilinum* [*L.*]) Kuhn) clones and certain periods of site history. *Acta For. Fenn.* 83:1–51.

———. (1967b) Summary: Sporal regeneration of ground pine (*Lycopodium complanatum L.*) in southern Finland in the light of the dimensions and the age of its clones. *Acta For. Fenn.* 83:76–85.

———. (1969) The time table of vegetative spreading of the lily-of-the-valley (*Convalaria majalis L.*) and the wood small-reed (*Calamagrostis epigeios* [*L.*] Roth) in southern Finland. *Acta For. Fenn.* 97:1–35.

Ojeda, S. R., Andrews, W. W., Advis, J. P., and White, S. S. (1980) Recent advances in the endocrinology of puberty. *Endocrine Rev.* 1:228–257.

Ojeda, S. R., Urbanski, H. F., and Ahmed, C. E. (1986) The onset of female puberty: Studies in the rat. *Rec. Prog. Horm. Res.* 42:385–442.

Oka, H. I. (1976) Mortality and adaptive mechanisms of *Oryza perennis* strains. *Evolution* 30:380–392.

Okamoto, K., Yamori, Y., and Nagaoka, A. (1974) Establishment of the stroke-prone spontaneously hypertensive rat (SHR). *Circ. Res.* 34 (suppl. 1): I143–I151.

Okulicz, W. C., Fournier, D. J., Esber, H., and Frederickson, T. N. (1985) Relationship of oestrogen and progesterone and their oviductal receptors in laying and non-laying 5-year-old hens. *J. Endocrinol.* 106:343–348.

Olashaw, N. E., Kress, E. D., and Cristofalo, V. J. (1983) Thymidine triphosphate synthesis in senescent WI-38 cells. *Exp. Cell Res.* 149:547–554.

Oliver, C. N., Levine, R. L., and Stadtman, E. R. (1987) A role of mixed-function oxidation reactions in the accumulation of altered enzyme forms during aging. *J. Am. Geriatr. Soc.* 35:947–956.

Ollason, J. C., and Dunnet, G. M. (1988) Variation in breeding success in fulmars. In *Reproductive Success. Studies of Individual Variation in Contrasting Breeding Systems*, T. H. Clutton-Brock (ed.), 263–278. Chicago: Univ. of Chicago Press.

Ollier, W., Spector, T., Silman, A., Perry, L., Ords, J., Thomson, W., and Festenstein, H. (1989) Are certain HLA haplotypes responsible for low testosterone levels in males? *Dis. Markers* 7:139–143.

Ollier, W., Venables, P. J., Mumford, P. A., Maini, R., Awad, J., Jaraquemada, D., D'Amaro, J., and Festenstein, H. (1984) HLA antigen association with extra articular rhematoid arthritis. *Tiss. Antigens* 24:279–291.

Olmo, E. (1983) Nucleotype and cell size in vertebrates: A review. *Bas. Appl. Histochem.* 27:227–256.

Olmo, E., and Morescalchi, A. (1975) Evolution of the genome and cell sizes in salamanders. *Experientia* 31:804–806.

Olsen, G. G., and Everitt, A. V. (1965) Retardation of the ageing process in the collagen fibres from the tail tendon of the old hypophysectomized rat. *Nature* 206:307.

Olson, S. L. (1985) The fossil record of birds. In *Avian Biology*, vol. 8, D. Farner and J. King (eds.), 79–238. New York: Academic Press.

Olzewski, J. (1950) On anatomical and functional organization of the spinal trigeminal nucleus. *J. Comp. Neurol.* 92:401–413.

O'Malley, B. W. (1989) Editorial: Did eucaryotic steroid receptors evolve from intracrine gene regulators? *Endocrinology* 125:1119–1120.

O'Meara, G. F., and Krasnick, G. J. (1970) Dietary and genetic control of the expression of autogenous reproduction in *Aedes atropalpus* (Coq.) (Diptera: Culicidae). *J. Med. Entomol.* 7:328–334.

O'Meara, G. F., and Lounibos, L. P. (1981) Reproductive maturation in the pitcher-plant mosquito, *Wyeomyia smithii. Physiol. Entomol.* 6:437–443.

152–year-old lake sturgeon caught in Ontario. (1954) *Commercial Fish. Rev.* 16:28.

Ono, I. (1933) The life history of the masu salmon of Hokkaido (in Japanese). *Salmon J.* 5(2): 15–26, 5(3): 13–25. Cited in Tanaka, 1965.

Ono, T., and Cutler, R. G. (1978) Age-dependent relaxation of gene repression: Increase of endogenous murine leukemia virus-related and globin-related RNA in brain and liver of mice. *Proc. Natl. Acad. Sci.* 75:4431–4435.

Ono, T., Okada, S., Kawakami, T., Honjo, T., and Getz, M. J. (1985) Absence of gross change in primary DNA sequence during aging process of mice. *Mech. Age Dev.* 32:227–234.

Ono, T., Okada, S., and Sugahara, T. (1976) Comparative studies of DNA size in various tissues of mice during the aging process. *Exp. Gerontol.* 11:127–132.

Opie, E. L, Lynch, C. J., and Tershakovec, M. (1970) Sclerosis of the mesenteric arteries of rats. *Arch. Pathol.* 89:306–313.

Orentreich, N., Brind, J. L., Rizer, R. L., and Vogelman, J. H. (1984) Age changes and sex differences in serum dehydroepiandrosterone sulfate concentrations throughout adulthood. *J. Clin. Endocrinol. Metab.* 59:551–555.

Orentreich, N., and Selmanowitz, V. J. (1969) Levels of biological functions with aging. *Trans. Acad. Sci. N.Y.,* ser. B, 31:992–1012.

Orentreich, N., and Sharp, N. J. (1967) Keratin replacement as an aging parameter. *J. Soc. Cosmetic Chem.* 18:537–547.

Orentreich Foundation for Advancement of Science. (1987) *Annual Report.* New York.

Orgel, L. E. (1963) The maintenance of accuracy of protein synthesis and its relevance to aging. *Proc. Natl. Acad. Sci.* 49:512–517.

———. (1970) The maintenance of the accuracy of protein synthesis and its relevance to ageing: A correction. *Proc. Natl. Acad. Sci.* 67:1476.

Orlander, J., and Aniansson, A. (1980) Effects of physical training on skeletal muscle metabolism and ultrastructure in 70 to 75–year-old men. *Acta Physiol. Scand.* 109:149–154.

Orren, A., and Dowdle, E. B. (1975) The ef-fects of sex and age on serum IgE concentration in three ethnic groups. *Int. Arch. Allergy Appl. Immunol.* 48:824–835.

Orton, J. H. (1929) Reproduction and death in invertebrates and fishes. *Nature* 123:14–15.

Orwoll, E. S., and Meier, D. E. (1986) Alterations in calcium, vitamin D, and parathyroid hormone physiology in normal men with aging: Relationship to the development of senile osteopenia. *J. Clin. Endocrinol. Metab.* 63:1262–1269.

Osani, M., and Kikuta, S. (1981) Age-related changes in amino acid pool sizes in the adult silkworm, *Bombyx mori. Exp. Gerontol.* 16:445–459.

Osani, M., and Yonezawa, Y. (1984) Age-related changes in amino acid pool sizes in the adult silkmoth, *Bombyx mori,* reared at low and high temperatures: A biochemical examination of the rate-of-living theory and urea accumulation when reared at high temperature. *Exp. Gerontol.* 19:37–51.

Osborn, J. W. (1973) The evolution of dentitions. *Am. Sci.* 61:548–559.

———. (1974) On the control of tooth replacement in reptiles and its relationship to growth. *J. Theoret. Biol.* 46:509–527.

Osborne, D. J. (1980) Senescence in seeds. In *Senescence in Plants,* K. V. Thimann (ed.), 13–37. Boca Raton, FL: CRC Press.

Osborne, T. B., Mendel, L. B., and Ferry, E. L. (1917) The effect of retardation of growth upon the breeding period and duration of life in rats. *Science* 45:294–295.

Oshima, M. (1934) Life-history and the distribution of the freshwater salmons found in the waters of Japan. *Proceedings of the Fifth Pacific Scientific Congress (1933),* 3751–3774. Cited in Vladykov, 1963.

Osiewacz, H. D., and Esser, K. (1983) DNA sequence analysis of the mitochondrial plasmid of *Podospora anserina. Curr. Genet.* 7:219–223.

———. (1984) The mitochondrial plasmid of *Podospora anserina.* A mobile intron of a mitochondrial gene. *Curr. Genet.* 8:299–305.

Oster, J., Mikkelsen, M., and Nielsen, A. (1975) Mortality and life table in Down's syndrome. *Acta Paediatr. Scand.* 64:322–326.

Otsuka, K., Sodek, J., and Limeback, H. (1984) Synthesis of collagenase and collagenase inhibitors by osteoblast-like cells in culture. *Eur. J. Biochem.* 145:123–129.

Otter, G. W. (1933) On the biology and life history of *Rhabditis pellio* (Nematoda). *Parasitology* 25:296–307.

Ottinger, M. A. (1983) Sexual behaviour and endocrine changes during reproductive maturation and aging in the avian male. In *Hormones and Behavior in Higher Vertebrates*, J. Balthazart, E. Prove, and R. Gilles (eds.), 350–367. Berlin: Springer-Verlag.

Ottinger, M. A., Adkins-Regan, E., Buntin, J., Cheng, M. F., DeVoogd, T., Harding, C., and Opel, H. (1984) Hormonal mediation of reproductive behavior. *J. Exp. Zool.* 232: 605–616.

Ottinger, M. A., and Balthazart, J. (1986) Altered endocrine and behavioral responses with reproductive aging in the male Japanese quail. *Horm. Behav.* 20:83–94.

Ottinger, M. A., Duchala, C. S., and Masson, M. (1983) Age-related reproductive decline in the male Japanese quail. *Horm. Behav.* 17:197–207.

Ottinger, M. A., Green, C., and Palmer, S. S. (1987) Effect of age on testis size, serum testosterone concentrations, and FSH receptor number and affinity in quail. *Biol. Reprod.*, suppl. 1, 36:158 (abstract).

Oudea, M. C., Collette, Z. M., Dedieu, P. H., and Oudea, P. (1973a) Morphometric study of the ultrastructure of human alcoholic fatty liver. *Biomedicine* 18:455–459.

Oudea, M. C., Collette, Z. M., and Oudea, P. (1973) Morphometric study of ultrastructural changes induced in rat liver by chronic alcohol intake. *Digest. Dis.* 18:398–402.

Ozato, K., and Wakamatsu, Y. (1983) Multistep genetic regulation of oncogene expression in fish hereditary melanoma. *Differentiation* 24:181–190.

Pacifici, R., Rifas, L., McCraken, R., Vered, I., McMurtry, C., Avioli, L. V., and Peck, W. A. (1989) Ovarian steroid treatment blocks a postmenopausal increase in blood monocyte interleukin 1 release. *Proc. Natl. Acad. Sci.* 86:2398–2402.

Padhi, S. L., and Patnaik, B. K. (1976) Regional distribution of RNA, DNA, and protein in the brain of male garden lizard during aging. *J. Neurochem.* 26:617–619.

Paffenholz, V. (1978) Correlations between DNA repair of embryonic fibroblasts and different life spans of 3 inbred mouse strains. *Mech. Age. Dev.* 7:131–150.

Paganini-Hill, A., Ross, R. K., and Henderson, B. E. (1988) Postmenopausal oestrogen treatment and stroke: A prospective study. *Brit. Med. J.* 297:519–523.

Page, L. B., Danion, A., and Moellering, R. D. (1974) Antecedents of cardiovascular disease in six Solomon Islands societies. *Circulation* 49:1132–1146.

Page, R. D., Kirkpatrick-Keller, D., and Butcher, R. L. (1983) Role of age and length of oestrous cycle in alteration of the oocyte and intrauterine environment in the rat. *J. Reprod. Fertil.* 69:23–28.

Pai, S. (1928) Die Phasen des Lebenzyklus der *Anguillula aceti* und ihre experimentellmorphologische Beeinflussung. *Z. Wiss. Zool.* 131:293.

———. (1934) Regenerationsversuche an Rotatorien. *Sci. Rep. Univ. Cheklang* 1:247–259.

Paigen, B., Holmes, P. A., Mitchell, D., and Albee, D. (1987) Comparison of atherosclerotic lesions and HDL-lipid levels in male, female, and testosterone-treated female mice from strains C57BL/6, BALB/c, and C3H. *Atherosclerosis* 64:215–221.

Paigen, B., Ishida, B. Y., Verstuyft, J., Waters, R. B., and Albee, D. (1990) Arteriosclerosis susceptibility differences among progenitors of recombinant inbred strains of mice. *Arteriosclerosis* 10:316–326.

Paigen, B., Mitchell, D., Reue, K., Morrow, A., Lusis, A. J., and LeBoeuf, R. C. (1987) *Ath-1*, a gene determining atherosclerosis susceptibility and high density lipoprotein levels in mice. *Proc. Natl. Acad. Sci.* 84:3763–3767.

Paigen, B., Morrow, A., Brandon, C., Mitchell, D., and Holmes, P. A. (1985) Variation in susceptibility to atherosclerosis among inbred strains of mice. *Atherosclerosis* 57: 65–73.

Paigen, B., Nesbitt, M. N., Mitchell, D., Albee, D., and LeBoeuf, R. C. (1989) *Ath-2, a* second gene determining atherosclerosis susceptibility and high density lipoprotein levels in mice. *Genetics* 122:163–168.

Palinski, W., Rosenfeld, M. E., Yla-Herttuala, S., Gurtner, G. C., Steinberg, D., Witztum, J. L., Socher, S. S., Butler, S. W., Parthasarathy, S., and Carew, T. E. (1989) Low density lipoprotein undergoes oxidative modification *in vivo. Proc. Natl. Acad. Sci.* 86:1372–1376.

Palmer, E. D. (1955) Course of egg output over a 15–year period in a case of experimentally induced *Necatoriasis americanus*, in the absence of hyperinfection. *Am. J. Trop. Med. Hyg.* 4:756–757.

Palmiter, R. D., Brinster, R. L., Hammer, R. E., Trumbauer, M. E., Rosenfeld, M. G., Birnberg, N. C., and Evans, R. M. (1982) Dramatic growth of mice that develop from eggs microinjected with metallothionein-growth horme fusion genes. *Nature* 300:611–616.

Palumbi, S. R., and Jackson, J. B. (1983) Aging in modular organisms: Ecology of zooid senescence in *Steginoporella* sp. (Bryozoa: Cheilostomata). *Biol. Bull.* 164:267–278.

Paniagua, R., Nistal, M., Amat, P., Rodriguez, M. C., and Martin, A. (1987) Seminiferous tubule involution in elderly men. *Biol. Reprod.* 36:939–947.

Papadaki, L., Beilby, J. O., Chowaniec, J., Coulson, W. F., Darby, A. J., Newman, J., O'Shea, A., and Wykes, J. R. (1979) Hormone replacement therapy in the menopause: A suitable animal model. *J. Endocrinol.* 83:67–77.

Parfitt, A. M., Mathews, C. H. E., Villanueva, A. R., and Kleerekoper, M. (1983) Relationships between surface, volume, and thickness of iliac trabecular bone in aging and in osteoporosis. *J. Clin. Invest.* 72:1396–1409.

Park, J.-W., and Ames, B. N. (1988) 7–methylguanine adducts in DNA are normally present at high levels and increase on aging: Analysis by HPLC with electrochemical detection. *Proc. Natl. Acad. Sci.* 85:7467–7470.

Parkening, T. A., Collins, T. J., and Au, W. W.

(1988) Paternal age and its effects on reproduction in C57BL/6NNia mice. *J. Gerontol.* 1–20.

Parker, J., Flanagan, J., Murphy, J., and Gallant, J. (1981) On the accuracy of protein synthesis in *Drosophila melanogaster. Mech. Age. Dev.* 16:127–139.

Parker, R. J., Bekowitz, B. A., Lee, C. H., and Denckla, W. D. (1978) Vascular relaxation, aging, and thyroid hormones. *Mech. Age. Dev.* 8:397–405.

Parker, S. P. (ed.) (1982) *Synopsis and Classification of Living Organisms.* 2 vols. New York: McGraw-Hill.

Parrington, F. R. (1971) On the Upper Triassic mammals. *Phil. Trans. Roy. Soc.*, ser. B, 261:231–272.

———. (1978) A further account of the Triassic mammals. *Phil. Trans. R. Soc.*, ser. B, 282:177–204.

Parsons, P. A. (1962) Maternal age and development variability. *J. Exp. Biol.* 39:251–260.

Partridge, L. (1988) Lifetime reproductive success in *Drosophila.* In *Reproductive Success. Studies of Individual Variation in Contrasting Breeding Systems*, T. H. Clutton-Brock (ed.), 11–23. Chicago: Univ. of Chicago Press.

Partridge, L., and Andrews, R. (1985) The effect of reproductive activity on the longevity of male *Drosophila melanogaster* is not caused by an acceleration of senescence. *J. Insect Physiol.* 31:393–395.

Partridge, L., and Farquhar, M. (1981) Sexual activity reduces lifespan of male fruitflies. *Nature* 294:580–582.

Partridge, L., Fowler, K., Trevitt, S., and Sharp, W. (1986) An examination of the effects of males on the survival and egg-production rates of female *Drosophila melanogaster. J. Insect Physiol.* 32:925–929.

Partridge, L., and Harvey, P. H. (1985) Costs of reproduction. *Nature* 316:20–21.

Partridge, L., Hoffman, A., and Jones, J. S. (1987) Male size and mating success in *Drosophila melanogaster* and *D. pseudoobscura* under field conditions. *Anim. Behav.* 35:468–76.

Pashko, L. L., Fairman, D. K., and Schwartz, A. G. (1986) Inhibition of proteinuria development in aging Sprague-Dawley rats and C57BL/6 mice by long-term treatment with dehydroepiandrosterone. *J. Gerontol.* 41:433–438.

Pashko, L. L., Hard, G. C., Rovito, R. J., Williams, J. R., Sobel, E. L., and Schwartz, A. G. (1985) Inhibition of 7,12-dimethyl-benz(*a*)anthracene-induced skin papillomas and carcinomas by dehydroepiandrosterone and 3β-methyl-androst-5–en-17–one in mice. *Canc. Res.* 45:164–166.

Pashko, L. L., Rovito, R. J., Williams, J. R., Sobel, E. L., and Schwartz, A. G. (1984) Dehydroepiandrosterone (DHEA) and 3β-methyl-androst-5–en-17–one: Inhibitors of 7,12–dimethylbenz(*a*)anthracene (DMBA)–initiated and 12–O-tetradecanoyl-phorbol-13–acetate (TPA)–promoted skin papilloma formation in mice. *Carcinogenesis* 5:463–466.

Pasinetti, G. M., Lerner, S. P., Johnson, S. A., Morgan, D. G., Telford, N. A., and Finch, C. E. (1989) Chronic lesions differentially decrease tyrosine hydroxylase mRNA in dopaminergic neurons and the substantia nigra. *Mol. Brain Res.* 5:203–209.

Patelka, L. F., and Ashmun, J. W. (1985) Physiology and integration of ramets in clonal plants. In *Population Biology and Evolution of Clonal Organisms*, J. B. Jackson, L. W. Buss, and R. E. Cook (eds.), 399–435. New Haven: Yale Univ. Press.

Patnaik, B. K., and Behera, H. N. (1981) Age-determination in the tropical agamid garden lizard, *Calotes versicolor* (Daudin), based on bone histology. *Exp. Gerontol.* 16:295–307.

Paton, J. A., and Nottebohm, F. N. (1984) Neurons generated in the adult brain are recruited into functional circuits. *Science* 225:1046–1048.

Patrick, J. S., Thorpe, S. R., and Baynes, J. W. (1990) Nonenzymatic glycosylation of protein does not increase with age in normal human lenses. *J. Gerontol.* 45:B18–B23.

Patzner, R. (1978) Cyclical changes in the ovary of the hagfish, *Eptatretus burgeri* (Cyclostomata). *Acta Zool.* (Stockholm) 59:57–62.

Paulweber, B., Friedl, W., Krempler, F., Humphries, S. E., and Sandhofer, F. (1988) Genetic variation in the apolipoprotein AI-CIII-

AIV gene cluster and coronary heart disease. *Atherosclerosis* 73:125–133.

Pavlov, E. P., Harman, S. M., Chrousos, G. P., Loriaux, D. L., and Blackman, M. R. (1986) Responses of plasma adrenocorticotropin, cortisol, and dehydroepiandrosterone to ovine corticotropin-releasing hormone in healthy aging men. *J. Clin. Endocrinol. Metab.* 62:767–772.

Payne, F. (1952) Cytological changes in the pituitary, thyroids, adrenals, and sex glands of the aging fowl. In *Cowdry's Problems of Ageing*, 3d ed., A. Lansing (ed.), 381–402. New York: Plenum.

Paysinger, J. T., Noblet, R., Adkins, T. R., Jr., and Vaughn, E. A. (1978) Scanning electron microscopy of the adult mouthparts of the horn fly and the face fly. *J. Ga. Entomol. Soc.* 13:28–39.

Pearce, L., and Brown, W. H. (1960a) Hereditary premature senescence of the rabbit. 1. Chronic form: General features. *J. Exp. Med.* 111:485–504.

———. (1960b) Hereditary premature senescence of the rabbit. 2. Acute form: General features. *J. Exp. Med.* 111:505–516.

Pearl, R. (1928) *The Rate of Living*. New York and London: Knopf.

Pearl, R., and Doering, C. R. (1923) A comparison of mortality of certain lower organisms with that of man. *Science* 57:209.

Pearl, R., Parker, S. L., and Gonzalez, B. M. (1923) Experimental studies on the duration of life. 7. The Mendelian inheritance of duration of life in crosses of wild type and quintuple stocks of *Drosophila melanogaster*. *Am. Nat.* 57:153–192.

Pearson, M. W., and Roberts, C. J. C. (1984) Drug induction of hepatic enzymes in the elderly. *Age and Aging* 13:313–316.

Pearson, O. P., and Enders, R. K. (1944) Duration of pregnancy in certain mustelids. *J. Exp. Zool.* 95:21–35.

Pearson, T. A., Solez, K., Dillman, J. M., and Heptinstall, T. H. (1980) Evidence for two populations of fatty streaks with different roles in the atherogenic of atherosclerosis. *Lancet* 2:496–498.

Pearson, T. G. (1935) A herring gull of great age. *Bird Lore* 37:412–413.

Pekkanen, J., Marti, B., Nissinen, A., and Tuomilehto, J. (1987) Reduction of premature mortality by high physical activity: A 20–year follow-up of middle-aged Finnish men. *Lancet* 1:1473–1477.

Pellika, P. A., Sullivan, W. P., Coulam, C. B., and Toft, D. O. (1983) Comparison of estrogen receptors in human premenopausal uteri using isoelectric focusing. *Obstet. Gynecol.* 62:430–434.

Pelton, J. (1953) Studies on the life-history of *Symphoricarpos occidentalis* Hook in Minnesota. *Ecol. Monographs* 23:17–39.

Peluso, J. J., England-Charlesworth, C., and Hutz, R. (1980) Effect of age and of follicular aging on the preovulatory oocyte. *Biol. Reprod.* 22:999–1005.

Peluso, J. J., Montgomery, M. K., Steger, R. W., Meites, J., and Sacher, G. (1980) Aging and ovarian function in the white-footed mouse (*Peromyscus leucopus*) with specific reference to the development of preovulatory follicles. *Exp. Aging Res.* 6:317– 328.

Pena de Grimaldo, E., and Lavoipierre, M. M. J. (1960a) Efecto de la fertilización sobre la ovopostura de los mosquitos *Aedes aegypti* variedad *queenslandensis,* con algunas observaciónes sobre anomalias y viabilidad de los huevos retenidos por los mosquitos estériles fecundizados a diferentes intervalos después de la comida de sangre. *Rev. Ibérica Parasitol.* 20:163–176.

———. (1960b) Longevidad de los mosquitos *Aedes aegypti* variedad *queenslandensis* fecundados y no fecundados, alimentados con sangre o privados de ella; y dejados en ayuno; o proveídos de agua; de solución de azucar; o de agua y solución de azucar. *Rev. Ibérica Parasitol.* 20:39–52.

Pener, M. P. (1976) The differential effect of the corpora allata on male sexual behavior in crowded and isolated adults of *Locusta migratoria migratoroides* (M. & F.). *Acrida* 5:189–206.

Penfield, W. J., Penfield, H. B., and Phillips, M. (1937) *Taenia saginata:* Its growth and propagation. *J. Helminthol.* 15:41–48.

Peng, M. T., and Hsu, H. K. (1982) No neuron loss from hypothalamic nuclei of old male rats. *Gerontology* 28:19–22.

Penrose, L. S. (1955) Parental age and mutation. *Lancet* 312–313.

———. (1961) Mongolism. *Brit. Med. Bull.* 17:184–189. Pentney, R. J. (1986) Quantitative analysis of dendritic networks of Purkinje neurons during aging. *Neurobiol. Aging* 7:241–248.

Pereira, M. A., Burns, F. J., and Albert, R.' E. (1979) Dose response for benzo(a)pyrene adducts in mouse epidermal DNA. *Canc. Res.* 39:2558–2559.

Pereira-Smith, O. M., and Smith, J. R. (1983) Evidence for the recessive nature of cellular immortality. *Science* 221:964–966.

———. (1988) Genetic analysis of indefinite division in human cells. Identification of four complementation groups. *Proc. Natl. Acad. Sci.* 85:6042–6046.

Pericak-Vance, M. A., Bebout, J. L., Yamaoka, L. A., Gaskell, P. C., Hung, W.-Y., Alberts, M. J., Clark, C. M., et al. (In press) Linkage studies in familial Alzheimer's disease: Application of the affected pedigree member (APM) method of linkage analysis. Proceedings of the International Symposium on Dementia: Molecular Biology Genetics of Alzheimer Disease. *Excerpta Medica* International Congress Series. Amsterdam: Elsevier.

Pericak-Vance, M. A., Yamaoka, L. H., Haynes, C. S., Speer, M. C., Haines, J. L., Gaskell, P. C., Hung, W.-Y., Clark, C. M., Heyman, A. L., Trofatter, J. A., Eisenmenger, J. P., Gilbert, J. R., Lee, J. E., Alberts, M. J., Dawson, D. B., Bartlett, R. J., Earl, N. L., Siddique, T., Vance, J. M., Conneally, P. M., and Roses, A. D. (1988) Genetic linkage studies in Alzheimer's disease families. *Exp. Neurol.* 102:271–279.

Perillo, N. L., Walford, R. L., Newman, M. A., and Effros, R. B. (1989) Human T lymphocytes possess a limited *in vitro* lifespan. *Exp. Gerontol.* 24:177–187.

Perkins, E. H., Massucci, J. M., and Glover, P. L. (1982) Antigen presentation by peritoneal macrophages from young adult and old mice. *Cell. Immunol.* 70:1–10.

Pero, R. W., and Ostlund, C. (1980) Direct comparison, in human resting lymphocytes, of the inter-individual variations in unscheduled DNA synthesis induced by N-acetyl-laminofluorene and ultraviolet irradiation. *Mutat. Res.* 73:349–361.

Perron, F. E. (1982) Inter- and intraspecific patterns of reproductive effort in four species of cone shells (*Conus* spp). *Mar. Biol.* 68:161–167.

Perry, R. E., Swamy, M. S., and Abraham, E. C. (1987) Progressive changes in lens crystallin glycation and high-molecular-weight aggregate formation leading to a cataract development in streptozotocin-diabetes rats. *Exp. Eye Res.* 44:269–282.

Persov, G. M. (1975) *Sex Differentiation in Fishes* (in Russian). Leningrad: Publishing House Leningrad Univ.

Perutz, M. F. (1983) Species adaptation in a protein molecule. *Evolution* 1:1–28.

Perzigian, A. J. (1973) Osteoporotic bone loss in two prehistoric Indian populations. *Am. J. Phys. Anthropol.* 39:87–95.

Pesonen, E., Hirvonen, J., Laaksonen, H., Mottonen, M., Raekallio, J., and Akerblom, H. K. (1982) Morphometry of coronary arteries. *Arch. Pathol. Lab. Med.* 106:381–384.

Pesonen, E., Norio, R., and Sarna, S. (1975) Thickenings in the coronary arteries in infancy as an indication of genetic factors in coronary heart disease. *Circulation* 51:218–225.

Peters, A., Harriman, K. M., and West, C. D. (1986) The effect of increased longevity, produced by dietary restriction of the neuronal population and area 17 in rat cerebral cortex. *Neurobiol. Aging* 8:7–20.

Peters, J. G., and Peters, W. L. (1986) Leg abscission and adult *Dolania* (Ephemeroptera: Behningiidae). *Fla. Entomol.* 69:245–250.

Peters, J. G., Peters, W. L., and Fink, T. J. (1987) Seasonal synchronization of emergence in *Dolania americana* (Ephemeroptera: Behningiidae). *Can. J. Zool.* 65:3177–3185.

Peters, R. L., Hartley, J. W., Spahn, G. J., Rabstein, L. S., Whitmire, C. E., Turner, H. C., and Huebner, R. J. (1972) Prevalence of the group-specific (gs) antigen and infectious virus expression of the murine C-type RNA viruses during the lifespan of BALB/cCr mice. *Int. J. Canc.* 10:283–289.

Peters, W. L., and Peters, J. G. (1977) Adult life and emergence of *Dolania americana* in northwestern Florida (Ephemeroptera: Behningiidae). *Int. Rev. Ges. Hydrobiol.* 62:409–438.

———. (1988) The secret swarm. In the pre-dawn mass mating of sand-burrowing mayflies, timing is everything. *Nat. Hist.*, May, 8–13.

Peterson, B. V. (1981) Simuliidae. In *Manual of Nearctic Diptera*, vol. 1, J. F. McAlpine, B. V. Peterson, G. E. Shewell, H. J. Teskey, J. R. Vockeroth, and D. M. Wood (eds.). *Res. Br. Agric. Can.* 27:355–391.

Peterson, C. C., Nagy, K. A., and Diamond, J. (1990) Sustained metabolic scope. *Proc. Natl. Acad. Sci.* 87:2324–2328.

Peterson, C. H. (1986) Quantitative allometry of gamete production by *Mercenaria mercenaria* into old age. *Mar. Ecol. Prog. Ser.* 29:93–97.

Peterson, C. R. D., Cryar, J. R., and Gaubatz, J. W. (1984) Constancy of ribosomal RNA genes during aging of mouse heart cells and during serial passage of WI-38 cells. *Arch. Gerontol. Geriatr.* 3:115–125.

Petkau, A., Chelack, W. S., Pleskach, S. D., Meeker, B. E., and Brady, C. M. (1975) Radioprotection of mice by superoxide dismutase. *Biochem. Biophys. Res. Commun.* 65:886–893.

Peto, R. (1977) Epidemiology, multistage models, and short-term mutagenicity tests. In *Origins of Human Cancer*, vol. C, H. H. Hiatt, J. D. Watson, and J. A. Winston (eds.), 1403–1428. Cold Spring Harbor, NY: Cold Spring Harbor Laboratory Press.

———. (1979) Detection of risk of cancer to man. *Proc. Roy. Soc. Lond.* ser. B, 205:111–120.

Peto, R., Roe, F. J., Lee, P. N., Levy, L., and Clack, J. (1975) Cancer and ageing in mice and men. *Brit. J. Canc.* 32:411–426.

Pettengill, N. E., and Martin, A. W. (1947) The amounts of connective tissue in various muscles of the dog with some interspecific comparisons. *Fed. Proc.* 6:179.

Petter-Rousseaux, A. (1953) Récherches sur la croissance et le cycle d'activité testiculaire de *Natrix natrix helvetica* (Lacepede). *Terre et Vie* 4:175–223.

Pettifor, R. A., Perrins, C. M., and McCleeny, R. H. (1988) Individual optimization of clutch size in great tits. *Nature* 336:160–162.

Peyer, H. B. (1968) *Comparative Odontology.* Translated by R. Zangerl (ed.). Chicago: Univ. of Chicago Press.

Phelan, J. P., and Austad, S. N. (1989) Natural selection, dietary restriction, and extended longevity. *Growth Dev. Aging* 53:4.

Phillips, D. P., and King, E. W. (1988) Death takes a holiday: Mortality surrounding major social occasions. *Lancet* 728–732.

Phillips, J. P., Campbell, S. D., Michaud, D., Charbonneau, M., and Hilliker, A. J. (1989) Null mutation of copper/zinc superoxide dismutase in *Drosophila* confers hypersensitivity to paraquat and reduced longevity. *Proc. Natl. Acad. Sci.* 86:2761–2765.

Phillips, M. L., Moule, M. L., Delovitch, T. L., and Yip, C. C. (1986) Class I histocompatibility antigens and insulin receptors: Evidence for interactions. *Proc. Natl. Acad. Sci.* 83:3474–3478.

Phillips, P. D., Nishikura, K., and Cristofalo, V. J. (1990) Transient transfection of senescent WI-38 cells with an inducible c-fos construct leads to a significant stimulation of DNA synthesis.

Phillips, P. D., Pignolo, R. J., and Cristofalo, V. J. (1987) Insulin-like growth factor. 1. Specific binding to high and low affinity sites and mitogenic action throughout the life span of WI-38 cells. *J. Cell Physiol.* 133:135–143.

Piacsek, B. E. (1985) Altered negative feedback responses to ovariectomy and estrogen in prepubertal restricted-diet rats. *Biol. Reprod.* 32:1062–1068.

Pianka, E. R. (1970) On r and K-selection. *Am. Nat.* 104:592–597.

Piatigorsky, J., and Wistow, G. J. (1989) Enzyme/crystallins: Gene sharing as an evolutionary strategy. *Cell* 57:197–199.

Piatt, J., and Piatt, M. (1958) Transaction of the spinal cord in the adult frog. *Anat. Rec.* 131:81–95.

Picard-Bennoun, M. (1985) Introns, protein syntheses, and aging. *FEBS Lett.* 184:1–5.

Pickering, A. D. (1976) Stimulation of intestinal degeneration by oestradiol and testoster-

one implantation in the migrating river lamprey, *Lampetra fluviatilis L. Gen. Comp. Endocrinol.* 30:340–346.

Pickles, A. (1931) On the metamorphosis of the alimentary canal in certain Emphemeroptera. *Trans. Roy. Entomol. Soc.* (London) 79:263–276.

Pierre, R. V., and Hoagland, H. C. (1972) Age-associated aneuploidy: Loss of Y chromosome from human bone marrow cells with aging. *Cancer* 30:889–894.

Pike, M. C., Krailo, M. D., Henderson, B. E., Casagrande, J. T., and Hoel, D. G. (1983) "Hormonal" risk factors, "breast tissue age," and the age-incidence of breast cancer. *Nature* 303:767–770.

Pikó, L., Bulpitt, K. J., and Meyer, R. (1984) Structural and replicative forms of mitochondrial DNA in tissues from adult and senescent BALB/c mice and Fischer 344 rats. *Mech. Age. Dev.* 26:113–131.

Pikó, L., Hougham, A. J., and Bulpitt, K. J. (1988) Studies of sequence heterogeneity of mitochondrial DNA from rat and mouse tissues: Evidence for an increased frequency of deletions/additions with aging. *Mech. Age. Dev.* 43:279–293.

Pikó, L., and Matsumoto, L. (1977) Complex forms and replicative intermediates of mitochondrial DNA in tissues from adult and senescent mice. *Nucleic Acids Res.* 4:1301–1314.

Pikó, L., Meyer, R., Eipe, J., and Costea, N. (1978) Structural and replicative forms of mitochondrial DNA from human leukocytes in relation to age. *Mech. Age. Dev.* 7:351–365.

Pisciotta, A. V., Westring, D. W., DePrey, C., and Walsh, B. (1967) Mitogenic effect of phytohaemagglutinin at different ages. *Nature* 215:193–194.

Pixell-Goodrich, E. L. M. (1919) Determination of age in honey-bees. *Q. J. Microscop. Sci.* 64:191–206.

Platt, K. A., Min, H. Y., Ross, S. R., and Spiegelman, B. M. (1989) Obesity-linked regulation of the adipsin gene promoter in transgenic mice. *Proc. Natl. Acad. Sci.* 86:7490–7494.

Plattner, H. (1968) Comparative quantitative

cytology of liver parenchymal cells. In *Fourth European Regional Conference on Electron Microscopy,* Rome, vol. 2, D. S. Bocciarelli (ed.), 469–470. Rome: Tipografia Poliglotta Vaticana.

Pohl, C. R., Richardson, D. W., Hutchinson, J. S., Germak, J. A., and Knobil, E. (1983) Hypophysiotropic signal frequency of the functioning of the pituitary-ovarian system in the rhesus monkey. *Endocrinology* 112: 2076–2080.

Pohley, H.-J. (1987) A formal mortality analysis for populations of unicellular organisms (*Saccharomyces cerevisiae*). *Mech. Age. Dev.* 38:231–243.

Poirier, J. (1987) Pathophysiology and biochemical mechanisms involved in MPTP-induced Parkinsonism. *J. Am. Geriatr. Soc.* 35:660–668.

Polani, P. E., and Jagiello, G. M. (1976) Chiasmata, meiotic univalents, and age in relation to aneuploid imbalance in mice. *Cytogenet. Cell. Genet.* 16:505–529.

Polenov, A. L., and Pavlovic, M. (1978) The hypothalamo-hypophysial system in Acipenseridae. *Cell Tiss. Res.* 186:559–570.

Pollack, M. L., Foster C., Knapp, D., Rod, J. L., and Schmidt, D. H. (1987) Effect of age and training on aerobic capacity and body composition of master athletes. *J. Appl. Physiol.* 62:725–731.

Pollard, M., Luckert, P. H., and Snyder, D. (1989) Prevention of prostate cancer and liver tumors in L-W rats by moderate dietary restriction. *Cancer* 64:686–690.

Poltorykhina, A. N. (1971) Metamorphosis of the arctic brook lamprey (*Lampetra japonica kessleri* [Anikin]) in the Upper Irtysh. *J. Ichthyol.* 11:281–285.

Pommerville, J. C., and Kochert, G. D. (1981) Changes in somatic cell structure during senescence of *Volvox carteri. Eur. J. Cell Biol.* 24:236–243.

———. (1982) Effects of senescence on somatic cell physiology in the green alga *Volvox carteri. Exp. Cell Res.* 140:39–45.

Pongor, S., Ulrich, P. C., Bencsath, A., and Cerami, A. (1984) Aging of proteins: Isolation and identification of a fluorescent chromophore from the reaction of polypeptides with glucose. *Proc. Natl. Acad. Sci.* 81:2684–2688.

Ponte, P., Gonzalez-De Whitt, P., Schilling, J., Miller, J., Hsu, D., Greenberg, B., Davis, K., Wallace, W., Lieberburg, I., Fuller, F., and Cordell, B. (1988) A new A4 amyloid mRNA contains a domain homologous to serine protease inhibitors *Nature* 331:525–527.

Ponten, J. (1971) Spontaneous and virus-induced transformation in cell culture. In *Handbook of Virus Research,* Virology Monographs, S. Gard, C. Hallauer, and K. F. Meyer (eds.), 1–253. New York: Springer-Verlag.

———. (1977) Abnormal cell growth (neoplasia) and aging. In *Handbook of the Biology of Aging,* 1st ed., C. E. Finch and L. Hayflick (eds.), 537–560. New York: Van Nostrand Reinhold.

Poole, A. R. (1986) Changes in the collagen and proteoglycan of articular cartilage in arthritis. *Rheumatology* 10:316–371.

Popp, D. M. (1978) Use of congenic mice to study the genetic basis of degenerative disease. In *Genetic Effects on Aging,* D. Bergsma and D. E. Harrison (eds.), 261–279. Birth Defects: Original Article Series, no. 14. New York: Alan R. Liss.

———. (1979) Basal serum immunoglobulin levels. 2. Integration of genetic loci and maternal effect on regulation of IgA. *Immunogenetics* 9:281–292.

———. (1982) Analysis of genetic factors regulating the lifespan in congenic mice. *Mech. Age. Dev.* 18:125–134.

Popp, D. M., Otten, J. A., and Popp, R. A. (1986) Interaction of *H-2* genotype and basal serum immunoglobulin A level influences longevity. *Mech. Age. Dev.* 36:79–93.

Popp, R. H., Bailiff, E. G., Hirsch, G. P., and Conard, R. A. (1976) Errors in human hemoglobin as a function of age. In *Interdisciplinary Topics in Gerontology,* vol. 9, R. G. Cutler and H. P. Von Hahn (eds.), 209–218. Basel: Karger.

Porsch, O. (1958) Alte Insektentypen als Blumenausbeuter. *Oster. Bot. Z.* 104:115–164.

Poskanzer, D. C., and Schwab, R. S. (1961) Studies in the epidemiology of Parkinson's disease predicting its disappearance as a major clinical entity by 1980. *Trans. Am. Neurol. Assoc.* 86:234.

Post, A., and Tesch, A.-W. (1982) Midwater trawl catches of adolescent and adult anguilliform fishes during the Sargasso Sea eel expedition 1979. *Helgoländer Meeruntersuchungen* 35:341–356.

Potter, V. R. (1978) Phenotypic diversity in experimental hepatomas: The concept of partially blocked ontogeny. *Brit. J. Canc.* 38:1–23.

———. (1980) Initiation and promotion in cancer formation: The importance of studies on intercellular communication. *Yale J. Biol. Med.* 53:367.

Pourriot, R., and Clément, P. (1975) Influence de la durée de l'éclairment quotidien sur le taux de femelles mictiques chez *Notommata copeus* Ehr (rotifere). *Oecologia* 22:67–77.

Prange, H. D., Anderson, J. F., and Rahn, H. (1979) The allometrics of rattlesnake skeletons. *Copeia* 3:524–545.

Preston, D. L., Kato, H., Kopecky, K. J., and Fujita, S. (1987) Studies of A-bomb survivors. 8. Cancer mortality, 1950–1982. *Rad. Res.* 111:151–178.

Prieur, M., Al Achkar, W., Aurias, A., Couturier, J., Dutrillaux, A. M., Dutrillaux, B., Flury-Herard, A., Gerbault-Seureau, M., Hoffschir, F., Lamoliatte, E., Lefrancois, D., Lombard, M., Muleris, M., Ricoul, M., Sabatier, L., and Viegas-Pequignot, E. (1988) Acquired chromosome rearrangements in human lymphocytes: Effect of aging. *Human Genet.* 79:147–150.

Prinz, P. N., Witzman, E. D., Cunningham, G. R., and Karacan, I. (1983) Plasma growth hormone during sleep in young and aged men. *J. Gerontol.* 38:519–524.

Probst, R. T., and Cooper, E. L. (1954) Age, growth, and production of the lake sturgeon (*Acipenser fulvescens*) in the Lake Winnebago region, Wisconsin. *Trans. Am. Fish. Soc.* 84:207–227.

Promislow, D. E., and Harvey, P. H. (1990) Living fast and dying young: A comparative analysis of life history variation among mammals. *J. Zool.* (London) 220:417–437.

Prothero, J. W. (1980) Scaling of blood parameters in mammals. *Comp. Biochem. Physiol.* 67A:649–657.

———. (1986) Methobiological aspects of scaling in biology. *J. Theoret. Biol.* 188:259–286.

Prothero, J. W., and Jürgens, K. D. (1987) Scaling of maximal lifespan in mammals: A review. In *Evolution of Longevity in Mammals. A Comparative Approach*, A. D. Woodhead and K. H. Thompson (eds.), 49–74. New York: Plenum.

Prout, T. (1964) Observations on structural reduction in evolution. *Am. Nat.* 98:239–249.

Prusiner, S. B. (1990) Prion diseases and aging. In *Molecular Biology of Aging*, C. E. Finch and T. E. Johnson (eds.). UCLA Symposia on Molecular and Cellular Biology, n.s. 123:311–338. New York: Wiley-Liss.

Pryor, W. A. (1986) Oxy-radicals and related species: Their formation, lifetimes, and reactions. *Ann. Rev. Physiol.* 48:657.

———. (1987) The free-radical theory of aging revisited: A critique and a suggested disease-specific theory. In *Modern Biological Theories of Aging*, H. R. Warner, R. N. Butler, R. L. Sprott, and E. L. Schneider (eds.), 89–112. New York: Raven Press.

Pugasek, B. H. (1981) Increased reproduction effort with age in the California gull (*Larus californicus*). *Science* 212:822–823.

———. (1983) The relationship between parental age and reproductive effort in the California gull (*Larus californicus*). *Behav. Ecol. Sociobiol.* 13:161–171.

Pugasek, B. H., and Diem, K. L. (1983) A multivariate study of the relationship of parental age to reproductive success in California gulls. *Ecology* 64:829–839.

Pugh, S. Y. R., and Fridovich, I. (1985) Induction of superoxide dismutases in *Escherichia coli* B by metal chelators. *J. Bacteriol.* 162:196–202

Purves, D., and Lichtman, J. W. (1985) Geometric differences among homologous neurons in mammals. *Science* 228:298–302.

Purves, D., Rubin, E., Snider, W. D., and Lichtman, J. W. (1986) Relation of animal size to convergence, divergence, and neuronal number in peripheral sympathetic pathways. *J. Neurosci.* 6:158–163.

Purvis, H. A. (1980) Effects of temperature on metamorphosis and the age and length at metamorphosis in sea lamprey (*Petromyzon marinus*) in the Great Lakes. *Can. J. Fish. Aquat. Sci.* 37:1827– 1834.

Putnam, P., and Shannon, R. C. (1934) The biology of *Stegomyia* under laboratory conditions. 2. Egg-laying capacity and longevity of adults. *Proc. Entomol. Soc. Wash.* 36:217– 242.

Quinn, L. S., Nameroff, M., and Holtzer, H. (1984) Age-dependent changes in myogenic precursor cell compartment sizes: Evidence for the existence of a stem cell. *Exp. Cell Res.* 154:65–82.

Quint, W., Quax, W., van der Putten, H., and Berns, A. (1981) Characterization of AKR murine leukemia virus sequences in AKR mouse substrains and structure of integrated recombinant genomes in tumor tissues. *J. Virol.* 39:1–10.

Raab, W., Stark, E., Macmillan, W. H., and Gigea, W. R. (1961) Sympathogenic origin and antiadrenergic prevention of stress-induced myocardial lesions. *Am. J. Cardiol.* 8:203–211.

Rabinovitch, P. S., and Martin, G. M. (1982) Encephalomyocarditis virus as a probe of errors in macromolecular synthesis in aging mice. *Mech. Age. Dev.* 20:155–163.

Rabinowich, H., Lyte, M., Steiner, Z., Klaiman, A., and Shinitzky, M. (1987) Augmentation of mitogen responsiveness in the aged by a special lipid diet AL 721. *Mech. Age. Dev.* 40:131– 138.

Racey, P. A. (1969) Diagnosis of pregnancy and experimental extension of gestation in the pipistrelle bat, *Pipistrellus pipistrellus. J. Reprod. Fertil.* 19:464–474.

Radman, M., Wagner, R., and Jego, P. (1982) DNA and time in carcinogenesis. In *Gene Expression in Normal and Transformed Cells,* J. E. Celis and R. Bravo (eds), 177–191. New York: Plenum.

Raes, M., Michiels, C., and Remacle, J. (1987) Comparative study of the enzymatic defense systems against oxygen-derived free radicals: The key role of glutathione peroxidase. *Free Rad. Biol. Med.* 3:3–7.

Raff, R. A. (1987) Constraint, flexibility, and phylogenetic history in the evolution of direct development in sea urchins. *Dev. Biol.* 119:6–19.

Ragland, S. S., and Sohal, R. S. (1968) Ambient temperature, physical activity, and aging in the housefly *Musca domestica. Exp. Gerontol.* 10:279–289.

———. (1973) Mating behavior, physical activity, and aging in the housefly, *Musca domestica. Exp. Gerontol.* 8:135–145.

Ragozzino, M. W., Melton, L. J., Kurland, L. T., Chu, C. P., and Perry, H. O. (1982) Population-based study of herpes zoster and its sequelae. *Medicine* 61:310–316.

Rahmani, Z., Blouin, J.-L., Creau-Goldberg, N., Watkins, P. C., Mattei, J.-F., Poissonnier, M., Prieur, M., Chettouh, Z., Nicole, A., Aurias, A., Sinet, P.-M., and Delabar, J.-M. (1989) Critical role of the D21S55 region on chromosome 21 in the pathogenesis of Down syndrome. *Proc. Natl. Acad. Sci.* 86:5958–5962.

Raikova, E. V. (1976) Evolution of the nucleolar apparatus during oogenesis in Acipenseridae. *Embryol. Exp. Morphol.* 35:667–687.

Raikow, R. (1975) The evolutionary reappearance of ancestral muscles as developemental anomalies in two species of birds. *Condor* 77:514–517.

Raisz, L. G. (1982) Osteoporosis. *J. Am. Geriatr. Soc.* 30:127–138.

Rakic, P. (1985) Limits of neurogenesis in primates. *Science* 227:1054–1056.

Ralls, K., Brownall, R. L., and Ballou, J. (1980) Differential mortality by sexual age in mammals, with specific reference to the sperm whale. *Rep. Int. Whal. Commn.,* special issue, 2:233–243.

Randall, P. K., Severson, J. A., and Finch, C. E. (1981) Aging and the regulation of striatal dopaminergic mechanisms in mice. *J. Pharmacol. Exp. Ther.* 219:690–700.

Randerath, K., Reddy, M. V., and Disher, R. M. (1986) Age- and tissue-related modifications in untreated rats: Detection by ^{32}P-postlabeling assay and possible significance for spontaneous tumor induction and aging. *Carcinogenesis* 7:1615–1617.

Randolph, P. A., Randolph, J. C., and Barlow, C. A. (1975) Age-specific energetics of the pea aphid, *Acyrthosiphon pisum*. *Ecology* 56:359–369.

Ranieri, R., and Levy, H. R. (1970) On the specificity of steroid interaction with mammalian glucose-6–phosphate dehydrogenase. *Biochemistry* 9:2233–2243.

Rao, G., Xia, E., and Richardson, A. (1990) Effect of age on the expression of antioxidant enzymes in male Fischer 344 rats. *Mech. Age. Dev.* 53:49–60.

Rao, T. R., and Slobin, L. I. (1987) Regulation of the utilization of mRNA for eukaryotic elongation factor Tu in Friend erythroleukemia cells. *Mol. Cell. Biol.* 7:687–697.

Rapoport, S. I. (1988) Brain evolution and Alzheimer's disease. *Rev. Neurol.* (Paris) 144: 79–90.

Rapp, J. P. (1973) Age-related pathologic changes, hypertension, and 18–hydroxydeoxycorticosterone in rats selectivity bred for high or low juxtaglomerular granularity. *Lab. Invest.* 28:343–351.

Rasheed, S., and Gardner, M. B. (1983) Resistance of fibroblasts and hematopoietic cells to ecotropic murine leukemia virus infection: An *Akvr-1ᴿ* gene effect. *Int. J. Canc.* 31:491–496.

Rasnitsyn, A. P. (1980a) Order Vespidae (in Russian). In *Historical Development of the Class Insecta,* B. B. Rohdendorf and A. P. Rasnitsyn (eds.). *Trudy Paleontol. Inst.* 175:122–127. [not read]

———. (1980b) Origin and evolution of the Hymenoptera (Insecta) (in Russian). *Trudy Paleontol. Inst.* 174:1–191. [not read]

Rasquin, P., and Hafter, E. (1951) Age changes in the testes of the teleost *Astyanax mexicanus*. *J. Morphol.* 89:397–407.

Rath, P. C., and Kanungo, M. S. (1988) Age-related changes in the expression of cytochrome P-450 (b+e) gene in the rat after phenobarbitone administration. *Biochem. Biophys. Res. Commun.* 157:1403–1409.

———. (1989) Methylation of repetitive DNA sequences in the brain during aging of the rat. *FEBS Lett.* 244:193–198.

Raven, P. H., Evert, R. F., and Eichhorn, S. E.

(1986) *Biology of Plants*. 4th ed. New York: Worth.

Rawlins, J. N. (1985) Associations across time: The hippocampus as a temporary memory store. *Behav. Brain Sci.* 8:479–496.

Read, A. F., and Harvey, P. H. (1989) Life history differences among the eutherian radiations. *J. Zool.* (London) 219:329–353.

Read, C. P. (1967) Longevity of the tapeworm, *Hymenolepis diminuta*. *J. Parasitol.* 53: 1055–1056.

Reaka, M. L. (1979) The evolutionary ecology of life history patterns in stomatopod Crustacea. In *Reproductive Ecology of Marine Invertebrates,* S. E. Stancyk (ed.), 235–260. Columbia: Univ. of South Carolina Press.

Reaven, G. M., and Reaven, E. P. (1980) Effects of age on various aspects of glucose and insulin metabolism. *Mol. Cell. Biochem.* 31:37–47.

Rebar, R. W. (1982) The thymus gland and reproduction: Do thymic peptides influence the reproductive lifespan in females? *J. Am. Geriatr. Soc.* 30:603–606.

Rebeitz, J. J., Moore, M. J., Holden, E. M., and Adams, R. D. (1972) Variations in muscle status with age and systemic diseases. *Acta Neuropathol.* (Berlin) 22:127–144.

Rechsteiner, M. (1988) Regulation of enzyme levels by proteolysis the role of PEST regions. *Adv. Enzyme Reg.* 27:135–151.

Reddy, M. V., and Randerath, K. (1986) Nucleas P1–mediated enhancement of sensitivity of ^{32}P-postlabeling test for structurally diverse DNA adducts. *Carcinogenesis* 7:1543–1551.

Reeds, P. J., Cadenhead, A., Fuller, M. F., Lobley, G. E., and McDonald, J. D. (1980) Protein turnover in growing pigs. Effects of age and food intake. *Brit. J. Nutrit.* 43:445–455.

Rees, T. S., and Duckert, L. G. (1990) Auditory and vestibular dysfunction in aging. In *Principles of Geriatric Medicine and Gerontology,* 2d ed., W. R. Hazzard, R. Andres, E. L. Bierman, and J. P. Blass (eds.), 432–444. New York: McGraw-Hill.

Reeve, R., Smith, E., and Wallace, B. (1990) Components of fitness become effectively neutral in equilibrium populations. *Proc. Natl. Acad. Sci.* 87:2018–2020.

Reff, M. E. (1985) RNA and protein metabolism. In *Handbook of the Biology of Aging,* 2d ed., C. E. Finch and E. L. Schneider (eds.), 225–254. New York: Van Nostrand.

Reff, M. E., and Schneider, E. L. (eds.) (1982) *Biological Markers of Aging.* NIH Publication 82–2221. Washington, DC: USDHHS.

Regal, P. J. (1977a) Ecology and evolution of flowering plant dominance. *Science* 196:622–629.

———. (1977b) Evolutionary loss of useless features: Is it molecular noise suppression? *Am. Nat.* 111:123–133.

Regoeczi, E., and Hatton, M. W. C. (1980) Transferrin catabolism in mammalian species of different body sizes. *Am. J. Physiol.* 238:R306–R310.

Reichel, W., Bailey, J. A., Zigel, S., Garcia-Bunuel, R., and Knox, G. (1971) Radiological findings in progeria. *J. Am. Geriatr. Soc.* 19:657–674.

Reichel, W., Hollander, J., Clark, J. H., and Strehler, B. L. (1968) Lipofuscin pigment accumulation as a function of age and distribution in rodent brain. *J. Gerontol.* 23:71–78.

Reid, A. H., and Maloney, N. F. (1974) Giant cell arteritis and arteriolitis associated with amyloid angiopathy in an elderly Mongol. *Acta Neuropathol.* (Berlin) 27:131–137.

Reid, D. H., and Parsons, P. A. (1963) Sex of parent and variation of recombination with age in the mouse. *Heredity* 18:107–108.

Reid, J. B. (1980) Apical senescence in *Pisum:* A direct or indirect role for the flowering genes? *Ann. Bot.* 45:195–201.

Reif, W. E. (1976) Morphogenesis, pattern formation, and function of the dentition of Heterodontus (Selachii). *Zoomorphologie* 83:1–47.

———. (1984) Pattern regulation in shark dentitions. In *Pattern Formation,* G. M. Malincinski and S. V. Bryant (eds.), 603–621. New York: Macmillan.

Reik, W., Collick, A., Norris, M. L., Barton, S. C. and Surani, M. A. (1987) Genomic imprinting determines methylation of parental alleles in transgenic mice. *Nature* 328:248–251.

Reimers, N. (1979) A history of a stunted brook trout population in an alpine lake: A lifespan of 24 years. *Calif. Fish and Game* 65:196–214.

Reiser, K. M., Hennessy, S. M., and Last, J. A. (1987) Analysis of age-associated changes in collagen crosslinking in the skin and lung in monkeys and rats. *Biochim. Biophys. Acta* 926:339–348.

Reiss, U., and Rothstein, M. (1975) Age-related changes in isocitrate lyase in the free-living nematode *Turbatrix aceti. J. Biol. Chem.* 250:826–830.

Reiter, E. O., and Grumbach, M. M. (1982) Neuroendocrine control mechanisms and the onset of puberty. *Ann. Rev. Physiol.* 44:595–613.

Remane, A. (1932) Rotatoria. *Klassen and Ordnungen des Tier-reiches Acoela und Rhabdocoelida,* vol. 4, H. G. Bronn (ed.), part 2, sec. 1:1–576. Leipzig: Winter'sche Verlags Ges.

Resnitzky, P., Segal, M., Barak, Y., and Dassa, C. (1987) Granulopoiesis in aged people: Inverse correlation between bone marrow cellularity and myeloid progenitor cell numbers. *Gerontology* 33:109–114.

Retallack, G. J., Dugas, D. P., and Bestland, E. A. (1990) Fossil soils and grasses of a middle Miocene East African grassland. *Science* 247:1325–1328.

Reyes, F. I., Koh, K. S., and Faiman, C. (1976) Fertility in women with gonadal dysgenesis. *Am. J. Obstet. Gynecol.* 126:668–670.

Reynolds, J. J. (1986) Inhibition of production and action of tissue metalloproteinases. *Ann. Biol. Clin.* 44:188.

Reynoldson, T. B. (1960) A quantitative study of the population biology of *Polycelis tenuis* (Ijama) (Turbellaria, Tricladida). *Oikos* 11:125–141.

———. (1966) The distribution and abundance of lake-dwelling triclads: Towards a hypothesis. *Adv. Ecol. Res.* 3:1–72.

Reznick, A. Z., Dovrat, A., Rosenfelder, L., Shpund, S., and Gershon, D. (1985b) Defective enzyme molecules in cells of aging animals are partially denatured, totally inactive normal degradation intermediates. In *Modifications of Proteins during Aging,* R. C. Adelman (ed.), 69–82. New York: Liss.

Reznick, A. Z., and Gershon, D. (1977) Age-related alterations in purified fructose-1,6–diphosphate aldolase in the nematode *Turbatrix aceti. Mech. Age. Dev.* 6:345–353.

———. (1979) The effect of age on the protein degradation system in the nematode *Turbatrix aceti. Mech. Age. Dev.* 1:403–415.

Reznick, A. Z., Lavie, L., Gershon, H., and Gershon, D. (1981) Age-associated accumulation of altered FDP aldolase B in mice. Conditions for detection and determination of aldolase half-life in young and old animals. *FEBS Lett.* 128:221–224.

Reznick, A. Z., Rosenfelder, L., Shpund, S., and Gershon, D. (1985) Identification of intracellular degradation intermediates of aldolase B by antiserum to the denatured enzyme. *Proc. Natl. Acad. Sci.* 82:6114–6118.

Reznick, D. (1985) Costs of reproduction: An evaluation of the empirical evidence. *Oikos* 44:257–267.

Ribbands, C. R. (1952) Division of labour in the honey bee community. *Proc. Roy. Soc.,* ser. B, 140:32–43.

Richards, O. W. (1965) Concluding remarks on the social organization of insect communities. *Symp. Zool. Soc. Lond.* 14:169–172.

Richards, O. W., and Davies, R. G. (1964) *A General Textbook of Entomology,* A. D. Imm (ed.). London: Butter and Tanner.

———. (1977) *Imms' General Textbook of Entomology.* Vol. 2, *Classification and Biology.* 10th ed. London: Chapman and Hall.

Richardson, A. (1981) The relationship between aging and protein synthesis. In *CRC Handbook of Biochemistry and of Aging,* J. R. Florini (ed.), 79–101. Boca Raton, FL: CRC Press.

Richardson, A., and Birchenall-Sparks, M. C. (1983) Age-related changes in protein systhesis. *Rev. Biol. Res. Aging* 1:225–273.

Richardson, A., Butler, J. A., Rutherford, M. S., Semsei, I., Gu, M. Z., Fernandes, G., and Chiang, W.-S. (1987) Effect of age and dietary restriction on the expression of alpha$_{2u}$-globulin. *J. Biol. Chem.* 262:12821–12825.

Richardson, A., Roberts, M. S., and Rutherford, M. S. (1985) Aging and gene expression. *Rev. Biol. Res. Aging* 2:395–419.

Richardson, D. M. (1972) Body weights of a few JAX inbred strains. *JAX Notes,* no. 412. Jackson Laboratory, Bar Harbor, ME.

Richardson, G. (1990) Cirdadian rhythms and aging. In *Handbook of the Biology of Aging,* 3d ed., E. L. Schneider and J. W. Rowe (eds.), 275–305. San Diego: Academic Press.

Richardson, S. J., Senikas, V., and Nelson, J. F. (1987) Follicular depletion during the menopausal transition: Evidence for accelerated loss and ultimate exhaustion. *J. Clin. Endocrinol. Metab.* 65:1231–1237.

Richey, J. E., Perkins, C. M., and Goldman, C. R. (1975) Effects of kokanee salmon (*Oncorhynchus nerka*) decomposition on the ecology of a subalpine stream. *J. Fish. Res. Bd. Can.* 32:817–820.

Richie, J. P., and Lang, C. A. (1988) A decrease in cysteine levels causes the glutathione deficiency of aging in the mosquito (42660). *Proc. Soc. Exp. Biol. Med.* 187:235–240.

Richie, J. P., Jr., Mills, B. J., and Lang, C. A. (1987) Correction of a glutathione deficiency in the aging mosquito increases its longevity. *Proc. Soc. Exp. Biol. Med.* 184:113–117.

Richter, C. P. (1922) A behavioristic study of the activity of the rat. *Comp. Psych. Monographs* 1:1–55.

Ricker, W. E. (1945) Natural mortality rates among Indiana bluegill sunfish. *Ecology* 26:111–121.

———. (1949) Mortality rates in some little-exploited populations of fresh-water fishes. *Amer. Fish. Soc. Trans.* 77:114–127.

———. (1979) Growth rates and models. *Fish Physiol.* 8:677–743.

Ricketts, W. G., Birchenall-Sparks, M. C., Hardwick, J. P., and Richardson, A. (1985) Effect of age and dietary restriction on protein synthesis by isolated kidney cells. *J. Cell Physiol.* 125:492–498.

Rickleffs, R. E. (1977) On the evolution of reproductive strategies in birds: Reproductive effort. *Am. Nat.* 111:453–478.

Riddle, D. L. (1988) The dauer larva. In *The Nematode Caenorhabditis Elegans,* W. B. Wood (ed.), 393–414. Cold Spring Harbor, New York: Cold Spring Harbor Laboratory.

Riddle, D. L., Swanson, M. M., and Albert, P. S. (1981) Interacting genes in nematode dauer larva formation. *Nature* 290:668–671.

Riederer, P., and Wuketich, S. T. (1976) Time

course of nigrostriatal degeneration in Parkinson's disease. *J. Neural. Transm.* 38:277–301.

Rifai, N. (1986) Lipoproteins and apolipoproteins. Composition, metabolism, and association with coronary heart disease. *Arch. Pathol. Lab. Med.* 110:694–701.

Riggs, A. (1960) The nature and significance of the Bohr effect in mammalian hemoglobins. *J. Gen. Physiol.* 43:737–752.

———. (1971) Mechanism of the enhancement of the Bohr effect in mammalian hemoglobins by diphosphoglycerate. *Proc. Natl. Acad. Sci.* 68:2062–2065.

Riggs, B. L., and Melton, J. L., III (1986) Involutional osteoporosis. *N. Engl. J. Med.* 314:1676–1685.

Riggs, B. L., Wahrner, H. W., Dunn, W. L., Mazess, R. B., Offord, K. P., and Melton, L. J., III (1981) Differential changes in bone mineral density of the appendicular and axial skeleton with aging: Relationship to spinal osteoporosis. *J. Clin. Invest.* 67:328–355.

Riggs, B. L, Wahrner, H. W., Melton, L. J., III, Richelson, L. S., Judd, H. L., and Offord, K. P. (1986) Rates of bone loss in the axial and appendicular skeletons of women: Evidence of substantial vertical bone loss prior to menopause. *J. Clin. Invest.* 77:1487–1491.

Rikans, L. E., and Notley, B. A. (1984) Effect of methyltestosterone administration on microsomal drug metabolism in aging rats. *Mech. Age. Dev.* 25:335–341.

Rines, J., and vom Saal, F. (1984) Fetal effects on sexual behavior and aggression in young and old female mice treated with estrogen and progesterone. *Horm. Behav.* 18:117–129.

Rinkevich, B., and Loya, Y. (1986) Senescence and dying signals in a reef-building coral. *Experientia* 42:320–322.

Rinkevich, B., and Weissman, I. L. (1987) A long-term study on fused subclones in the ascidian *Botryllus schlosseri:* The resorption phenomenon. *J. Zool.* 213:717–733.

Ritossa, F. M., Atwood, K. D., and Spiegelman, S. (1966) A molecular explanation of the bobbed mutants of *Drosophila* as partial deficiencies of ribosomal DNA. *Genetics* 54:818–834.

Ritter, M. A. (1977) Enhancing effect of corti-

costerone at physiological levels. *Immunology* 33:241.

Rittling, S. R., Brooks, K. M., Cristofalo, V. J., and Baserga, R. (1986) Expression of cell cycle-dependent genes in young and senescent WI-38 fibroblasts. *Proc. Natl. Acad. Sci.* 83:3316–3320.

Rivnay, B., Orbital-Harel, T., Shinitzky, M., and Globerson, A. (1983) Enhancement of the response of aging mouse lymphocytes by *in vivo* treatment with lecithin. *Mech. Age. Dev.* 23:329–336.

Rizet, G. (1953) Sur la longévité des souches de *Podospora anserina. Compt. Rend. Acad. Sci. Paris* 237:1106–1109.

Robakis, K., Ramakrishna, N., Wolfe, G., and Wisniewski, H. M. (1987) Molecular cloning and characterization of a cDNA encoding the cerebrovascular and the neuritic plaque amyloid peptides. *Proc. Natl. Acad. Sci.* 84:4190–4194.

Robbins, J. H., Kraemer, K. H., Lutzner, M. A., Festoff, B. W., and Coon, H. G. (1974) Xeroderma pigmentosum. An inherited disease with sun sensitivity, multiple cutaneous neoplasms, and abnormal DNA repair. *Ann. Int. Med.* 80:221–248.

Robbins, W. J. (1957) Physiological aspects of aging in plants. *Am. J. Bot.* 44:289–294.

Roberts, P. A., and Iredale, R. B. (1985) Can mutagenesis reveal major genes affecting senescence? *Exp. Gerontol.* 20:119–121.

Roberts, T. R. (1967) Tooth formation and replacement in characoid fishes. *Stanford Ichthyol. Bull.* 8:231–243.

———. (1981) Sundasalangidae, a new family of minute freshwater salmoniform fishes from Southeast Asia. *Proc. Calif. Acad. Sci.* 42:295–302.

Robertson, G. L., and Rowe, J. (1980) The effect of aging on neurohypophyseal function. *Peptides* 1:159–162.

Robertson, O. H. (1957) Survival of precociously mature king salmon (*Oncorhynchus tshawytscha*) after spawning. *Calif. Fish and Game* 43:119–130.

———. (1961) Prolongation of the lifespan of kokanee salmon (*O. nerka kennerlyi*) by castration before beginning development. *Proc. Natl. Acad. Sci.* 47:609–621.

Robertson, O. H., Hane, S., Wexler, B. C.,

and Rinfret, A. P. (1963) The effect of hydrocortisone on immature rainbow trout (*Salmo gairdnerii*). *Gen. Comp. Endocrinol.* 3:422–436.

Robertson, O. H., Krupp, M. A., Thomas, S. F., Favour, C. B., Hane, S., and Wexler, B. C. (1961) Hypoadrenocorticism in spawning migratory and non-migratory rainbow trout: *Salmo gairdnerii. Gen. Comp. Endocrinol.* 1:473–484.

Robertson, O. H., and Wexler, B. C. (1960) Histological changes in the organs and tissues of senile castrated kokanee salmon: *Oncorhynchus nerka kennerlyi. Gen. Comp. Endocrinol.* 2:458–472.

Robertson, O. H., Wexler, B. C., and Miller, B. F. (1961) Degenerative changes in the cardiovascular system of the spawning Pacific salmon (*Oncorhynchus tshawytscha*). *Circ. Res.* 9:826–834.

Roberts-Thompson, I. C., Whittingham, S., Youngchalyud, U., and Mackay, I. R. (1974) Ageing, immune response, and mortality. *Lancet* 2:368–370.

Robinson, A. B., McKerrow, J. H., and Cary, P. (1970) Controlled deamidation of peptides and proteins: An experimental hazard and a possible biological timer. *Proc. Natl. Acad. Sci.* 66:753–757.

Robinson, A. B., and Rudd, C. J. (1974) Deamidation of glutaminyl and asparginyl residues in peptides and proteins. *Curr. Top. Cell. Regul.* 8:247–295.

Robinson, D. W., and Shipton, M. S. (1973) Tables for the estimation of noise-induced hearing loss. N.P.L. Acoustics Report Ac61. Toddington, U.K.: National Physical Laboratory.

Robinson, G. E. (1987a) Hormonal regulation of age polyethism in *Apis mellifera.* In *Chemistry and Biology of Social Insects,* J. Eder and H. Rembold (eds.), 124–125. Munich: Verlag J. Peperny.

———. (1987b) Regulation of honey bee age polyethism by juvenile hormone. *Behav. Ecol. Sociobiol.* 20:329–338.

Robinson, G. E., and Ratnieks, F. L. (1987) Induction of premature honey bee (Hymenoptera: Apidae) flight activity with juvenile hormone analogs administered orally or topically. *J. Econ. Entomol.* 80:784–788.

Rockstein, M. (1950) The relation of cholinesterase activity to change in cell number with age in the brain of the adult worker honeybee. *J. Cell. Comp. Physiol.* 35:11–24.

———. (1959) The biology of aging in insects. In *The Lifespan of Animals,* Wolstenholm and O'Conner (eds.), 247–264. *CIBA Foundation Colloquium on Aging,* vol. 5.

Rockstein, M., and Lieberman, H. M. (1958) Survival curves for male and female houseflies (*Musca domestica*). *Nature* 181:787–788.

Rockstein, M., and Miquel, J. (1973) Aging in insects. In *The Physiology of Insecta,* 2d ed., M. Rockstein (ed.), 371–478. New York: Academic Press.

Rodaniche, A. F. (1984) Iteroparity in the lesser Pacific striped octopus *Octopus chierchiae* (Jatta, 1889). *Bull. Mar. Sci.* 35:99–104.

Rodeheffer, F. J., Gerstenblith, G., Becker, L. C., Fleg, J. L., Weisfeldt, M., and Lakatta, E. G. (1984) Exercise cardiac output is maintained with advancing age with healthy human subjects: Cardiac dilitation and increased stroke volume compensate for a diminished heart rate. *Circulation* 69:203–213.

Roderick, T. H., Staats, J., and Womack, J. E. (1981) Strain distribution of polymorphic variants. In *Genetic Variants and Strains of the Laboratory Mouse,* M. C. Green (ed.), 377–396. New York: Gustav Fischer.

Roff, D. A. (1981) Reproductive uncertainty and the evolution of iteroparity: Why don't flatfish put all their eggs in one basket? *Can. J. Fish. Aquat. Sci.* 38:968–977.

Rogers, M. B., and Karrer, K. M. (1985) Adolescence in *Tetrahymena thermophila. Proc. Natl. Acad. Sci.* 82:436–439.

Rogers, S., Wells, R., and Rechsteiner, M. (1986) Amino acid sequences common to rapidly degraded proteins: The PEST hypothesis. *Science* 234:364–368.

Roghmann, K. J., Tabloski, P. A., Bentley, D. W., and Schiffman, G. (1987) Immune response of elderly adults to pneumococcus: Variation by age, sex, and functional impairment. *J. Gerontol.* 42:265–270.

Rohdendorf, B. (1974) *The Historical Development of Diptera.* Translated from Russian by J. E. Moore and I. Thiele. Calgary: Univ. of Alberta Press.

Rohme, D. (1981) Evidence for a relationship between longevity of mammalian species and lifespans of normal fibroblasts in vitro and erythrocytes in vivo. *Proc. Natl. Acad. Sci.* 78:5009–5013.

Romer, A. S. (1966) *Vertebrate Paleontology.* Chicago: Univ. of Chicago Press.

Rook, A., and Dowber, R. (1982) In *Diseases of the Hair and Scalp,* London: Blackwell. [not read]

Roozendaal, B., van Gool, W. A., Swaab, D. F., Hoogendijk, J. E., and Mirmiran, M. (1987) Changes in vasopressin cells of the rat suprachiasmatic nucleus with aging. *Brain Res.* 409:259–264.

Rose, A. M., and Baillie, D. L. (1979) The effect of temperature and parental age on recombination and nondisjunction in *Caenorhabditis elegans. Genetics* 92:409–418.

Rose, M. R. (1983) Theories of life-history evolution. *Am. Zool.* 23:15–23.

———. (1984a) The evolution of animal senescence. *Can. J. Zool.* 62:1661–1667.

———. (1984b) Genetic covariation in *Drosophila* life history: Untangling the data. *Am. Nat.* 123:565–569.

———. (1984c) Laboratory evolution of postponed senescence in *Drosophila melanogaster. Evolution* 38:1004–1010.

———. (1985) Life-history evolution with antagonistic pleiotropy and overlapping generations. *Theor. Pop. Biol.* 28:342–358.

———. (1991) *The Evolutionary Biology of Aging.* Oxford: Oxford Univ. Press.

Rose, M. R., and Charlesworth, B. (1980) A test of evolutionary theories of senscence. *Nature* 287:141–142.

———. (1981) Genetics of life-history of *Drosophila melanogaster.* 1. Sib. analysis of adult females. *Genetics* 97:173–186.

Rose, M. R., Dorey, M. L., Coyle, A. M., and Service, P. M. (1984) The morphology of postponed senescence in *Drosophila melanogaster. Can. J. Zool.* 62:1576–1580.

Rose, M. R., and Graves, J. L. (1989) Minireview: What evolutionary biology can do for gerontology. *J. Gerontol.* 44:B27–B29.

———. (1990) Evolution of aging. In *Review of Biological Research in Aging,* vol. 4, M. Rothstein (ed.), 3–14. New York: Liss.

Rosen, D. R., Martin-Morris, L., Luo, L., and

White, K. (1989) A *Drosophila* gene encoding a protein resembling the human β-amyloid protein precursor. *Proc. Natl. Acad. Sci.* 86:2478–2482.

Rosen, S., Bergman, M., Plester, D., El-Mofty, A., and Satti, M. H. (1962) Presbycusis study of a relatively noise-free population in the Sudan. *Ann. Otol. Rhinol. Laryngol.* 71:727–743.

Rosen, S., Plester, D., El-Mofty, A., and Rosen, H. V. (1964) Relation of hearing loss to cardiovascular disease. *Trans. Am. Acad. Opthal. Otolaryng.* 68:433–444.

Rosen, S., and Rosen, H. V. (1971) High frequency studies in school children in nine countries. *Laryngoscope* 81:1007–1013.

Rosenberg, B., and Juckett, D. A. (1987) Preliminary studies of a new stochastic human death function involving small integers. *Mech. Age. Dev.* 40:223–241.

Rosenberg, B., Kemeny, G., Smith, L. G., Skurnick, I. D., and Bandurski, M. J. (1973) The kinetics and thermodynamics of death in multicellular organisms. *Mech. Age. Dev.* 2:275–293.

Rosendaal, M., Hodgson, G. S., and Bradley, T. R. (1976) Hemopoietic stem cells are organised for use on the basis of their generation age. *Nature* 264:68–70.

Rosenfeld, M. E., Faggiotto, A., and Ross, R. (1985) The role of the mononuclear phagocyte in primate and rabbit models of atherosclerosis. *Proceedings of the Fourth Leiden Conference on Mononuclear Phagocytes,* 795–802. The Hague: Martinus Nijhoff.

Rosenfeld, M. E., Tsukada, T., Gown, A. M., and Ross, R. (1987) Fatty streak initiation in Watanabe heritable hyperlipemic and comparably hypercholesterolemic fat-fed rabbits. *Arteriosclerosis* 7:9–23.

Rosenthal, I. M., and Weiss, E. B. (1980) Precocious puberty and sexual immaturity. In *Gynecologic Endocrinology,* J. J. Gold and J. B. Josimovich (eds.), 625–641. San Francisco: Harper and Row.

Rosenzweig, M. R., Bennett, E. L., and Diamond, M. C. (1971) Chemical and anatomical plasticity of brain: Replications and extensions, 1970. In *Macromolecules and Behavior,* 2d ed., J. Gaito (ed.), 205–277. New York: Appleton Century Crofts.

Ross, D. H., and Merritt, R. W. (1987) Factors affecting larval blackfly distributions and population dynamics. In *Black Flies. Ecology, Population Management, and Annotated World List*, K. C. Kim and R. W. Merritt (eds.), 90–108. University Park: Pennsylvania State Univ. Press.

Ross, M. H. (1959) Protein, calories, and life expectancy. *Fed. Proc.* 18:1190–1207.

———. (1961) Length of life and nutrition in the rat. *J. Nutrit.* 75:197–210.

———. (1972) Length of life and caloric intake. *Am. J. Clin. Nutrit.* 25:834–838.

Ross, M. H., and Bras, G. (1971) Lasting influence of early caloric restriction on prevalence of neoplasms in the rat. *J. Natl. Canc. Inst.* 47:1095–1113.

———. (1973) Influence of protein under- and overnutrition on spontaneous tumor prevalence in the rat. *J. Nutrit.* 103:944–963 (abstract).

Ross, M. H., Lustbader, E. D., and Bras, G. (1982) Dietary practices of early life and spontaneous tumors of the rat. *Nutrit. Canc.* 3:150–167.

Ross, R. (1986) The pathogenesis of atherosclerosis: An update. *N. Engl. J. Med.* 314:488–500.

Rossman, I. (1942) On the lipin and pigment of the corpus luteum of the rhesus monkey. *Contributions to Embryology*, no. 193:97–109. Carnegie Institute of Washington Publication 541. Washington, DC.

Roth, G. S. (1986) Effects of aging on the mechanisms of estrogen action in the rat uterus. In *Aging, Reproduction, and the Climacteric*, L. Mastroianni and C. A. Paulsen (eds.), 97–116. New York: Plenum.

Roth, G. S., and Hess, G. D. (1982) Changes in the mechanisms of hormone and adipocytes during aging. *Endocrinology* 99:831–839.

Roth, G. S., Ingram, D. K., and Joseph, J. A. (1984) Delayed loss of striatal dopamine receptors during aging of dietarily restricted rats. *Brain Res.* 300:27–32.

Roth, G. S., and Livingston, J. N. (1976) Reductions in glucocorticoid inhibition of glucose oxidation and presumptive glucocorticoid receptor content in rat adipocytes during aging. *Endocrinology* 99:831–839.

Roth, M., Tomlinson, B. E., and Blessed, G. (1967) The relationship between quantitative measures of dementia and of degenerative changes in the cerebral grey matter of elderly subjects. *Proc. Roy. Soc. Med.* 60:254–260.

Rothman, F. G., and Lewis, J. L. (In prep.) Age-dependent decline of feeding in *C. elegans*. Personal communication.

Rothstein, M. (1982) *Biochemical Approaches to Aging*. New York: Academic Press.

———. (1985) The alteration of enzymes in aging. *Modern Aging Res.* 7:53–67.

Rouchy, R. (1964) Squelette et musculature céphaliques de deux types d'imagos de lepidopteres hétéroneures. *Trab. Lab. Zool. Univ. Dijon.* 61:1–59.

Roudier, J., Petersen, J., Rhodes, G. H., Luka, J., and Carson, D. A. (1989) Susceptibility to rheumatoid arthritis maps to a T-cell epitope shared by the HLA-Dw4 DR B-1 chain and the Epstein-Barr virus glycoprotein gp110. *Proc. Natl. Acad. Sci.* 86:5104–5108.

Roule, L. (1933) *Fishes, Their Journeys, and Migration*. London: Norton.

Rounsefell, G. A. (1957) Fecundity of North American Salmonidae. *Fish. Bull.* 57:451–468.

Rowe, J. W., and Minaker, K. L. (1985) Geriatric medicine. In *Handbook of the Biology of Aging*, 2d ed., C. E. Finch and E. L. Schneider (eds.), 932–960. New York: Van Nostrand.

Rowe, J. W., and Troen, B. R. (1980) Sympathetic nervous system and aging in man. *Endocrine Rev.* 1:167–179.

Rowe, R. W. (1969) The effect of senility on skeletal muscles in the mouse. *Exp. Gerontol.* 4:119–126.

Rowe, W. P., Pugh, W. E., and Hartley, J. W. (1970) Plaque assay techniques for murine leukemia virus. *Virology* 42:1136–1139.

Rowenkamp, W., Hofer, E., and Sekeris, C. E.(1976) Translation of mRNA from rat-liver polysomes into tyrosine aminotransferase and tryptophan oxygenase in a protein synthesizing system from wheat germ. Effect of cortisol on the translatable levels of mRNA for these two enzymes. *EMBO J.* 1:1287–1293.

Rowley, W. A., and Graham, C. L. (1968) The effect of age on the flight performance of female *Aedes aegypti* mosquitoes. *J. Insect Physiol.* 14:719–728.

Roy, A. K., McMinn, D. M., and Biswas, N. M. (1975) Estrogenic inhibition of the hepatic synthesis of alpha$_{2u}$-globulin in the rat. *Endocrinology* 97:1501–1508.

Roy, A. K., Milin, B. S., and McMinn, D. M. (1974) Androgen receptor in rat liver: Hormonal and developmental regulation of the cytoplasmic receptor and its correlation with the androgen-dependent synthesis of alpha^{2}u-globulin. *Biochim. Biophys. Acta* 354:213–232.

Roy, A. K., Nath, T. S., Motwani, N. M., and Chatterjee, B. (1983) Age-dependent regulation of the polymorphic forms of α_{2u}-globulin. *J. Biol. Chem.* 258:10123–10127.

Roy, A. K., Sarkar, F. H., Nag, A. C., and Mancini, M. A. (1986) Role of cytodifferentiation and cell-cell communication in the androgen dependent expression of α_{2u}-globulin gene in the rat liver. In *Cellular Endocrinology: Hormonal Control of Embryonic and Cellular Differentiation*, 401–415. New York: Liss.

Roy, S., Sala, R., Cagliero, E., and Lorenzi, M. (1990) Overexpression of fibronectin induced by diabetes or high glucose: Phenomenon with a memory. *Med. Sci.* 87:404–408.

Rubner, M. (1908) *Das Problem der Lebensdauer und seine Beziehungen zum Wachstum und Ernahrung*. Munich: Oldenbourg. [not read]

Rucker, R. B., and Tinker, D. (1977) Structure and metabolism of arterial elastin. *Int. Rev. Exp. Pathol.* 17:1–47.

Rudman, D., Kutner, M. H., Rogers, C. M., Lubin, M. F., Flemming, G. A., and Barin, R. P. (1981) Impaired growth hormone secretion in the adult population: Relation to age and adiposity. *J. Clin. Invest.* 67:1361–1369.

Rudzinska, M. A. (1951) The influence of amount of food on the reproduction rate and longevity of a suctorian (*Tokophrya infusionum*). *Science* 113:10–11.

———. (1952) Overfeeding and life span in *Tokophyra infusionum. J. Gerontol.* 7:544–550.

———. (1961a) The use of a protozoan for studies on ageing. 1. Differences between young and old organisms of *Tokophrya infusionum* as revealed in light and electron microscopy. *J. Gerontol.* 16:213–223.

———. (1961b) The use of a protozoan for studies on ageing. 2. Macronucleus in young and old organisms of *Tokophrya infusionum*: Light and electron microscopy. *J. Gerontol.* 16:326–334.

———. (1962) The use of a protozoan for studies on ageing. 3. Similarities between young overfed and old normally fed *Tokophyra infusionum*: A light and electron microscope study. *Gerontologia* (Basel) 6:206–212.

Ruffner, D. E., and Dugaiczyk, A. (1988) Splicing mutation in human hereditary analbuminemia. *Proc. Natl. Acad. Sci.* 85:2125–2129.

Ruggeri, B. A., Klurfeld, D. M., and Kritchevesky, D. (1987) Biochemical alterations in 7,12–dimethyl-benz[*a*]anthracene-induced mammary tumors from rats subjected to caloric restriction. *Biochim. Biophys. Acta.* 929:239–246.

Rumsey, W. L., Kendrick, Z. V., and Starnes, J. W. (1987) Bioenergetics in the aging Fischer 344 rat: Effects of exercise and food restriction. *Exp. Gerontol.* 22:271–287.

Russ, S., and Scott, G. M. (1937) Some biological effects of continuous gamma irradiation, with a note on protection. *Brit. J. Radiol.* 10:619–625.

Russell, E. M. (1974) The biology of kangaroos (Marsupialia: Macropodia). *Mammal. Rev.* 4:1–59.

Russell, E. S. (1966) Lifespan and aging patterns. In *Biology of the Laboratory Mouse*, 2d ed., E. L. Green (ed.), 511–519. New York: McGraw-Hill.

———. (1978) Genetic origins and some research uses of C57BL/6, DBA/2, and B6D2F1 mice. In *Development of the Rodent as a Model System of Aging*, D. C. Gibson, R. C. Adelman, and C. E. Finch (eds.), 37–44. NIH Publication 79–161. Bethesda, MD: DHEW.

Russell, E. S., and Meier, H. (1966) Constitutional diseases. In *Biology of the Laboratory Mouse*, 2d ed., E. L. Green (ed.), 571– 587. New York: McGraw-Hill.

Russell, R. L., and Jacobson, L. A. (1985) Some aspects of aging can be studied early in mammals. In *Handbook of the Biology of Aging*, 2d ed., C. E. Finch and E. L. Schneider (eds.), 128–146. New York: Van Nostrand.

Rutherford, M. S., Baehler, C. S., and Richardson, A. (1986) Genetic expression of complement factors and alpha₁-glycoprotein by liver tissue during senescence. *Mech. Age. Dev.* 35:245–254.

Rutz, W., Imboden, H., Jaycox, E. R., Wille, H., Gerig, L., and Luscher, M. (1977) Juvenile hormone and polyethism in adult worker honeybees (*Apis mellifica*). *Proceedings of the Eighth International Congress IUSSI, Wageningen*.

Ruvkun, G., and Giusto, J. (1989) The *Caenorhabditis elegans* heterochronic gene lin-14 encodes a nuclear protein that forms a temporal developmental switch. *Nature* 338:313–319.

Ryan, J. M., Ostrow, D. G., Breakefield, X. O., Gershon, E. S., and Upchurch, L. (1981) A comparison of the proliferative and replicative life span kinetics of cell cultures derived from monozygotic twins. *In Vitro* 17:20–27.

Rydzewski, W. (1962) Longevity of ringed birds. *Ring* 33:147–152.

———. (1963) Longevity records 2. *Ring* 34:177–181.

Sabbadin, A. (1979) Colonial structure and genetic patterns in ascidians. In *Biology and Systematics of Colonial Organisms*, G. Larwood and B. R. Rosen (eds.), 433–444. New York: Academic Press.

Sacher, G. A. (1956) On the statistical nature of mortality, with especial reference to chronic radiation mortality. *Radiology* 67:250–257.

———. (1959) Relation of lifespan to brain weight and body weight in mammals. *CIBA Found. Colloq. Aging* 5:115–133.

———. (1966) The Gompertz transformation in the study of the injury-mortality relationship: Application to late radiation effects and

aging. In *Radiation and Aging*, P. J. Lindop and G. A. Sacher (eds.), 411–441. London: Taylor and Francis.

———. (1967) The complementarity of entropy terms for the temperature-dependence of development and aging. *Ann. N.Y. Acad. Sci.* 138:680–712.

———. (1975a) Maturation and longevity in relation to cranial capacity in hormonal evolution. In *Primate Functional Morphology and Evolution*, R. Tuttle (ed.), 417–441. The Hague: Mouton.

———. (1975b) Use of zoo animals for research on longevity and aging. In *A Symposium Held at the Meeting of the American Association of Zoological Parks and Aquariums, Houston, TX*, 191–198. Washington, DC: National Academy of Sciences.

———. (1977) Life table modification and life prolongation. In *Handbook of the Biology of Aging*, 1st ed., C. E. Finch and L. Hayflick (eds.), 582–638. New York: Van Nostrand.

———. (1978) Longevity and ageing in vertebrate evolution. *Bio. Sci.* 28:497–501.

Sacher, G. A., and Hart, R. W. (1978) Longevity, aging, and comparative cellular and molecular biology of the house mouse, *Mus musculus,* and the white-footed mouse, *Peromyscus leucopus*. In *Genetic Effects on Aging*, D. Bergsma and D. E. Harrison (eds.), 71–96. New York: Liss.

Sacher, J. A. (1957) Relationship between auxin and membrane-integrity in tissue senescence and abscission. *Science* 125:1199–1200.

———. (1965) Senescence: Hormonal control of RNA and protein synthesis in excised pod tissue. *Am. J. Bot.* 52:841–848.

Sadoglu, P. (1967) The selective value of eye and pigment loss in Mexican cave fish. *Evolution* 21:541–549.

Sahu, A., Kalra, P. S., Crowley, W., and Kalra, S. P. (1988) Evidence that hypothalamic neuropeptide Y secretion decreases in aged male rats: Implications for reproductive aging. *Endocrinology* 122:2199–2203.

Saidler, R. M. S. (1969) *The Ecology of Reproduction in Wild and Domestic Animals*. London: Methuen. [not read]

St. George-Hyslop, P., Tanzi, R., Polinsky, R.,

Haines, J. L., Nee, L., Watkins, P. C., Myers, R. H., Feldman, R. G., Pollen, D., Drachman, D., Growden, J., Burni, A., Fancin, J., Samon, D., Formmelt, P., Amaducci, L., Sorbi, S., Piacentini, S., Stewart, G. D., Hobbs, W., Conneally, P. M., and Guisella, J. F. (1987) The genetic defect causing familial Alzheimer's disease maps on chromosome 21. *Science* 235:885–890.

Saito, F. (1984) A pedigree of homozygous familial hyperalphalipoproteinemia. *Metabolism* 33:629–633.

Sakagami, S. F., and Fukuda, H. (1968) Life tables for worker honeybees. *Res. Pop. Ecol.* 10:127–139.

Sakaguchi, H. (1968) Studies on the digestive enzymes of devilfish (in Japanese, with abstract and tables in English). *Bull. Jpn. Soc. Sci. Fish.* 34:716–632.

Salk, D. (1982) Werner syndrome: A review of recent research with an analysis of connective tissue metabolism, growth control of cultured cells, and chromosomal aberrations. *Human Genet.* 62:1–15.

Salk, D., Bryant, E., Au, K., Hoehn, H., and Martin, G. M. (1981) Systematic growth studies, cocultivation, and cell hybridization studies of Werner's syndrome cultured skin fibroblasts. *Cytogenet. Cell Genet.* 30:108–117.

Salmony, D., and Weale, R. A. (1961) *The Aging Eye.* R. A. Weale (ed.). London: W. K. Lewis. Cited in Fozard et al., 1985.

Salthouse, T. A., and Prill, K. (1987) Inferences about age impairments in inferential reasoning. *Psychol. Aging* 2:43–51.

Saltin, B. (1969) Physiological effects of physical conditioning. *Med. Sci. Sports* 1:50–56.

Samis, H. V., Erk, F. C., and Baird, M. B. (1971) Senescence in *Drosophila.* 1. Sex differences in nucleic acid, protein, and glycogen levels as a function of age. *Exp. Gerontol.* 6:9–18.

Sanadi, D. R. (1977) Metabolic changes and their significance in aging. In *Handbook of the Biology of Aging,* 1st ed., C. E. Finch and L. Hayflick (eds.), 73–100. New York: Van Nostrand Reinhold.

Sandberg, A. A. (1984) Chromosomal changes and cancer causation. In *Accomplishments in Cancer Research,* J. G. Fortner and J. E. Rhoads (eds.), 157–169. Philadelphia: J. B. Lippincott.

Sandberg, L. B. (1975) Elastin structure in health and disease. *Int. Rev. Connect. Tiss. Res.* 7:159–199.

Sandground, J. H. (1936) On the potential longevity of various helminths with a record for a species of *Trichostrozeylus* in man. *J. Parasitol.* 22:464–470.

Sandler, L., and Hiraizumi, Y. (1961) Meiotic drive in natural populations of *Drosophila melanogaster.* 8. A heritable aging effect on the phenomenon of segregation-distortion. *Can. J. Genet. Cytol.* 3:34–46.

Sanford, C. (1987) Phylogenetic relationships of salmonid fishes. Ph.D. diss., University of London. [not read]

Sankaranarayanan, K. (1979) The role of nondisjunction in aneuploidy in man. An overview. *Mutat. Res.* 61:1–28.

Santa Maria, C., and Machada, A. (1988). Changes in some hepatic enzyme activities related to phase II drug metabolism in male and female rats as a function of age. *Mech. Age. Dev.* 44:115–125.

Sapolsky, R. M. (1985) A mechanism for glucocorticoid toxicity in the hippocampus: Increased neuronal vulnerability to metabolic insults. *J. Neurosci.* 5:1228–1232

———. (1990) The adrenocortical axis. In *Handbook of the Biology of Aging,* 3d ed., E. L. Schneider and J. W. Rowe (eds.), 330–348. San Diego: Academic Press.

Sapolsky, R. M., Krey, L., and McEwen, B. (1983) Corticosterone receptors decline in a site-specific manner in the aged rat brain. *Brain Res.* 289:235–241.

———. (1984a) Glucocorticoid-sensitive hippocampal neurons are involved in terminating the adrenocortical stress response. *Proc. Natl. Acad. Sci.* 81:6174–6178.

———. (1984b) Stress down-regulates corticosterone receptors in a site-specific manner in the brain. *Endocrinology* 114:287–292.

———. (1985) Prolonged glucocorticoid exposure reduces hippocampal neuron number: Implications for aging. *J. Neurosci.* 5:1222–1227.

———. (1986) The neuroendocrinology of

stress and aging: The glucocorticoid cascade hypothesis. *Endocrine Rev.* 7:284–301.

Sapolsky, R. M., Krey, L. C., McEwen, B. S., and Rainbow, T. C. (1984) Do vasopressin-related peptides induce hippocampal corticosterone receptors? Implications for aging. *J. Neurosci.* 4:1479–1485.

Sapolsky, R. M., Packan, D. R., and Vale, W. W. (1988) Glucocorticoid toxicity in the hippocampus: In vitro demonstration. *Brain Res.* 453:367–371.

Sapolsky, R. M., Rivier, C., Yamamoto, G., Plotsky, P., and Vale, W. (1987) Interleukin-1 stimulates the secretion of hypothalamic corticotropin-releasing factor. *Science* 238:522–526.

Sapolsky, R. M., Uno, H., Rebert, C. S., and Finch, C. E. (1990) Hippocampal damage associated with prolonged glucocorticoid exposure in primates. *J. Neurosci.* In press.

Sargent, C. A., Dunham, I., Trowsdale, J., and Campbell, R. D. (1989) Human major histocompatibility complex contains genes for the major heat shock protein HSP70. *Proc. Natl. Acad. Sci.* 86:1968–1972.

Sarkar, D. K., Gottschall, P. E., and Meites, J. (1984) Reduced tuberoinfundibular dopaminergic neural function in rats with in situ prolactin-secreting tumors. *Neuroendocrinology* 38:498–503.

Satchell, J. E. (1967) Lumbricidae. In *Soil Biology*, F. Raw (ed.), 259–322. New York: Academic Press.

Saul, R. L., Gee, P., and Ames, B. N. (1987) Free radicals, DNA damage, and aging. In *Modern Biological Theories of Aging*, H. R. Warner, R. N. Butler, R. L. Sprott, and E. L. Schneider (eds.), 113–130. New York: Raven Press.

Saunders, D. S. (1962a) Age determination for female tsetse flies and the age compositions of *Glossina pallipides* Aust., *G. palpis fuscipes* Newst., and *G. brevipalpis* Newst. *Bull. Entomol. Res.* 53:579–596.

———. (1962b) The effect of the age of female *Nasonia vitripennis* (Walker) (Hymenoptera, Pteromalidae) upon the incidence of larval diapause. *J. Insect Physiol.* 8:309–318.

Saunders, J. W. (1966) Death in embryonic systems. *Science* 154:604–612.

Saunders, R. L., and Farrell, A. P. (1988) Cor-

onary arteriosclerosis in Atlantic salmon. *Arteriosclerosis* 8:378–384.

Savory, T. (1977) *Arachnida*. New York: Academic Press.

Sax, K. (1962) Aspects of aging in plants. *Ann. Rev. Plant Physiol.* 13:489–506.

Saxton, J. A., Jr., Barness, L. L., and Sperling, G. (1946) A study of the pathogenesis of chronic pulmonary disease (bronchiectasis) of old rats. *J. Gerontol.* 1:165.

Scammon, R. E., and Hesdorffer, M. B. (1937) Growth in man and volume of the human lens in postnatal life. *Arch. Ophthalmol.* 17:104–112. Cited in Fozard et al., 1985.

Scarbrough, K., and Wise, P. M. (1990) Age-related changes in pulsatile luteinizing hormone release precede the transition to estrous acyclicity and depend upon estrous cycle history. *Endocrinology* 126:884–890.

Schach von Wittenau, M., and Gans, D. J. (1981) Aging as a confounding factor in carcinogenicity bioassays. *Drug Chem. Toxicol.* 4:307–310.

Schaffalitzky de Muckadell, M. (1954) Juvenile stages in woody plants. *Physiol. Plant.* 7:782–796.

———. (1959) Investigations on aging of apical meristems in woody plants and its importance in silviculture. *Forstl. Forsgsvaesen Danmark* 25:309–455.

Schaffer, W. M. (1974a) Optimal reproductive effort in fluctuating environments. *Am. Nat.* 108:783–790.

———. (1974b) Selection for optimal life histories: The effects of age structure. *Ecology* 55:291–303.

Schaie, K. W., and Hertzog, C. (1985) Measurement in the psychology of adulthood and aging. In *Handbook of the Psychology of Aging*, J. E. Birren and K. W. Schaie (eds.), 61–94. New York: Van Nostrand Reinhold.

Schaller, G. B. (1972) *The Serengeti Lion: A Study in Predator-Prey Relations*. Chicago: Univ. of Chicago Press.

Schapira, F., Weber, A., Guilluozo, C., and Dreyfus, J. C. (1978) Search for alterations of three enzymes with a difference turnover rate in the liver of senescent rats. In *Liver and Aging*, K. Kitani (ed.), 47–54. Amsterdam: Elsevier–North-Holland Biomedical Press.

Schapiro, H., Hotta, S. S., Outten, W. E., and

Klein, A. W. (1982) The effect of aging on rat liver regeneration. *Experientia* 38:1075–1076.

Schapiro, M. B., Kumar, A., White, B., Grady, C. L., Friedland, R. P., and Rapoport, S. I. (1989) Alzheimer's disease (AD) in mosaic/translocation Down's syndrome (Ds) without mental retardation. *Neurology,* suppl. 1, 39:169.

Schapiro, S., and Percin, C. J. (1966) Thyroid hormone induction of α-glycerophosphate dehydrogenase in rats of different ages. *Endocrinology* 79:1075–1078.

Schechter, J. E., Felicio, L. S., Nelson, J. F., and Finch, C. E. (1981) Pituitary tumorigenesis in aging female C57BL/6J mice: A light and electron microscopic study. *Anat. Rec.* 199:423–432.

Scheckler, S. E. (1978) Ontogeny of progymnosperms. 2. Shoots of Upper Devonian Archaeopteridales. *Can. J. Bot.* 56:3136–3170.

Scheibel, A. B. (1978) Structural aspects of the aging brain: Spine systems and the dendritic arbor. In *Alzheimer's Disease, Senile Dementia, and Related Disorders,* R. Katzman, R. D. Terry, and K. L. Bick (eds.). *Aging* 7:353–373. New York: Raven Press.

Scheiner, S. M., Caplan, R. L., and Lyman, R. F. (1989) A search for trade-offs among life history traits in *Drosophila melanogaster. Evol. Ecol.* 3:51–63.

Schellenberg, G. D., Bird, T. D., Wijsman, E. M., Moore, D. K., Boenke, M., Bryant, E. M., Lampe, T. H., Nochlin, D., Sumi, S. M., and Deeb, S. S. (1988) Absence of linkage of chromosome 21q21 markers to familial Alzheimer's disease. *Science* 241:1507–1510.

Schellenberg, G. D., Moore, D. K., Bird, T. D., Bryant, E. M., and Martin, G. M. (1990) Down syndrome in familial Alzheimer disease kindreds: Origin of non-dysjunction and inheritance of chromosome 21. In prep.

Schellenberg, G. D., Pericak-Vance, M. A., Wijsman, E. M., Moore, D. K., Gaskell, P. C., Yamaoka, L. A., BeBout, J. L., Anderson, L., Welsh, K. A., Clark, C. M., Martin, G. M., Roses, A. D., and Bird, T. D. (1990) Linkage analysis of familial Alzheimer's disease using chromosome 21 markers. In prep.

Schellenberg, G. D., Wijsman, E. M., Bird, T. D., Moore, D. K., and Martin, G. M. (In press) Linkage analysis of familial Alzheimer's disease: Genetic heterogeneity and the search for the gene. Proceedings of the International Symposium on Dementia: Molecular Biology and Genetics of Alzheimer Disease. *Excerpta Medica.* International Congress Series. Amsterdam: Elsevier.

Schmale, H., and Richter, D. (1984) Single base deletion in the vasopressin gene is the cause of diabetes insipidus in Brattleboro rats. *Nature* 308:705–709.

Schmid, C. W., and Jelinek, W. R. (1982) The Alu family of dispersed repetitive sequences. *Science* 216:1065–1070.

Schmidt, A. J. (1968) *Cellular Biology of Vertebrate Regeneration and Repair.* Chicago: Univ. of Chicago Press. [not read].

Schmidt, B. (1986) Eight pound salmon in Great Lakes future? *Fishing Facts Magazine* (June):56.

Schmidt, M. (1978) Elephants. In *Zoo and Wild Animal Medicine,* 1st ed., M. E. Fowler (ed.), 709. Philadelphia: Saunders.

Schmidt-Nielsen, K. (1984) *Scaling. Why Is Animal Size So Important?* Cambridge: Cambridge Univ. Press.

Schmidt-Nielsen, K., and Larimer, J. L. (1958) Oxygen dissociation curves of mammalian blood in relation to body size. *Am. J. Physiol.* 195:424–428.

Schmieder, R. G. (1933) The polymorphic forms of *Melittobia chalybii* and the determining factors involved in their production. *Biol. Bull.* (Woods Hole) 65:338–354.

———. (1939) The significance of two types of larvae in *Sphecophaga burra* and the factors conditioning them. *Entomol. News* 50:125–131.

Schmucker, D. L., Mooney, J. S., and Jones, A. L. (1977) Age-related changes in the hepatic endoplasmic reticulum: A quantitative analysis. *Science* 197:1005–1008.

Schmucker, D. L., and Wang, R. K. (1981) Effects pf aging and phenobarbital on the rat liver microsoma drug-metabolizing system. *Mech. Age. Dev.* 15:189–202.

Schneider, D., Naryshkin, S., and Lang, C. A.

(1982) Blood glutathione, a biochemical index of aging women. *Fed. Proc.* 41:7671.

Schneider, E. L. (1985) Cytogenetics of aging. In *Handbook of the Biology of Aging,* 2d ed., C. E. Finch and E. L. Schneider (eds.), 357–373. New York: Van Nostrand.

Schneider, E. L., and Mitsui, Y. (1976) The relationship between *in vitro* cellular aging and *in vivo* human age. *Proc. Natl. Acad. Sci.* 73:3584–3588.

Schneider, E. L., and Reed, J. D. (1985) Modulations of aging processes. In *Handbook of the Biology of Aging,* 2d ed., C. E. Finch and E. L. Schneider (eds.), 45–76. New York: Van Nostrand.

Schneider, E. L., and Rowe, J. W. (eds.) (1990) *Handbook of the Biology of Aging,* 3d ed. San Diego: Academic Press.

Schneiderman, H. A., and Horwitz, J. (1958) The induction and termination of facultative diapause in the chalcid wasps *Mormoniella vitripennis* (Walker) and *Trineptis klugi* (Ratzeburg). *J. Exp. Biol.* 35:520–551.

Schnider, S. L., and Kohn, R. R. (1980) Glucosylation of human collagen in aging and diabetes mellitus. *J. Clin. Invest.* 66:1179–1181.

———. (1981) Effects of age and diabetes mellitus on the solubility and nonenzymatic glucosylation of human skin collagen. *J. Clin. Invest.* 67:1630–1635.

———. (1982) Effects of age and diabetes mellitus on the solubility of collagen from human skin, tracheal cartilage, and dura mater. *Exp. Gerontol.* 17:185–194.

Schreibman, M. P., and Kallman, K. D. (1977) The genetic control of the pituitary-gonadal axis in the platyfish, *Xiphophorus maculatus. J. Exp. Zool.* 200:277–294.

Schreibman, M. P., Margolis-Kazen, H., Bloom, J. L., and Kallman, K. D. (1983) Continued reproductive potential in aging platyfish as demonstrated by the persistence of gonadotropin, luteinizing hormone releasing hormone, and spermatogenesis. *Mech. Age. Dev.* 22:105–112.

Schreibman, M. P., and Margolis-Nunno, H. (1989) The brain-pituitary-gonad axis in poikilotherms. In *Development, Maturation, and Senescence of Neuroendocrine Systems:* *A Comparative Approach,* M. P. Schreibman and C. G. Scanes (eds.), 97–133. New York: Academic Press.

Schreibman, M. P., Margolis-Nunno, H., and Halpern-Sebold, L. (1987) Aging in the neuroendocrine system. In *Hormones and Reproduction in Fishes, Amphibians, and Reptiles,* D. O. Norris and R. E. Jones (eds.), 563–584. New York: Plenum.

Schroeder, H. C., Messer, R., Breter, H.-J., and Muller, H. E. (1985) Evidence for age-dependent impairment of ovalbumin heterogeneous nuclear RNA (HnRNA) processing in hen oviduct. *Mech. Age. Dev.* 30:319–324.

Schroeder, P. C., and Hermans, C. O. (1975) Annelida: Polychaeta. In *Reproduction of Marine Invertebrates,* vol. 3, *Annelids and Echiurans,* A. C. Giese and J. S. Pearse (eds.), 1–213. New York: Academic Press.

Schroots, J. J., and Birren, J. E. (1988) The nature of time: Implications for research on aging. *Comprehensive Gerontol. C.* 2:1–29.

Schuhmacher, H., and Hoffmann, H. (1982) Zur Funktion der Mundwerkzeuge von Schwebfliegen bei der Nahrungsaufnahme (Diptera: Syrphidae). *Entomol. Gen.* 7:327–342.

Schuknecht, H. F. (1974) *Pathology of the Ear.* Cambridge, MA: Harvard Univ. Press.

Schulman, E. (1954) Longevity under adversity in conifers. *Science* 19:396–399.

Schultz, E., and Lipton, B. H. (1982) Skeletal muscle satellite cells: Changes in proliferative potential as a function of age. *Mech. Age. Dev.* 20:377–383.

Schultz, J. J., and Shain, S. A. (1988) Effect of aging on AXC/SSh rat ventral and dorsolateral prostate S-adenosyl-L-methionine decarboxylase and L-ornithine decarboxylase messenger RNA content. *Endocrinology* 122:120–126.

Schultz, R. J. (1971) Special adaptive problems associated with unisexual fishes. *Am. Zool.* 11:351–360.

Schulze-Robbecke, G. (1951) Untersuchungen über Lebensdauer, Altern, und Tod bei Arthropoden. *Zool. Jb.* 62:366–394.

Schuster, R. M. (1979) The phylogeny of the Hepaticae. In *Bryophyte Systematics,* G. C.

S. Clarke and J. F. Duckett (eds.), 41–82. New York: Academic Press.

———. (1981) Paleoecology, origin, and distribution through time and evolution of Hepaticae and Anthocerotae. In *Paleobotany, Paleoecology, and Evolution*, vol. 2, K. J. Niklas (ed.), 129–192. New York: Praeger.

Schwarcz, R., Okuno, E., White, R. J., Bird, E. D., and Whetsell, W. O. (1988) 3-hydroxyanthranilate oxygenase activity is increased in the brains of Huntington disease victims. *Proc. Natl. Acad. Sci.* 85:4079–4081.

Schwartz, A. G. (1975) Correlation between species lifespan and capacity to activate 7,12-dimethylbenz(a)anthracene to a form mutagenic to a mammalian cell. *Exp. Cell Res.* 1:293–346.

———. (1979) Inhibition of spontaneous breast cancer formation in female C3H (A^{vy}/a) mice by long-term treatment with dehydroepiandrosterone. *Cancer* 39:1129–1132.

Schwartz, A. G., Fairman, D. K., Polansky, M., Lewbart, M. L., and Pashko, L. L. (1989) Inhibition of 7,12-dimethylbenz (α)anthracene-iniated and tetradecanoylphorbol–13-acetate-promoted skin papilloma formation in mice by dehydroepiandrosterone and two synthetic analogs. *Carcinogenesis* 10:1809–1813.

Schwartz, A. G., Lewbart, M. L., and Pashko, L. L. (1988) Novel dehydroepiandrosterone analogues with enhanced biological activity and reduced side effects in mice and rats. *Canc. Res.* 48:4817–4822.

Schwartz, A. G., and Pashko, L. L. (1986) Food restriction inhibits (3H) 7,12-dimethylbenz(a)anthracene binding to mouse skin DNA and tetradecanoylphobol–13-acetate stimulation of epidermal (³H) thymidine incorporation. *Anticanc. Res.* 6:1279–1282.

Schwartz, A. G., and Tannen, R. H. (1981) Inhibition of 7,12-dimethylbenz(a)-anthracene and urethan-induced lung tumor formation in A/J mice by long-term treatment with dehydroepiandrosterone. *Carcinogenesis* (London) 2:1335–1337.

Schwartz, A. G., Whitcomb, J. M., Nyce, J. W., Lewbart, M. L., and Pashko, L. L. (1988) Dehydroepiandrosterone and structural analogs: A new class of cancer chemopreventive agents. *Adv. Canc. Res.* 51:391–423.

Schweitzer, P., and Bodenstein, D. (1975) Aging and its relation to cell growth and differentiation in *Drosophila* imaginal discs: Developmental response to growth restricting conditions. *Proc. Natl. Acad. Sci.* 72:4674–4678.

Scofield, V. L., Schlumpberger, J. M., West, L. A., and Weissman, I. L. (1982) Protochordate allorecognition is controlled by a MHC-like gene system. *Nature* 295:499–502.

Scollay, R., Butcher, E. C., and Weissman, I. L. (1980) Thymus cell migration. Quantitative aspects of cellular traffic from the thymus to the periphery in mice. *Eur. J. Immunol.* 10:210–218.

Scott, A. (1941) Reversal of sex production in *Micromalthus*. *Biol. Bull.* 81:420–431.

Scott, A. C. (1937) Paedogenesis in the Coleoptera. *Z. Morphol. Okol. Tiere* 33:633–653.

Scott, A. C., and Chaloner, W. G. (1983) The earliest fossil conifer from the Westphalian-B of Yorkshire. *Proc. Roy. Soc.* (London), ser. B, 220:163.

Scott, A. F., Bunn, H. F., and Brush, H. A. (1977) The phylogenetic distribution of red cell 2,3-diphosphoglycerate and its interaction with mammalian hemoglobins. *J. Exp. Zool.* 201:269–288.

Scott, D. K. (1988) Reproductive success in Bewick's swans. In *Reproductive Success. Studies of Individual Variation in Contrasting Breeding Systems*, T. H. Clutton-Brock (ed.), 220–236. Chicago: Univ. of Chicago Press.

Scott, M., Foster, D., Mirenda, C., Serban, D., Coufal, F., Walchli, M., Torchia, M., Groth, D., Carlson, G., DeArmond, S. J., Westaway, D., and Prusiner, S. B. (1989) Transgenic mice expressing hamster prion protein produce species-specific scrapie infectivity and amyloid plaques. *Cell* 59:847–857.

Scott, W., Bolla, R., and Denckla, W. D. (1979) Age-related changes in immune function of rats and the effect of long-term hypophysectomy. *Mech. Age. Dev.* 11:127–136.

Scott, W. B., and Crossman, E. J. (1973)

Freshwater fishes of Canada. *Bull. Fish. Res. Bd. Can.* 184:1–966. [not read]

Sedlis, S. P., Schechtman, K. B., Ludbrook, P. A., Sobel, B. E., and Schonfeld, G. (1986) Plasma apoproteins and the severity of coronary artery disease. *Circulation* 73:978–986.

Segall, P. E., and Timiras, P. S. (1976) Pathophysiologic findings after chronic tryptophan deficiency in rats: A model for delayed growth and aging. *Mech. Age. Dev.* 5:109–124.

———. (1983) Low tryptophan diets delay reproductive aging. *Mech. Age. Dev.* 23:245–253.

Seilacher, A. (1984) Late Precambrian Metazoa: Preservational or real extinctions? In *Patterns of Change in Earth Evolution*, H. D. Holland and A. F. Trendall (eds.), 159–168. Berlin: Springer-Verlag. [not read]

Seip, M. (1971) Generalized lipodystrophy. In *Ergebnisse der Inneron Medizes und Kinderheilkunde*, vol. 31, 59–95. New York: Springer-Verlag. [not read]

Selkoe, D. J., Bell, D. S., Podlisny, M. B., Price, D. L., and Cork, L. C. (1987) Conservation of brain amyloid proteins in aged mammals and humans with Alzheimer's disease. *Science* 235:873–877.

Sell, D. R., and Monnier, V. M. (1990) Structure elucidation of a senescence cross-link from human extracellular matrix. *J. Biol. Chem.* 266:21597–21602.

Sell, S., Nichols, M., Becker, F. F., and Leffert, H. L. (1974) Hepatocyte proliferation and α_1-fetoprotein in pregnant, neonatal, and partially hepatectomized rats. *Canc. Res.* 34:865–871.

Selmanowitz, V. J., Rizer, R. L., and Orentreich, N. (1977) Aging of the skin and its appendages. In *Handbook of the Biology of Aging*, 1st ed., C. E. Finch and L. Hayflick (eds.), 496–512. New York: Van Nostrand Reinhold.

Selye, H. (1976) *Stress in Health and Disease.* London: Butterworths.

Sergeant, D. E. (1962) The biology of the pilot whale, *Globicephala melaena* (Traill) in Newfoundland waters. *Bull. Fish. Res. Bd. Can.* 132:1–84.

Service, P. M. (1989) The effect of mating status on lifespan, egg laying, and starvation resistance in *Drosophila melanogaster* in relation to selection on longevity. *J. Insect Physiol.* 35:447–452.

Service, P. M., Hutchinson, E. W., Mackinley, M. D., and Rose, M. R. (1985) Resistance to environmental stress in *Drosophila melanogaster* selected for postponed senescence. *Physiol. Zool.* 58:380–389.

Service, P. M., Hutchinson, E. W., and Rose, M. R. (1988) Multiple genetic mechanisms for the evolution of senescence in *Drosophila melanogaster. Evolution* 42:708–716.

Seshadri, T., and Campisi, J. (1990) Repression of c-fos transcription and an altered genetic program in senescent human fibroblasts. *Science* 247:205–209.

Sessoms, A. H., and Huskey, R. J. (1973) Genetic control of development in *Volvox.* Isolation and characterization of morphogenetic mutants. *Proc. Natl. Acad. Sci.* 70:1355–1338.

Setlow, R. B., and Stelow, J. K. (1972) Effects of radiation on polynucleotides. *Ann. Rev. Biophys. Bioeng.* 1:293–346.

Sevanian, A., Davies, K. J. A., and Hockstein, P. (1985) Conservation of vitamin C by uric acid in blood. *J. Free Rad. Biol. Med.* 1:117–124.

Severinghaus, C. W. (1949) Tooth development and wear as criteria of age in white-tailed deer. *J. Wildlife Mgt.* 13:195–216.

Severson, J. A., and Finch, C. E. (1990) Reduced dopaminergic binding during aging in the rodent striatum. *Brain Res.* 192:147–162.

Seymour, F. I., Duffy, C., and Koerner, A. (1935) A case of authenticated fertility in a man of 94. *J.A.M.A.* 105:1423–1424.

Shah, B. G., Krishna Rao, C. V. C., and Draper, H. H. (1967) The relationship of Ca and P nutrition during adult life and osteoporosis in aged mice. *J. Nutrit.* 92:30–42.

Shain, S. A., Schultz, J. J., and Lancaster, C. M. (1986) Aging in the AXC/SSh rat: Diminished prostate L-ornithine decarboxylase (ODC) activity reflects diminished prostate ODC protein and transcript content. *Endocrinology* 119:1830–1838.

Shantz, L. M., Talalay, P., and Gordon, G. B.

(1989) Mechanism of inhibition of growth of 3T3-L1 fibroblasts and their differentiation to adipocytes by dehydroepiandrosterone and related steroids: Role of glucose–6-phosphate dehydrogenase. *Proc. Natl. Acad. Sci.* 86:3852–3856.

Shapiro, A. R. (1976) Seasonal polyphenism. *Evol. Biol.* 9:259–333.

Sharma, H. K., Prasanna, H. R., Lane, R. S., and Rothstein, M. (1979) The effect of age on enolase turnover in the free-living nematode, *Turbatrix aceti. Arch. Biochem. Biophys.* 194:275–282.

Sharma, H. K., and Rothstein, M. (1980a) Altered enolase in aged *Turbatrix aceti* results from conformational changes in the enzyme. *Proc. Natl. Acad. Sci.* 77:5865–5868.

———. (1980b) Altered phosphoglycerate kinase in aging rats. *J. Biol. Chem.* 255:5043–5050.

Sharma, S. P., Jit, I., and Sharma, G. (1982) Age-related lipid changes in *Callosobruchus maculatus* (Coleoptera) and *Zaprionus paravittiger* (Diptera). *Acta Entomol. Bohemoslov.* 80:336–340.

Sharma, V. P., Hollingworth, R. M., and Paschke, J. D. (1970) Incorporation of tritiated thymidine in male and female mosquitos, *Culex pipiens,* with particular reference to spermatogenesis. *J. Insect Physiol.* 16:429–436.

Sharp, A., Zipori, D., Toledo, J., Tal, S., Resnitzky, P., and Globerson, A. (1989) Age related changes in hemopoietic capacity of bone marrow cells. *Mech. Age. Dev.* 48:91–99.

Shaskan, E. G. (1977) Brain regional spermidine and spermine levels in relation to RNA and DNA in aging rat brain. *J. Neurochem.* 28:509–516.

Shattuck, G. C., and Hilferty, M. M. (1932) Sunstroke and allied conditions in the United States. *Am. J. Trop. Med. Hyg.* 12:223–245.

Shaw, P. H., Held, W. A., and Hastie, N. D. (1983) The gene family for major urinary proteins: Expression in several secretory tissues of the mouse. *Cell* 32:755–761.

Shea, B. T., Hammer, R. E., and Brinster, R. L. (1987) Growth allometry of the organs in giant transgenic mice. *Endocrinology* 121:1924–1930.

Sheldon, W. G., and Greenman, D. L. (1980) Spontaneous lesions in control BALB/c female mice. *J. Environ. Pathol. Toxicol.* 3:155.

Shepherd, J. C., Walldorf, U., Hug, P., and Gehring, W. J. (1989) Fruit flies with additional expression of the elongation factor EF–1α live longer. *Proc. Natl. Acad. Sci.* 86:7520–7521.

Sherman, B. M., West, J. H., and Korenman, S. G. (1976) The menopausal transition: Analysis of LH, FSH, estradiol, and progesterone concentrations during menstrual cycles of older women. *J. Clin. Endocrinol. Metab.* 42:629–636.

Sherwood, S. W., Rush, D., Ellsworth, J. L., and Schimke, R. T. (1988) Defining cellular senescence in IMR-90 cells: A flow cytometric analysis. *Proc. Natl. Acad. Sci.* 85:9086–9090.

Shick, J. M., Hoffman, R. J., and Lamb, A. N. (1979) Asexual reproduction, population structure, and genotype: Environment interactions in sea anemones. *Am. Zool.* 19:699–713.

Shick, J. M., and Lamb, A. N. (1977) Asexual reproduction and genetic population structure in the colonizing sea anemone *Haliplanella luciae. Biol. Bull.* 153:604–617.

Shields, W. G., and Bockheim, J. G. (1981) Deterioration of trembling aspen clones in the Great Lakes region. *Can. J. For. Res.* 11:530–537.

Shigo, A. L. (1979) Tree decay. An expanded concept. *Agricultural Information Bulletin* no. 419.

Shima, A., and Egami, N. (1978) Absence of systematic polyploidization of hepatocyte nuclei during the aging process of the male medaka, *Oryzias latipes. Exp. Gerontol.* 13:51–55.

Shima, A., Egami, N., and Sugahara, T. (1978) Comparative study on the polyploidizatiion of hepatocytes of mouse and fish during aging. In *Liver and Aging,* K. Kitani (ed.), 93–102. Amsterdam: Elsevier–North-Holland Biomedical Press.

Shima, A., and Sugahara, T. (1976) Age-dependent ploidy class changes in mouse hepatocyte nuclei as revealed by Feulgen-DNA cytofluorometry. *Exp. Gerontol.* 11:193–203.

Shimazu, T. (1978) Impairment of neural regulation of liver enzymes in aged animals. In *Liver and Aging,* K. Kitani (ed.), 263–276. Amsterdam: Elsevier–North-Holland Biomedical Press.

Shimazu, T., Matshushita, H., and Ishikawa, K. (1978) Hypothalamic control of liver glycogen metabolism in adult and aged rats. *Brain Res.* 144:343–352.

Shimokata, H., Andres, R., Coon, P. J., Elahi, D., Muller, D. C., and Tobin, J. D. (1989) Studies in the distribution of body fat. 2. Longitudinal effects of change in weight. *Int. J. Obesity* 13:455–464.

Shirai, T., and Mellors, R. C. (1971) Natural thymocytotoxic autoantibody and reactive antigen in New Zealand black and other mice. *Proc. Natl. Acad. Sci.* 68:1412–1415.

Shock, N. W. (1961) Physiological aspects of aging in man. *Ann. Rev. Physiol.* 23:97–122.

———. (1985) Longitudinal studies of aging in humans. In *Handbook of the Biology of Aging,* 2d ed., C. E. Finch and E. L. Schneider (eds.), 721–743. New York: Van Nostrand.

Shock, N. W., Greulich, R. C., Andres, R., Arneberg, D., Costa, P. T., Lakatta, E. G., and Tobin, J. D. (1984) *Normal Human Aging: The Baltimore Longitudinal Study.* NIH Publication 84–2450. Washington, DC: USDHHS.

Short, R., Williams, D. D., and Bowden, D. M. (1987) Cross-sectional evaluation of potential biological markers of aging in pig-tailed macaques: Effects of age, sex, and diet. *J. Gerontol.* 42:644–654.

Shostak, S. (1981) Hydra and cancer: Immortality and budding. In *Phyletic Approaches to Cancer,* C. J. Dawe (ed.), 275–286. Tokyo: Japan Science Society Press.

Siakatos, A. N., and Koppang, N. (1973) Procedures for the isolation of lipopigments from brain, heart, and liver, and their properties: A review. *Mech. Age. Dev.* 2:177–200.

Sibly, R. M., and Calow, P. (1982) Asexual reproduction in protozoa and invertebrates. *J. Theoret. Biol.* 96:401–424.

———. (1986) *Physiological Ecology of Animals. An Evolutionary Approach.* Boston: Blackwell.

Sicard, R. E. (1988) *Regulation of Vertebrate Limb Regeneration.* Oxford: Oxford Univ. Press. [not read]

Sidenius, P., and Jakobsen, J. (1982) Reversibility and preventability of the decrease is slow axonal-transport velocity in experimental diabetes. *Diabetes* 31:689–693.

Siesjo, B. K. (1981) Cell damage in the brain: A speculative synthesis (review). *J. Cerebral Blood Flow Metab.* 1:155–185.

Siiteri, P. K., Wilson, J. D., and Mayfield, J. A. (1970) Dihydrotestosterone in prostatic hypertrophy. 1. The formation and content of dihydrotestosterone in the hypertrophic prostate of man. *J. Clin. Invest.* 49:1737–1745.

Sikes, S. K. (1971) *The Natural History of the African Elephant.* New York: American Elsevier.

Silander, J. A. (1985) Microevolution in clonal plants. In *Population Biology and Evolution of Clonal Organisms,* J. B. Jackson, L. W. Buss, and R. E. Cook (eds.), 107–152. New Haven: Yale Univ. Press.

Silberberg, M., and Silberberg, R. (1941) Age changes of bones and joints in various strains of mice. *Am. J. Anat.* 68:69–90.

———. (1962) Osteoarthrosis and osteoporosis in senile mice. *Gerontologia* (Basel) 6:19–101.

———. (1965) Aging changes and osteoarthritis in castrate mice receiving progesterone. *J. Gerontol.* 20:228–232.

Silber, J. R., Fry, M., Martin, G. M., and Loeb, L. A. (1985) Fidelity of DNA polymerases isolated from regenerating liver chromatin of aging *Mus musculus. J. Biol. Chem.* 260:1304–1310.

Silverberg, E., and Lubera, J. (1986) Cancer statistics, 1986. *CA Canc. J. Clin.* 36:9–25.

Silvestri, F. (1905) Descrizione di un nuovo genere di Rhipiphoridae. *Redia* 3:315–324.

Siman, R., Gall, C., Perlmutter, L., Christian, C., Baudry, M., and Lynch, G. (1985) Distribution of calpain I, an enzyme associated with degenerative activity, in rat brain. *Brain Res.* 347:399–403.

Simmonds, F. J. (1948) The influence of maternal physiology on the incidence of diapause. *Proc. Roy. Soc.* (London), ser. B, 233:385–414.

Simmonds, N. W. (1979) *Principles of Crop Management.* London: Longman.

Simms, H. S. (1942) The use of measurable cause of death (hemorrhage) for the evaluation of aging. *J. Gen. Physiol.* 26:169–178.

——. (1945) Logarithmic increase of mortality as a manifestation of aging. *J. Gerontol.* 1:13–25.

Simms, H. S., and Berg, B. N. (1957) Longevity and the onset of lesions in male rats. *J. Gerontol.* 12:244–252.

Simpson, E., and Cantor, H. (1975) Regulation of the immune response by subclasses of T lymphocytes. 2. The effect of adult thymectomy upon humoral and cellular responses in mice. *Eur. J. Immunol.* 5:337–343.

Simpson, L. O. (1976) An NZB virus or NZB mice with viral infections? *Lab. Anim. Sci.* 10:249.

Simpson, S. B., and Rausch, D. M. (1989) Cells from the lizard *Anolis* do not exhibit *in vitro* senescence. *Gerontologist,* special issue, 29:248A (abstract).

Sing, C. F., Boerwinkle, E., Moll, P. P., and Davignon, J. (1985) Apolipoproteins and cardiovascular risk: Genetics and epidemiology. *Ann. Biol. Clin.* 43:407–417.

Singer, M. (1973) *Limb Regeneration in Vertebrates.* Addison-Wesley.

Singh, S. M., Toles, J. F., and Reaume, J. (1985) Genotype- and age-associated in vivo cytogenetic alterations following mutagenic exposures in mice. *Can. J. Genet. Cytol.* 28:286–293.

Singhal, R. P., Mays-Hoopes, L. L., and Eichhorn, G. L. (1987) DNA methylation in aging of mice. *Mech. Age. Dev.* 41:199–210.

Sivori, E., Nakayama, F., and Cigliano, E. (1968) Germination of *Achira* seed (*Canna* sp.) approximately 550 years old. *Nature* 219:1269–1270.

Skadsen, R. W., and Cherry, J. H. (1983) Quantitative changes in *in vitro* and *in vivo* protein synthesis in aging and rejuvenated soybean cotyledons. *Plant Physiol.* 71:861–868.

Skaife, S. H. (1954) The black-mound termite of the cape, *Amitermes atlanticus* Fuller. *Roy. Soc. S. Africa Trans.* 34:251–271.

Sladek, J. R., and Sladek, C. D. (1978) Relative quantitation of monamine histofluoresc-

ence in young and old non-human primates. In *Parkinson's Diseases,* vol. 2, *Aging and Neuroendocrine Relationships,* C. E. Finch, D. E. Potter, and A. D. Kenny (eds.). *Adv. Exp. Biol. Med.* 113:321–239.

Slatis, H. M. (1964) A study of normal variation in man. 2. Polymorphism and pleiotropy. *Cold Spring Harbor Symp. Quant. Biol.* 29:64–67.

Slatis, H. M., and Apelbaum, A. (1963) Hairy pinna of the ear in Israeli populations. *Am. J. Human Genet.* 15:74–85.

Slatkin, D. N., Friedman, A. P., Irsa, A. P., and Micca, P. L. (1985) The stability of DNA in human cerebellar neurons. *Science* 228:1002–1004.

Slavens, F. L. (1988) Inventory, longevity, and breeding notes: Reptiles and amphibians in captivity. Seattle: Woodland Park Zoo. See Chapter 3, n. 2.

Slavkin, H. C., Graham, E., Zeichner-David, M., and Hildemann, W. (1983) Enamel-like antigens in hagfish: Possible evolutionary significance. *Evolution* 37:404–412.

Slob, W., and Janse, C. (1988) A quantitative method to evaluate the quality of interrupted animal cultures in aging studies. *Mech. Age. Dev.* 42:275–290.

Sloviter, R. S. (1987) Decreased hippocampal inhibition and a selective loss on interneurons in experimental epilepsy. *Science* 235:73–76.

——. (1989) Calcium-binding protein (calbindin-D_{28k}) and parvalbumin immunocytochemistry: Localization in the rat hippocampus with specific reference to the selective vulnerability of hippocampal neurons to seizure activity. *J. Comp. Neurol.* 280:183–196.

Sloviter, R. S., Barbaro, N. M., and Laxer, K. D. (1989) Calcium-binding protein (calbindin) and parvalbumin-like immunoreactivity (LI) in the "normal" and "epileptic" human hippocampus. *Epilepsia* 30:719.

Smith, C. B., Crane, A. M., Kadekaro, M., Agranoff, B. W., and Sokoloff, L. (1984) Stimulation of protein synthesis and glucose utilization in the hypoglossal nucleus induced by axotomy. *J. Neurosci.* 4:2489–2496.

Smith, D. M., Nance, W. E., Kang, K.-W., Christian, J. C., and Johnston, C. C. (1973) Genetic factors in determining bone mass. *J. Clin. Invest.* 52:2800–2808.

Smith, D. W. (1989) Is greater female longevity a general finding among animals? *Biol. Rev.* 64:1–12.

———. (1991) *Human Longevity*. Baltimore: Johns Hopkins Univ. Press. In press.

Smith, D. W., and Warner, H. R. (1989) Does genotypic sex have a direct effect on longevity? *Exp. Gerontol.* 24:277–288.

Smith, E. R. (1971) Protein metabolism in the aging mosquito. Ph.D. thesis, University of Louisville, Dept. Biochemistry.

Smith, G. R., and Stearley, R. F. (1989) The classification and scientific names of rainbow and cutthroat trouts. *Fisheries* 14:4–10.

Smith, G. S., Carew, M., and Walford, R. L. (1990) *Peromyscus* as a gerontologic animal: Aging and the MHC. In *Genetic Effects on Aging II*, D. E. Harrison (ed.). Caldwell, NJ: Telford Press. In press.

Smith, G. S., and Walford, R. L. (1977) Influence of the main histocompatibility complex on ageing in mice. *Nature* 270:727–729.

———. (1978) Influence of the H–2 and H–1 histocompatibility systems upon life span and spontaneous cancer incidences in congenic mice. In *Genetic Effects on Aging*, D. Bergsma and D. E. Harrison (eds.), 281–312. Birth Defects: Original Article Series, 14. New York: Alan R. Liss.

———. (1990) *Peromyscus*. In *Genetics of Aging*, vol. 2, D. Harrison (ed.). Caldwell, NJ: Telford Press.

Smith, G. S., Walford, R. L., and Mickey, M. R. (1973) Lifespan and incidence of cancer and other diseases in selected long-lived inbred mice and their F_1 hybrids. *J. Natl. Canc. Inst.* 50:1195–1213.

Smith, J. M. (1981) Does high fecundity reduce survival in song sparrows? *Evolution* 35:1142–1148.

Smith, J. R., and Rubenstein, I. (1973) The development of "senescence" in *Podospora anserina*. *J. Gen. Microbiol.* 76:283–296.

Smith, J. R., and Whitney, R. G. (1980) Intraclonal variation in proliferative potential of human diploid fibroblasts: Stochastic mechanism for cellular aging. *Science* 207:82–84.

Smith, R. E. (1956) Quantitative relations between liver mitochondria metabolism and total body weight in mammals. *Ann. N.Y. Acad. Sci.* 62:405–422.

Smith, S. E. (1984) Timing of vertebral band deposition in tetracycline injected leopard sharks. *Trans. Am. Fish. Soc.* 113:308–313.

Smith-Sonneborn, J. (1981) Genetics and aging in Protozoa. *Int. Rev. Cytol.* 73:319–354.

———. (1985) Aging in unicellular organisms. In *Handbook of the Biology of Aging*, 2d ed., C. E. Finch and E. L. Schneider (eds.), 79–104. New York: Van Nostrand.

———. (1990) Aging in Protozoa. In *Handbook of the Biology of Aging*, 3d ed., E. L. Schneider and J. W. Rowe (eds.), 24–44. San Diego: Academic Press.

Snider, W. D. (1987) The dendritic complexity and innervation of submandibular neurons in five species of mammals. *J. Neurosci.* 7:1760–1768.

Snodgrass, R. E. (1956) *Anatomy of the Honey Bee*. Ithaca: Cornell Univ. Press.

Snow, E. C. (1987) An evaluation of antigen-driven expansion and differentiation of hapten-specific B lymphocytes purified from aged mice. *J. Immunol.* 139:1758–1762.

Snowdon, D. A., Kane, R. L., Beeson, W. L., Burke, G. L., Sprafka, M., Potter, J., Iso, H., Jacobs, D. R., and Phillips, R. L. (1989) Is early natural menopause a biologic marker of health and aging? *Am. J. Publ. Hlth.* 79:709–714.

Society of Actuaries. (1951) Impairment study, 1951. New York: Peter F. Mallon. [not read]

Soderstrom, T. R., and Calderon, C. E. (1979) A commentary on the bamboos (Poaceae: Bambusoideae) *Biotropica* 11:161–172.

Soderwall, A. L., Kent, H. A., Turbyfill, C. L., and Britenbaker, A. L. (1960) Variation in gestation length and litter size of the golden hamster. *Mesocricetus auratus*. *J. Gerontol.* 15:246–248.

Sohal, R. S. (1975) Mitochondrial changes in flight muscles of normal and flightless *Drosophila melanogaster* with age. *J. Morphol.* 145:337–345.

———. (1976) Metabolic rate and lifespan. *Interdiscipl. Top. Gerontol.* 9:25–40.

Sohal, R. S., Farmer, K. J., Allen, R. G., and Cohen, N. R. (1983) Effect of age on oxygen consumption, superoxide dismutase, catalase, glutathione, inorganic peroxides, and chloroform soluble antioxidants in the adult male housefly, *Musca domestica. Mech. Age. Dev.* 24:185–195.

Sohal, R. S., Svensson, I., and Brunk, U. T. (1990) Hydrogen peroxide production by liver mitochondria of different species. *Mech. Age. Dev.*

Sohal, R. S., Svensson, I., Sohal, B. H., and Brunk, U. T. (1989) Superoxide anion radical products in different animal species. *Mech. Age. Dev.* 49:121–135.

Sojar, H. T., and Rothstein, M. (1986) Protein synthesis by liver ribosomes from aged rats. *Mech. Age. Dev.* 35:47–57.

Sokal, R. R. (1970) Senescence and genetic load: Evidence from *Tribolium. Science* 167:1733–1734.

Sokoloff, L. (1969) *The Biology of Degenerative Joint Disease*. Chicago: Univ. of Chicago Press.

———. (1981) Localization of functional activity in the central nervous system by measurement of glucose utilization with radioactive deoxyglucose. *J. Cerebral Blood Flow Metab.* 1:7–36.

Sokoloff, L., and Varma, A. A. (1988) Chondrocalcinosis in surgically resected joints. *Arthrit. Rheum.* 31:750–756.

Solano, A. R., Sanchez, M. L., Sardanons, M. L., Dada, L., and Podesta, E. J. (1988) Luteinizing hormone triggers a molecular association between its receptor and the major histocompatibility complex class I antigen to produce cell activation. *Endocrinology* 122: 2080–2083.

Sonneborn, T. M. (1930) Genetic studies of *Stenostomum incaudatus* (nov. sp.). 1. Origin and nature of differences among individuals formed during vegetative reproduction. *J. Exp. Zool.* 57:57–108.

———. (1947) Recent advances in the genetics of *Paramecium* and *Euplotes. Adv. Genet.* 1:263.

———. (1954) The relation of autogamy to senescence and rejuvenescence in *P. aurelia. J. Protozool.* 1:36–53.

———. (1957) Breeding systems, reproductive methods, and species problems in Protozoa. In *The Species Problem*, E. Mayr (ed.). AAAS Publication 50. Washington, DC: AAAS.

———. (1960) Enormous differences in length of life of closely related ciliates and their significance. In *The Biology of Aging*, B. L. Strehler, J. Ebert, H. B. Glass, and N. W. Shock (eds.), 289. American Institute of Biolological Sciences Symposium 6. Baltimore: Waverly.

Sonneborn, T. M., and Schneller, M. V. (1960) Age induced mutations in *Paramecium*. In *The Biology of Aging*, B. L. Strehler, J. Ebert, H. B. Glass, and N. W. Shock (eds.), 286–287. American Institute of Biological Sciences Symposium 6. Baltimore: Waverly.

Sonntag, W. E., Forman, L. J., Miki, N., Trapp, J. M., Gottschall, P. E., and Meites, J. (1982) L-DOPA restores amplitude of growth hormone pulses in old male rats to that observed in young male rats. *Neuroendocrinology* 34:163–168.

Sonntag, W. E., and Gough, M. A. (1988) Growth hormone releasing hormone induced release of growth hormone in aging male rats: Dependence on pharmacological manipulation and endogenous somastostatin release. *Neuroendocrinology* 47:482–488.

Sonntag, W. E., Hylka, V. W., and Meites, J. (1985) Growth hormone restores protein synthesis in skeletal muscle of old male rats. *J. Gerontol.* 40:689–694.

Sonntag, W. E., Steger, R. W., Forman, L. J., and Meites, J. (1980) Decreased pulsatile release of growth hormone in old male rats. *Endocrinology* 107:1875–1879.

Sopelak, V. M., and Butcher, R. L. (1982) Decreased amount of ovarian tissue and maternal age affect embryonic development in old rats. *Biol. Reprod.* 27:449–455.

Sørensen, A. B., Weinert, F. E., and Sherrod, L. R. (eds.) (1986) *Human Development and the Life Course: Multidisciplinary Perspectives*. Hillsdale, NJ: Lawrence Erlbaum Associates.

Spector, I. M. (1974) Animal longevity and plasma turnover rate. *Nature* 240:66.

Spector, W. S. (1961) *Handbook of Biological Data*. Philadelphia: Saunders.

Speed, R. M. (1977) The effects of ageing of the meiotic chromosomes of male and female mice. *Chromosoma* 64:241–254.

———. (1985) The prophase stages in human foetal oocytes studied by light and electron microscopy. *Human Genet.* 69:69–75.

Spemann, F. W. (1924) Über Lebensdauer, Altern, und andere Fragen der Rotatorien-biologie. Nach Beobachtungen an Rotifer vulgaris. *Z. Wiss. Zool.* 123:1–36.

Spence, A. M., and Herman, M. M. (1973) Critical re-examination of the premature aging concept in progeria: A light and electron microscopic study. *Mech. Age. Dev.* 2:211–227.

Spence, M. A., Heyman, A., Marazita, M. L., Sparkes, R. S., and Weinberg, T. (1986) Genetic linkage studies in Alzheimer's disease. *Neurology* 36:581–584.

Spicer, G. S. (1988) The effects of culture media on the two-dimensional electrophoretic protein pattern of *Drosophila virilis. Drosophila Info. Service.* Research notes, 67:74–75.

Spielman, A. (1964) Studies on autogeny in *Culex pipiens* populations in nature. 1. Reproductive isolation between autogenous and anautogenous populations. *Am. J. Hyg.* 80:175–183.

———. (1971) Bionomics of autogenous mosquitos. *Ann. Rev. Entomol.* 16:231–248.

Spight, T. M. (1979) Environment and life history: The case of two marine snails. In *Reproductive Ecology of Marine Invertebrates,* S. E. Stancyk (ed.), 135–143. Columbia: Univ. South Carolina Press.

Spinage, C. A. (1972) African ungulate life tables. *Ecology* 53:645–652.

Spoendlin, H., and Brun, J. P. (1973) Relation of structural damage to exposure time and intensity in acoustic trauma. *Acta Otolaryng.* 75:220–226.

Sprules, W. M. (1952) The arctic char of the west coast of Hudson Bay. *J. Fish. Res. Bd. Can.* 9:1–15.

Squire, L. R. (1986) Mechanisms of memory. *Science* 232:1612–1619.

Srivastava, P. N., Gao, Y., Levesque, J., and Auclair, J. L. (1985) Differences in amino acid requirements between two biotypes of the pea aphid, *Acyrthosiphon pisum. Can. J. Zool.* 63:603–606.

Staats, J. (1976) Standardized nomenclature for inbred strains of mice: Sixth listing. *Canc. Res.* 36:4333–4377.

———. (1980) Standardized nomenclature for inbred strains of mice: Seventh listing. *Canc. Res.* 40:2083–2128.

Stadtman, E. R. (1988) Minireview: Protein modification in aging. *J. Gerontol.* 43:B112–B120.

Stahl, U., Lemke, P. A., Tudzynski, P., Kuck, U., and Esser, K. (1978) Evidence for plasmid-like DNA in a filamentous fungus, the ascomycete *Podospora anserina. Mol. Gen. Genet.* 162:341–343.

Stahl, W. R. (1962) Similarity and dimensional methods in biology. *Science* 137:205–212.

———. (1963) Similarity analysis of physiological systems. *Persp. Biol. Med.* 6:291–321.

———. (1967) Scaling of respiratory variables in mammals. *J. Appl. Physiol.* 22:453–460.

Staiano-Coico, L., Darzynkiewicz, Z., Hefton, J. M., Dutrowski, R., Darlington, G. J., and Weksler, M. E. (1982) Increased sensitivity of lymphocytes from people over 65 years to cell cycle arrest and chromosomal damage. *Science* 219:1335–1337.

Stampfer, M. J., Willett, W. C., Colditz, G. A., Rosner, B., Speizer, F. E., and Hennekens, C. H. (1985) A prospective study of postmenopausal estrogen therapy and coronary heart disease. *N. Engl. J. Med.* 313:1044–1049.

Stanley, J. F., Pye, D., and MacGregor, A. (1975) Comparison of doubling numbers attained by cultured animal cells with life span of species. *Nature* 255:158–159.

Stanley, R. D. (1987) A comparison of age estimates derived from the surface and cross-section methods of otolith reading for Pacific ocean perch (*Sebastes alutus*). In *Proceedings of the International Rockfish Symposium,* 187–196. University of Alaska Sea Grant Rep. 87–2. [not read]

Stanley, S. M. (1973) An explanation for Cope's rule. *Evolution* 27:1–26.

Stanulis-Praeger, B. M. (1987) Cellular senescence revisited: A review. *Mech. Age. Dev.* 38:1–48.

Stanulis-Praeger, B. M., and Gilchrest, B. A. (1989) Effect of donor age and prior sun exposure on growth inhibition of cultured human dermal fibroblasts by all trans-retinoic acid. *J. Cell Physiol.* 139:116–124.

Starr, M., Himmelman, J. H., and Therriault, J. C. (1989) Direct coupling of marine invertebrate spawning with phytoplankton blooms. *Science* 247:1071–1074.

Stastny, P. (1978) Alloantigen DRw4 with rheumatoid arthritis. *N. Engl. J. Med.* 298:869–871.

Stavenhagen, J. B., and Robins, D. M. (1988) An ancient provirus has imposed androgen regulation on the adjacent mouse sex-limited protein gene. *Cell* 55:247–254.

Stearns, S. C. (1976) Life-history tactics: A review of the ideas. *Q. Rev. Biol.* 51:4–47.

——. (1977) The evolution of life-history traits: A critique of the theory and a review of the data. *Ann. Rev. Ecol. Sys.* 8:145–171.

——. (1983a) The genetic basis of differences in life-history traits among six populations of mosquitofish (*Gambusia affinis*) that shared ancestors in 1905. *Evolution* 37:618–627.

——. (1983b) The influence of size and phylogeny on patterns of covariation among life-history traits in the mammals. *Oikos* 41:173–187.

——. (1984) The effects of size and phylogeny on patterns of covariation in the life history traits of lizards and snakes. *Am. Nat.* 123:56–72.

Stebbins, G. L. (1950) *Variation and Evolution in Plants*. New York: Columbia Univ. Press.

Steenbergh, W. F., and Lowe, C. H. (1977) Reproduction, germination, establishment, growth, and survival of the young plant. *Ecology of the Saguaro: II*. Monograph series, no. 8. Washington, DC: U.S. Government Printing Office.

——. (1983) Growth and demography. *Ecology of the Saguaro: III*. Monograph series, no. 17. Washington, DC: U.S. Government Printing Office.

Stefano, G. B. (1981) Decrease in the number of high affinity opiate binding sites during the aging process in *Mytilus edulis* (Bivalvia). *Cell. Mol. Neurobiol.* 1:343–350.

——. (1982) Aging: Variations in opiate binding characteristics and dopamine responsiveness in subtidal and intertidal *Mytilus edulis* visceral ganglia. *Comp. Biochem. Physiol.* 72:349–352.

Stefano, G. B., Stanec, A., and Catapane, E. J. (1982) Aging: decline of dopamine-stimulated adenylate cyclase activity in *Mytilus edulis* (Bivalvia). *Cell. Mol. Neurobiol.* 2:249–253.

Steger, R. W., Huang, H., and Meites, J. (1981) Reproduction. In *CRC Handbook of Physiology in Aging*, E. J. Masoro (ed.), 333–382. Boca Raton, FL: CRC Press.

Stein, D. G., and Mufson, E. J. (1979) Tumor-induced brain damage in rats: Implications for behavioral and anatomical studies with aging animals. *Exp. Aging Res.* 5:537–547.

Steinberg, A. D., Baron, S., and Talal, N. (1969) The pathogenesis of autoimmunity in New Zealand mice. 1. Induction of antinucleic acid antibodies by polyinosinic polycytidylic acid. *Proc. Natl. Acad. Sci.* 63:1102.

Steinberg, D., Parthasarathy, S., Carew, T. E., Khoo, J. C., and Witztum, J. L. (1989) Beyond cholesterol. Modifications of low-density lipoprotein that increase its atherogenicity. *N. Engl. J. Med.* 320:915–924.

Steinfeld, A. D., and Reinhardt, C. (1980) Male climacteric after orchiectomy in patient with prostatic cancer. *Urology* 16:620–623.

Stenback, F., Peto, R., and Shubik, P. (1981) Initiation and promotion at different ages and doses in 2,200 mice. 1. Methods, and the apparent persistence of initiated cells. *Brit. J. Canc.* 44:1–14.

Stepan, J. S., Lachman, M., Zverince, J., Pacousky, V., and Baylink, D. J. (1989) Castrated men exhibit bone loss: Effect of calcitonin treatment on biochemical indices of bone remodeling. *J. Clin. Endocrinol. Metab.* 69:523–527.

Sterba, V. G. (1955) Das Adrenal- und Interrenalsystem im Lebensablauf von *Petromyzon planeri* Bloch. 1. Morphologie und Histologie einschliesslich Histogenese. *Zool. Anz.* 155:151–168.

Stern, C. (1926) An effect of temperature and age on crossing-over in the first chromosome of *Drosophila melanogaster*. *Proc. Natl. Acad. Sci.* 12:530–533.

Stern, C., Centerwall, W. R., and Sarker, S. S. (1964) New data on the problem of Y-linkage of Hairy Pinnae. *Am. J. Hum. Gen.* 16:455–471.

Sternberg, E. M., Young, W. S., Bernadini, R., Calogero, A. E., Chrousos, G. P., Gold, P. W., and Wilder, R. L. (1989) A central nervous system defect in biosynthesis of corticotropin-releasing hormone is associated with susceptibility to streptococcal cell wall–induced arthritis in Lewis rats. *Proc. Natl. Acad. Sci.* 86:4771–4775.

Stevens, J. G., Wagner, E. K., Devi-Rao, G. B., Cook, M. L., and Feldman, L. T. (1987) RNA complementary to a herpes virus a gene mRNA is prominent in latently infected neurons. *Science* 235:1056–1059.

Steward, F. C., and Mohan Ram, H. Y. (1961) Determining factors in cell growth: Some implications for morphogenesis in plants. *Adv. Morphogen.* 1:189–265.

Stewart, W. N. (1983) *Paleobotany and the Evolution of Plants.* Cambridge: Cambridge Univ. Press.

Stöcker, E. (1976) Zur DNS-Synthese in Zellkernen der regenierenden Rattenleber also Funktion des Alters. In *Alternstheoriern,* D. Platt (ed.), 121–156. Stuttgart: Schattauer Verlag.

Stöcker, E., and Heine, W.-D. (1971) Regeneration of liver parenchyma under normal and pathological conditions. *Beitr. Path. Bd.* 144:400–408.

Stocker, R., Glazer, A. N., and Ames, B. N. (1987) Antioxidant activity of albumin-bound bilirubin. *Proc. Natl. Acad. Sci.* 84:5918–5922.

Stone, L. S. (1960) Regeneration of the lens, iris, and neural retina in a vertebrate. *Yale J. Biol. Med.* 32:464–473.

Storer, J. (1966) Longevity and gross pathology at death in 22 inbred mouse strains. *J. Gerontol.* 21:404–409.

Storer, J. B. (1978) Effect of aging and radiation in mice of different genotypes. In *Genetic Effects on Aging,* D. Bergsma and D. E. Harrison (eds.), 55–71. New York: Liss.

Storer, J. B., and Graham, D. (1960) Vertebrate radiobiology: Late effects. *Ann. Rev. Nucl. Sci.* 10:561–582.

Storrs, E. E., Burchfield, H. P., and Rees, R. J. W. (1988) Superdelayed parturition in armadillos: A new mammalian survival strategy. *Lepr. Rev.* 59:11–15.

————. (1989) Reproduction delay in the common long-nosed armadillo, *Dasypus novemcintus* L. In *Advances in Neotropical Mammalogy,* K. H. Redford and J. F. Eisenberg (eds.), 535–548. Gainesville, FL: Sandhill Crane.

Stoye, J. P., and Coffin, J. M. (1988) Polymorphism of murine endogenous proviruses revealed by using virus class-specific oligonucleiotide probes. *J. Virol.* 62:168–175.

Stoye, J. P., Fenner, S., Greenoak, G. E., Moran, C., and Coffin, J. M. (1988) Role of endogenous retroviruses as mutagens: The hairless mutation of mice. *Cell* 54:383–391.

Strathern, J. N., Jones, E. W., and Broach, J. R. (eds.) (1981) *The Molecular Biology of the Yeast Saccharomyces: Life Cycle and Inheritance.* Cold Spring Harbor, NY: Cold Spring Harbor Laboratory Press.

Strathern, J. N., Jones, E. W., and Broach, J. R. (eds.) (1982) *The Molecular Biology of the Yeast Saccharomyces: Metabolism and Gene Expression.* Cold Spring Harbor, NY: Cold Spring Harbor Laboratory Press.

Strauss, R. H., and Albertini, R. J. (1979) Enumeration of 6–thioguanine-resistant peripheral blood lymphocytes in man as a potential test for somatic cell mutations arising in vivo. *Mutat. Res.* 61:353–379.

Strehler, B. L. (1961) Studies on the comparative physiology of aging. 2. On the mechanism of temperature life shortening in *Drosophila melanogaster. J. Gerontol.* 16:2–12.

————. (1977) *Time, Cells, and Aging.* 2d ed. New York: Academic Press.

————. (1986) Genetic instability as the primary cause of human aging. *Exp. Gerontol.* 21:283–319.

Strehler, B. L., and Chang, M. P. (1979a) Loss of hybridizable ribosomal DNA from human post-mitotic tissues during aging. 1. Age-dependent loss in human myocardium. *Mech. Age. Dev.* 11:371–378.

————. (1979b) Loss of hybridizable ribosomal DNA from human post-mitotic tissues during aging. 2. Age-dependent loss in human cerebral cortex: Hippocampal and somatosensory cortex comparison. *Mech. Age. Dev.* 11:379–382.

Strehler, B. L., and Crowell, S. (1961) Studies on comparative physiology of aging. 1. Function vs. age of *Campanularia flexuosa. Gerontologia* (Basel) 5:1–8.

Strehler, B. L., Mark D. D., Mildvan, A. S., and Gee, M. V. (1959) Rate and magnitude of age pigment accumulation in the human myocardium. *J. Gerontol.* 14:430–439.

Strehler, B. L., and Mildvan, A. S. (1960) General theory of mortality and aging. A stochastic model relates observations on aging, physiologic decline, mortality, and radiation. *Science* 132:14–21.

Strehler, B. L., Shock, N. W., Ebert, J., and Glass, H. B. (eds.) (1960) *The Biology of Aging,* American Institute of Biological Sciences Symposium 6. Baltimore: Waverly.

Strong, J. P., Restrepo, C., and Guzman, M. (1978) Coronary and aortic atherosclerosis in New Orleans. 2. Comparison of lesions by age, sex, and race. *Lab. Invest.* 39:364–369.

Strong, L. C. (1951) Litter seriation phenomena in fibrosarcoma susceptibility. A contribution to the subject of cancer susceptibility in relation to age. *J. Gerontol.* 6:339–357.

———. (1954) The cumulative effect of litter seriation on fibrosarcoma susceptibility in mice. *Ann. N.Y. Acad. Sci.* 57:507–516.

———. (1968) *Biological Aspects of Cancer and Aging.* Oxford: Pergamon Press.

Strong, L. C., and Johnson, F. (1962) Oncology and gerontology-genetic implications. In *The Morphological Precursors of Cancer,* L. Severi (ed.), 119–151. Perugia, Italy: Division of Cancer Research.

———. (1965) The effects of fifty-four generations of inbreeding on age of first litters. Selective influences of early maternal age and polydactylism. *J. Gerontol.* 20:405–409.

Strother, S. (1988) The role of free radicals in leaf senescence. *Gerontology* 34:151–156.

Stuchliková, E., Juricobá-Horaková, M., and Deyl, Z. (1975) New aspect of the dietary effect of the life prolongation in rodents. What is the role of obesity in aging? *Exp. Gerontol.* 10:141–144.

Stunkard, H. W. (1937) The physiology and life cycles of the parasitic flatworms. *Am. Mus. Novit.* 908:1–27.

———. (1954) The life-history and systematic relations of the Mesozoa. *Q. Rev. Biol.* 29:230–244.

———. (1962) The organization, ontogeny, and orientation of the Cestoda. *Q. Rev. Biol.* 37:23–34.

Suarez, G., Rajaram, R., Bhuyan, K. C., Oronsky, A. L., and Goidl, J. A. (1988) Administration of an aldose reductase inhibitor induces a decrease of collagen fluorescence in diabetic rats. *J. Clin. Invest.* 82:624–627.

Subba Rao, K. S., Martin, G. M., and Loeb, L. A. (1985) Fidelity of DNA polymerase-B in neurons from young and very aged mice. *J. Neurochem.* 45:1273–1278.

Suda, Y., Suzuki, M., Ikawa, Y., and Aizawa, S. (1987) Mouse embryonic stem cells exhibit indefinite proliferative potential. *J. Cell Physiol.* 133:197–201.

Sugawara, O., Oshimura, M., Koi, M., Annab, L. A., and Barett, J. C. (1990) Induction of cellular senescence in immortalized cells by human chromosome I. *Science* 247:707–710.

Sugawara, S., and Mikamo, K. (1983) Absence of correlation between univalent formation and meiotic nondisjunction in aged female Chinese hamsters. *Cytogenet. Cell. Genet.* 35:34–40.

Sullivan, C. B., Visscher, B. R., and Detels, R. (1984) Multiple sclerosis and age at exposure to childhood diseases and animals: Cases and their friends. *Neurology* 34:1144–1148.

Sullivan, J. M., Vander Zwaag, R., Lemp, G. F., Hughes, J. P., Maddock, V., Kroetz, F. W., Ramanathan, K. B., and Mirvis, D. M. (1988) Postmenopausal estrogen use and coronary atherosclerosis. *Ann. Intern. Med.* 108:358–363.

Suntzeff, V., Cowdry, E., and Hixon, B. (1962) Influence of maternal age on offspring in mice. *J. Gerontol.* 17:2–7.

Suomalainen, E., Saura, A., and Lokki, J. (1976) Evolution of parthenogenetic insects. *Evol. Biol.* 9:209–257.

Suominen, H., Heikkinen, E., Liesen, H., Michel, D., and Hollmann, W. (1977) Effects of eight weeks' endurance training on skeletal muscle metabolism in 70-year-old sedentary men. *Eur. J. Appl. Physiol.* 37:173–180.

Suominen, H., Heikkinen, E., and Parkatti, T. (1977) Effects of eight weeks' physical training on muscle and connective tissue of the m. vastus lateralis in 69-year-old men and women. *J. Gerontol.* 32:33–37.

Sussex, I., and Rosenthal, D. (1973) Differential ³H-thymidine labeling of nuclei in the shoot apical meristem of *Nichotiana. Bot. Gaz.* 134:295–301.

Sutcliffe, J. G. (1988) mRNA in the mammalian central nervous system. *Ann. Rev. Neurosci.* 11:157–198.

Svardson, G. (1979) Speciation of Scandinavian *Corgonus. Rep. Inst. Freshwater Res. Drottningholm* 57:3–95.

Svare, B., Kinsley, C., Miele, J., Konen, C., and Ghiraldi, L. (1986) Intrauterine contiguity influences regulatory activity in adult female and male mice. *Horm. Behav* 20:7–19.

Svare, B., Kinsley, C., Miele, J., Konen, C., Ghiraldi, L., and Broida, J. (1986) Prior intrauterine position influences body weight in male and female mice. *Horm. Behav.* 20:201–211.

Sved, J. A., and Sandler, L. (1981) Relation of maternal age effect in Down syndrome to nondisjunction. In *Trisomy 21 (Down Syndrome) Research Perspective,* F. F. del la Cruz and P. D. Gerald (eds.), 95–98. Baltimore: University Park Press.

Svejgaard, A., Platz, P., and Ryder, L. P. (1983) HLA and disease 1982: A survey. *Immunol. Rev.* 70:193–218.

Svoboda, J. A., and Thompson, M. J. (1983) Comparative sterol metabolism in insects. In *Metabolic Aspects of Lipid Nutrition in Insects,* T. E. Mittler and R. H. Dadd (eds.), 1–16. Boulder, CO: Westview Press.

Swaab, D. F., Fliers, E., and Partiman, T. S. (1985) The suprachiasmatic nucleus of the human brain in relation to sex, age, and senile dementia. *Brain Res.* 342:37–44.

Swanson, C. P. (1957) *Cytology and Cytogenetics.* Englewood Cliffs, NJ: Prentice-Hall.

Swanson, M. M., and Riddle, P. L. (1981) Critical periods in the development of the *Caenorhabditis elegans* dauer larva. *Dev. Biol.* 84:27–40.

Swedmark, M. (1957) Travaux de la station biologique de roscoff. 2. Sur la variation géographique de *Gobius minutus* Pallas. I. Biol-

ogie et croissance. *Arch. Zool. Exp. Gen.* 95:32–51.

Swick, R. W., Koch, A. L., and Handa, D. T. (1956) The measurement of nucleic acid turnover in rat liver. *Arch. Biochem. Biophys.* 63:226–242.

Swisshelm, K., Disteche, C. M., Thorvaldsen, J., Nelson, A., and Salk, D. (1990) Age-related increase in methylation of ribosomal genes and inactivation of chromosome-specific rRNA gene clusters in mouse. *Mutat. Res.*

Swisshelm, K., Nelson, A., and Salk, D. (In prep.) Evidence for rRNA dosage stability in aging mice. Typescript.

Switzer-Dunlap, M., and Hadfield, M. G. (1979) Reproductive patterns of Hawaiian aplysiid gastropods. In *Reproductive Ecology of Marine Invertebrates,* S. E. Stancyk (ed.), 199–210. Columbia: Univ. of South Carolina Press.

Szabó, I. (1931) The three types of mortality curve. *Q. Rev. Biol.* 6:462–463.

———. (1935) Senescence and death in invertebrate animals. *Riv. Biol.* 19:378–436.

Szalay, F. S., and Delson, E. (1979) *Evolutionary History of the Primates.* New York: Academic Press.

Szarski, H. (1976) Cell size and nuclear DNA content in vertebrates. *Int. Rev. Cytol.* 44:93–111.

Szilard, L. (1959) On the nature of the aging process. *Proc. Natl. Acad. Sci.* 45:30–45.

Szmant, A. M. (1986) Reproductive ecology of Caribbean reef corals. *Coral Reefs* 5:43–54.

Szmant-Froelich, A. (1985) The effect of colony size on the reproductive ability of the Caribbean coral *Montastrea annularis* (Ellis and Solander). In *International Association of Biological Oceanography,* vol. 4, *Symposia and Seminars,* 295–300.

Taber, S., III, and Poole, H. K. (1973) Caste determination in honeybees: The production of queen-worker intermediates. *J. Apicult. Res.* 12:111–116.

Tajima, T., Watanabe, T., Iijima, K., Ohshika, Y., and Yamaguchi, H. (1981) The increase of glycosaminoglycans synthesis and accumulation on the cell surface of cultured skin fibroblasts in Werner's syndrome. *Exp. Pathol.* 20:221–229.

Takagi, Y. (1987) Aging. In *Paramecium,* H.-D. Gortz (ed.), 131–140. Berlin and Heidelberg: Springer-Verlag.

Takagi, Y., and Kanazawa, N. (1982) Age associated change in macronuclear DNA content in *Paramecium caudatum. J. Cell Sci.* 54:137–147.

Takahashi, R., and Goto, S. (1987a) Age-associated accumulation of heat-labile aminoacyl-tRNA synthetases in mice and rats. *Arch. Gerontol. Geriatr.* 6:73–82.

———. (1987b) Influence of dietary restriction on accumulation of heat-labile enzyme molecules in the liver and brain of mice. *Arch. Biochem. Biophys.* 257:200–206.

Takahashi, R., Mori, N., and Goto, S. (1985a) Accumulation of heat-labile elongation factor 2 in the liver of mice and rats. *Exp. Gerontol.* 20:325–331.

———. (1985b) Alteration of aminoacyl tRNA synthetases with age: Accumulation of heat-labile enzyme molecules in rat liver, kidney, and brain. *Mech. Age. Dev.* 33:67–75.

Takahashi, S., and Kawashima, S. (1983) Age-related changes in prolactin cells in male and female rats of the Wistar/Tw strain. *J. Sci. Hiroshima Univ.* ser. B, div. 1 (31): 185–195.

Takayama, S., Ishikawa, T., Masahito, P., and Matsumoto, J. (1981) Overview of biological characterization of tumors in fish. In *Phyletic Approaches to Cancer,* C. J. Dawe (ed.), 3–17. Tokyo: Japan Scientific Society Press.

Takeda, T., Hosokawa, M., Takeshita, S., Irino, M., Higuchi, K., Matsushita, T., Tomita, Y., Yashuira, K., Hamamoto, H., Shimizu, K., Ishii, M., and Yamamuro, T. (1981) A new murine model of accelerated senescence. *Mech. Age. Dev.* 17:183–194.

Takimoto, Y., Hagino, S., Yamada, H., and Miyamoto, J. (1984) The acute toxicity of fenitrothion to killifish (*Oryzias latipes*) at twelve different stages of its life history. *J. Pesticide Sci.* 9:463–470.

Takunaga, M., Futami, T., Wakamatsu, E., Endo, M., and Yosizawa, Z. (1975) Werner's syndrome as "hyaluronuria." *Clin. Chim.* 62:89–96.

Takunaga, M., Wakamatsu, E., Soto, K., Satake, S., Aoyama, K., Saito, K., Sugawara, M., and Yosizawa, Z. (1978) Hyaluronuria in a case of progeria (Hutchinson-Gilford syndrome) *J. Am. Geriatr. Soc.* 26:296.

Talan, M. I., Engel, B. T., and Whitaker, J. R. (1985) A longitudinal study of tolerance to cold stress among C57BL/6J mice. *J. Gerontol.* 40:8–14.

Talan, M. I., and Ingram, D. K. (1985) Effect of intermittent feeding on thermoregulatory abilities of young and aged C57BL/6J mice. *Arch. Gerontol. Geriatr.* 4:251–259.

Talman, W. T., Snyder, D., and Reis, D. J. (1980) Chronic lability of arterial pressure produced by destruction of A2 catecholamine neurons in rat brain stem. *Circ. Res.* 46:842–853.

Tam, S.-P., Haché, R. J., and Deeley, R. G. (1986) Estrogen memory effect in human hepatocytes during repeated cell division without hormone. *Science* 234:1234–1237.

Tamm, C. O. (1956) Further observations on the survival and flowering of some perennial herbs. *Oikos* 7:273–292.

Tamura, E., and Honma, Y. (1969) Histological changes in the organs and tissues of the gobiid fishes throughout the lifespan. 1. Hypothalamic-hypophyseal neurosecretory system of the ice-goby, *Leucopsarion petersi* Hilgendorf. *Bull. Jpn. Soc. Sci. Fish.* 35:19–28.

———. (1970) Histological changes in the organs and tissues of the gobiid fishes throughout the life-span. 3. Hemopoietic organs in the ice-goby, *Leucopsarion petersi* Hilgendorf. *Bull. Jpn. Soc. Sci. Fish.* 36:661–669.

Tan, Y. H., Chou, E. L., and Lundh, N. (1975) Regulation of chromosome 21–directed antiviral gene(s) as a consequence of age. *Nature* 257:310–312.

Tanabe, Y. (1985) Effect of ageing on plasma LH, testosterone, estradiol, and progesterone levels in embryonic and post-embryonic male and female ducks. In *Current Trends in Comparative Endocrinology,* B. Lofts and W. N. Holmes (eds.), 617–618. Hong Kong: Hong Kong Univ. Press.

Tanaka, S. (1965) Salmon of the North Pacific Ocean. 9. Coho, chinook, and masu salmon in offshore waters. A review of the biological information on masu salmon (*Oncorhynchus masou*). International North Pacific Fisheries Commission, Vancouver, B.C., bulletin 16:75–135.

Tannen, R. H., and Schwartz, A. G. (1982) Reduced weight gain and delay of Coomb's positive hemolytic anemia in NZB mice treated with dehydroepiandrosterone. *Fed. Proc.* 41:463 (abstract).

Tanner, J. M. (1962) *Growth at Adolescence.* 2d ed. Oxford: Blackwell.

———. (1969) Growth and endocrinology of the adolescent. In *Endocrine and Genetic Diseases of Childhood and Adolescence,* 2d ed., L. I. Gardner (ed.), 14–63. Philadelphia: Saunders.

Tannreuther, G. W. (1919) Studies on the rotifer *Asplanchna ebbesbornii,* with special reference to the male. *Biol. Bull.* 36:194–208.

Tanzi, R. E., Gusella, J. F., Watkins, P. C., Bruns, G. A. P., St. George-Hyslop, P. H., van Keuren, D., Patterson, D., Pagan, S., Kurnit, D. M., and Neve, R. L. (1987) Amyloid β protein gene: cDNA, mRNA distribution, and genetic linkage near the Alzheimer locus. *Science* 235:880–884.

Tanzi, R. E., Haines, J. L., Watkins, P. C., Stewart, G. D., Wallace, M. R., Hallewell, R., Wong, C., Wexler, N. S., Conneally, P. M., and Gusella, J. F. (1988) Genetic linkage map of human chromosome 21. *Genomics* 3:129–136.

Tanzi, R. E., St. George-Hyslop, P. H., and Gusella, J. F. (1989) Molecular genetic approaches to Alzheimer's disease. *Trends in Neuroscience* 12:152–163.

Tappel, A., Fletcher, B., and Deamer, D. (1973) Effect of antioxidants and nutrients on lipid peroxidation fluorescent products and aging parameters in the mouse. *J. Gerontol.* 28:415–424.

Tappel, A. L. (1973) Lipid peroxidation damage to cell components. *Fed. Proc.* 32:1870–1876.

Tassin, J., Malaise, E., and Courtois, Y. (1979) Human lens cells have an in vitro proliferative capacity inversely proportional to the donor age. *Exp. Cell Res.* 123:388–392.

Taubold, R. D., Siakotos, A. N., and Perkins, E. G. (1975) Studies on the chemical nature of lipofuscin (age pigment) isolated from normal human brain. *Lipids* 10:383–390.

Taylor, A., Zuliani, A. M., Hopkins, R. E., Dallal, G. E., Treglia, P., Kuck, J. F., and Kuck, K. (1989) Moderate caloric restriction delays cataract formation in the Emory mouse. *FASEB J.* 3:1741–1746.

Taylor, C. E., and Condra, C. (1980) r- and K-selection in *Drosophila pseudoobscura. Evolution* 34:1183–1193.

Taylor, C. R. (1980) Evolution of mammalian homeothermy: A two-step process? In *Comparative Physiology: Primitive Mammals,* R. Schmidt-Nielsen, L. Bolis, and C. R. Taylor (eds.), 100–109. Cambridge: Cambridge Univ. Press.

Taylor, C. R., Maloiy, G. M. O., Weibel, E. R., Langman, V. A., Kamau, J. M., Seherman, H. J., and Heglund, N. C. (1981) Design of the mammalian respiratory system. 3. Scaling maximum aerobic capacity to body mass: Wild and domestic animals. *Resp. Physiol.* 44:25–37.

Taylor, F. (1989) Optimal switching to diapause in relation to the onset of winter. *Theor. Pop. Biol.* 18:125–133.

Teague, P. O., Friou, G. J., and Myers, L. L. (1968) Antinuclear antibodies in mice. 1. Influence of age and possible genetic factors on spontaneous and induced responses. *J. Immunol.* 101:791–798.

Teague, P. O., Yunis, E. J., Rodey, G., Fish, A. J., Stutman, O., and Good, R. A. (1970) Autoimmune phenomena and renal disease in mice. *Lab. Invest.* 22:121–130.

Tease, C., and Fisher, G. (1986) Oocytes from young and old mice respond differently to colchicine. *Mutat. Res.* 173:31–34.

Technau, G. M. (1984) Fibre number in the mushroom bodies of adult *Drosophila melanogaster* depends on age, sex, and experience. *J. Neurogenet.* 1:113–126.

Teissier, G. (1939) Biométrique de la cellule animale et végétale. *Tabulae Biologique* part 1, 19:1–64. [not read]

Telford, N., Mobbs, C. V., Sinha, Y. N., and Finch, C. E. (1986) The increase of anterior pituitary dopamine in aging C57BL/6J female mice is caused by ovarian steroids, not intrinsic pituitary aging. *Neuroendocrinology* 43:135–142.

Temple, S. A., and Wallace, M. P. (1989) Survivorship patterns in a population of Andean condors *Vultur gryphus.* In *Raptors in the Modern World,* B.-U. Meyburg and R. D. Chancellor (eds.), 247–251. Berlin: WWGBP.

Templeton, A. R. (1982) The prophecies of parthenogenesis. In *Evolution and Genetics of Life Histories,* H. Dingle and J. P. Hegmann (eds.), 75–101. Berlin: Springer-Verlag.

———. (1983) Natural and experimental parthenogenesis. In *The Genetics and Biology of Drosophila,* vol. 3C, M. Ashburner, H. L. Carson, and J. N. Thompson (eds.), 343–398. New York: Academic Press.

Templeton, A. R., Grease, T. J., and Shah, F. (1985) The molecular through ecological genetics of abnormal abdomen in *Drosophila mercatorum. Genetics* 111:805–818.

Templeton, A. R., Hollocher, H., Lawler, S., and Johnston, J. S. (1989) Natural selection and ribosomal DNA in *Drosophila. Genome* 31:296–303.

———. (1990) The ecological genetics of abnormal abdomen in *Drosophila melanogaster.* In *Ecological and Evolutionary Genetics of Drosophila,* J. S. Barker and W. T. Starmer (eds.). San Diego: Academic Press. In press.

Templeton, A. R., and Johnston, J. S. (1982) Life history evolution under pleiotropy and K-selection in a natural population of *Drosophila mercatorum.* In *Ecological Genetics and Evolution: The Cactus-Yeast-Drosophila Model System,* J. S. Barker and W. T. Starmer (eds.) 225–239. New York: Academic Press.

Templeton, A. R., Johnston, J. S., and Sing, C. F. (1987) The proximate and ultimate control of aging in *Drosophila* and humans. In *Evolution of Longevity in Animals: A Comparative Approach,* A. D. Woodhead and K. H. Thompson (eds.), 123–135. New York: Plenum.

Templeton, A. R., and Rankin, M. A. (1978) Genetic revolutions and control of insect populations. In *The Screwworm Problem,* R. H. Richardson (ed.), 81–111. Austin: Univ. of Texas Press.

Terao, A. (1928) Growth of the lobster, *Homarus americanus. Proc. Soc. Exp. Biol. Med.* 25:353–355.

Terasawa, E. (1985) Developmental changes in the positive feedback effect of estrogen on luteinizing hormone release in ovariectomized female rhesus monkeys. *Endocrinology* 117:2490–2497.

Terasawa, E., Bridson, W. E., Nass, T. E., Noonan, J. J., and Dierschke, D. J. (1984) Developmental changes in the luteinizing hormone secretory pattern in peripubertal female rhesus monkeys: Comparisons between gonadally intact and ovariectomized animals. *Endocrinology* 115:2233–2240.

Terpstra, G. K., and Raaijmaker, J. A. (1976) Loss of an adrenergic effect in Swiss mice. *Eur. J. Pharmacol.* 38:373–376.

Terry, R. D., DeTeresa, R., and Hansen, L. A. (1987) Neocortical cell counts in normal human adult aging. *Ann. Neurol.* 21:530–539.

Tesch, F. W. (1977) *The Eel. Biology and Management of Anguillid Eels.* London: Chapman and Hale.

Thane, A. (1974) Rotifera. In *Reproduction of Marine Invertebrates,* vol. 1, A. C. Giese and J. S. Pearse (eds.), 471–484. New York: Academic Press.

Thiele, H.-U. (1977) Carabid beetles in their environments. A study on habitat selection by adaptations in physiology and behavior. In *Zoophysiology and Ecology,* vol. 10. New York: Springer-Verlag.

Thimann, K. V. (1980) The senescence of leaves. In *Senescence in Plants,* K. V. Thiman (ed.), 85–115. Boca Raton, FL: CRC Press.

Thimann, K. V., and Giese, A. C. (1981) A glance at senescence in plants. *Biological Mechanisms in Aging,* R. T. Schimke (ed.), 702–709. NIH Publ. 81-2194. Bethesda, MD: USDHEW.

Thoman, M. L., and Weigle, W. O. (1981) Lymphokines and aging: Interleukin-2 production and activity in aged animals. *J. Immunol.* 127:2102–2106.

Thomas, C. S., and Coulson, J. C. (1988) Reproductive success of kittiwake gulls, *Rissa tridactyla.* In *Reproductive Success. Studies of Individual Variation in Contrasting Breeding Systems,* T. H. Clutton-Brock (ed.), 251–262. Chicago: Univ. Chicago Press.

Thomas, H. (1976) Delayed senescence in leaves treated with the protein synthesis inhibitor MDMP. *Plant Sci. Lett.* 6:369–377.

Thomas, H., Luthy, B., and Matile, P. (1985) Leaf senescence in a non-yellowing mutant of *Festuca pratensis* Huds. *Planta* 164:400–405.

Thomas, H., and Stoddart, J. L. (1975) Separation of chlorophyll degradation from other senescence processes in leaves of a mutant genotype of meadow fescue (*Festuca pratensis* L.) *Physiology* 56:438–441.

———. (1980) Leaf senescence. *Ann. Rev. Plant Physiol.* 31:83–111.

Thomas, P. K. (1987) Vascular factors in the causation of diabetic neuropathy. *Trends in Neuroscience* 10:6–7.

Thomas, W. K., Withler, R. E., and Beckenbach, A. T. (1986) Mitochondrial DNA analysis of Pacific salmonid evolution. *Can. J. Zool.* 64:1058–1064.

Thompson, I., Jones, D. S., and Dreibelbis, D. (1980) Annual internal growth banding and life history of the ocean quahog *Artica islandica* (Mollusca: Bivalvia) *Mar. Biol.* 57:25–34.

Thompson, J. S., Saxena, S. K., and Sharp, J. G. (1990) Effect of age on intestinal regeneration in the rabbit. *Mech. Age. Dev.* 52:305–312.

Thompson, M., and Bywaters, E. G. L. (1962) Unilateral rheumatoid arthritis following hemiplegia. *Ann. Rheum. Dis.* 21:370.

Thompson, M. J. A. (1987) Longevity and survival of female pipistrelle bats (*Pipistrellus pipistrellus*) on the vale of York, England. *J. Zool.* 211:209–214.

Thompson, S. D., and Nicoll, M. E. (1986) Basal metabolic rate and energies of reproduction in theian mammals. *Nature* 321:690–693.

Thompson, T., Straus, R., and Tietz, N. (1984) The immune status of healthy centenarians. *J. Am. Geriatr. Soc.* 32:274–281.

Thomsen, O., and Kettel, K. (1929) Die Stärke der menschlichen Isoagglutinine und entsprechenden Blutkörperchenrezeptoren in verschiedenen Lebensaltern. *Z. Immunitätsforsch.* 63:67–93.

Thorneycroft, I. H., and Soderwall, A. L. (1969) The nature of litter size loss in senescent hamsters. *Anat. Rec.* 165:343–348.

Thorpe, J. E., Miles, M. S., and Keay, D. S. (1984) Developmental rate, fecundity, and egg size in Atlantic salmon, *Salmo salar L. Aquaculture* 43:289–305.

Thorpe, J. E., and Mitchell, K. A. (1981) Stocks of Atlantic salmon (*Salmo salar*) in Britain and Ireland: Discreteness and current management. *Can. J. Fish. Aquat. Sci.* 38:1576–1590.

Thorpe, J. E., and Morgan, R. I. (1978) Parental influence on growth rate, smolting rate, and survival in hatchery reared juvenile Atlantic salmon, *Salmo salar. J. Fish. Biol.* 13:549–556.

Thorpe, J. E., Talbot, C., and Villarreal, C. (1982) Bimodality of growth and smolting in Atlantic salmon, *Salmo salar L. Aquaculture* 28:123–132.

Thorsen, T. B., and Lacy, E. J., Jr. (1982) Age, growth rate, and longevity of *Carcharhinus leucas* estimated from tagging and vertebral rings. *Copeia* 1:110–116.

Thrasher, J. D., and Greulich, R. C. (1965a) The duodenal cell population. 1. Age-related increase in the duration of the cryptal progenitor cell cycle. *J. Exp. Zool.* 159:39–46.

———. (1965b) The duodenal cell population. 2. Age-related increases in size and distribution. *J. Exp. Zool.* 159:385–396.

Thung, P. J. (1957) The relation between amyloid and ageing in comparative pathology. *Gerontologia* (Basel) 1:234–254.

Tice, R. R., and Setlow, R. B. (1985) DNA repair and replication in aging organisms and cells. In *Handbook of the Biology of Aging,* 2d ed., C. E. Finch and E. L. Schneider (eds.), 173–224. New York: Van Nostrand.

Tiffney, B. H., and Niklas, K. J. (1985) Clonal growth in land plants: A paleobotanical perspective. In *Population Biology and Evolution of Clonal Organisms,* J. B. Jackson, L. W. Buss, R. E. Cook (eds.), 35–66. New Haven: Yale Univ. Press.

Tigges, J., Gordon, T. P., McClure, H. M., Hall, E. C., and Peters, A. (1988) Survival rate and life span of rhesus monkeys at the Yerkes Regional Primate Research Center. *Am. J. Primatol.* 15:263–273.

Tilley, S. G. (1980) Life histories and comparative demography of two salamander populations. *Copeia* 4:806–821.

Timms, M., Westerman, M., and Düring, B. (1982) Studies on the DNA of some species of *Antechinus* (Dasyuridae, Marsupialia). In *Carnivorous Marsupials,* M. Archer (ed.), 715–721. Sydney: Royal Zoological Society of New South Wales.

Tinkle, D. W., Wilbur, H. M., and Tilley, S. G. (1970) Evolutionary strategies in lizard reproduction. *Evolution* 24:55–74.

Tiwari, J. L., and Terasaki, P. I. (1985) *HLA and Disease Associations.* New York: Springer-Verlag.

Tobari, Y. N., Matsuda, M., Tomimura, Y., and Moriwaki, D. (1983) Discriminative effects of male age on male recombination and minute mutation frequencies in *Drosophila ananassae. Jpn. J. Genet.* 58:173–179.

Todaro, G. J., and Green, H. (1963) Quantitative studies of the growth of mouse embryo cells in culture and their development into established lines. *J. Cell Biol.* 17:299–313.

Todd, P. R. (1980) Size and age of migrating New Zealand freshwater eels (*Anguilla* spp.) *New Zealand J. Mar. Freshwater Res.* 14:283–293.

Tohmé, G., and Tohmé, H. (1978) Accroissement de la société et longévité de la reine et des ouvrières chez *Messor semirufus* (Andre) (Hym.: Formicoidea) *Compt. Rend. Acad. Sci. Paris* 286:961–963.

Tokarz, R. R. (1978) Oogonial proliferation, oogenesis, and folliculogenesis in nonmammalian vertebrates. In *The Vertebrate Ovary: Comparative Biology and Evolution,* R. E. Jones (ed.), 145–179. New York: Plenum.

Tokunaga, C. (1970) The effects of low temperature and aging on nondisjunction in *Drosophila. Genetics* 65:75–94.

Tolmasoff, J. M., Ono, T., and Cutler, R. G. (1980) Superoxide dismutase: Correlation with life span and specific metabolic rate in primate species. *Proc. Natl. Acad. Sci.* 77:2777–2781

Tomita, S., and Riggs, A. (1971) Studies of the interaction of 2,3–diphosphoglycerate and carbon dioxide with hemoglobins from mouse, man, and elephant. *J. Biol. Chem.* 246:547–554.

Tomkins, G. A., Stanbridge, E. J., and Hayflick, L. (1974) Vireal probes of ageing in the human diploid strain, WI–38. *Proc. Soc. Exp. Biol. Med.* 146:385–390.

Tomlinson, P. B. (1983) Tree architecture. *Am. Sci.* 71:141–150.

Toth, S. E. (1968) The origin of lipofuscin age pigments. *Exp. Gerontol.* 3:19–30.

Tower, D. B. (1954) Structural and functional organization of mammalian cerebral cortex: The correlation of neuron density with brain size. *J. Comp. Neurol.* 101:19–51.

Tower, D. B., and Young, O. M. (1973) The activities of butyrylcholinesterase and carbonic anhydrase, the rate of anaerobic glycolysis, and the question of a constant density of glial cells in cerebral cortices of various mammalian species from mouse to whale. *J. Neurochem.* 20:269–278.

Tracey, K. M., Sr. (1958) Effects of parental age on the life cycle of the mealworm, *Tenebrio molitor* Linnaeus. *Ann. Entomol. Soc. Am.* 51:429–432.

Traina, V. L., Taylor, B. A., and Cohen, J. C. (1981) Genetic mapping of endogenous mouse mammary tumor viruses: Locus characterization, segregation, and chromosomal distribution. *J. Virol.* 40:735–744.

Trainor, K. J., Wigmore, D. J., Chrysostomou, A., Dempsey, J. L., Seshadri, R., and Morley, A. (1984) Mutation frequency in human lymphocytes increases with age. *Mech. Age. Dev.* 27:83–86.

Treloar, A. E., Boynton, R. E., Behn, B. G., and Brown, B. W. (1967) Variation of the human menstrual cycle through reproductive life. *Int. J. Fertil.* 12:77–126.

Trenton, J. A., and Courtois, Y. (1982) Correlation between DNA excision repair and mammalian lifespan in lens epithelial cells. *Cell Biol. Int. Rep.* 6:253–260.

Tribe, M. A., and Ashhurst, D. E. (1972) Biochemical and structural variations in the flight muscle mitochondria of ageing butterflies, *Calliphora erythrocephala. J. Cell Sci.* 10:443–469.

Trippi, V.S., and Brulfert, J. (1973) Organization of morphophysiologic unit in *Anagallis arvensis* and its relation with perpetuation mechanism in senescence. *Am. J. Bot.* 60:641–647.

Trippi, V. S., and Montaldi, E. (1960) The aging of sugarcane clones. *Phyton* (Buenos Aires) 14:79–91.

Trosko, J. E., and Chang, C. C. (1980) An integrative hypothesis linking cancer, diabetes, and atherosclerosis: The role of mutations and epigenetic changes. *Med. Hypotheses* 6: 455.

Trounce, I., Byrne, E., and Marzuki, S. (1989) Decline in skeletal muscle mitochondrial respiratory chain function: Possible factor in ageing. *Lancet* 637–639.

Troup, G. M., Smith, G. S., and Walford, R. L. (1960) Life span, chronologic disease patterns, and age-related changes in spleen weights for the Mongolian gerbil. *Exp. Gerontol.* 4:139–144.

Trout, W. E., and Kaplan, W. D. (1981) Mosaic mapping of foci associated with longevity in the neurological mutants Hk and Sh of *Drosophila melanogaster. Exp. Gerontol.* 16:461–474.

Trpis, M. (1978) Genetics of hematophagy and autogeny in the *Aedes scutellaris* complex (Diptera: Culicidae). *J. Med. Entomol.* 15: 73–80.

Tsai, K. S., Heath, H., III, Kumar, R., and Riggs, B. L. (1984) Impaired vitamin D metabolism with aging in women. *J. Clin. Invest.* 73:1668–1672.

Tsepkin, Y. A., and Sokolov, L. I. (1971) The maximum size and age of some sturgeons. *J. Ichthyol.* 11:444– 446.

Tsien, H. C., and Wattiaux, M. (1971) Effect of maternal age on DNA and RNA content of *Drosophila* eggs. *Nat. New Biol.* 230:147–148.

Tsitouras, P. D., Martin, C. E., and Harman, S. M. (1982) Relationship of serum testosterone to sexual activity in healthy elderly men. *J. Gerontol.* 37:288–293.

Tsuda, T., Kim, Y. T., Siskind, G. W., and Weksler, M. E. (1988) Old mice recover the ability to produce IgG and high-avidity antibody following irradiation with partial bone marrow shielding. *Proc. Natl. Acad. Sci.* 85:1169–1173.

Tsuji, T. (1987) Ultrastructure of deep wrinkles in the elderly. *J. Cutan. Pathol.* 14:158–164.

Tsuneki, K., Ouji, M., and Saito, H. (1983) Seasonal migration and gonadal changes in the hagfish *Eptatretus burgeri. Jpn. J. Ichthyol.* 29:429–440.

Tucker, S. M., Mason, R. L., and Beauchene, R. E. (1976) Influence of diet and food restriction on kidney function in aging male rats. *J. Gerontol.* 31:264–270.

Tudzynski, P., and Esser, K. (1979) Chromosomal and extrachromosomal control of senescence in the ascomycete *Podospora anserina. Mol. Gen. Genet.* 173:71–84.

———. (1982) Extrachromosomal genetics of *Cephalosporium acremonium.* 2. Development of mitochondrial DNA hybrid vector replicating in *Saccharomyces cerevisiae. Curr. Genet.* 6:153–158.

Tudzynski, P., Stahl, U., and Esser, K. (1980) Transformation to senescence with plasmid-like DNA in the ascomycete *Podospora anserina. Curr. Genet.* 2:181–184.

Turker, M. S., Domenico, J. M., and Cummings, J. D. (1987) A novel family of mitochondrial plasmids associated with longevity mutants of *Podospora anserina. J. Biol. Chem.* 262:2250–2256.

Turker, M. S., Monnat, R. J., Fukuchi, K.-I., Johnston, P. A., Ogburn, C. E., Weller, R. E., Park, J. F., and Martin, G. M. (1988) A novel class of unstable 6–thioguanine-resistant cells from dog and human kidneys. *Cell Biol. Toxicol.* 4:211–223.

Turker, M. S., Nelson, J. G., and Cummings, D. J. (1987) A *Podospora anserina* longevity mutant with a temperature-sensitive phenotype for senescence. *Mol. Cell. Biol.* 7:3199–3204.

Turnbull, W. D. (1970) Mammalian mastica-tory apparatus. *Fieldiana Geol.* 18:149–356.

Turner, B. J. (1964) An introduction to the fishes of the genus *Nothobranchius*. *African Wildlife* 18:117–124.

Turner, B. M. (1975) Posttranslational altera-tions of human erythrocyte enzymes. *Iso-zyme*, vol. 1, *Molecular Structure*, C. L. Markert (ed.), 781–795. New York: Aca-demic Press.

Turner, C. L. (1950) The reproductive potential of a single clone of *Pelmatohydra oligactis*. *Biol. Bull.* 99:285–299.

Turturro, A., and Shafiq, S. A. (1979) Quanti-tative morphological analysis of age-related changes in the flight muscle of *Musca domes-tica*. *J. Gerontol.* 43:823–833.

Tyan, M. L. (1980) Marrow colony-forming units: Age-related changes in responses to anti-*O*-sensitive helper/suppressor stimuli. *Proc. Soc. Exp. Biol. Med.* 165:354–360.

Tyndale-Biscoe, M. (1984) Age-grading meth-ods in adult insects: A review. *Bull. Entomol. Res.* 74:341–377.

Uemura, E. (1980) Age-related changes in neu-ronal RNA content in rhesus monkeys (*Ma-caca mulatta*). *Brain Res. Bull.* 5:117–119.

Uemura, E., and Hartmann, H. A. (1979) RNA content and volume of nerve cell bodies in hu-man brain. 2. Subiculum in aging normal pa-tients. *Exp. Neurol.* 65:107–117.

Úlehlová, L. (1975) Ageing and the loss of au-ditory neuroepithelium in the guinea pig. *Adv. Exp. Med. Biol.* 53:257–264.

Ulrich, P., Pongor, S., Chang, J. C., Bencsath, F. A., and Cerami, A. (1985) Aging of pro-teins. The furoyl furanyl imidazole crosslink as a key advanced glycosylation event. In *Modifications of Proteins during Aging*, R. C. Adelman (ed.), 83–92. New York: Liss.

Umminger, B. L. (1975) Body size and whole blood sugar concentrations in mammals. *Comp. Biochem. Physiol.* 52A:455–458.

Uno, H., Adachi, K., and Montagna, W. (1969) Baldness of the red uacari (*Carajao rubicun-dus*): Histological properties and enzyme ac-tivities of hair follicles. *J. Gerontol.* 24:23–27.

Uno, H., Allegra, F., Adachi, K., and Mon-tagna, W. (1967) Studies of common baldness of the stump-tailed macaque. 1. Distribution of hair follicles. *J. Invest. Dermatol.* 49:88–296.

Uno, H., Tarara, R., Else, J., Suleman, M., and Sapolsky, R. (1989) Hippocampal damage as-sociated with prolonged and fatal stress in pri-mates. *J. Neurosci.* 9:1705–1711.

Upadhyay, S., and Zamboni, L. (1982) Ectopic germ cells: Natural model for the study of germ cell differentiation. *Proc. Natl. Acad. Sci.* 79:6584–6588.

Upton, A. C. (1957) Ionizing radiation and the aging process. *J. Gerontol.* 12:306–311.

Upton, A. C., Kimbal, A. W., Furth, J., Chris-tenberry, K. W., and Benedict, W. H. (1960) Some delayed effects of atom-bomb radia-tions in mice. *Canc. Res.* 20:1–60.

Urbano-Marquez, A., Estruch, R., Navarro-Lopez, F., Grau, J. M., Mont, L., and Rubin, E. (1989) The effects of alcoholism on skeletal and cardiac muscle. *N. Engl. J. Med.* 320:409–415.

Utermann, G. (1989) The mysteries of lipopro-tein(a). *Science* 246:904–910.

Utida, S. (1972) Density dependent polymorph-ism observed in the adult of *Callosobruchus maculatus* (Coleoptera, Bruchidae) *J. Stored Prod. Res.* 8:111–126.

Valentin, J. (1972) Effect of maternal age on re-combination in X in *Drosophila melanogas-ter*. *Drosophila Info. Services* 48:127. Re-search notes.

van Broeckhoven, D., Genthe, A. M., Vanden-berghe, A., Horsthemke, H., Backhovens, H., Raeymaekers, P., Van Hul, W., Wehnet, A., Gheuens, J., Cras, P., Bruyland, M., Martin, J. J., Salbaum, M., Mullan, M. J., Holland, A., Barton, A., Irving, N., Wil-liamson, R., Richards, S. J., and Hardy, J. A. (1987) Failure of familial Alzheimer's disease to segregate with the A4–amyloid gene in several European families. *Nature* 329:153–155.

van Dijk, S. (1979) On the relationship between

reproduction, age, and survival in two carabid beetles: *Calathus melanocephalus L.* and *Pterostichus coerulescens L.* (Coleoptera, Carabidae). *Oecologia* (Berlin) 40:63–80.

Van Heukelem, W. F. (1976) Growth, bioenergetics, and life-span of *Octopus cyanea* and *Octopus maya.* Ph.D. diss., University of Hawaii, Honolulu.

———. (1977) Laboratory maintenance, breeding, rearing, and biomedical research potential of the Yucatan octopus (*Octopus maya*). *Lab. Anim. Sci.* 27:852–859.

———. (1978) Aging in lower mammals. In *The Biology of Aging,* J. A. Behnke, C. E. Finch, and G. B. Moment (eds.), 115–130. New York: Plenum.

———. (1979) Environmental control of reproduction and lifespan in octopus: An hypothesis. In *Reproductive Ecology of Some Marine Invertebrates,* S. E. Stancyk (ed.), 123–133. Columbia: Univ. South Carolina Press.

Van Hoesen, G., Damasio, A. R., and Barnes, C. L. (1984) Alzheimer's disease: Cell-specific pathology isolates the hippocampal formation. *Science* 225:1168–1170.

Van Kleef, F. S., Willems-Thijssen, W., and Hoenders, H. J. (1976) Intracellular degradation and deamidation of a-crystallin subunits. *Eur. J. Biochem.* 66:477–483.

van Leeuwen, F., van der Beek, E., Seger, M., Burbach, P., and Ivell, R. (1989) Age-related development of a heterozygous phenotype in solitary neurons of the homozygous Brattleboro rat. *Proc. Natl. Acad. Sci.* 86:6417–6420.

van Nie, R., and Verstraeten, A. A. (1975) Studies of genetic transmission of mammary tumor virus by C3Hf mice. *Int. J. Canc.* 16:922–931.

van Overbeeke, A. P., and McBride, J. R. (1971) Histological effects of 11–ketotestosterone, 17α-methyltestosterone, estradiol cyprionate, and cortisol on the interrenal tissue, thyroid gland, and pituitary gland of gonadectomized sockeye salmon (*Oncorhynchus nerka*) *J. Can. Res. Bd. Can.* 28:477–484.

Van Peenen, H. J., and Gerstl, B. (1962) Arteriosclerotic massive hypertrophy of the heart. *J. Am. Geriatr. Soc.* 10:505–508.

Van Staden, J., Cook, E. L., and Noodén, L. D. (1988) Cytokinins and senescence. In *Senescence and Aging in Plants,* L. D. Noodén and A. C. Leopold (eds.), 281–328. San Diego: Academic Press.

Van't Hof, J. (1965) Relationships between mitotic cycle time duration, S period duration, and the average ratio of DNA synthesis in the root-tip meristem cells of several plants. *Exp. Cell Res.* 39:48–58.

Van't Hof, J., and Sparrow, A. H. (1963) A relationship between DNA content, nuclear volume, and minimum mitotic cycle time. *Proc. Natl. Acad. Sci.* 49:897–902.

Van Valen, L. (1960) A functional index of hypsodonty. *Evolution* 14:531–532.

Vanyushin, B., Nemirovsky, L. E., Klimenko, V. V., Vasiliev, V. K., and Belozersky, A. N. (1973) The 5–methylcytosine in DNA of rats. *Gerontologia* (Basel) 19:138–152.

Varma, S. K. (1970) Morphology of ovarian changes in the garden lizard, *Calotes versicolor. J. Morphol.* 131:195–210.

Vasek, F. C. (1980) Creosote bush: Long-lived clones in the Mojave Desert. *Am. J. Bot.* 67:246–255.

Vaysse, J., Gattegno, L., Bladier, D., and Aminoff, D. (1986) Adhesion and erythrophagocytosis of human senescent erythrocytes by autologous monocytes and their inhibition by β-galactosyl derivatives. *Proc. Natl. Acad. Sci.* 83:1339–1343.

Vermeulen, A., Reuteus, R., and Verdonck, L. (1972) Testosterone secretion and metabolism in male senescence. *J. Clin. Endocrinol. Metab.* 34:730–735.

Verzar, F. (1956) Das Altern des Kollagens. *Helv. Physiol. Acta* 14:207–221.

Verzar, F., and Spichtin, H. (1966) The role of the pituitary in the aging of collagen. *Gerontologia* (Basel) 12:48.

Vestal, R. E., and Dawson, G. W. (1985) Pharmacology and aging. In *Handbook of the Biology of Aging,* 2d ed., C. E. Finch and E. L. Schneider (eds.), 744–819. New York: Van Nostrand.

Vierny, C., Keller, A., Begel, O., and Belcour, L. (1982) A sequence of mitochondrial DNA is associated with the onset of senescence in a fungus. *Nature* (London) 297:157–159.

Vijayalaxmi and Evans, H. J. (1984) Measurement of spontaneous and X-irradiation–induced 6–thioguanine-resistant human blood lymphocytes using a T-cell cloning technique. *Mutat. Res.* 125:87–94.

Vijayashankar, N., and Brody, H. (1979) A quantitative study of the pigmented neurons in the nuclei locus coeruleus and subcoeruleus in man as related to aging. *J. Neuropathol. Exp. Neurol.* 38:490–497.

Vijg, J., and Uitterlinden, A. G. (1987) A search for DNA alterations in the aging mammalian genome: An experimental strategy. *Mech. Age. Dev.* 41:47–63.

Vladykov, V. D. (1956) *Poissons du Québec, no. 6 (The eel).* Quebec: Dépt. des Pêcheries Province du Québec, Canada.

———. (1963) A review of salmonid genera and their broad geographic distribution. *Trans. Roy. Soc. Can.,* ser. 4, 1:459–504.

Vlassara, H., Brownlee, M., and Cerami, A. (1984) Accumulation of diabetic rat peripheral nerve myelin by macrophages increases with the presence of advanced glycosylation endproducts. *J. Exp. Med.* 160:197–207.

———. (1985a) High-affinity-receptor-mediated uptake and degradation of glucose-modified proteins: A potential mechanism for the removal of senescent macromolecules. *Proc. Natl. Acad. Sci.* 82:5588–5592.

———. (1985b) Recognition and uptake of human diabetic peripheral nerve myelin by macrophages. *Diabetes* 34:553–557.

———. (1986) Novel macrophage receptor for glucose-modified proteins is distinct from previously described scavenger receptors. *J. Exp. Med.* 164:1301–1309.

Vlassara, H., Brownlee, M., Manogue, K. R., Dinarello, C. A., and Pasagian, A. (1988) Cachectin/TNF and IL–1 induced by glucose-modified proteins: Role in normal tissue remodeling. *Science* 240:1546–1548.

Vlijm, L., Kessler, A., and Richter, C. J. J. (1963) The life history of *Pardosa amentata* (CI) (Araneae, Lycosidae) *Entomol. Bericht.* 23:75–80.

Vlodaver, Z., Kahn, H. A., and Neufeld, H. N. (1969) The coronary arteries in early life in three different ethnic groups. *Circulation* 39:541–550.

Vogel, F. (1983) Mutation in man. In *Principles and Practices of Medical Genetics,* A. E. H. Emery and D. Rimoin (eds.), 20–46. London: Churchill and Livingston.

Vogel, F., and Motulsky, A. G. (1986) *Human Genetics: Problems and Approaches,* 301–311. Berlin: Springer-Verlag.

Voitenko, V. P. (1980) Aging, diseases, and X-chromatin. *Z. Gerontol.* 13:18–23.

Volk, E. C. (1986) Use of calcareous elements (statoliths) to determine age of sea lamprey (*Petromyzon marinus*). *Can. J. Fish. Aquat. Sci.* 43:718–722.

Vollmer, W. M., Wahl, P. W., and Blagg, C. R. (1983) Survival with dialysis and transplantation in patients with end-stage renal disease. *N. Engl. J. Med.* 308:1553–1558.

vom Saal, F. S., and Bronson, F. H. (1980) Sexual characteristics of adult female mice are correlated with their blood testosterone levels during prenatal development. *Science* 208:597–599.

vom Saal, F. S., Coquelin, A., Schoonmaker, J., Shryne, J., and Gorski, R. (1984) Sexual activity and sexually dimorphic nucleus volume in male rats are correlated with prior intrauterine position. *Abs. Soc. Neurosci.* 10:927 (abstract).

vom Saal, F. S., and Finch, C. E. (1988) Reproductive senescence: Phenomena and mechanisms in mammals and selected vertebrates. In *The Physiology of Reproduction,* vol. 2, E. Knobil, J. Neill, L. L. Ewing, G. S. Greenwald, C. L. Markart, and D. W. Pfaff (eds.), 2351–2413. New York: Raven Press.

vom Saal, F. S., and Moyer, C. L. (1985) Prenatal effects on reproductive capacity during aging in female mice. *Biol. Reprod.* 32:1111–1126.

vom Saal, F. S., Quadagno, D., Even, M., Keisler, L., Keisler, D., and Khan, S. (1990) The correlation between serum testosterone and estradiol during fetal life and postnatal reproductive traits in female mice is eliminated by maternal stress. *Biol. Reprod.* In press.

vom Saal, F. S., and Rines, J. (1984) Fetal effects on sexual behavior and aggression in young and old female mice treated with estrogen and progesterone. *Horm. Behav* 18:117–129.

Vorontsova, M. A., and Liosner, L. D. (1960) *Asexual Propagation and Regeneration.* London: Pergamon Press.

Vracko, R., and McFarland, B. H. (1980) Life-spans of diabetic and non-diabetic fibroblasts in vitro. *Exp. Cell Res.* 129:345–350.

Waddington, C. H. (1940) *Organizers and Genes.* Cambridge: Cambridge Univ. Press.

Waddy, S. L., and Aiken, D. E. (1985) Multiple fertilization and consecutive spawning in large American lobsters, *Homarus americanus. Can. J. Fish. Aquat. Sci.* 43:2291–2294.

Wade, A. W., and Szewczuk, M. R. (1984) Aging, idiotype repertoire shifts, and compartmentalization of the mucosa-associated lymphoid system. *Adv. Immunol.* 36:143–188.

Wadsworth, J. R., and Hill, W. C. O. (1956) Selected tumors from the London Zoo menagerie. *Univ. Pa. Vet. Ext. Q.* 141:70–73.

Wagner, G. F., and McKeown, B. A. (1985) The purification, partial characterization and bioassay of growth hormone from two species of Pacific Salmon. In *Current Trends in Comparative Endocrinology,* B. Lofts and W. N. Holmes (eds.), 557–561. Hong Kong: Hong Kong Univ. Press.

Wahle, C. M. (1983) Regeneration of injuries among Jamaican gorgonians: The roles of colony physiology and environment. *Biol. Bull.* (Wood's Hole) 165:778–790.

Wajc, E., and Pener, M. P. (1971) The effect of the corpora allata on the flight activity of the male African migratory locust, *Locust migratoria migratorioides* (R. & F.) *Gen. Comp. Endocrinol.* 17:327–333.

Walburg, H. E. (1975) Radiation-induced life-shortening and premature aging. *Adv. Radiat. Biol.* 5:145–179.

Waldron, I. (1987) Causes of the sex differential in longevity. *J. Am. Gerontol. Soc.* 35:365–370.

Walford, R. L. (1969) *The Immunological Theory of Aging.* Williams and Wilkins.

———. (1986) Maximum lifespan and the im-munological theory of aging. *Immunopathol. Immunol. Ther. Lett.* 1:3–11.

———. (1987) MHC-regulation of aging: An extension of the immunologic theory of aging. In *Modern Biological Theories of Aging,* H. R. Warner, R. N. Butler, R. L. Sprott, and E. L. Schneider (eds.), 243–260. New York: Raven Press.

Walford, R. L., and Bergmann, K. (1979) Influence of genes associated with the main histocompatibility complex on deoxyribonucleic acid excision repair capacity and bleomycin sensitivity in mouse lymphocytes. *Tiss. Antigens* 14:336–342.

Walford, R. L., and Liu, R. K. (1965) Husbandry, life span, and growth rate of the annual fish, *Cynolebias adolffi* E. Ahl. *Exp. Gerontol.* 1:161–171.

Walford, R. L., Liu, R. K., Troup, G. M., and Hsiu, J. (1969) Alteration in soluble/insoluble collagen ratios in the annual fish, *Cynolebias bellottii,* in relation to age and environment temperatures. *Exp. Gerontol.* 4:103–109.

Walker, B. W. (1952) A guide to the grunion. *Calif. Fish and Game* 38:409–420.

Walker, E. P. (1975) *Mammals of the World,* 2d ed. Baltimore: Johns Hopkins Univ. Press.

Walker, M. V. (1938) Evidence of Triassic insects in the Petrified Forest National Monument. *Proc. U.S. Natl. Mus.* 85:137–141.

Walker, T. J. (1987) Wing dimorphism in *Gryllus rubens* (Orthoptera: Gryllidae). *Ann. Entomol. Soc. Am.* 80:547–560.

Wallace, J. E., Krauter, E. E., and Campbell, B. A. (1980) Motor and reflexive behavior in the aging rat. *J. Gerontol.* 35:364–370.

Wallentin, L., and Larsson-Cohn, U. (1977) Metabolic and hormonal effects of post-menopausal oestrogen replacement treatment. 2. Plasma lipids. *Acta Endocrinol.* (Copenhagen) 86:597–607.

Waller, B. F. (1988) Hearts of the "oldest old." *Mayo Clin. Proc.* 63:625–27.

Waloff, N. (1949) Observations on larvae of *Ephestia elutella* during diapause. *Trans. Roy. Entomol. Soc.* (London) 100:147–159.

Waloff, N., Norris, M. J., and Broadhead, E. C. (1948) Fecundity and longevity of

Ephestia elutella. Trans. Roy. Entomol. Soc. (London) 99:245–268.

Walsh, S. J., and Burr, B. M. (1981) Distribution, morphology, and life history of the "least brook lamprey," *Lampetra aegyptera* (Pisces: Petromyzontidae), in Kentucky. *Brimleyana* 6:83–100.

Wang, R. K., and Mays, L. L. (1978) Isozymes of glucose–6–phosphate dehydrogenase in livers of aging rats. *Age* 1:2–7.

Wang, S. Y., Halban, P. A., and Rowe, J. W. (1988) Effects of aging on insulin synthesis and secretion. Differential effects on pre-proinsulin messenger RNA levels, proinsulin biosynthesis, and secretion of newly made and preformed insulin in the rat. *J. Clin. Invest.* 81:176–185.

Wang, S. Y., and Rowe, J. W. (1988) Age-related impairment in the short term regulation of insulin biosynthesis by glucose in rat pancreatic islets. *Endocrinology* 123:1008–1013.

Wangermann, E. (1965) Longevity and ageing in plants and plant organs. *Encyclopedia Plant Physiol.* 15:1026–1057.

Warburton, D., and Henderson, A. S. (1979) Sequential silver staining and hybridization *in situ* on nucleolus organizing regions in human cells. *Cytogenet. Cell Genet.* 24:168–175.

Ward, W. F. (1988) Enhancement by food restriction of liver protein synthesis in the aging Fischer 344 rat. *J. Gerontol.* 43:B50–B53.

Wardle, J. A. (1984) *The New Zealand Beeches. Ecology, Utilisation, and Management.* New Zealand Forest Service. Christchurch, New Zealand: Caxton Press.

Wareham, K. A., Lyon, M. F., Glenister, P. H., and Williams, E. D. (1987) Age-related reactivation of an X-linked gene. *Nature* 327:725–727.

Wass, W. M., Thompson, J. R., Moss, E. W., Kunesh, J. P., and Eness, P. G. (1981) Examination of the teeth and diagnosis and treatment of dental diseases. In *Current Veterinary Therapy and Food Animal Practice*, J. L. Howard (ed.), 874–875. Philadelphia: Saunders.

Waterlow, J. C., and Jackson, A. A. (1981) Nutrition and protein turnover in man. *Brit. Med. Bull.* 37:5–10.

Watkinson, A. R., and White, J. (1985) Some life-history consequences of modular construction in plants. *Phil. Trans. Roy. Soc.* (London), ser. B, 313:31–51.

Watt, F., and Molloy, P. L. (1988) Cytosine methylation prevents binding to DNA of a HeLa cell transcription factor required for optimal expression of the adenovirus major late promoter. *Genes Dev.* 2:1136–1143.

Wattiaux, J. M. (1966) Cumulative parental age effects in *Drosophila subobscura. Evolution* 22:406–421.

———. (1967) Influence de l'age sur le fonctionnement ovarien chez *Dropsophila melanogaster. J. Insect Physiol.* 13:1279–1282.

———. (1968) Parental age effects in *Drosophila subobscura. Evolution* 22:406.

Wattiaux, J. M., Libion-Mannaert, M., and Delcour, J. (1971) Protein turnover and protein synthesis following actinomycin-D injection as a function of age in *Drosophila melanogaster. Gerontologia* (Basel) 17:289–299.

Wattiaux, J. M., and Tsien, H. C. (1971) Age effects on the variation of RNA synthesis in the nurse cells in *Drosophila. Exp. Gerontol.* 6:235–247.

Waugh, S. M., and Low, P. S. (1985) Hemichrome binding to band 3: Nucleation of Heinz bodies on the erythrocyte membrane. *Biochemistry* 24:34–39.

Weber, A., Gugen-Guillouzo, C., Szart, C., Beck, M. F., and Schapira, F. (1980) Tyrosine aminotransferase in senescent rat liver. *Gerontology* 26:9–15.

Weber, M. (1927–1928) *Die Säugetiere. Einführung in die Anatomie und Systematik der rezenten und fossilen Mammalia.* 2 vols. Jena: Gustav Fischer. [not read]

Webster, G. C. (1986) Effect of aging on the components of the protein synthesis system. In *Insect Aging*, K.-G. Collatz and R. S. Sohal (eds.), 207–216. Heidelberg: Springer-Verlag.

Webster, G. C., Beachell, V. T., and Webster, S. L. (1980) Differential decrease in protein synthesis by microsomes from aging *Drosophila melanogaster. Exp. Gerontol.* 15:495–497.

Webster, G. C., and Webster, S. L. (1981)

Aminoacylation of tRNA by cell-free preparations from aging *Drosophila melanogaster*. *Exp. Gerontol.* 16:487–494.

———. (1982) Effects of age on the postinitiation stages of protein synthesis. *Mech. Age. Dev.* 16:369–378.

———. (1983) Decline in synthesis of elongation factor one (EF–1) precedes the decreased synthesis of total protein in aging *Drosophila melanogaster*. *Mech. Age. Dev.* 22:121–128.

———. (1984) Specific disappearance of translatable messenger RNA for elongation factor one in aging *Drosophila melanogaster*. *Mech. Age. Dev.* 24:335–342.

Webster, G. C., Webster, S. L., and Landis, W. A. (1981) The effect of age on the initiation of protein synthesis in *Drosophila melanogaster Mech. Age. Dev.* 16:71–79.

Weinbach, E. C., and Garbus, J. (1959) Oxidative phosphorylation in mitochondria of aged rats. *J. Biol. Chem.* 234:412–417.

Weinberg, R. A. (1985) The action of oncogenes in the cytoplasm and nucleus. *Science* 230:770–776.

Weindruch, R. H., Cheung, M. K., Verity, M. A., and Walford, R. L. (1980) Modification of mitochondrial respiration by aging and dietary restriction. *Mech. Age. Dev.* 12:375–392.

Weindruch, R. H., Gottesman, S. R. S., and Walford, R. L. (1982) Modification of age-related immune decline in mice dietarily restricted from or after midadulthood. *Proc. Natl. Acad. Sci.* 79:898–902.

Weindruch, R. H., Kristie, J. A., Cheney, K. E., and Walford, R. L. (1979) Influence of controlled dietary restriction on immunologic function and aging. *Fed. Proc.* 38:2007–2016.

Weindruch, R. H., Kristie, J. A., Naeim, F., Mullen, B. G., and Walford, R. L. (1982) Influence of weaning-initiated dietary restriction on responses to T cell mitogens and on splenic T cell levels in a long-lived F_1-hybrid mouse strain. *Exp. Gerontol.* 17:49–64.

Weindruch, R. H., McFeeters, G., and Walford, R. L. (1984) Food intake reduction and immunologic alterations in mice fed dehydroepiandrosterone. *Exp. Gerontol.* 19:297–304.

Weindruch, R. H., Naylor, P. H., Goldstein, A. L., and Walford, R. L. (1988) Influences of aging and dietary restriction on serum thymosin levels in mice. *J. Gerontol.* 43:B40–B42.

Weindruch, R. H., and Suffin, S. C. (1980) Quantitative histologic effects on mouse thymus of controlled dietary restriction. *J. Gerontol.* 35:525–531.

Weindruch, R. H., and Walford, R. L. (1982) Dietary restriction in mice beginning at 1 year of age: Effect on life-span and spontaneous cancer incidence. *Science* 215:1415–1418.

———. (1988) *The Retardation of Aging and Disease by Dietary Restriction*. Springfield, IL: C. C. Thomas.

Weindruch, R. H., Walford, R. L., Fligiel, S., and Guthrie, D. (1986) The retardation of aging in mice by dietary restriction: Longevity, cancer, immunity, and lifetime energy intake. *J. Nutrit.* 116:641–654.

Weiner, M., Davis, B., Mohs, R., and Davis, K. (1987) Influence of age and relative weight on cortisol suppression in normal subjects. *Am. J. Psychiatry* 144:646–648.

Weinrich, D. H. (1954) Sex in protozoa: A comparative review. In *Sex in Microorganisms*, 134–207. Washington, DC: American Association for the Advancement of Science.

Weinstein, R., Stemmerman, M. B., and Maciag, T. (1981) Hormonal requirements for growth of arterial smooth muscle cells in vitro: An endocrine approach to atherosclerosis. *Science* 212:818–820.

Weintraub, J. A., and Burt, B. A. (1985) Oral health status in the United States: Tooth loss and edentulism. *J. Dental Educ.* 49:368–376.

Weisbart, M., Youson, J. H., and Wiebe, J. P. (1978) Biochemical, histochemical, and ultrastructural analyses of presumed steroid-producing tissues in the sexually mature sea lamprey, *Petromyzon marinus* L. *Gen. Comp. Endocrinol.* 34:26–37.

Weismann, A. (1889a) The duration of life (a paper presented in 1881). In *Essays upon Heredity and Kindred Biological Problems*, E. B. Poulton, S. Schonland, and A. E. Shipley (eds.), 1–66. Oxford: Clarendon Press. Reprint. Oceanside, NY: Dabor Science.

————. (1889b) Life and death (a paper presented in 1883). In *Essays upon Heredity and Kindred Biological Problems,* E. B. Poulton, S. Schonland, and A. E. Shipley (eds.), 111–157. Oxford: Clarendon Press. Reprint. Oceanside, NY: Dabor Science.

Weiss, K. M. (1972) A general measure of human population growth regulation. *Am. J. Phys. Anthropol.* 37:337–344.

————. (1973) Demographic models for anthropology. *Am. Antiq.* vol. 38, part 2. [not read]

Weiss, R. A. (1986) The oncogene concept. *Canc. Rev.* 2:1–17.

Weissenberg, R. (1927) Beitrage zur Kenntniss der Biologie und Morphologie der Neuenaugen. 2. Das Reifewachstum der Gonaden bei *Lampetra fluviatilis* und *planeri. Z. Mikr. Anat. Forsch.* 8:193–249.

Weksler, M. (1985) Changes in the immune response with age. In *Handbook of the Biology of Aging,* 2d ed., C. E. Finch and E. L. Schneider (eds.), 414–432. New York: Van Nostrand.

Weksler, M. E., Innes, J. B., and Goldstein, G. (1978) Immunological studies of aging. 4. The contribution of thymic involution to the immune deficiencies of aging mice and reversal with thymopoietin 32–36. *J. Exp. Med.* 148:996–1006.

Welden, B. A., Calliet, G. M., and Flegal, A. R. (1987) Comparison of radiometric with vertebral band age estimates in four California elasmobranchs. In *The Age and Growth of Fish,* R. C. Summerfelt and G. E. Hall (eds.), 301–315. Ames: Iowa State Univ. Press.

Welford, A. T. (1977) Motor performance. In *Handbook of the Psychology of Aging,* J. E. Birren and K. W. Schaie (eds.), 450–496. New York: Van Nostrand Reinhold.

————. (1980) Sensory, perceptual, and motor processes in older adults. In *Handbook of Mental Health and Aging,* J. E. Birren and R. B. Sloane (eds.), 192–213. Englewood Cliffs, NJ: Prentice Hall.

Welford, A. T., Norris, A. H., and Shock, N. W. (1969) Speed and accuracy of movement and their changes with age. *Acta Psychol.* 30:3–15.

Wellinsick, S. J. (1952) Rejuvenation of woody plants by formation of sphaeroblasts. *Proc. Kon. Ned. Akad. Wet.,* ser. C, 55:567–573.

Wellinger, R., and Guigoz, Y. (1986) The effect of age on the induction of tyrosine aminotransferase and tryptophan oxygenase genes by physiological stress. *Mech. Age. Dev.* 34:203–217.

Wensel, L. C., and Krumland, B. (1986) A site index system for redwood and Douglas fir in California's north coast system. *Hilgardia* 54:1–14.

Werner, B. (1979) Coloniality in the Scyphozoa: Cnidaria. In *Biology and Systematics of Colonial Organisms,* G. Larwood and B. R. Rosen (eds.), 81–103. New York: Academic Press.

Wertheim, I., Jagiello, G. M., and Ducayern, M. B. (1986) Aging and aneuploidy in human oocytes and follicular cells. *J. Gerontol.* 41:567–573.

Wesenberg-Lund, G. (1923) Contributions to the biology of the Protifera. 1. The males of the Rotifera. *Konigl. Danske Vidensk. Selsk. Skrifter Naturvidensk.* og Mathem. Afd. 8, 4(3): 191–345. [not read]

West, M. D., Pereira-Smith, O. M., and Smith, J. R. (1989) Replicative senescence of human skin fibroblasts correlates with a loss of regulation and overexpression of collagenase activity. *Exp. Cell Res.* 184:138–147.

Westaway, D., Goodman, P. A., Mirenda, C. A., McKinley, M. P., Carlson, G. A., and Prusiner, S. B. (1987) Distinct prion proteins in short and long scrapie incubation period mice. *Cell* 51:651–662.

West-Eberhard, M. J. (1986) Alternative adaptations, speciation, and phylogeny (a review). *Proc. Natl. Acad. Sci.* 83:1388–1392.

Westerberg, E., Akesson, B., Rehncrom, S., Smith, D. S., and Siesjo, B. K. (1979) Lipid peroxidation in brain *in vitro*: Effects on phospholipids and fatty acids. *Acta Physiol. Scand.* 105:524–526.

Western, D. (1979) Size, life history, and ecology in mammals. *African J. Ecol.* 17:185–204.

Western, D., and Ssemakula, J. (1982) Life history patterns in birds and mammals and their evolutionary interpretation. *Oecologia* (Berlin) 54:281–290.

Westing, A. H. (1964) The longevity and aging of trees. *Gerontologist* 4:10–15.

Wexler, B. C. (1964a) Correlation of adrenocortical histopathology with arteriosclerosis in breeder rats. *Acta Endocrinol.* 46:613–631.

———. (1964b) Spontaneous arteriosclerosis in repeatedly bred male and female rats. *J. Atheroscl. Res.* 4:57–80.

———. (1964c) Spontaneous arteriosclerosis of the mesenteric renal and peripheral arteries of repeatedly bred rats. *Circ. Res.* 15:485–496.

———. (1964d) Spontaneous coronary arteriosclerosis in repeatedly bred male and female rats. *Circ. Res.* 14:32–43.

———. (1976a) Comparative aspects of hyperadrenocorticism and aging. In *Hypothalamus, Pituitary, and Aging,* A. V. Everitt and J. A. Burgess (eds.), 333–361. Springfield, IL: C. C. Thomas.

———. (1976b) Effects of hypophysectomy on alloxan-diabetic, arteriosclerotic, breeder vs. non-arteriosclerotic, virgin rats. *Atherosclerosis* 25:13–30.

Wexler, B. C., Antony, C. D., and Kittinger, G. W. (1964) Serum lipoprotein and lipid changes in arteriosclerotic breeder rats. *J. Atheroscl. Res.* 4:131–143.

Wexler, B. C., and Fischer, C. W. (1963a) Abnormal glucose tolerance in repeatedly bred rats with arteriosclerosis. *Nature* 200:133–136.

———. (1963b) Hyperplasia of the islets of Langerhans in arteriosclerotic rats. *Nature* 200:33–37.

Wexler, B. C., Iams, S. G., and Judd, J. T. (1977) Comparative effects of adrenal regeneration hypertension on non-arteriosclerotic and arteriosclerotic Sprague-Dawley vs. spontaneously hypertensive rats. *Atherosclerosis* 26:1.

Wexler, B. C., and Kittinger, G. W. (1965) Adrenocortical function in arteriosclerotic female breeds rats. *J. Atheroscl. Res.* 5:317–329.

Wexler, B. C., and True, C. W. (1963) Carotid

and cerebral arteriosclerosis in the rat. *Circ. Res.* 12:659–666.

Whalley, P. E., and Jarzembowski, E. A. (1985) Fossil insects from the lithographic limestone of Montsech (Late Jurassic–Early Cretaceous), Lerida Province, Spain. *Bull. Brit. Mus. Nat. Hist. Geol.* 38:381–412.

Wheeler, A. (1969) *The Fishes of the British Isles and Northwest Europe.* East Lansing: Michigan State Univ. Press.

Wheeler, K. T., and Lett, J. T. (1974) On the possibility that DNA repair is related to age in non-dividing cells. *Proc. Natl. Acad. Sci.* 71:1862–1865.

Wherry, E. T. (1972) Box-huckleberry as the oldest living protoplasm. *Castanea* 37:94–95.

White, A. C., and Gershbein, L. L. (1987) Steroid modulation of liver regeneration and hepatic microsomal enzymes in rats of either sex. *Res. Comm. Chem. Pathol. Pharmacol.* 55:317–334.

White, J. (1973) Viable hybrid young from crossmated periodical cicadas. *Ecology* 54:573–580.

———. (1980) Resource partitioning in oviposition strategies in periodical cicadas. *Am. Nat.* 115:1–28.

White, J., and Lloyd, M. (1975) Growth rates of 17– and 13–year periodical cicadas. *Am. Midl. Nat.* 94:127–143.

White, J., Lloyd, M., and Zar, J. H. (1979) Faulty eclosion in crowded suburban periodical cicadas: Populations out of control. *Ecology* 60:305–315.

White, J., and Strehl, C. (1978) Xylem feeding by periodical cicada nymphs on tree roots. *Ecol. Entomol.* 3:323–327.

White, P. C., Chaplin, D. D., Weis, J. H., Dupont, B., New, M. I., and Seidman, J. G. (1984) Two steroid 21–hydroxylase genes are located in the murine S region. *Nature* 312:465–467.

White, P. C., Grossberger, D., Onufer, B. J., Chaplin, D. D., New, M. I., Dupont, B., and Strominger, J. L. (1985) Two genes encoding steroid 21–hydroxylase are located near the genes encoding the forth component of complement in man. *Proc. Natl. Acad. Sci.* 82:1089–1093.

White, P. R. (1934) Potentially unlimited growth of excised tomato root tips in a liquid medium. *Plant Physiol.* 9:585–600.

———. (1939) Potentially unlimited growth of excised plant calus in an artificial medium. *Am. J. Bot.* 26:59–64.

Whitehead, H., and Carscadden, J. E. (1985) Predicting inshore whale abundance: Whales and capelin off the Newfoundland coast. *Can. J. Fish. Aquat. Sci.* 42:976–981.

Whitfield, P. J., and Evans, N. A. (1983) Parthenogenesis and asexual multiplication among parasitic platyhelminths. *Parasitology* 86:121–160.

Whiting, P. W. (1926) Influence of age of mother on appearance of an hereditary variation in *Habrobracon. Biol. Bull.* 50:371–385.

Whitten, P. L., and Naftolin, F. (1987) Estrogenic substances in commercial rodent chow could affect biomarkers of estrogen action. PNAS National Research Council Panel on Reproductive and Developmental Toxicology.

Whittington, H. B. (1985) *The Burgess Shale.* New Haven: Yale Univ. Press.

Whittow, G. C. (1970) *Comparative Physiology of Thermoregulation.* Vol. 1, *Invertebrates and Nonmammalian Vertebrates.* G. C. Whittow (ed.). New York: Academic Press.

———. (1971) *Comparative Physiology of Thermoregulation.* Vol. 2, *Mammals.* G. C. Whittow (ed.). New York: Academic Press.

———. (1973) *Comparative Physiology of Thermoregulation.* Vol. 3, *Special Aspects of Thermoregulation.* G. C. Whittow (ed.). New York: Academic Press.

Wichman, H. A., Potter, S. S., and Pine, D. S. (1985) Mys, a family of mammalian transposable elements isolated by phylogenetic screening. *Nature* 317:77–81.

Widdowson, E. M., and Kennedy, G. C. (1962) Rate of growth, mature weight, and life-span. *Proc. Roy. Soc.,* ser. B, 156:96–108.

Wielckens, K., and Delfs, T. (1986) Glucocorticoid-induced cell death and poly[adenosine diphosphate (ADP)-ribosyl]ation: Increased toxicity of dexamethasone on mouse S49.1 lymphoma cells with the poly(ADP-ribosyl)ation inhibitor benzamide. *Endocrinology* 119:2383–2392.

Wigglesworth, V. B. (1934) The physiology of ecdysis in *Rhodnius prolixus* (Hemiptera). 2. Factors controlling moulting and metamorphosis. *Q. J. Microscop. Sci.* 77:193–222.

———. (1972) *The Principles of Insect Physiology.* 7th ed. London: Chapman and Hall.

Wiklund, J. A., and Gorski, J. (1982) Genetic differences in estrogen-induced deoxyribonucleic acid synthesis in the rat pituitary: Correlations with pituitary tumor susceptibility. *Endocrinology* 111:1140–1149.

Wilcock, G. K., and Esiri, M. M. (1982) Plaques, tangles, and dementia: A quantitative study. *J. Neurol. Sci.* 56:343–356.

Wilkens, H. (1971) Genetic interpretations of regressive evolutionary processes: Studies on hybrid eyes of two *Astyanax* cave populations (Characidae, Pisces) *Evolution* 25:530–544.

Wilks, A. F., Cozens, P. J., Mattaj, J. W., and Jost, J.-P. (1982) Estrogen induces a demethylation at the 5′ end region of the chicken vitellogenin gene. *Proc. Natl. Acad. Sci.* 79:4252–4255.

William, F., Beamish, H., and Medland, T. E. (1988) Metamorphosis of the mountain brook lamprey *Ichthyomyzon greeleyi. Environ. Biol. Fishes* 23:45–54.

Williams, C. M., Barness, L. A., and Sawyer, W. H. (1943) The utilization of glycogen by flies during flight and some aspects of the physiological aging of *Drosophila. Biol. Bull.* 84:263–272.

Williams, D. H., and Anderson, D. T. (1975) The reproductive system, embryonic development, larval development, and metamorphosis of the sea urchin *Heliocidaris erythrogramma* (Vol.) (Echinoidae: Echinometridae). *Austral. J. Zool.* 23:371–403.

Williams, G. C. (1957) Pleiotropy, natural selection, and the evolution of senescence. *Evolution* 11:398–411.

———. (1966a) *Adaptations and Natural Selection. A Critique of Some Current Evolutionary Thought.* Princeton, NJ: Princeton Univ. Press.

———. (1966b) Natural selection, the costs of reproduction, and a refinement of Lack's principle. *Am. Nat.* 100:687–690.

———. (1975) *Sex and Evolution.* Princeton, NJ: Princton Univ. Press.

Williams, G. C., and Koehn, R. K. (1984) Population genetics of North Atlantic catadromous eels (*Anguilla*). In *Evolutionary Genetics of Fishes,* B. J. Turner (ed.), 529–560. New York: Plenum.

Williams, G. C., and Taylor, P. D. (1987) Demographic consequences of natural selection. In *Evolution of Longevity of Animals. A Comparative Approach,* A. D. Woodhead and K. H. Thompson (eds.), 235–246. New York: Plenum.

Williams, J. B. (1961) The dimorphism of *Polystoma integerrimum* (Frolich) rudolphi and its bearing on relationships within the Polystomatidae: Part 3. *J. Helminthol.* 35:181–202.

Williams, J. B., and Sharp, P. J. (1978) Age-dependent changes in the hypothalamo-pituitary-ovarian axis of the laying hen. *J. Reprod. Fertil.* 53:141–146.

Williams, J. R., and Dearfield, K. L. (1985) Nonhuman fibroblast-like cells in culture. In *CRC Handbook of the Cell Biology of Aging,* V. J. Cristofalo, R. C. Adelman, and G.S. Roth (eds.), 443–451. Boca Raton, FL: CRC Press.

Williams, M. (1990) Feeding behaviour in Cleveland shale fishes. In *Evolutionary Paleobiology of Behavior and Coevolution,* A. Boucot (ed.), 273–287. Amsterdam: Elsevier.

Williams, R. M., Kraus, L. J., Lavin, P. T., Steele, L. L., and Yunis, E. J. (1981) Genetics of survival in mice: Localization of dominant effects to subregion of the major histocompatibility complex. In *Immunological Aspects of Aging,* D. Segre and L. Smith (eds.), 247–253. New York: Marcel Dekker.

Williamson, A. R., and Askonas, B. A. (1972) Senescence of an antibody-forming clone. *Nature* 238:337–339.

Williamson, H., and Fennell, D. J. (1981) Nonrandom assortment of sister chromatids in yeast mitosis. In *Molecular Genetics in Yeast,* D. Von Wettstein, A. Stenderup, M. Kielland-Brandt, and J. Friis (eds.), 89–100. Copenhagen: Munksgaard.

Willis, E. R., Riser, G. R., and Roth, L. M. (1958) Observations on reproduction and development in cockroaches. *Ann. Entomol. Soc. Am.* 51:53–69.

Willott, J. F., Pankow, D., Hunter, K. P., and Kordyban, M. (1985) Projections from the anterior ventral cochlear nucleus to the central nucleus of the inferior colliculus in young and aging C57BL/6J mice. *J. Comp. Neurol.* 237:545–551.

Wilmut, I., Sales, D. I., and Ashworth, C. J. (1986) Maternal and embryonic factors associated with prenatal loss in mammals. *J. Reprod. Fertil.* 76:851–864.

Wilson, D. E., Hall, M. E., and Stone, G. C. (1978) Test of some aging hypotheses using two-dimensional protein mapping. *Gerontology* 24:426–433.

Wilson, E. B. (1925) *The Cell in Development and Heredity.* 3d ed. New York: Macmillan.

Wilson, E. O. (1971) *The Insect Societies.* Cambridge, MA: Belknap Press.

———. (1975) *Sociobiology. The New Synthesis.* Cambridge, MA: Harvard Univ. Press.

———. (1987) The earliest known ants: An analysis of the Cretaceous species and an inference concerning their social organization. *Palaeontology* 13:44–53.

Wilson, J. D., and Goldstein, J. L. (1975) Classification of hereditary disorders of sexual development. *Birth Defects* 11:1–16.

Wilson, M. C., and Harrison, D. E. (1983) Decline in male mouse pheromone with age. *Biol. Reprod.* 29:81–86.

Wilson, P. D. (1973) Enzyme changes in ageing mammals. *Gerontologia* (Basel) 19:79–125.

Wilson, P. H., and Coursen, B. W. (1987) Further studies on the induction of o-pyrocatechuic acid carboxylyase in aging cultures of *Aspergillus ornatus. Mech. Age. Dev.* 40:31–40.

Wilson, P. W., Castelli, W. P., and Kannel, W. B. (1987) Coronary risk prediction in adults (the Framingham Heart Study). *Am. J. Cardiol.* 59:91G–94G.

Wilson, T. G., Landers, M. H., and Happ, G. M. (1983) Precocene I and II inhibition of vitellogenic oocyte development in *Drosophila melanogaster. J. Insect Physiol.* 29:249–254.

Wilson, V., and Jones, P. (1983). DNA methylation decreases in aging but not in immortal cells. *Science* 220:1055–1057.

Wilson, V. L., Smith, R. A., Ma, S., and Cutler, R. G. (1987) Genomic–5–methyldeoxycytidine decreases with age. *J. Biol. Chem.* 262:9948–9951.

Wing, S. L., and Tiffney, B. H. (1987) The reciprocal interaction of angiosperm evolution and tetrapod herbivory. *Rev. Paleobot. Palynol.* 50:179–210

Winklbauer, R., and Hausen, P. (1983) Development of the lateral line system in *Xenopus laevis*. 2. Cell multiplication and organ formation in the supraorbital system. *J. Embryol. Exp. Morphol.* (U.K.) 76:283–296.

Winston, J. E., and Jackson, J. B. (1984) Ecology of cryptic coral reef communities. 4. Community development and life histories of encrusting cheilostome Bryozoa. *J. Exp. Mar. Biol. Ecol.* 76:1–21.

Winston, M. L. (1987) *The Biology of the Honey Bee*. Cambridge, MA: Harvard Univ. Press.

Winston, M. L., and Katz, S. J. (1981) Longevity of cross-fostered honey bee workers (*Apis mellifera*) of European and Africanized races. *Can. J. Zool.* 59:1571–1575.

Wise, P. M. (1983) Aging of the female reproductive system. *Adv. Biol. Res. Aging* 1:195–224.

Wise, P. M., and Parsons, B. (1984) Nuclear estradiol and cytosol progestin receptor concentrations in the brain and pituitary gland and sexual behaviour in ovariectomized estradiol-treated middle-aged rats. *Endocrinology* 115:810–816.

Wise, P. M., Walovitch, R. C., Cohen, I. R., Weiland, N. G., and London, E. D. (1987) Diurnal rhythmicity and hypothalamic deficits in glucose utilization in aged ovariectomized rats. *J. Neurosci.* 7:3469–3473.

Wise, T. H. (1988) The thymus: Old gland, new perspectives. *Dom. Anim. Endocrinol.* 5:109–128.

Wiske, P. S., Epstein, S., Bell, N. H., Queener, S. F., Edmondson, J., and Johnston, C. C., Jr. (1979) Increases in immunoreactive parathyroid hormone with age. *N. Engl. J. Med.* 300:1419–1421.

Wisniewski, K., Howe, J., Williams, D. G., and Wisniewski, H. M. (1978) Precocious aging and dementia in patients with Down's syndrome. *Biol. Psychiatry* 13:619.

Wisniewski, K. E., and Maslinska, D. (1989) Immunoreactivity of ceroid lipofuscin storage pigment in batten disease with monoclonal antibodies to the amyloid beta-protein. *N. Engl. J. Med.* 320:256–257.

Withler, I. L. (1966) Variability in life history characteristics of steelhead trout (*Salmo gairdneri*) along the Pacific coast of North America. *J. Fish. Res. Bd. Can.* 23:365–392.

Witten, M. (1983) A return to time, cells, systems, and aging: Rethinking the concept of senescence in mammalian organisms. *Mech. Age. Dev.* 21:69–81.

———. (1984) A return to time, cells, systems, and aging. 2. Relational and reliability theoretic approaches to the study of senescence in living systems. *Mech. Age. Dev.* 27:323–340.

———. (1985) A return to time, cells, systems, and aging. 3. Gompertzian models of biological aging and some possible roles for critical elements. *Mech. Age. Dev.* 32:141–177.

———. (1986) A return to time, cells, systems, and aging. 4. Further thoughts on Gompertzian survival dynamics: The neonatal years. *Mech. Age. Dev.* 33:177–190.

———. (1987) Information content of biological survival curves arising in aging experiments: Some further thoughts. In *Evolution of Longevity in Animals*, A. Woodhead and K. Thompson (eds.), 295–317. *Basic Life Sciences*, vol. 42. New York: Plenum.

———. (1988) A return to time, cells, systems, and aging. 5. Further thoughts on Gompertzian survival dynamics: The geriatric years. *Mech. Age. Dev.* 46:175–200.

Witztum, J. L., and Koschinsky, T. (1989) Metabolic and immunological consequences of glycation of low density lipoproteins. In *The Maillard Reaction in Aging Diabetes*, J. W. Baynes and V. M. Monnier. Proceedings of the NIH Conference, Bethesda, Maryland, September 1988. *Prog. Clin. Biol. Res.* 304:219–234.

Wodinsky, J. (1977) Hormonal inhibition of feeding of death in octopus. Control by optic gland secretion. *Science* 198:948–951.

Wodsedalek, J. E. (1917) Five years of starvation of larvae. *Science* 46:366–367.

Woke, P. A., Ally, M. S., and Rosenberger, C. R. (1956) The numbers of eggs developed related to the quantities of human blood ingested in *Aedes aegypti* (L.) (Diptera: Culicidae). *Ann. Entomol. Soc. Am.* 49:435–441.

Wolf, N. S., Giddens, W. E., and Martin, G. M. (1988) Life table analysis and pathologic observations in male mice of long-lived hybrid stain (A_f × C57BL/6)F_1. *J. Gerontol.* 43:B71–B78.

Wolfe, L. S., Ng Ying Kin, N. M., Palo, J., and Haltia, M. (1983) Dolichols in brain and urinary sediment in neuronal ceroid lipofuscinosis. *Neurology* 33:103.

Wolff, S. P., Garna, A., and Dean, R. T. (1986) Free radicals, lipids, and protein degradation. *Trends in Biochem. Sci.* 11:27–43.

Wolff, T. (1978) Maximum size of lobsters (*Homarus*) (Decapoda, Nephrodpidae). *Crustacea* 34:1–14.

Wolters, E. C., and Calne, D. B. (1989) Is Parkinson's disease related to aging? In *Parkinsonism and Aging*, D. B. Calne (ed.), 125–132. *Aging* 35. New York: Raven Press.

Wong, D. F., Wagner, H. N., Jr., Dannals, R. F., Links, J. M., Frost, J. J., Ravert, H. T., Wilson, A. A., Rosenbaum, A. E., Gjedde, A., Douglass, K. H., Petronis, J. D., Folstein, M. F., Toung, J. K., Burns, H. D., and Kuhar, M. J. (1984) Effects of age on dopamine and serotonin receptors measured by positron tomography in the living human brain. *Science* 226:1393–1396.

Wood, D. M. (1978) Taxonomy of the nearctic species of *Twinnia* and *Gymnopais* (Diptera: Simuliidae) and discussion of the ancestry of the Simuliidae. *Can. Entomol.* 110:1297–1337.

———. (1987) Oestridae. In *Manual of Nearctic Diptera*, vol. 2, J. F. McAlpine (ed.). *Res. Br. Agric. Can.* 28:1147–1158.

Wood, D. M., and Borkent, A. (1989) Phylogeny and classification of the Nematocera. In *Manual of Nearctic Diptera*, vol. 3, J. F. McAlpine and D. M. Wood (eds.). *Res. Br. Agric. Can.* 32:1333–1370.

Wood, A. E. (1962) The early Tertiary rodents of the family Paramyidae. *Trans. Am. Phil. Soc.*, n.s. 52:1–261.

Wood, W. B. (ed.) (1988) *The Nematode Caenorhabditis Elegans*. Cold Spring Harbor, New York: Cold Spring Harbor Laboratory.

Woodard, A. E., and Abplanalp, H. (1971) Longevity and reproduction in Japanese quail. *Poultry Sci.* 50:688–694.

Woodhead, A. D. (1974) Ageing changes in the Siamese fighting fish, *Betta splendens:* 1. The testis. *Exp. Gerontol.* 9:75–81.

———. (1978) Fish in studies of aging. *Exp. Gerontol.* 13:125–140.

Woodhead, A. D., and Ellett, S. (1969a) Endocrine aspects of ageing in the guppy, *Lebistes reticulatus* (Peters). 3. The testis. *Exp. Gerontol.* 4:17–25.

———. (1969b) Aspects of aging in the guppy, *Lebistes reticulatus*. 4. The Ovary. *Exp. Gerontol.* 4:197–205.

Woodhead, A. D., Settow, R. B., and Grist, E. (1980) DNA repair and longevity in three species of cold-blooded vertebrates. *Exp. Gerontol.* 15:301–304.

Woodley, N. E. (1989) Phylogeny and classification of the "orthorrhaphous" Diptera. In *Manual of Nearctic Diptera*, vol. 3, J. F. McAlpine and D. M. Wood (eds.). *Res. Br. Agric. Can.* 32:1371–1396.

Woodruff, L. L. (1943) The pedigreed culture of *Paramecium aurelia*. *Proc. Natl. Acad. Sci.* 29:135–136.

Woodward, A. E., and Abplanalp, H. (1971) Longevity and reproduction in Japanese quail maintained under stimulatory lighting. *Poultry Sci.* 50:688–692.

Woolhouse, H. W. (1978) Cellular and metabolic aspects of senescence is higher in plants. In *The Biology of Aging*, J. A. Behnke, C. E. Finch, and G. B. Moment (eds.), 83–99. New York: Plenum.

———. (1983) The general biology of plant senescence and the role of nucleic acids in protein turnover in the control of senescence processes which are genetically programmed. In *Post Harvest Physiology and Crop Production*, NATO Advanced Study Institute, M. Lieberman (ed.), 1–45. New York: Plenum.

———. (1984) The biochemistry and regulation of senescence in chloroplasts. *Can. J. Bot.* 62:2934–2942.

Woolley, P. (1966) Reproduction in *Antechinus* spp. and other dasyurid marsupials. *Symp. Zool. Soc. Lond.* 15:281–294.

———. (1971) Differential mortality of *Antechinus stuartii* (Macleay): Nitrogen balance of somatic changes. *Austral. J. Zool.* 19:347–353.

———. (1981) *Antechinus bellus,* another dasyurid marsupial with post-mating mortality of males. *J. Mammal.* 62:381–382.

Wourms, J. P. (1972) The developmental biology of annual fishes. 2. Naturally occuring dispersion and reaggregation of blastomeres during the development of annual fish eggs. *J. Exp. Zool.* 182:169–200.

———. (1973) The developmental biology of annual fishes. 3. Pre-embryonic and embryonic diapause of variable duration in the eggs of annual fishes. *J. Exp. Zool.* 182:389–414.

———. (1977) Reproductive and development in chondrichthyan fishes. *Am. Zool.* 17:379–410.

Wright, R. M., and Cummings, D. J. (1983) Integration of mitochondrial gene sequences within the nuclear genome during senescence in a fungus. *Nature* 302:86–88.

Wright, R. M., Horrum, M. A., and Cummings, D. J. (1982) Are mitochondrial structural genes selectively amplified during senescence in *Podospora anserina? Cell* 29:505–515.

Wright, W. E., Pereira-Smith, O. M., and Shay, J. W. (1989) Reversible cellular senescence: Implications for immortalization of normal human diploid fibroblasts. *Mol. Cell. Biol.* 9:3088–3092.

Wronski, T. J., Lowry, P. L., Walsh, C. C., and Ignaszewski, L. A. (1985) Skeletal alterations in ovariectomized rats. *Calc. Tiss. Int.* 37:324–328.

Wu, C. I., and Li, W.-H. (1985) Evidence for higher rates of nucleotide substitution in rodents than in man. *Proc. Natl. Acad. Sci.* 82:1741–1745.

Wu, W., Pahlavani, M., Cheung, H. T., and Richardson, A. (1986) The effect of aging on the expression of interleukin-2 messenger ribonucleic acid. *Cell. Immunol.* 100:224–231.

Wundsch, H. H. (1953) Das Vorkommen von Aalen in "vorgeschrittenem Reifezustand" in einem markischen Binnengewassen. *Z. Fisch.* 2:1–18.

Wyllie, A. H., Kerr, J. F., and Currie, A. R. (1980) Cell death: The significance of apoptosis. *Int. Rev. Cytol.* 68:251–306.

Yagi, H., Katoh, S., Akiguchi, I., and Takeda, T. (1988) Age-related deterioration of ability of acquisition in memory and learning in senescence accelerated mouse: SAM-P/8 as an animal model of disturbances in recent memory. *Brain Res.* 474:86–93.

Yalden, D. W. (1985) Feeding mechanisms as evidence for cyclostome monophyly. *J. Linn. Soc. Zool.* (London) 84:291–300.

Yamada, M., Tsukagoshi, H., Otomo, E., and Hayakawa, M. (1987) Cerebral amyloid angiopathy in the aged. *J. Neurol.* 234:371–376.

Yamada, T. (1968) Cellular and subcellular events in Wolffian lens regeneration. *Curr. Top. Dev. Biol.* 2:247–283.

Yamagishi, H. (1962) Growth relation in some small experimental populations of rainbow trout fry, *Salmo gairdneri* Richardson, with special reference to social relations among individuals. *Jpn. J. Ecol.* 12:43.

Yamamoto, K. (1962) Origin of the yearly crop of eggs in the medaka, *Oryzias latipes. Annot. Zool. Jpn.* 35:218–222.

Yamamoto, K. R. (1985) Steroid receptor regulated transcription of specific genes and gene networks. *Ann. Rev. Genet.* 19:209–232.

Yamamoto, O., Fuji, I., Yoshida, T., Cox, A. B., and Lett, J. T. (1988) Age dependency of base modification in rabbit liver DNA. *J. Gerontol.* 43:B132–B136.

Yamauchi, M., Woodley, D. T., and Mechanic, G. L. (1988) Aging and cross-linking of skin collagen. *Biochem. Biophys. Res. Commun.* 152:898–903.

Yamazaki, K., Beauchamp, G. K., Wysocki, C. J., Bard, J., Thomas, L., and Boyse, E. A. (1983) Recognition of H–2 types in relation to the blocking of pregnancy in mice. *Science* 221:186–188.

Yang, Y. J., Hope, I. D., Ader, M., and Bergman, R. N. (1989) Insulin transport across
capillaries is rate limiting for insulin action in
dogs. *J. Clin. Invest.* 84:1620–1628.

Yanishevsky, R., Mendelsohn, M. L., Mayall,
B. H., and Cristofalo, V. J. (1974) Proliferative capacity and DNA content of aging human diploid cells in culture: A cytophotometric and autoradiographic analysis. *J. Cell
Physiol.* 84:164–170.

Yarwood, E. A., and Hansen, E. L. (1969)
Dauerlarvae of *Caenorhabditis briggsae* in
axenic culture. *Nematology* 1:184–189.

Yates, F. E. (1987) Senescence from the aspect
of physical stability. Typescript.

Yates, F. E., and Kugler, P. N. (1986) Similarity principles and intrinsic geometries: Contrasting approaches to interspecies scaling. *J.
Pharmacol. Sci.* 75:1019–1027.

Yeargers, E. (1981) Effect of gamma-radiation
of dauer larvae *Caenorhabditis elegans*. *Nematology* 13:235–237.

Yen, I. I., Allan, J. V., Pearson, D. V., Acton,
J. M., and Greenberg, M. M. (1977) Prevention of obesity in *Ar/a* mice by dehydroepiandrosterone. *Lipids* 12:409–413.

Yen, S. S. C. (1977) The biology of menopause. *J. Reprod. Med.* 18:287.

Yetter, R. A., Hartley, J. W., and Morse, H. C.
(1983) *H–2*-linked regulation of xenotropic
murine leukemia virus expression. *Proc.
Natl. Acad. Sci.* 80:505–509.

Yonezu, T., Tsunasawa, S., Higuchi, K., Ko-
gishi, K., Naiki, H., Hanada, K., Sakiyama,
F., and Takeda, T. (1987) A molecular-path-
ologic approach to murine senile amyloidosis. Serum precursor-apo A-II variant
(pro5-glu) presents only in amyloidosis prone
SAM-P/1 and -P/2 mice. *Lab. Invest.* 57:65–
70.

Yoshida, H., Kohno, A., Ohta, K., Hirose, S.,
Maruyama, N., and Shirai, T. (1981) Genetic
studies of autoimmunity in New Zealand
mice. 3. Associations among anti-DNA antibodies, NTA, and renal disease in (NZB ×
NZW)F$_1$ × NZW backcross mice. *J. Immunol.* 127:433–437.

Yoshida, Y. (1961) Nuclear control of chloroplast activity in *Elodea* leaf cells. *Protoplasma* 54:476–492.

Young, J. B., Rowe, J. W., Pallotta, J. A.,
Spen, D., and Landsberg, L. (1980) Enhanced plasma and epinephrine response to
upright posture and oral glucose administration in elderly human subjects. *Metabolism*
29:532–539.

Young, J. O., and Reynoldson, T. B. (1964) A
quantitative study of the population biology
of *Dendrocoelum lacteum* (Muller) (Turbellaria, Tricladia) *Oikos* 15:237–264.

Young, T. P. (1984) The comparison demography of semelparous *Lobelia telekii* and iteroparous *keniensis* on Mount Kenya. *J. Ecol.*
72:637–650.

Youson, J. H. (1980) Morphology and physiology of lamprey metamorphosis. *Can. J. Fish.
Aquat. Sci.* 37:1687–1710.

Yu, B. P. (1987) Update on food restriction and
aging. *Rev. Biol. Res. Aging* 3:495–505.

Yu, B. P., Masoro, E. J., and McMahan, C. A.
(1985) Nutritional influences on aging in
Fischer 344 rats. 1. Physical metabolic and
longevity characteristics. *J. Gerontol.*
40:657–670.

Yu, B. P., Masoro, E. J., Murata, I., Bertrand,
H. A., and Lynd, F. T. (1982) Life span
study of SPF Fischer 344 male rats fed *ad libitum* or restricted diets: Longevity, growth,
lean body mass, and disease. *J. Gerontol.*
37:130–141.

Yunis, E. J., Watson, A. L. M., Gelman,
R. S., Sylvia, S. J., Bronson, R., and Dorf,
M. E. (1984) Traits that influence longevity
in mice. *Genetics* 108:999–1011.

Yunis, J. J. (1983) The chromosomal basis of
human neoplasia. *Science* 221:227–236.

Zahn, R. F. (1983) Measurement of the molecular weight distributions in human muscular
deoxyribonucleic acid. *Mech. Age. Dev.*
22:355–379.

Zahn, R. F., Reinmuller, J., Beyer, R., and
Pondeljak, V. (1987) Age-correlated DNA
damage in human muscle tissue. *Mech. Age.
Dev.* 41:73–114.

Zamenhof, S., Van Marthens, E., and Grauel,
L. (1972) DNA (cell number) and protein in
rat brain. Second generation (F2) alteration
by maternal (F0) dietary restriction. *Nutrit.
Metab.* 14:262–270.

Zammuto, R. M. (1986) Life histories of birds: Clutch size, longevity, and body mass among North American game birds. *Can. J. Zool.* 64:2739–2749.

Zanandrea, G. (1954) Corrispondenza tra forma parassite nei generi *Ichthyomyzon é Lampetra* (problemi di speciazione) *Boll. Zool.* 21:461–466.

———. (1957) Neoteny in a lamprey. *Nature* 179:925–926.

———. (1961) Studies on European lampreys. *Evolution* 15:523–534.

Zar, J. H. (1969) The use of the allometric model for avian standard metabolism-body weight relationships. *Comp. Biochem. Physiol.* 29:227–234.

Zarins, C. K., Giddens, D. P., and Glagov, S. (1983) Atherosclerotic plaque distribution and flow velocity profiles in the carotid bifurcation. In *Cerebrovascular Insufficiency,* J. J. Bergan and J. S. Yao (eds.), 19–30. New York: Grune and Stratton.

Zauber, N. P., and Zauber,.A. G. (1987) Hematologic data of healthy very old people. *J.A.M.A.* 257:2181–2184.

Zebrower, M., Kieras, F. J., and Brown, W. T. (1985) Glycosaminoglycan and hyaluronic acid elevation in genetic aging syndromes. *Am. J. Human Genet.* 36:84S.

Zeelon, P., Gershon, H., and Gershon, D. (1973) Inactive enzyme molecules in aging organisms. Nematode fructose–1,6–diphosphate aldolase. *Biochemistry* 12:1743–1750.

Zeman, F. D. (1962) Pathologic anatomy of old age. *Arch. Pathol.* 73:126–145.

Zerwekh, J. E., Sakhaec, K., Glass, K., and Pak, C. Y. C. (1983) Long-term 25–hydroxyvitamin D3 therapy in postmenopausal osteoporosis: Demonstration of responsive and nonresponsive subgroups. *J. Clin. Endocrinol. Metab.* 56:410–413.

Ziegler, M. G., Lake, C. R., and Kopin, I. J. (1978) Plasma noradrenaline increases with age. *Nature* 361:333–334.

Zigler, J. S., Jr., and Goosey, J. (1981) Aging of protein molecules: Lens crystallins as a model system. *Trends in Biochem. Sci.* 6:133–136.

Zihlman, A. L., Morbeck, M. E., and Sumner, D. R. (1989) Tales of Gombe chimps as told in their bones. In *Anthroquest* (Leakey Foundation News), no. 40 (Summer):20–22.

Zilles, K., Armstrong, E., Schlaug, G., and Schlechter, A. (1986) Quantitative cytoarchetectonics of the posterior cingulate cortex in primates. *J. Comp. Neurol.* 253:514–524.

Zimmerman, M. H. (1983) *Xylem Structure and the Ascent of Sap.* Berlin: Springer-Verlag.

Zola-Morgan, S., Squire, L. R., and Amaral, D. G. (1986) Human amnesia and the medial temporal region: Enduring memory impairment following a bilateral lesion limited to field CA_1 of the hippocampus. *J. Neurosci.* 6:2950–2967.

ZS.-Nagy, I. (ed.) (1988) Lipofuscin–1987 State of the Art. *Excerpta Medica.* International Congress Series 782.

Zuckerman, S., and Baker, T. G. (1977) The development of the ovary and the process of oogenesis. In *The Ovary,* vol. 1, *General Aspects,* 2d ed., S. Zuckerman and B. J. Weir (eds.), 41–67. New York: Academic Press.

Zuckerman, S., and Groome, J. R. (1937) The aetiology of benign enlargement of the prostate in the dog. *J. Pathol. Bacteriol.* 44:113.

Zurcher, C., van Zwieten, M. J., Solleveld, H. A., and Hollander, C. F. (1982) Aging Research. In *The Mouse in Biomedical Research,* vol. 4, H. L. Foster, J. D. Small, and J. G. Fox (eds.), 11–35. New York: Academic Press.

Zurcher, C., van Zwieten, M. J., Solleveld, H. A., van Bezooijen, C. F. A., and Hollander, C. F. (1982) Possible influence of multiple pathological changes in aging rats on studies of organ aging, with emphasis on the liver. In *Liver and Aging: 1982: Liver and Drugs,* K. Kitani (ed.), 19–36. Amsterdam: Elsevier.

Zweifel, R. G. (1980) Aspects of the biology of a laboratory population of kingsnakes. In *Reproductive Biology and Diseases of Captive Reptiles,* J. B. Murphy and J. T. Collins (eds.), 145–158. SSAR Contributions to Herpetology 1. Society for the Study of Amphibians and Reptiles.

Author Index

Species Index

Accipiter nisus (sparrow hawk), 148

Acetabularia mediterranea (bluegreen algae), 112

Achatinella mustellina (Hawaiian tree snail), 131–32

Acipenser (sturgeon), 591–92

Acipenser fulvescens (lake sturgeon), 122–23, 134–35, 663

Acipenser güldenstädti (Russian sturgeon), 134–35, 592, 594

Acipenser nudiventris (schip sturgeon), 272

Acipenser ruthenus (sterlet), 135

Acipenser stellatus (sevrjuga), 134, 592

Acmaea insessa. See Lottia insessa

Acropora, 233

Acyrthosiphon kondo (pea aphid), 394–95

Acyrthosiphon pisum (pea aphid), 394–95

Aedes, 54, 394

Aedes aegypti (yellow fever mosquito), 54, 64, 394, 511

Aedes aegypti var. *queenslandiensis*, 511

Aedes atropalpus, 312

Aedes taeniorhynchus, 56, 585

Aelosoma hemprichii, 234

Aepyceros (impala), 35

African elephant. *See Loxodonta africana*

African green monkey. *See Cercopithecus aethiops*

Agamermis decaudata, 60

Agaricia agaricites, 234

Agaricia lamarcki, 234

Agathis australis (Kauri), 209

Agave americana (century plant), 101, 208

Agnatha, 80–81, 84, 115, 216, 572, 588–91, 597

Agouti paca, 608

Agraulis vanillae, 52

Alectura lathami (brush turkey), 26–27, 122, 124, 147, 605, 664

Alerce. *See Fitzroya cupressoides*

Alewife. *See Alosa pseudoharengus*

Alligator mississippiensis, 144, 664

Allolobophora (= *Helodrilus*) *longa*, 128

Alosa pseudoharengus (alewife), 84, 597

Alosa sapidissima (American shad), 84, 92–93, 595, 597, 630

Alouatta seniculus (red howler monkey), 268, 286

Amago. *See Oncorhynchus rhodurus*

Ambloplites rupestris (rock bass), 136

Ambrosia beetle. *See Xyleborus ferrugineus*

Ambystoma maculata (axolotl), 219, 664

American badger. *See Taxidea*

American shad. *See Alosa sapidissima*

Amphioxus, 572

Anagallis arvensis (scarlet pimpernel), 230

Anas platyrhynchos (domestic duck), 150

Anchovy. *See Engraulis mordax*

Ancyclostoma duodenale (hookworm), 215

Andean condor. *See Vultur gryphus*

Aneurophyton, 577

Anguilla anguilla (European freshwater eels), 93, 593, 663

Anguilla dieffenbachii, 94, 593

Anguilla rostrata (American eel), 593

Anguillula aceti. See Turbatrix aceti

Anodonta cygnea, 132

Anolis (lizard), 291, 417

Anopheles maculopensis, 56–57

Anoplopoma fimbria (sablefish), 217

Ant. *See Stenamma westwoodi*

Antechinus (marsupial mice), 95–98, 116, 151, 548, 605–6

Antechinus bellus, 96

Antechinus bilarni, 95–96

Antechinus flavipes, 96

Antechinus stuartii (brown antechinus), 95–97

Antechinus swainsonii (dusky antechinus), 95–96

Antrozous pallidus, 288

Aotus trivirgatus (night monkey), 286

Aphaenogaster picea, 71

Aphelocoma coerulescens coerulescens (Florida scrub jay), 148

Aphia minuta, 86, 599

Apis mellifera (honeybee), 46, 65, 68, 663

Aplochiton taeniatus, 84

Carabid beetle. _See Carabus coerulescens_

Carabus, 60, 203, 244

Carabus coerulescens (carabid beetle), 122–23, 130, 663

Carabus glabratus, 66, 122–23, 130, 663

Carajao rubicundus (red uacari), 617

Carcharodon carcharias (white shark), 133

Carp. _See Cyprinus carpio_

Castor (beaver), 35

Cat. _See Felis catus_

Cavia porcellus (guinea pig), 289

Cebus capucinus (capuchin), 286

Cedar of Lebanon. _See Cedrus libani_

Cedrus libani (Cedar of Lebanon), 209

Cellepora pumicosa (bryozoan), 233

Century plant. _See Agave americana_

Cephalobus dubius, 113

Cercocebus albigena, 268

Cercopithecus aethiops (vervet or African green monkey), 286–87, 546

Cercopithecus jacchus, 268

Cercopithecus talapoin, 268

Cereas pedunculatus, 232

Cereus giganteus (giant saguaro cactus), 224, 666

Cervus (deer), 35

Cervus elaphus (red deer), 548–49

Cetorhinus maximus, 134

Char. _See Salvelinus_ sp.

Cheirogaleus major, 268

Chelonus annulipes (wasp), 470

Chenopodium album, 212

Chicken. _See Gallus gallus domesticus_

Chimpanzee. _See Pan troglodytes_

Chipmunk. _See Tamias_

Chiton tuberculatus, 224, 244, 665

Chondrosteus, 573, 591

Chortoicetes terminifera, 470

Chub. _See Leucichthyus kiyi_

Chum. _See Oncorhynchus keta_

Cicada. _See Magicicada_

Citellus (ground squirrels, gophers), 271

Cladoselache, 588

Clethrionomys (vole), 35

Club moss. _See Lycopodium_

Clupea harengus (Atlantic herring), 272

Clupea pallasi (Pacific herring), 136

Clupeidae (shad), 116

Clytia johnstone, 112

Cnephia eremites (blackfly), 56, 586

Cnephia mutata, 56, 586

Cocklebur. _See Xanthium pennsylvanicum_

Cockroach. _See Blaberus giganteus_

Cod. _See Gadus morhua_

Colias erytheme (pierid butterfly), 52

Common dragonet. _See Callionymus lyra_

Common tree shrew. _See Tupaia glis_

Condylarth, 612

Conger eel. _See Conger vulgaris_

Conger vulgaris (conger eels), 84, 94, 593

Connochaetes (gnu), 35

Conus, 131

Convallaria majalis (lily of the valley), 229

Cooksonia, 572, 576

Cordaites, 580

Coregonus clupeaformis (whitefish), 136

Cornborer. _See Pyraustra nubilis_

Cottontail rabbit. _See Sylvilagus_

Coturnix coturnix japonica (Japanese quail), 27, 122, 124, 146–47, 149, 384, 605, 664

Crab-eating macaque. _See Macaca irus_

Creosote bush. _See Larrea tridentata_

Cricetulus griseus (Chinese hamster), 289

Cricetulus triton (rat-like hamster), 289

Cricket. _See Gryllus rubens_

Crocus sativa (saffron crocus), 229

Crowned pigeon. _See Goura netoria_

Cryptus inortus (wasp), 482

Crystallogobius linearis, 86, 599

Crytophria, 77

Ctenodrilus monostylos, 235

Culex pipiens (mosquito), 63, 312

Cyanea capillata, 232

Cyclostigma, 579

Cygnus columbianus bewickii (Bewick's swan), 148

Cynolebias, 137, 139, 143, 222

Cynolebias adolffi, 122–23, 499, 664

Cynolebias bellottii, 137, 139, 470

Cyprinidontiformes, 603

Cyprinodon macularis (desert pupfish), 500

Cyprinus carpio (domestic carp), 139

Cyrthosiphon, 395

Dacrydium franklinii (Australian conifer), 209

Daphnia, 59, 250

Daphnia longispina (water flea), 46, 474, 663

Daphnia magna (water flea), 488, 501

Dasypus novemcintus (armadillo), 472

Daucus carota, 212

Deer. _See Cervus_ sp.

Deer mouse. _See Peromyscus_

Dendrobaena subrubicunda, 129

Dendrocoelum lacteum, 512

Dendrosoma, 59

Desert pupfish. _See Cyprinodon macularis_

Desmognathus ochrophaeus (salamander), 219–20

Dicrobezzia venusta (biting midge), 53, 55

Didelphis marsupialis (southern opossum), 151, 606

Didelphis virginiana marsupialis (common opossum), 151, 166, 606

Didelphoidea, 606

Diphyllobothrium latum, 235

Diplodocus, 615

Diploptera punctata, 130

Dirphia eumedide, 46, 663

Dirphia hircia, 46, 663

Subject Index